CHEMISTRY FOR DEGREE STUDENTS

B.Sc. First Year

For all Indian Universities

As per UGC Model Curriculum

- **Part I : Inorganic Chemistry**
- **Part II : Organic Chemistry**
- **Part III : Physical Chemistry**

Dr. R.L. MADAN
M.Sc., Ph.D., H.E.S. (I)
Former Head of the Chemistry Department
Government Postgraduate College, Faridabad
and Principal, Government College, Panchkula

S. CHAND
PUBLISHING

S Chand And Company Limited
(ISO 9001 Certified Company)

AF646752

S Chand And Company Limited

(ISO 9001 Certified Company)

Head Office: D-92, Sector–2, Noida – 201301, U.P. (India), Ph. 91-120-4682700

Registered Office: A-27, 2nd Floor, Mohan Co-operative Industrial Estate, New Delhi – 110 044, Phone: 011-49731800

S. CHAND PUBLISHING

www.**schandpublishing.com;** e-mail: **info@schandpublishing.com**

Marketing Offices:

Chennai	:	Ph: 23632120; chennai@schandpublishing.com
Guwahati	:	Ph: 2738811, 2735640; guwahati@schandpublishing.com
Hyderabad	:	Ph: 40186018; hyderabad@schandpublishing.com
Jalandhar	:	Ph: 4645630; jalandhar@schandpublishing.com
Kolkata	:	Ph: 23357458, 23353914; kolkata@schandpublishing.com
Lucknow	:	Ph: 4003633; lucknow@schandpublishing.com
Mumbai	:	Ph: 25000297; mumbai@schandpublishing.com
Patna	:	Ph: 4011400; patna@schandpublishing.com

© S Chand And Company Limited, 2008

All rights reserved. No part of this publication may be reproduced or copied in any material form (including photocopying or storing it in any medium in form of graphics, electronic or mechanical means and whether or not transient or incidental to some other use of this publication) without written permission of the copyright owner. Any breach of this will entail legal action and prosecution without further notice.

Jurisdiction: *All disputes with respect to this publication shall be subject to the jurisdiction of the Courts, Tribunals and Forums of New Delhi, India only.*

First Edition 2010

Second Revised Edition 2011

Reprints 2012, 2013 (Twice), 2014 (Thrice), 2015 (Twice), 2016 (Twice), 2018 (Twice), 2019, 2020

Low Priced Students' Paperback Edition (LPSPE), 2021; Reprint 2022

Reprint 2026

ISBN: 978-81-947-7178-4 **Product Code:** H6CDYCHEM6810ENAA21L

PRINTED IN INDIA

By Shri Shyamlal Printing Press Private Limited, Plot 20/4, Site-IV, Industrial Area Sahibabad, Ghaziabad – 201 010 and Published by S Chand And Company Limited, A-27, 2nd Floor, Mohan Co-operative Industrial Estate, New Delhi – 110 044.

PREFACE TO THE SECOND EDITION

Overwhelming response to the first edition of the book has encouraged me to make some additions and revisions in the book. Notable changes that have been made are as under:

- Matter on inclusion compounds, charge transfer complexes and clatherates in Chapter 1 of organic chemistry has been rewritten to cover them thoroughly.
- Determination of mechanism of organic reaction has been dealt with elaborately in Chapter 2 of organic chemistry.
- Objective questions which form an essential part of the question paper in university examination have been added before the subjective questions in the exercises in each chapter.
- A new chapter on Thermodynamics– I containing first law of thermodynamics and thermochemistry, which forms a part of syllabus for B.Sc.– I in some universities, has been added as Chapter 8 in physical chemistry section.

I hope that the changes made in the book will be found to be even more useful by the students. Patronage received by the teachers is gratefully acknowledged. Suggestions regarding further improvement in the book are welcome.

I am thankful to the management and the editorial team of S. Chand & Company Pvt. Ltd., New Delhi for help and support in publication of this revised edition.

Dr. R.L. Madan
Mobile : 9971775666

Disclaimer : While the author of this book has made every effort to avoid any mistakes or omissions and has used his skill, expertise and knowledge to the best of his capacity to provide accurate and updated information, the author and S. Chand do not give any representation or warranty with respect to the accuracy or completeness of the contents of this publication and are selling this publication on the condition and understanding that they shall not be made liable in any manner whatsoever. S.Chand and the author expressly disclaim all and any liability/responsibility to any person, whether a purchaser or reader of this publication or not, in respect of anything and everything forming part of the contents of this publication. S. Chand shall not be responsible for any errors, omissions or damages arising out of the use of the information contained in this publication.
Further, the appearance of the personal name, location, place and incidence, if any; in the illustrations used herein is purely coincidental and work of imagination. Thus the same should in no manner be termed as defamatory to any individual.

PREFACE TO THE FIRST EDITION

I feel pleasure in presenting this book to the teachers and students of B.Sc.– I year of all the Indian universities. The book has been prepared keeping in view the syllabi prepared by different universities on the basis of Model UGC curriculum. This book is an outgrowth of my more than three decades of classroom teaching experience at B.Sc. level.

A long span of interaction with students has guided me to give special attention to dark areas which students find difficult to understand. Mathematics background required is simple and the necessary mathematical techniques have been developed within the text. Some salient features of the book are :

- The book has been written in a simple and comprehensible language.
- A large number of illustrations, pictures and interesting examples have been provided to make the reading interesting and understandable.
- The important laws, concepts and principles have been given either in italics or in bold type to help the students to focus on them.
- A large number of solved examples which have appeared in the examination papers of various universities have been given in each chapter, for the benefit of students.
- The questions that have been provided in the Exercise are in tune with the latest pattern of examination.

I believe that the book contains all that is needed to understand the subject in a systematic manner. However, no work is perfect. I would welcome suggestions from the teaching fraternity and students for further improvement of the book.

I express my gratitude to the staff of S.Chand & Co., New Delhi for providing all assistance and bringing out the book in the present shape.

Dr. R.L. Madan
Mobile : 9971775666

UGC MODEL SYLLABUS

CHEMISTRY, B.Sc. FIRST YEAR

CH-101 Inorganic Chemistry

I Atomic Structure

Idea of de Broglie matter waves, Heisenberg uncertainty principle, atomic orbitals, Schrodinger wave equation, significance of ψ and ψ^2, quantum numbers, radial and angular wave functions and probability distribution curves, shapes of *s*, *p*, *d* orbitals. Aufbau and Pauli exclusion principles, Hund's multiplicity rule. Electronic configurations of the elements, effective nuclear charge

II Periodic Properties

Atomic and ionic radii, ionization energy, electron affinity and electronegativity – definition, methods of determination or evaluation, trends in periodic table and applications in predicting and explaining the chemical behaviour.

III Chemical Bonding

(A) Covalent Bond – Valence bond theory and its limitations, directional characteristics of covalent bond, various types of hybridization and shapes of simple inorganic molecules and ions. Valence shell electron pair repulsion (VSEPR) theory to NH_3, H_3O^+, SF_4, ClF_3, ICl_2^- and H_2O. MO theory, homonuclear and heteronuclear (CO and NO) diatomic molecules, multicenter bonding in electron deficient molecules, bond strength and bond energy, percentage ionic character from dipole moment and electronegativity difference.

(B) Ionic Solids – Ionic structures, radius ratio effect and coordination number, limitation of radius ratio rule, lattice defects, semiconductors, lattice energy and Born-Haber cycle, solvation energy and solubility of ionic solids, polarizing power and polarisability of ions, Fajan's rule. Metallic bond-free electron, valence bond and band theories.

(C) Weak Interactions – Hydrogen bonding, van der Waals forces.

IV *s*-Block Elements

Comparative study, diagonal relationships, salient features of hydrides, solvation and complexation tendencies including their function in biosystems, an introduction to alkyls and aryls.

V *p*-Block Elements

Comparative study (including diagonal relationship) of groups 13-17 elements, compounds like hydrides, oxides, oxyacids and halides of groups 13-16, hydrides of boron-diborane and higher boranes, borazine, borohydrides, fullerenes, carbides, fluorocarbons, silicates (structural principle), tetrasulphur, tetranitride, basic properties of halogens, interhalogens and polyhalides.

VI Chemistry of Noble Gases

Chemical properties of the noble gases, chemistry of xenon, structure and bonding in xenon compounds.

CH-102 Organic Chemistry

I Structure and Bonding

Hybridization, bond lengths and bond angles, bond energy, localized and delocalized chemical bond, van der Waals interactions, inclusion compounds, clatherates, charge transfer complexes, resonance, hyperconjugation, aromaticity, inductive and field effects, hydrogen bonding.

II Mechanism of Organic Reactions

Curved arrow notation, drawing electron movements with arrows, half-headed and double-headed arrows, homolytic and heterolytic bond breaking. Types of reagents – electrophiles and nucleophiles. Types of organic reactions. Energy considerations.

Reactive intermediates – carbocations, carbanions, free radicals, carbenes, arynes and nitrenes (with examples). Assigning formal charges on intermediates and other ionic species.

Methods of determination of reaction mechanism (product analysis, intermediates, isotope effects, kinetic and stereochemical studies).

III Stereochemistry of Organic Compounds

Concept of isomerism. Types of isomerism.

Optical isomerism – elements of symmetry, molecular chirality, enantiomers, stereogenic centre, optical activity, properties of enantiomers, chiral and achiral molecules with two stereogenic centres, diastereomers, threo and erythro diastereomers, meso compounds, resolution of enantiomers, inversion, retention and racemization.

Relative and absolute configuration, sequence rules, D & L and R & S systems of nomenclature.

Geometric isomerism – determination of configuration of geometric isomers. E & Z system of nomenclature, geometric isomerism in oximes and alicyclic compounds.

Conformational isomerism – conformational analysis of ethane and *n*-butane; conformations of cyclohexane, axial and equatorial bonds, conformation of mono substituted cyclohexane derivatives. Newman projection and Sawhorse formulae, Fischer and flying wedge formulae.

Difference between configuration and conformation.

IV Alkanes and Cycloalkanes

IUPAC nomenclature of branched and unbranched alkanes, the alkyl group, classification of carbon atoms in alkanes. Isomerism in alkanes, sources, methods of formation (with special reference to Wurtz reaction, Kolbe reaction, Corey-House reaction and, decarboxylation of carboxylic acids), physical properties and chemical reactions' of alkanes.

Mechanism of free radical halogenation of alkanes: orientation, reactivity and selectivity.

Cycloalkanes – nomenclature, methods of formation, chemical reactions, Baeyer's strain theory and its limitations. Ring strain in small rings (cyclopropane and cyclobutane), theory of strainless rings. The case of cyclopropane ring: banana bonds.

V Alkenes, Cycloalkenes, Dienes and Alkynes

Nomenclature of alkenes, methods of formation, mechanisms of dehydration of alcohols and dehydrohalogenation of alkyl halides, regioselectivity in alcohol dehydration. The Saytzeff rule, Hofmann elimination, physical properties and relative stabilities of alkenes.

Chemical reactions of alkenes – mechanisms involved in hydrogenation, electrophilic and free radical additions, Markownikoff's rule, hydroboration–oxidation, oxymercuration-reduction. Epoxidation, ozonolysis, hydration, hydroxylation and oxidation with $KMnO_4$, Polymerization of alkenes. Substitution at the allylic and vinylic positions of alkenes. Industrial applications of ethylene and propene.

Methods of formation, conformation and chemical reactions of cycloalkenes.

Nomenclature and classification of dienes: isolated, conjugated and cumulated dienes. Structure of allenes and butadiene, methods of formation, polymerization. Chemical reactions – 1,2 and 1,4 additions, Diels-Alder reaction

Nomenclature, structure and bonding in alkynes. Methods of formation. Chemical reactions of alkynes, acidity of alkynes. Mechanism of electrophilic and nucleophilic addition reactions, hydroboration-oxidation, metal-ammonia reductions, oxidation and polymerization.

VI Arenes and Aromaticity

Nomenclature of benzene derivatives. The aryl group. Aromatic nucleus and side chain. Structure of benzene: molecular formula and Kekule structure. Stability and carbon-carbon bond lengths of benzene, resonance structure, MO picture.

Aromaticity: the Huckel rule, aromatic ions.

Aromatic electrophilic substitution – general pattern of the mechanism, role of σ- and π- complexes. Mechanism of nitration, halogenation, sulphonation, mercuration and Friedel-Crafts reaction. Energy profile diagrams. Activating and deactivating substituents, orientation and ortho/para ratio. Side chain reactions of benzene derivatives. Birch reduction.

Methods of formation and chemical reactions of alkylbenzenes, alkynylbenzenes and biphenyl.

VII Alkyl and Aryl Halides

Nomenclature and classes of alkyl halides, methods of formation, chemical reactions. Mechanisms of nucleophilic substitution reactions of alkyl halides, S_N2 and S_N1 reactions with energy profile diagrams.

Polyhalogen compounds: chloroform, carbon tetrachloride.

Methods of formation of aryl halides, nuclear and side chain reactions. The addition-elimination and the elimination-addition mechanisms of nucleophilic aromatic substitution reactions.

Relative reactivities of alkyl halides vs allyl, vinyl and aryl halides. Synthesis and uses of DDT and BHC.

CH-103 Physical Chemistry

I Mathematical Concepts and Computers

(A) Mathematical Concepts

Logarithmic relations, curve sketching, linear graphs and calculation of slopes, differentiation of functions like k_x, e^x, x^n, sin x, log x; maxima and minima, partial differentiation and reciprocity relations. Integration of some useful/relevant functions; permutations and combinations. Factorials. Probability.

(B) Computers

General introduction to computers, different components of a computer, hardware and software, input-output devices; binary numbers and arithmatic; introduction to computer languages. Programming, operating systems.

II Gaseous States

Postulates of kinetic theory of gases, deviation from ideal behavior, van der Waals equation of state.

Critical Phenomena : PV isotherms of real gases, continuity of states, the isotherms of van der Waals equation, relationship between critical constants and van der Waals constants, the law of corresponding states, reduced equation of state.

Molecular velocities : Root mean square, average and most probable velocities. Qualitative discussion of the Maxwell's distribution of molecular velocities, collision number, mean free path and collision diameter. Liquefaction of gases (based on Joule-Thomson effect).

III Liquid State

Intermolecular forces, structure of liquids (a qualitative description).

Structural differences between solids, liquids and gases.

Liquid crystals: Difference between liquid crystal, solid and liquid. Classification, structure of nematic and cholestric phases. Thermography and seven segment cell.

IV Solid State

Definition of space lattice, unit cell.

Laws of crystallography – (*i*) Law of constancy of interfacial angles (*ii*) Law of rationality of indices (*iii*) Law of symmetry. Symmetry elements in crystals.

X-ray diffraction by crystals. Derivation of Bragg equation. Determination of crystal structure of NaCl, KCl and CsCl (Laue's method and powder method).

V Colloidal State

Definition of colloids, classification of colloids.

Solids in liquids (sols): properties – kinetic, optical and electrical, stability of colloids, protective action, Hardy-Schulze law, gold number.

Liquids in liquids (emulsions) : types of emulsions, preparation. Emulsifier.

Liquids in solids (gels): classification, preparation and properties, inhibition, general applications of colloids.

VI Chemical Kinetics and Catalysis

Chemical kinetics and its scope, rate of a reaction, factors influencing the rate of a reaction – concentration, temperature, pressure, solvent, light, catalyst. Concentration dependence of rates, mathematical characteristics of simple chemical reactions – zero order, first order, second order, pseudo order, half life and mean life. Determination of the order of reaction –

differential method, method of integration, method of half life period and isolation method. Radioactive decay as a first order phenomenon.

Experimental methods of chemical kinetics: conductometric potentiometric, optical methods, polarimetry and spectrophotometer.

Theories of chemical kinetics: effect of temperature on rate of reaction, Arrhenius equation, concept of activation energy.

Simple collision theory based on hard sphere model, transition state theory (equilibrium hypothesis). Expression for the rate constant based on equilibrium constant and thermodynamic aspects.

Catalysis, characteristics of catalysed reactions, classification of catalysis, miscellaneous examples.

CONTENTS OF B.Sc. II AND B.Sc. III BOOKS

B.Sc. II

PART I: INORGANIC CHEMISTRY

1. Chemistry of Elements of First Transition Series
2. Chemistry of Elements of Second and Third Transition Series
3. Oxidation and Reduction
4. Coordination Compounds
5. Chemistry of Lanthanide Elements
6. Chemistry of Actinides
7. Acids and Bases
8. Non-Aqueous Solvents

PART II: ORGANIC CHEMISTRY

1. Electromagnetic Spectrum : Absorption spectra
2. Alcohols
3. Phenols
4. Ethers and Epoxides
5. Aldehydes and Ketones
6. Carboxylic Acids
7. Carboxylic Acid Derivatives
8. Organic Compounds of Nitrogen

PART III: PHYSICAL CHEMISTRY

1. Thermodynamics-I (Thermochemistry and First Law)
2. Thermodynamics-II (Second and Third Laws)
3. Chemical Equilibrium
4. Phase Rule
5. Liquid-Liquid Mixures
6. Distribution Law
7. Electrochemistry-I
8. Electrochemistry-II

B.Sc. III

PART I: INORGANIC CHEMISTRY

1. Hard and Soft Acids and Bases
2. Metal Ligand Bonding in Transition Metal Complexes
3. Magnetic Properties of Transition Metal Complexes
4. Electronic Spectra of Transition Metal Complexes
5. Thermodynamic and Kinetic Aspects of Metal Complexes
6. Organometallic Chemistry
7. Bioinorganic Chemistry
8. Silicones and Phosphazenes

PART II: ORGANIC CHEMISTRY

1. Spectroscopy
2. Organometallic Compounds
3. Organosulphur Compounds
4. Heterocyclic Compounds
5. Organic Synthesis via Enolates (Acetoacetic Ester, Malonic Esters, 1,3-Dithianes, Enamines)
6. Carbohydrates
7. Amino Acids, Peptides, Proteins and Nucleic Acids
8. Fats, Oils, Soaps and Detergents
9. Synthetic Polymers
10. Synthetic Dyes

PART III: PHYSICAL CHEMISTRY

1. Elementary Quantum Mechanics
2. Spectroscopy
3. Photochemistry
4. Physical Properties and Molecular Structure
5. Solutions, Dilute Solutions and Colligative Properties
6. Statistical Mechanics

CONTENTS

PART I: INORGANIC CHEMISTRY

PART II: ORGANIC CHEMISTRY

PART III: PHYSICAL CHEMISTRY

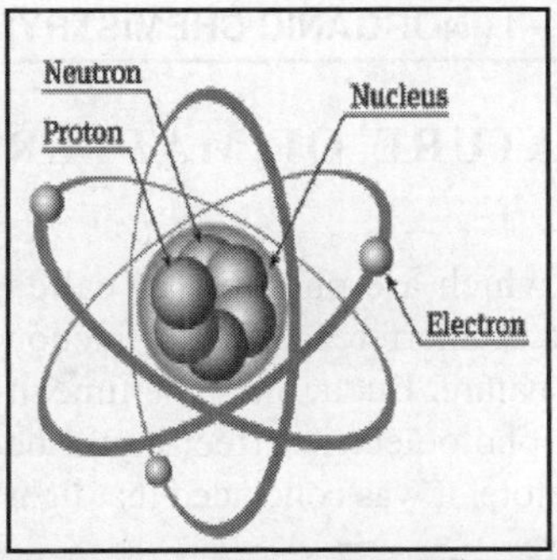

PART – I
INORGANIC
CHEMISTRY

1

Atomic Structure

1.1 INTRODUCTION

Bohr achieved the major breakthrough in the understanding of structure of atom, by presenting his model of atom. He gave the concept of discrete energy levels (or orbits) in which the electrons can revolve. He arrived at these energy levels using Planck's quantum theory. According to Bohr's model of atom, as long as the electron moves in a particular orbit, it does not gain or lose energy. Bohr postulated that the angular momentum of an electron revolving around the nucleus has definite values which are whole number multiples of $h/2\pi$.

The appearance of various spectral lines in the emission spectra of hydrogen can be explained in terms of Bohr's theory of structure of atom. The Lyman series in the hydrogen spectrum are produced when the electrons jump from second, third, fourth or a higher energy level to the first energy level. The Balmer series in the hydrogen spectrum are obtained when the electrons jump from third, fourth or a higher energy level to the second energy level. Similarly, Paschen series in the hydrogen spectrum are obtained when the electrons jump from fourth, fifth or higher energy levels to the third energy level.

Niels Bohr (1885-1962) Danish Physicist 1922 Nobel Prize winner for contribution to atomic structure and quantum theory.

But, Bohr theory had to be abandoned because of many shortcomings. The main objection to Bohr's theory came from new ideas of de Broglie and Heisenberg.

Shortcomings of Bohr's theory

The main shortcomings of Bohr's theory are:

1. Bohr's atomic model failed to account for the effect of magnetic field on the spectra of atoms or ions. It was observed that when the source of a spectrum is placed in a strong magnetic field, each spectral line is further split into a number of lines. This is called **Zeeman effect**. This observation could not be explained on the basis of Bohr's model.

2. Bohr's theory of an atom was successful in accounting for the spectra of hydrogen and hydrogen-like atoms (*e.g.*, He^+, Li^{2+}). But, it failed to explain the spectra of multi-electron atoms. The spectroscopic measurements with high precisian revealed the fine structure in the line spectra of atoms. Bohr's theory could not explain the occurrence of these fine spectral lines.

3. Another objection of Bohr's theory came from Heisenberg's uncertainty principle. According to this principle, it is impossible to determine simultaneously the exact position and momentum of a small moving particle like an electron. The postulate of Bohr that electrons revolve in well defined orbits around the nucleus is, thus not tenable.

4. In 1923, de Broglie suggested that electron like light has a dual character. It has particle as well as wave character. Bohr treated the electron only as a particle.

1.1.1. DE BROGLIE HYPOTHESIS (DUAL NATURE OF MATTER AND RADIATION)

Earlier, it was thought that light is a stream of particles which are photons (or called corpuscles). However, this concept failed to explain the phenomena of interference and diffraction which could be explained only if light is considered to have wave nature. But at the same time, it was observed that the phenomena of black body radiation and photoelectric effect could be explained only if light is considered to have particle character. Therefore, it was concluded that light has a particle nature as well as wave nature *i.e.*, it has a dual nature.

Louis de Broglie, a French physicist, in 1924, advanced *the idea that like photons, all material particles such as electron, proton, atom, molecule, a piece of chalk, a piece of stone or an iron ball* (*i.e. microscopic as well as macroscopic objects*) *also possessed dual character.* The wave associated with a particle is called a **matter wave** or **de Broglie wave.**

The de Broglie Relation. The wavelength of the wave associated with any material particle was calculated by analogy with photon as follows :—

In case of a photon, if it is assumed to have wave character, its energy is given by

$$E = h\nu \qquad \text{... }(i)$$

(according to the *Planck's quantum theory*)

where ν is the frequency of the wave and h is Planck's constant.

If the photon is supposed to have particle character, its energy is given by

$$E = mc^2 \qquad \text{... }(ii)$$

(according to *Einstein equation*)

where m is the mass of photon and c is the velocity of light.

From equations (*i*) and (*ii*), we get

$$h\nu = mc^2$$

But $\quad \nu = c/\lambda$

$\therefore \quad h \cdot c/\lambda = mc^2 \qquad \text{... }(iii)$

or $\quad \lambda = h/mc$

de Broglie pointed out that the above equation is applicable to any material particle big or small. The mass of the photon is replaced by the mass of the material particle and the velocity c of the photon is replaced by the velocity v of the moving particle. Thus, for any material particle like electron, we may write

$$\lambda = h / mv$$

or $\quad \lambda = h / p \qquad \text{... }(iv)$

where $mv = p$ is the momentum of the particle.

The above equation is called **de Broglie equation.**

Significance of de Broglie equation. *Although the Broglie equation is applicable to all material objects but it has significance only in case of microscopic particles.* This is because the wavelength produced by a bigger particle (say a ball) using eq. (*iv*) come out to be too small to be observed. Only particles like electrons, atoms, etc. give an observable value of λ according to eq. (*iv*).

1.1.2. Heisenberg's Uncertainty Principle

Werner Heisenberg, a German physicist, in 1927 gave a principle about the uncertainties in simultaneous measurements of position and momentum of small particles. It is known as Heisenberg's uncertainty principle and it states as follows :

It is impossible to measure simultaneously the position and momentum of a small particle with absolute accuracy or certainty. If an attempt is made to measure any one of these two quantities with greater accuracy, the other becomes less accurate. The product of the uncertainty in the position (Δx) and the uncertainty in the momentum ($\Delta p = m.\ \Delta v$ where m is the mass of the particle and Δv is the uncertainty in velocity) is always constant and is equal to $h/4\pi$, where h is Planck's constant i.e.

$$\Delta x \cdot \Delta p \approx \frac{h}{4\pi} \qquad \text{... }(i)$$

Putting $\Delta p = m \times \Delta v$ eqn. (*i*) becomes

$$\Delta x \cdot (m\,\Delta v) = \frac{h}{4\pi} \qquad \text{... }(ii)$$

or

$$\Delta x \cdot \Delta v = \frac{h}{4\pi m} \qquad \text{... }(iii)$$

This implies that the *position and velocity of a particle cannot be measured simultaneously with certainty.*

Explanation of Heisenberg's uncertainty principle. Suppose we attempt to measure both the position and momentum of an electron. To locate the position of the electron, we have to use light so that the photon of light strikes the electron and the reflected photon is seen in the microscope (Fig. 1.1). As a result of the hitting by the photon, the position as well as the velocity of the electron are altered.

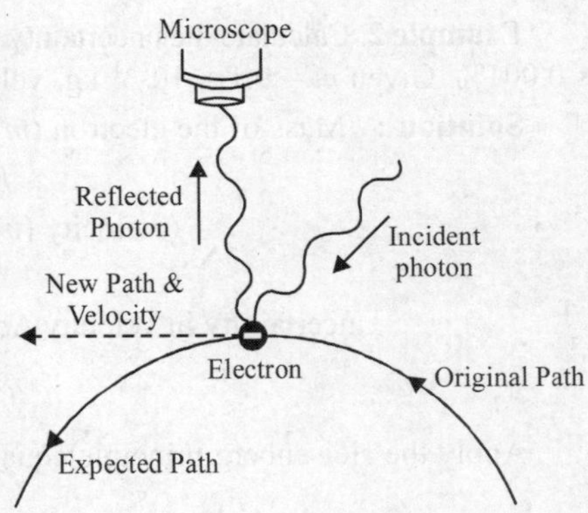

Fig. 1.1. Change of momentum and position of electron on impact with a photon

Significance of Heisenberg's uncertainty principle. Although Heisenberg's uncertainty principle holds good for all objects but *it is of significance only for microscopic particles.* Obviously the energy of the photon is insufficient to change the position and velocity of bigger bodies when it collides with them. For example, the light from a torch falling on a running ball in a dark room neither changes the speed of the ball nor its direction *i.e.* position.

This may be further illustrated by comparing $\Delta x \cdot \Delta v$ values for a particle of mass 1 *g* and a microscopic particle.

For a particle of mass 1g, we have

$$\Delta x \cdot \Delta v = \frac{h}{4\pi m} = \frac{6.626\times10^{-34}\ \text{kg}\,m^2 s^{-1}}{4\times3.1416\times(10^{-3}\ \text{kg})} \simeq 10^{-25}\ \text{m}^2\text{s}^{-1}$$

Thus the product of Δx and Δv is extremely small. Hence these values are negligible.

For a microscopic particle like an electron, we have

$$\Delta x \cdot \Delta v = \frac{h}{4\pi m} = \frac{6.626\times10^{-34}\ \text{kg}\,m^2 s^{-1}}{4\times3.1416\times(9.11\times10^{-31}\ \text{kg})} \simeq 10^{-4}\ \text{m}^2\text{s}^{-1}.$$

Thus if uncertainty in position is 10^{-8} m, uncertainty in velocity will be = 10^4 ms^{-1} which is quite significant.

Example 1. Calculate the wavelength associated with an electron of mass = 9.11×10^{-28} g travelling at 50% of the velocity of light. (Given $h = 6.63 \times 10^{-34}$ kg m^{2}s^{-1})

Solution : Use de Broglie relation

$$\lambda = \frac{h}{mv}$$

$$\text{Given that } m = 9.11 \times 10^{-31} \text{ kg}$$
$$h = 6.63 \times 10^{-34} \text{ kg m}^2\text{s}^{-1}$$
$$v = 0.5 \times 10^8 \text{ ms}^{-1}$$

Substituting the values in the de Broglie equation

$$\lambda = \frac{6.63\times10^{-34}\,\text{kg m}^2\text{s}^{-1}}{9.11\times10^{-31}\,\text{kg}\times0.5\times10^8\,\text{ms}^{-1}}$$

$$= \frac{6.63\times10^{-11}}{0.5\times9.11}\text{m} = 1.46 \times 10^{-11} \text{ m.}$$

Example 2. Calculate the uncertainty in the position of the electron if uncertainty in its velocity is 0.001%. Given $m = 9.1 \times 10^{-31}$ kg, velocity = 300 ms^{-1}. ($h = 6.6 \times 10^{-34}$ kg m^{2}s^{-1})

Solution : Mass of the electron (m) $= 9.1 \times 10^{-31}$ kg

$$h = 6.6 \times 10^{-34} \text{ kg m}^2\text{s}^{-1}$$
$$\text{velocity } (v) = 300 \text{ ms}^{-1}$$
$$\text{Uncertainity in velocity } \Delta v = \frac{0.001}{100}\times300\,\text{ms}^{-1}$$
$$= 3 \times 10^{-3} \text{ ms}^{-1}$$

Apply the Heisenberg principle equation and substitute the values

$$\Delta x \,.\, \Delta v = \frac{h}{4\pi m}$$

$$\Delta x \times 3\times10^{-3}\text{ ms}^{-1} = \frac{6.6\times10^{-34}\,\text{kg m}^2\text{s}^{-1}}{4\times3.14\times9.1\times10^{-31}\,\text{kg}}$$

or

$$\Delta x = \frac{6.6\times10^{-34}\,\text{kg m}^2\text{s}^{-1}}{3\times10^{-3}\,\text{ms}^{-1}\times4\times3.14\times9.1\times10^{-31}\,\text{kg}}$$

$$= \frac{6.6}{3\times4\times3.14\times9.1}\text{m}$$

$$= 0.019 \text{ m.}$$

Example 3. Find the number of waves made by a Bohr electron in one complete revolution in the third orbit.

Solution : According to Bohr

$$m\,v\,r = n\frac{h}{2\pi}$$

$$\text{or } n = \frac{mvr}{h/2\pi}$$

on rearranging the above equation

$$n = \frac{2\pi r}{h/mv} = \frac{2\pi r}{\lambda} = \frac{\text{Circumference}}{\lambda} = \text{No. of waves}$$

or No. of waves in one complete revolution = n

= 3 (Given n = 3, third orbit)

PROBLEMS FOR PRACTICE

1. A moving electron has 4.55×10^{-25} J of K.E. Calculate its wavelength. Given mass of electron = 9.1×10^{-31} kg and $h = 6.6 \times 10^{-34}$ kg m^{2}s^{-1}. (**Ans.** 7.25×10^{-7} m)
2. Calculate the wavelength associated with a neutron having a mass of 1.675×10^{-24} g and kinetic energy equal to 6.21×10^{-21} J. (**Ans.** 1.45×10^{-10} m)
3. Calculate the momentum of a particle having a wavelength of 0.1 nm. (**Ans.** 6.62×10^{-24} kg ms^{-1})
4. Calculate the minimum uncertainty in velocity of a particle of mass 1.1×10^{-27} kg if uncertainty in its position is 3×10^{-10} cm.
   Given $h = 6.6 \times 10^{-34}$ kg m^{2}s^{-1}. (**Ans.** 1.59×10^4 ms^{-1})
5. Calculate the uncertainty in velocity of an electron if uncertainity in its position is 1Å. (Given $m = 9.1 \times 10^{-31}$ kg, $h = 6.6 \times 10^{-34}$ kg m^{2}s^{-1}).

   Hint : 1 Å = 10^{-10} m. (**Ans.** 5.77×10^5 ms^{-1})

1.1.3 Derivation of Schrodinger wave equation

Schrodinger developed a mathematical wave equation to describe the wave motion of the electron in a hydrogen atom. Here the nucleus is surrounded by a vibrating electron wave which may be mathematically compared with stationary as standing wave formed by a vibrating string fixed between two points as shown in the figure. The electron in a hydrogen atom is under constraint imposed by the attraction of the nucleus. An electron motion can be described by an equation analogous to that used to describe a "standing or stationary" wave system whose solution in one dimension is given by the sine wave equation (i).

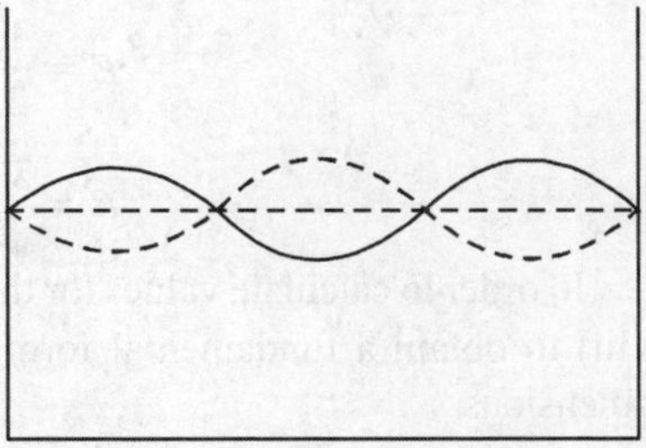

A standing wave.

$$\Psi = A \sin \frac{2\pi x}{\lambda} \quad \text{...(i)}$$

where Ψ = amplitude of the wave of wavelength λ at a distance x from the origin A is a constant.

On double differentiation w.r.t. x, we get

$$\frac{d\Psi}{dx} = A . \frac{2\pi}{\lambda} . \cos \frac{2\pi x}{\lambda} \quad \text{...(ii)}$$

$$\frac{d^2\Psi}{dx^2} = A . \frac{2\pi}{\lambda} . \frac{2\pi}{\lambda} \left[\frac{-\sin 2\pi x}{\lambda} \right]$$

$$= \frac{-4\pi^2}{\lambda^2} A \sin \frac{2\pi x}{\lambda} \quad \text{...(iii)}$$

Substituting $\Psi = A \sin\frac{2\pi x}{\lambda}$ from Eq. (i) into Eq. (iii), we get

$$\frac{d^2\Psi}{dx^2} = \frac{-4\pi^2}{\lambda^2}\Psi \qquad \text{...(iv)}$$

or
$$\frac{d^2\Psi}{dx^2} + \frac{4\pi^2}{\lambda^2}\Psi = 0 \qquad \text{...(v)}$$

This is the equation for one dimensional standing wave which can be extended to describe motion of an electron in three dimensions by a second order partial differential equation (vi)

$$\frac{\partial^2\Psi}{\partial x^2} + \frac{\partial^2\Psi}{\partial y^2} + \frac{\partial^2\Psi}{\partial z^2} + \frac{4\pi^2}{\lambda^2}\Psi = 0 \qquad \text{...(vi)}$$

where Ψ is a function of Cartesian coordinates *x, y, z.* This may be written more precisely as

$$\nabla^2\Psi + \frac{4\pi^2}{\lambda^2}\Psi = 0 \qquad \text{...(vii)}$$

where ∇^2 stands for $\frac{\partial^2\Psi}{\partial x^2} + \frac{\partial^2\Psi}{\partial y^2} + \frac{\partial^2\Psi}{\partial z^2}$

Combining de Broglie equation $\lambda = h/mv$ and eq. (vii), we have

$$\nabla^2\Psi + \frac{4\pi^2 m^2 v^2}{h^2}\Psi = 0 \qquad \text{...(viii)}$$

Considering the electron as a particle, the total energy (E) of the system is sum of kinetic ($1/2\ mv^2$) and potential (V) energies, *i.e.*,

$$E = \frac{1}{2}mv^2 + V$$

or
$$v^2 = \frac{2(E-V)}{m} \qquad \text{...(ix)}$$

In order to calculate values for the energy states value of v^2 (Eq. (ix)) is substituted into equation (viii) to obtain a fundamental form of Schrodinger wave equation for a single particle in three dimensions.

$$\nabla^2\Psi + \frac{8\pi^2 m}{h^2}(E-V)\Psi = 0 \qquad \text{...(x)}$$

or
$$\frac{\partial^2\Psi}{\partial x^2} + \frac{\partial^2\Psi}{\partial y^2} + \frac{\partial^2\Psi}{\partial z^2} + \frac{8\pi^2 m}{h^2}(E-V)\Psi = 0 \qquad \text{...(xi)}$$

where, Ψ = amplitude of the wave function associated with the electron

x, y, z = Cartesian coordinates
m = mass of the electron
h = Planck's constant
E = total energy of the system
V = potential energy of the system

Erwin Schrodinger

The potential energy term V in Schrodinger wave equation (x) is given by the following equation

$$V = \frac{-Ze^2}{r} \quad \text{...(xii)}$$

where $+Ze$ = charge of the nucleus

Z = atomic number

$-e$ = charge on the electron

Consequently, the equation (xi) for one electron atomic system in three dimensions may be written as

$$\frac{\partial^2\Psi}{\partial x^2}+\frac{\partial^2\Psi}{\partial y^2}+\frac{\partial^2\Psi}{\partial z^2}+\frac{8\pi^2 m}{h^2}\left[E+\frac{Ze^2}{r}\right]\Psi = 0 \quad \text{...(xiii)}$$

1.1.4 Application of Schrodinger wave equation to hydrogen atom

Schrodinger wave equation for one electron atomic system is

$$\frac{\partial^2\Psi}{\partial x^2}+\frac{\partial^2\Psi}{\partial y^2}+\frac{\partial^2\Psi}{\partial z^2}+\frac{8\pi^2 m}{h^2}\left[E+\frac{Ze^2}{r}\right]\Psi = 0$$

For hydrogen atom, Z= 1; therefore $V=\frac{-e^2}{r}$ and Schrodinger wave equation for H atom may be written as

$$\frac{\partial^2\Psi}{\partial x^2}+\frac{\partial^2\Psi}{\partial y^2}+\frac{\partial^2\Psi}{\partial z^2}+\frac{8\pi^2 m}{h^2}\left[E+\frac{e^2}{r}\right]\Psi = 0 \quad \text{...(xiv)}$$

Equation (xiv) results from the assumptions:

(*i*) that the behaviour of an electron in an atom is analogous to a system of standing wave; and

(*ii*) de Broglie relationship $\lambda=\frac{h}{mv}$ describes the wavelength of an electron.

Justification for the above equation for hydrogen atom is found in the fact that the solution of equation (xiv) gives values of energy which agree well with those obtained experimentally from atomic spectra or calculated from the equation (xv) given by Bohr for his model of the atom.

$$E = \frac{-2\pi^2 me^4 Z^2}{h^2 n^2} \quad \text{...(xv)}$$

That is, Schrodinger wave equation leads to exactly the same allowed energies as those deduced from the Bohr model. Besides that the "most probable" radius for 1*s* electron in H atom from wave mechanics comes out to be the same as determined by Bohr (0.529 Å) from equation (xvi) from Bohr model

$$r = \frac{h^2 n^2}{4\pi^2 me^2 Z^2} \quad \text{...(xvi)}$$

1.1.5 Physical significance of Ψ and Ψ^2

In all wave equations, the square of the wave function is the property which has physical significance. For example, in case of light waves, square of the amplitude of the wave at a point is proportional to the intensity of the light wave. Similarly in electron wave motion, square of Ψ (ψ^2) should represent intensity. According to uncertainty principle, the position of electron cannot be determined with certainty. All we can say is that greater the intensity of wave function at a particular point, greater is the probability of locating the electron at that point. Thus ψ^2 may be interpreted as being proportional to electron density. When ψ^2 is high, electron density is high, that is, the probability of finding an electron is high. If ψ^2 is low, probability is low. In order to avoid imaginary values, $\psi\,\psi^*$ is used for ψ^2. ψ^* is a complex conjugate of ψ. (See below) We expect that if ψ extends over a finite volume in space $dx.\,dy.\,dz\,(= d\tau)$, then the electron would be found in that volume. So $\psi\psi^*\,d\tau$ gives the probability of finding the electron in the volume $d\tau$. The wave function ψ may be positive, negative or imaginary but the probability density $\psi\psi^*$ will always be positive and real. The probability of fmding the electron is finite, which means $\psi\psi^*\,d\tau$ must give real values. If ψ has real value, then $\psi = \psi^*$ and we can use $\psi^2\,d\tau$ in place of $\psi\psi^*\,d\tau$.

1.1.6 Main conditions which Ψ must satisfy to give meaningful solution (Eigen function)

Ψ as a wave function must obey the following conditions if it is to give meaningful results:

(*i*) Ψ must be single valued, *i.e.*, it must have only one definite value at a particular point in space.

(*ii*) Ψ must be continuous at all points in space, *i.e.*, there must not be sudden changes in the values of ψ when its variables are changed.

(*iii*) Ψ must have a finite value over the space $d\tau$ that the electron can occupy.

(*iv*) Ψ must be normalized, *i.e.*, the probability of finding the electron in the space under consideration must be unity.

(*v*) Ψ must become zero at infinity.

This means that of all the values of Ψ obtained by solving Schrodinger wave equation, only those values of Ψ are accepted which obey the above conditions. Thus only acceptable values of Ψ have significance.

The significant values of Ψ are called **Eigen functions** or **wave functions**. Eigen functions give significant values of the total energy of electron (E) called Eigen values.

1.1.7 Physical significance of Ψ^*

Solution of Schrodinger wave equation gives Ψ values which can be real as well as imaginary but Ψ^2 is always real and has a positive value. This is because the square of even negative values is positive. If Ψ has imaginary values, then it can be expressed as

$$\Psi = a + ib$$

where a and b are real numbers and i (known as iota) $= \sqrt{-1}$. In such a case, Ψ is made real by multiplying it by its own complex conjugate (Ψ^*) which is expressed as

$$\Psi^* = a - ib$$

The product obtained by multiplying Ψ and Ψ^* for real values of Ψ is as:

$$\Psi\,\Psi^* = (a + ib)\,(a - ib) = a^2 - (i^2\,b^2)$$

$$= a^2 - (-1 \times b^2) \qquad (\because i = \sqrt{-1})$$

$$= a^2 + b^2 \text{ (which is real)}$$

Thus $\Psi\,\Psi^*$ becomes a real quantity.

However, if Ψ is real, then $\Psi = \Psi^*$, then $\Psi\Psi^* = \Psi^2$.

1.1.8 Origin of quantum numbers *n, l* and *m*

The three quantum numbers *n, l* and *m* and their allowed values arise as a consequence of the solution of the wave equation for hydrogen atom. These quantum numbers and their allowed values are discussed below:

(*i*) Principal or Radial Quantum Number, *n*

It arises from the solution of the radial part of Ψ and identifies the shell or level to which the electron belongs. These shells are regions in space indicating the "most probable" position of the electron from the nucleus. The quantum number n is restricted to have any positive integral values from 1 to infinity. The larger the value of n, the further is the shell from the nucleus.

$$n = 1, 2, 3, 4, 5 \ldots\ldots \infty$$

Theoretically there is no limit to the number of shells or levels but seventh shell ($n = 7$) is the highest shell occupied by the known elements.

Principal quantum number n alone does not give complete energy but designates the main energy level to which an electron belongs. The lowest energy state of an atom corresponds to $n = 1$ and the energy increases with increasing n.

(*ii*) Azimuthal or Subsidiary Quantum Number, *l*

It refers to the subshell or sublevel to which an electron belongs and describes the motion of the electron. The magnitude of the angular momentum of an electron (angular velocity × moment of inertia) is related to l by the expression:

$$\text{Angular momentum} = \sqrt{l(l+1)}\,\frac{h}{2\pi} \quad \ldots(1.17)$$

where $l = 0,\ 1,\ 2,\ 3,\ \ldots.\ (n-1)$

This means that the values of l for the subshells of a shell are governed by n, when $n = 1$, there is only one subshell, with the value $l = 0$, when $n = 2$, there are two subshells with the values $l = 0, 1$. When $n = 3$, there are three subshells that have the values $l = 0, 1, 2$. Similarly when $n = 4$, number of shells is 4 with values, $l = 0, 1, 2, 3$. For every value of l, there is a separate notation to indicate subshell. For example,

Quantum No.	$l = 0$	1	2	3	4	5 ......
Subshell Notation	$= s$	p	d	f	g	h.....

The first four notations : *s, p, d* and *f* are adjectives used earlier to identify spectral lines; **sharp, principal, diffuse and fundamental** but subsequent notations used for $l > 3$ proceed alphabetically *e.g., g, h ,i , j* ..... so on.

Knowing the value of n and finding out l, we designate a subshell. For example, subshell with $n = 1$, $l = 0$ is called 1*s* subshell and others are given as under:

n	*l*	*Subshell notation*
2	0	2 *s*
	1	2 *p*
3	0	3 *s*
	1	3 *p*
	2	3 *d*
4	0	4 *s*
	1	4 *p*
	2	4 *d*
	3	4 *f*

(*iii*) Magnetic Quantum Number, *m*

Each subshell is comprised of one or more orbitals (regions with maximum probability of finding the electron) whose number is equal to the number of ways the electrons in a subshell can orient themselves in space. The number of orientations is given by $2l + 1$. Thus, the number of orientations or orbitals is one when $l = 0$ (*s*-subshell), three for $l = 1$ (*p* -subshell), five for $l = 2$ (*d*-subshell) and seven for $l = 3$ (f-subshell). In other words, s-subshell has one s-orbital, p-subshell has three p-orbitals, d-subshell has five d-orbitals and so on. In the absence of electric or magnetic field, these orbitals in a subshell are degenerate, that is, they have identical energy. For example, for $l = 1$, the three orbitals P_x, P_y, and P_z are degenerate but this degeneracy is split in the presence of a magnetic field.

In this way three p-orbitals of different energies with the lowest energy orbital aligned with the magnetic field ($m = +1$) and five d-orbitals of different energies with the lowest aligned with magnetic field ($m = +2$) are obtained. Each orbital within a subshell is identified in term of the component of the angular momentum $m\dfrac{h}{2\pi}$ where m = magnetic quantum number. In other words, m is associated with different orientations of the orbital with reference to some defined directions.

For a given subshell, m can have values

$m = -l, -(l-1), -(l-2) \ldots.0\ldots.. (l-2), (l-1), l$

Thus for $l = 0$, $m = 0$, we get an s-orbital.

For $l = 1$, m can have three values $m = -1, 0, +1$ corresponding to three p-orbitals; p_x, p_y, p_z.

For $l = 2$, m can have five values $m = -2, -1, 0, +1, +2$ corresponding to five d-orbitals; $d_{x^2-y^2}, d_{xy}, d_{yz}, d_{zx}, d_{z^2}$.

Some salient features of these quantum numbers are given below:

(*i*) Radial quantum number n and its allowed values arise from the solution of radial part of ψ, whereas l and m quantum numbers arise from solution of angular part of ψ.

(*ii*) The lowest energy state of an atom corresponds to $n = 1$ and energy increases as n increases.

(*iii*) The number of subshells in a shell is equal to the value of n where $n = 2$, there are 2 subshells ($l = 0, 1$); for $n = 3$, there are 3 subshells ($l = 0, 1, 2$) and so on.

(*iv*) The number of orbitals of a type in a given subshell is given by $2l + 1$. That is, we have one s-type orbital of $l = 0$; three p-type orbitals for $l = 1$; five d-type orbitals for $l = 2$ and seven f-type orbitals for $l = 3$.

(*v*) The total number of orbitals in a given shell = n^2 (n = principal quantum number). For example for $n = 2$, the total number of orbitals = $2^2 = 4$.

Spin Quantum Number m_s

In 1896, Zeeman observed that spectral lines in the atomic spectrum get split under the influence of a strong magnetic field. This phenomenon, called the **Zeeman effect**, is attributed to the orientation of the components of the angular momentum with respect to the external magnetic field and is given by $m\dfrac{h}{2\pi}$, where m = magnetic quantum number.

Later on it was observed in the atomic spectra of alkali metals that the spectral lines which were earlier considered to be single lines are actually narrow doublets (two lines quite close together). An explanation to these doublet lines was offered by Uhlenbeck and Goudsmit in 1925. They proposed that the electron is also associated with rotation about its own axis (spinning). Axial spin of the electron is associated with spin angular momentum whose magnitude is given by $m_s\dfrac{h}{2\pi}$, where m_s is called the spin quantum number. The spin quantum number can have values $+\dfrac{1}{2}$ or $-\dfrac{1}{2}$ which arise from the direction of spin (clockwise or anticlockwise). It is also designated as ↑ or ↓. Thus, the description of an electron in an atom in terms of three quantum numbers (described earlier) is not complete and, in fact, four quantum numbers, *n, l, m* and m_s: are needed to describe the electron in an atom completely.

An electron is a spinning negative charge and a spinning charge is magnetic, therefore, a spinning electron can be considered to behave like a tiny magnet. Two electrons with opposite spins in an orbital will behave as two tiny magnets with opposite poles towards each other and thus attract each other.

***Example 1.* Which sets of quantum numbers are not permissible and why?**

(*i*) $n = 5,\ l = 4,\quad m = 0,\ m_s = \frac{1}{2}$

(*ii*) $n = 3,\ l = 0,\quad m = -1,\ m_s = -\frac{1}{2}$

(*iii*) $n = 3,\ l = 1,\quad m = 2,\ m_s = \frac{1}{2}$

(*iv*) $n = 2,\ l = 2,\quad m = 0,\ m_s = \frac{1}{2}$

Solution. (*i*) This is allowed.

(*ii*) It is not allowed because value of *m* (+ 1, ... 0 ... –1) are governed by *l* and when *l*=0, *m* cannot have values other than zero (*m* = 0).

(*iii*) It is not allowed because when *l* = 1, *m* has allowed values = + 1, 0, –1 only.

(*iv*) It is not allowed because values of *l* are governed by *n*. For *n* = 2, 1 can have values equal = 0, 1 only. Therefore for *n* = 2, *l* = 2 is not permissible.

***Example 2.* Explain why orbitals like 1*p*, 2*d* and 3*f* are not possible.**

Solution. This is because the sets of quantum numbers required for these orbitals are not allowed. 1*p* refers to *n* = 1, *l* = 1 which is not allowed because values of *l* are governed by *n,* that is *l*= 0, 1, 2..... (*n* – 1). Thus, for *n* = 1, *l* cannot have a value equal to one.

Similarly, 2*d* refers to *n* = 2, *l* = 2 which again is not possible. For *n* = 2; *l* can have values 0, 1 only.

Also 3 *f* is not possible because for 3 *f,* the combination *n* = 3, *l* = 3 is again not allowed. For *n* = 3, *l* can't have a value higher than 2.

***Example 3.* Calculate and identify all the individual orbitals for (*i*) *n* = 3 (*ii*) *n* = 4.**

Solution. (*i*) *n* = 3; *l* = 0, 1, 2; Total no. of orbitals = 9

These are: $3s;\ 3p_x,\ 3p_y;\ 3p_z;\ 3d_{xy},\ 3d_{yz},\ 3d_{zx},\ 3dz^2,\ 3d_{x^2-y^2}$

(*ii*) For *n* = 4; *l* = 0, 1, 2, 3. Individual orbitals are:

l	*m*	*Orbitals*
0	0	$4s$
1	± 1, 0	$4p_x,\ 4p_y,\ 4p_z$
2	± 2, ± 1, 0	$4d_{xy},\ 4d_{yz},\ 4d_{xz},\ 4dz^2,\ 4d_{x^2-y^2}$
3	± 3, ± 2, ± 1, 0	$4f_{x^3}, 4f_{y^3}, 4f_{z^3}, 4f_{xyz}\ 4f_{(z^2-y^2)}, 4f_{(z^2-x^2)}\ 4f_{(x^2-y^2)}$

For *n* = 4; total number of orbitals = 1 + 3 + 5 + 7 = 16.

Example 4. **Draw the shapes of each of the following orbitals $2p_y, 3d_{z^2}, 3d_{x^2-y^2}, 3d_{xy}$.**

Solution. Shapes of the orbitals are as shown in the figure below.

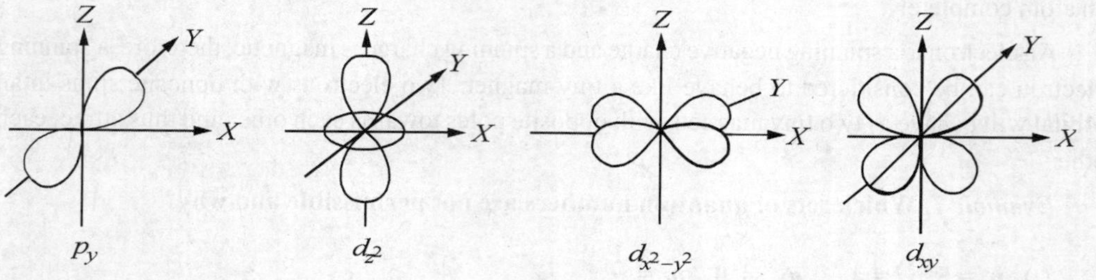

Fig. 1.2

Example 5. **Which of the following sets of quantum numbers are not allowable and why?**

(*i*) $n = 2 \quad l = 1 \quad m = 0 \quad s = +\frac{1}{2}$

(*ii*) $n = 2 \quad l = 2 \quad m = -1 \quad s = -\frac{1}{2}$

(*iii*) $n = 2 \quad l = 1 \quad m = +1 \quad s = -\frac{1}{2}$

(*iv*) $n = 3 \quad l = 2 \quad m = +1 \quad s = 0$

(*v*) $n = 2 \quad l = 0 \quad m = 0 \quad s = +\frac{1}{2}$

Solution. Quantum numbers in (*ii*) and (*iv*) are not allowable because the values of l exceeds the maximum permissible value of $(n-1)$.

1.1.9 Schrodinger wave equation in terms of polar coordinates

Schrodinger wave equation as given by the following equation is in terms of cartesian coordinates (expressed in terms of x, y, z coordinates).

$$\frac{\partial^2 \psi}{\partial x^2} + \frac{\partial^2 \psi}{dy^2} + \frac{\partial^2 \psi}{\partial z^2} + \frac{8\pi^2 m}{h^2}(E - V)\,\psi = 0$$

However, for the solution of this equation, it is more convenient to transform the equation in terms of spherical polar co-ordinates (expressed in terms of r, θ and ϕ).

The spherical polar co-ordinates are related to cartesian co-ordinates as shown in Fig. 1.3. The location of a point P with respect to the point O may be given in terms of polar co-ordinates (r, θ and ϕ). Here r is the distance from the origin (O) called radial distance, θ is the angle which radial line OP makes with the Z-axis and ϕ is the angle which the projection of the radial line OP on XY plane makes with the X-axis. The relationship between the cartesian co-ordinates and polar co-ordinates can be obtained by simple geometry. The co-ordinates x, y and z are related to r, θ and ϕ as:

$x = r \sin \theta$ and $\cos \phi$

$y = r \sin \theta \sin \phi$

and $z = r \cos \theta$

Substituting the values of x, y and z in the Schrodinger wave equation (Eq. 1.13), we get

Schrodinger wave equation in terms of polar coordinates (r, θ and φ) is obtained as under :

$$\frac{1}{r^2}\frac{\partial}{\partial r}\left(r^2\frac{\partial\psi}{\partial r}\right)+\frac{1}{r^2\sin\theta}\frac{\partial}{\partial\theta}\left(\sin\theta\frac{\partial\psi}{\partial\theta}\right)$$

$$+\frac{1}{r^2\sin^2\theta}\frac{\partial^2\psi}{\partial\phi^2}+\frac{8\pi^2\mu}{h^2}(E-V)\,\psi=0 \qquad ...(1.18)$$

where μ is reduced mass and is equal to $\dfrac{m_1 m_2}{m_1+m_2}$

Here m_1 = mass of the electron, m_2 = mass of the nucleus.

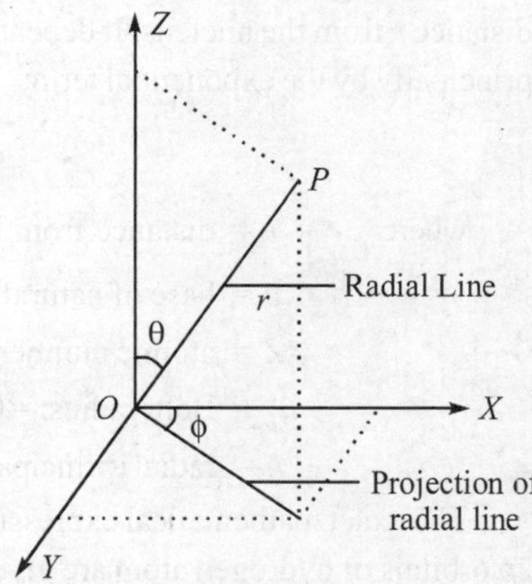

Fig. 1.3. The relation of cartesian coordinates (x, y and z) and polar co-ordinates (r, θ and φ).

The above equation may be solved by separating the variables r, θ and φ. The solutions $\psi(r, \theta, \phi)$ may be assumed to be product of three functions: $R(r)$ which depends upon distance r alone, $\Theta(\theta)$ which depends upon angle θ alone and $\Phi(\phi)$ which depends upon angle φ alone. Thus,

$$\psi(r,\theta,\phi)=\mathrm{R}(r)\,\Theta(\theta)\,\Phi(\phi)$$

We obtain three simpler equations each having one variable only and can be easily solved.

During the solution of Schrodinger wave equation, certain constants are required, known as quantum numbers. For example, the wave function $\mathrm{R}(r)$ depends on two numbers n and l, the wave function, $\Theta(\theta)$ depends upon two numbers l and m while the wave function $\Phi(\phi)$ depends on m only. This may be expressed as:

$$\psi(r,\theta,\phi)=\mathrm{R}_{n,l}(r)\,\Theta_{l,m}(\theta)\,\Phi_m(\phi)$$

The above equation shows that the wave function may be represented as the product of three functions, one of which depends only on distance (r) from the nucleus and is determined by two quantum numbers n and l; and other two which depend on the direction (angles θ, φ) and are determined by quantum numbers l and m. The first is called *radial wave function* and the second is called *angular wave function*. Thus, the solution of wave function may be expressed in terms of two parts:

1. **Radial wave function.** *The radial wave function gives the distribution of the electron as a function of the distance* (r) *from the nucleus*. It depends on the values of two quantum numbers n and l and may be expressed as:

Radial wave function = $\mathrm{R}_{n,l}(r)$

2. **Angular wave function.** The product $\Theta(\theta)$ and $\Phi(\phi)$ is called angular wave function. It depends upon the direction given by angles θ and φ. Therefore, *it determines the shape of the electron cloud or orbital and depends on the values of quantum numbers l and m.* The angular wave function may be expressed as:

$$\Theta(\theta)\,\Phi(\phi) \text{ or simply as } \mathrm{Y}(\theta,\phi)$$

Thus, the complete wave function may be expressed as:

$$\psi(r,\theta,\phi)=\mathrm{R}(r)\,\mathrm{Y}(\theta,\phi)$$

The complete wave function represents an orbital in which the radial part gives its size and the angular part determines its shape.

1.1.10 Distribution curves for radial wave function of 1*s*, 2*s* and 3*s* orbitals

The radial part of the wave function $R(r)$ gives the distribution of the electron with respect to its distance r from the nucleus. It depends only on the value of quantum numbers n and l. It is governed principally by the exponential term:

$$e^{-Zr/na_0}$$

where r = distance from the nucleus

e = base of natural logarithm

Z = atomic numbers; $Z = 1$ for H

a_o = Bohr radius; = 0.529 Å (52.9 pm) for 1s in H atom

n = radial (principal) quantum number

The exact mathematical expressions (1.19 –1.20) for radial part of wave functions of 1s, 2s, and 2p orbitals of hydrogen atom are given in Table 1.1.

Table 1.1. Radial part of Ψ for orbitals of hydrogen atom.

Quantum Nos. n	l	*Orbitals*	*Radial part of the wave function*
1	0	1s	$R_{(r)} = 2\left(\frac{Z}{a_0}\right)^{3/2} e^{-Zr/a_0}$ ...(1.19)
2	0	2s	$R_{(r)} = \left(\frac{Z}{2a_0}\right)^{3/2}\left(2-\frac{Zr}{a_0}\right)e^{-Zr/2a_0}$ ...(1.19(a))
2	1	2p	$R_{(r)} = \frac{1}{2\sqrt{6}}\left(\frac{Z}{a_0}\right)^{3/2}\left(\frac{Zr}{a_0}\right)e^{-Zr/2a_0}$ ...(1.20)

The radial wave function may be represented as under:

1. $R_{(r)}$, which is r dependent part of the wave function Ψ, is plotted against distance r from the nucleus as shown for 1s, 2s, 2p, 3s and 3p orbitals for the hydrogen atom in Figs. 1.4 to 1.8.

A look at expression 1.19 (Table 1.1) indicates that $R_{(s)}$ of Ist orbital depends on the exponential term.

$$e^{-Zr/2a_0}$$

(because other terms in expression 1.19 are constant; $z = 1$ for H and $a_0 = 52.9$ pm)

This shows that $R_{(s)}$ should decrease as r increases. This is reflected in Fig.1.4 showing a decrease in radial wave function $R_{(s)}$ with increasing distance (r) from the nucleus for 1s orbital of H atom.

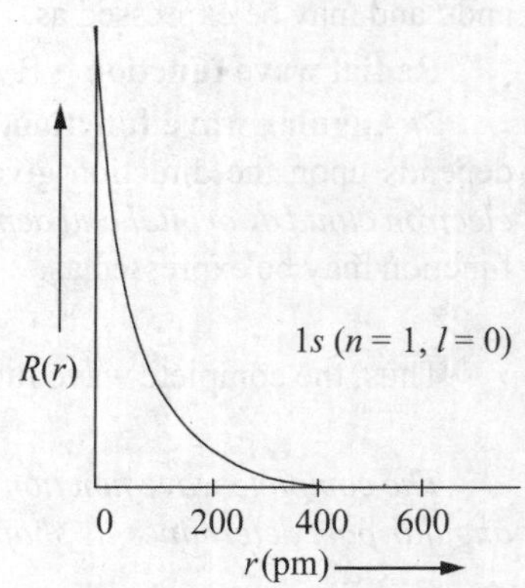

Fig. 1.4. Radial part of the wave function for 1s of H atm.

Expression 1.19a indicates that the radial wave function for 2s orbital is governed primarily by the term

$$\left[2-\frac{Zr}{a_0}\right]e^{-Zr/2a_0}$$

As r increases, the value of $R_{(s)}$ for 2s (Fig. 1.5) decreases and becomes zero at $r = \frac{2a_0}{Z}$ as calculated below:

$$\left[2-\frac{Z}{a_0}\cdot\frac{2a_0}{z}\right]e^{-\frac{z}{2a_0}\cdot\frac{2a_0}{z}}$$

$$R_{(s)} = (2-2)\,e^{-1} = 0$$

The point at which $R_{(s)} = 0$ is called the **node**. That is, the $2s$ orbital has a radial node which is a spherical surface on which $R_{(s)} = 0$. It is at a distance $= \frac{2a_0}{z}z$ from the nucleus.

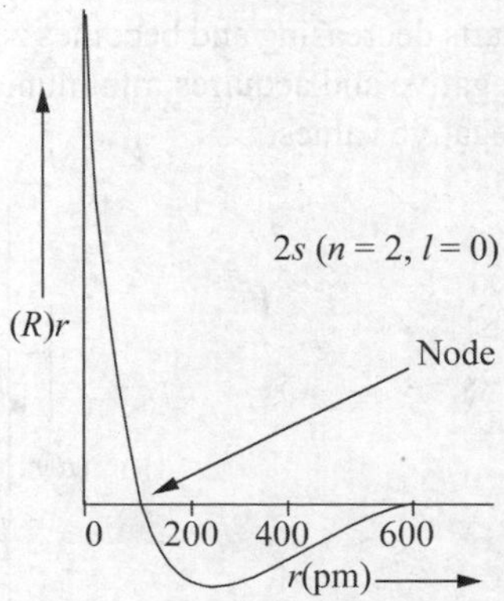

Fig. 1.5. Radial part of the wave function for 2s of H atom.

At r greater than $2a_0$ ($Z = 1$ for H atom), the radial wave function acquires a negative value and decreases to a minimum value before increasing again to an infinitesimally small negative value as shown in Fig. 1.5 for $2s$. The $2s$ orbital has one node.

The radial dependence of $3s$ is shown in Fig. 1.6. The radial function acquires zero value at two points and gets a positive value beyond second node. It acquires a maximum value before decreasing to an infinitesimally small positive value. Fig. 1.6 orbital has two nodes.

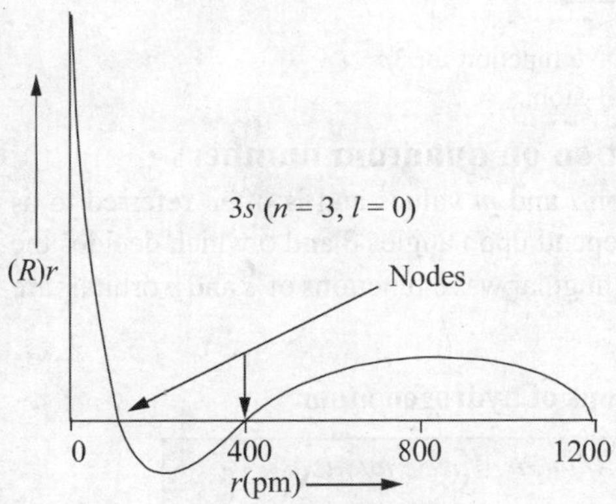

Fig. 1.6. Radial part of the wave function for 3s atomic orbital of H-atom.

Orbital **1*s*** has **no** nodes, **2*s*** has **one** node, **3*s*** has **2** nodes and so on. In general s-orbitals have $(n - 1)$ nodes.

1.1.11 Distribution curves for radial wave function of $2p$ and $3p$ orbitals

Expression 1.20 (Table 1.1) shows that the radial wave function for $2p$ orbital of H-atom ($Z = 1$) depends mainly on the term.

$$\left[\frac{r}{a_0}\right]e^{-\frac{r}{2a_0}}$$

The above expression does not become zero (except at $r = 0$) and hence the curve has no nodes as shown in Fig. 1.12. As r increases, the radial wave function $R_{(s)}$ increases to a maximum and then decreases before becoming infinitesimally small. It is positive throughout as shown in Fig. 1.7.

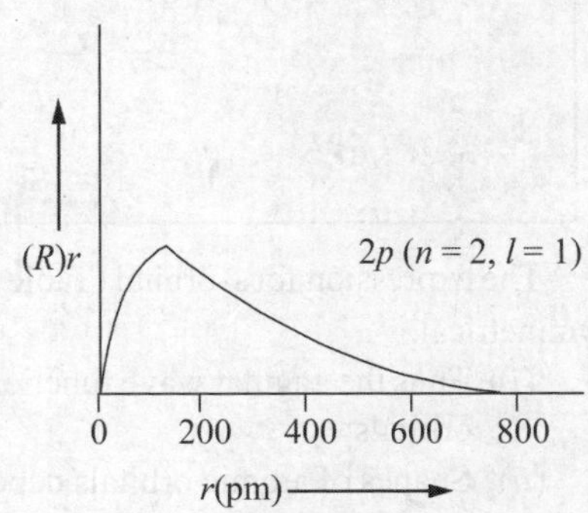

Fig. 1.7. Radial part of the wave function for $2p$atomic orbital of H-atom.

The radial wave function of $3p$ is shown in Fig. 1.8. The value of radial wave function for $3p$ shows an increase in the beginning and after acquiring maximum value, it starts decreasing and becomes zero to indicate a radial node. Beyond that, $R_{(s)}$ decreases to become negative and acquires minimum negative value before increasing again to an infinitesimally small negative values.

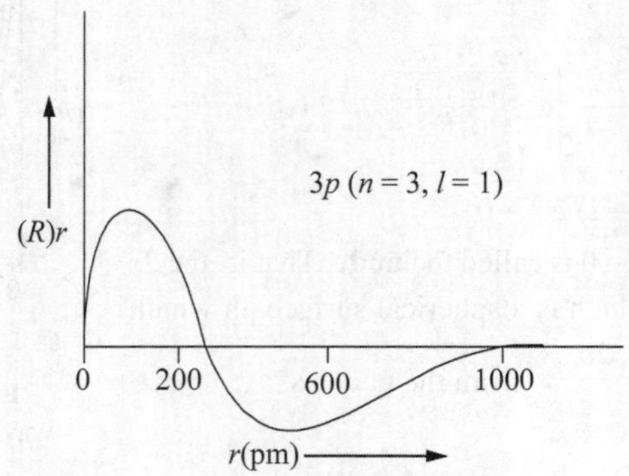

Fig. 1.8. Radial part of the wave function for 3*p* atomic orbital of H-atom.

1.1.12 Dependence of angular wave function on quantum numbers

The angular part of the wave function depends on *l* and *m* values and is often referred to as atomic orbitals. The shapes of these atom is orbitals depend upon angles θ and ϕ which decides the geometry of the orbitals. Mathematical expressions for angular wave functions of *s* and *p* orbitals are given in Table 1.2.

Table 1.2. Angular wave functions of hydrogen atom.

Quantum *l*	*Nos.* *m*	*Orbitals*	*Angular Wave Functions*
0	0	*s*	$\left(\frac{1}{4\pi}\right)^{1/2}$
1	0	p_z	$\left(\frac{3}{4\pi}\right)^{1/2} \cos\theta$
1	1	p_x	$\left(\frac{3}{4\pi}\right)^{1/2} \sin\theta \cos\phi$
1	–1	p_y	$\left(\frac{3}{4\pi}\right)^{1/2} \sin\theta \cos\phi$

The expression for *s*-orbital (Table 1.2) is independent of angles θ and ϕ and hence is spherically symmetrical.

(*i*) Thus the angular wave function depends on *l* and *m* values and is often referred to as atomic orbitals;

(*ii*) Shapes of atomic orbitals depends on angles θ and ϕ and determine geometry of the atom; and

(*iii*) *s*-orbitals have no angular dependence and hence spherically symmetrical.

***Example 6.* State the number of nodes in $R_{(r)}$ plots of**

(i)* 3*s ***(ii)* 2*p*** ***(iii)* 4*d*** ***(iv)* 5*f***

Solution. *s* orbitals have $(n-1)$ nodes, *p* orbitals have $(n-2)$ nodes, *d* orbitals have $(n-3)$ nodes and *f* orbitals have $(n-4)$ nodes. Based upon this generalisation

3*s* orbital has $(3-1)$ *i.e.*, 2 nodes.

2*p* orbital has $(2-2)$ *i.e.*, no nodes.

4*d* orbital has $(4-3)$ *i.e.*, 1 node.

5*f* orbital has $(5-4)$ *i.e.*, 1 node.

***Example 7.* Draw the plot of Ψ^2 and *r* for *s* and *p* orbitals.**

Solution. *R*(*r*) plots of orbitals like 2*s*, 3*s*, and 3*p* show negative values of radial distribution. To avoid negative values, variation of Ψ^2 is plotted as a function of *r*. Here negative part of radial distribution becomes positive in Ψ^2 plots.

Such plots known as probability electron distribution curves are given for 1*s*, 2*s* and 2*p* orbitals of hydrogen atom in Fig. 1.9.

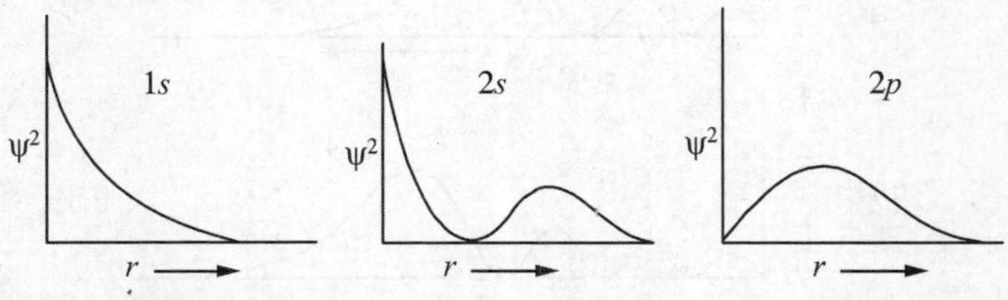

Fig. 1.9. Variation of Ψ^2 (probability of finding an electron) in 1*s*, 2*s* and 2*p* orbitals of H-atom.

1.1.13 Radial probability distribution curves of 1*s*, 2*s*, 3*s* and 4*s* orbitals

These are obtained by evaluating $\Psi^2\, d\tau$ in thin spherical shell, at different distance (*r*) from the nucleus. Considering the orbital to consist of layers of thin concentric spherical shells, one such shell of thickness $d\tau$ between distance *r* and $r + dr$ from origin is shown in Fig. 1.10. The volume of such a spherical shell is $4\pi r^2 dr$ and the probability of finding the electron in this volume element would be equal to $\Psi^2 \times 4\pi r^2\, dr$. A radial probability distribution curve of 1*s*, obtained by plotting relative probabilities of finding electron in this spherical shells at various distances from the nucleus, is shown in Fig. 1.11. The probability of finding the electron is highest close to the nucleus as shown in Fig. 1.9 but the volume element $4\pi r^2 dr$ is very small as *r* is very small and hence the product $4\pi r^2 dr \Psi^2$ is small. On the other hand, as distance increases, the volume element $4\pi r^2 dr$ increases but the probability

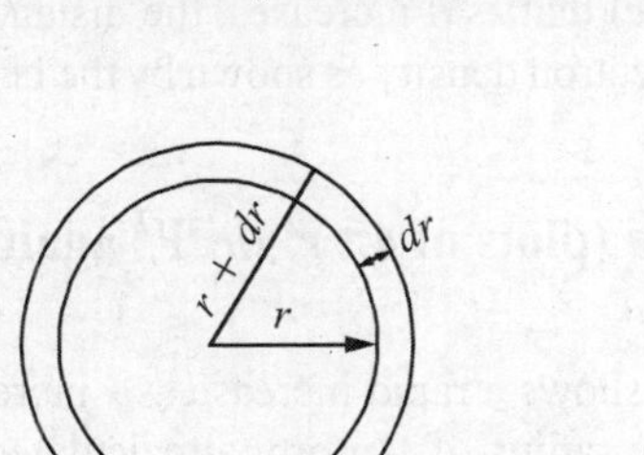

Fig. 1.10. Volume of shell of thickness *dr* around the nucleus = 4p*r*²*dr*.

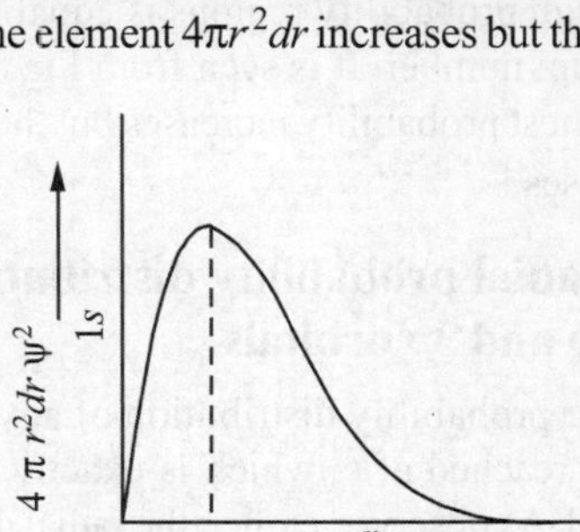

Fig. 1.11. Radial probability distribution curve of 1*s*.

of finding the electron Ψ^2 decreases and becomes negligible at greater distances. As shown in Fig. 1.11, the radial probability distribution of 1*s* is zero at the nucleus but increases gradually reaching a maximum at $r = r_0$ which is 52.9 pm, equal to Bohr radius for a hydrogen atom. The distance r_0 is called radius of maximum probability. Beyond r_0, the radial probability distribution of 1*s* shows a gradual decrease with value becoming infinitesimally small at larger distances from nucleus.

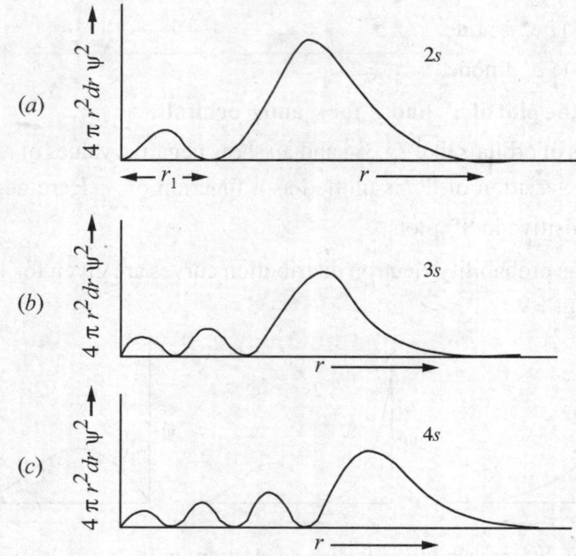

Fig. 1.12. Radial probability distribution curves for 2*s*, 3*s* and 4*s* orbitals.

Solution of the Schrodinger wave equation for $n = 2$, $l = 0$, gives an expression Ψ_{2s} whose probability distribution plot is drawn in Fig. 1.12 (*a*). Plots of $4\pi r^2 dr \Psi^2$ against *r* for 3*s* and 4*s* are given in Fig. 1.12(*b*) and (*c*). Plot for 2*s* has two regions of higher probability separated by a node.

For $n = 2$, the number of high probability regions = 2 and number of node = 1

Similarly, for $n = 3$, $l = 0$ the probability plot is as shown in Fig. 1.12 (*b*). There are three regions of high probability and two nodes. For 4*s*-orbital, there are four regions of high probability and three nodes as shown in Fig. 1.12 (*c*). It may therefore be inferred that for an *s*-orbital, the number of high probability regions is equal to *n* and number of nodes is equal to $n - 1$, where *n* is radial quantum number. It is seen from Fig. 1.12 (*a, b, c*) that as *n* increases, the distance *r* of the region of highest probability increases but the average electron density as shown by the height of the hump decreases.

1.1.14 Radial probability distribution curves (plots of $4\pi r^2 dr \Psi^2$ against *r*) for 2*p* and 3*p* orbitals

Radial or probability distribution of any 2*p*-orbital shows a rapid increase as *r* increases till a maximum is reached at r_0 which is exactly equal to the radius of 2*p*-orbital calculated by Bohr [Fig. 1.13 (*a*)]. As observed earlier, the radial distribution curve for 3*p*-orbital shows a node separating the regions of high probability [Fig. 1.13 (*b*)]. It may be stated that for any *p*-orbital, the number of regions of high probability is equal to $n - 1$ and number of nodes is equal to $n - 2$.

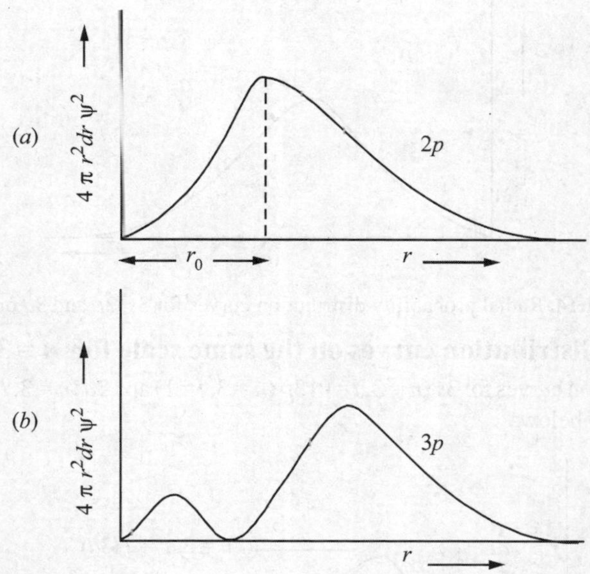

Fig. 1.13. Radial probability distribution curves for $2p$ and $3p$ orbitals.

1.1.15 Comparison of radial probability distribution curves for 3*s*, 3*p* and 3*d* in hydrogen atom

Probability distribution curves for $n = 3$ in a hydrogen atom are given in Fig. 1.14 (*a*), 1.14 (*b*) and (*c*). Figure shows that the largest maximum for $3s$ is more extended in space than that of $3p$ which in turn is more extended than $3d$. That is, the most probable radius decreases in the order $3s > 3p > 3d$.

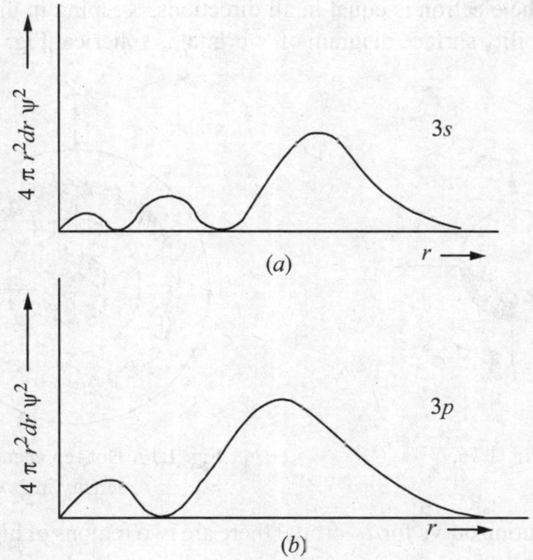

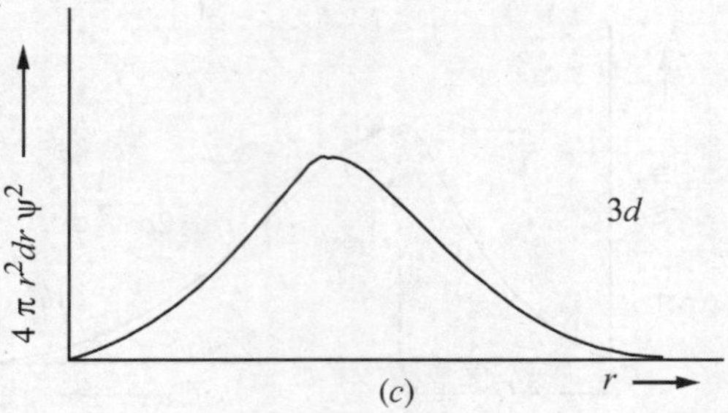

Fig. 1.14. Radial probability distribution curves for 3*s*, 3*p* and 3*d* orbitals.

1.1.16 Radial distribution curves on the same scale for $n = 3, l = 0, 1, 2$

Radial distribution curves for 3*s* ($n = 3, l = 0$) 3*p* ($n = 3, l = 1$) and 3*d* ($n = 3, l = 2$) on the same scale are given in Fig. 1.15 below.

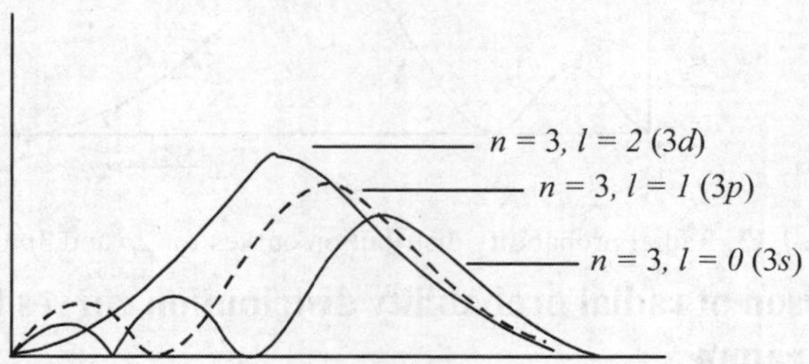

Fig. 1.15. Radial distribution curves for 3*s*, 3*p* and 3*d* on the same scale.

1.1.17 Shape of *s*-orbitals, node and nodal surface

The expression for angular wave function for *s*-orbital is not dependent on either of the angles θ or ϕ. In other words, the angular part of the wave function is spherically symmetrical. That is, the probability of finding the electron is equal in all directions. Keeping in mind that atoms are three dimensional, the probability surface diagram of *s*-orbital is spherical Fig. 1.16.

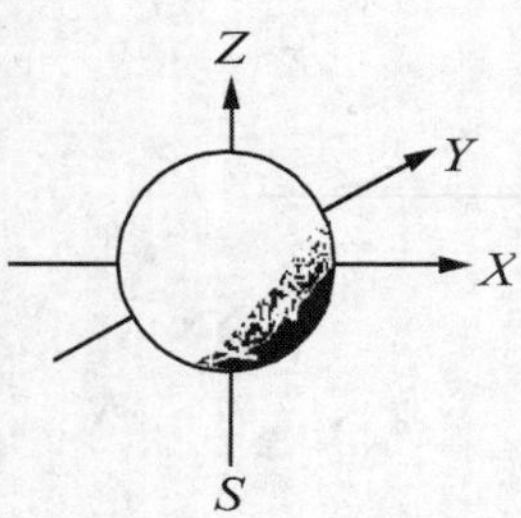

Fig. 1.16.

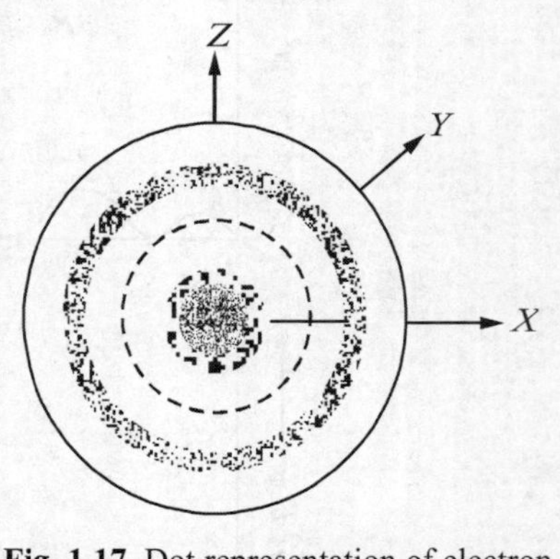

Fig. 1.17. Dot representation of electron density in 2*s* orbital.

In the radial distribution curve for 2*s*-orbital there are two regions of high probability separated by a node at $r = r_1$. Fig. 1.17 for 2*s* shows two regions of higher dot density separated by a spherical nodal surface (shown as dotted circle) at distance r_1 from the nucleus.

Such diagrams showing nodal surfaces will be avoided because by doing so we are extending the mathematical model to too much of simplification. The wave model prevents us firstly to locate the electron as a particle in one region and then explain its motion to the other region through an intervening nodal surface. For such a visualization, the electron as standing wave would be more appropriate because such a wave can exist simultaneously on both sides of nodes.

1.1.18 Shape of *p*-orbitals

For $l = 1$, m has three values, ± 1, 0 representing the three orientations of *p*-orbitals, p_x, p_y, p_z.

Angular wave function for $l = 1$, $m = 0$ is labelled as p_z, and the expression $\left[\frac{3}{4}\pi\right]^{1/2}$ cos θ gives two tangent dumb-bells symmetrical about Z-axis. In other words, the angular part of the probability distribution is concentrated along Z-axis and the resulting surface of p_z, electron has the form of a dumb-bell along the Z-axis. The dumb-bell shaped lobes of high probability are separated by a nodal surface in XY-plane. Other two orbitals p_x and p_y have similarly been found to have dumb-bell shapes along X and Y-axis, respectively as shown in Fig. 1.18. The electron in a *p*-orbital is found with equal probability in either of the lobes. The three *p*-orbitals p_x, p_y and p_z, together having an electron in each one of these give a spherical distribution of electron density.

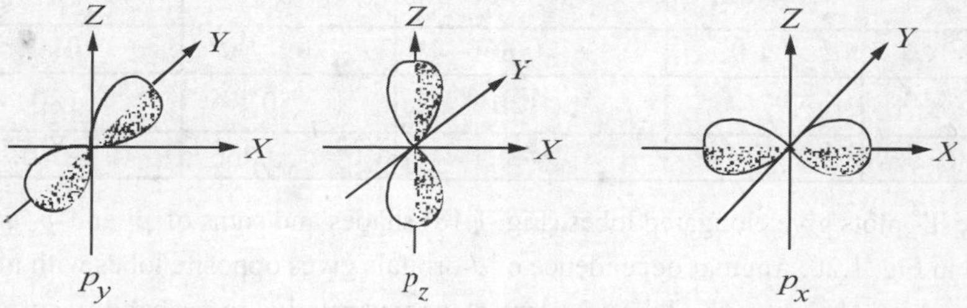

Fig. 1.18. Shape of *p*-orbitals.

1.1.19 Shape of *d*-orbitals

For $l = 2$, m has five values 0, ± 1, ± 2 representing the five orientations of *d*-orbitals designated as $d_{z^2}, d_{x^2-y^2}, d_{xy}, d_{yz}, d_{zx}$, as shown in Fig. 1.19. Their angular dependence is more complicated than that of *p*-orbital. Probability surface diagrams of three orbitals d_{xy}, d_{yz} and d_{xz} have four lobes of high electron density in *xy, yz* and *zx* planes respectively with each lobe bisecting the principal axis into two equal halves.

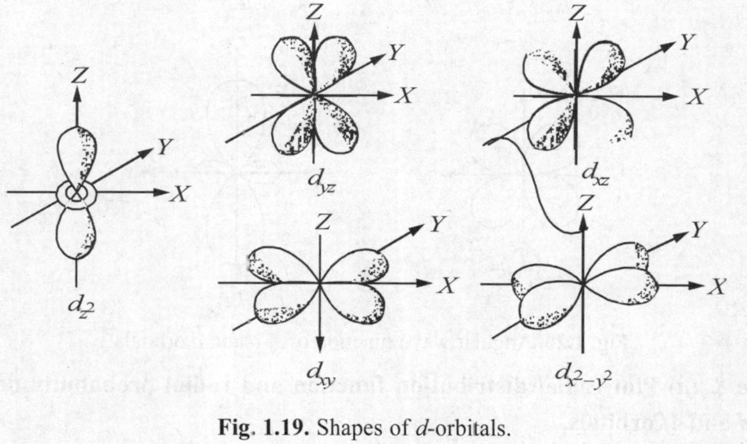

Fig. 1.19. Shapes of *d*-orbitals.

The fourth *d*-orbital $d_{x^2-y^2}$ has four lobes of electron density in *xy* plane along the *x* and *y* axes.

The fifth *d*-orbital d_{z^2} has a different shape. It has two lobes along the *z*-axis with a ring of electron density around the nucleus in *xy*-plane. The five orbitals together having an electron in each one of these give a spherical distribution of electron density. Each of the *d*-orbitals possesses two nodal planes.

1.1.20 Sign of angular wave function

It may be pointed out that the angular part of wave function Ψ may be positive or negative depending on θ and φ. Angular part of the wave function for hydrogen like orbitals is given in Fig. 1.20. The *s*-orbital is positive all through because it is independent of θ and φ and there is no change in its shape which remains spherical. The *p*-orbital has one lobe +ve and the other –ve because the angular part of, say p_z, depends on cosθ which has +ve sign between θ – 90° and 270 – 360° and –ve sign between 90 –180 and 180 –270 along *z*-axis. The *p*-orbitals (Fig. 1.20) have spherical lobes;

θ	0	30	60	90
cosθ	1.0	0.866	0.5	0
θ		120	150	180
cosθ		– 0.5	– 0.866	–1.0

whereas the Ψ^2 plots give elongated lobes (Fig. 1.18) shapes and signs of p_y and p_x are similar to p_x as shown in Fig. 1.20. Angular dependence of *d*-orbitals gives opposite lobes with identical sign (Fig. 1.20) and the shapes of the lobes are less elongated relative to probability surface diagram (Fig. 1.19). The true significance of signs is realised when orbitals combine to form bonds in molecules.

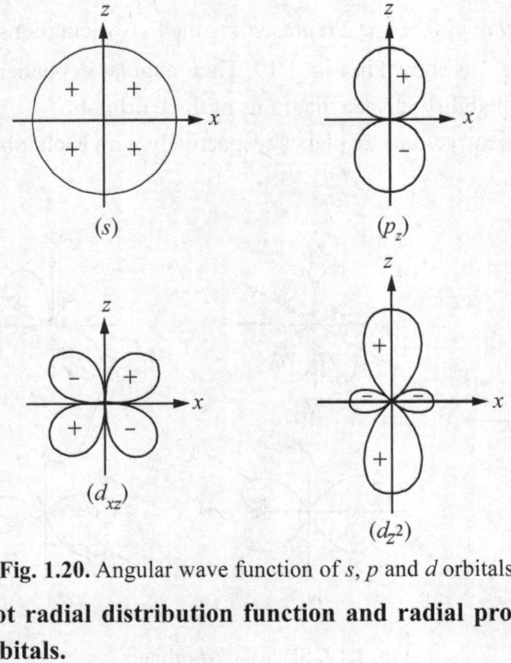

Fig. 1.20. Angular wave function of *s*, *p* and *d* orbitals.

Example 8. **(*a*) Plot radial distribution function and radial probability density curves for 4*s*, 4*p*, 4*d* and 4*f* orbitals.**

(b) Explain as to why only 's' orbitals are spherical.

(c) Draw the shapes of 5'd' orbital indicating sign of wave function 4, 4, 2, 2½.

Solution. (b) Radial distribution function for 4*s*, 4*p*, 4*d* and 4*f* orbitals may be shown as below:

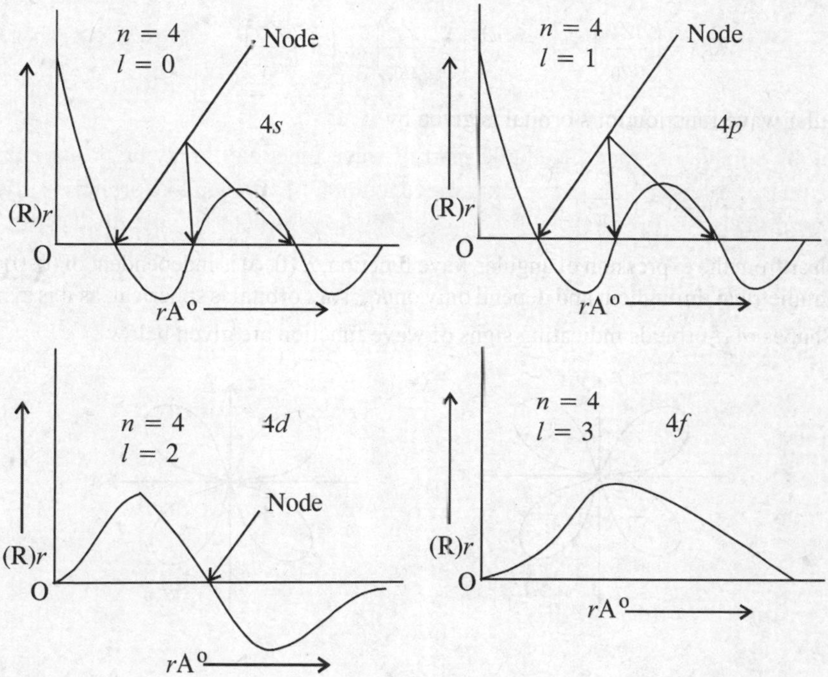

Radial probability density of curves of 4*s*, 4*p*, 4*d* and 4*f* may be shown as below:

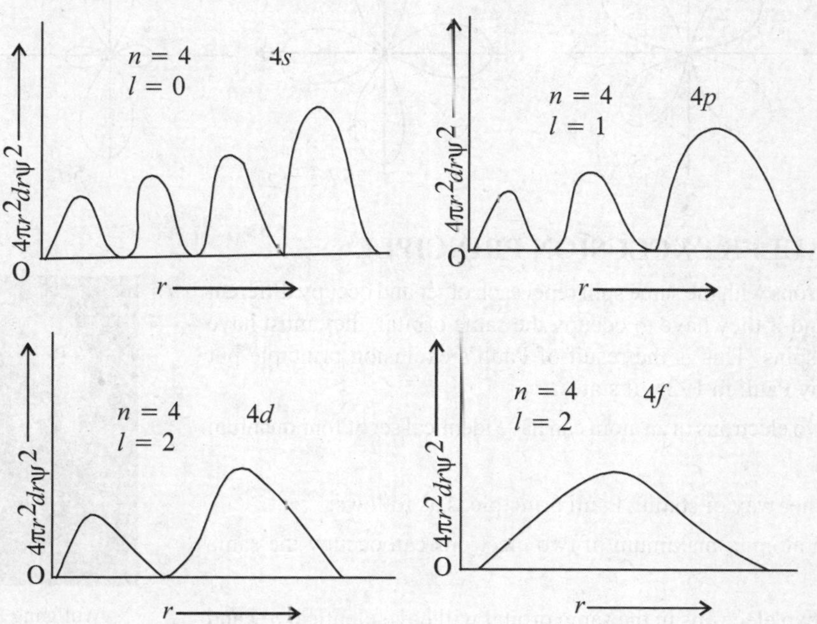

(*b*) Shape of an orbital depends on two factors *viz* radial wave function and angular wave function.

Radial wave function for s orbital is given by

$$R_{(r)} = \left(\frac{Z}{na_0}\right)^{3/2} e^{-zr/na_0}\left(2-\frac{Zr}{a_0}\right)......\left(n-\frac{Zr}{a_0}\right)$$

Angular wave function for s orbital is given by

$$R_{(\theta+\phi)} = \left(\frac{1}{4\pi}\right)^{1/2}$$

As clear from the expression of angular wave function, $R(\theta, \phi)$ is independent of (ϕ, θ) so, orbital will be symmetrical throughout and depend only on $R_{(r)}$ so, *s* orbital is spherical, as it is symmetrical.

(*c*) Shapes of *d*-orbitals indicating signs of wave function are given below :

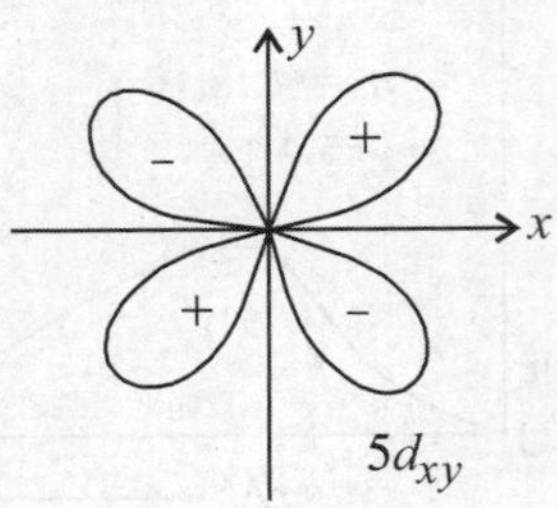

$5d_{xy}$

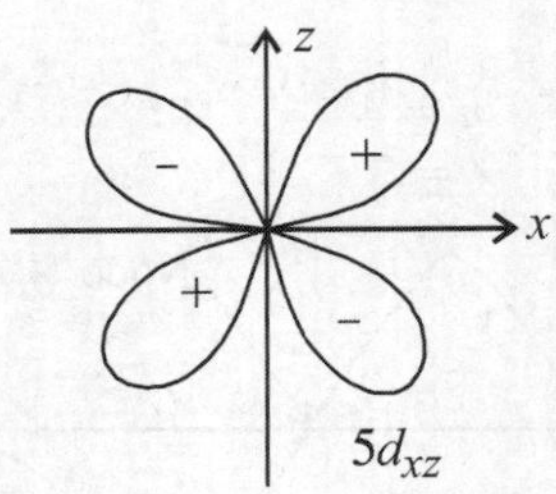

$5d_{xz}$

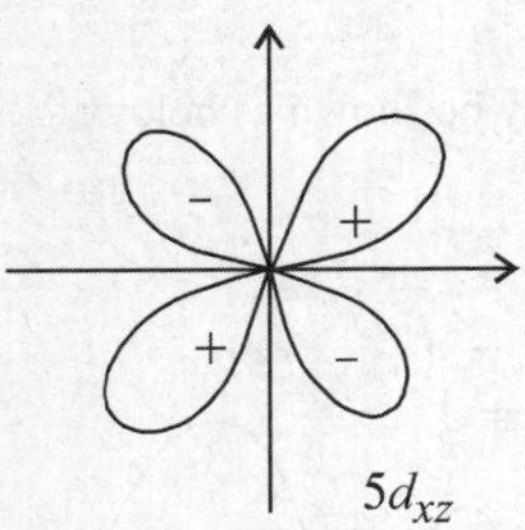

$5d_{xz}$

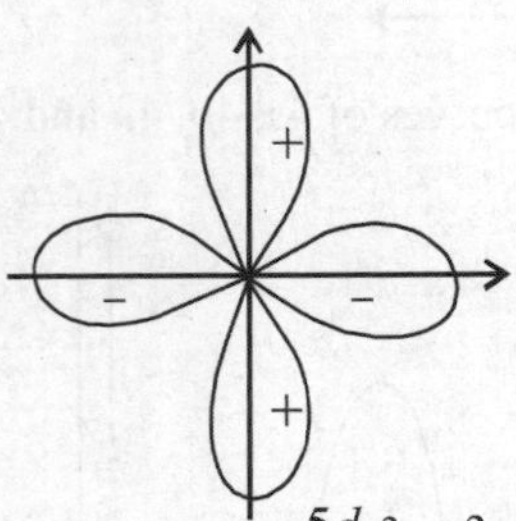

$5d_{x^2-y^2}$

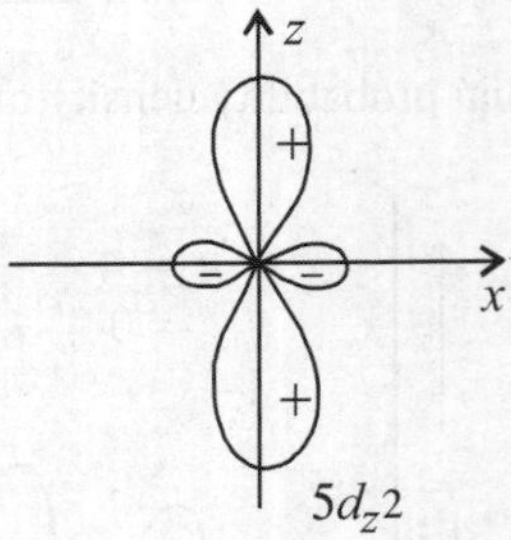

$5d_{z^2}$

1.2 PAULI'S EXCLUSION PRINCIPLE

Electrons with the same spin repel each other and occupy different orbitals and if they have to occupy the same orbital, they must have opposite spins. This is the result of Pauli's exclusion principle put forward by Pauli in 1925. It states:

No two electrons in an atom can have identical set of four quantum numbers.

Another way of stating Pauli principle is as follows:

In an atom, a maximum of two electrons can occupy the same orbital.

Wolfgang Pauli

The two electrons in the same orbital will have identical *n, l* and *m* but must have different spins (+ 1/2 and – 1/2).

Pauli exclusion principle has no mathematical explanation but in an experimental fact. Although no proof of the exclusion principle exists, there has been no contradiction to this.

1.3 HUND'S RULE OF MAXIMUM MULTIPLICITY

It can be stated in two parts as given below:

(*a*) In an atom, no electron pairing takes place in the *p* and *d* or *f* orbitals respectively until each orbital of the given subshell contains one electron.

(*b*) The unpaired electrons present in the various orbitals of the same subshell should have parallel spins.

For example, each of three *p*-orbitals of the 2*p*-subshell gets one electron of parallel spin before any one of them receives a second electron of opposite spin.

This is because presence of the electron each in two orbitals of the same subshell involves lesser inter-electronic repulsion as compared to the electrons in the same orbital. Therefore, it leads to lower energy state.

Moreover, electrons in half-filled orbitals of the same subshell experience less inter-electronic repulsion in space when they have parallel spins than when their spins are opposite.

1.4 ORDER OF FILLING OF VARIOUS ORBITALS

The energy state of an orbital depends not only upon the principal quantum number, *n,* but also, to some extent, on the angular momentum quantum numbers, *l.* In other words, the orbitals having the same principal quantum number can have different energy values.

This happens because

(*i*) The increased nuclear charge leads to somewhat contraction of all the inner orbitals.

(*ii*) The shielding of nucleus by the intervening core of electrons affects the energy of the outer electrons.

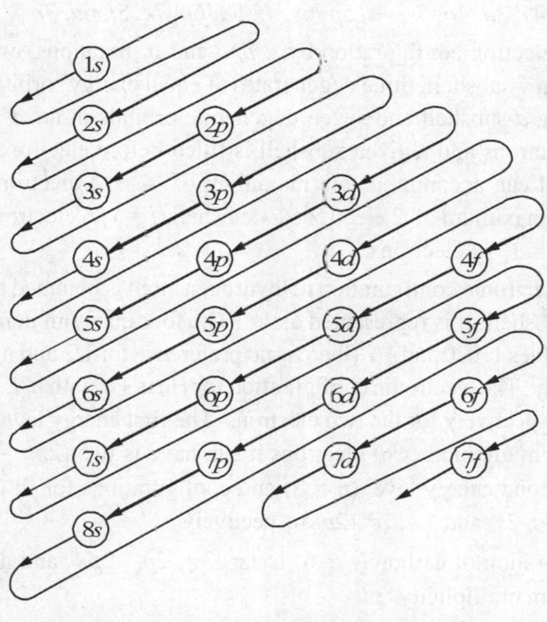

Fig. 1.21. An aid to memory for order of filling orbitals.

It has been found that within a given main energy level (n), the energy states do not remain degenerate but increase in the order $s < p < d < f$. Since the energy of a given orbital depends on the nuclear charge (z) different types of orbitals are affected to different extents. The following sequence is considered to be the most useful for assigning outermost or valence electrons for all the elements:

$$1s < 2s < 2p < 3s < 3p < 4s < 3d < 4p < 5s < 4d < 5p < 6s < 4f < 5d < 6p < 7s$$

An easy way to remember the order of filling of energy states is shown in Fig. 1.19. Represent a given orbital by a circle. Then draw orbitals of each principal quantum number on the same horizontal level. Draw parallel arrows and follow these as shown in Fig. 1.26, starting from 1*s*, 2*s*, 2*p* $\longrightarrow$ 3*s*, 3*p* $\longrightarrow$ 4*s* and so on.

Example 9. **Arrange the following orbitals in order of increasing energy for a multielectron atom 1*s*, 3*s*, 5*s*, 2*p*, 4*p*, 3*d*, 4*d*, and 4*f*.**

Solution. Follow Fig. 1.21. The order comes out to be: 1*s*, 2*p*, 3*s*, 3*d*, 4*p*, 5*s*, 4*d*, 4*f*

1.5 AUFBAU PRINCIPLE ($n + l$) RULE

According to Aufbau principle: "In the ground state of an atom the electrons occupy the lowest energy orbitals available to them". This means that the electrons enter the orbitals in the order of increasing energies. The lowest energy available orbitals being filled up first.

Since the energy of orbitals is determined by the quantum numbers n and l, the sequence of filling the orbitals proceeds according to the following rules:

(*i*) Orbitals are filled in order of increasing value of $n + l$. For instance, 4*s* ($n + l = 4 + 0 = 4$) is filled before 3*d* ($n + l = 3 + 2 = 5$).

(*ii*) For the orbitals having the same value of $n + l$ the orbital having lower value of n is filled up first. For example, 2*p* ($n + l = 2 + 1 = 3$) is filled before 3*s* ($n + l = 3 + 0 = 3$) because 2*p* has lower value of n.

The general order of increasing energies of the orbitals is:

1 *s*, 2*s*, 2*p*, 3*s*, 3*p*, 4*s*, 3*d*, 4*p*, 5*s*, 4*d*, 5*p*, 6*s*, 4*f*, 5*d*, 6*p*, 7*s*, 5*f*, 6*d*, 7*p*

In describing the electron configuration by *s*, *p*, *d* and *f* notations, one must remember that there is one orbital in an *s*-subshell, three degenerate (of equal energy) orbitals in a *p*-subshell, five degenerate orbitals in a *d*-subshell and seven degenerate orbitals in an *f*- subshell. Each orbital accommodates two electrons and a given subshell is filled before electrons are added to the next subshell. Each subshell can accommodate a maximum of $4l + 2$ electrons. As for example, an *s*-subshell ($l = 0$) has a maximum of 2 electrons, *p*-subshell ($l = 1$) 6 electrons *d*-subshell ($l = 2$) 10 electrons, *f*-subshell ($l = 3$) 14 electrons.

The ground state electronic configuration of hydrogen atom is obtained by placing the electrons in the lowest energy subshell. It is represented as $1s^1$. The four quantum numbers n, l, m and m_s for hydrogen atom have values 1, 0, 0 and 1/2 (there is no preference for 1/2 and it may be any orientation 1/2 or – 1/2) respectively. The electronic configuration for He is $1s^2$ with n, l, m and m_s values 1, 0, 0, 1/2 and 1, 0, 0, –1/2 respectively for the two electrons. The first energy level gets completely filled at He because the maximum number of electrons it can have is two ($2n^2 = 2 \times 1^2 = 2$). The next electron goes to the second energy level ($n = 2$) and configurations for Li ($z = 3$), Be ($z = 4$) and B ($z = 5$) are $1s^2$, $2s^1$ $1s^2$, $2s^2$ and $1s^2$, $2s^2$, $2p^1$ respectively.

Electronic configuration of carbon ($z = 6$) is $1s^2$, $2s^2$, $2p_x^1$, $2p_y^1$, and not $1s^2$,$2s^2$, $2p_x^2$ as per Hund's rule of maximum multiplicity.

Thus, the electronic configuration of nitrogen ($z = 7$) would be $1s^2$, $2s^2$, $2p_x^{\ 1}$, $2p_y^{\ 1}$, $2p_z^{\ 1}$ and not $1s^1$, $2s^2$, $2p_x^{\ 2}$, $2p_y^{\ 1}$ and $2p_z^{\ 0}$. Once all the three *p*-orbitals in the subshell are half-filled, pairing will start at oxygen ($z = 8$) to give $1s^2$,$2s^2$, $2P^2_{\ x}$,$2p_y^{\ 1}$, $2p_z^{\ 1}$. The pattern continues and nothing unusual is observed up to Argon ($z = 18$), $1s^2$, $2s^2$, $2p^6$, $3s^2$ $3p^6$ as given in Table 1.3.

Although 3*d*-orbitals are vacant, yet the ground state configurations of potassium ($z = 19$) and calcium ($z = 20$) are $1s^2$, $2s^2$, $2p^6$, $3s^2$, $3p^6$, $4s^1$ and $1s^2$, $2s^2$, $2p^6$, $3s^2$, $3p^6$, $4s^2$ respectively because 4*s* subshell is lower in energy than the 3*d* subshell. Because of greater penetration, 4*s* electron faces higher nuclear charge and this effect is so pronounced that in spite of higher principal quantum number, the energy of the 4*s* subshell is lower than that of 3*d* subshell. At the next element scandium ($z = 21$), the 3*d* orbitals start filling up and this continues till 3*d* subshell is completely filled at zinc ($z = 30$).

Electronic configuration of the elements are given in Table 1.3.

Table 1.3. Ground state electronic configurations of gaseous atoms.

Z	*Symbol*	*Configuration as [Core] plus "Outermost" Orbitals*	*Z*	*Symbol*	*Configuration as [Core] plus "Outermost" Orbitals*
1	H	$1s^1$	21	Sc	[Ar] $3d^1 4s^2$
2	He	$1s^2$, or [He]	22	Ti	[Ar] $3d^2 4s^2$
3	Li	[He] $2s^1$	23	V	[Ar] $3d^3 4s^2$
4	Be	[He] $2s^2$	24	Cr	[Ar] $3d^5 4s^1$
5	B	[He] $2s^2 2p^1$	25	Mn	[Ar] $3d^5 4s^2$
6	C	[He] $2s^2 2p^2$	26	Fe	[Ar] $3d^6 4s^2$
7	N	[He] $2s^2 2p^3$	27	Co	[Ar] $3d^7 s^2$
8	O	[He] $2s^2 2p^4$	28	Ni	[Ar] $3d^8 4s^2$
9	F	[He] $2s^2 2p^5$	29	Cu	[Ar] $3d^{10} 4s^1$
10	Ne	[He] $2s^2 2p^6$ or [Ne]	30	Zn	[Ar] $3d^{10} 4s^2$
11	Na	[He] $3s^1$	31	Ga	[Ar] $3d^{10} 4s^2 4p^1$
12	Mg	[Ne] $3s^2$	32	Ge	[Ar] $3d^{10} 4s^2 4p^2$
13	Al	[Ne] $3s^2 3p^1$	33	As	[Ar] $3d^{10} 4s^2 4p^3$
14	Si	[Ne] $3s^2 3s^2$	34	Se	[Ar] $3d^{10} 4s^2 4p^4$
15	P	[Ne] $3s^2 3p^3$	35	Br	[Ar] $3d^{10} 4s^2 4p^5$
16	S	[Ne] $3s^2 sp^4$	36	Kr	[Ar] $3d^{10} 4s^2 4p^6$ or [Kr]
17	Cl	[Ne] $3s^2 3p^5$	37	Rb	[Kr] $5s^1$
18	Ar	[Ne] $3s^2 3p^5$ or [Ar]	38	Sr	[Kr] $5s^2$
19	K	[Ar] $4s^1$	39	Y	[Kr] $4d^1 5s^2$
20	Ca	[Ar] $4s^2$	40	Zr	[Kr] $4d^2 5s^2$

(Contd.)

Z	*Symbol*	*Configuration as [Core] plus "Outermost" Orbitals*	*Z*	*Symbol*	*Configuration as [Core] plus "Outermost" Orbitals*
41	Nb	[Kr] $4d^4 5s^1$	69	Tm	[Xe] $4f^{13} 6s^2$
42	Mo	[Kr] $4d^5 5s^1$	70	Yb	[Xe] $4f^{14} 6s^2$
43	Tc	[Kr] $4d^6 5s^1$	71	Lu	[Xe] $4f^{14} 5d^1 6s^2$
44	Ru	[Kr] $4d^7 5s^1$	72	Hf	[Xe] $4f^{14} 5d^2 6s^2$
45	Rh	[Kr] $4d^8 5s^1$	73	Ta	[Xe] $4f^{14} 5d^3 6s^2$
46	Pd	[Kr] $4d^{10}$	74	W	[Xe] $4f^{14} 5d^4 6s^2$
47	Ag	[Kr] $4d^{10} 5s^1$	75	Re	[Xe] $4f^{14} 5d^5 6s^2$
48	Cd	[Kr] $4d^{10} 5s^2$	76	Os	[Xe] $4f^{14} 5d^6 6s^2$
49	In	[Kr] $4d^{10} 5s^2 5p^1$	77	Ir	[Xe] $4f^{14} 5d^7 6s^2$
50	Sn	[Kr] $4d^{10} 5s^2 5p^2$,	78	Pt	[Xe] $4f^{14} 5d^9 6s^1$
51	Sb	[Kr] $4d^{10} 5s^2 5p^3$	79	Au	[Xe] $4f^{14} 5d^{10} 6s^1$
52	Te	[Kr] $4d^{10} 5s^2 5p^4$	80	Hg	[Xe] $4f^{14} 5d^{10} 6s^2$
53	I	[Kr] $4d^{10} 5s^2 5p^5$	81	Tl	[Xe] $4f^{14} 5d^{10} 6s^2 6p^1$
54	Xe	[Kr] $4d^{10} 5s^2 p^6$ or [Xr]	82	Pb	[Xe] $4f^{14} 5d^{10} 6s^2 6p^2$
55	Cs	[Xe] $6s^1$	83	Bi	[Xe] $4f^{14} 5d^{10} 6s^2 6p^3$
56	Ba	[Xe] $6s^2$	84	Po	[Xe] $4f^{14} 5d^{19} 6s^2 6p^4$
57	La	[Xe] $5d^1 6s^2$	85	At	[Xe] $4f^{14} 5d^{10} 6s^2 6p^5$
58	Ce	[Xe] $4f^2 5d^{10} 6s^2$	86	Rn	[Xe] $4f^{14} 5d^{10} 6s^2 6p^6$ or [Rn]
59	Pr	[Xe] $4f^3 6s^2$	87	Fr	[Rn] $7s^1$
60	Nd	[Xe] $4f^4 6s^2$	88	Ra	[Rn] $7s^2$
61	Pm	[Xe] $4f^5 6s^2$	89	Ac	[Rn] $6d^1 7s^2$
62	Sm	[Xe] $4f^6 6s^2$	90	Th	[Rn] $6d^2 7s^2$
63	Eu	[Xe] $4f^7 6s^2$	91	Pa	[Rn] $5f^2 6d^1 7s^2$
64	Gd	[Xe] $4f^7 5d^1 6s^2$	92	U	[Rn] $5f^3 6d^1 7s^2$
65	Tb	[Xe] $4f^9 6s^2$	93	Np	[Rn] $5f^5 6d^0 7s^2$
66	Dy	[Xe] $4f^{10} 6s^2$	94	Pu	[Rn] $5f^6 7s^2$
67	Ho	[Xe] $4f^{11} 6s^2$	95	Am	[Rn] $5f^7 7s^2$
68	Er	[Xe] $4f^{12} 6s^2$	96	Cm	[Rn] $5f^7 6d^1 7s^2$

(*Contd.*)

Z	Symbol	Configuration as [Core] plus "Outermost" Orbitals	Z	Symbol	Configuration as [Core] plus "Outermost" Orbitals
97	Bk	[Rn] $5f^8 6d^1 7s^2$	103	Lr	[Rn] $5f^{14} 6d^1 7s^2$
98	Cf	[Rn] $5f^{10} 7s^2$	104	Rf	[Rn] $5f^{14} 6d^2 7s^2$ (?)
99	Es	[Rn] $5f^{11} 7s^2$	105	Ha	[Rn] $5f^{14} 6d^3 7s^2$ (?)
100	Fm	[Rn] $5f^{12} 7s^2$	106	?	[Rn] $5f^{14} 6d^4 7s^2$ (?)
101	Md	[Rn] $5f^{13} 7s^2$			
102	No	[Rn] $5f^{14} 7s^2$			

Example 10. **Why is 4*s* orbital lower in energy than 3*d* orbital?**

Solution. According to Aufbau Rule, an orbital having a lower value for $(n + l)$ is lower in energy.

For 4*s* orbital $n = 4$ and $l = 0$ $\quad\therefore\quad n + l = 4 + 0 = 4$

For 3*d* orbital $n = 3$ and $l = 2$ $\quad\therefore\quad n + l = 3 + 2 = 5$

Thus 4*s* has lower energy than 3*d*.

Example 11. **Describe the following orbitals in terms of their corresponding $(n + l)$ value.**

3*s*, 4*d* and 5*p*

Solution. For 3*s* :

$n = 3, \quad l = 0 \qquad \therefore\ n + l = 3$

For 4*d* :

$n = 4, \quad l = 2 \qquad \therefore\ n + l = 6$

For 5*p*:

$n = 5, \quad l = 1 \qquad \therefore\ n + l = 6$

Example 12. **Write electronic configuration of $_{25}Mn^{2+}$, $_{30}Zn^{2+}$, $_{64}Cd^{3+}$ and $_{13}Al^{3+}$**

Solution. Mn^{2+} [Ar] $3d^5$

Zn^{2+} [Ar]$3d^{10}$

Cd^{3+} [Xe] $5d^1$

Al^{3+} [Ne]

1.5.1 Limitations of Aufbau principle

Although electron configurations based on Aufbau order of filling agree mostly with those derived experimentally, yet there are quite a few exceptions where errors in Aufbau configurations arise due to one (rarely two) misplaced electron. Some of the limitations of Aufbau order are listed here.

(*i*) The Aufbau cannot be used to predict electron configurations of atoms on ionization. In other words, it does not tell us which electrons are to be removed when an ion is formed from an atom. For example, the configuration of Fe based on Aufbau principle is $1s^2$, $2s^2$, $2p^6$, $3s^2$,

$3p^6, 4s^2, 3d^6$. It has been experimentally confirmed by spectral and magnetic studies that Fe^{2+} has configuration $1s^2, 2s^2, 2p^6, 3s^2, 3p^6, 3d^6$ and not $1s^2, 2s^2, 2p^6, 3s^2, 3p^6, 4s^2, 3d^4$. It means that ionization results in the loss of $4s$ electrons in preference to $3d$ electrons even though the $3d$ were the last to be added in building up the configuration of Fe atom. It means that in Fe^{2+}, $3d$ has lower energy than $4s$ which is contrary to Aufbau order of filling. This can be said of any transition elements on ionization.

In general, in the process of ionization, the first electrons are lost from the subshell with the higher n and if n is the same then from subshell with the highest l.

(*ii*) Although $(n-1)\,d$ subshells and ns subshells lie quite close together, yet the former is somewhat higher in energy. Accordingly, the predicted Aufbau outer configuration of Cr (Z = 34) is ..... $3d^4, 4s^2$ but the experimentally supported configuration is ..... .$3d^5, 4s^1$.

Similarly the experimentally observed configuration of Cu, Ag, Au, are different from Aufbau configurations as shown in Table 1.4.

Table 1.4. Limitations of Aufbau order.

Element	*At. No.*	*Aufbau Config.*	*Experimental Config.*
Copper	29	...$3d^9 4s^2$	...$3d^{10}\ 4s^1$
Silver	47	...$4d^9\ 5s^2$	...$4d^{10}\ 5s^1$
Gold	79	...$5d^9\ 6s^2$	...$5d^{10}\ 6s^1$

The reasons for these deviations lie in the extra stability associated with half filled or completely filled subshells.

(*iii*) In the sixth period, $4f$ and $5d$ subshells are exceedingly close in energy. At lanthanum ($Z = 57$), the last electron does not go to $4f$ as predicted in Aufbau order but is added to $5d$ subshell to give it a configuration ....$4d^{10}\ 5s^2, 5p^6, 5d^1, 6s^2$. However, the next electron for cerium (Z = 58) is added to the $4f$ subshell and the previous electron in $5d$ in lanthanum is also shifted to $4f$ to give it a configuration ....$4d^{10}\ 5s^2\ 5p^6\ 4f^2\ 5d^0\ 6s^2$.

(*iv*) The Aufbau order configuration of zirconium (Z = 40) is ..... $5s^2\ 4d^2$ and that the predicted order for the next element niobium Nb (Z = 41) would be ...... $5s^2\ 4d^3$. However, the observed configuration of Nb is ... $5s^1\ 4d^4$. It is understandable that the next element molybedenum Mo (Z = 42) acquire ..... $5s^1\ 4d^5$ configuration due to higher stability associated with half-filled configuration. However, the next element technitium Tc (Z = 43) with observed configuration $5s^1\ 4d^6$ in place of predicted $5s^2\ 4d^5$ shows that the preference for half-filled or completely filled configurations is not the only controlling factor but some complex nuclear-electron and electron-electron forces also have some role to play in determining these configurations.

(*v*) In case of palladium Pd (Z = 46), the predicted Aufbau configuration is $5s^2\ 4d^8$ but the experimentally observed configuration is $5s^0\ 4d^{10}$. This is perhaps the only case where the observed configuration has two misplaced electrons.

1.5.2 Factors governing special stability associated with half-filled and completely filled subshells (concept of exchange energy)

The Aufbau order fails to predict sometimes the correct electronic configurations in cases where energies of neighbouring subshells (*e.g.*, $4s$, $3d$; $5s$, $4d$; $4f$, $5d$; etc.) are quite close together and comparable. As for example, the observed configurations, $3d^5 4s^1$ and $3d^{10}\ 4s^1$ for Cr (Z = 24) and Cu (Z = 29) respectively, differ from respective Aufbau predicted configurations, $3d^4 4s^2$ and $3d^9 4s^2$.

This is explained on the basis of **"special stability associated with half-filled and completely-filled subshells".**

According to Hund's rule, electrons distribute themselves in degenerate orbitals so as to keep like spins as far apart as possible. For example, the most stable carbon configuration $1s^2\ 2s^2\ 2p_x^{\ 1}\ 2p_y^{\ 1}\ 2p_z^{\ 0}$, has both the *p*-electrons in degenerate orbitals with like spins. Similarly the lowest energy configurations for nitrogen $1s^2\ 2s^2\ 2p_x^{\ 1}\ 2p_y^{\ 1}\ 2p_z^{\ 1}$ has all the three *p*-electrons with the same spin. This happens because electrons as charged particles repel one another and electron-electron repulsion is minimum when electrons with like spins in a subshell keep as far apart as possible. This situation may be visualized in two steps shown in Fig. 1.22. The first step called the charge correlation means interactions among electrons by virtue of their charges (coulomb forces) and refers to lower energy situation for two electrons having different m and m_s values. The next step called the spin correlation refers to further lowering of energy for like (parallel) spins. Spin correlation leading to decrease in repulsion for electrons of like spins than for electrons of different spins is called the **exchange** energy. Spin correlation and its exchange energy of stabilization are thought to be the basis for special stability of half-filled subshells. Placing two electrons in the same orbital will lead to higher repulsion than if they occupy degenerate orbitals as shown in Fig. 1.22 for configuration.

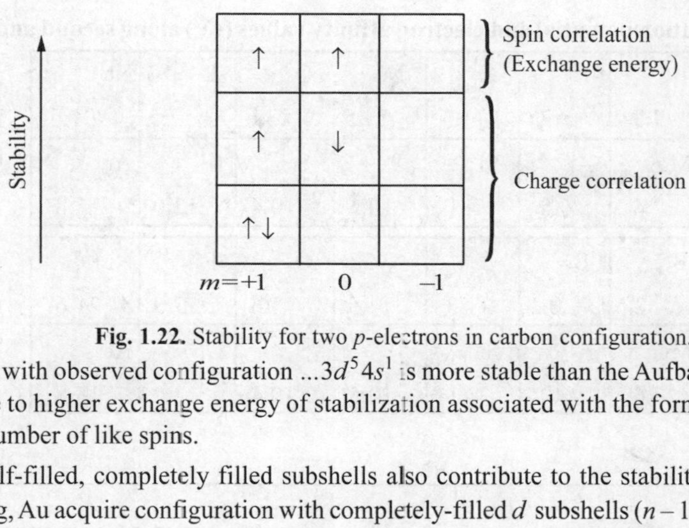

Fig. 1.22. Stability for two *p*-electrons in carbon configuration.

Chromium with observed configuration ...$3d^5\,4s^1$ is more stable than the Aufbau configuration ..... $3d^4\ 4s^2$ due to higher exchange energy of stabilization associated with the former arrangement having larger number of like spins.

Besides half-filled, completely filled subshells also contribute to the stability of atoms. For example, Cu, Ag, Au acquire configuration with completely-filled *d* subshells $(n-1)\,d^{10}\ ns^1$, instead

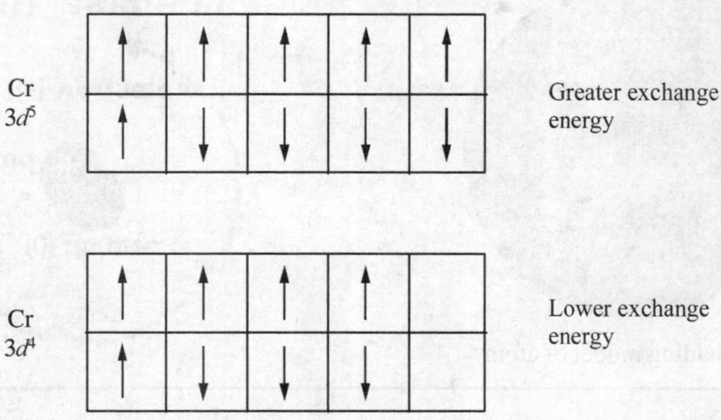

Fig. 1.23. Configuration of chromium

of the predicted configuration $(n-1)\, d^9\, ns^2$. The observed configuration of palladium $Pd - 4d^{10}, 5s^0$ involves a transfer of two electrons from predicted configuration $4d^8\, 5s^2$ to acquire a completely-filled d sub-shell. The stability of noble gases is the most important example of special stability associated with completely-filled subshells.

1.5.3 Special stability of filled and half-filled configuration related to ionisation potential and electron affinity

Special stability associated with half-filled and completely filled subshells illustrated in the trends of ionization potential and electron affinity (eV) is given in Table 1.5. Ionization potentials and electron affinities of Be and Mg which have filled s subshell in the outermost shell have exceptional values that fall out of trends along a period (Table 1.5). The trend of increasing ionization potential along a period for p-block elements with valence configuration $ns^2\, np^x$ ($x = 1 - 6$) is followed smoothly up to nitrogen s^2p^3. The break occurs at s^2p^4, when the fourth electron with opposite spin is added to a p-orbital which already has an electron. This additional electron does not benefit from the exchange energy of stabilization possible for C and N. Special stability of half-filled and completely filled configuration is supported by abnormal values of electron affinity for atoms with s^2p^3 and s^2p^6 configuration.

Table 1.5. Ionization potential and electron affinity values (eV) along second and third periods.

Ionization potential trend

Li	Be	B	C	N	O	F	Ne
5.4	9.3	8.3	11.3	14.5	13.6	17.5	21.6
Na	Mg	Al	Si	P	S	Cl	Ar
5.1	7.6	6.0	8.1	10.5	10.4	13.0	15.8

Electron affinity trend

Li	Be	B	C	N	O	F	Ne
0.62	– 2.5	0.24	11.3	14.5	13.6	17.5	21.6
Na	Mg	Al	Si	P	S	Cl	Ar
5.1	7.6	6.0	8.1	10.5	10.4	13.0	15.8

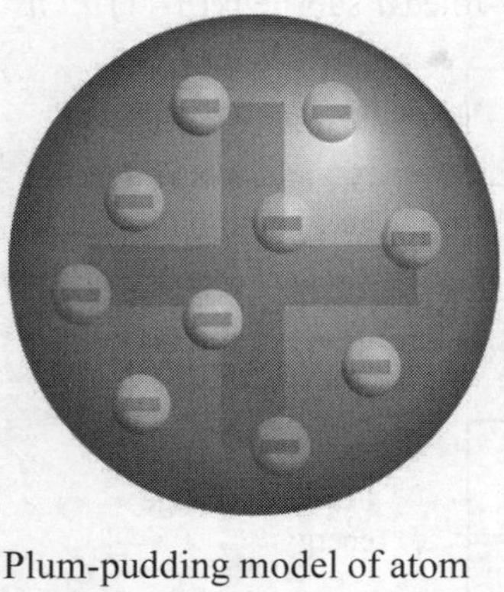

Plum-pudding model of atom

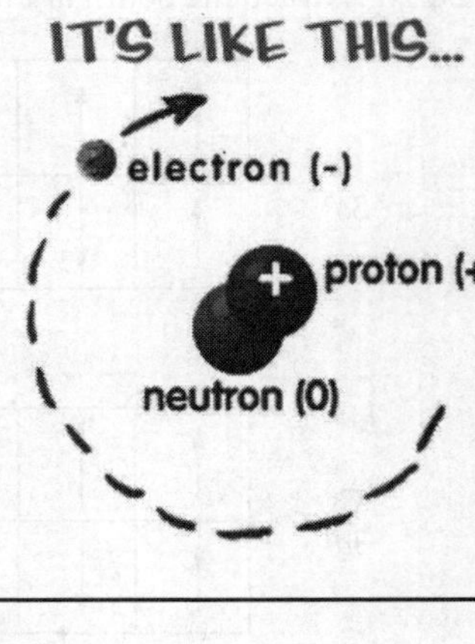

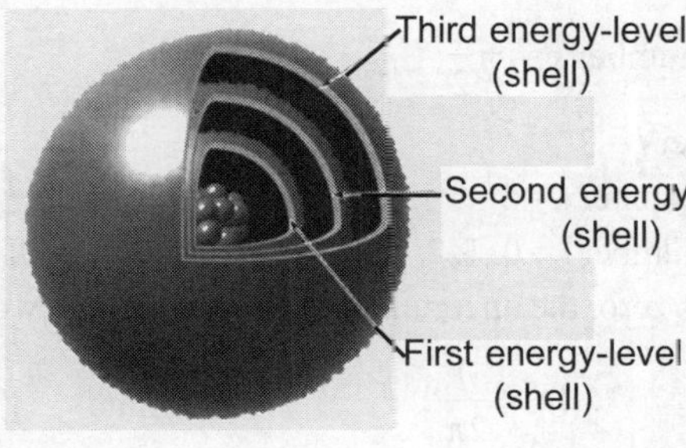

Orbitals

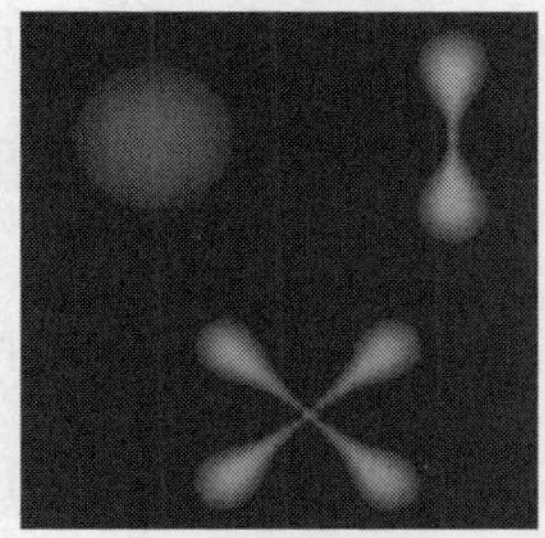

s, p and d orbitals

Example 13. Give two examples where observed configuration has no preference for half-filled and completely filled configuration.

Solution. In Tungsten W (Z = 74) the configuration is $5d^4 6s^2$.
In Technicium Tc (Z = 43) the configuration is $4d^6 5s^1$.
The above two elements show no preference for half-filled configuration.
Note : For effective nuclear charge, refer to Periodic Properties chapter 2.

EXERCISES
(INCLUDING QUESTIONS FROM DIFFERENT UNIVERSITY PAPERS)

Objective Questions

Fill in the blanks.

1. The energy change in electron jump is given by ....
2. If the principal quantum number is 3, the azimuthal quantum number can have values ....
3. According to .... principle, no two electrons in an atom will have all the four quantum numbers ....
4. The outermost electronic configuration of Cr is ....
5. Orbital is the term given to the .... where probability of finding an electron is ....
6. Wave functions of electrons in atoms and molecules are called..............................
7. The $2p_x$, $2p_y$ and $2p_z$ orbitals of atoms have identical shapes, but differ in their ....
8. The uncertainty principle and the concept of wave nature of matter were proposed by .... and .... respectively.
9. The number of protons and neutrons in $^{226}_{88}\text{Ra}$ are .... and ....
10. $\frac{h}{2\pi}$ is the angular momentum of the electron in the .... orbit of He^+

Multiple Choice Questions

Choose the correct option

1. Which of the following statements is not correct for an electron that has the quantum number $n = 4$ and $m = 2$?

(*a*) The electron may have the quantum number $s = +\frac{1}{2}$

(*b*) The electron may have quantum numer $l = 2$

(*c*) The electron may have the quantum number $l = 3$

(*d*) The electron may have the quantum number $l = 0, 1, 2, 3$

2. If uncertainty in position of an electron is zero, the uncertainty in its momentium would be

(*a*) zero (*b*) $> \frac{h}{2\pi}$ (*c*) $< \frac{h}{2\pi}$ (*d*) infinite

3. Electromagnetic radiation with highest wavelength is

(*a*) IR (*b*) Radiowaves (*c*) X-rays (*d*) U.V.

4. Which of the following does not characterise X-rays ?

(*a*) The radiation can ionize gases

(*b*) It causes ZnS to fluoresce

(*c*) Deflected by electric and magnetic field

(*d*) Have the wavelength shorter than U,V, rays

5. The element with a non-spherical outermost orbital is

(*a*) B (*b*) Li (*c*) Be (*d*) He

6. Hund's Rule of maximum multiplicity indicates that

(*a*) one orbital cannot contain more than two electrons of opposite spin.

(*b*) one orbital can contain more than two electrons.

(*c*) one orbital can contain two electrons of same spin.

(*d*) none of these.

7. Bohr model postulate is

(*a*) $mvr = \frac{mn}{2\pi}$ (*b*) $mvr = \frac{hm}{2\pi}$ (*c*) $mvr = \frac{nh}{2\pi}$ (*d*) none

8. The presence of three unpaired electrons in the nitrogen atom can be explained by

(*a*) Pauli's exclusion principle (*b*) Aufbau principle

(*c*) Uncertainty principle (*d*) Hund's rule

9. Which of the following electronic transitions in a hydrogen atom will require the largest amount of energy ?

(*a*) From n = 1 to n = 2 (*b*) From n = 2 to n = 1

(*c*) From n = 0 to h = 1 (*d*) From n = 3 to n = 3

10. The equation $\overline{v} = R\left(\frac{1}{n_1^2} - \frac{1}{n_2^2}\right) m^{-1}$ is known as

(*a*) Moseley's equation (*b*) Estein's equation

(*c*) Ritz-combination principle (*d*) Pauling equation

11. Which of the following is an energy particle ?

(*a*) Photon (*b*) Neutrino (*c*) Antiproton (*d*) Positron

12. The electron is

(*a*) an α-particle (*b*) a β-particle (*c*) hydrogen ion (*d*) positron.

13. The wave nature of an electron was first given by

(*a*) de Broglie (*b*) Sommerfeld (*c*) Heisenberg (*d*) Bohr

14. Magnetic quantum number signifies

(*a*) orientation of orbitals in space (*b*) size of orbitals
(*c*) shape of orbitals (*d*) nuclear stability

15. Correct set of four quantum numbers for the valance (outermost) electron of rubidium (Z = 37) is

(*a*) 5, 0, 0, – 1/2 (*b*) 5, 1, 0, + 1/2 (*c*) 5, 1, 1,+ 1/2 (*d*) 5, 0, 0, + 1/2

16. Angular momentum of an electron depends upon

(*a*) principal quantum number (*b*) magnetic quantum number
(*c*) azimuthal quantum number (*d*) all the above

17. Transition of electron from n = 3 level to n = 1 level results in

(*a*) x-ray spectrum (*b*) emission spectrum
(*c*) band spectrum (*d*) infrared spectrum

18. Pauli's exclusion principle states that

(*a*) two electrons in the same atom cannot have the same energy.
(*b*) two electrons in an orbital cannot have same spin.
(*c*) the electrons tend to occupy different orbitals as far as possible.
(*d*) electrons tend to occupy lower energy orbitals preferentially.

19. That the position and velocity of small particles e.g. electron cannot be simultaneously determined, is a statement of

(*a*) Heisenberg's uncertainty principle
(*b*) Aufbau principle
(*c*) Pauli exculsion principle
(*d*) de Broglie's wave concept of electron.

20. The quantum numbers for the electrons of an element are given below : n = 2, l = 0, m = 0, s = + 1/2. The atom is

(*a*) lithium (*b*) beryllium (*c*) hydrogen (*d*) boron

ANSWERS

Fill in the blanks

1. Planck equation **2.** 0, 1 and 2 **3.** Pauli's exclusion **4.** $3s^2\ 3p^6\ 3d^5\ 4s^1$
5. region, maximum **6.** orbitals **7.** orientation **8.** Heisenberg, de Broglie
9. 88, 138 **10.** second

Multiple choice questions

1. (*d*)	**2.** (*d*)	**3.** (*b*)	**4.** (*c*)	**5.** (*a*)
6. (*d*)	**7.** (*c*)	**8.** (*d*)	**9.** (*a*)	**10.** (*c*)
11. (*a*)	**12.** (*b*)	**13.** (*a*)	**14.** (*a*)	**15.** (*d*)
16. (*d*)	**17.** (*b*)	**18.** (*b*)	**19.** (*a*)	**20.** (*a*)

General Questions

1. (*a*) The basic equation which describes the standing motion of a stretched string is:

$$\psi = A \sin \frac{2\pi x}{\lambda}$$

in one of the directions (*x*-axis). Formulate Schrodinger wave equation from this.

(*b*) Discuss the significance of ψ and ψ^2.

(*c*) What do you understand by the terms eigen functions and eigen values as used while solving Schrodinger wave equation?

2. (*a*) How does the wave mechanical model of an atom lead to the concept of orbital?
(*b*) List the important differences between an orbital and an orbit.
3. (*a*) Write Schrodinger wave equation for hydrogen atom. What are the various parameters used in the equation?
(*b*) What physical significance is attached to ψ?
4. (*a*) What are quantum numbers? How do they arise in the wave mechanical model of an atom? Name these.
(*b*) Discuss the physical significance of different quantum numbers.
(*c*) An electron is in 4*d*-orbital. What possible values for the quantum numbers *n*, *l* and *m* can it have?
5. (*a*) What are the main conditions which ψ must satisfy to give meaningful solutions (Eigen functions?
(*b*) Write Schrodinger wave equation. Name the three quantum numbers obtained from it and informations conveyed by them.
6. (*a*) What do you understand by radial and angular wave functions? How will you represent the complete wave function in terms of radial and angular wave functions?
(*b*) The complete wave function for 1*s*-orbital of hydrogen atom is:

$$\psi_{1s} = \frac{1}{\sqrt{\pi}}\left(\frac{Z}{a_0}\right)^{3/2} e^{-Zr/a_0}$$

where a_0 = Bohr radius = 0.529 Å. How does this equation lead to conclusion that 1*s*-orbital is spherically symmetrical?
(*c*) Write the relationship between *x*, *y*, *z* and *r*, θ, ϕ.
7. (*a*) How many variables are there in polar co-ordinate equation? Mention these.
(*b*) What is the significance of radial part of ψ? Is it same for all the three 2*p*-orbitals or not?
(*c*) Why is 1*s*-orbital spherically symmetrical?
8. (*a*) Which of the following sets of quantum numbers are not permissible?

1. $n = 3$	$l = 2$	$m = 0$	$s = +\frac{1}{2}$
2. $n = 4$	$l = 3$	$m = -3$	$s = -\frac{1}{2}$
3. $n = 5$	$l = 0$	$m = 0$	$s = -\frac{1}{2}$
4. $n = 2$	$l = 2$	$m = +2$	$s = +\frac{1}{2}$
5. $n = 3$	$l = -2$	$m = +2$	$s = -\frac{1}{2}$

9. (*a*) Formulate Schrodinger wave equation.
(*b*) What are angular and radial wave functions? On what quantum numbers do these depend?
(*c*) Why do we use $\psi\psi^*$ in place of ψ^2?
10. (*a*) Discuss the shapes of 2p-orbitals.
(*b*) What is the symmetry of an orbital with
(*i*) $l = 0$ (*ii*) $l = 1$?
(*c*) Do 2*d* and 3*f*-orbitals make sense according to wave mechanics ?

11. (*a*) Describe briefly the various methods of giving the electron probability distribution of an electron in an atom.
(*b*) Name the various orbitals possible for
(*i*) $n = 3, l = 0$ (*ii*) $n = 3, l = 1$
(*c*) Name the part of wave function which governs the shapes of the orbitals.
(*d*) Draw radial probability distribution curves for $n = 3, l = 0; n = 3, l = 1$ and $n = 3, l = 2$ for hydrogen atom.
12. (*a*) What is the significance of ψ and ψ^2?
(*b*) What do you mean by the term $4\pi r^2 dr\psi^2$? Discuss.
(*c*) Describe the shapes of 1*s* and 2*s*-orbitals. What is the basic difference between their shapes?
13. (*a*) What do you understand by radial probability distribution? Draw radial probability distribution curves for an orbital with $n = 3$ and $l = 0$. What information does this provide?
(*b*) Write expressions for radial part of wave function of 2*s* and 2*p*.
(*c*) Write expressions for total angular part of wave functions of 2*p*-orbitals.
(*d*) Draw the shape of an orbital corresponding to $n = 2, l = 0$.
14. (*a*) How many orientations are possible for *d*-orbitals? Name these orbitals. Discuss the shapes of 3*d*-orbitals.
(*b*) Which of the following statements are correct and which are incorrect? Correct the incorrect statements.
(*i*) The shape of an orbital is determined by the angular part of the wave function.
(*ii*) For $l = 0$, *m* does not have any value.
(*iii*) ψ^2 gives the probability of finding the electron in a given volume.
(*iv*) The wave functions for all *s*-orbitals are independent of angles θ and ϕ.
(*v*) Spin quantum number did not originate from the solution of Schrodinger wave equation.
(*vi*) For $n = 2$, the total values of *m* are equal to three.
(*c*) Fill in the blanks:
(*i*) Radial part of wave function depends upon quantum numbers ....
(*ii*) For $l = 2$, the values of *m* are equal to ....
(*iii*) The radial probability distribution curve for 3*s*-orbital have .... nodes.
(*iv*) The $3d_{x^2-y^2}$ orbital consists of .... lobes lying in .... the coordinates axes *x* and *y*.
(*v*) ψ is called ....
15. (*a*) Draw the shapes of the following orbitals:
(*i*) $2p_x$ (*ii*) 3*s* (*iii*) $3d_{xy}$ (*iv*) 3*p*
(*b*) How many nodes are present in each of the following:
(*i*) 3*s*-orbital (*ii*) $2p_z$-orbital (*iii*) $3d_{x^2-y^2}$ orbital
(*c*) Why is 4*s*-orbital lower in energy than 3*d*-orbital?
16. (*a*) Arrange the following orbitals in the order of increasing energy for a multielectron atom: 1*s*, 3*s*, 5*s*, 2*p*, 4*p*, 5*p*, 3*d*, 5*d* and 4*f*.
(*b*) What do you understand by Hund's rule of maximum multiplicity? Discuss this rule by taking into examples.

(*c*) What is Pauli's exclusion principle? How many electrons may exist in the M shell of an atom? How are these electrons distributed within the shell.

(*d*) What are the anticipated electronic configurations of elements with atomic numbers 106 and 107, which have been prepared artificially.

17. (*a*) Write the Schrodinger wave equation for the wave mechanical model of an atom.

(*b*) The wave mechanical model of an atom has over-ruled the concept of circular orbits proposed by Bohr. Discuss.

(*c*) Name the three quantum numbers of an electron which are yielded during the solution of Schrodinger wave equation.

(*d*) Draw the shapes of $3d_{x^2-y^2}$ and $3d_{z^2}$ orbitals. How do they differ?

18. (*a*) What is Aufbau principle? What are its limitations?

(*b*) Why are half-filled and completely filled orbitals more stable?

(*c*) The energies of 4*s*, 4*p*, 4*d* and 4*f*-orbitals are same for H-atom but different for multielectron atoms. Discuss.

19. (*a*) Define the following:

(*i*) eigen function

(*ii*) radial wave function

(*b*) (*i*) Which of the orbitals 1*p*, 2*s*, 2*p*, 3*f* are not possible. Give reasons.

(*iii*) What can be the maximum number of electrons in the following orbitals?

(*a*) 3*d* (*b*) 4*d* (*c*) 5*d*

20. (*a*) State and explain the rules which govern the filling of various orbitals in an atom.

(*b*) Write the ground state electronic configurations of the following:

(*i*) Cd (Z = 48) (*ii*) Sn (Z = 50) (*ii*) Cu (Z = 29) (*iv*) I (Z = 53)

(*c*) Calculate the number of unpaired electrons in each of the following:

(*i*) Fe^{3+} (Z = 26) (*ii*) Se^{2-} (Z = 34) (*iii*) Ag^{+} (Z = 47) (*iv*) Cu^{2+} (Z = 29)

(*d*) Which of the following ions are dimagnetic or paramagnetic?

(*i*) Al^{3+} (Z = 13) (*ii*) Ni^{2+} (Z = 28) (*iii*) Ti^{4+} (Z = 22) (*iv*) Co^{3+} (Z = 27)

21. (*a*) What is the order of energy of 4*s*, 4*p*, 4*d* and 4*f*-orbitals for H-atom? Is the order same or different for Na-atom? Explain.

(*b*) What are the factors responsible for the stability of a half-filled orbital?

(*c*) Draw the shapes for each of the following orbitals: $2p_x, 3d_{z^2}, 3d_{x^2-y^2}, 3d_{xy}$

(*d*) Discuss in detail Pauli's exclusion principle.

(*e*) Discuss the physical significance of various quantum numbers.

(*f*) What is the radial probability of an electron at the nucleus?

22. (*a*) Explain radial wave function of H-atom.

(*b*) Give radial probability distribution curves for 2*p* and 3*p*-orbitals and give their characteristics.

(*c*) Give the shapes of 3*d*-orbitals. How many maximum number of electrons this subshell can occupy and why?

(*d*) How do principal quantum number, azimuthal quantum number and magnetic quantum number explain the relative energies of the orbitals in a given subshell?

23. (*i*) What are the factors responsible for the stability of half-filled and completely filled orbitals?

(*ii*) Draw the shapes of 2*s*, 2*p* and $3d_{x^2-y^2}$ orbitals.

(*iii*) Give the actual electronic configuration of Pt (Z = 78) and Cu (Z = 29).

(*iv*) Give the relationship between polar and cartesian co-ordinates. Give the value of angular wave function for *s*-orbitals.

26. Give reasons for the following:

(*i*) In forming Fe^{2+} from Fe (Z = 26), the electrons are removed from 4*s* and not from 3*d*-orbitals.

(*ii*) The electronic configuration of copper is $[Ar]3d^{10}4s^1$ and not $[Ar]3d^9 4s^2$.

(*iii*) Though the maximum of 2*p*-orbitals in radial probability distribution curve is slightly closer to the nucleus than that of 2*s*-orbital, yet the energy of 2*s*-orbital is less than that of 2*p*-orbital in multielectron atom.

(*iv*) 4*s*-orbital has less energy than 3*d*-orbital.

27. (*a*) Write down the electronic configurations of Ag (Z = 47) (both actual and theoretical) and give suitable reasons for the difference.

(*b*) Write down the actual electronic configurations of elements with atomic numbers 57 and 79 in terms of their preceding noble gas configuration.

28. (*a*) Draw the radial probability distribution curves for 2*s* and 2*p*-electrons and explain the difference.

(*b*) Give the shapes of *d*-orbitals.

(*c*) State and explain Aufbau principle and give its limitations.

(*d*) Give the significance of *l* and *m*.

29. (*a*) What are the main conditions which ψ must satisfy to give meaningful solution (eigen function).

(*b*) Sketch the angular wave function diagram of *s*, *p* and *d*-orbitals.

(*c*) Which of the orbitals are not possible? 1*p*, 2*s*, 3*p* and 3*d*. Give reasons.

30. (*a*) Discuss the physical significance of four quantum numbers.

(*b*) Draw the shapes of $3d_{x^2-y^2}$ and $3d_{z^2}$ orbitals. How do they differ?

(*c*) In forming Fe^{2+} from Fe (Z = 26), the electrons are removed from 4*s* and not from 3*d* orbitals. Explain.

(*d*) Draw orbital probability functions for *n* = 3 for hydrogen atom.

31. (*a*) Draw the radial probability distribution curves for the orbitals with (*i*) *n* = 2, *l* = 0 and (*ii*) *n* = 2, *l* = 1

(*b*) Give the shapes of *d*-orbitals.

(*c*) How will you apply Schrodinger wave equation to hydrogen atom?

(*d*) Differentiate between the quantum numbers of all the electrons of 3*p* and 3*d*-orbitals.

32. (*a*) Draw radial probability distribution curves for 2*s*, 3*s* and 4*s* orbitals.

(*b*) What are quantum numbers? How do they arise in the wave mechanical model of an atom? Name these.

(*c*) Discuss the limitations of Aufbau principle and anomalous configurations with suitable examples.

(*d*) 4*s*-subshell is filled first than 3*d* while 2*p*-subshell is filled first than 3*p*. Explain the reasons.

(*e*) Draw the shapes of $3d_{x^2-y^2}$ and $3d_{z^2}$ orbitals. How do they differ?

(*f*) Which of the following ions are diamagnetic or paramagnetic?

Al^{3+} (Z = 13), Ni^{2+} (Z = 28), Ti^{4+} (Z = 22), Co^{3+} (Z = 27)

(*g*) Arrange the following orbitals in order of increasing energy for multielectron atom:

1*s*, 3*s*, 5*s*, 2*p*, 4*p*, 5*p*, 3*d*, 4*d* and 4*f*

33. Discuss the role of Pauli's exclusion principle and Hund's rule in electron configuration determination.

34. (*a*) Describe Schrodinger wave equation. Show how the wave equation supports Bohr's theory of hydrogen atom.

(*b*) What is $(n + l)$ rule? How do it help in buildings up of the elements?

(*c*) Describe the following orbitals in terms of their corresponding $(n + l)$ values: 3*s*, 4*d*, 4*p*.

(*d*) Show the relationship between *n*, *l* and *m* when *n* = 3. Give the shape of d_{z^2} and $d_{x^2-y^2}$ orbitals.

35. (*a*) What part of ψ govern shape of orbitals? Give expression for angular wave function of *s*-orbital and p_z orbital in the hydrogen atom.

(*b*) How do the $d_{x^2-y^2}$ and d_{xy} orbitals differ in their orientation in space?

(*c*) What physical significance is attached to ψ and ψ^2?

(*d*) Find the values of *n*, *l*, *m* and *s* for an electron in 4*f* orbital of an atom.

36. (a) What do you understand by orbital and angular wave functions? How will you represent the complete wave function in terms of radial and angular wave functions?

(*b*) Write the electronic configurations of palladium (At. No. 46) and Cu^{2+} (At. No. 29) ion.

(*c*) Why is 1*s* orbital spherically symmetrical?

37. (*a*) Explain the terms, E, V, eigen values and eigen functions involved in Schrodinger wave equation.

(*b*) Find the values of *n*, *l*, *m* and *s* for an electron in 4*f*-orbital of an atom.

(*c*) What do you understand by Pauli's exclusion principle?

38. (*a*) Write the expression for the nontime wave equation for a subatomic particle in three dimensions and explain the meaning of various terms involved in it.

(*b*) Give two examples where observed configuration has no preference for half-filled and completely filled configurations.

39. (*a*) Discuss Aufbau principle. What are its various limitations?

(*b*) Define exchange energy and discuss its importance in influencing electronic configuration.

(*c*) What is $(n + l)$ rule? Explain with example.

40. (*a*) What are the values of all four quantum numbers of ^{46}Pd for outermost shell?

(*b*) Draw radial probability distribution curves for 3*s* and 3*p*-orbitals.

(*c*) Draw the shapes of $3d_{z^2}$, $3d_{x^2-y^2}$ and 3*s* orbital.

(*d*) What are eigen values?

(*e*) Give three examples which have completely filled stability in a group of periodic table and give their electronic configuration.

41. (*a*) Draw the shapes of orbitals as represented by

(*i*) d_{xy} (*ii*) $d_{x^2-y^2}$ (*iii*) d_{z^2} (*iv*) 2*s*

(*b*) Write the electronic configurations of Ti^{2+} (Z = 22) and Cr^{3+} (Z = 24).

(*c*) What is the ordinate for a radial probability distribution curve? Draw radial probability distribution curves for 3*s* and 3*p*-orbitals and state the following in each:

(*i*) Number of high probability regions

(*ii*) Number of nodes.

42. (*a*) What is (*n* + *l*) rule? Explain with examples.

(*b*) How do $d_{x^2-y^2}$ and d_{xy} orbitals differ in their orientation in space?

(*c*) Cu (Z = 29) shows the oxidation states Cu^+ and Cu^{2+}. Give the electronic configuration of all the three. Which oxidation state is more stable and why?

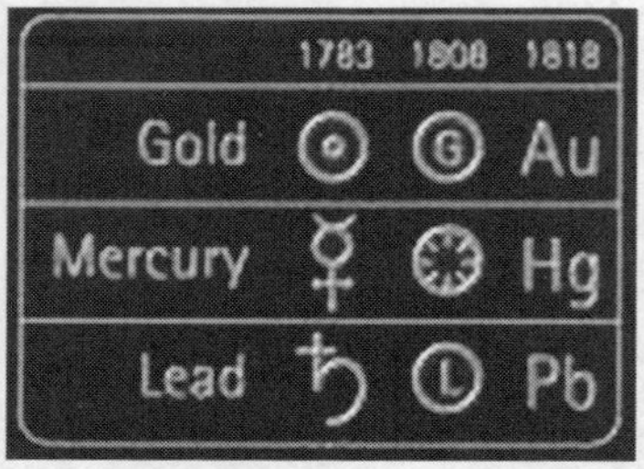

Earlier symbols of elements

2

Periodic Properties

2.1 INTRODUCTION

If the elements are arranged in order of increasing atomic numbers, then their properties get repeated after regular intervals. The repetition of similar properties of the element after regular intervals is called **periodicity.** It is possible to explain the cause of periodicity in the light of the structure of the atom.

It has now been established that chemical properties of the elements are governed by the number of valence electrons in their atom. Therefore it is expected that atoms having similar valence shell electronic configuration will have similar properties. This is actually true. The arrangement of elements in order of increasing atomic number results in the repetition of similar electronic configuration after regular intervals.

Consider the electronic configuration of alkali metals (Table 2.1). The alkali metals have one electron in the *s* orbital of their valence shell. In other words, their electronic configuration can be written as ns^1 where n is the number of outmost energy level. It is due to the similar electronic configuration in the outermost orbit that all the alkali metals have similar properties.

Table 2.1. Electronic configuration of alkali metals.

Element	*At. No.*	*Electronic configuration*
Li	3	$1s^2;2s^1$
Na	11	$1s^2$; $2s^2\,2p^6$; $3s^1$
K	19	$1s^2$; $2s^2\,2p^6$; $3s^2 3p^6$; $4s^1$
Rb	37	$1,s^2$; $2s^2\,2p^6$; $3s^2\,3p^6\,3d^{10}$; $4s^2\,4p^6$; $5s^1$
Cs	55	$1s^2$; $2s^2\,2p^6$; $3s^2\,3p^6\,3d^{10}$; $4s^2\,4\mathrm{p}^6\,4d^{10}$; $5s^2 5p^6$; $6\mathrm{s}^1$
Fr	87	$1s^2$; $2s^2\,2p^6$; $3s^2\,3p^6\,3d^{10}$; $4s^2 4p^6\,4d^{10}\,4f^{14}$; $5s^2\,5p^6\,5d^{10}$; $6s^2 6p^6$; $7s^1$

Hence, the cause of periodicity of the properties of elements is the repetition of similar electronic configurations after certain regular intervals.

2.1.1 Features of long form of periodic table

The long form periodic table is arranged in seven horizontal rows called *periods* and eighteen vertical columns called *groups* (See Table 2.2).

Periods

The seven periods are numbered from 1 to 7 starting from top and going downwards. Each period begins with an alkali metal and ends with an inert gas element. The number of elements in each period is as given in Table 2.3:

Table 2.2. Periodic Table of the Elements.

Alkali Metals	Alkaline earths	Transition metals								Coinage metals					Chal-cogens	Halo-gens	Noble gases
1	2	3	4	5	6	7	8	9	10	11	12	13	14	15	16	17	18
1008 **H** 1																	4.003 **He** 2
6.939 **Li** 3	9.012 **Be** 4											10.811 **B** 5	12.011 **C** 6	14.007 **N** 7	15.999 **O** 8	18.998 **F** 9	20.183 **Ne** 10
22.990 **Na** 11	24.312 **Mg** 12											26.982 **Al** 13	28.085 **Si** 14	30.974 **P** 15	32.064 **S** 16	35.453 **Cl** 17	39.94 **Ar** 18
39.102 **K** 19	40.08 **Ca** 20	44.956 **Sc** 21	47.90 **Ti** 22	50.942 **V** 23	51.996 **Cr** 24	54.938 **Mn** 25	55.847 **Fe** 26	58.933 **Co** 27	58.71 **Ni** 28	63.546 **Cu** 29	65.37 **Zn** 30	69.72 **Ga** 31	72.59 **Ge** 32	74.922 **As** 33	78.96 **Se** 34	79.904 **Br** 35	83.80 **Kr** 36
85.47 **Rb** 37	87.62 **Sr** 38	88.905 **Y** 39	91.22 **Zr** 40	92.906 **Nb** 41	95.94 **Mo** 42	(99) **Tc** 43	101.07 **Ru** 44	102.905 **Rh** 45	106.4 **Pd** 46	107.868 **Ag** 47	112.40 **Cd** 48	114.82 **In** 49	118.69 **Sn** 50	121.75 **Sb** 51	127.60 **Te** 52	126.904 **I** 53	83.80 **Xe** 54
138.905 **Cs** 55	137.34 **Ba** 56	138.91 **La** 57	178.49 **Hf** 72	180.948 **Ta** 73	183.85 **W** 74	186.2 **Re** 75	190.2 **Os** 76	192.2 **Ir** 77	195.09 **Pt** 78	196.967 **Au** 79	200.59 **Hg** 80	204.37 **Tl** 81	207.19 **Pb** 82	208.980 **Bi** 83	(210) **Po** 84	(210) **At** 85	(222) **Rn** 86
(223) **Fr** 87	(226) **Ra** 88	(227) **Ac** 89	(257) **Unq** 104	(260) **Unp** 105	(263) **Unh** 106	(265) **Uns** 107	(265) **Uno** 108	(267) **Une** 109									

Lanthanide series	140.12 **Ce** 58	140.907 **Pr** 59	144.24 **Nd** 60	(147) **Pm** 61	150.35 **Sm** 62	151.96 **Eu** 63	157.25 **Gd** 64	158.924 **Tb** 65	162.50 **Dy** 66	164.930 **Ho** 67	167.26 **Er** 68	168.934 **Tm** 69	173.04 **Yb** 70	174.97 **Lu** 71
Actinde series	232.038 **Th** 90	(231) **Pa** 91	238.03 **U** 92	(237) **Np** 93	(242) **Pu** 94	(243) **Am** 95	(247) **Cm** 96	(247) **Bk** 97	(249) **Cf** 98	(254) **Es** 99	(253) **Fm** 100	(256) **Md** 101	(253) **No** 102	(257) **Lr** 103

Table 2.3. Periods with Number of Elements.

Period	*No. of elements*
1	2
2	8
3	8
4	18
5	18
6	32
7	19

The first three periods are called *short* periods because they contain small number (2 or 8) of elements. The next four periods are called *long* periods because they contain large number (18 or 32) of elements. The seventh period is a long period but is *incomplete.*

This period consists of 26 elements, 14 of them being placed in *f*-block and 12 in *d*-block.

Elements from atomic numbers 104 to 112 have been reported recently. But these heavy elements are highly unstable. Names of these elements as suggested by American Chemical Society and IUPAC are given in Table 2.4. American chemical society has given the names only for elements up to 109.

Table 2.4. American chemical society and IUPAC names of some elements.

Atomic number	American chemical society names	IUPAC names
104	Rutherfordium (Rf)	Unnilquadium (Unq)
105	Hahnium (Ha)	Unnilpentium (Unp)
106	Seaborgium (Sg)	Unnilhexium (Unh)
107	Bohrium (Bh)	Unnilseptium (Uns)
108	Hassium (Hs)	Unniloctium (Uno)
109	Meitnerium (Mt)	Unnilennium (Une)
110	—	Ununnilium (Uun)
111	—	Unununilium (Uuu)
112	—	Unundodium (Uud)

Thus this period is incomplete and requires 6 more elements to be discovered to complete $7p^{1-6}$.

Groups

There are 18 vertical columns in the periodic table which constitute 18 groups and are numbered as 1, 2, 3, 4, 5, 6, 7, 8, 9, 10, 11, 12, 13, 14, 15, 16, 17, and 18.

The elements belonging to groups 1, 2, 13, 14, 15, 16, 17 and 18 are known as *normal or representative elements.* Elements in between groups 2 and 13 are called *transition elements* because they represent a transition (change) from electropositive elements on the left-hand side to electronegative elements on the right-hand side of the periodic table.

Two rows of elements (each containing 14 elements) are known as *lanthanides and actinides* respectively. They are placed separately in the periodic table to avoid uncontrolled sideway expansion of the periodic table. Names of elements in alphabetical order, their symbols and at. numbers are given in Table 2.5.

Table 2.5. Name of Elements, their Symbols and Atomic Numbers.

Element	Symbol	Atomic No.	Element	Symbol	Atomic No.
Actinium	Ac	89	Mercury	Hg	80
Aluminium	Al	13	Molybdenum	Mo	42
Americium	Am	95	Neodymium	Nd	60
Antimony	Sb	51	Neon	Ne	10
Argon	Ar	18	Neptunium	Np	93
Arsenic	As	33	Nickel	Ni	28
Astatine	At	85	Niobium	Nb	41
Barium	Ba	56	Nitrogen	N	7
Berkelium	Bk	97	Nobelium	No	102
Beryllium	Be	4	Osmium	Os	76
Bismuth	Bi	83	Oxygen	0	8
Boron	B	5	Palladium	Pd	46
Bromine	Br	35	Phosphorus	P	15
Cadmium	Cd	48	Platinum	Pt	78
Calcium	Ca	20	Plutonium	Pu	94
Californium	Cf	98	Polonium	Po	84
Carbon	C	6	Potassium	K	19
Cerium	Ce	58	Praseodymium	Pr	59
Cesium	Cs	55	Promethium	Pm	61
Chlorine	Cl	17	Protactinium	Pa	91
Chromium	Cr	24	Radium	Ra	88
Cobalt	Co	27	Radon	Rn	86
Copper	Cu	29	Rhenium	Re	75
Curium	Cm	96	Rhodium	Rh	45
Dysprosium	Dy	66	Rubidium	Rb	37
Einsteinium	Es	99	Ruthenium	Ru	44
Erbium	Er	68	Samarium	Sm	62
Europium	Eu	63	Scandium	Sc	21
Fermium	Fm	100	Selenium	Se	34
Fluorine	F	9	Silicon	Si	14
Francium	Fr	87	Silver	Ag	47
Gadolinium	Gd	64	Sodium	Na	11
Gallium	Ga	31	Strontium	Sr	38
Germanium	Ge	32	Sulphur	S	16
Gold	Au	79	Tantalum	Ta	73
Hafnium	Hf	72	Technetium	Tc	43
Helium	He	2	Tellurium	Te	52
Holmium	Ho	67	Terbium	Tb	65
Hydrogen	H	1	Thallium	Tl	81
Indium	In	49	Thorium	Th	90
Iodine	I	53	Thulium	Tm	69
Iridium	Ir	77	Tin	Sn	50
Iron	Fe	26	Titanium	Ti	22
Krypton	Kr	36	Tungsten	W	74
Lanthanum	La	57	Uranium	U	92
Lawrencium	Lr	103	Vanadium	V	23
Lead	Pb	82	Xenon	Xe	54
Lithium	Li	3	Ytterbium	Yb	70
Lutetium	Lu	71	Yttrium	Y	39
Magnesium	Mg	12	Zinc	Zn	30
Manganese	Mn	25	Zirconium	Zr	40
Mendelevium	Md	101			

2.1.2 Different blocks constituting the periodic table

To simplify the study of elements the periodic table has been divided into four blocks known as *s, p, d* and *f* blocks. The division depends upon the type of atomic orbital which receives the last electron in an atom *e.g.*, magnesium ($1s^2 2s^2 2p^6 3s^2$) which receives its last electron in *s*-orbital belongs to *s*-block. The division of periodic table into four blocks is depicted in Fig. 2.1.

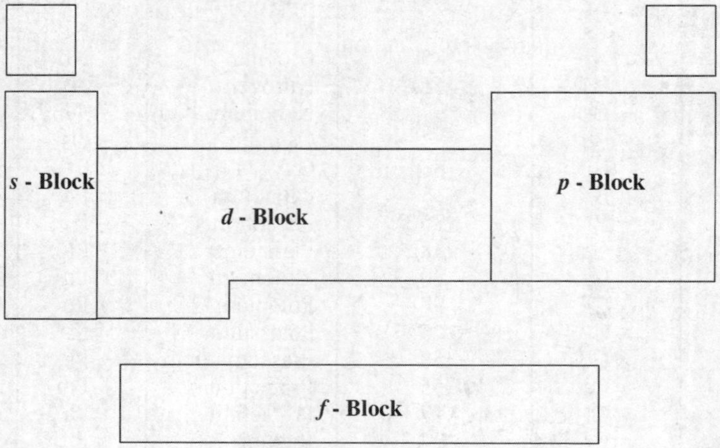

Fig. 2.1. Division of periodic table into four blocks.

s-block elements are those elements whose last electron enters the *s*-orbital of the valence shell. Therefore, elements of group 1 and group 2 constitute *s*-block elements. General electronic configuration of s-block elements is: ns^{1-2}.

p-block elements are those elements whose last electron enters the *p*-orbital of the valence shell. Therefore, elements of groups 13, 14, 15, 16, 17 and 18 constitute *p*-block. General electronic configuration of *p*-block elements is ns^2np^{1-6}.

d-block elements are those elements whose last electron enters *d*-orbitals. Therefore, elements of groups 3, 4, 5, 6, 7, 8, 9, 10, 11 and 12 constitute *d*-block. These elements are called transition elements. General electronic configuration of *d*-block elements is: $(n-1)\,d^{1-10}ns^{1,\,2}$.

f-block elements are those whose last electron enters *f*-orbital. Two series of elements at the bottom of the periodic table known as lanthanides and actinides constitute *f*-block elements. These elements are also known as inner transition elements because they constitute transition series within the transition elements. General electronic configuration of *f*-block elements is: $(n-2)\,f^{1-14}$ $(n-1)\,d^{0-1}ns^2$.

Periodic Properties

There are many physical and chemical properties of elements which show periodic variation with atomic number. Some of these properties such as valency, atomic size, ionisation energy, electron affinity and electronegativity are directly related to the electronic configuration of the atom while some properties such as melting point, density, etc., are indirectly related to electronic configurations.

2.2 ATOMIC RADII

Generally, the term atomic radius means *the distance from the centre of the nucleus to the outermost shell of electrons*. If the atom is assumed to be spherical, it is the radius of the sphere. However, the absolute size of an atom is difficult to be defined because of the following reasons:

(*i*) According to the probability picture of electrons, an atom does not have well defined boundary. The probability of finding the electron is not zero even at large distances from the nucleus.

(*ii*) It is not possible to isolate an atom and measure its radius. The probability distribution of an atom is also affected by the presence of other atoms in its neighbourhood. Therefore, the size of atom may change from one set of environment to another.

Thus, we can only arbitrarily define atomic radius as the effective size which means the distance of closest approach of one atom to another atom in a given bonding situation. Therefore, the atomic radii may be called as covalent radius, van der Waal's radius, ionic or crystal radius, etc., depending upon the types of bonding prevalent.

2.2.1 Covalent Radius

The radius of an atom is calculated by measuring the distance between the centres of two atoms (internuclear distance) in molecules with the help of X-ray diffraction or spectral studies. In the case of homonuclear diatomic molecules, the atomic radius is equal to one half of the distance between the centres of nuclei of two similar atoms bonded together in a molecule. The radius is also called covalent radius because the atoms in a homonuclear diatomic molecule are held together by covalent bonds. For example, the internuclear distance between the two atoms in F_2 molecule is 1.28Å or 128 pm (Fig. 2.2).

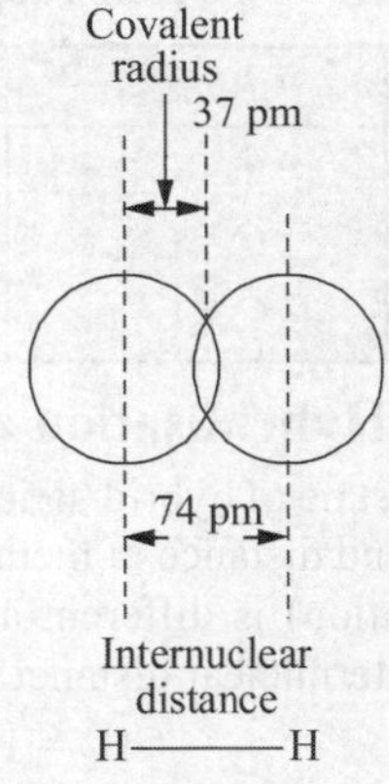

Fig. 2.2. Covalent radius of hydrogen atom.

Therefore, Atomic radius of fluorine $= \dfrac{128 \text{ pm}}{2} = 64$ pm.

It may be mentioned that 1Å = 100 pm.

Similarly, the atomic radii of chlorine and bromine are 99 pm and 114 pm respectively because the internuclear distances in Cl_2 and Br_2 are 198 pm and 228 pm respectively.

In the case of heteronuclear diatomic molecules (containing different atoms) it has been observed that when the covalent radii of different atoms are added, the resulting value agrees fairly well with the experimentally determined internuclear distance of that molecule. For example, the internuclear distance in HCl molecule can be obtained by adding the covalent radii of H and Cl.

$$r\,(\text{HCl}) = r\,(\text{H}) + r\,(\text{Cl}) = 37 + 99 = 136 \text{ pm}$$

The calculated value of internuclear distance is in fair agreement with the experimentally observed internuclear distance of 131 pm in HCl molecule. Again, the atomic radius of H atom may be calculated by subtracting the atomic radius of Cl from the internuclear distance of HCl molecule. In general, the atomic radius of an atom B in AB molecule can be calculated by measuring the internuclear distance (r_{AB}) of AB molecule and subtracting the atomic radius of atom A. Thus,

$$r_B = r_{AB} - r_A$$

The atomic radii of atoms which do not form diatomic molecules with single bonds are calculated by indirect methods. For example, the single bond radii of nitrogen and oxygen are not obtained from the bond lengths in N_2 ($N \equiv N$) and O_2 ($O = O$) because these molecules contain multiple bonds. One half of the internuclear distance of these molecules will correspond to multiple bond radii. The single bond atomic radius of nitrogen is obtained by indirect method. Compounds in which N atom is bonded to another atom by a single bond is selected. For example, the internuclear distance between C – N in H_3 CNH_2 has been found to be 147 pm. Since the atomic radius of carbon is 77 pm (obtained from C – C bond distance in diamond), the atomic radius of nitrogen is (147 – 77) = 70 pm.

2.2.2 Multiple Bond Radii

For multiple bonded atoms, the corresponding radii are called multiple bond radii. For example, carbon has double bonded radius of 66 pm in ethylene ($H_2C = CH_2$) and triple bonded radius of 60 pm in acetylene ($HC \equiv CH$). The carbon has single bond radius of 77 pm (as in ethane, $CH_3 - CH_3$). Similarly, nitrogen has single, double and triple bond radius of 70 pm, 60 pm and 55 pm respectively. As the number of bonds increases, the attraction between the atoms also increases and, therefore, the atoms approach closer and atomic radius decreases.

Bond	*Radius*	*Bond*	*Radius*
C – C	77 pm	N – N	70 pm
C = C	66 pm	N = N	60 pm
C ≡ C	60 pm	N ≡ N	55 pm

2.2.3 Hybridisation and Atomic Radius

The state of hybridisation also affects the radius of an atom. For example, the carbon-hydrogen single bond distance in methane (sp^3 hybridisation), ethylene (sp^2 hybridisation) and acetylene (*sp* hybridisation) is different and therefore, the atomic radius of carbon is different. For example, C – H, internuclear distances are:

$$CH_4 = 114 \text{ pm}, H_2C = CH_2 = 111 \text{ pm}, HC \equiv CH = 107 \text{ pm}$$

Hybridisation	*s-character*	*Example*	*C – H distance*	*Radius of C*
sp^3	25%	CH_4	114 pm	77 pm
sp^2	33.3%	$H_2C = CH_2$	111 pm	74 pm
sp	50%	$HC \equiv CH$	107 pm	70 pm

Since the atomic radius of hydrogen is 37 pm, the atomic radius of carbon in CH_4, $H_2C = CH_2$ and $HC \equiv CH$ is 77 pm, 74 pm and 70 pm respectively. The decrease in atomic radius of carbon or carbon-hydrogen internuclear distance may be explained in terms of increase in *s*-character in going from sp^3 (25%) to sp^2 (33%) to *sp* (50%). With increase in *s*-character, the electron pair between C and H becomes closer to the nucleus resulting in greater attraction. Therefore, atomic radius decreases.

The atomic radii of elements of *s*- and *p*-block elements are given in Table 2.6.

Table 2.6. Atomic radii (in pm) of main group (*s* and *p*-block) elements.

Li	Be	B	C	N	O	F	Ne*
123	89	80	77	70	66	64	(160)
Na	Mg	Al	Si	P	S	Cl	Ar
157	136	125	117	110	104	99	(191)
K	Ca	Ga	Ge	As	Se	Br	Kr
203	174	125	122	121	117	114	(200)

Rb	Sr	In	Sn	Sb	Te	I	Xe
216	191	150	141	137	133	(220)	
Cs	Ba	Tl	Pb	Bi	Po	At	Rn
237	198	155	154	152	153	—	—

* The values for atomic radii of noble gas atoms are van der Waal's radii rather than their covalent bonded radii.

2.2.4 Metallic or crystal radius

In case of metals, a large number of atoms are closely packed and are held together by means of metallic bonds. The close packing of the metal atoms is known as *crystal lattice* or *metallic lattice* and this packing is different in different metals. The **metallic radius** may be defined as: *one half of the distance between the nuclei of two adjacent atoms of a metal in a metallic lattice.*

The internuclear distance between the nearest neighbouring atoms in a metallic crystal is determined by X-ray diffraction This is divided by 2 to obtain the metallic radius.

$$r_{\text{metallic}} = \frac{\textbf{Internuclear distance between two adjacent atoms in metallic lattice}}{\mathbf{2}}$$

The metallic radii are larger than the corresponding single bond covalent radii. For example, the covalent radius of sodium is 154 pm, while its metallic radius is 186 pm. This is because metallic bond is weaker than a covalent bond and, therefore, the metal atoms in a metallic lattice are not drawn as close to each other as in a covalent bond.

2.2.5 van der Waals' Radius

van der Waals' radius is defined as *one half of the internuclear distance between the adjacent atoms belonging to two nearest neighbouring molecules of the same substance in the solid state.*

It has been observed that there are some weak forces of attraction between the neutral atoms of noble gas atoms and non-polar molecules of various substances. These are known as van der Waals' forces. Thus, the van der Waals' radius corresponds to the closest approach of two atoms of noble gases or non-polar molecules. The measurement of van der Waals' radius for chlorine is shown in Fig. 2.3.

Thus,

$$r_{\text{van der Waals}} = \frac{\textbf{Internuclear distance between two non-bonded nearest neighbouring atoms}}{\mathbf{2}}$$

Internuclear distance between two chlorine atoms of nearest neighbouring molecules = 360 pm.

$$\therefore \quad \text{van der Waals' radius of chlorine} = \frac{360}{2} = 180 \text{ pm}$$

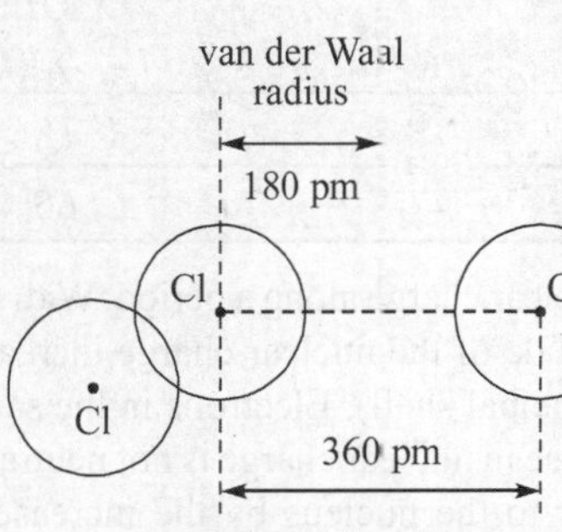

Fig. 2.3. van der Waals' radius of chlorine.

van der Waals

The van der Waals' radii of some atoms are given in Table 2.7 along with their covalent radii. The van der Waals' radii are larger than the covalent radii. This is obvious because in the formation of covalent bond, the atoms come close to each other due to the overlapping of orbitals. On the other hand, van der Waals' forces existing between atoms or molecules (non-bonded) are weaker and the atoms are at larger distances. Therefore, the internuclear distances in case of atoms held by van der Waals' forces is more than that between covalently bonded atoms. Consequently, van der Waals' radii are larger than covalent radii.

Table 2.7. Covalent and van der Waals' radii in pm of some atoms.

Element	N	O	F	Ne	P	S	Cl	Ar
Covalent	77	74	72	—	110	104	99	—
van der Waals'	150	140	135	131	180	190	180	174

2.2.6 Bragg slater radii

Bragg and Slater proposed a set of internally consistent atomic radii. These radii can be added, irrespective of the bond type involved in the interaction between two atoms to give good correlations with observed internuclear distances. In general terms, the Bragg Slater radii represent the radii of inner cores of electrons, which in turn determine how closely two atomic species can approach each other. The Bragg Slater radii of elements of second period are given in Table 2.8.

William Bragg

Table 2.8. Covalent and Bragg-Slater radii.

Element	Li	Be	B	C	N	O	F
Covalent radius (pm)	123	89	80	77	70	66	64
Bragg Slater radius (pm)	145	105	85	70	65	60	50

2.2.7 Periodic Trends in Size

1. Variation of atomic radii in a period. In general, the atomic radii decrease as we move from left to right in a period in the periodic table. For example, in the second period, the atomic radii decrease from Li to F through Be, B, C, N and O as given in Table 2.9.

Table 2.9. Decrease in atomic radii in second period.

Element and its atomic no.	Li 3	Be 4	B 5	C 6	N 7	O 8	F 9
Nuclear charge	+3	+4	+5	+6	+7	+8	+9
Atomic radius (pm)	123	89	80	77	70	66	64

This may be explained on the basis of increasing nuclear charge along a period. With the increase in atomic number from lithium to flourine, the magnitude of the nuclear charge increases and the addition of electrons takes place in the same orbit (principal shell). Electrons in the same shell do not screen each other from the nucleus, therefore, increase in nuclear charge is not neutralised by the extra electrons added. The electrons are pulled closer to the nucleus by the increased effective nuclear charge. Hence, we observe a decrease in the size of the atom.

2. Variation of atomic radii in a group. The atomic radius increases from top to bottom in a group. Table 2.10 below gives the variation of atomic radius of alkali metals group.

Table 2.10. Atomic radii of alkali metals.

Element	*Atomic No.*	*Outermost Electronic Configuration*	*Atomic radius (pm)*
Li	3	$2s^1$	123
Na	11	$3s^1$	157
K	19	$4s^1$	203
Rb	37	$5s^1$	216
Cs	55	$6s^1$	237

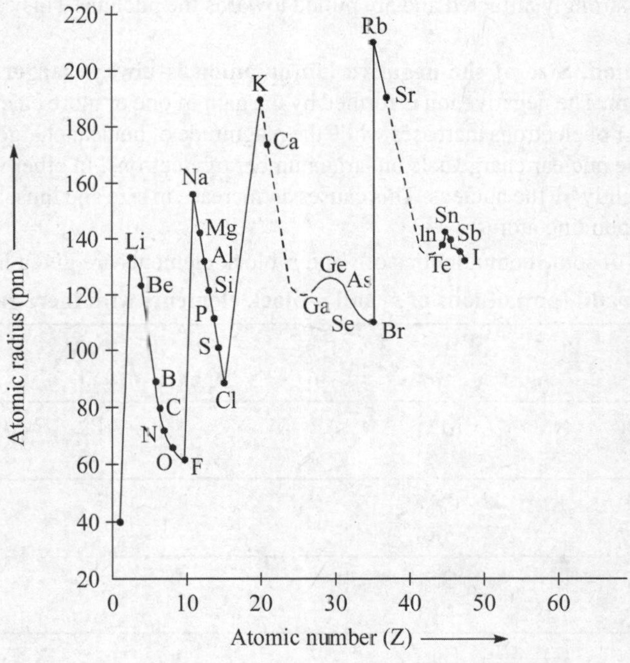

Fig. 2.4. Variation of atomic radii with atomic number.

As we move down the group, the nuclear charge increases and it is expected that atomic size should decrease. However, we observe the reverse. This happens because the extra electrons that are added go to the next shell. The effect of increase in size is more pronounced than the effect of increased nuclear charge. Consequently, the distance of outermost electron from the nucleus gradually increases down the group. In other words, atomic size increases as we move down the group.

If we plot atomic radius *vs*. atomic number, we obtain a graph as shown in Fig. 2.4.

It is evident from the figure that alkali metals ocupy the maxima and halogen atoms occupy the minima in the graph. We can infer that alkali metal is the largest and the halogen is the smallest atom in any period.

Noble gases have not been considered in the graph because their radii correspond to the van der Waals' radii.

2.2.8 Ionic and crystal radii

The ions are formed as a result of addition or removal of electrons from the outermost shells of atoms. The ions formed by the loss of electrons acquire positive charge and are called **cations** while the ions formed by the gain of electrons acquire negative charge and are called **anions. Ionic radius may be defined as the effective distance from the nucleus of the ion up to which it has an influence in the ionic bond.**

The internuclear distance between the nuclei of adjacent positive and negative ions in a crystal is obtained by X-ray studies or spectroscopic studies. By knowing the radius of one ion, that of the other ions can be calculated.

Size of the cation. Size of a cation is smaller than a neutral atom. This is due to the fact that with the removal of electrons from an atom, the magnitude of nuclear charge remains same while the number of electrons decreases. As a result, the nuclear charge now acts on lesser number of electrons. Electrons are more strongly attracted and are pulled towards the nucleus. This causes a decrease in the size of the atom.

Size of the anion. Size of the negative ion or anion is always larger than that of the corresponding atom. The negative ion is formed by the gain of one or more electrons in the neutral atom and the number of electrons increases while the magnitude of nuclear charge remains the same. As a result, the same nuclear charge acts on large number of electrons. In other words, the electron cloud is held less tightly by the nucleus. This causes an increase in size and thus, anions are larger in size than the corresponding atoms.

The ionic radii of some common ions of *s* and *p*-block elements are given in Table 2.11.

Table 2.11. Ionic radii (pm) of ions of *s*- and *p*-block elements with inert gas configuration.

	H 154	Li^+ 60	Be^{2+} 39	B^{3+} 20	C^{4+} 15	N^{5+} 11	O^{6+} 9	
O^{2-} 140	F^- 136	Na^+ 95	Mg^{2+} 65	Al^{3+} 50	Si^{4+} 41	P^{5+} 34	S^{6+} 29	Cl^{7+} 26
S^{2-} 184	Cl^{-} 181	K^+ 133	Ca^{2+} 99					
Se^{2-} 198	Br^- 195	Rb^+ 148	Sr^{2+} 113					
Te^{2-} 221	I^- 216	Cs^+ 169	Ba^{2+} 135					

Periodic Trends in Ionic Radii

1. **Variation of ionic radii in a period.** For an isoelectronic series (ions having same number of electrons), the ionic radius decreases as the nuclear charge increases.

In an isoelectronic series, there is an increase in nuclear charge and, therefore, the attraction for the same number of electrons increases. As a result, the electrons are pulled more strongly and thus ionic radius decreases. This is given in Table 2.12.

Table 2.12. Variation of ionic radii in an isoelectronic series.

Ion	N^{3-}	O^{2-}	F^-	Na^+	Mg^{2+}	Al^{3+}
Nuclear charge	+7	+8	+9	+11	+12	+13
Radius (pm)	172	140	126	95	65	50

2. **Variation of ionic radii in a group.** As we go down a group, the ionic radii increase with increase in atomic number.

This is mainly due to increase in the principal quantum levels. The effect of increase in principal quantum level is more pronounced than the effect of increased nuclear charge.

Example 1. **Explain: Increasing order of size in case of following species should be**

$$\mathbf{Fe^{3-} < Fe^{2+} < Fe}$$

Solution. When two electrons are removed from Fe atom, Fe^{2+} ion is obtained. As the positive charge on the nucleus outnumbers negative charge on the electrons, the nucleus attracts the electrons with a much greater force resulting in the shrinking of size.

On removal to a further electron to convert Fe^{2+} into Fe^{3+}, difference in the positive charge on nucleus and negative charge on the electron further increases, this increases further the pull of the nucleus on the electrons. Hence the size decreases further. Increasing order of size is therefore

$$Fe^{3+} < Fe^{2+} < Fe$$

Example 2. **Ionic radius of K^+ is smaller than that of Cl^-, both are isoelectronic. Explain.**

Solution. Electronic configurations of K^+ and Cl^- are

$$K^+ : 1s^2, 2s^2, 2p^6, 3s^2, 3p^6 \text{ (At. No. 19)}$$

$$Cl^- : 1s^2, 2s^2, 2p^6, 3s^2, 3p^6 \text{ (At. No. 17)}$$

Thus both configurations are isoelectronic, *i.e.*, same configuration. However K^+ is smaller in size compared to Cl^-. This is because K^+ has a greater nuclear charge (No. of protons) compared to Cl^-. Therefore electron cloud in K^+ is pulled to a greater extent than in Cl^-. Thus K^+ is smaller in size compared to Cl^-.

Example 3. **Radius of a cation is invariably smaller than that of an atom and that of an anion invariably lower than that of the corresponding atom. Explain.**

Solution. After an electron is removed from an atom, the cation is obtained.

$$\underset{\text{Atom}}{M} \longrightarrow \underset{\text{Cation}}{M^+} + e$$

Positive charge in M^+ outweighs the negative charge on electron. The pull of the nucleus on the electron, therefore increases. That is why a cation is smaller than the atom. However, when an electron is added to an atom, an anion is obtained.

$$M + e^- \longrightarrow \underset{\text{Anion}}{M^-}$$

Now the electronic charge is greater than the nuclear charge. Moreover repulsion among the electrons will increase thereby increasing the size of the anions compared to that of the neutral atom.

2.3 IONISATION ENERGY AND SUCCESSIVE IONISATION ENERGY

The ionisation energy of an element is defined as amount of energy required to remove the most loosely bound electron from an isolated gaseous atom.

$$M\,(g) + \text{Ionisation Energy} \longrightarrow M^+\,(g) + e^-$$

The experimental values of I.E. of elements are determined from their spectra and are reported either in electron volts or in kcal or kJ (1 kJ = 4.184 kcal). Ionisation energy is also known as ionisation potential or ionisation enthalpy.

Successive ionisation energies. After the first electron has been removed from a neutral atom, it is possible to remove second, third or even more electrons. The energies required for the removal of subsequent electrons are known as second **IE_2, third IE_3, fourth IE_4 or in general successive ionisation energies.** This may be expressed as

$$A\,(g) + IE_1 \longrightarrow A^+\,(g) + e^- \qquad \text{(first ionisation energy)}$$

$$A^+ (g) + IE_2 \longrightarrow A^{2+} (g) + e^- \quad \text{(second ionisation energy)}$$

$$A^{2+} (g) + IE_3 \longrightarrow A^{3+} (g) + e^- \quad \text{(third ionisation energy)}$$

It is observed that successive ionisation energies go on increasing progressively because it becomes increasingly difficult to remove each successive electron. This is because of the following reasons. (*i*) After the removal of first electron, the atom changes into a positive ion in which the electrons are held more tightly by the nucleus. This is because the positive ion contains smaller number of electrons while the nuclear charge remains the same. Similarly, the removal of the second electron would further increase the attractive force of the nucleus on the remaining electrons and so on. This naturally leads to increases in ionisation energies. (*ii*) Another situation that may arise in many cases is that after the removal of one, two or more electrons from the atom, the next one may be present in a lower energy level. This electron would naturally require much more energy for its removal *e.g.*, after 2 electrons have been removed from *2s* orbital of Be to form Be^{2+}, the next electron lies in 1*s* orbital. Therefore, the third ionisation energy of beryllium which involves the removals of 1*s* electron has a much higher value than second ionisation energy.

2.3.1 Factors that govern ionisation energy of an element

The ionisation energy depends upon the following factors:

(*i*) *Magnitude of nuclear charge.* This attractive force of the nucleus on the electron is directly proportional to the product of the nuclear and electronic charges. Therefore as the nuclear charge increases, the electron would be held more strongly by the nucleus. In other words, more energy would have to be supplied to remove the electron from the atom. Thus the ionisation energy increases as the nuclear charge increases.

(*ii*) *Size of the atom.* As the size of the atom increases, the outer electrons are less strongly attracted by the nucleus, because the force of attraction is inversely proportional to the square of the distance between the charged particles. With the decrease in attractive force, the outer electron is less tightly held and is, therefore, easier to remove. Thus the ionisation energy decreases as the size of the atom increases.

(*iii*) *Screening effect of the inner electrons.* The inner electronic shells act as a screen between the nucleus and outer shell. This reduces the pull of the outer shell towards the nucleus. In other words, the electrons that lie between the nucleus and outermost shell have a shielding or screening effect on the electrons of outermost shell. The larger the number of electrons in the inner shells, the greater will be the screening effect and the lesser will be nuclear attraction on the outermost electrons. Thus increase in the number of inner electrons will trend to decrease I.E. It should be carefully noted that electrons in the same or similar type of orbital do not shield each other. Thus the electrons in *s*-orbital will not shield each other. Similarly, electrons in p_x, p_y and p_z orbital of the same subshell do not shield each other. It has also been found that the screening effect of *s*-electron is more than that of *p*-electron whose effect is in turn more than that of a *d*-electron and so on.

Actual charge felt by an electron is termed as **effective nuclear charge** (Z_{eff}).

$$Z_{eff} = \text{Total nuclear charge (Z)} - \text{Screening constant (S)}$$

(*iv*) *Penetration effect of the electrons.* Electrons in the *s*-orbitals have the maximum probability of being nearer to the nucleus while this probability goes on decreasing in case of *p, d* and *f* electrons. In other words *s*-electrons have the greatest penetration towards the nucleus and the penetrating effect of the electron with a given principal quantum number, *n* is in the order : $s > p > d > f$. Greater the penetration of electron, greater is the ionisation energy since the electron would be bound more tightly to the nucleus in that case. Thus, for the same value of *n,* the ionisation energy would be more for *s*-electron compared to that for a *p*-electron which is turn will be greater than for *d*-electron and so on.

(*v*) *Electronic arrangement.* (*a*) Noble gases have the most stable electronic configuration having s^2p^6 arrangement of the outermost shell while all the inner shells are also complete.

(*b*) The arrangement in which all the occupied orbitals are completely filled as, for instance, in the case of Be ($1s^2$, $2s^2$) and Mg ($1s^2$, $2s^2$, $2p^6$, $3s^2$).

(*c*) The configuration in which all the orbitals of the same sublevel are exactly half filled also represents a very stable arrangement such as in the case of N ($1s^2$, $2s^2$, $2p^1_x 2p^1_y$ and $2p^1_z$) and *P* ($1s^2$, $2s^2$, $2p^6$, $3s^2$, $3p^1_x 3p^1_y$ $3p^1_z$).

The more stable the electronic arrangement the greater is the ionisation energy. That is why each noble gas has the highest ionisation energy of the element in the same period.

***Example 3.* First ionisation energy of Al is lower than that of Mg. Explain.**

Solution. Electric configuration of Al is $1s^2$, $2s^2$, $2p^6$, $3s^2$, $3p^1$) and that of Mg is $1s^2$,$2s^2$, $2p^6$ $3s^2$. Although we expect the I.E. to increase as we move form left to right in the periodic table, here the situation is different. Fully filled orbitals are more stable compared to half filled orbital and therefore greater energy is needed to remove the electron. Therefore Mg has a fully filled 3s orbital and it is more difficult to remove the electron from the atom. In Al, it is the electron from half filled 3*p* orbital which is to be removed, this will require comparatively less energy.

2.3.2 Variation in ionisation energy (*i*) along a period and (*ii*) down a group

Variation of ionisation energy with atomic number is shown in Fig. 2.5.

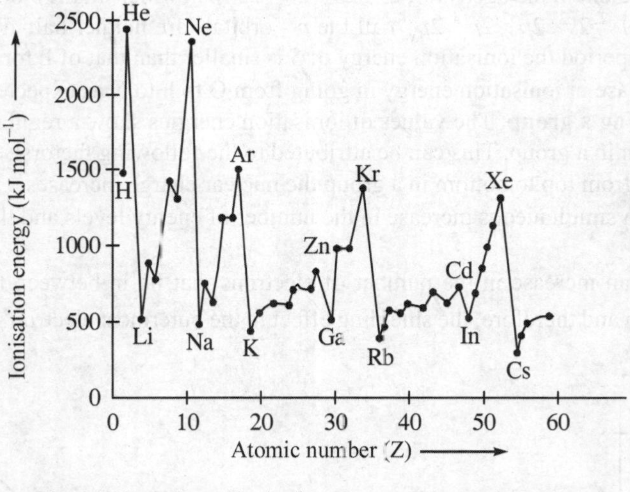

Fig. 2.5. Variation of ionisation energy with atomic number.

Figure 2.5 shows the variation of first ionisation energy of elements with atomic number upto atomic number 60. It can be seen that in each period, the maxima are found at the noble gases while minima are found at alkali metals. Thus, the metals of group 1 with one electron in outermost *s*-orbital are easy to ionise while the noble gases (group 18) with ns^2np^6 configuration are the most difficult to ionise. The following periodic trends have been observed.

Variation along a period. In general, the ionisation energy increases with increasing atomic number as we move from left to right along a period. This is because of the following reasons:

(*i*) The nuclear charge increases from left to right.

(*ii*) The number of main energy levels remains the same while the atomic size decreases. Due to the gradual increase in the nuclear charge and simultaneous decrease in the atomic size, the electrons get more and more firmly bound to the nucleus. Therefore it becomes increasingly difficult to remove an electron and hence the ionisation energy keeps on increasing. Some of the exceptional values observed within a period can be explained in terms of the role played by the other factors governing ionisation energy. To illustrate this let us consider the changes in the second period.

As we move from Li to Be there is an increase in the ionisation energy. This may be easily explained in terms of increased nuclear charge and smaller atomic size while other factors do not make a significant contribution.

On passing from Be to B, we observe that in spite of the increased nuclear charge and smaller atomic size, there is a slight decrease in ionisation energy.

This can be explained in terms of following factors:

(*a*) The electronic arrangement of B ($1s^2$, $2s^2$, $p_x^{\ 1}$) is less stable that than of Be ($1s^2$, $2s^2$).

(*b*) The outermost electron in B is present in $2p$ orbital which is less penetrating than $2s$ orbital which contains the outermost electron of Be.

(*c*) The outermost electron in B is shielded by 4 electrons while that in Be is shielded only by 2 electrons.

As a result of the above factors, the outermost electron of B is not so firmly held by the nucleus and is easily removed in comparison. For similar reasons there is a slight decrease in energy in moving from Mg to Al in third period.

As we pass from B to C to N the ionisation energy keeps on increasing primarily due to increasing nuclear charge and decreasing atomic size, the combined effect of other factors not being appreciable.

Oxygen has a slightly smaller ionisation energy than N. This is primarily due to the more stable electronic arrangement of nitrogen ($1s^2$, $2s^2$, $2p_x^{\ 1}$ $2p_x^{\ 1}$ $2p_x^{\ 1}$) in which all the p orbitals are half-filled while in oxygen ($1s^2$ $2s^2$, $2p_x^{\ 2}$ $2p_y^{\ 1}$ $2p_x^{\ 1}$) all the p - orbitals are neither half-filled not completely filled. In the third period the ionisation energy of S is smaller than that of P for similar reasons.

There is increase in ionisation energy in going from O to F to N as expected.

Variation along a group. The values of ionisation energies show a regular decrease in going from top to bottom in a group. This can be attributed to the following factors:

(*i*) In going from top to bottom in a group the nuclear charge increases.

(*ii*) There is a simultaneous increase in the number of energy levels and therefore, the atomic size increases.

(*iii*) There is an increase in the number of electrons that lie in between the nucleus and the outermost electron and therefore, the shielding effect in the outermost electron gradually increases.

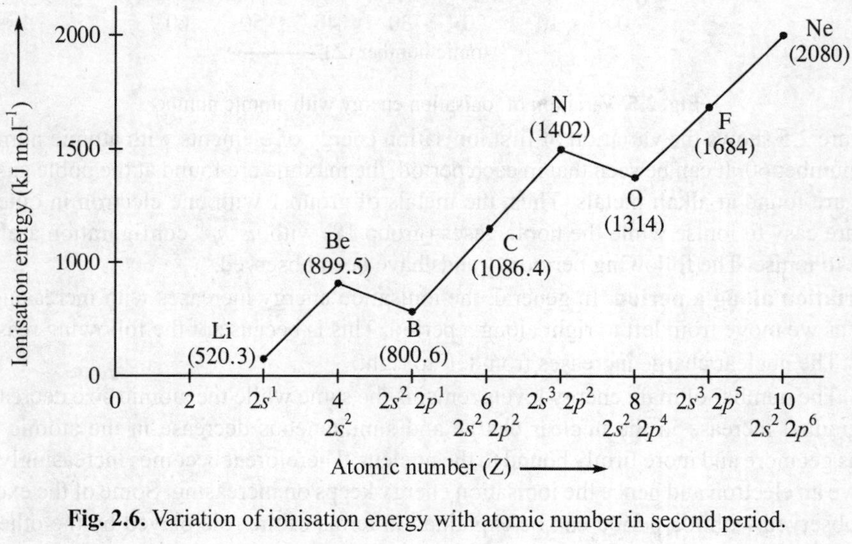

Fig. 2.6. Variation of ionisation energy with atomic number in second period.

The increase in atomic size and the shielding effect more than neutralise the effect of the increased charge. As such the electron in the higher orbital is less firmly bound to the nucleus. Consequently there is gradual decrease in the ionisation energy.

Example 4. **The Be and N in second period and Mg and P in the third period of the table have slightly higher ionisation energies than expected. Explain.**

Solution. Configurations in which there are fully filled and half filled orbitals are extraordinarily more stable than others. So more energy is needed to take out electrons from the outermost shells. In Be and Mg, there are fully filled orbitals in the outermost shell. Similarly in nitrogen and phosphorus, the outermost shells have half-filled orbitals.

$_7N: 1s^2, 2s^2, 2p^2{}_x\, 2p_y{}^1\, 2p_z{}^1,$ $\quad _{15}P: 1s^2, 2s^2\, 2p^6\, 3s^2\, 3p_x{}^1\, 3p_y{}^1\, 3p_z{}^1$

Therefore these atoms have higher ionisation energy than expected.

Example 5. **Explain: Nitrogen has higher value of I.E. than oxygen.**

Solution. Nitrogen has more stable electronic configuration in which the orbitals are all half-filled.

$_7N : 1s^2, 2s^2\; 2p_x{}^1\, 2p_y{}^1\, 2p_z{}^1$

$8O : 1s^2, 2s^2\; 2p_x{}^2\, 2p_y{}^1\, 2p_z{}^1$

Half-filled configuration is more stable than others. Hence greater energy is needed to extract the electron from them.

Example 6. **Calculate the amount of energy required to convert 1.5 g of potassium atoms in the gaseous state to form potassium ions. Ionisation energy of potassium is 419 kJ mole^{-1} and atomic mass of potassium is 39 a.m.u.**

Solution. $\underset{(1\text{mole})}{K(g)} + 419 \text{ kJ (IE)} \longrightarrow \underset{(1\text{mole})}{K^+(g)} + \underset{(1\text{mole})}{\text{electron}}$

Thus 39 g of potassium atoms require = 419 kJ energy for ionisation

$\therefore \quad 1.5 \text{ g of potassium atoms required} = \frac{419}{39} \times 1.5.$

$= 16.1$ **kJ**

2.4 ELECTRON AFFINITY

The amount of energy released when a neutral atom in the gaseous state acquires an extra electron to form a negatively charged ion is known as electron affinity. The process may be depicted as

$$A(g) + e^-(g) \longrightarrow A^-(g) + \text{Energy (Electron affinity)}$$

The electron affinity determines the tendency of an atom to add an extra electron and the firmness with which the added electron is bound to the nucleus.

If an atom has greater tendency to accept an electron, larger amount of energy will be released. Consequently, the electron affinity will be high. On the other hand, if an electron has less tendency to hold the electron, smaller amount of energy will be released. Consequently, electron affinity will be less.

2.4.1 Factors determining electron affinity

Electron affinity of an element depends mainly upon the following factors:

(*i*) *Nuclear charge.* More the nuclear charge of the atom, more strongly will it attract additional electron. Therefore, as electron affinity increases, the nuclear charge increases.

(*ii*) *Atomic size.* The smaller the size of atom, smaller will be the distance between the extra electron and the nucleus. Therefore, electrostatic force of attraction will be more and the electron affinity will be higher.

(*iii*) *Electronic configuration.* Atoms having stable electronic configuration (*i.e.*, those having completely filled or exactly half-filled outer orbitals) do not show much tendency to add extra electron and have either zero or very low electron affinities.

2.4.2 Variations of electron affinity in the periodic table (*i*) down the group and (*ii*) along the period

Variation of electron affinity down the group and along a period in the periodic table is discussed as under: (also see Table 2.13)

Down a group. The atomic size as well as the nuclear charge increase on moving down the group. But the overall effect of these factors is that the effective nuclear attraction for the electron decreases. As such atom would have less tendency to attract additional electron towards itself so that its electron affinity would decrease.

Across a period. The atomic size decreases while the nuclear charge increases. Naturally, therefore, the nucleus would have increased attraction for additional electron and electron affinity would be more. But there are many exceptions to this general trend which are mostly due to the relative stabilities of electronic configurations. Electron affinities of some elements are given in Table 2.13.

Table 2.13. Electron affinities of some elements (kJ mol^{-1}).

H 72.8							He 0
Li 59.8	Be ≈ 0	B 83.0	C 127.5	N ≈ 0	O 140.9	F 327.9	Ne 0
Na 53.1	Mg ≈ 0	Al 50.2	Si 119.6	P 74.3	S 200.7	Cl 348.3	Ar 0
K 48.4	Ca ≈ 0	Ga 35.7	Ge 115.8	As 77.2	Se 195.0	Br 324.2	Kr 0
Rb 47.0						I 295.3	Xe 0

2.4.3 Some important features of electron affinities of different elements

(*i*) *Electron affinity of fluorine is smaller than that of chlorine.* It has been pointed out above that in general electron affinities go on decreasing in moving down a group. Therefore, amongst the halogens the electron affinity of fluorine should be higher than that of chlorine. Actually the electron affmity of fluorine is lesser than that of chlorine (Fig. 2.7).

This is due to the small size of fluorine atom. In fluorine the second shell is the outermost shell. Due to the small volume of this shell the addition of an extra electron results in strong electron-electron repulsions. As a result, attractive interaction between the incoming electron and the nucleus of fluorine atom is considerably depressed. This result is low electron affinity of fluorine.

(*ii*) *Electron affinities of halogens are very high.* This is because halogens have the general electronic configuration $n^2\ np^5$. As such they have a strong tendency to acquire an extra electron to change into ions having the stable s^2p^6 configuration of the next noble gas. Therefore their electron affinities are very high.

(*iii*) *Electron affinities of noble gases are zero.* This can be attributed to the fact that their electronic configurations are very stable and they have absolutely no tendency to add an extra electron in the higher energy level. Similarly, the low value of electron affinity in case of elements like Be,

Mg. N and P can be explained in terms of their stable electronic arrangements marked by the presence of either only completely filled or exactly half filled orbitals of the same type. Therefore it is not easy to add an electron in their case.

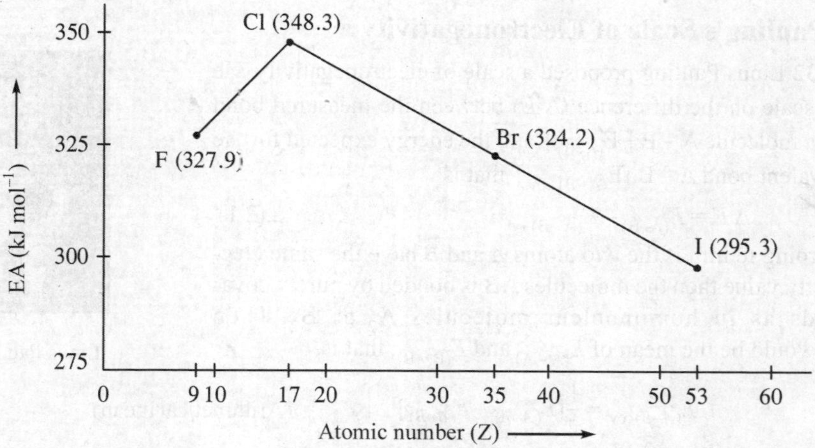

Fig. 2.7. Electron affinity values of the halogens.

***Example 7.* Successive electron affinities have negative value. Explain.**

Solution. Successive electron affinity means affinity for the addition of second, third or next electrons to the anion after one electron has been added to the neutral atom.

Actually there is a lot of resistance put up by the anions (uninegative) to the addition of more electrons because of repulsive forces. That is why successive electron affinities have negative values.

***Example 8.* How much energy in kJ is released when 1.0 g of chlorine atom are converted completely to Cl^- ions in the gaseous state, electron affinity of Cl (g) = 349 kJ and atomic mass of chlorine is 35.5 a.m.u.**

Solution. $\underset{\text{1 mole}}{Cl(g)} + \underset{\text{1mole}}{\text{electron}} \longrightarrow \underset{\text{1mole}}{Cl^-(g)} + 349\,kJ$

35.5 g of chlorine on conversion to Cl^- ions release 349 kJ of energy.

$\therefore$ 1 g of chlorine on conversion to Cl^- releases $= \frac{349}{35.5} \times 1\,kJ = 9.83\,kJ$

2.5 ELECTRONEGATIVITY

In a covalent bond between two homonuclear atoms, the shared pair of the covalent bond lies exactly midway between the nuclei of the two atoms because both the atoms may have same attraction for these electrons. For example, in molecules such as H_2, Cl_2, F_2, the electron pair remains exactly at half distance between the two nuclei. However, in the case of covalent bond between dissimilar atoms, the electron pair may be closer to one atom than to the other. This is due to the fact that different atoms have different tendencies to attract the bonded pair of electrons towards itself. *The tendency of an atom to attract the shared electron pair in a molecule towards itself is called* **electronegativity**.

Different atoms have different tendencies to attract the bonding electron pair towards themselves. Greater the ability of an atom to attract electrons in a bond, larger is the value of its electronegativity. It must be remembered that unlike other atomic properties such as ionisation energy and electron affinity which are related to the individual gaseous atoms, the electronegativity is related to the atoms in their bonded state. Various attempts have been made to derive a scale of electronegativity for the elements. Some of these are discussed below:

2.5.1 Pauling's Scale of Electronegativity

Linus Pauling

In 1932 Linus Pauling proposed a scale of electronegativity. He based his scale on the difference (ΔE) between the measured bond energy of a molecule A – B [$E_{(A-B)}$] and the energy expected for the purely covalent bond A – B [$E_{(A-B)\,cov.}$] that is

$$\Delta E = E_{(A-B)} - E_{(A-B)\,cov} \quad ...(2.1)$$

According to him if the two atoms A and B have the same electronegativity value then the molecules AB is bonded by purely covalent bonds as in homonucleur molecules A_2 or B_2. Then $E_{(A-B)\,cov}$ would be the mean of $E_{(A-A)}$ and $E_{(B-B)}$, that is,

$$E_{(A-B)\text{cov}} = \frac{1}{2}[E_{(A-A)} + E_{(B-B)}] \quad \text{(Arithmetical mean)}$$

or

$$E_{|A-B|\text{cov}} = [E_{(A-A)} \times E_{(B-B)}]^{1/2} \quad \text{(Geometrical mean)} \quad ...(2.2)$$

Pauling used geometric mean for his calculations. Substituting from (2.2) into (2.1),

$$\Delta E = E_{A-B} - [\, E_{A-A} \cdot E_{B-B}\,]^{1/2} \quad ...(2.3)$$

When molecules AB is purely covalent then

$$E_{(A-B)} = E_{(A-B)\text{cov}} \quad \text{and } \Delta E = 0$$

$$\Delta E = E_{(A-B)} - E_{(A-B)\text{cov}} = 0$$

$$\Delta E = E_{(A-B)} - [E_{(A-B)} \cdot E_{B-B}]^{1/2} = 0$$

However, if the two atoms A and B have different electronegativities the bond A – B will no longer be purely covalent and the energy $E_{(A-B)}$ would be greater than the mean of $E_{(A-A)}$ and $E_{(B-B)}$.

$$E_{(A-B)} > [E_{(A-A)} \cdot E_{(B-B)}]^{1/2}$$

In other words, ΔE would be greater than zero. This excess bond energy ΔE is called ionic-covalent resonance energy and arises due to the resonance stabilization of the sort

$$\underset{\text{covalent}}{A-B} \rightleftharpoons \underset{\text{ionic}}{A^+B^-} \quad \text{(B is more electronegative than A)}$$

Such a resonance hybrid of ionic and covalent characters gives rise to a strong bond than a purely covalent or purely ionic bond. The value of ΔE increases as the difference in electronegativity between the two atoms increases. This is clearly indicated on hydrohalides where the value of ΔE decreases as the electronegativity difference between the two atoms (H and $X = F$ to I) decreases.

Halide	HF	HCl	HBr	HI
ΔE (kJ/ mol^{-1})	268	42	46	5

This also indicates decrease in ionic character from HF to HI. In order to assign numerical values for the electronegativity of an element, Pauling proposed the following empirical expression to correlate ΔE with electronegativity difference.

$$0.18\,(\Delta E)^{1/2} = \chi_A - \chi_B \qquad \text{... (2.4)}$$

where χ_A is electronegativity of element A and χ_B is the electronegativity of element B and ΔE is expressed in kcal mole^{-1}.

However if ΔE is expressed in eV per molecule then the above expression becomes

$$0.208\,(\Delta E)^{1/2} = \chi_A - \chi_B \qquad \text{... (2.5)}$$

The equation valid for S.I. unit of energy is

$$0.088\,(\Delta E)^{1/2} = \chi_A - \chi_B \qquad \text{... (2.6)}$$

Assigning an arbitrary χ value of 2.1 for H, Pauling got a suitable range of numerical value for the element C to F in the range 2.5 to 4.0. This procedure allowed the evaluation of electronegativity of 14 elements H, C, N, O, F, Si, P, S, Cl, Ge, As, Se, Br and I. Values for others were then calculated from thermochemical data of compounds containing one of these 14 elements. Pauling values of electronegativity are given in Table 2.14. The determination of ΔE is subject to considerable errors and hence numerical values of electronegativity are subject to variation. This exposes the drawback of Pauling's method.

Table 2.14. Electronegativities of some elements.

Element	*Revised Pauling*	*Alfred-Rochow*
H	2.2	2.20
Li.	0.98	0.97
Be	1.57	1.47
B	2.04	2.01
C	2.55	2.50
N	3.04	3.07
O	3.44	3.50
F	3.98	4.10
Na	0.93	1.01
Mg	1.31	1.23
Al	1.61	1.47
Si	1.90	1.74
P	2.19	2.06
S	2.58	2.44
CI	3.16	2.83
Ar	......	......
K	0.82	0.91
Ca	1.00	1.04
Zn	1.65	1.66
Ga	1.81	1.82
Ge	2.01	2.02
As	2.18	2.20

(Contd.)

Element	*Revised Pauling*	*Alfred-Rochow*
Se	2.55	2.48
Br	2.96	2.74
Kr	2.9	......
Rb	0.82	0.89
Sr	0.95	0.99
Y	1.22	1.11
Cd	1.69	1.46
In	1.78	1.49
Sn	1.96	1.72
Sb	2.05	1.82
Te	2.1	2.01
I	2.66	2.21
Xe	2.6	.....
Cs	0.79	0.86
Ba	0.89	0.97
Hg	2.00	1.44
TI	2.04	1.44
Pb	2.33	1.55
Bi	2.02	1.67

Example 9. **Calculate electronegativity of C in C – H bond if E_{C-H}, E_{H-H} and E_{C-C} are 98.8, 104 and 83 kcal/mole respectively.**

Solution. Pauling equation relating ΔE and electronegativity difference between the two elements

$$0.18 \times (\Delta E)^{1/2} = \chi_C - \chi_H$$

ΔE in the present case is obtained as under :

$$\Delta E = E_{C-H} - [E_{C-C} \cdot E_{H-H}]^{1/2}$$
$$= 98.8 - (83 \times 104)^{1/2}$$
$$= 98.8 - 92.9$$
$$= 5.9$$

Now $\qquad 0.18\,(5.9)^{1/2} = X_C - 2.1$

$$X_C = 2.1 + 0.18 \times 5.9^{1/2}$$
$$= 2.1 + 0.18 \times 2.43$$
$$= 2.1 + 0.44 = 2.54$$

Example 10. **Electronegativities of hydrogen and fluorine are 2.1 and 4.0 respectively. Calculate the percentage ionic character in HF.**

Solution. According to Pauling, when the electronegativity difference between two elements is 1.7 percentage of ionic character is 50.

Therefore, per cent of ionic character if the electronegativity difference is (4.0 – 2.1) or 1.9

$$= \frac{1.9}{1.7} \times 50 = 56$$

2.5.2 Mulliken's Scale of electronegativity

Mulliken Scale of Electronegativity. According to Mulliken, electronegativity of an element is the arithmetic mean of its ionization potential and electron affinity.

$$\text{Thus electronegativity} = \frac{\text{Ionization Potential} + \text{Electron affinity}}{2}$$

The physical picture of Mulliken concept is reasonable because the tendency of an atom to attract the shared electrons in a bond should be average of the tendency of the atom to hold its own electrons (IE) and the tendency to attract the additional electron (EA).

Mulliken also suggested the following empirical equation for calculating electronegativity of an element:

$$\chi = 0.374\left[\frac{\text{IE} + \text{EA}}{2}\right] + 0.17$$

$$= 0.187\,[\text{IE} + \text{EA}] + 0.17$$

Here IE and EA values are in electron-volts.

It is possible to calculate the electronegativity of an element on Pauling scale if its electronegativity is known on Mulliken scale by using the relationship

$$\text{Electronegativity on Pauling scale (in eV)} = \frac{\text{Electronegativity on Mulliken scale (in eV)}}{2.8}$$

Disadvantage of Mulliken Scale is that IE and EA values change with the valance state of the element.

2.5.3 Alfred-Rochow electronegativity

Alfred and Rochow proposed an empirical approach to electronegativity based on electrostatic force between the nucleus and electron. They defined electronegativity as electrostatic force (F) exerted by the effective nuclear charge (Z_{eff}) on the outermost electron at a distance r from the nucleus. That is

$$F \propto \frac{Z_{eff}}{r^2}$$

where r = covalent radius of the atom in Å,

Z_{eff} is calculated according to Slater rules based on screening effect of intervening electrons. The attractive force (F) is then correlated to electronegativity value on the Pauling scale using an empirical relationship to obtain Alfred-Rochow electronegativities χ_{AR} (AR for Alfred-Rochow).

$$\chi_{AR} = 0.359 \times \frac{Z_{eff}}{r^2} + 0.744$$

The electronegativity values so obtained agree quite closely with those of Pauling's (Table 5.4).

Advantages

(*i*) Alfred-Rochow method depends on measuring covalent distance which are obtained with greater accuracy by X-ray diffraction. Therefore the X_{AR} values are quite accurate.

(*ii*) The Alfred-Rochow approach is based on the more fundamental and reasonable dependence of coulombic force on distance.

Disadvantages

(*i*) Slater rules used for calculated Z_{eff} are empirical and may not give accurate values.

(*ii*) Covalent radii are inaccurate when the bond order is not known.

2.5.4 Mulliken-Jaffee electronegativity concept

Jaffee and his co-workers revised the Mulliken definition and suggested that electronegativity of an atom is different in different environment. Some of the factors which affect electronegativity values are as follows:

(*a*) Oxidation state. Electronegativity of an atom varies from one bonding situation to another depending upon the oxidation number of the atom in the molecule, for example, Pauling electronegativity value is 1.62 for TI(I) whereas it is around 2.04 for T1(II). Similarly, electronegativity of S in SF_6 is higher than that in $SC1_2$ because the former has S in higher oxidation state. The higher the oxidation state, the greater the attraction for electron and higher the electronegativity.

(*b*) Hybridization. It affects electronegativity value considerably because *s*-orbital due to its higher penetration effect is more tightly held than other orbitals and hence has low energy. Consequently it will have greater tendency to attract electrons and hence show higher electronegativity values, for example, carbon in CH_4 is sp^3 hybridized which has 25% *s*-character. The electronegativity of C in CH_4 is 2.53. In ethylene $H_2C = CH_2$, carbon is sp^2 hybridized which has 33% *s*-character. The electronegativity of C in C_2H_4 is 2.75. In case of acetylene $HC \equiv CH$, carbon is *sp* hybridized which has 50% *s*-character. Thus electronegativity of C in C_2H_2 is raised and its value is referred to be 3.3.

(*c*) Charge. An atom with a positive charge will attract electrons more readily than a neutral atom while a negatively charged atom will attract electron less readily than a neutral atom.

(*d*) Bond order. As bond order of the atom whose electronegativity is under discussion increases, its electronegativity increases, for example, C in C_2H_2 is more electronegative than C in C_2H_4.

Periodic Variation of Electronegativity

The electronegativity generally increases on moving across a period from left to right. This is due to decrease in atomic size and increase in effective nuclear charge. As a result of increase in effective nuclear charge, the attraction for the electrons increases.

Electronegativity generally decreases from top to bottom in a group as atomic size increases and the bonding electrons become away from the nucleus.

Thus, the alkali metals of group 1 have the lowest electronegativities and the halogens of group 17 have the highest electronegativities. Further, inspection of Table 2.14 shows that the non-metals on the right-hand side of the periodic table have high electronegativities while the metals on the left-hand side have low electronegativities. As an approximation, elements with electronegativity 2.0 or greater are non-metals while those with electronegativity less than 2.0 are metals.

It is clear from Table 2.14 that the electronegativity generally increases on moving across a period from left to right and decreases on moving down a group.

Applications of Electronegativities

Some of the important applications of electronegativity values are described below:

1. **Types of bonds.** The type of bond formed between two atoms would depend upon the difference in their electronegativities. If this difference is zero or very small, the bond formed would be covalent and if the difference exceeds 2.5, the bond formed would be electrovalent. If, however, the electronegativity difference is less than 2.5 but otherwise quite appreciable, the bond formed would be a polar bond.

2. **Explanation of diagonal relationship.** The electronegativity increases as we go from Li to Be (variation of electronegativity in a period) but it decreases as we move from Be to Mg (variation of electronegativity in a group). As a result of these two opposite changes (one along the period and the other down the group), as we move diagonally, these two effects partly cancel each other and there is no appreciable change in electronegativity. This is the reason why Li (1.0) and Mg (1.2) have close values of electronegativities.

Because of similar values of electronegativities, Li and Mg have similar chemical properties.

3. **Explanation for bond angles.** The lesser the electronegativity of the central atom in a polyatomic molecule, the lesser would be the magnitude of the bond angle. If the electronegativity of the central atom is less, it would not be able to hold the bonding electron pairs towards itself. Therefore, the bonding electron pairs would shift more towards the other atoms attached to the central atom. This would result in a decrease in the bond pair–bond pair repulsion and hence a decrease in the bond angle will be observed.

This is illustrated by the bond angles and electronegativities of trifluorides of the fifth group elements as given below. The decrease in electronegativity of the central atom results in a decrease of the bond angle.

Compound	*Electronegativity of Central Atom*	*Bond Angle*
NF_3	3.0	102°
PF_3	2.1	97°
AsF_3	2.0	96°
SbF_3	1.9	88°

4. **Calculation of bond length.** If the two atoms A and B bonded together through a covalent bond differ in their electronegativities, then the covalent bond would acquire some ionic character. In other words, the bond acquires polarity. The greater the polarity, the shorter would be the length of the bond formed between A and B. Shoemaker and Stevenson proposed the following empirical equation for calculating the bond length in such cases:

$$d_{A-B} = r_A + r_B - 0.09\ (\chi_A - \chi_B)$$

where d_{A-B} is the actual bond distance between A and B; r_A and r_B are the covalent radii of A and B and $\chi_A - \chi_B$ is the electronegativity difference. The normal length of the covalent bond between A and B would have been equal to $r_A + r_B$. Evidently, the shortening of the bond between A and B due to electronegativity difference is equal to $0.09\ (\chi_A - \chi_B)$. Due to reduction in length, the bond would be more stable.

We can calculate the bond distance N – O with the help of above equation. Covalent radii of nitrogen and oxygen atoms are, respectively, 0.75 and 0.74 Å and the electronegativities of N and O are 3.0 and 3.5, respectively. Thus,

$$d_{N-O} = 0.75 + 0.74 - 0.09\ (3.5 - 3.0) = 1.445\ \text{Å}$$

The actual N – O bond length is 1.46 Å.

5. **Calculation of partial ionic character of a covalent bond.** The ionic character in a covalent bond between two atoms, say, A and B, is due to the difference in the electronegativities of A and B. Greater the difference in the electronegativities, greater would be the ionic character and consequently the higher would be the stability of the resulting bond.

Pauling suggested the following correlation between the electronegativity difference $\chi_A - \chi_B$ and the per cent ionic character of the covalent bond:

$$\text{Per cent ionic character} = [1 - e^{-0.25(\chi_A - \chi_B)}] \times 100$$

Let us calculate the per cent ionic character in HF molecule using Pauling's equation. Electronegativity values (Alfred-Rochow) of H and F are 2.2 and 4.1 respectively.

According to Pauling's equation,

$$\left.\begin{array}{r}\text{Per cent ionic character} \\ \text{of H} - \text{F bond}\end{array}\right] = [1 - e^{-0.25(1.9)}] \times 100$$

$$= 37.77$$

Pauling suggested an alternative empirical equation also for calculating the per cent ionic character, according to which

$$\text{Per cent ionic character} = 18\,(\chi_A - \chi_B)^{1.4}$$

According to this equation, the per cent ionic character of HF comes out to be 44.12. This value is quite close to the one (*viz.*, 44.8) obtained by Pauling from dipole moment of HF.

2.6 EFFECTIVE NUCLEAR CHARGE AND SCREENING EFFECT

The relation between energy and energy shell for an electron in hydrogen atom is given by expression

$$E = -R\frac{Z^2}{n^2} \quad \text{... (2.9)}$$

For hydrogen atom ($Z = 1, n = 1$), the energy of the outermost electron calculated from equation (2.9) agrees with the observed I.E. 1312 kJ mole^{-1}. However the calculated and observed value for atom with $Z > 1$ do not agree fairly. The observed ionization energy of say Li is considerably lower than that calculated from equation (2.9) because of following reasons:

(*i*) The expression does not hold good for multi-electron atom because it does not take inter-electronic repulsions into account.

(*ii*) The $2s^1$ electron does not face the nuclear charge directly. There are intervening $1s^2$ electrons which not only shield the nucleus but also repel the $2s^1$ electron resulting in removal of the electron.

Thus the inner core of electrons shield or screen the nucleus from outer electron. Consequently the outer electrons experience only a part of the nuclear charge or the effective nuclear charge faced by the outer electron is lower than that expected from total nuclear charge. In case of Li, the effective nuclear charge faced by $2s^1$ electron is not 3 but between 1 and 2 (actual value is 1.59). However if we recall the radical probability distribution curve of 2*s* orbital in hydrogen like atoms, it is apparent that 2*s* electron has definite probability of finding itself near the nucleus and hence would penetrate the inner 1*s* orbital core. This means that although 2*s* electron is shielded by the 1*s* core yet 2*s* electron penetrates it to some extent and experiences somewhat higher nuclear charge. In fact the energy of outer electron is determined by the effective nuclear charge (Z_{eff}) which is less than *Z* (nuclear charge) by an amount shielding constant *(S)* of the intervening core of the electrons

$$Z_{eff} = Z - S$$

Due to greater penetration, *s*-electrons tend to shield the nucleus more effectively than *p*-electrons which in turn shield the nucleus more effectively than *d*-electrons. That is, for a given *n,* the screening or shielding effect varies as

$$s > p > d > f$$

2.6.1 Slater Rules for determining effective nuclear charge

Slater gave a set of empirical rules for calculating the effective nuclear charge (Z_{eff}) experienced by electrons in different orbitals. As stated above, the Z_{eff} acting on a given electron is calculated by subtracting the screening constant (S) from the atomic number Z (nuclear charge).

That is: $Z_{eff} = Z - S$

The 'Slater rules' for calculating screening on shielding constant (S) are as follows:

(1) The electronic configuration of the element is first written in the following order and groupings: (1*s*); (2*s*, 2*p*); (3*s*, 3*p*); (3*d*); (4*s*, 4*p*); (4*d*); (4*f*); (5*s*, 5*p*); (5*d*); (5*f*) and so on.

(2) For an electron considered in a group of *s*, *p* electrons the shielding constant (S) is the sum of the following contributions:

(*a*) No contribution for electrons in groups beyond the one considered.

(*b*) An amount of 0.35 for each electron in the group considered (except in 1*s* group where 0.30 is used instead of 0.35).

(*c*) An amount of 0.85 for each electron in the next inner shell ($n - 1$).

(*d*) An amount of 1.00 for each of all other inner shell electrons.

(3) For an electron considered in *d* or *f* group, rules 2*a* and 2*b* apply as such; however rules 2*c* and 2*d* are replaced with the rule that all other electrons lying to the left of *d* or *f* group contribute 1.00 each.

***Example 11*. Calculate Z_{eff} faced by a 3*s* or 3*p* electron in phosphorus atom ($Z = 15$, $1s^2$ $2s^2$ $2p^6$ $3s^2$ $3p^3$)**

***Solution*. Group configuration** = ($1s^2$) ($2s^2$, $2p^6$) ($3s^2$, $3p^3$). Four other electrons in the group ($3s^2$, $3p^3$) contribute 4 × 0.35 and next inner shell contributes 8 × 0.85 and then the next 2 × 1.

For a 3*s* or 3*p* electron, $S = (2 \times 1.00) + (8 \times 0.85) + (4 \times 0.35)$

$$S = 2.00 + 6.80 + 1.40 = 10.20$$

$$Z_{eff} = 15 - 10.20 = 4.80$$

***Example 12*. Calculate Z_{eff} experienced by a 2*p* electron in oxygen atom Z (8, $1s^2$, $2s^2$ $2p^4$)**

***Solution*.** Group configuration = ($1s^2$); ($2s^2$ $2p^4$)

There are 5 other electrons in group ($2s^2$, $2p^4$) under consideration which contribute 5 × 0.35 towards S. Next inner shell contributes 2 × 0.85. Thus for a 2*p* electron

$$S = (2 \text{ x } 0.85) + 5 \text{ x } (0.35) = 3.45$$

$$Z_{eff} = (8 - 3.45) = 4.55$$

***Example 13*. Calculate Z_{eff} experienced by (*i*) 5*s* electron (*ii*) 4*d* electron in Ag atom (Z = 47, $1s^2$, $2s^2$, $2p^6$, $3s^2$, $3p^6$, $3d^{10}$, $4s^2$, $4p^6$, $4d^{10}$, $5s^1$).**

***Solution*.** Group configuration = ($1s^2$), ($2s^2$, $2p^6$); ($3s^2$, $3p^6$); ($3d^{10}$) ($4s^2$, $4p^6$), ($4d^{10}$) ($5s^1$)

(*i*) **For 5*s* electron,** $S = (28 \times 1.00) + (18 \times 0.85) + (0 \times 0.35)$

$= 43.30$

$Z_{eff} = 47 - 43.30 = 3.70$

(*ii*) **For 4*d* electron,** $S = (36 \times 1.00) + (9 \times 0.35) + 0$

$= 39.15$

$Z_{eff} = 47 - 39.15 = 7.85$

Example 14. **Calculate Z_{eff} for (*i*) 4s and (*ii*) 3*d* electron in iron atom**

$$Z = (26, 1s^2, 2s^2, 2p^6, 3s^2, 3p^6, 3d^6, 4s^2)$$

Solution. **Group configuration** = $(1s^2);(2s^2, 2p^6); (3s^2\ 3p^6); (3d^6); (4s^2)$

(*i*) **For 4s electron,** $S = (10 \times 1.00) + (14 \times 0.85) + (1 \times 0.35) = 22.25$

$$Z_{eff} = 26 - 22.25 = 3.75$$

(*ii*) In case of 3*d* electron, there is no contribution from outer 4*s* group electrons inner of 3*d* contribute one each.

Example 15. **Calculate the effective nuclear charge for**

(*i*) an electron in 3*d* orbital, and

(*ii*) an electron in 4s orbital of chromium (*Z* = 24) with electronic configuration =$1s^2, 2s^2, 2p^6, 3s^2, 3p^6, 3d^5, 4s^1$

Solution. **Group configuration** $(1s^2), (2s^2\ 2p^6)\ (3s^2\ 3p^6)\ (3d^5), (4s^1)$

(*i*) For 3*d* electron. There is no contribution from outer 4*s*. Remaining 4 electrons in 3*d* contribute 4 × 0.35 and all others contribute one each.

$$S = (18 \times 1) + (4 \times 0.35) + (4 \times 0)$$
$$= 18 + 1.40 + 0 = 19.40$$
$$Z_{eff} = 24 - 19.40 = 4.60$$

(*ii*) For 4*s* electron. There are no other electron in 4*s*. Consequently there is no contribution and we have 0 × 0.35 = 0.

Next inner shell (n = 3) has 13 electrons contributing 13 × 0.85, others contribute each and we have

$$S = (10 \times 1) + (13 \times 0.85) + (0 \times 0.35)$$
$$= 10 + 11.45 + 0 = 21.45$$
$$Z_{eff} = 24 - 21.45 = 2.55$$

Example 16. **Calculate Z_{eff} for valence electrons in Ca and Br.**

Solution. (*a*) Electronic configuration for Ca is

$$1s^2\ 2s^2\ 2p^6\ 3s^2\ 3p^6\ 4s^2$$

This can be grouped as

$$1s^2, (2s^2, 2p^6), (3s^2, 3p^6), 4s^2$$

Screening constant of 4*s* electron

$$= (1 \times 0.35) + (8 \times 0.85) + (10 \times 1)$$
$$= 17.15$$

Hence $Z_{eff} = 20 - 17.15 = 2.85$

(*b*) Electronic configuration for Br is

$$1s^2\ 2s^2\ 2p^6\ 3s^2\ 3p^6\ 4s^2\ 3d^{10}\ 4p^5$$

This can be grouped as

$$1s^2, (2s^2, 2p^6), (3s^2, 3p^6), 3d^{10}, (4s^2, 4p^5)$$

Screening constant of 4*p* electron

$$= (4 \times 0.35) + (10 \times 0.85) + (18 \times 1)$$
$$= 27.9$$

Hence $Z_{eff} = 35 - 27.9 = 7.1$

2.6.2 Important applications of Slater rules

1. It justifies the filling of 4*s* orbitals in preference to (*n* –1) *d* orbitals. For example, potassium has electronic configuration,

$$1s^2\ 2s^2\ 2p^6\ 3s^2\ 3p^6\ 4s^1$$

Here Z_{eff} for $3d = 19 - (10 \times 1.00) + (8 \times 0.85) = 2.20$

If the electronic configuration were

$$1s^2\ 2s^2\ 2p^6\ 3p^6\ 3d^1$$

Then Z_{eff} for $3d = 19 - [(18 \times 1.00) + 0 \times 0.35)] = 1.00$

This shows that a 4*s* electron experiences a higher Z_{eff} and hence a configuration in which 4*s* is occupied in preference to 3*d* would be more stable.

2. Although Slater rules do not allow the exact calculations of the energies of electron, yet these are useful in understanding variation in size, ionization potential and electronegativity. For example, for an outer electron in elements along the 3rd period, we notice the following trend in Z_{eff} value.

Element	P	S	Cl	Ar	H
Z_{eff}	4.80	5.45	6.10	6.75	(2.2)

This shows that Z_{eff} increases along the period and attains maximum value at the noble gas which accounts for increasing ionization potential along the period with maximum at the noble gas. Ionization potential shows sudden drop after that as evidenced in Z_{eff} value of *H*.

***Example 17.* 4*s* subshell is filled first and 3*d* later on while 2*p* subshell is filled first and 3*p* later on. Explain the reason.**

Solution. (*i*) The problem can be discussed by taking the case of potassium (K).

It has an electronic configuration:

$1s^2\ 2s^2\ 2p^6\ 3p^6\ 4s^1$ and not $1s^2\ 2s^2\ 2p^6\ 3s^2\ 3p^6\ 3d^1$

Let us calculate effective nuclear charge on the outermost electron based on the two configurations.

Configuration I

E.N.C. for 4*s* electron $= [19 - (10 \times 1) + (8 \times 0.85)] = 2.20$

Configuration II

E.N.C. for 4*s* electron $= 19 - [(18 \times 1) + (0 \times 0.35)] = 1$

Greater E.N.C means greater stability and lower energy. Hence the configuration I is observed and 4*s* subshell has lower energy and filled first.

(*ii*) 2*p* is filled first and 3*p* later on.

This can be studied by calculating the E.A.N. of last electron in atom with atomic No. 5 viz., boron. The two posibilities are:

$1s^2\ 2s^2\ 2p^1$ (Configuration I)

$1s^2\ 2s^2\ 3p^1$ (Configuration II)

E.A.N. of last *i.e.*, $2p^1$ electron (Configuration I) according to Slater Rule is

$$5 - [(2 \times 0.35) + (2 \times 0.85)]$$
$$= 5 - [0.70 + 1.70]$$
$$= 5 - (2.40) = 2.6$$

E.A.N. of last $3p^1$ electron (Configuration II)

$$= 5 - (2 \times 0.85) + (2 \times 1)]$$
$$= 5 - 3.7 = 1.3$$

Thus in configuration I, E.A.N of the $2p^1$ electron is higher indicating lower energy and greater stability of $2p$ orbital compared to $3p$ orbital.

2.6.3 Limitations of Slater Rules

(*i*) Both *s* and *p* orbitals are grouped together for calculating Z_{eff} whereas the radial probability distribution curves for a given *n* indicate a higher penetration or shielding effect of *s* as compared to *p*-orbital.

(*ii*) All electrons irrespective of *s, p, d,* or *f* orbitals at positions lower than the $(n-1)$ shell shield the *n* shell electron with equal contribution of 1.00. This is unjustified because different orbitals cannot have the same contribution on energy consideration.

(*iii*) Penetration by higher occupied orbitals (than the one considered) into inner orbitals is well known but even then their contribution is ignored as per Slater Rule 2*a*.

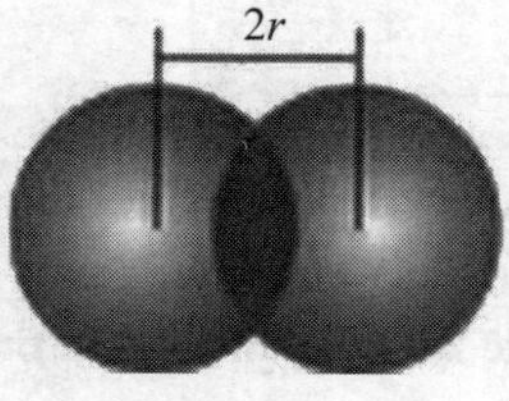

Covalent radius

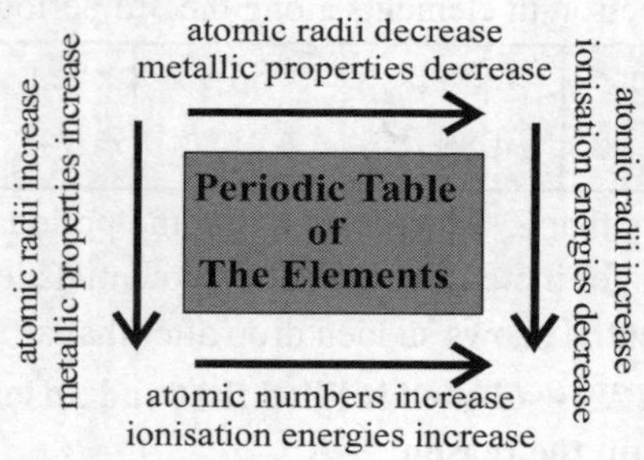

Periodic trends

EXERCISES
(INCLUDING QUESTIONS FROM DIFFERENT UNIVERSITY PAPERS)

Objective Questions

Fill in the blanks

1. Elements of *s* and *p* block are also called ....
2. The third ionisation potential of aluminium is .... than its second ionisation Potential.
3. On Mulliken scale, the average of ionisation potential and electron affinity is known as ....
4. Elements of second period are known as ....
5. Electronegativity of an atom is given by the average of electron affinity and ....
6. Electronegativity .... from left to right across a period and .... on descending a group.
7. Elements of third period are known as ....
8. The size of positive ion is .... while that of negative ion is .... than that of the parent atom .
9. The number of transition series in the periodic table is ....
10. The cause of periodicity in properties is the recurrence of .... at .... intervals.

Multiple Choice Questions

Choose the correct option

1. The most electronegative elements have electronic configuration
(*a*) $ns^2\ np^3$ (*b*) $ns^2\ np^4$ (*c*) $ns^2\ np^5$ (*d*) $ns^2\ np^6$

2. In this order of elements: C, N, O, F the electronegativity
(*a*) decreases (*b*) increases
(*c*) remains constant (*d*) none of these

3. Highest electron affinity is of the element
(*a*) P (*b*) Cl (*c*) Br (*d*) I

4. $1s^2\ 2s^2\ 2p^6\ 3s^2$ is the electronic configuration of the element. The element is a
(*a*) metal (*b*) non-metal (*c*) inert gas (*d*) metalloid

5. Which ion is colourless ?
(*a*) Cu^+(*b*) Co^{2+} (*c*) Ni^{2+} (*d*) Fe^{3+}

6. General electronic configuration of transition elements is
(*a*) $ns^2\ nd^{1-10}$ (*b*) $ns^2\ np^6\ nd^{1-10}$
(*c*) $(n-1)\ d^{1-10}\ np^6$ (*d*) $(n-1)\ d^{1-10}\ ns^{1-2}$

7. Increasing order of electronegativity of the elements is
(*a*) C, N, Si, P (*b*) N, Si, C, P (*c*) Si, P, C, N (*d*) P, Si, C, N

8. Which of the *s* block elements is liquid
(*a*) Na (*b*) K (*c*) Mg (*d*) Cs

9. Lowest ionisation potential will be of
(*a*) halogens (*b*) inert gases (*c*) d-block (*d*) alkali metals

10. In modern periodic table, how the elements are arranged ?
(*a*) Increasing mass (*b*) Decreasing volume
(*c*) Increasing atomic number (*d*) Alphabetically

11. Which ion has smallest size ?
(*a*) N^{-3} (*b*) O^{-2} (*c*) F^- (*d*) Na^+

12. Which of the following alkaline earth metals has lowest ionisation potential ?
(*a*) Mg (*b*) Ca (*c*) Sr (*d*) Ba

13. Which one of the following statement is not correct for transition elements ?
(*a*) They form coloured compounds
(*b*) All compounds of them are paramagnetic
(*c*) They show variable oxidation states
(*d*) They have incomplete *d*-orbital

14. Which element is radioactive ?
(*a*) Tc (*b*) Fe (*c*) Os (*d*) W

15. Most reactive element is
(*a*) Na (*b*) Fe (*c*) Hg (*d*) Cs

16. The electronic configuration of an elements is [Ar] $3d^2\ 4s^2$. The element belongs to
(*a*) *s*-block (*b*) *p*-block (*c*) *d*-block (*d*) *f*-block

17. Magic number is
(*a*) 3 (*b*) 54 (*c*) 20 (*d*) 19

18. Which element has lowest melting point and boiling point ?
(*a*) Hydrogen (*b*) Helium (*c*) Neon (*d*) None

19. Elements of atomic number 10,18,36, 54, 86 are called
(*a*) light element (*b*) inert gases (*c*) halogens (*d*) rare earth elements

20. Which element has highest electron affinity ?
(*a*) Cl (*b*) Br (*c*) N (*d*) Na

ANSWERS

Fill in the blanks

1. normal or representative **2.** higher **3.** electronegativity
4. bridge elements **5.** ionisation potential **6.** increases, decreases **7.** typical elements
8. smaller, greater **9.** three **10.** similar electronic config., regular

Multiple choice questions

1. (*c*)	**2.** (*b*)	**3.** (*b*)	**4.** (*a*)	**5.** (*a*)
6. (*a*)	**7.** (*d*)	**8.** (*d*)	**9.** (*a*)	**10.** (*c*)
11. (*d*)	**12.** (*d*)	**13.** (*b*)	**14.** (*a*)	**15.** (*d*)
16. (*c*)	**17.** (*c*)	**18.** (*b*)	**19.** (*b*)	**20.** (*a*)

General Questions

1. Define atomic radius. How do the atomic radii of elements vary when we move (*i*) down a group (*ii*) across a period? How would you account for this conclusion?
2. What is ionisation energy? On what factors does it depend? How does the ionisation energy of the elements vary as we move down a group and along a period?
3. Give reasons for each of the following:
   (*a*) The size of an anion is larger than its parent atom.
   (*b*) Electron affinities of halogens are by far the highest.
   (*c*) The second ionisation energy of an atom is always than the first ionisation energy of an atom.
   (*d*) The second electron affinity of oxygen is negative.
   (*e*) The size of an atom cannot be defined precisely.
4. (*a*) Define ionisation energy and electron affinity. How do these vary in the periodic table as the atomic number increases?
   (*b*) Account for the following trend in electron affinity values of halogens?

F	Cl	Br	I
332.6 kJ mol^{-1}	341.5 kJ mol^{-1}	324.7 kJ mol^{-1}	295.5 kJ mol^{-1}

   (*c*) The electron affinities of noble gases are zero. Why?
5. (*a*) The radii of positive ions are always smaller than the radii of the corresponding atoms. Discuss.
   (*b*) What are isoelectronic ions? Account for the decrease in size of the following isoelectronic ions:

$$O^{2-} > F^{-} > Na^{+} > Mg^{2+}$$

6. (*a*) Which of the following electronic configurations would you expect to have the lowest ionisation energy? Explain.
   (*i*) $1s^2\ 2s^2\ 2p^6$ (*ii*) $1s^2\ 2s^2\ 2p^5$ (*iii*) $1s^2\ 2s^2\ 2p^6\ 3s^1$
   (*b*) Mg^{2+} ion is smaller than O^{2-} ion although both have same electronic structure. Explain.
   (*c*) The three ionisation energies of boron atom are 800, 2427 and 3658 kJ mol^{-1}. Explain these values.
7. (*a*) Explain the term electron affinity. Why are electron affinities of halogens so high?
   (*b*) Predict which element in each of the following pairs has higher ionisation energy and why?
   (*i*) Be, B (*ii*) N, C (*iii*) Ne, F (*iv*) I, Br (*v*) Na, K

8. (*a*) What is screening effect? How does it govern the ionisation energy of an atom?
(*b*) Account for the difference in size of Fe^{2+} and Fe^{3+} as

$$Fe^{2+} = 0.076 \text{ nm}$$
$$Fe^{3+} = 0.064 \text{ nm}$$

9. (*a*) What do you understand by successive ionisation energy?
(*b*) Why is ionisation energy of B less than that of Be and of O is less than that of N?
(*c*) Among the elements Li, Na, Mg, S and Xe, which one has the lowest first ionisation energy and which has the highest first ionisation energy?

10. Give reasons for the following?
(*a*) The second ionisation energy of sodium is very high as compared to its first ionisation energy.
(*b*) Electron affinity of fluorine is less than that of chlorine.
(*c*) Ionic radius of K^+ is smaller than that of Cl^-.
(*d*) Electron affinity of N is almost zero while that of F is very high.
(*e*) Size of Cl^- is greater than that of Cl atom while that of Na^+ is less than that of Na atom.
(*f*) Electronegativity increases along a period.
(g) Ionisation energies of C, N and O follow the order C < N > O.

11. (*a*) What are isoelectronic ions? Account for the decrease in size of the following isoelectronic ions:

$$O^{2-} > F^- > Na^+ > Mg^{2+}$$

(*b*) Account for the fact that the 4th period has eighteen and not eight elements.

12. (*a*) What is electronegativity? How does it vary in period and a group in the periodic table?
(*b*) Write notes on the following scales of electronegativity:
(*i*) Pauling scale
(*ii*) Mulliken's scale
(*iii*) Allred and Rochow scale

13. (*a*) For each of the following pairs, predict which one has greater ionisation energy and greater electron affinity:
(*i*) I, I^- (*ii*) Br, Cl (*iii*) Li, Li^+
(*b*) How do the following properties change on moving down a group
(*i*) ionisation energy
(*ii*) electron affinity
(*iii*) ionic radius
(*c*) Distinguish between electron affinity and electronegativity.

14. (*a*) Define the following and explain their trends in a period and in a group:
(*i*) Electron affinity (*ii*) Electronegativity
(*iii*) Atomic radius (*iv*) Ionisation energy
(*b*) Answer the following with reasons:
(*i*) Which element has higher value of electron affinity, F or Cl?
(*ii*) Which element has larger atomic volume, Na or K?
(*iii*) Which element is more metallic, Na or Cl?
(*iv*) Which element has smaller size, O or F?
(*v*) Which element has larger second ionisation energy, Na or F?

15. (*a*) Explain the term electronegativity. How is it related to electron affinity and ionisation energy?

(*b*) Arrange the following species in the order of increasing size:

(*i*) Ra^{2+}, Sr^{2+}, Ca^{2+}, Ba^{2+}

(*ii*) Si, P, S, Cl

(*iii*) Ca^{2+}, Cl^{-}, S^{2-}, K^{+}

(*c*) Why is the radius of an atom not defined precisely?

16. (*a*) What do you understand by effective nuclear charge?

(*b*) Give Slater rules for calculating shielding constant.

(*c*) Calculate the effective nuclear charge for

(*i*) a 3*d*-electron in zinc (Z = 30).

(*ii*) a 4*s*-electron in zinc (Z = 30).

(*d*) The electronic configuration of Ag atom is:

$$1s^2\ 2s^2\ 2p^6\ 3s^2\ 3p^6\ 3d^{10}\ 4s^2\ 4p^6\ 4d^{10}\ 5s^1$$

Compare the effective nuclear charge experienced by

(*i*) a 5*s*-electron (*ii*) a 4*d*-electron

What conclusion do you make from this?

(*e*) Give some important applications of Slater rules.

17. Give reasons for the following:

(*a*) Electron affinity of fluorine is less than that of chlorine.

(*b*) The electron affinity of Be and N are almost zero while that of Ne is zero.

(*c*) Electron affinities of halogens are the highest.

(*d*) Ionisation potential decreases down the group but tends to increase across a period.

18. (*a*) Define electronegativity. How does it vary along a period and down the group in a periodic table?

(*b*) Explain:

(*i*) Electron affinities of halogens are very high.

(*ii*) Electron affinities of noble gases are zero.

(*iii*) Successive electron affinities have negative values.

(*c*) Calculate effective nuclear charge for an electron in 4*s* orbital of Cr (Z = 24).

19. (*a*) Define electronegativity. Give its applications.

(*b*) Name the factors upon which I.E. depend.

(*c*) (*i*) Which has the largest I.E. and why (B or Be)?

(*ii*) Which has the highest E.N. and why (F or Cl)?

(*iii*) Which has the largest size and why (N or O)?

20. (*a*) Give Slater rules for calculating shielding constant of Z_{eff}.

(*b*) Calculate Z_{eff} experienced by

(*i*) 5*s* electrons

(*ii*) 4*d* electrons in Ag atom Z = 47, $1s^2$, $2s^2\ 2p^6\ 3s^2\ 3p^6\ 3d^{10}\ 4s^2\ 4p^6\ 4d^{10}\ 5s^1$.

21. (*a*) State and explain ionisation energy.

(*b*) Discuss the various factors which govern ionisation energies of atoms.

22. (*a*) Give Slater rules for calculating shielding constant for Z effective.

(*b*) Calculate effective nuclear charge for an electron in ^{19}K.

23. Given $E(H_2) = 436\ kJ\ mol^{-1}$, $E(Cl_2) = 244\ kJ\ mol^{-1}$ and $E(HCl) = 432\ kJ\ mol^{-1}$. Calculate X_{Cl} given $X_H = 2.2$.

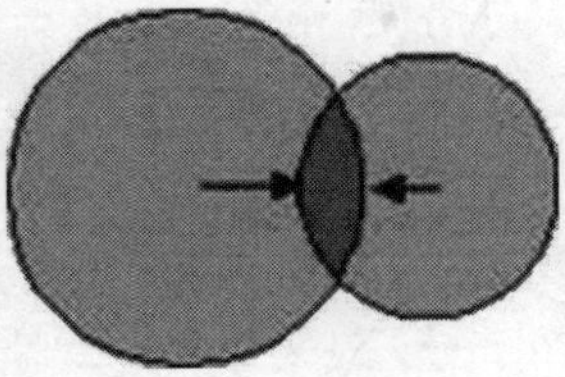

3

Covalent Bond

3.1 VALENCE BOND THEORY

The main points of valence bond approach are as under:

(*i*) Atoms retain their individuality in the molecule.

(*ii*) The bond is formed due to the interaction of valence electrons as the atoms approach each other, *i.e.,* the inner atomic orbits from each atom forming the bond are undisturbed.

(*iii*) When they form a bond, only valence electrons from each bonded atom lose their identity.

(*iv*) The stability of the bond is explained by the fact that electrons forming the bond between two atoms undergo exchange between them.

This method is based on two **theorems** which are stated as:

(*i*) If Ψ_A and Ψ_B are the wave functions for two isolated independent atoms, with energies E_A and E_B, respectively, then the total wave function ψ of the system can be written as a product of the wave functions of the two atoms

$$\Psi = \Psi_A \Psi_B \qquad ...(3.1)$$

and energy of the system by

$$E = E_A \pm E_B \qquad ...(3.2)$$

(*ii*) If a system can be represented by a number of wave functions such as Ψ_1, Ψ_2, Ψ_3, ..., then the true wave function Ψ can be obtained by taking a linear combination of all these wave functions.

$$\Psi = N(C_1\Psi_1 + C_2\Psi_2 + C_3\Psi_3 + ...) \qquad ...(3.3)$$

where N is the normalization constant and C_1, C_2, C_3 .... are the coefficients indicating the weight of each of the Ψ_n's. They are so adjusted as to give a state of lowest energy. The squares of the coefficients may the taken as a measure of the weights of each wave function to the total wave function.

3.1.1 Formation of the hydrogen molecule

Let us now apply the concept of valence bond theory of the formation of hydrogen molecule from hydrogen atoms and consider the various forces of attraction and repulsions which may come into play.

For two hydrogen atoms which are infinitely apart when there is no interaction between them the wave function of system is:

$$\Psi = \Psi_a \Psi_b \qquad ...(3.4)$$

where Ψ_a and Ψ_b represent the wave functions for two hydrogen atoms H_a and H_b respectively. Now if we bring them close together, *i.e.*, distance close enough for interaction, the wave function for the system I, $H_a^{\ 1} \cdot H_b^{\ 2}$ can be written as :

$$\Psi_1 = \Psi_a\ (1) \cdot \Psi_b(2) \qquad ...(3.5)$$

where the hydrogen nuclei are represented as H_a and H_b and the electrons as 1 and 2 (Fig. 3.1).

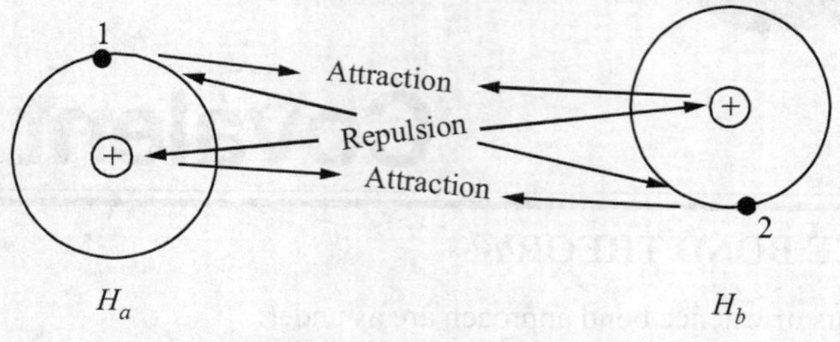

Fig. 3.1. Various interactions between two H atoms.

With the help of wave function Ψ_1, as given by eq. (3.5), the energy of the system consisting of two hydrogen atoms may be calculated as a function of internuclear distance r_{ab} by applying Schrodinger wave equation. The energy changes as a function of distance r_{ab} may be shown graphically in Fig. 3.2 (curve I).

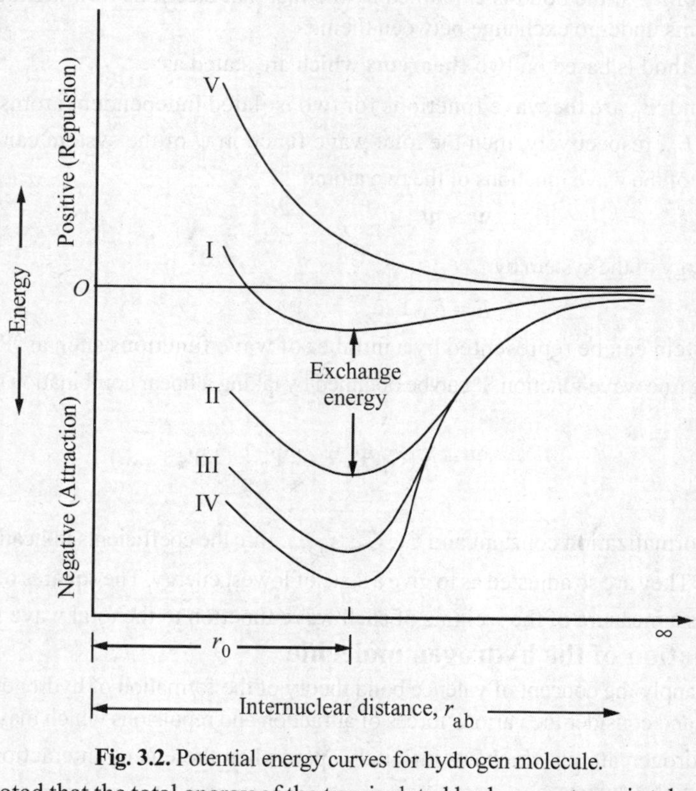

Fig. 3.2. Potential energy curves for hydrogen molecule.

It may be noted that the total energy of the two isolated hydrogen atoms is taken as zero. As the two atoms approach one another [without exchange of electrons, eq. (3.5)], the electron 1 of H_a is attracted by the nucleus of H_b and electron 2 of H_b is similarly attracted to the nucleus of H_2. As a result there is weak attraction and consequently energy of the system starts decreasing. With decrease in distance between the two atoms, forces of attraction increase rapidly. Ultimately an equilibrium internuclear distance (r_0) is reached where the system has lowest energy. At distances closer than this, stronger repulsion due to close proximity between the two nuclei sets in and consequently the energy of the system becomes positive (curve I). The minima in curve I represents that a molecule is formed but the bond energy represented by wave function Ψ_1 (about 25.2 kJ/mole) compares very poorly with the experimental value of 431.4 kJ/mole (curve IV). One may thus conclude that the wave function Ψ_1 representing structure I is not correct. Certain improvements are therefore necessary to match the theoretical and observed values.

***(a)* First Improvement : Exchange Energy.** A careful examination of the wave function Ψ_1 reveals that electron 1 is associated with atom H_a and electron 2 is associated with atom H_b. When a molecule is formed the two atoms come close together, their atomic orbitals overlap and we are no longer definite that the atoms retain their individual electrons. We cannot trace any one electron in a system according to uncertainty principle. It clearly states that we cannot exactly locate an electron but only the probability of finding the electron is given. Hence the system of two hydrogen atoms may as well be represented by structure II.

$$\text{II} \qquad H_a^{\ 2} \cdot H_b^{\ 1}$$

The wave function for this may be represented as

$$\Psi_{II} = \Psi_a(2) \cdot \Psi_b(1)$$

So if both wave functions Ψ_I and Ψ_{II} are equally probable representations then the real wave function is likely to be a linear combination of these two wave functions.

$$\Psi = C_1 \Psi_I + C_2 \Psi_{II}$$

or

$$\Psi = C_1 \Psi_a(1) \Psi_b(2) + C_2 \Psi_a(2) \Psi_b(1) \qquad ...(3.6)$$

Since structures I and II are equally probable, the contribution of wave functions, Ψ_I and Ψ_{II} to the true wave function must be equal. As stated earlier in theorem II, the weight of each wave function is proportional to the squares of its coefficients. So we have

$$C_1^{\ 2} = C_2^{\ 2} \text{ or } C_I = \pm C_2$$

and if for simplicity we assume $C_1 = 1$, then $C_2 = \pm 1$.

Equation (3.6) may be represented by two equations

$$\Psi_+ = \Psi_a(1) \Psi_b(2) + \Psi_a(2) \Psi_b(1) \qquad ...(3.7)$$

$$\Psi_- = \Psi_a(1) \Psi_b(2) - \Psi_a(2) \Psi_b(1) \qquad ...(3.8)$$

Equation (3.7) represented by Ψ_+ is a symmetric combination as there is no change of sign on exchange of electrons. The energy of the system calculated as a function of internuclear distance, r_{ab} by using wave function Ψ_+ results in a significant lowering of energy and is represented by curve II. The bonding energy is shown in curve II and is approximately 302.4 kJ/mole and compares well with the experimental curve IV. The additional lowering in energy *i.e.*, 302.4 – 25.2 = 277.2 kJ/mole has been obtained by virtue of exchange of electrons between the two hydrogen atoms and is known as **exchange energy.** If we plot the energy of the system using Ψ_- eq. (3.8) as function of distance r_{ab} we obtain curve V which shows that the energy curve represented by Ψ_- is an antisymmetric combination since there is a change of sign on electron interchange. In this case the two atoms approach each other with their electrons having parallel spins, Fig. 3.3 (*b*). It is clear that Ψ_+ is the wave function which represents the H_2 molecule, *i.e.*, electrons approach each other with opposite spins, Fig. 3.3 (*a*). Ψ_+ is also referred to as Ψ_{cov} as each nuclei is always in possession of one electron at all times.

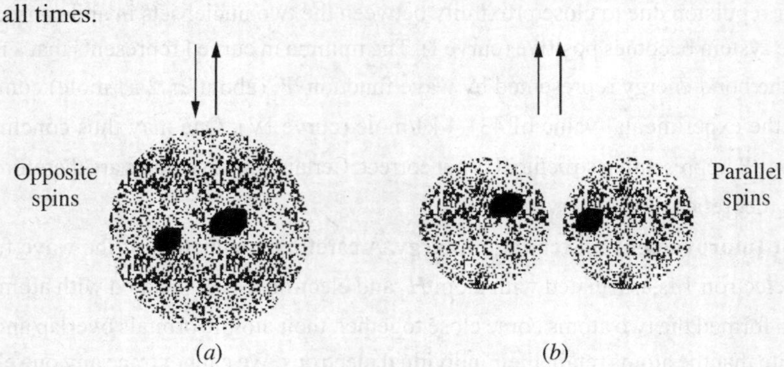

Fig. 3.3. Electrons approaching each other.

(*b*) Second Improvement : Shielding Effect. The wave function can be improved by considering the screening effect of electrons. Since electrons shield one another by varying extents from the nuclear charge, so we must use, Z_{eff} (effective nuclear charge) rather than Z. When Z_{eff} is used, bonding energy of 339.4 kJ/mole is obtained.

(*c*) Third Improvement : Ionicity. By the use of wave function Ψ_+ (or Ψ_{cov}) in eq. (3.7), we have made a significant improvement in the choice of our wave function. But so far we have assumed that one electron is always associated with one atom. Such a situation may not always be true. There is the possibility that both electrons 1 and 2 are either on atom H_a or H_b. The wave function describing such a state would be a combination of eqs. 3.9 and 3.10.

III $\quad H_a^{1,\,2}$..........H_b

Its wave function may be written as

$$\Psi_{III} = \Psi_a(1)\,\Psi_a(2) \qquad ...(3.9)$$

and $\quad$ IV $\quad H_b^{1,\,2}$ ....H_a

is represented as

$$\Psi_{IV} = \Psi_b(1)\,\Psi_b(2) \qquad ...(3.10)$$

The resulting wave function representing both electrons on one nucleus may be summed as

$$\Psi_{ion} = \Psi_a(1)\Psi_a(2) + \Psi_b(1)\,\Psi_b(2) \qquad ...(3.11)$$

and is called Ψ_{ion}. Since it represents a state in which ionic forms also contribute to the structure. If we take wave function Ψ_{cov} and Ψ_{ion} together then the real wave function may be written as

$$\Psi = \Psi_{cov} + \lambda\,\Psi_{ion}$$

or

$$\Psi = \Psi_a(1)\,\Psi_b(2) + \Psi_a(2)\,\Psi_b(1) + \lambda[\Psi_a(1)\,\Psi_a(2) + \Psi_b(1)\,\Psi_b(2)] \qquad ...(3.12)$$

where λ represents the mixing coefficient. It tells us how much Ψ_{ion} wave function is to be introduced in the real wave function Ψ. By applying the variation method, we can adjust the value of λ so as to obtain the lowest energy. This way we can measure the extent of ionic character in the covalent bond. For H_2, the value of λ is about 0.17, and the ionic contribution to the bonding energy is about

23.0 kJ/mole. If we now calculate the energy of the system using wave function as given by eq. (3.12) we get a new energy wave which is marked III. The energy associated with curve III is 362.4 kJ/mol and is an improvement over our earlier results and is quite close to the experimental value.

This treatment which we have discussed for the formation of H_2 is essentially the pairing of two electrons, one from each bonded atom. The valence bond method is also referred to as method of electron pairing. It is obvious even from eq. (3.7) that the atoms must approach one another with the spins of their electrons in opposite collection. It follows that the formation of covalent bond is explained by the lowering of the potential energy of the electrons, resulting from the increased electron density in the internuclear space.

3.1.2 Resonance and resonance energy.

Sometimes it is not possible to describe the properties of a substance on the basis of a single structure. We explain it by saying that the substance exists in a number of structures called **resonating** structures. The actual substance is then hybrid of these resonating structures. This phenomenon is called resonance. Each resonating structure makes a definite contribution, big or small, to the actual structure of the substance. This is explained with the help of resonating structures I, II, III of H_2.

$$\underset{\substack{\text{(covalent)}\\ \text{I}}}{\text{H : H}} \leftrightarrow \underset{\substack{\text{(ionic)}\\ \text{II}}}{\text{H}^+ \ \ :\text{H}^-} \leftrightarrow \underset{\substack{\text{(ionic)}\\ \text{III}}}{\text{H}\overset{-}{:}\ \ \text{H}^+}$$

These structures can be represented by wave functions defined by eqs. 3.7, 3.9, and 3.10. The physical significance of eq. (3.12) is that hydrogen molecule cannot be represented either by pure covalent structure eq. (3.7) or by the ionic structures eqs. (3.9 and 3.10). The double pointing arrows ($\leftrightarrow$) between these structures indicate that they are in resonance with one another. However it is not to be inferred that H_2 exhibits all of these three structures, now can it be said that hydrogen exists for a certain fraction of time in one form and for another fraction of time in some other form. The molecules have single structures which cannot be described in terms of electron dot diagram on the paper. The difficulty lies in the description and not in the molecule. The real structure is a resonance hybrid of these three structures. A resonance hybrid can be equated to a mule which exhibits the characteristics of both the donkey and the horse, but at no moment it is either of these. The resonance hybrid is more stable than any one of the contributing structures. This extra stability of the resonance hybrid as compared with the most stable of the resonating structures is known as resonance energy. It is the greatest when all the resonating structures are of the same energy.

Resonance energy = Actual bond energy – Energy of the pure covalent bond

It may be pointed out that if a molecule or ions cannot be represented by a single structure, then the real structure is expressed in terms of resonance hybrid of all possible structures.

Necessary conditions for writing resonating structures

The necessary conditions for writing resonating structures are as follows :

(*i*) The nuclei of the atoms is each structure must remain in fixed positions. Only the arrangement of electrons may change. The bond angles and bond lengths should be closer to the real structure. Only then the contribution of that structure to the hybrid will be significant. For example, the Dewar structures of benzene are less important than the Kekule structures because in Dewar structures there is one very long bond across the ring.

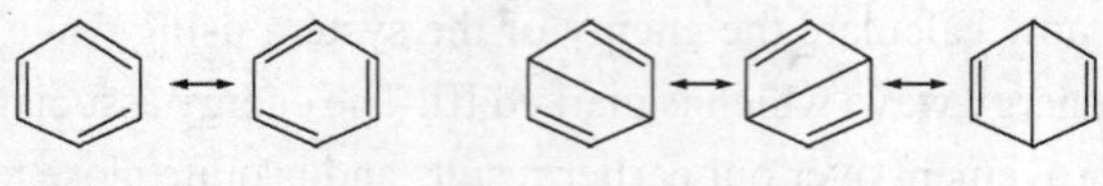

Kekule forms   Dewar forms

The following two structures of phosphorus acid (H_3PO_3) are not resonance structures because of the displacement in the position of atoms.

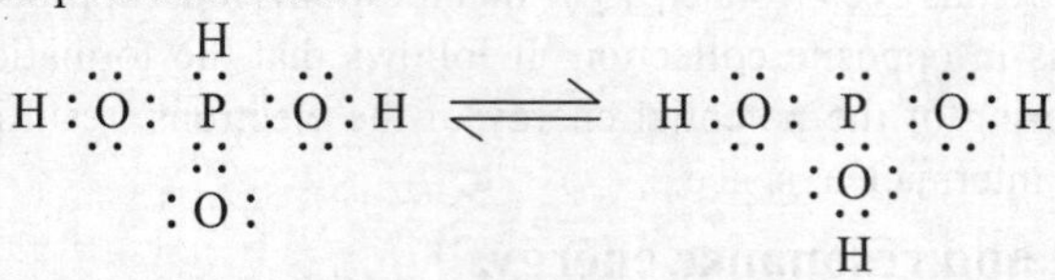

The two structures represent two different chemical species (tautomers) which are in equilibrium, but are not resonance structures.

(*ii*) Each structure must have the same number of unpaired electrons. We should not draw structures in which one form contains a single unpaired electron and another contains two unpaired electrons. For example:

$$:\dot{N}=\ddot{O}: \longleftrightarrow :\ddot{N}=\dot{O}: \quad \text{and not} \quad :\dot{N}=\dot{O}:$$

(*iii*) The contributing structures should all have similar energies. This rule may be illustrated by taking the example of benzene. The Dewar structures have higher energy and are, therefore, less important than the more stable Kekule forms in the combination that represents the true structure of benzene.

(*iv*) Those structures in which negative charge resides on the most electronegative atoms and positive charge on the most electropositive atom will be closest to the true structure. For example, we may write the following resonating structures of HC1.

$$\underset{\text{I}}{\underset{\text{(covalent)}}{H:\ddot{Cl}:}} \longleftrightarrow \underset{\text{II}}{\underset{\text{(ionic)}}{H^+ \; :\ddot{Cl}:^-}} \longleftrightarrow \underset{\text{III}}{\underset{\text{(ionic)}}{\bar{H}: \; \ddot{Cl}:^+}}$$

There is significant contribution from structure II (because of greater electronegative of chlorine). However, any contribution by structure III is not possible.

(*v*) The contributing structures should be so written that unlike charges reside on neighbouring atoms. Consider the resonance structures of HNO_3 . The last structure does not contribute because of +ve charge on adjacent N atoms and double –ve charge on the third nitrogen atom.

$$H-\ddot{N}=\overset{+}{N}=\overset{-}{\ddot{N}}:$$
$$H-\overset{-}{\ddot{N}}-\overset{+}{N}\equiv N:$$
$$H-\overset{+}{N}=\overset{+}{N}-\ddot{N}:^{2-}$$

(*vi*) The greater the number of covalent bonds, greater is the contribution of the resonating structures. For example, the double bonded structures of BF_3 are important because of the formation of an additional covalent bond.

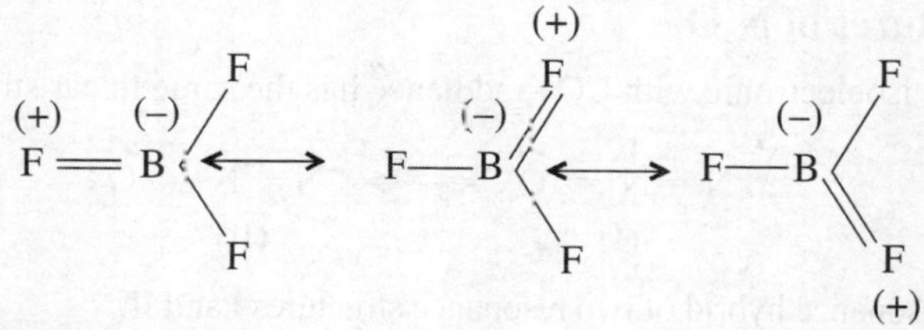

Resonating structures for CO

Following resonance structures can be written for this molecule.

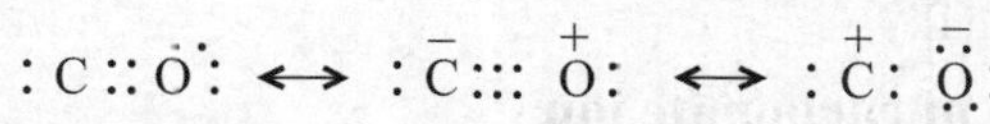

(covalent) (ionic) (ionic)
I II III

***(i)* Bond length.** The observed bond length between carbon and oxygen in CO is 113 pm. The bond length of a normal carbon to oxygen double bond as in ketone is 122 pm and that observed for carbon to oxygen triple bond is 110 pm. It is, therefore, obvious that the bond length in carbon monoxide is shorter that that of for a carbon to oxygen double bond but slightly longer than that for a triple bond. However, bond length in carbon monoxide is closer to carbon oxygen triple bond distance. Bond length measurements suggest that contribution from resonance structure II dominates over other structures, I and III.

***(ii)* Dipole moment.** On the basis of bond length, we may say that structure II makes the maximum contribution. However, we may note that this structure suggest a partial positive charge on oxygen and partial negative change on carbon. If the electronegativities of carbon and oxygen were the same, carbon monoxide would have an appreciable dipole moment in the direction shown.

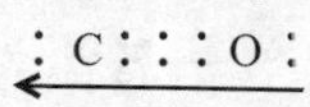

But since oxygen is more electronegative than carbon it, therefore, draws electron density back from carbon to the oxygen atom. This force works in the opposite direction and hence carbon monoxide has very low dipole moment (0.12 D).

Resonating structures for carbon dioxide

It may be represented as a resonance hybrid of the following three structures :

***(i)* Bond length.** On the basis of structure I the bond length in CO_2 should be equal to carbon-

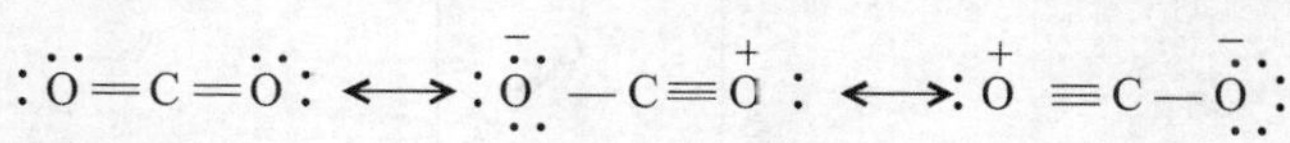

(covalent) (ionic) (ionic)
I II III

oxygen double bond distance of 122 pm. The observed bond distance is shorter than this value and is 115 pm. Therefore, there must be some contribution from structures II and III.

***(ii)* Dipole moment.** Contributions from structures II and III must be equal. If their contributions are unequal, the molecule will have dipole moment. But CO_2 has zero dipole moment.

***(iii)* Bond energy.** The energy of a carbon - oxygen double bond is 731.5 kJ/mole. Therefore, heat of formation of the molecule O = C = O (structure I) should be about 1463.0 kJ/mole. The experimental value of heat of formation is 1601 kJ/mole. The difference between these two values, *i.e.*, 138 kJ/mole suggests the extra stability acquired due to contribution by structures II and III. It may be noted here that structure I is most dominant form due to its symmetrical nature.

Resonance structures of N_2O

The molecule is isoelectronic with CO_2 and hence has the same linear structure. It can be

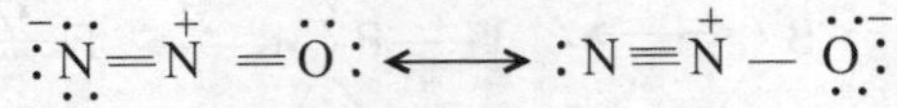

(I) (II)

represented as resonance hybrid of two resonance structures I and II.

Each of the structures I and II separately predicts a sizeable dipole moment for the molecule N_2O. The low value (0.12 D) is explained on the cancelling effect of dipole moments in opposite directions in structures I and II.

Resonance structure of carbonate ion

The Lewis dot representation of carbonate ion contains a double bond and two single bonds between central C and oxygen atoms as shown below. This structure indicates that one of the carbon oxygen bonds (C = O) should be shorter than the other two bonds. Also the angle α should be greater than the angle β. But this is contrary to observations.

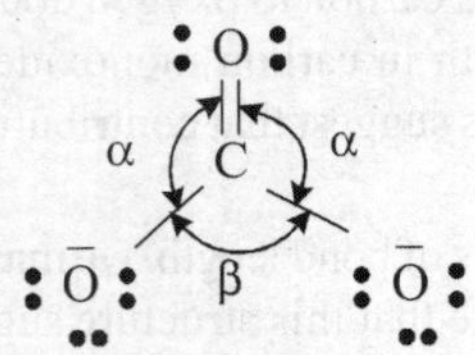

X-ray studies show that all the three bonds have the same bond length and each angle is 120°.

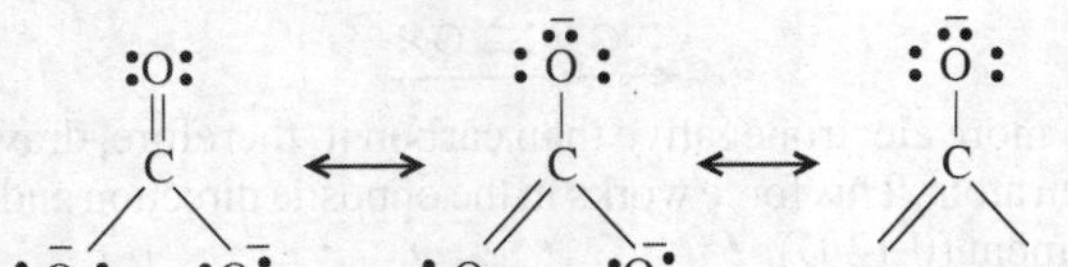

I II III

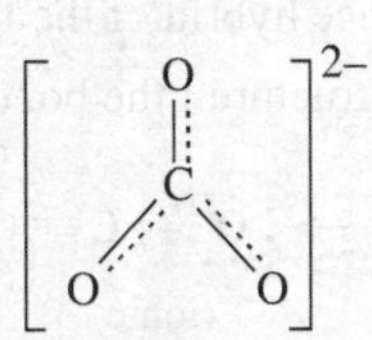

IV

This has been explained by a combination of resonance structures I, II and III for carbon dioxide. Thus carbonate ion may be represented by structure IV which is combination of these three structures.

Example 1: **Write down the resonance structures of**

(*i*) Cl_2O

(*ii*) N_3^-

(*iii*) HNO_2

(*iv*) SCN^-

(*v*) NO_2^-

***Solution*:**

(*i*) Cl_2O

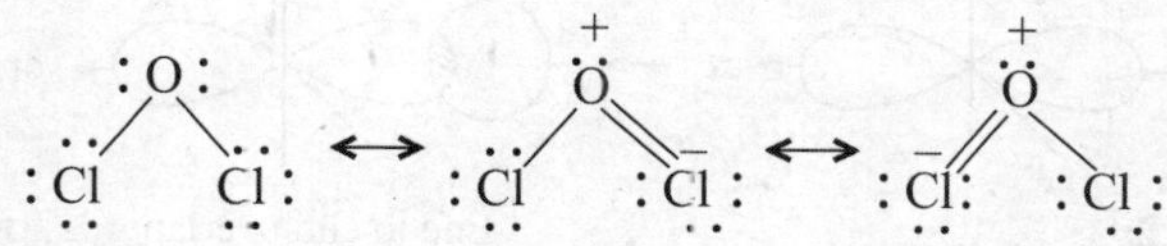

(*ii*) N_3^-

The three structures are equally important and result in the same bond order ($\cong 2$) for each bond in N_3^-.

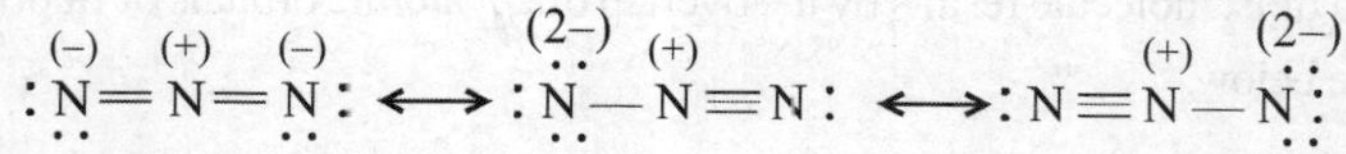

(*iii*) HNO_2

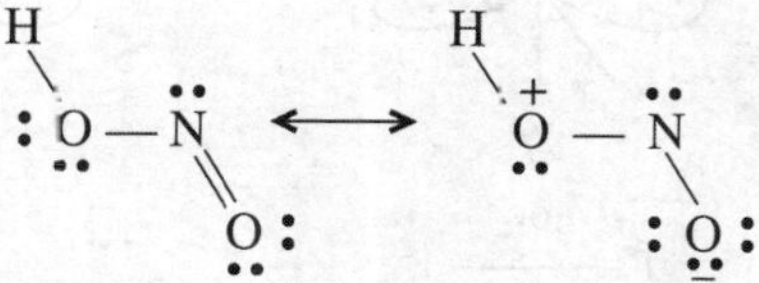

(*iv*) SCN^-

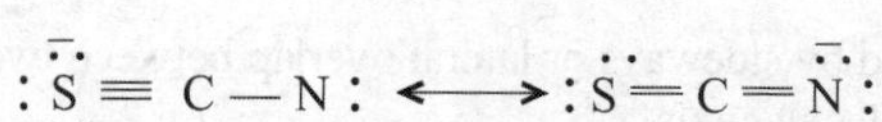

S — C bond order is between 1.5 and 2 and C—N bond order is between 2 and 3.

(*v*) NO_2^-

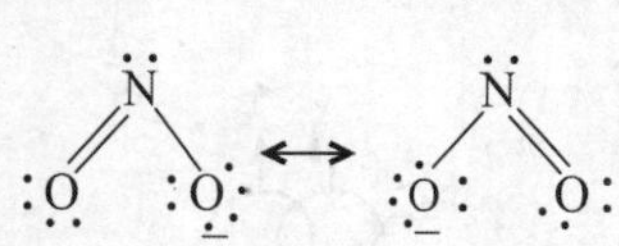

3.1.3 σ (sigma) and π (pi) bonds

σ-covalent bond. Sigma bond is formed by end-to-end overlap of any one of these combinations of atomic orbitals. This type of overlapping results in a strong covalent bond. In these cases, bond is formed by electron cloud having maximum density on the line connecting the centres of atoms.

(*a*) $s - s$ overlap (*b*) $s - p_z$ overlap (*c*) $p_z - p_z$ overlap

Bond formation between two hydrogen atoms to give H_2 results due to overlap of two 1*s* atomic orbitals. The covalent bond formed by the *s*-orbital electron cloud below is called a σ-bond.

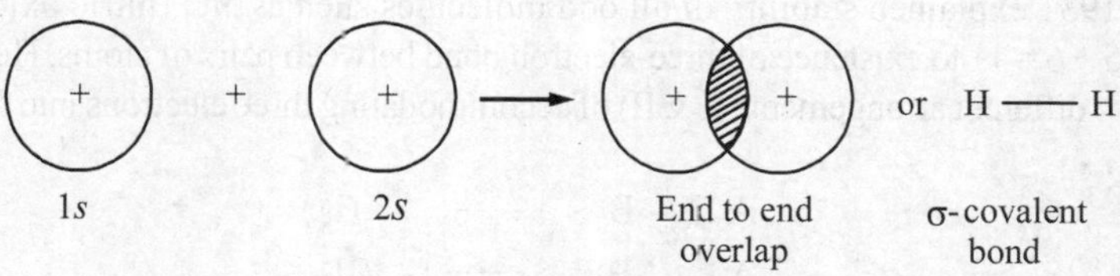

Formation of H_2 molecule.

The formation of the HF may be visualized as an example of the overlap between 1*s* and $2p_z$ atomic orbitals as shown below.

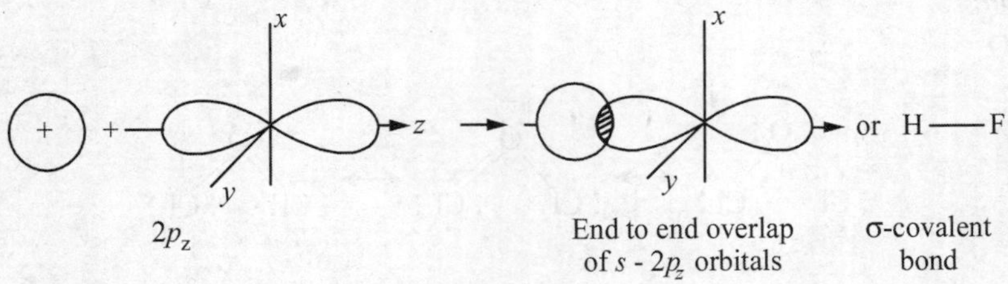

Formation of HF.

The formation of F_2 molecule results by the overlap of $2p_z$ atomic orbitals of fluorine atoms. This is shown in the figure below.

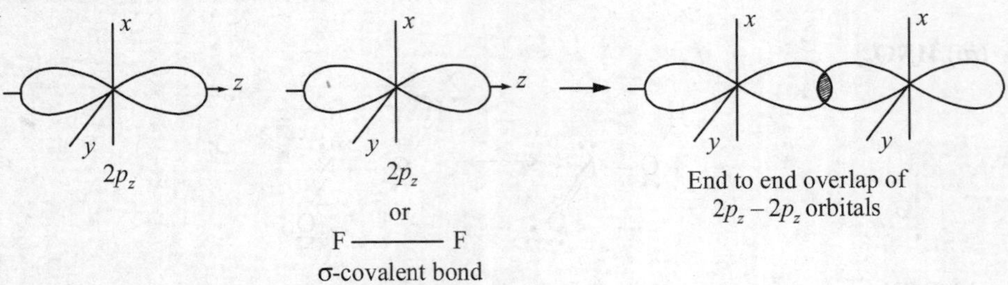

Formation of F_2

π bond. π bond is formed by sideways or lateral overlap between two *p*-orbitals which have their lobes perpendicular to the molecular axis.

The sideways overlap is not efficient and results in a weak bond. Sigma bond is always stronger than a pi-bond.

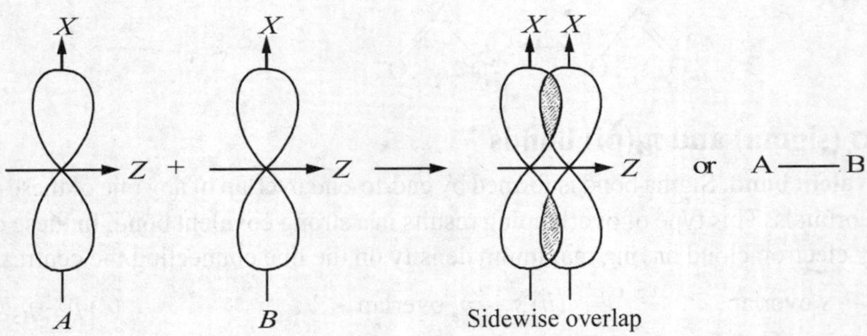

Sidewise overlap of $p_x - p_x$ orbits

3.1.4 Three-electron (odd-electron) bond

Pauling in 1931 explained stability of all odd molecules such as NO (nitric oxide) having odd no. of electrons 5 + 6 = 11 to existence of three-electron bond between pairs of atoms. He suggested that there are only two different arrangements (I & II) of accommodating three electrons into the two orbitals.

A **:** · B ------ (I)

A · **:** B ------ (II)

He suggested that calculation of energy for structure I alone shows repulsion between the two atoms (+ve potential energy) and so structure (I) does not correspond to the formation of a stable bond. Similarly, structure (II) alone also leads to an unstable arrangement. Pauling suggested that if two atoms A and B are identical or are very similar, so that the two structures I and II have very close energies, then resonance may occur between them. This resonance energy will lead to attraction between two atoms (and lowering in energy) and will lead to the formation of a stable three-electron bond between the two atoms. The resonance between structures I & II may be represented as

$$A\colon\ .B \rightleftharpoons A.\ \colon B$$

This type of a three-electron bond is represented as A· · · B. It is shown by Pauling that the three-electron bond is about one-half as strong as an electron pair bond i.e. it corresponds to half bond.

Paramagnetic nature of oxygen molecule can be explained with the help of three-electron bond. Normally, we represent the bond between two oxygen atoms as

:Ö::Ö: or O ═ O

But, this does not explain the paramagnetive nature of oxygen molecule. In the above structure of oxygen, all the electrons are in pairs. There is no unpaired electrons. Hence, paramagnetic nature remains unexplained. Pauling explained the paramagnetic nature of oxygen by proposing two three-electrons bonds between two oxygen atoms as shown below:

:O⁚⁚O: or O∸O

As a three-electron bond corresponds to a half bond. Total number of bonds between two oxygen atoms remains 2, or $[1 + (2 \times 1/2)]$.

This structure accounts for the presence of two unpaired electrons, one from each three-electron bond, hence the molecule of oxygen is paramagnetic.

Conditions for the formation of a three-electron bond

Pauling suggested the following conditions for the existence of a three-electron bond between atoms A and B.

(*i*) Atoms A and B must be identical or very similar.

A: ·B ------ (I)

A· :B ------ (II)

(*ii*) Structures (I) and (II) between atoms A and B must have nearly the same energy, so that resonance stabilization is provided to the system.

(*iii*) Three-electron bond is formed between two different atoms A and B if the electronegativity difference between A and B does not exceed by 0.5. Thus, three-electron bonds have been found to exist between oxygen and fluorine, nitrogen and oxygen and chlorine and oxygen.

Some examples of molecules containing three-electron bond

***(i)* Nitric oxide** is the most stable of various odd molecules. The valence bond representation of NO is given below. Structure (II) is expected to be less stable than structure (I) because of distribution of unfavourable charges, *i.e.*, we are placing positive charge on electronegative oxygen atom.

$$:\dot{N} = \ddot{O}: \text{ ------ (I)}$$

$$:\overset{-}{\ddot{N}} = \overset{+}{\dot{O}}: \text{ ------ (II)}$$

However, structures I and (II) are nearly of equal stability and thus there is complete resonance between the two structures.

$$:\dot{N} = \ddot{O}: \longleftrightarrow :\overset{-}{\ddot{N}} = \overset{+}{\dot{O}}:$$

The molecule can thus be assigned the following structure consisting of a double bond and

a three-electron bond. This structure is more stable than either of the two structures I and II. Thus molecule is described as having $2\frac{1}{2}$ bond. It is expected to show bond length between that for a double and a triple bond. The experimental value for bond length for NO is 115 pm, somewhat larger than the triple bond value of 106 pm for NO^+.

***(ii)* Nitrogen dioxide (NO_2).** It may be assigned the following resonance structures.

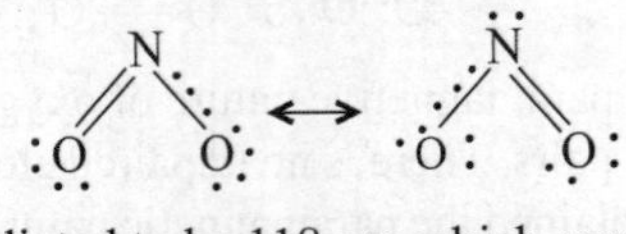

The NO bond distance was predicted to be 118 pm which agrees well with the experimentally determined value of 119 pm by infrared spectroscopy.

3.1.5 Bond Energy

For a diatomic molecule, say AB, the bond dissociation energy is defined as the energy needed (absorbed) to break the molecule AB into its atoms A & B. The process of bond rupture in a molecule is called **dissociation.** Therefore, the bond energy of a diatomic molecule which is identical with bond dissociation energy (D) may also be defined as the enthalpy change for dissociation of gaseous molecule at 25°C into gaseous atoms at the same temperature. As for example,

$$H_2\,(g) \longrightarrow 2H\,(g) \qquad \Delta\,(H-H) = \Delta\,H = 435.9\text{ kJ / mole}^{-1}.$$

It is observed that there is a definite trend in bond energy among successive numbers in a group. The heavier analogs generally show lower values of dissociation energy (F_2 is an exception). As far as range of dissociation energies is concerned, the diatomic alkali metals have low values and hydrogen halides have comparatively large values of dissociation energy. There is no trend along a period. It is found that for similar molecules the dissociation energy decreases as bond order decreases. In other words, for related molecules, the bond dissociation energy decreases as bond length increases. Bond dissociation energy of F_2 is exceptionally low because of high repulsion between the non-bonding electrons on the similar fluorine atoms (other halogens have longer bonds).

For polyatomic molecules, the bond dissociation energy and bond energy terms are not identical and should not be used synonymously. Bond dissociation energy (D) of a polyatomic molecule AB_n is defined as the energy required to break just one bond to form atom B and a group of atoms (radical) AB_{n-1} . Bond dissociation energies for a given bond are different for different structural and electronic arrangements in the radical.

Although the bond dissociation energy of a given bond is not independent of the environments but the variation is not very large. The bond dissociation energies may not change significantly with change in environments around a given bond. Thus, it should be possible to characterise a bond by an average bond energy (E) which is defined as the approximate energy required for breaking of the bond in any gaseous molecule. It does not include energies necessary for vaporization or sublimation.

The distinction between bond dissociation energy (D) and average bond energy (E) shall be clear from the following examples showing stepwise dissociation of gaseous water.

$$H_2O\,(g) \longrightarrow H\,(g) + OH\,(g) \qquad D_1\,(H-OH) = 500\text{ kJ}$$

$$OH\,(g) \longrightarrow H\,(g) + O\,(g) \qquad D_2\,(H-O) = 424\text{ kJ}$$

The average bond energy (E) for O – H bond is the mean of the two separate bond dissociation energies $(D_1 + D_2) / 2$, that is , (500 + 424)/2 = 462 kJ. The enthalpy change for the formation of water molecule by the reaction is –926 kJ mol^{-1}.

$$2H\ (g) + O\ (g) \longrightarrow H_2O\ (g)$$

It is negative because heat is evolved in the formation of gaseous water. As expected, it is almost twice the average bond energy (E = 462) for an O–H bond and hence in close agreement with the presence of two O–H bonds in H_2O. However, the average bond energy (E) is quite different from the bond dissociation energy D_1 or D_2.

Example 2. The bond energy of Li_2 (113 kJ mol^{-1}) is less than that of H_2 (431 kJ mol^{-1}) though both contain a σ-bond.

Solution. The bond energy of a bond depends upon the bond length. Greater the bond length, smaller the bond energy. Li—Li bond is larger than H—H bond. Hence bond energy of Li_2 is smaller than that of H_2 .

3.2 HYDROGEN BOND

(1) Hydrogen Bond. When hydrogen is joined to a highly electronegative element, it can form an additional bond to another electronegative element. This additional bond is known as hydrogen bond. For example, in the compound KHF_2, the anion HF_2^- is believed to have structure $(FHF)^-$ in which hydrogen acts as bridge between the two fluorine atoms. The hydrogen bond is considerably less stable than the usual valence bond. In order to distinguish a hydrogen bond from other valence bond, it is written as dotted line so that HF_2^- ion becomes $(F.......H — F)^-$. In general, when a hydrogen bond links two electronegative atoms A and B as a resonance hybrid between A – H.....B and A.....H – B, the more electronegative the atom to which hydrogen is joined , and the smaller it is, the stronger is the hydrogen bond. Hydrogen bonds are formed only between highly electronegative atoms like F, O and N. The energy of the hydrogen bond varies depending on the state and the nature of the species to which hydrogen is joined. The energy of the hydrogen bond H ..... F is about 40 kJ while the energies of the O.....H and N.....H bonds are about 20 and 8 kJ respectively. Evidence in support of the existence of hydrogen bonding is obtained, if we compare some of the physical properties of compounds containing hydrogen bonds.

3.2.1 Intermolecular hydrogen bond

We find that the boiling points of HF and H_2O are abnormally high as compared with other members of each series (Fig. 3.4). In the series HF, HCl, HBr and HI boiling points are expected to rise due to increasing number of electrons (because of increased van der Waals forces). The unexpectedly high boiling point of H_2O in the series H_2O, H_2S, H_2Se and H_2Te is also attributed to hydrogen boiling.

Hydrogen fluoride consists of $(HF)_n$ molecules which have the following structure.

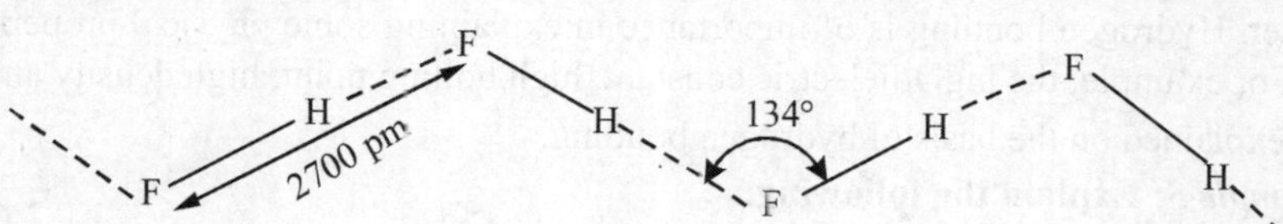

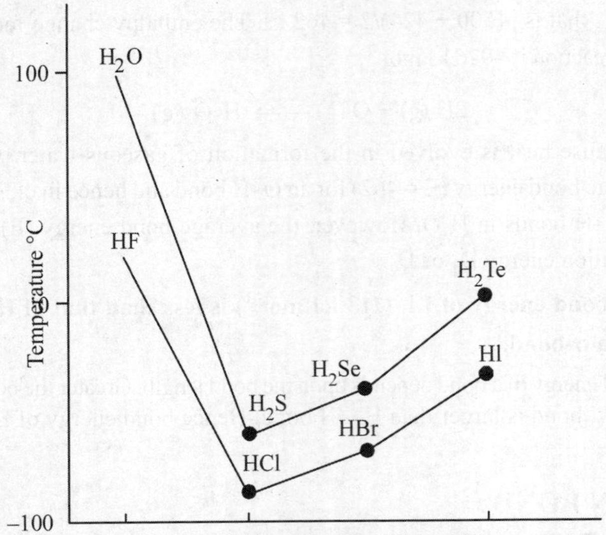

Fig. 3.4. Comparison of boiling points of hydrides of 6th and 7th group elements.

3.2.2 Intramolecular hydrogen bonding

In many organic compounds, often the hydrogen bond joins part of a single molecule (*i.e.*, within the molecule) and in such cases intramolecular hydrogen bonding is favoured. Substances which are intramolecularly hydrogen bonded are more soluble. In nonpolar solvents (such as $CC1_4$, C_6H_6 etc.) and have lower melting and boiling points than their other isomers which are intermolecularly hydrogen bonded.

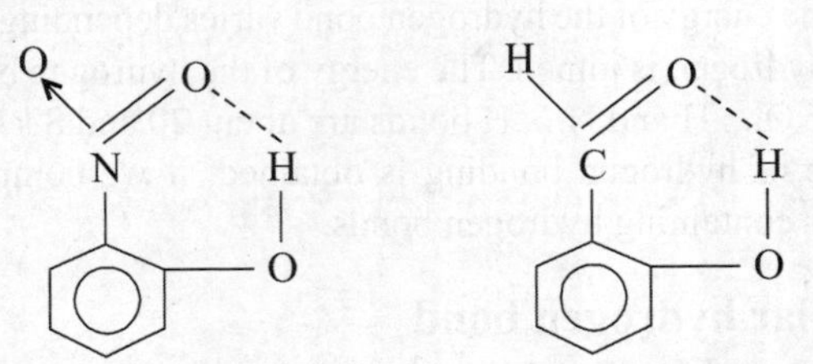

Fig. 3.5. Intramolecular hydrogen bonding.

The hydrogen bond plays important part in many processes. For example, the solubility of an uncharged organic compound in water depends mainly on the ability of a substance to form hydrogen bonds with the solvent. Thus dimethyl ether is completely miscible with water whereas dimethyl sulphide has slight solubility; sugar, an organic compound, is water soluble due to hydrogen bonding with water. Hydrogen bonding is of importance in explaining some physical properties of certain liquids. For example, the high dielectric constant, high boiling point, high density and viscosity of water is explained on the basis of hydrogen bonding.

***Example 3*: Explain the following:**

'H_2O has a higher boiling point while H_2S and H_2Te are gases'.

Solution. Oxygen is more electronegative than sulphur, selenium and tellurium. Therefore oxygen forms hydrogen bond more effectively. The hydrogen bonding in water may be depicted as follows.

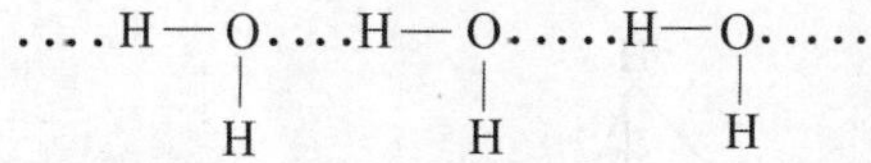

Substances having hydrogen bonding boil at higher temperature.

***Example 4*: Which of the following will have higher boiling point or melting point? Justify your answer:**

(*i*) H_2O or H_2S

(*ii*) Nitrophenol or nitrobenzene

***Solution*:** (*i*) H_2O boils at a high temperature. It has effective hydrogen bonding than H_2S because oxygen is more electronegative than sulphur.

(*ii*) *Nitrophenol* has a higher boiling point compared to nitrobenzene.

***Example 5.* Methane is very slightly soluble in water, but methanol is highly soluble. Explain.**

Solution. Methanol forms hydrogen bond with water. No such bonding takes place between methane and water.

***Example 6*: Why is boiling point of NH_3 higher than that of PH_3 ?**

Solution. There is greater electronegativity difference between N and H than between P and H. Thus due to hydrogen bonding, although small, NH_3 molecule is associated and boils at a higher temperature.

***Example 7*: Explain the following :**

(*i*) Alcohol is a better drying agent than acetone.

(*ii*) Density of ice is less than that of water.

Solution. (*i*) Alcohol can combine with water more than with acetone due to hydrogen bonding. Therefore, alcohol is a better drying agent.

(*ii*) Ice has a cage like structure comprising hydrogen bonds. There is a lot of empty space in the structure due to which its density is much less than that of water.

3.3 HYBRIDIZATION OF ORBITALS

Ground state of configuration of carbon is

$$1s^2\, 2s^2\, 2p_x^1\, 2p_y^1$$

To account for tetravalency of carbon it is assumed that one electron from 2*s* orbital is excited to $2p_z$ orbital so that the configuration in excited state is

$$1s^2\, 2s^1\, 2p_x^1\, 2p_y^1\; 2p_z^1$$

The structure of methane consists of four equivalent bonds arranged about the carbon atom at the centre in a manner that all the angles between carbon and hydrogen are equal, its value is 109° 28'. The geometric arrangement which fulfils these requirements is a tetrahedron. In such a structure, carbon atom occupies a position at the centre of the tetrahedron and four hydrogen atoms occupy the corner positions. This arrangement is shown in Fig. 3.6.

The equivalence of four bonds can be explained by assuming that the available orbitals of carbon, the 2*s* and three 2*p* orbitals are intermixed or hybridized in a manner as to result in four equivalent orbitals which are oriented towards the four corners of regular tetrahedron. *This mixing of atomic orbitals to give equal number of symmetrically oriented orbitals of equal energy is referred to as hybridization.*

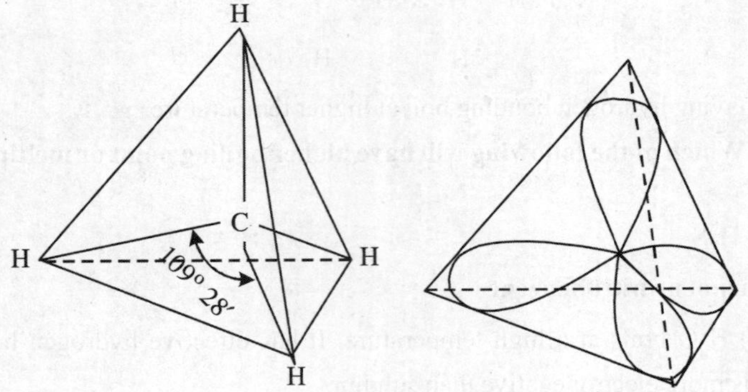

Fig. 3.6. Structure of methane.

The carbon in this case is said to be sp^3 hybridized and carbon hydrogen bonds are referred to as hybrid bonds.

Different types of hybridization are:

(*i*) sp^3 type
(*ii*) sp^2 type
(*iii*) *sp* type
(*iv*) sp^3d type
(*v*) sp^3d^2 or d^2sp^3 type
(*vi*) sp^3d^3 type

3.3.1 *sp* hybridization or diagonal hybridization and the shape of beryllium fluoride

Electronic configuration of beryllium in ground state is $1s^2\ 2s^2$. One of the 2*s* electrons gets promoted to next available empty orbital (*i.e.*, $2p_x$) and therefore electronic configuration of beryllium in excited state is $1s^2, 2s^1, 2p_x^1$.

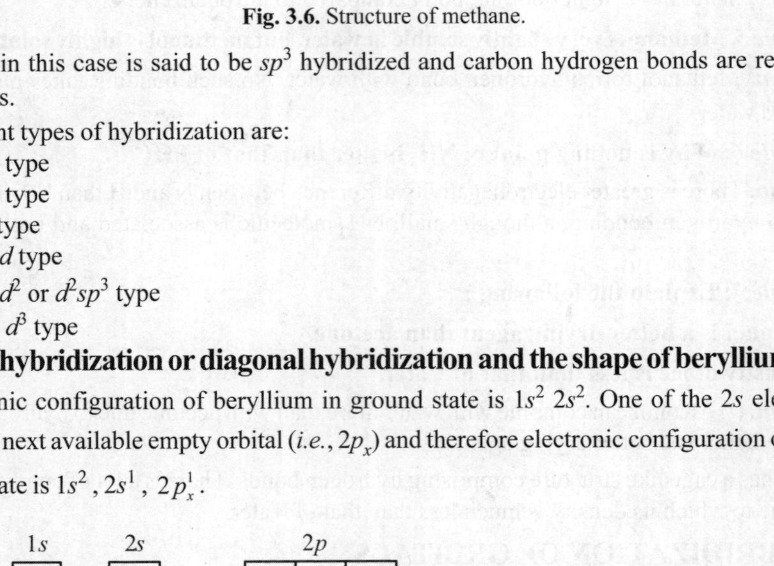

Now there are two half-filled orbitals, each having single unpaired electron. But they belong to two different orbitals. One belongs to *s* and the other two *p*-orbital. Therefore, to get equivalent bonds, these two orbitals hybridise or get mixed up to form two new equivalent orbitals each of which has half *s* and half *p* character.

In the above case one *s* and one *p*-orbitals are involved to form hybridized orbitals. This type of hybridization is therefore known as *sp hybridization* or *diagonal hybridization.* The hybridized

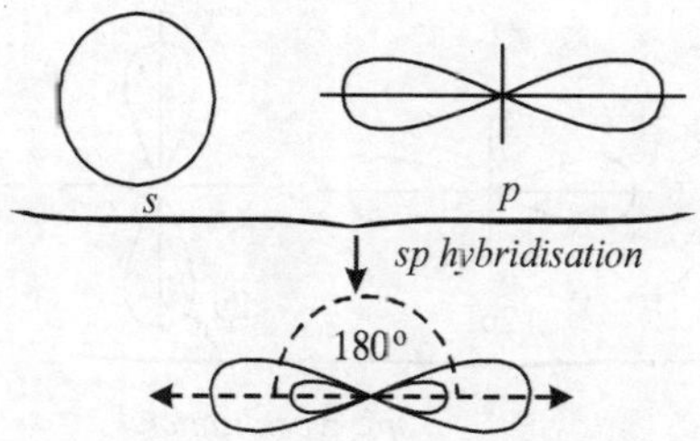

Fig. 3.7. *sp* hybridisation.

orbitals are oriented at an angle of 180° to each other along the same axis (Fig. 3.7).

These two hybridized orbitals overlap separately with the *p*-orbitals of fluorine to form two Be-F bonds. Shape of BeF_2 molecule can be represented as shown in Fig. 3.8. There are two bonded

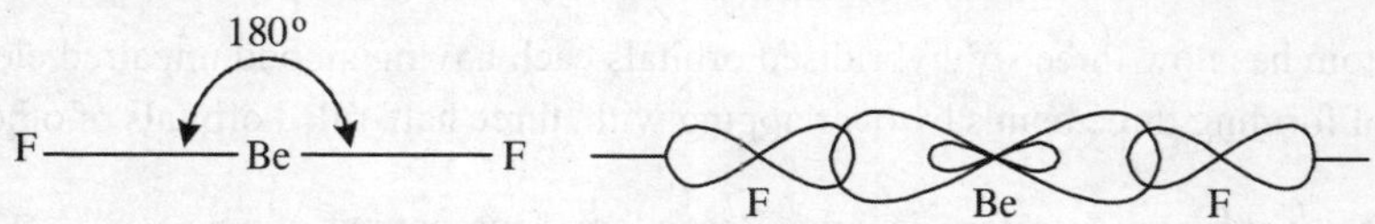

Fig. 3.8. Pictorial representation of BeF_2.

electron pairs. Therefore molecule is linear with two orbitals at an angle of 180° to each other.

3.3.2 *sp*^{2} hybridization or trigonal hybridization and the shape of BF_3 molecule

Electronic configuration of boron in ground state is $1s^2\,2s^2\,2p_x^1$. One of the two 2*s* electrons gets promoted to next available vacant orbital ($2p_y$). The electronic configuration of boron in the excited state is $1s^2\,2s^1\,2p_x^1\,2p_y^1$.

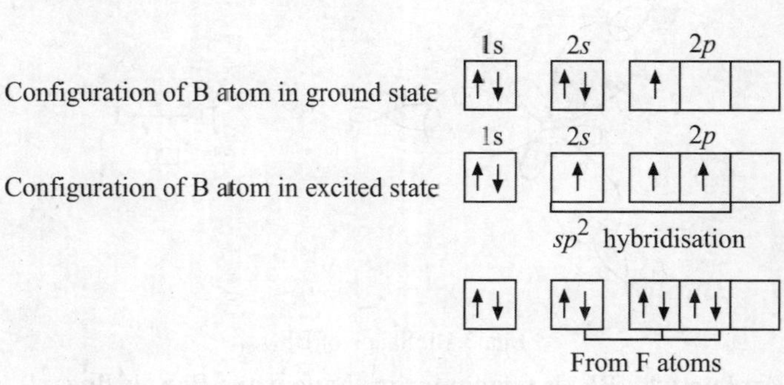

Now there are three half-filled orbitals each having odd unpaired electrons. But they belong to different orbitals. To get equivalent bonds these three orbitals hybridize to form three new equivalent hybridised orbitals. In the above case one *s* and two *p* orbitals are involved to form three hybridised orbitals. This type of hybridization is, therefore, known as *sp*^{2} *hybridization or trigonal hybridization.*

Boron forms *sp*^{2} hybridised orbitals. These three orbitals lie in one plane (*i.e.*, boron atom is planer) and are directed towards the three corners of an equivalent triangle, so that the angle between them is 120°. (Fig. 3.9)

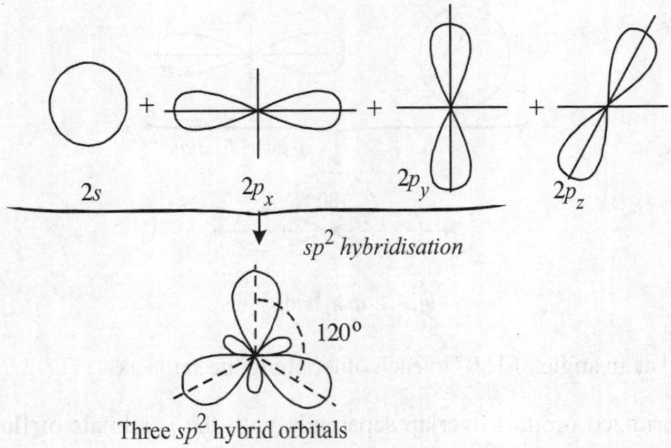

Three sp^2 hybrid orbitals

Fig. 3.9. Orientation of sp^2 hybrid orbitals.

Boron atom has now three sp^2 hybridised orbitals each having an odd unpaired electron. They are capable of forming three bonds by overlapping with, three half-filled orbitals of other atoms.

Bonding capacity of boron, therefore, is three. Boron forms compounds of type X — B(X)X where X = H or Cl

Shape of BF_3

Shape of BF_3 molecule can be represented as shown in Fig. 3.10. There are three bonded electron pairs. The three bonds are equivalent and 120° apart. The molecule is, therefore, triangular and planar in shape.

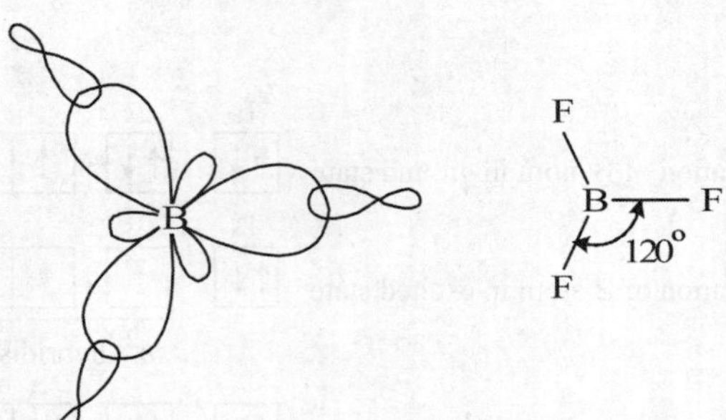

Fig. 3.10. Shape of BF_3.

Example 8. **Explain why BF_3 is triangular in shape while BeF_2 is linear.**

Solution. In BeF_2, Be atom is *sp* hybridized. The *sp* hybrid orbitals are linearly oriented. That is why the molecule of BeF_2 is linear.

In BeF_3, B atom is sp^2 hybridized. The sp^2 hybrid orbitals are oriented at an angle of 120° towards the corners of a triangle. That is why the shape of BF_3 is triangular.

3.3.2 Shapes of certain molecules

(*i*) Shape of stannous chloride molecule.

Atomic no. of Sn is 50. In stannous chloride the tin (Sn) atom has electronic configuration of outermost shell as : $5s^2\ 5p_x^{\ 1}\ p_y^{\ 1}$. Two electrons in the 5*s* orbital and two electrons in *p*-orbitals participate in hybridisation. It may be noted that an orbital with paired electrons can participate in hybridisation but not in bond formation. Thus it is a case of sp^2 hybridisation.

In $SnCl_2$, two unpaired electrons in outermost shell of tin are used for bond formation with two atoms of chlorine, while one electron pair is left unused as lone pair. Thus there are three electron pairs, two bond pairs and a lone pair. According to its hybridisation it is expected to be trigonal planar (Fig. 3.11) one of the three positions is occupied by lone pair of electrons. The expected angle is 120°. But the actual angle is less than 120°. This is because of a greater repulsion between the bond pair and lone pair than between the bond pair and bond pair. (VSEPR Theory, See ahead).

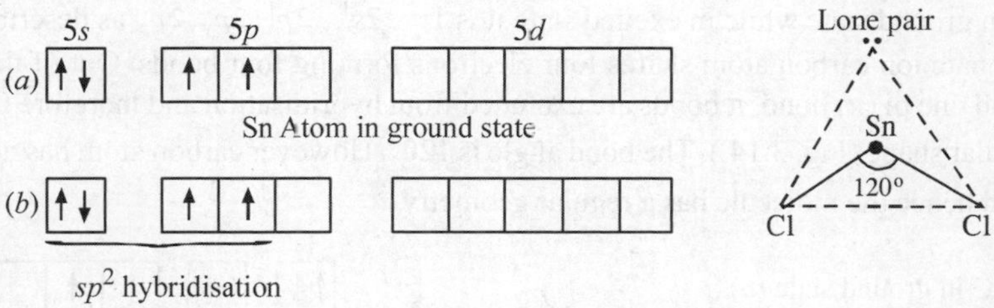

Fig. 3.11. Formation of stannous chloride.

(*ii*) Shape of nitrate ion.

It is a case of sp^2 hybridization. Atomic no. of 'N' is 7. In nitrate ion nitrogen is central atom and it has 5 electrons in outermost shell ($2s^2$, $2p^3$) while oxygen atoms have six electrons ($2s^2$, $2p^4$). The nitrate ion carries a unit negative charge. It means that one of the oxygen atoms carries an extra electron.

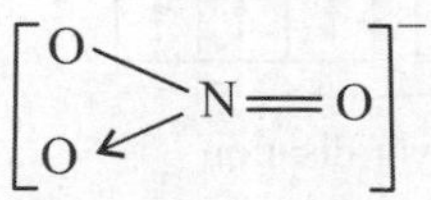

The bond formation in nitrate ion can be shown in Fig. 3.12.

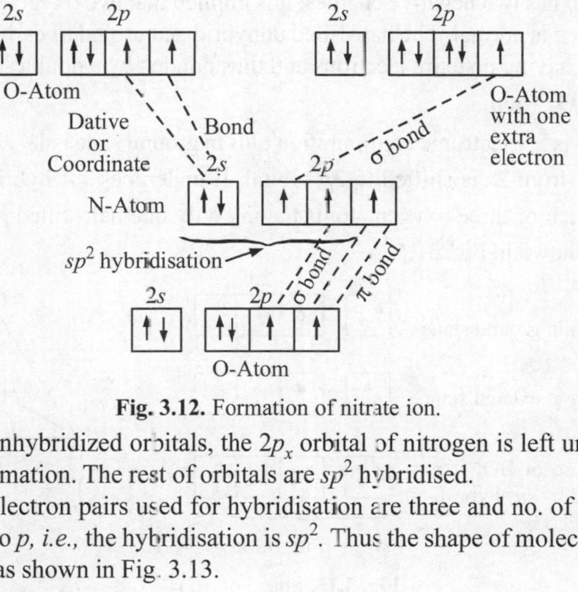

Fig. 3.12. Formation of nitrate ion.

π bond is formed by unhybridized orbitals, the $2p_x$ orbital of nitrogen is left unhybridized and it is involved in π bond formation. The rest of orbitals are sp^2 hybridised.

Total number of electron pairs used for hybridisation are three and no. of orbitals required are three *viz.*, one *s* and two *p, i.e.,* the hybridisation is sp^2. Thus the shape of molecule is trigonal planar with bond angle 120° as shown in Fig. 3.13.

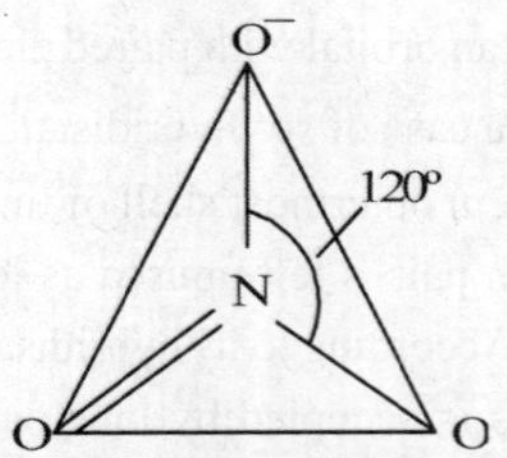

Fig. 3.13. Shape of NO_3^- ion.

(*iii*) Shape of carbonate ion.

It is a case of sp^2 hybridisation. Atomic no. of carbon is 6. Electronic configuration of carbon is $1s^2\ 2s^2\ 2p^2$ in ground state while in excited state it is $1s^2$, $2s^1$, $2p_x^1, 2p_y^1, 2p_z^1$ as described below.

In carbonate ion carbon atom shares four electrons forming four bonds. Out of these three are sigma (σ) and one pi (π) bond. π bonds are excluded from hybridisation and therefore $CO^2{}_3^-$ ion has plane triangular shape (Fig. 3.14.). The bond angle is 120°. However carbon atom has no lone pair of electrons and hence the molecule has a regular geometry.

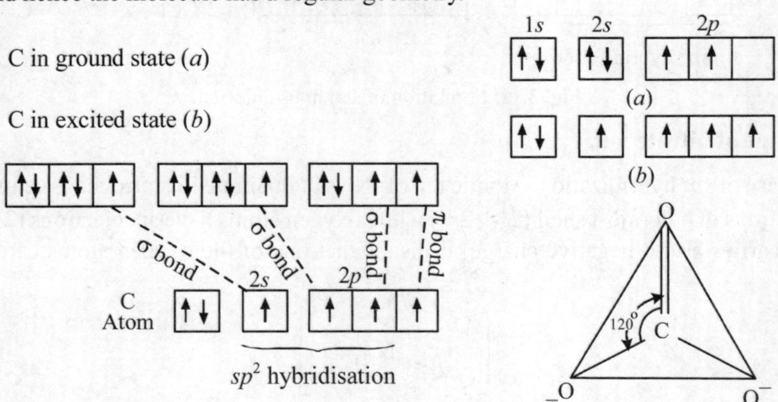

Fig. 3.14. Shape of CO_3^{2-} ion.

As carbonate ion has two negative charges, it is implied that two oxygens have an extra electron while the third oxygen is normal. The half-filled unhybridised orbital of carbon forms a π bond with that of oxygen atom having no extra electrons and thus bonded by a double-bond to carbon atom.

(*iv*) Shape of BO_3^{3-} ion.

Atomic no. of B is 5. Electronic configuration of B in ground state is $1s^2$, $2s^2$, $2p^1$. In excited state one of the electrons from $2s$ is shifted to $2p$ orbital. It undergoes sp^2 hybridisation forming three sigma bonds with each of three oxygen atoms having only one half–filled p-orbital giving a plane triangular shape as shown in Fig. 3.15.

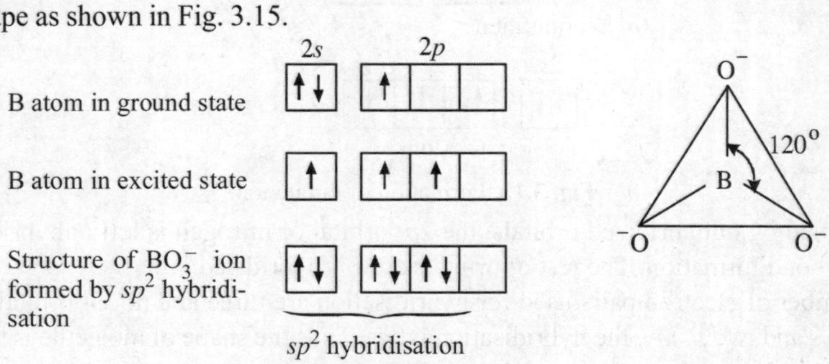

Fig. 3.15. Shape of BO_3^{3-} .

3.3.3 sp^3 hybridisation or tetrahedral hybridisation

When one s orbital and three p orbitals combine together to produce equivalent type of orbitals, sp^3 hybridisation takes place. Four sp^3 hybridised orbitals are produced which have the same energy and shape and are oriented at an angle of 109°28' to one another. This is also known as tetrahedral hybridisation because the hybridised orbitals are oriented towards the four corners of tetrahedron with the nucleus occupying the centre of tetrahedron. In this position, hybridised orbitals experience minimum repulsive forces. This is illustrated in Fig. 3.16.

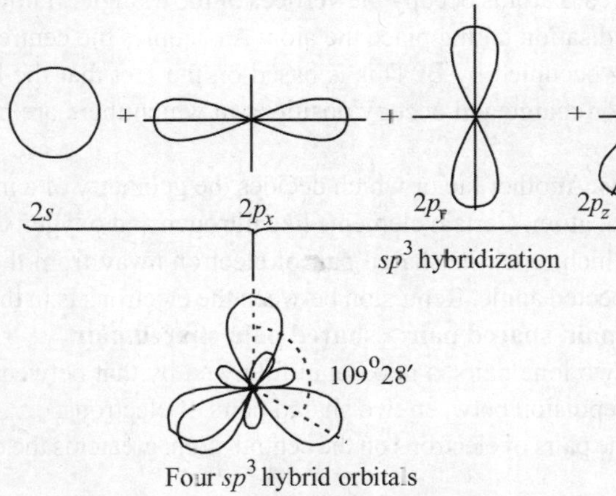

Fig. 3.16. sp^3 hybridisation.

Shape of methane molecule

At no. of carbon = 6

Electronic configuration of C in ground state

$$= 1s^2\ 2s^2\ 2p_x^1\ 2p_y^1$$

Electronic configuration of C in excited state

$$= 1s^2\ 2s^1\ 2p_x^1\ 2p_y^1\ 2p_z^1$$

There are four orbitals having unpaired electrons. These $2s$, $2p_x$, $2p_y$ and $2p_z$ orbitals will first hybridise. So it is a case of sp^3 hybridisation. Four hybridised orbitals tetrahedrally are shown in Fig. 3.17.

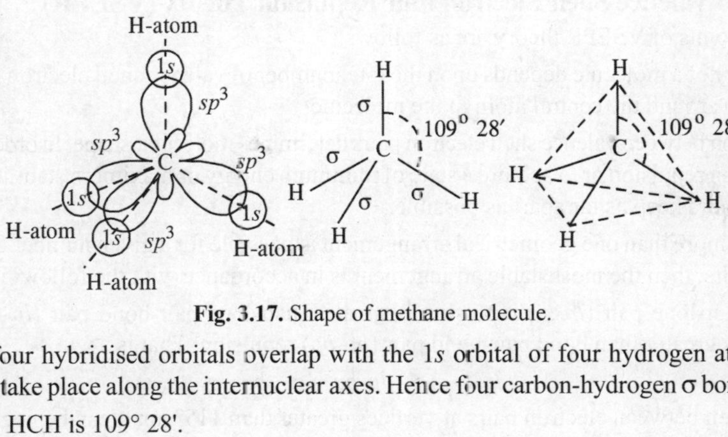

Fig. 3.17. Shape of methane molecule.

All the four hybridised orbitals overlap with the $1s$ orbital of four hydrogen atoms. All these overlappings take place along the internuclear axes. Hence four carbon-hydrogen σ bonds are formed. The angle ∠ HCH is 109° 28'.

3.3.4 Factors that determine the shape of molecules

The shapes of molecules depend upon the following factors:

1. Types of hybridisation. Type of hybridisation decides the shape of molecules. If *sp* hybridisation takes place and the molecule is of the type AB_2 where the atom A is central and surrounded by two B atoms then molecule will have a linear shape B — A — B with the ∠ BAB equal to 180°.

Similarly if the molecule is of the type AB_3 with sp^2 hybridisation the atom A lies in the centre of an equilateral traingle and three B atoms occupy the vertices of the triangle. If however the molecule is of type AB_4 with sp^3 hybridisation taking place the atom A occupies the centre of tetrahedral and vertices of tetrahedron being occupied by B. This is based on the fact that the hybridised orbitals repel each other due to similar charge and occupy positions in which there are minimum repulsive forces.

2. Lone pair of electrons. Another factor which decides the geometry of a molecule is the lone pair of electrons on the central atom. Certain elements like nitrogen and oxygen contain lone pair or pairs of electrons on them which repel the shared pair of electron away from them. In such a case bond angle changes from expected angle. Repulsion between the electrons is in the following order :
Lone pair, lone pair > lone pair, shared pair > shared pair, shared pair.

Repulsion between the two lone pairs is maximum followed by that between a lone pair and a shared pair followed by the repulsion between two shared pairs of electrons.

Greater the number of lone pairs of electrons on the central atom, greater is the distortion produced in the normal angle or shape.

3.3.5 Valence Shell Electron Pair Repulsion (VSEPR) Theory

Nevil Sidgwick

Valence shell electron pair repulsion theory (in short VSEPR theory) is the simplest theory for predicting the geometries of molecules. It is based on the repulsive interactions between the electron pairs in the valence shell. This theory was proposed by Sidgwick and Powell in 1940 and given the final shape by Gillespie and Nyholm (1957). According to this theory, the shapes of molecules can be determined by the number of electron pairs in the valence shell of the central atoms. These electron pairs experience electrostatic repulsions from one another and the bonded atoms in a molecule adopt that particular arrangement in space around the central atom which keeps them on the average as far apart as possible.

Main poins of Valence Shell Electron Pair Repulsion Theory (VSEPR)

The main points of VSEPR theory are as follows:

(*i*) Geometry of a molecule depends upon the total number of valence shell electron pairs (bonded or non-bonded) around the central atom in the molecule.

(*ii*) Repulsion between valence shell electron pairs determines molecular shape. In order to minimise electron-electron repulsion or to acquire a state of minimum energy or maximum stability the valence shell electron pairs keep as far apart as possible.

(*iii*) In case more than one geometrical arrangement is possible for a given number of bonded and non-bonded pairs, then the most stable arrangement is in accordance with the following:

(*a*) Lone pair-lone pair (*lp-lp*) repulsion is greater than lone pair-bond pair (*lp-bp*) repulsion which in turn is greater than bond pair-bond pair (*bp-bp*) repulsion. That is

$$lp\text{-}lp > lp\text{-}bp > bp\text{-}bp$$

(*b*) Repulsion between electron pairs at vertices greater than 115° apart can be neglected.

In the formation of a bond, the central atoms shares its valence electrons with the surrounding atoms. However, in certain cases, all the valence shell electrons may not take part in the bond formation. The electrons left in the valence shell without forming bonds exist as lone pairs. For example, in methane, CH_4, carbon uses all the four valence electrons in forming four bond pairs. In water, H_2O, the central oxygen atoms has ($1s^2$, $2s^2$ $2p^4$) six valence electrons. The hydrogen atom shares two of these six valence electrons in bond formation leaving four electrons as two lone pairs.

$\cdot\dot{C}\cdot$	$:\ddot{O}\cdot$
(Four valence electrons)	(Six valence electrons)

```
    H                 ..
    ×·               :O×H
 H×C×H                ·×
    ·×                 H
    H
```

(Four bond pairs)   (Two bond pairs and two lone pairs)

Thus, in methane, there are four pairs around carbon, while oxygen in water has two bond pairs and two lone pairs around it. According to VSEPR theory, it is the total number of electron pairs (bonded as well as lone pairs) that determines the geometry of molecule.

Regular and irregular geometries

The molecules in which the central atom is surrounded only by similar bonded electron pairs have ***regular geometries*** *while those in which the central atom is surrounded by bond pairs as well as lone pairs have* ***irregular geometries.***

Molecules such as CH_4, BF_3, PF_5, SF_6, etc., have regular geometries because the central atom in these molecules are bonded to similar atoms and there are no lone pairs in these molecules.

Molecules such as CH_2Cl_2, $CHCl_3$ etc., in which the central atom is bonded to different atoms and the molecules such as NH_3 and H_2O which have lone pairs as well as bond pairs, have irregular geometries.

VSEPR theory provides a simple approach to predict the shapes of molecues. The results obtained on the basis of this theory are quite in agreement with those obtained from hybridisation concept.

This theory is very simple to use. We make no distinction between *s* and *p*-electrons. We take into account only the number of electron pairs present in the valence shell of the central atom. The basic point is that the electron pairs (bonded or lone pairs) try to remain as far apart as possible to get a state of stability.

Limitations of VSEPR theory

Important limitations of VSEPR theory are:

(*i*) The theory is applicable to covalent compounds only and cannot be applied to ionic compounds.

(*ii*) The theory does not make any distinction between *s*, *p*, *d* and *f* orbitals.

(*iii*) The theory does not help in the prediction of shapes of coordination compounds.

(*iv*) The theory simply predicts that molecules such as NH_3, H_2O, $CHCl_3$, etc. have distinction from the regular geometrical shapes but does not quantify the extent of distortion.

Shape of ammonia molecule

Atomic no. of nitrogen = 7

Electronic configuration of nitrogen, $1s^2\ 2s^2\ 2p_x^1\ 2p_y^1\ 2p_z^1$

Here, there is no possibility of shifting of electron from $2s$ level to $2p$ level as there is no empty $2p$ orbital. $2s$ orbital, although incapable of overlapping or forming the bond or forming the bond, participates in hybridisation. This will give rise to tetrahedral shape. Hybridised orbitals are oriented towards the corners of a tetrahedron with one corner occupied by the lone pair of electrons from $2s$ orbital. Three sp^3 hybridised orbitals overlap with the $1s$ orbitals of three hydrogen atoms (Fig. 3.18.). We expect an angle of 109°28' between H — N — H in tetrahedral geometry. Actually this angle is 107°. This contraction in angle takes place because the lone pair of electrons repels the shared pair of electrons or bonded pair of electrons with a greater force (Fig. 3.18.) thereby reducing the angle. (Valence Shell Electron Pair Repulsion, VSEPR Theory).

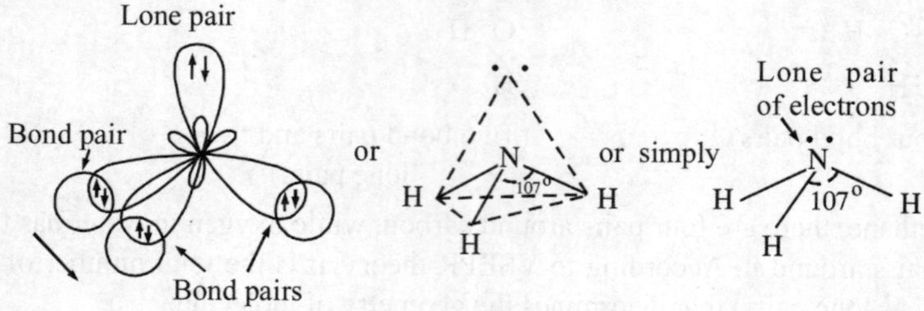

Fig. 3.18. Shape of ammonia molecule.

Thus it has pyramidal shape.

Shape of water molecule

At. no. of oxygen = 8

Electronic configuration = $1s^2\ 2s^2\ 2p_x^2\ 2p_y^1\ 2p_z^1$

In the 2nd orbit there are four orbitals : $2s$, $2p_x$ $2p_y$ and $2p_z$. Out of these $2s$ and $2p_x$ cannot form the bonds as they are paired. But they can participate in hybridisation. Hence it is case of sp^3 hybridisation. The hybridised orbitals are oriented along the corners of a tetrahedron. Two of the positions on the tetrahedron are occupied by lone pair of electrons from $2s$ and $2p_x$ orbitals. The other two hybridised orbitals will overlap with the $1s$ orbital of two hydrogen atoms to form a water molecule as shown in Fig. 3.19. We expect an angle of 109°28' in HOH. Actually the angle is 105°. This is due to stronger repulsion of two lone pairs of electrons with shared pair of electrons (VSEPR Theory). Thus water is a V-shaped molecule.

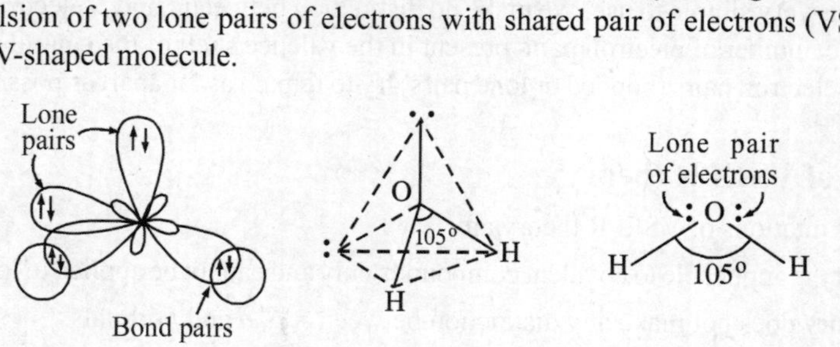

Fig. 3.19. Shape of water molecule.

Structure of hydronium ion $[H_3O]^+$

In the formation of hydronium ion, coordinate, bond is involved. Here, oxygen atom in the water molecule is the donour while hydrogen ion is the acceptor. The formation of coordinate bond in H_3O^+ may be represented as below pointing from the donor to the acceptor, as shown below in the formation of hydronium ion.

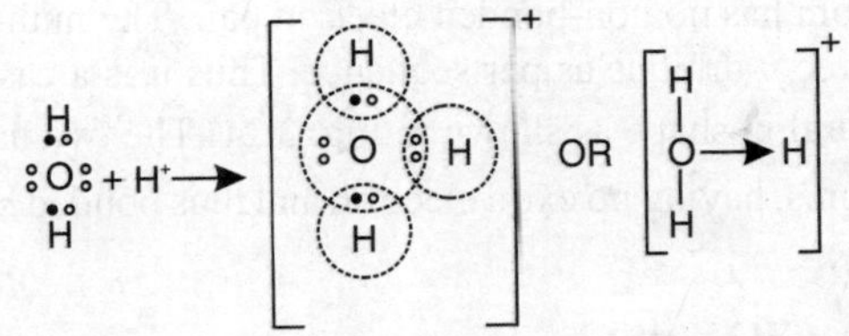

It is important to note that the coordinate covalent bond once formed is indistinguishable from a covalent bond.

According to the atomic orbital overlap concept of covalency, a coordinate covalent bond results when an empty orbital present in the other atom. The central atom, oxygen, in water molecule has four orbitals. Two of these orbitals contain bond pairs of electrons while the other two (shaded ones) contain lone pairs of electrons, as shown in Fig. 3.20. The H^+ ion has an empty s orbital which overlaps with one of the orbitals of oxygen atom containing a lone pair of electrons forming hydronium ion, H_3O^+, as illustrated in Fig. 3.20.

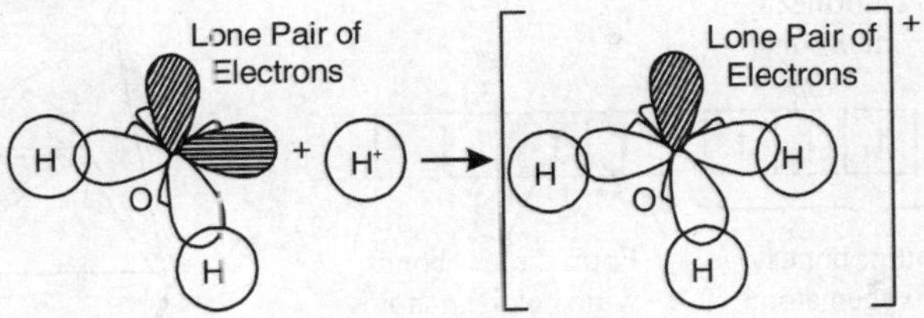

Fig. 3.20. Formation of hydronium ion.

It may be noted that the shape of $[H_3O]^+$ will continue to remain tetrahedral because the hybridisation involved remains sp^3. The difference between H_2O and H_3O^+ is: while in H_2O, two tetrahedral positions are occupied by lone pairs of electrons, it is only one tetrahedral position that is occupied by loans pair of electrons in H_3O^+.

Shape of sulphate ion

Atomic no. of sulphur is 16. Its outer electronic configuration is $3s^2\ 3p^4$. The shape of molecule suggests that it is a case of sp^3 hybridisation. To have the requisite no. of half-filled orbitals one electron from *s* and one electron from *p* orbitals are excited to *d*-orbitals as shown in Fig. 3.21. It

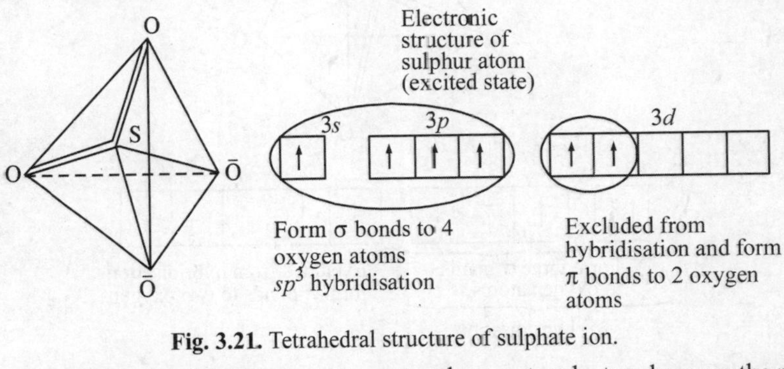

Fig. 3.21. Tetrahedral structure of sulphate ion.

may be remembered that out of four oxygen atoms two have extra electron because the net charge on SO_4^{2-} is –2. Sulphur combines with four oxygen atoms forming four bonds, thus using four pairs of electrons. Also sulphur atom has no non-bonded electron pair. The number of hybrid orbitals used are four *viz.*, one *s* and three *p* orbitals as per sequence. Thus it is a case of sp^3 hybridisation and therefore SO_4^{2-} is tetrahedral in shape as shown in Fig. 3.21. The two half-filled *d*-orbitals form π bonds with two oxygen atoms, having no extra electron and thus bonded by double bonds to sulphur.

Shape of perchlorate ClO_4^- ion

This is a case of sp^3 hybridisation At. no. of Cl is 17. Its outer electronic configuration is $3s^2$ $3p^5$. One electron each from 3*s*, $3p_x$ and $3p_y$ are excited to 3*d* orbitals. Single electron in 3*s*, $3p_x$ $3p_y$ and $3p_z$ hybridise sp^3 to give tetrahedral shape. Electrons in 3*d* orbital form π bonds. ClO_4^- ion is tetrahedral. It is because chlorine makes use of four bond orbitals and these are one *s* and three *p*-orbitals and hybridisation is sp^2 and shape is tetrahedral. Chlorine has no non-bonded electron pair in the excited state. Three half-filled *d*-orbitals form π bonds with three oxygen atoms without extra electron and thus bonded by double bonds to chlorine as shown in Fig. 3.22.

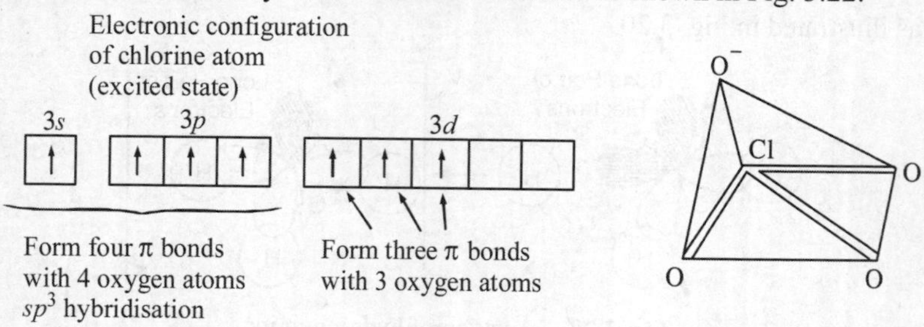

Fig. 3.22. Tetrahedral shape of perchlorate ion.

Shape of chlorate ion (ClO_3^-)

At. no. of Cl is 17. Its outer configuration is $3s^2\ 3p_x^2\ 3p_y^2\ 3p_z^1$. It combines with three atoms of oxygen using three of its electrons. Thus it forms three single σ bonds with oxygen using three of its

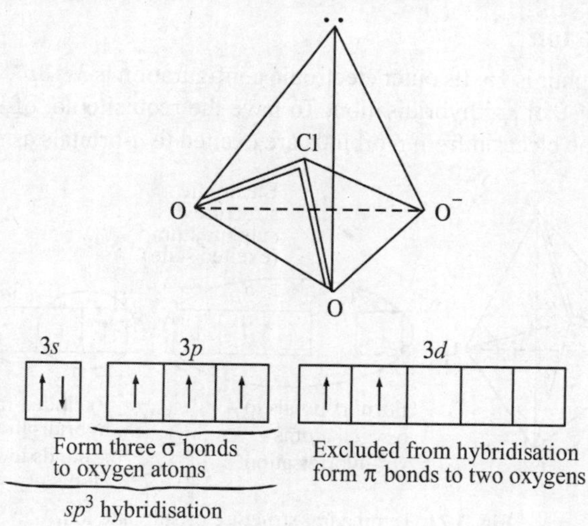

Fig. 3.23. Structure of ClO_3^-.

electrons. Here two electrons from $3p_x$ and $3p_y$ are excited to $3d$ orbitals which don't participate in hybridisation but form π bonds with two oxygen atoms without extra electron. $3s$ and $3p$ orbitals (sp^3) hybridise. It forms three single σ bonds with oxygen using three orbitals. Its actual shape is pyramidal with one corner position occupied by a lone pair of electrons. Its expected geometry is tetrahedral as sp^3 hybridisation is involved. The two half-filled d-orbitals form π bonds with two oxygen atoms and thus bonded to chlorine by double bonds as shown in Fig. 3.23.

Example 9. **ClO_3^- and ClO_4^- ions have same number of electron pairs around central chlorine but their geometry is different. Why?**

Solution. sp^3 hybridisation takes place in the formation of both molecules. In ClO_3^- one tetrahedral position is occupied by a lone pair, therefore the ion gets pyramidal shape.

As expected from sp^3 hybridisation, ClO_4^- achieves tetrahedral shape.

Structure and shape of carbonate

Ground state electronic configuration of carbon atom is $1s^2$, $2s^2$, $2p^2$ *i.e.*, it contains only two unpaired electrons. In order to form bonds to three oxygen atoms, carbon requires three unpaired electrons which may be obtained by promoting one of the $2s$ electrons to the vacant $2p$ orbitals. (Fig. 3.24).

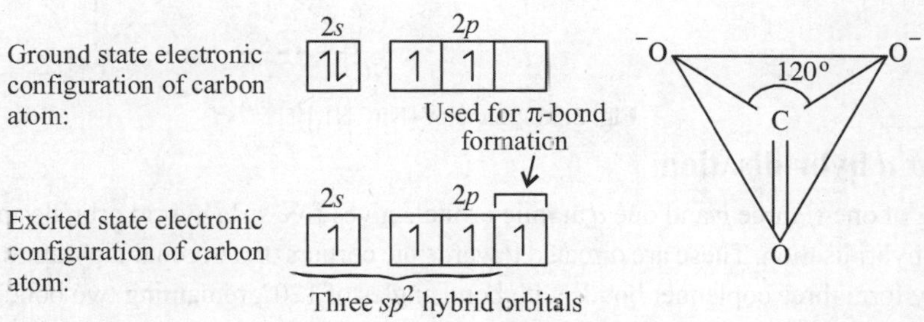

Fig. 3.24. Structure of carbonate ion.

The three sp^2 hybridised orbitals are obtained by mixing $2s$ and two $2p$ atomic orbitals. Each sp^2 hybridised orbital contains only one electron and is used to form three σ covalent bonds with three oxygen atoms. This accounts for the planner geometry of carbonate ion.

The third $2p$ orbital containing one electron overlaps with a vacant $2p$ orbitals of an oxygen atom to give a π bond between carbon and one of the three oxygen atoms.

3.3.4 dsp^2 square planar hybridisation

This type of hybridisation takes place when one s, two p (p_x, p_y) and one d ($d_{x^2-y^2}$) orbitals mix together to give four equivalent hybrid orbitals. These four hybrid orbitals keep their lobes directed towards the corners of a square in the XY plane, *i.e.*, these hybrid orbitals point away from the origin in the $\pm X$ and $\pm Y$ direction of X and Y axes. The angle of maximum repulsion between their orbitals is 90° as shown in Fig. 3.25.

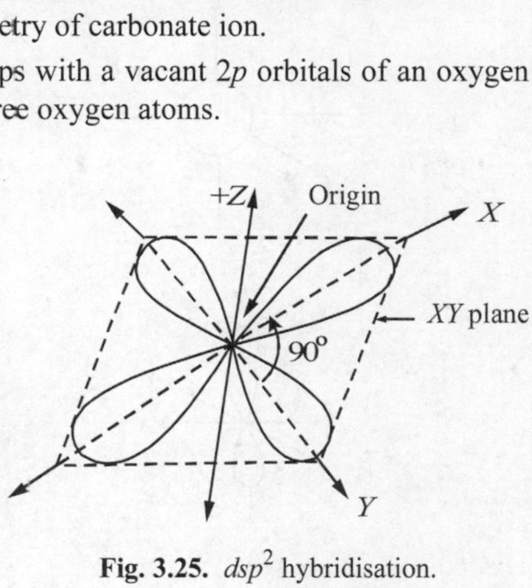

Fig. 3.25. dsp^2 hybridisation.

Shape of $[Ni(CN)_4]^{2-}$

Pairing of single electrons in 3*d* orbitals in Ni^{2+} takes place thereby creating one vacant 3*d* orbital.

Electron pairs from four CN^- ions are filled up in four hybrid orbitals obtained as a result of dsp^2 hybridisation as shown in Fig. 3.26.

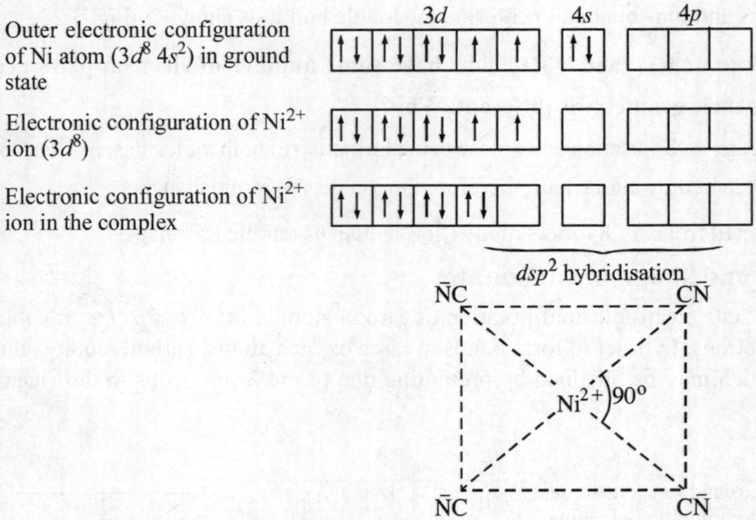

Fig. 3.26. Shape of $[Ni(CN)_4]^{2-}$]

3.3.5 sp^3d hybridisation

Mixing of one *s,* three *p* and one *d* atomic orbitals gives five sp^3d hybrid orbitals and process is called sp^3d hybridisation. These are directed towards the corners of a trigonal bipyramid as shown in figure. They form three coplanner bonds with bond angles of 120°, remaining two bonds are at right angles to this plane one above and one below.

Formation of five sp^3 *d* hybrid orbitals is shown in Fig. 3.27.

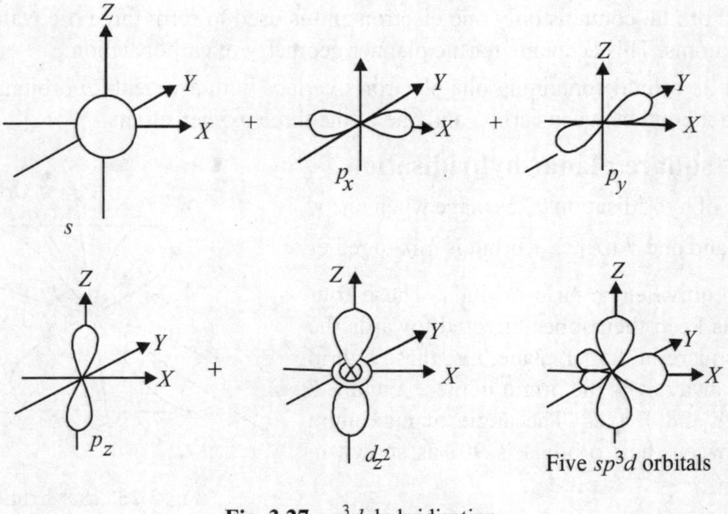

Fig. 3.27. sp^3d hybridisation.

Shapes of molecules involving sp^3d hybridisation

(*i*) Shape of PF_5

Here sp^3d type hybridisation is involved. At. no. of phosphorus atom is 15. Its outer electronic configuration in the ground state is $3s^2\ sp^3$ as shown in Fig. 3.28.

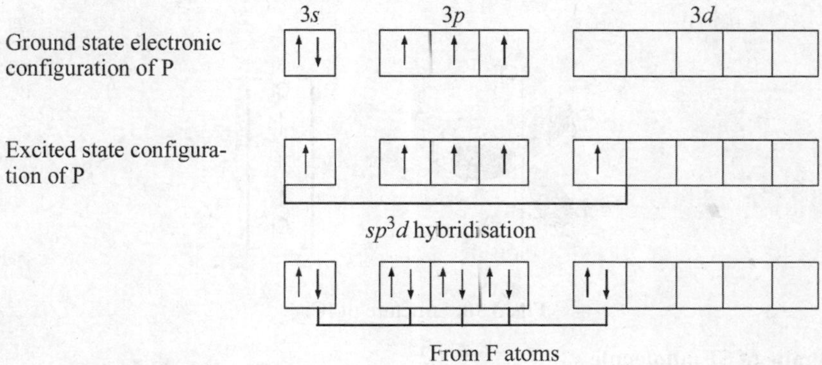

Fig. 3.28. Electronic configuration in P.

In order to have the requisite number of half-filled orbitals available for hybridisation purpose one of the 3*s* electrons is made to jump to one of the 3*d* empty orbitals, thereby giving rise to 5 half-filled orbitals for bonding purposes. These five orbitals hybridise (sp^3d). Now five orbitals are available for combination which permit the formation of five covalent bonds with five fluorine atoms. But all the hybridised orbitals are not equivalent. Three of these are oriented towards the three corners of an equilateral triangle making an angle of 120° between them. The bonds formed by these orbitals are known as equitorial bonds. Remaining two orbitals are oriented at right angles to the plane of first set of three orbitals. Thus PF_5 has trigonal bipyramidal shape as shown in Fig. 3.29.

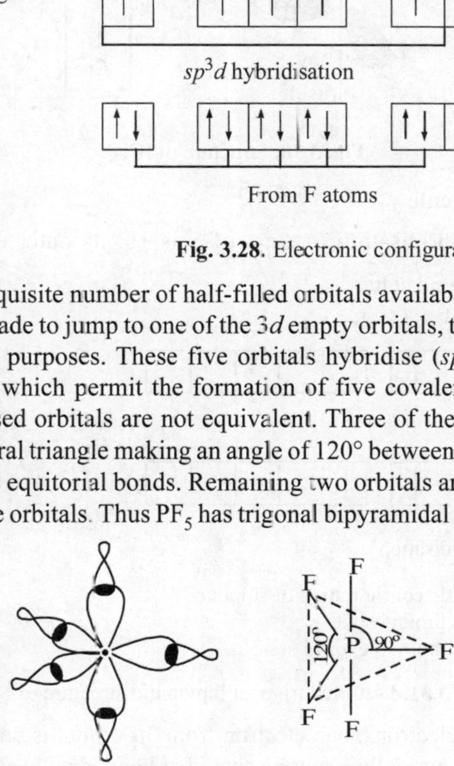

Fig. 3.29. Formation of trigonal bipyramidal PF_5 molecule.

(*ii*) Shape of ICl_2^-

ICl_2^- molecule has been found to be linear in shape. This can be explained by assuming sp^3d hybridisation of the orbitals of iodine (the bigger of I and Cl). sp^3d hybridisation leads to trigonal

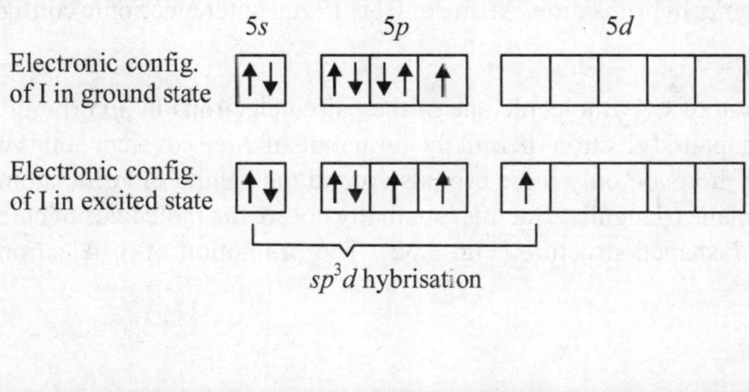

bipyramidal shape. Two chlorine atoms occupy axial positions (perpendicular to the plane of the triangle base) because the repulsions will the minimum in this arrangement. One electron on the chlorine atom (negative charge) will enter one of the sp^3d hybrid orbitals. Two sp^3d hybrid orbitals containing single electrons will overlap separately with the *p*-orbitals of two chlorine atoms to form two σ bonds. Thus, in actual appearance the ICl_2^- ion will have T-shape because the three base position of the triangular bipyramid will be occupied by lone pairs of electrons as shown in Fig. 3.30.

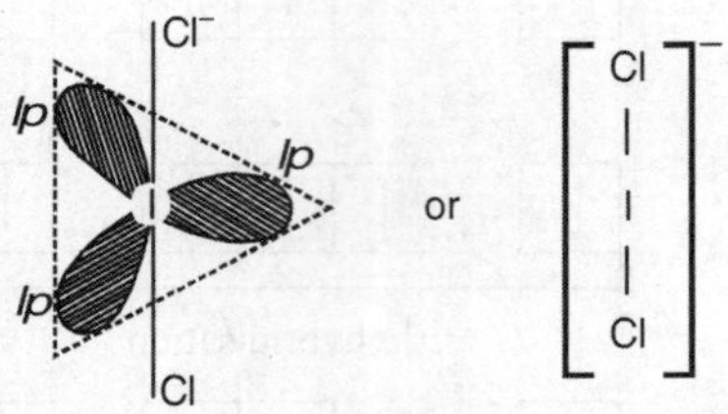

Fig. 3.30. Structure of ICl_2^-.

(*iii*) Shape of SF_4 molecule

It is a case of sp^3d hybridisation; At. no. of S is 16, its outer electronic configuration is $3s^2\ 3p_x^2\ 3p_y^1\ 3p_z^1$ in the ground state.

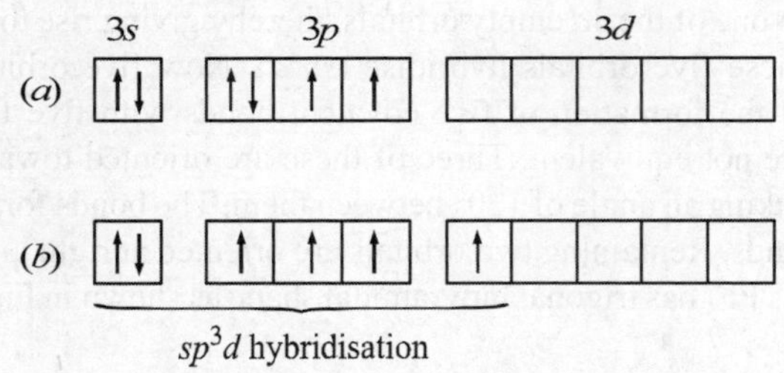

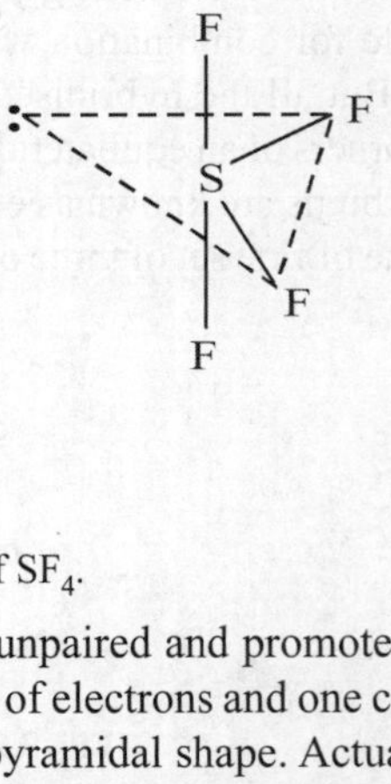

(a) Electronic configuration of sulphur atom in ground state.
(b) Sulphur atom in excited state.

Fig. 3.31. Distorted trigonal bipyramid structure of SF_4.

To have four unpaired electrons, one electron from 3*p* orbital is unpaired and promoted to 3*d* orbital. Thus five orbitals are used, four contain shared or bonded pairs of electrons and one contains a lone pair, sp^3d hybridisation occurs, giving an expected trigonal bipyramidal shape. Actually the four fluorine atoms are joined tetrahedrally to sulphur atom in an irregular manner due to presence of a lone pair (Fig. 3.31).

(*iv*) Shape of ClF_3 molecule

It is a case of sp^3d hybridisation. At. no. of Cl is 17. Its outer electronic configuration in ground state is $3s^2\ 3p_x^2\ 3p_y^2\ 3p_z^1$.

In the formation of ClF_3 molecule one of the paired electrons in 3*p* orbital is promoted to 3*d* orbital. The three unpaired electrons permit the formation of three covalent bonds with three fluorine atoms. If however there are only three orbitals around the central chlorine atom $C1F_3$ molecules should have been plane triangular. But this is actually not so, the molecules of chlorine trifluoride is known to have a *T*-shaped structure (Fig. 3.32). The promotion of 3*p* electrons is followed by

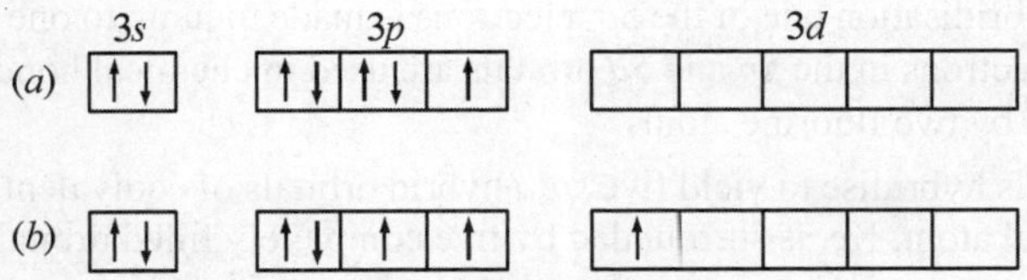

sp^3d hybridisation
(a) Electronic configuration of chlorine atom–ground state.
(b) Chlorine atom–excited state.

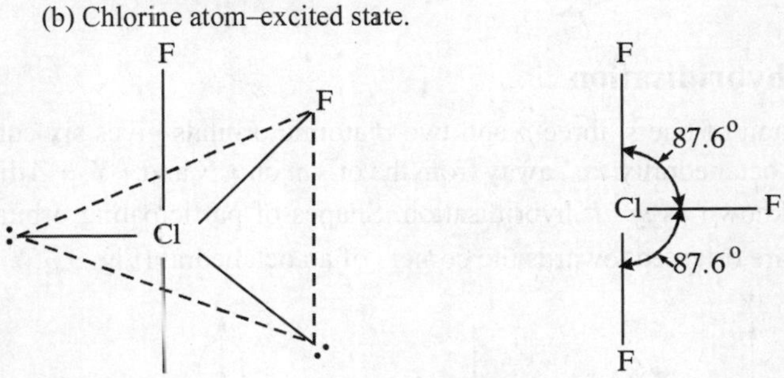

(*a*) (*b*)

Fig. 3.32. Shape of ClF_3 molecule.

hybridisation involving one *s*, three *p* and one *d* orbitals. This is thus a case of sp^3d hybridisation. The chlorine atom is therefore surrounded by five orbitals, three containing a pair of electrons while the other two contain lone pair of electrons. As two positions of the trigonal, bipyramid are occupied by lone pairs, the molecule has a *T*-shape. The angle F — Cl —F expected from this geometry is 90°. However actual angle is 87.6°. This is due to repulsion from lone pairs which result in contraction of angle.

Example 10. **In XeF_2, Xe undergoes sp^3d hybridisation but the molecule is linear. Explain.**

Solution. It is a case of sp^3d hybridisation. Atomic number of Xe is 54. Its outer electronic configuration is $5s^2\ 5p_x^2\ 5p_y^2\ 5p_z^2$ as shown in Fig. 3.33. In order to have the requisite number of

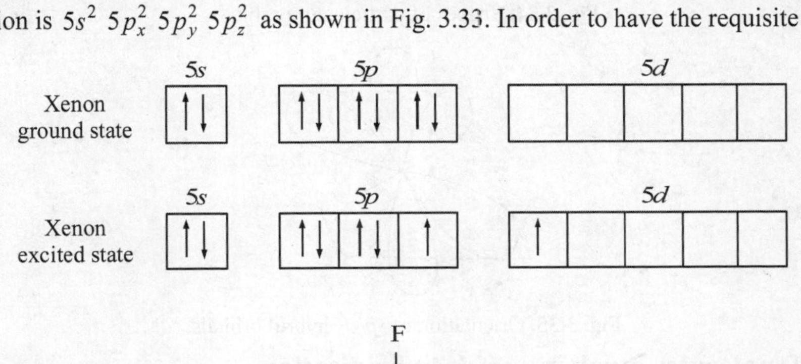

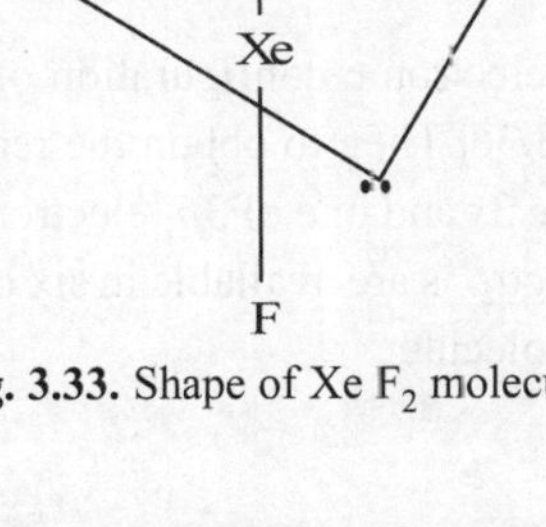

Fig. 3.33. Shape of Xe F_2 molecule.

half-filled orbitals for hybridisation one of the $5p_z$ electrons is made to jump to one of the $5d$ empty orbitals. The unpaired electrons in the $5p$ and $5d$ orbitals are used in chemical bonding, the pairing electrons being provided by two fluorine atoms.

The *s, p* and *d* orbitals hybridise to yield five sp^3d hybrid orbitals of equivalent energy. Thus in XeF_2 molecule the central atom, Xe, is surrounded by five completely filled orbitals, two of which contain shared pair of electrons and remaining three contain lone pair of electrons. Three positions in trigonal bipyramid are occupied by lone pairs because minimum repulsive forces between the lone pairs and bond pairs are involved in this position. Thus actual structure of XeF_2 is linear (Fig. 3.33).

3.3.6 sp^3d^2 hybridisation

A combination of one *s,* three *p* and two *d* atomic orbitals gives six equivalent sp^3d^2 hybrid orbitals directed octahedrally, *i.e.*, away from the origin on ± X and ± Y, ± Z directions. This type of combination is known as sp^3d^2 hybridisation. Shapes of participating orbitals is shown in 3.34. Hybrid orbitals are oriented towards the corners of an octahedral (Fig. 3.35).

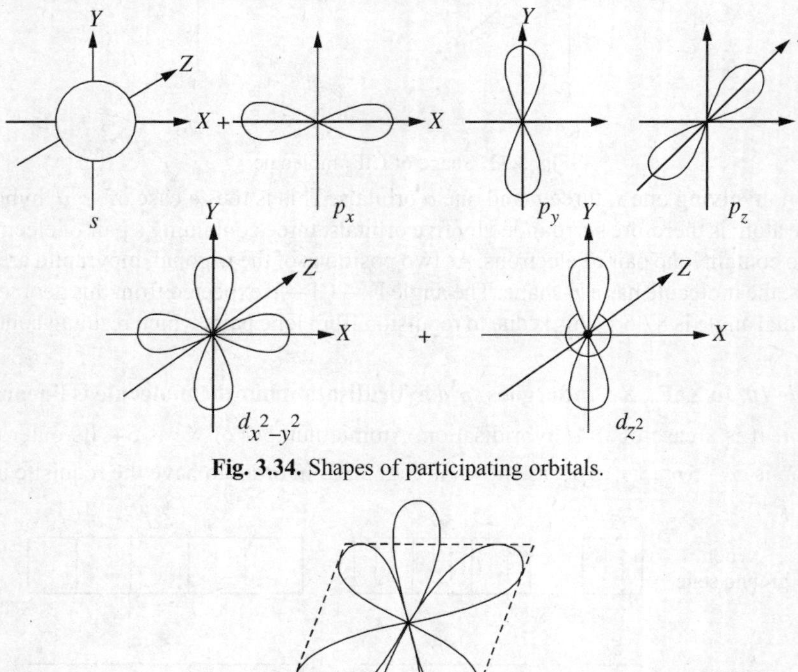

Fig. 3.34. Shapes of participating orbitals.

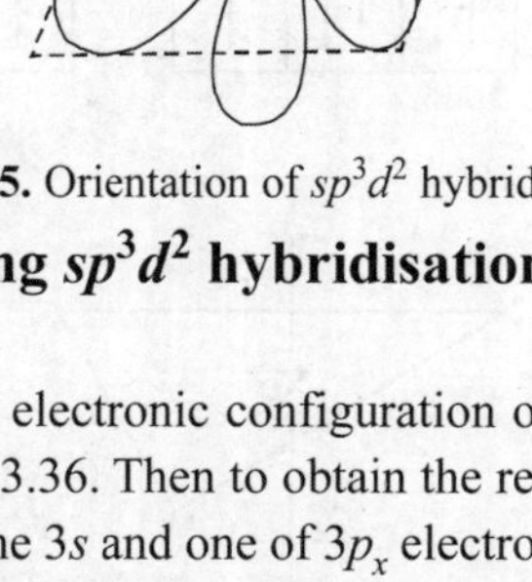

Fig. 3.35. Orientation of sp^3d^2 hybrid orbitals.

Shapes of molecules involving sp^3d^2 hybridisation

(*i*) Shape of SF_6

At. no. of sulphur is 16. Outer electronic configuration of central atom in the ground state is $3s^2\ 3p_x^2\ 3p_y\ 3p_z$ as shown in Fig. 3.36. Then to obtain the requisite number of half-filled orbitals for hybridisation purposes, one of the $3s$ and one of $3p_x$ electrons are made to jump to two of the $3d$ empty orbitals. Now six unpaired electrons are available in six orbitals (one *s,* three *p* and two *d*), for bond formation to give rise to SF_6 molecule.

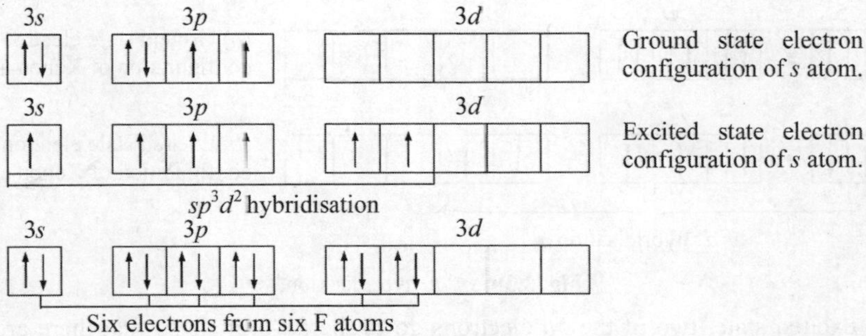

Fig. 3.36. Electronic configuration of S.

These six orbitals hybridise in sp^3d^2 type. The six hybrid orbitals are now directed towards the corners of a regular octahedron. These six hybrid orbitals overlap with the p-orbitals of six fluorine atoms. So SF_6 molecule has an octahedral structure as shown in Fig. 3.37.

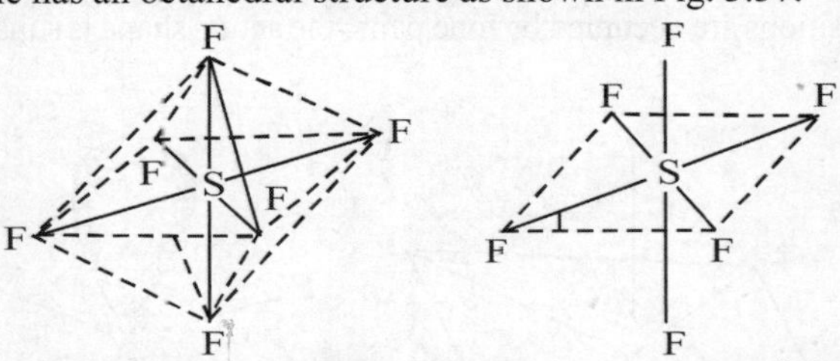

Fig. 3.37. Octahedral molecule of sulphur hexafluoride formed by sp^3d^2 hybridisation.

(*ii*) Shape of iodine pentafluoride IF_5.

It is a case of sp^3d^2 hybridisation. At. no. of iodine is 53. Its outer electronic configuration is

$$5s^2\ 5p_x^2\ 5p_y^2\ 5p_z^1$$

In the excited state two electrons from $5p$ orbitals are excited to $5d$ orbitals, thereby giving five single electrons for the formation of five bonds with fluorine. $5s$, $5p_x$, $5p_y$, $5p_z$ and two d-orbitals (sp^3d^2) hybridise to give equivalent orbitals oriented octahedrally.

However one position is occupied by a lone pair. Therefore actual shape is square pyramidal (Fig. 3.38).

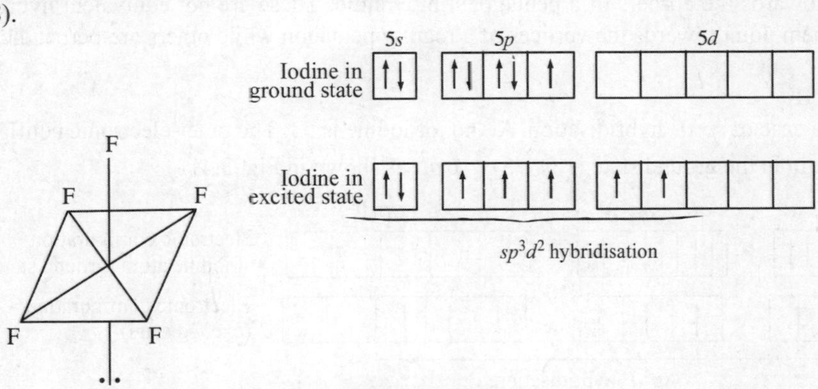

Fig. 3.38. Shape of IF_5 molecule.

(*iii*) Shape of XeF_4.

It is a case of sp^3d^2 hybridisation. At. no. of Xe is 54. Its outer electronic configuration in the ground state is $5s^2\ sp^6$ (Fig. 3.39).

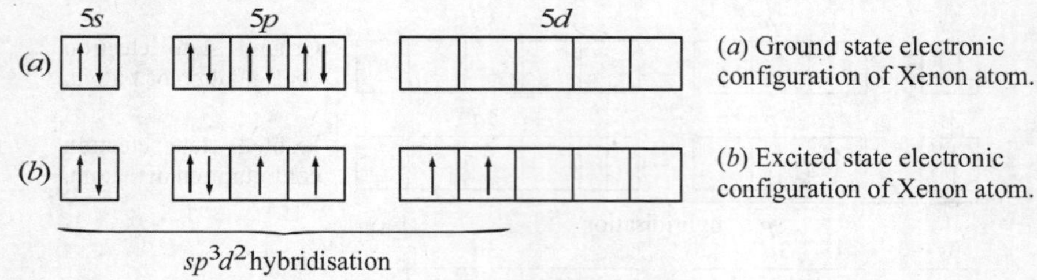

Fig. 3.39. sp^3d^2 hybridisation.

In the excited state, two of the *5p* electrons go to the *5d*-orbitals so that there are now four unpaired electrons (two in *5p* and two in *5d* orbitals). These are used in bonding with four atoms of fluorine as shown in figure. The central atom, Xe, is surrounded by six orbitals resulting from sp^3d^2 hybridisation. The hybrid orbitals are oriented octahedrally. Four of these orbitals contain shared pairs while the remaining two contain lone pair of electrons.

As the vertex positions are occupied by lone pairs, the actual shape is square planer as represented in Fig. 3.40.

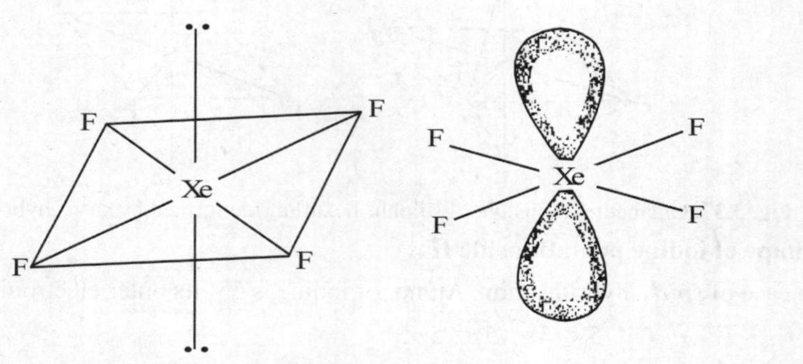

Fig. 3.40. Shape of XeF_4 molecule.

3.3.7 sp^3d^3 hybridisation

Mixing of one *s*, three *p* and three *d* atomic orbitals gives seven sp^3d^3 hybrid orbitals which are directed towards the corners of a pentagonal bipyramid. These are not equivalent hybrid orbitals. Five of them point towards the vertices of a regular pentagon while others are perpendicular to the plane.

Shape of IF_7

It is a case of sp^3d^3 hybridisation. At. no. of iodine is 53. The outer electronic configuration of central atom in the ground state is $5s^2\ 5p_x^2\ 6p_z^2$ as shown in Fig. 3.41.

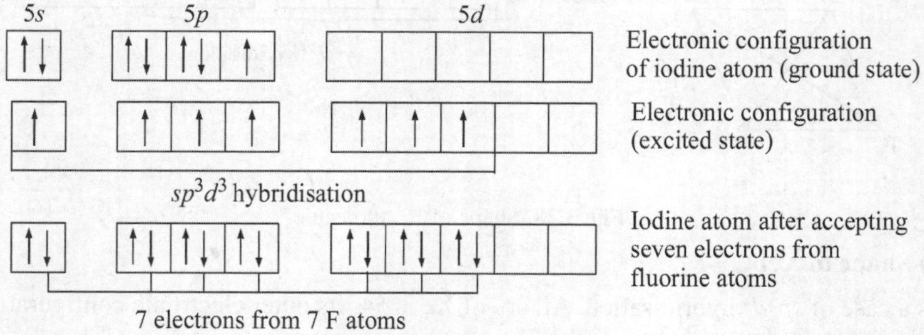

Fig. 3.41. sp^3d^3 hybridisation.

To have the requisite number of half-filled orbitals for hybridisation purposes, unpairing of 5*s* as well as of $5p_x$ and $5p_y$ electrons takes place. Thus seven unpaired electrons are available for bond formation to give rise to IF_7 molecule. The resulting molecule acquires pentagonal bipyramidal shape shown in Fig. 3.42.

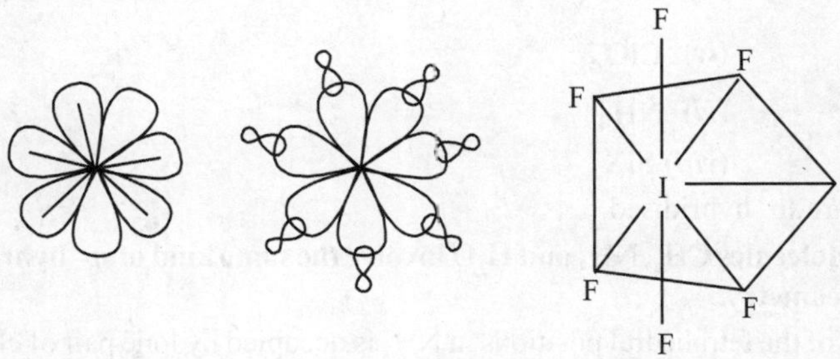

Fig 3.42. Formation of IF_7 molecule.

In this case all the seven hybrid orbitals are not equivalent. Five of them are directed towards the vertices of a regular pentagon giving an angle 72° with one another, the rest of two are oriented at right angle to the plane of the first set of five orbitals. Thus the bond angles in the molecule are of 72° and 90°.

***Example 11*. Pick up species with sp^3d^3 hybridisation.**

(*i*) XeF_6 (*ii*) IF_7 (*iii*) BrF_5 (*iv*) $XeOF_4$

Solution. XeF_6, IF_7.

***Example 12*. Give shapes and hybridisations of following molecules or ions:**

(*i*) $(Me)_2Be$ (*ii*) CS_2

(*iii*) BH_4^- (*iv*) $Ni(CO)_4$

(*v*) Hg_2Cl_2 (*vi*) GaI_3

(*vii*) Al_2Cl_5 (*viii*) $SeBr_2$

(*ix*) ICl_2^- (*x*) XeF_4

(*xi*) PF_3Cl_2 (*xii*) NF_3

Solution. (*i*), (*ii*), (*v*) : linear, *sp*

(*iii*), (*iv*), (*vii*) : tetrahedral sp^3

(*vi*) Plane triangle sp^2

(*viii*) Angular (V-shaped), sp^3

(*xii*) Pyramidal, sp^3

(*ix*) Linear sp^3d

(*x*) Square planner, sp^3d

(*xi*) Trigonal bipyramidal, sp^3d^2

***Example 13*: Give the shapes acquired by the following gaseous molecules or ions.**

(*i*) SO_3 (*ii*) XeF_2

(*iii*) CS_2 (*iv*) ICl_4^-

(*v*) SCl_2 (*vi*) NH_4^+

***Solution*:** (*i*) Plane triangle (*ii*) Linear

(*iii*) Linear (*iv*) Square planer

(*v*) Angular (*vi*) Tetrahedral

***Example 14*: Indicate the species with sp^3 hybridisation of central atom.**

(*i*) SO_4^{2-}	**(*ii*) $AsCl_3$**
(*iii*) BF_4^-	**(*iv*) ClO_4^-**
(*v*) ClO_2^-	**(*vi*) NH_4^+**
(*vii*) OCl_2	**(*viii*) SiX_4**

***Solution*:** All are sp^3 hybridised.

***Example 15*: Molecules CH_4, NH_3 and H_2O involve the same kind of sp^3 hybridisation, yet they possess different geometry.**

***Solution*:** One of the tetrahedral positions in NH_3 is occupied by lone pair of electrons. Similarly in the case of water two positions are occupied by lone pair of electrons.

The repulsion between lone pair-bond pair is greater than that between bond pair and bond pair. As a result $\angle$HNH is decreased to 107°. There are still greater repulsion in the case of water. Here the $\angle$HOH is further decreased to 105°.

***Example 16*. Explain: *lp-lp* repulsion > *lp-bp* repulsion > *bp-bp* repulsion.**

***Solution*:** A bond pair is under the influence of two nuclei of bonding atoms whereas the lone pair is under the influence of only one nucleus. Hence lone pair is more active and mobile in causing repulsion with other orbitals. As a result, *lp-lp* repulsion is greater than *lp-bp* repulsion which in turn is greater than *bp-bp* repulsion.

***Example 17*. Giving reason arrange the following molecules in order of increasing bond angle**

$$H_2O, NH_3, PH_3$$

***Solution*.** The order of increasing bond angle in

$$H_2O < PH_3 < NH_3$$

In case of water, there are two lone pair of electrons, therefore repulsion between lone pair and bond pair will be maximum and angle will be minimum.

Out of NH_3 and PH_3 repulsion between bond pair and bond pair will be smaller in case of PH_3 because of smaller electronegativity of phosphorus and larger P-H bond length and hence $\angle$HPH will be smaller than $\angle$HNH.

3.4 MOLECULAR ORBITAL THEORY

Main points of molecular orbital theory of covalent bonding.

The main points of molecular orbital theory put forward by F. Hund and R.S. Mulliken in 1930 are:

R.S. Mulliken

(*i*) Electrons belonging to two bonding atoms are controlled by the entire molecule. The electrons will occupy molecular orbitals that are analogous to atomic orbitals. These molecular orbitals belong to the entire molecule.

(*ii*) In an atom there are *s, p, d* .... atomic orbitals determined by various values of quantum numbers. In a molecule we have a set of molecular orbitals σ, π, δ ..... determined by various quantum numbers.

(*iii*) Just as an atomic orbital can contain a maximum of two electrons, molecular orbitals can also accommodate only a maximum of two electrons as per Pauli's exclusion principle.

(*iv*) The various electrons occupy different orbitals according to Hund's rule of maximum multiplicity as in the case of atomic orbitals.

3.4.1 Linear combination of atomic orbitals (LCAO)

Molecular orbitals can be visualised as being made up of appropriate atomic orbitals provided by the atoms joined together in the molecule. A very simple, reasonable and satisfactory model is obtained if it is assumed that atomic orbitals of constituents combine in a linear additive fashion to form molecular orbitals. Consider two atoms *A* and *B* each possessing a single valence electron. Let ψ_A and ψ_B represent the wave functions of atoms *A* and *B*. Let atomic orbitals A and B accommodating valence electrons overlap. The valence electrons will be in region where the orbitals overlap and therefore, they can be associated with either nucleus. Consequently, the valence electrons have a much larger volume available for their movement as compared with that available in each isolated atom. The electrons can be thought of as moving in a molecular orbital whose total volume is approximately equal to the sum of the volume of atomic orbitals of *A* and *B*. In this fashion, the wave function of molecular orbitals can be expressed as a linear combination of atomic orbitals (LCAO). ψ_{AB} is the wave function of molecular orbital.

$$\psi_{AB} = C_A \psi_A - C_B \psi_B$$

and C_A and C_B are mixing coefficients. Value of C_A and C_B is so chosen as to get a molecular orbital of lowest energy.

In case we are considering the bonding in two similar atoms (homonuclear) then the magnitude of the coefficients C_A and C_B must be same.

In that case

$$C_A^2 = C_B^2 \text{ or } C_A = \pm C_B$$

There will be two molecular orbitals as a result of two types of combinations as given below.

$$\psi^b = \psi_A + \psi_B$$

$$\psi^* = \psi_A - \psi_B$$

Values of ψ_A and ψ_B have been taken as unity.

Thus one molecular orbital (ψ^b) can be obtained by addition of wave function of two atomic orbitals and then the other molecular orbital (ψ^*) can be obtained by the subtraction of wave function of the atomic orbitals. This is shown in Fig. 3.43.

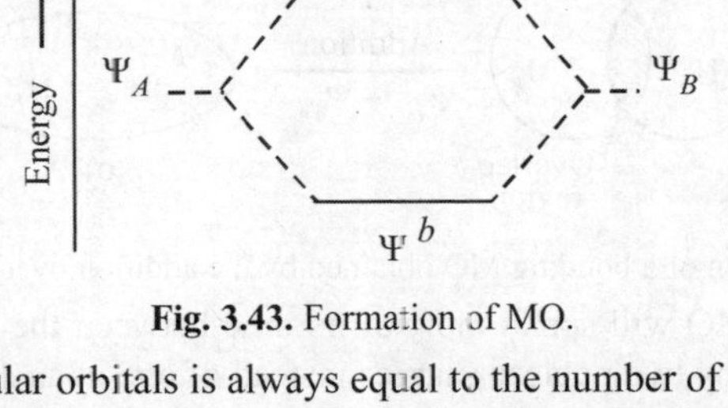

Fig. 3.43. Formation of MO.

The number of molecular orbitals is always equal to the number of atomic orbitals involved in bonding. Effective combination of atomic orbitals to give rise to stable molecular orbitals requires:

(*i*) Symmetry and relative energies of atomic orbitals ψ_A and ψ_B must be close to each other.

(*ii*) Extent of overlap between the charge clouds of ψ_A and ψ_B should be large.

(*iii*) Atomic orbitals of same sign should overlap.

3.4.2 Bonding and antibonding molecular orbitals

Bonding orbitals. The molecular orbital that is obtained by the addition overlap of two atomic orbitals is termed bonding molecular and is represented as ψ^b

$$\psi^b = \psi_A + \psi_B$$

Some features of bonding molecular orbitals are:

(*i*) It possesses lower energy than the atomic orbitals from which it is formed.

(*ii*) It has high electron density in the region between the two nuclei and this accounts for stability of bond.

(*iii*) It is formed when lobes of the combining atomic orbitals have been same sign. The magnitude of ψ^b is numerically bigger than ψ_A or ψ_B.

(*iv*) Every electron in a bonding M.O. contributes to attraction between two atoms.

Antibonding M.O. The other molecular orbital which is obtained by the subtraction overlap of two atomic orbitals is termed the antibonding molecular orbital and is represented as ψ^*.

$$\psi^* = \psi_A - \psi_B$$

Some features of antibonding molecular orbitals are:

(*i*) It possesses higher energy than the atomic orbitals from which it is formed.

(*ii*) The probability of finding the electron in between the nuclei is negligible. The electron density is concentrated only on one nucleus at any particular moment.

(*iii*) It is formed when the lobes of combining atomic orbitals have opposite signs. ψ^* is numerically smaller than ψ_A or ψ_B.

(*iv*) Every electron in an antibonding molecular orbital contributes to repulsion between two atoms.

Combination of *s* orbitals to give bonding and antibonding molecular orbitals

Let us consider the combination of two 1*s* atomic orbitals belonging to say two hydrogen atoms. Using LCAO approach the two atomic orbitals will overlap to give two molecular orbitals. The first combination is obtained by the addition of two atomic orbitals. The pictorial representation is shown in Fig. 3.44.

ψ^b MO $= N(\Psi_{1s_a} + \Psi_{1s_b})$ where N is a normalizing constant.

For simplicity its value may be taken as 1

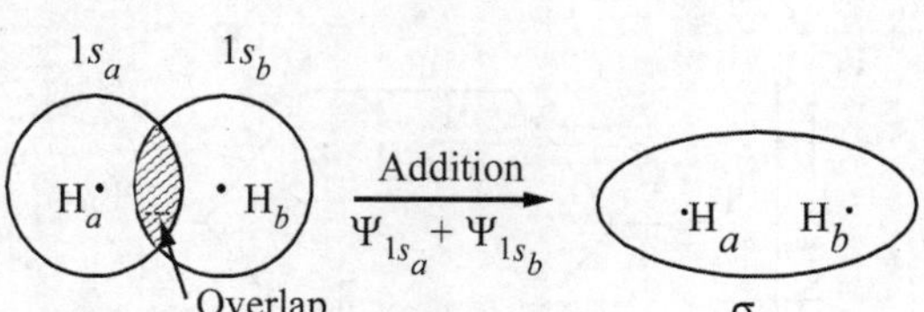

Fig. 3.44. Formation of a bonding MO obtained by the addition overlap of 1*s* atomic orbitals.

The electron in this MO will spend most of its time between the two nuclei H_a and H_b. The electron is now stabilized by two nuclear charges instead of one in an atomic orbital. The resulting MO is of lower energy than the 1*s* orbital. That is why it is called a bonding MO.

The other linear combination is obtained by subtracting the two atomic orbitals and is shown in Fig. 3.45.

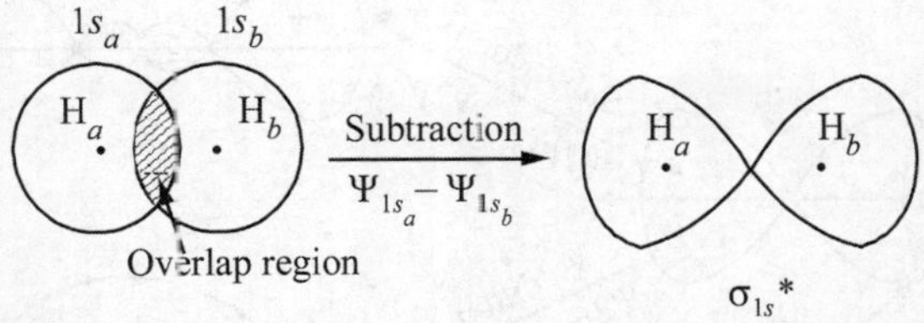

Fig. 3.45. Formation of an antibonding MO formed by the subtraction overlap of 1*s* atomic orbitals.

$\overset{*}{\psi}_{MO} + N^* (\psi_{1sa} - \psi_{1sb})$ where N^* is a normalizing constant.

The resulting molecular orbital has zero electron density (Nodal Plane) between the nuclei. The electron in this orbital is concentrated in regions away from each nuclei. The net effect is to push the nuclei apart rather than to hold them together. The resulting molecular orbital is of higher energy than the isolated 1*s* atomic orbital. The resulting MO is called an antibonding molecular orbital. The molecular orbitals discussed above have cylindrical symmetry and are called σ (sigma) molecular orbitals.

The combination of 2*s* atomic orbitals also gives rise to σ-molecular orbitals. The situation is exactly like the overlapping of 1*s* atomic orbitals. The only difference is that their sizes are different. Their overlap is illustrated in a diagram below (Fig. 3.46).

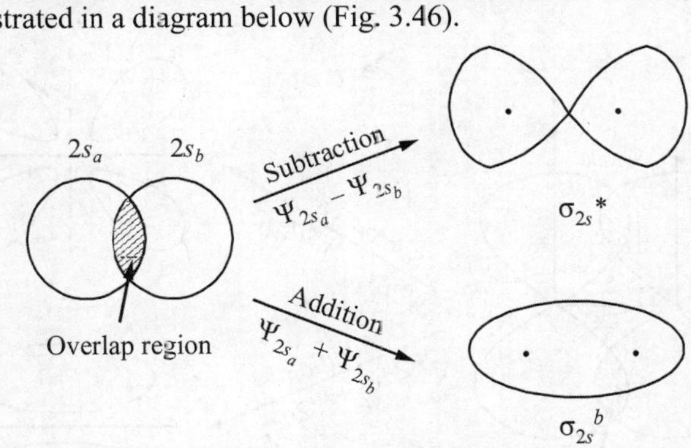

Fig. 3.46. Formation of bonding and antibonding MO's obtained by the overlap of 2*s* atomic orbitals.

3.4.4 Formation of σ molecular orbitals resulting from the overlapping of *p*-orbitals along the internuclear axis

σ-type molecular orbitals are also formed when p_z atomic orbitals of two atoms combine with each other. We are now assuming that the internuclear axis is the *z*-axis. This is illustrated in a diagram shown in Fig. 3.47.

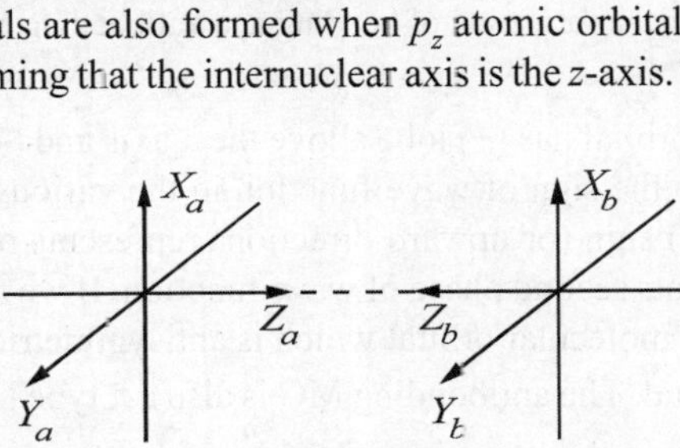

Fig. 3.47. Co-ordinate system for the overlap of 2*p* atomic orbitals.

The molecular orbital formed by the addition, overlap of $2p_z$ atomic orbital is designated as σ^b_{2pz} or simply as σ^b_z while the molecular orbital formed by the subtraction of $2p_z$ atomic orbitals is designated as $\sigma^*_{2p_z}$ or simply σ^*_z. These orbitals also have cylindrical symmetry around the molecular axis and hence they are also classified as σ-molecular orbitals (Fig. 3.48.)

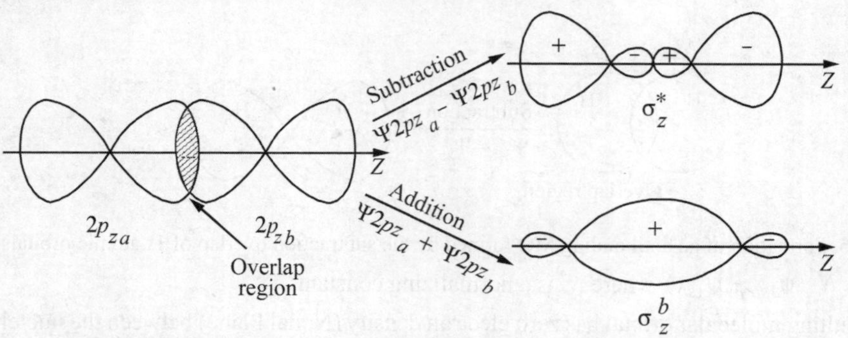

Fig. 3.48. Diagram showing the formation of bonding and antibonding MO's by the overlap of $2p_z$ atomic orbitals.

3.4.5 Formation of π orbitals by the lateral overlap of *p*-atomic orbitals

Molecular orbitals which are obtained by the lateral (or sidewise) overlap of p_x or p_y orbitals are designated as π (pi) molecular orbitals. This is illustrated by showing overlap of $2p_x$ atomic orbitals.

Overlaps containing the same algebraic signs yield bonding MO and those of opposite sign give antibonding M.O. (Fig. 3.49)

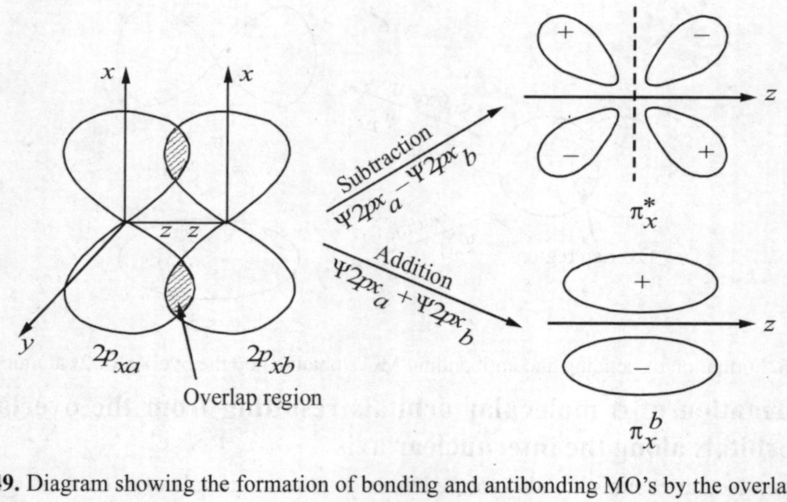

Fig. 3.49. Diagram showing the formation of bonding and antibonding MO's by the overlap of $2p_x$ atomic orbitals.

The bonding molecular orbital has (+) lobe above the *z*-axis and (–) lobe below the z-axis. The plus and minus signs refer to the sign of wave function in the various lobes and not to nuclear or electronic charges. The (+ve) sign (or upward direction) represents one phase and (–ve) sign (or downward direction) represents second phase of wave function. If we rotate this MO through 180°, it changes sign (Fig. 3.50). A molecular orbital which is antisymmetric about the molecular axis is called a π (pi) molecular orbital. The antibonding MO is also a π type M.O. They are labelled as π^b_x and π^*_x orbitals.

The two $2p_y$ atomic orbitals will overlap in a similar fashion to give π^b_y and π^*_y. Both π_x and π_y molecular orbitals are equivalent and have the same energy. These orbitals are said to be degenerate, *i.e.*, orbitals with same energy.

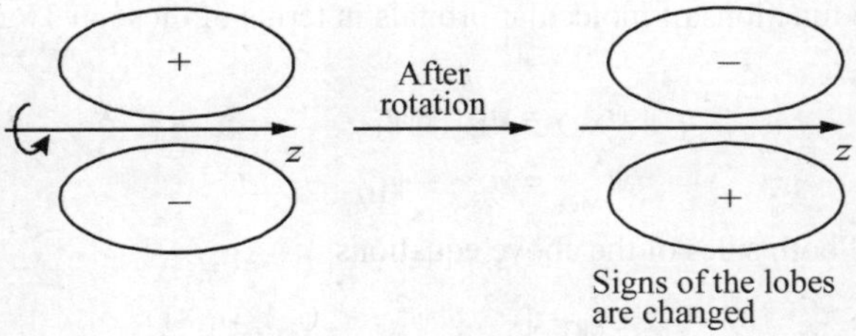

Fig. 3.50. Rotation of bonding MO.

3.4.6 Comparison of electron overlap in σ and π molecular orbitals

In a σ-overlap, the overlap region is much greater than in a π–overlap. The difference arises because in a σ-overlap, two orbitals involve head-on overlap whereas in a π–overlap, the two orbitals overlap laterally (or sidewise) as shown in diagram (Fig. 3.51). It is assumed that z-axis is the molecular axis.

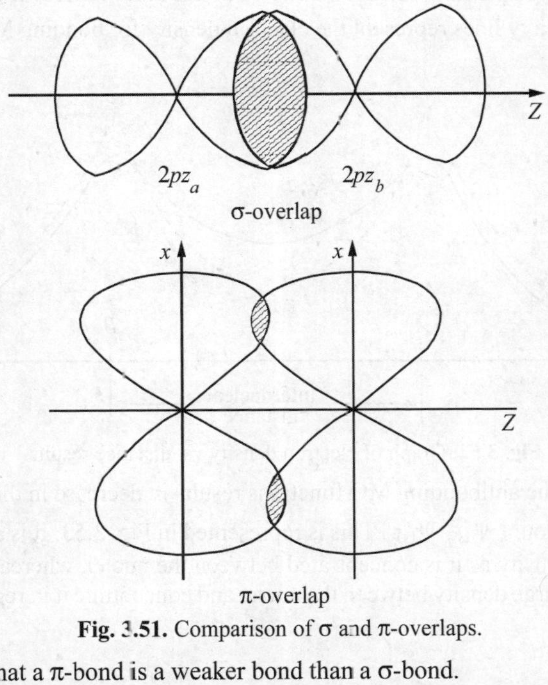

Fig. 3.51. Comparison of σ and π-overlaps.

The net result is that a π-bond is a weaker bond than a σ-bond.

Example 18. **Show by LCAO method as to which of these pairs of AO's result in positive overlap**

$$s + p_x$$
$$s + p_z$$
$$p_x + p_z$$
$$p_z + p_z$$

z-axis may be assumed as a molecular axis.

***Solution*:** $s + p_z$ and $p_z + p_z$ will result in positive overlap.

3.4.7 Variation of electronic charge with internuclear distance

From the molecular orbital wave functions ψ_{MO} and υ^*_{MO}, we can determine the distribution of electron charge in the molecule.

Writing the wave functions of molecular orbitals in terms of those of 1*s* atomic orbitals of two atoms *a* and *b*.

$$\psi_{MO} = \psi_{1s_a} + \psi_{1s_b}$$

$$\psi^*_{MO} = \psi_{1s_a} - \psi_{1s_b}$$

Taking squares of both sides of the above equations

$$\psi^2_{MO} = [\psi^2_{1s_a} + \psi^2_{1s_b} + 2\psi_{1s_a}\ \psi_{1s_b}]$$

$$\psi^{*2}_{MO} = [\psi^2_{1s_a} + \psi\psi^2_{1s_b} - 2\psi_{1s_a} \cdot \psi_{1s_b}]$$

It is observed that the bonding MO functions shows an increase in the electron charge density in the overlap region by an amount $2\psi_{1s_a}\ \psi_{1s_b}$ over that of the individual atoms as described by the functions $\psi^2_{1s_a}$ and $\psi^2_{1s_b}$.

This is represented graphically in Fig. 3.52. The dotted lines represent the charge densities of individual atoms and heavy lines represent the electron density for bonding MO.

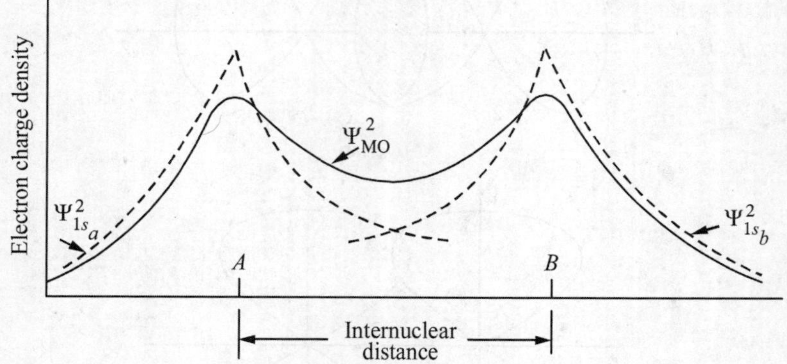

Fig. 3.52. Graph of electron density vs. distance results.

On the other hand the antibonding MO functions results in decrease in the charge density in the overlap region by an amount $\psi_{1s_a}\ \psi_{1s_b}$. This is represented in Fig. 3.53. It is also clear that bonding MO favours electron density as it is concentrated between the nuclei, whereas the antibonding MO tends to decrease the charge density between the nuclei and concentrate it in regions on the individual atoms.

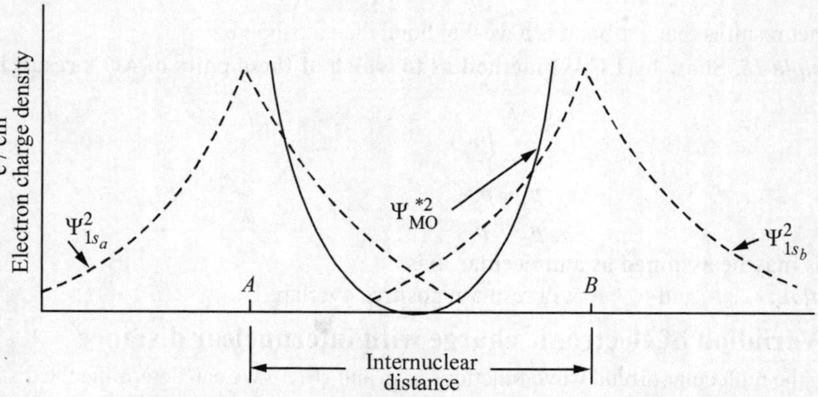

Fig. 3.53. Graph of electron density vs. internuclear density.

Electron density maps for bonding and antibonding (σ_{1s}^{b} and σ_{1s}^{*}) MO's are shown in Fig. 3.54.

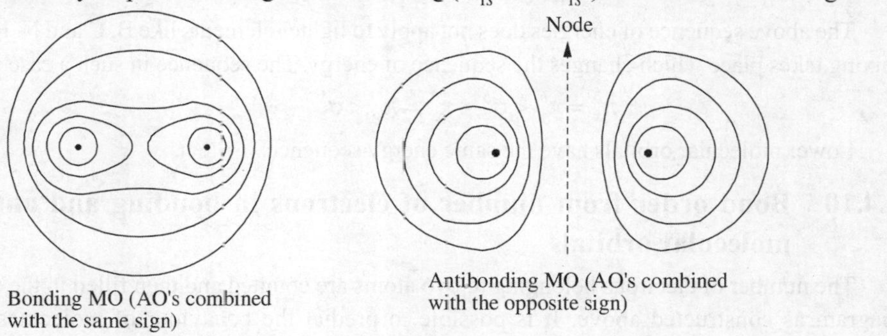

Fig. 3.54. Electron density map of σ_{1s}^{b} and σ_{1s}^{*} MO's.

3.4.8 Rules for filling of electrons in molecular orbitals

The filling of electrons in different molecular orbitals is done according to following rules:

(*i*) Electrons enter molecular orbitals in the order of their increasing energies, *i.e.*, the electron will first enter the lowest energy molecular orbital (Aufbau Principle).

(*ii*) Each molecular orbital can accommodate a maximum of two electrons with their spins in opposite direction in accordance with Pauli's exclusion principle.

(*iii*) In agreement with Hund's rule whenever two equal energy molecular orbitals are available, electrons prefer to remain unpaired.

(*iv*) Electrons (equal to the sum of the electrons actually present in combining atoms) are placed one at a time into molecular orbital to obtain the electronic configuration of the molecule.

3.4.9 Energy level diagram of molecular orbitals

Energy level diagram of MO's is shown in Fig. 3.55.

In this type of diagram the combining atomic orbitals are shown on two extremes and the resulting molecular orbitals are represented in the middle column. Atomic orbitals of same energy are shown at the same level. The difference between the energies of atomic orbitals and the bonding MO's depend upon the extent of overlap of atomic orbitals. A large overlap will result in greater lowering of bonding MO (and a corresponding large raise in the antibonding MO). Larger the overlap, greater is the overlapping and hence stronger will be resulting bond.

According to above diagram the sequence of energy levels in which they are filled up is:

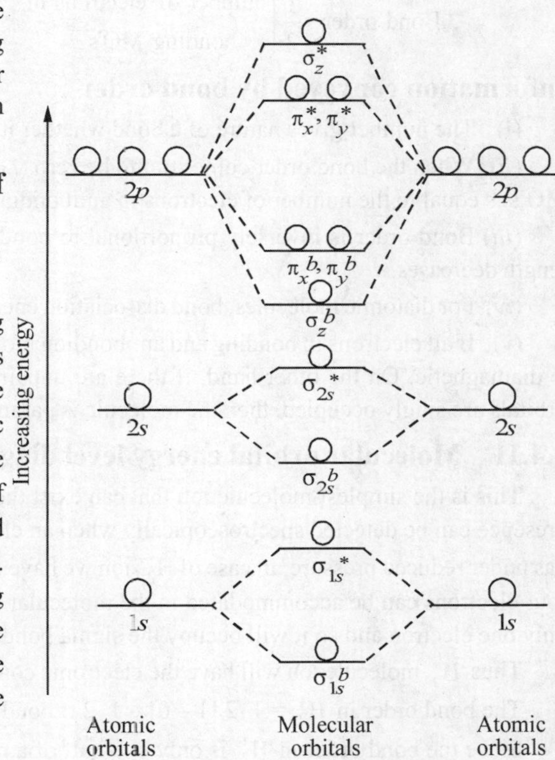

Fig. 3.55. Energy level diagram.

$$\sigma_{1s}^b < \sigma_{1s}^* < \sigma_{2s}^b < \sigma_{2s}^* < \sigma_z^b < \pi_x^b = \pi_y^b < \pi_x^* = \pi_y^* < \sigma_z^*$$

The above sequence of energies does not apply to lighter elements like B, C and N. Here $2s - 2p_z$ mixing takes place which changes the sequence of energy. The sequence in such a case is

$$\pi_x^b = \pi_y^b < \sigma_z^b < \pi_x^* = \pi_y^* < \sigma_z^*$$

Lower molecular orbitals have the same energy sequence.

3.4.10 Bond order from number of electrons in bonding and antibonding molecular orbitals

The number of electrons belonging to two atoms are counted and then filled in the energy level diagram as constructed above. It is possible to predict the behaviour of molecules from their configuration. All the electrons occupying bonding molecular orbitals (N^b) and antibonding molecular orbitals (N^*) are counted separately. The following cases arise.

(*a*) $N^b > N^*$. If the number of electrons in bonding MO's are greater than the number of electrons in antibonding MO's, then the molecule would be stable.

(*b*) $N^b = N^*$. If the number of electrons in bonding MO's are equal to the number of electrons in antibonding MO's, then the molecule is unstable and it does not exist. This is because an electron in an antibonding MO cancels out the stability gained by electron in the bonding MO.

Bond order. It is defined as the number of covalent bonds present in a molecule. It is calculated as

$$\text{Bond order} = \frac{1}{2}\left\{\begin{matrix}\text{number of electrons in} \\ \text{bonding MO's}\end{matrix} - \begin{matrix}\text{Number of electrons in} \\ \text{antibonding MO's}\end{matrix}\right\}$$

Information conveyed by bond order

(*i*) The number gives nature of a bond whether it is a single, double or a triple bond.

(*ii*) When the bond order comes out to be zero (*i.e.*, when the number of electrons in bonding MO's is equal to the number of electrons in antibonding MO's), then the molecule does not exist.

(*iii*) Bond order is inversely proportional to bond length, *i.e.*, as bond order increases, bond length decreases.

(*iv*) For diatomic molecules, bond dissociation energy is directly proportional to the bond order.

(*v*) If all electrons in bonding and antibonding molecular orbitals are paired, then the molecule is diamagnetic. On the other hand, if there are unpaired electrons (or electrons), *i.e.,* molecular orbitals are singly occupied, then the molecule is paramagnetic.

3.4.11 Molecular orbital energy level diagram for H_2^+

This is the simplest molecule ion that can exist though it does not have any real existence. Its presence can be detected spectroscopically when an electric discharge is passed through hydrogen gas under reduced pressure. In case of H_2^+ ion we have two molecular orbitals σ_{1s}^b and σ_{1s}^*. In total four electrons can be accommodated in the molecular orbitals. In hydrogen molecule ion we have only one electron and so it will occupy the sigma bonding molecular orbital as shown in Fig. 3.56.

Thus H_2^+ molecule ion will have the electronic configuration $[(\sigma_{1s}^b)^1]$

The bond order in $H_2^+ = 1/2\,[1 - 0] = 1/2\ \sigma$ bond

Since the bond order in H_2^+ is only one-half of a normal covalent bond, it will have low bond dissociation energy and large bond length.

Bond dissociation energy of H_2^+ = 255 kJ/mole. Bond length = 106 pm.

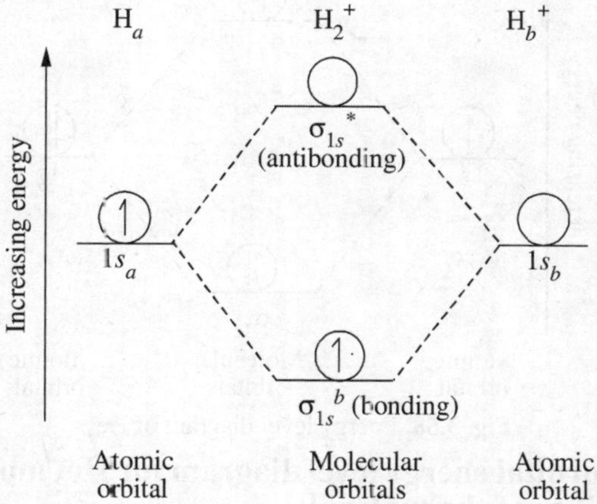

Fig. 3.56. Energy level diagram of H_2^+ ion.

3.4.12 Molecular energy level diagram for hydrogen molecule (H_2)

The hydrogen molecule is formed by the overlap of 1*s* atomic orbitals. It shall have two molecular orbitals, one bonding (σ_{1s}^b) and the other antibonding σ_{1s}^*). The H_2 molecule has two electrons and these can be accommodated in the σ_{1s}^b molecular orbital. In accordance with Pauli exclusion principle, these electrons must have opposite spins. The energy level diagram of H_2 is shown in Fig. 3.57.

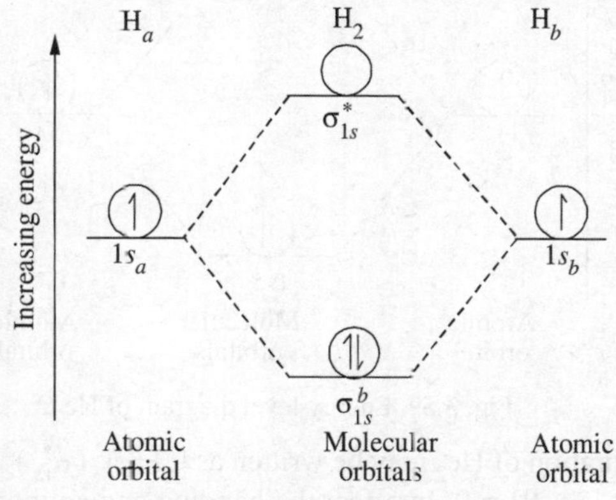

Fig. 3.57. Energy level diagram of H_2.

The ground state of H_2 is represented as $[(\sigma_{1s}^b)^2]$ and the bond order is 1/2 (2 – 0) = 1, *i.e.*, one σ-bond. Bond dissociation energy in H_2 is 431.4 kJ/mole and bond length is 77 pm.

3.4.13 Molecular energy level diagram in He_2^+

Here three electrons are to be filled in the molecular orbitals (Fig. 3.58). The configuration is $(\sigma_{1s}^b)^2 (\sigma_{1s}^*)$.

The bond order is: $\frac{1}{2}[2-1]=\frac{1}{2}$. This is an hypothetical ion.

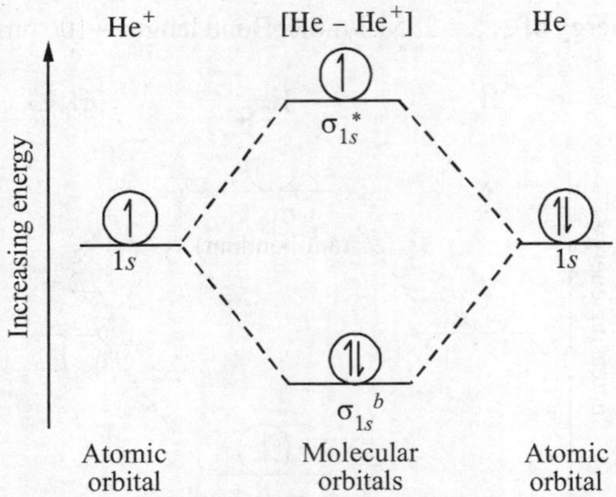

Fig. 3.58. Energy level diagram of He_2^+.

3.4.14 Molecular orbital energy level diagram for He_2 molecule and comment on the stability of the molecule

If a He_2 molecule is to be formed by the combination of helium ions, a total of four electrons are involved. From a combination of two 1*s* atomic orbitals we have two molecular orbitals σ_{1s}^b and σ_{1s}^*. Two electrons would go in the bonding σ_{1s}^b MO and the other two would have to go into the antibonding σ_{1s}^* MO (Fig. 3.59).

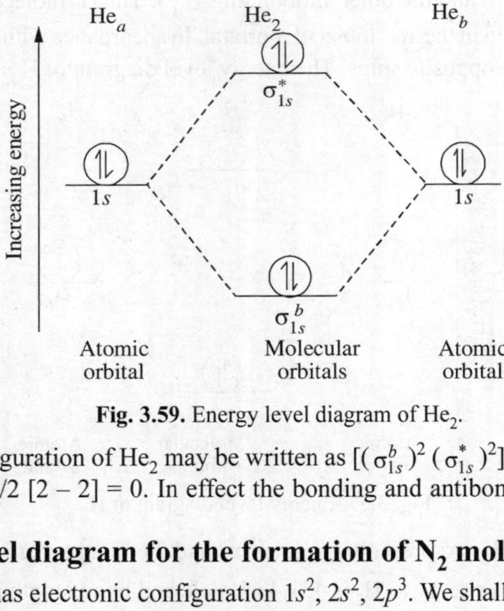

Fig. 3.59. Energy level diagram of He_2.

The electronic configuration of He_2 may be written as $[(\sigma_{1s}^b)^2 (\sigma_{1s}^*)^2]$.

The bond order is 1/2 [2 – 2] = 0. In effect the bonding and antibonding orbitals cancel each other.

3.4.15 Energy level diagram for the formation of N_2 molecule

The nitrogen atom has electronic configuration $1s^2, 2s^2, 2p^3$. We shall not take into account the electrons in σ_{1s} and σ_{1s}^* molecular orbitals. Electrons in inner M.O. do not contribute to bonding. The N_2 molecule is well known. Its molecular orbital configuration may be written as $[KK (\sigma_{2s}^b)^2 (\sigma_{2s}^*)^2 (\pi_x^b)^2 (\pi_y^b)^2 (\sigma_z^b)^2]$. KK signifies that the K shells of the combining atoms are full. Not taking into account σ_{1s} and σ_{1s}^* orbitals it has eight electrons in bonding and two electrons in antibonding orbitals. Since there are six electrons in excess in bonding orbitals so the bond order in N_2 is three. The molecule has 1σ and 2π bonds. This is in conformity with its very high bond energy of 940 kJ/mole and very short bond length of 110 pm. The energy level diagram is shown in Fig. 3.60.

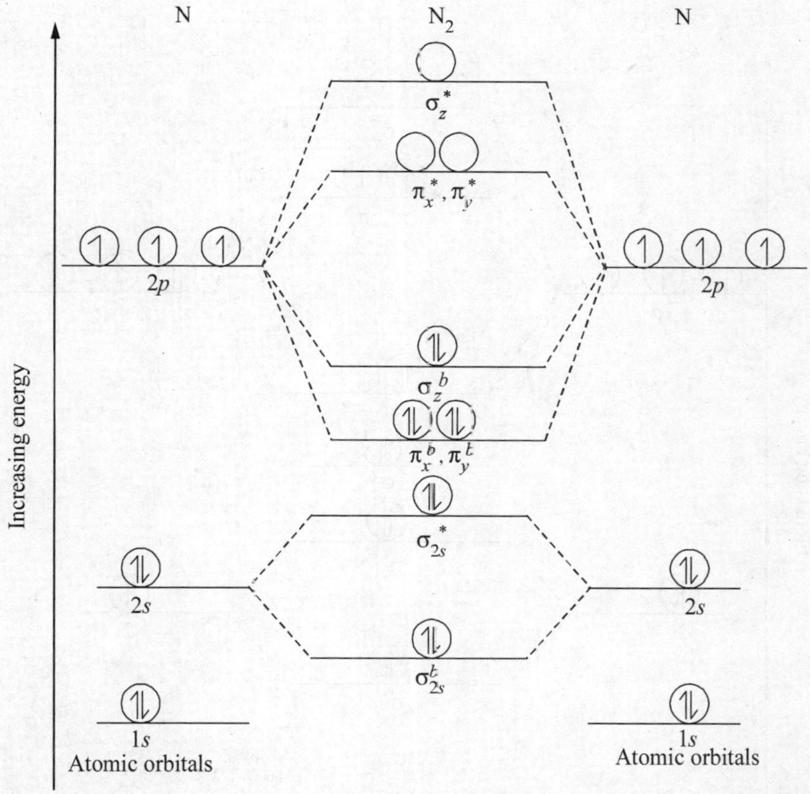

Fig. 3.60. Molecular orbital energy level diagram of hypothetical N_2 molecule.

3.4.16 Molecular orbital energy level diagram for oxygen molecule (O_2)

The oxygen has electronic configuration $1s^2, 2s^2, 2p^4$. The O_2 molecule is well known and shows paramagnetism (Fig. 3.61), equivalent to that of two unpaired electrons. The molecule has the following configuration:

$$[\text{KK}\,(\sigma_{2s}^b)^2(\sigma_{2s}^*)^2(\sigma_z^b)^2(\pi_x^b)^2(\pi_y^b)^2(\pi_y^*)^1(\pi_y^*)^1]$$

$$[\text{Bond order} = \frac{1}{2}(8-4) = \frac{1}{2}\times 4 = 2$$

The bond order is two. In accordance with Hund's rule the two electrons in the π^* orbitals have parallel spins. The molecular orbital method explains the presence of two unpaired electrons in the molecule which accounts for its paramagnetism (Fig. 3.61). The valence bond theory had failed to predict the presence of unpaired electrons in O_2 molecule.

This is in accordance with the experimental results that O_2 consists of a double bond (one σ and one π bond).

Bond dissociation energy = 493.2 kJ/mol

Bond length = 121 pm.

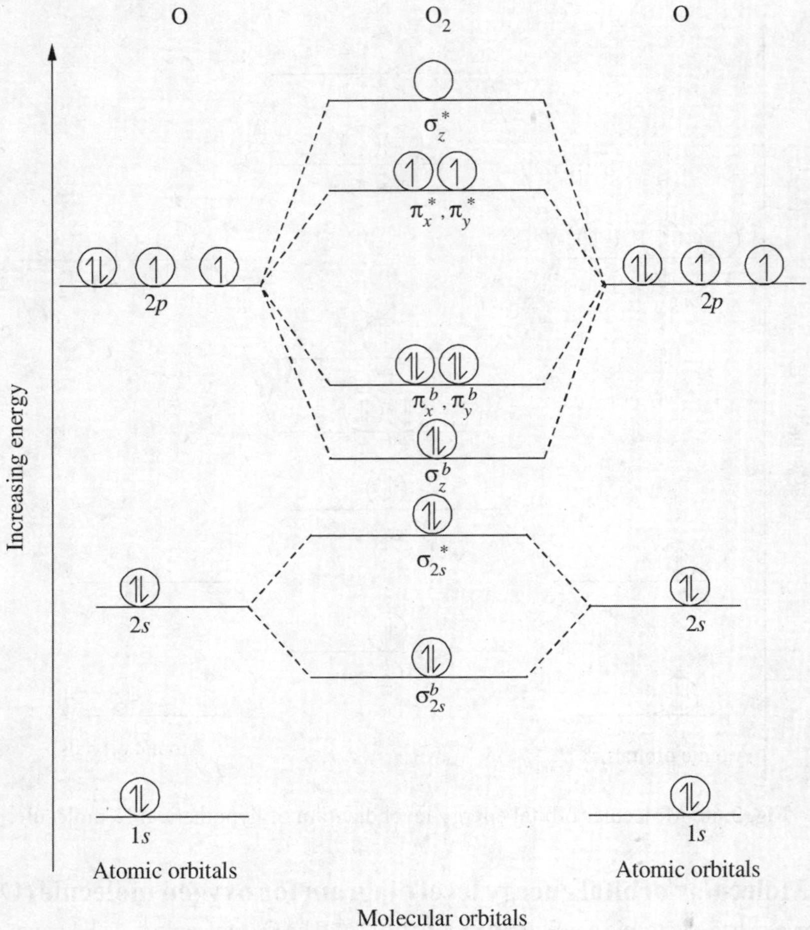

Fig. 3.61. Molecular orbital energy level diagram of O_2.

3.4.17 MO energy level diagram for O_2^+ ion

There are 12 – 1 = 11 electrons to be filled in the LL shells.

The bond order is = 1/2 (8 – 3) = $2\frac{1}{2}$. Thus it has a stronger bond compared to oxygen. O_2^+ is paramagnetic.

The O_2^+ ion will have the configuration:

$$[KK(\sigma_{2s}^b)^2\ (\sigma_{2s}^*)^2\ (\sigma_z^b)^2\ \ (\pi_x^b)^2\ \ (\pi_y^b)^2\ (\pi_y^*)^1]$$

3.4.18 MO energy level diagram of fluorine molecule

On adding two more electrons to the O_2 configuration we get configuration for fluorine molecule. This is shown in Fig. 3.62. The molecular orbital configuration of F_2 is:

$$[KK(\sigma_{2s}^b)^2\ (\sigma_{2s}^*)^2(\sigma_z^b)^2\ (\pi_x^b)^2\ (\pi_y^b)^2(\pi_x^*)^2(\pi_y^*)^2]$$

and its bond order is 1/2 [10 – 8] = 1. It is diamagnetic since it has no unpaired electrons.

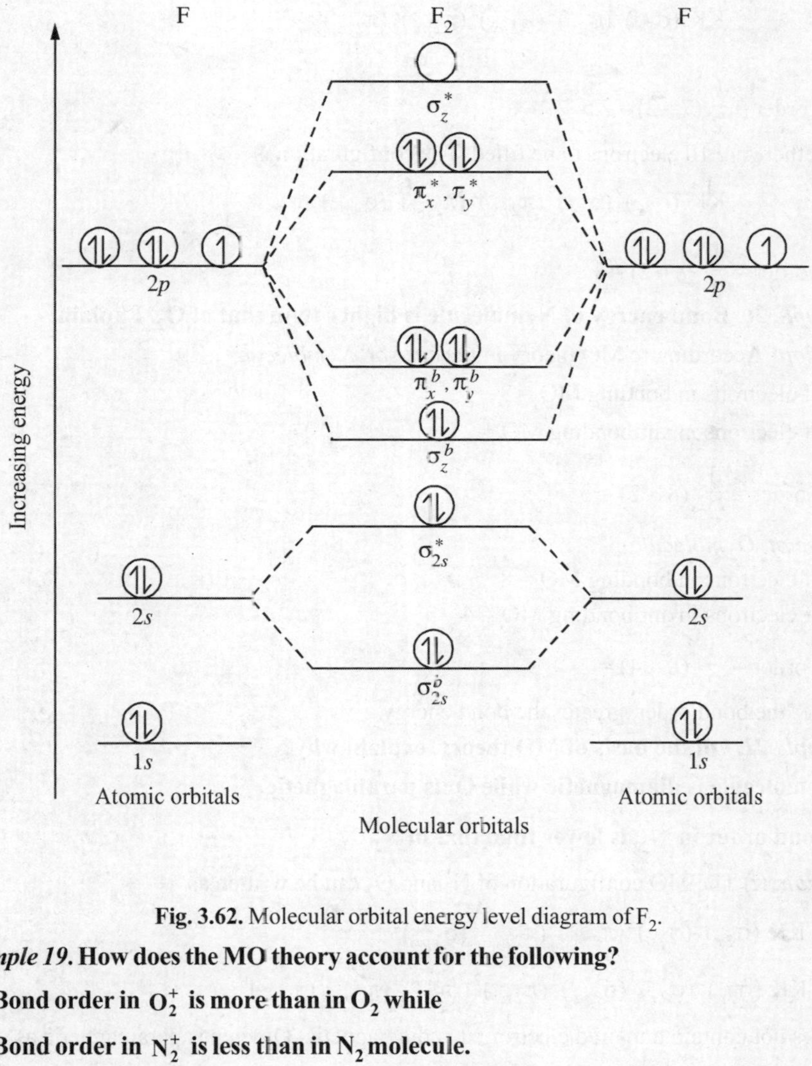

Fig. 3.62. Molecular orbital energy level diagram of F_2.

Example 19. **How does the MO theory account for the following?**

(*i*) Bond order in O_2^+ is more than in O_2 while

(*ii*) Bond order in N_2^+ is less than in N_2 molecule.

Solution. (*i*) In O_2^+, there are 11 valence electrons to be filled. The outer shell molecular orbital configuration is

$$(\sigma_{2s})^2\ (\sigma_{2s}^*)^2\ (\sigma_{2pz})^2\ (\pi_{2px})^2\ (\pi_{2py})^2\ (\pi_{2px}^*)^1$$

$$\text{Bond order} = \frac{1}{2}\ (8 - 3) = 2.5$$

In O_2, there are 12 valence electrons to be filled. The outer shells MO configuration is

$$(\sigma_{2s})^2 (\sigma_{2s}^*)^2 (\sigma_{2pz})^2 (\pi_{2px})^2 (\pi_{2py})^2 (\pi_{2px}^*)^1 (\pi_{2py}^*)^1$$

$$\text{Bond order} = \frac{1}{2}\ (8 - 4) = 2$$

(*ii*) In N_2^+, there are 9 valence electrons to be filled. The configuration is

$$KK\ (\sigma_{2s})^2(\sigma_{2s}^*)^2(\pi_{2px})^2(\pi_{2py})^2(\pi_{2pz})^1$$

$$\text{Bond order} = \frac{1}{2}(7-2) = 2.5$$

In N_2, there are 10 electrons to be filled. The configuration is

$$KK\ (\sigma_{2s})^2(\sigma_{2s}^*)^2(\pi_{2px})^2(\pi_{2py})^2(\sigma_{2pz})^2$$

$$\text{Bond order} = \frac{1}{2}(8-2) = 3$$

***Example 20.* Bond energy of N_2 molecule is higher than that of O_2. Explain.**

***Solution*:** According to MO theory *in the case of N_2 molecule*

No. of electrons in bonding MO = 8

No. of electrons in antibonding MO = 2

$$\text{Bond order} = \frac{1}{2}(8-2) = 3$$

In case of O_2 molecule

No. of electrons in bonding MO = 8

No. of electrons in antibonding MO = 4

$$\text{Bond order} = \frac{1}{2}(8-4) = 2$$

Greater the bond order, greater the bond energy.

***Example 21.* On the basis of MO theory, explain why?**

(*i*) N_2 molecule is diamagnetic while O_2 is paramagnetic.

(*ii*) Bond order in N_2^+ is lower than that in N_2.

Solution. (*i*) The MO configuration of N_2 and O_2 can be written as

$$N_2 : KK\ (\sigma_{2s})^2(\sigma_{2s}^*)^2(\pi_{2px})^2(\pi_{2py})^2(\sigma_{2pz})^2$$

$$O_2 : KK\ (\sigma_{2s})^2(\sigma_{2s}^*)^2(\sigma_{2pz})^2(\pi_{2px})^2(\sigma_{2py})^2(\pi_{2px}^*)^1(\pi_{2py}^*)^1$$

N_2 does not contain unpaired electron, it is diamagnetic. Oxygen is paramagnetic as it contains two unpaired electrons.

(*ii*) Bond order in $N_2^+ = \frac{1}{2}(7-2) = 2.5$

Bond order in $N_2 = \frac{1}{2}(8-2) = 3$

3.4.19 MO energy level diagram for CO molecule

Solution. In CO, the oxygen atomic orbitals are lower in energy than the carbon atomic orbitals. This is because oxygen is more electronegative than carbon. The 2*s* and 2*p* orbitals of oxygen have lower energy than the corresponding orbitals of carbon. Energy levels of different M.O are shown in the figure. Due to this difference, the σ_{2s}^* orbital is higher in energy than the σ_{2px} orbital (Fig. 3.63). Carbon monoxide possesses 10 valence electrons and, therefore, has a valence configuration :

$$(\sigma_{2s}^b)^2(\sigma_{2pz}^b)^2(\pi_x^b\,\pi_y^b)^4(\sigma_{2s}^*)^2$$

There are three net bonding pairs (one σ and two π) describing the C ≡ O molecule.

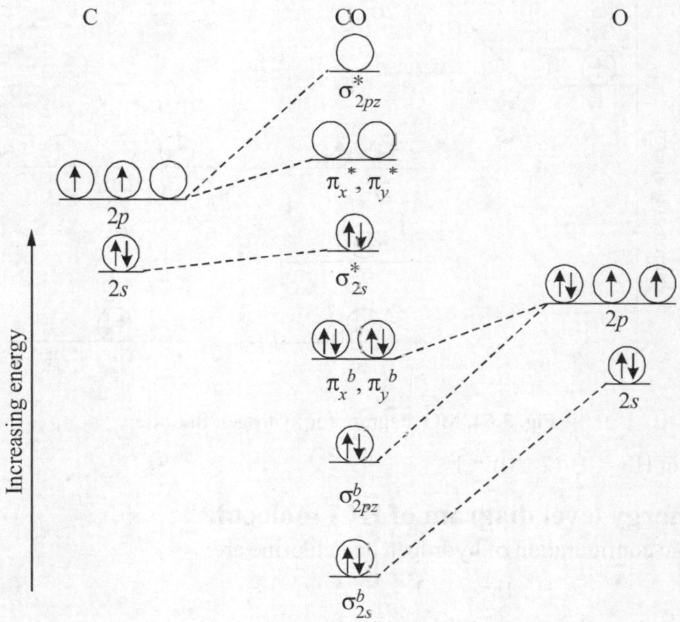

Fig. 3.63. MO diagram for carbon monoxide.

3.4.20 MO electrons configuration for nitrogen oxide (NO) molecule

NO (11 valence electrons) has an electron configuration :

$$(\sigma^b_{2s})^2(\sigma^b_{2pz})^2(\pi^b_x.\pi^b_y)^4(\sigma^*_{2s})^2(\pi^*_x)^1$$

$$\text{Bond order in NO} = \frac{1}{2}(8-3) = 2.5$$

It is paramagnetic with a single unpaired electron. This is manifested by its easy oxidation to NO^+ (by loss of π^* electron). This is associated with an increase in bond order from $2\frac{1}{2}$ in NO to 3 in NO^+, and a decrease in bond length

3.4.21 MO energy level diagram for hydrogen fluoride molecule

The 2*s* atomic orbital of the fluorine atom does not interact with the 1*s* orbital of hydrogen atom because it is of low energy and is retained as atomic orbital. Only the $2p_z$ orbital of fluorine atom has proper symmetry and energy to interact with the 1*s* orbital of the hydrogen atom, giving one sigma bonding molecular orbital and one sigma antibonding molecular orbital (Fig. 3.64).

The molecular orbitals formed are named as σ_{sp} (bonding) and σ^*_{sp} (non-bonding) because they result from the combination of 1s orbital of hydgrogen and $2p_z$ orbital of fluorine. The $2p_x$ and $2p_y$ orbitals of fluorine yield non-bonding molecular orbitals at the same energies because of their symmetry because they yield zero overlap with 1*s* orbital.

It is obvious from Fig. 3.64 that the atomic orbitals of electromagnetic fluorine are much lower in energy than the 1*s* atomic orbital of hydrogen. M.O. configuration of HF molecule may the written as:

$$kk\,(2s)^2\,(\sigma_{sp})^2\,(2p_x)^2\,(2p_y)^2$$

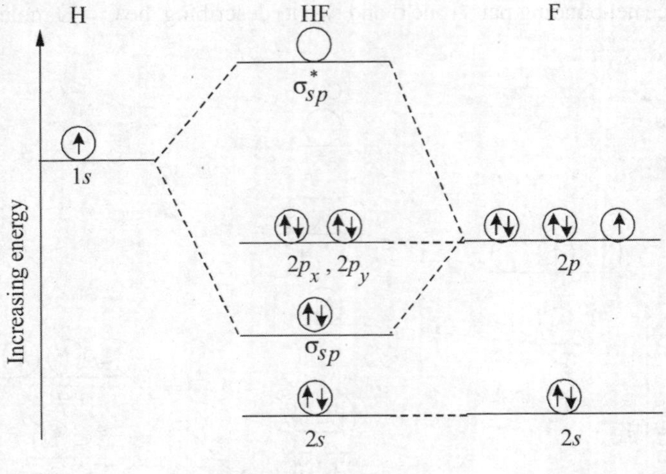

Fig. 3.64. MO diagram for hydrogen fluoride.

Bond under in HF $= \frac{1}{2}(2-0) = 1$

3.4.22 MO energy level diagram of HCl molecule

The electronic configuration of hydrogen and chlorine are:

$$H : 1s^1$$

$$Cl : 1s^2\,2s^2\,2p^6 3s^2\,3p^5$$

Molecular orbital energy level diagram of hydrogen chloride (Fig. 3.65) is similar to that of hydrogen fluoride. The energy of 3*s* and 3*p* atomic orbitals of the chlorine atom is lower than that of 1*s* atomic orbitals of the hydrogen atom. The 3*s* atomic orbitals of chlorine atom does not interact with the 1*s* atomic orbital of H atom because the 3*s* has a very low energy and is, therefore, retained as atomic orbital. The $3p_z$ atomic orbital of chlorine atom has proper symmetry and energy to interact with the 1*s* atomic orbital of hydrogen atom giving one sigma bonding molecular orbital and one sigma antibonding molecular orbital. The $3p_x$ and $3p_y$ orbitals of chlorine yield non-bonding molecular orbitals which are actually unchanged atomic orbitals of chlorine atom. These represent the two *p* orbitals on chlorine atom which lie perpendicular to the bond axis and hence do not overlap with 1*s* atomic orbital of the hydrogen atom.

We name the molecular orbitals as σ_{spz} and σ^*_{spz} because these are obtained by the combination of 1*s* orbital of H and $3p_z$ orabital of Cl. M.O. energy diagram of HCl is given in fig. 3.65.

Bond order in HCl $= \frac{1}{2}(2-0) = 1$

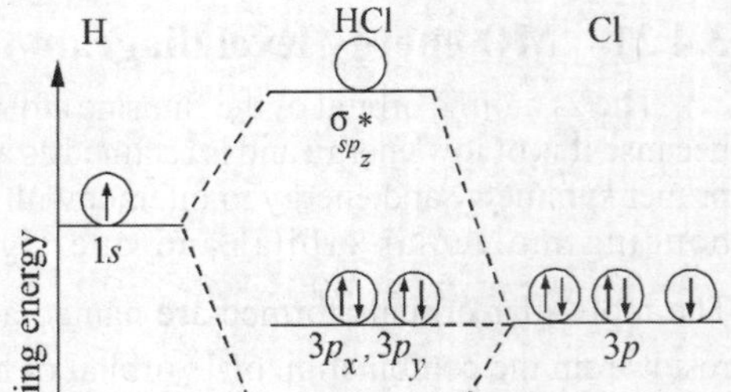

Fig. 3.65. Molecular orbital energy level diagram of HCl.

Example 22. **On the basis of MO theory, explain the magnetic character of N_2, O_2, NO and CO molecules. Also give the bond order in each case.**

Solution: (*i*) Molecular orbital configuration of N_2 is

$$KK\ (\sigma_{2s})^2(\sigma^*_{2s})^2(\pi_{2px})^2(\pi^*_{2py})^2(\sigma_{2pz})^2$$

As there is no unpaired electron, the molecule is diamagnetic

$$\text{Bond order} = \frac{1}{2}(n_b - n_a)$$

$$= \frac{1}{2}(8-2) = 3$$

(*ii*) Molecular orbital configuration of O_2 is

$$KK\ (\sigma_{2s})^2(\sigma^*_{2s})^2(\sigma_{2pz})^2(\pi_{2py})^2(\pi_{2py})^2(\pi^*_{2px})^1(\pi^*_{2py})^1$$

As there are two unpaired electrons, the molecule is paramagnetic.

$$\text{Bond order} = \frac{1}{2}(8-4) = 2$$

(*iii*) Molecular orbital configuration of NO is

$$KK\ (\sigma_{2s})^2(\sigma_{2pz})^2(\pi_{2px})^2(\pi_{2py})^2(\sigma^*_{2s})^2(\pi^*_{2px})^1$$

The molecule is paramagnetic due to presence of an unpaired electron

$$\text{Bond order} = \frac{1}{2}(8-3) = 2.5$$

(*iv*) Molecular orbital configuration of CO is

$$KK\ (\sigma_{2s})^2(\sigma_{2pz})^2(\pi_{2px})^2(\pi_{2py})^2(\sigma^*_{2s})^2$$

As there are no unpaired electrons, the molecule is diamagnetic.

$$\text{Bond order} = \frac{1}{2}(8-2) = 3.$$

Example 23. **Draw MO diagram for O_2^+, O_2^- and O_2^{2-} and give the value of bond order in each case.**

Solution. Molecular orbital configuration of O_2^+, O_2^- and O_2^{2-} are written as under:

$$O_2^+ :\ KK\ (\sigma_{2s})^2(\sigma^*_{2s})^2(\sigma_{2pz})^2(\pi_{2px})^2(\pi_{2py})^2(\pi^*_{2px})^1$$

$$O_2^- :\ KK\ (\sigma_{2s})^2(\sigma^*_{2s})^2(\sigma_{2pz})^2(\pi_{2px})^2(\pi_{2py})^2(\pi^*_{2px})^2(\pi^*_{2py})^1$$

$$O_2^{2-} :\ KK\ (\sigma_{2s})^2(\sigma^*_{2s})^2(\sigma_{2pz})^2(\pi_{2px})^2(\pi_{2py})^2(\pi^*_{2px})^2(\pi^*_{2py})^2$$

Bond order

(*i*) in $O_2^+ = \frac{1}{2}(8-3) = 2.5$

(*ii*) in $O_2^- = \frac{1}{2}(8-5) = 1.5$

(*iii*) in $O_2^{2-} = \frac{1}{2}(8-6) = 1.0$

Example 24. **The bond energy of NO is 623 kJ/mol whereas that of NO^+ molecule ion is 1049 kJ/mol. Justify.**

***Solution*:** NO : KK $(\sigma_{2s})^2(\sigma_z)^2(\pi_x)^2(\pi_y)^2(\sigma^*_{2s})^2(\pi^*_x)^1$

$$\text{Bond order} = \frac{8-3}{2} = 2\frac{1}{2}$$

NO^+ : KK $(\sigma_{2s})^2(\sigma_z)^2(\pi_x)^2(\pi_y)^2(\sigma^*_{2s})^2$

$$\text{Bond order} = \frac{8-2}{2} = 3$$

Higher bond order of NO^+ justifies its higher bond energy than NO.

***Example 25*. Using molecular orbital configurations indicate paramagnetic nature of B_2 and non-existence of He_2.**

***Solution*.** He_2 : $(\sigma_{1s})^2(\sigma^*_{1s})^2$

$$\text{Bond order} = \frac{2-2}{2} = 0$$

Bond order zero supports non-existence of helium molecule.

B_2 : KK $(\sigma^2_{2s})^2(\sigma^*_{2s})^2(\pi_x)^1(\pi_y)^1$

$$\text{Bond order} = \frac{4-2}{2} = 1$$

***Example 26*. Give molecular orbital configuration of O_2, O_2^+ and O_2^-. Give order of stability with appropriate reasons.**

***Solution*:** O_2 : KK $(\sigma_{2s})^2(\sigma^*_{2s})^2(\sigma^b_z)^2(\pi^b_x)^2(\pi^b_y)^2(\pi^*_x)^1(\pi^*_y)^1$

$$\text{Bond order} = \frac{8-4}{2} = 2$$

O_2^+ : KK $(\sigma^b_{2s})^2(\sigma^*_{2s})^2(\sigma^b_z)^2(\pi^b_x)^2(\pi^b_y)^2(\pi^*_x)^1$

$$\text{Bond order} = \frac{8-3}{2} = 2\frac{1}{2}$$

O_2^- : KK $(\sigma^b_{2s})^2(\sigma^*_{2s})^2(\sigma^b_2)^2(\pi^b_x)^2(\pi^b_y)^2(\pi^*_x)^2(\pi^*_y)^1$

$$\text{Bond order} = \frac{8-5}{2} = 1.5$$

Because of decreasing bond order, the stability decreases as

$$O_2^+ > O_2 > O_2^-$$

3.4.23 Points of similarity and difference between valence-bond and molecular orbital theory

Points which are common to both theories:

(*i*) Both these theories are approximations.
(*ii*) Both interpret covalent bonds as orbitals embracing two atomic nuclei.
(*iii*) Both require that the combining atomic orbitals should be of nearly equal energy : they should be capable of overlapping and the combining orbitals must have the same symmetry.
(*iv*) Both predict the concentration of electron density between the nuclei.
(*v*) Both can account for directed valency.
(*vi*) Both predict the non-existence of helium molecule.

Some points of difference between the two theories:

Valence bond theory	*Molecular orbital theory*
(*i*) Electrons in molecules are localized as if they are in isolated atoms.	It treats the nuclei of the molecule as polycentric and then constructs one electron molecular orbital.
(*ii*) The total wave function in the valence bond treatment takes the contribution due to ionic terms.	Molecular orbital theory overlapping estimates the role of ionic terms in its total wave functions.
(*iii*) It could not explain the paramagnetism of O_2.	It can easily explain the paramagnetism of O_2.
(*iv*) Resonance is an integral part of valence bond theory.	Resonance has no role in molecular orbital theory.
It is simple to apply.	It is more difficult to apply.

Example 27. **Why a σ^b_{2s} orbital has a higher energy than σ^*_{1s} orbital? Why is the former MO called a bonding orbital while the latter is called an antibonding MO?**

Solution. The relative energies of σ^b_{2s} and σ^*_{2s} are determined by the energies of the atomic orbitals from which they originate. The σ^b_{2s} MO is lower in energy compared to the 2*s* atomic orbital from which it is originated and is, therefore, called a bonding orbital while the σ^*_{1s} MO is higher in energy than the 1*s* atomic orbital from which it originates and thus called an antibonding MO.

Example 28. **Draw a diagram showing the formation of a σ_{2p} bonding MO from atomic orbitals.**

Solution. Formation of σ_{2p} bonding MO from atomic orbitals is shown in Fig. 3.66.

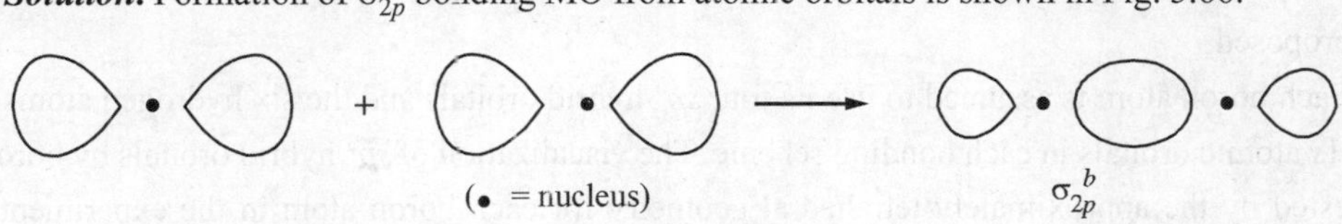

Fig. 3.66. Formation of σ_{2p} orbital.

3.5 ELECTRON - DEFICIENT COMPOUNDS

An electron -deficient compound is one in which there is an insufficient number of electrons to complete the octet of central atom. These compounds contain insufficient numbers of electrons to form normal electron pairs bonds, between each pair of bonded atoms. For example, boron which has electronic configuration [He], $2s^2\ 2p^1$ forms compounds which are described as electron-deficient compounds. Boron has only three valence electrons but has four available orbitals (2*s*, $2p_x$, $2p_y$ and $2p_z$) to accommodate these electrons. Other members of boron family, namely, aluminium, gallium, indium and thallium, show tendency to lose their valence electrons to form ionic M^{3+} species in their compounds. However, boron is an exception. Due to its very small size the ionization energies are very high and the formation of ionic B^{3+} species does not take places. Boron fills its fourth available valence orbital in the following different ways:

(*i*) $p\pi - p\pi$ donation as observed in boron halides (BF_3 etc.).
(*ii*) donor-acceptor donation as in $Me_3N \rightarrow BF_3$.
(*iii*) Formation of electron-deficient compounds as in diborane (B_2H_6).

3.5.1 Structure and bonding in electron-deficient diborane, B_2H_6

Diborane B_2H_6 does not possess sufficient number of electrons to allow a bonding similar to ethane, C_2H_6 in which each bond consists of a shared electron pair (Fig. 3.67).

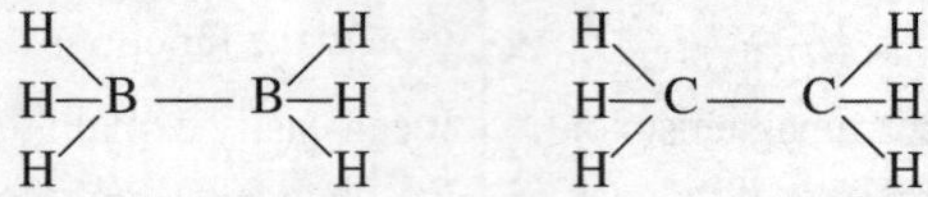

Not possible Possible

Fig. 3.67

There are twelve valency electrons in B_2H_6 (three per boron atom and one for each H atom) and for molecule to have ethane type structure a minimum of fourteen electrons would be required, since seven bonds would require seven electron pairs (*i.e.*, 2-centre, 2-electrons) bonds.

The experimentally determined structure of diborane suggests the following bonding scheme as shown in Fig. 3.68.

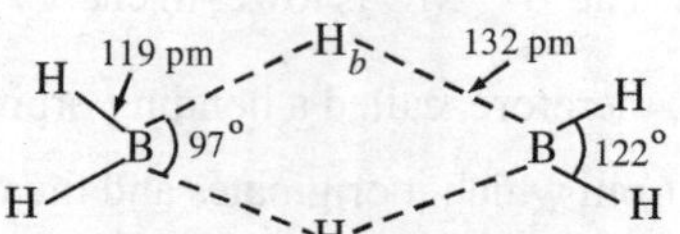

3.68. Structure of diborane.

A 3-centre, 2-electron bonds in addition to the normal electron pair (2-centre, 2-electron) bonds was proposed.

Each boron atom is assumed to use its four sp^3 hybrid orbitals and the six hydrogen atoms use their 1*s* atomic orbitals in each bonding scheme. The visualization of sp^3 hybrid orbitals by boron is suggested by the approximately tetrahedral geometry for each boron atom in the experimentally determined structure of diborane. Each of the two boron atoms then uses two of its sp^3 hybrid orbitals to overlap with the 1*s* atomic orbitals of each of the two hydrogen atoms (terminal B —H bonds). These terminal B — H bonds are formed by the overlap of one electron from boron and one electron from hydrogen to form normal electron pair (2-centre, 2-electron, 2c -2e) bonds.

The remaining two sp^3 hybrid orbitals of each boron atoms are then arranged in a manner so that an sp^3 hybrid orbital from each boron atom and ls orbital from a bridging H atom overlap to form two multicentre orbitals as shown below in Fig. 3.69.

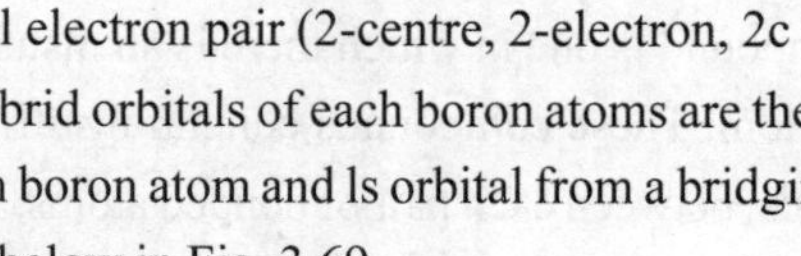

Fig. 3.69.

Their overlap results in the formation of two bonding B—H—B orbitals, two non-bonding B—H—B orbitals and two antibonding B—H—B orbitals as depicted below in Fig. 3.70.

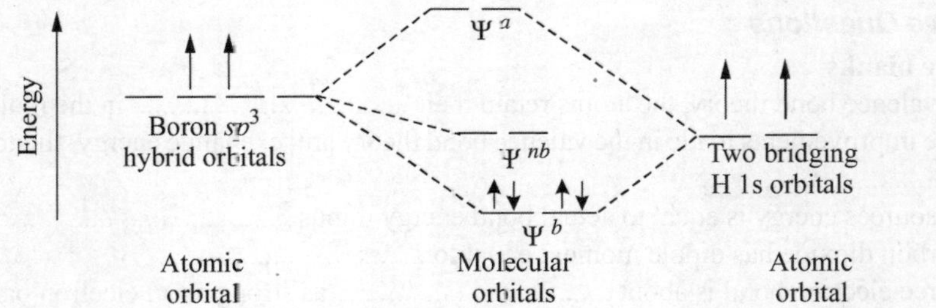

Fig. 3.70.

The two lowest energy bonding B—H—B molecular orbitals are then occupied by a total of 4 electrons (one from each boron and hydrogen atom) to result in the formation of two 3-centre, 2-electron B—H—B bridge bonds. Since only two electrons are available for the formation of two B—H_b bonds among the three B—H_b—B atoms, the bond order for each B—H_b bonds comes out to be equivalent to 0.5 σ bond. Such electron-deficient 3-centre, 2-electron bonds are weak and get readily cleaved.

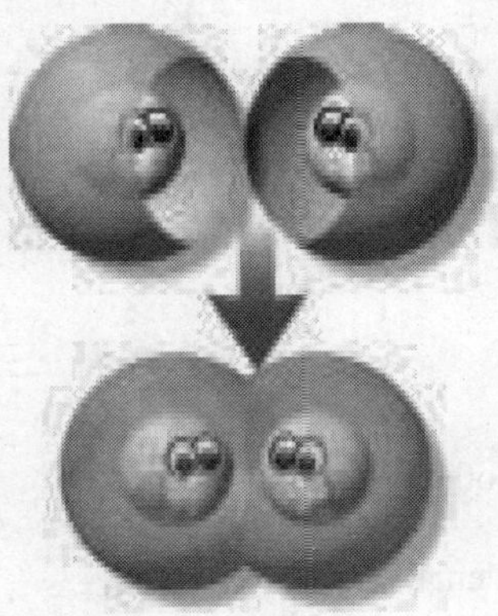

Covalent bond formation

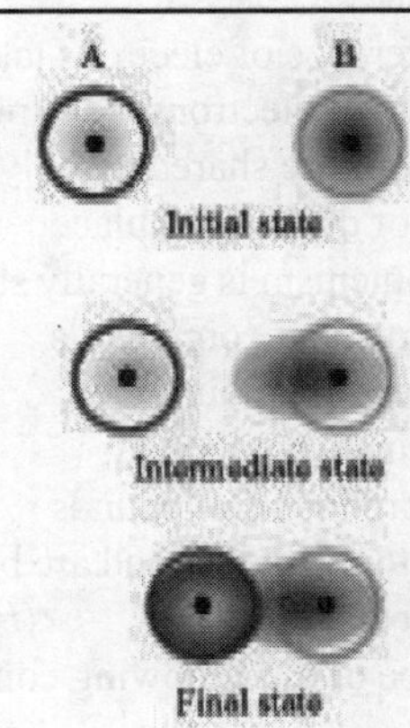

Formation of bond

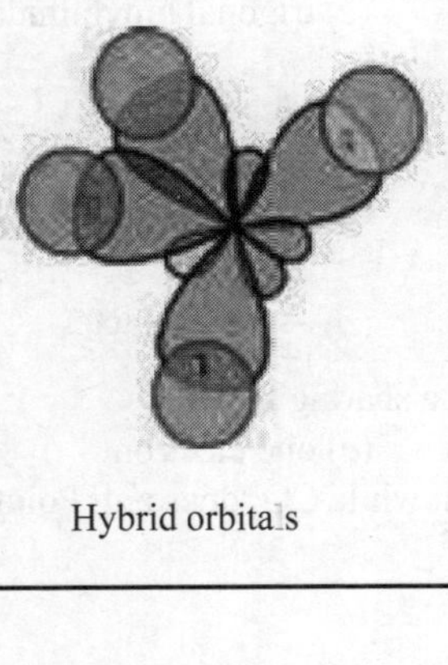

Hybrid orbitals

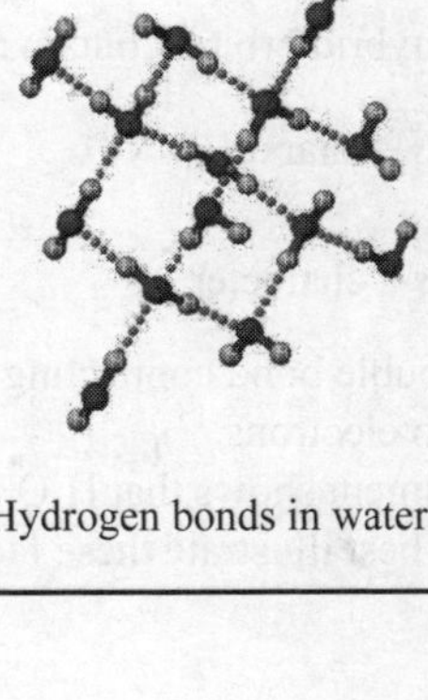

Hydrogen bonds in water

EXERCISES

(INCLUDING QUESTIONS FROM DIFFERENT UNIVERSITY PAPERS)

Objective Questions

Fill in the blanks

1. In valence bond theory, the atoms retain their .... in the molecule.
2. The improvements made in the valence bond theory are: exchange energy, shielding effect and ....
3. Resources energy is equal to actual bond energy minus ....
4. Carbon dioxide has dipole moment equal to ....
5. Three electron bond is about .... as strong as an electron pair bond.
6. Bond order in a molecule of NO is ....
7. Bond energy of H – F bond is ....
8. Intramolecular hydrogen bonding .... the melting and boiling points.
9. BF_3 involves .... hybridisation and the shape of the molecule is ....
10. The order of repulsion between lone pair (*lp*) and bond pair (*bp*) of electrons is ....

Multiple Choice Questions

Choose the correct option

1. In covalency
   (*a*) transference of electrons takes place
   (*b*) sharing of electrons takes place
   (*c*) electrons are shared by only one atom
   (*d*) none of these takes place
2. Stereoisomenism is generally shown by
   (*a*) electrovalent compounds
   (*b*) covalent compounds
   (*c*) coordinate compounds
   (*d*) isomorphous compounds
3. Carbon atoms in diamond are bonded to one another in configuration
   (*a*) planar (*b*) tetrahedral (*c*) linear (*d*) octahedral
4. Which one of the following compounds is covalent ?
   (*a*) H_2 (*b*) CaO (*c*) KCl (*d*) Na_2S
5. The shape of the sulphate ion is
   (*a*) tetrahedral (*b*) square planar (*c*) trigonal bipyramidal (*d*) hexagonal
6. A sp^3 hybrid orbital contains
   (*a*) $\frac{1}{4}$ *s* – character (*b*) $\frac{1}{2}$ *s* – character
   (*c*) $\frac{2}{3}$ *s* – character (*d*) $\frac{3}{4}$ *s* – character
7. In a double bond connecting two atoms, there is a sharing of
   (*a*) two electrons (*b*) four electrons (*c*) one electron (*d*) all electrons
8. Experiment shows that H_2O has a dipole moment while CO_2 does not. Point out the structures which best illustrate these facts.

(*a*) 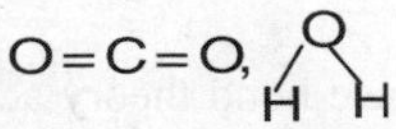

(*b*) O=C=O H–O–H

(*c*) 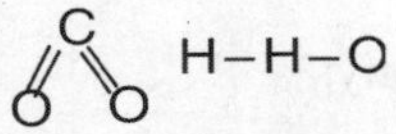

(*d*) 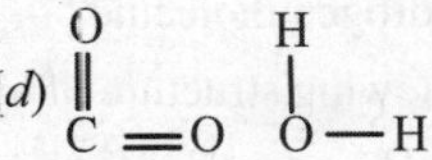

9. The octet rule is not valid for

(*a*) CO_2 (*b*) H_2O (*c*) CO (*d*) O_2

10. Noble gases do not react with other atoms because they

(*a*) are monoatomic (*b*) are not found in abundance

(*c*) have small sizes (*d*) are completely paired up

11. Which among the following has a tendency to form covalant compounds?

(*a*) Ba (*b*) Be (*c*) Mg (*d*) Ca

12. The compound formed by sp^3d hybridisation will have the structure

(*a*) planar (*b*) pyramidal

(*c*) angualr (*d*) trigonal bipyramidal

13. Which of the following can form hydrogen bonds?

(*a*) CH_4 (*b*) H_2O (*c*) NaCl (*d*) $CHCl_3$

14. Which of the following compounds of boron will be more covalent in nature ?

(*a*) BCl_3 (*b*) BF_3 (*c*) BBr_3 (*d*) BI_3

15. Bond order of a molecule is given by

(*a*) The difference between the number of electrons in bonding and antibonding orbitals

(*b*) total number of electrons in bonding and antibonding orbitals

(*c*) twice the differences between the number of electrons in bonding and anti bonding orbitals

(*d*) half the differnce between the number of electrons in bonding and non-bonding orbitals.

16. Which one of the following will have a tetrahedral disposition?

(*a*) SO_2 (*b*) SO_3 (*c*) SO_4^{2-} (*d*) SO_3^{2-}

17. Resonance structure of a molecule should have

(*a*) identical arrangement of atoms (*b*) nearly same energy contant

(*c*) the same number of paired elelctrons (*d*) identical bonding

18. CO_2 is isostructural with

(*a*) $Hg\,Cl_2$ (*b*) $SnCl_2$ (*c*) C_2H_6 (*d*) NO_2

19. Which of the following is paramagnetic?

(*a*) O_2^- (*b*) CN^- (*c*) CO (*d*) NO^+

20. Covalant compounds generally show the following properties:

(*a*) They are soluble in water (*b*) They are insoluble in water

(*c*) They ionise in water (*d*) They hydrolyse in water

ANSWERS

Fill in the blanks

1. individuality **2.** ionicity **3.** energy of the pure covalent bond

4. zero **5.** one half **6.** 2.5 **7.** 10 *k* cal

8. lowers **9.** sp^2, triangular planar

10. $lp - lp > lp - bp > bp - bp$

Multiple choice questions

1. (*b*)	**2.** (*b*)	**3.** (*b*)	**4.** (*a*)	**5.** (*a*)
6. (*a*)	**7.** (*b*)	**8.** (*a*)	**9.** (*c*)	**10.** (*d*)
11. (*b*)	**12.** (*d*)	**13.** (*b*)	**14.** (*d*)	**15.** (*d*)
16. (*c*)	**17.** (*a*), (*b*), (*c*)	**18.** (*a*)	**19.** (*a*)	**20.** (*b*)

General Questions

1. (*a*) What is valence bond theory of bonding? How does valence bond theory account for bonding in a hydrogen molecule?

   (*b*) What do the following structure of hydrogen molecule signify:

   $H—H \longleftrightarrow H^+ H^- \longleftrightarrow H^- H^+$

   (*c*) Write the various resonating structures for carbon monoxide molecule.

2. (*a*) What do you understand by the terms resonance and resonance energy?

   (*b*) What are the essential conditions for writing resonating structures?

   (*c*) Draw all the resonating structures of nitrous oxide.

   (*d*) Why does nitrous acid (HNO_2) show *cis* and *trans* isomers?

3. Give the salient points of valence bond theory.

4. In the potential energy diagram, the combined wave function ψ, for two hydrogen atoms. $\psi = \psi_A(1)\,\psi_B(2)$ (where $\psi_A(1)$ and $\psi_B(2)$ represent wave functions of two hydrogen atoms H_A and H_B associated with electrons 1 and 2 respectively, results in decrease in energy of the system by 24 kJ mol^{-1}. But the experimental bond energy of H_2 is 458 kJ mol^{-1}. What improvement in the wave function of the system are made to account for this difference?

5. (*a*) Explain what is meant by the terms exchange energy of hydrogen molecule. How would you account for it ?

   (*b*) What is resonance? Discuss this by taking the example of CO_3^{2-} ion.

   (*c*) Draw the various resonating structures of CO and CO_2 molecules. Justify the significance of each structure on the basis of their physical properties.

6. (*a*) What do you understand by resonance and resonance energy?

   (*b*) Discuss resonance structure of CO_3^{2-} ion.

7. Write a short note on resonance structure of CO and cause of its low dipole moment.

8. What is a three electron bond? Give examples.

9. (*a*) What do you mean by resonance? Write the resonance structures of CO and explain the cause of its low dipole moment.

10. (*a*) Write short note on:

    Resonance structure of CO and cause of its low dipole moment.

11. (*a*) Give two examples of a three electron bond.

    (*b*) N_2O is linear structure. Explain with the help of resonating suructures.

    (*c*) Why is CO_2 linear? Explain with the help of resonating structures.

12. (*a*) Boiling point of H_2O is higher than H_2S.

    (*b*) HF is less volatile than HCl.

    (*c*) Boiling point of PH_3 is less than NH_3.

    (*d*) Formic acid exists as dimer.

13. What is hydrogen bond? What are the essential conditions for hydrogen bond formation: Give example.

14. Use valence bond theory to explain bonding in N_2 and CO.

15. What is hybridisation? What shapes are associated with the following hybrid orbitals:

    sp^2, sp^3d and sp^3d^2

16. On the basis of hybridisation, explain the shapes of the following molecules/ions:

    SF_6, NH_3, BF_3, IF_7, CO_3^{2-}, NO_3^-.

17. (*a*) Discuss the geometry of PF_3 on the basis of VSEPR theory. Compare its bond angle with PCl_3 and PBr_3.
(*b*) Using VSEPR theory, how will you show that
(*i*) XeF_4 is square planar
(*ii*) ClF_3 is T-shaped
(*iii*) SF_4 is see-saw shaped?

18. (*a*) Why is the use of hybrid orbitals preferred than atomic orbitals? Illustrate your answer.
(*b*) Give the hybridisation of the central atom and expected as well as actual geometry of the following molecules/ions.

$$NO_3^-, PF_5, NH_3, SF_4, ClO_3^-$$

19. What are the main postulates of VSEPR theory? How does this account for the geometries of the following molecules?
(*i*) PF_5 (*ii*) H_2O (*iii*) $SnCl_2$
(*iv*) BF_3 (*v*) SF_6 (*vi*) ClF_3

20. (*a*) Use VSEPR theory to predict the shapes of the following molecules:
(*i*) NH_3 (*ii*) SF_4 (*iii*) XeF_4 (*iv*) PF_5,
(*b*) Account for the following order of bond angles:
(*i*) $PF_3 < PBr_3 < PCl_3$ (*ii*) $H_2O > H_2S > H_2Se$ (*iii*) $PF_3 > PH_3$,
(*c*) What are the limitations of VSEPR theory.

21. (*a*) What is hybridisation? What are the important characteristics of hybridisation?
(*b*) Beryllium (Z = 4) has no unpaired electron in its ground state. However, in all its compounds, beryllium shows divalency. Explain
(*c*) The central atom in the molecules of CH_4, NH_3 and H_2O involves sp^3 hybridisation. But the bond angles in these molecules are 109.5° 107° and 104.5° respectively. Account for this.
(*d*) What shapes are associated with the molecules involving sp^3d and sp^3d^2 hybridisations? Which *d*-orbitals are specifically required for each set?

22. (*a*) Boron has only one unpaired electron in its ground state and is expected to show monovalency. However, in all its compounds, boron shows trivalency. Discuss.
(*b*) The ground state electronic configuration of chlorine is $1s^2\ 2s^2\ 2p^6\ 3s^2\ 3p^5$. Account for the geometry of following molecules/ions of chlorine:
(*i*) ClF_3^- (*ii*) ClO_3^-
(*c*) In XeF_2, xenon undergoes sp^3d hybridisation but the molecule is linear. Explain.

23. Discuss the geometry of the following ions:
(*a*) NO_3^- (*b*) CO_3^{2-} (*c*) SO_4^{2-} (*d*) PO_4^{3-} (*e*) ClO_4^-

24. Account for the following:
(*a*) All the bond lengths in SF_6 are equal but these are not equal in IF_5 molecule.
(*b*) BeF_2 molecule is linear while SF_2 is angular, though both are triatomic.
(*c*) The geometics of H_2O and H_2S are same, but their bond angles are $H_2O > H_2S$.
(*d*) BH_3 is trigonal planar while NH_3 is pyramidal.

25. (*a*) What is hybridisation? Discuss the stereochemistry of the following molecules on the basis of hybridisation:

$PF_5, SnCl_2, NH_3, ClF_3$.

(*b*) Which of the following molecules has largest bond angle?

H_2O, NH_3, CH_4, BH_3

26. (*a*) What do you understand by hybridisation? Explain with one example sp^3 hybridisation.

(*b*) What type of geometry is possible for molecules undergoing the following types of hybridisations?

(*i*) sp (*ii*) sp^2 (*iii*) sp^3 (*iv*) sp^3d (*v*) sp^3d^2 (*vi*) sp^3d^3.

(*c*) Name the four important types of hybrid orbitals which can be formed by an atom with *s*, *p* and *d*-orbitals in its valence shell.

27. One the basis of hybridisation, discuss the geometry but the molecules is linear. Explain.

$SF_4, IF_5, SnCl_2, XeF_4, SF_6, H_2O, PCl_5$ and NH_3.

28. (*a*) Give the hybridisation of the central atom of the following molecules/ions:

IF_7, ClO_3^- and SF_6.

(*b*) Giving examples explain the concept of hybidisation with reference to:

(*i*) sp^3 and (*ii*) sp^3d.

29. (*a*) What is LCAO molecular orbital approach?

(*b*) What do you understand by bonding and antibonding molecular orbitals?

(*c*) Draw the molecular ortbitals obtained by the combination of two 2*s*-atomic orbitals.

30. (*a*) Discuss the conditions necessary for the combination of atomic orbitals. Sketch and discuss the combination of *p*-atomic orbitals to form molecular orbitals.

(*b*) What is meant by bond order? Calculate the bond order in He_2^+, O_2 and N_2 molecules.

(*c*) Is B_2 molecule paramagnetic or diamagnetic? Discuss.

31. (*a*) Sketch the shapes of MO obtained by the overlap of:

(*i*) two *s*-orbitals

(*ii*) end on overlap of two *p*-orbitals

(*iii*) side wise overlap of two *p*-orbitals

(*b*) How does molecular orbital theory account for the following:

(*i*) bond order in O_2^+ is more than in O_2 molecule while bond order in N_2^+ is less than in N_2 molecule.

(*ii*) He_2 does not exist.

32. (*a*) Is it correct to say that bond energy in a diatomic molecule always increases when an electron is lost?

(*b*) Why is the bond energy of Li_2 less than that of H_2?

(*c*) Draw the molecular orbital diagram for HCl molecule.

33. (*a*) Discuss the shapes of SF_6 and NO_3^-.

(*b*) What do you understand by linear combination of atomic orbitals? Explain.

(*c*) Discuss the resonance structures of CO_3^{2-}.

34. The internuclear distance is some O—O bonds are as follows:

O_2^+	O_2	O_2^-	O_2^{2-}
1.12Å	1.21 Å	1.30Å	1.49 Å

(*a*) Draw the molecular orbital diagram for O_2 molecule and account for the gradation of the internuclear distance in the above series.

(*b*) Which of these species are paramagnetic?

35. (*a*) Write molecular orbital configurations for O_2, O_2^+, O_2^- and O_2^{2-} and arrange them in increasing order of their bond lengths.

(*b*) With the help of MO diagrams explain

(*i*) that the bond energy of NO^+ is larger than that of NO.

(*ii*) that the bond length of CO^+ is larger than that of CO.

36. (*a*) Give the postulates of molecular orbital theory and discuss the distribution of electron charge density in bonding as well as antibonding molecular orbitals.

(*b*) What is meant by *dsp*^{2} hybridisation? How are these hybrid orbitals formed? Discuss the formation of $[Ni(CN)_4]^{2-}$ ion on the basis of hybridisation.

(*c*) PF_5 is more reactive than SF_6. How will you account for this?

37. (*a*) Distinguish two aspects of the following:

(*i*) atomic and molecular orbitals

(*ii*) bonding and antibonding orbitals

(*iii*) sigma and pi bonds.

(*b*) Do you expect bond order in superoxide ion (O_2^-) more or less than in peroxide ion (O_2^{2-})?

(*c*) Give an account of the relationship between electronic configuration and molecular behaviour.

38. (*a*) Explain the following:

(*i*) lone pair—lone pair repulsion > lone pair — bond pair repulsion > bond pair—bond pair repulsion.

(*ii*) He_2 does not exist. (Use MO theory).

39. (*a*) Bring out the differences between the valence bond and molecular orbital theories.

(*b*) Distinguish between σ and π molecular orbitals.

(*c*) (*i*) Why is the force of repulsion between lone pairs different from that of the bond pairs of electrons?

(*ii*) All P–F bonds in PF_5 are not equivalent in length. Explain.

40. (*a*) Discuss the structure of the following on the basis of hybridisation

(*i*) IF_5, (*ii*) CO_3^{2-} ion.

(*b*) Define bonding and antibonding molecular orbitals. Describe salient features of these orbitals.

41. (*a*) Which of the following combinations of orbitals are not allowed in LCAO method (consider Z-axis to be molecular axis):

(*i*) $s + p_z$ (*ii*) $p_x + p_x$ (*iii*) $p_y + p_z$ (*iv*) $s + p_x$.

(*b*) What is resonance? What are the rules which have resctricted our choice of resonating structures?

(*c*) State the geometrical arrangements possible in sp^3d and sp^3d^2 hybridisations. Name the *d*-orbitals involved in these.

42. (*a*) Explain the following:

(*i*) PF_5 is more reactive than SF_6.

(*ii*) BeF_2 molecule is linear while BF_3 molecule is trigonal.

(*iii*) Bond angle in NH_3 is 107° while in H_2O it is 104°.

(*iv*) He_2^+ ion is less stable than H_2^+ ion.

(*b*) Distinguish between molecular orbital theory and valence bond theory.

43. (*a*) How is molecular orbital wave function (ψ_{MO}) obtained from atomic orbital wave functions (ψ_A and ψ_B) for a molecule AB?

(*b*) How will you account for the fact that bond strength of nitrogen molecule is more than that of N_2^+ ion whereas bond strength of O_2 is less than that of O_2^+ ion. Explain your answer on the basis of molecular orbital theory.

(*c*) Is B_2 molecule paramagnetic or diamagnetic ? Discuss.

44. (*a*) Give the hybridisation of the central atom and expected as well as actual geometry of the following molecules/ions :

H_2O, NO_3^-, XeF_4, $SnCl_2$.

(*b*) Explain the following :

(*i*) Nitrogen molecule is diamagnetic while oxygen molecule is paramagnetic.

(*ii*) Boron molecule is stable while beryllium molecule is extremely unstable.

(*c*) Explain what is meant by the term exchange energy of hydrogen molecule. How would you account for it ?

45. (*a*) Which of the following molecules/ions have sp^2 hybridized central atom and draw their structures:

(*i*) ClO^- (*ii*) NO_2 (*iii*) $SnCl_2$ (*iv*) SO_3^{2-} (*v*) CO_3^{2-}

(*b*) Draw an MO diagram of CO and show its bond order. How does this description compare with its valence bond strucuture.

46. (*a*) How does molecular orbital theory account for the following:

Bond order in O_2^+ is more than in O_2 molecule while bond order in N_2^+ is less than in N_2 molecule?

(*b*) Why is the force of repulsion between lone pairs different from that of the bond pairs of electrons.

(*c*) Distinguish the two aspects of the following:

(*i*) bonding and antibonding orbitals

(*ii*) sigma and pi bonds.

47. (*a*) Use valence bond theory to explain bonding in N_2 and CO.

(*b*) Explain on the basis of V.B.T. and hybridisation that

(*i*) BeF_2 molecule is linear while H_2O is bent.

(*ii*) Phosphorus forms two chlorides (PCl_3 and PCl_5) while nitrogen forms only one *i.e.* NCl_3.

(*c*) Chlorate and perchlorate ions have the same number of electron pairs around the chlorine atom but their geometry is different.

(*d*) Explain on the basis of M.O.T.

(*i*) N_2 molecule is diamagnetic while O_2 molecule is paramagnetic.

(*ii*) Bond order of N_2^+ is lower than that of N_2.

48. (*a*) Account for the decreasing bond angle in the order $NO_2^+ > NO_2 > NO_2^-$.

(*b*) Use VSEPR theory to predict the shapes and hybridisation of the following molecules:

$$NH_3, SF_4, PF_5, IF_7$$

(*c*) Account for the following bond angles:

(*i*) $PF_3 < PBr_3 < PCl_3$ (*ii*) $H_2O > H_2S > H_2Se$ (*iii*) $PF_3 > PH_3$.

(*d*) In XeF_2, xenon undergoes sp^3d hybridisation, but the molecule is linear. Explain.

49. Discuss M.O. theory. How does this theory differ from VB theory? Discuss the diamagnatic nature of N_2 molecule with the help of M.O. theory.

50. (a) On the basis of hybridisation, discuss thegeometry of the following molecules:

$$IF_5, XeF_4, PCl_5, NH_3$$

(*b*) Draw the molecular orbital diagram for HCl molecule.

(*c*) Draw the M.O. energy level diagram for fluorine and predict

(*i*) bond order

(*ii*) magnetic properties, therein.

51. Explain the following:

(*i*) Oxygen molecule is paramagnetic

(*ii*) Bond order in NO^+ is 3.

(*iii*) Bond distance in O_2^+ is smaller than in O_2 molecule.

(*iv*) Bond strength of N_2 molecule is more as compared to O_2 molecule.

(*v*) Non-existence of He_2 molecule.

52. (*a*) All P-F bonds in PF_5 are not equivalent length. Explain.

(*b*) Write down the differences between valence bond and molecular orbital theory?

(*c*) Give the hybridisation of the central atom and expected as well as actual geometry of the following molecules/ions:

$$H_2O, NO_3^-, XeF_4, SnCl_2.$$

53. (*a*) Draw MO diagram of CO and calculate its bond order.

(*b*) Discuss the geometry shapes of molecules involving sp^3d and sp^3d^2 hybridisation with at least one example each.

(*c*) PF_5 is more reactive than SF_6, Explain.

54. (*a*) Discuss the main features and limitations of VB theory.

(*b*) Discuss the shapes of

(*i*) NO_3^- and CO_3^{2-}

(*ii*) ClF_3 and IF_5

(*iii*) ClO_3^- and ClO_4^-

55. What do you understand by bonding and antibonding molecular orbitals? Give the number of electrons which occupy the bonding and antibonding molecular orbitals of

(*i*) H_2^+ (*ii*) H_2 (*iii*) He_2 molecule.

56. (*a*) How will you explain that the geometries of H_2O and H_2S are the same but their bond angles are $H_2O > H_2S$.

(*b*) What is bond order? Calcualte the bond order of He_2^+.

(*c*) Explain why phosphorus forms two chlorides (PCl_5 and PCl_3) while nitrogen forms only NCl_3?

(*d*) What are the postulates of VSEPR theory?

57. Discuss the shapes and types of hybridisation in the following:

$$ClO_3^-, ClO_4^-, PCl_5, SF_6, NO_3^-$$

58. (*a*) On the basis of M.O. theory calculate bond order in

(*i*) O_2^{2-} ion (*ii*) N_2^+ ion.

(*b*) Draw the actual geometry and type of hybridisation:

(*i*) NO_3^- (*ii*) IF_7 (*iii*) ClO_3^-

(*c*) On the basis of molecular orbital theory explain why ionisation energy of O_2 molecules is less than that of O_2^{2+} ion.

(*d*) Using VSEPR theory, describe the shape of

(*i*) $SnCl_2$, (*ii*) $XeOF_4$, (*iii*) ClF_3

59. Explain on the basis of M.O. theory

(*i*) N_2 molecule is diamagnetic while O_2 molecule is paramagnetic.

(*ii*) Bond order of N_2^+ is lower than that of N_2.

60. (*a*) Give the structure and bonding in electron deficient diborane, B_2H_6.

(*b*) What are molecular orbitals? Sketch sigma and pi molecular orbitals that result from the combination of the p-orbitals. One of each atom in a diatomic molecule.

61. (*a*) Name the actual geometry and type of hybridisation of the following:

(*i*) CO_3^-, (*ii*) ClO_4^-, (*iii*) ClO_3^-, (*iv*) $SnCl_2$

(b) Draw and name the shapes which are associated with the following hybrid orbitals:

(*i*) sp^3d (*ii*) sp^3d^2 (*iii*) sp^3d^3

(*c*) Discuss the structure of diborane. Name two compounds which show bonding similar to that of diborane.

(*d*) Discuss the points of similarities and differences between V.B. theory and M.O. theory of covalent bonding.

4

Ionic Bond

4.1 IONIC BOND

An ionic bond is formed by the transference of electron from one atom to the other. Hence this process will be facilitated when one of the constituent atoms has a smaller ionization energy, so that the electron could be removed from the outermost orbit. And the second atom should have a high value of electron affinity so that it could easily accept the electron released by the other atom.

4.1.1 General characteristics of ionic compounds

Some general characteristics of ionic compounds are described as under:

(*i*) Melting and Boiling Points

Melting and boiling points are dependent on the nature of binding forces in the crystalline solid. Melting involves breaking of the crystal lattice. Since electrostatic interactions between positive and negative ions in the crystal are generally strong, therefore their breakdown requires higher energies and hence the ionic compounds have higher melting and boiling points.

Significant variations in melting and boiling points within ionic compounds are attributed primarily to (*i*) charges on the ions, (*ii*) ionic radii.

For example, NaC1 and BaO have similar internuclear distances (NaC1 = 2.81 Å and BaO = 2.7 Å) but the former has a melting point of 800°C only while the latter melts at 1921°C. The significantly higher melting point of BaO is attributed to the stronger cation-anion interaction between the doubly charged Ba^{2+} and O^{2-} ions as compared to singly charged ions in NaC1 crystal. Melting point also depends upon the size of the ions. The larger the ions, the weaker the attractive forces and lower would be the melting points. This is illustrated if we compare melting points of NaF and NaI. The force of attraction between Na^+ and larger I^- ions will be lower than that between Na^+ and small F^- ions. As a result, NaI lattice is less stable than NaF lattice. Hence melting points of NaI (925°C) is lower than that of NaF (1270°C).

(*ii*) Hardness

The electrostatic interaction between oppositely charged ions determines hardness of an ionic compound. The electrostatic interaction between oppositely charged ions depends on:

(*i*) ionic radii (*ii*) charges on the ions

For a given ionic charge, hardness increases with decrease in interionic distance for crystals with similar structure. For example, alkaline earth oxides CaO, SrO and BaO have similar NaC1 type of structure but their interionic distance increases along the group from CaO to SrO to BaO. Consequently, their hardness decreases down the group. In short, more compact the structure, harder is the substance.

Hardness increases with increase in charge on the ions for crystals with similar interionic distances for ionic compounds having similar structure. For example, LiCl and SrO have the same NaC1 type of structure and same interionic distance of 2.57 Å. Their hardness increases as we move from singly charged Li^+ and Cl^- ions to doubly charged Sr^{2+} and O_2^-. Similarly, NaC1 and CaS (with similar structure and interionic distance) show an increase in hardness from NaC1 to CaS because of increase in charge on ions.

***(iii)* Brittleness**

The constituent units which make up ionic solids are oppositely charged ions [Fig. 4.1 (*a*)]. It may be observed that if requisite energy is supplied to the crystal layers, the movement of layers of ions brings ions of like charges near each other [Fig. 4.1 (*b*)] and this causes strong repulsions which lead to the breakdown of the crystal. For this reason ionic crystals, in spite of their hardness, are brittle and can be powdered easily.

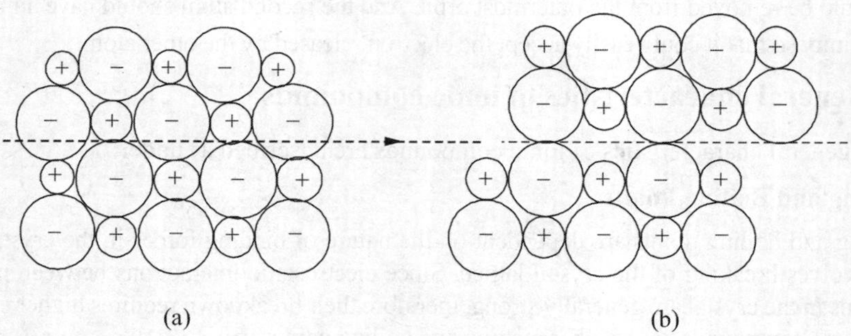

Fig. 4.1. Brittleness.

***(iv)* Cleavage**

The phenomenon of cleavage is observed in ionic crystals. That is, in the crystals the splitting of layers is easier in some directions than others. For example, in a crystal of sodium chloride, cleavage can be achieved only along planes parallel to cube faces. Any attempt to cleave such a crystal along an arbitrary plane will shatter the structure as shown in Fig. 4.2.

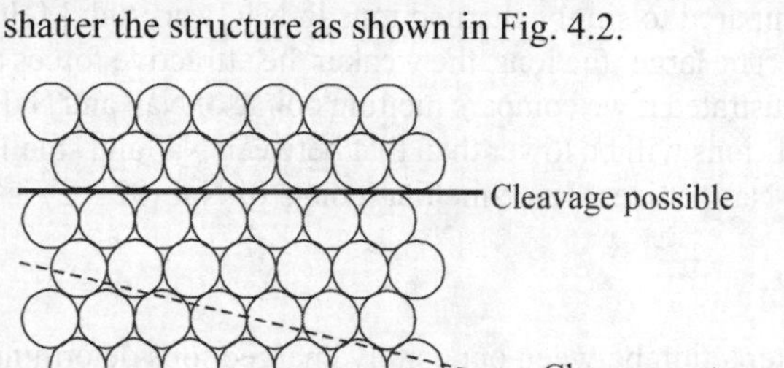

Fig. 4.2. Cleavage.

***(v)* Solubility**

Ionic compounds are soluble in highly polar solvents. Water is a good solvent for dissolving ionic compounds because

(*a*) It has a larger dipole moment which provides high hydration energy due to strong ion-solvent dipole interactions. If hydration energy is greater than the crystal energy, the lattice breaks.

(*b*) It has high dielectric constants (78.5 at 25°C) which is responsible for reducing attractive force between cations and anions in solutions. This keeps the unlike charges apart from each other.

Suitable solvents for dissolving ionic compounds should have high dipole moment and high dielectric constant.

(*vi*) Conductance

In an ionic crystal, the ions are situated in fixed positions in the lattice, that is, they are not mobile. Therefore, ionic solids do not conduct electricity in the solid state. However, in molten state or in solution the crystal lattice breaks and ions are free to move in an electric field. Thus, they conduct electricity due to increase in relative mobility of ions in fused state or in solution.

(*vii*) Sterochemistry

The constituent ions of an ionic solid possess an orderly arrangement in a definite geometric pattern which repeats itself throughout the crystal. Crystals are always bound by plane faces and these faces always meet at some fixed angles. For any particular substance, the angle between corresponding pair of faces is always the same in all crystals.

***Example 1.* Give trends in m.p. and hardness for the following having similar tructures:**

(*i*) LiCl and SrO (*ii*) NaCl and BaO

Solution. Melting points and hardness have similar trends. These two properties decrease in the following order : SrO > LiC1 and BaO > NaCl.

This is because of higher charges on cations and anions in alkaline earth metal oxides.

***Example 2.* AgF, AgCI and AgBr have similar structure. Explain their trends in hardness.**

Solution. Charge on cation remains the same but size increases from AgF to AgC1 to AgBr. Consequently hardness decreases in the same order AgF > AgCl > AgBr.

***Example 3.* Why is the radius of a cation smaller and that of an anion bigger than that of the atom?**

Solution. When a neutral atom gains an electron to become an anion, its radius increases. This is expected due to the increase in the number of electrons which the same nuclear charge has to hold. The force of attraction on the electron decreases and so the radius increases. On the other hand, when a neutral atom loses an electron to form a cation, the radius decreases, because the nuclear charge is now acting on a fewer number of electrons. It is also expected that the size of the ions decreases in an isoelectronic series with increase in nuclear charge.

***Example 4.* Though LiCl and MgS have similar structure and nearly same internuclear distance, MgS is harder than LiCl.**

***Solution*:** Mg^{2+} and S^{2-} ions have greater charges compared to Li^+ and Cl^-. Therefore packing is more compact in MgS lattice. There are greater attractive fores between the cations and anions. Hence MgS is harder than LiCl.

***Example 5.* Melting point of NaCl is higher than that of $AlC1_3$. Explain.**

***Solution*:** Na^+ are smaller than $A1^{3+}$ ions. In sodium chloride, therefore, cation-anion interactions are stronger. The lattice is more compact and thus melting point of NaC1 is higher than that of $A1C1_3$. This can also be explained on the basis of covalent character in $A1Cl_3$. (See Fajan's rules ahead)

***Example 6.* CsCl lattice is less stable than NaCl structure. Explain.**

***Solution*:** Distance between Cs and Cl in CsCl is more than that between Na and Cl in NaC1. Greater distance mean smaller force of attraction between the cation and anion. Hence CsC1 will have a relatively weaker lattice or less stable structure than NaCl.

4.1.2 Pauling's method of calculation of ionic radius

The radii calculated by Pauling are based on the observed internuclear distances in four crystals namely, NaF, KC1, RbBr and CsI. The set of radii calculated by this method are given in Table 4.1.

Table 4.1. Pauling set of ionic radii (pm).

Li^+	60	Be^{2+}	31	B^{3+}	20	C^{4+}	15	O^{2-}	140	F^-	136
Na^+	95	Mg^{2+}	65	Al^{3+}	50	Si^{4+}	41	S^{2-}	184	Cl^-	181
K^+	133	Ca^{2+}	99					Se^{2-}	198	Br^-	195
Rb^+	148	Sr^{2+}	113	Zn^{2+}	74			Te^{2-}	221	I^-	216
Cs^+	169	Ba^{2+}	135					–	–	–	–

Some of the factors which affect interionic distances in crystals are (*i*) electron distribution of ions, (*ii*) charges on the ions, (*iii*) crystal structure, and (*iv*) cation to anion radius ratio. The four compounds mentioned above were so chosen that all these factors remain constant. In such a situation, it was proposed by Pauling that the radius of an ion is inversely proportional to the effective nuclear charge, Z_{eff} of the ion.

$$R_{ion} \propto \frac{1}{Z_{eff}}$$

$$\propto \frac{1}{Z-S} \text{ (where } Z = \text{nuclear charge, } S = \text{screening constant)}$$

$$= \frac{C}{Z-S} \text{ (where } C \text{ is proportionality constant)}$$

Let us calculate the radii of Na^+ and F^- from the measured internuclear separation at NaF of 231 pm.

Both sodium and fluoride ions are isoelectronic and have the same value of screening constant, *i.e.*, $S = 4.5$.

Effective nuclear charge will be calculated as under:

The effective nuclear of $Na^+ = 11 - 4.5 = 6.5$

Similarly, effective nuclear of $F^- = 9 - 4.5 = 4.5$

The proportionality constant C may be calculated as

$$95 = \frac{C}{6.5} \text{ for } Na^+$$

and $$136 = \frac{C}{4.5} \text{ for } F^- \qquad \text{(See Table 4.1)}$$

Combining the above two equations

$$\frac{C}{6.5} + \frac{C}{4.5} = 231$$

or $$C = 641$$

Using the above value of C, radius of each ion may be found out as:

$$r_{Na^+} = \frac{C}{Z_{eff}} = \frac{641}{6.5} = 95\,\text{pm}$$

and $$r_{F^-} = \frac{641}{4.5} = 136\,\text{pm}$$

4.2 RADIUS RATIO

The ratio of radius of the cation (r^+) and radius of anion (r^-) is called radius ratio.

Their radius ratio = $\frac{r^+}{r^-}$

Ionic compounds consist of −ve and −ve ions held together by electrostatic interactions in such a way that the distance between cations and anions is minimum. If each ion is considered as a hard sphere the optimum arrangement of ions for the most stable structure would be one that involves touching of the largest number of oppositely charged spheres without any crowding. Such a situation leads to maximum attraction between opposite charges and minimum repulsion between like charges. Such an arrangement of ions in a crystal is greatly influenced by the radius ratio; it determines how many ions of a given size can be held in contact with a single ion of smaller size.

The relation between radius ratio and structure is given in Table 4.2.

Table 4.2. Relation between radius ratio and structure.

Radius ratio r^+/r^-	*Coordination number*	*Structural arrangement*	*Ionic lattices*
0.155 – 0.225	3	Trigonal planar	Boron oxide
0.225 – 0.414	4	Tetrahedral	ZnS(Wurtzite, Zinc blende)
0.414–0.732	6	Octahedral	NaC1, TiO_2(Rutile)
0.732–1.00	8	Cubic	CsCl,CaF_2 (Fluorite)

A cation will fit into an octahedral arrangement of anions for radius ratio $\frac{r^+}{r^-}$ in the range 0.414 - 0.732. That is, only in such a situation a cation will be able to keep six anions in contact with each other. Smaller cations will prefer to fit into tetrahedral holes in the lattice. For radius ratio $\frac{r^+}{r^-}$ ranging between 0.225 and 0.414, tetrahedral sites will be preferred. A cubic lattice with coordination number 8 is preferred for radius ratio $\frac{r^+}{r^-}$ in the range 0.732 – 1.

Coordination number is the number of anions that surrounded a cation to give a stable lattice.

In zinc sulphide, with a radius ratio of 74/184 *i.e.*, 0.4, zinc will occupy tetrahedral holes in the close packed lattice of sulphide ions whereas in sodium chloride (radius ratio 5.2), Na^+ will be surrounded by six chloride *i.e.*, it will have octahedral arrangement.

By using radius ratio rule, it is possible to predict the cation/anion coordination number in any compound. So radius ratio is a useful measure in establishing the structure of ionic solids. In Fig. 4.3, it is shown how the arrangement of ions in a single layer is effected by the differences in the sizes of the ions.

Figure 4.3 (*a*) shows the ideal situation where M^+ fits into X^- anions.

In Fig. 4.3 (b), if the anions move closer to the size of M^+, then they would crowd each other. There would be repulsion amongst them and the energy of the system would be raised. If they hold each other apart then they are not very close to the oppositely charged M^+. Fig. 4.3 (*c*) does not allow the X^- ions to come close enough to feel any $X^- – X^-$ repulsions. Therefore, if M^+ is too large

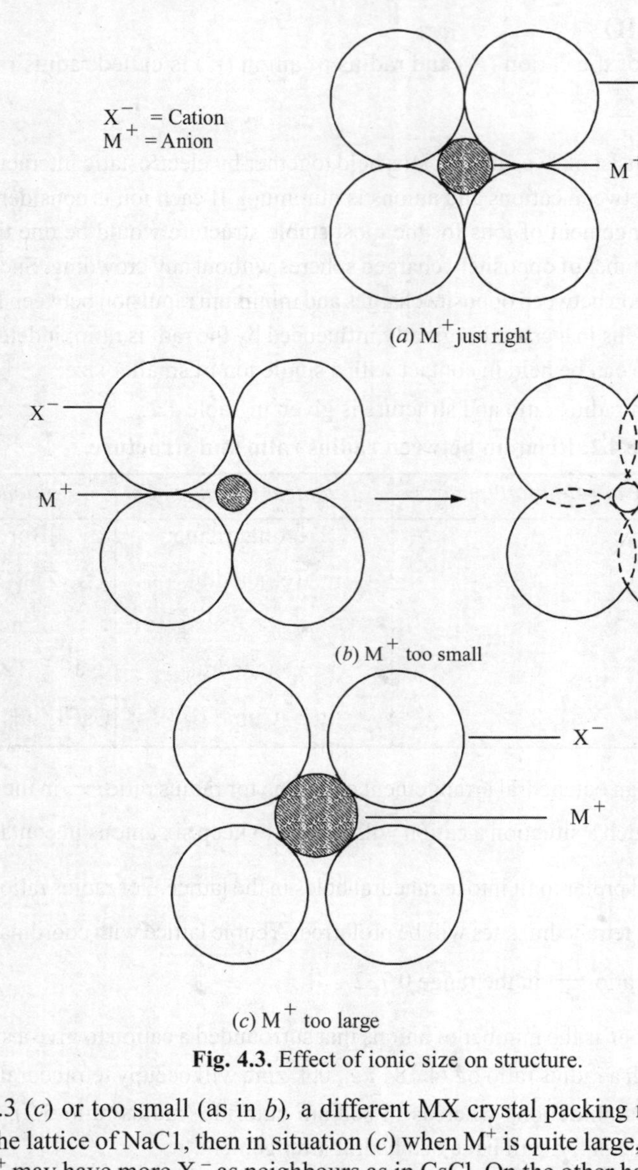

Fig. 4.3. Effect of ionic size on structure.

as in Fig. 4.3 (*c*) or too small (as in *b*), a different MX crystal packing might be obtained. If (*a*) represents the lattice of NaC1, then in situation (*c*) when M^+ is quite large, we have an arrangement in which M^+ may have more X^- as neighbours as in CsCl. On the other hand, if M^+ is too small, as in (*b*), only four negative ions pack around the positive ions. ZnS provides an example of this type. The change in arrangement occurs in order to attain stability by touching maximum number of anions.

4.2.1 Limitations of radius-ratio rule

Limitations of radius ratio rule are:

(*i*) An ion is considered as a sphere for determining the optimum arrangements of ions in the crystal lattice. This is far from reality and serious errors can be made if anions get polarized and the bonding has some partial covalent character.

For example, radius ratio r^+/r^- for AgI is 0.52 which is indicative of coordination number 6 but actually it is 4 coordinated. Similarly, ZnO, HgS and CdS have radius ratios greater than 0.5 but have tetrahedral lattice.

(*ii*) Some compounds may exist in more than one modifications with different coordination number. In such cases, anion-anion repulsions and hence internuclear distances would be different. For example, TIC1 crystallises out in NaC1 structure (coordination number 6) with internuclear distance 3.15Å but can also adopt CsC1 structure (coordination number 8) with internuclear distance 3.32 Å.

(*iii*) Effective radius of a cation is greatly influenced by the anions with the consequence that the radius ratio changes. For example, AgF and NaF crystallises out in NaC1 type of structure with co-ordination number 6 and if we assume that the size of F^- remains constant then Ag^+ is bigger than Na^+. On the other hand, for a given size of chloride ion in AgC1 and NaC1, the sizes of cations are reversed, that is, Na^+ is bigger than Ag^+. It is also true for their bromides. This is understandable because Ag^+ is softer than Na^+ and introduces relatively more covalent character with polarizable anions like Cl^- and Br^-.

4.3 LATTICE ENERGY

An ionic bond is likely to be formed most readily between an atom with the lowest ionization potential and an atom with the highest electron affinity. However, even the combination of the most electropositive cesium atom with the lowest ionization energy of 357.3 kJ/mol (energy absorbed : endothermic) and the electronegative chlorine with the highest electron affinity of 347.36 kJ/mol (energy released : exothermic) is energetically unfavourable. This indicates that even for the most ionic bond *i.e.*, Cs^+ Cl^- (g), there is insufficient energy available for its formation. However, the shortfall is more than compensated by the energy gained through electrostatic attraction in the formation of an ionic crystal. This energy is known as the **crystal energy** or the **lattice energy** and is sufficient to permit the formation of ions.

The stability of an ionic solid is measured in terms of its lattice energy (U_0). **It is defined as the energy released when one mole of an ionic crystal is formed from one mole of gaseous positive and one mole of gaseous negative ions, when these are separated from each other by infinite distance.** Greater the value of lattice energy, more stable is the ionic solid. The lattice energy of sodium chloride is – 774 kJ $mo1^{-1}$. The negative sign indicates that the process is exothermic and that of the energy of the system is lowered as the solid is formed.

The force of attraction (F) between oppositely charged ion is governed by

(*i*) their ionic radii

(*ii*) the charges on the ions.

It is given by the equation : $F \propto \frac{q_1 q_2}{r^2}$

where q_1 and q_2 are the points charges separated by distance r is vacuum.

It may be observed that if the radii of the ions are large, the force of attraction will be less. This is better illustrated if we compare the stability of NaC1 and CsCl. The force of attraction between the large Cs^+ ions and Cl^- ions will be less than that between the smaller Na^+ ions and Cl^- ions. As a result, CsC1 is less stable than NaCl. We may say that the stability of an ionic solid increases with decreasing interionic distance for crystals with similar ionic charges.

The role played by the charges on the ions demonstrated if we compare the lattice energy of BaO and NaC1, since both have very similar internuclear distances (NaC1 = 281 and BaO = 277 pm). Since BaO consists of doubly charged Ba^{2+} and O^{2-} ions, the force of attraction is much larger than that in NaC1 which contains singly charged ions. Hence BaO is more stable than NaC1 (melting points NaCl = 800°C, BaO = 1911°C). The stability of an ionic solid increases with increasing nuclear charge for crystals with similar interionic distances.

4.3.1 Calculation of lattice energy of ionic solids

The lattice energy of an ionic crystal is determined by coulombic interaction between all of its ions; attractive forces between oppositely charged ions and repulsive forces because of the interpenetration of electron clouds. The potential energy of a pair of oppositely charged ions representing force of attraction varies inversely with distance, *r*, between the two ions. (Fig. 4.4).

$$PE_{(attraction)} = -\frac{Z_1 Z_2 e^2}{r} \qquad ...(i)$$

where Z_1 and Z_2 are charges on the ions and *e* is electronic charge. For NaCl, $Z_1 = Z_2 = 1$ and for CaO, $Z_1 = Z_2 = 2$ but for $BaCl_2$, $Z_1 = 2$, $Z_2 = 1$.

As the ions come very close to each other, they repel one another because of interpenetration of the electron clouds. The repulsion energy is inversely proportional to the nth power of the distance *r* between the ions.

$$PE_{(repulsion)} = \frac{be^2}{r^n} \qquad ...(ii)$$

where *e* is the electronic charge and *n* is the constant called the Born repulsion exponent and *b* is a constant called a repulsion coefficient. The Born equation gives the net potential energy for a pair of ions as the summation of the two terms, *i.e.*, (*i*) and (*ii*) when these ions are brought together from infinite distance.

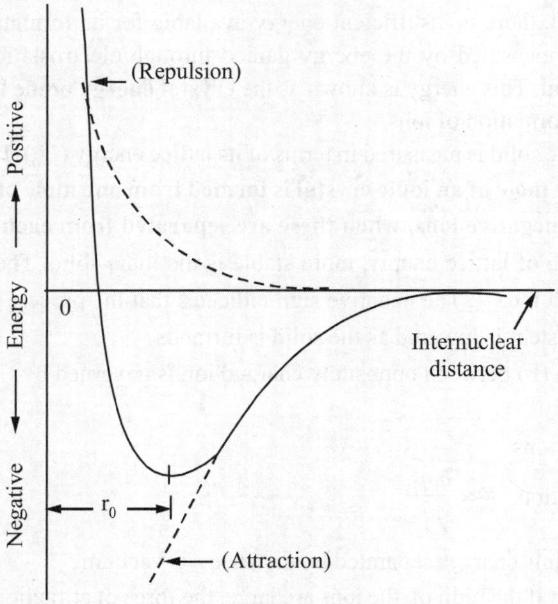

Erwin Madelung

Fig. 4.4. Potential energy diagram of an ionic solid.

$$PE = -\frac{Z_1 Z_2 e^2}{r} + \frac{be^2}{r^n} \qquad ...(iii)$$

The negative sign on attraction term indicates release of energy. It is clear from Fig. 4.4 that repulsion term increases very rapidly as the separation between the ions decreases beyond the internuclear separation. The potential energy is at a minimum at the equilibrium separation, r_0.

Modification in Born equations.

In ionic crystals, no two separate ions can be considered, the entire crystal must be considered as a single entity. The force on any one ion will be determined not only by the oppositely charged ions which are directly surrounding it but also somewhat due to other ions (both positive and negative) at greater distances. To consider the forces due to all the ions, we may take a specific example, say sodium chloride. The potential energy of an ion will depend on the crystal structure of the solid.

In a crystal of sodium chloride (Fig. 4.5) the potential at a sodium ion results from:

(*a*) Six nearest Cl^- ions at a distance r ;

(*b*) Twelve next nearest Na^+ ions at a distance $\sqrt{2}r$;

(*c*) Eight $(next)^2$ nearest Cl^- ions at a distance $\sqrt{3}r$;

(*d*) Six $(next)^3$ nearest Na^+ ions at a distance $2r$;

(*e*) Twenty-four $(next)^4$ nearest Cl^- ions at a distance $\sqrt{5}r$.

These distances from a sodium ion are clearly shown in Fig. 4.5.

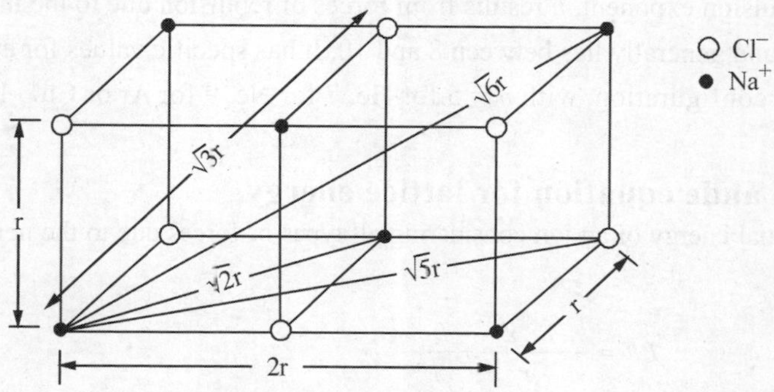

Fig. 4.5. Various forces acting in NaCl crystal.

The potential energy expression for the energy released (attraction) in bringing a sodium ion in a crystal lattice of NaC1 for infinity is given by the summation of all such terms:

$$PE = -\frac{6Z_1Z_2e^2}{r} + \frac{12Z_1Z_2e^2}{\sqrt{2}\,r} - \frac{8Z_1Z_2e^2}{\sqrt{3}\,r} + \frac{6Z_1Z_2e^2}{2\,r} - \frac{24Z_1Z_2e^2}{\sqrt{5}\,r} + \ldots$$

or $$PE = -\frac{Z_1Z_2e^2}{r}\left(6 - \frac{12}{\sqrt{2}} + \frac{8}{\sqrt{3}} - \frac{6}{2} + \frac{24}{\sqrt{5}} \ldots\right) \qquad \ldots\text{(iv)}$$

The quantity in the bracket in equation (iv) is the summation of an infinite series which is called the **Madelung constant**. The summation value for sodium chloride is given as 1.747558. Since the value of this constant depends only on the geometry of the crystal, the value will be the same for all other solids which exhibit the NaC1 structure. The Madelung constant is represented by the symbol A. This quantity can be calculated in the same way for other crystal lattices. Incorporating A, equation (iv) becomes.

$$PE = -\frac{AZ_1Z_2e^2}{r}$$

Values of Madelung constant for some common crystal structure type are given in Table 4.3.

Table 4.3. Madelung constant (A) for some common crystal lattices.

Structure	*Formula*	*Coordination number*	*A*
Sodium chloride	NaC1	Na 6; Cl 6	1.7475
Cesium chloride	CsC1	Cs 8; Cl 8	1.7627
Zinc blende	ZnS	Zn 4; S 4	2.6081
Wurtzite	ZnS	Zn 4; S 4	1.6413
Fluorite	CaF_2	Ca 8; F 4	5.0388

As per convention, potential energy is negative when interaction between oppositely charged ions is involved (energy is released). On the other hand potential energy is positive when interaction between similarly charged ions is involved (energy is absorbed).

Similarly, the second term in the potential energy is modified. The repulsive energy term $\frac{be^2}{r^n}$ increases very steeply as interionic distance decreases after ions come into contact. The repulsion coefficient (*b*) is modified in view of its approximate proportionality to the number of nearest neighbours. In case of sodium chloride, six Cl^- ions are the immediate neighbours surrounding the Na^+ ion. Therefore instead of (*b*), the repulsion coefficient is taken as (6*b*) for NaCl. Thus, the expression for repulsion term becomes $\frac{6be^2}{r^n}$. This is expressed in a more general form: $\frac{Be^2}{r^n}$, where $B = 6b$ for NaCl.

The Born repulsion exponent, *n* results from forces of repulsion due to the interpenetration of ionic atmosphere and generally lies between 8 and 10. It has specific values for each noble gas, or pseudo noble gas configuration, with $n = 5$ for He, 7 for Ne, 9 for Ar or Cu^+, 10 for Kr or Ag^+, 12 for Xe or Au^+.

4.3.2 Born-Lande equation for lattice energy

The net potential energy of an ion considering all types of forces due to the neighbouring ion is given by

$$PE = -\frac{Ae^2Z^2}{r} + \frac{Be^2}{r^n} \qquad \text{...(v)}$$

where *Z* replaces Z_1 and Z_2. This is called Born-Lande equation. At the equilibrium internuclear distance, r_0 the attractive and repulsive forces are exactly balanced. The potential energy is minimum at r_0. Differentiation of Eq. (v) for $r = r_0$ and equating it to zero we obtain the value of r_0 and *B* and then eliminate the latter from the potential energy expression.

Equation (v) may be written as

$$PE = -Ae^2Z^2r^{-1} + Be^2r^{-n}$$

On differentiation, we get

$$\frac{d(PE)}{dr} = (-1)(-)Ae^2Z^2r^{-1-1} + Be^2(-n)r^{-n-1}$$

or $$Ae^2Z^2r^{-2} - nBe^2r^{-n-1}$$

On applying the necessary conditions, that $\frac{d}{dr}(PE) = 0$ for $r = r_0$, we have

$$0 = Ae^2Z^2r_0^{-2} - nBe^2r_0^{-n-1}$$

or $$Ae^2Z^2r_0^{-2} = nBe^2r_0^{-n-1}$$

$\therefore$ $$B = \frac{AZ^2}{n}r_0^{-2+n+1}$$

or $$B = \frac{AZ^2}{n}r_0^{n-1} \qquad \text{...(vi)}$$

and similarly, we can get

$$r_0 = \left(\frac{nB}{AZ^2}\right)^{\frac{1}{n-1}} \qquad \text{...(vii)}$$

By substituting the value of B in expression (v), we get

$$(PE)_0 = \frac{-Ae^2Z^2}{r_0} + \frac{AZ^2e^2}{nr_0^n} r_0^{n-1}$$

$$= \frac{-Ae^2Z^2}{r_0} + \frac{AZ^2e^2}{n} r_0^{n-1-n}$$

$$= \frac{Ae^2Z^2}{r_0}\left[\frac{1}{n} - 1\right]$$

$$= \frac{-Ae^2Z^2}{r_0}\left[1 - \frac{1}{n}\right] \quad \text{...(viii)}$$

The lattice energy U_0 is the amount of energy released when one mole of NaC1 crystal is formed from gaseous ions which are at infinite separation.

Lattice energies of ionic compounds are always negative (being energy released).

For one mole of the ion-pairs the said expression (viii) is multiplied by Avogadro's number (N)

$$U_0 = \frac{-NAe^2Z^2}{r_0}\left[1 - \frac{1}{n}\right] \quad \text{... (ix)}$$

This expression is known as **Born-Lande** equation for lattice energy. The charge Z is taken as unity for NaC1 and other similar uni-univalent systems and two for BaO and other divalent ionic compounds. Care should be taken in cases like say CaF_2 where the value of charge on the cation is 2 and the value of charge on the anion is 1. Consequently, Z^2 in CaF_2 would be 2.

***Example 7.* Calculate the value of Madelung constant A for MgO, given**

$$\mathbf{r_0 = 2.10\ Å,\ n = 7,\ N = 6.02 \times 10^{23}}$$

$$\mathbf{U_0 = -3940\ kJ/mol,\ e = 4.8 \times 10^{-10}\ e.s.u.}$$

Solution. In case of MgO, the values of charges on cation or anion are 2 each. Therefore $Z^2 = 4$.

Substituting values and taking care of conversion of erg/mol to kJ/mole, we have

$$-3940 = \frac{-6.02\times10^{23}\times2^2\left(4.8\times10^{-10}\right)^2\times A}{2.10\times10^{-8}\times10^{10}}\left(1-\frac{1}{7}\right)$$

$$3940 = 2264.5\ A$$

or $\quad A = 1.74$

***Example 8.* Calculate the value of Z^2 in CaF_2, given**

$$\mathbf{A = 2.519,\ r_0 = 2.35\ A,\ n = 7}$$

$$\mathbf{N = 6.02 \times 10^{23},\ e = 4.8 \times 10^{-10}\ e.s.u.}$$

and $\quad \mathbf{U_0 = -2550\ kJ/mol}$

Solution. Convert R.H.S. expression (eq. 4.7) into kJ/mol by dividing by a factor 10^{10} because 10^{10} erg = 1 kJ

$$-2550 = \frac{-6.02\times10^{23}\times2.519\times\left(4.8\times10^{-10}\right)^2\times Z^2}{2.35\times10^{-8}\times10^{10}}\left[1-\frac{1}{7}\right]$$

$$2550 = 1274\ Z^2$$

or $\quad Z^2 = 2$

Example 9. **Calculate lattice energy of sodium chloride (in kJ/mol) from the following data:**

$$A = 1.75,\ r_0 = 2.8\text{Å},\ n = 9$$
$$N = 6.02 \times 10^{23},\quad e = 4.8 \times 10^{10}\text{e.s.u.}$$

Solution.
$$U = -\frac{NAe^2Z^2}{r_0}\left(1-\frac{1}{n}\right)$$

$$= -\frac{6.02\times10^{23}\times1.75\times\left(4.8\times10^{-10}\right)^2\times1^2}{2.8\times10^{-8}}\left[1-\frac{1}{9}\right]$$

$$= -770.56 \times 10^{10}\ \text{ergs/mol}$$

$$U_0 = -770.56 \times 10^{10}\ \text{ergs/mol.}$$

Since 10^7 ergs $= 1$ J

or 10^{10} ergs $= 1$ kJ

∴ $V = -770.56$ kJ/mol.

4.3.3 Born-Haber cycle

The formation of an ionic solid MX from its elements may be formulated by two different paths interrelating different thermochemical quantities, as shown in Fig. 4.6.

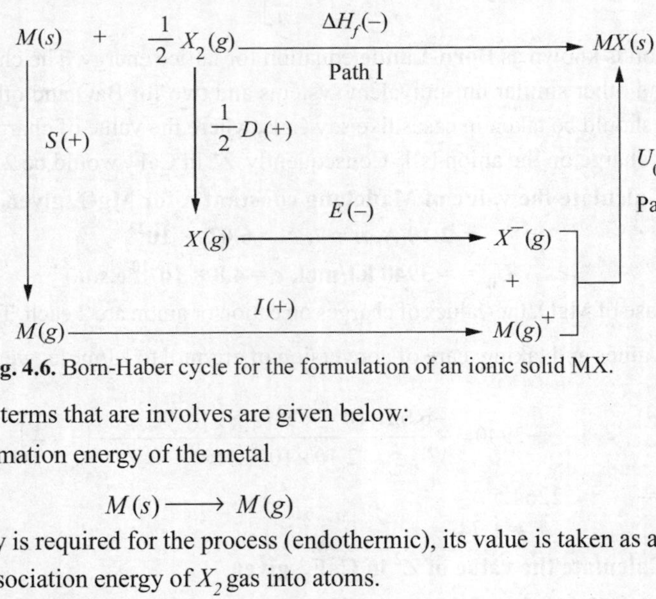

Fig. 4.6. Born-Haber cycle for the formulation of an ionic solid MX.

The various terms that are involves are given below:

(*i*) S = sublimation energy of the metal

$$M(s) \longrightarrow M(g)$$

Since energy is required for the process (endothermic), its value is taken as a positive quantity.

(*ii*) D = Dissociation energy of X_2 gas into atoms.

$$X_2(g) \longrightarrow 2X(g)$$

This process is endothermic and its value is taken as a positive quantity.

(*iii*) I = Ionization energy, *i.e.*, energy required for the removal of an electron from an isolated gaseous atom, M. The process is endothermic and is taken as a positive quantity.

$$M(g) \longrightarrow \text{M}^+(g) + e^-$$

(*iv*) E = Electron affinity, *i.e.*, amount of energy released when a neutral gaseous atom gains an electron. The process is exothermic and its value is taken as a negative quantity.

$$X(g) + e \longrightarrow X^-(g)$$

(*v*) U_0 = Lattice energy; it is amount of energy released when one mole of gaseous positive and negative ion condense to form an ionic solid.

$$M^+(g) + X^-(g) \longrightarrow MX(s)$$

The process is exothermic and its value is always negative.

(*vi*) ΔH_f = Heat of formation, it is defined as the amount of energy released when one mole of a substance is formed from its element.

$$M(s) + \frac{1}{2}X_2(g) \longrightarrow MX(s)$$

Energy is released in the process and its value is taken as a negative quantity. This is a single step reaction equivalent to sum of all other steps (*i*) to (*v*). There are two paths leading to the formation of MX(s) from M(*s*) and $\frac{1}{2}$ $X_2(g)$

Since the two processes are independent of each other, the two can be equated, *i.e.*, heat of formation of *MX* must be equal to the summation of all other terms. That is energy change via path 1 must be equal to the energy change via path 2.

$$\Delta H_f = S + \frac{1}{2}D + I + E + U_0$$

$$U_0 = \Delta H_f - S - \frac{1}{2}D - I - E \quad \text{... (4.8)}$$

Thus, the lattice energy of a solid can be calculated with the help of other thermochemical quantities which can be measured experimentally. The quantitative value of each term along with their sign are substituted in eq. 4.8.

***Example 10.* Calculate the heat of formation (ΔH_f) of KF from its elements from the following data by the use of Born-Haber cycle.**

Sublimation energy of potassium (S) = 87.8 kJ mol^{-1}
Dissociation energy of F_2 (D) = 158.9 kJ mol^{-1}
Ionization energy of K (g) (I) = 414.2 kJ mol^{-1}
Electron affinity for F (g) (E) = – 334.7 kJ mol^{-1}
Lattice energy of KF (U_0) = – 807.5 kJ mol^{-1}.

Solution. Substituting the values in eq. 2.8

$$\Delta H_f = S + \frac{1}{2}D + I + E + U_0$$

$$\Delta H_f = 87.8 + \frac{1}{2}(158.9) + 414.2 + (-334.7) + (-807.5)$$

$$= -560.8 \text{ kJ mol}^{-1}$$

This can be represented as:

$$K(s) + \frac{1}{2}F_2(g) \longrightarrow KF(s) + 560.8 \text{ kJ}$$

***Example 11.* Calculate the heat of formation (ΔH_f) of MgF_2 from its elements using the Born-Haber cycle.**

The thermochemical data are as follows:
Sublimation energy of magnesium, (S) = 146.4 kJ mol^{-1}
Dissociation of F_2, (D) = 158.9 kJ mol^{-1}
Ionization energy of Mg (g), I (Mg^{2+}) = 2184.0 kJ mol^{-1}

Electron affinity for F (g), E(F^-) = – 334.7 kJ mol^{-1}

Lattice energy of MgF_2 (U_0) = – 2922.5 kJ mol^{-1}

Solution. Substituting these values in the expression for MgF_2:

$$Mg\,(s) + F_2\,(g) \longrightarrow MgF_2 + 1102.6\text{ kJ}$$

$$\Delta H_f = S + D + I + 2E + U_0$$

Here *D* will be used instead of $\frac{1}{2}D$ because of the following process

$$F_2\,(g) + D \longrightarrow 2F\,(g)$$

Also 2*E* is taken in place of *E* because of the following change

$$2\,F\,(g) + 2e \longrightarrow 2F^-\,(g)$$

$$\Delta H_f = 146.4 + 158.9 + 2184.0 + 2\,(-334.7) - 2922.5$$

$$= -1102.6\text{ kJ mol}^{-1}$$

***Example 12.* Calculate lattice energy of NaCl crystal from the following data by use of Born-Haber cycle.**

Sublimation energy (S) = 108.7 kJ mol^{-1}

Dissociation energy for Cl_2 (D) = 225.9 kJ mol^{-1}

Ionization energy for Na (g) (I) = 489.5 kJ mol^{-1}

Electron affinity for Cl (g) E = – 351.4 kJ mol^{-1}

Heat of formation of NaCl (ΔH_f) = – 414.2 kJ mol^{-1}

Solution. Substituting these values in equation 4.8

$$U_0 = (-414.2) - (+108.7) - \frac{1}{2}\,(225.9) - (+489.5) - (-315.4)$$

$$= -414.2 - 108.7 - 112.9 - 489.5 + 351.4$$

$$= -773.9\text{ kJ mol}^{-1}$$

The lattice energy of NaCl is –773.9 KJ mol^{-1}.

4.3.4 Calculation of electron affinity with the help of lattice energy

Electron affinities can be measured directly only for the conversion of an isolated gaseous atom into its singly charged anion. That is,

$$O\,(g) + e \longrightarrow O^-\,(g) + E$$

However, in order to find the sum of the first and second electron affinities of the oxygen atom, that is, to know ΔH for the reaction

$$O\,(g) + 2e \longrightarrow O^{2-}\,(g)$$

we can apply Born-Haber cycle to any metal oxide whose lattice energy can be calculated. For example, the lattice energy of CaO, which has sodium chloride structure with Madelung constant A = 1.7475, interionic distance r_0= 240 pm, z = 2 and fairly accurate value of Born exponent has been calculated to be about –3389 kJ mol^{-1} using expression 4.7. The heat of formation of CaO is – 635.9 kJ mol^{-1}. The values of *S,* 1/2 D, and I (Ca^{2+}) are 153.9, 227.1 and 1735.1 kJ mol^{-1}; respectively. From this data we can calculate electron affinity.

Using Born-Haber cycle, $E(O^{2-})$ can be calculated as follows :

$$\Delta H_f = S + \frac{1}{2}D + I + E + U_0$$

Substituting these values, we get

$$-635.9 = 153.9 + 227.1 + 1735.1 + E(O^{2-}) - 3389$$

or

$$E(O^{2-}) = 3389 - 635.9 - 153.99 - 227.1 - 1735.1$$
$$= +637 \text{ kJ mol}^{-1}$$

In other words, ΔH for the reaction

$$O(g) + 2e \longrightarrow O^{2-}(g)$$

is + 637 kJ mol^{-1}.

The first electron affinity of oxygen is –118.0 kJ mol^{-1} but the addition of the second electron absorbs 774.8 kJ mol^{-1}. Their sum is + 636.8 kJ /mol. This shows that the formation of oxide ion is not favoured but the only reason why an oxide ion (O^{2-}) exists, is its high lattice energy in contact with metal ions.

4.3.5 Estimation of heat of formation of an unknown compound from lattice energy measurement

Heats of formation of an unknown compound can be estimated by making a reasonable guess about its lattice energy. As for example, in order to find out if the formation of argon chloride ArC1 is favourable or not, we make use of Born-Haber cycle. If argon chloride were to exist as a salt $Ar^+ Cl^-$, it would have NaCl structure and lattice energy close to the nearest alkali metal chloride, KC1.

or

$$U_0 = -722 \text{ kJ mol}^{-1}$$

Other thermochemical values are given in the cycle shown in Fig. 4.7.

$$\Delta H_f = \frac{1}{2}D + I + E + U_0$$
$$= 113.4 + 1526.3 - 352.8 - 693$$
$$= 593.9 \text{ kJ / mol}$$

This indicates that the formation of hypothetical solid ArC1 is highly endothermic, that is, strongly unfavourable.

It is clear that the very high ionization energy of argon (1526.3 kJ/mol) is responsible for making the reaction highly endothermic.

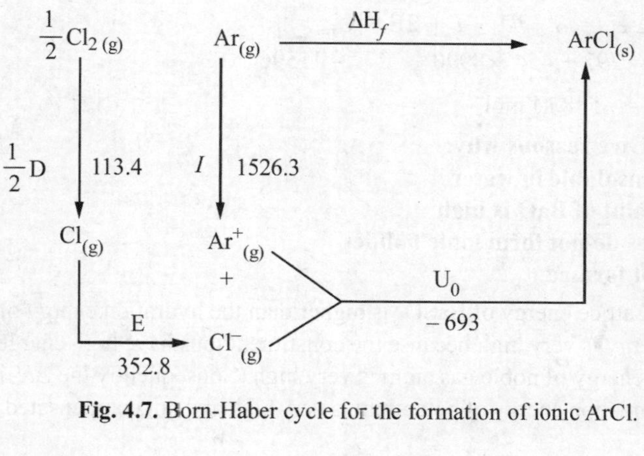

Fig. 4.7. Born-Haber cycle for the formation of ionic ArCl.

4.4 LATTICE ENERGY AND HYDRATION (SOLVATION) ENERGY

Solubility of a solid in a solvent depends upon two factors: lattice energy and solvation energy. The role of the two quantities is discussed as under:

(*i*) Greater is the lattice energy, greater the stability of an ionic solid. Because more energy will be needed to pull apart the +ve and –ve ions.

(*ii*) Salts with high lattice energies require a greater input of thermal energy to break down the crystal lattice. Consequently, such salts have high melting points.

(*iii*) The magnitude of lattice energy of a solid may give us an idea about the solubility of that substance in different solvents. Ionic solids in general are insoluble in non-polar solvents such as carbon tetrachloride, whereas these are soluble in polar solvents, like water. For a solid to dissolve in water, the strong forces of attraction between its ions (lattice energy) must be overcome. This energy can be overcome by ion-solvent interactions. The solvation of ions is referred to in terms of solvation energy, which is always negative, *i.e.*, energy is released in the process. The amount of solvation energy depends upon the nature of the solvent.

The solvation energy increases if the solvent has high dipole moment/or high polarizability. As a general rule, we may say that for a solid to dissolve in water its hydration energy should be greater than the lattice energy of that solid. Therefore if

Hydration energy > Lattice energy ... (salt dissolves in water)

Hydration energy < Lattice energy ... (salt does not dissolve in water)

Example 13. **Calculate the heat of formation of ionic solid neon monochloride $Ne^+ Cl^-$ and indicate whether this hypothetical solid is favoured or not. Given:**

$$U_0 = -840 \text{ kJ mol}^{-1} \quad D(Cl_2) = 226 \text{ kJ mol}^{-1}$$
$$I_0 = 2080 \text{ kJ mol}^{-1} \quad E(Cl^-) = -351 \text{ kJ mol}^{-1}$$

Solution. $\Delta H_f = \frac{1}{2}D + I + E + U_0$ Substitute the values,

$$\Delta H_f = 113 + 2080 - 351 - 840$$
$$= 1002 \text{ kJ mol}^{-1}$$

Formation of NaCl is endothermic and hence not favoured.

Example 14. **The experimental lattice energy of SnO_2 is –11596 kJ mol^{-1}. Calculate, ΔH_f of SnO_2. Given: *S* (Sn) = 292 kJ mol^{-1}, *D* (O_2) = 454 kJ mol^{-1}, *I* (Sn^{4+}) = 8990 kJ mol^{-1} and $E(O^{2-})$ = + 636 kJ mol^{-1}.**

Solution. $\Delta H_f = S + D + I + 2E + U_0$

$$= 292 + 454 + 8990 + 1272 - 11596$$

$$\Delta H_f = -588 \text{ kJ mol}^{-1}$$

Example 15. **Give reasons why:**

(*a*) $BaSO_4$ is insoluble in water

(*b*) Melting point of BaO is high

(*c*) Noble gases do not form ionic halides

(*d*) KCl_2 is not formed.

Solution. (*a*) Lattice energy of $BaSO_4$ is higher than the hydration energy of the ions.

(*b*) Lattice energy is very high because the constituent ions have high charges.

(*c*) Ionization energy of noble gas atom is very high. Consequently the ΔH_f is not favoured.

(*d*) Second ionization energy of potassium is very high and not compensated by lattice energy of KCl_2.

4.5 POLARIZATION, POLARIZING POWER AND POLARIZABILITY

Anions are larger in size than cations and, therefore, their electron clouds are less lightly held. A small highly charged cation shall, therefore, distort the electron cloud of the large anions in a manner that it increases the electron density between the nuclei. For example, the large iodide ion by itself is perfectly symmetrical. However when a small positively charged lithium ion comes close to the iodide ion, the electron cloud on the anion is pulled towards the positive lithium ion. The iodide ion is said to be polarized and the process is called polarization. The polarization effect produces covalence in LiI molecule because there is some electron density in between the nuclei as in the sharing of electrons in a covalent bond, The power of an ion to distort the other ion is known as its **polarizing power** and the tendency of the anion to distort is known as its **polarizability**. The polarizing power of a cation is proportional to its charge/radius ratio. This is illustrated in Fig. 4.8 when a large and soft anion comes under the influence of a small cation, the cation is able to polarize the anion (Fig. 4.8 (*b*)).

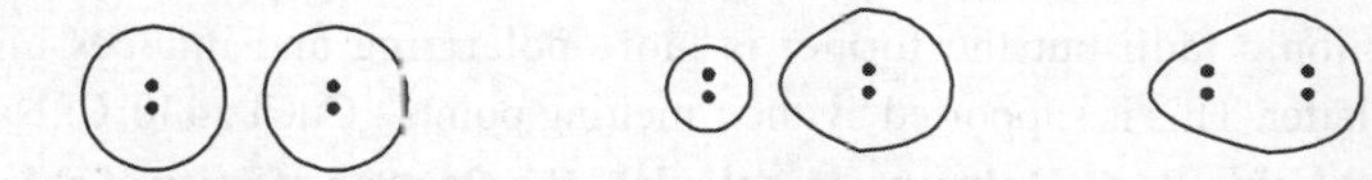

Fig. 4.8. Interaction of + ve and – ve ions.

This results in deformation of electron cloud on anion, called the polarization of anion. Due to small size of the cation, polarization of the cation is not much. Extensive polarization would be witnessed when cation penetrates the anionic electron cloud giving a covalent bond (Fig. 4.8). In compound consisting of large negative ions and small positive ions, the polarization leads to partial covalent character in ionic compounds. Examples of such ionic covalent compounds are $FeCl_3$, $AlCl_3$, LiBr etc.

4.5.1 Fajans's rules and covalent character in ionic compounds

Fajans explained the variation of nonpolar character in ionic compounds in terms of polarization effects. He gave qualitative ideas to offer simple approach to the problem of partial covalent character in ionic compounds. Fajans proposed the following rules to determine the extent of polarization of an anion by a cation, that is, the extent of covalent character induced in an ionic compound.

***(i)* Small, highly charged cations.** Small, highly charged cations have higher polarizing effects on anions than large, lowly charged cations. In other words, induced covalent character in ionic compounds increases with decrease in size of the cation or increase in charge on the cation. This is because very small cations or cations with higher charge have higher charge density which tends to distort or polarize the electron cloud around the anion. The greater the polarization, the higher the induced covalency. For example, Na^+ and Ca^{2+} have almost similar ionic radii but $CaCl_2$ (m.p. = 772°C) has higher covalent character than NaCl (m.p. = 800 °C) because of higher charge on Ca^{2+}. With identical charges on cations, $BeCl_2$ (m.p. = 405°C) is relatively more covalent than $CaCl_2$ (m.p. = 772°C) because of smaller size of Be^{2+} as compared to Ca^{2+}.

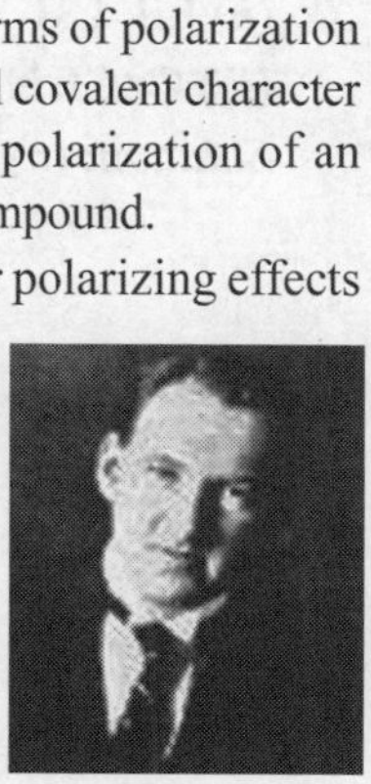

Kasimir Fajan

***(ii)* Large, highly charged anions.** Covalent character induced in ionic compound due to polarization of the anion increases with an increase in size and charge on the anion. The electron cloud on the large-sized anion is less strongly held by its nucleus and hence it is easily polarized by the approaching cation. In other words, the larger is the anion charge, the higher is its polarizability. For example, anions like Br^-, I^-, S^{2-}, Te^{2-}, As^{3-}, Sb^{3-} etc., are easily polarizable and have tendency to induce covalent character in ionic compounds. For a given cation, say Li^+, as the size of the anion, increases from F^- to I^-, the covalent character increases and consequently melting point decreases: LiF = 870 °C, LiCl = 613 °C, LiBr = 574 °C, LiI = 446 °C.

***(iii)* Cation configuration.** Cations with pseudo noble gas configuration (*i.e.*, 18 electron outer-shell) have higher polarizing power than cations with noble gas configurations. In other words, cations with $ns^2\ np^6$ configuration cause less distortion than those with $ns^2\ np^6\ nd^{10}$ configuration. This is because *d*-electrons shield the nucleus poorly with the consequence that such a cation will exert a higher nuclear charge on the interacting anion than a cation with 8 electron configuration. In other words, a cation with 18 electron configuration will lead to higher polarization of the anion. For example, Hg^{2+} (Xe) $4f^{14}\ 5d^{10}$ and Ca^{2+} (.... $3s^2\ 3p^6$) have similar ionic radii but the former is more polarizing and its compounds are more covalent than those of Ca^{2+}. For example, for a given anion (Cl^-), the melting point of $HgCl_2$ = 276°C and that of $CaCl_2$ = 772°C. Also Cu^+ (..... $3s^2\ 3p^6\ 3d^{10}$) and Na^+ (..... $2s^2\ 2p^6$) have similar ionic radii but the former is more polarizing and induces higher covalent character than the latter. This is supported by their melting points : CuCl : 430°C; NaCl= 800°C.

***Example 16* (*a*) Using Born-Haber cycle, calculate the electron affinity of chlorine from the following data :**

Bond enthalpy of Cl_2 = + 240.0 kJ mole^{-1}
Enthalpy of formation of NaCl(s) = – 440.0 kJ mole^{-1}
Enthalpy of sublimation of Na(s) = + 110.0 kJ mole^{-1}
Enthalpy of ionization of Na(g) = + 480.0 kJ mole^{-1}
Enthalpy of lattice formation of NaCl(s) = – 810.0 kJ mole^{-1}

(*b*) Describe Fajan's rule. How does it explain the relative covalent/ionic nature of the following pairs of compounds :

(*i*) $FeCl_2$, $FeCl_3$;
(*ii*) LiI, CsI
(*iii*) CuCI, NaCl.

***Solution*:** (*a*) Electron affinity of chlorine may be calculated as under :

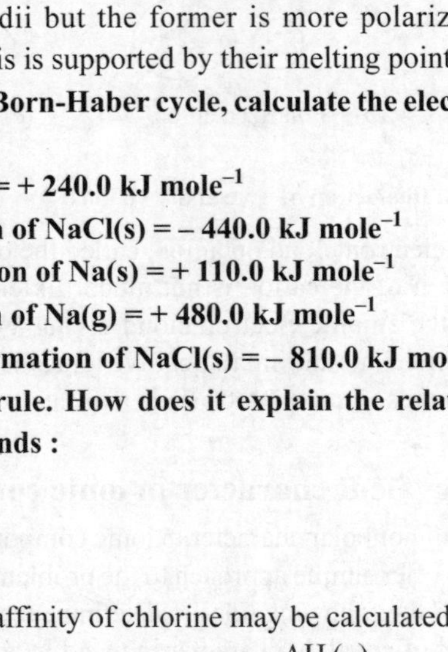

$$\Delta H_f = S + \tfrac{1}{2} D + I + E + U_0$$

Using data

$$-440 = 110 + (\tfrac{1}{2} \times 240) + 480 + E - 810$$

or $\quad E = -340$ kJ/mole.

(*b*) The Fajan's rule may be stated as

(*i*) Induced covalent character on ionic compound increases with decrease in size of the cation or increase in charge on the cation.

(*ii*) Covalent character induced in ionic compounds due to polarization of anion increases with an increase in size and charge of the anion.

(*iii*) Cations with pseudo noble gas configuration (*i.e.*, 18 e^- in outer shell) have higher polarizing power than cation with noble gas configuration.

Covalent/ionic nature of the pairs of compounds may be compared as under:

(*i*) $FeCl_2$, $FeCl_3$

As both Fe^{2+} and Fe^{3+} have almost same ionic radii but Fe^{3+} has more charge. So according to Fajan's rule $FeCl_3$ will be more covalent or less ionic than $FeCl_2$.

(*ii*) LiI, CsI

As both Li and Cs have some charge but radius of Cs is much greater than that of Li so according to Fajan' rule LiI will be more covalent.

(*iii*) CuCl, NaCl

As Cu and Na have some charge but Na is much smaller than Cu in radius so according to Fajan's rule NaCl will be more covalent.

4.5.2 Per cent ionic character of a polar covalent bond and (*i*) electronegativity difference, (*ii*) Dipole moment

1. Electronegativity Difference and Per cent Ionic Character. A pure covalent bond is formed between atoms *A* and *B* when the electronegativities of the two atoms are the same. If *A* and *B* have different electronegativities, the resulting bond would be a polar covalent bond because the bonding electron-pair is pulled towards the more electronegative atom causing some excess negative charge on it. As a result, some excess positive charge is developed on the less electronegative atom. If *A* is more electronegative than *B*, a polar covalent bond $\overset{-\delta}{A} \bullet \overset{+\delta}{B}$ would be formed between *A* and *B*. Larger the difference in the electronegativities *of A* and *B*, greater would be the magnitude of δ and the higher would be the polarity of the bond.

Some empirical equations have been proposed to calculate the per cent ionic character of a covalent bond from the electronegativities of the bonding atoms. Two of these equations are the *Pauling equation* and the *Hannay-Smith* equation.

The Pauling Equation. Pauling proposed the following equation for determining the per cent ionic character of a covalent bond:

$$\text{Per cent Ionic Character} = 1 - e^{-1/4(\chi_A - \chi_B)} \quad .....(1)$$

Where χ_A and χ_B are the electronegativities of *A* and *B*. He established the following empirical relationship between $\chi_A - \chi_B$ and the per cent ionic character :

$\chi_A - \chi_B$	Per cent ionic character
0.6	9
1.0	22
1.4	39
2.0	63
2.4	76
3.0	89

The Hannay-Smith Equation. Hannay and Smith proposed the following equation for the purpose :

$$\text{Per cent Ionic Character} = 16(\chi_A - \chi_B) + 3.5\,(\chi_A - \chi_B)^2 \quad .....(2)$$

2. Dipole Moment and Per cent Ionic Character. Consider the bond H – X where X is a halogen atom. If the bond is purely covalent, its dipole moment would be zero. If the bond is ionic, *viz.*, H^+X^-, it implies that complete transfer of an electron has taken place from H to X and the bond would have a high value of dipole moment which is obtained by the product of the electronic charge and the internuclear distance between H and X. For a polar covalent bond which is intermediate between a pure covalent and a pure ionic bond, the dipole moment would be given by the product of the actual charge (δ) caused by polarisation and the internuclear distance. The magnitude of δ will depend upon the extent of the ionic character developed in the bond.

Thus,

$$\text{Per cent Ionic Character} = \frac{\text{Actual dipole moment of the bond}}{\text{Dipole moment of pure ionic bond}} \times 100$$

For example, in **HF**, the internuclear distance is 0.92 Å and the dipole moment for the ionic structure $\mathbf{H^+F^-}$ when an electron is completely transferred from hydrogen atom to fluorine atom is given by 4.8×10^{-10} esu $\times\ 0.92 \times 10^{-8}$ cm, *i.e.,* 4.42 D. The actual dipole moment of HF is found to be 1.98 D. Per cent ionic character is calculated as under:

$$\text{Per cent Ionic Character} = \frac{1.98\,\text{D} \times 100}{4.42\,\text{D}} = 44.8$$

***Example 17.* Silver halides are least soluble in water, though their lattice energy is almost the same as that of highly soluble alkali metal halides. Explain.**

OR

Ag I is covalent while NaI is covalent. Explain.

Solution. Ag^+ can cause greater polarization of the halide because of its electronic configuration. There are *d*-electrons which cannot screen the nucleus effectively and thus pull the electrons of halide more than expected. This brings about covalency leading to the substance being insoluble in water.

***Example 18.* Explain why NaCl is soluble in water but $BaSO_4$ is not.**

Solution. In the case of $BaSO_4$, lattice energy is much more than hydration energy. Hence it does not dissolve.

With NaCl, hydration energy exceeds the lattice energy. So it is soluble.

***Example 19.* Give the trends of thermostability of alkaline earth metal carbonates.**

Solution. The thermal stability increases in the order

$$BeCO_3 < MgCO_3 < CaCO_3 < SrCO_3 < BaCO_3$$

This can be explained on the basis of decreasing polarizing power of the cation as we move from L.H.S. to R.H.S. making the substance more and more ionic or thermally stable.

***Example 20.* Explain the sequence of solubility of lithium halide in ethyl alcohol.**

Solution. The halides to be considered are:

LiF

LiCl

LiBr

LiI

The cation is same in all the cases. Hence the nature of the anion will decide solubility. Larger the anion, greater the polarizability hence greater the covalent character. And greater the covalent character greater the solubility in ethyl alcohol. On this basis solubility increases from LiF to LiI.

***Example 21.* State, giving reasons, which cation will have greater polarizing power in the following:**

(*i*) Na^+ or Mg^{2+} (*ii*) Cu^{2+} or Ca^{2+} (*iii*) Pb^{2+} or Pb^{4+}

Solution. (*i*) Mg^{2+} has greater polarizing power because of greater charge.

(*ii*) Ca^{2+} has greater polarizing power because of smaller size.

(*iii*) Pb^{4+} has greater polarizing power because of greater charge.

Example 22. **$AlCl_3$ anhydrous is covalent but $AlCl_3 . 6H_2O$ is ionic in nature. How would you account for this behaviour?**

Solution. $AlCl_3$ is covalent in nature because of small and highly charged Al^{3+} ion. This polarizes the chloride ion causing covalency.

In the hydrated sample, the Al^{3+} ion get surrounding by water molecules. A lot of energy called hydration energy is released and thus Al^{3+} lose their polarizing power and remain in ionic state.

Example 23. **Give reasons why $ZnCl_2$ is soluble in organic solvents while $MgCl_2$ is insoluble.**

Solution. Zn^{2+} ions have greater polarizing power, because of the presence of low shielding *d*-orbitals, as compared to Mg^{2+} ions. Thus Zn^{2+} polarizes the chloride ion and brings about covalent character and therefore $ZnCl_2$ is soluble in organic solvents.

Example 24. **Highly charged cations are rare. Explain.**

Solution. Highly charged cations strongly polarize the anion and bring about covalent character in the molecule. Thus the charge on the cation is neutralized and a covalent bond involving sharing of electrons or overlapping of orbitals is formed.

Example 25. **Explain the following:**

(*i*) $SnCl_4$ is more covalent than $SnCl_2$

(*ii*) NaCl is more ionic than CuCl.

Solution. (*i*) $SnCl_4$ is more covalent because Sn^{4+} has greater polarizing power than Sn^{2+}.

(*ii*) NaCl is more ionic than CuCl because Cu^+ has greater polarizing power than Na^+. This is because there are *d*-electrons which poorly shield the nucleus.

The cation thus can polarize the anion more powerfully.

Example 26. **The inter-nuclear distance in HI is 1.62 Å. Assuming complete transfer of electron from H to I, calculate % ionic character and fractional charge on each ion. If the measured dipole moment is 0.38 D, also calculate the % ionic character of HI.**

Given *e* = (electronic charge) = 4.8×10^{-10} e.s.u.

Solution. Assuming complete transfer of electron from H to I, the charge on H and I would be $+e$ and $-e$ respectively.

The dipole moment of HI in such case,

$$= q \times r$$
$$= e \times r$$
$$= (4.8 \times 10^{-10} \text{ e.s.u.})(1.62 \times 10^{-8} \text{ cm})$$
$$= 7.776 \text{ D}$$

Observed dipole moment = 0.38 D

$$\% \text{ ionic character} = \frac{0.38}{7.776} \times 100 = 4.88.$$

4.6 CLOSE PACKING OF PARTICLES (ATOMS OR IONS) IN CRYSTALS

In most of the solids, the constituent atoms are closely packed so as to utilize the maximum available space. For simplicity, we shall consider the constituents as spheres. These spheres are in contact with one another in various layers. Naturally, there will be some empty space in between these spheres. This empty space is called the *void*. Depending upon the size of the spheres and their arrangement, the voids will have different sizes. Then within the same solid, there are different types of voids like *tetrahedral* void or *octahedral* void.

Packing of particles in a crystal

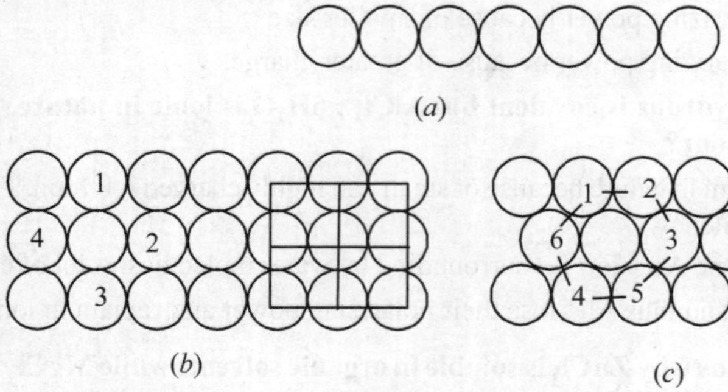

Fig. 4.9. Packing of spheres for efficient use of space. (*a*) Edge formation (*b*) Square close packing (*c*) Hexagonal close packing.

Considering the particles as spheres of equal size, we can visualize the formation of a solid as shown in Fig. 4.9.

One Dimensional Packing

We can arrange the spheres in a row [see Fig. 4.9 (*a*)], such that they touch one another. This is to utilize the space economically and to account for the high density of the solids.

Two Dimensional Packing

Then we place another row of spheres by the side of the first row. This can be done in two ways. One way is depicted in Fig. 4.9 (*b*). Second method is shown in Fig. 4.9 (*c*). The first type [Fig. 4.9 (*b*)] is called *Square Close Packing* and the second type [Fig. 4.9 (*c*)] is called *Hexagonal Close Packing.*

In square close packing, a sphere is surrounded by four spheres in the plane whereas in hexagonal close packing, a sphere is surrounded by six spheres in the plane. Thus the empty space in square close packing is more as compared to that in hexagonal close packing. Or in other words, Hexagonal Close packing is more dense than the Square Close packing.

Three Dimensional Packing

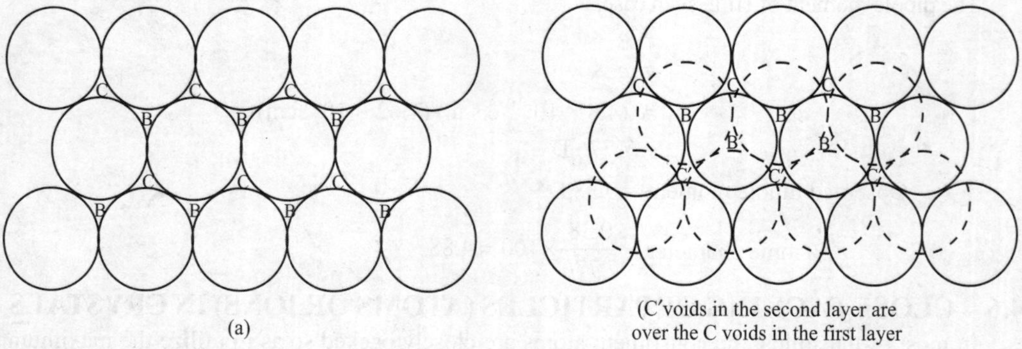

Fig. 4.10. Layers in close packing of spheres (*a*) Hexagonal close packed layer (*b*) Formation of the second layer.

Now let us form a second layer of spheres on the first layer in hexagonal close packing (see Fig. 4.10).

Spheres in the first layer are shown as full circles while those in the second layer are shown as dotted circles. There are voids in between them. These voids are *C* voids and *B* voids as indicated in Fig. 4.10 (*a*). We could place the spheres of the second layer on the first layer in such a way that every sphere in the second layer lies vertically over the sphere in the first layer. But this won't be an efficient packing as this would leave lots of gaps in between.

So let us place the spheres in the second layer over the voids *B* in the first layer as shown in Fig. 4.10 (*b*). It will then not be possible to place spheres over the voids *C* because of space limitation. Now in the second layer of spheres there are *B′* voids and there are second *C′* voids, which are in fact over the *C* voids of the first layers.

Let us now place a third layer of spheres over the second layer. There are two possibilities. Either we place spheres on the *B′* voids or on the *C′* voids. If we place the spheres on the *B′* voids in the third layer, these spheres in fact would be seated over the spheres of the first layer. This arrangement would be *ABABAB* ...arrangement. This type of packing is called *hexagonal close packing*.

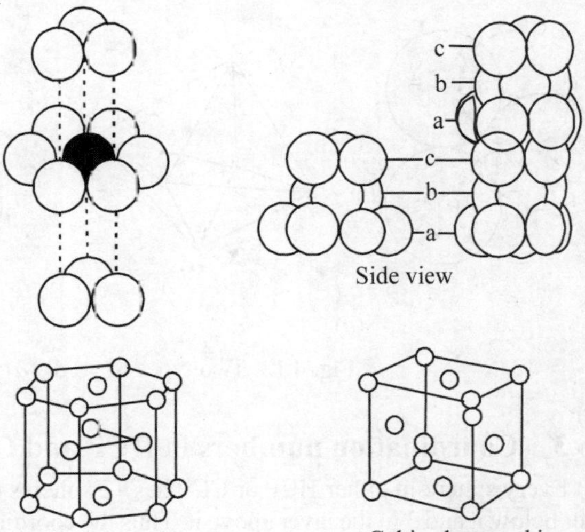

Fig 4.11. Three dimensional view of hexagonal and cubic close packed lattices.

If we place the spheres in the third layer over the *C′* voids then the spheres in the fourth layer will be found seated over the spheres in the first layer. This will give rise to *ABCABC*... arrangement. This type of packing is called *cubic close packing.* This is similar to face centred cubic system. Three dimensional views of hexagonal and cubic close packed lattices are shown in Fig. 4.11.

4.6.1 Body-centred cubic packing (BCC structure)

This is deviation from close packing. If the spheres of the first layer are loosened up, we obtain body-centred Cubic Structure. Each sphere in the second layer is in contact with four spheres of the first layer (see Fig. 4.12). The spheres in the second layer have the same packing as those in the first layer. The spheres in the third layer, having the same packing, will be lying exactly over the corresponding spheres in the first layer. Thus every sphere will be in contact with 8 spheres, four of the layer above and four of the layer below it. The coordination number is thus 8.

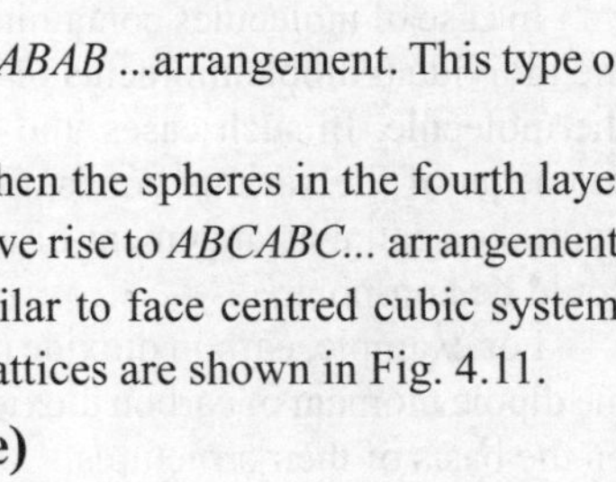

Fig. 4.12. Body centred cubic packing.

Alkali metals like Li, Na, K etc. crystallise in BCC structure. The structure is less closely packed than the CCP and HCP structures and the empty space is about 32%.

4.6.2 Tetrahedral and octahedral voids

A void or empty space in a lattice which is surrounded by four spheres is called a tetrahedral void. For examples, the voids *B* in Fig.4.10 are tetrahedral voids because they are surrounded by four spheres three in the first layer and the fourth in the second layer.

A void which is surrounded by six spheres is called the octahedral void. The voids C' in Fig. 2.10 are octahedral voids because they are surrounded by six spheres, three in the first layer and three in the second layer. These voids are shown in Fig. 4.13.

(*a*) (*b*)

Fig. 4.13. Two types of voids (*a*) octahedral (*b*) tetrahedral.

4.6.3 Coordination numbers in HCP and CCP

Every sphere in either HCP or CCP has 12 spheres directly touching it, 6 in its own layer, 3 in layer below it and 3 in the layer above it. Thus the coordination number of a sphere in HCP or CCP is 12. In other types of crystals, coordination number of 4, 6, 8 is quite common.

4.7 STRUCTURE OF MOLECULES FROM DIPOLE MOMENT DATA

In case of molecules containing more than two atoms, the dipole moment not only depends upon the individual dipole moments of the bonds but also on the spatial arrangement of various bonds in the molecule. In such cases, the dipole moment of the molecule is the vector sum of the dipole moments of the various bonds. Obviously, the polar or non-polar nature of molecule will depend upon the spatial arrangement of atoms in the molecule. The bonds may be polar yet the molecule could be non-polar.

For example, carbon dioxide (CO_2) and water (H_2O) molecules are both triatomic molecules but the dipole moment of carbon dioxide is zero while that of water is 1.84 D. This can be easily explained on the basis of their structures.

In *carbon dioxide* molecule, both the carbon-oxygen bonds are polar. Since the molecule is linear, the dipole moment of C = O bond on one side of the molecule is cancelled by that on the other side of the molecule giving net zero dipole moment. On the other hand, *water* molecule is not linear, it has a *bent structure* as shown. Therefore, the dipole moment of water molecule is not zero. It is the resultant of the dipole moments of the two O — H bonds inclined at an angle of 104.5°. It has been found to be 1.84 D.

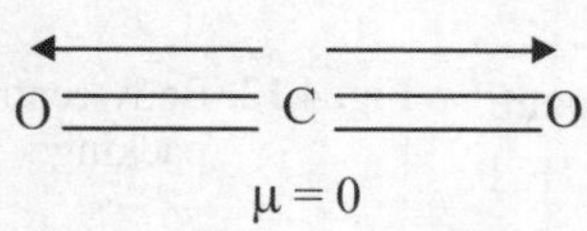

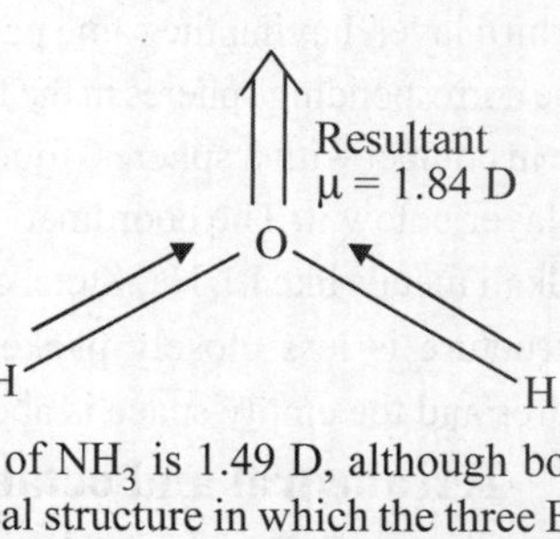

Similarly, the dipole moment of BH_3 is zero while that of NH_3 is 1.49 D, although both the molecules are tetra-atomic. This is because BH_3 has symmetrical structure in which the three B — H bonds are oriented at an angle of 120° to one another. The three bonds lie in one plane and the dipole moments of these bonds cancel one another giving net dipole moment equal to zero. On the other hand, the ammonia molecule has *pyramidal structure.* The dipole moments of three N—H bonds do not cancel one another (as they are in different planes). The molecule of ammonia shows a net value of dipole moment.

4.8 METALLIC BOND

Metals are characterised by high electrical and thermal conductivity, bright lustre, malleability, ductility and high tenstile strength. It has been observed that metals, generally, have low ionisation energies because the valency electrons can be taken out relatively easily.

The following three models have been proposed to account for the nature of bonding in metals :

1. *The Free Electron Model. The 'Electron Sea' Model.*
2. *The Valence Bond Model.*
3. *The Band Model : Molecular Orbital Approach.*

4.8.1 Free electron model or electron sea model of metallic bond

Ans. According to this model, a metal behaves as if *it is an assemblage of positive ions (kernels) mmersed in a 'sea' of mobile electrons* (Fig. 4.14). Thus *each electron belongs to a number of positive ions and each positive ion belongs to a number of electrons.*

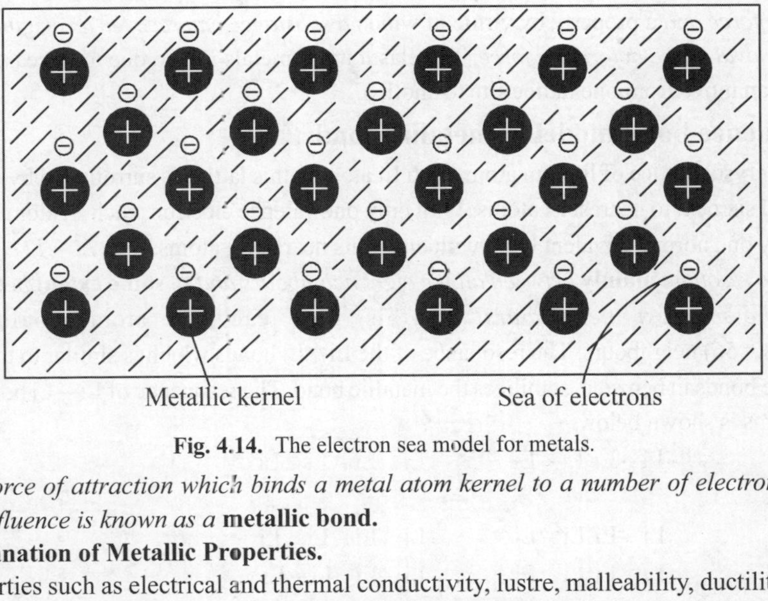

Fig. 4.14. The electron sea model for metals.

The force of attraction which binds a metal atom kernel to a number of electrons within its sphere of influence is known as a **metallic bond.**

Explanation of Metallic Properties.

Properties such as electrical and thermal conductivity, lustre, malleability, ductility, elasticity, etc., can be explained satisfactorily on the basis of free electron model.

1. The high electrical conductivity of metals, for example, is due to the presence of the mobile valency electrons. They move readily in an electric field and thus conduct electricity throughout the metal from one end to the other.

2. The high thermal conductivity is again due to the presence of these mobile electrons. If one part of a metal is heated, the electrons in that part acquire a large amount of kinetic energy. These electrons move rapidly through the crystal and pass on heat to other parts of the metal.

3. The metallic lustre can be explained as due to the presence of the highly mobile electrons. As a beam of light comprising of electromagnetic waves falls on the surface of a metal, *the electric field associated with light waves sets the electrons present on the surface of the metal into to and fro oscillations.* A moving charge always emits electromagnetic energy. Hence, oscillating electrons emit electromagnetic energy in the form of light. Thus, when light falls on a metal surface, it appears as if light is being reflected.

4. Malleability, ductility and softness. The metallic bond holding the positive ions (say, M^+) and the valency electrons is *non-directional.* In other words, the force of attraction between the M^+ ions and the valency electrons is uniform in all directions. Also the bonds holding the crystal lattice in metals are not rigid. The result is that M^+ ions can be easily moved from one lattice site to another.

The nearest neighbours can thus be changed easily and new metal bonds can be formed readily. This explains why metals are malleable, *i.e.,* they can be flattened out into thin sheets when hammered.

The ease with which the metal ions can be moved from one lattice site to another is also responsible for the fact that the metals are *ductile, i.e.,* they can be drawn into wires by very little expenditure of energy.

5. Metals have high tensile strength, *i.e., they can resist stretching without breaking.* This is due to the existence of strong electrostatic attraction between the positively charged metal ions and the 'sea' of negative electrons surrounding them.

6. Metals possess elasticity. *Elasticity is a property by virtue of which a substance can resist a deforming force and* a property by virtue of which *a substance can recover its original form soon after the removal of the deforming force.* The elasticity of metals is also due to the ease with which metal ions can move from one lattice site to another.

4.8.2 Valence bond model of metallic bond

In the crystal lattice of lithium atom, each Li atom in this lattice is surrounded by eight nearest Li atoms and six next to nearest Li atoms. With only one valence electron, each Li atom is capable of forming only one normal covalent bond with one of its nearest Li atoms. *This Li— Li covalent bond is assumed to resonate* **mainly** *between all the eight neighbours and* **to some extent** *between the six next to the nearest neighbours.* In this way, each Li atom would bind its 56 ($8 + 6 \times 8 = 56$) neighbours. The resonance of the Li – Li bonds which is similar to the resonance of the double bonds in benzene, stabilizes the metallic bond. The resonance of Li—Li bonds between a few Li atoms is shown below.

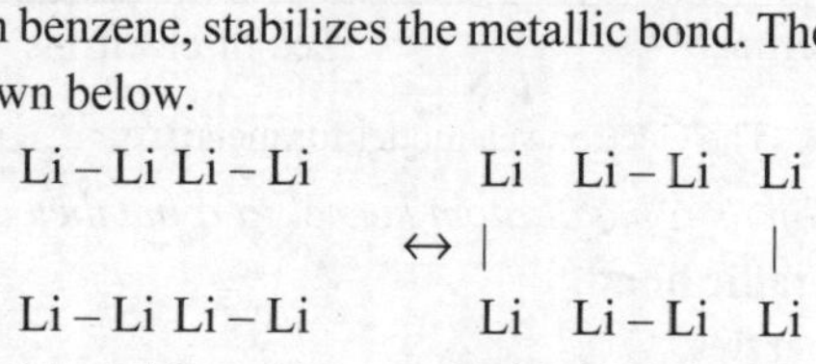

Greater stabilisation results from resonance of the following type :

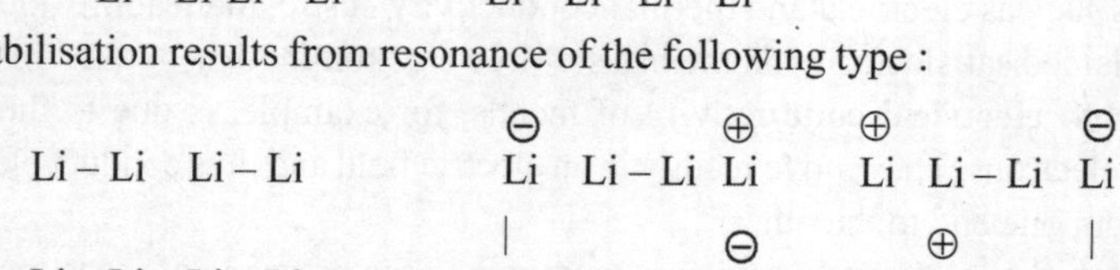

Although the valence bond model does meet some success in explaining some of the characteristics of metals, yet it is only qualitative in nature.

4.8.3 Band model (Molecular Orbital Approach) of metallic bonding

When two atomic orbitals combine to form two molecular orbitals, the energy of one of the resulting molecular orbitals is higher and that of the other is lower than the energy of either of the combining atomic orbitals. The energies of the two molecular orbitals are represented in the form of two energy levels whose magnitudes vary with the internuclear distance between the two atoms, as shown in Fig. 4.15.

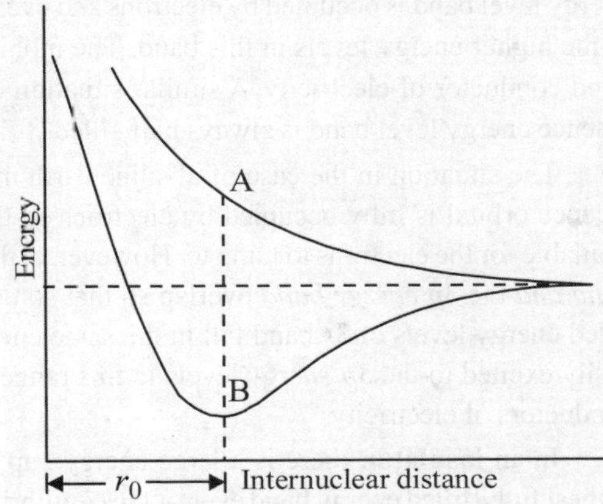

Fig. 4.15. Variation of the two molecular orbital energy levels with internuclear distance in the case of H_2 molecule. r_0 is the equilibrium internuclear distance.

The energy separation between the two molecular orbital energy levels at equilibrium internuclear distance r_0 is presented by *AB*.

If there are *n* atomic orbitals, one on each atom, in a metallic crystal, they may combine to give *n* molecular orbitals corresponding to *n* energy levels. It is essential, however, that the atomic orbitals have the same symmetry and are associated similar amount of energy. Thus, larger the number of atomic orbitals satisfying the above conditions, larger would be the number of molecular orbitals each of which can be represented by a separate energy level. The energy separation between any two adjacent energy levels would decrease with increase in the number of overlapping orbitals. Consequently, if we have an exceedingly large number of overlapping orbitals, the energy separation between any two adjacent energy levels would be exceedingly small so that we get an almost *continuous energy band,* as shown in Fig. 4.16

A similar situation exists in case of metals. A metal lattice has an extremely large number of atoms. The atomic orbitals on these atoms with same symmetry and same energy overlap resulting in an almost continuous energy band of extremely large number of energy levels of molecular orbitals. Each energy level in this band is capable of holding two electrons. The overlaping of *3s* and *3p* orbitals (valence orbitals) in the crystal lattice of **sodium metal** is shown in Fig. 4.17.

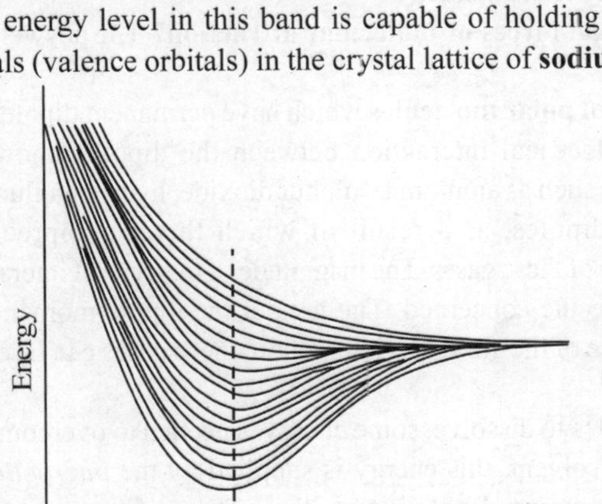

Fig. 4.16. An almost continuous energy band of energy levels of molecular orbitals obtained by overlapping of an exceedingly large number of atomic orbitals.

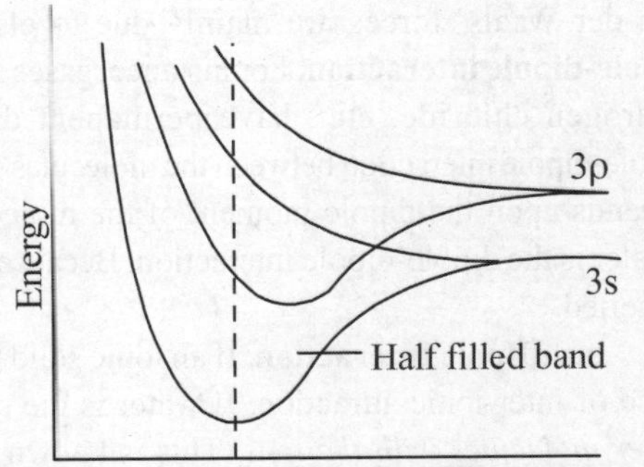

Fig. 4.17. The valence band of sodium metal. The dotted line indicates equilibrium internuclear distance.

Since sodium atom has only a single electron in the *3s* orbital, therefore, only half of the 3*s* energy level band is occupied by electrons and even on slightest excitation, these electrons can move to the higher energy levels in this band. The mobility of these electrons makes the sodium metal a good conductor of electricity. A similar situation occurs in all the other alkali metals in which the valence energy level band is always half-filled.

The situation in the case of alkaline earth metals is, however, different. Here, the outermost valence orbital is fully occupied by electrons so that no vacant energy level of the energy band is available for the electrons to jump to. However, at the equilibrium internuclear distance, the 3*s energy band and* the 3*p energy band* overlap so that some of the vacant energy levels of 3*p band and* the filled energy levels of 3*s* band fall in the same energy range. Consequently, the 3*s* electrons can be easily excited to the 3*p energy* levels in this range. This makes the alkaline earth metals also good conductors of electricity.

In an **insulator,** there is a large energy gap between the lowest vacant energy band and the highest fully-filled energy band even at the equilibrium internuclear distance. As a result, the electrons cannot be easily excited. This makes the material non-conducting.

In **a semi-conductor,** there is a lesser energy gap between the two energy bands. Although the electrons are not very easily excited from the filled to the vacant energy bands yet these can be excited if an appropriate amount of energy is made available.

4.9 VAN DER WAALS' FORCES

Forces of attraction exist between *neutral molecules* as well. The magnitude of these forces is maximum in case of solids and decreases on passing from solid to liquid and from liquid to gaseous states. The forces existing between the molecules of a substance are known as **intermolecular forces.** These are also designated as **van der Waals' forces.** The attractive forces which keep the molecules in position in a molecular crystal are van der Waals' forces.

The question arises, how can there be electrical attraction between neutral molecules lying close to one another ? The existence of electrical repulsion between such molecules, on the other hand, can be more readily understood as these may arise from the interaction of the electron clouds of the molecules when they are in close proximity to each other.

Thus, what could be the cause of **physical forces or molecular attraction?** The answer is as follows:

Dipole-Dipole Interaction. In the case of **polar molecules** which have permanent dipoles, the van der Waals' forces are mainly due to electrical interaction between the dipoles known as **dipole-dipole interaction.** For instance, gases such as ammonia, sulphur dioxide, hydrogen fluoride, hydrogen chloride, etc., have permanent dipoles, as a result of which there is appreciable dipole-dipole interaction between the molecules of these gases. The magnitude of this type of interaction depends upon the dipole moment of the molecule concerned. The greater the dipole moment, the greater is the dipole-dipole interaction. Because of the attractive interaction, these gases can be easily liquefied.

Ion-Dipole Interaction. If an ionic solid is to dissolve, some energy is needed to overcome the force of inter-ionic attraction. If water is the solvent, this energy is supplied by the *interaction of water molecules with the ions.* This is known as **ion-dipole interaction.** The positive pole of the water molecules reacts with the negative ion while the negative pole reacts with the positive ion.The ion-dipole interaction results in the release of a considerable amount of energy which becomes available for the separation of the oppositely charged ions. This results in the dissolution of the solute.

London Forces. But van der Waals' forces exist even in non-polar molecules such as F_2, O_2 and N_2 and even in non-polar mono-atomic molecules such as He, Ne, Ar, etc. This attraction is evident from the condensation of these gases into liquids at sufficiently high pressures and low temperatures. This is explained by London forces.

In 1930, F. London provided an explanation for the existence of forces of attraction between *non-polar* molecules, based on quantum mechanics. According to this view, *electrons of a neutral molecule keep on oscillating with respect to the nuclei of the atoms. As a result of this, at a given instant, positive charge may be concentrated in one region and negative charge in another region of the same molecule. Thus, a non-polar molecule may become momentarily self-polarised.* This polarised molecule may induce a dipole moment in a neighbouring molecule, as shown in Fig. 4.18.

The electrostatic forces of attraction between induced dipoles and the original dipoles (due to electron oscillation) are known as **London forces**. These forces are also called **dispersive forces** because the well known phenomenon of dispersion of light is also connected with these dipoles. For a pair of adjacent molecules, London forces vary inversely as the *seventh power* of the distance between them, *i.e.*, F ∝ $1/r^7$. The magnitude of these forces, evidently, would be very small.

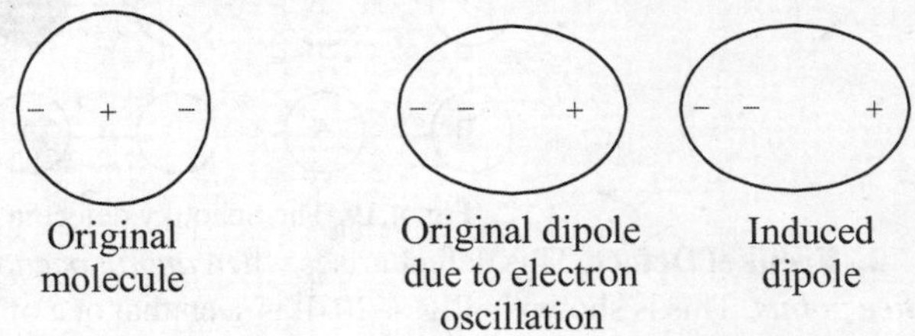

Fig. 4.18. Origin of cohesive forces in non-polar molecules.

The van der Waals' attraction in *non-polar molecules* is thus *exclusively* due to London forces. In the case of polar molecules having permanent dipoles, however, other electrostatic forces due to dipole-dipole interaction also contribute towards the van der Waals' attraction.

Factors determining van der Waals' forces

van der Waals' forces depend largely on the following factors:

1. The Number of Electrons Present in a Molecule. van der Waals' forces increase with increase in the number of electrons in the molecule. Thus, boiling points (T_b) of the halogens increase as we move from fluorine to chlorine, bromine and iodine. Similarly, boiling points of noble gases increase as we move from helium to radon, as shown below:

Noble gas	He	Ne	Ar	Kr	Xe	Rn
Boiling pt. (K)	4.1	27	67.2	121	165	211.2

2. Molecular Size. The larger the molecular size (or roughly the larger the molar mass), the stronger the van der Waals' forces. Thus, the boiling point of ethane, C_2H_6 (185 K) is higher than that of methane, CH_4 (112 K). In compounds having similar structures, viz. CH_4, SiH_4, GeH_4, the boiling points increase with increase in the number of electrons and size of the nolecule, as shown below:

Substance	CH_4	SiH_4	GeH_4
No. of electrons	10	18	36
Boiling point (K)	112	161	183

4.10 CRYSTAL DEFECTS

Crystals in which the number of positive and negative ions are exactly in the ratios indicated by their chemical formulae are called stoichiometric crystals. For simplicity, we can consider the crystals of the type *AB* having equal number of A^+ and B^- ions.

4.10.1 Defects in stoichiometric crystals

1. Schottky Defect. This defect arises if *some of the lattice points are unoccupied.* The points which are unoccupied are called *lattice vacancies* or *'holes'*. The existence of two holes, one due to a missing positive ion and the other due to a missing negative ion in a crystal lattice, is illustrated in Fig. 4.19. It may be noted that the crystal, as a whole, remains neutral because the number of missing positive and negative ions is the same.

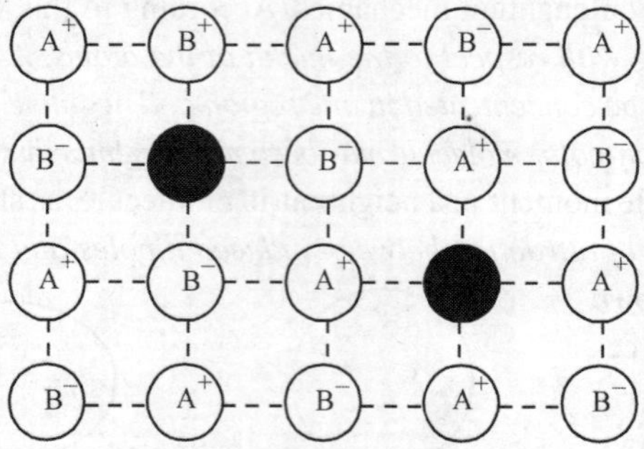

Fig. 4.19. The Schottky defect in crystals.

2. Frenkel Defect. This defect arises when *an ion occupies an interstitial position between the lattice points.* This is shown in Fig. 4.20. It is seen that one of the positive ions occupies a position in the interstitial space rather than at its own appropriate site in the lattice. A 'hole' is thus created in the lattice as shown. It may be noted again that the crystal remains neutral since the number of positive ions is the same as that of the negative ions. Positive ions being smaller are more likely to occupy interstitial positions than the negative *ions*.

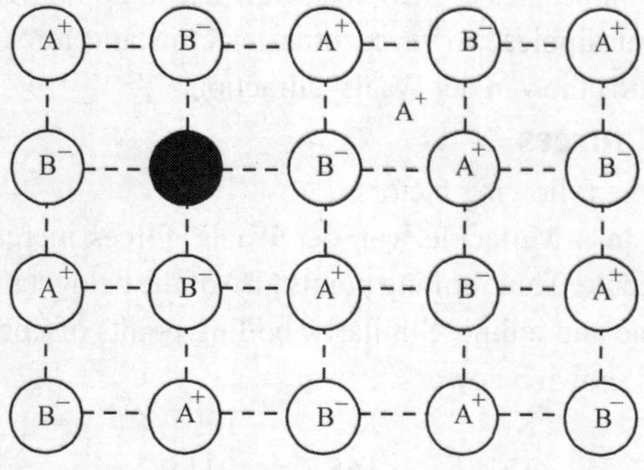

Fig. 4.20. The Frenkel defect in crystals.

Yakov Llich Frenkel

In silver bromide, AgBr, some of the Ag^+ ions are generally missing from their regular positions and occupy positions in between the other ions in the lattice. The presence of Ag^+ ions in the interstitial spaces of the silver bromide crystal is responsible for the formation of a photographic image on exposure of silver bromide crystals *(i.e.* photographic plate) to light.

Lattice vacancies ('holes') occur in almost all types of ionic crystals. However, Schottky defect appears more often than Frenkel defect. The reason is that the *energy needed to form a Schottky defect is much less than that needed to form a Frenkel defect.*

4.10.2 Defects in non-stoichrometric crystals

These crystal do not obey strictly the law of constant proportions. The numbers of positive and negative ions present in such compounds are different from those expected from their ideal chemical formulae. In other words, there is excess of either positive charge or negative charge in the crystal. However, the crystal as a whole is neutral. If positive charge is in excess, it is balanced by the presence of *extra electrons.* If negative charge is in excess, it is balanced by the presence of *extra positive ions.* The crystal structure, therefore, becomes irregular and this implies the presence of defects. Defects in such crystals are called **non-stoichiometric defects.**

Metal Excess Defects. In these defects, positive ions are in excess. These defects may manifest in two ways.

1. *A negative ion may be missing from its lattice site leaving a 'hole' which is occupied by an extra electron to maintain the elecrrical balance,* as shown in Fig. 4.21. There is, evidently, *an excess of positive (metal) ions,* although the crystal, as a whole, is neutral.

2. *An extra positive ion occupies an interstitial position in the lattice and to maintain electrical neutrality, an electron is also presence in the interstitial space,* as shown in Fig. 4.22.

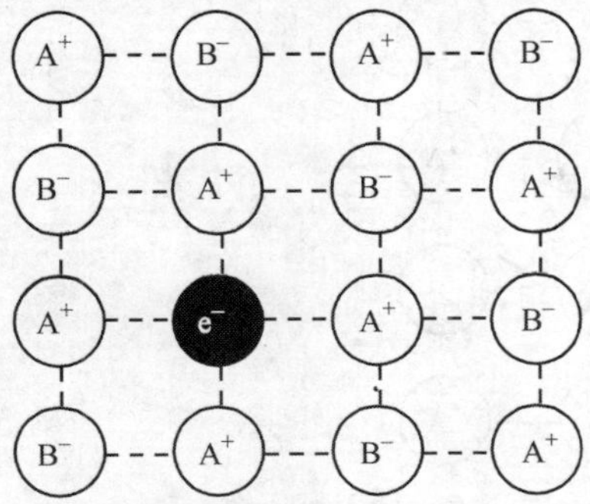

Fig. 4.21. Metal excess defect caused by a missing negative ion.

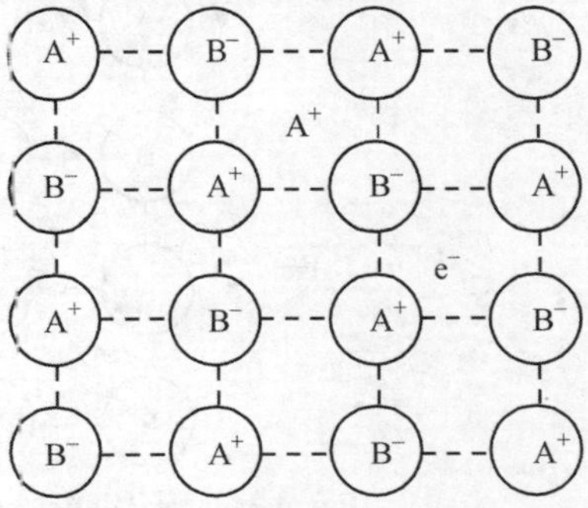

Fig. 4.22. Metal excess defect due to extra positive ion.

Consequences of Metal Excess Defects. 1. The crystals associated with metal excess defects of the first kind or second kind, *contain free electrons* (Fig 4.21 and 4.22). Since these electrons can migrate in an electric field, *such crystals can conduct electricity.* However, since the number of defects and, therefore, the number of electrons is very small, these crystals can conduct only small amounts of current, much smaller than the current conducted by metals, fused salts or salts dissolved in water. These materials are, therefore, termed as **semiconductors**. Since current, is carried by electrons in *normal way,* these substances are called ***n*-type semiconductors.**

2. The crystals with metal excess defects are generally coloured. This, again, is due to the presence of free electrons. These electrons get excited easily to higher energy levels by absorption of certain wavelengths from the visible white light and, therefore, the compounds appear coloured.

Metal Deficiency Defects. There are *two possible ways* in which metal deficiency defects can occur.

1. *One of the positive ions may be missing from its lattice site. The extra negative charge may be balanced by some nearby metal ion acquiring two charges instead of one.* This is represented in Fig. 4.23. *The metal,* obviously, *must be in a position to exhibit variable valency.* This type of defect, therefore, is generally found amongst the compounds of transition metals.

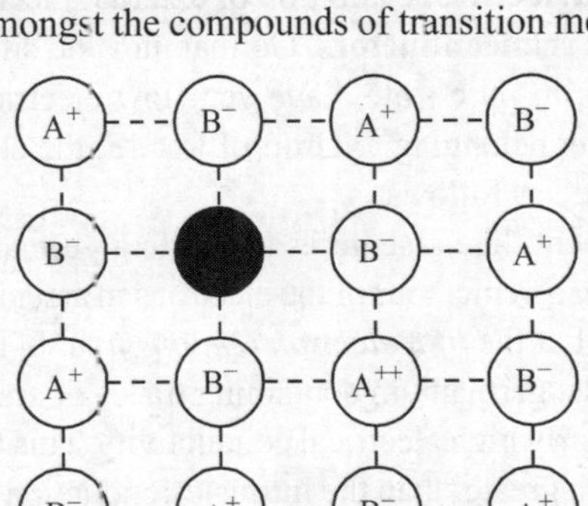

Fig. 4.23. Metal deficiency defect due to a missing positive ion.

2. An *extra negative ion may occupy an interstitial position. The extra negative charge is balanced again by means of an extra charge on one of the adjacent metal ions,* as shown in Fig. 4.24. *The metal* in this case also, *should be capable of exhibiting variable valency, i.e.,* it should be one of the transition metals. However, since the negative ions are usually very large, it is not expected that they would be able to fit into the interstitial positions. As a matter of fact, no crystal showing this type of metal deficiency defect is known.

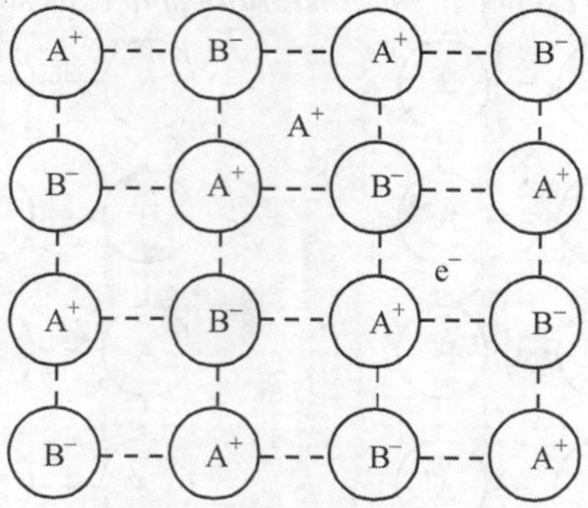

Fig. 4.24. Metal deficiency defect due to an extra negative ion.

Consequences of Metal Deficiency Defects. Crystals with metal deficiency defects can also be semiconductors. This property arises from the movement of an electron from one ion to another ion. In this way, the ion, say, A^+, changes into A^{++}. Thus, the movement of an electron from A^+ ion is *an apparent movement of* A^{++} *ion.* This is called *movement of positive 'hole'* and the substances showing this type of movement are known as ***p-type*** **semiconductors.**

Thermal Defects. Intrinsic Conduction. Atoms of both germanium and silicon have four electrons in the outer shell. Each atom, therefore, is covalently bonded with four neighbouring atoms through sp^3 hybrid bonds, giving rise to a highly stable tetrahedral structure. There are no free or conducting electrons. The electrical conductance, therefore, is very low. Now, suppose a sufficient amount of energy, is supplied to the crystal as a result of which *one of the covalent bonds gets broken and the electrons are set free.* The electrons, thus released, can migrate leaving behind a 'positive hole', at the site of the missing bond. *The crystal will now be able to conduct electricity* because, when electric field is applied, the electrons migrate in one direction and the 'positive holes' in the other. This type of conduction in known as **intrinsic conduction** because if happens without adding an external substance.

Impurity Defects. Certain defects in crystals happen due to the presence of chemical impurities. These are known as **impurity defects.** One addition of external metal impurities to germanium and silicon crystals, the latter act as **semiconductors.** Germanium and silicon belong to Group 14 of the periodic table. These elements, in pure state, have very low electrical conductivity. However, on adding even traces of an element belonging to Group 13 or 15, the electrical conductivity is greatly enhanced. This may be explained as follows:

Suppose, a Group 15 element, like arsenic, is added to a germanium crystal. As a germanium atom is substituted by an atom of asrenic, four of the electrons in arsenic form covalent bonds with the surrounding germanium atoms but the *fifth electron remains free.* In this way, an extra electron gets introduced into the crystal. Thus, germanium containing traces of arsenic (known as *arsenic-doped germanium)* begins to exhibit fairly high electrical conductivity. This type of conduction is known as **extrinsic conduction**. It is much greater than the intrinsic conduction discussed above. Since, in this type of conduction, current is carried by excess electrons in the *normal way,* it is called ***n*-type semiconduction.**

Now, suppose a Group III element, like indium, having only three electrons in the outer shell, is added in small traces to, say pure germanium. The atoms of indium, evidently, are not able to complete tetrahedral covalent structures because they have one electron short of the requirement. This gives rise to *electron-vacancies*.

The electron-vacant sites are known as 'positive holes' because the net charge at these sites is positive. When electric field is applied, adjacent electrons move into the positive holes and in this way other electron-vacancies (or positive holes) are formed. The migration of positive holes thus continues and current is carried thereby throughout the crystal.

Thus, doping of germanium with traces of indium increases the electrical conductivity of germanium crystal. In the present case, the current is carried *by positive holes,* therefore, this type of conduction is called ***p*-type semiconduction.**

4.10.3 Effect of temperature on the conductivity of semiconductors

The conductivity of semiconductors, unlike that of metals, *increases with rise in temperature.* This is due to the fact that more extra electrons or positive holes, as the case may be, are released from the crystal lattice for the conduction of electricity.

Applications of semiconductors

The combination of *n* and *p-type* semiconductors (known as ***n-p* junctions**), finds interesting applications in the **fabrication of transistors**. This device can conduct electric current more easily in one particular direction than in the reverse direction and, therefore, can be used as **rectifier** for changing alternating current into direct current. The device is schematically shown in Fig. 4.25. The material on the left side of each junction is *n-type* conductor obtained by doping germanium with, say, arsenic. The minus signs represent extra electrons. The material on the right side of the junction is *p*-type conductor obtained by doping germanium with, say indium. The positive signs represent 'positive holes' arising from the deficiency of electrons at the indium impurity centres. When an external voltage is applied in such a way as to cause motion of electrons (*n*-current) from left to right and motion of positive holes (*p*-current) from right to left (Fig. 4.25), current is readily conducted. If direction of voltage is reversed so that there is separation of electrons and positive holes as shown in Fig. 4.25, there is cancellation of *n* and *p*-currents and hence the conduction stops. Thus, the *n-p* junction permits the current from an outside source to flow *in one direction only.*

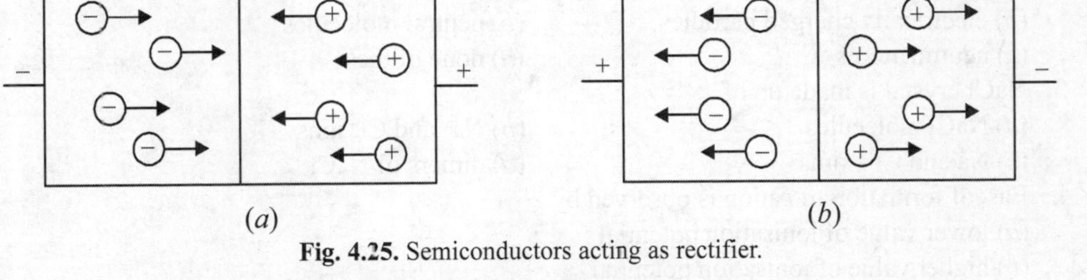

Fig. 4.25. Semiconductors acting as rectifier.

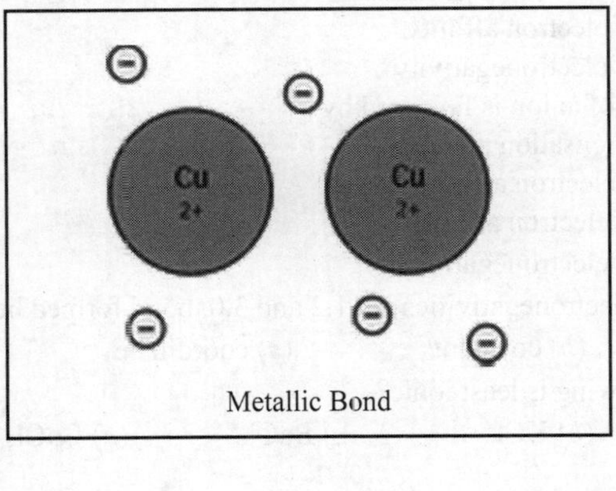

Metallic Bond

EXERCISES
(INCLUDING QUETIONS FROM DIFFERENT UNIVERSITY PAPERS)

Objective Questions

Fill in the blanks

1. Ionic compounds have .... melting and boiling points.
2. The structural arrangement in solids with radius ratio in the range 0.155 – 0.225 is ....
3. Coordination number is the number of anions that surround a .... to give a stable lattice
4. The stability of an ionic solid is measured in terms of its ....
5. The Born-Haber cycle equation is ....
6. A salt dissolves in water if hydration energy is .... than lattice energy
7. For a given cation, as the size of the anion increases, the covalent character .... and melting point ....
8. Barium sulphate is insoluble in water because the .... is much more than ....
9. If some of the lattice points in a solid are unoccupied, it causes .... defects.
10. The conductivity of a semiconductor .... with the rise of temperature.

Multiple Choice Question

Choose the correct option

1. In electrovalency
   (*a*) The transference of electrons takes place
   (*b*) The sharing of electrons takes place
   (*c*) The electrons are shared by one atom only
   (*d*) none of these takes place
2. An electrovalent compound is made up of
   (*a*) electrically charged particles (*b*) neutral molecules
   (*c*) neutral atoms (*d*) none of these
3. NaCl crystal is made up of
   (*a*) NaCl molecules (*b*) Na^+ and Cl^- ions
   (*c*) Na and Cl atoms (*d*) dimers of NaCl
4. Easeof formation of cation is observed by
   (*a*) lower value of ionisation potential
   (*b*) higher value of ionisation potential
   (*c*) higher value of electron affinity
   (*d*) higher value of electronegativity
5. Ease of formation of anion is favoured by
   (*a*) lower value of ionsation potential
   (*b*) lower value of electron affinity
   (*c*) higher value of electron affinity
   (*d*) higher value of electronegativity
6. Elements whose electronegativities are 1.2 and 3.0, bond formed between them would be
   (*a*) ionic (*b*) covalent (*c*) coordinate (*d*) metallic
7. Which of the following is least ionic?
   (*a*) AgCl (*b*) KCl (*c*) $BaCl_2$ (*d*) $CoCl_2$

8. Which of the following has maximum polar character?
(*a*) C – I (*b*) C – Br (*c*) C – F (*d*) C – S
9. The compound which contains both ionic and covalent bond is
(*a*) CH_4 (*b*) H_2 (*c*) KCN (*d*) KCl
10. Which one is electron deficient compound ?
(*a*) ICl (*b*) NH_3 (*c*) BCl_3 (*d*) PCl_3
11. Element X is strongly electropositive and element Y is strongly electronegative, both are univalent. the compound formed should be
(*a*) X^+Y^- (*b*) X^-Y^+ (*c*) X – Y (*d*) Y – X
12. In a solid lattice, cation has left a lattice site and is located at an interstitial position, the lattice defect is
(*a*) interstitial defect (*b*) vacancy defect
(c) Frenkel defect (*d*) Schottky defect
13. The molecule which has the largest depole moment among the following is
(*a*) CH_4 (*b*) $CHCl_3$ (*c*) CCl_4 (*d*) CHI_3
14. Both ionic and covalant bonds are present in
(*a*) CH_4 (*b*) KCl (*c*) SO_2 (*d*) NaOH
15. The molecule which has zero dipole moment is
(*a*) CH_2Cl_2 (*b*) BF_3 (*c*) NF_3 (*d*) ClO_2^-
16. Which is the hardest substance ?
(*a*) Ag F (*b*) Ag Cl (*c*) Ag Br (*d*) Ag I
17. In trigonal planar arrangement, the ratio is
(*a*) 0.155 – 0.225 (*b*) 0.225 – 0.414
(*c*) 0.414 – 0.732 (*d*) 0.732 – 1.00
18. A substance is soluble in water if
(*a*) lattice energy > hydration energy (*b*) hydration energy > lattice energy
(*c*) hydration energy = lattice energy (*d*) no relation
19. The rule to determine the extent of polarization of an anion by a cation was given by
(*a*) Pouling (*b*) Born (*c*) Haber (*d*) Fajan
20. In a certain solid a negative ion is missing from its lattice site. The type of defect in the solid is
(*a*) Schottky defect (*b*) Frenkel defect
(*c*) Metal excess defect (*d*) Metal deficiency defect

ANSWERS

Fill in the blanks

1. high **2.** trigonal planar **3.** cation **4.** lattice energy

5. $U_0 = \Delta H_f - S - \frac{1}{2} D - I - E$ **6.** more **7.** decreases, decreases

8. lattice energy, hydration energy **9.** Schottky **10.** increases

Multiple choice questions

1. (*a*) **2.** (*a*) **3.** (*b*) **4.** (*a*) **5.** (*c*)
6. (*a*) **7.** (*d*) **8.** (*c*) **9.** (*c*) **10.** (*c*)
11. (*a*) **12.** (*c*) **13.** (*b*) **14.** (*d*) **15.** (*b*)
16. (*a*) **17.** (*a*) **18.** (*b*) **19.** (*d*) **20.** (*c*)

General Questions

1. (*a*) Explain the factors which influence the formation of ionic compounds.
   (*b*) "$AlCl_3$ anhydrous is covalent while $AlCl_3 . 6H_2O$ is ionic in nature." How would you account for this behaviour?
   (*c*) Trivalent and tetravalent ions are less common. Why?
2. (*a*) Discuss the radius ratio rule for determining the structures of ionic crystals.
   (*b*) Ionic radius of K^+ is smaller than that of Cl^- ion although both possess isoelectronic configurations. Explain.
   (*c*) What is lattice energy? How does the solubility of an ionic solid depend upon its lattice energy?
3. (*a*) What is an ionic bond? What factors influence the formation of an ionic bond? List four characteristic properties of ionic compounds?
   (*b*) Melting point of NaCl is higher than that of $AlCl_3$. Explain.
4. (*a*) The internuclear distance in KCl is 3.14 Å. Calculate the ionic raddi of K^+ and Cl^- ions using Pauling method.
   (*b*) Describe Born–Haber cycle for calculating the lattice energy of sodium chloride.
   (*c*) Give three uses of lattice energy.
   (*d*) What is Madelung constant ? Is the Madelung constant same or different for NaCl and CsCl structures ?
5. (*a*) How do melting points vary within ionic compounds? Illustrate your answer with suitable examples.
   (*b*) Write note on the hardness and brittleness of ionic solids.
6. (*a*) Describe briefly with one example the Pauling's method to calculate the ionic radii of a univalent cation and anion.
   (*b*) What is the size of tetrahedral and octahedral holes in closed packed structure.
   (*c*) In a close packed array of N spheres, how many (*i*) octahedral and (*ii*) tetrahedral holes are present?
   (*d*) A compound consisting of cations (*i*) of radius 0.97 Å and anions (*ii*) of radius 2.51 Å will probably crystallise in the NaCl form. Is this statement correct? Justify.
7. (*a*) What do you understand by the terms polarizing power and polarizability? How do these determine ionic character of a compound?
   (*b*) Discuss the Fajan rules.
   (*c*) The lattice energies of silver halides are almost the same as that of alkali metal halides, yet, they are insoluble in water. Explain.
8. (*a*) Derive Born-Lande equation for the lattice energy of sodium chloride.
   (*b*) Why do NaCl and MgO have the same Madelung constant?
9. (*a*) What is Born-Haber cycle? How does it explain the stability of ionic compounds?
   (*b*) Write a note on polarization and polarizing power.
   (*c*) Write brief notes on:
   (*i*) Radius ratio rule
   (*ii*) Born-Haber cycle for calculating lattice energy.

10. Explain the following:

(*a*) Though LiCl and MgS have similar structurs and nearly the same internuclear distance yet MgS is harder than LiCl.

(*b*) HF has less ionic character than HI.

(*c*) CuCl and AgCl are insoluble in water whereas NaCl is highly soluble in water.

(*d*) $ZnCl_2$ is soluble in organic solvents while $MgCl_2$ is insoluble.

11. (*a*) What is Born-Haber cycle? Discuss its usefulness by explaining the stability of solids.

(*b*) Calculate the lattice energy of sodium chloride (in kJ/mol) from the following data:

A = 1.75 r_0 = 2.8 Å, n = 9

N = 6.02×10^{23} $e = 4.8 \times 10^{-10}$ *e.s.u.*

(*c*) Discuss any two rules proposed by Fajan.

12. (*a*) What do you mean by lattice energy? Mention the factors on which it depends.

(*b*) Calculate the lattice energy of Na^+Cl^-, when the Madelung constant for NaCl is 1.74759, Born exponent is 9 and the radii of Na^+, Cl^- ions are 0.96 Å and 1.76 Å respectively.

13. (*a*) State Fajan's rule. Explain with the help of this rule, which compound of each of the following is more covalent.

(*i*) AgCl or AgI

(*ii*) LiCl or KCl

(*b*) Explain the term lattice energy.

14. (*a*) Write briefly about polarization and polarizability.

(*b*) Explain the term lattice energy.

15. (*a*) Explain how Born-Haber cycle is used to calculate the lattice energy of NaCl.

(*b*) What is meant by radius ratio? Explain its role in ionic structures.

16. (*a*) Explain why melting point of NaCl is higher than that of $AlCl_3$.

(*b*) $AlCl_3$ anhydrous is covalent but $AlCl_3 . 6H_2O$ is ionic in nature. How will you account for it?

(*c*) Ionic radius of K^+ is smaller than that of Cl^- ion although both possess isoelectronic configuration. Explain.

(*d*) The lattice energies of silver halides are almost the same as that of alkali metal halides, yet they are insoluble in water. Explain.

17. (*a*) Discuss radius ratio rule for the prediction of structures of ionic compounds.

(*b*) Highly charged ions are rare. Why?

(*c*) Explain the following:

(*i*) Ionic bonds are non-directional in nature.

(*ii*) LiCl has higher boiling point than HCl.

(*d*) What is meant by radius ratio? How will you determine the ionic radius ratio of ionic crystals in which the central metal ion, A^+ is surrounded octahedrally by six, B^- ions?

18. (*a*) Describe Born-Haber cycle for finding the crystal energy of NaCl. "The lattice energy calculated by Born-Haber cycle does not agree with the observed value." Comment.

(*b*) Explain the following:

(*i*) Highly charged cations are rare.

(*ii*) Silver halides are least soluble in water through their lattice energy is almost of the same order as that of highly soluble alkali metal halides.

19. (*a*) Describe briefly with one example the Pauling method to calculate the ionic radii of an univalent cation and anion.

(*b*) What is Born-Haber cycle? How does it explain the stability of ionic compounds?

(*c*) Explain polarization and polarizing powder.

(*d*) Give reasons that $ZnCl_2$ is soluble in organic solvents while $MgCl_2$ is insoluble.

20. (*a*) Define ionic bond and factors which favour formation of ionic bond, what is the general position of elements in the periodic table forming ionic bond?

(*b*) Both NaCl and CaS have similar structure and ionic distances, which one is harder and why?

(*c*) AgF, AgCl and AgBr have similar structure. Explain their hardness.

21. (*a*) What is radius ratio rule? Discuss the radius ratio rule for predicting the structures of ionic crystals.

(*b*) The experimental lattice energy of SnO_2 is –1159.5 kJ mol^{-1}. Calculate the heat of formation of SnO_2. Given:

S (Sn) = 291.6 kJ mol^{-1}; I (Sn^{2+}) = 8991.4 kJ mol^{-1}.

D (O_2) = 454.3 kJ mol^{-1}, E (O^{2-}) = + 635.9 kJ mol^{-1}.

22. (*a*) Describe Born-Haber cycle for calculating lattice energy of an ionic solid MX.

(*b*) What is radius ratio rule? How can it help to predict the structure of an ionic crystal? What are its limitations?

23. (*a*) Define lattice energy.

(*b*) Silver halides have lattice energy of the order of alkali metal halides, even then they are soluble in water, explain.

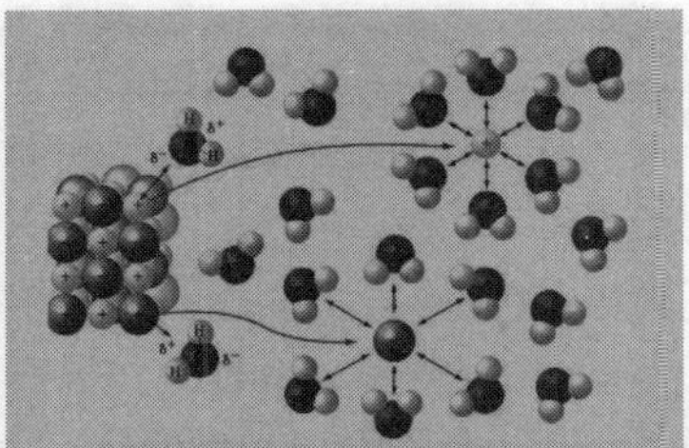
Hydration of ions

5 s–Block Elements

5.1 INTRODUCTION

Groups 1 and 2 constitute *s*-block of the periodic table.

Names of elements of group 1 (alkali metals) along with their electronic configurations are given as under:

Element	*Electronic Configurations*
Lithium	[He] $2s^1$
Sodium	[Ne] $3s^1$
Potassium	[Ar] $4s^1$
Rubidium	[Kr] $5s^1$
Cesium	[Xe] $6s^1$
Francium	[Rn] $7s^1$

Names of alkaline earth elements (group 2) along with their electronic configurations are given below:

Element	*Electronic configurations*
Beryllium	[He] $2s^2$
Magnesium	[Ne] $3s^2$
Calcium	[Ar] $4s^2$
Strontium	[Kr] $5s^2$
Barium	[Xe] $6s^2$
Radium	[Rn] $7s^2$

Alkali metals do not form bivalent cation because second ionisation energy (*i.e.*, energy required to remove second electron) is much higher than the first ionisation energy.

Some important physical properties of group 1 (alkali metals) and group 2 (alkaline earth metals) are given in Table 5.1 an 5.2 respectively.

Table 5.1. Some important physical properties of the elements of group 1.

Properties	*Lithium*	*Sodium*	*Potassium*	*Rubidium*	*Caesium*
Electronic configuration	2, 1	2, 8, 1	2, 8, 8, 1	2, 8, 18, 8, 1	2, 8, 18, 18, 8, 1
Atomic mass	6.94	22.9	39.10	85.48	132.91

Properties	*Lithium*	*Sodium*	*Potassium*	*Rubedium*	*Caesium*
Single bond covalent radius (Å)	1.23	1.54	2.03	2.16	2.35
Radius of M^+ion (Å)	0.68	0.95	1.33	1.48	1.69
Density (g cm^{-3})	0.53	0.97	0.86	1.53	1.90
Melting point (°C)	181	97.8	64.0	39.0	28.5
Boiling point (°C)	1330	888	756.0	688.0	669.0
Heat of atomisation (kJ mol^{-1})	159.0	108.1	90.2	82.1	78.2
Ionisation energy (kJ mol^{-1})	520.3	495.8	418.9	403.0	375.7
Oxidation state	+1	+1	+1	+1	+1
Flame colouration	Crimson red	Golden yellow	Violet	Violet	Violet
Hydration energies of M^+ cations (kJ mol^{-1})	531.4	422.6	338.9	318.9	280.3
Size of hydrated M^+ ions, i.e., size[of $[M(aq)]^+$ ions (Å)	3.40	2.76	2.32	2.28	2.28
Ionic conductance of $[M\ (aq)]^+$ ions (ohm^{-1} cm^2 $equiv^{-1}$)	33.5	43.5	64.6	67.5	68.0
Standard reduction potential for $M^+\ (aq) + e^- \rightarrow M(s)$ (volts)	–3.05	–2.71	–2.93	–2.93	–2.93
Electronegativity	1.0	0.9	0.8	0.8	0.7
Electron affinity (kJ mol^{-1})	59.8	52.7	48.36	46.89	45.50

Table 5.2. Some important physical properties of the element of group 2.

Properties	*Beryllium*	*Magnesium*	*Calcium*	*Strontium*	*Barium*
Electronic configuration	2,2	2,8,2	2,8,8,2	2,8,18,8,2	2,8,18,18,8,2
Atomic mass	9.01	24.31	40.08	87.67	137.34
Density (g cm^{-3})	1.84	1.74	1.55	2.54	3.75
Melting point (°C)	1277	650	838	768	714
Boiling point (°C)	2770	1107	1440	1380	1640
Single bond covalent radius (Å)	0.90	1.36	1.74	1.91	1.98
Radius of M^{2+} cation (Å)	0.31	0.65	0.99	1.13	1.35
1st ionisation energy (kJ mol^{-1})	899.5	737.7	589.8	549.5	502.9

Properties	*Beryllium*	*Magnesium*	*Calcium*	*Strontium*	*Barium*
2nd ionisation energy (kJ mol^{-1})	1757.1	1450.1	1146.0	1046.0	965.0
Hydration energy of M^{2+} ion (kJ mol^{-1})	2494.0	1953.9	1577.0	1443.0	1352.0
Standard reduction potential (volts) for $M^{2+}(aq) + 2e^- \rightleftharpoons M(s)$	–1.70	–2.38	–2.76	–2.89	–2.90
Electronegativity	1.5	1.2	1.0	1.0	0.9
Heat of atomisation (kJ mol^{-1})	78.2	35.2	42.5	39.1	42.2
Flame colouration	None	None	Brick red	Crimson	Grassy green
Oxidation state	+2	+2	+2	+2	+2

5.2 LATTICE ENERGY

When a large number of cations and anions are brought nearer to each other to form an ionic lattice, they arrange themselves in a regular pattern. In the process of attaining this type of structure, energy is released and the system gets *stabilised* and ionic crystal is formed. The amount of energy released is called lattice energy of the ionic lattice or solid.

Thus the *lattice energy of an ionic solid can be defined as the amount of energy released when one mole of the gaseous cations and one mole of the gaseous anions are brought closer to each other from an infinite distance to combine together to form one mole of the ionic crystal or solid. Lattice energy is represented as* H_{lat}. *Since the energy is released*, it is denoted with a negative sign. Thus, for Na^+Cl^- (*s*) ionic crystal:

$$Na^+ (g) + Cl^- (g) \longrightarrow Na^+ Cl^- (s), \text{ Energy released} = -(H_{lat})_{NaCl (s)}$$

Due to the release of energy, the formation of Na^+ Cl^- (*s*) by the combination Na^+ (*g*) and Cl^- (*g*) is an *exothermic process*.

It follows from the above discussion that when one mole of Na^+ Cl^- (*s*) ionic crystal is broken up into its constituent gaseous ions *viz.*, Na^+ (*g*) and Cl^-(*g*), energy will be required. The energy required in the above reaction has been found to be numerically equal to the lattice energy of Na^+Cl^-(*s*). Thus lattice energy of an ionic crystal can also be defined *as the amount of energy needed to break one mole of the ionic solid into its constituent gaseous ions*. Thus:

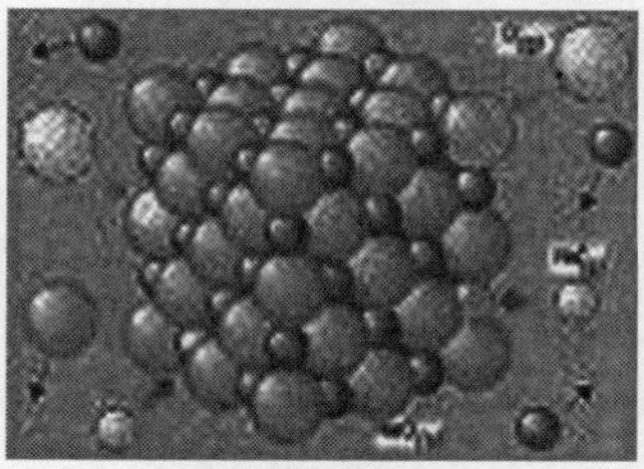

Lattice Energy

$$Na^+ Cl^- (s) \longrightarrow Na^+ (g) + Cl^- (g); \text{ Energy required} = + (H_{lat})_{NaCl (s)}$$

A compound having negative value for its heat of formation is stable while that having positive value is unstable. Thus, since NaCl (*s*) has negative value for its heat of formation (= – 118.2 kcal mol^{-1}) it is stable.

5.2.1 Comparison between lattice energy and heat of formation

Lattice energy of an ionic compound is the energy released (negative sign) when one molecule of the ionic compound in the solid state is formed by the combination of its constituent ions in the gaseous state, while heat of formation of a compound is defined as the energy released (negative sign) or absorbed (positive sign) when one molecule of that compound (in any state) is formed by the combination of the constituent elements (in any state).

5.2.2 Factors affecting the lattice energy of ionic solids

Lattice energy depends upon the following factors:

(*i*) *Charge on the cation (z^+) or on the anion (z^-) or the product of charges ($z^+ z^-$)*. Greater the charges on the cation or the anion or the product of charges on the ions, greater is the extent of lattice energy. Thus, the lattice energies of ionic solids containing monovalent-monovalent ions, divalent-monovalent ions and divalent-divalent ions are in the increasing order, e.g., (values of lattice energies are in kJ/mole).

$Na^+ F^-$	<	$Mg^{2+}F_2^-$	<	$Mg^{2+} O^{2-}$
914.2		2882.0		3895.0
K^+F^-	<	$Ca^{2+}F_2^-$	<	$Ca^{2+}O^{2-}$
812.1		2581.0		3520.0

(*ii*) *Inter-ionic distance (r_0) between the ions or the size of the cation (r_+) or of the anion (r_-).* Lattice energy is inversely proportional to the inter-ionic distance (r_0) between the ions. Since r_0 is equal to the sum of r_+ and r_-, smaller the size of the ions, smaller would be the value of r_0 and hence higher would be the value of lattice energy. Thus smaller the size of a cation for a given anion or of an anion for a given cation, smaller is the value of inter-nuclear distance (d) and hence greater is the force of attraction existing between the ions and consequently greater is the magnitude of lattice energy.

(*a*) Since the size of monovalent alkali metal cations (M^+) increases on moving down the group from Li^+ to Cs^+, the lattice energy of their fluorides (or of any other halides) decreases in the order LiF > NaF > KF > RbF > CsF.

Size of M^+ ions: $Li^+ < Na^+ < K^+ < Rb^+ < Cs^+$

Increasing ⟶

Lattice energy of fluorides (kJ/mole)/of alkali metals: 1034 > 914.2 > 812.1 > 780.3 > 743.9

Decreasing ⟶

(*b*) Since the size of monovalent halide ions (X^-) increases on moving down the group from F^- to I^-, the lattice energy of the halides of a given alkali metal (say lithium) *decreases* in the order LiF > LiCl > LiBr > LiI.

Size of X ions: $F^- < Cl^- < Br^- < I^-$

Increasing ⟶

Lattice energy of halides of Li (kJ/ mole): 1034 > 840.1 > 781.2 > 718.2

Decreasing ⟶

The results in (*a*) and (*b*) above, show that in the series of a particular halide of various alkali metals or in the series of halides of a given alkali metals, the lattice energy of the halides decreases with the increase in internuclear distance ($M^+ - X^-$ distance). The lattice energies of salts of alkali metals, with a given anion, decreases with the increase in atomic number of the alkali metals.

(*c*) The lattice energies of the oxides of alkaline earth metals also decrease on moving down group, since the size of Be^{2+}, Mg^{2+} etc., ions increase in the same direction. Thus :

Oxides:	BeO	MgO	CaO	SrO	BaO
Lattice energies of oxides (kJ/mole)	4541	3895	3520	3325	3108
	——— Decreasing ——→				
Size of M^{2+} ions (Å):	Be^{2+} 3.1	Mg^{2+} 6.5	Ca^{2+} 9.9	Sr^{2+} 1.13	Ba^{2+} 1.35
	——— Increasing ——→				

(*d*) Since the monovalent cations of alkali metals (M^+ ions) are larger in size than the divalent cations of alkaline earth metals (M^{2+} ions), the lattice energies of the salts of M^+ ions are lower in magnitude than those of the corresponding salts of M^{2+} ions, e.g.,

NaF	<	MgF_2	:	KF	<	CaF_2
914.2		2882.0	;	812.1		2581.0

(values in kJ/mole)

Conclusion. In order to get high value of lattice energy, the size of the cation and (or) anion should be small and electric charges on the cation/anion should be high.

Lattice energy of the sulphates, chlorides, nitrates etc., of alkaline earth metals

Since the size of the divalent cations of alkaline earth metals increase as we move down the group from Be^{2+} to Ba^{2+} ions, the lattice energy of their sulphates, chlorides, nitrates etc., should decrease in the same direction. But it has been observed that their lattice energies are almost the same. The similarity in their lattice energy values is due to the fact that the anions viz., sulphates, chlorides, nitrates etc., present in these salts are very large in size. compared to the size of the cation.

5.2.3 Explanation of some properties of ionic crystals on the basis of their lattice energy values

1. Solubility/insolubility of ionic solids in polar (e.g., H_2O) and non-polar (e.g. C_6H_6, CCl_4 etc.) solvents

Whether a given ionic solid is soluble or insoluble in a polar (ionic) or a non-polar solvent depends on two factors *viz*., (*a*) *lattice energy of the ionic solid* which keeps the constituent ions (*i.e.*, cations and anions) of the ionic solid held together tightly in the ionic crystal and (*b*) *hydration energy of the ionic solid*, which is the energy released when the ions of the ionic solid obtained by dissolution of the ionic solid in ionic solvent (e.g., H_2O) interact with water dipoles to form hydrated ions. Higher the magnitude of lattice energy, more tightly are the ions held together in the ionic crystal and hence smaller is the tendency of the ionic solid to split into ions by dissolution and to interact with water dipoles to form hydrated ions, i.e., smaller is the solubility of the ionic solid. Higher the magnitude of hydrated energy, greater will be the tendency of the ionic solid to break into ions in water and then to get hydrated by H_2O molecules, and consequently higher will be the solubility of the solid. Thus we see that lattice energy tends to oppose the dissolution of the ionic solid, while the hydration energy tends to dissolve the ionic solid into water. We can therefore say that if the hydration energy is greater than the lattice energy, the ionic solid will be soluble in polar solvent like water and *vice-versa*.

Ionic solids are not soluble in non-polar solvents, (*e.g.*, C_6H_6, CCl_4 etc.) since the ions of ionic solid donot interact with the solvent molecules (due to the non-polar nature of the solvent molecules) to form the hydrated ions and hence no hydration energies are obtained.

(*a*) LiCl, Li Br, NaCl, NaBr etc., are the examples of icnic solids which are soluble in water, since their hydration energies are greater than their lattice energies as shown below:

Hydration energies: LiCl = 883, Li Br = 854, NaCl = 775, NaBr = 741 (kJ/mole)

Lattice energies: LiCl = 840, Li Br = 781, NaCl = 770, NaBr = 728 (kJ/mole)

(*b*) The ionic solids like KCl, KBr and KI are insoluble in water, since their hydration energies are smaller than their lattice energies.

High value of dielectric constant of the solvent also favours the dissolution of the ionic solids, because it weakens the electrostatic force of attraction existing between the oppositely charged ions present in the ionic solids.

Comparision between the solubility of halides of a given metal. Let us compare the solubility of the halides given by Li. As discussed above, higher the magnitude of lattice energy, smallar is the solubility of the ionic solid in water. Since the lattice energies of the halides of Li decrease from LiF to LiI, the solubility of these halides moves in the increasing order. Thus:

Halides of Li:	LiF	LiCl	LiBr	LiI
Values of lattice energy (kJ/mole):	1034	840.1	781.1	718.2

Decreasing ⟶

Solubility: Increasing ⟶

(LiF < LiCl < LiBr < LiI)

Infact LiF is insoluble in water while LiI is soluble in this solvent. In general we can say that the order of the solubility of the halides of a given metal can be written as:

fluorides < chlorides < bromides < iodides.

Comparison between the solubility of hydroxides and fluorides of a alkaline earth metals. Since the size of the divalent alkaline earth metal cations (M^{2+} ions) increases on moving down the group from Be^{2+} to Ba^{2+}, the lattice energy of the hydroxides and fluorides of these metals decreases in the same direction. With the decrease in the lattice energy value, the solubility of these compounds increases. Thus $Be(OH)_2$ and $Mg(OH)_2$ are almost insoluble in water while hydroxides of other alkaline earth metals are slightly soluble. Similarly MgF_2 and CaF_2 are also insoluble in water.

Comparison between the solubility of M^+ salts with ions such as SO_4^{2-}, $C_2O_4^{2-}$, PO_4^{3-} etc. and that of M^{2+} salts with the same ions. Due to the smaller size of alkaline earth metal ions (M^{2+} ions) as compared to those of alkali metal cations (M^+ ions) of the same period, the lattice energies of M^{2+} salts with ions such as SO_4^{2-}, $C_2O_4^{2-}$, PO_4^{3-} etc. are higher than those of the corresponding to M^+ salts. Consequently M^{2+} salts are less soluble than M^+ salts. For example, $CaCO_3$ is less soluble in water than K_2CO_3.

2. Melting points and boiling points of ionic solids.

Since the electrostatic force of attraction existing between the cations and anions present in an ionic crystal are very strong, the lattice energies of ionic solids are very high. It is because of high values of their lattice energies that large amount of energy is required to separate the ions of the ionic solid. Hence melting and boiling points of ionic solids are very high. The melting points of ionic solids, therefore increase with the increase of their lattice energy and *vice-versa.*

(*a*) Since the lattice energies of the halides of Li decrease as LiF > LiCl > LiBr > LiI, the melting points of these halides also decrease in the same direction.

Halides of Li	:	LiF	LiCl	LiBr	LiI
Lattice energy (kJ/mole)	:	1034	840.1	781.2	718.2
		Decreasing ⟶			
Melting points (°C)	:	870	613	597	446
		Decreasing ⟶			

(*b*) Melting poitns of the chlorides of alkali metals also decrease as NaCl > KCl > RbCl > CsCl, since the lattice energy values of these chlorides also decrease in the same direction. LiCl is an exception because it is a covalent compound.

Chlorides of alkali metals	:	LiCl	NaCl	KCl	RbCl	CsCl
Lattice energy (kJ/mole)	:	840.1	770.3	701.2	682.4	629.7
		Decreasing ⟶				
Melting points (°C)		613	800	776	710	625
		Decreasing ⟶				

(*c*) The melting point of AgCl (= 455°C) in higher than that of AgBr (= 434°C), since AgCl has higher value of its lattice energy. Lattice energy of AgCl = 895 kJ/mole and that of AgBr = 883 kJ/mole.

5.3 SOLVATION ENERGY

When a gaseous cation or a gaseous anion reacts with a solvent, solvated cation or solvated anion is obtained and in this process energy is liberated. This liberated energy is called *solvation energy*. Thus:

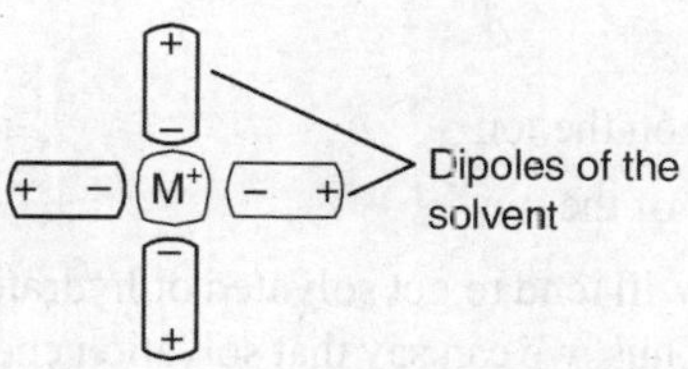

Fig. 5.1. Representation of solvation process.

$M^+(g)$ + solvent ⟶ M^+ (*sol*) + solvation energy
solvated cation

X^- (*g*) + solvent ⟶ X^- (*sol*) + solvation energy
solvated anion

Due to the liberated of energy solvation process is an *exothermic process*.

Solvation process is illustrated by Fig. 5.1.

5.3.1 Hydration energy

When a gaseous cation or a gaseous anion reacts with water, hydrated cation or hydrated anion is produced and in this process energy is released. This released energy is called *hydration energy*. Thus:

M^+ (*g*) + H_2O (*l*) or *aq* ⟶ M^+ (*aq*) + Hydration energy
Hydrated cation

$$X^-(g) + H_2O\,(l) \text{ or } aq \longrightarrow \underset{\text{Hydrated anion}}{X^-(aq)} + \text{Hydration energy}$$

Obviously, hydration process is an exothermic process.

Hydration energies (heat of hydration) of some ions and some salts are given below (in kJ/mole).

Ions

H^+ = 1091

Li^+ = 531.4	Be^{2+} = 2494.0		F^- = 515.0
Na^+ = 422.6	Mg^{2+} = 1953.9	Al^{3+} = 4694.5	Cl^- = 381.0
K^+ = 338.9	Ca^{2+} = 1577.0		Br^- = 347.0
Rb^+ = 318.9	Sr^{2+} = 1443.0		I^- = 305.0
CS^+ = 280.3	Ba^{2+} = 1352.0		

Salts

LiCl = 883,	LiBr = 854	
NaCl = 775,	NaBr = 741	
KCl = 686,	KBr = 657,	KI = 619

5.3.2 Factors affecting solvation (or hydration) energy

Following factors are important in the determination of hydration energy.

1. Ionic potential (ϕ) of the ion. The charge-to-radius ratio of an ion is called its ionic potential (ϕ). Thus:

$$\text{Ionic potential } (\phi) = \frac{\text{Charge on the ion}}{\text{Radius of the ion}}$$

An ion with high value of ϕ will tend to get solvated or hydrated more easily than the ion with a smaller value of ionic potential. Thus, we can say that solvation energy or hydration energy increases with increasing ionic potential and *vice versa*. Hydration energy increases with increase in the charge on the ion and decreases with increase in the radius of the ion.

(*i*) Since the size of alkali metal cations increases as $Li^+ < Na^+ < K^+ < Rb^+ < Cs^+$, the hydration energy of these cations are in the decreasing order as shown below:

Alkali metal cations :	Li^+	Na^+	K^+	Rb^+	Cs^+
Radii (Å) :	0.60	0.95	1.33	1.48	1.69
	Increasing ⟶				
Hydration energy (kJ/mole) :	531.4	422.6	338.9	318.8	280.3
	Decreasing ⟶				

(*ii*) On the basis of above generalisation it can be shown that with the increase of the size of divalent cations of alkaline earth metals from Be^{2+} to Ba^{2+}, the hydration energies of these cations decrease down the group. The values of hydration energies (in kJ/mol) are:

Be^{2+} = 2494, Mg^{2+} = 1953.9, Ca^{2+} = 1577, Sr^{2+} = 1443 and Ba^{2+} = 1352.

(*iii*) Similarly it can be explained why hydration energy of halide ions decreases down the group as shown below:

Halide ions	:	F^-	Cl^-	Br^-	I^-
Radii (Å)	:	1.36	1.81	1.95	2.16
		Increasing →			
Hydration energy (kJ/mole)	:	515	381	347	305
		Decreasing →			

Hydration energies of K^+ and F^- ions. Although both these ions have the same size (K^+ = 1.33 Å, F^- = 1.36 Å), there is a considerable difference in the value of their hydration energy (K^+ = 338.9 kJ/mole, F^- = 515.0 kJ/mole). This difference is due to the fact that F^- ion interacts with the small $H^{\delta+}$ atom while K^+ ion has to interact with the much larger $O^{\delta-}$ atom of water dipoles. Thus greater number of H–atoms can get around a F^- ion compared to O-atoms around the K^+ ion. Greater the dipole-dipole interaction, larger is the energy, released.

5.3.3 Variation of hydration energies of ions in the periodic table

As already said the tendency of ions to get hydrated decreases with the increase of their size, *i.e.*, the magnitude of hydration energy of ions decrease with the increase of their size (radius) and *vice versa*.

(*a*) **Variation of hydration energy in a group**. Radii of the ions derived from the atoms of the elements of a given group of *s*- and *p*-block elements increase on descendig the group, therefore hydration energy of the ions decreases down the group.

Examples. (*i*) Since the radii of alkali metal cations increase down the group, their hydration energy decreases, *i.e.* the relative order of the hydration energies of alkali metal cations is as:

$Li^+ > Na^+ > K^+ > Rb^+ > Cs^+$ (*Decreasing order*)

(*ii*) Since the size of Mg^{2+} ion is less than that of Ba^{2+}, Mg^{2+} ion has higher value of its hydration energy than Ba^{2+} ion.

(*iii*) Since the radii of halide ions increase down the group, their hydration energy decrease in the same direction, or the relative order of the hydration energies of halide ions is as:

$F^- > Cl^- > Br^-\ I^-$ (*Decreasing order*)

(*b*) **Variation of hydration energy in a period**. Since the radii of the ions derived from the atoms of the elements of a given period *s*- and *p*-block elements decrease on moving from left to right, the magnitude of hydration energy of the ions increases.

Example. Since the radii of Na^+, Mg^{2+} and Al^{3+} ions (period 3) decrease as $Na^+ > Mg^{2+} > Al^{3+}$, the hydration energies of these ions are in the increasing order, *i.e.*, the relative order of hydration energies of the these ions is as:

$Na^+ < Mg^{2+} < Al^{3+}$ (*Increasing order*)

5.3.4 Explanation of some properties on the basis of hydration values

1. Size and electrical conductivity of hydrated alkali metal cations, $[M (aq)]^+$. Smaller is the size of an ion, greater is the degree of its hydration. Thus as we move down the group of alkali metals from Li^+ to Cs^+ ion, the size of the ion increases and hence the degree to which the ion gets hydrated decreases, i.e., the degree of hydration of alkali metal ions is in the order:

$Li^+ > Na^+ > K^+ > Rb^+ > Cs^+$

The above order shows that Li^+ ion is hydrated to the maximum extent and Cs^+ ion gets hydrated to the minimum extent. Thus, Li^+ ion is surrounded by the maximum number of H_2O molecules and hence has the largest size when it is in the *aqueous solution*. This means that $[Li (aq)]^+$ is the *largest in size*. On the other hand Cs^+ ion is surrounded by the minimum number of H_2O molecules and hence has the smallest size in *aqueous solution*. This means that $[Cs(aq)]^+$ is the smallest in size. Thus we see that the size of alkali metal hydrated cations decreases down the group as shown below:

$[Li(aq)]^+ > [Na(aq)]^+ > [K(aq)]^+ > [Rb(aq)]^+ > [Cs(aq)]^+$ (*Decreasing order*)

The radii (Å) of the hydrated ions, M^+ (*aq*) have been found to be equal to Li^+ (*aq*) = 3.40, Na^+ (*aq*) = 2.76, K^+ (*aq*) = 2.32, Rb^+ (*aq*) = 2.28, Cs^+ (*aq*) = 2.28.

As the tendency of alkali metal cations to get hydrated decreases down the group, the electrical conductivity (ionic conductance) of these cations in solution of infinite dilution increases in the order as shown below:

$[Li(aq)]^+ < [Na(aq)]^+ \; [K(aq)]^+ < [Rb(aq)]^+ < [Cs(aq)]^+$ (*Increasing order*)

This is because a smaller hydrated ion will show greater mobility or electrical conductivity.

2. Formation of hydrated salts by alkali metals. Since the size of alkali metal monovalent cations (M^+ ions) increases down the group from Li^+ to Cs^+ with the increases in their atomic number, the extent to which these cations get hydrated decrease in the same direction. Thus all salts of Li^+ ion which get hydrated to the maximum extent are hydrated, many salts of Na^+ ion are also hydrated while those of Rb^+ and Cs^+ ions are very rarely hydrated.

3. Formation of hydrated salts by alkaline earth metals. Since the alkline earth metal divalent cations (M^{2+} ions) are smaller in size than alkali metal monovalent cations (M^+ ions), the extent to which M^{2+} ions get hydrated is greater than that of M^+ ions. For example Mg^{2+}, Ca^{2+} and Ba^{2+} ions give a large number of hydrated salts which are more extensively hydrated than those given by M^+ ions. Example of such salts are $MgCl_2.6H_2O$, $Mg(NO_3)_2.6H_2O$, $MgSO_4.\ 7H_2O$, $CaCl_2.6H_2O$, $BaCl_2.6H_2O$. On the other hand Na^+ ion gives a limited number of hydrated salts and Rb^+ gives hydrated salts very rarely.

4. Solubility of the nitrates, sulphates, chlorides etc., of alkaline earth metals. As told earlier, the solubility of an ionc salt in water depends on the difference in the hydration energy and lattice energy of the ionic salt. Since the lattice energies of nitrates, chlorides etc., of divalent alkaline earth metal cations (M^{2+}) are almost the same, the relative order of the solubility of these salts cannot be predicted on the basis of the difference in the magnitude of hydration energy and lattice energy. The relative order is determined by the magnitude of hydration energies of the gaseous M^{2+} ions only. Greater is the magnitude of hydration energy of M^{2+} cation, greater is the solubility of the salt of this ion. Since the hydration energies of M^{2+} ion decrease down the group, the solubility of the sulphates, nitrates, chlorides etc., also decreases in the same direction. $MgSO_4$ is soluble in H_2O while $BaSO_4$ is insoluble or has very low solubility.

5.4 HYDRIDES OF ALKALI AND ALKALINE EARTH METALS

Two types of hydrides are formed by alkali and alkaline earth metals

1. Ionic or Salt-like Hydrides
2. Molecular or Covalent Hydrides

1. Ionic Hydrides. These are formed by metals which are more electropositive than hydrogen. Such metals are located towards the left of the Periodic table. The examples are alkali metals and alkaline earth metals except Be and Mg. These are formed by *transfer of electrons from metals to hydrogen atoms* and contain hydrogen as hydride ion, e.g., lithium hydride (Li^+H^-) and calcium hydride ($Ca^{2+}\ 2H^-$). These are white crystalline substances and behave like salts. They have high melting and boiling points and conduct electricity in the fused state liberating hydrogen at the anode. The reaction

$$\frac{1}{2}H_2(g) + e^- \longrightarrow H^-(g);\ \Delta H = +146.4 \text{kJ mol}^{-1}$$

is evidently *endothermic*. Therefore, unless the element with which hydrogen combines is highly electropositive, the ionic hydride formed would be unstable. This is the reason why only alkali and alkaline earth metals, which can readily lose electrons, form stable ionic hydrides.

Ionic hydrides are obtained by direct combination of the metals with hydrogen at high temperature, usually close to 970 K. They are strong reducing agents, as illustrated by the following reactions:

$$2CO + NaH \longrightarrow HCOONa + C$$

$$SiCl_4 + 4NaH \longrightarrow SiH_4 + 4\,NaCl$$

$$PbSO_4 + 2CaH_2 \longrightarrow PbS + 2Ca(OH)_2$$

They burn spontaneously in air on heating as hydrogen is produced which is inflammable.

$$CaH_2 \longrightarrow Ca + H_2$$

They are strong bases. When dissolved in water they form hydroxides and evolve hydrogen.

$$NaH + H_2O \longrightarrow NaOH + H_2$$

They form complexes, like lithium aluminium hydride, Li $[AlH_4]$ and sodium borohydride, $Na[BH_4]$, as shown below:

$$LiH + AlH_3 \longrightarrow Li[AlH_4]$$

$$NaH + BH_3 \longrightarrow Na[BH_4]$$

The complexes are also good reducing agents.

Uses. 1. The ionic hydrides and their complexes are used as reducing agents.

2. They evolve hydrogen when heated and, therefore, they are used as solid fuels.

2. Molecular or Covalent Hydrides. Beryllium and magnesium of Group 2 form hydrides which have considerable covalent character in their M–H bonds. Magnesium hydride has lower enthalpy of formation and is thermally less stable than the true saline (ionic) hydrides. In fact MgH_2 lies intermediate between pure ionic hydrides and pure covalent hydrides. BeH_2, is predominantly covalent. It has a polymeric structure with hydrogen atoms forming bridges between the Be atoms as in boranes. Both BeH_2 and MgH_2 are prepared by direct action of the metal with hydrogen at elevated temperatures.

$$Mg + H_2 \xrightarrow[\text{temp.}]{\text{High}} Mg\,H_2$$

The hydrides formed by elements of Groups III to VII are predominantly covalent although in some cases, as for example, in HF, the bonds are partly ionic.

5.5 VARIATION IN PROPERTIES OF *s*-BLOCK ELEMENTS

5.5.1 Change of atomic radius (*i*) along the period (*ii*) down in group in *s*-block elements

The alkali metals have the largest atomic radii in their respective periods and those of alkaline earths are smaller than the preceding alkali metals. For example, in the second period, the atomic radius of Na is maximum, that of Mg comes next while those of Al, Si, P, Cl ar still smaller. This is explained on the basis of greater nuclear pull on the valency electrons with increasing atomic number. As we move from group 1 to 2, group 2 is one unit higher in molecular charge than that of the preceding element of group 1. With the increase in the nuclear charge, the electrons in the outer shell are pulled closer towards the nucleus. This results in a decrease in the atomic radius. On moving down the group the atomic radii increase. This is because with every succeeding element in the group, a new energy level is added up.

5.5.2 Variation in ionisation energy along a period and down the group

The ionization energy is defined as the amount of energy required to remove the most loosely bound electron from an isolated gaseous atom to form a gaseous positive ion. The most loosely bound electron must lie in the outermost shell. The single electron in the valency shall of each alkali metal can be removed relatively easily. Hence the ionization energy of these elements are relatively low. On moving down the group as the atomic radius increase, the outer electron gets farther and farther away from the nucleus and therefore, the ionization energy decreases on moving down from lithium to cesium. In the case of alkaline earth metals, since there are two electrons in the *s*-orbital of outer level, two ionization energies are involved.

The first, I.E. of beryllium is 895 kJ/mole,

$$Be \longrightarrow Be^{+} + e^{-}, E_1 = 895 \text{ kJ/mole}$$

$$Be^{+} \longrightarrow Be^{2+} + e^{-}, E_2 = 1557 \text{ kJ/mole}$$

As is clear, E_2 is invariably higher than E_1. This is due to the fact that it is more difficult to remove an electron from a positively charged particle than from a neutral particle. The *s*-block elements have relatively low I.E.

It appears that the alkaline earth metals would prefer to form +1 ions (M^{+}) rather than M^{2+} ions e.g., Mg^{2+}, Ca^{2+} etc. However, it is the M^{2+} ions that are obtained. This is explained due to greater *lattice energy* associated with M^{2+} ion compared to that with M^{+} ions. Also +2 ions are excessively hydrated and a large amount of energy, known as *hydration energy*, is released in the process which more than counterbalances the higher value of second ionization energy.

5.5.3 Alkali metals are good reducing agents

A substances that can easily impart its electron can act as a good reducing agent. Electron donating tendency of a metal depends upon its ionisation energy. Smallar the I.E., greater is the reducing power. Since alkali metals possess smaller values of I.E., they are good reducing agents.

***Exmaple 1.* Potassium is more reactive than calcium. Explain.**

Solution. Reactivity of an element depends upon the I.E. Smaller the I.E. greater is the reactivity. As potassium has smaller I.E. than calcium, the former is more reactive than the latter.

***Example 2.* Radius of a cation is smaller than that of its parent atom. Explain.**

***Solution*:** A cation is obtaned when an electron or electrons are removed from the outermost orbit of the atom.

$$M \longrightarrow M^{n+} + ne^{-}$$

After the removal of the electrons, the nuclear charge becomes greater than the electronic charge. The nuclear charge pulls the outer electrons with greater force, thereby reducing the size of cation. In some cases (1 and 2 group), the outer orbit is completely removed thereby reducing the size of cation compared to parent atom.

***Example 3*: First member of each group differs from other members of the same group. Explain.**

***Solution*:** First member of every group is smallest in size and has extraordinary large value of I.E. and small value for electron affinity. These characteristics are responsible for difference in behaviour from other members of the group. Non-availability of *d*-orbitals in the first member also accounts for difference in properties from other members.

***Example 4*: Calcium forms CaF_2 and not CaF or CaF_3; why?**

***Solution*:** Calcium could from CaF because first I.E. of Ca is much smaller than second I.E. But Ca^{2+} is formed in the preference because the lattice energy of CaF_2 would be much higher than that of CaF. Thus CaF_2 forms a more stable lattice. Formation of Ca^{3+} (i.e., CaF_3) does not take place because it involves a great amount of energy in removing the third electron from the inner shell.

Example 5. **I.E. of Be is larger than that of Mg. Explain.**

Solution. Beryllium is a smaller atom than magnesium. The outer electrons in Be are held more strongly than in magnesium. Hence I.E. of Be is greater than that of Mg.

Example 6. **Why are alkali metals used in photoelectric cells?**

Solution**:** A metal which can release electrons on irradation forms a suitable substance for photoelectric cells. Since I.E. of alkali metals is small, they can easily release electrons on irradiation.

5.5.4 Colours imparted to the flame by *s*-block elements

s-block elements impart characteristic flame colourations to the non-luminous flame. Thus all alkali metals impart a colour to the flame while all alkaline earth elements except Be and Mg also impart colourations to the flame. This is explained by the fact that in the flame, the extra energy excities the electrons in these elements to higher energy levels. When these electrons return to their original orbitals, the absorbed energy is now emitted as visible light. Thus, lithium imparts a crimson red colouration while sodium a yellow, potassium a violet and both rubidium and calcium impart a reddish violet colour to the flame. Similarly, the colour of the flame is brick-red in case of calcium, crimson in the case of strontium and apple green in the case of barium.

The colour imparted by an element depends upon the I.E. Higher the I.E., higher will be the frequency of radiation absorbed, and consequently lower will be frequency of radiation emitted i.e., the colour imparted. Since Li has higher I.E., it emits radiation of lower frequency i.e., red colour.

The same explanation also holds good for alkaline earths and the colours are also in accordance with the I.E. of these metals.

Beryllium and magnesium atoms are comparatively smaller and their I.E. are very high. Hence their electrons are more strongly bound and are not excited by the energy of the flame to higher energy states. These elements, therefore, do not give any colour in Bunsen flame.

Why only *s*-block elements give colouration in a flame and not other atoms?

Colour can be imparted to a flame if the I.E. of the element is low and the outermost electron absorbs radiation in the visible range of light. These conditions are satisfied by *s*-block elements and not others.

5.5.5 Variation of metallic character in the period and down the group in *s*-block elements

Alkali metals are the highly electropositive elements. For this reason, they form ionic compounds. Same argument applies to the alkaline earth metals. Since the I.E. and electronegativity decrease as we go down a group, the electropositive character and the metallic character increase as we go down groups 1 and 2. Thus Cs is more electropositive than Li and Ba is more electropositive than Mg. In fact Cs is the most electropositive element in the periodic table. All the *s*-block elements are soft and malleable metals. They are good conductros of heat and electricity. These are the characteristics of a metallic nature. Along the period, metallic properties decrease.

5.6 COMBINATION OF *s*-BLOCK ELEMENTS WITH WATER AND OXYGEN

Reaction of alkali metals with water

With water, hydrogen is liberated, *e.g.*,

$$2Na + 2H_2O \longrightarrow 2NaOH + H_2$$

The reactivity with water increases on going down from Li to Cs. This is because electropositive character increases in this order. With potassium, rubidium and caesium, the reaction is so vigorous that the evolved hydrogen catches fire. Alkaline earth metals react with water slowly. It is not certain if Be reacts with water at all.

$$Ca + 2H_2O \longrightarrow Ca(OH)_2 + H_2$$

Formation of oxides

Li forms oxide, Na forms peroxide and potassium form superoxide.

$$4Li + O_2 \longrightarrow \underset{\text{Lithium oxide}}{2Li_2O}$$

$$2Na + O_2 \longrightarrow \underset{\text{Sod. peroxide}}{Na_2O_2}$$

$$K + O_2 \longrightarrow \underset{\text{Pot. superoxide}}{KO_2}$$

Lithium forms only monoxide, Li_2O. In excess of oxygen sodium forms peroxide; Na_2O_2 while others form superoxides of the general formula MO_2. The superoxides contain the ion, O_2^- presumably, possessing the structure $(:\ddot{O} \text{ ⋯ } \ddot{O}:)^-$ which has a three electron bond. Due to the presence of one unpaired electron, the superoxides are generally coloured and paramagnetic.

The stability of superoxides increases with increasing atomic number of size. This is due to the fact that packing of larger cations and anions gives stable crystal lattices. The smaller Li^+ ion has a strong positive field around it which attracts the negative charge of O^{2-} anion so strongly that the latter iond does not bind another oxygen to form peroxide ion O_2^{2-}. On the other hand, the weak positive field around sodium and potassium ion allows the dispersal of the negative charge to form peroxide and superoxide ion respectively. This can be explained by the following reactions:

$$\underset{\text{Monoxide ion}}{O^{2-}} \xrightarrow{\frac{1}{2}O_2} \underset{\text{Peroxide ion}}{O_2^{2-}} \xrightarrow{O_2} \underset{\text{Superoxide ion}}{2O_2^-}$$

Example 7. **Why do alkaline earth metals form bivalent ions though the second ionisation energy (E_2) is much higher than the first ionisation energy (E_1)?**

Solution: Though E_2 is almost twice E_1 for alkaline earths yet there are no univalent positive ion $[M^+]$ compounds of alkaline earths. This is explained on the basis of the lattice energy accompanying the formation of compounds with dispositive ions which is much greater than the lattice energy of compounds of unipositive ions. The higher lattice energy of compounds of dispositive ions more than compensates for additional energy required to remove the second electron. Moreover in aqueous solution, M^{+2} ions have more hydration energy than that of M^+ ions. The large amount of hydration energy more than counterbalances the higher value of the second ionisation energy.

Example 8. **Lithium, an alkali metal, forms chloride having covalent nature. Explain.**

Solution. Lithium atom being small in size holds its outer electron firmly. Thus transfer of electron to chloride to produce an ionic compound is difficult.

It prefers to combine with chlorine by sharing of electrons and forms a covalent bond.

5.7 ALKALI METAL–LIQUID AMMONIA SOLUTIONS

All the alkali and alkaline earth metals except Be and Mg dissolve in liquid ammonia to give deep blue solutions which are good conductors of electricity. The dissolution of the metal is accompanied by its ionisation which yields ammoniated cation and ammoniated electron. For example,

$$Na \longrightarrow Na^+ + e^-$$

$$Na^+ + xNH_3 \longrightarrow Na(NH_3)_x^+$$
Ammoniated sodium cation

$$e^- - yNH_3 \longrightarrow e(NH_3)_y^-$$
Ammoniated electron

As the electropositive nature of the metal ion increases, the stability of ammoniated metal cation decreases down the group. The blue colour of the solution is due to the ammoniated cation as well as the ammoniated electrons present in the solution.

Reducing nature of alkali metals

The *s*-block elements show strong reducing properties. This is because of the ease with which they can lose their valency electrons because of low ionisation energy.

$$Na \longrightarrow Na^+ + e^-$$

$$Ca \longrightarrow Ca^{2+} + 2e^-$$

Some of the reducing properties of alkali metals are given below:

$$B_2O_3 + 6K \longrightarrow 2B + 3K_2O$$

$$Al\,Cl_3 + 3Na \longrightarrow 3NaCl + Al$$

Reducing character is also related to high oxidation potential which, in fact, represents the tendency to release electrons and get oxidised. This property increases as we go down the group.

Group 1 elements have higher oxidation potential than group 2 elements. The former are therefore stronger reducing agents than the latter. The difference is more striking among the top elements of the groups. Great reducing power of lithium is due to its large heat of hydration because of small size.

5.8 BASIC STRENGTH OF HYDROXIDE OF ALKALI AND ALKALINE EARTH METALS

The hydroxides of alkali and alkaline earth metals are basic. Alkali metal hydroxides such as NaOH are strong bases while the alkaline earth metal hydroxides such as $Ca(OH)_2$ are weak bases. This fact can be explained on the basis of I.E. of the metal, say M forming the hydroxide M–O–H. The compound M–O–H will act as an acid if the I.E. of M is high and as a base if the ionization energy of M is low.

The ionization energies of alkali metals are low, therefore, their hydroxides are basic in character. As the I.E. decreases from top to bottom in a group, the basic properties of their hydroxides increase as we move down the group. Thus, KOH is stronger than NaOH and Sr $(OH)_2$ is stronger than Mg $(OH)_2$. The ionization energies of alkalien earth metals are higher then those of alkali metals. The hydroxides of alkaline earth metals are, therefore, weaker than those of alkali metals.

If the elements M is less electropositive, it would have then high I.E. instead of releasing the electrons, it would tend to hold the electrons strongly and the bond M–O–H would not furnish OH– and would not act as a base. If the element M is more electropositive, it would have then low I.E. It will have a strong tendency to lose electron and furnish OH^- ions.

5.8.1 How does the electronegativity of M in M–OH classify it to be an acid or base?

If M is quite electropositive, it will help in the breaking of M–O bond in the hydroxide leading to the release of hydroxide ions.

$$M - O - H \longrightarrow M^+ + OH^-$$

Consequently the hydroxide will behave as a base. If on the other hand M is electronegative, it will attract electrons on oxygen and the bond O–H will be broken giving rise to protons.

$$M - O - H \longrightarrow M - O^- + H^+$$

Cosequently the hydroxide will behave as an acid.

5.9 DIAGONAL RELATIONSHIP

The first element in any given family of the periodic table tends to resemble in certain respect the second element in the family to the right. This resemblance is known as *diagonal relationship* which is largely explained by Fajans theory. Thus lithium and magnesium have much in common and the chemistry of Be and Al is surprisingly similar in spite of the difference in preferred oxidation state. Elements behave similarly if they have similar polarising power. Polarising power of the cation is determined by the ionic charge to ionic radius ratio. This property of the cation increases as we move to the right owing to the increase in the ionic charge and decreases when we move downwards owing to increase in the ionic radius. If we move diagonally from Li to Mg; from Be to Al or from B to Si, the two changes will more or less balance or another and we will have about the same polarising power and hence similar properties.

5.9.1 Diagonal relationship between lithium and magnesium

Lithium in many respects differs from the other alkali metals and resembles magnesium with which it is diagonally related. The resemblances are usually due to a similarity in ionic size and high polarising power of Li^+ and Mg^{2+} ions. The following points illustrate the abnormal behaviour of lithium in the alkali metal group and its similarity to magnesium.

(*i*) Both lithium and magnesium are harder metals having higher melting point than other metals in their respective groups.

(*ii*) Lithium is usually the least reactive element among the alkali metals. It is slowly attacked by water at 298 K whereas other alkali metals react vigorously. Magnesium also reacts slowly with water.

(*iii*) Lithium is the only alkali metal which combines direct with nitrogen to form the nitride Li_3 N. Similarly magnesium also combines with nitrogen yielding $Mg_3 N_2$. These are decomposed by water giving ammonia.

$$Li_3 N + 3H_2O \longrightarrow 3 LiOH + NH_3$$

$$Mg_3 N_2 + 6H_2O \longrightarrow 3 Mg(OH)_2 + 2NH_3$$

(*iv*) Lithium, when treated in air or oxygen, gives the normal oxides, Li_2O. The other alkali metals first yield the normal oxide M_2O which combines further with oxygen to form peroxides, M_2O_2 and superoxies, MO_2. Magnesium under similar conditions forms a normal oxide MgO only.

(*v*) Lithium hydroxide, like magnesium hydroxide, is sparingly soluble in water and decomposes at red heat to form the oxide. The other alkali metal hydroxides are quite soluble and remain unaffected on heating.

$$2Li\,OH \xrightarrow{\text{Red heat}} Li_2O + H_2O$$

$$Mg(OH)_2 \xrightarrow{\text{Red heat}} MgO + H_2O$$

(*vi*) The solubilities of lithium salts resemble those of the corresponds magnesium salts and differ markedly from those of the other alkali metals. Thus the fluorides, carbonates and phosphates of lithium and magnesium are sparingly soluble in water but the corresponding compounds of the other alkali metals are freely soluble.

(*vii*) Because of strong polarising power of Li^+ and Mg^{2+}, these ions have great tendency to form hydrates. Thus a number of lithium and magnesium salts crystallize from water with definite number of water molecules of crystallization. For example,

$$LiCl, 2H_2O \quad \text{and} \quad LiCO_3\ 3H_2O$$

$$MgCl_2.6H_2O \quad \text{and} \quad Mg(NO_3)_2\ 6H_2O$$

The salts of other alkali metals are not hydrated.

(*viii*) Again, because of the strong distorting effect of Li^+ and Mg^{2+} ions on the charge cloud of the anions, their oxy-compounds such as carbonates and nitrate possess low thermal stability and decompose to form the respective oxides.

$$Li_2CO_3 \xrightarrow{\text{heat}} L1_2O + CO_2$$

$$MgCO_3 \longrightarrow MgO + CO_2$$

$$4LiNO_3 \longrightarrow 2Li_2O + 4NO_2 + O_2$$

$$2Mg\ (NO_3)_2 \longrightarrow 2MgO + 4NO_2 + O_2$$

The carbonates of other alkali metals are thermally stable but the nitrates decompose at elevated temperatures yielding nitrites and oxygen.

$$2MNO_3 \longrightarrow 2MNO_2 + O_2$$

where M stands for alkali metal.

5.9.2 Diagonal relationship between Be and Al

Beryllium and aluminium show diagonal relationship because of the following reasons:

(1) Polarizing powers of Be^{2+} and Al^{3+} are of similar magnitude.

(2) The standard oxidation potentials of beryllium and aluminium are quite close to each other.

Beryllium resembles aluminium in the following respects:

(*i*) Both Be and Al metals dissolve rapidly in $3MH_2SO_4$ but very slowly in nitric acid in which the metals are rendered passive due to formation of an inert film of oxide on the surface of metals.

(*ii*) Be and Al both react with strong alkalies liberating hydrogen and forming beryllate and aluminate ions.

$$Be + 2OH^- \longrightarrow Be\,O_2^{2-} + H_2$$

$$2Al + 2OH^- + 2H_2O \longrightarrow 2AlO_2^- + 3H_2$$

(*iii*) Oxides and hydroxides of both the metals are amphoteric in nature. They dissolve in acids as well as in alkalies giving their corresponding salts and beryllate and aluminate ions respectively.

$$BeO + 2HCl \longrightarrow Be\ Cl_2 + H_2O$$

$$BeO + 2NaOH \longrightarrow Na_2BeO_2 + H_2O$$

$$Al_2O_3 + 6HCl \longrightarrow 2AlCl_3 + 3H_2O$$

$$Al_2O_3 + 2NaOH \longrightarrow 2\ Na\ AlO_2 + H_2O$$

$$Be\ (OH)_2 + 2HCl \longrightarrow Be\ Cl_2 + 2\ H_2O$$

$$Be\ (OH)_2 + 2NaOH \longrightarrow Na_2BeO_2 + 2H_2O$$

$$Al\ (OH)_3 + 3HCl \longrightarrow AlCl_3 + 3H_2O$$

$$Al\ (OH)_3 + NaOH \longrightarrow Na\ AlO_2 + 2H_2O$$

(*iv*) The anhydrous halides (except fluorides) of both Be and Al are covalent in nature and as such dissolve in organic solvents and fume in moist air.

(*v*) The carbides of Be and Al yeild methane on hydrolysis.

$$Be_2C + 4H_2O \longrightarrow 2\,Be(OH)_2 + CH_4$$
$$Al_4C_3 + 12H_2O \longrightarrow 4\,Al(OH)_3 + 2CH_4$$

***Example 9.* Why do lithium compounds show covalent character?**

***Solution*:** Because of the small size of the atom, the charge density of lithium is maximum. Lithium ion, therefore, possesses the greatest polarizing power out of all the alkali metal ions. It exerts a great distorting effect on a negative ion. Consequently the ion has a remarkably great tendency towards the covalent bond formation.

5.9.3 Points of difference between lithium and other alkali metals

Points of difference in the behaviour of lithium and the other alkali metals are given below:

(1) Lithium possesses higher melting and boiling points.

(2) Lithium is much harder than the other alkali metals.

(3) As compared to the other alkali metals, lithium is the least reactive.

(4) It directly combines with nitrogen to form nitride while other alkali metals do not.

(5) When burnt in air or oxygen, it only forms the monoxide Li_2O while other give higher oxides also. The monoxide is less basic than the oxides of other metals. Also it dissolves in water less readily.

(6) Lithium nitrate like magnesium nitrate evolves nitrogen dioxide and oxygen on heating.

$$4\,LiNO_3 \longrightarrow 2Li_2O + 4NO_2 + O_2$$

Sodium and potassium nitrates evolve only oxygen, leaving nitrites, e.g.

$$2Na\,NO_3 \longrightarrow 2Na\,NO_2 + O_2$$

(7) Its hydroxide decomposes at red heat to Li_2O. Also it is a much weaker base and is much less soluble as compared to the hydroxides of other alkali metals. Hydroxides of others sublime unchanged.

(8) Lithium carbonate is not stable to heat and gives CO_2 while carbonates of other alkali metals are very stable towards heat.

$$Li_2\,CO_3 \longrightarrow Li_2O + CO_2$$

(9) *Behaviour of chlorides*: (i) Lithium chloride crystallizes out as $LiCl.2H_2O$. Sodium and other alkali metal chlorides do not form hydrates at ordinary temperature.

(*ii*) Lithium chloride like $MgCl_2$ is deliquescent.

(*iii*) Lithium chloride hydrolyses in hot water to a small extent.

(10) *Solubiltity of salts*: (*i*) The cabonate, phosphate, oxalate and fluoride of lithium are only sparingly soluble in water. The corresponding salts of sodium and potassium are fairly soluble.

(*ii*) Lithium chloride and nitride are soluble in alcohol.

(*iii*) Lithium chloride and nitride are soluble in organic solvent.

***Example 10.* Conductivity of Li^+ in aqueous solution is smaller than that of Cs^+ ion though the former is much lighter than the latter. Explain.**

***Solution*:** Li^+ having a high charge density gets excessively hydrated and thus becomes heavy, that is why its conductivity is much smaller compared to Cs^+.

***Example 11.* Lithium is most extensively hydrated among alkali metals. Explain.**

Solution. All the alkali metal ions have unit positive charge but the atomic volume goes on increasing from Li to Cs. Hence charge density which is the ratio of charge to volume decreases from Li to Cs. Due to maximum value of charge density of Li, it gets hydrated to the maximum extent.

5.9.4 Points of difference between beryllium and other members of group 2

Points of difference between beryllium and other members of group 2 are given below:

(*i*) Melting point of beryllium is higher than those of rest of the members.

(*ii*) Beryllium is very hard metal while other alkaline earth metals are soft.

(*iii*) Unlike other alkaline earth metals, beryllium is not easily affected by dry air.

(*iv*) Beryllium does not liberate hydrogen from acids readily while others do so readily.

(*v*) Oxide and hydroxide of beryllium are amphoteric while others oxides are basic.

(*vi*) Beryllium does not decompose even in boiling water, while other members react with normal water.

(*vii*) Beryllium sulphate is soluble in water while sulphates of other metals are either sparingly soluble in water or insoluble in water.

(*viii*) In general, compounds of beryllium are covalent while compounds of other alkaline earth metals are ionic.

5.10 CHARACTERISTICS OF OXIDES OF *s*-BLOCK ELEMENTS

Alkali metal oxides are formed by burning the metals in air. The oxides are of three types. Monoxide Li_2O is formed by only lithium, peroxide Na_2O_2 is formed by sodium while the superoxides, MO_2, are formed by potassium and cesium. The oxides are related to each other as follows:

$$\underset{\text{Monoxide ion}}{O^{2-}} \xrightarrow{\frac{1}{2}O_2} \underset{\text{Peroxide ion}}{O_2^{2-}} \xrightarrow{O_2} \underset{\text{Superoxide ion}}{2(O_2^-)}$$

Formation of superoxides is favoured by heavier alkali members. This is due to the comparable sizes of the large cations and O_2^- anions, the packing of which gives stable crystal lattices. The superoxide contains the ion $(:\ddot{O} \cdots \ddot{O}:)^-$ which has a three-electron bond. Its presence in superoxides is supported by the X-ray data and paramagnetism of the compounds. The monoxides are ionic, strongly basic and highly soluble in water. Their solutions are strongly alkaline and possess M^+ and OH^- ions.

Higher oxides such as peroxides and superoxides behave as strong oxidisign agents. On treatment with dilute acids they form hydrogen peroxide and oxygen respectively as shown below:

$$\underset{\text{Sodium peroxide}}{Na_2O_2} + H_2SO_4 \longrightarrow Na_2SO_4 + H_2O_2$$

$$\underset{\text{Potassium superoxide}}{4KO_2} + 2H_2SO_4 \longrightarrow 2K_2SO_4 + 2H_2O + 3O_2$$

Alkaline earth metals give two types of oxides. Normally oxide, MO, is formed either by heating the metal in oxygen or by decomposing the carbonates at high temperature.

$$2Ca + O_2 \xrightarrow{\text{Heat}} 2\,CaO$$

$$CaCO_3 \longrightarrow CaO + CO_2$$

Normal oxides are ionic and basic in nature except beryllium oxide which is amphoteric. They dissolve in water giving hydroxides. All of them have lattice of NaCl type.

5.10.1 Comparison of basicity of hydroxides of alkali metals and alkaline earth metals

Alkali metal hydroxides with the general formula M – O – H are obtained when metals react with water. These are highly soluble in water.

$$2M + 2H_2O \longrightarrow 2MOH + H_2$$

Since I.E. of alkali metals is very low, their hydroxides act as strong bases. The basic character of the alkali metal hydroxides increases on moving down the group as the I.E. of the metals decreases in the same order. LiOH is the weakest and CsOH is the strongest base. Alkali metal hydroxides are not decomposed on heating.

Alkaline earth hydroxides are obtained when metal monoxides (MO) react with water.

$$CaO + H_2O \longrightarrow Ca(OH)_2$$

Magnesium oxide reacts only slowly with water than those of alkali metal hydroxides. Alkaline earth hydroxides are less basic. The solubility of the alkaline earth hydroxides increases with increasing size of the metals down the group. In fact, none of them is freely soluble in water. Magnesium hydroxide is very little soluble whereas calcium hydroxide is sparingly soluble. Consequently Mg $(OH)_2$ may be prepared by the addition of an alkali to an aqueous solution of its salts.

$$\underset{\text{Solution}}{Mg\,Cl_2} + 2NaOH \longrightarrow Mg\,(OH)_2 + 2NaCl$$

Solid calcium hydroxide is known as **slaked lime**, a dilute solution as **lime water** and the milky suspension as **milk of lime**.

***Example 12.* Be$(OH)_2$ is insoluble in water but Ba $(OH)_2$ is soluble. Explain.**

Solution. Beryllium ion has a small size and a high positive charge density. According to Fajan's rule, it will exhibit covalent character and is therefore insoluble in water in contrast to Ba$(OH)_2$.

5.10.2 Hydrides of alkali and alkaline earth metals

Alkali and alkaline earth metals form hydrides of the type $M^+ H^-$ and $M^{2+} (H^-)_2$ respectively. These are obtained by the direct combination of metals with hydrogen at high temperatures. The stability of the hydrides decreases down the group because of increasing difference between cation and hydride sizes and consequent decreasing stability of lattices.

$$2Na + H_2 \longrightarrow 2NaH$$

$$Mg + H_2 \longrightarrow MgH_2$$

The hydrides react with water forming the corresponding hydroxides with liberation of hydrogen.

$$MH + H_2O \longrightarrow MOH + H_2$$

$$CaH_2 + 2H_2O \longrightarrow Ca(OH)_2 + 2H_2$$

All hydrides (except those of Li and Be) are ionic in nature. Hydrogen in these compounds gets liberated at the anode on hydrolysis, showing the presence of hydride ion H^- in them.

Magnesium hydride is more conveniently prepared by the action of lithium-aluminium hydride on magnesium in ethernal solution.

$$2Mg\,Cl_2 + LiAlH_4 \rightarrow 2\,MgH_2 + LiCl + AlCl_3$$

5.10.3 Halides of alkali and alkaline earth metals

(1) All the alkali metals form ionic halides of the type MX where X is F, Cl, Br or I.

Alkali metals form halides readily while alkaline earth metals require elevated temperatures.

$$2Na + H_2 \longrightarrow 2\,NaH \qquad Ca + H_2 \longrightarrow CaH_2$$

(2) Lithium halides, except fluoride, have some covalent character also. This is due to the polarizing effect of small lithium ion on large halogen atoms.

(3) Cesium halides crystallize in a body centred cubic lattice where the coordination number is eight while all other halides form cubic crystals and show a coordination number of six.

(4) The thermal stability of halides increases with increasing electropositive character down the group. They are all soluble in water. Lithium halides except fluoride are also soluble in organic solvents such as alcohol and ether.

(5) Alkaline earth metals form halides of the type MX_2 where X is F, Cl, Br or I. Fluorides of Ca, Sr and Ba are insoluble and precipitate from aqueous solution of the metal salt by adding a soluble fluoride. For example.

$$CaCl_2 + 2\,NaF \longrightarrow \underset{\text{Precipitate}}{CaF_2} + 2\,NaCl$$

Magnesium fluoride is obtained by dissolving magnesium in aqueous hydrofluoric acid. Hydrated chlorides $MCl_2.\ 6H_2O$ where M is Mg, Ca and Sr and $BaCl_2.\ 2H_2O$ crystallize from solution when metal carbonates are dissolved in dilute hydrochloride acid.

$$MCO_3 + 2HCl \longrightarrow MCl_2 + CO_2 + H_2O$$

(6) The fluorides of Mg, Ca, Sr and Ba are ionic character whereas their chlorides, bromides and iodides are appreciably covalent. Covalent character of these halides increases with decreasing size of the cation and increasing size of the anions.

5.10.4 Carbonates and bicarbonates of alkali and alkaline earth metals

(1) Alkali metals form carbonates of the general formula $M_2\,CO_3$. All the carbonates are white solids, soluble in water and thermally stable. However, lithium carbonate is an exception. It is sparingly soluble and thermally unstable. On decomposition, it gives lithium oxide and carbon dioxide. Due to strong polarizing nature of small Li^+ ions, it distorts the electron cloud of the oxygen atom in $CO^2{}_3{}^-$ ion so that Li – O bond becomes stronger as compared to C – O bond. It can also be explained in terms of more stable lattice of Li_2O than $Li_2\,CO_3$

$$Li_2\,CO_3 \xrightarrow{\text{Heat}} Li_2O + CO_2$$

(2) The carbonates of K, Rb and Cs are deliquescent. All carbonates are decomposed by acids liberating carbon dioxide. Their aqueous solutions are alkaline due to hydrolysis.

$$CO_3^{2-} + 2H^+ \longrightarrow CO_2 + H_2O$$

$$CO_3^{2-} + H_2O \xrightarrow{\text{Hydrolysis}} HCO_3^{2-} + OH^-$$

(3) Solid bicarbonates are known for all the alkali metals except lithium. The stability of the bicarbonates increases with the electropositive nature of the metals down the group. All bicarbonates are white solids, soluble in water and decompose on gently heating to give carbonates and liberate carbon dioxide.

$$2MHCO_3 \xrightarrow{\text{Heat}} M_2CO_3 + H_2O + CO_2$$

The normal carbonates of calcium, strontium and barium can be precipitated by the addition of sodium carbonate solution to the solution of a salt of these metals.

$$MCl_2 + Na_2\,CO_3 \longrightarrow MCO_3 + 2\,NaCl$$

On heating, all carbonates decompose to give metal oxides and liberate carbon dioxide.

$$MCO_3 \xrightarrow{\text{Heat}} MO + CO_2$$

The thermal stability of the carbonates increases with increasing size of the metal cation down the group. Alkaline earth metal carbonates are generally insoluble but dissolve in water in presence of carbon dioxide forming bicarbonates. For example,

$$CaCO_3 + H_2O + CO_2 \longrightarrow Ca(HCO_3)_2$$

***Example 13.* BeCO$_3$ is less stable than MgCO$_3$. Explain.**

Solution. Be^{2+} ion is a small ion. It does not form a stable lattice with carbonate ion. That is why on heating it decomposes to BeO.

$$BeCO_3 \longrightarrow BeO + CO_2$$

Mg^{+2} forms a stable compound.

5.11 ALKYLS AND ARYLS OF ALKALI METALS (ORGANOMETALLIC COMPOUNDS)

Alkali metals, especially lithium, form a number of organometallic compounds which are frequently employed in the synthesis of a variety of organic compounds.

Organolithium compounds aer generally synthesised as follows:

(*i*) $Li(n\text{–}C_3H_7) + C_6H_5I \xrightarrow[\text{ether}]{\text{Solvent}} Li\ (C_6H_5) + n\text{–}C_3H_7I$

(*ii*) $4Li\ (C_6H_5) + Sn(CH_2 - CH = CH_2)_4 \xrightarrow[\text{ether}]{\text{Solvent}} 4Li\ (CH_2 - CH = CH_2) + Sn\ (C_6\ H_5)_4$

(*iii*) $2Li + C_2H_5I \xrightarrow[\text{ether}]{\text{Solvent}} LiI + Li(C_2H_5)$

Organometallic compounds of other alkali metals such as Na $(C_6\ H_5)$ Ph_2 C(Na)C(Na)Ph_2, $CsCH_2CH_2Cs$, Cs, are also known. These are generally prepared from the corresponding mercury compounds. For example,

$$2K + HgR_2 \longrightarrow Hg + 2KR$$

5.11.1 Applications of alkali metal alkyls and aryls

Lithium alkylis are extensively used in the synthesis of organic compounds and,organometallic compounds of other elements. For example,

(*i*) $3LiR + BCl_3 \longrightarrow 3LiCl + BR_3$

(*ii*) $Li(C_6H_5) + (CH_3C_6H_4)_2CO \longrightarrow (CH_3C_6H_4)_2\ C\ (C_6H_5)OLi$

$$\downarrow H_2O$$

$$(CH_3C_6H_4)_2\ C(OH)C_6H_5 + LiOH$$

(Tertiary alcohols)

(*iii*) $Li(C_6H_5) + CO_2 \xrightarrow[\text{solvent}]{\text{Nonpolar}} [C_6\ H_5CO_2Li]$

$$\downarrow H_2O$$

$$C_6H_5COOH + LiOH$$

Carboxylic acid

Alkyl and aryl compounds of lithium are polymeric having considerable covalent character in their Li–C bonds whereas such organometallic compounds of other alkali metals have predominantly ionic character so that these can be represented as M^+R^-. The alkyl and aryl of sodium, potassium, etc., are much more reactive than the corresponding alkyls and aryls of lithium.

5.11.2 Structure of beryllium dialkyls

X-ray analysis reveals a chain polymeric structure for dimethyl beryllium ($BeMe_2$) as depicted below in Fig. 5.2.

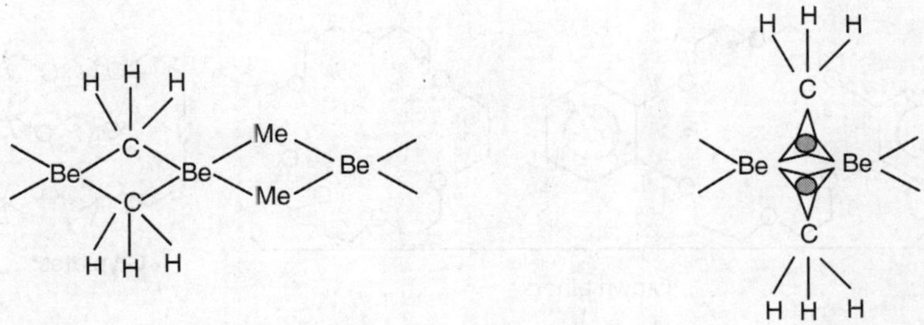

Fig. 5.2. Structure of Be $(Me)_2$.

The bonding can be explained in terms of 3-centre, 2-electron bridge bonds involving Be (CH_3) Be bonds. Each Be atom is sp^3 hybridized and has two electrons. Each bridging methyl carbon atom contributes one electron making a total of 4–electrons for $[BeMe_2]$ group. Each Be —(CH_3)—Be may then be described in terms of 3–centre, 2–electron (3c–2e) bonding as in electron-deficient diborane molecule.

5.12 COMPLEXES OF ALKALI METALS

Alkali metal ions with the exception of Li^+ have very little polarising power due their large size and only a unit positive charge. As a result, they have only a week tendency to form complexes with unidentate ligands. There is concrete evidence for the existence of $[M(H_2O)_4]^+$ (where M is Li^+, Na^+, and K^+) and $[M'(H_2O)_6]^+$ (where M′ is Rb^+ and Cs^-) in aqueous solutions. Also, lithium iodide reacts with liquid ammonia to produce $[Li(NH_3)_4]^+$.

However, alkali metal ions form reasonably stable complexes with some of the bidentate and polydentate ligands such as salicylaldehyde, β-diketones, nitrophenols, α-nitroso-β-naphthol, etc. Thus, when NaOH is added to a solution of salicylaldehyde, a neutral complex is formed which is soluble in organic solvents.

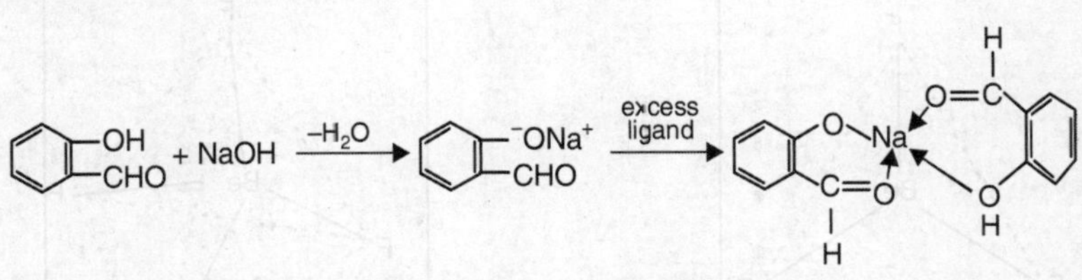

Similarly, the anhydrous sodium derivative of acetylacetone takes up water to produce a hydrated compelx which is also soluble in organic solvents.

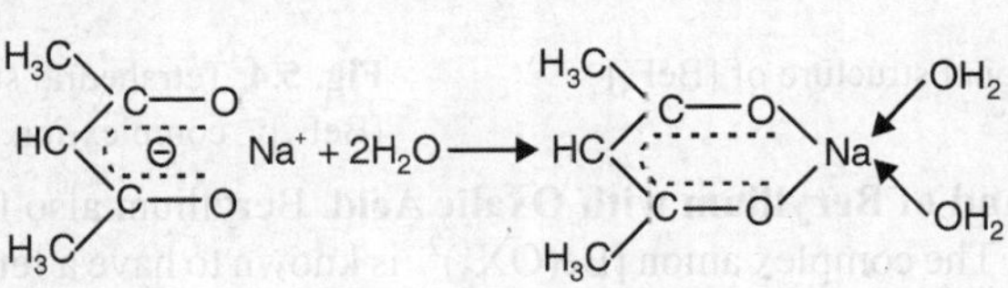

The 'Wrap Around' Complexes. Polydentate ligands such as *crown ethers* and cryptates are the most important ligands which form *highly stable complexes with alkali metal ions*. These ligands are also highly relective in complex formation. These ligands warp the metal ions completely and protect them from coming into contact with the solvent.

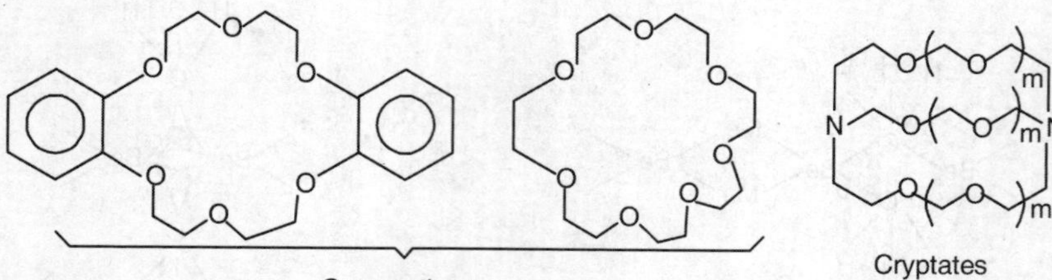

The cause of selectivity lies in the size of the 'hole' available in polydentate ligand of the type mentioned above. Only that alkali metal ion which *exactly fits into* the 'hole', forms stable complex. The cations which are either too small or too big to fit in the hole, would not form stable complexes. For example, the 'hole' in cryptate-222 is just of the right size for the K^+ ion to fit in whereas Li^+ and Na^+ ions are *too small* to make contact with the oxygen atoms of the ligand. On the other hand, Rb^+ and Cs^+ ions are *too big* in size to fit in the hole without causing steric strain. The K^+ complex of cryptate-222 is thus most stable. Similarly, Na^+ complex of crystal-221 is most stable out of all the alkali metal ion complexes formed with this ligand.

5.12.1 Complex formation by alkaline earth metals

Be, because of its small size, forms stable complexes with halogens, as, for example, $[BeF_3]^-$ and $[BeF_4]^{2-}$.

The complex ion, $[BeF_3]^-$, involves sp^2 hybridisation and therefore, has a **planar trigonal** structure, as shown in Fig. 5.3.

The complex ion $[BeF_4]^{2-}$ involves sp^3 hybridisation and hence has a **tetrahedral** structure (Fig. 5.4).

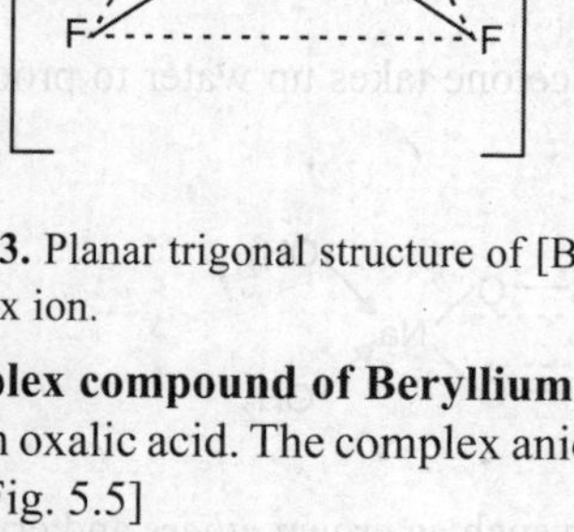

Fig. 5.3. Planar trigonal structure of $[BeF_3]^-$ complex ion.

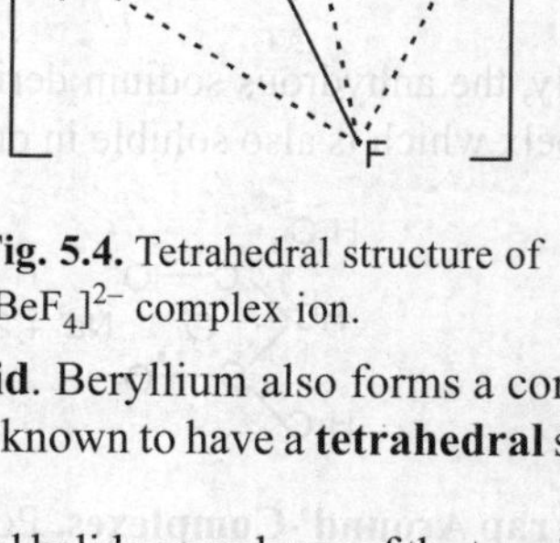

Fig. 5.4. Tetrahedral structure of $[BeF_4]^{2-}$ complex ion.

Complex compound of Beryllium with Oxalic Acid. Beryllium also forms a complex compound with oxalic acid. The complex anion $[Be(OX_2)]^{2-}$ is known to have a **tetrahedral** structure, as shown in Fig. 5.5]

Complex of Magnesium. Magnesium forms tetrahedral halide complexes of the type $[(C_2H_5)_4N]_2[MgCl_4]$. The chelates of magnesium with EDTA and pyrophosphates are also known. However, the most important complex of magnesium is the naturally occuring **chlorophyll**, the green pigment present in plants, which photosensitises the synthesis of carbohydrates in nature from CO_2 and H_2O as, for example,

$$6CO_2 + 6H_2O \xrightarrow[\text{chlorophyll}]{h\nu} C_6H_{12}O_6$$

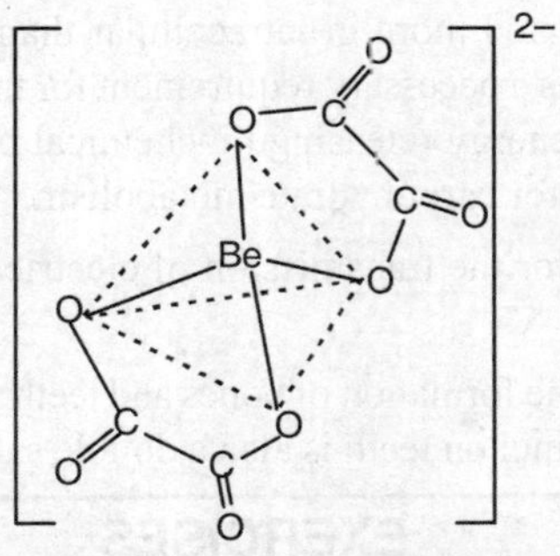

Fig. 5.5. Structure of beryllium oxalate, $[Be(OX)_2]^{2-}$ ion.

The macrocyclic ligand involved is derived from the parent tetrapyrrole molecule porphylin. The structure of chlorophyll is shown in Fig. 5.6

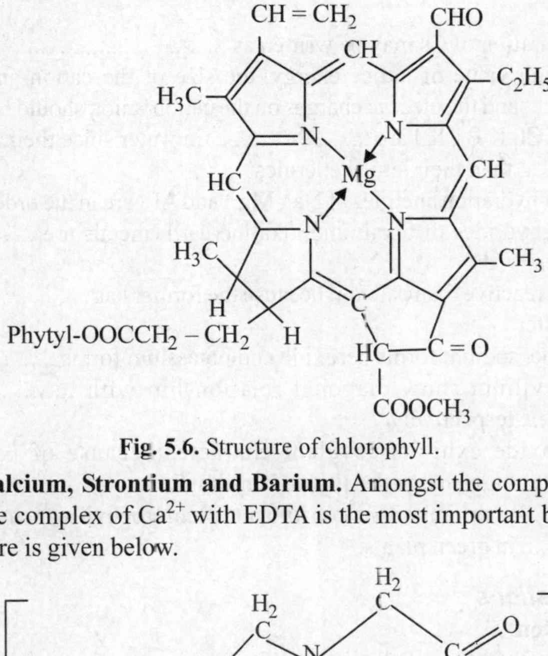

Fig. 5.6. Structure of chlorophyll.

Complexes of Calcium, Strontium and Barium. Amongst the complexes of Ca^{2+}, Sr^{2+} and Ba^{2+}, the water soluble complex of Ca^{2+} with EDTA is the most important because of its analytical application. Its structure is given below.

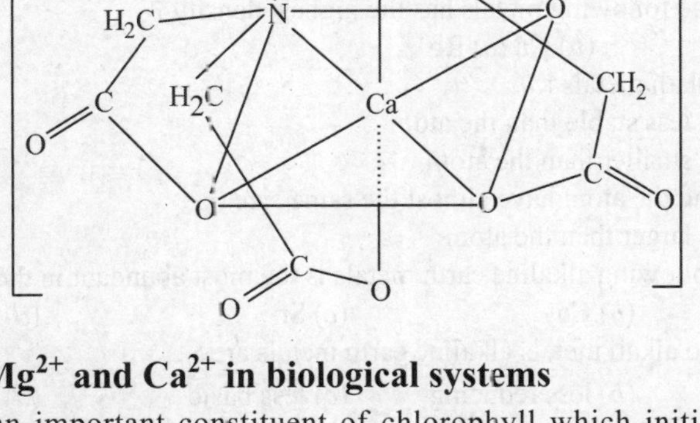

5.12.2 Role of Mg^{2+} and Ca^{2+} in biological systems

Magnesium is an important constituent of chlorophyll which initiates the process of photosynthesis in green plants.

Magnesium ions are concentrated more in intracellular than in extracellular fluids in animal bodies. The presence of Mg^{2+} ions is a necessary requirement for the activation of phosphate transfer enzymes which participate in the energy releasing biochemical process occurring in animal body. Magnesium ions are also required for carbohydrate metabolism.

Mg^{2+} and Ca^{2+} ions are vital for the transmission of electrical impulses along the nerve fibre and for the contraction of muscles.

Calcium ions are essential for the formation of bones and teeth and for maintaining rythm of heart and for clotting for blood. The enamel on teeth is also a double salt of calcium, 3 $Ca_3(PO_4)_2$, CaF_2.

EXERCISES
(INCLUDING QUESTIONS FROM DIFFERENT UNIVERSITY PAPERS)

Objective Questions

Fill in the blanks

1. Electronic configuration of Cs may be written as ....
2. In order to get high value of lattice energy, the size of the cation and/or anion should be .... and the electric charges on the cation/anion should be ....
3. Ionic solids like KCl, K Br, K I are .... in water since their hydration energies are .... than their lattice energies
4. Increasing order of hydration energies of Na^+, Mg^{2+} and Al^{3+} are in the order ....
5. The two types of hydrides of alkali and alkaline earth metals are .... and ....
6. Potassium is more reactive than calcium because the former has .... ionisation energy than the latter
7. Lithium form oxide, sodium forms peroxide and potassium forms ....
8. Lithium and beryllium show diagonal relationship with .... and .... respectively.
9. Beryllium hydroxide exhibits covalent character because of .... size and .... high charge density on beryllium ion.
10. .... is an important constituent of chlorophyll which initials the process of .... in green plants.

Multiple Choice Questions

Choose the correct option

1. The outoumost configuration of alkali metals is :
   (*a*) ns^1 (*b*) np^1 (*c*) ns^2 (*d*) none of the above
2. Which one of the following metals has the highest density ?
   (*a*) Li (*b*) Na (*c*) Rb (*d*) Cs
3. In the case of alkali metals :
   (*a*) the cation is less stable than the atom
   (*b*) the cation is smaller than the atom
   (*c*) the cation and the atom have almost the same size
   (*d*) the cation is larger than the atom
4. Which one of following alkaline earth metals is the most abundant in the earth's crust ?
   (*a*) Be (*b*) Ca (*c*) Sr (*d*) Ra
5. In comparison to alkali metals alkaline earth metals are :
   (*a*) less reactive (*b*) less reducing (*c*) less basic (*d*) all the three
6. Which one of the following alkali metals is used in photoelectric cells ?
   (*a*) Li (*b*) Na (*c*) Cs (*d*) Fr

7. Which of the following is not an alkali metal?
(a) Rh (b) Cs (c) Fr (d) Rb
8. Alkali metals contain :
(a) 3 valency electrons (b) 1 valency electron
(c) 4 valency electrons (d) 2 valency electrons
9. Of the four alkali metals given below which has the smallest size ?
(a) Rb (b) K (c) Ma (d) Li
10. Density is the highest for :
(a) Mg (b) Ca (c) Sr (d) Ba
11. Which of the following alkaline earth metals shows some properties similar to aluminium ?
(a) Be (b) Ca (c) Sr (d) Ba
12. Which of the following alkali metals is the most abundent in the earth's crust ?
(a) Li (b) K (c) Rb (d) Cs
13. The correct sequence of the alkali metals in the group is :
(a) Fr, Na, K, Rb,CS, Li (b) Li, Na, K, Rb, Cs, Fr
(c) Ma, K, Rb, Cs, Fr, Li (d) Rb, Cs, Li, Na, K, Fr
14. Which of the following is a liqid at room temperature (37°C) ?
(a) Na (b) Fr (c) Os (d) Cs
15. The most electropositive amongst the alkaline earth metals is :
(a) beryllium (b) magnesium (c) calcium (d) barium
16. Which one of the alkaline earth metals shows some anomalous behaviour and has the same electronegativity as aluminium :
(a) Ba (b) Sr (c) Ca (d) Be
17. The wire of flash bulb is made of
(a) Mg (b) Cu (c) Ba (d) Ag
18. Magnesium is present in
(a) ascorbic acid (Vitamin C) (b) haemoglobin
(c) chlorophyll (d) vitamin B_{12}
19. Which of the alkaline earth metal sulphates is the least soluble ?
(a) Be SO_4 (b) Mg SO_4
(c) Sr SO_4 (d) $BaSO_4$
20. Which of the following imparts brick red colour to the flame ?
(a) Ca (b) Ba (c) Sr (d) none of these

ANSWERS

Fill in the blanks

1. [Xe] $6s^1$ **2.** small, high **3.** insoluble, smaller **4.** $Na^+ < Ba^{2+} < Al^{3+}$
5. ionic hydrides, molecular hydrides **6.** smaller **7.** superoxide
8. magnesium, aluminium **9.** small, high **10.** Mg, photosynthesis

Multiple choice questions

1. (a)	**2.** (d)	**3.** (b)	**4.** (b)	**5.** (d)
6. (c)	**7.** (a)	**8.** (b)	**9.** (d)	**10.** (d)
11. (a)	**12.** (b)	**13.** (b)	**14.** (b)	**15.** (d)
16. (d)	**17.** (a)	**18.** (c)	**19.** (a)	**20.** (a)

General Questions

1. How do you account for the gradation in the following properties of the alkali metals?
   (*a*) Electropositive character
   (*b*) Ionisation energy
   (*c*) Hydration energies
   (*d*) Strength of covalent bond in M_2 molecules.
2. (*a*) How does lithium differ from its congeners?
   (*b*) Show that lithium resembles more with magnesium than with its congeners.
3. (*a*) How do the following properties of alkali metals vary with rise in atomic number?
   (*i*) Ionization energy (*ii*) Reactivity towards air and water.
   (*b*) Which of the following compounds will be paramagnetic?
   K_2O, K_2O_2, KO_2, K_3N
4. (*a*) Why would you not expect to find oxides and sulphides of alkali metals in nature as is common with many other metals?
   (*b*) Discuss that the conductivity of lithium ion in aqueous solution is less than that of cesium although the sizes of Li^+ and Cs^+ suggest an opposite order.
5. (*a*) The lattice energies of LiF, NaF, KF, RbF and CsF are 1000, 890, 795, 760, 725 kcal $mole^{-1}$ respectively.
   (*i*) How can you account for the gradually diminishing value?
   (*ii*) Which of the above fluoride do you expect to be least soluble?
   (*b*) Lithium has the lowest reduction potential of the alkali metals. How can this be reconciled with the fact that ionisation energy of lithium is also the highest in group 1.
6. Explain the following:
   (*a*) Why does Li forms normal oxide, Na the peroxide and K, Rb and Cs the superoxides?
   (*b*) Why do solutions of alkali metals in liquid ammonia have high conductivity, are paramagnetic and possess blue colour in dilute solution whereas are copper coloured in concentrated solutions.
   (*c*) Electrolysis of an aqueous solution of sodium chloride does not yield metallic sodium.
   (*d*) Inspite of fact that Li is an alkali metal, its halides have covalent character.
   (*e*) Potassium hydroxide is a stronger base than barium hydroxide.
   (*f*) Lithium ion has the maximum hydration energy.
   (*g*) Why does the chemical reactivity of alkali metals increase from Li to Cs?
   (*h*) How does the reactivity and nature of products vary from Li to Cs when the alkali metals react with oxygen?
   (*i*) What is the active reducing agent in solutions of alkali metals in liquid ammonia? What prevents the formation of stable solutions of alkali metals in water?
   (*j*) How would you extinguish a sodium fire in the laboratory?
   [**Hint:** Since sodium reacts vigorously with water, sodium fire should not be extinguished with water. Carbon tetrachloride (pyrene) should be used to extinguish sodium fire in the laboratory.]
7. (*a*) What are the important consequences of loose binding of *s*-electron in alkali metals?
   (*b*) The hydration energies of Group 1 ions have been estimated to be Li^+ 515, Na^+ 450, K^+ 322, Rb^+ 293 and Cs^+ 263 K cal. $mole^{-1}$. Suggest a reason for the trend which is observed.

8. Give reasons as to why alkali metals

(*a*) have low ionization energies,

(*b*) are strong reducing energies,

(*c*) are poor complexing agents,

(*d*) lose electrons even when exposed to light,

(*e*) have low melting and boiling points,

(*f*) are soft, volatile and highly electropositive.

9. What are the general trends in the solubilities of the following alkaline earth salts:

(*a*) Sulphates, (*b*) Hydroxides, (*c*) Iodates

Give reasons for your answer.

10. Account for the following:

(*a*) The conductivity of Li^+ ion in equeous solution is less than that of Cs^+ ion.

(*b*) Group 2 elements form compounds in which they exhibit +2 oxidation state.

11. Explain, why

(*a*) Calcium forms CaF_2 but neither CaF_3 nor CaF.

(*b*) Group 2 salts are generally less soluble than Group 1 salts

(*c*) Beryllium forms covalent compounds inspite of its having no unpaired electrons.

(*d*) Magnesium does not impart any colour to the flame while calcium does.

(*e*) Linear molecules, X—Be—X exist only in the gas phase.

(*f*) Alkaline earth metals only form compounds in which they exhibit +2 oxidation state.

12. Compare the alkaline earth metals with alkali metals with regards to the following properties:

(*a*) Ionisation energies, (*b*) Reducing behaviour, (*c*) Flame spectra, (*d*) Ionic radii, (*e*) Melting and boiling points, (*f*) Density.

13. (*a*) To what do we attribute the diagonal relationship that exists between Li and Mg? List three ways in which lithium and magnesium are similar to one another.

(*b*) Beryllium resembles more with aluminium than with alkaline earth metals. Explain how and why.

14. (*a*) How will you account for the fact that the first ionisation energies of alkaline earth metals are very low as compared to the second?

(*b*) Which ion in group 2 would you expect to have the highest ionic mobility?

(*c*) Why do alkaline earth metals have higher melting points than alkali metals?

(*d*) How does the basic strength of alkaline earth metal hydroxides vary within the group and why?

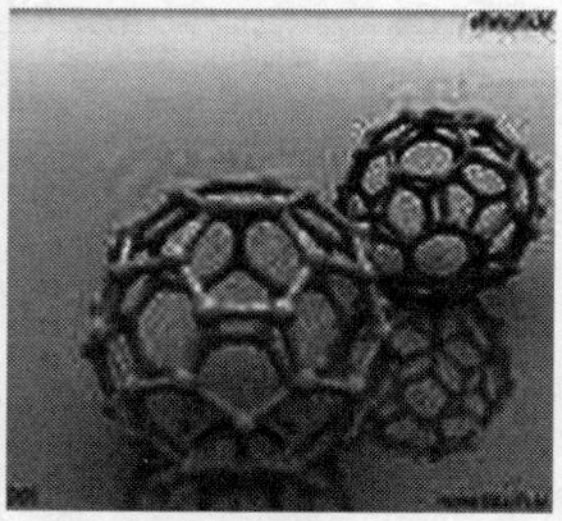

Fullerenes

6

p–Block Elements

p-Block Elements

Group 13	Group 14	Group 15	Group 16	Group 17
Boron (B)	Carbon (C)	Nitrogen (N)	Oxygen (O)	Fluorine (F)
Aluminium (Al)	Silicon (Si)	Phosphorus (P)	Sulphur (S)	Chlorine (Cl)
Gallium (Ga)	Germanium (Ge)	Arsenic (As)	Selenium (Se)	Bromine (Br)
Indium (In)	Tin (Sn)	Antimony (Sb)	Tellurium (Te)	Iodine (I)
Thallium (Tl)	Lead (Pb)	Bismuth (Bi)	Polonium (Po)	Astatine (At)

6.1 INTRODUCTION

All the elements belonging to groups 13, 14, 15, 16, 17 and 18 in which *p*-orbitals are being progressively filled are collectivley called *p*-block elements since their properties depend upon the electrons present in *p*-orbitals.

Variations of physical and chemical properties. The physical properties of elements in a group show gradual variations while there prevails similarity in their chemical properties. Again, the properties of elements in a row show no similarity, but show gradual variations of the various properties, with the increasing atomic number.

6.1.1 Variation of valency of *p*-block elements in the period

As we pass from left to right in a period, the number of electrons in the outermost orbital increases from 3 to 8. It can be shown that the valency of an element in a group is either the number of electrons in the outermost shell or eight minus the valence electrons. Thus the elements of the 13 group which have three electrons in the outermost shell have valency equal to three. Elements of 15 group have valency equal to (8–5) i.e., 3. Phosphorus also exhibits a valency of 5 in some cases when it shares all its 5 valency electrons with other atoms. Oxygen is in group 16 and its valency is 8–6, i.e., 2. Fluorine is in gruop 17. It needs to gain one electron more either by electrovalent or covalent bond formation. Thus its valency is $8 - 7 = 1$.

6.1.2 Variation of atomic radius in a period and in a group in *p*-block elements

Variation in a period. Atomic radii decrease from left to right across a period in the periodic table. This is explained on the basis of increasing nuclear charge which pulls the electrons inside and thus decreases the atomic size and radius.

Variation in a group. On descending a group in a periodic table, such as moving from B to Al to Ga to In and finally to Tl the size of the atoms increases due to the effect of extra shells of electrons being added. This effect is so large that it outweighs the effect of increased nuclear charge. Hence the radius of the atom increases in moving from top to bottom in a group.

6.1.3 Variation of ionization energy in a group and along a period in *p*-block elements

(*a*) **Variation in a group.** The values of I.E. reveal that there is a regular decrease in going from top to bottom in a group.

(*i*) There is simultaneous increase in the number of energy levels and therefore, the atomic size increases.

(*ii*) There is an increase in the number of electrons that lie in between the nucleus and the outermost electron and therefore, the shielding effect on the outermost electron gradually increases.

The increase in the atomic size and the shielding effect dominate the effect of the increased nuclear charge and the electron in the outer orbital is rendered less firmly bound to the nucleus. Consequently, there is gradual decrease in ionization energy.

(*b*) **Variation in a period.** It is observed that in general, the ionization energy increases with increasing atomic number, as we move from left to right along a period. We have the following observations:

(*i*) The nuclear charge increases from left to right.

(*ii*) The number of main energy levels remains the same while the atomic size decreases. Both these factors tend to increase the ionization energy.

As we move from B to C and to N, ionization energies keep on increasing due to increasing nuclear charge and decreasing atomic size, the combined effect of other factors not being appreciable. Oxygen has a slightly smaller ionization energy than N. This is primarily due to the more stable electronic arrangement of nitrogen in which all the orbitals are half-filled while the next *p*-electron in oxygens are somewhat less tightly held. Half-filled and fully filled electronic configurations are more stable than others.

The expected increase in ionization energy is observed in going from O to F and to Ne.

6.1.4 Variation of electron affinity in a group and in a period in *p*-block elements

Energy released in the process in which an electron is added to an isolated neutral gaseous is called electron affinity.

$$\underset{\text{Neutral atom}}{X} + e \longrightarrow \underset{\text{Anion}}{X^-} + \text{Energy}$$

So electron affinity is a measure of tightness with which an additional electron is bound to atom. The greater the energy released in the process of taking up the extra electron, the greater is the electron affinity.

It is observed that:

(*i*) Electron affinity generally decreases on moving down the group. This is expected on account of increase in the size of the atom on moving down the group. Due to increase in the size of atoms, the effective nuclear attraction for electrons decreases and hence there is less tendency to attract additional electrons with increase in atomic number.

(*ii*) Electron affinity generally increases as we move from left to right in a period. This is due to the increase in the nuclear charge which results in greater attraction for electrons. In the second period, fluorine has a maximum value of electron affinity. As the electron affinity increases the element becomes less metallic and more non-metallic, *i e.*, the electron affinities of metals are low while those of non-metals are high.

(*iii*) The halogens have high electron affinity. This is due to the fact that the addition of an electron leads to a stable electronic configuration of $s^2\ p^6$ of the valence shell.

(*iv*) Chlorine possesses high value of E.A. than fluorine which is contrary to expectation. The unexpected low value of fluorine as compared with chlorine may be associated with strong inter-electronic repulsive force between the electron charge cloud of fluorine atom and the added electron in the relatively small and compact $2p$ sub-level. Thus due to the small size and non-availability of d-orbitals in fluorine atom, there is a high electron density in the p sub-level which exerts a strong electron-electron repulsion on the incoming electron. Hence, the electron affinity of fluorine is lower than that of chlorine.

(*v*) The electron affinities of noble gases have zero values since their outermost shells have completed ns^2 and np^6 orbitals and the electron added has to go to a new orbital with a higher value of n (principal quantum number).

Similarly the E.A. of beryllium and magnesium also have zero values since the $2s$ and $3s$ orbitals are completely filled respectively and the atoms do not accept an additional electron. Hence also the added electron will have to go to the sub-shell p of considerably high energy. Sometimes their E.A. are given a negative sign.

(*vi*) Nitrogen and phosphorus have lower value of electron affinities. This is due to the fact that their p-subshells are half-filled and so they are extra stable. This decreases the possibility of adding an electron.

6.1.5 Variation of electronegativity in a group and in a period

The tendency of an atom to attract the shared electrons in a molecule towards itself is called the electronegativity of the atom. It is different from electron affinity which is a measure of the tendency of an isolated atom to attract electrons. Electron affinity is thus property of an isolated atom, whereas electronegativity is a property of a bonded atom. It therefore varies with the change in the oxidation state of the element.

For example Pb (II) has electronegativity 1.87 and Pb (IV) has electronegativity value 2.33.

This is clear that the electronegativity increases with the increase in oxidation number or formal charge on an atom. An increase in the oxidation number increases the attraction for electrons and hence, there is an increase in electronegativity also.

Electronegativity increases on moving from left to right in a period. This is due to the progressive decrease in the atomic size and increase in the effective nuclear charge as a result of which the binding electrons are attracted more strongly by the nucleus.

Electronegativity decreases on moving down the group. This is due to the increase in size of the atom so that the binding electron is removed further away from the nucleus and therefore is held less tightly. With the help of the electronegativity values, it is possible to predict the type of bond that may result between any two atoms. If two atoms have similar electronegativities, *i.e.*, there is a similar tendency to attract electrons, the bond between them will be covalent. Similarly, a large difference in electronegativity between two elements leads to an ionic bond.

When the electronegativity difference is about 1.9 the bond is 50% ionic. For a large difference, the bond is more ionic than covalent. If the difference is less than 1.9, bond is more covalent than ionic.

***Example 1*. Electron affinity of Cl is greater than that F but reverse is true for their electronegativity. Explain.**

Solution. Electronegativity of an element is different from E.A. It is the tendency of an atom to attract the shared pair of electrons towards itself. Its value decreases down the group. In the group of halogens, since F is at the top, it has the maximum electronegativity. In a normal case, even the E.A. would also be maximum for the topmost element in a group. There is a special reason why F has a lower value of electron affinity. This is because in small fluorine atom, strong repulsive forces come into play between the electrons of F and the extra electron that would come to it.

***Example 2.* Explain why second electron affinities are negative.**

Solution. Second E.A. means energy released when a mononegative ion gains another electron to produce dinegative ion.

$$X^- + e^- \longrightarrow X^{2-} + \text{Energy}$$

The second electron faces strong repulsion from the electrons of mononegative ion. Therefore instead of release of energy, absorption of energy takes place. Thus second E.A. is negative.

***Example 3.* Al forms an ion $[Al\,F_6]^{3-}$ but B does not form $[BF_6]^{3-}$ ion**

Solution. $$Al^3 + 6F^- \longrightarrow [Al\,F_6]^{3-}$$

Al forms the ion $[Al\,F_6]^{3-}$ because there are empty *d*-orbitals in Al to accommodate the F^- ligands.

B does not form $[BF_6]^{3-}$ because there are no empty *d*-orbitals in B to accommodate the F^- ligands.

6.1.6 Variation in metallic character of *p*-block elements (*i*) in a period and (*ii*) down the group

Metals have a tendency to lose electrons, if supplied with energy.

$$M + \text{Energy} \longrightarrow M^+ + e^-$$

The stronger is this tendency, the more electropositive and more metallic an element is. Since ionisation energy increases from left to right across the periodic table, the possibility of delocalization of valence electrons becomes less and less. Accordingly, the metallic character of elements from left to right across the periodic table changes to non-metallic character. On moving down a group, the I.E. and the electronegativity of elements generally decreases which accounts for the gradual change from non-metallic to metallic character down a group. Fig. 6.1. illustrates the gradual transition from metallic to non-metallic character. The elements B, Si, Ge, As and Te are metalloids which have metals on their left and non-metals to their right. The noble gases have completed outermost shells with explains their chemical inactivity. They have very high ionisation energy and low electron affinity. It is for the above reasons that these elements exist in atomic state. The atoms of these noble gases are held together by van der Waals forces. These are monoatomic elements.

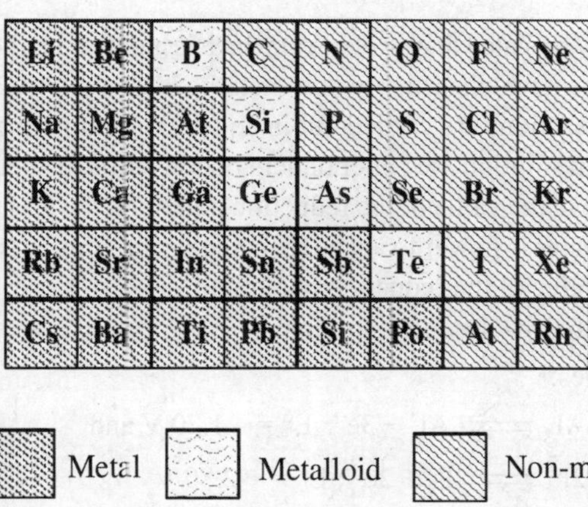

Fig. 6.1. Transition from metallic to non-metallic character.

***Example 4.* Graphite is a conductor of electricity but diamond is not. Explain.**

Solution. This can be explained in terms of structure of graphite and diamond

Graphite conducts electricity because of loose electrons binding different layers. In diamond, there are no such loose electrons. There are four covalent bonds formed by each carbon atom in a network structure. Therefore diamond does not conduct electricity.

***Example 5.* SiO_2 is a solid but CO_2 is a gas. Explain.**

Solution. Silicon is tetravalent. It forms a network solid with oxygen having structure :

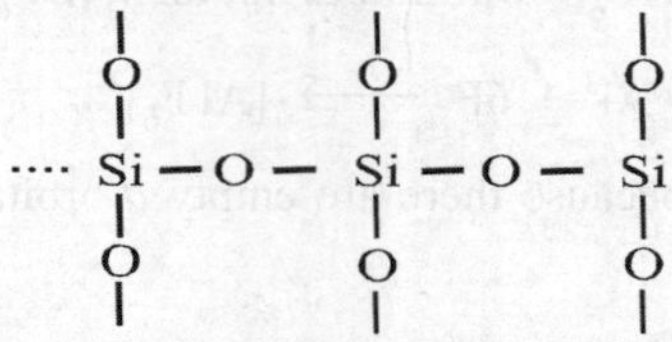

Various atoms are bound to each other by covalent bonds which require a large amount of energy to break. Carbon dioxide molecules are held to one another by weak van der Waals forces. It is a gas at room temperature.

6.1.7 Variation in oxidising and reducing properties of *p*-block elements

Oxidation is a process in which electrons are lost and reduction is process in which electrons are gained Thus an oxidizing agent is a substance which can take up electrons while a reducing agent is a substance which supplies electrons.

It can be understood that elements with low ionisation energy will have greater tendencies to lose electrons than those with higher I.E.

Since the I.E. increases and oxidation potential of *p*-block elements decreases with rising atomic number in a period, the reducing properties decrease from group 13 to group 17 and their oxidizing properties naturally increase in the same order.

In the third period, for instance, while the first element sodium of group 1 is a strong reducing agent, the last element, chlorine of group 17, is a strong oxidizing agent, the intermediate elements, silicon, phosphorus, sulphur, have only an intermediate character.

In a group, the I.E. decreases in moving from the top to bottom. Hence the reducing character increases and oxidizing character decreases as we move down a column. Thus, amongst the elements of group 14, while the first element, carbon, is a weak reducing agent, the last element, lead, is a strong reducing agent.

As a general rule, it may be stated that an element which is highly metallic in nature is also a strong reducing agent and an element which is highly non-metallic in character is a strong oxidizing agent.

As most of chemical reactions are carried out in aqueous solution, the oxidizing and reducing properties of the elements in solution are better understood in terms of standard electrode potential, E° (volts). The more negative the value of the reduction potential, the greater is the tendency of the metal to ionize, *i.e.*, the greater is power to act as reducing agent. Similarly, the powerful oxidizing agent has a large positive value of the reduction potential.

Note. Oxidation potential = – Reduction potential

In the following changes:

$$Al \rightleftharpoons Al^3 + 3e^-;\ E^\circ = -1.70\ V \text{ and}$$

$$2F^- \rightleftharpoons F_2 + 2e^-;\ E^\circ = +2.80\ V$$

E° values in the above reactions refer to reduction potentials.

Aluminium acts as a powerful reducing agent whereas fluorine acts as powerful oxidizing agent.

Example 6. **Al is a good reducing agent. Explain.**

Solution. A substance that can easily donate electrons can acts as a reducing agent. Al has a lower I.E. and a negative reduction potential, therefore it is a good reducing agent.

6.1.8 Variation in basic and acidic character of hydroxides of *p*-block elements

Let us consider an element M and its compound M–O–H. The following observation can be made:

(*i*) If the ionisation energy of the element M is low, it tends to lose electron/electrons to the oxygen atom, resulting in the weakening of the M–O bond as

$$M - O - H \longrightarrow M^+ + OH^-$$

So the element 'M' in this case has basic nature.

(*ii*) If the I.E. of element M is high, it draws electrons from the OH group and holds the oxygen atom firmly to itself. Consequently the OH bond becomes weak and breaks up as:

$$M - O - H \longrightarrow M - O^- + H^+$$

Element 'M' in this case has acidic character. It may be noted that

(*a*) There is a gradual decrease in basic character and increase in acidic character of the hydroxides of second and third row of elements.

(*b*) The intermediate elements form acidic or basic solutions of intermediate strength.

(*c*) Halogens (except F) form strongly acidic solutions of oxy-acids.

Example 7. **Why is $(CH_3)_3$ SiOH a stronger acid than $(CH_3)_3$ COH?**

Solution. $(CH_3)_3$ SiO^- is stabilized due to the presence of vacant 3*d*-orbitals on silicon. Negative charge is easily dispersed into the vacant 3*d*-orbitals. This stabilization is not possible in $(CH_3)_3$ COH because of the absence of *d*-orbitals in carbon.

6.1.9 Diagonal relationship between B and Si

Boron, the first member of group 13, differs from other members in certain respects and resembles silicon, the second member of group 14, to which it is diagonally related. The various points of similarities are:

(*i*) Both boron and silicon are semiconductors of electricity in their crystalline form and possess high melting points.

(*ii*) Both boron and silicon form oxides such as B_2O_3 and SiO_2 respectively which are acidic in nature. They dissolve in alkalies to form borates and silicates.

$$B_2O_3 + 2\,NaOH \longrightarrow \underset{\text{Sodium metaborate}}{2NaBO_2} + H_2O$$

$$SiO_2 + 2NaOH \longrightarrow \underset{\text{Sodium metasilicate}}{Na_2SiO_3} + H_2O$$

(*iii*) Boron and silicon form hydroxides which are weak acids with no basic properties.

(*iv*) Boron and silicon halides (except BF_3) readily hydrolyse to give their corresponding acids.

$$BCl_3 + 3H_2O \xrightarrow{\text{Hydrolysis}} \underset{\text{Boric acid}}{H_3BO_3} + 3HCl$$

$$SiCl_4 + 4H_2O \xrightarrow{\text{Hydrolysis}} \underset{\text{Orthosilicic acid}}{H_4SiO_4} + 4HCl$$

(*v*) Hydrides of boron and silicon known as boranes and silanes, respectively, are covalent in nature and have similar methods of preparation.

$$\underset{\text{Magnesium bromide}}{Mg_3B_2} + 6HCl \longrightarrow 3MgCl_2 + \underset{\text{Borane}}{B_2H_6}$$

$$\underset{\text{Magnesium silicide}}{Mg_2Si} + 4HCl \longrightarrow 2MgCl_2 + \underset{\text{Silane}}{SiH_4}$$

(*vi*) The nitrides of both boron and silicon decompose steam and liberate ammonia.

$$\underset{\text{Boron nitride}}{BN} + \underset{\text{Steam}}{3H_2O} \longrightarrow H_3BO_3 + NH_3$$

$$\underset{\text{Silicon nitride}}{Si_3N_4} + \underset{\text{Steam}}{12H_2O} \longrightarrow 3H_4SiO_4 + 4NH_3$$

(*vii*) The alkoxides $B(OR)_3$ and $Si(OR)_4$ may be prepared similarly by refluxing benzene solution of the anhydrous chlorides with alcohol as:

$$BCl_3 + \underset{\text{Anhydrous ethanol}}{3C_2H_5OH} \xrightleftharpoons{\text{Re flux}} B(OC_2H_5)_3 + 3HCl$$

$$SiCl_4 + 4C_2H_5OH \xrightleftharpoons{\text{Re flux}} Si(OC_2H_5)_4 + 4HCl$$

6.1.10 Variation in melting point and boiling point along the period and down the group in *p*-block element

The melting and boiling points depend on the nature of chemical bonding between atoms of these elements. It may be noted that the elements on the extreme left of the row are metals and involve metallic bonds. The atoms for their separation require small energy to separate the metallic bonds which are not as strong as covalent bonds and hence have lower melting and boiling points.

The network solids having two-dimensional or three-dimensional giant molecules, *e.g.*, C, Si, etc., on the other hand involve covalent bonds. Each atom may have two or more than two bonds to break away, which explains the high melting and boiling points of network solids. To the right of these elements are N, P, S and Cl. These elements form molecular solids. In these molecules, the atoms are strongly covalent bonded but the molecules are attracted towards each other by weak van der Waals forces. Since it is the molecule which breaks away from the solid, it requires little thermal energy to overcome the van der Waals forces. Accordingly, the melting points and boiling points of these elements on the extreme right of the periodic table are lowest.

In a group the metallic character increases and therefore the m.p. and b.p. decrease on moving down the group.

6.1.11 Fajan's Rules (Polarizing power and Polarizibility)

If two atoms A^+ and B^- are brought together to the equilibrium distance, the type of bond between them depends on interaction between ions. The +ve ion attracts the electrons on the –ve ion and repels the nucleus, thus distorting or polarizing the –ve ion. If the polarization is quite small, then the ionic bond is formed. If it is more, electrons are drawn from the –ve ion to the +ve ion and appreciable covalent bond formation takes place.

The extent of ion distortion depends on both the power of an ion to distort the other, i.e. *polarizing power*, and susceptibility of an ion to distortion, i.e. *polarizability*. It is the cation which causes polarization and it is the anion which gets polarized. Generally speaking, the polarizing power increases as ions become smaller and more highly charged. The polarizability of a –ve ion is greater than that of a +ve ion since the additional electrons are less firmly bound to the nucleus because of the difference in the effective nuclear charge. This is illustrated in Fig. 6.2.

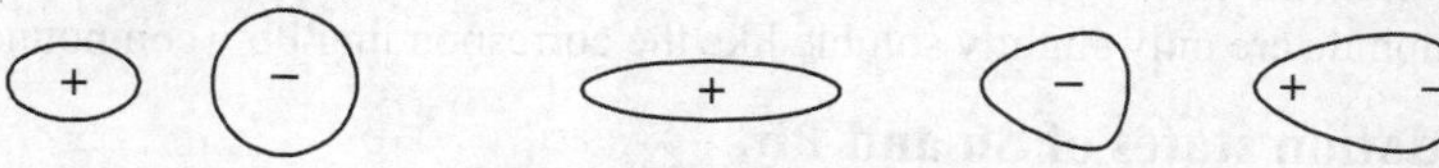

Fig. 6.2. Polarization of ions.

The above observation form the basis of *Fajan's rules*. **Fajan's rules state that covalent bonding is favoured by (*i*) small +ve ion, (*ii*) large –ve ion and (*iii*) large charge on either ion.** In moving along a row of elements, the size of the cations decreases and due to greater effective charge, the polarizing power increases.

For example, degree of covalence increases in the order NaCl, $MgCl_2$ and $AlCl_3$.

On moving down a group the size of the cation increases and in accordance with the Fajan's Rule, the polarizing power decreases. For example, degree of covalence decreases in the order $BeCl_2$, $MgCl_2$ and $CaCl_2$. However, with the increase in the size of the anion on moving down a group, its polarizability and hence the covalent character in the compounds increases. Thus, among the halides of a given cation, bromides and iodides are generally more covalent than the corresponding chlorides and fluorides.

6.1.12 Inert pair effect

Where several +ve oxidation states occur among a family of representative elements, they normally differ by two units and the lower oxidation states tend to be become more stable as one goes down a group in the periodic table. For example, in group 13, thallium, the heaviest member in the family, is quite unstable in the +3 sate and is readily reduced to the more stable +1 state. Likewise Pb^{2+} ion is more stable than Pb^{4+} ion in group 14. The greater stability of lower oxidation state among the heavier *p*-block elements is attributed to tendency of ns^2 electrons to remain paired. That is, these electrons do not take part in bond formation. This is known as **Inert Pair Effect.**

6.1.13 Stability of In^+ and Tl^+ ions

Indium and thallium belong to group 13 containing B, Al, Ga, In and Tl. Both indium and Tl form trivalent ions, In^{3+} and Tl^{3+}, but they also form monovalent ions, In^+ and Tl^+, respectively. Loss of electrons from an atom of an element can take place in two ways. Look at the electronic configuration of In given below:

	In	2,	8,	18,	18,	3	$(5s^2p^1)$
I	In^{3+}	2,	8,	18,	18		
II	In^+	2,	8,	18,	18,	2	

Either all its *s* and *p* electrons $(5s^2p^1)$ may be lost as in I forming an atom in which its oxidation number equals three or the outermost *s*-electrons are not lost, only *p*-electron is lost and the oxidation number equals 1 (as in II). In this case, the pair of *s*-electrons is said to have become inert.

Similarly for thallium, the formation of ions can be shown as follows:

Tl = 2, 8, 18, 32, 18 , 3 $(6s^2p^1)$

(At. no. = 81)

Tl^{3+} = 2, 8, 18, 32, 18

Tl^+ = 2, 8, 18, 32, 18, 2

It has been observed that compounds of Tl^+ are more stable than those of Tl^{3+}. In fact Tl^{3+} salts are oxidizing agents. On the other hand, In^+ ions are immediately converted into In^{3+} ions and the metal (In) in presence of water. Gallium forms no monovalent ions. This explains the tendency of the elements in a group to show more inert pair effect as we move down the group.

As a matter of fact Tl^+ ion has the same electronic configuration as that of the Pb^{2+} ion and this explains the similarities between certain Tl^+ and Pb^{2+} compounds. Tl^+ chloride, bromide, iodide, sulphate and sulphide are only slightly soluble like the corresponding Pb^{2+} compounds.

6.1.14 Oxidation states of Sn and Pb

Sn and Pb both form the expected tetravalent ions but they also form divalent ions as shown below:

$$\begin{aligned}
Sn &= 2, 8, 18, 18, 4\ (5s^2p^2) \\
&\quad\ \text{(At. no. = 50)} \\
Sn^{4+} &= 2, 8, 18, 18 \\
Sn^{2+} &= 2, 8, 18, 18, 2\ (5s^2) \\
Pb &= 2, 8, 18, 32, 18, 4\ (6s^2p^2) \\
&\quad\ \text{(At. no. = 82)} \\
Pb^{4+} &= 2, 8, 18, 32, 18 \\
Pb^{2+} &= 2, 8, 18, 32, 18, 2\ (6s^2)
\end{aligned}$$

As the inert pair effect is more marked in the heavier element, *i.e.*, Pb, the Pb^{2+} ion is more stable than the Pb^{4+}. On the other hand, Sn^{4+} ion is more stable than the Sn^{2+} ion. This explain *why SnO and $SnCl_2$ are reducing agents whereas PbO_2 and $PbCl_4$ are oxidizing agents*. Moreover, $PbBr_4$ and PbI_4 do not exist but corresponding compounds of Sn can be obtained. The non-existence of $PbBr_4$ and PbI_4 is due to the oxidation of Br^- and I^- ions by Pb (IV) ion to Br_2 and I_2 and in the process Pb (IV) is reduced to Pb (II) ion.

$$PbBr_4 \longrightarrow PbBr_2 + Br_2$$

$$PbI_4 \longrightarrow PbI_2 + I_2$$

The lower oxidation states are more ionic since the radius of M^{2+} is greater than M^{4+} and according to Fajan's rules, the smaller the ion, the greater is the tendency to covalency. The reduction potential values indicate that Pb (IV) is a stronger oxidant than Sn (IV), Pb^{4+} / Pb^{2+}, + 1.46 volts. Sn^{4+} / Sn^{2+}, + 0.188 volt) and conversely, Sn (II) is superior to Pb (II) as a reducing agent.

6.1.15 Oxidation states of Sb and Bi

Both Sb and Bi form trivalent as well as pentavalent ions.

$$\begin{aligned}
Sb &= 2, 8, 18, 18, 5\ (5s^2p^3) \\
&\quad\ \text{(At. no. = 51)} \\
Sb^{5+} &= 2, 8, 18, 18 \\
Sb^{3+} &= 2, 8, 18, 18, 2\ (5s^2) \\
Bi &= 2, 8, 18, 32, 18, 5\ (6s^2p^3) \\
&\quad\ \text{(At. no. = 83)} \\
Bi^{5+} &= 2, 8, 18, 32, 18 \\
Bi^{3+} &= 2, 8, 18, 32, 18, 2\ (6s^2)
\end{aligned}$$

The inert pair effect, however, is more marked in the case of Bi which is the heaviest element of this group. Thus, Bi forms only the trichloride, $BiCl_3$, but not the pentachloride, $BiCl_5$, whereas P, As and Sb are trivalent as well as pentavalent.

6.1.16 Similarity and dissimilarity between boron and aluminium

Points of similarity. (*i*) *Covalent character.* Both B and Al form covalent compounds, *e.g.*, BCl_3 and $AlCl_3$. Their chlorides are volatile and hygroscopic. They are hydrolysed by water.

$$BCl_3 + 3H_2O \longrightarrow B(OH)_3 + 3HCl$$

$$AlCl_3 + 3\,H_3O \longrightarrow Al(OH)_3 + 3HCl$$

(*ii*) *Nitrides.* Both B and Al form nitrides which react with water to give ammonia.

$$BN + 3H_2O \longrightarrow H_3BO_3 + NH_3$$

$$AlN + 3H_2O \longrightarrow Al(OH)_3 + NH_3$$

(*iii*) *Oxides.* Both B and Al form oxides when they are heated in air.

$$6B + 3O_3 \longrightarrow 3B_2O_3$$

$$4Al + 3O_2 \longrightarrow 2Al_2O_3$$

(*iv*) *Action of alkalis.* Both B and Al react with alkalies to evolve hydrogen.

$$2B + 6NaOH \longrightarrow \underset{\text{Sodium borate}}{2Na_3BO_3} + 3H_2$$

$$2Al + 2H_2O + 2NaOH \longrightarrow \underset{\text{Sod. meta-aluminate}}{2NaAlO_2} + 3H_2$$

(*v*) *Halides as Lewis acids.* Halides of both elements such as BCl_3 and $AlCl_3$ behave as Lewis acids, because boron and aluminium atoms are electron deficient.

Points of Dissimilarity

Due to small size, high electronegativity and high I.E., boron differs from aluminium and other members of the group in the following respects:

(*i*) *Melting point.* Boron is quite hard with high m.p. while other members of the group are soft with low m.p.

(*ii*) *Electrical conductivity.* Boron is a bad conductor while other members are good conductors of electricity.

(*iii*) *Metallic character.* Because of smaller size and higher I.E. boron is non-metal while other members of the group are metals.

(*iv*) *Nature of oxides and hydroxides.* Oxides and hydroxides of boron are acidic in nature.

$$B_2O_3 + 3H_2O \longrightarrow \underset{\text{Boric acid}}{2H_3BO_3}$$

Hydroxides and oxides of Al are amphoteric in nature.

$$Al_2O_3 + 6\,HCl \longrightarrow 2\,AlCl_3 + 3H_2O$$

$$Al_2O_3 + 2NaOH \longrightarrow \underset{\text{Sod. meta-aluminate}}{2NaAlO_2} + H_2O$$

On the other hand, other members form basic oxides and hydroxides.

(*v*) *Action of acids.* Boron does not displace hydrogen from acids but other members do. For example,

$$2\,Al + 6HCl \longrightarrow 2\,AlCl_3 + 3H_2$$

With conc. HNO_3 boron is oxidized to boric acid while Al becomes passive.

(*vi*) *Action of steam.* Water or steam has no action on boron but other members decompose steam to give out hydrogen.

$$2Al + 6H_2O \longrightarrow 2\,Al\,(OH)_3 + 3H_2$$

6.1.17 Trends in acidic nature and basic nature of the oxides of *p*-block elements in a group and in a period

As we move in a period, in *p*-block, from left to right, the basic character of the oxides decreases (as the metallic character of the elements decreases) while acid character increases. On the other hand, in a group the basic character of oxides increases as we descend a group. Metallic character of elements increases.

The behaviour of oxides of third row and group 13 elements is given below.

III Row	Oxides :	Na_2O	MgO	Al_2O_3	SiO_2	P_2O_5	SO_2	Cl_2O_7
	Nature :	Strongly basic	Basic	Amphoteric	Weakly acidic	Acidic	More acidic	Most acidic
13 Group	Oxides :	B_2O_3	Al_2O_3	Ga_2O_3	In_2O_3	Tl_2O		
	Nature :	Acidic	Amphoteric	Weakly basic	Basic	Strongly basic		

We observe that acid character of 3rd row elements increases from Na to Cl and basic character of oxides of group 13 increases from B to Tl.

Example 7. **Categorise the following oxide in terms of acidic, basic, and amphoteric character:**

$$\mathbf{BeO,\ B_2O_3,\ MgO,\ SiO_2\ and\ N_2O_5}$$

***Solution*:** Basic oxide—MgO

Acidic oxides – B_2O_3, SiO_2, N_2O_5

Amphoteric oxide – BeO

6.2 HYDRIDES OF *p*-BLOCK ELEMENTS

The hydrides of Group 13 elements (*i.e.*, BH_3 and AlH_3) are electron-deficient compounds and exist in polymerised forms [e.g., B_2H_6, B_4H_{10} and $(AlH_3)_n$]. Hydrides of B and Al are employed as efficient reducing agents.

Carbon of Group 14 forms polynuclear hydrides in which two or more atoms of the element are directly bonded to each other as well as to hydrogen thus giving numerous hydrocarbons. Silicon and some other non-metal elements of Group 14 also show this property but to a very limited extent. Silicon, for example, gives hydrides such as Si_2H_6, Si_3H_8, Si_4H_{10}, etc. Sn and Pb also form hydrides which are rather unstable.

The covalent hydrides of As, Sb, Bi are also known.

The hydrides of elements after Group 13 become increasingly acidic in character on moving from left to right along a given row in the Periodic table. Thus, while NH_3 is a weak base, H_2O is neutral and HF is acidic. Similarly, in the next row, while PH_3 is a weak base, H_2S is a weak acid and HCl is highly acidic. The acidity can also be related to the increasing stability of their anions as shown below:

$$\xrightarrow[\text{Increasing stability}]{CH_3^- < NH_2^- < OH^- < F^-}$$

In covalent hydrides, the H–X bond energies generally decrease with increase in atomic number (and hence atomic size) of the element within a group. The lesser the bond energy, the lesser is the stability of the H–X bond. This readily accounts for the instability of the hydrides of heavier elements of Groups 14 and 15. In fact, PbH_4 and BiH_5 are so unstable that these can be prepared only in traces.

Within a group, the boiling points of covalent hydrides (barring the hydride of the first element) increase regularly with increase in atomic number of the element and the molar mass of the hydride. Thus, boiling points of hydrides of elements of Group 15, 16, and 17 (with the exception of top elements) increase as we move down the groups, as shown below.

Group 15		Group 16		Group 17	
Hydride	**b.p. (°C)**	**Hydride**	**b.p. (°C)**	**Hydride**	**b.p. (°C)**
NH_3	– 33.0	H_2O	+100	HF	+ 19.4
PH_3	– 84.6	H_2S	– 59.6	HCl	– 85.0
AsH_3	– 55.1	H_2Se	– 42.0	HBr	– 67.1
SbH_3	– 17.0	H_2Te	– 1.8	HI	– 35.5

The boiling points of the hydrides of top elements, *viz*., NH_3, H_2O and HF are abnormally high. This is because of the association of these molecules through hydrogen bonding. These molecules, thus, have long polymeric structures, as shown below.

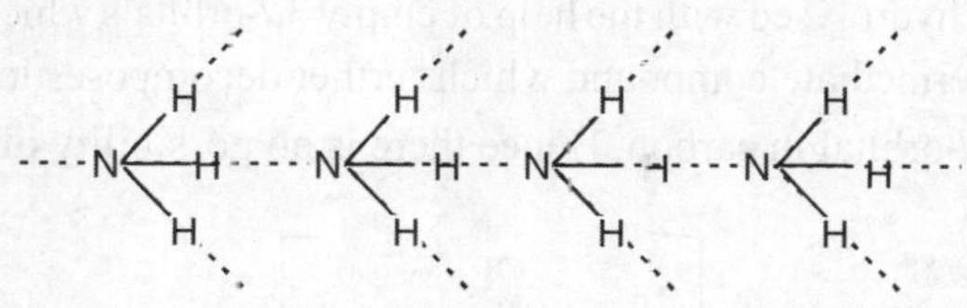

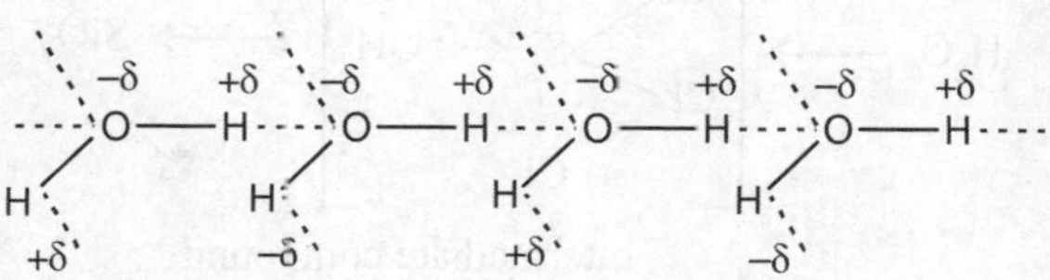

$$\overset{+\delta}{H}-\overset{-\delta}{F}\cdots\overset{+\delta}{H}-\overset{-\delta}{F}\cdots\overset{+\delta}{H}-\overset{-\delta}{F}\cdots\overset{+\delta}{H}-\overset{-\delta}{F}$$

The electronegativity difference in carbon (the first element of Group IV) and hydrogen is very small. There is thus no asociation of CH_4 molecules through hydrogen bonding. It has, therefore, the lowest boiling point among the hydrides of group 16 elements. The boiling point of these hydrides increase in the order:

$$CH_4 < SiH_4 < Ge\ H_4 < Sn\ H_4$$

6.3 HALIDES OF *p*-BLOCK ELEMENTS

All the *p*-block elements form halides. Some of the halides show strong ionic character. There is a gradation from purely ionic halides through those of intermediate character to those which are clearly covalent. The halides of the alkali metal (group 1) are ionic in character. However, as we move along a period with decrease in ionic radius, *i.e.*, with increase in the charge radius ratio, the covalent character of the halides increases. Thus amongst chlorides of 3rd period elements, *viz.*, NaCl, $MgCl_2$, $AlCl_3$ and $SiCl_4$, the first one is completely ionic while the last one is essentially covalent. On moving down a group, the electropositive nature of the element increases and thus the ionic nature of halides increases.

Some of the metals form halides in several oxidation states. In such cases, the halides formed in the lower oxidation states are ionic while those in the higher oxidation states tend to be covalent. Thus while $PbCl_2$ is largely ionic, $PbCl_4$ is largely covalent.

The size of the halide ion is another important factor in determining the character of the halides of elements. Thus, while AlF_3 is essentially ionic, $AlCl_3$ has an intermediate character and $AlBr_3$ and AlI_3 exist as covalent dimers. This behaviour can be explained by Fajans's rules.

Most of the ionic halides of *p*-block elements are soluble in water. There are, however, a few exceptions as for example $PbCl_2$. The covalent halides do not show any trend in their solubility. If a given metal forms all the four halides, MF_n, MCl_n, MBr_n, MI_n, solubility decreases in the order

$$I^- > Br^- > Cl^- > F^-$$

This is also because ionic character decreases from iodide to fluoride.

***Example 8.* Which of the following is ionic? Justify your answer:**

(*i*) $SnCl_2$ (*ii*) $SnCl_4$.

Solution. $SnCl_2$ is ionic and $SnCl_4$ is covalent because Sn^{4+} has large polarizing power. It polarizes chloride ions and induces covalent character according to Fajan's rules.

***Example 9.* CCl_4 is not hydrolysed with water while $SiCl_4$ is readily hydrolysed. Explain.**

Solution. $SiCl_4$ gets hydrolysed with the help of empty 3*d*-orbitals which accommodate electron pairs of H_2O to form intermediate compound which further decomposes into SiO_2.

There is no empty *d*-orbital in carbon. Hence there is no possibility of such a reaction.

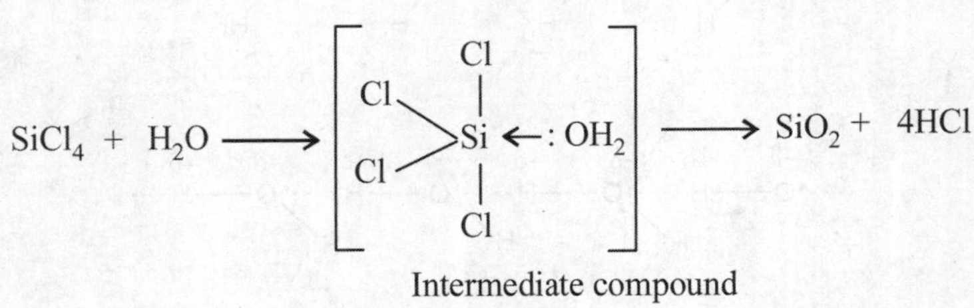

Intermediate compound

***Example 10.* A metallic element M forms two stable chlorides with formulae MCl_3 and MCl_5 respectively. Where the metal M is to be placed in the periodic table?**

Solution. Metal M belongs to group 15. The configuration in the outermost orbit is $ns^2\ np^3$.

MCl_5 will be obtained when all the five electrons are removed whereas MCl_3 will be obtained when *s* electrons remain inert and only the electrons belonging to *p*-orbital are removed.

***Example 11.* A metallic element M forms two stable chlorides with formulae MCl and MCl_3 respectively. Where the element M is to be placed in the periodic table?**

Solution. Metal M is to be placed in group 13. These elements have the configuration $ns^2\ np^1$ in the outermost orbit. MCl and MCl_3 are obtained when one electron in the *p*-orbital and all the three electrons respectively are removed.

***Example 12.* PbI_4 is unstable while SnI_4 is stable. Explain.**

Solution. In Sn, + 4 oxidation state is more stable. That is why SnI_4 is stable.

In Pb, + 2 oxidation state is more stable because cf inert pair effect in lead. It decomposes to more stable compound PbI_2 as follows:

$$PbI_4 \longrightarrow PbI_2 + I_2$$

***Example 13.* Fluorine cannot be liberated from fluoride by a hemical reagent. Explain.**

Solution. Oxidation potential for the reaction

$$2F^- \longrightarrow F_2 \text{ is } -2.80 \text{ volts}$$

This is the maximum oxidation potential in the periodic table. There is no reagent which has oxidation potential more than this. Hence liberation of fluorine from fluoride is not possible.

***Example 14.* HF is the weakest acid among hydrogen halides. Explain.**

Solution. Bond dissociation energy in H – X, where X stands for halogen, decreases from F to I. Thus HF has the strongest bond and most difficult to break. Therefore HF is the weakest acid.

*d*π – *p*π bond

A bond that is formed between lateral overlap of a *p*-and *d*-orbital is called *d* π – *p* π bond. For example, 3*d* orbitals are involved in *d*π – *p*π multiple bonding between sulphur and oxygen in SO_2 and SO_3 as shown in Fig. 6.3.

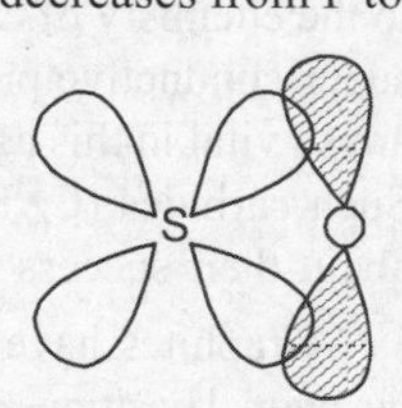

Fig. 6.3. *d*π -*p* π bonding.

6.3.1 Discuss relative strength of BF_3, BCl_3 and B Br_3 as Lewis acids (Back bonding)

The relative strengths of trihalides of boron to accept a pair of electrons and thus to act as Lewis acids have been determined by measuring their heats of formation as well as by studying their infra-red spectra. The power to accpet an electron pair has been found to decreases in the order:

$$BF_3 < BCl_3 < BBr_3$$

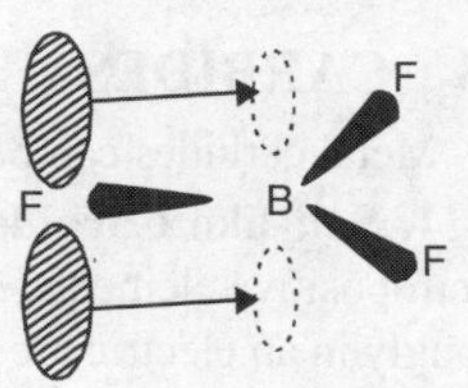

Fig. 6.4. Back bonding.

This order is just the reverse of what is expected from the electronegativities of hologens. We can explain it on the basis of **back bonding**.

Halogen have a tendency to back-donate their electron to the boron atom.

In BF_3, 2*p* orbitals of all fluorine atom are fully-filled while one of the 2*p* orbitals of B is vacant. The 2*p* orbitals of F involved in the formation of B—F bond overlap laterally thus transferring electrons of fluorine into the vacant 2*p* orbital of B. Thus an additional pπ–pπ bond is established between B and F atoms, as shown in Fig. 6.4.

As a result of back donation of electrons from F to B, electron deficiency of boron atom gets compensated and Lewis character (tendency to attract electrons) of BF_3 decreases.

As we move from BF_3 to BCl_3 and $B\,Br_3$, tendency to form $p\pi - p\pi$ bond decreases. The reason for this is the energy difference between the 2p orbital of B and 3*p* orbital of C*l* and between the 2*p* orbital of B and 4*p* orbital of Br. In the case of BF_3, it is the 2p orbitals of B and F of equivalent energies.

Therefore trendency to attract electrons or Lewis acid character shows the gradation.

$$BF_3 < BCl_3 < BBr3$$

6.4 FULLERENES

Prof. Kroto of Susex University (U.K.) and Prof. Smalley of Rice University (USA), while conducting experiments on carbon, vaporised graphite with the help of lasers. To their surprise, some atoms which were vaporised by lasers, regrouped together to form hollow shell-like structures. In such structures, sixty or even more carbon atoms joined together to produce cage-like structures. This form of carbon (allotrope) was given the name **Buckminster fullerene** after the name of a famous architect **Buckminster fuller,** who shaped famous Geodesic dome.

Prof. H.W. Kroto

This discovery has opened new vistas in the chemistry of carbon compounds. It has been found that *Buckminster fullerene* agglomerate which is of the shape of a base-ball is big enough to accommodate atoms of potassium to form $C_{60}K$, which is grey in colour. It will give new dimensions to the chemisry of carbides (alkali metal and others). Such carbides show super-conductive properties (no resistance to the flow of current) which are so vital in this age of power-crisis. Lot of further research is going on. Such carbides ($C_{60}K$) are highly coloured and thus paint a rosy picture about their success as *materials.*

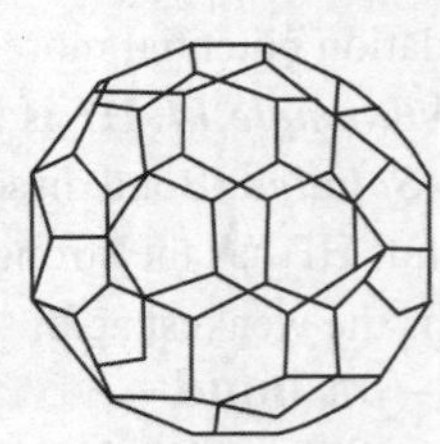

Fig. 6.5. Fullerene

Graphites have sheet structures, in which the carbon atoms occupy hexagonal positions, different sheets being held together by weak van der Waals' forces. On evaporation by lasers, the carbon atoms arrange themselves in mixed pentagons and hexagons as shown in Fig. 6.5 maintaining the basic sp^2 hybridisation scheme and producing ball-like hollow shells. More recently even bigger fullerenes containing 70, 76 and 78 carbons have been reported.

6.5 CARBIDES

Metal carbides can be discussed under the following heads:

1. Salt-like Carbides. Carbides of this type are simple compounds of carbon with more electropositive elements, *i.e.,* with metals of Groups 1, 2, 3. The metal, its oxide or hydride is heated strongly in an electric arc with carbon to get the required carbide. For example,

Be_2C is obtained by heating BeO and C at 1900-2000°C.

$$2BeO + 3C \xrightarrow{1900-2000°C} Be_2C + 2CO$$

Calcium carbide is obtained by heating lime and coke at 2200 — 2250°C.

$$CaO + 3C \longrightarrow CaC_2 + CO$$

Aluminium carbide is prepared by heating Al and C in an electric furnace.

$$4A1 + 3C \longrightarrow A1_4C_3$$

In general, these compounds exist as transparent crystals and in the solid state they are non-conductors of electricity.

They are readily hydrolysed in water yielding different hydrocarbons. These carbides are further classified according to the hydrocarbon which they produce on hydrolysis.

1. *Acetylides*. The carbides which yield **acetylene** on hydrolysis are known as **acetylides.** MgC_2, BeC_2 and CaC_2 fall in this category.

$$CaC_2 + 2H_2O \longrightarrow Ca(OH)_2 + C_2H_2$$

They contain discrete C_2^{2-}, *i.e.*, $(C \equiv C)^{2-}$ units.

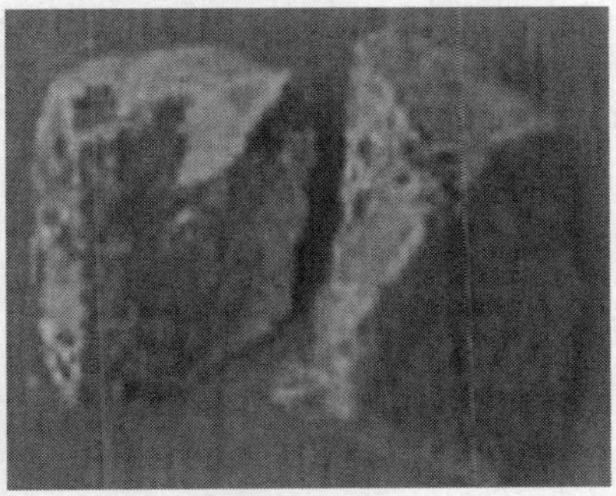

Calcium carbide

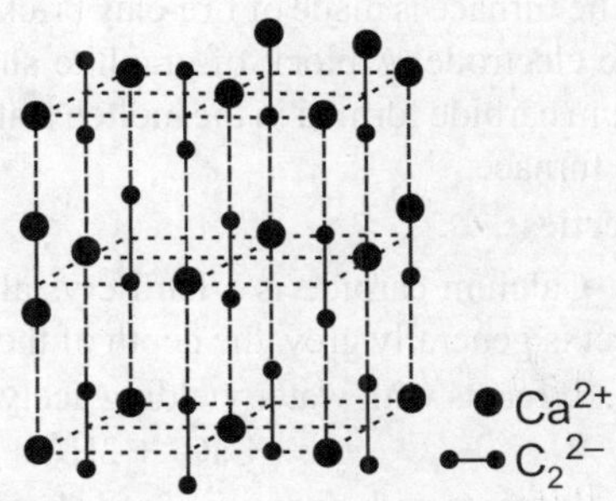

Structure of calcium carbide

2. *Methanides*. The carbides such as aluminium and beryllium carbides, Al_4C_3 and Be_2C, which give **methane** on hydrolysis, are known as **methanides.**

$$Al_4C_3 + 12H_2O \longrightarrow 4Al(OH)_3 + 3CH_4$$

These carbides contain C^{4-} ions.

3. *Allylides*. Magnesium carbide, Mg_2C_3, is the only carbide which yields **allylene** on hydrolysis.

$$Mg_2C_3 + 4H_2O \longrightarrow 2Mg(OH)_2 + \underset{\text{Allylene}}{CH_3C \equiv CH}$$

2. Covalent Carbides. These carbides contain carbon covalently bonded with elements of comparable electronegativity such as hydrogen, sulphur and chlorine. The covalent carbides can be further divided into two categories.

(*i*) The first category consists of covalent carbides which contain *small discrete molecules*. These are formed by combination of carbon with hydrogen or the elements of Groups 16 and 17. The examples are CH_4, CS_2 and CCl_4. These carbides are either gases or volatile liquids and are not thermally stable.

(*ii*) The second category consists of giant molecules. The common examples are boron carbide, B_4C and silicon carbide, SiC. These are thermally stable, hard and chemically inert solids.

3. Interstitial Carbides. These carbides are obtained, by the direct combination of metal and carbon or by the reduction of metal oxide with carbon at high temperature close to 2000°C. Transition metals like Ti, Zr, Hf, V, Nb, Ta and Mo form these carbides. The metals retain their close-packed lattice strcutures and *incorporate carbon atom in their octahedral interstices* to give MC and M_2C type carbides. The common examples are WC, TaC, V_2C, Nb_2C, Ta_2C, W_2C. These carbides melt at very high temperatures. They are extremely hard refractory materials and have melting points in the range of 3000–4000°C. They are very hard and brittle. They are good conductors of electricity and are chemically inert. Due to their extensive hardness and chemical inertness WC and TaC are used for the manufacture of high speed cutting tools.

Uses of Carbides in Industry

1. Silicon carbide is used as *abrasive* under the name *carborundum* and also as a *refractory material*. It is also very resistant to abrasion and corrosion. It also finds applications in atomic reactors as it provides a good material for making containers for fuel elements at high temperatures.

2. Tungsten carbide, WC, is used in making tools and dies.

3. Boron carbide, B_4C, which is harder than silicon carbide, is being used now increasingly as an abrasive. It is also used as a shield against radioactive radiations.

6.5.1 Calcium Carbide, CaC_2

Preparation. It is manufactured by heating lime with powdered coke at about 3000°C in an electric furnace.

$$CaO + 3C \xrightarrow{3000°C} CaC_2 + CO$$

The furnace is made of fire-clay bricks and has a thick layer of graphite at the bottom. This serves as one electrode. A block of graphite suspended from the middle serves as the second electrode. Calcium carbide formed in the molten state is taken out from a tapping hole provided near the bottom of the furnace.

Properties

1. Calcium carbide is a hard crystalline solid. The pure product is white but the commercial product is generally grey, the depth of the colour increases with the extent of impurities.
2. It reacts with water yielding acetylene.

$$CaC_2 + 2H_2O \longrightarrow Ca(OH)_2 + C_2H_2$$

3. It reacts with nitrogen at red heat to yield calcium cyanamide.

$$CaC_2 + N_2 \longrightarrow CaNCN + C$$

The mixture of calcium cyanamide and carbon is used as such as a fertilizer under the name **nitrolim.** It gets hydrolysed supplying ammonia to the soil :

$$CaNCN + 3H_2O \longrightarrow CaCO_3 + 2NH_3$$

Uses.

Calcium carbide is used for the preparation of acetylene gas and for the manufacture of calcium cyanamide which is used as a fertilizer.

Structure. Calcium carbide is an ionic solid consisting of Ca^{2+} and $(C \equiv C)^{2-}$ ions. Its structure is similar to that of sodium chloride because of similar radius ratio (0.52). Thus each Ca^{2+} is surrounded by six carbide ions which lie at the corners of a regular octahedron.

6.5.2. Silicon carbide or carbon silicide, Si C (also called carborundum)

Preparation. 1. It is prepared by heating a mixture of coke and SiO_2 in an electric furnace.

$$SiO_2 + 3C \longrightarrow SiC + 2CO$$

2. It may also be prepared by passing acetylene over heated silicon.

$$2Si + C_2H_2 \longrightarrow 2SiC + H_2$$

Manufacture (Acheson's process)

A mixture of sand (54%), coke (34%), sawdust (10%) and salt (2%) is heated in a furnace made of fire-bricks at 1500–2200ºC. The furnace is provided with carbon electrodes consisting of sixty carbon rods at either end. A heavy current is passed for 36 hours. The following reaction takes place.

$$SiO_2 + 3C \xrightarrow{1500\text{-}2000ºC} SiC + 2CO \uparrow$$

At the end of operation, black crystals of SiC are removed, crushed and washed successively with H_2SO_4 and NaOH solutions to remove impurities.

The salt acts as flux and saw dust increases the porosity of the charge.

Uses. 1. Silicon carbide is extremely hard and hence used as an abrasive for cutting and grinding glasses.

2. SiC has a low coefficient of thermal expansion and high thermal stability. It is used for making crucibles for melting different metals.

3. Because of high melting point, it is used in furnace lining.

4. SiC is used as a high temperature semi-conductor.

5. It is used in the fabrication of grinding wheels for sharpening metallic tools.

Structure : There is an infinite three dimensional array of silicon and carbon in which each carbon is tetrahedrally surrounded by four silicon atoms and each silicon is tetrahreadely surrounded by four carbon atoms.

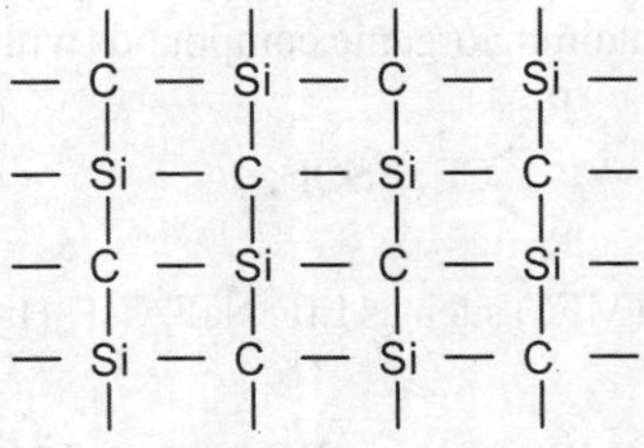

Structure of SiC

6.5.3 Tetrasulphur tetranitride, S_4N_4

Preparation. It can be prepared by one of the following reactions:

1. $10S + 4NH_3 \longrightarrow S_4N_4 + 6H_2S$

2. $4NH_4Cl + 6S_2Cl_2 \xrightarrow{160°C} S_4N_4 + 8S + 16HCl$

3. $16NH_3 + 6S_2Cl_2 \xrightarrow[CCl_4]{50°C} S_4N_4 + 8S + 12NH_4Cl$

Properties

1. S_4N_4 is an orange compound with a melting point of 178°C.
2. It explodes when struck or heated.
3. It forms addition compounds S_4N_4. BF_3 and $S_4N_4.SbCl_5$ with BF_3 and $SbCl_5$, respectively.
4. S_4N_4 gets hydrolysed by dilute and concentrated alkali solutions.

$$2S_4N_4 + 6OH^- \text{ (dilute)} + 9H_2O \longrightarrow S_2O_3^{2-} + 2S_3O_6^{2-} + 8NH_3$$

$$S_4N_4 + 6OH^- \text{ (concentrated)} + 3H_2O \longrightarrow S_2O_6^{2-} + 2SO_3^{2-} + 4NH_3$$

Structure of S_4N_4. Tetrasulphur tetranitride has a cradle-like structure as shown in Fig. 6.6.

All the S–N bond distances are equal (1.60Å) which are shorter than the calculated S – N single bond distance of 1.75 Å. This indicates some double bond character in S—N bonds. The trans annular S – S distances are shorter than twice the van der Waals' radius for S indicating that there may be a weak association between the trans annular S atoms. The actual structure of S_4N_4 is a resonance hybrid of a number of the following types of structures :

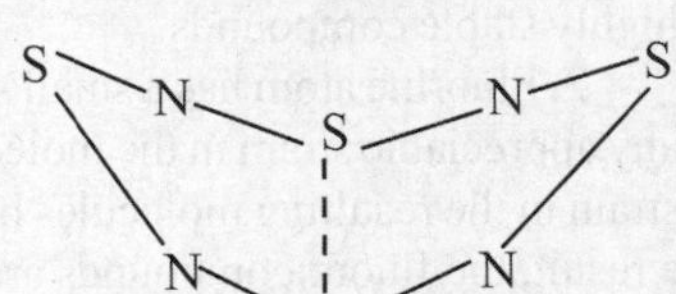

Fig. 6.6. Structure of S_4N_4.

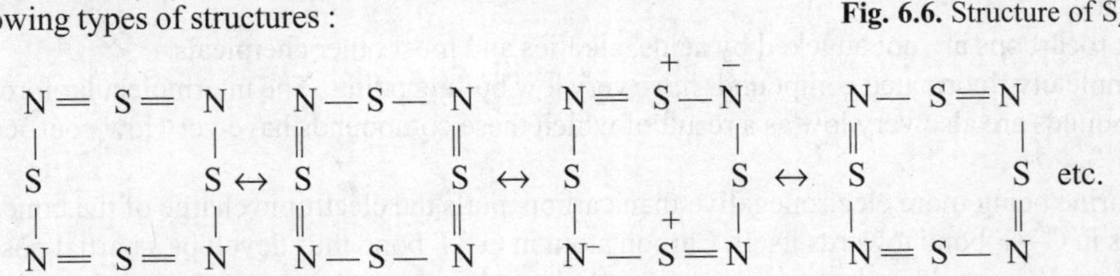

6.6 FLUOROCARBONS

Fluorocarbons are hydrocarbons or other organic molecules in which most of the hydrogen atoms are replaced by fluorine atoms. These compounds have wide industrial applications.

Preparation. Fluorocarbons can be prepared by the following methods.

1. By heating organic halides with anhydrous HF in the presence of SbF_5 at 50°-100°C temperature and 50-500 atm pressure.

$$2CCl_4 + 3HF \xrightarrow[50-100°C \atop 50-500\,atm]{SbF_5} CCl_2F_2 + CCl_3F + 3HCl$$

2. By fluorinating oxygen-containing organic compounds with SF_4.

$$\begin{matrix} R \\ \quad \diagdown \\ & CO + SF_4 \\ \quad \diagup \\ R' \end{matrix} \longrightarrow \begin{matrix} R \\ \quad \diagdown \\ & CF_2 + SOF_2 \\ \quad \diagup \\ R' \end{matrix}$$

3. By heating metal fluorides (MF_X) such as LiF, NaF, CsF, HgF_2, ZnF_2, SbF_3, etc., with organic halides.

$$RCl + MF \xrightarrow{\Delta} \underset{\text{Fluorocarbon}}{RF} + MCl$$

The difference in the lattice energies of MF and RCl pushes the reaction forward. Within a group, the above difference increases with increase in the size of the metal M. Hence, within a group, the fluorides with bigger metal ions make the reaction more facile. Accordingly, CsF is a better fluorinating agent than NaF.

4. By electrolysing organic compounds in liquid HF in steel cells with Ni anodes and steel cathodes. For example,

$$C_8H_{18} \xrightarrow[\text{in liquid HF}]{\text{Electrolysis}} C_8F_{18}$$

$$(C_4H_9)_3N \xrightarrow[\text{in liquid HF}]{\text{Electrolysis}} (C_4F_9)_3N$$

5. By reacting organic compounds with fluorine in the presence of a catalyst such as copper gauze. The reaction is carried out in the presence of an inert gas such as helium or nitrogen.

$$C_6H_6 + 9F_2 \xrightarrow[\text{Cu at } 256°C]{N_2 \text{ or He}} C_6F_{12} + 6HF$$

Properties. 1. Since bond energy of C – F bond (486 kJ mol^{-1}) is very high fluorocarbons are highly stable compounds.

2. Fluorine atom has a small size so that the replacement of H atoms by F atoms does not produce any appreciable strain in the molecule. Replacement of H atoms by other halogens causes appreciable strain in the resulting molecules because of larger differences in the sizes of H and halogen atoms. As a result, the fluoro compounds are more stable than the other halogen compounds.

3. Fluorocarbons are stable towards hydrolysis. This is because carbon does not have the *d* orbitals which could permit the formation of the 5-coordinated intermediate by necleophilic attack of H_2O molecule.

4. Fluorocarbons are not attacked by acids, alkalies and most other chemicals.

5. Completely fluorinated compounds have very low boiling points. The intermolecular forces in these compounds are also very low as a result of which these compounds have very low coefficients of friction.

6. Fluorine being more electronegative than carbon, pulls the electronic charge of the bond pair of electrons in C—F bond towards itself. Carbon atom in C—F bond thus develops a partial positive charge and can be considered as *oxidised.* The fluorocarbons are not oxidised at all because the partial positive charge on the carbon atom of fluorocarbons makes further loss of electrons from it extremely difficult.

7. The fluorine atoms in fluorocarbons protect the carbon atoms effectively making them less vulnerable to attack by other reagents. The fluoro compounds are thus not very reactive.

Uses. 1. $CF_3CHBrCl$ and some similar other compounds are used as anaesthetics.

2. $CHClF_2$ is used for preparing tetrafluoro ethylene which is further converted into teflon by polymerisation.

$$2CHClF_2 \xrightarrow[\text{Polymerisation}]{500-1000°C} CF_2=CF_2+2HCl$$

$$nCF_2=CF_2 \longrightarrow \underset{\text{Teflon}}{-(CF_2-CF_2)_n-}$$

3. Fluorocarbons are used as lubricants because of their very low coefficients of friction.

4. Chlorofluorocarbons such as CCl_3F, $CClF_3$ and CF_2Cl_2 are used as refrigerants.

***Example 15*: Which allotrope of carbon is a non-conductor of electricity and why?**

***Solution*:** Diamond is non-conductor of electricity. Its structure is an infinite network of carbon atoms in which each carbon is tetrahedrally surrounded by four carbon atoms. No free electrons are available. Thus, there is no scope for conductance of electricity.

***Example 16*: Nitrogen cannot form nitrogen penta chloride while phosphorus can. Explain.**

Or

Nitrogen forms NCl_3 whereas phosphorus forms PCl_3 and PCl_5 both. Explain.

Solution. Electronic configurations of N and P are given below:

N $\quad 1s^2 2s^2 2p_x^1 2p_y^1 2p_z^1$

P $\quad 1s^2 2s^2 2p^6 3s^2 3p_x^1 3p_y^1 3p_z^1$

In the case of N, only three half filled orbitals are available for bonding. Of course, 2s orbital which is paired also participates in hybridisation (sp^3). Thus N forms three bonds to give NCl_3.

In the case of P, *d* orbitals are also available. One of the electrons in 3*s* orbital is excited to 3*d* orbitals. Now there are five orbitals which are half filled. They hybridise (sp^3d) to give five equivalent orbitals. Thus, P can form PCl_5.

6.7 HYDRIDES OF BORON

Preparation. Some common methods of preparation of hydride are given as under.

***(i)* By pyrolysis of diborane.** Diborane is a source for obtaining the higher hydrides mainly B_4H_{10}, B_5H_{11}, B_5H_9 and $B_{10}H_{14}$. Normally diborane required for the purpose is made by treating sodium borohydride, $NaBH_4$, with concentrated H_2SO_4.

$$2NaBH_4 + H_2SO_4 \longrightarrow Na_2SO_4 + B_2H_6$$

$$B_2H_6 \xrightarrow{373K} B_4H_{10}$$

$$B_2H_6 \xrightarrow{393K} B_5H_{11}$$

$$B_2H_6 \xrightarrow{453K} B_5H_9 + B_{10}H_{14}$$

***(ii)* By reduction of boron halide.**

$$LiAlH_4 + BF_3 \xrightarrow{\text{Ether}} [LiBH_4 + AlF_3] \times 3$$

$$3\,LiBH_4 + BF_3 \longrightarrow 2B_2H_6 + 3LiF$$

$$4BF_3 + 3LiAlH_4 \longrightarrow 2B_2H_6 + 3AlF_3 + 3LiF$$

***(iii)* By hydrolysis of magnesium boride.** Magnesium boride on dissolving in aqueous acids yields mainly B_4H_{10}, B_5H_9 and B_6H_{10}.

$$\underset{\text{Magnesium boride}}{2Mg_3B_2} + 12HCl \longrightarrow 6MgCl_2 + B_4H_{10} + H_2$$

***(iv)* Diborane can also be obtained by action of hydrogen on boric oxide.**

$$B_2O_3 + 6H_2 \xrightarrow[Al_2Cl_6]{Al\ or} B_2H_6 + 3H_2O$$

***(v)* By passing electric discharge either on lower boranes or on a mixture of boron halide and hydrogen.**

$$2B_5H_9 \xrightarrow{Discharge} 2B_5H_8 + H_2$$

$$2BCl_3 + 6H_2 \longrightarrow B_2H_6 + 6HCl$$

Properties of diborane.

***(i)* Combustion.** It bums in air or oxygen producing large amount of heat.

$$B_2H_6 + 3O_2 \longrightarrow B_2O_3 + 3H_2O;\ \Delta H = -201.7\ kJ\ mol^{-1}$$

***(ii)* Action of alkalies.** Diborane undergoes hydrolysis in the presence of water or alkalies.

$$B_2H_6 + 2KOH + 2H_2O \longrightarrow 2KBO_2 + 6H_2$$

(iii) $B_2H_6 \xrightarrow[High\ temp.]{Excess\ NH_3} BN$

(iv) $B_2H_6 \xrightarrow{HCl} B_2H_5Cl$

(v) $B_2H_6 \xrightarrow{BCl_3} B_2H_5Cl$

(vi) $B_2H_6 \xrightarrow[298K]{Cl_2} BCl_3$

(vii) $B_2H_6 \xrightarrow{H_2O} B(OH)_3 + H_2$

(viii) $B_2H_6 \xrightarrow{CO} COBH_3$

(ix) $B_2H_6 \xrightarrow[373-523K]{H_2}$ Higher hydrides

(x) $B_2H_6 \xrightarrow[473K]{NH_3}$ Borozole

***Example 17.* Complete the following reaction and write the structure of the product formed:**

$$\mathbf{B_2H_6 \xrightarrow[High\ temp.]{Excess\ NH_3}}$$

Solution.

$B_2H_6 + NH_3 \longrightarrow$

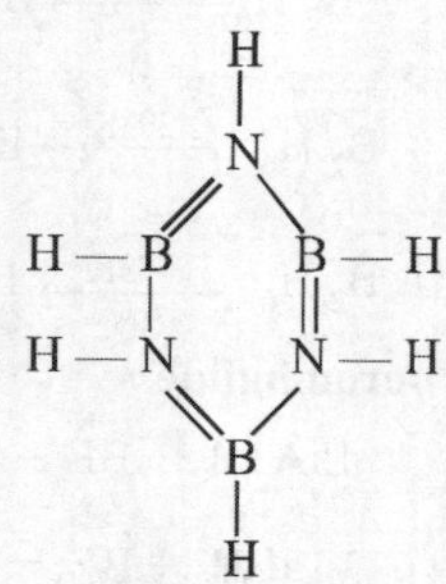

Borazine

6.7.1 Structure of diborane

Structure of diborane presents an interesting case since there are not enough valence electrons to form the expected number of covalent bonds. This is an *electron deficient compound.*

Since boron atom has three electrons in the outermost orbit in the excited state, therefore, it can link itself with three hydrogen atoms. Thus while each of the boron atoms in diborane can link itself to three hydrogen atoms, there are no electrons left to form a bond between the two boron atoms.

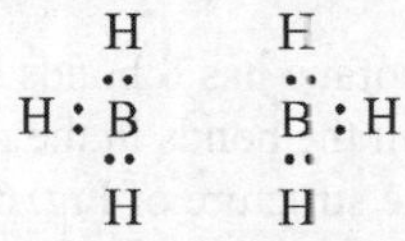

To account for this anomaly, various structures for diborane have been proposed. These can be discussed under two main types:

1. Ethane-like structures. It is assumed that all the six hydrogen atoms in B_2H_6 are equivalent as in case of ethane, C_2H_6. Such structures can be visualised by taking into account the phenomenon of resonance. The following three different possibilities of resonance are:

***(i)* Covalent resonance.** This gives two possible structures.

(*a*) (*b*)

***(ii)* Covalent-ionic resonance.** It gives the following structure.

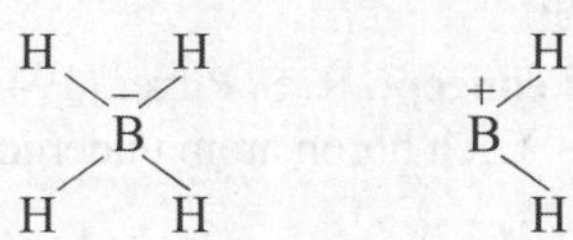

***(iii)* One-electron bond and two-electron bond resonance.** Here two possible structures obtained are:

(*a*) (*b*)

Experimental evidence has shown that in B_2H_6 molecule, two of the six hydrogen atoms behave differently from the others and also there is a hindered rotation around *B-B* linkage in the molecule. These facts rule out the possibility of an ethane-like structure and indicates the existence of an ethylene-like structure.

2. *Hydrogen bridge structure* (Ethylene-type structure). Diborane, B_2H_6, has been found to prossess a bridge structure in which each B atom is bonded to two *H* atoms (called terminal H atoms) by regular electron-pair bonds. The resulting BH_2 fragments are co-planar and are bridged by two H atoms. The bridging H atoms are in a plane perpendicular to

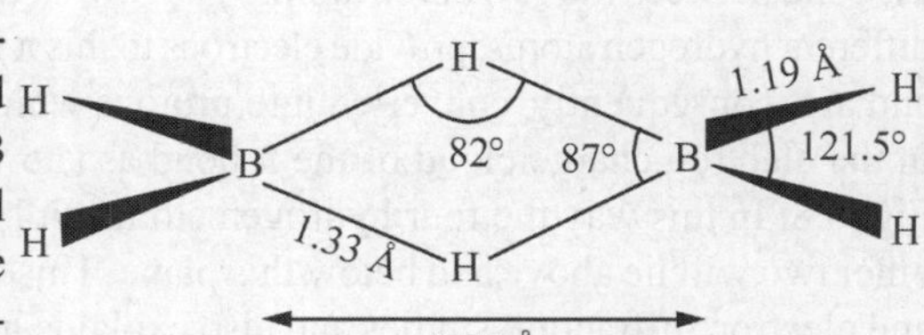

Fig. 6.7. Bridge structure of diborane.

the plane of BH_2 fragments, *i.e.*, one lies above and one below the plane to the rest of the molecule and prevent rotation between the two boron atoms. The various bond distances and angles are shown in Fig. 6.7.

In view of the bridge structure, diborane has 8 bonds but there are 12 electrons available for bonding (3 per B and 1 per H), hence, all the bonds in the molecule cannot be electron-pair bonds which would require 16 electrons for the structure of Fig. 6.7. Four of the terminal H-B bonds are normal single bonds as evidenced by bond angles and infrared stretching frequencies. The other two hydrogen atoms and four electrons are in a different environment.

The bond lengths between boron and hydrogen at the bridges is 1.33 Å while that of terminal positions is 1.19 Å. The terminal B-H distances are the same as the bond lengths measured in non-electron deficient compounds. So the electron deficiency is associated with the bridge groups.

A number of bonding theories have been proposed to account for the nature of the bridge in diborane structure. These are:

(*a*) Normal hydrogen bond concept. This concept was proposed by Buraway that the bridge between the boron atoms involves normal hydrogen bonds as shown below:

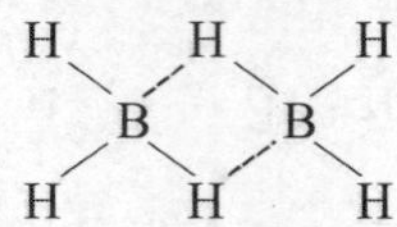

However, this concept is not acceptable due to a very small difference in the electronegativity values of B (1.00) and H (2.1) atoms.

(*b*) Protonated double bond concept. R. S. Pitzer (1945) suggested the concept of a double bond between the two boron atoms. Each boron atom undergoes sp^2 hybridization as follows:

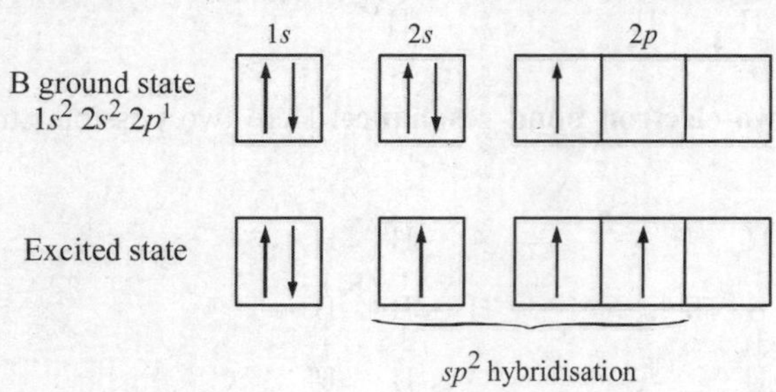

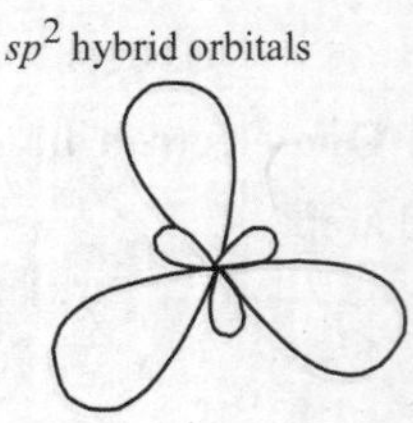

One of the sp^2 hybrid orbitals of each boron atoms overlaps to form one B-B σ bond. There is one unhybridized empty *p* orbital associated with each boron atom which lies in a plane perpendicular to the sp^2 hybrid orbitals. Sideways overlapping of these *p* orbitals gives a π bond between the two boron atoms. The two, so called different hydrogen atoms, provide electrons to this π bond and are, consequently, converted into protons which lie in the electron charge cloud of the π bond as shown in Fig. 6.8. In this way the four hydrogen atoms will be coplanar with the two boron atoms while the other two will lie above and below this plane. This structure is consistent with the results of infrared and electron diffraction studies and also explains hindered rotation around B-B axis. This structure envisages acidic properties which are, however, not given by the compound.

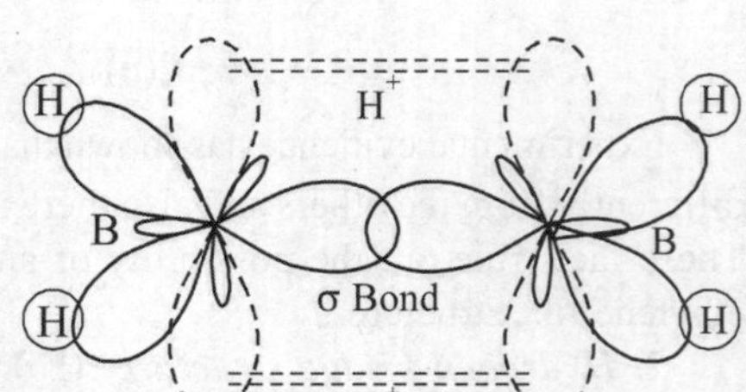

Fig. 6.8. Structure of diborane.

(c) Three-centre bond concept. This concept, also known as the molecular orbital concept, was proposed by Longuet-Higgins and Professor William N. Lipscomb. They modified Pitzer' s concept in such a way that the double bond no longer involved protons, rather it involved two similar 'three centre bonds'.

It is supposed that boron atoms undergo sp^3 hybridisation. Each boron then utilizes two such orbitals containing one electron each for bonding with terminal hydrogens. There are now two sp^3 hybrid orbitals and one electron left with each boron atom. Also there are two 1*s* orbitals containing one electron each belonging to two hydrogen atoms not yet bonded. Thus, in all, there are six atomic orbitals and four electrons available which can form two three centre bonds known as the hydrogen bridges between the boron atoms as shown in Fig. 6.9.

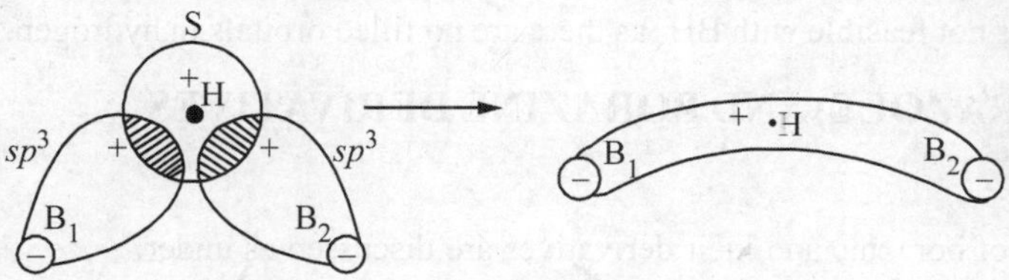

Fig. 6.9. Three-centred bonds in diborane.

Christopher Longuet-Higgins

This is known as three-centre electron pair bond B-H-B. The molecule of diborane contains two such bonds. Due to repulsion between the two hydrogen nuclei, the delocalised orbitals or bridges are bent away from each other in the middle. This is why the hydrogen bridge orbitals is shifted upwards in the middle. The other bridge orbital will be shifted downwards. This is the reason that these three atom orbitals are frequently referred to as *banana bonds*.

The molecule in a more simple form may be represented as in Fig. 6.10 (*a*) and (*b*).

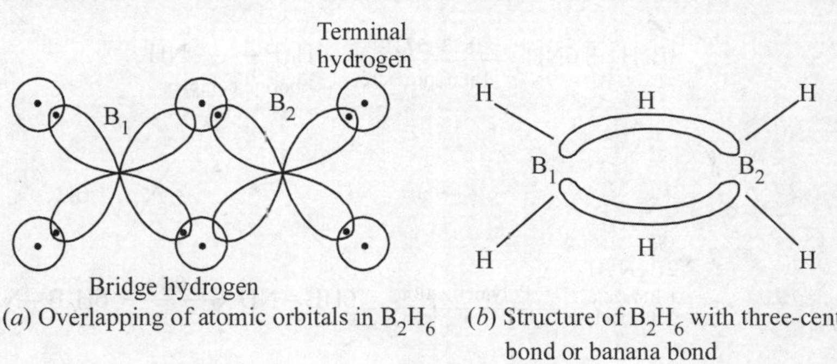

(*a*) Overlapping of atomic orbitals in B_2H_6 (*b*) Structure of B_2H_6 with three-centre bond or banana bond

Fig. 6.10. Representation of structure of diborane.

***Example 18.* BF_3 exists but BH_3 does not exist. Assign reason.**

Solution. BH_3 does not exist as a molecule because boron in BH_3 has a vacant 2*p* orbital. Two BH_3 molecules combine to give a B_2H_6 molecule and boron thus removes its electron deficiency.

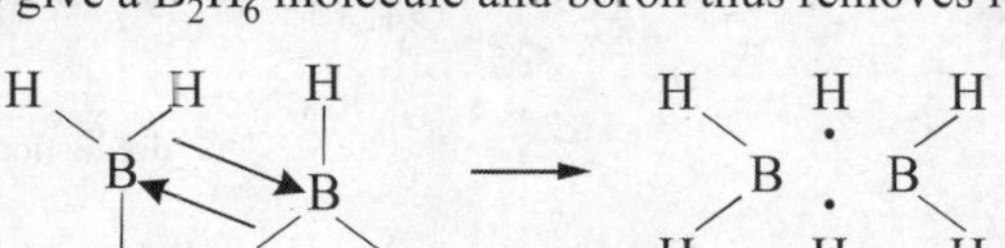

In BF_3 the vacant 2*p* orbital of boron forms π bond by overlapping with filled 2*p* orbital of fluorine. There is a possibility of overlap with all the three fluorine atoms as depicted below.

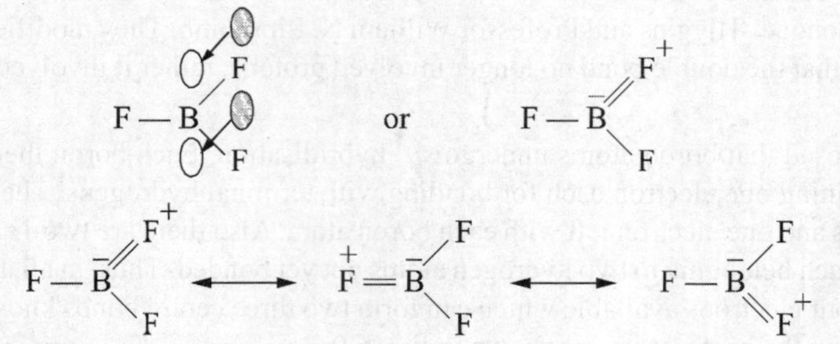

Such π bond formation is not feasible with BH_3 as there are no filled orbitals in hydrogen.

6.8 BORAZINE (BORAZOLE) AND BORAZINE DERIVATIVES

Preparation

Methods of preparation of borazine and their derivatives are discussed as under:

(*i*) Borazine was first synthesized by pyrolysis of the adduct of diborane and ammonia at 453-473 K.

$$3B_2H_6 + 6NH_3 \longrightarrow \underset{\text{Adduct}}{3(B_2H_6)(NH_3)_2} \xrightarrow[453-473K]{\text{Heat}} \underset{\text{Borazine}}{2B_3N_3H_6} + 12H_2$$

(*ii*) Borazole can also be prepared by the direct reaction of diborane with ammonia at 523-573 K under one atmospheric pressure. The yield is only 50%. The reaction involves the formation of the borine complex $H_3B \leftarrow NH_3$ which loses hydrogen and polymerises as shown below:

$$3B_2H_6 + 6NH_3 \xrightarrow[\text{1 atm-pressure}]{523-573\text{K}} \underset{\text{borine complex}}{6H_3B \longleftarrow NH_3}$$

$$\downarrow -6H_2$$

$$\underset{\substack{\text{Borazole}\\(50\%\text{ yield})}}{2B_3N_3H_6} \xleftarrow{\text{Polymerization}} 6HB=NH \xleftarrow{-6H_2} 6H_2B-NH_2$$

(*iii*) It can be obtained by heating together ammonium chloride and boron trichloride at 413-423 K in presence of chlorobenzene. The initial product is B-chloroborazine which is reduced with sodium borohydride in polyether to give borazole or borazine.

$$3NH_4Cl + 3BCl_3 \xrightarrow[C_6H_5Cl]{413-423K} 9HCl + \underset{\text{Trichloroborazine}}{Cl_3B_3N_3H_3}$$

$$\downarrow \text{Reduction} \quad NaBH_4 \text{ in polymer}$$

$$\underset{\text{Borazole}}{B_3N_3N_6} + \frac{3}{2}B_2H_6 + 3NaCl$$

Preparation of derivatives. B or N-substituted borazines may be prepared by appropriate substitution on the starting materials.

(*a*) Preparation of tri-B-methyl borazoles. (*i*) Tri-B-methyl borazole can be prepared by the reaction of boron trimethyl with ammonia followed by heating the addition compound at 603 K and 30 atmospheric pressure.

$$2B(CH_3)_3 + 3NH_3 \longrightarrow 3[(CH_3)_3 B \leftarrow NH_3] \xrightarrow[\text{20 atom pressure}]{\text{Heat 603 K}} \underset{\text{B·B'B''–trimethy borazole}}{(CH_3)_3B_3N_3H_3}$$

(*ii*) These are also formed when borazole is treated with boron trimethyl.

$$B_3N_3H_6 + B(CH_3)_3 \longrightarrow \underset{\text{B · B' B''– trimethyl borazole}}{(CH_3)_3 B_3N_3H_3}$$

(*b*) Preparation of tri-N-methyl borazoles. (*i*) Tri-*N*-methyl borazole can be prepared by reaction of monomethyl ammonium chloride with $LiBH_4$.

$$3CH_3NH_3Cl + 3LiBH_4 \longrightarrow \underset{\text{N, N',N''-trimethyl borazole}}{(CH_3)_3N_3H_3B_3} + 3LiCl + 9H_2$$

(*ii*) Tri-N-alkyl borazines can be prepared by the following method in good yield.

$$3RNH_3Cl + 3BCl_3 \longrightarrow Cl_3B_3NR_3 + 9HCl$$

$$\downarrow \text{Reduction Na}BH_4$$

$$(RNBH)_3 + 3B_2H_6 + 3NaCl$$

Properties of borazine

(*i*) At ordinary temperature, both air and oxygen have no action on borazine.

(*ii*) Reaction with water. It hydrolyses in excess of water giving H_2, H_3BO_3 and NH_3. Hydrolysis becomes fast above 423 K.

$$B_3N_3H_6 + 3H_2O \longrightarrow 3B(OH)_3 + NH_3 + 3H_2$$

(*iii*) Addition reactions. (*a*) Borazine reacts with two moles of bromine at 273 K, the resulting compound decomposes at 333 K to give dibromoborazine.

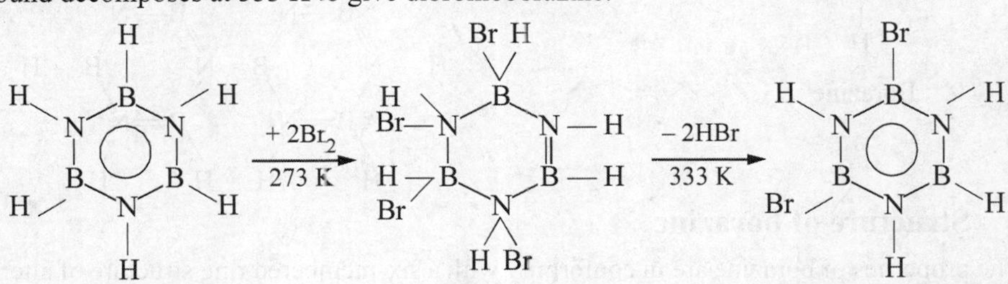

Borazine B-dibromo-borazine

(*b*) Borazine, in contrast to benzene, readily undergoes addition reactions with hydrogen halides, methyl iodide, methanol and water. In the last two cases very mild conditions must be used or disruption of the ring occurs. One mole of borazine adds three moles of the above molecules. The more negative group in the adding molecule normally goes to boron atom.

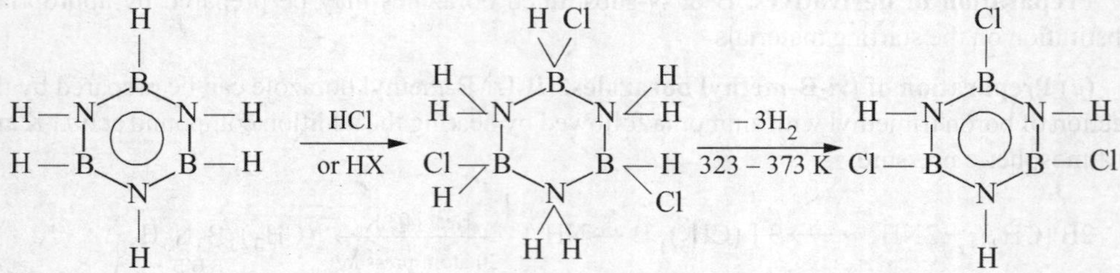

Borazine B-trichloroborazole

***(iv)* Substitution reactions.** Borazine reacts with boron trimethyl, B $(CH_3)_3$ B give to trimethyl borazine (about 60% yield) together with small quantities (about 5%) of the mono and dimethyl derivatives and a non-volatile product of composition BNH

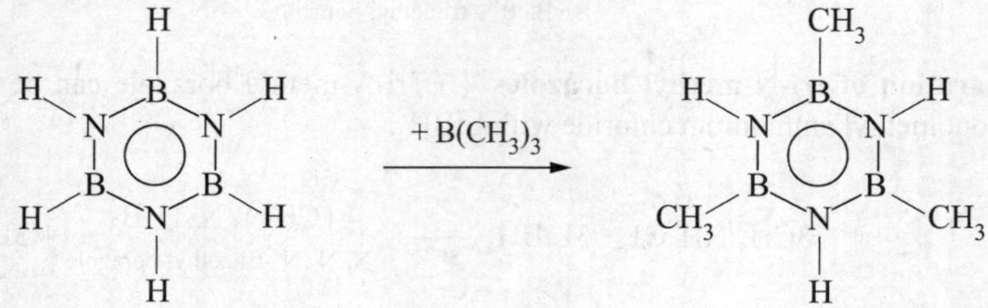

Borazine B-trimethyl borazine

***(v)* Pyrolysis of borazine** by passing a silent electric discharge through it yields compounds similar to naphthalene.

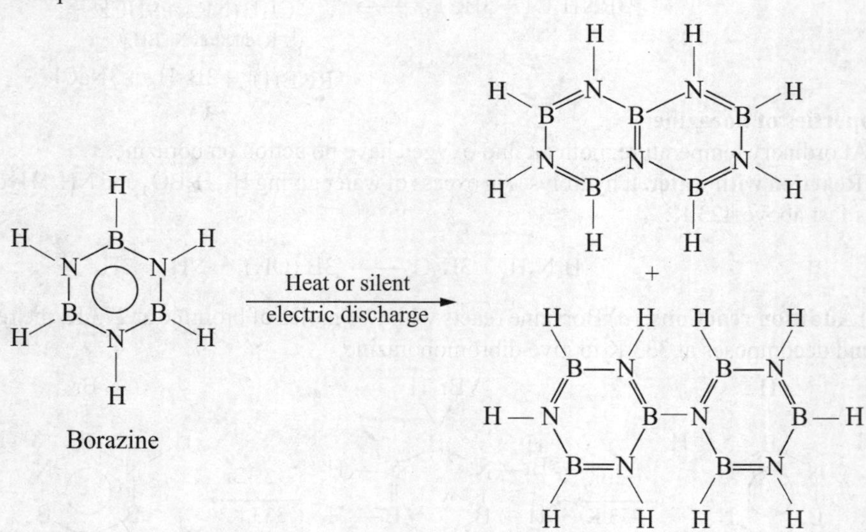

Borazine

6.8.1 Structure of borazine

The properties of borazine are in conformity with a six-membered ring structure of alternating boron and nitrogen atoms.

(*i*) Structural studies of borazine including electron diffraction show the planar ring structure exactly comparable to that found in benzenes with equal B-N bond lengths (144 Å) and equal angles (120°).

(*ii*) Infrared and Raman spectra studies also confirm the planar structure. The bond distance (1.44 Å) lies between those expected for single (1.54 Å) and double (1.36 Å) B-N bonds. This suggests resonance in the structure of the Kekule type as in benzene ring. Such a structure may be fomulated as

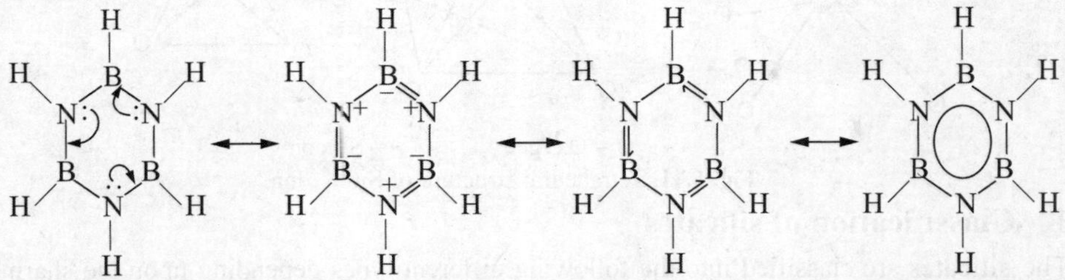

(*iii*) This ring structure is related to the layer lattice of boron nitride as the ring structure of benzene is related to the layer lattice of graphite. That is the reason borazine is also called 'inorganic benzene'. However, in benzene the C = C bonds are nonpolar and the nucleus is very resistant to addition reactions whereas the borazine nucleus, because of its polarities, is fairly reactive.

***Example 19.* Why is borazine called inorganic benzene? Explain with reasons.**

Solution. Borazine is called inorganic benzene because of the following reasons:

(*i*) Borazine is a six-membered cyclic compound like benzene where the ring positions are occupied by boron and nitrogen in alternate positions. Like benzene, there is a system of conjugated single and double bonds.

(*ii*) Borazine undergoes addition and substitution reactions with HC1 and Br_2 etc., like in the case of benzene.

(*iii*) The ring structure of borazine is related to layer structure of BN like the structure of benzene is related to layer structure of graphite.

6.9 SILICATES

Structure of SiO_4^{4-} ion

Consider the electronic configuration of silicon in the ground state and excited state.

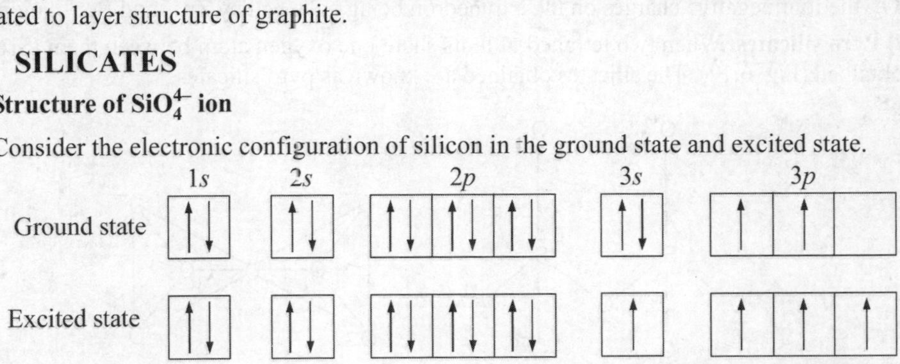

The four unpaired electrons in the four sp^3 hybrid orbitals form covalent bonds with oxygen ions to give SiO_4^{4-} tetrahedron. The negative charge on SiO_4^{4-} is due to the fact that any oxygen (which is not a bridge atom) has to pick up an electron from some metal to complete its octet. The negative charge is being balanced by the metal cations such as, Li^+, K^+ ,Ca^{2+}, Al^{3+}, etc. These serve to hold the solid together by ionic attraction.

The tetrahedral structure of the SiO_4^{4-} ion can also be represented by projection as shown on the extreme right of Fig. 6.11. This is obtained when the tetrahedron is viewed from above. The oxygen is shown as a small open circle and the silicon a small dot. The silicon is shown inside and the oxygen at the apex. Thus in silicates the basic unit of structure is SiO_4 tetrahedron. These tetrahedra occur singly or by sharing oxygen atoms in small groups, in small cyclic groups, infinite chain or infinite sheets.

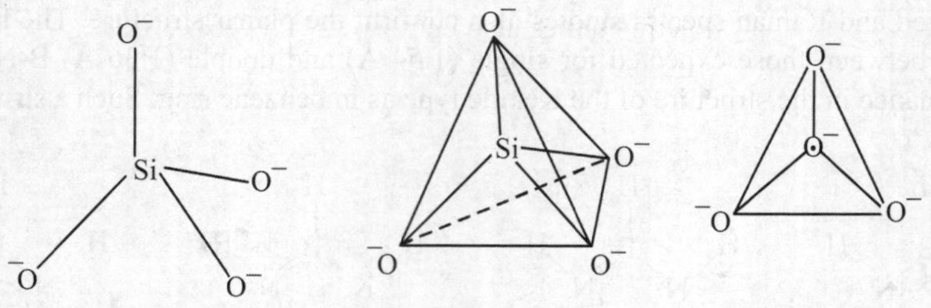

Fig. 6.11. Tetrahedral structure of SiO_4^{4-} ion.

6.9.1 Classification of silicates

The silicates are classified into the following different types depending upon the sharing of oxygen atoms between the adjacent SiO_4^{4-} tetrahedra and the substitution of Al^{3+} for Si^{4+} ion in the structure:

(*i*) Orthosilicates
(*ii*) Pyrosilicates
(*iii*) Cyclic silicates
(*iv*) Chain silicates
(*v*) Sheet silicates
(*vi*) Three-dimensional silicates.

Each type is discussed separately as under:

(*i*) Orthosilicates. In the ortho silicates, discrete individual SiO_4^{4-} tetrahedra (Fig. 6.11) are present and there is no sharing of oxygen atoms between adjacent tetrahedra but the oxygens present are coordinated to the metal cations. The important examples are willemite, Zn_2SiO_4 and phenacite, Be_2SiO_4, the four negative charges on the tetrahedron being balanced by Zn^{2+} and Be^{2+} cations.

(*ii*) Pyro silicates. When two tetrahedral units share one oxygen atom between them $(Si_2O_7)^{6-}$ ion is obtained (Fig. 6.12). The silicates obtained are known as pyro silicates.

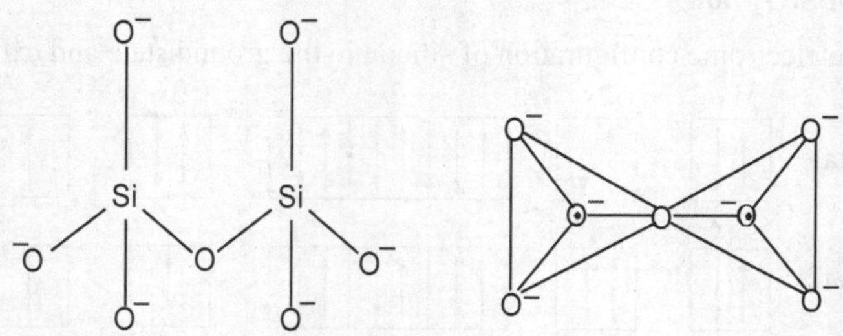

Fig. 6.12. Structure of pyrosilicates.

This is the simplest of the condensed silicate ions. Thortveitite, $Sc_2(Si_2O_7)$ is an example of this type of silicates.

(*iii*) Cyclic silicates. If two oxygens per tetrahedron are shared ring structures (Fig. 6.13) may be formed with the general formula $(SiO_3)_n^{2n-}$. The cyclic ion SiO_9^{6-} where $n = 3$ occurs in wollastonite, $Ca_3(Si_3O_9)$ and benitoite $BaTi(Si_3O_9)$. Another example is Beryl $Be_3Al_2(Si_6O_{18})$. It contains $Si_6O_{18}^{12-}$ cyclic ion where $n = 6$. The twelve negative charges are being balanced by Be^{2+} and Al^{3+} cations. The Si_6O_{18} units are superimposed one above the other leaving channels, hence, the mineral is permeable to gases which have small molecules. In benitoite, Ti^{4+} and Ba^{2+} ions are 6-coordinate but in beryl the Be^{2+} ions are 4-coordinate and Al^{3+} ions are 6-coordinate.

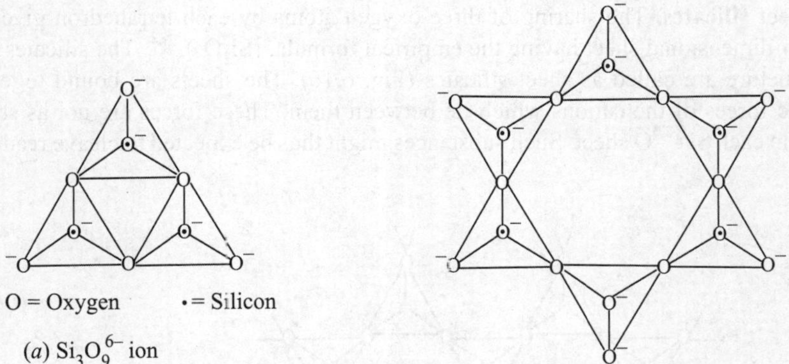

Fig. 6.13. Structures of cyclic silicates.

***(iv)* Chain silicates.** The sharing of two oxygen atoms per tetrahedron can also result in the formation of linear simple chains of indefinite length with two types of compositions, *i.e.,*

(*a*) Metasilicate having the general formula $(SiO_3)_n^{2n-}$ where the chains are single and occur in pyroxenes, and

(*b*) Metatetrasilicate having the general formula $(Si_4O_{11})_n^{6n-}$ where the chains are double and occur in amphiboles. The single chains can be represented by Fig. 6.14. Important examples of pyroxenes are spodumene LiAl $(SiO_3)_2$ enstatite, $MgSiO_3$, diopside. CaMg $(SiO_3)_2$ and jadeite, NaAl $(SiO_3)_2$.

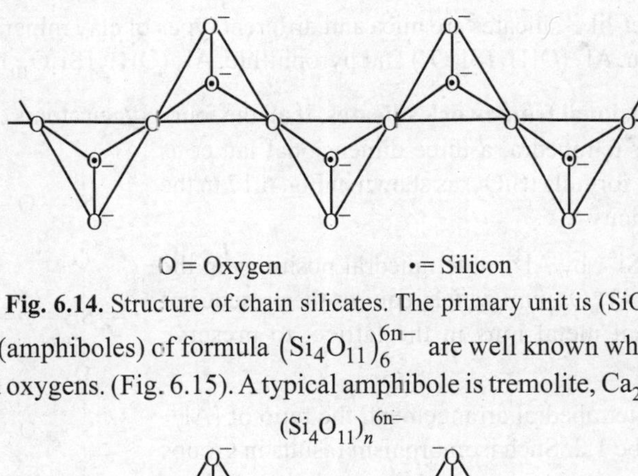

Fig. 6.14. Structure of chain silicates. The primary unit is $(SiO_3)_n^{2n-}$.

Double chains (amphiboles) of formula $(Si_4O_{11})_6^{6n-}$ are well known where two simple chains are joined by shared oxygens. (Fig. 6.15). A typical amphibole is tremolite, $Ca_2Mg_5(Si_4O_{11}) \cdot (OH)_2$.

$(Si_4O_{11})_n^{6n-}$

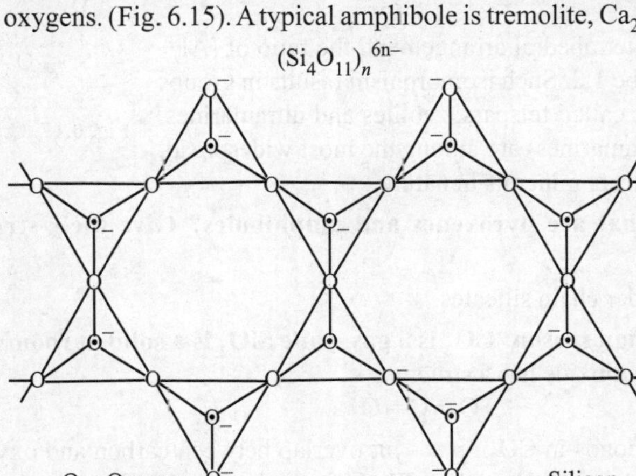

Fig. 6.15. Double chain silicates (Amphiboles).

(*v*) Sheet silicates. The sharing of three oxygen atoms by each tetrahedron gives rise to an infinite two-dimensional sheet having the empirical formula, $(Si_2O_5)_n^{2n-}$. The silicates having this type of structure are called as sheet silicates (Fig. 6.16). The sheets are bound together by the electrostatic forces of the cations which lie between them. These forces are not as strong as the bonds within each Si — O sheet. Such substances might thus be expected to cleave readily into thin sheets.

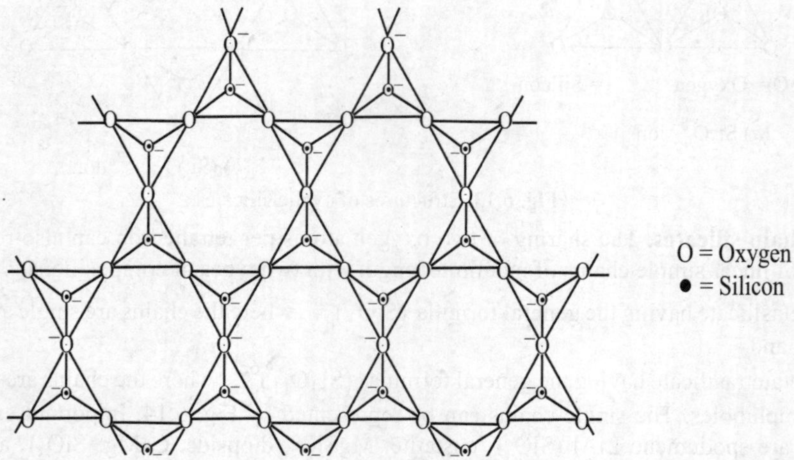

Fig. 6.16. Sheet silicates

Examples of sheet-like-silicates are mica and different types of clay minerals such as talc. $Mg_3(OH)_2(Si_4O_{10})$ Kaolin, $Al_2(OH)_4(Si_2O_5)$ and pyrophillite, $Al_2(OH)_2(Si_4O_{10})$.

(*vi*) Three-dimensional framework silicates. If all the four oxygen atoms of SiO_4^{4-} tetrahedron are shared with other tetrahedra, a three-dimensional lattice is formed with a general formula (SiO_2) as shown in Fig. 6.17 in the planer form for simplicity.

Replacement of Si^{4+} by Al^{3+} in tetrahedral positions in the SiO_2 lattice gives $(Si, Al)_nO_{2n}$ unit which requires the presence of additional monovalent metal ions in the lattice, to preserve electrical neutrality.

Also to preserve tetrahedral arrangement, the ratio of (Al + Si) to O must always be 1:2. Such isomorphism results in groups of important minerals called felspars, zeolites and ultramarines, which (excepting ultramarines) are among the most widespread, diverse and useful silicate minerals in nature.

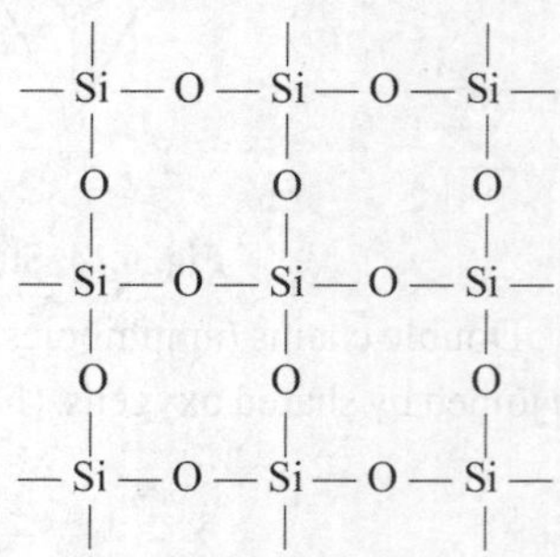

Fig. 6.17. Crystal structure of SiO_2

Example 20. **What are pyroxenes and amphiboles? Give their structural units with examples.**

Solution. See under chain silicates.

Example 21. **Assign reason: CO_2 is a gas while SiO_2 is a solid at room temperature.**

Solution. Carbon dioxide has a structure

$$O = C = O$$

There are two π bonds in $CO_2 \cdot p\pi - p\pi$ overlap between carbon and oxygen is efficient and quite strong. Now there are weak van der Waals forces between different molecules of CO_2. Hence it is a gas at room temperature.

In SiO_2, silicon cannot form double bond with oxygen as $p\pi - p\pi$ overlap is not efficient because of size difference between Si and O. Hence it cannot give molecules of type O = Si = O. It completes its octet by network of single bonds as shown here under:

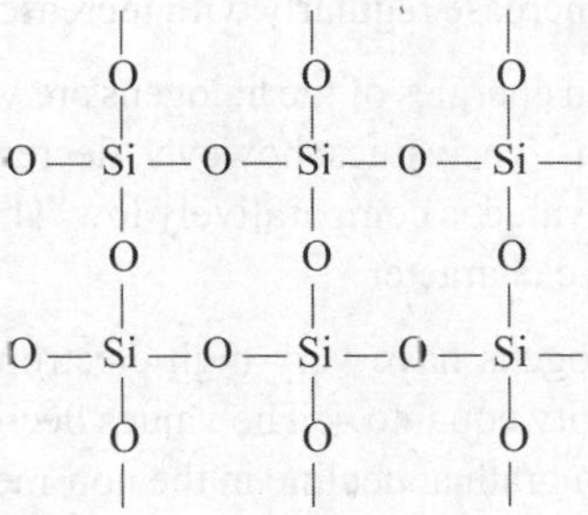

As a result, it is a high-boiling solid.

6.9.2 (*i*) Felspar, (*ii*) Zeolites, (*iii*) Ultramarine

(*i*) Felspar. Felspars are silicates in which Si^{4+} is partly replaced by Al^{3+} in the tetrahedron position giving rise to units $(Si, Al)_n O_{2n}$.

Felspars constitute some two-thirds of the igneous rocks and are, in fact, the most abundant of all the rock forming minerals. They are divided into two groups according to the symmetry of their structure.

(*a*) *Orthoclase felspars, e.g.*, orthoclase, $KAlSi_3O_8$ and

(*b*) *Plagioclase felspars, e.g.*, albite, $NaAlSi_3O_8$.

In orthoclase, K^+ ion has the appropriate size to fit into the lattice while in plagioclase, Na^+ is smaller and causes distortion. Therefore, orthoclase are more symmetrical.

(*ii*) Zeolites are the most important three-dimensional silicates. They have a much open structure than the felspar which permits them to take up or lose water quite readily from the channels in the honeycomb-like structure. A variety of other molecules such as CO_2, NH_3, C_2H_2OH may also be absorbed instead of water. Zeolites act as molecular sieves by absorbing molecules small enough to enter cavities but not those which are too big to enter. Zeolites also act as natural ion exchangers. For example, permutit, water softeners, are sodium zeolites. They take Ca^{2+} ions from hard water and replace them by Na^+ thereby softening the water. The permutit softener is regenerated by treatment with brine (NaCl-solution), when the reverse process takes place.

$$\underset{\text{Permutit}}{Na_2Z} + \underset{\text{from hard water}}{CaSO_4} \longrightarrow \underset{\text{Softening of water}}{CaZ + Na_2SO_4} \qquad \text{where Z = Zeolite ion}$$

$$CaZ + \underset{\text{(Brine)}}{2NaCl} \longrightarrow \underset{\substack{\text{Regeneration}\\ \text{of permutit}}}{Na_2Z + CaCl_2}$$

(*iii*) Ultramarines are produced synthetically. Apart from $(Si, Al)_n O_{2n}$, frameworks and balancing cations, there are additional negative ions such as Cl^-, SO_4^{2-} or S_2^{2-} (persulphide ion) also present in these materials. Most of the ultramarines are coloured and are used as pigments. Examples are, ultramarine $[Na_8 (Al_6 Si_6 O_{24}) S_2]$, sodalite, $[Na_8 (Al_6 Si_6 O_{24}) Cl_2]$ and noselite $[Na_8 (Al_6 Si_6 O_{24}) SO_4]$. Ultramarines have basket-like structures and all baskets are joined to each other. The various cations and anions occupy the cavities in the lattice and can exchange their positions.

6.10 BASIC PROPERTIES OF HALOGENS

Some important physical properties of the halogens ar given in Table 6.1.

1. Atomic radius and density increase regularly with increase in atomic number, as expected.

2. Ionisation energy. Ionisation energies of the halogens are very high indicating that they have very small tendency to lose electrons. The values, however, decrease on moving down the group, as expected. In the case of iodine, the value is comparatively low. This enables the iodine atom to lose an electron and show electropositive character.

3. Electronegativity. The halogens have very high electronegativities. Fluorine is the most electronegative with electronegativity equal to 4. The values decrease as we move down the group from fluorine to iodine, indicating a gradual decline in the non-metallic character. The last element iodine, with an electronegativity of 2.5, shows some metallic character also. For instance, it forms in a few cases a positive ion and has a metallic lustre.

Table 6.1. Physical properties of the Halogens

Property	Fluorine	Chlorine	Bromine	Iodine
Atomic Number	9	17	35	53
Atomic mass (a.m.u.)	18.998	35.453	79.909	126.904
Physical state	Gas	Gas	Liquid	Solid
M.P. (° C)	– 218.6	–101.0	–7.2	113.9
B.P. (° C)	–188.1	–34.6	59.5	185.2
Density in liquid state (g cm^{-3})	1.108	1.1557	2.948	3.76 (in solid state)
Colour of vapour	Pale yellow	Greenish yellow	Orange red	Violet
Atomic radius (Å)	0.72	0.99	1.14	1.33
Ionic radius of X^- ion (Å)	1.33	1.84	1.96	2.20
Electronegativity	4.0	3.0	2.8	2.5
Electron affinity (kJ mol^{-1})	332.6	348.5	324.7	295.5
Ionisation energy (kJ mol^{-1})	1680.8	1255.5	1142.8	1008.8
Oxidation states	–1	–1, +1, +3, +4, +5, +6, +7	–1, +1, +3 +4, +5, +6	–1, +1, +3 +4,+5,+6+7
Bond energy (kJ mol^{-1})	158.8	242.6	192.8	151.1
Standard reduction potential (volts)	+2.8	+1.36	+1.08	+0.54

4. Non-metallic character. On account of high electronegativities, all the halogens are non-metallic in character. The non-metallic character shows, decreasing trend as we move down the group. The last element, iodine, has comparatively very small electronegativity and exhibits metalloidal character. It has a metallic lustre and has a tendency to lose an electron to form a positive ion, I^+.

5. Electron affinity. All the elements, have high electron affinities. Therefore, they have a strong tendency to gain an electron to acquire noble gas configuration, *viz.*, s^2p^6. Chlorine has the highest electron affinity which decreases gradually as we move down the group from Cl to I. The electron affinity of fluorine however is lower than that of chlorine. The unexpectedly lower value is due to the small size of the atom. The addition of an extra electron in fluorine atom produces, a high electron density in a relatively compact size and, therefore, there is a strong electron repulsion. Consequently, fluorine has lesser tendency to accept extra electron, that is, its electron affinity is low.

6. Colour. Al the halogens are coloured. This is because of the absorption of visible light by their molecules resulting in the excitation of electrons from the highest filled molecular orbitals to the lowest unoccupied molecular orbitals. The excitation energy depends upon the size of the halogen atom. Smaller the atom, the greater is the force of attraction between the nucleus and the outer electrons and hence greater is the amount of the energy required excite the electrons to higher levels. Consequently, the excitation energy is very large in the case of fluorine and comparatively very small in the case of iodine. Hence, fluorine absorbs violet portion of the light which is associated with higher energy and thus appears yellow. Iodine, on the other extreme, absorbs radiations which are associated with lower energy and thus appears violet. Similarly, we can explain the greenish yellow colour of chlorine and orange red colour of bromine.

7. Oxidation states. All the halogens have the configuration s^2p^5 in the outer shell in the ground state. These elements, therefore, tend to gain or share one electron to acquire noble gas configuration. Hence they show an oxidation state of –1 or +1, depending upon whether the element with which they combine is less electronegative or more electronegative than the halogen (transference and sharing of electrons).

Since fluorine is the most electronegative element, it always exhibits an oxidation state of –1. It does not exhibit any positive oxidation state. Also, fluorine has no *d* orbitals in its valence shell and hence it cannot have any excited states. Consequently, it does not exhibit any higher oxidation state.

The other halogens exhibit higher positive oxidation states of +3, +4, +5, +6 and +7 also because they have vacant *d* orbitals in their valency shells which allow for excitation of *p* and *s* electrons, as illustrated below.

Various oxidation states exhibited by the halogens.

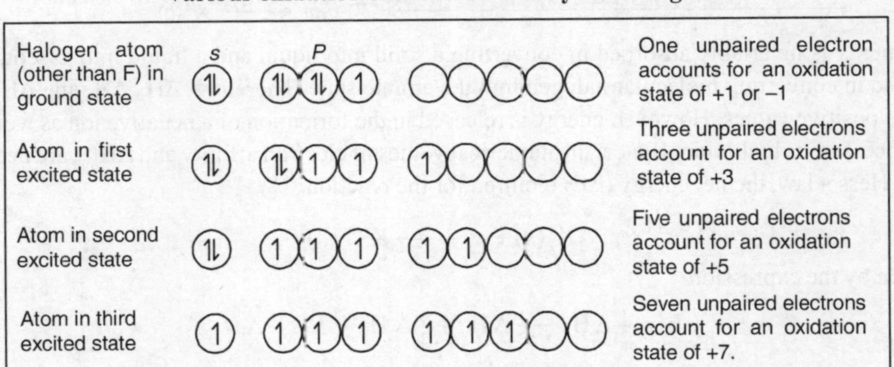

F. Haber

8. Oxidising power. Halogens, on account of high electron affinities, have a strong tendency to pick up electrons. Thus they, act as strong oxidising agents. The oxidising character, however, decreases on moving down the group. This can be explained with the help of the Born-Haber cycle. The oxidising action of the halogen can be represented by the following reaction:

$$\frac{1}{2}X_2\,(s, l \text{ or } g) + e^- + aq \xrightarrow{\Delta E} X^-\,(aq)$$

The following steps are involved in the process :

1. The solid halogen (e.g. I_2) changes into the liquid state. The energy required in this process is known as the heat of fusion, ΔH_f. Thus,

$$\tfrac{1}{2}X_2\,(s) \xrightarrow{1/2\,\Delta H_f} \tfrac{2}{1}X_2\,(l)$$

2. The liquid halogen changes into the gaseous state. The energy required in this process is known as the heat of vaporisation. ΔH_v. Thus,

$$\tfrac{1}{2}X_2\,(l) \xrightarrow{1/2\,\Delta H_v} X\,(g)$$

3. The gaseous molecular halogen changes into the atomic halogen. The energy required in this case is knwon as the heat of dissociation, ΔH_d. Thus,

$$\tfrac{1}{2}X_2\,(g) \xrightarrow{1/2\,\Delta H_d} X\,(g)$$

4. The gaseous atom changes into the gaseous ion. The energy involved in this process is known as electron affinity. Thus,

$$X(g) + e^- \xrightarrow{EA} X^-\,(g)$$

5. The gaseous ion finally changes into the hydrated ion. The energy involved in this process is known as the heat of hydration of the ion, ΔH_{hyd}.

$$X^-\,(g) + aq \xrightarrow{\Delta H_{hyd}} X^-\,(aq)$$

The value of ΔE for the halogen which is a measure of its oxidising strength, can be obtained as under:

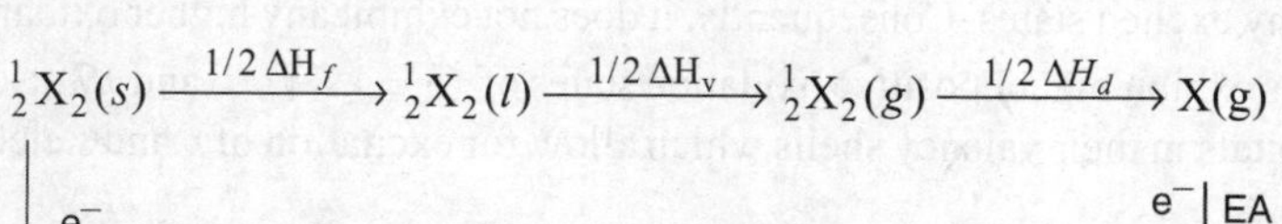

Energy is invariably absorbed in converting a solid into liquid and a liquid into gaseous state and also in converting molecular halogen into the atomic state. Therefore, ΔH_f, ΔH_v and ΔH_d have always positive values. However, energy is released in the formation of a negative ion as well as in the process of its hydration. Hence, the numerical values of electron affinity and ΔH_{hyd} are negative. Using Hess's law, the net energy (ΔE) required for the reaction

$$\tfrac{1}{2}X_2(s) + e^- \longrightarrow X^-\,(aq)$$

is given by the expression

$$\Delta E = \tfrac{1}{2}\Delta H_f + \tfrac{1}{2}\Delta H_v + \tfrac{1}{2}\Delta H_d - EA - \Delta H_{hyd}$$

ΔE measures the oxidising potential, *i.e.*, the tendency of the halogen to oxidise.

The values of ΔE calculated as above for the various halogens are given below. Values of ΔH_f, ΔH_v, ΔH_d, EA and ΔH_{hyd} for different halogen are also provided.

Halogen	$\frac{1}{2}$ Heat of fusion (kJ)	$\frac{1}{2}$ Heat of vaporisationg (kJ)	$\frac{1}{2}$ Heat of dissociation (kJ)	Electron affinity (kJ)	Heat of hydration (kJ)	ΔE (kJ)
F_2	–	–	+ 79	– 332	– 515	– 768
Cl_2	–	–	+ 122	– 348	– 381	– 607
Br_2	–	+ 15	+ 96	– 324	– 347	– 575
I_2	+ 8	+ 22	+ 76	– 295	– 305	– 524

Since fluorine and chlorine are gaseous at room temperature, the heats of fusion and vaporisation are not involved. Similarly, in the case of bromine, which is liquid at ordinary temperatures, the heat of fusion has not been taken into consideration.

The magnitude of ΔE is maximum in the case of fluorine and gradually decreases as we move down the group. Thus, fluorine is the strongest oxidising agent and iodine is the weakest oxidising agent. We can also explain why chlorine, which has the maximum electron affinity, is not the strongest oxidising agent. Inspection of the above table shows that the stronger oxidising character of F_2 in comparison to Cl_2 is mainly due to its relatively lower bond dissociation energy and higher heat of hydration. It is the oxidising potential and not the electron affinity alone which determines the strength of an oxidising agent in aqueous medium.

9. Combination with metals and non-metals. All the halogens react with metals and also with many non-metals to form halides. Fluorine is the most reactive. The reactive decreases as we move down the group. With sulphur for examples, fluorine forms the hexafluoride (SF_6), chlorine forms the tetrachloride (SCl_4), bromine forms the dibromide (SBr_2) while iodine does not react at all.

10. Reaction with water. The interaction of halogens with water decreases with increase in atomic number. For example, fluorine decomposes water very energetically; chlorine and bromine do so much less readily according to the equation.

$$X_2 + H_2O \longrightarrow HOX + HX$$

while iodine hardly reacts with water.

11. Reaction with hydrocarbons. All the halogens react with hydrocarbons. The reactivity decreases with increase in atomic number. Fluorine is the most reactive and brings about decomposition of the hydrocarbon into carbon.

$$CH_4 + 2F_2 \longrightarrow C + 4HF$$

Chlorine brings about the substitution of hydrogen atoms.

$$CH_4 + 4Cl_2 \longrightarrow CCl_4 + 4HCl$$

Bromine also brings about substitution but at a slower rate. Iodine has little action.

6.10.1 Peculiarities of Fluorine

The first element, fluorine, differs considerably from the other members of the halogen family. This difference is due to the following characteristics of fluorine:

(*i*) Small size

(*ii*) Very high electronegativity (fluorine, in fact, is the most electronegative of all the elements).

(*iii*) Non-availability of *d* orbitals in the valency shell.

Different behaviour of fluorine is observed in the following

1. Since fluorine is the most electronegative element, it exhibits only a negative oxidation stat of –1. It does not show any positive oxidation state. The other members, on the other hand, show positive oxidation states also. Thus, oxidation states of +1, +3, +5, +6 and +7 are quite frequent amongst halogens others than fluorine.

2. Fluorine is most reactive amongst the halogen. This is due to the low bond enegy of the F – F bond, the value for which is 158.8 kJ per mole. The bond is weak because of the repulsion between the non-bonding electrons in the molecule.

The X – X bond is much stronger in Cl_2, Br_2 and I_2. This is due to the hybridisation of *p* and *d* orbitals.

3. On account of high electronegativity, fluorine extensively forms hydrogen bond. Other elements, on account of comparatively small electronegativities, do not form hydrogen bonds. Thus, while HCl, HBr and HI are gaseous under ordinary conditions, HF is a liquid (b.p. 19.5°C).

4. Maximum covalency of fluorine is one because it cannot expand its valency shell beyond the octet as there are no *d* orbitals in the valency shell. Other elements, can observe covalencies upto 7 because of the availability of vacant *d* orbitals.

5. Most of the fluorides are ionic; these contain the F^- ion. This is due to small size of F^- ion, its radius being 1.36 Å. Chlorides, bromides and iodides form frequently molecular lattices.

6. The fluoride ion has great tendency to form complex ions, e.g., $[AlF_6]^{3-}$, $[FeF_6]^{3-}$, in which the metal atom exhibits a higher covalency. Aluminium and iron exhibit coordination number equal to 6 in these complexes. The other halide ions show this tendency to a much smaller extent.

7. Fluorine, because of the absence of *d* orbitals in its valency shell, does not form any polyhalides. Other halogens frequently associate with the halide ions forming polyhalide ions because of the vacant *d* orbitals in their valency shell. Thus, we have I_3^-, Br_3^-, Cl_3^- ions but no F_3^- ion.

8. The fluoride ion, by a process of hydrogen bonding forms the anion, HF_2^-. This tendency is missing with the other halides. Thus, the compound KHF_2, written as K^+ (F...H–F)$^-$, is well known but the corresponding compounds of other halogens (*e.g.,* $KHCl_2$ and $KHBr_2$) are unknown.

6.10.2 Interhalogen compounds

The binary compounds formed by halogens among themselves are known as interhalogen compounds. The halogens form various compounds which are binary combination of halogens among themselves. With the exception of BrCI, ICI, $IC1_3$ and IBr the compounds are all halogen fluorides. It will be noted that all the interhalogen compounds are of the type A X_n. where *n* is an odd number and X is always the lighter and smaller halogen than A. The value of *n* is always odd. This is odd number of unpaired electrons available in the ground state, first, second and third excited states of the halogen atoms forming the interhalogen compounds.

These compounds can be divided into the following four types.

AX type	*AX_3 type*	*AX_5 type*	*AX_7 type*
ClF	ClF_3	BrF_5	IF_7
BrF	BrF_3	IF_5	
BrCl	ICl_3		
ICl.			
IBr			
IF (unstable)			

[Here *A* represents the larger halogen atom and X the smaller one]

The bonds in these compounds are essentially covalent because of the small electronegativity difference. The compounds formed in the AX and AX_3 group are those where the electronegativity difference is not too great; the higher types AX_5 and AX_7 are shown by large atoms such as Br and I combining with small atoms such as F, because it is possible to pack more small atoms round a larger one.

Example 22. **Why are interhalogens more reactive than halogens?**

Solution. Interhalogen compounds, say of the type ClF, BrF and ClF_3, have polarity in the bonds whereas halogens, say F_2 or Cl_2, have non-polar bonds. A polar bond is easier to break compared to a nonpolar bond, that is the reason why interhalogen compounds are more reactive. The bonds in such compounds can be easily broken.

Interhalogens of AX type

Iodine monochloride

Preparation. (*i*) It can be prepared by the direct combination of liquid chlorine with solid iodine under controlled conditions of temperature and pressure.

$$\underset{\text{Solid}}{I_2} + \underset{\text{Liquid}}{Cl_2} \longrightarrow 2ICl$$

(*ii*) It may also be obtained by heating iodine with potassium chlorate.

$$KClO_3 + I_2 \longrightarrow KIO_3 + ICl$$

Properties. (*i*) It is a reddish oil and is readily hydrolyzed by water.

$$ICl + H_2O \rightleftharpoons HCl + HOI$$

(*ii*) It is also decomposed by sodium hydroxide.

$$ICl + NaOH \longrightarrow NaCl + HOI$$

(*iii*) When electricity is passed through iodine monochloride, iodine is liberated at the cathode and chlorine at the anode, indicating that the compound is a chloride of monovalent positively charged iodine.

$$2ICl \longrightarrow I^+ + [ICl_2]^-$$

(*iv*) With alkali metal halides it gives polyhalides.

$$KCl + ICl \longrightarrow KICl_2$$

Uses. (*i*) It is used to iodinate organic compounds, though chlorination may also occur depending on the condition of reaction.

$$\text{Salicylic acid} \xrightarrow{\text{ICl vapour}} \text{Chlorination}$$

$$\text{Salicylic acid} \xrightarrow[\text{nitrobenzene}]{\text{ICl in}} \text{Iodination}$$

(*ii*) It is used as Wij's reagent in the estimation of iodine number which is a measure of unsaturation of fats and oils.

$$-CH=CH- + ICl \longrightarrow -\underset{\displaystyle I}{\underset{|}{CH}}-\underset{\displaystyle Cl}{\underset{|}{CH}}-$$

Iodine monobromide, IBr

Preparation. It is prepared by direct combination of bromine with iodine.

$$I_2 + Br_2 \longrightarrow 2IBr$$

Properties. (*i*) It crystallizes at 315°C to a black solid.

(*ii*) Iodine monobromide undergoes decomposition when heated.

$$2IBr \rightleftharpoons I_2 + Br_2$$

Chlorine monofluoride, ClF

Preparation. (*i*) It is best prepared by mixing Cl and $C1F_3$ in equimolar ratios.

$$Cl_2 + C1F_3 \longrightarrow 3ClF$$

(*ii*) It is also obtained by heating fluorine and chlorine together in a copper vessel at 425 K.

(*iii*) It is obtained when chlorine reacts with hydrogen flouride at liquid air temperature.

$$Cl_2 + 2HF \longrightarrow 2ClF + H_2$$

Properties. (*i*) Chlorine monofluoride dissociates on strong heating

$$2ClF \longrightarrow Cl_2 + F_2$$

(*ii*) It reacts with a number of metals forming their fluorides even more vigorously than fluorine.

$$2Na + 2ClF \longrightarrow NaF + Cl_2$$

Bromine monofluoride

Preparation. It is obtained when bromine trifluoride and bromine vapours are mixed together.

$$BrF_3 + Br_2 \longrightarrow 3BrF$$

Properties. (*i*) It is a reddish brown liquid b.p. 296 K.

(*ii*) It is unstable and decomposes readily into bromine trifluoride and bromine.

$$3BrF \longrightarrow BrF_3 + Br_2$$

Bromine monochloride, BrCl

Preparation. BrC1 is obtained in traces when chlorine and bromine vapours are mixed together. However when chlorine gas is passed into liquid bromine covered with water, the formation of a solid hydrate, $BrC14H_2O$, has been reported.

Properties. With CsI it forms polyhalide as

$$CsI + BrC1 \longrightarrow CsIBrCl$$

Interhalogens of AX_3 type

Iodine trichloride, ICl_3

Preparation. It is obtained by action of excess of chlorine on iodine or by treating the monochloride with chlorine.

$$I_2 + 3Cl_2 \longrightarrow 2ICl_3$$

$$ICl + Cl_2 \rightleftharpoons IC1_3$$

Properties. (*i*) It forms fluffy orange powder which is deliquescent in air and has a penetrating pungent smell.

(*ii*) With alkali metal halides it gives polyhalides.

$$KCl + ICl_3 \longrightarrow KICl_4$$

(*iii*) It dissociates at 341 K into ICl and Cl_2.

$$ICl_3 \rightleftharpoons ICl + Cl_2$$

Structure. In the solid state, two molecules combine to form the dimer $(ICl_3)_2$ which is a flat molecule.

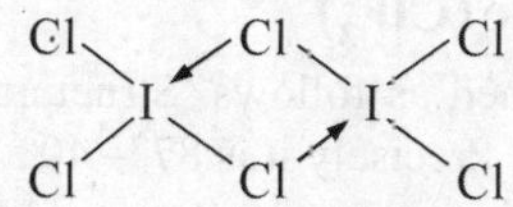

Bromine trifluoride, BrF_3

Preparation. It is obtained by mixing bromine and fluorine vapours or by the action of ClF_3 on bromine at 238 K.

$$Br_2 + 3F_2 \longrightarrow 2BrF_3$$

$$2ClF_3 + Br_2 \longrightarrow 2BrF_3 + Cl_2$$

Properties. (*i*) Bromine trifluoride is a greenish-yellow fuming liquid having b.p. 400 K.

(*ii*) With alkali fluorides it forms polyhalides.

$$KF + BrF_3 \longrightarrow KBrF_4$$

(*iii*) It converts many metal oxides, halides and even metals to their respective fluorides.

$$6MO + 4BrF_3 \longrightarrow 6MF_2 + 2Br_2 + 3O_2$$

where M is a divalent metal.

(*iv*) It ionizes considerably as

$$2BrF_3 \rightleftharpoons BrF_2^+ + BrF_4^-$$

Structure of BrF_3. This molecule involves sp^3d hybridisation. The T shape structure can be illustrated as

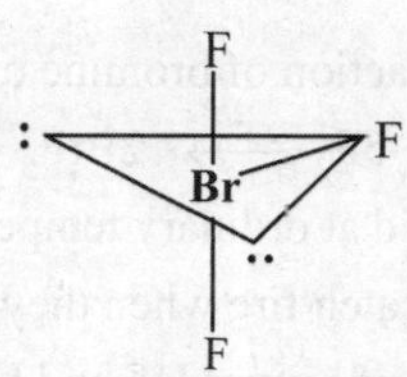

Structure of BrF_5. This molecule involves sp^3d^2 hybridisation. Square pyramidal structure can be represented as

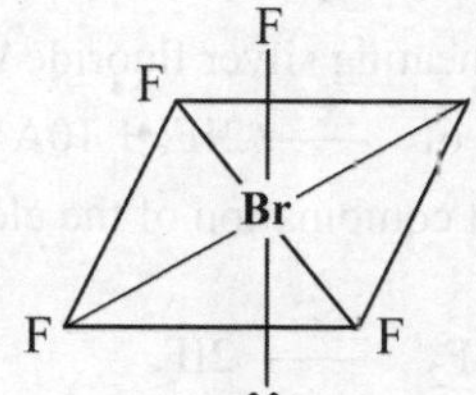

Chlorine trifluoride ClF_3

Preparation. It is prepared by passing a mixture of chlorine monofluoride and fluorine (excess) through a nickel tube maintained at 523 K.

$$ClF + F_2 \longrightarrow ClF_3$$

Properties. It is colourless gas which condenses to a green liquid and freezes to a white solid.

Structure of chlorine trifluoride (ClF_3)

The structure of ClF_3 is explained as follows. Structural studies show that the molecule is *T*-shaped with bond angle about 90°. Precisely it is 87°– 40'. The distortion is due to the repulsion between lone pairs of electrons. Two bond lengths are equal but different from the third, which is to be expected because a trigonal bipyramid is not a regular shape (Fig. 6.18).

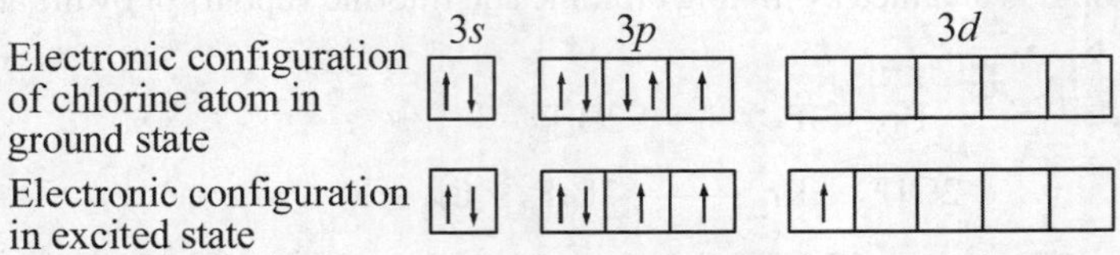

Three unpaired electrons form bonds with 3 fluorine atoms

sp^3d hybridization trigonal bipyramid with two positions occupied by lone pairs

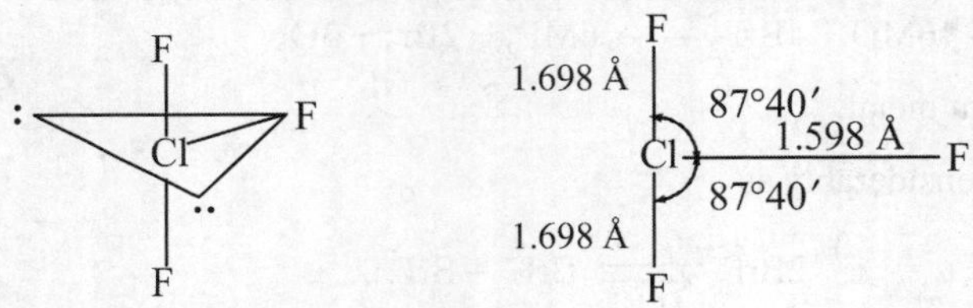

Fig. 6.18. Structure of ClF_3.

Bromine pentafluoride, BrF_5

Preparation. It is obtained by the action of bromine on excess of fluorine (1:5 molar ratio)

$$Br_2 + 5F_2 \longrightarrow 2BrF_5$$

Properties. (*i*) It is colourless liquid at ordinary temperature and is highly reactive.

(*ii*) Many metals and non-metals catch fire when they are dropped into the liquid.

(*iii*) Hydrolysis of bromopentafluoride gives HF and $HBrO_3$.

$$BrF_5 + 3H_2O \longrightarrow 5HF + HBrO_3$$

Iodine pentafluoride IF_5

Preparation. (*i*) It is obtained by heating silver fluoride with iodine.

$$10AgF + 6I_2 \longrightarrow 2IF_5 + 10AgI$$

(*ii*) It is also obtained by the direct combination of the elements under controlled condition or by the action of fluorine pentoxide.

$$I_2 + 5F_2 \longrightarrow 2IF_5$$

$$2I_2O_5 + 10F_2 \longrightarrow 4IF_5 + 5O_2$$

Properties. (*i*) Liquid FI_5 conducts electricity and self-ionizes.

$$2IF_5 \rightleftharpoons IF_4^+ + IF_6^-$$

(*ii*) It undergoes hydrolysis as

$$IF_5 + 3H_2O \longrightarrow 5HF + HIO_3$$

(*iii*) Complex salts are formed by heating IF_5 with alkali halides.

$$KI + IF_5 \longrightarrow K[IF_6]$$

Structure of iodine pentafluoride

IF_5 is formed by sp^3d^2 hybridisation and hence is octahedral in structure as shown below (Fig. 6.19).

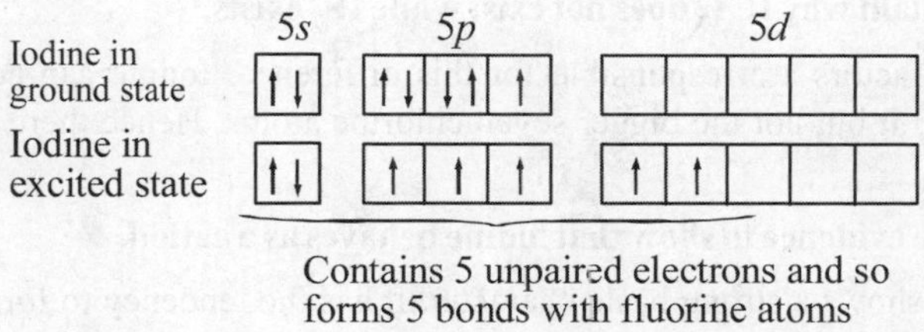

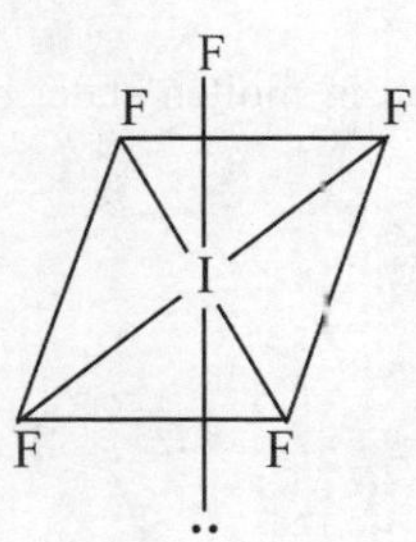

$sp^3 d$ hybridization octahedral with one position occupied by lone pair

Square pyramid

Fig. 6.19. Structure of IF_5.

The presence of one lone pair in the molecule converts its shape from regular octahedron to square pyramid.

Interhalogens of AX_7, type

Iodine heptafluoride, IF_7

Preparation. It is formed by combination of the pentafluoride with fluorine at 525-550 K.

$$IF_5 + F_2 \longrightarrow IF_7$$

Properties. (*i*) It undergoes hydrolysis giving HF and H_5IO_6.

$$IF_7 + 6H_2O \rightarrow 7HF + H_5IO_6$$

(*ii*) It is a powerful fluorinating agent.

Structure of IF_7. This compound has a pentagonal bipyramidal structure as it is formed by sp^3d^3 hybridisation as shown in Fig. 6.20.

Electronic configuration of iodine in excited state

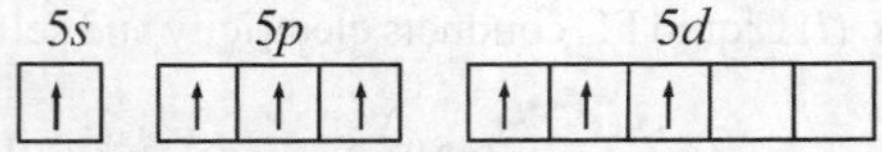

Iodine atom in excited state

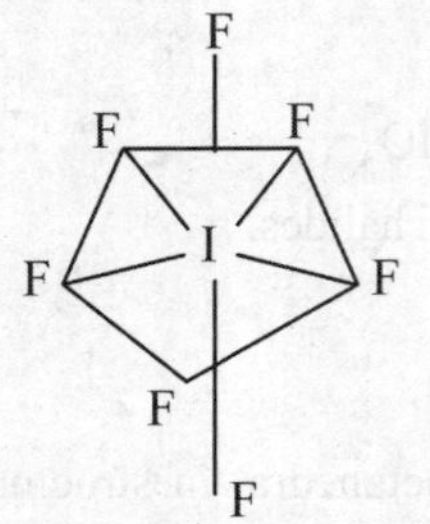

Fig. 6.20. Structure of IF_7.

***Example 23.* Explain why ICl_7 does not exist while IF_7 exists.**

Solution. Steric factors are responsible for this difference. Iodine can accommodate seven fluorine atoms around it but not the bigger seven chlorine atoms. Hence there is no possibility of formation of ICl_7.

***Example 24.* Give evidence to show that iodine behaves as a cation.**

Solution. Iodine shows a strong basic character. It has the tendency to form cations. Positive iodine compounds are many and their structures are well established. There are various evidences for existence of I^+ and I^{3+}.

(*a*) **Evidence for I^+.** (*i*) ICl conducts electricity when in molten state, both I_2 and Cl_2 are liberated at the anode but only I_2 at the cathode.

The ionization is probably of following type:

$$2ICl \rightleftharpoons I^+ + ICl_2^-$$

At cathode

$$2I^+ + 2e^- \longrightarrow I_2$$

At anode

$$2ICl_2^- - 2e^- \longrightarrow I_2 + 2Cl_2$$

(*ii*) If the electrolysis of molten ICN or a solution of ICN in pyridine is carried out, iodine is liberated at the cathode to confirm that iodine forms positive ions. Thus, this suggests the more simple type of ionization of ICN as

$$ICN \rightleftharpoons I^+ + CN^-$$

(*b*) **Evidence for I^{3+}.** (*i*) In the case of ICl_3, I_2 and Cl_2 are liberated at both electrodes. This is explained by ionisation of ICl_3 into ICl_2^+ and ICl_4^-.

(*ii*) If I_2 is treated with fuming HNO_3 and acetic anhydride, iodine triacetate, $I(CH_3COO)_3$ is formed. If electricity is passed through the solution of $I(CH_3COO)_3$ using silver electrodes, one equivalent of AgI is deposited at the cathode for the passage of every three Faradays of electricity. This proves the ionization :

$$I\,(\text{acetate})_3 \rightleftharpoons I^{3+} + 3\,(\text{acetate}^-)$$

(*iii*) If I_2 is oxidized with conc. HNO_3 in the presence of acetic anhydride and phosphoric acid, iodide phosphate IPO_4 is formed.

$$I_2 + Ac_2O + H_3PO_4 \xrightarrow{\text{Conc. } HNO_3} IPO_4 + \text{Acetic acid}$$

If iodine phosphate ionizes, I^{3+} could be expected.

6.10.3 Polyhalides

Halides often associate with molecules of halogens or interhalogens to form polyhalide ions, *e.g.*, the solubility of iodine in water is greatly increased by the addition of iodide ions, due to the formation of the tri-iodide ion, I_3^-

$$I_2 + I^- \longrightarrow I_3^-$$

More complex ions such as I_5^-, I_7^-, I_9^- combined with large metal ion are known.

A number of polyhalides are known which contain 2 or 3 different halogens, *e.g.*, $K[ICl_2]$, $K[ICl_4]$, $Cs[IBrCl]$. These are formed from metal halides and interhalogens.

$$KCl + ICl \longrightarrow K[ICl_2]$$

$$KCl + ICl_3 \longrightarrow K[ICl_4]$$

The structures of polyhalides are known. The trihalides, *e.g.*, $K[ICl_2]$ contain a linear trihalide ion ICl_2^-. This is shown in Fig. 6.21 Structure of pentahalide ion $[ICl_4]^-$ is illustrated in Fig. 6.22. Although sp^3d^2 hybridisation of iodine orbitals takes place giving rise to octahedral geometry, the actual shape is square planar because two positions are occupied by lone pairs.

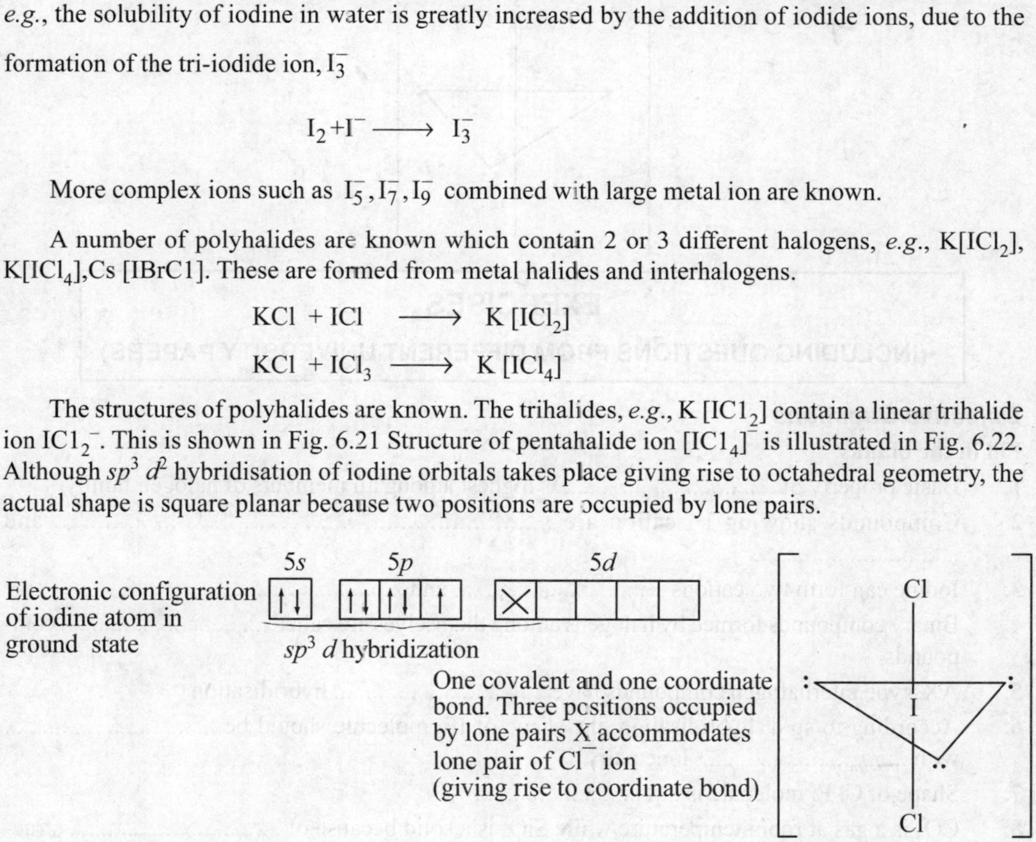

Fig. 6.21 Structure of $[ICl_2]^-$

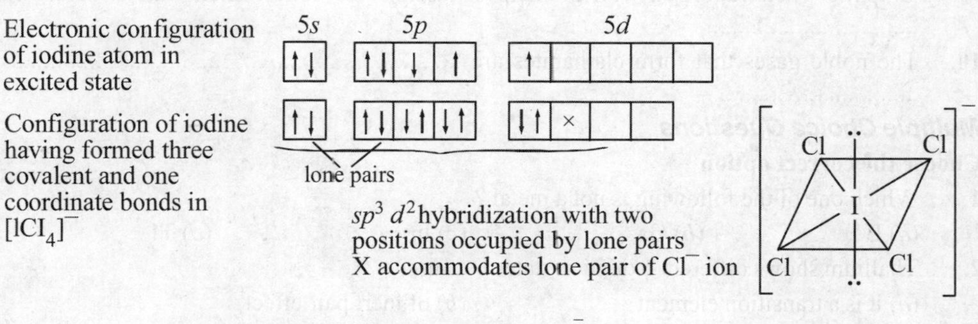

Fig. 6.22. Structure of $[ICl_4]^-$ ion.

***Example 25.* I_2 is usually insoluble in water. It dissolves readily in a solution of KI. Discuss the structure of product in this solution.**

Solution. Iodine dissolves in a solution of potassium iodide due to the following reaction:

$$I_2 + KI \rightleftharpoons KI_3$$

The structure of I_3^- is as follows:

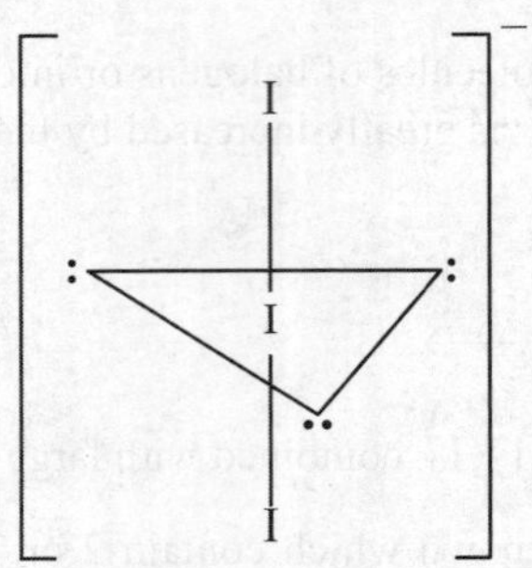

EXERCISES
(INCLUDING QUESTIONS FROM DIFFERENT UNIVERSITY PAPERS)

Objective Questions

Fill in the blanks

1. Basic property of .... is highest among all members of halogen family.

2. Compounds showing I^{3+} cation are ...., ... and ....

3. Iodine can form two cations .... and ....

4. Binary compounds formed by halogens among themselves are called .... compounds

5. AX_3 type interhalogen compounds have .... hybridisation

6. According to sp^3d^2 hybridisation, the shape of IF_5 molecule should be .... with .... lone pair

7. Shape of ClF_3 molecule is ....

8. CO_2 is a gas at room temperature while SiO_2 is a solid because of .... structure.

9. Graphite when vaporised with the help of lasers form hollow shell- like structure called ....

10. The noble gases that form clatharates are ...., ... and ....

Multiple Choice Questions

Choose the correct option

1. Which one of the following is not a metal ?

(*a*) B (*b*) Ga (*c*) In (*d*) TI

2. Thallium shows different oxidation states because:

(*a*) it is a transition element (*b*) of inert pair effect

(*c*) of its amphoteric character (*d*) of its higher reactivity

3. In which of the following elements + 1 oxidation state is more stable than + 3?
(*a*) B (*b*) Al (*c*) Ga (*d*) TI
4. What is general electronic configuration of group 14 elements :
(*a*) ns^2np^2 (*b*) ns^2np^6 (*c*) ns^2 (*d*) None
5. Which one of the following has the lowest melting point ?
(*a*) B (*b*) Al (*c*) Ga (*d*) Tl
6. Which of the following is the smallest cation ?
(*a*) Na^+ (*b*) Mg^{+2} (*c*) Ca^{+2} (*d*) Al^{3+}
7. Which of the following mineral does not contain Al ?
(*a*) Cryolite (*b*) Mica (*c*) Feldspar (*d*) Felspar
8. Which one of the following resembles silicon in some of its properties ?
(*a*) B (*b*) Ga (*c*) In (*d*) TI
9. Which one of the following is a metalloid ?
(*a*) C (*b*) Si (*c*) Ge (*d*) Sn
10. Lead shows oxidation states :
(*a*) + 3, + 4 (*b*) + 1, + 2 (*c*) + 2, + 4 (*d*) + 4
11. Which one of the following elements occur free in nature ?
(*a*) Nitrogen (*b*) Phosphorus (*c*) Arsenic (*d*) Antimony
12. Which does not form complex ?
(*a*) N (*b*) P (*c*) As (*d*) Bi
13. Which of the following is most electronegative ?
(*a*) O (*b*) S (*c*) Te (*d*) Se
14. Halogen atoms molecules are
(*a*) monoatomic and form X_2^{2-} ions (*b*) diatomic and form X^- ions
(*c*) diatomic and form X_2^{2-} ions (*d*) monoatomic and form X^- ions
15. The outermost elelctronic configuration of the most electronegative element is :
(*a*) ns^2np^5 (*b*) ns^2np^3 (*c*) ns^2 (*d*) ns^2np^4
16. Most abundant element on earth is :
(*a*) O (*b*) S (*c*) Te (*d*) Se
17. N, P, As, Sb, Bi elements belong to group
(*a*) 16 (*b*) 15 (*c*) 17 (*d*) 18
18. The molecule of which element is octaatomic ?
(*a*) O (*b*) S (*c*) Te (*d*) Se
19. Which of the following is a metalliod ?
(*a*) S (*b*) Se (*c*) Te (*d*) At
20. As the atomic number of halogen increases the halogens
(*a*) lose the outermost electrons less readily
(*b*) become lighter in colour
(*c*) become less denser
(*d*) gain electrons less readily

Answers

Fill in the blanks

1. iodine	**2.** IPO_4, $I(CH_3COO)_3$ ICl_3		**3.** I^+, I^{3+}
4. interhalogen	**5.** sp^3d	**6.** octahedral, one	**7.** T-shaped
8. network	**9.** fullerenes	**10.** Ar, Kr, Xe	

Multiple choice questions

1. (*a*)	**2.** (*b*)	**3.** (*d*)	**4.** (*d*)	**5.** (*c*)
6. (*d*)	**7.** (*d*)	**8.** (*a*)	**9.** (*c*)	**10.** (*c*)
11. (*a*)	**12.** (*a*)	**13.** (*a*)	**14.** (*b*)	**15.** (*a*)
16. (*a*)	**17.** (*a*)	**18.** (*b*)	**19.** (*c*)	**20.** (*d*)

General Questions

1. What are *p*-block elements? Discuss their characteristics.

2. Explain the following:
(*i*) Radius of chloride ion is larger than that of chlorine.
(*ii*) Ionization potential decreases from nitrogen to oxygen
(*iii*) Electron affinity of chlorine is more than that of fluorine
(*iv*) Atomic radii decrease in moving from B to F in the period and increase in moving from B to TI in the group.

3. How would you account for the diagonal relationship in elements arranged in the periodic table?

4. Explain how the type of bonding between two atoms can be predicted from their electronegativity values?

5. Explain the term electronegativity? How does this property change as we move down in the group and from left to right in a period in case of *p*-block elements?

6. Discuss the trends in the following properties of elements, group and period wise:
(*i*) Non-metallic character.
(*ii*) Melting and boiling points.
(*iii*) Density.
(*iv*) Oxidizing and reducing nature.
(*v*) Oxidation states.
(*vi*) Nature of oxides.
(*vii*) Ionization energy.
(*viii*) Electronegativity.
(*ix*) Electron affinity.

7. "Metallic character increases considerably as we move down a group". Why is it so?

8. Name a few important compounds of *p*-block elements. What is the treand in their properties?

9. (*a*) Why do elements exhibit different oxidation states? Account for the stability of lower oxidation states with rise of atomic number in a group.
(*b*) Discuss the trends in the following aspects of the chemistry of the *p*-block elements giving suitable explanation in each case.
(*i*) Acidic/basic nature of the oxides in a group and in a period.
(*ii*) Nature of bonding in the hydrides in terms of the electronegativity of the central atom.

10. What do you mean by "inert-pair effect"? Give two examples in support of your answer.

11. (*a*) Suppose A and B are the first and the last elements respectively (except inert gas elements) in a group of *p*-block. Which of the following statements are correct and justify your answer?

(*i*) Ionization energy decreases from A to B.

(*ii*) Lower oxidation state of B will be more stable than that of A.

(*iii*) Oxides of A will be more acidic than that of B.

(*iv*) Reducing power of their hydrides decrease from A to B.

(*b*) A metallic element M forms two stable chlorides with formulae MCl and MCl_3 respectively, where the element M is to be placed in the periodic table?

(*c*) Be resembles Al although they belong to two different groups in the periodic table. Assign reasons.

12. Discuss in detail the reasons as to why the first member of each group and the elements of the second short period differ from rest of the elements in their respective groups.

13. (*a*) Explain the following observations:

(*i*) H_2O is a liquid while H_2S is a gas.

(*ii*) Addition of excess of NaOH solution to $Al_2(SO_4)_3$ solution gives a clear solution.

14. Give reasons for the following:

(*i*) $PbBr_4$ and PbI_4 do not exist.

(*ii*) CCl_4 is not hydrolysed with water while $SiCl_4$ is readily hydrolysed.

(*iii*) Metallic character of the elements increases considerably as we move down a group.

15. (*a*) In what respect, the first member in each group of *p*-block elements differs from the other members of the same group? What are the consequences of such differences?

(*b*) Categorize the following oxides in terms of acidic, basic and amphoteric character:

BeO, B_2O_3, MgO, SiO_2, N_2O_5

(*c*) Which of the following is ionic?

Justify your answer:

(i) $SnCl_2$, (ii) $SnCl_4$

16. Explain: electron affinity of chlorine is more than that of fluorine but fluorine is a better oxidising agent than chlorine.

17. (*a*) Discuss three-central bonds in diborane.

(*b*) BF_3 exists but BH_3 does not. Explain.

18. Describe preparation, properties and structure of borazine.

19. (*a*) Classify silicates into different types.

(*b*) Discuss the structure of silicate ion.

(*c*) Write down the structure of silica.

20. (*a*) Discuss the variation in the following properties of halogens

(*i*) oxidation states, (*ii*) Electron affinity, (*iii*) Ionization potential.

(*b*) Why are halogens coloured elements.

21. (*a*) Sketch the Born-Haber cycle to arrive at the oxidation potential of a halogen.

(*b*) Fluorine has stronger oxidising character than chlorine. Explain.

22. (*a*) What are interhalogen compounds?

(*b*) Give two methods of preparation of iodine monochloride.

(*c*) Discuss the structure of chlorine trifluoride.

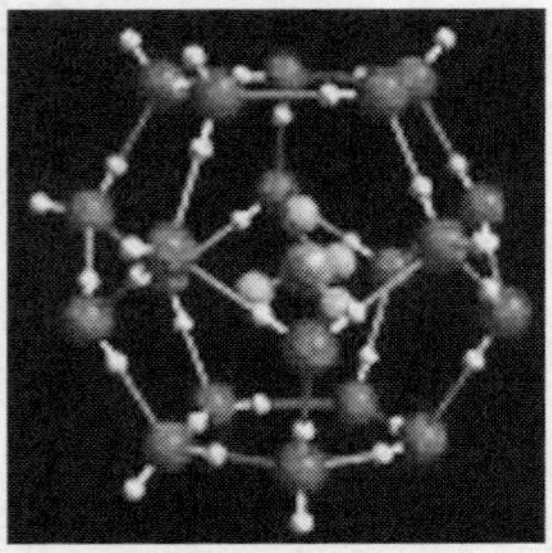

Semi-conducting clathrates

7 Chemistry of Noble Gases

7.1 INTRODUCTION

The zero group of the periodic table consists of six *gaseous elements* namely *helium* (He), *neon* (Ne), *argon* (Ar), *krypton* (Kr), *xenon* (Xe) and *radon* (Rn). All of them, except Radon, (which is a radioactive substance), are present in small amounts in the *gases* because they did not seem to enter into any chemical reaction. For the same reason, they are regarded as zero- valent and therefore called *zero group elements*. On account of their rare occurrence in nature, they are also called *rare gases* and *noble gases*.

7.2 POSITION IN THE PERIODIC TABLE

William Ramsay

The original Mendeleef's table had not left any blank space for noble gases because he could not imagine the existence of such elements which were devoid of chemical reactivity. However, Ramsay, the discoverer of these gases, introduced a new group in the periodic table to accomodate these elements. Based on their atomic weights and inertness, he classed them as a separate group of 'no valency' or 'zero valency' and called it 'Zero group' and placed it after the group 17 and before group 1. They, thus, form a bridge between the most electronegative halogens and the most electropositive alkali metals.

7.3 ELECTRONIC CONFIGURATION AND GENERAL PROPERTIES

The electronic configuration of noble gases are given in the following table 7.1.

Table 7.1. Electronic configuration of noble gases.

Element	*At. No.*	*Electronic configuration*
He	2	$1s^2$
Ne	10	$1s^2; 2s^2, 2p^6$
Ar	18	$1s^2; 2s^2, 2p^6; 3s^2, 3p^6$
Kr	36	$1s^2; 2s^2; 2p^6; 3s^2, 3p^6, 3d^{10}; 4s^2, 4p^6$
Xe	54	$1s^2; 2s^2; 2p^6; 3s^2, 3p^6, 3d^{10}; 4s^2, 4p^6, 4d^{10}, 5s^2, 5p^6$
Rn	86	$1s^2; 2s^2; 2p^6; 3s^2, 3p^6, 3d^{10}; 4s^2, 4p^6, 4d^{10}, 4f^{11}$ $5s^2, 5p^6, 5d^{10}; 6s^2, 6p^6$

All the noble gases except helium (which has a stable, $1s^2$ configuration) have a stable ns^2, np^6 configuration (complete octet) of their outermost orbit, *n*. Because of this configuration, they are chemically so unreactive and possess zero valency. Some of their physical constants are given in Table 7.2.

Table 7.2. Physical constants of noble gases.

	Property	*He*	*Ne*	*Ar*	*Kr*	*Xe*	*Rn*
1.	At. No.	2	10	18	36	54	86
2.	At. mass	4.003	20.18	39.94	83.80	131.30	222.0
3.	B.P. (K)	4.2	27.2	87.3	120.0	165.0	211
4.	M.P. (K)	—	24.6	83.9	116	161	202
5.	Ionization energy (kJ/mole)	2372	2079	1514	1351	1171	1037
6.	Atomic radius (°A)	1.2	1.6	1.9	2.0	20	—
7.	Density	0.1785	0.89909	1.7836	37.48	5.806	9.07
8.	*Cp/Cv*	1.652	1.642	1.65	1680	1.66	—

7.4 DISCUSSION OF PHYSICAL PROPERTIES

(*i*) Boiling and melting points. Due to very weak intermolecular van der Waals forces between them, they possess very low b.p. and m.p. However, they show a general increase with increase in the atomic mass. Small gaps in melting and boiling points show that solids change to vapours easily and they remain gases at temperatures when other elements are liquids and solids.

(*ii*) Ionization energies. They possess much higher ionization energies than other elements because their outermost shells are complete. Ionization energies decrease down the group.

The ionisation energy of xenon (1171 kJ/mole) comparable to that of molecular oxygen, 1167 kJ/mole. Thus, there is a striking resemblance between the two.

(*iii*) Electron affinities. Their electron affinity values are almost zero. This is because of their stable outer shell configurations. This stability comes from the complete filling of their orbitals. Thus, they show no tendency to acquire an electron.

(*iv*) Atomic radius. It increases with increase in atomic number on moving down a group. This is because every time, on stepping down, a new energy level is added.

(*v*) Monatomic nature. As their atoms do not gain or lose electrons due to complete octet under normal conditions, they do not form many bonds. They, therefore exist as single atoms. i.e., they are monatomic gases. It is also supported from their c_p/c_v ratio which is close to 1.667.

(*vi*) State. They are all colourless, odourless and tasteless gases.

(*vii*) Adsorption by charcoal. All of them except helium are readily adsorbed by wood charcoal at low temperature.

The degree of adsorption of these gases increases with atomic mass. Helium thus shows minimum degree of adsorption. Fractional adsorption followed by fractional desorption is employed for separation of noble gases because of difference in the degree of their adsoption on wood charcoal at low temperatures.

(*viii*) Electrical conductivity. They are better conductors of electric current than nitrogen or hydrogen. They are fairly good concuctors of electricity.

(*ix*) Characterisitc spectra. These gases are identified from the characteristic spectra shown by them. For example, helium is identified by spectrum cf green and yellow lines and neon gives spectrum of orange and red lines.

(*x*) Solubility. These gases are slightly soluble in water. Argon is comparatively more soluble in water as compared to nitrogen and cxygen. Solubility in water increases with atomic mass. Their solubility in water is due to dipole-induced dipole interaction. An induced dipole is created in noble gas atoms by water molecule. The polarizing water molecules distort their electron cloud to a great extent with the increase in atomic size, the attraction of the electron cloud by the nucleus decreases. Thus the dipole induced dipole interactions increase with increasing size of the gas and thus the solubility increases.

(*xi*) Liquefaction of these gases. The low heats of vaporization of noble gases show that they have very weak inter-atomic forces in them. Because of low inter-atomic forces, their liquefaction is very difficult. The van der Waals' forces increase with the size of their electron cloud, hence the ease of liquefaction of the gases increases with their atomic mass.

7.5 OCCURRENCE OF NOBLE GASES

Noble gases exist as under:

(*a*) In the earth's atmosphere. The non-radioactive gases are present in the atmosphere to the extent of 1% by volume. The composition of these gases in air is as follows:

100 litres of air contains 9.34 ml of argon, 1.82 ml of neon, 0.524 ml of helium, 0.114 ml of krypton and 0.0087 ml of xenon.

(*b*) In the stars and interstellar space. It has been established that the main constituents of the sun are hydrogen and helium. The masses of the major planets Jupiter and Saturn are estimated to be due to 23% and 37% of helium respectively. The atmosphere of Mars consists mostly of nitrogen and argon.

(*c*) In minerals and springwaters. Noble gases occur in minerals in the occluded form. Except helium, the concentration of noble gases in minerals is quite low. The concentration of helium in thorianite has been found to be 10.5 ml/g.

(*d*) In natural gas. Helium along with traces of neon are found to occur in natural gas in the wells of certain oil fields of U.S.A. In the natural gas, helium concentration varies from 2 to 7%. At present most of the helium is commercially recovered from natural gas which is a much cheaper source than atmosphere.

7.6. ISOLATION OF NOBLE GASES

By fractionation of liquid air

Principle. This method is based on the difference in the boiling points of noble gases. The gases are separated from one another and also from oxygen and nitrogen (which are the other major constituents of air) by fractional distillation. The boiling points of noble gases along with those of oxygen and nitrogen are given in Table 7.3.

Table 7.3. Boiling points of some gases.

GAS	He	Ne	N_2	Ar	O_2	Kr	Xe
Boiling point, K	4.2	27.2	77	87.3	90.3	121	165

Some important conclusions from the above table are:

(*i*) Helium and neon have very low boiling points. In other words, during the process of liquefying, air, helium and neon remain in gaseous state while the other constituents change into liquid.

(*ii*) There is only a difference of +3 K in the boiling points of oxygen and argon. So these two gases are collected more or less as a single fraction.

(*iii*) Krypton and xenon have fairly high boiling points as compared to those of nitrogen, oxygen and argon. So krypton and xenon are always found in the less volatile portion of liquid air and are always easy to separate from other gases.

The plant used for isolation of noble gases is shown in Fig. 7.1.

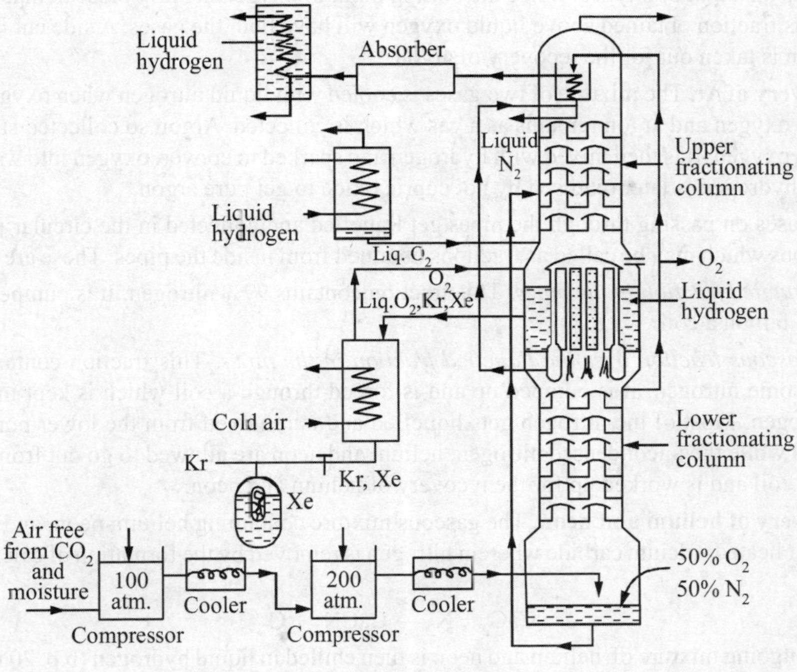

Fig. 7.1. Isolation of noble gases from air.

Procedure. Air free from carbon dioxide and moisture is compressed to 100 atmospheric pressure. Air is then cooled to remove heat produced by compression. The air is again compressed to about 200 atmospheric pressure and again cooled by a mixture of salt and ice. It is then passed into liquefier in which it escapes through a valve and expands to a pressure of about 20 atmospheres. Then air cools down. The cold compressed air enters the lower fractionating column near its bottom (Fig. 7.1). The air expands and as a result cools down further to an extent that oxygen and nitrogen change into liquid and collect at the bottom of the fractionating column. This liquid fraction consists of equal proportions of oxygen and nitrogen. It is pumped up and is dropped from the middle of the upper fractionating column.

The uncondensed gases rise up and reach the middle portion where they pass through pipes surrounded by liquid oxygen. Four different fractions are obtained, here two fractions are obtained from outside the pipes and two fractions are obtained from condensed material in the pipes. These fractions are worked up as follows:

Two fractions outside the pipes are:

(*i*) Liquefied fraction in liquid oxygen;

(*ii*) Gaseous fraction over liquid oxygen.

(*i*) Krypton and xenon get cooled and are dissolved in liquid oxygen. This liquid fraction of O_2, Kr and Xe is taken out for recovery of Kr and Xe.

(*i*) Recovery of Kr and Xe. The liquid is taken to another rectifying column and is dropped from above while cold air is being circulated through it. Consequently, liquid oxygen changes into vapours and it is collected from the top of this column while krypton and xenon still remain liquids and are collected from the bottom of this column. From the residual liquid containing krypton and xenon, the two gases are separated either by selective low temperature absorption or by fractionation in a separate rectifier.

***(ii)* Gaseous fraction over the liquid oxygen.** Argon remains in the gaseous state as its boiling point is less than that of oxygen. Since the boiling point of oxygen is fairly close to that of argon so the gaseous fraction obtained above liquid oxygen will have both the gases. A side cut is made and the fraction is taken out for the recovery of argon.

Recovery of Ar. The mixture of two gases is cooled with liquid nitrogen when oxygen changes into liquid oxygen and argon remains as a gas which is collected. Argon so collected still contains about 20% oxygen. It is then mixed with hydrogen and sparked to convert oxygen into water and the unreacted hydrogen is later oxidised by hot cupric oxide to get pure argon.

The gases on passing through the pipes get liquefied and collected in the circular jacket from two fractions which may be called as fractions obtained from inside the pipes. These are as follows:

(*i*) *Liquefied fraction in the pipes.* This fraction contains 99% nitrogen. It is pumped up and is allowed to fall on a coil.

(*ii*) *Gaseous fraction over the liquefied fraction in the pipes.* This fraction contains helium, neon and some nitrogen. It is pumped up and is forced through a coil which is kept immersed in liquid nitrogen. Most of the nitrogen gets liquefied and is removed from the lower portion of the cooled coil while the uncondensed nitrogen, helium and neon are allowed to go out from the top of the cooled coil and is worked up for the recovery of helium and neon.

Recovery of helium and neon. The gaseous mixture containing helium, neon, and nitrogen is passed over heated calcium carbide wherein nitrogen is removed by the formation of calcium cyanamide.

$$CaC_2 + N_2 \rightarrow CaCN_2 + C$$

The outgoing mixture of helium and neon is then chilled in liquid hydrogen (b.p. 20 K) wherein neon changes into a liquid while helium still remains a gas and the two are collected.

By Chemical Method

The methods involves the following two steps:

(*a*) Removal of nitrogen and oxygen from air. Any of the following method may be used to remove nitrogen and oxygen from pure dry air.

(*i*) Rayleigh's method. This apparatus is shown Fig. 7.2. It consists of a 50-litre round-bottomed flask and is fitted with a cork having five holes. Leads for platinum are fitted though two holes. A mixture of pure dry air and oxygen in the ratio of 9:11 is introduced through a tube from the third hole. Long glass tubes are fitted through the remaining two holes, one of these tubes is used as an inlet for sodium hydroxide solution while the other is used as an exit for sodium hydroxide solution.

Mixture of air and oxygen is introduced. Sodium hydroxide solution is forced into the flask. An electric discharge from an induction coil of 6000–8000 volts between the platinum electrodes is passed. Due to combination of nitrogen and oxygen, two oxides of nitrogen are formed.

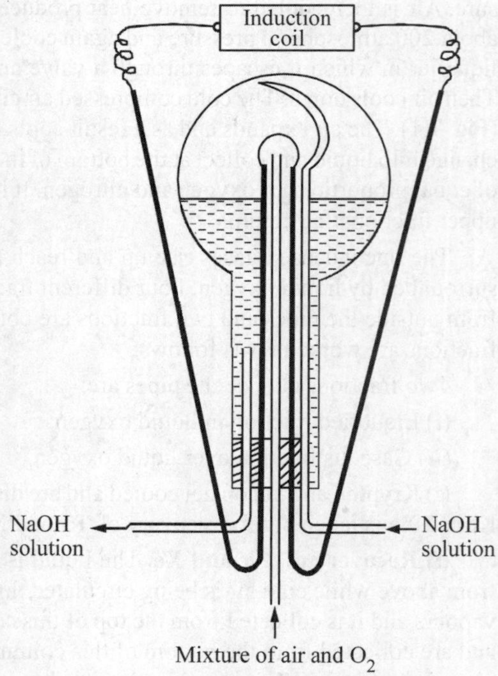

Fig. 7.2. Isolation of noble gases from air.

Lord Rayleigh

$$N_2 + O_2 \longrightarrow 2NO$$

$$2NO + O_2 \longrightarrow 2NO_2$$

These oxides of nitrogen get adsorbed in concentrated sodium hydroxide solution.

$$2NaOH + 2NO_2 \longrightarrow NaNO_2 + NaNO_3 + H_2O$$

$$2NaOH + NO + NO_2 \longrightarrow 2NaNO_2 + N_2O$$

As a result, most of oxygen and nitrogen are removed from the air in the flask. The residual gases contain noble gases with traces of oxygen. The mixture of residual gases is passed through tubes containing alkaline pyrogallol where oxygen gets adsorbed and the mixture of noble gases, free from oxygen, comes out.

***(ii)* Ramsay's method.** In this method, pure and dry air is passed slowly over hot finely divided copper contained in a long tube to remove oxygen.

$$2Cu + O_2 \longrightarrow 2CuO$$

Thereafter, the air (free from oxygen) is passed over heated magnesium ribbon when nitrogen is removed due to the formation of magnesium nitride.

$$3Mg + N_2 \longrightarrow Mg_3N_2$$

Though this method gives very satisfactory results but it takes a very long time to complete.

***(iii)* Fischer-Ringe's method.** In this method, pure dry air is passed over a mixture of 90% calcium carbide and 10% calcium chloride at 800°C when oxygen and nitrogen are removed according to the following reactions:

$$CaC_2 + N_2 \longrightarrow CaNCN + C$$

$$C + O_2 \longrightarrow CO_2$$

$$2C + O_2 \longrightarrow 2CO$$

The resulting gases are passed over hot cupric oxide to oxidise carbon monoxide to carbon dioxide.

$$CO + CuO \longrightarrow Cu + CO_2$$

Therefore, the residual gases are passed through potash bulbs to absorb carbon dioxide.

$$2KOH + CO_2 \longrightarrow K_2CO_3 + H_2O$$

The gases are dried by bubbling them through concentrated sulphuric acid and phosphorus pentoxide. The gaseous mixture now obtained contains only the non-radioactive noble gases.

***(b)* Separation of Noble Gases by Dewar's Charcoal Method. Principle.** This method is based on the principle that the adsorption capacity of these gases at low temperatures by coconut charcoal increases with increase in their atomic weights. Thus, helium has the lowest adsorption capacity while xenon has the maxmum adsorption capacity. Moreover, adsorption capacity depends on temperature, *i.e.*, the lower the atomic weight of the noble gas, the lower is the temperature needed to adsorb it.

Procedure. The mixture of noble gases is brought in contact with coconut charcoal packed in a bulb maintained at –100°C in a Dewar's flask for one hour (Fig. 7.3). Xenon, Krypton and argon are adsorbed by the coconut charcoal while helium and neon remain unadsorbed and are pumped out of the bulb. The mixture of helium and neon is then introduced into another bulb containing coconut charcoal and maintained at –180°C when neon gets adsorbed while helium is pumped out. Thereafter, this charcoal is warmed to recover neon.

The first charcoal in which xenon, krypton and argon had been absorbed is brought in contact with another charcoal sample which had earlier been cooled to the liquid air temperature. Argon being a lighter gas diffuses into third charcoal unit and is recovered by warming it.

Krypton and xenon remain absorbed on the first charcoal which is slowly warmed so that its temperature rises to –90°C. At this temperature krypton does not remain adsorbed and so it starts coming out of the charcoal and is collected. Xenon remains adsorbed on the charcoal which is released on further warming and is collected.

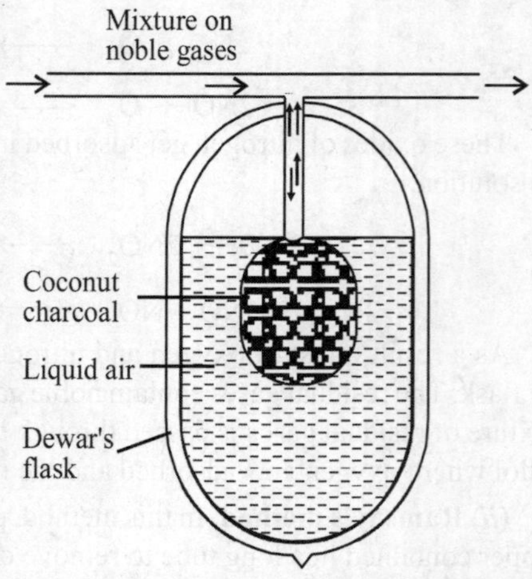

Fig. 7.3. Fractional adsorption of noble gases by activated charcoal.

7.7 WHY ARE NOBLE GASES UNREACTIVE?

Noble gases are not very reactive. This may be ascribed to the following reasons:

(*a*) They have high ionization energies.

(*b*) Their electron affinities are almost zero.

Both these factors are a direct result of their electronic configuraion. At each noble gas the electronic sub-level is just complete ($ns^2\ np^6$) and this particular electronic configuration resists change or disturbance of any kind. Therefore, all noble gases are extremely unreactive as compared to the other elements.

7.8 FORMATION OF FIRST NOBLE GAS COMPOUND BY BARTLETT

The first chemical compound of the noble gases was reported by N. Bartlett in 1962. The preparation of this compound, $XePtF_6$, was based on the chance discovery of another compound formed by the reaction between oxygen gas and platinum hexafluoride (deep red vapours).

$$O_2(g) + PtF_6(g) \longrightarrow \underset{\text{Dioxygen hexafluoroplatinate(V)}}{O_2PtF_6(s)}$$

The above compound is formulated as $O_2^+(PtF_6)^-$ .This word means that platinum hexafluoride is a very strong oxidising material, and could even remove electron from molecular oxygen. Since the ionization energy of oxygen gas (1165.6 kJ mol^{-1}) is comparable to that of xenon gas (1170 $kJmol^{-1}$) and as their molecular diameters are also similar, Bartett thought that xenon would also be susceptible to oxidation with platinum hexafluoride. Therefore, he proceeded to mix xenon and platinum hexafluoride (both gases) at room temperature and obtained on his first attempt an orange yellow crystalline compound of composition, $XePtF_6$.

Niel Bartlett

$$Xe(g) + PtF_6(g) \longrightarrow Xe^+(PtF_6)^-(s)$$

***Example 1.* Xenon is known to form some compounds but not He and Ar. Explain why?**

Or

Xenon compounds are more stable than those of other noble gases. Explain.

Or

Prove that the noble gases are not so inert.

***Solution*:** We may expect only the heavier noble gases which have low ionization energy values to show some reactivity as a result of electron loss. Ionization energies of noble gases are given in Table 7.4.

Table 7.4. Comparison of ionization energy values of noble gases with other reactive elements.

Element	Ionization energy value. *kJ mol*$^{-1}$
Helium	2372.3
Neon	2080.7
Argon	1520.5
Krypton	1350.7
Xenon	1170.4
Radon	1037.0

From the above ionization energy values, it is obvious that we may expect radon to go into chemical combination with the greatest ease. But radon is radioactive and we know very little about its ability to form chemical compounds. Xenon forms the largest number of compounds as compared to any other noble gas because xenon is the stable element which has the smallest ionization energy. Krypton tends to form compounds but with difficulty as compared to xenon. The lighter noble gases, helium, neon and argon which have very high ionisation energies, are expected to show very little reactivity. So far experimental evidence also suggests that no chemical compounds of these gases have been reported.

***Example 2*. Why most of noble gas compounds involve only fluorine and oxygen?**

***Solution*:** Any chemical reactivity shown by these gases may be attributed to their tendency to lose electrons. For this reason, the combining atom must be highly electronegative, so as to accept these electrons. Xenon forms stable compounds only with the most electronegative elements; fluorine ($X = 4.0$) and oxygen ($X = 3.5$) or with highly electronegative groups such as $OSeF_5$ and $OTeF_5$ that contain both oxygen and fluorine.

7.9 XENON DIFLUORIDE, XeF_2

Preparation. (*i*) It can be prepared quantitatively by photosynthesis on exposing to sunlight mixtures of xenon and oxygen difluoride (OF_2) or xenon and excess of oxygen monofluoride (O_2F_2) at –118°C.

(*ii*) It is best prepared in the laboratory in high yields and purity by heating xenon and fluorine in the molar ratio of 2:1 (excess of xenon is taken to avoid the formation of any xenon tetrafluoride).

The reaction is carried out in a nickel or Monel vessel at 400°C. The reaction products are quenched to room temperature and XeF_2 is isolated by vacuum sublimation.

$$Xe + F_2 \xrightarrow[\text{1 atm.}]{400°C} XeF_2 \text{ (Xe in excess)}$$

Properties. (*i*) It is a colourless crystalline solid melting at 140°C.

(*ii*) The fluorine of XeF_2 may be substituted by highly electronegative ligands by reacting it with protonic acids.

$$XeF_2 + HX \longrightarrow FXeX + HF$$

$$FXeX + HX \longrightarrow XeX_2 + HF$$

(*iii*) The xenon-F bond in XeF_2 has almost the lowest bond energy among the known fluorides. Due to this reason XeF_2 is a strong oxidising and fluorinating agent.

(*iv*) It reacts with sulphur trioxide to give Xe.

$$XeF_2 + 2SO_3 \longrightarrow Xe\uparrow + S_2O_6F_2$$

(*v*) It is reduced by hydrogen to xenon

$$XeF_2 + H_2 \longrightarrow Xe\uparrow + 2HF$$

(*vi*) It reacts with NO to give nitrosyl fluoride.

$$2NO + XeF_2 \longrightarrow 2NOF + Xe\uparrow$$

It appears that in all of the above fluorination and oxidation reactions, the first step involves the ionization of XeF_2 to XeF^+ followed by electron transfer to give XeF. The XeF species is very weakly bound and is a potential source of fluorine atoms.

(*vii*) It dissolves in HF (hydrofluoric acid) without any reaction. The solution does not show any change in conductivity.

(*viii*) It reacts with SbF_5 or AsF_5 to give solid compounds in which it reacts as a fluoride ion donor.

$$XeF_2 + SbF_5 \longrightarrow XeF_2 . SbF_5$$

$$XeF_2 + AsF_5 \longrightarrow XeF_2 . AsF_5$$

It has now been established that the compounds contain ionic species, $XeF^+.AsF_6^-$ and XeF^+ SbF_6^-. The reaction can also take place as follows:

$$Xe F_2 + 2SbF_5 \longrightarrow XeF_2 . 2SbF_5$$

XeF_2 also reacts with AsF_3 in a different stoichiometry:

$$2XeF_2 + AsF_5 \longrightarrow 2XeF_2 . AsF_5 \longrightarrow (Xe_2F_3)^+ (AsF_6)^-$$

The ion $Xe_2F_3^+$ has the expected linear geometry around xenon and an angular geometry at the central fluorine.

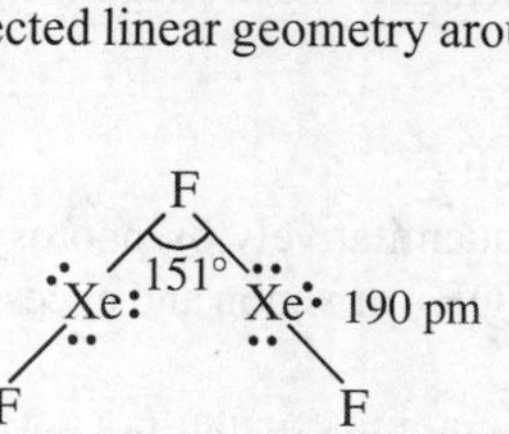

$[Xe_2F_3]^+$

(*ix*) It oxidizes iodine in the presence of fluoride ion acceptor to give IF.

$$I_2 + XeF_2 \xrightarrow{BF_3} 2IF + Xe\uparrow$$

$$IF + XeF_2 \xrightarrow{BF_3} IF_3 + Xe\uparrow$$

Structure and bonding in XeF_2, xenon difluoride

(*i*) Infrared and Raman spectra show that XeF_2 is a linear molecule in the vapour state.

F—Xe—F

(*ii*) Crystal structure of solid XeF_2 shows that XeF_2 consists of linear units.

Bonding. Xenon difluoride may be represented in terms of Lewis electron dot representation as:

The actual structure is a resonance hybrid of the following two structures:

Valence bond treatment. In the valence bond approach, the valence shell of xenon is expanded by promotion of an electron into the higher energy vacant 5*d* orbitals:

Ground state of Xe : [Kr] $5s^2\,4d^{10}\,5p^6$

Valence state of Xe : [Kr] $5s^2\,4d^{10}\,5p^5\,5d^1$

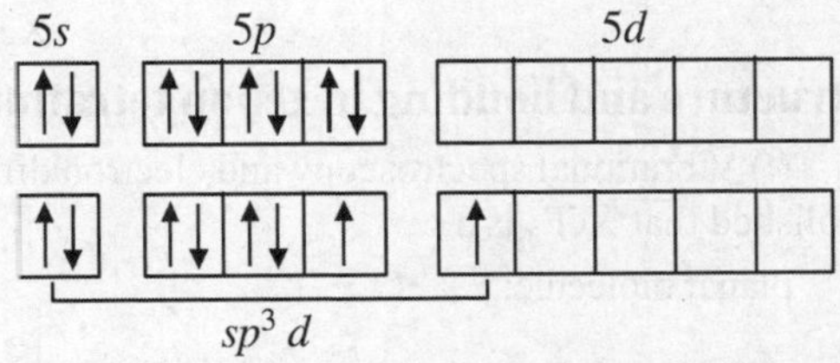

The five hybridised orbitals are oriented towards the corners of an equivalent triangle and along the two vertices to give the shape of trigonal bipyramid (Fig. 7.4). The orbitals along the triangle accommodate lone pairs and those along the vertices contain single electrons. These two orbitals form bonds with fluorine. The shape of XeF_2 is given below in Fig. 7.4.

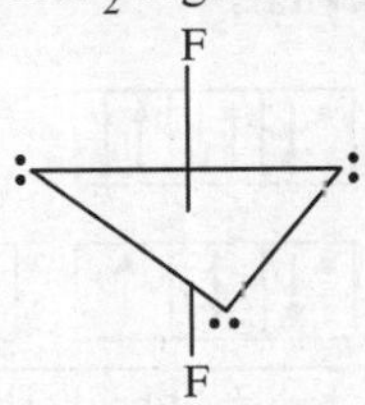

Fig. 7.4. Shape of XeF_2.

7.10 XENON TETRAFLUORIDE, XeF_4

Preparation. (*i*) It is prepared by heating a mixture of xenon and fluorine in a molar ratio 1:5 at 400°C and 5-6 atmosphere in an enclosed nickel vessel.

$$Xe + 2F_2 \xrightarrow[6\text{ atm}]{400°C} XeF_4$$

(*ii*) By passing an electric discharge in a reaction vessel maintained at –80°C containing xenon and fluorine in the molar ratio 1:2. The pressure is maintained between 2-15 mm of Hg.

Properties. (*i*) It is a colourless crystalline solid, melting at 117.1°C.

(*ii*) It is reduced by mercury.

$$XeF_4 + 2Hg \longrightarrow Xe + 2HgF_2$$

(*iii*) It is also reduced by hydrogen gas at 130°C.

$$XeF_4 + 2H_2 \longrightarrow Xe + 4HF$$

(*iv*) It forms an addition compound with antimony pentafluoride.

$$XeF_4 + SbF_5 \longrightarrow XeF_4SbF_5 \longrightarrow (XeF_3)^+\,(SbF_6)^-$$

$$XeF_4 + 2SbF_5 \longrightarrow XeF_4.2SbF_5$$

$$\longrightarrow (Xe\,F_3)^+\,(Sb_2F_{11})^-$$

The XeF_3^+ ions have the expected T-shaped structure as shown in Fig. 7.5.

(*v*) It reacts with boron trichloride to give boron trifluoride.

$$3XeF_4 + 4BCl_3 \longrightarrow 4BF_3 + 3Xe + 6Cl_2$$

(*vi*) It is readily hydrolysed by water. The reaction can be controlled to yield different products.

$$XeF_4 + H_2O \xrightarrow{-80°C} \underset{\text{Xenon oxyfluoride}}{XeOF_2} + 2HF$$

$$XeF_4 + H_2O \longrightarrow \underset{\text{Xenon trioxide}}{XeO_3}$$

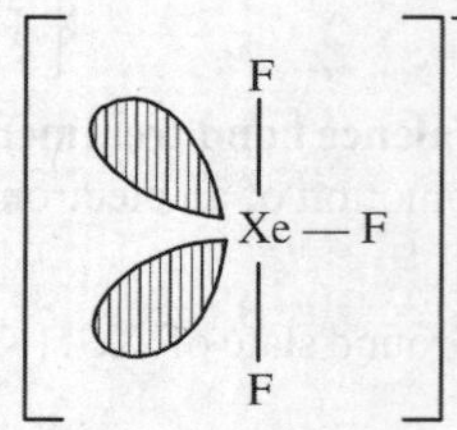

Fig. 7.5. Structure of $(XeF_3)^+$

Structure and bonding in xenon tetrafluoride, XeF_4

(*i*) Vibrational spectroscopy and electron diffraction data have established that XeF_4 is a planar molecule:

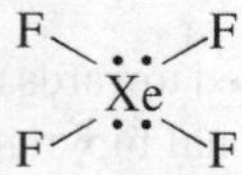

(*ii*) The valence bond representation of XeF_4 can be explained by the promotion of electron from the 5*p* sub-level to high energy 5*d* orbitals.

Ground state configuration of Xe :

Excited state configuration of Xe :

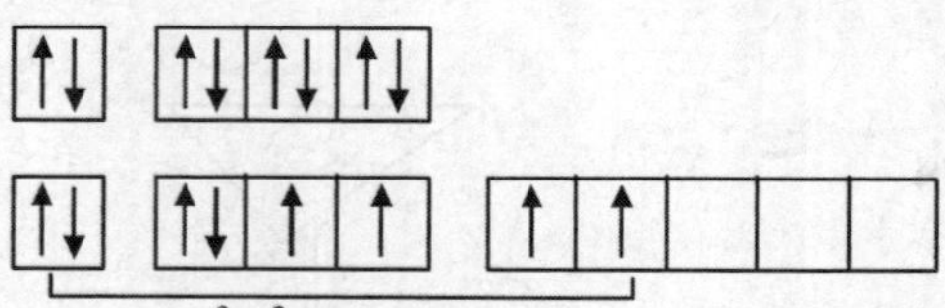

Six sp^3d^2 hybrid orbitals

The 5*s*, three 5*p* and two 5*d* atomic orbitals of xenon hybridize to produce six sp^3d^2 hybridized orbitals. The four singly occupied hybridized orbitals are used by fluorine for bond formation to xenon. The other two orbitals contain lone pairs. The molecule has octahedral geometry with two transpositions occupied by lone pairs (Fig. 7.6).

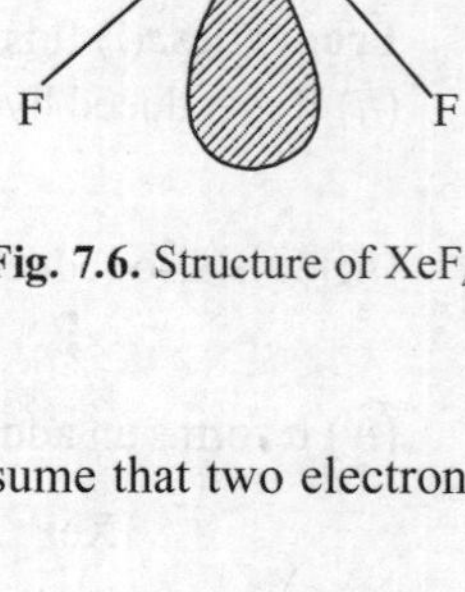

Fig. 7.6. Structure of XeF_4.

7.11 XENON OXIDE DIFLUORIDE, $XeOF_2$

Preparation. It is prepared by the hydrolysis of XeF_4 at –80°C. $XeF_4 + H_2O \xrightarrow{-80°C} XeOF_2 + 2HF$

Structure. (*i*) It has T-shaped structure. (Fig. 7.7)

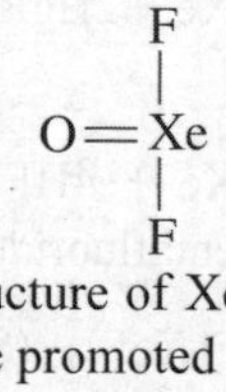

(*ii*) The valence bond structure of $XeOF_2$ may be explained if we assume that two electrons from 5*p* sub-level of xenon are promoted to 5*d* sub-level.

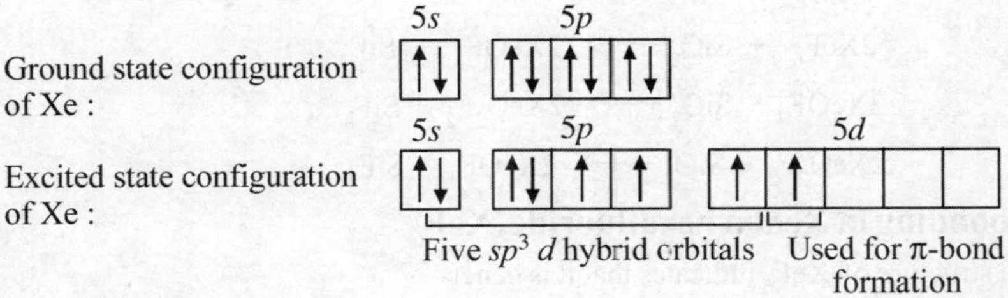

The 5*s*, three 5*p* and 5*d* orbitals of xenon hybridize to form five sp^3d hybridized orbitals. The three singly occupied orbitals are used for σ-bond formation of two fluorine atoms and one oxygen atom. The two positions of trigonal bipyramidal structure are occupied by lone pairs. One of the electrons in the 5*d* orbital which is not used in the hybridization scheme is used for π-bond formation to oxygen atom. It may be mentioned that two of orbitals lying along the corners of a triangle are occupied by lone pairs.

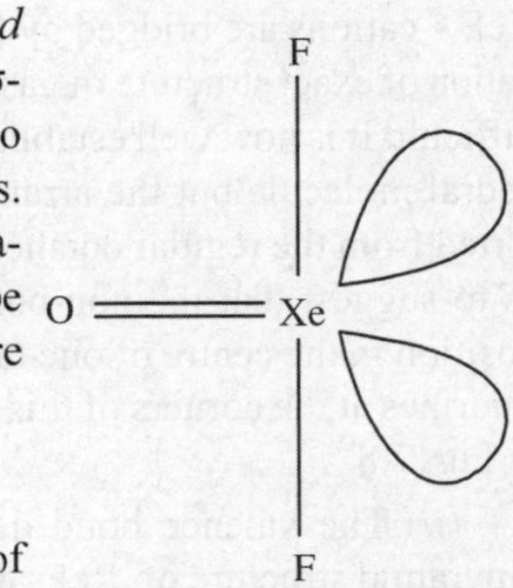

Fig. 7.7. Structure of $XeOF_2$.

7.12 XENON HEXAFLUORIDE, XeF_6

Preparation. (*i*) XeF_6 may also be prepared by the oxidation of XeF_4 with O_2F_2 at temperatures below –80°C.

(*ii*) It may also be prepared by the electric discharge of a mixture of fluorine (3 parts) and xenon (1 part) at low temperature.

(*iii*) It is best prepared by reacting xenon and fluorine in the ratio 1 : 20 in a nickel vessel. The reaction is carried out at 300°C and at 60 atmospheric pressure.

$$Xe + 3F_2 \xrightarrow[60\text{ atm.}]{300°C} XeF_6$$

Properties. (*i*) It is a colourless solid melting at 49.5°C.

(*ii*) It is very rapidly hydrolysed.

(*iii*) It is a stronger oxidation and fluorinating agent than both XeF_2 and XeF_4.

(*iv*) It reacts with fluoride ion acceptors to produce $(XeF_5)^+$ species as

$$XeF_6 + PtF_5 \longrightarrow XeF_6 . PtF_5 \longrightarrow (XeF_5)^+ (PtF_6)^-$$

$$XeF_6 + AsF_5 \longrightarrow XeF_6 . AsF_5 \longrightarrow (XeF_5)^+ (AsF_6)^-$$

$$XeF_6 + SbF_5 \longrightarrow XeF_6 . SbF_5 \longrightarrow (XeF_5)^+ (SbF_6)^-$$

$$XeF_6 + BF_3 \longrightarrow XeF_6 . BF_3 \longrightarrow (XeF_5)^+ (BF_4)^-$$

(*v*) It reacts with water to give xenon trioxide which is highly explosive and makes the handling of XeF_6 extremely difficult.

$$XeF_6 + 3H_2O \longrightarrow XeO_3 - 6\,HF$$

(*vi*) It is reduced to xenon by the action of H_2, HCl and NH_3.

$$XeF_6 + 3H_2 \longrightarrow Xe + 6HF$$

$$XeF_6 + 6\,HCl \longrightarrow Xe + 3Cl_2 + 6HF$$

$$XeF_6 + 8NH_3 \longrightarrow Xe + 6NH_4F + N_2$$

(*vii*) It cannot be stored in glass or quartz vessels owing to the following reactions.

$$2XeF_6 + SiO_2 \longrightarrow 2XeOF_4 + SiF_4$$
$$2XeOF_4 + SiO_2 \longrightarrow 2XeO_2F_2 + SiF_4$$
$$2XeO_2F_2 + SiO_2 \longrightarrow 2XeOF_3 + SiF_4$$

Structure and bonding in Xenon hexafluoride, XeF_6

(*i*) The crystal structure of XeF_6 indicates that it is generally present as $(XeF_5)^+F^-$ in the solid state. The pyramidal XeF_5^+ cations are bridged by free fluoride ions. The determination of exact structure of gaseous XeF_6 molecule has proved difficult. It is now well established that it is not a regular octahedral molecule but the arrangement of the fluorines is distorted from the regular octahedral arrangement in such a way as to suggest that the non-bonding electron pair occupies a position in the centre of one face of an octahedron so that the fluorines at the corners of this face are forced apart, as shown in Fig. 7.8.

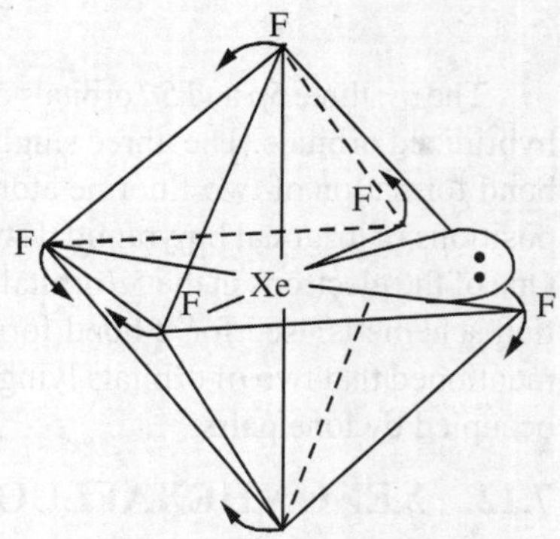

Fig. 7.8. Distorted octahedral structure of XeF_6.

(*ii*) The valence bond description predicts pentagonal bipyramid structure of XeF_6 with one transposition occupied by a lone pair. It may be assumed that three electrons from the 5*p* sublevel are promoted to 5*d* sublevel as:

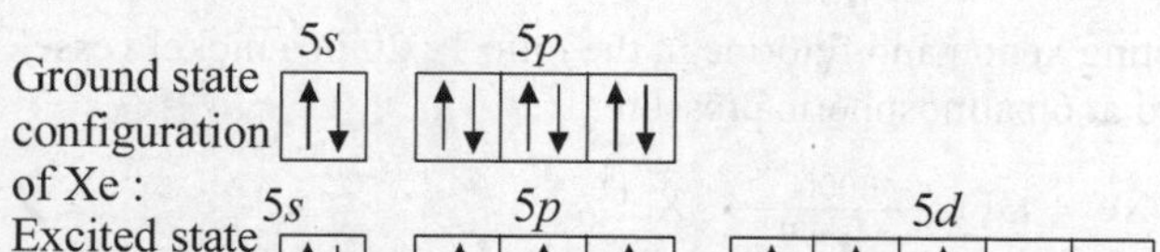

The 5*s*, three 5*p* and three 5*d* orbitals hybridize to produce seven sp^3d^3 hybrid orbitals. The singly occupied orbitals are used for bond formation to six fluorine atoms. The seventh orbital contains a lone pair. (Fig. 7.9).

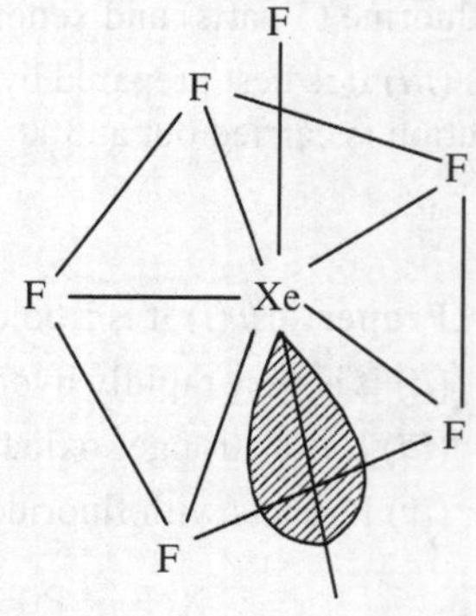

Fig. 7.9. Pentagonal bipyramidal structure of XeF_6.

7.13 XENON OXIDE TETRAFLUORIDE, $XeOF_4$

Preparation. (*i*) It can be prepared by the reaction of SiO_2 with XeF_6.

$$2XeF_6 + SiO_2 \longrightarrow 2\,XeOF_4 + SiF_4$$

(*ii*) It may be prepared by the partial hydrolysis of XeF_6.

$$XeF_6 + H_2O \longrightarrow XeOF_4 + 2HF$$

(*iii*) By heating a mixture of xenon (1 part) and fluorine (4 parts) with large excess of oxygen round 230°C.

$$Xe + 2F_2 + O \xrightarrow{230°C} XeOF_4$$

(*b*) **Properties.** (*i*) It is a colourless substance melting at –46°C.

(*ii*) It is reduced by hydrogen gas to xenon.

$$XeOF_4 + 3H_2 \longrightarrow Xe + H_2O + 4HF$$

(*iii*) It reacts with water or SiO_2 to give XeO_2F_2 and XeO_3.

$$XeOF_4 + H_2O \longrightarrow XeO_2F_2 + 2HF$$

$$XeO_2F_2 + H_2O \longrightarrow XeO_3 + 2HF$$

$$2XeOF_4 + SiO_2 \longrightarrow XeO_2F_2 + SiF_4$$

$$2XeO_2F_2 + SiO_2 \longrightarrow 2XeO_3 + SiF_4$$

Structure and bonding in $XeOF_4$

(*i*) The structure of molecule $XeOF_4$ is square pyramid as shown below:

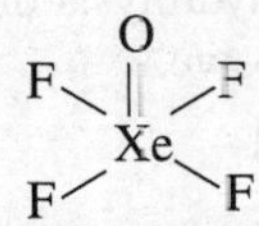

(*ii*) The valence bond description of $XeOF_4$ may be given by assuming that three 5*p* electrons are promoted to the 5*d* sub-level.

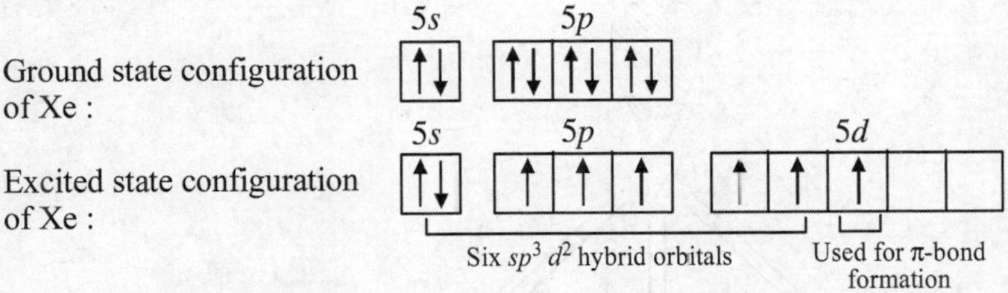

The 5*s*, three 5*p* and two 5*d* atomic orbitals hybridize to give six sp^3d^2 hybrid orbitals. The five singly occupied orbitals are used for s-bond formation to four fluorine and one oxygen atom. The sixth hybrid orbital is occupied by the lone pair. The single electron in a *d*-orbital is used for p-bond formation to oxygen atom. This explains the required geometry of $XeOF_4$ (Fig. 7.10).

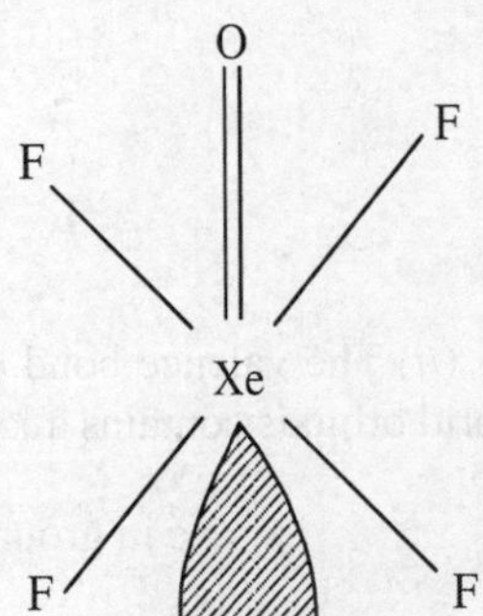

Fig. 7.10. Structure of $XeOF_4$.

7.14 XENON DIOXIDE DIFLUORIDE, XeO_2F_2

Preparation. It is prepared by mixing XeO_3 and $XeOF_4$ at low temperatures (–78°C).

Properties. (*i*) It is a colourless solid which melts at 30°C.

(*ii*) It is easily hydrolysed to give XeO_3.

Structure and bonding. (*i*) It has a see-saw structure.

(*ii*) The valence bond structure suggests trigonal bipyramid geometry with one equatorial position occupied by a lone pair.

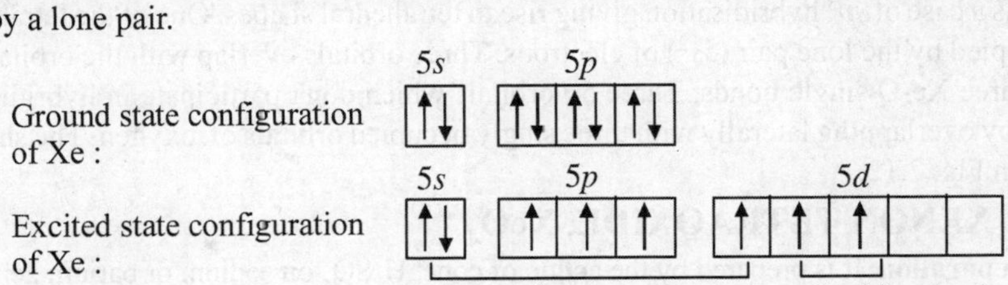

The three $5p$ electrons are promoted to higher energy $5d$ sub-level. The $5s$, three $5p$ and one $5d$ orbitals hybridize to give five sp^3d hybrid orbitals. The four singly occupied orbitals are used for bond formation to two fluorine and two oxygen atoms. The fifth hybrid orbital contains the lone pair. The other two $5d$ electrons of xenon which do not take part in the hybridization scheme are involved in p-bond formation to two oxygen atoms. The structure of XeO_2F_2 is represented in Fig. 7.11.

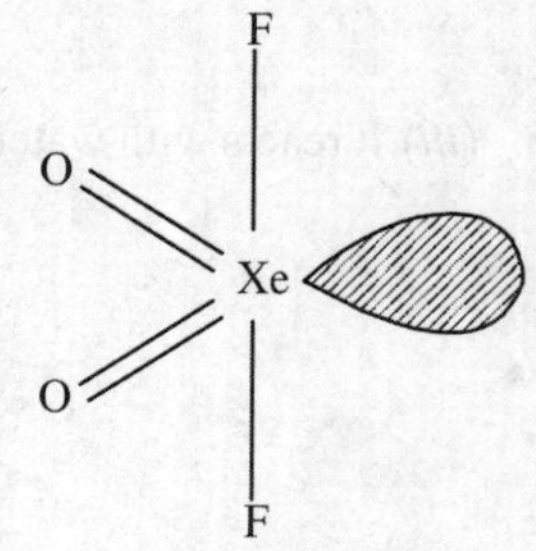

Fig. 7.11. Structure of XeO_2F_2

7.15 XENON TRIOXIDE, XeO_3

Preparation. It is prepared by the hydrolysis of XeF_6 and XeF_4. Great care is to be taken during its preparation as it is highly explosive.

Properties. (*i*) It is a colourless solid.

(*ii*) It is a highly explosive substance.

Structure. (*i*) The molecule is pyramidal, as shown in Fig. 7.12.

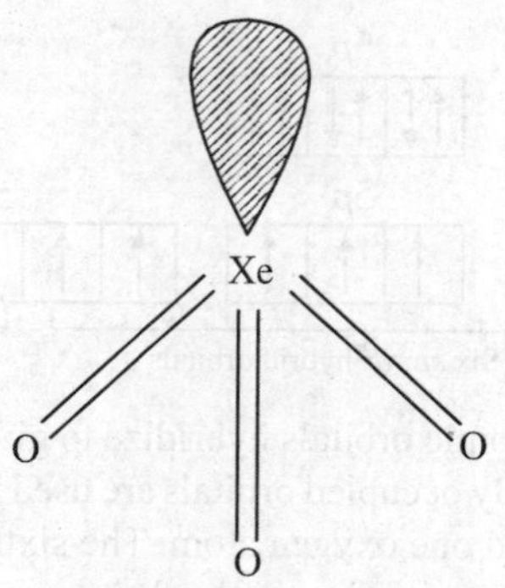

Fig. 7.12. Structure of XeO_3

(*ii*) The valence bond description suggests xenon to acquire sp^3 hybridization and one of the hybrid orbitals contains a lone pair.

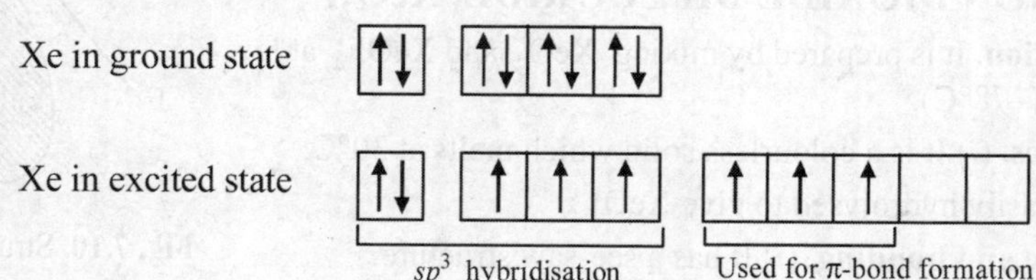

It is a case of sp^3 hybridisation giving rise to tetrahedral shapes. One of the tetrahedral positions is occupied by the lone pair ($5s^2$) of electrons. Three orbitals overlap with the orbitals of oxygen to form three Xe-O single bonds. Three $5d$ orbitals which don't participate in hybridisation form p-bonds by overlapping laterally with three singly occupied orbitals of oxygen. The shape of XeO_3 is given in Fig. 7.12.

7.16 XENON TETRAOXIDE, XeO_4

Preparation. It is prepared by the action of conc. H_2SO_4 on sodium or barium xenate (Na_4XeO_6 ; Ba_2XeO_6) at room temperature. It may be purified by vacuum sublimation at $-78°C$.

Properties. It is quite unstable and decomposes to xenon and oxygen.

Structure. The molecule is tetrahedral. In this case, the electrons from 5*s* and 5*p* orbitals are excited to 5*d* orbital, thus giving rise to eight orbitals having single electrons. Out of these, one *s* and three *p* orbitals hybridise. Hybridised orbitals form four σ-bonds with oxygens. Unhybridised orbitals (4 in no.) form four π-bonds with oxygen to give XeO_4 as shown in Fig. 7.13.

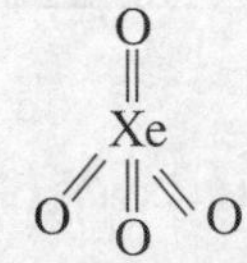

Fig. 7.13 Shape of XeO_4

7.17 COMPREHENSIVE LIST OF HYBRIDISATION AND SHAPE OF XENON COMPOUNDS

The hybridisation and shape of xenon compounds are given in the table below:

Molecule	*Shape*	*Description*
1. XeF_2	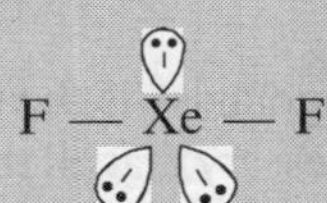	Linear; three nonbonding electron pairs at the points of an equilateral triangle. sp^3d hybridisation.
2. XeF_4	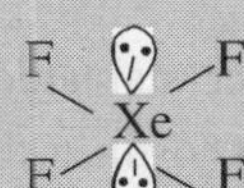	Square planar; two nonbonding electron pairs one above and one below the plane of the molecule. sp^3d^2 hybridisation.
3. XeF_6	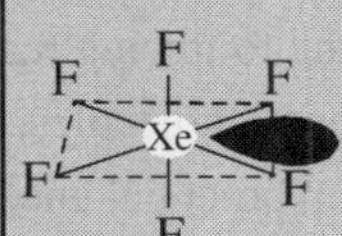 or 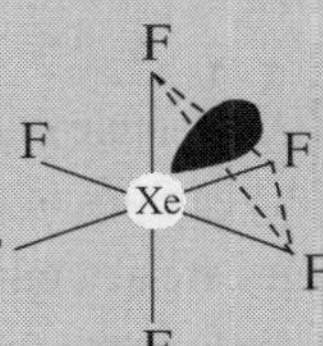	Distorted octahedron; one nonbonding electron pair either at the centre of a face or the midpoint of an edge. sp^3d^3 hybridisation.
4. Xe OF_4	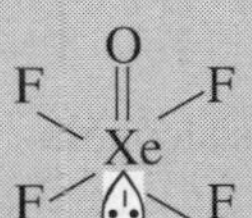	Square pyramidal; one nonbonding electron pair coming out from the base of the pyramid. sp^3d^2 hybridisation.
5. $XeOF_2$	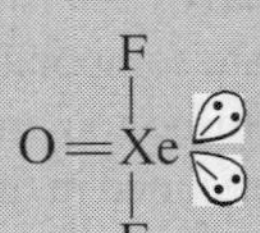	T-shaped; two nonbonding electron pairs pointing towards the corners of an equilateral triangle. sp^3d hybridisation.
6. XeO_2F_2	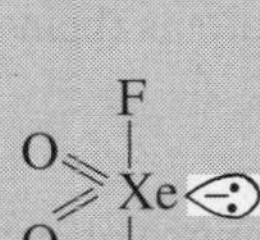	See-saw; one nonbonding electron pair pointing towards the corner of an equilateral triangle. sp^3d hybridisation.
7. XeO_3	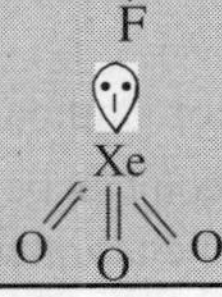	Trigonal pyramidal; one nonbonding electron pair pointing out from the apex of the pyramid. sp^3 hybridisation.

(Contd.)

Molecule	*Shape*	*Description*
8. XeO_4	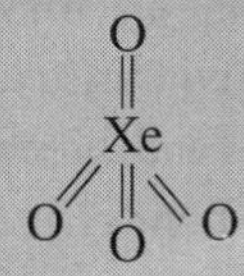	Tetrahedron sp^3 hybridisation.
9. XeF_5^-	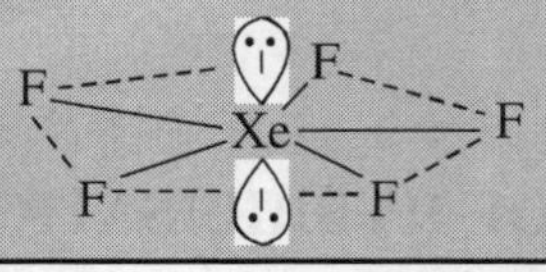	Pentagonal planar; two nonbonding electron pairs one above the plane and the other below the plane. sp^3d^3 hybridisation.

7.18 CLATHRATES

According to Powell, clathrate compounds consist of atoms or molecules of one substance called the 'guest' held in the crystalline cages formed by the molecules of a second substance called the 'host'.

The only type of interaction in these "cage compounds" is the relatively weak van der Waals forces and there is no chemical bonding. It is not necessary that all the cavities in the host structure are filled. As a result, the clathrates are normally non-stoichiometric compounds.

Essential Conditions for Clathrate Formation

(*a*) Presence of cavities of appropriate size in the crystal lattice of the 'host'.

(*b*) Size of the 'guest' particle should be such as to fit in the cavities of the 'host' without bringing any atoms closer together than which correspond to the van der Waals radii of the atoms.

Stability of Clathrate Compounds

Clathrate compounds once formed are quite stable due to the following reasons:

(*a*) The 'guest' molecules tightly fit in the cavities of the 'host' molecules.

(*b*) The 'guest' molecules within the cages are at a minimum potential energy.

The 'guest' can escape from the 'host' only when the forces holding the molecular cages together are overcome. This can be achieved by the following two methods:

(*i*) By heating the crystals.

(*ii*) By dissolving them in a suitable solvent such as alcohol.

Preparation of Clathrates

Clathrate compounds of argon, krypton and xenon have been isolated by locking up these gases in the cages of water, phenol or quinol. However, clathrate compounds of helium and neon have not yet been prepared because helium atoms are thought to be too small and difficult to deform or polarize as required by van der Waals bonding to be imprisoned in the cages. The neon is borderline element and its clathrate could possibly be formed at pressures in excess of 160 atmospheres. It is predicted that in clathrate of neon, if formed, the neon atoms (being so small) will diffuse through the crystal from one cavity to another.

Uses of Clathrates

(*i*) Krypton-85 clathrate provides a safe and useful source of beta radiation which are useful for measuring thickness of gauges.

(*ii*) Xenon-133 clathrate provides a compact source for gamma radiation.

(*iii*) In the separation of noble gases. For example, neon can be separated from argon, krypton and xenon because neon is the only gas that does not form clathrates with quinol.

(*iv*) Clathrates play an important role in some physiological actions, *e.g.*, it is thought that the anaesthetic action of xenon and many other anaesthetics is due to aqueous clathrate in physiologically strategic spots. When the anaesthetic is no longer administered, the clathrate equilibrium is destroyed, the clathrate decomposes and the consciousness returns.

***Example 1.* Which of the following compounds is most difficult to prepare and why?**

$$XeF_2, XeF_4, XeF_6$$

Solution. XeF_6 is most difficult to prepare. Maximum number of electrons are required to be excited from $5p$ to $5d$ orbitals to prepare this compound, hence large amount of energy is needed.

***Example 2.* Why xenon fluorides cannot be stored in quartz vessel?**

Solution. Xenon fluorides react with silicon dioxide present in quartz to form SiF_4.

$$2XeF_6 + SiO_2 \longrightarrow 2XeOF_4 + SiF_4$$

EXERCISES
(INCLUDING QUESTIONS FROM DIFFERENT UNIVERSITY PAPERS)

Objective Questions

Fill in the blanks

1. The original Mendeleef's Table did not leave any blank space for .... gases
2. Noble gases show very low m.p. and b.p. due to weak .... forces between them
3. Helium gives a spectrum of green and yellow lines while neon gives a spectrum of .... and ....lines
4. Because of low .... forces, the liquefaction of noble gases is difficult
5. $XeOF_2$ has .... structure.
6. XeO_3 involves .... hybridisation
7. Xenon flurides cannot be stored in quartz vessel because of the reaction leading to the formation of ....
8. Helium and neon do not form clathrates because of the .... of these noble gases
9. Rayleigh's method can be used to remove .... and .... from pure dry air
10. Dewar's charcoal method for the separation of noble gases uses the principle of ....

Multiple Choice Questions

Choose the correct option

1. Argon gas was discovered by :
   (*a*) Ramsay (*b*) Raylaigh (*c*) Rutherford (*d*) None
2. Which one of the following configuration represents the noble gas ?
   (*a*) $1s^2, 2s^22p^6, 3s^2$ (*b*) $1s^2, 2s^2, 2p^6, 3s^1$
   (*c*) $1s^2, 2s^2, 2p^6$ (*d*) $1s^2, 2s^2, 2p^6, 3p^6, 4s^2$
3. Which of the following represents outer electronic configuration of argon ?
   (*a*) ns^2 (*b*) $ns^2\ np^3$ (*c*) $ns^2\ np^6$ (*d*) $ns^2\ np^4$
4. The following has zero valency :
   (*a*) Na (*b*) Br (*c*) Al (*d*) Kr
5. Which of the following noble gases is present in atmosphere ?
   (*a*) He (*b*) Ne (*c*) Ar (*d*) Rn

6. The greatest contribution to the chemistry of noble gases was made by
(*a*) Lord Rayleigh (*b*) Sir Cavendish (*c*) William Ramsay (*d*) None

7. Which of the following gases exists in monoatomic state ?
(*a*) Chlorine (*b*) Neon (*c*) Oxygen (*d*) Nitrogen

8. The forces acting between noble gas atoms are :
(*a*) van der waals forces (*b*) ion-dipole forces
(*c*) London – dispersion force (*d*) magenetic forces

9. Which of the following rare gases is abundant in atmosphere ?
(*a*) He (*b*) Ne (*c*) Ar (*d*) Kr

10. The spectrum of He is expected to be similar to that of:
(*a*) H (*b*) Li^+ (*c*) Na (*d*) He^+

11. Which one of the following gases has the least tendency of form compounds ?
(*a*) He (*b*) Ne (*c*) Kr (*d*) Xe

12. The coloured discharge tubes for advertisement mainly contain :
(*a*) xenon (*b*) helium (*c*) neon (*d*) argon

13. Which of the flourides of xenon is impossible?
(*a*) XeF_2 (*b*) XeF_3 (*c*) XeF_4 (*d*) XeF_6

14. Noble gases do not react with other elements because :
(*a*) they are monoatomic
(*b*) they are found in abundance
(*c*) they size of their atom is very small
(*d*) they have complete paired up and stable electron shells

15. The noble gas forming maximum number of compounds is
(*a*) Ne (*b*) Xe (*c*) Kr (*d*) Rn

16. Percentage of argon in air is about :
(*a*) 10% (*b*) 0.1% (*c*) < 0.1% (*d*) 1%

17. Least chemical activity is known in
(*a*) ammonia (*b*) methane (*c*) argon (*d*) sulphuric acid

18. Deep sea divers are used to respirate in a mixture of :
(*a*) oxygen and argon (*b*) oxygen and helium
(*c*) oxygen and nitrogen (*d*) oxygen and hydrogen

19. The separation of inert gases from air is possible due to difference in their:
(*a*) density (*b*) boiling point (*c*) specific heat (*d*) solubility

20. The gas that surrounds a radioactive element is :
(*a*) He (*b*) Ne (*c*) N_2 (*d*) O_2

ANSWERS

Fill in the blanks

1. noble	**2.** ven der waals	**3.** orange, red	**4.** inter-atomic
5. T-shaped	**6.** sp^3	**7.** SiF_4	**8.** smaller size
9. nitrogen, oxygen	**10.** adsorption		

Multiple choice questions

1. (*a*)	**2.** (*c*)	**3.** (*b*)	**4.** (*d*)	**5.** (*d*)
6. (*c*)	**7.** (*b*)	**8.** (*a*)	**9.** (*c*)	**10.** (*b*)
11. (*a*)	**12.** (*b*)	**13.** (*b*)	**14.** (*d*)	**15.** (*b*)
16. (*d*)	**17.** (*c*)	**18.** (*b*)	**19.** (*b*)	**20.** (*a*)

General Questions

1. (*a*) Discuss the position of noble gases in the periodic table.
   (*b*) Why is it more appropriate to call the zero group elements as noble gases instead of inert gases?
2. (*a*) How can noble gases be separated from one another?
   (*b*) Compare the buoyancy of helium with that of hydrogen.
3. (*a*) Describe the methods employed, indicating the physico-chemical principles involved in them, for the separation of rare gases of the atmosphere. Indicate their importance in industry.
4. What prompted Bartlett to conjecture that the fluorides of xenon are possible. Give the preparation and properties of one fluoride of xenon. Discuss its structure.
5. (*a*) Give methods of preparation, properties and structure of the following:
   (*i*) XeF_2, (*ii*) $XeOF_4$, (*iii*) XeF_6
   (*b*) Complete the following reactions:
   (*i*) $XeF_4 + H_2O \longrightarrow$ (*ii*) $XeF_2 + H_2 \longrightarrow$ (*iii*) $XeF_6 + SiO_2 \longrightarrow$
   (*iv*) $XeF_6 + NH_3 \longrightarrow$ (*v*) $Xe\,F_6 + 3H_2O \longrightarrow$
6. (*a*) How is XeF_4 prepared? Describe its reactions with (*i*) Hg (*ii*) F_2 (*iii*) BCl_3 (*iv*) SF_4.
7. Discuss the methods of preparations, properties and structures of fluorides and oxofluorides of xenon in VI oxidation state.
8. Comment on the statement that there is little likelihood of the formation of compounds of helium and neon.
9. Give the methods of preparation, properties and structures of the following:
   (*i*) XeO_3, (*ii*) XeF_4, (*iii*) XeO_2F_2
10. Explain why
    (*i*) Electron affinity of noble gases is zero.
    (*ii*) He_2 molecule is unknown, while H_2 is known.
    (*iii*) Xenon compounds are more stable than those of other noble gases.
11. Write the preparation and stereochemistry of two oxides and three fluorides of xenon.
12. Explain why
    (*i*) Melting points of noble gases increase with increase in atomic number of the gas.
    (*ii*) Solubility of noble gases in water is greater than that of hydrogen.
13. Discuss the structures of Xe F_2 and Xe O_3.

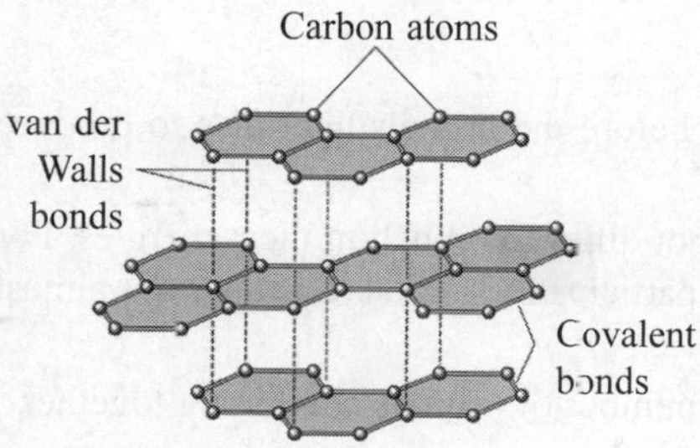

PART – II ORGANIC CHEMISTRY

1

Structure and Bonding

1.1 HYBRIDISATION OF ORBITALS

The electronic configuration of carbon is $1s^2\ 2s^2\ 2p_x^1\ 2p_y^1$ as its At. No. is 6. As the bonds are formed by unpaired electrons, carbon should show a valency of 2. But, we observe that carbon shows tetravalency in its compounds like methane or carbon dioxide. This was explained by saying that the above-mentioned configuration is the configuration in the ground state. During the formation of bonds with other atoms, some of the electrons in lower energy levels are excited to higher levels. Thus, in the case of carbon, one electron from 2*s* level is excited to empty $2p_z$ level. (Fig. 1.1)

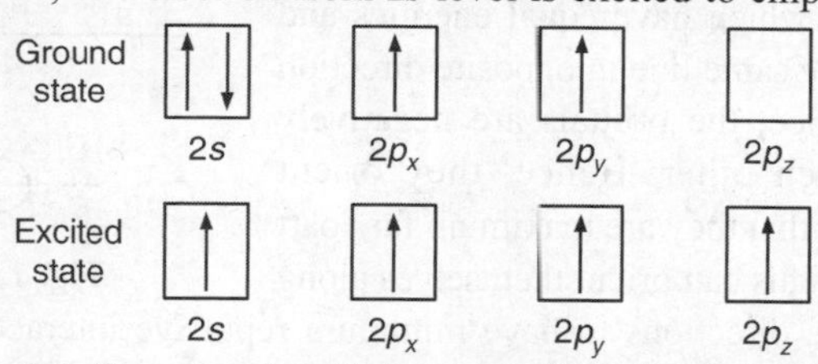

Fig. 1.1. Ground and excited state configurations of carbon.

Now, there are 4 electrons which are unpaired. These four electrons will form four bonds by overlapping with the orbitals of other atoms. For example, they combine with four hydrogen atoms to form methane (CH_4).

However, there is one thing which is surprising. We find that in methane, all the C—H bonds have the same bond energy and all the angles H—C—H are the same, *i.e.*, 109°– 28′. Looking back at the orbitals, which participate in bonding, *i.e.*, one *s*-orbital and three *p*-orbital which have different shapes and energies, we expect one of the bonds formed by *s*-orbital to be different from the three bonds formed by *p*-orbitals. Moreover, as the *p*-orbitals are inclined at an angle of 90° to each other, we expect the bond angles of 90° in methane also. But this actually is not so. To explain this anomaly, the concept of *hybridisation* was introduced.

It is assumed that the 2*s*-orbital and three 2*p*-orbitals which are associated with different shapes and energies mix with each other and produce four equivalent types of orbitals with the same shape and energy and are oriented symmetrically in space. These are called **hybrid** or **hybridised** orbitals and this phenomenon is called **hybridisation.** These hybridised orbitals then overlap with the orbitals of other atoms to form bonds.

Thus, hybridisation is the process of mixing of orbitals of different shapes and energies, to produce orbitals of equivalent shape and energy oriented symmetrically with respect to one another.

This phenomenon explains satisfactorily why the bond energies and bond angles are equal in the case of methane and many other molecules.

Necessary conditions for hybridisation

There are certain conditions which must be satisfied before the orbitals hybridise to produce equivalent type of orbitals. These are :

(*i*) The orbitals taking part in hybridisation should not differ too much in their energies. For example, 1*s* and 2*p* orbitals cannot hybridise. The participating orbitals must possess almost similar energies.

(*ii*) As many hybrid orbitals will be obtained as the number of orbitals combining together.

(*iii*) Whereas the paired orbitals do not take part in bonding, they can take part in hybridisation. In such a case, one of the positions in the geometrical shape is occupied by the lone pair of electrons. This happens in the case of ammonia and water molecules.

(*iv*) It is not essential for all the unpaired orbitals to participate in hybridisation. Those unpaired orbitals which do not take part in hybridisation form π-bonds by lateral overlapping.

1.2 TYPES OF HYBRIDISATION

Depending upon which atomic orbitals are combining together, we have different types of hybridisation, which are given below:

1. *sp* or diagonal hybridisation. This type of hybridisation occurs when one *s* orbital and one *p* orbital combine together to give two hybridised orbitals known as *sp* hybridised orbitals which have equal energies and they are oriented along the same line in opposite direction at an angle of 180°. Since, the orbitals are negatively charged, they repel each other. Hence, they orient themselves in such a way that they are maximum far apart from each other. Two orbitals can orient themselves along the same line in opposite directions to have minimum repulsive interactions. The shapes of the hybridised orbitals in *sp* hybridisation is given in Fig. 1.2.

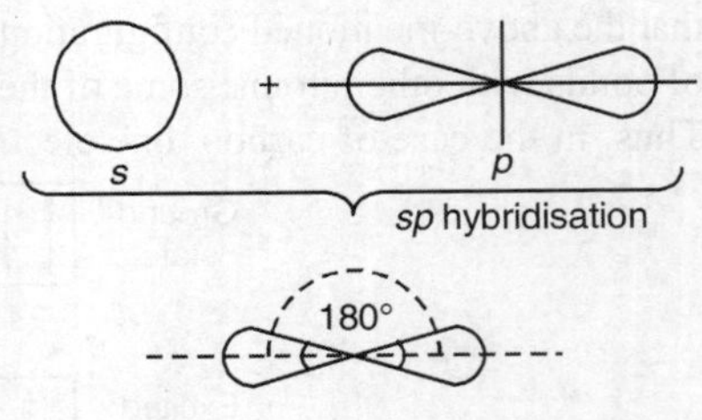

Fig. 1.2. *sp* hybridisation.

2. sp^2 or trigonal hybridisation. This type of hybridisation takes place when one *s* orbital and two 2*p* orbitals combine together to give three sp^2 hybridised orbitals of equal energies. These hybrid orbitals are oriented at an angle of 120° in the same plane. The three orbitals point towards the corners of an equilateral triangle with the nucleus occupying the centre of the triangle. This is illustrated in Fig. 1.3.

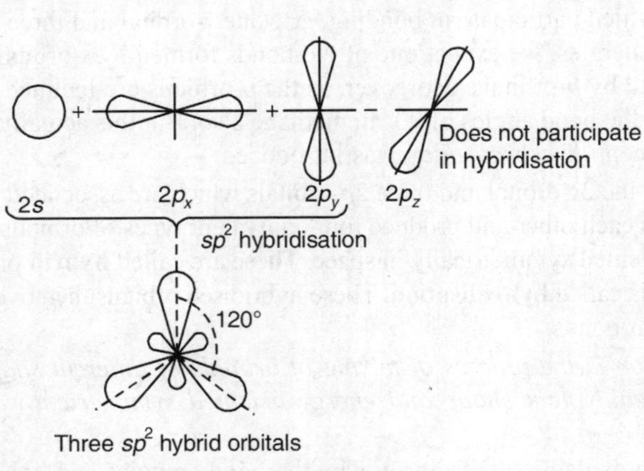

Fig. 1.3. sp^2 hybridisation.

3. sp^3 or tetrahedral hybridisation. When one *s*-orbital and three *p*-orbitals combine together to produce equivalent type of orbitals, sp^3 hybridisation takes place. Four sp^3 hybridised orbitals are produced which have the same energy and shape and are oriented at an angle of 109°28′ to one another. This is also known as *tetrahedral* hybridisation because the hybridised orbitals are pointing towards four corners of a tetrahedron with the nucleus occupying the centre of the tetrahedron. In this position, the hybridised orbitals experience minimum repulsive forces. This is illustrated in Fig. 1.4.

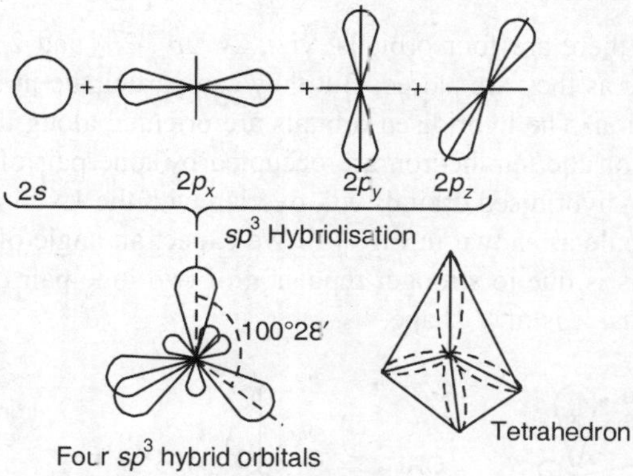

Fig. 1.4. sp^3 hybridisation.

The geometry and shape of molecules associated with different types of hybridisation is summarised in Table 1.1.

S. No.	Type of Hybridisation	Combination of Orbitals	Bond Angle	Geometry
1.	sp	$s + p_x$	180°	Linear
2.	sp^2	$s + p_x + p_y$	120°	Trigonal
3.	sp^3	$s + p_x + p_y + p_z$	109°28′	Tetrahedral
4.	dsp^2	$d + s + p_x + p_y$	90°	Square planer
5.	sp^3d	$s + p_x + p_y + p_z + d$	90°, 120°	Trigonal bipyramidal
6.	sp^3d^2	$s + p_x + p_y + p_z + (2\times d)$	90°	Octahedral

1.3 SHAPES OF CERTAIN MOLECULES

1. Shape of methane molecule (CH$_4$)

At. No. of carbon = 6. Electronic configuration of C in ground state = $1s^2\ 2s^2\ 2p_x^1\ 2p_y^1$.

Electronic configuration of C in excited state = $1s^2\ 2s^1\ 2p_x^1\ 2p_y^1\ 2p_z^1$.

There are four orbitals having unpaired electrons. These are $2s$, $2p_x$, $2p_y$, and $2p_z$ orbitals. They will first hybridise. So, it is a case of sp^3 hybridisation. Four hybridised orbitals are oriented tetrahedrally as shown in Fig. 1.5.

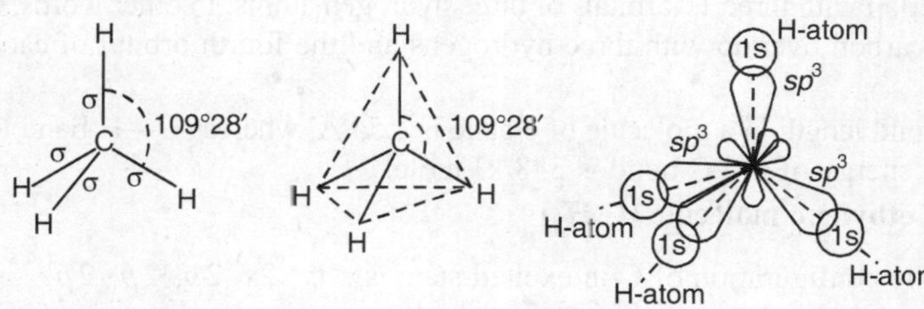

Fig. 1.5. Shape of methane molecule.

All the four hybridised orbitals overlap with the 1*s*-orbital of four hydrogen atoms. All these overlappings take place along the internuclear axes. Hence, four carbon-hydrogen σ-bonds are formed. The angle ∠HCH is 109°28′.

2. Shape of water molecule H_2O

Atomic number of oxygen = 8

Electronic configuration = $1s^2\ 2s^2\ 2p_x^2\ 2p_y^1\ 2p_z^1$

In the 2nd orbit, there are four orbitals, viz., 2*s*, $2p_x$, $2p_y$ and $2p_z$. Out of these 2*s* and $2p_x$ cannot form the bonds as they are paired. But they can participate in hybridisation. Hence, it is a case of sp^3 hybridisation. The hybridised orbitals are oriented along the corners of a tetrahedron. Two of the positions on the tetrahedron are occupied by lone pair of electrons from 2*s* and $2p_x$ orbitals. The other two hybridised orbitals will overlap with the 1*s*-orbitals of two hydrogen atoms to form a water molecule as shown in Fig. 1.6. We expect an angle of 109°28′ in HOH. Actually the angle is 105°. This is due to stronger repulsion of two lone pair of electrons with the shared pairs of electrons. Thus, it is of V shape.

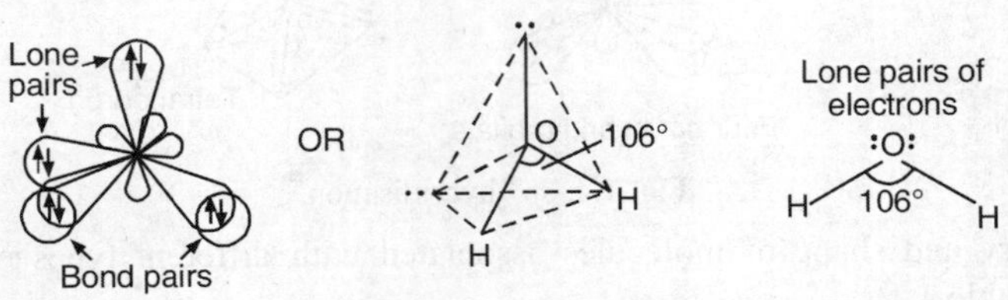

Fig. 1.6. Shape of water molecule.

3. Shape of ethane (C_2H_6) molecule

In the molecule of ethane two carbon atoms are linked to each other by a σ-bond and each carbon is linked to three hydrogens (Fig. 1.7)

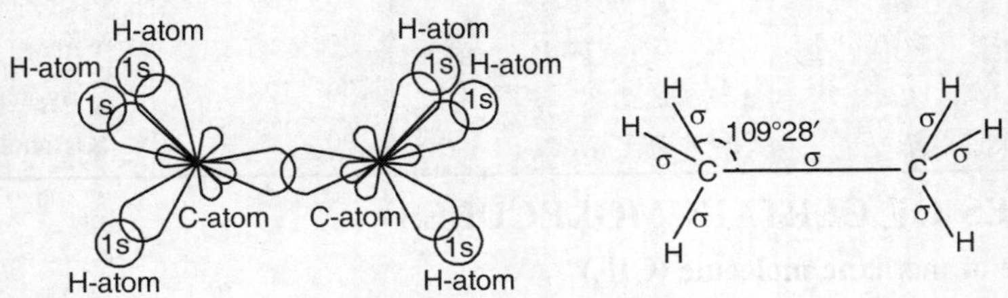

Fig. 1.7. Shape of ethane molecule.

Both the carbon atoms are sp^3 hybridised. Out of the four sp^3 hybridised orbitals of one carbon, one orbital overlaps with one of the hybridised orbitals of the second carbon making a C—C bond. The other three hybridised orbitals overlap with the 1*s*-orbitals of three hydrogen atoms. Similarly, one hybridised orbital of the second carbon overlaps with one of the orbitals of the first carbon and three orbitals overlap with three 1*s*-orbitals of three hydrogen atoms. In other words, three hybridised orbitals of each carbon overlap with three hydrogens and the fourth orbital of each overlaps with each other.

The C—C bond length in a molecule of ethane is 1.54 Å, whereas C—H bond length is 1.10 Å. The dissociation energy of C—C bond is 348 kJ/mole.

4. Shape of ethylene molecule (C_2H_4)

The electronic configuration of C in excited state is : $1s^2\ 2s^1\ 2p_x^1\ 2p_y^1\ 2p_z^1$

Here, all the four unpaired orbitals do not participate in hybridization. Out of four, only three orbitals viz. $2s$, $2p_x$ and $2p_y$ hybridise. The fourth one $2p_z$ remains unhybridised. Hence, it is a case of sp^2 hybridisation. The hybridised orbitals are oriented in the same plane at an angle of 120°. We have two such carbons. One orbital of each carbon overlaps with each other to form a σ-bond. Two orbitals of each carbon overlap with two 1*s*-orbitals of hydrogen to form four C—H bonds. We are left with 2 unhybridised orbitals of two carbons which are oriented perpendicular to the plane of the rest of the molecule. They will overlap laterally giving rise to a weak π-bond between carbon-carbon. Hence, there will be a double bond, one σ-bond and one π-bond between carbon-carbon (Fig. 1.8)

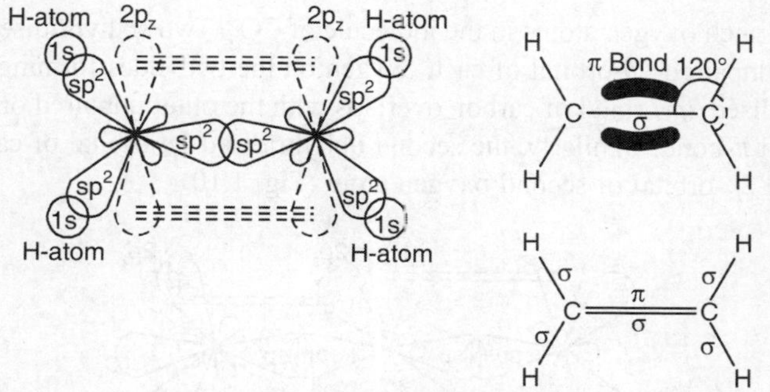

Fig. 1.8. Shape of ethylene molecule.

A double bond brings the carbon atoms together. Hence, the C = C bond length in C_2H_4 is shortened to 1.34 Å. The angle ∠HCH or ∠HCC is 120°, the trigonal angle. Bond dissociation energy of C = C bond is 614 kJ/mole.

5. Shape of acetylene molecule (C_2H_2)

Coming again to the electronic configuration of carbon :

$$1s^2\ 2s^1\ 2p_x^1\ 2p_y^1\ 2p_z^1 \quad \text{(Excited state).}$$

Here, only 2*s* and $2p_x$ orbitals take part in hybridisation. $2p_y$ and $2p_z$ orbitals remain unhybridised. Hence, it is a case of *sp* hybridisation. We have seen earlier that *sp* hybridised orbital are oriented along the same axis at an angle of 180°. One *sp* hybridised orbital of each carbon overlaps with the 1*s*-orbital of hydrogen atom on the two sides. One *sp* hybridised orbital of each carbon overlaps with each other forming a C—C sigma bond. The unhybridised $2p_y$ orbital of one carbon overlaps laterally with the $2p_y$ orbital of the other carbon forming a π-bond. Similarly, $2p_z$ orbital of one carbon overlaps with $2p_z$ orbital of the other carbon laterally forming another π-bond. The molecule of acetylene has a linear shape (Fig. 1.9). There is a cylinder of negative charge around C—C bond.

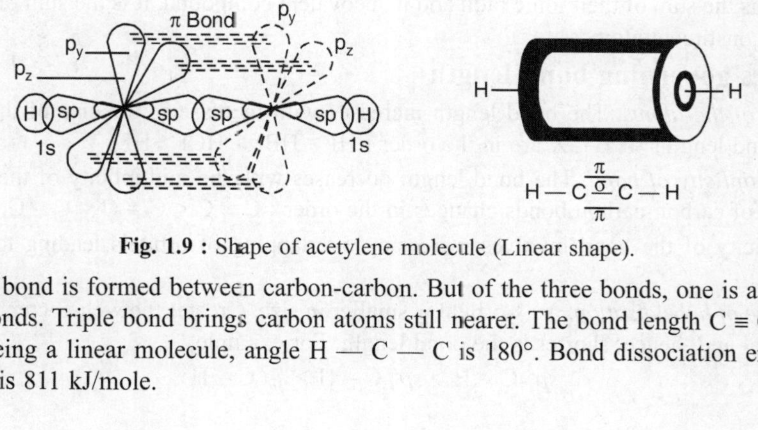

Fig. 1.9 : Shape of acetylene molecule (Linear shape).

A triple bond is formed between carbon-carbon. But of the three bonds, one is a σ-bond and two are π-bonds. Triple bond brings carbon atoms still nearer. The bond length C ≡ C is 1.20 Å. Acetylene being a linear molecule, angle H — C — C is 180°. Bond dissociation energy of the C ≡ C bond is 811 kJ/mole.

6. Shape of carbon dioxide molecule. At. No. of carbon is 6 and its electronic configuration in the excited state is :

$$1s^2\ 2s^1\ 2p_x^1\ 2p_y^1\ 2p_z^1.$$

sp hybridisation takes place here. The hybridised orbitals are oriented along the same line at an angle of 180°. $2p_y$ and $2p_z$ orbitals do not participate in the hybridisation.

Coming to oxygen, its electronic configuration is :

$$1s^2\ 2s^2\ 2p_x^2\ 2p_y^1\ 2p_z^1$$

There are two such oxygen atoms in the molecule of CO_2. Two *sp* hybridised orbitals of carbon overlap with one unpaired 2*p*-orbital of each oxygen on the two sides forming two σ-bonds. And then, one unhybridised 2*p*-orbital of carbon overlaps with the other unpaired orbital of one oxygen laterally forming a π-bond. Similarly, the second unhybridised 2*p*-orbital of carbon overlaps with the other unpaired 2*p*-orbital of second oxygen atom (Fig. 1.10)

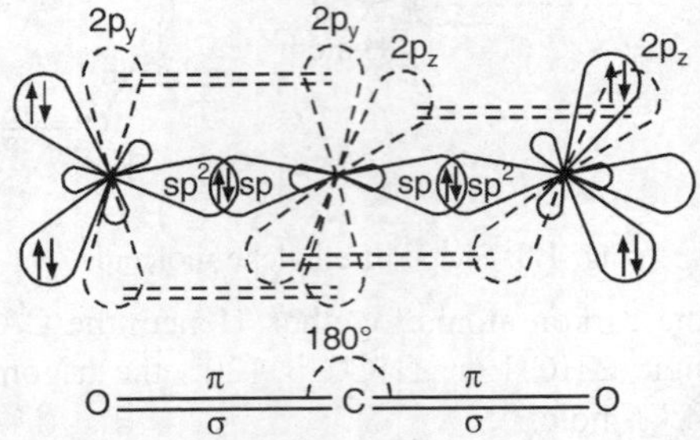

Fig. 1.10. Shape of CO_2 molecule.

The molecule of carbon dioxide has a linear shape.

1.4 SOME IMPORTANT BOND PARAMETERS

1. Bond Length

When atoms are brought close to each other, attraction takes place between them and, therefore, the potential energy of the system keeps on decreasing till at a particular distance, the potential energy is minimum. If the atoms are further brought closer, the repulsion starts and the potential energy of the system begins to increase.

The average distance between the centres of the nuclei of the two bonded atoms is called its **bond length.**

Bond length is expressed in Angstrom (1Å = 10^{-10}m) or picameter (1 pm = 10^{-12} m). It is determined experimentally by X-ray diffraction methods or spectroscopic methods. In an ionic compound, the bond length is the sum of their ionic radii and in a covalent compound, it is the sum of the covalent radii of the constituent atoms.

Factors governing bond length

(*i*) *Size of the atoms*. The bond length increases with increase in the size of the atoms. For example, bond lengths of H—X are in the order : HI > HBr > HCl > HF.

(*ii*) *Multiplicity of bond*. The bond length decreases with the multiplicity of the bond. Thus, bond length of carbon-carbon bonds changes in the order : C ≡ C < C = C < C – C.

Multiplicity of the bond involves increasing overlapping of orbitals leading to decrease in bond length.

(*iii*) *Type of hybridisation*. An *s*-orbital is smaller in size. Greater the *s*-character, shorter is the hybrid orbital and, hence, shorter is the bond length. For example,

$$sp^3\ \text{C – H} > sp^2\ \text{C – H} > sp\ \text{C – H}$$

sp hybrid orbital has 50% *s* character (one *s* and one *p*). This percentage of *s* decreases as we move to sp^2 and sp^3.

The bond lengths of a few common bonds are given in the Table below.

Some Bond Lengths

Bond	*Bond length (Å)*	*Bond*	*Bond length (Å)*
C – H	1.09	C – O	1.43
C – C	1.54	C = O	1.21
C = C	1.34	C – N	1.47
C ≡ C	1.20	C – Cl	1.77
O – H	0.96	C – Br	1.91

2. Bond Energy

Energy is released when atoms come close together resulting in the formation of bond between them. Conversely, the energy is required to break the bond to get the separated atoms.

The amount of energy required to break one mole of bonds of a particular type so as to separate them into gaseous atoms is called **bond dissociation energy** or simply **bond energy.**

Bond energy is expressed in kJ mol^{-1}. Greater is the bond dissociation energy, stronger is the bond.

Since, a particular type of bond present in different molecules (e.g., O – H bond in H – O – H) or even in the same molecule (*e.g.*, four C – H bonds in CH_4) do not possess the same bond energy, bond energies are usually the *average values*.

Table 1.2 gives the average bond energies of a few common bonds.

Table 1.2. Some Bond Energies.

Bond	*Bond energy (kJ/mole)*	*Bond*	*Bond energy (kJ/mole)*
C – H	414.2	C – O	359.8
C – C	347.3	C = O	748.9
C = C	610.9	C – N	305.4
C ≡ C	836.8	C – Cl	338.9
O – H	464.4	C – Br	284.5
N – H	389.1	C – I	213.4

Factors governing bond energy

(*i*) *Size of the atoms.* Greater the size of the atoms, greater is the bond length and smaller is the bond energy.

(*ii*) *Multiplicity of bonds.* Greater the multiplicity of the bond, greater is the bond dissociation energy.

(*iii*) *Number of lone pairs of electrons present.* Greater the number of lone pairs of electrons present on the bonded atoms, greater is the repulsion between the atoms and, hence, smaller is the bond dissociation energy.

Bond	C – C	N – N	O – O	F – F
Lone pairs of electrons	0	1	2	3
Bond energy (kJ mol^{-1})	348	163	146	139

3. Bond Angle

A bond is formed by the overlap of atomic orbitals of different atoms. The direction of overlap gives direction of the bond.

The angle between the lines representing the directions of the bonds is called the **bond angle.**

It is expressed in degrees, minutes and seconds. H—C—H bond angle in CH_4, H—N—H bond angle in NH_3 and H—O—H bond angle in H_2O are shown in the Fig. 1.11.

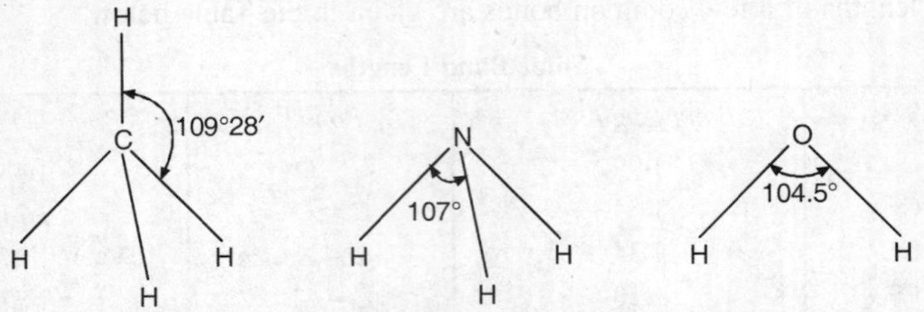

Fig. 1.11. Bond angles in CH_4, NH_3 and H_2O

1.5 LOCALIZED AND DELOCALIZED BONDS

Bonds in which the overlapping orbitals are almost in fixed positions are called localized bonds. Sigma(σ) bonds are localized bonds. Consider again the case of methane molecule. There are four carbon–hydrogen sigma bonds.

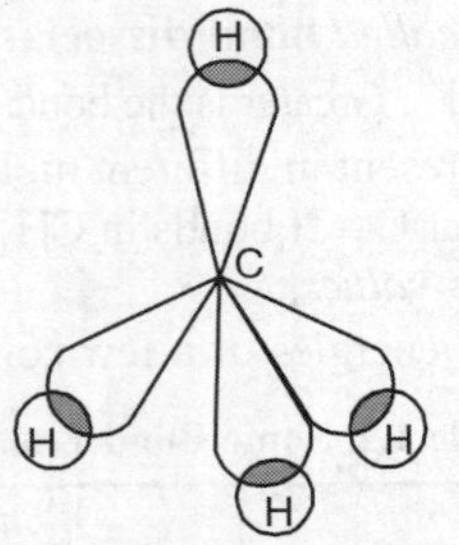

These are all localized bonds since the positions of the overlapping orbitals are fixed.

Now consider the case of 1, 3-Butadiene. This is a conjugated diene.

All four carbon atoms in 1, 3-butadiene are sp^2 hybridized. The sp^2 hybrid orbitals overlap with each other and with *s* orbitals of the hydrogen atoms to form C–C and C–H σ bonds. Since the bonds result from the overlap of trigonal sp^2 orbitals, all carbon and hydrogen atoms lie in one plane.

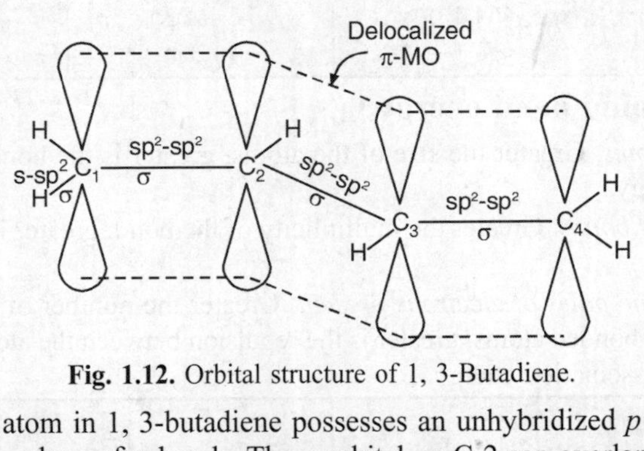

Fig. 1.12. Orbital structure of 1, 3-Butadiene.

Also each carbon atom in 1, 3-butadiene possesses an unhybridized *p* orbital. The *p* orbitals are perpendicular to the plane of σ bonds. The *p* orbital on C-2 can overlap with the *p* orbitals on C-1 and C-3. The *p* orbital on C-3 can overlap with the *p* orbitals on C-2 and C-4. That is, all four *p* orbitals overlap to form a large π molecular orbital (Fig. 1.12). Each pair of π electrons is thus attracted, not by two, but all four carbons.

The overlap of *p* orbitals of C-2 and C-3 in both directions, which allows the π electrons to be spread over a larger area, is referred to as **delocalization**. This delocalization of π electrons is responsible for greater stability of 1, 3-butadiene.

The C–C single bonds in 1, 3-butadiene are shorter (1.48 Å) than normal carbon—carbon single bonds. The C–C double bonds are longer (1.37 Å) than normal isolated carbon—carbon double bonds.

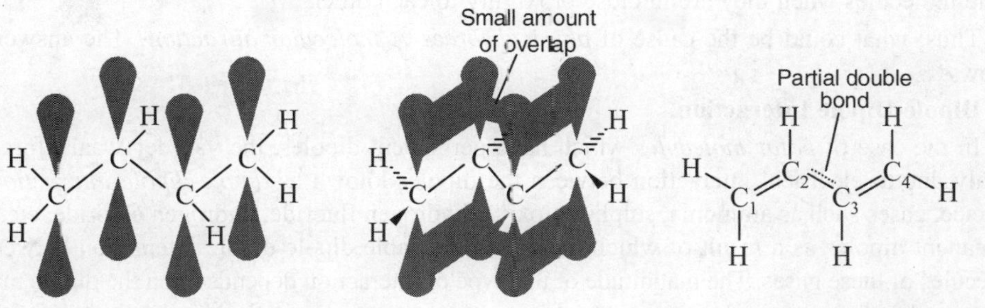

The delocalized molecular orbital contains four electrons. The central carbon-carbon single bond is shorter than the 1.54 Å bonds typical of alkanes because of its partial double bond character.

The electrons forming two π bonds in the molecule of 1, 3-butadiene are thus delocalized chemical bonds.

Another example of delocalized bonds is provided by the structure of benzene. It has a hexagonal structure in which the six carbon atoms occupy the corners of a regular hexagon. All the carbon atoms are linked to each other by sigma bonds. Also each carbon is linked to one hydrogen by sigma bond. It may be noted that carbon atoms in benzene are sp^2 hybridised. Thus, all carbons lie in the same plane. There is one orbital in each carbon, which does not participate in hybridisation.

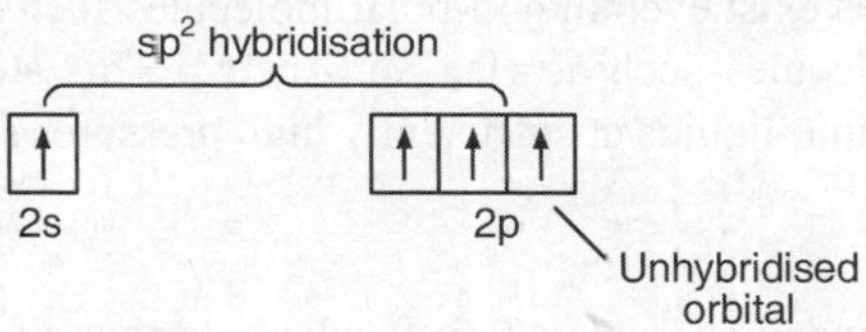

There are six such unhybridised orbitals these can overlap laterally in pairs to form three π bonds as shown in Fig. 1.13 (*a*) and (*b*). But the same situation exists as in 1, 3-butadiene. All the six orbitals overlap together forming a circular electron cloud above and below the benzene ring (Fig. 1.13)

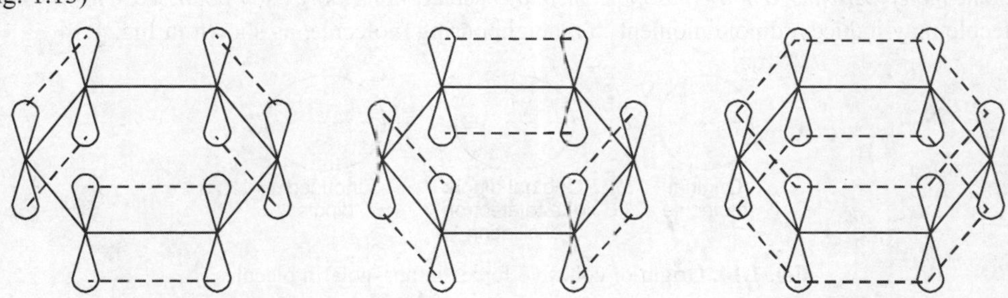

(*a*) (*b*) (*c*)

Fig. 1.13

1.6 VAN DER WAALS' FORCES

Forces of attraction exist between *neutral molecules* as well. The magnitude of these forces is maximum in case of solids and decreases on passing from solid to liquid and from liquid to gaseous states. The forces existing between the molecules of a substance are known as **intermoleculer forces.** These are also designated as **van der Waals' forces.** The attractive forces which keep the molecules in position in a molecular crystal are van der Waals' forces.

The question arises, how can there be electrical attraction between neutral molecules lying close to one another? The existence of electrical repulsion between such molecules, on the other hand, can be more readily understood as these may arise from the interaction of the electron clouds of the molecules when they are in close proximity to each other.

Thus, what could be the cause of *physical forces* or *molecular attraction*? The answer is as follows:

Dipole-Dipole Interaction.

In the case of *polar molecules* which have permanent dipoles, the van der Waals' forces are mainly due to electrical interaction between the dipoles known as *dipole-dipole interaction.* For instance, gases such as ammonia, sulphur dioxide, hydrogen fluoride, hydrogen chloride, etc., have permanent dipoles as a result of which there is appreciable dipole-dipole interaction between the molecules of these gases. The magnitude of this type of interaction depends upon the dipole moment of the molecule concerned. Greater the dipole moment, greater is the dipole-dipole interaction. Because of the attractive interaction, these gases can be easily liquefied.

Ion-Dipole Interaction.

If an ionic solid is to dissolve, some energy is needed to overcome the force of inner-ionic attraction. If water is the solvent, this energy is supplied by the *interaction of water molecules with the ions*. This is known as ion-dipole interaction. The positive pole of the water molecules reacts with the negative ion while the negative pole reacts with the positive ion. The ion-dipole interaction results in the release of a considerable amount of energy which becomes available for the separation of the oppositely charged ions. This results in the dissolution of the solute.

But van der Waals' forces exist even in non-polar molecules such as F_2, O_2 and N_2 and even in non-polar mono-atomic molecules such as He, Ne, Ar, etc. This attraction is evident from the condensation of these gases into liquids at sufficiently high pressures and low temperatures. This is explained by London forces.

London Forces.

In 1930, F. London provided an explanation for the existence of forces of attraction between *non-polar* molecules, based on quantum mechanics. According to his view, *electrons of a neutral molecule keep on oscillating with respect to the nuclei of the atoms. As a result of this, at a given instant, positive charge may be concentrated in one region and negative charge in another region of the same molecule. Thus, a non-polar molecule may become momentarily self-polarised.* This polarised molecule may induce a dipole moment in a neighbouring molecule, as shown in Fig. 1.14.

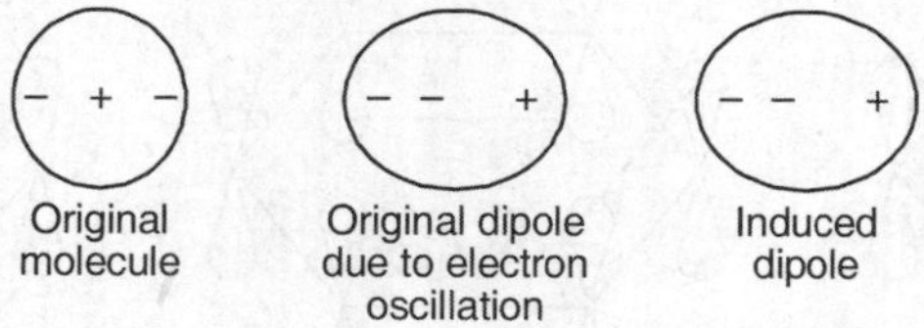

Fig. 1.14. Origin of cohesive forces in non-polar molecules.

The electrostatic forces of attraction between induced dipoles and the original dipoles (*due to electron oscillation*) *are known as* **London forces.** These foreces are also called dispersive forces because the well known phenomenon of dispersion of light is also connected with these dipoles. For a pair of adjacent molecules, London forces vary inversely as the *seventh power* of the distance between them, *i.e.*, $F \propto 1 / r^7$. The magnitude of these forces, evidently, would be very small.

The van der Waals' attraction in *non-polar molecules* is thus *exclusively* due to London forces. In the case of polar molecules having permanent dipoles, however, other electrostatic forces due to dipole-dipole interaction also contribute towards the van der Waals' attraction.

Factors determining van der Waals' forces

Van der Waals' forces depend largely on the following factors:

1. The Number of Electrons present in a Molecule. van der Waals' forces increase with increase in the number of electrons in the molecule. Thus, boiling points (T_b) of the halogens increase as we move from fluorine to chlorine, bromine and iodine. Similarly, boiling points of noble gases increase as we move from helium to radon, as shown below:

Noble gas	He	Ne	Ar	Kr	Xe	Rn
Boiling pt. (K)	4.1	27	67.2	121	165	211.2

2. Molecular Size. The larger the molecular size (or roughly the larger the molar mass), the stronger the van der Waals' forces. Thus, the boiling point of ethane, C_2H_6 (185 K) is higher than that of methane, CH_4 (112 K). In compounds having similar structures, *viz.*, CH_4, SiH_4 and GeH_4, the boiling points increase with increase in the number of electrons and the size of the molecule, as shown below:

Substance	CH_4	SiH_4	GeH_4
No. of electrons	10	18	36
Boiling point (K)	112	161	183

1.7 INCLUSION COMPOUNDS

An inclusion compound is a complex in which one chemical compound (the 'host') form a cavity in which molecules of a second 'guest' compound are located. The definition of inclusion compound is very broad, extending to channels formed between molecules in a crystal lattice in which guest molecules can fit. If the spaces in the host lattice are enclosed on all sides so that the guest species is 'trapped' as in a cage, the compound is known as a clathrate. In molecular encapsulation, a guest molecule is actually trapped inside another molecule.

Cyclodextrin inclusion compounds

Inclusion compounds have been reported between cyclodextrins and ferrocene. When a solution of both compounds in a 2 : 1 ratio in water is boiled for 2 days and then allowed to rest for 10 hours at room temperature, orange-yellow crystals are obtained. X-ray diffraction analysis of these crystals reveal a 4 : 5 inclusion complex with 4 molecules of ferrocene in the cavity of 4 cyclodextrin molecules with the fifth ferrocene molecule sandwiched between two stacks, of ferrocene – cyclodextrin dimers.

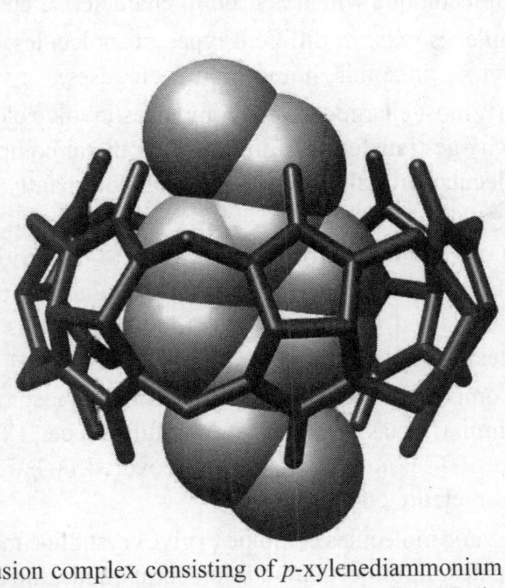

Example of an inclusion complex consisting of *p*-xylenediammonium bound within a cucurbituri reported by Freeman in Acta. Crystallogr. B. in 1984

There is no chemical affinity between the host and guest molecules in the inclusion compounds. Formation of such compounds depends upon the size and shape of the guest molecules and the size of the cavities or channels created in the host molecules.

Urea forms an inclusion compound with straight chain alkanes. Urea crystallises in hexagonal lattice, in the presence of guest molecules and traps them in the cavity. This is represented as under:

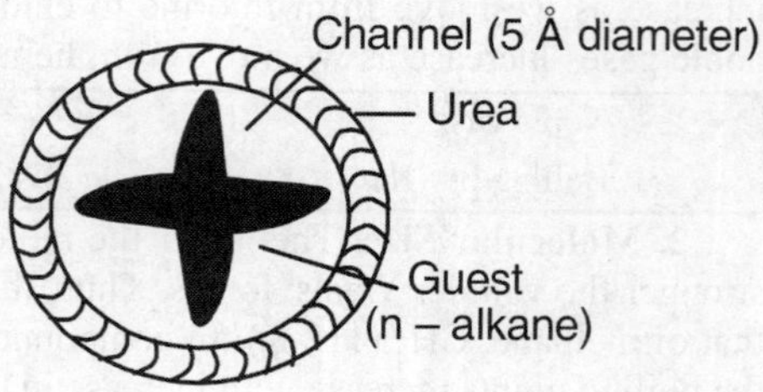

The channels in crystallised urea are of the dimensions 5 A which can trap only the straight chain hydrocarbons. Branched-chain hydrocarbons which are of bigger dimensions do not form inclusion compounds with urea. However thiourea which crystallises with bigger cavities can from inclusion compounds with branched chain hydrocharbon, the dimension of thiourea channel is appropriate in size and shape to trap the molecules of branched alkanes, cycloalkanes and chloroform etc.

1.8 CHARGE TRANSFER COMPLEXES

A charge transfer complex (or CT complex, electron-donor-acceptor-complex is a chemical association of two or more molecules, or of different parts of one very large molecule, in which attraction between the molecules (or parts) occurs due to an **electronic transition** into an excited electronic state, such that a fraction of electron charge is transferred between the molecules. The resulting **electrostatic** attraction stabilizes the moleculer complex. The source molecule from which the charge is transferred is called the **electrons donor** and the receiving molecule is called the **electron acceptor,** hence the alternate name, electron-donor-acceptor-complex.

The attraction in a charge-transfer complex is not equivalent of a stable chemical bond and is much weaker than **covalent forces**. It is better characterised as a weak electron resonance. As a result excitation energy of this resonance occurs very frequently in the visible region of the **electromagnetic spectrum.** This produces the visually intense colours characteristic for these complexes. These optical absorption bands are referred to as **charge-transfer bands** or **CT bands**. Optical spectrocopy is a powerful technique which is used to characterise charge-transfer bands

Charge transfer complexes exist in different types of molecules, organic as well as inorganic and in all phases of matter *i.e.* in solids, liquids and even gases.

In inorganic chemistry, most charge-transfer complexes involve electron transfer between metal atoms and **ligands**. The charge-transfer band in transition metal complexes result from movement of electrons between molecular orbitals (MO) that are predominantly metal in character and those that are predominantly ligand in character.

The classic example of charge-transfer complex is that between iodine and starch to form an intense purple colour. This is widely used for screening counterfeit currency. Unlike most paper, paper used in currency is not sized with starch, thus formation of this purple colour on application of iodine solution indicates a counterfeit.

Coming to organic compounds, in 1962, the well known acceptor tetracyanoquinodimethane (TCNQ) was reported. Similarly classic donor tetrathiafulvalence (TTF) was synthesised in 1970. A CT complex composed of TTF and TCNQ was discovered in 1973. This was the first organic conductor to show almost metallic conductance.

Some electron rich organic molecules combine to give crystalline molecular addition compounds like picrates, trinitrobenzoates. In molecular addition compounds, the donor molecule is capable of donating non bonding or π electrons to the acceptor molecule. The binding force between the donor and acceptor is a kind of weak coulombic force. This is illustrated by taking the example of anthracene and picric acid and toluene and 1, 3, 5 trinitrobenzene forming charge transfer complexes as shown below:

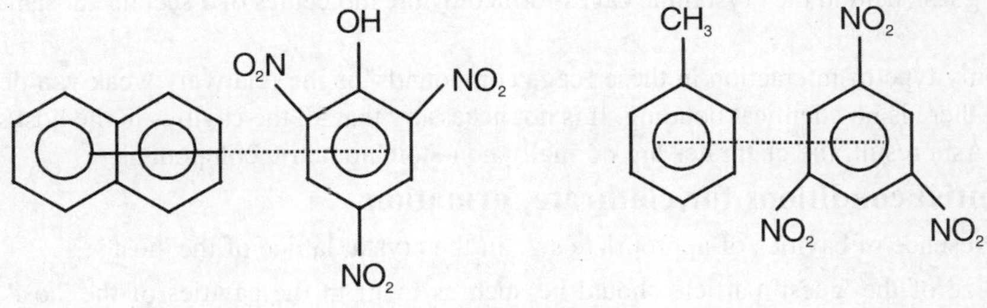

Charge transfer complexes

If we represent the donor by D and acceptor by A, then the addition compound may be represented by the resonance hybrid of the following three structures I, II and III

$$\underset{\text{I}}{(A-D)} \longleftrightarrow \underset{\text{II}}{(\bar{A}-\overset{+}{D})} \longleftrightarrow \underset{\text{III}}{(\overset{+}{A}-\bar{D})}$$

However form I contributes to the actual compound by the greatest extent. The forces that are involved in charge-transfer complexed are van der Waals, dipole-dipole, ion-dipole, dipole- induced dipole forces. The formation of picrate using picric acid and trimethylbenzene (mesitylene) is represented as under

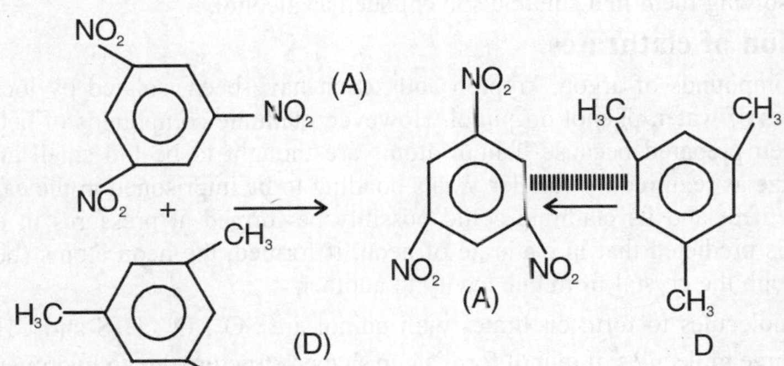

It may be noted that presence of strong electron donating groups in the donor molecule (D) and presence of strong electron withdrawing groups in acceptor molecule (A) favours the formation of charge transfer complexes (C.T.C). The C.T.C as given above has electron donating methyl groups in the donor moelcule and electron withdrawing $-NO_2$ groups in the acceptor molecule.

The distance between donor and acceptor molecules is close to 3Å which is greater than a normal covalent bond length, indicating that the π orbitals of both rings in the complex overlap poorly. That is why the binding forces are not as strong as in a covalent bond. The resonance that occurs in the formation of the C T C is shown as under taking the examples of toluene and trinitrobenzene.

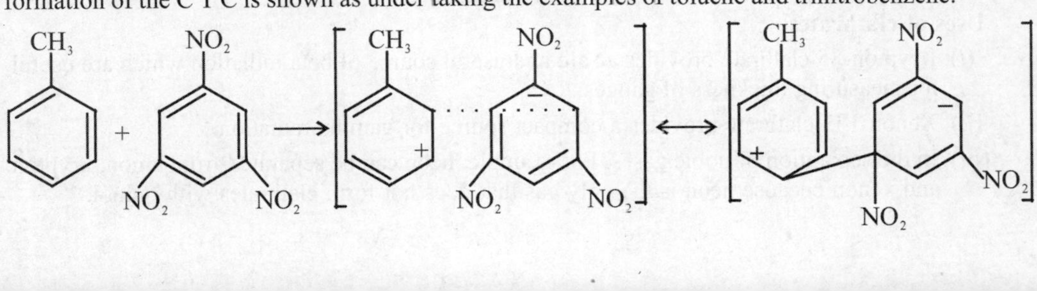

1.9 CLATHRATES

When it comes to inclusion compounds of noble gases, we use the term clathrates.

According to Powell, clathrate compounds consist of atoms or molecules of one substance called the 'guest' held in the crystalline cages formed by the molecules of a second substance called the 'host'.

The only type of interaction in these "cage compounds" is the relatively weak van der Waals' forces and there is no chemical bonding. It is not necessary that all the cavities in the host structure are filled. As a result, the clathrates are normally non-stoichiometric compounds.

Essential conditions for clathrate formation

(*a*) Presence of cavities of appropriate size in the crystal lattice of the 'host'.

(*b*) Size of the 'guest' particle should be such as to fit in the cavities of the 'host' without bringing any atoms closer together than corresponding to the van der Waals' radii of the atoms.

Stability of clathrate compounds

Clathrate compounds once formed are quite stable due to the following reasons:

(*a*) The 'guest' molecules tightly fit in the cavities of the 'host' molecules.

(*b*) The 'guest' molecules within the cages are at a minimum potential energy.

The 'guest' can escape from the 'host' only when the forces holding the molecular cages together are overcome. This can be achieved by the following two methods:

(*i*) By heating the crystals

(*ii*) By dissolving them in a suitable solvent such as alcohol.

Preparation of clathrates

Clathrate compounds of argon, krypton and xenon have been isolated by locking up these gases in the cages of water, phenol or quinol. However, clathrate compounds of helium and neon have not yet been prepared because helium atoms are thought to be too small and difficult to deform or polarize as required by van der Waals bonding to be imprisoned in the cages. The neon is borderline element and its clathrate could possibly be formed at pressures in excess of 160 atmospheres. It is predicted that in clathrate of neon, if formed, the neon atoms (being so small) will diffuse through the crystal from one cavity to another.

The other molecules to form clathrates with quinol are SO_2, O_2, H_2S and CH_3OH. During crystallisation, three molecules of quinol form a cup shaped structure due to hydrogen bonding and two such cups give rise to a cage like structure for the guest molecule. The crystals are quite stable and exist for a pretty long time.

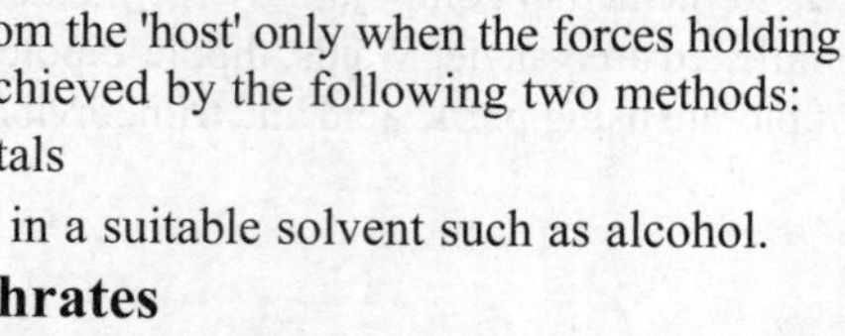

Quinol Clathrate

Uses of clathrates

(*i*) Krypton-85 clathrate provides a safe and useful source of beta radiation which are useful for measuring thickness of gauges.

(*ii*) Xenon-133 clathrate provides a compact source for gamma radiation.

(*iii*) In the separation of noble gases. For example, neon can be separated from argon, krypton and xenon because neon is the only gas that does not form clathrates with quinol.

(*iv*) Clathrates play an important role in some physiological actions, e.g., it is thought that the anaesthetic action of xenon and many other anaesthetics is due to aqueous clathrate in physiologically strategic spots. When the anaesthetic is no longer administered, the clathrate equilibrium is destroyed, the clathrate decomposes and the consciousness returns.

1.10 RESONANCE EFFECT

If two double bonds in a molecule are separated by a single bond, they are said to be in **conjugation** and the molecule having such bonds are called conjugated molecules e.g., 1, 3-butadiene.

$$H_2C{=}CH{-}CH{=}CH_2 \qquad\qquad H_2C{=}CH{-}\ddot{\underset{..}{C}}l{:}$$

1, 3-Butadiene   Vinyl chloride

Another example of conjugation is given by benzene which is a hexagonal ring of carbon atoms with three double bonds in the alternate positions. Alternatively, a **double** bond or a **triple** bond may also be in conjugation with a lone electron pair e.g., in *vinyl chloride.*

The conjugated molecules do not exhibit the character of pure double bond or triple bond, for example, benzene is expected to be highly reactive since it has three double bonds in the ring but is actually quite stable. In order to explain the difference in the expected and the actual behaviour of the conjugated molecules, *Robinson* and *Ingold* stated that such compounds exist in two or more forms none of which can explain all the properties of the molecule under investigation. They called this concept as **mesomerism** or **mesomeric effect.** *Heisenberg* studied the same mathematically and named it as **resonance** or **resonance effect.** According to the concept,

If a compound having a certain molecular formula can be represented by different structural formulae which differ only in the arrangement of the electron pairs and not of the atoms, such like structures are called **resonating** or **contributing** or **canonical structures** *and the phenomenon is known as* **resonance.** *The compound cannot be represented completely by any of the resonating or contributing structures but by a mixture of all of them which is called* **resonance hybrid** (cannot be actually represented). Resonance is indicated by the sign $\longleftrightarrow$.

It may be noted that the various resonating or contributing structures do not actually exist i.e., they are all hypothetical structures. They have been given simply to explain certain properties of the compounds which cannot be otherwise explained.

Examples of Resonance

A common analogy of a resonance hybrid is that of a mule which is a hybrid of horse and donkey.

1. Monocarboxylic acid. The acidic character of monocarboxylic acid is explained with the help of resonance. The following contributing structures for the monocarboxylic acid are possible which differ in the position of the electron pairs.

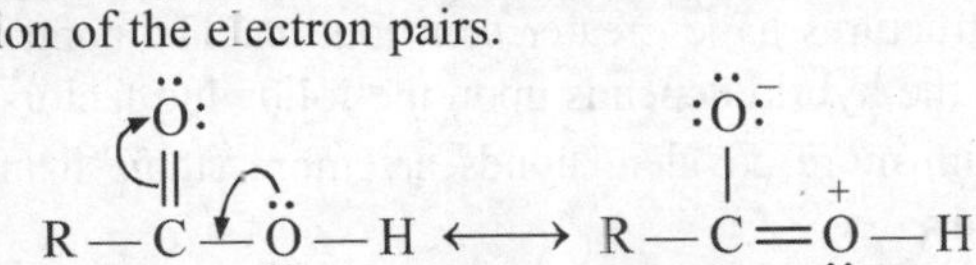

As a result of resonance, the oxygen atom of the O — H group acquires a positive charge *i.e.*, it draws the electrons pair towards itself resulting in the release of protons.

2. Benzene. Benzene is a hybrid of two equivalent contributing structures (I and II) which differ in the position of the π-electron pairs representing double bonds. These were suggested by Kekule.

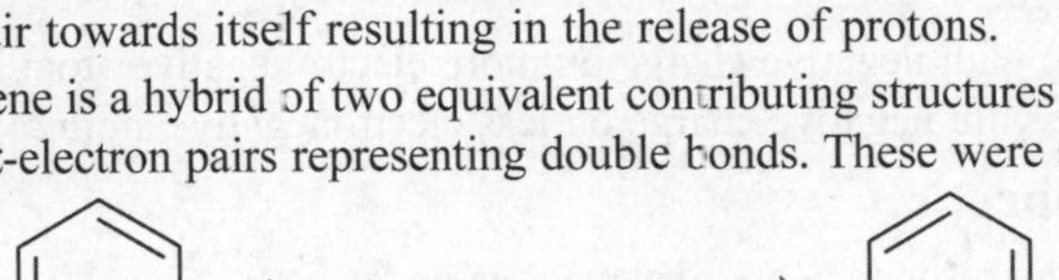

I   II

Three more contributing structures (III, IV and V) have been proposed by **Dewar**. Being less symmetrical they have small contribution (28% only) towards the hybrid. Benzene exists mainly in the form of **Kekule structures** (I and II). Benzene is, in fact, quite stable and its stability is explained

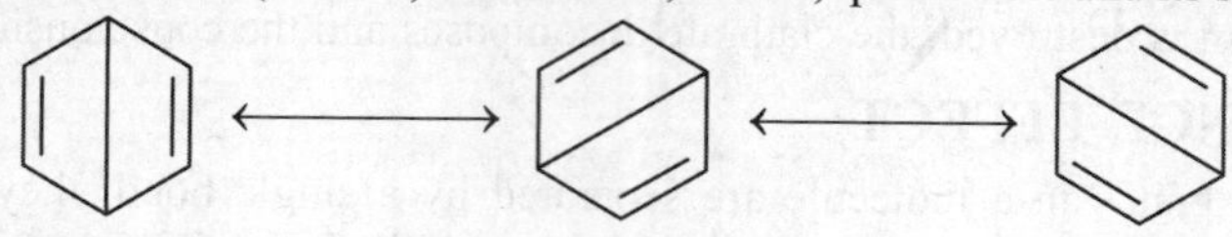

III IV V

with the help of resonance. As the π-electron charge is distributed over greater area *i.e.*, it gets delocalised.

3. 1, 3- Butadiene. The diene can be regarded as the hybrid of the following contributing structures :

$$\underset{\text{(I)}}{CH_2{=}CH-CH{=}CH_2} \longleftrightarrow \underset{\text{(II)}}{\overset{\oplus}{C}H_2-CH{=}CH_2-\overset{\ominus}{C}H_2} \longleftrightarrow \underset{\text{(III)}}{\overset{\ominus}{C}H_2-CH{=}CH-\overset{\oplus}{C}H_2}$$

The charged structures II and III make lesser contribution towards the hybrid. However, they do explain the 1, 4-addition in the conjugated dienes.

4. Carbon dioxide. The structural formula of carbon dioxide (CO_2) molecule is O = C = O. The standard C = O bond length is 122 pm but the bond length which is obtained for the molecule from spectroscopic studies is 115 pm. The difference in the bond length values can be explained by considering the following structures for carbon dioxide which are known as the *resonating structures,* or *contributing structures* or *canonical structures.*

$$\underset{\text{(I)}}{:\ddot{O}{=}C{=}\ddot{O}:} \qquad \underset{\text{(II)}}{:\overset{-}{\ddot{\underset{..}{O}}}-C{\equiv}\overset{+}{O}:} \qquad \underset{\text{(III)}}{:\overset{+}{O}{\equiv}C-\overset{-}{\ddot{\underset{..}{O}}}:}$$

The carbon-oxygen bond length is the mean of all the bond length values. Carbon dioxide cannot be represented by any of the contributing structures but by an average of all. Carbon dioxide is said to be a resonance hybrid of these structures.

Conditions of resonance

Wheland has suggested the following conditions for resonance :

(*i*) The resonance or the contributing structures must differ only in the position of the electron pairs and not of the atomic nuclei.

(*ii*) The resonating structures must have the same number of paired and unpaired electrons.

(*iii*) The energies of the various resonating structures must be either same or nearly the same.

(*iv*) All the contributing or resonating structures do not contribute equally towards the hybrid. The equivalent structures have greater contribution. The contribution of any resonating structure towards the hybrid depends upon the following factors :

(*a*) Structures with more covalent bonds are more stable than the structures with less covalent bonds.

(*b*) The charged contributing structure is less stable as compared to the structure without any charge.

(*c*) Structure with negative charge on more electronegative atom is more stable than the structure with negative charge on less electronegative atom.

Effects of resonance

Some effects of the resonance are explained as under :

1. Stability. As a result of resonance, the energy of the hybrid decreases and its stability, therefore, increases. Greater the number of contributing structures, greater will be the stability of the hybrid. The stability also depends upon the equivalence of the contributing structures. Moreover, the charged structures have less contribution than the uncharged structures.

2. Resonance energy. The relative stabilities of two different resonance hybrids may be compared in terms of **resonance energy.** *It may be defined as the difference in the energy content of the hybrid and its most stable contributing structure.* It may be expressed either as the difference in the heat of combustion of the contributing structures and the hybrid or in terms of the difference in the heat of hydrogenation values. This is discussed as follows :

(*a*) Heat of combustion value. In benzene, the heat of combustion of the contributing structure is calculated by determining the values for different bonds mathematically.

Heat of combustion of 6 C — H bonds = 6 × 225.5 = 1355 kJ / mole
Heat of combustion of 3 C — C bonds = 3 × 206.0 = 618 kJ / mole
Heat of combustion of 3 C == C bonds = 3 × 491.0 = 1473 kJ / mole
3446 kJ / mole.

The value as determined experimentally for the hybrid is 3300 kJ / mole. Thus, the resonance energy for benzene is (3446 – 3300) kJ or **146 kJ / mole**.

Similarly, resonance energies for some typical molecules are as follows :

1, 3-Pentadiene (17.6 kJ / mole); 1, 3-Butadiene (14.6 kJ/mole) and 1, 3-Cyclohexadiene (7.5 kJ / mole).

(*b*) Heat of hydrogenation value. In benzene, the expected heat of hydrogenation value is 376 kJ/mole. But the value for the hybrid as determined experimentally is 230 kJ/mole *i.e.* the resonance energy is the same *i.e.* (376 – 230) kJ or (146 kJ / mole).

The relative stabilities of two hybrids can be compared in terms of the resonance energy value.

3. Bond length values. The resonance causes a change in bond length values. For example, the standard C — C bond length is 154 pm and C == C bond length is 134 pm. But the value as determined for benzene is 139 pm which does not coincide with any of the two values and is intermediate between the two values.

Types of resonance or mesomeric effect

Depending upon the nature of the functional group present adjacent to the multiple bond, the resonance effect (R) or mesomeric effect (M) are of two types :

(*i*) Groups having electron-withdrawing resonance effect. Groups such as — NO_2, > C == O, — C ≡≡ N, — COOH etc. tend to withdraw the electrons from the multiple bond through resonance and are said to have – M or – R effect. For example, in the following compound, the aldehyde group (— CH == O) has – R effect.

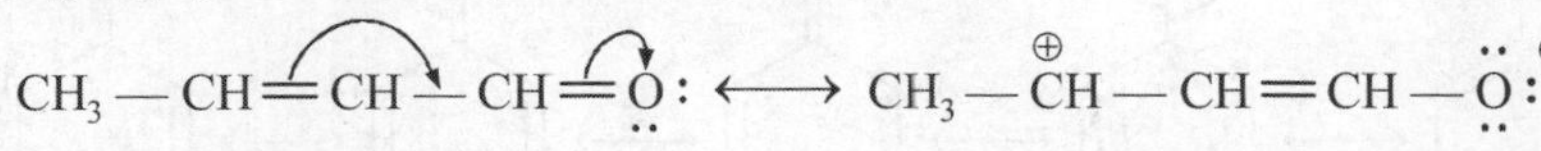

(*ii*) Groups having electron-releasing resonance effect. Groups such as — $\ddot{N}H_2$, — $\ddot{N}HR$, — $\ddot{N}R_2$, — $\ddot{O}$ — H, — $\ddot{O}$ — R etc. which can release electrons through resonance are said to have + M or + R effect. For example, the amine group (– NH_2) in the following compound has + R effect

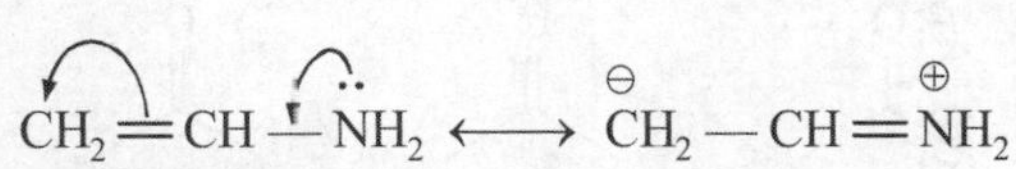

Applications of resonance effect

Resonance effect or mesomeric effect is quite useful in explaining many observations in organic compounds. A few typical applications are being listed as follows :

1. Explanation for the acidic character of carboxylic acids. The acidic character of caboxylic acids is due to release of H^+ ion in aqueous solution.

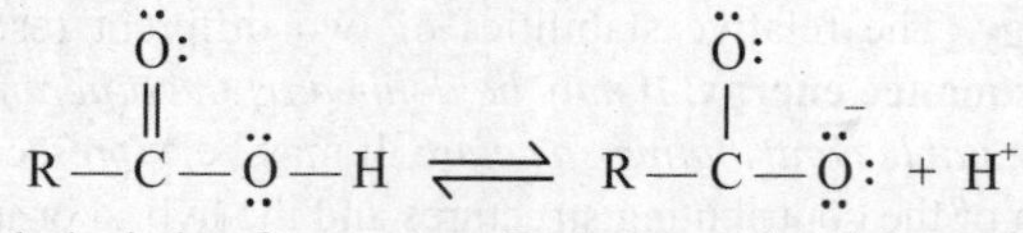

This is explained with the help of resonance. Both the acid and the carboxylate ion are resonance stabilised as shown below :

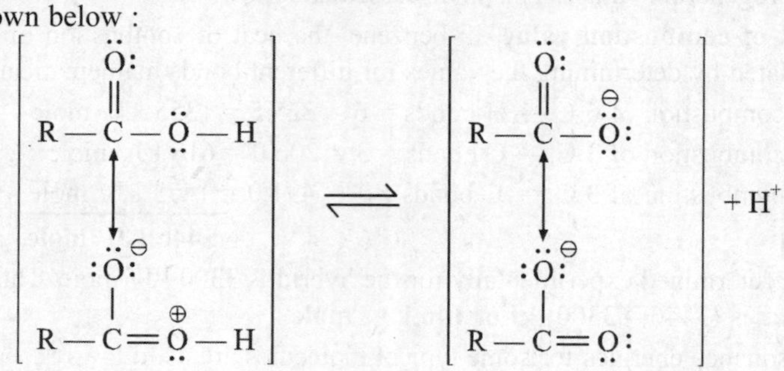

Acid   Carboxylate ion

While the contributing structures for the carboxylate ion are exactly equivalent, they are not so for the acid. *Therefore, hybrid for the ion is more stable than hybrid for the acid. The acid changes to a more stable ion by releasing a proton and this accounts for the acidic character of carboxylic acid.*

2. Explanation for acidic character of phenols. The acidic nature of phenols is due to the release of H^+ ion in aqueous solution.

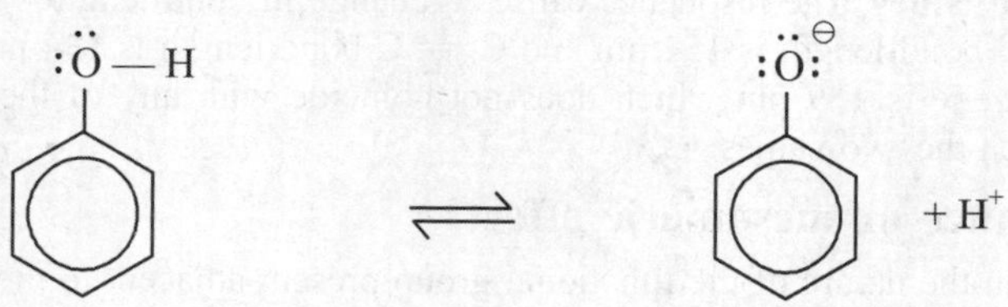

Phenol   Phenate ion

The acidic character in phenol is exhibited on the basis of resonance both in phenol and phenate ion. They exist as hybrid of a number of contributing structures as follows :

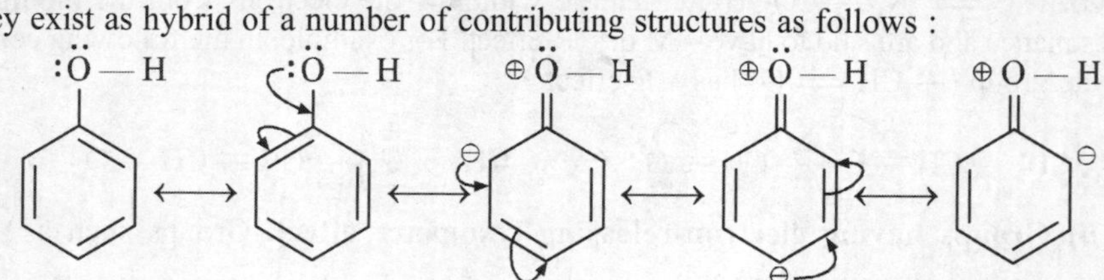

I   II   III   IV   V

Resonating structures of phenol

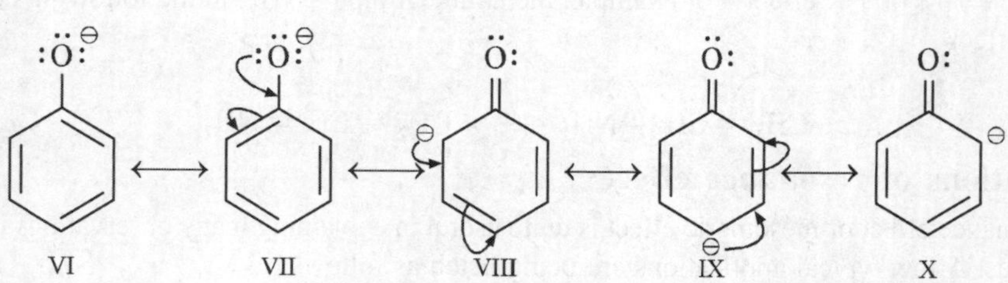

VI   VII   VIII   IX   X

Resonating structures for phenate ion

In case of phenol, contributing structures (III), (IV) and (V) involve charge separation i.e. they have both positive and negative charges. Energy is needed to separate the opposite charges. But no such structures are noticed in case of phenate ion. Therefore, the hybrid for the phenate ion is more stable and phenol changes to phenate ion by releasing proton. This accounts for the acidic character of phenol.

3. Comparison of relative basic strength of ethyl amine and acetamide. Ethyl amine is a stronger base than acetamide. In ethyl amine, the ethyl group with + I effect increases the electron density on the nitrogen atom. As a result, its electron releasing tendency is more. In acetamide the carbonyl group is an electron withdrawing group and takes away the electron pair from the nitrogen atom because of resonance as shown below :

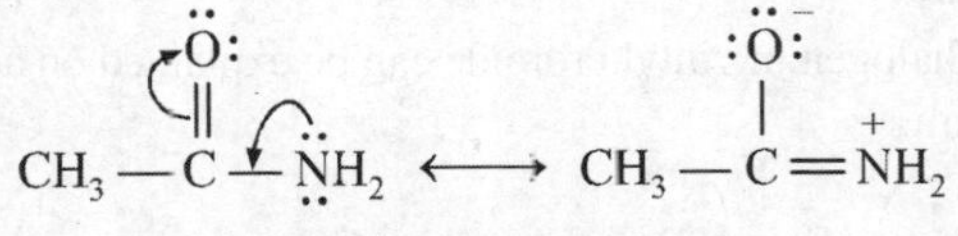

Resonance hybrids of acetamide

4. Comparison of relative basic strength of ethyl amine and aniline. Ethyl amine ($K_b = 5.6 \times 10^{-4}$) is much more basic than aniline ($K_b = 3.8 \times 10^{-10}$). In the former, the ethyl group with + I

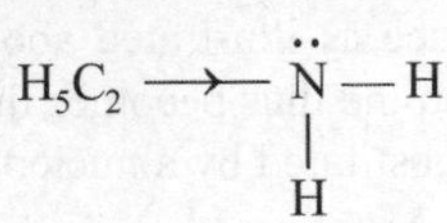

Ethyl amine

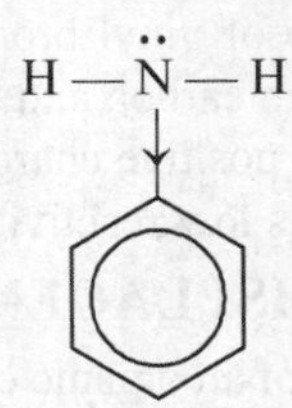

Aniline

effect increases the electron density on the nitrogen atom and, thus, increases electron releasing tendency or basic character. In the latter, the phenyl group with – I effect tends to decrease the electron density on the nitrogen atom resulting in a decreased basic strength. Aniline is a hybrid of following contributing structures :

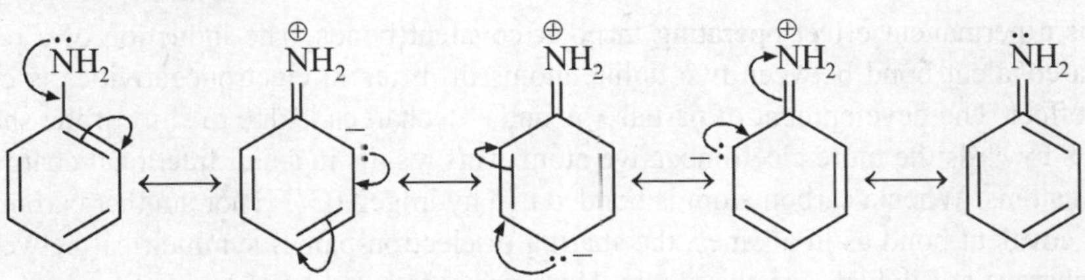

The electron density on the nitrogen atom decreases, consequently, the electron releasing tendency or basic strength decreases.

Example 1. **Which of the following canonical forms would contribute most towards resonance? Explain.**

$$\underset{(a)}{CH_2{=}CH{-}CH{=}\overset{-}{C}H_2} \longleftrightarrow \underset{(b)}{\overset{-}{C}H_2{-}CH{=}CH{-}\overset{+}{C}H_2}$$
$$\longleftrightarrow \underset{(c)}{\overset{-}{C}H_2{-}\overset{+}{C}H{-}CH{=}CH_2}$$

Solution. The resonance structures fulfilling the following conditions are more stable.

1. Greater number of covalent bonds
2. Less separation of charges

3. Negative charge on electronegative atom
4. Positive charge on electropositive atom
5. More dispersal of charges

Structure (*b*) takes care of dispersal of charge.

Hence, it is more stable and contributes more towards resonance.

Example 2. **By what electronic effect can you explain the low reactivity of halogen atom in vinyl bromide.**

or

On the basis of resonance, how will you explain low reactivity of vinyl bromide as compared to ethyl bromide?

Solution. Low reactivity of halogen in vinlyl bromide can be explained on account of the phenomenon of resonance or mesomerism.

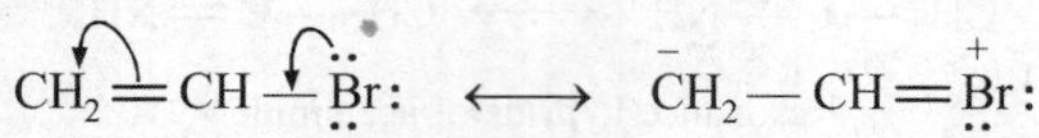

Halogen compounds generally give nucleophilic substitution reactions, in which the halogen is removed as halide ion.

But in the case of vinyl bromide, resonance takes place as illustrated above. This creates a double bond between carbon and bromine. Removal of bromine thus becomes difficult. Moreover, bromine, acquires a positive charge and hence cannot be substituted by a nucleophile. That is why vinyl bromide shows low reactivity.

ELECTRON DISPLACEMENT IN ORGANIC COMPOUNDS

The behaviour of an organic compound is influenced by the electron displacement taking place in its covalent bonds. These displacements may be permanent or temporary which take place in presence of another species in the molecule. The acidity and basicity of organic compounds, their stability, their reactivity towards other substances, etc. can be predicted based on such electronic displacements.

1.11 INDUCTIVE EFFECT

This is a permanent effect operating in polar covalent bonds. The induction of a permanent dipole in a covalent bond between two unlike atoms of different electronegativities is called the inductive effect. The development of partial +ve and –ve charges is due to shift of the shared pair of electrons towards the more electronegative atom. This results in small fractional charges on the constituent atoms. When a carbon atom is bonded to a hydrogen (C – H) or another carbon (C – C) atom by a covalent bond as in alkanes, the sharing of electron-pair is symmetrical between them. Thus, no charges are induced on the atoms. However, when carbon is bonded to a halogen (X), charges are created.

$-C-H$ or $-C:H$ (Symmetrical sharing of electron pair) Electronegativities of C and H are almost the same.

$-C-C$ or $-C:C-$ (Symmetrical sharing of electron pair)

$-C:X$ or $-\overset{\delta^+}{C}-X^{\delta^-}$ (Fractional charges due to greater electronegativity of X)

The direction of displacement is shown by placing an arrow head midway along the line representing the sigma bond.

$$\overset{}{C_4} \rightarrow C_3 \rightarrow C_2 \rightarrow \overset{\delta^+}{C_1} \rightarrow \overset{\delta^-}{Cl}$$

The inductive effect of an atom or a group of atoms diminishes rapidly with distance. Infact, the inductive effect is almost negligible beyond two carbon atoms from the active atom or group.

Inductive effect does not involve actual transfer of electrons from one atom to another but simply helps in displacing them permanently.

Groups with –I effect:

$$NO_2 > CN > F > COOH > Cl > Br > I > OCH_3 > C_6H_5$$

Decreasing order of – I effect ⟶

Groups with + I effect :

$$(CH_3)_3\,C > (CH_3)_2\,CH > C_2H_5 > CH_3 > -H$$

Decreasing order of + I effect ⟶

Applications

***(i)* Reactivity of alkyl halides :** The presence of halogen atoms in the molecule of alkyl halide creates a centre of low electron density on adjacent carbon which is readily attacked by the negatively charged reagents.

$$CH_3 \rightarrow Cl \quad \text{or} \quad \overset{\delta^+}{CH_3} - \overset{\delta^-}{Cl}$$

↑ Site for the attack of electron rich reagent

***(ii)* Dipole moment :** As the inductive effect increases, the dipole moment increases. This is because the dipole moment is the product of distance and charge.

$CH_3 \rightarrow I$	$CH_3 \rightarrow Br$	$CH_3 \rightarrow Cl$
1.64 D	1.79 D	1.83 D

Inductive effect increases ⟶

***(iii)* *(a)* Acidity of acetic acid and halogen substituted acetic acid**

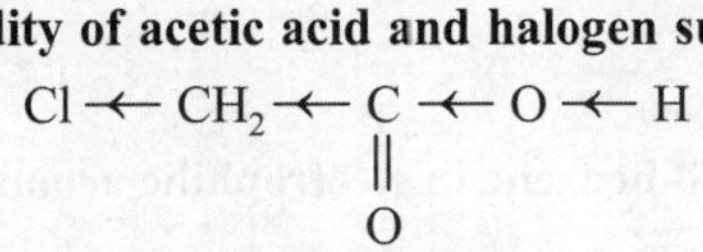

Chloroacetic acid
$pK_a = 2.86$

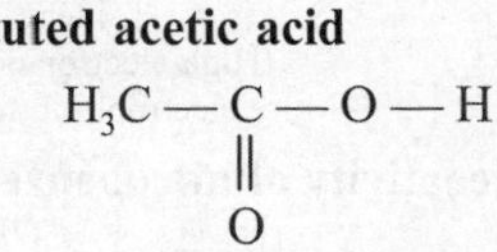

Acetic acid
$pK_a = 4.76$

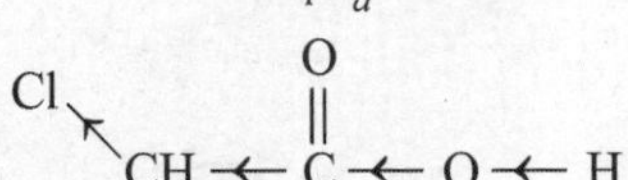

Dichloroacetic acid
$pK_a = 1.25$

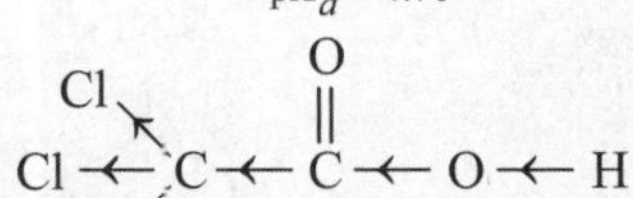

Trichloroacetic acid
$pK_a = 0.65$

The decreasing order of acid strength

$$Cl_3CCOOH > Cl_2CHCOOH > ClCH_2COOH > CH_3COOH$$

On the other hand, electron withdrawing –I effect decreases electron availability on carbonyl carbon and therefore increases the rate of nucleophilic addition.

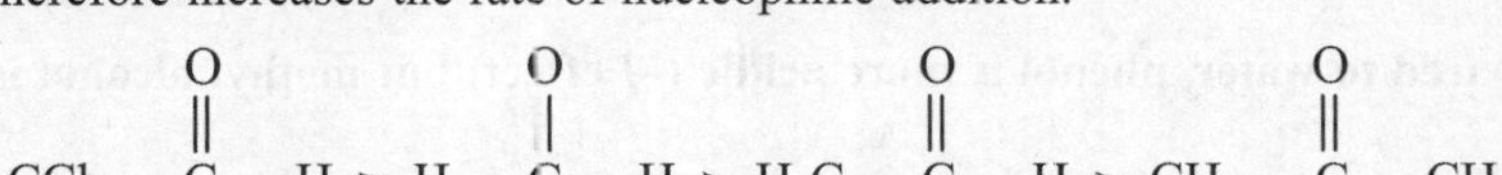

Decreasing order of nucleophilic addition ⟶

(b) Relative acid strength of formic acid and acetic acid :

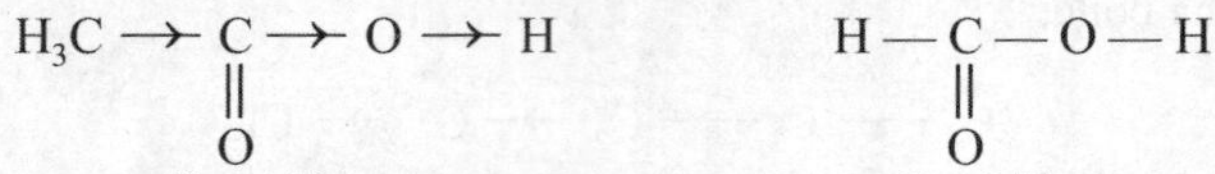

$pK_a = 4.79$ $pK_a = 3.77$

Methyl group has an electron-releasing inductive effect (+*I* effect). It reduces the release of protons. Therefore acetic acid is a weaker acid than formic acid.

(c) Relative acid strength of fluoroacetic acid, chloroacetic acid, bromoacetic acid and iodoacetic acid

Halogenated acids are much stronger acids than the parent acid and the acidity increases with the increase in electronegativity of the halogen present which helps in the release of protons. Thus, the strength of halogenated acids follows the order :

$$FCH_2COOH > ClCH_2COOH > BrCH_2COOH > ICH_2COOH$$

Inductive effect decreases with increase in distance of halogen atom from the carboxylic group, hence the strength of the acid is proportionally decreased. Thus,

$$\underset{\alpha\text{-Chlorobutyric acid}}{CH_3CH_2CH(Cl)COOH} > \underset{\beta\text{-Chlorobutyric acid}}{CH_3CH(Cl)CH_2COOH} > \underset{\gamma\text{-Chlorobutyric acid}}{CH_2ClCH_2CH_2COOH} > \underset{n\text{-Butyric acid}}{CH_3CH_2CH_2COOH}$$

(iv) (a) Relative reactivity of toluene and benzene in aromatic substitution reactions

Aromatic substitution reactions are electrophilic in nature. Methyl has an electron-releasing inductive effect (+*I* effect). Therefore, toluene with higher electron density than benzene has greater reactivity in electrophilic substitution reactions.

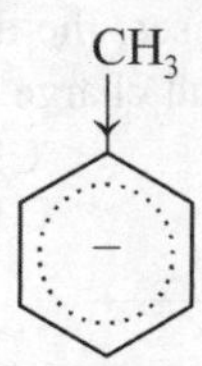

Toluene
(High electron density)
(More electrophilic substitution)

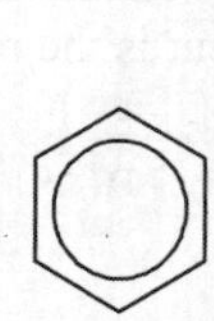

Benzene

(b) Relative reactivity of nitrobenzene and benzene in electrophilic aromatic substitution reactions

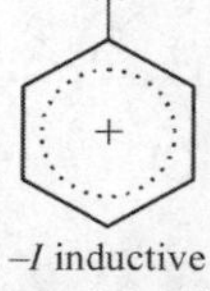

NO_2 group has –*I* inductive effect. It draws electrons from the benzene ring and decreases electron density on the ring
(Less reactive towards electrophilic substitution reactions)

Benzene

(v) Compared to water, phenol is more acidic (–*I* effect) but methyl alcohol is less acidic (+*I* effect)

This is because phenyl group has –*I* effect while methyl group has +*I* effect.

 $\leftarrow$ OH > H — OH > $CH_3 \rightarrow$ OH

Phenol   Water   Methyl alcohol

(*vi*) Inductive effect is dependent on state of hybridisation of the atoms linked by covalent bond

Relative acidity of hydrocarbons may be given as

$$\underset{sp}{HC \equiv CH} > \underset{sp^2}{CH_2 = CH_2} > \underset{sp^3}{CH_3 - CH_3}$$

Electron density on *s*-orbital is more than on, *p*-orbital. Percentage of *s* character decrease from left to right.

Relative basicity of corresponding carbanions is as :

$$CH_3 - \bar{C}H_2 > CH_2 = \bar{C}H > CH \equiv \bar{C}$$

(*vii*) Strength of base : Base strength is defined as the tendency to donate an electron pair for sharing. Strength of a base can be explained on the basis of inductive effect.

(*a*) As compared to ammonia, methyl amine is more basic (+*I* effect) but aniline is less basic and diphenyl amine is a still weaker base (–*I* effect).

$$CH_3NH_2 > NH_3 > C_6H_5NH_2 > (C_6H_5)_2NH$$

(*b*) The decreasing order of base strength in alcohols from tertiary to primary is due to +I effect of alkyl groups.

$$\underset{3°}{(CH_3)_3COH} > \underset{2°}{(CH_3)_2CHOH} > \underset{1°}{CH_3CH_2OH} > \underset{methyl}{CH_3OH}$$

(*c*) Greater the tendency to donate electron pair for coordination with proton, more is the basic nature, or more the negative charge on nitrogen atom (due to +*I* effect of alkyl group), higher is the basic strength.

Thus, the basic nature decreases in the order,

Alkyl group	Relative basic strength
CH_3	$R_2NH > RNH_2 > R_3N > NH_3$
C_2H_5	$R_2NH > RNH_2 > NH_3 > R_3N$
$(CH_3)_2CH$	$RNH_2 > NH_3 > R_2NH > R_3N$
$(CH_3)_3C$	$NH_3 > RNH_2 > R_2NH > R_3N$

(*viii*) Relative stabilities of carbocations :

Greater the number of alkyl groups attached to positive carbon, greater the dispersal of charge and hence greater is the stability.

$$(CH_3)_3\overset{\oplus}{C} > (CH_3)_2\overset{\oplus}{C}H > (CH_3)\overset{\oplus}{C}H_2 > \overset{\oplus}{C}H_3$$

(*ix*) +*I* group increases electron availability on carbonyl carbon :

This therefore decreases the rate of nucleophilic addition.

1.12 FIELD EFFECT

Inductive effect is a permanent effect in the ground state of the molecule and usually operates through single bonds. However, when the inductive effect is transmitted through space or solvent molecules, it is called as field effect. Consider the two isomers I and II of the following molecule:

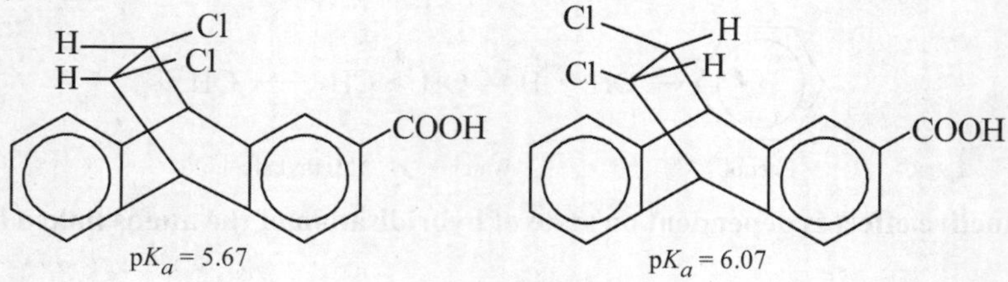

pK_a = 5.67 I   pK_a = 6.07 II

The two chlorine atoms in I and II exert the same inductive effect with respect to the position of electrons associated with the – COOH group and therefore the two compounds should exhibit the same acid strength. However, the two chlorine atoms in the molecule I are closer to the acid group as compared to II, and therefore will exert greater electron withdrawing effect in I as compared to II. Hence the isomer I is a stronger acid than II.

1.13 ELECTROMERIC EFFECT

This is a temporary effect operating in unsaturated compounds only at the demand of a reagent. As soon as this attacking reagent is removed, the original condition is restored.

It involves the complete transfer of π-electrons of multiple bond, because π-bonds are loosely held and are easily polarisable.

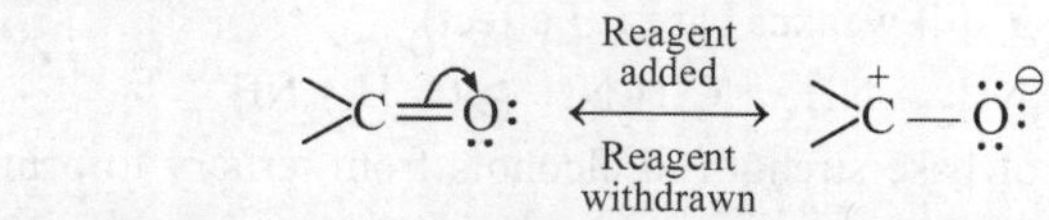

The complete transfer of shared pair of π-electrons of a multiple bond to the more electronegative atom of the bonded atoms at the requirement of an attacking reagent is called electromeric effect (*E*-effect).

When the transfer of π-electrons takes place towards the attacking reagent (electrophile), the effect is called +*E*-effect.

For example,

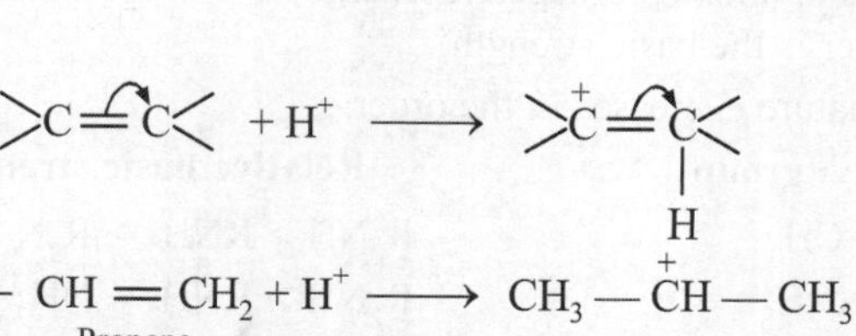

$$CH_3 \rightarrow CH=CH_2 + H^+ \longrightarrow CH_3-\overset{+}{C}H-CH_3$$

Propene

When the transfer of electrons takes place away from the attacking reagent, the effect is called –*E* effect, For example,

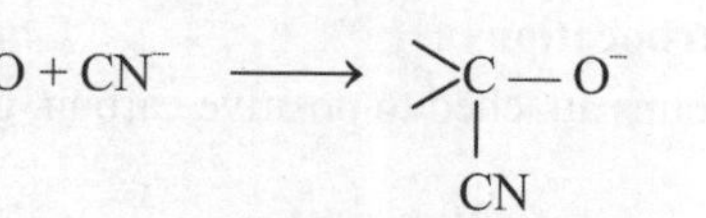

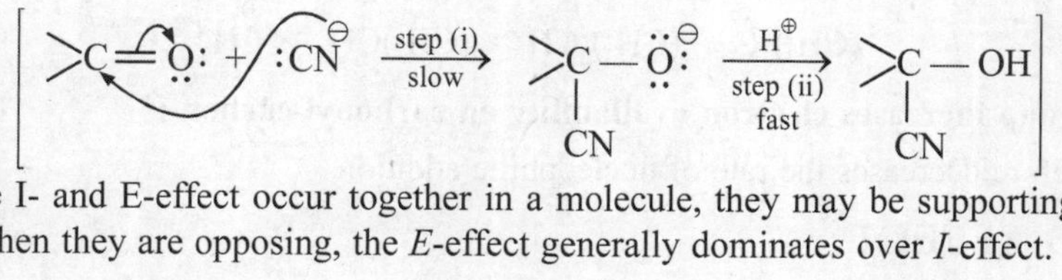

When the I- and E-effect occur together in a molecule, they may be supporting or opposing each other. When they are opposing, the *E*-effect generally dominates over *I*-effect.

Applications

(*i*) Electrophilic addition reactions of unsaturated compounds involve electromeric effect or the polarisation of the carbon-carbon double bond in the presence of attacking electrophiles like H^+.

(*ii*) Nucleophilic addition reactions of carbonyl compounds involve electromeric effect or polarisation through $-E$ electromeric effect of the carbon-oxygen double bond in the presence of a nucleophile.

(*iii*) Electrophilic substitution reactions of benzenoids involve polarisation through electromeric effect of the benzene ring when an electrophile approaches them. For example,

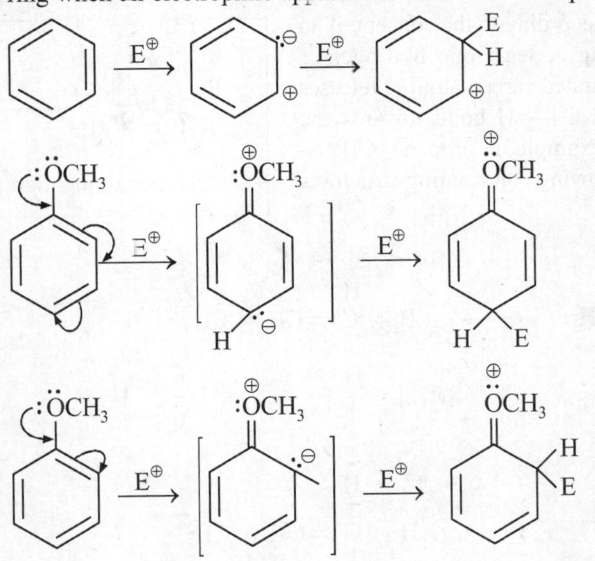

Comparison of Inductive and Electromeric Effects

Inductive effect	Electromeric effect
1. It involves a permanent displacement of the electron pairs in the molecule.	It involves a temporary transference of the electron pair towards more electronegative atom in the molecule.
2. The displaced electron pair does not leave its molecular orbital. There is only a distortion in shape of the molecular orbital.	The electron pair which gets transferred, completely leaves its molecular orbital and takes up a new position.
3. The presence of the outside attacking reagent is not needed.	The presence of the outside attacking reagent is essential.
4. There is a partial separation of the charges.	There is a complete separation of the charges.
5. No ions are formed.	Ions are formed.
6. The presence of multiple bond is not essential.	The presence of multiple bond is essential.

Comparison of Resonance and Inductive Effects

Resonance Effect	Inductive Effect
1. It occurs in conjugated systems.	It occurs in saturated compounds.
2. It involves the transference of the π-electrons.	It involves the displacement of the σ-electrons.
3. It involves delocalisation of electrons.	It does not involve any delocalisation of electrons.
4. It does not undergo any change in charge density with distance.	It decreases as we move away from the atom involved in the initial polar bond.

1.14 HYPERCONJUGATION OR NO BOND RESONANCE

In the study of inductive effect, we have observed that the alkyl group has +I effect and the order of +I effects of different alkyl groups follow the order :

$$-CH_3 < -C_2H_5 < -CH(CH_3)_2 < -C(CH_3)_3$$

But when the alkyl group is attached to an unsaturated system such as — $CH = CH_2$ group, the order of the inductive effect simply gets reversed. This anomaly in the electron releasing tendency of the alkyl groups is explained on the basis of **hyperconjugation** which was developed by *Baker* and *Nathan.* It is also called *Baker-Nathan effect.*

Explanation. According to the concept, if an alkyl group carrying at least one hydrogen is attached to an unsaturated carbon atom, it releases the electrons of the C — H bond towards the multiple bond. For example, in propene (CH_3 — $CH = CH_2$), the following contributing structures are involved.

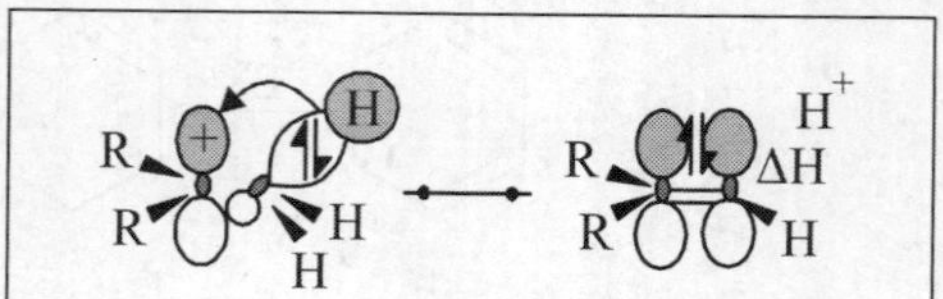

Hyperconjugation phenomenon

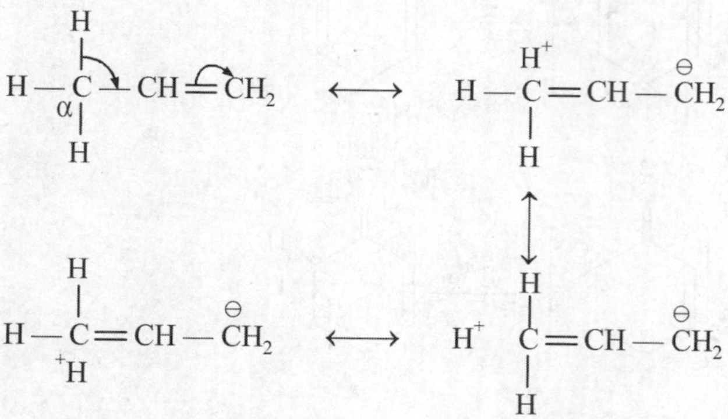

Since there is no bond between the α-carbon atom and one of the hydrogen atoms, the hyper conjugation is also called **no-bond resonance.** Now, hyperconjugation due to CH_3— group is expected to be more than due to CH_3CH_2 — group since the latter has only two hydrogen atoms attached to the α-carbon atom directly linked to the double bonded carbon atom. Therefore, the number of contributing structures for ethyl group will be less. Similarly, the electron releasing tendency in $(CH_3)_2CH$ — group will be still less as it has only one hydrogen atom attached to the α-carbon atom and this tendency will be the least when the $(CH_3)_3C$ — group is attached to the $CH = CH_2$ group. Thus, the order of the electron releasing tendency or inductive effect of the alkyl groups get reversed because of the phenomenon of hyperconjugation.

Orbital concept of hyperconjugation

The orbital concept of hyperconjugation is illustrated with the help of propene ($H_3C — CH = CH_2$). As mentioned earlier the electron pair of C — H bond (σ-bond) is involved in conjugation with the π-electron pair of the double bond. This may be shown with the help of Fig. 1.15.

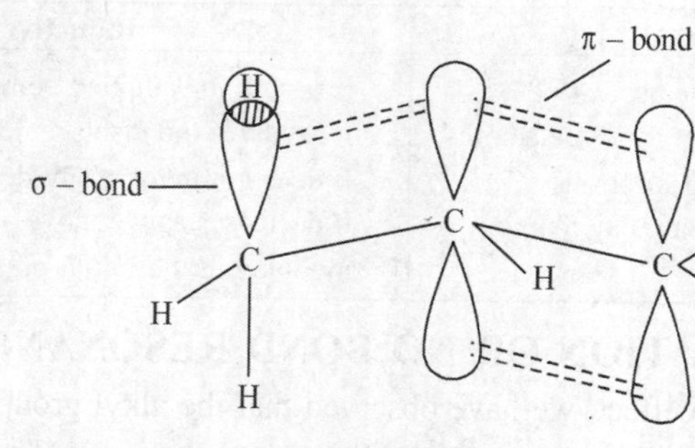

Fig. 1.15. Hyperconjugation in propene.

The σ — π conjugation is called hyperconjugation. Though hydrogen atom of the methyl group is not forming any real bond with the double bonded carbon atom but it is not actually free to leave its original position. From the orbital structure, it is evident that there is interaction between the electrons of the C — H bond and the π-electrons of the adjoining double bond. This will bring about a change in the bond lengths of C — H and C = C bonds. This has been supported by the experimental determination of the bond lengths.

Applications of hyperconjugation

The concept of hyperconjugation is quite useful in explaining relative stabilities of a number of species such as alkyl free radicals, carbocations and alkenes. These are briefly discussed as follows :

1. Relative Stabilities of Alkyl Free Radicals

The order of the relative stabilities of different alkyl free radicals is :

$$(CH_3)_3\dot{C} > (CH_3)_2\dot{C}H > CH_3\dot{C}H_2 > \dot{C}H_3$$

Tert-butyl	Isopropyl	Ethyl	Methyl
(3°)	(2°)	(1°)	(1°)

Ethyl free radical is regarded as a hybrid of four contributing structures.

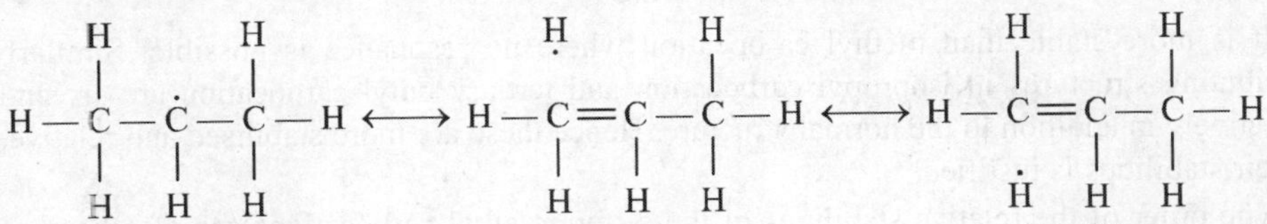

(I) (II) (III)

But there is no possibily of any hyperconjugation in methyl free radical. Therefore, ethyl free radical is more stable than the methyl free radical.

In isopropyl free radical, there are two methyl groups attached to the carbon atom which has single electron. As a result, six contributing structures in addition to normal structure are possible as shown below :

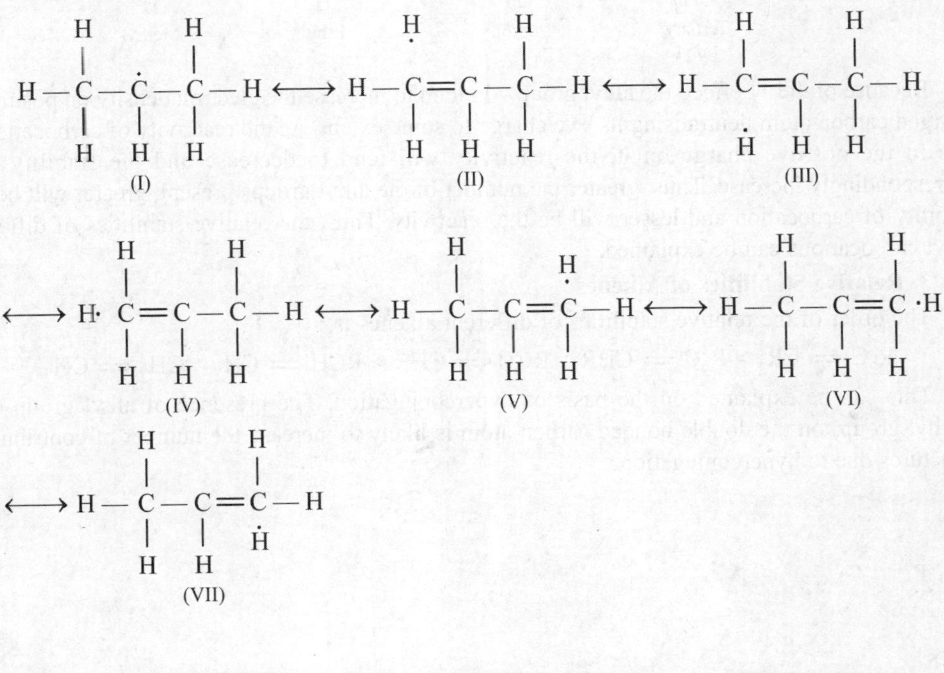

(I) (II) (III) (IV) (V) (VI) (VII)

Similarly, *nine contributing structures* are possible for the tertiary butyl free radical in addition to its normal structure. Therefore, it is still more stable and thus, the order of the relative stabilities of the different alkyl free radicals can be explained.

2. Relative Stabilities of Alkyl Carbocations

The order of relative stabilities of the alkyl carbocations is

$$\underset{\text{Tert-butyl}}{(CH_3)_3C^{\oplus}} > \underset{\text{Isopropyl}}{(CH_3)_2CH^{\oplus}} > \underset{\text{Ethyl}}{CH_3\overset{\oplus}{C}H_2} > \underset{\text{Methyl}}{\overset{\oplus}{C}H_3}$$

and the order is the same as in case of free radicals. This can also be explained with the help of hyperconjugation. The ethyl carbocation (primary) has three contributing structures in addition to its normal structure based upon hyperconjugation.

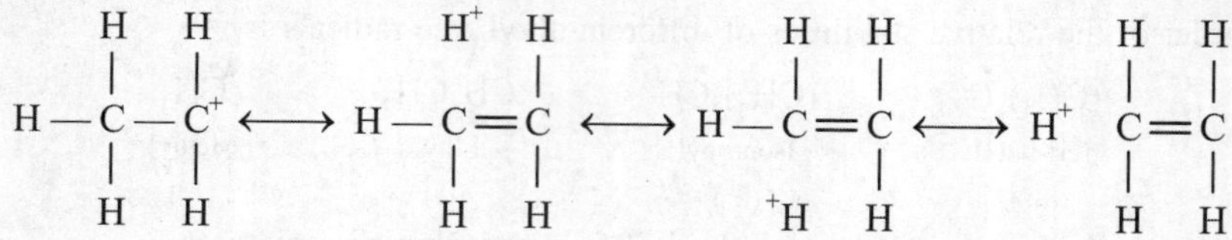

(I) (II) (III) (IV)

It is more stable than methyl carbocation where no resonance is possible. Similarly, the contributing structures in isopropyl carbocation and tertiary butyl carbocation are six and nine respectively in addition to the normal structure. Hence these are more stabilised and relative order of their stabilities is justified.

The order of the relative stabilities of the different alkyl carbocations can also be explained with the help of inductive effect (+*I* effect).

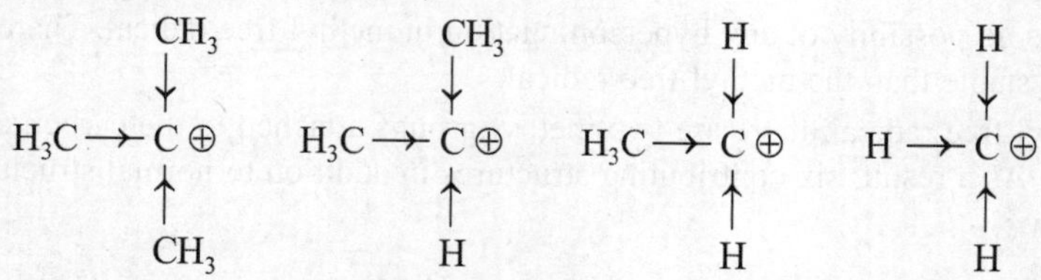

Tertiary butyl Isopropyl Ethyl Methyl

Because of the +I effect, the alkyl group will tend to increase the electron density on positively charged carbon atom neutralising its +ve charge to some extent. As the reactivity of carbocation is due to the positive charge on it, the reactivity will tend to decrease and the stability will correspondingly increase. Thus, greater the number of the alkyl groups present, greater will be the stability of carbocation and lesser will be the reactivity. Thus, the relative stabilities of different alkyl carbocations can be explained.

3. Relative Stabilities of Alkenes

The order of the relative stabilities of different alkenes is :

$$R_2C = CR_2 > R_2C = CHR > RCH = CHR > RCH = CH_2 > CH_2 = CH_2$$

This can be explained on the basis of hyperconjugation. The presence of alkyl group (say methyl group) on the double bonded carbon atom is likely to increase the number of contributing structures due to hyperconjugation.

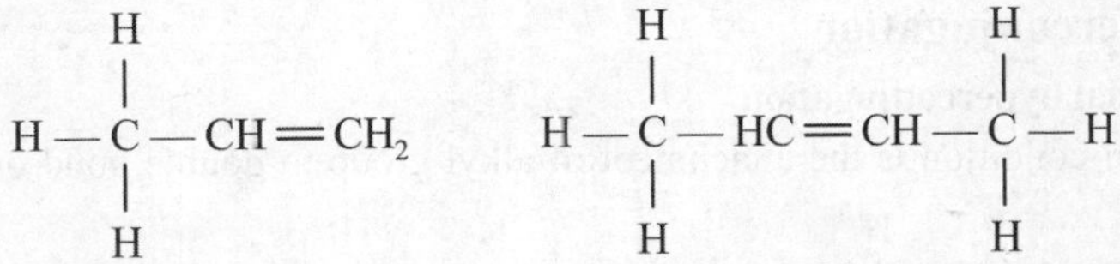

Propene (Three contributing structures) 2-Butene (Six contributing structures)

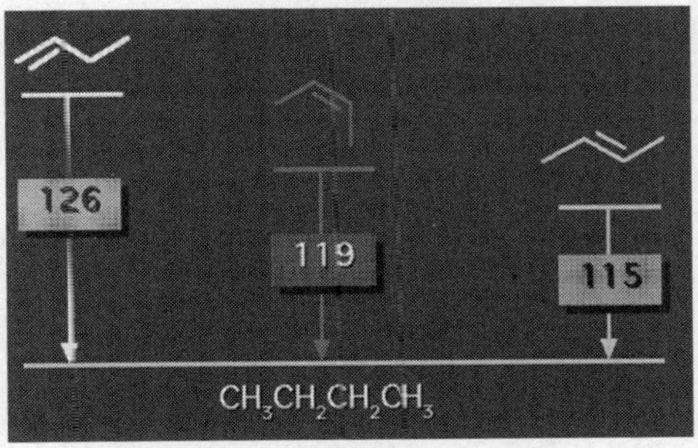

Stability of isomeric alkenes

Thus, greater the number of the contributing structures, greater will be the stability of the alkene.

4. Bond length

Hyperconjugation, like conjugation and resonance, also affects bond length.

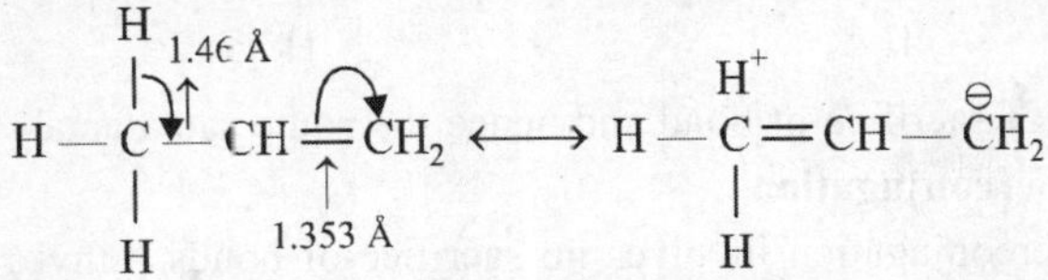

Bond length in propene is 1.46 Å in contrast to normal 1.54 Å (in propane). It is due to the partial double bond character acquired and hence a little shorter.

5. Dipole moment

Since hyperconjugation causes the separation of charges, it also affects the dipole moment in the molecule.

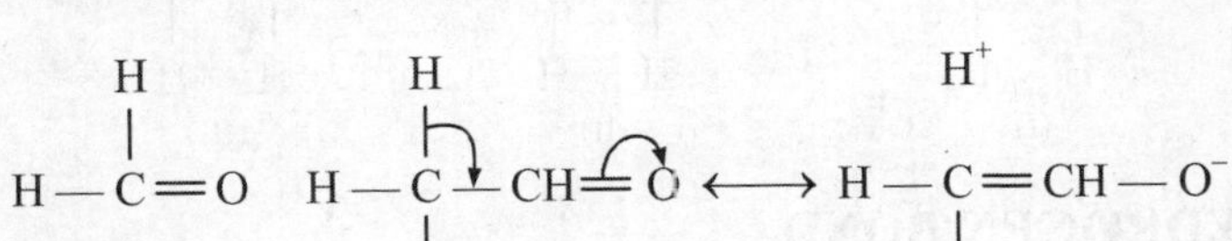

6. Ortho-para directing property of methyl group in toluene is partly due to +I effect and partly due to hyperconjugation.

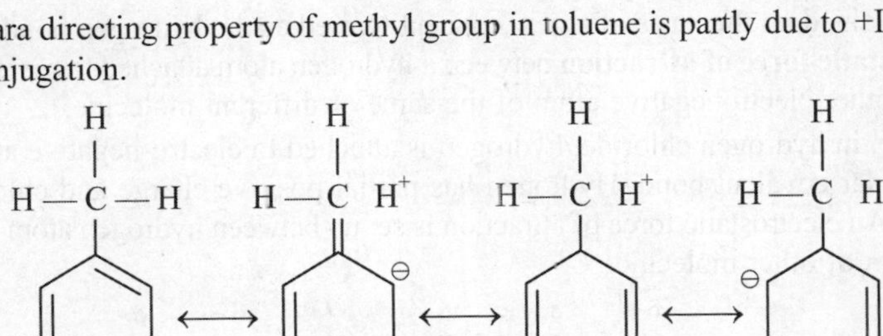

Types of Hyperconjugation

(*a*) Sacrafacial hyperconjugation

The essential condition is the attachment of alkyl group to double bond or triple bond.

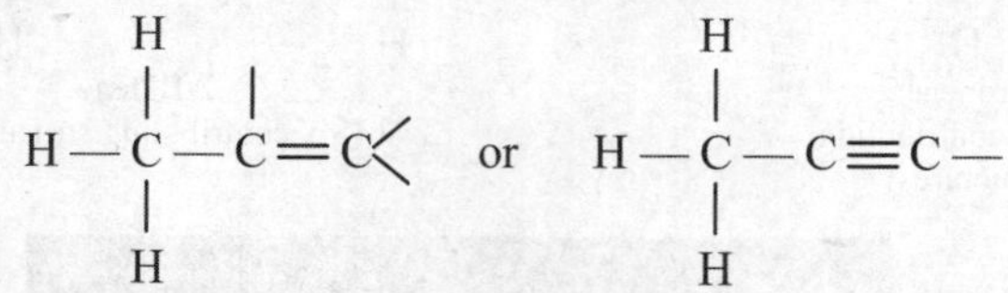

Carbon atom of alkyl group attached to double bond must contain atleast one hydrogen atom in hyperconjugation.

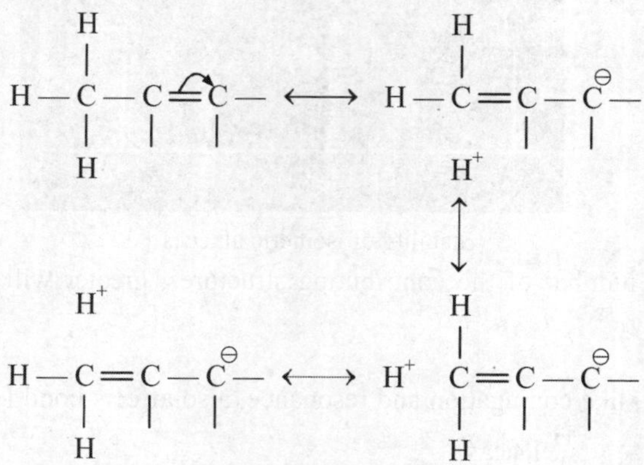

It involves a sort of sacrifice of bond and hence the name sacrafacial.

(*b*) Isovalent hyperconjugation

This kind of hyperconjugation involves no sacrifice of bonds. Ethyl radicals have the same number of real bonds as the classical structure.

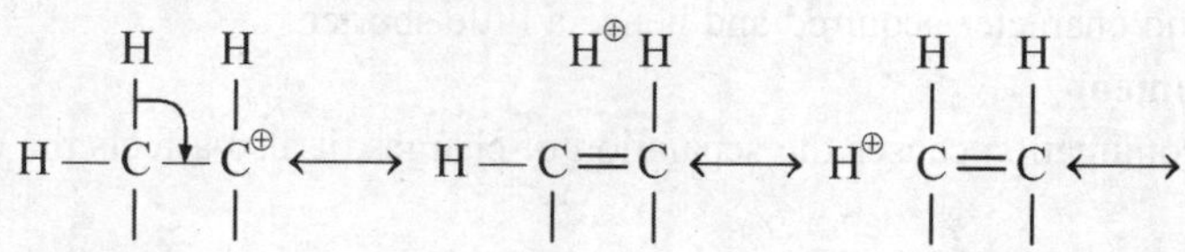

(I) (II) (III)

1.15 THE HYDROGEN BOND

The compounds, in which hydrogen is attached to some electronegative atom, general exhibit unexpected properties like relatively high melting and boiling points, high solubility in water. The behaviour is explained on the basis of a new concept termed as the hydrogen bond. **The hydrogen bond an electrostatic force of attraction** between a hydrogen atom attached to more electronegative atom and some other electronegative atom of the same or different molecule.

For example, in hydrogen chloride, hydrogen is attached to electro-negative atom of chlorine. The bond is a polar covalent bond. Hydrogen has partial positive charge and chlorine has partial negative charge. An electrostatic force of attraction is set up between hydrogen atom of one molecule and chlorine atom of other molecule.

$$\overset{\delta+}{H} — \overset{\delta-}{Cl} \cdots\cdots \overset{\delta+}{H} — \overset{\delta-}{Cl} \cdots\cdots \overset{\delta+}{H} — \overset{\delta-}{Cl}$$

The force of attraction is a hydrogen bond and is represented by a dotted line.

Thus, as a result of hydrogen bonding, a H-atom links the two electronegative atoms simultaneously, one by a covalent bond and the other by a hydrogen bond. Hence, it is said to form a **hydrogen bridge.**

Conditions for hydrogen bonding

The following conditions must be fulfilled for the formation of hydrogen bond :

(*i*) *The molecule must contain a highly electronegative atom linked to H-atom,* Higher the electronegativity, greater is the polarization of the molecule.

(*ii*) *The size of the electronegative atom should be small.* Smaller the size of the electronegative atom, greater is the electrostatic attraction.

Thus, only F, O and N atoms satisfy these conditions and form hydrogen bonds of high electronegativities.

Chlorine having the same electronegativity as that of nitrogen does not form hydrogen bond due to its large size.

Examples of hydrogen bonding

(*i*) Water (H_2O) : Water molecule contains highly electronegative oxygen atom linked to hydrogen atom. Due to large electronegativity difference, oxygen end of the molecule becomes negative, whereas the H-atom becomes positive. The negative end of one molecule attracts the positive end of the other and hydrogen bonding takes place as shown below :

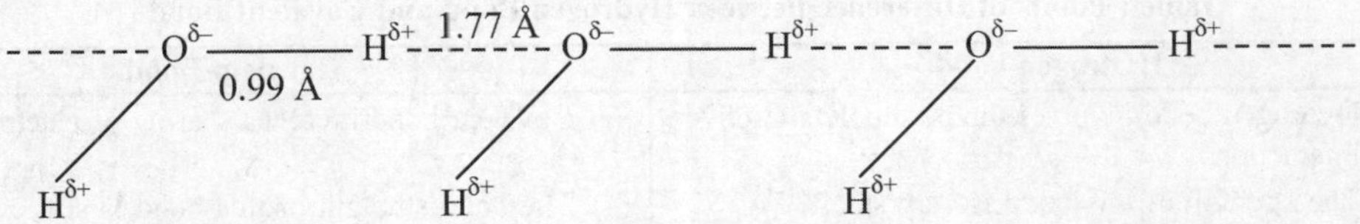

Parameters of the two kinds of O — H bonds are also indicated.

(*ii*) Ammonia (NH_3) : It contains the highly electronegative atom N linked to H-atoms. Hence, hydrogen bonding takes place as follows :

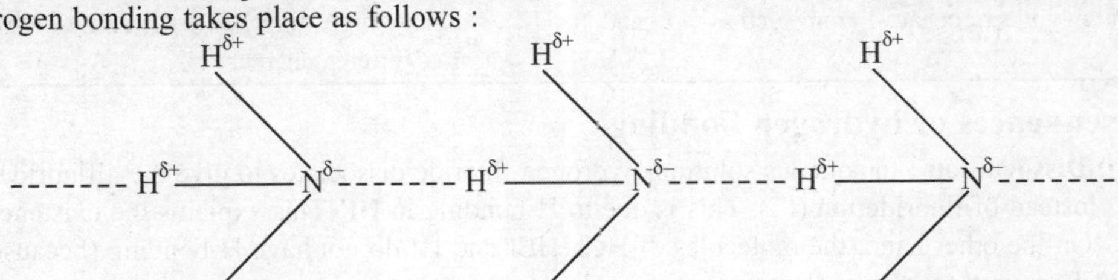

(*iii*) Hydrogen fluoride (HF) : Fluorine, having the highest value of electronegativity forms the strongest hydrogen bonds. Orientation of various atoms and bond lengths are indicated below :

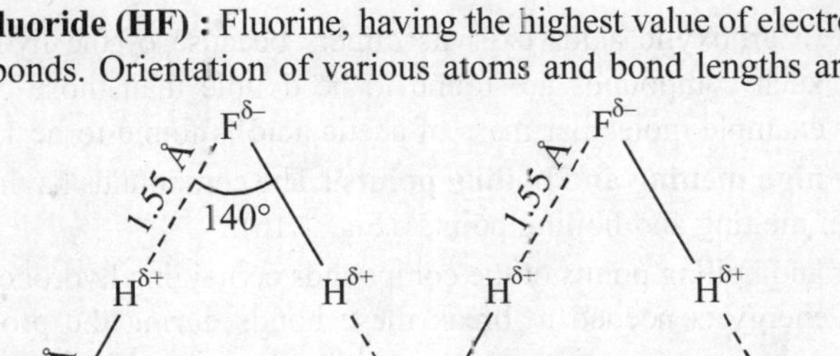

(*iv*) Alcohols (ROH) and Carboxylic acids (RCOOH) : These molecules contain the highly electronegative oxygen atom linked to H-atom and, hence, form associated molecules as follows :

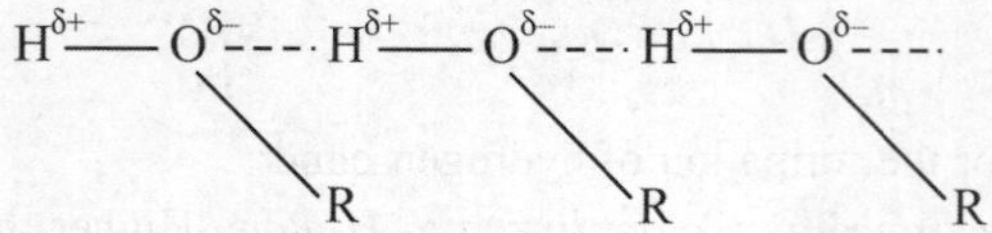

Alcohols

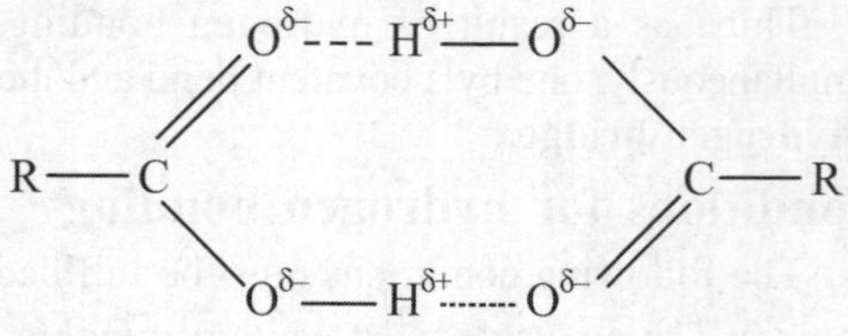

Carboxylic acids

In carboxylic acids, the hydrogen bonding causes association of two molecules only.

Strength of hydrogen bond

Hydrogen bond is a very weak bond. The strength of hydrogen bonds is intermediate between the weak van der Waals' forces and the strong covalent bonds. Thus, whereas the bond dissociation energy of a covalent bond is 209 – 418 kJ mol^{-1}, that for H-bond is only 12.6 – 41.8 kJ mol^{-1}. The dissociation energy of the H-bond depends upon the attraction of the shared pair of electrons and, hence, on the electronegativity of the atom. The bond energies in case of three elements are as follows :

H .... F bond dissociation energy is 41.8 kJ $mole^{-1}$.
H .... O bond dissociation energy is 29.3 kJ $mole^{-1}$.
H .... N bond dissociation energy is 12.6 kJ $mole^{-1}$.

Table : Points of Difference between Hydrogen Bond and Covalent Bond

Hydrogen bond	Covalent Bond
(*i*) Hydrogen bond involves dipole-dipole attractive interactions.	(*i*) Covalent bond involves sharing of electrons.
(*ii*) The strength of this bond is very small, *e.g.*, bond strength of H .... F bond is 41.8 kJ mol^{-1}.	(*ii*) The bond strength of this bond is sufficiently high, *e.g.*, bond strength of H — H bond is 433 kJ mol^{-1}.
(*iii*) It is formed between a hydrogen atom and a highly electronegative atom such as F, O and N.	(*iii*) It is formed between two electronegative atoms which may be of the same element or of different elements.

Consequences of hydrogen bonding

***(i)* Dissociation :** In aqueous solution, hydrogen fluoride dissociates to give the difluoride ion (HF_2^- instead of fluoride ion (F^-). This is due to H-bonding in HF. This explains the existence of KHF_2. On the other hand, the molecules of HCl, HBr and HI do not have H-bonding (because Cl Br and I are not so highly electronegative). This explains the non-existence of compounds like $KHCl_2$, $KHBr_2$ or KHI_2.

***(ii)* Association :** Carboxylic acids exist as dimers because of the hydrogen bonding. The molecular masses of such compounds are found to be double than those calculated from their simple formulae. For example molecular mass of acetic acid is found to be 120.

***(iii)* Abnormally high melting and boiling points :** The compounds having hydrogen bonding show abnormally high melting and boiling points. (Fig. 1.16).

The high melting and boiling points of the compounds containing hydrogen bonds is due to the fact that some extra energy is needed to break these bonds during the process of melting and boiling.

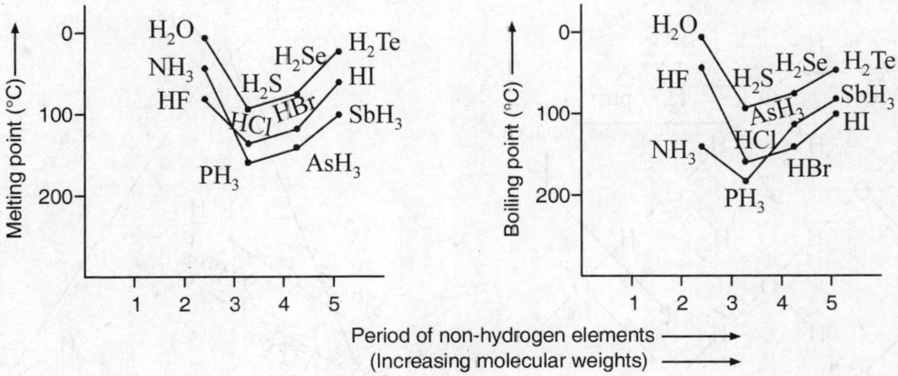

Fig. 1.16

(*a*) *The unusually high boiling point of HF (292.4 K) among the halogen acids* is due to the existence of hydrogen bonding in HF. Hydrogen bond exists to a much smaller extent in HCl and is non-existent in HBr and HI.

(*b*) *H_2O is a liquid whereas H_2S, H_2Se and H_2Te are all gases at room temperature.* This is because hydrogen bonding causes association of the H_2O molecules with the result that the boiling point of water is more than that of the other compounds. On the other hand, there is no such hydrogen bonding in H_2S, H_2Se and H_2Te.

(*c*) *Ethanol has higher boiling point than diethyl ether or diethylamine* because there is hydrogen bonding in the former but there is no hydrogen bonding in the later.

It is interesting to note that a stronger H-bond exists in HF than in water, yet water boils at higher temperature than HF. It is probably due to the fact that water forms two H-bonds as compared to one in HF.

***(iv)* Solubility :** Compounds that can form hydrogen bonds with the solvent are soluble in such solvents. For example, lower alcohols are soluble in water because of the hydrogen bonding which can take place between water and alcohol molecules as shown below :

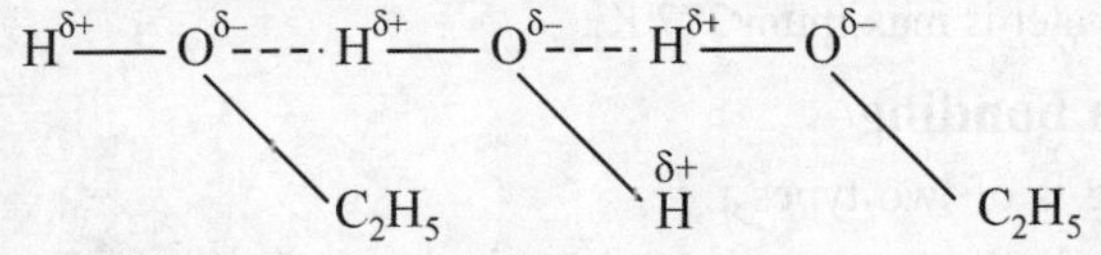

Similarly, ammonia (NH_3) is soluble in water because of hydrogen bonding as represented below :

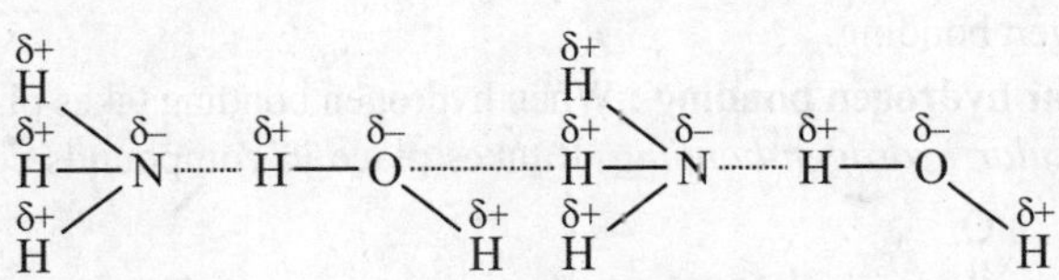

***(v)* Viscocity and surface tension :** The substances which contain hydrogen bonding exist as associated molecules. So, their flow becomes comparatively difficult. In other words, they have higher viscosity and high surface tension.

***(vi)* Explanation of lower density of ice than water :** Hydrogen bonding gives a cage like structure to solid ice as shown in Fig. 1.17 (*a*). As a matter of fact, each water molecule is linked tetrahedrally to four water molecules tetrahedrally as shown in Fig. 1.17 (*b*).

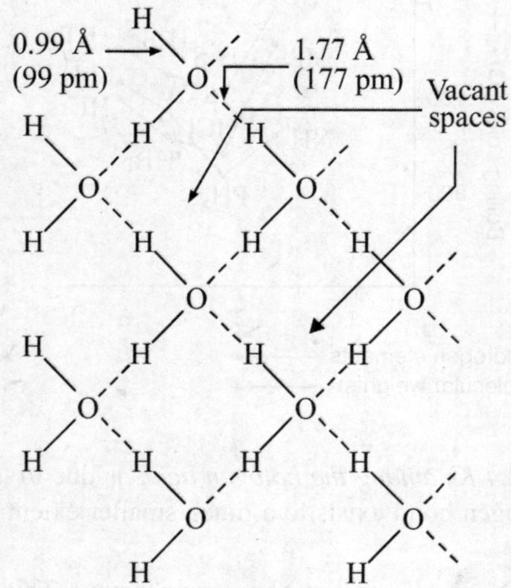

Fig. 1.17 (*a*) Cage like structure of H_2O in the ice.

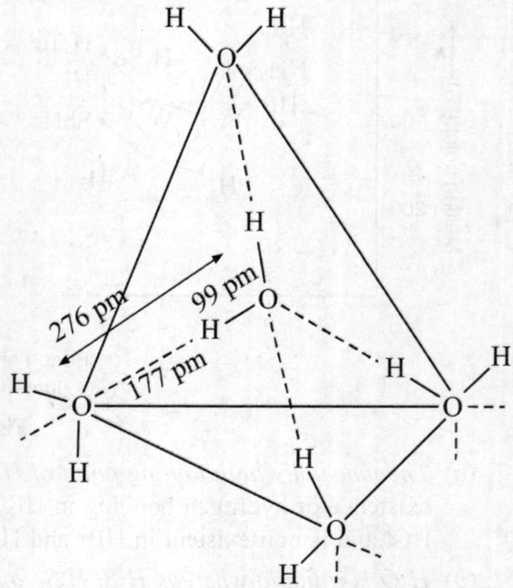

Fig. 1.17 (*b*) Tetrahedral linkage in solid ice.

Ice possesses low density because the molecules are not so closely packed as they are in the liquid state. When ice melts, this cage like structure collapses and the molecules come closer to each other. Thus, for the same mass of water, the volume decreases and, hence, density increases. Therefore, ice has lower density than water at 273 K and it floats on water.

(*vii*) Explanation of maximum density of water at 277 K : At 273 K, complete hydrogen bonding exists in ice. Thus, the open cage like structure exists which has larger volume and, thus, makes the density low. On heating, the hydrogen onds starts collapsing and, thus, the molecules start coming together resulting in the decrease of volume and, hence, increase of density. This goes on upto 277 K. After 277 K, normal increase in volume of a liquid with temperature takes place. Thus, after 277 K, there is net increase of volume on heating which means derease in density. Hence, the density of water is maximum 277 K.

Types of hydrogen bonding

Hydrogen bonding is of two types :

(*i*) Intermolecular hydrogen bonding : When hydrogen bonding takes place between different molecules of the same or different compounds, it is called *intermolecular hydrogen bonding.* Hydrogen fluoride, water, alcohols and solutions of alcohol or ammonia in water involve intermolecular hydrogen bonding.

(*ii*) Intramolecular hydrogen bonding : When hydrogen bonding takes place within a molecule, it is called *intramolecular hydrogen bonding.* It takes place in compounds containing two groups

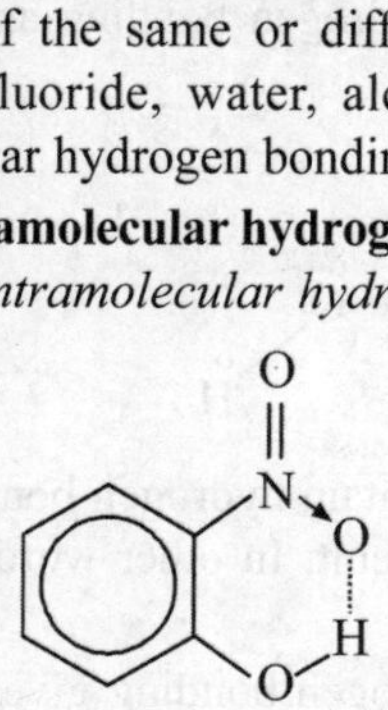

o-Nitrophenol

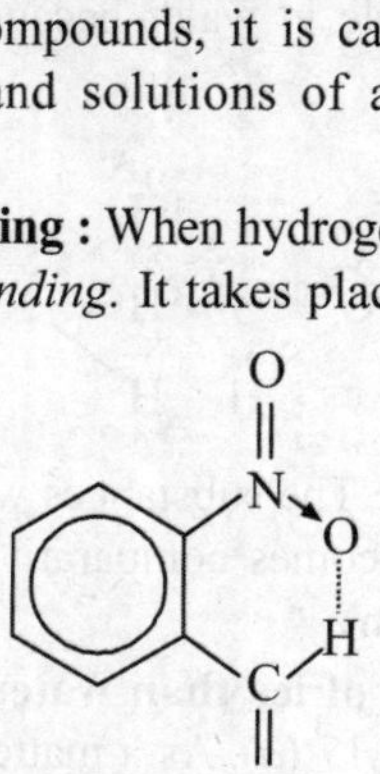

o-Nitrobenzaldehyde

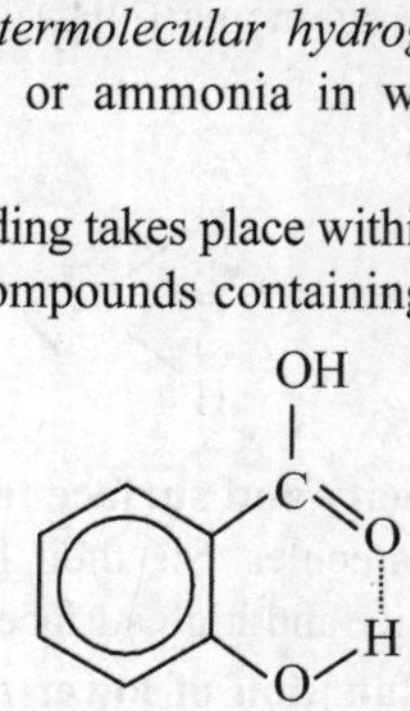

Salicylic acid

such that one group contains a H – atom linked to an electronegative atom and other group contains a highly electronegative atom linked to a lesser electronegative atom. The bond is formed between the H-atom of one group with the more electronegative atom of the other group. Intramolecular hydrogen bonding is involved in compounds like *n*-nitrophenol, *o*-nitrobenzaldehyde and salicylic acid.

Effects of intramolecular hydrogen bonding

The intramolecular hydrogen bonding prevents association of molecules, causes a contraction of size of the molecules and, hence, decreases its surface area. Consequently, intramolecular hydrogen bonding results in lowering their melting and boiling points. It also decreases their solubility and increases their vapour pressure. Intramolecular hydrogen bonding produces results opposite to that of intermolecular hydrogen bonding.

1.16 AROMATICITY OR AROMATIC CHARACTER

Benzene and arenes are aromatic in nature. The property of exhibiting aromatic character is called **aromaticity**. In order that a compound may be aromatic a nature, it must fulfil the following characteristics :

(*i*) The compounds must be cyclic in nature.

(*ii*) The molecular formula might suggest high degree of unsaturation due to the presence of one or more double bonds in the ring but it must behave as an saturated compound.

(*iii*) The compounds must be flat in nature. This has been confirmed by the X-ray and electron diffraction methods.

(*iv*) The compounds must resist addition but should readily take part in the electrophilic substitution reactions.

(*v*) They must have a tendency to show resonance with high value of resonance energy.

(*vi*) The most important criteria for the resonance is that the compounds must obey Huckel's rule.

It may be noted that the aromatic character is not confined merely to benzene and the compounds containing one or more benzene rings. There are many examples in which the ring does not have at six carbon atoms, it could also be heterocyclic in nature. In general the aromatic compounds are of two types.

(*a*) Those which contain a benzene ring and are called **benzenoids.**

(*b*) Those which do not contain a benzene ring and are called **non-benzenoids.**

Theories of Aromatic Character

The following theories have been put forward to explain the aromatic character of cyclic compounds :

1. Aromatic sextet theory

This theory was proposed by *Robinson* and co-workers in 1925. According to this theory, each carbon atom in benzene contributes one electron (not involved in hybridisation). These six electrons constitute a 'closed group' or *'closed sextet'* of electrons which is responsible for the stability of benzene and its aromatic character.

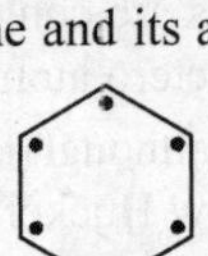

Benzene

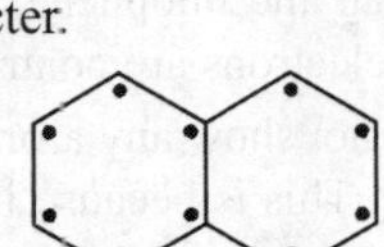

Naphthalene

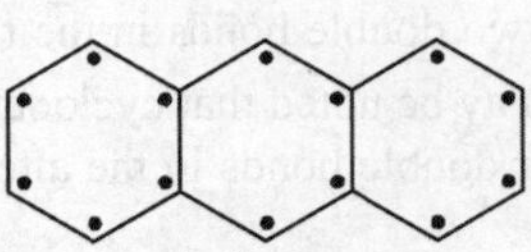

Anthracene

Similarly, in naphthalene which has two condensed benzene rings, there are ten electrons out of which two re common to both the rings. In the same way, in anthracene, there are fourteen electrons and four electrons are shared by three rings.

Some heterocyclic componds such as pyrrole, furan, pyridine etc. are aromatic in character because the heterocyclic ring has a sextet of electrons.

Pyrrole

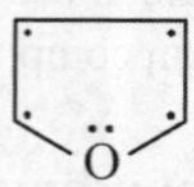

Furan

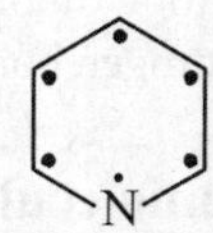

Pyridine

2. Resonance Theory

The theory of resonance explains high stability of aromatic compounds. The stability of these compounds is expressed in terms of **resonance energy** which is the difference in the energy contents of contributing structures and of resonance hybrid. Aromatic compounds possess such a large resonance energy that M.J.S. Dewar defined the aromatic compounds as the *cyclic compounds with definite resonance energy where all the atoms take part in single conjugated system.*

The resonance energy of benzene is 150 kJ mol^{-1}. It may be noted that the relative stabilities of any two compounds can be compared in terms of resonance energy.

3. Huckel's Rule

The aromatic character of cyclic compounds can be best explained in terms of *Huckel's rule* which has been derived from the molecular orbital theory. According to this rule,

> *monocyclic planar systems of trigonally hybridised atoms which contain (4n + 2) π-electrons possess considerable stability and hence, aromatic character where n may vary from 0, 1, 2, 3, 4...etc.*

Thus, cyclic compounds having 2, 6, 10, 14, 18 or 22 π-electrons are expected to show aromatic character. But '*n*' does not denote the number of rings in the cyclic compound.

Erich Huckel

Aromatic character of some cyclic systems

***(i)* Monocyclic systems.** All the monocyclic compounds shown below have a number of π-electrons in the ring according to the Huckel's rule. *i.e.*, all of them have six π-electrons in the ring. They are, therefore, aromatic.

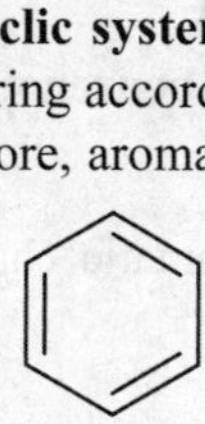

Benzene

Pyrrole

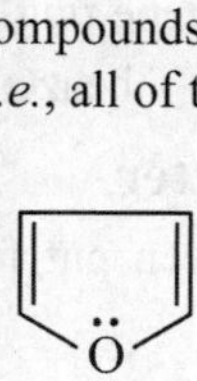

Furan

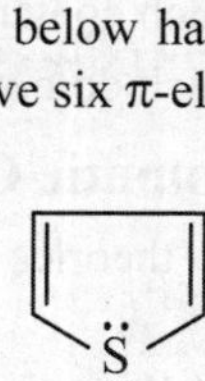

Thiophene

In the heterocyclic compounds pyrrole, furan and thiophene, four π-electrons are contributed by the two double bonds in the ring while two electrons are contributed by the hetero atom.

It may be noted that cyclooctatetraene does not show any aromatic character although the ring has four double bonds in the alternate positions. This is because it does not follow Huckel's rule.

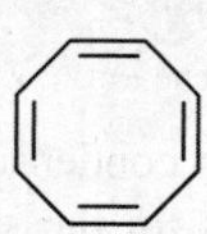

The number of π-electrons (8) is not according tc Huckel's rule.

***(ii)* Fused ring systems.** The polynuclear hydrocarbonds such as naphthalene, anthracene and phenanthrene are aromatic in nature according to Huckel's rule. (6 and 14 π electrons)

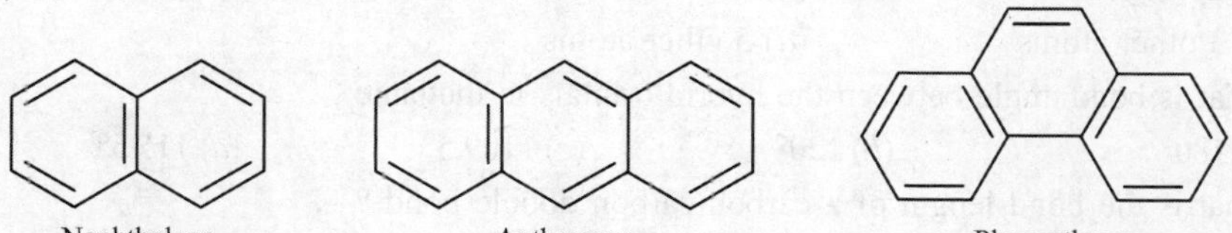

Naphthalene (10 π-Electrons) Anthracene (14 π-Electrons) Phenanthrene (14 π-Electrons)

***(iii)* Aromatic ions system.** It may be noted that not only the molecules but even some cyclic ions exhibit aromatic character. A few such ions are listed :

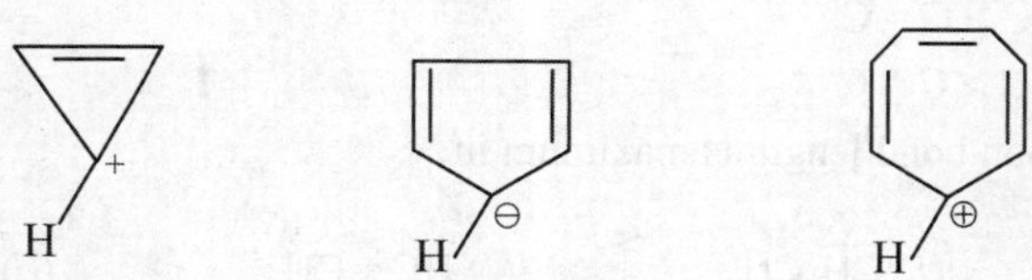

Cyclopropene cation (2 π-Electrons) Cyclopentadiene anion (6 π-Electrons) Cycloheptatriene cation (2 π-Electrons)

EXERCISES
(INCLUDING QUESTIONS FROM DIFFERENT UNIVERSITY PAPERS)

Objective Questions

Fill in the blanks

1. For hydrogen bonding, hydrogen should be attached to ....

2. Compounds which have spaces in the shape of long channels or tunnels are called ....

3. With the increase in carbon chain length, the inductive effect ....

4. H_2S is a gas but H_2O is a liquid at room temperature because of ....

5. Out of *sp*, sp^2 and sp^3 hybridisation, ... hybridisation gives trigonal planar geometry

6. Highest b.p out of H_2S, NH_3, PH_3 and CH_4, is that of ....

7. Out of $\overset{+}{N}O_2$, NH_3, $AlCl_3$ and SO_3, ... is not an electrophile.

8. Number of σ and π bonds in 1-pentene-3-yne are .... and .... respectively

9. The change $C_2H_5 \cdot + \cdot C_2H_5 \longrightarrow C_2H_6 + C_2H_4$ is called

10. Hyperconjugation is also called no – bond ....

Multiple Choice Questions

Choose the correct option

1. When the carbon atom is sp^2 hybridized in a compound, it is bonded to
   (*a*) 2 other atoms (*b*) 4 other atoms
   (*c*) 3 other atoms (*d*) 5 other atoms
2. What is bond angle between the hybrid orbitals in methane ?
   (*a*) 180° (*b*) 120° (*c*) 109.5° (*d*) 115.5°
3. What is the bond length of a carbon-carbon double bond ?
   (*a*) 1.20 Å (*b*) 1.34 Å (*c*) 1.54 Å (*d*) 1.68 Å
4. Which of the following is the correct order of bond lengths : -
   (*a*) $C-C < C=C < C\equiv C$
   (*b*) $C-C > C\equiv C > C=C$
   (*c*) $C\equiv C > C-C > C=C$
   (*d*) $C\equiv C < C-C > C=C$
5. The carbon-carbon bond length is maximum in
   (*a*) $CH_2=CH_2$ (*b*) CH_3CH_3 (*c*) $HC\equiv CH$ (*d*) 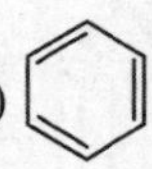
6. Compound in which carbons use only sp^2 hybrid orbitals for bond formation is
   (*a*) 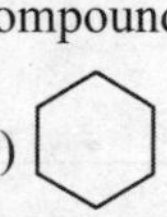 (*b*) 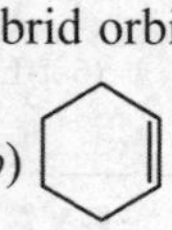
   (*c*) $CH_2=CH-CH=CH_2$ (*d*) $CH_3CH=C=CH_2$
7. The H-C-C bond angle in ethane is
   (*a*) 60° (*b*) 109°28′ (*c*) 120° (*d*) 118°28′
8. The carbon-carbon bond lengths in rank of increasing bond length is:
   (*a*) triple, double, single
   (*b*) single, double, triple
   (*c*) single, triple, double
   (*d*) triple, single, double
9. Which of the following hydrocarbons has the shortest C-C bond length ?
   (*a*) $CH_2=CH_2$ (*b*) CH_3CH_3 (*c*) $HC=CH$ (*d*) 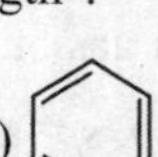
10. Which of the following is a planar molecule ?
    (*a*) Formaldehyde (*b*) Acetone (*c*) Formic acid (*d*) Acetic acid
11. The H-C-H bond angle in ethylene is
    (*a*) 60° (*b*) 120° (*c*) 90° (*d*) 180°
12. Which statement is true ?
    (*a*) Resonance hybrids are inherently unstable.
    (*b*) Resonance hybrids are more stable than any individual resonance form.
    (*c*) Resonance hybrids are averages of all resonance forms resembling .the less stable forms.
    (*d*) Resonance hybrids are averages of all resonance forms resembling the more stable forms.

13. The order of stability of carbonium ions is
(*a*) tertiary > secondary > primary
(*b*) secondary > tertiary > primary
(*c*) primary > secondary > tertiary
(*d*) primary > tertiary > secondary

14. Which of the following carbonium ions will be most stable ?

(*a*) $\overset{+}{C}H_3$ (*b*) $CH_3\overset{+}{C}H_2$ (*c*) $(CH_3)_2\overset{+}{C}H$ (*d*) $CH_2{=}CH\overset{+}{C}H_2$

15. Resonance structures of a molecule have
(*a*) same arrangement of atoms
(*b*) different arrangement of atoms
(*c*) same number of paired electrons
(*d*) different number of paired electrons

16. Which of the following compounds will show dipole moment ?
(*a*) *Cis*-l, 2-dichloroethylene (*b*) *o*-dichlorobenzene
(*c*) *trans.* l, 2-dichloroethylene (*d*) *p*-dichlorobenzene

17. The order of stability of carbanions is ‘
(*a*) primary > secondary > tertiary
(*b*) secondary > tertiary > primary
(*c*) tertiary > secondary > primary
(*d*) tertiary > primary > secondary

18. Which of the following compounds have planar molecules ?
(*a*) ethyl alcohol (*b*) formaldehyde (*c*) diethyl ether (*d*) 1,3-butadiene

19. In stable organic compounds, carbon will always form
(*a*) 2 bonds (*b*) 4 bonds (*c*) 3 bonds (*d*) 5 bonds

20. Compound in which carbons use only sp^3 hybrid orbitals for bond formation is
(*a*) $CH_3CH_2CH_3$ (*b*) $CH_3C \equiv CH$
(*c*) $CH_3CH = CH_2$ (*d*) $CH_2 = CH— CH = CH_2$

21. Which molecule has a nonzero dipole moment ?
(*a*) $C1_2$ (*b*) CO_2 (*c*) $CC1_4$ (*d*) $CHCl_3$

ANSWERS

Fill in the blanks

1. hetero atom viz. N, S, O, etc. **2.** inclusion compounds **3.** decreases
4. H-bonding in H_2O **5.** sp^2 **6.** ammonia
7. NH_3 **8.** 10, 3 **9.** disproportionation **10.** resonance

Multiple choice questions

1. (*c*) **2.** (*c*) **3.** (*b*) **4.** (*a*) **5.** *b*) **6.** (*c*)
7. (*b*) **8.** (*a*) **9.** (*c*) **10.** (*a*) **11.** (*b*) **12.** (*d*)
13. (*a*) **14.** (*d*) **15.** (*a*)and (*c*) **16.** (*a*) and (*b*) **17.** (*a*) **18.** (*b*) and (*d*)
19. (*b*) **20.** (*a*) **21.** (*d*)

General Questions

1. What are + I and – I effects ? Illustrate with two examples in each case.
2. How will you account for the polarisation of a covalent bond ?
3. Define resonance. What is the difference between resonance and isomerism ?
4. Why is phenol a stronger acid than aliphatic alcohol ?
5. Show how the carbanion $\overset{\ominus}{C}H_2-\overset{O}{\overset{\|}{C}}-CH_3$ is stabilised due to resonance ?
6. Arrange the following in decreasing order of – I effect.
   $-C_6H_5, -NO_2, -COOH, -CN, -I, -F, -Br, -Cl$
7. Compare the relative acidic strengths of the following on the basis of inductive effect.
   $HC\equiv CH, CH_3C\equiv CH, C_2H_5C\equiv CH$
8. The electromeric effect is only temporary in nature. How will you account for this ?
9. Give the decreasing order of basicity of the following carbanions :
   $$\overset{\ominus}{C}_2H_5 > \overset{\ominus}{H}C = CH_2 > \overset{\ominus}{C}\equiv CH$$
10. Electromeric effect is only a temporary effect. Comment.
11. Out of ethylamine and acetamide, which is more basic and why ?
12. How will you distinguish between resonance and inductive effect ?
13. Out of $(CH_3)_3\ C^+$ and $(CH_3)_2\ \overset{+}{C}H$ carbocations, which is more stable and why ?
14. The dipole moment of *m*-dichlorobenzene is 1.48 D while that of the *o*-dichlorobenzene is 2.30 D. How will you account for this ?
15. How does hyperconjugation help in comparing the relative stabilities of alkenes ?
16. Why is ethylamine more basic than ammonia ?
17. Give the names and definitions of two effects involving electron displacement.
18. Enlist the various conditions for a compound to exhibit resonance.
19. How will you account for the acidic character of phenol ?
20. Differentiate between inclusion compounds and clathrates.
21. What are charge transfer complexes ? Give an example.
22. What is meant by field effects ? Give an example.
23. What is meant by localized and delocalized chemical bond ? Explain by taking example of 1, 3-butadiene
24. What is hybridisation ? What are the distinguishing features of sp^3 and sp^2 orbitals ? Explain with examples.
25. Explain why four covalent bonds in methane are equivalent.
26. Discuss the orbital structure of methane.
27. Write a note on hydrogen bonding.
28. Discuss the effect of hydrogen bonding on the boiling points and water solubility of organic compounds.
29. Explain clearly (*a*) bond length (*b*) bond angle and (*c*) bond energy.

30. What is the difference between inductive and electromeric effects ?

31. In what way a covalent bond is fissioned ? What are its results ?

32. Write a short note on resonance effect.

33. Write note on electromeric effect.

34. Define inductive effect and mesomeric effect.

35. What is meant by inductive effect ? Describe with the help of inductive effect that dichloro acetic acid is a stronger acid than acetic acid.

36. (*a*) Explain what do you understand by inductive effect.

(*b*) How does inductive effect help in explaining.

(*i*) Relative acidic strengths of acetic acid and formic acid ?

(*ii*) Relative basic strengths of primary, secondary and tertiary amines ?

37. Explain the following :

(*i*) Hyperconjugation effect. (*ii*) Electromeric effect.

38. Write notes on :

(*i*) Inductive effect (*ii*) Electromeric effect (*iii*) Hyperconjugation.

39. What is hyperconjugation ? With its help, explain which of the following free radicals is the most stable :

$CH_3 - \dot{C}H_2$, $\dot{C}H_3$, $(CH_3)_3\dot{C}$, $(CH_3)_2\dot{C}H$

40. Give the main points of difference between inductive and electromeric effects.

41. Give a brief account of :

(*i*) Resonance effect (*ii*) Inductive effect.

42. What is inductive effect ? With its help, show that dichloro acetic acid is a stronger acid than acetic acid.

43. What is inductive effect ? How does inductive effect help in explaining the relative strengths of organic acids.

44. Write a short note on resonance.

45. Indicate the type of hybridization of each carbon atom in the following structures :

(*a*) CH_3CH_3 (*b*) $CH_3C \equiv CCH_3$ (*c*) $CH_2 = C = CH_2$ (*d*) $CH_3C \equiv N$

Ans.

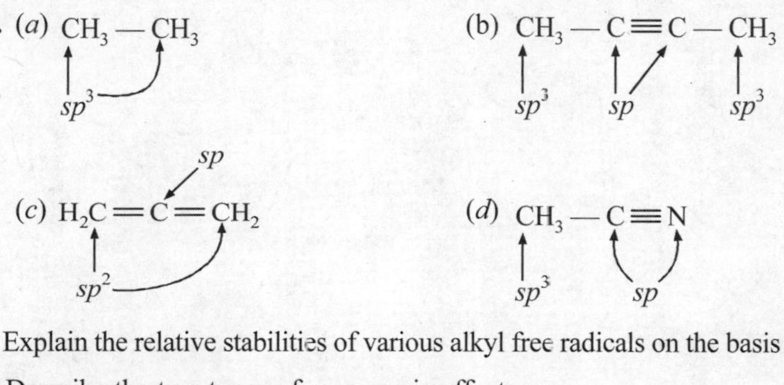

46. (*a*) Explain the relative stabilities of various alkyl free radicals on the basis of hyperconjugation.

(*b*) Describe the two types of mesomeric effects.

(*c*) Differentiate between inductive and electromeric effects.

47. Indicate the type of hybridization of each atom in the following structures :

(*a*) $H—C(=O)—H$ (*b*) $CH_3—O—CH_3$ (*c*) $H—C \equiv N$ (*d*) CH_3NH_2

Ans.

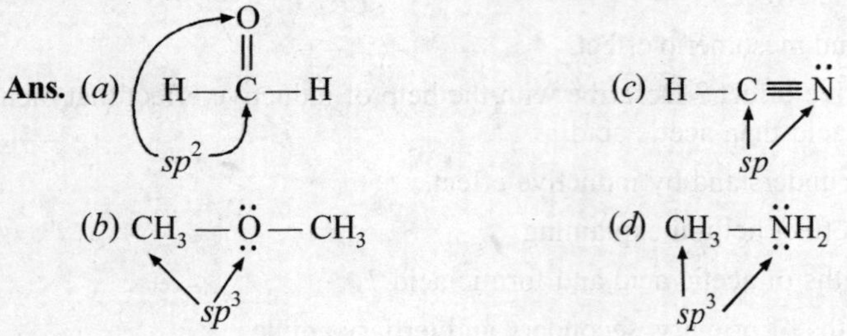

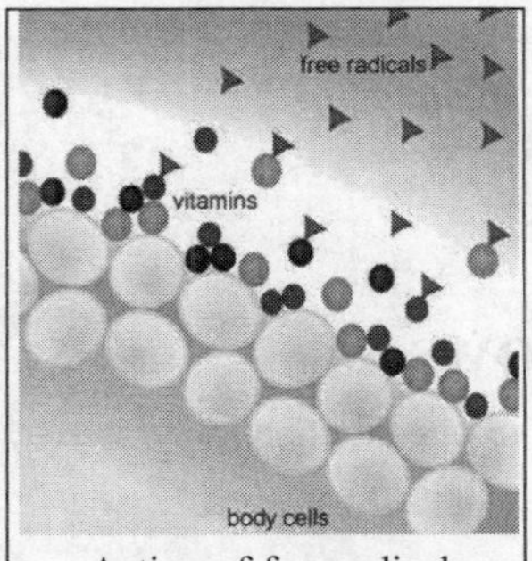

Action of free radicals

2 Mechanism of Organic Reactions

2.1 CURVED ARROW NOTATION

In the study of organic reaction mechanisms, we come across movement of electrons from one atom to the other. This movement of electron is represented with the help of curved arrows. This method is known as curved arrow notation. For example, when the atom B is more electronegative than atom A in the following compound, then in the presence of a reagent, the flow of electrons that takes place may be represented with the help of **curved arrows**.

$$A{=}B \xrightarrow{\text{Reagent added}} \overset{\delta+}{A} - \overset{\delta-}{B}$$

Electronegativity plays an important role in knowing the direction of transfer of the shared pair of electron to one of the atoms.

(*i*) In a carbonyl group, >C = O, present in aldehydes and ketones, the displacement is towards the oxygen atom which is represented by a curved arrow as under

$$>C{=}O \longrightarrow >\overset{+}{C} - \overset{-}{O}$$

(*ii*) In propylene, the displacement of shared pair of electrons is towards the carbon atom which is away from the methyl group due to the reason that methyl group is electron repelling.

$$CH_3 - CH{=}CH_2 \longrightarrow CH_3 - \overset{+}{C}H - \overset{-}{C}H_2$$

2.2 DRAWING ELECTRON MOVEMENTS WITH ARROWS

There are instances in organic molecules where the separation of charges (complete + ve or – ve) does not take place. But, due to the presence of electron-attracting or electron-repelling groups, some kind of polarization or movement of electrons takes place. We represent it with the help of arrows along the bonds. For example, in the molecule of ethyl chloride, the arrows would be drawn as :

$$CH_3 \longrightarrow CH_2 \longrightarrow Cl$$

Presence of a halogen group in a carboxy acid helps in the release of protons. This happens because the electron movement takes place in the direction of the halogen group. Greater the number of halogen groups, greater will be the charge movement away from the carboxylic group and towards the halogen groups. This can be illustrated as under:

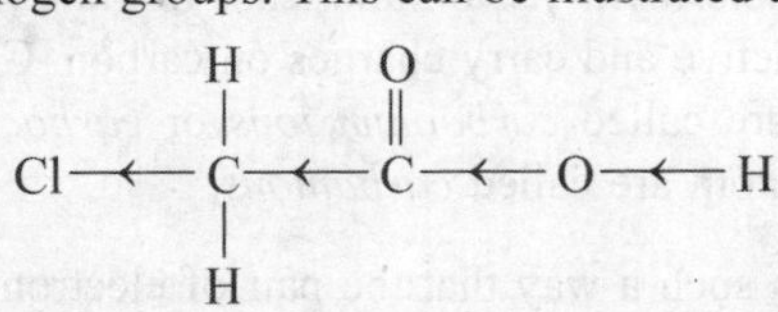
Chloroacetic acid

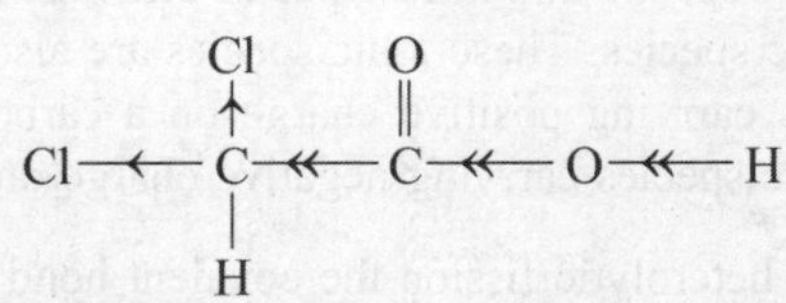
Dichloroacetic acid

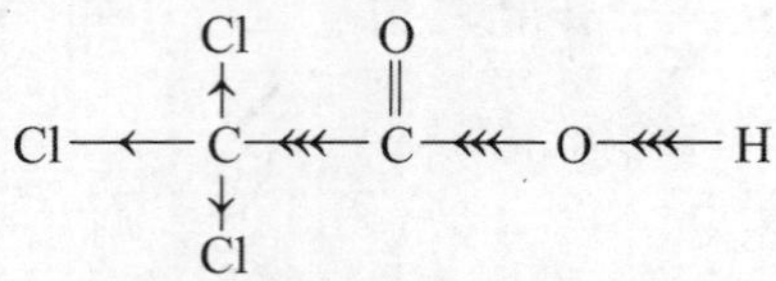

Trichloracetic acid

2.3 HALF-HEADED AND DOUBLE HEADED ARROWS

The curved arrows that we use to represent electron movement from one site to another can be classified as under:

(*i*) Half-headed arrows represented as  or 

(*ii*) Double-headed arrows represented as 

Half-headed or single headed arrows show the movement of a single electron. These are used in the discussion of radical chemistry mechanisms in particular. For example, we can depict the breaking of a bromine molecule into two bromine radicals as under:

Double-headed arrows show the movement of electron pair from one species to another. For example

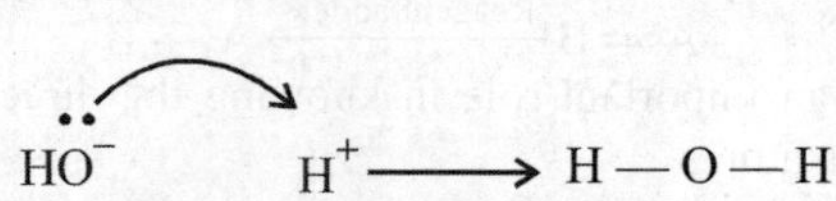

The movement takes place from the tail of the arrow and goes to head.

2.4 HOMOLYTIC AND HETEROLYTIC BOND BREAKING (FISSION)

Consider a covalent bond between atoms A and B.

A : B or A — B

The cleavage (or breaking) of this bond can take place in three possible ways depending upon the relative electronegativities of A and B.

(*i*) A : B $\longrightarrow$ A• + B• (A• and B• of equal electronegativity)

(*ii*) A : B $\longrightarrow$ $\overset{-}{A}: + \overset{+}{B}$ (A more electronegative than B)

(*iii*) A : B $\longrightarrow$ $\overset{+}{A} + :\overset{-}{B}$ (B more electronegative than A)

The first type of cleavage is called *homolytic fission* or *homolysis* and leads to the formation of very reactive species called '*free radicals*' (atoms or groups of atoms containing odd or unpaired electrons).

In homolytic fission the covalent bond breaks in such a way that each fragment carries one unpaired electron.

The second and third types of cleavage is called *heterolytic fission* and leads to the formation of ionic species. These ionic species are also very reactive and carry charges on carbon. Cationic species carrying positive charge on a carbon atom are called *carbonium ions* or *carbocations*. Anionic species carrying negative charge on carbon atom are called *carbanions*.

In heterolytic fission the covalent bond breaks in such a way that the pair of electrons stays on the more electronegative atom.

2.5 TYPES OF ORGANIC REACTIONS

Organic reactions may be classified into four main types:

(*a*) Substitution reactions

(*b*) Addition reactions

(*c*) Elimination reactions

(*d*) Rearrangements.

These are separately described as under.

(*a*) Substitution reactions. A substitution reaction is one in which a part of one molecule is replaced by other atom or group without causing a change in the rest of the molecule. Following are some examples of substitution reactions.

(*i*) $\underset{\text{Methane}}{CH_4} + Cl_2 \longrightarrow \underset{\text{Methyl chloride}}{CH_3Cl} + HCl$

(*ii*) $C_2H_5Br + KOH \longrightarrow C_2H_5OH + KBr$

(*iii*) 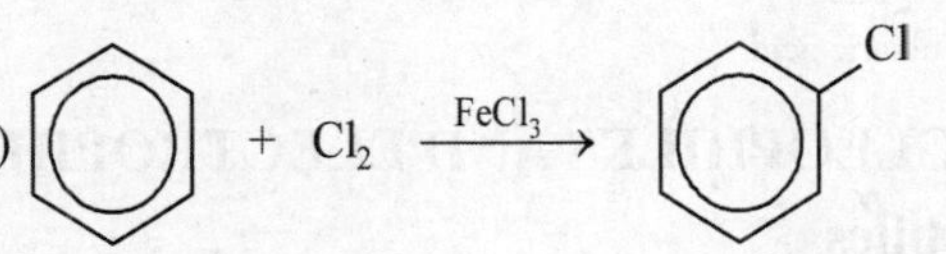

Benzene Chlorobenzene

The substitution reactions may be brought about by free-radicals, nucleophilic or electrophilic reagents. Thus, there are free radical substitution reactions, nucleophilic substitution reactions and electrophilic substitution reactions.

(*b*) Addition reactions. When two molecules of same or different substances combine together giving rising to a new product, it is an addition reaction. Examples of addition reactions are :

$$\underset{\text{Ethane}}{CH_2 = CH_2} + Br_2 \longrightarrow \underset{\text{Ethylene dibromide}}{CH_2BrCH_2Br}$$

$$\underset{\text{Acetaldehyde}}{CH_3CHO} + HCN \longrightarrow \underset{\text{Acetaldehyde cyanohydrin}}{CH_3CH(OH)CN}$$

Addition reactions could be brought about by free-radical, electrophilic or nucleophilic reagents.

As in the case of substitution, there are free radical, electrophilic and nucleophilic substitution reactions.

(*c*) Elimination reactions. These reactions involve the removal of atoms or groups from a molecule to form a new compounds containing multiple bonds. Dehydrohalogenation of alkyl halides is a common example of this type reaction.

$$\underset{n\text{-Propyl chloride}}{CH_3 - CH_2CH_2Cl} + KOH \xrightarrow{\text{Alc}} \underset{\text{Propene}}{CH_3CH = CH_2} + KCl + H_2O$$

Elimination reactions are of two types. Bimolecular eliminations or E_2 reactions and monomolecular or E_1 elimination reactions.

In E_2 reactions, rate determining step involves two reacting species. Dehydrohalogenation of tertiary alkyl halides with alc. KOH proceeds by E_2 mechanism. In E_1 reactions, the rate determining step involves only one reacting species. Dehydrohalogenation of primary alkyl halides with alc. KOH usually proceeds through E_1 mechanism.

(*d*) Rearrangement reactions. Rearrangement reactions involve the migration of an atom or a group from one atom to the other within the same molecule.

It is interesting to note that the first organic compound, *i.e.*, urea synthesized in the laboratory by Wohler actually involved a rearrangement reaction.

$$\underset{\text{Amm. cyanate}}{NH_4CNO} \xrightarrow{\Delta} \underset{\text{Urea}}{NH_2 — CO — NH_2}$$

Another important example of such reactions is *Hoffmann bromamide reaction.* This reaction involves the migration of an alkyl groups from the carbon to the nitrogen atom of an amide with the simultaneous elimination of CO as carbonate ion under the influence of Br_2/KOH.

$$\underset{\text{Acetamide}}{CH_3 — CONH_2} + Br_2 + 4KOH \longrightarrow \underset{\text{Methylamine}}{CH_3 — NH_2} + K_2CO_3 + 2KBr + 2H_2O$$

Similarly maleic acid, when heated in a sealed tube, is converted into fumaric acid.

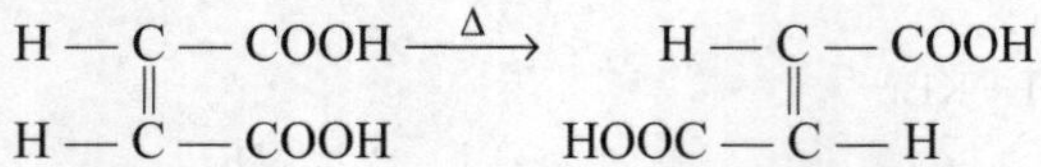

Maleic acid Fumaric acid

2.6 TYPES OF REAGENTS—NUCLEOPHILES AND ELECTROPHILES

Nucleophilic reagents or nucleophiles

A nucleophilic reagent is a reagent with an atom having an unshared or lone pair of electrons. Such a reagent is in search of a point where it can share these electrons to form a bond. Nucleophiles are of two types:

***(i)* Neutral nucleophiles.** These are the nucleophiles which are neutral in charge. But they carry some unshared electrons which they like to share with some positive centre or electron deficient centre. Ammonia $\ddot{N}H_3$, water $H_2\ddot{O}$ and alcohols $R — \ddot{O} — H$ are examples of neutral nucleophiles.

***(ii)* Negative nucleophiles.** These are the nucleophiles which carry negative charge. Examples of this type of nucleophiles are hydroxyl ions (OH^-), halide ion (X^-), alkoxide ion (RO^-) and cyanide ion (CN^-). Carbanions also come in the category of negative nucleophiles.

Electrophilic reagents or electrophiles

An electrophile is a reagent containing electron deficient atoms. Such species have a tendency to attach themselves to centres of high electron density. There are two types of electrophiles:

***(i)* Neutral electrophiles.** These electrophile do not carry any net charge. Lewis acids like $AlCl_3$, $FeCl_3$ and BF_3 belong to this category of electrophiles. Sulphonium ion (SO_3) carries no net charge, but it acts as an electrophile for sulphonation in benzene rings. This is because of its structure.

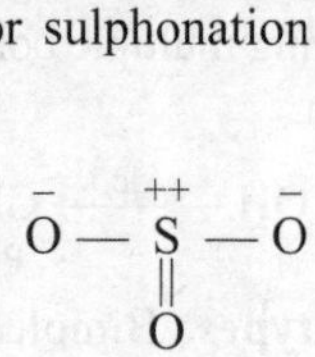

As the positive charge is concentrated and the negative charge is scattered, it acts as an electrophile. Substances like $SnCl_4$ which have vacant *d*-orbitals would like to accommodate electrons in them. Thus such substances also act as electrophiles.

***(ii)* Positive electrophiles.** These electrophiles carry a net positive charge. Examples of this category of electrophiles are hydrogen ion (H^+), hydronium ion (H_3O^+), nitronium ion (NO_2^+), and chloronium ion (Cl^+). In the halogenation and nitration of aromatic systems, these electrophiles are involved.

Example 1. **Classify the following as nucleophiles and electrophiles: H_3O^+, NH_3, $AlCl_3$, ROH, BF_3, CN^-, SO_3.**

Solution.	**Nucleophiles**	**Electrophiles**
	NH_3	H_3O^+
	ROH	$AlCl_3$
	CN^-	BF_3
		SO_3 (Sulphonium ion)

Example 2. **Pick up from the following, the electrophiles and nucleophiles:**

$AlCl_3$, PH_3, HNO_3, R_3N

Solution. Electrophiles: $AlCl_3$, HNO_3.
Nucleophiles: PH_3, R_3N.

Example 3. **Pick up from the following, the electrophiles and nucleophiles:**

BF_3, NH_3, ROH, $SnCl_4$.

Solution. Electrophiles: BF_3, $SnCl_4$.
Nucleophiles: NH_3, ROH.

2.7 REACTIVE INTERMEDIATES

In organic reactions, reactants do not change into products in one step. The change normally takes place via an intermediate product, which is short-lived. From the intermediate product, the reaction passes on to the products. Various reaction intermediates that we come across in the study of organic reactions are:

(*i*) Carbonium ion (or carbocation)
(*ii*) Carbanion
(*iii*) Free radical
(*iv*) Carbene
(*v*) Nitrene
(*vi*) Arynes (benzyne)

2.8 CARBOCATIONS

These are defined as the species in which the positive charge is carried by the carbon atom with six electrons in its valence shell. These are formed by the heterolytic fission in which an atom or group along with its pair of electrons leaves the carbon. In heterolytic fission, the shared pair of electrons between two atoms goes to one atom only.

$$R-CH_2-X \xrightarrow[\text{Fission}]{\text{Heterolytic}} R\overset{\oplus}{C}H_2 + \overset{\ominus}{\ddot{X}}$$

Stability of Carbocations

1. The order of stability among simple alkyl carbocations is: tertiary > secondary > primary. In most of the reactions, primary and secondary carbocations get rearranged to tertiary carbocations. Both *n*-propyl fluoride and isopropyl fluoride form the same isopropyl cation (2° carbocation). Similarly all the four butyl fluorides *viz.*, *n*-, iso-, sec. and tertiary butyl fluorides form the same tert-butyl cation. There are two factors which determine the stability:

(*i*) *Hyperconjugation or resonance.*
(*ii*) *Field effect or inductive effect of groups.*

Hyperconjugation

A large number of canonical forms can be written for tertiary carbocation compared to those for primary carbocation. Consider hyperconjugation in primary and tert-carbocation.

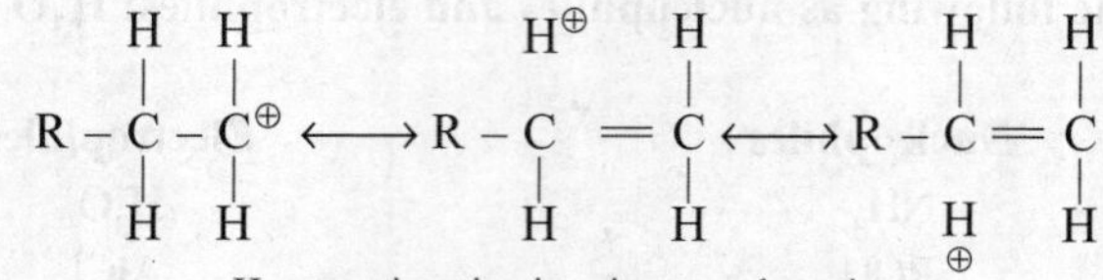

Hyperconjugation in primary carbocation

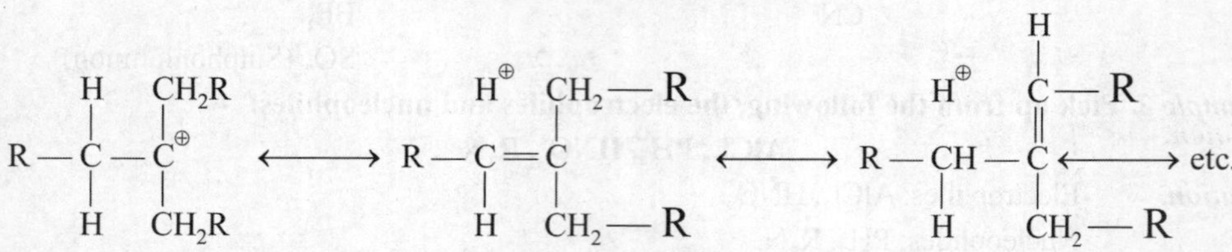

Hyperconjugation in tertiary cation

Greater the number of canonical forms, greater is the stability.

Field effect

Electron donating effect of alkyl groups increases the electron density around the +vely charged carbon. This results in reducing the magnitude of positive charge on it and thus the charge is delocalised on α-carbon. Dispersal of positive charge increases the stability. Of carbocation of all the simple cation, *tert*-butyl cation being most stable. Even *tert*-pentyl and *tert*-hexyl cations produce *tert*-butyl cation at high temperature. Lower alkanes like methane, ethane and propane when treated with super acid also yield *tert*-butyl cation as the main product. Salts of *tert*-butyl cation, like $(CH_3)_3\ C^+\ SbF_3^-$ have been prepared from super acid solutions.

2. The stability of carbocation containing a conjugated double bond is usually greater due to increased delocalisation by resonance. In such carbocations, the positive charge is dispersed on at least two carbon atoms. Consider the following carbocation.

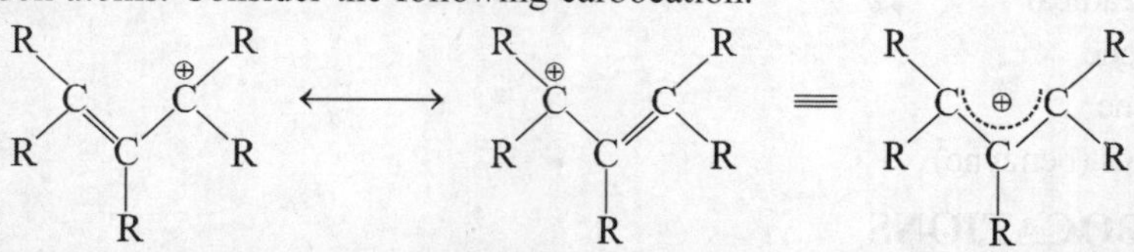

3. Allylic type cocations have been prepared from the solution of conjugated diene in conc. sulphuric acid and are found to be most stable.

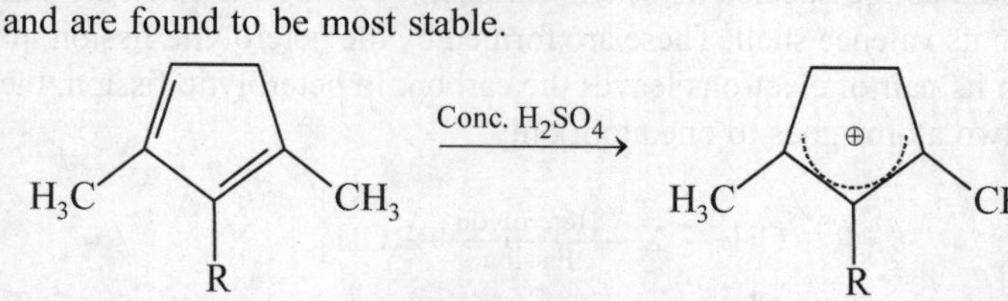

4. Carbocations can be stabilised through aromatisation also. 1-Bromocyclohepta-2, 4, 6-triene tropylium bromide is a crystalline solid. It is highly soluble in water and forms bromide ions in solution. Clearly, the compound is not covalent in nature. The reason for such a behaviour is the stability of tropylium cation which follows Huckel's rule (6 π electrons) for aromaticitiy.

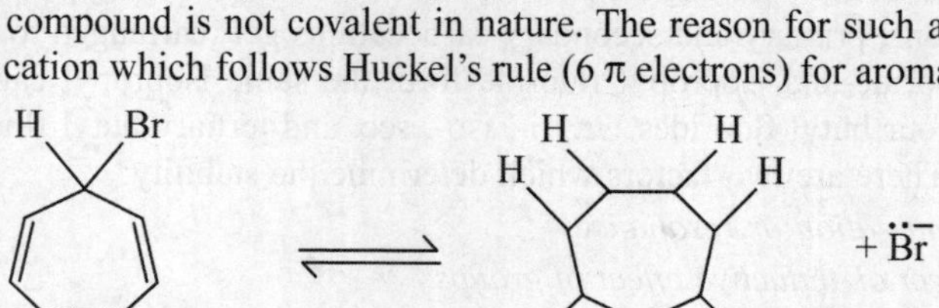

1, Bromo-2, 4, 6-Triene tropylium bromide

Tropylium cation

Its stability is also explained in terms of canonical structures which stabilise the tropylium cation.

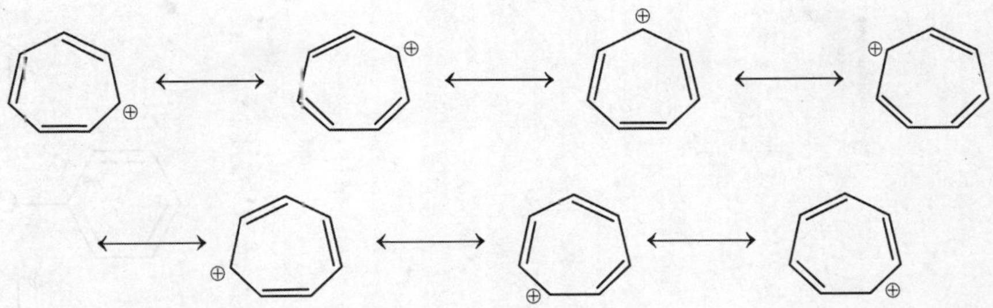

5. Substituted cyclopropenyl cation possesses even more aromatic stabilisation ($n = 0$).

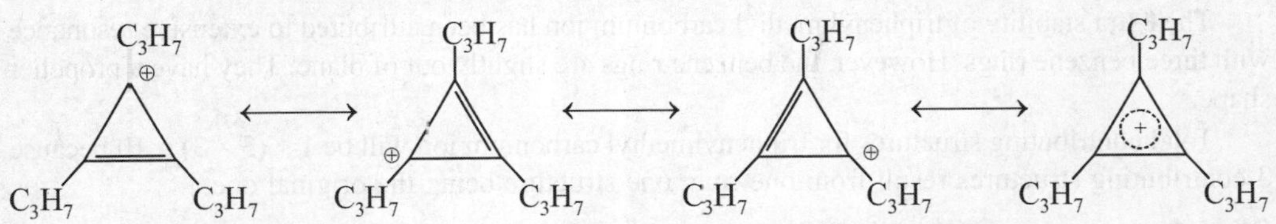

In the above tripropyl substituted cation, all the carbons are sp^2 hybridised. Thus all the p-orbitals including the vacant p-orbital of the positively charged carbon overlap forming a delocalised π molecular orbital, leading to stabilization.

6. Benzyl cation can be written in the following canonical forms:

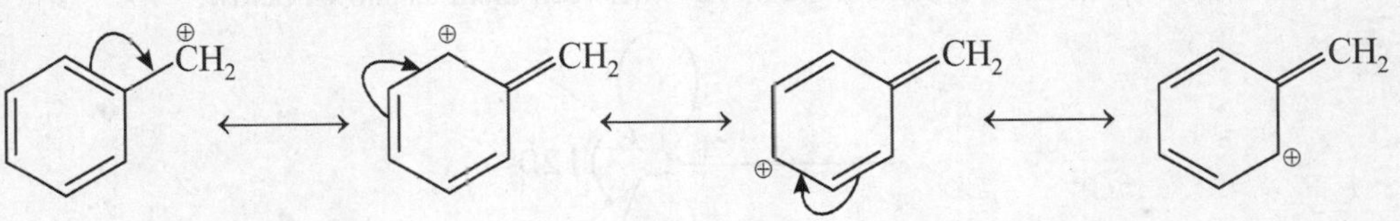

Due to larger number of canonical forms and greater dispersal of positive change, benzyl carbocation is still more stable.

7. Triphenyl chloromethane ionises in SO_2 as

$$(C_6H_5)_3\,C \cdot Cl \rightleftharpoons (C_6H_5)_3\,C^{\oplus} + Cl^-$$

Triphenyl methyl cation has been isolated as solid salt as $(C_6H_5)_3\ C^+.\ BF_4^-$. The stability of triarylmethyl cation is further increased if the rings are substituted at *ortho* and *para* positions by electron donating groups.

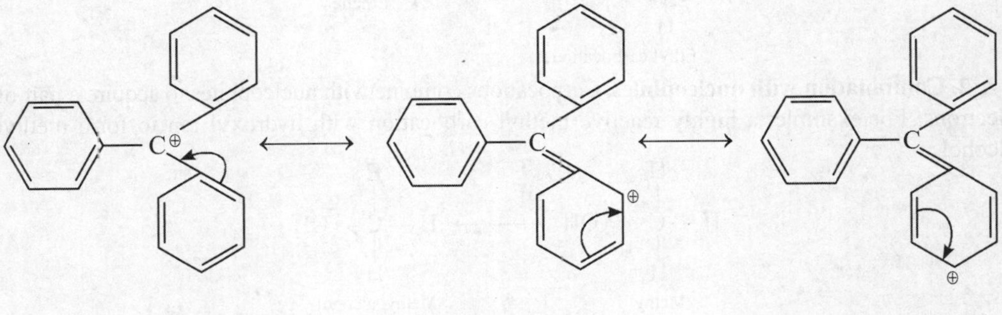

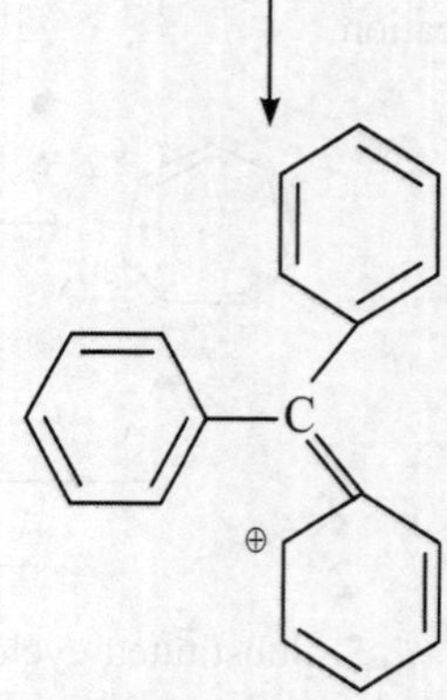

The extra stability of triphenyl methyl carbonium ion has been attributed to extensive resonance with three benzene rings. However, the benzene rings are slightly out of plane. They have a propeller shape.

Total contributing structures for triphenylmethyl carbonium ion will be $1 + (3 \times 3) = 10$ because 3 contributing structures result from one ring, one structure being the original one.

Structure and reactions of carbocations

Structure. The carbon atom bearing the positive charge in a carbocation is sp^2 hybridised. The three sp^2 hybrid orbitals are utilised in the formation of sigma bonds with three atoms or groups. The third unused *p*-orbital remains vacant. Thus, the carbocation is a flat species having all the three bonds in one plane with a bond angle of 120° between them as shown below:

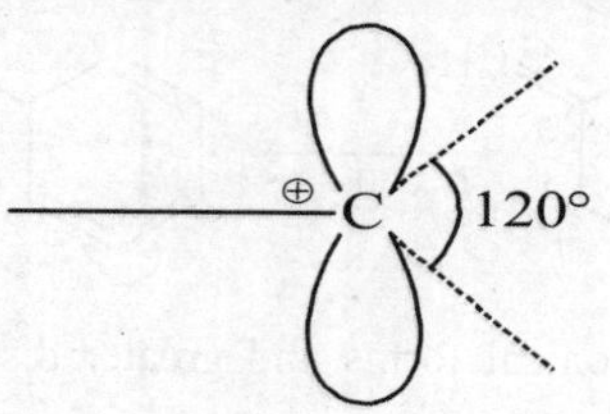

Reactions

Carbocations, which are shortlived species and very reactive give the following reactions:

1. Proton loss. Carbocation may lose a proton to form an alkene. An ethyl carbocation loses a proton to form ethene.

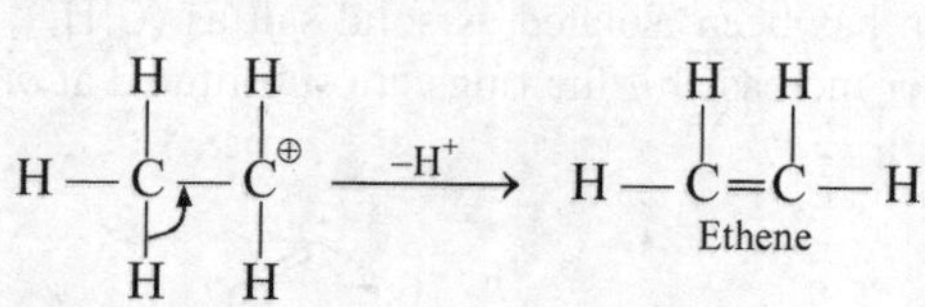

Ethyl carbocation

2. Combination with nucleophiles. Carbocations combine with nucleophiles to acquire a pair of electrons. For example, a highly reactive methyl carbocation with hydroxyl ion to form methyl alcohol.

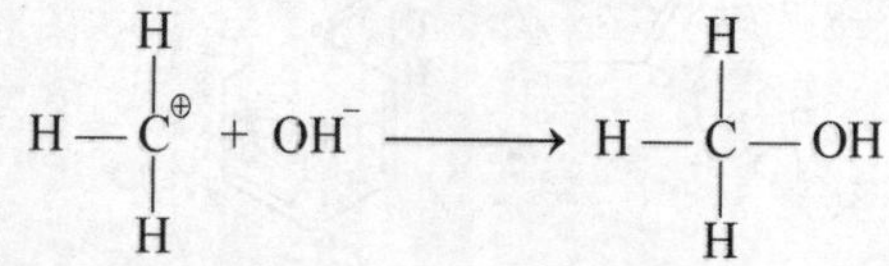

Methyl carbocation   Methyl alcohol

3. Addition to alkene. A carbocation may add to an alkene to produce another carbocation possessing higher molecular weight.

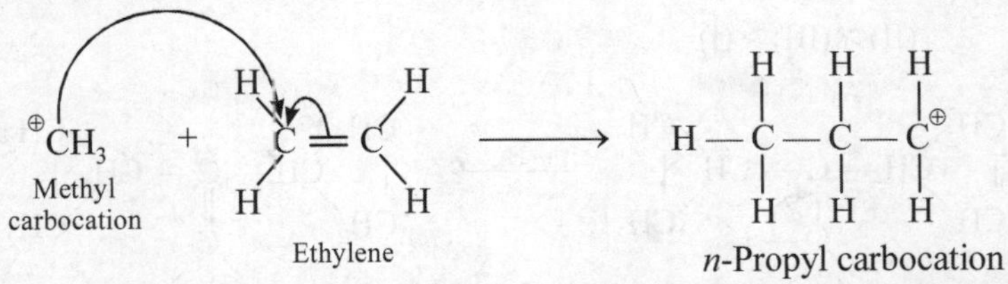

4. Rearrangement. The migration of alkyl or aryl or hydrogen along with an electron pair to the positive centre takes place which results in the formation of more stable carbocation. Some examples are given below:

$$\underset{(1°)}{CH_3-\overset{H}{\overset{|}{C}}H}-\overset{\oplus}{C}H_2 \xrightarrow{\text{Hydride shift}} CH_3-\underset{(2°)}{\overset{\oplus}{C}H}-CH_3$$

$$C_6H_5-CH_2-\overset{OH}{\overset{|}{C}}(CH_3)_2 \xrightarrow{FSO_3H} C_6H_5-\overset{H}{\overset{|}{C}}H-\overset{\oplus}{C}(CH_3)_2 \xrightarrow[\text{shift}]{\text{Hydride}}$$

$$\underset{\text{(More stable)}}{C_6H_5-\overset{\oplus}{C}H-\overset{H}{\overset{|}{C}}(CH_3)_2}$$

Such a rearrangement takes place because a tertiary carbocation is more stable than secondary carbocation, which in turn is more stable then primary cabocation. Similarly, a carbocation attached to a phenyl group (benzene ring) is more stable than others.

***Example 1.* Arrange the following in order of decreasing stability and give suitable explanation for it.**

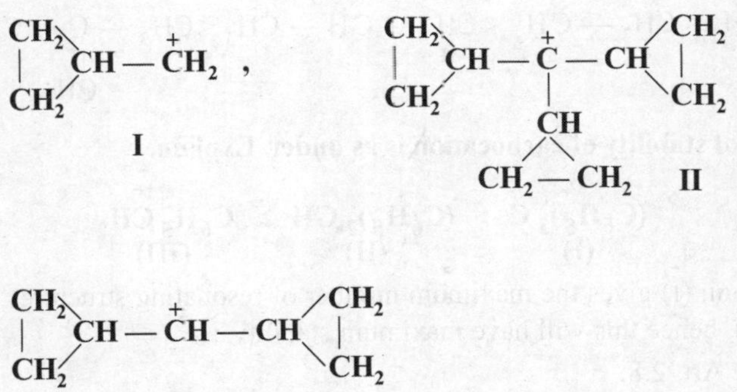

I II III

***Solution*.** The stability of a carbocation depends upon two factors:

(*i*) Hyperconjugation

(*ii*) Inductive effect of groups attached.

In this problem, it is hyperconjugation or resonance which is important. Structure II can form maximum number of resonating structures and hence this is most stable. It can form resonating structures with the help of hydrogen from three sides (from three cyclopropyl groups) whereas carbocations III and I can form resonating structures from two and one side respectively. Thus the order of stability in decreasing order is:

(II) > (III) > (I)

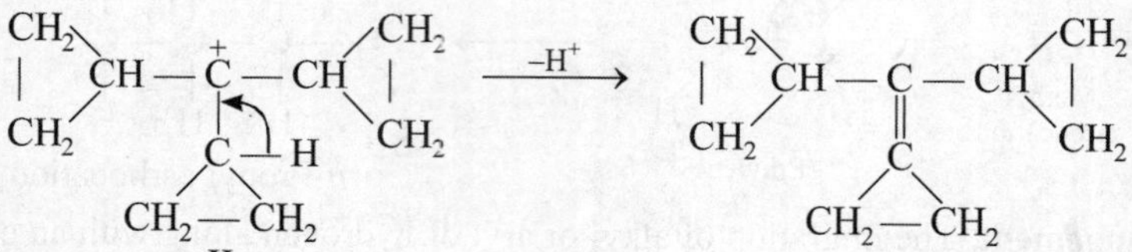

II

Example 2. **Arrange the following in order of increasing stability and explain the order on the basis of hyprconjugation.**

$$\mathbf{CH_3-\overset{+}{C}H_2,\ CH_3-\underset{\underset{CH_3}{|}}{\overset{+}{C}}-CH_3,\ \overset{+}{C}H_3,\ CH_3-\overset{+}{C}H-CH_3}$$

(I) **(II)** **(III)** **(IV)**

Solution. Greater the number of resonating structures, greater is the stability attached to a carbocation. Carbocation (II) which is a tertiary carbocation will give the maximum number of resonating structures involving hydride shift, like

$$\overset{+}{C}H_2-\underset{\underset{CH_3}{|}}{CH}-CH_3 \quad CH_3-\underset{\underset{CH_3}{|}}{CH}-\overset{+}{C}H_2 \quad CH_3-\underset{\underset{^{+}CH_2}{|}}{CH}-CH_3$$

Greater the number of hydrogens on carbon atoms in the immediate neighbourhood of carbocation, greater the number of resonating structures and hence greater the stability.

This is followed by secondary carbocation IV, primary carbocation I and methyl carbocation (III). Hence the order of stability in increasing order is:

$$\overset{+}{C}H_3 < CH_3-\overset{+}{C}H_2 < CH_3-\overset{+}{C}H-CH_3 < CH_3-\underset{\underset{CH_3}{|}}{\overset{+}{C}}-CH_3$$

Example 3. **Order of stability of carbocation is as under. Explain.**

$$\mathbf{\underset{(I)}{(C_6H_5)_3\overset{+}{C}} > \underset{(II)}{(C_6H_5)_2\overset{+}{C}H} > \underset{(III)}{C_6H_5\overset{+}{C}H_2}}$$

Solution. Carbocation (I) gives the maximum number of resonating structures, as this is linked to three benzene rings, hence this will have maximum stability.

For details, see Art. 2.8.

The order of stability in decreasing order is:

I > II > III

Example 4. **Why tropyllium bromide gives precipitate with $AgNO_3$ and CH_3Br does not?**

Solution. This is because tropyllium bromide behaves like an ionic compound whereas methyl bromide does not. Ionic nature of tropyllium ion is because of extra stability of tropyllium ion. For details refer to Art. 2.8.

2.8 CARBANIONS

These are defined as the species with an unshared pair of electrons and a negative charge on the central carbon atom. A carbanion may be formed in one of the following ways:

(*i*) An atom or group leaves carbon without the electron pair (Heterolytic fission).

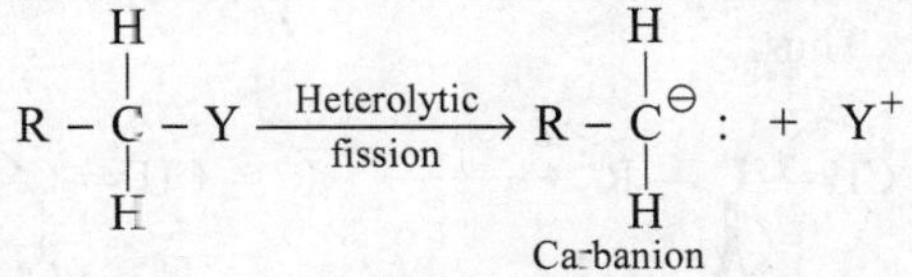

(*ii*) An anion adds to a carbon-carbon double or triple bond

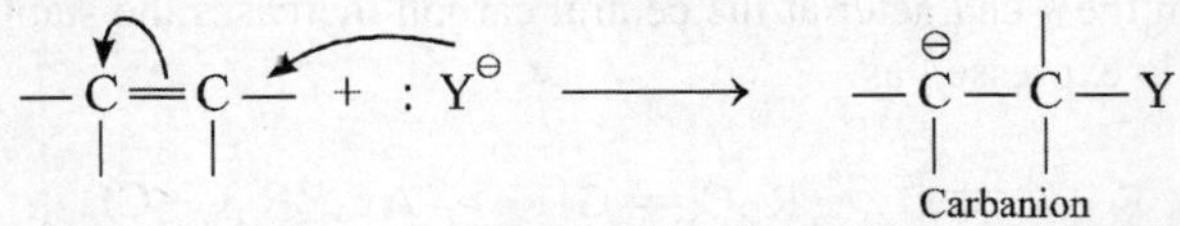

Every carbanion possesses unshared pair of electrons and is therefore a base. When the carbanion accepts a proton, it gives a conjugate acid. The stability of the carbanion is directly related with the strength of the conjugated acid.

$$\underset{\text{(Base)}}{\text{Carbanion}} + H^+ \longrightarrow \text{Conjugate acid}$$

Stability of carbanions

1. It is important to note that weaker the acid, stronger is the base and hence lower is the stability of the carbanion. Clearly, the order of stability of carbanion can be determined from the order of the strength of conjugate acids. Carbanions are highly unstable in solution as compared to carbocations.

2. The order of stability of carbanions is:

Vinyl > Phenyl > Cyclopropyl > Ethyl > *n*-propyl > Isobutyl

Also it has been found that the stability of carbanions decreases in the order methyl > prim-carbanion > sec-carbanion.

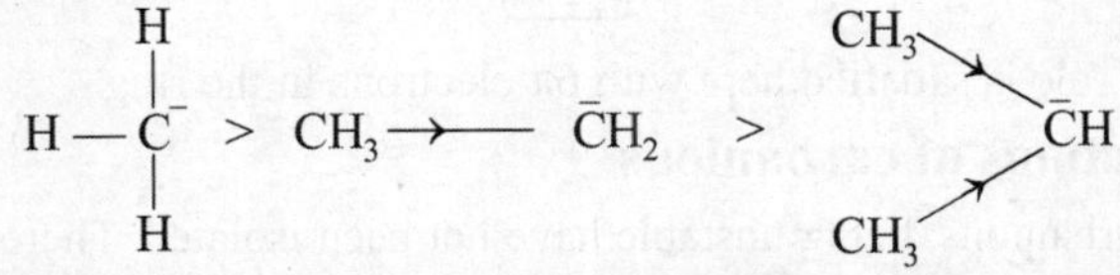

This stability order can be explained simply by the field effects. The presence of electron donating alkyl groups in secondary (Isopropyl) carbanion results in greater localisation of negative charge on the central carbon atom and hence the stability falls. Cyclopropyl carbanion has greater stability due to greater *s*-character on the carbanionic carbon.

3. Vinyl and phenyl carbanions are more stable due to resonance. In cases where a double or triple bond is located at α-position to the carbanionic carbon, the ion is stabilised by resonance in which the unshared pair overlaps with π-electrons of the double bond. The stability of allylic and benzylic carbanions can be illustrated as under.

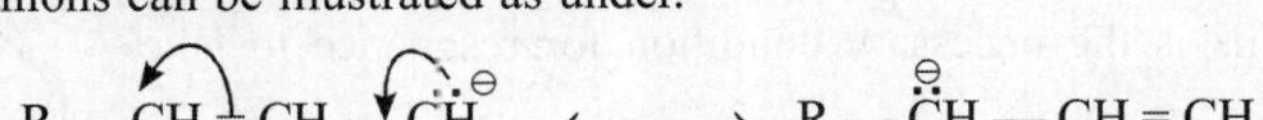

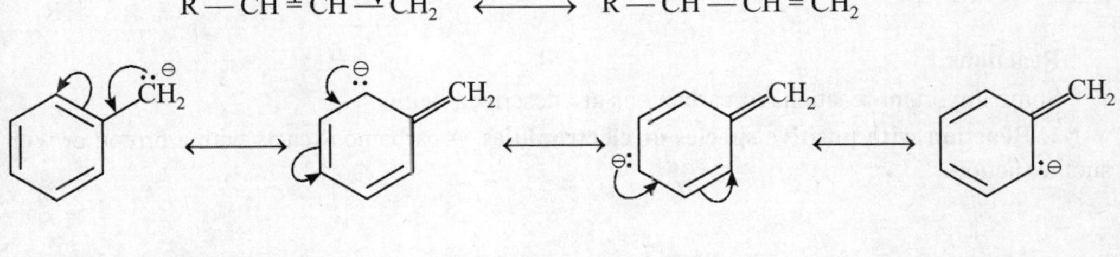

4. Diphenyl methyl and triphenyl methyl carbanions are more stable then even benzylic and can be kept in solution for a longer time if water is completely excluded.

5. When the carbanionic carbon is conjugated with a carbon-oxygen or carbon-nitrogen multiple bond, then the stability of carbanion is increased. The reason is that the presence of electronegative atoms helps in the dispersal of negative charge and thus such carbanions are better capable of bearing the negative charge. Thus,

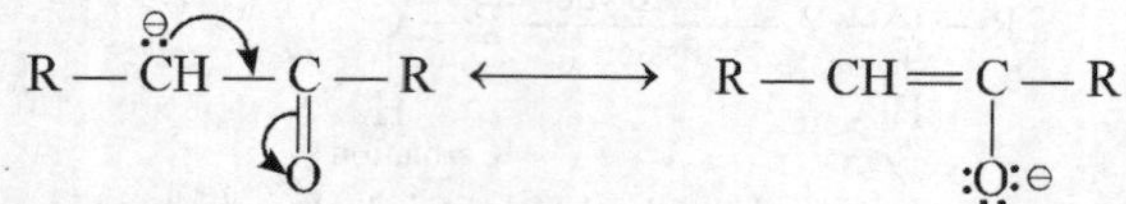

6. The increase in the *s*-character at the central carbon increases the stability of carbon. Thus, the order of stability is expressed as:

$$R-C\equiv\overset{\ominus}{\ddot{C}} > R_2C=\overset{\ominus}{\ddot{C}}H > \overset{\ominus}{Ar} > R_3C-\overset{\ominus}{\ddot{C}}H_2$$

Acetylenic carbon contains 50% *s*-character being *sp*-hybridised.

7. Stabilisation by Sulphur or Phosphorus: The bonding of sulphur or phosphorus atom with carbanionic carbon increases the stability. It is probably due to the overlap of unshared pair with an empty *d*-orbital. Consider a carbanion containing sulphur atom:

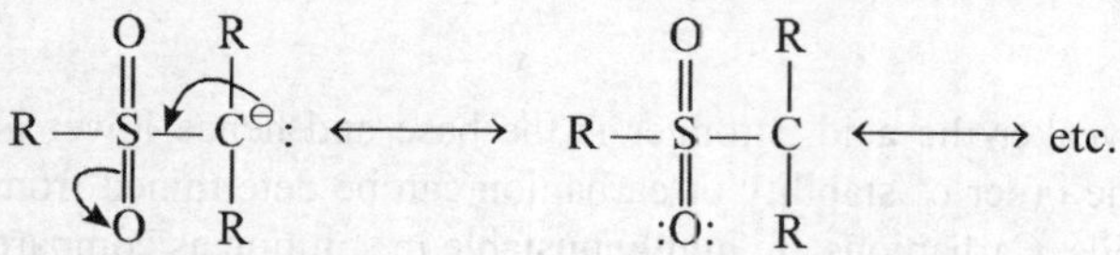

8. Aromatic nature. Some carbanions are stable as they are aromatic. *e.g.*, cyclopentadienyl anion is a stable carbanion.

Huckel's (4*n* + 2) rule is satisfied here with 6π electrons in the ring.

Structure and reactions of carbanions

Structure. The carbanions, being unstable have not been isolated. Therefore, their structure is not known with certainty. Most likely, the central carbon atom with an unshared pair of electrons (and negative charge) is sp^3 hybridised. The three sp^3 hybrid orbitals are used in forming three sigma bonds with the other atoms. The fourth sp^3 -orbital is occupied by a lone pair of electron. In fact, the structure of carbanion is quite similar to that of ammonia as shown in the figure.

It different alkyl groups are attached to the central carbon, the carbanion will be chiral. Reactions in which such a carbanion is formed, give a product with retention of configuration. Carbanions which get stabilized by resonance involving the lone pair of electrons and the π electrons of multiple bonds, should be planar as this is the necessary condition for resonance to take place.

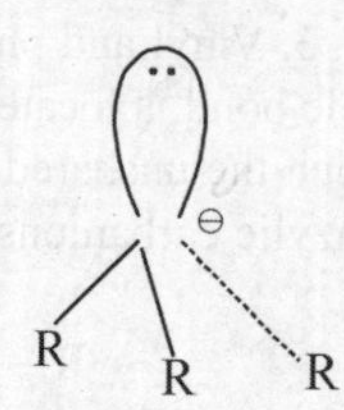

Reactions

Some important reactions of carbanions are described below:

1. Reaction with positive species or electrophiles. A carbanion reacts with a proton or with another action.

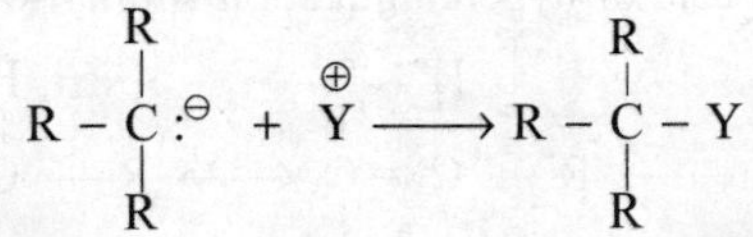

2. Reactions involving the displacement of an atom or a group. Such reactions are nucleophilic substitution (SN^2) reactions observed in alkyl halides.

3. Addition to carbonyl compounds. Carbanions attack an electron deficient carbon in the carbonyl compounds. Such reactions are called nucleophilic addition reactions.

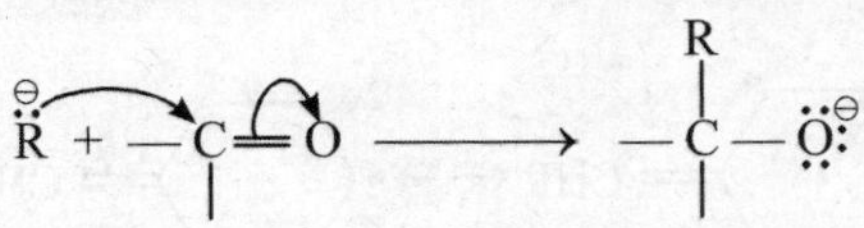

4. Rearrangement. In some cases, carbanions may rearrange to form more stable species. Consider the rearrangement in triphenylmethyl carbanion.

$$(C_6H_5)_3C-\overset{\ominus}{\ddot{C}}H_2 \longrightarrow (C_6H_5)_2\overset{\ominus}{\ddot{C}}-CH_2-C_6H_5$$

Here the two phenyl groups directly attached to the central carbon help in dispersing the negative charge.

2.9 FREE RADICALS

Free radicals are produced by the homolytic fission of a covalent bond. These are odd electron neutral species which are formed by the homolytic fission of a covalent bond. Free radicals are paramagnetic due to the presence of unpaired electron. Formation of free radicals is favoured by the presence of UV light, heat and organic peroxides. Reactions involving radicals widely occur in the gas phase. Such reactions also occur in solutions, particularly if carried in non-polar solvents. An important characteristics of free radical reactions is that, once initiated, they proceed very fast. The free radicals can be detected by magnetic susceptibility measurements.

$$R - R \xrightarrow[\text{Fission}]{\text{Homolytic}} \underset{\text{Free radical}}{2R\bullet}$$

$$Cl - Cl \xrightarrow{hv} \underset{\text{Free radical}}{2Cl\bullet}$$

$$\underset{\text{Acyl peroxide}}{RCO - O - OCOR} \xrightarrow{\text{Thermal}} \underset{\text{Free radical}}{2RCOO\bullet}$$

Stability of free radicals

1. Simplest alkyl free radicals are highly reactive like carbocations and carbanions. Their life time is extremely short in solution. The relative stability of simple alkyl radicals has the order:

$$\underset{3°}{R_3\dot{C}} > \underset{2°}{R_2\dot{C}H} > \underset{1°}{R\dot{C}H_2} > \underset{\text{Methyl}}{\dot{C}H_3}$$

It can be explained on the basic of hyperconjugation similar to that in carbocations.

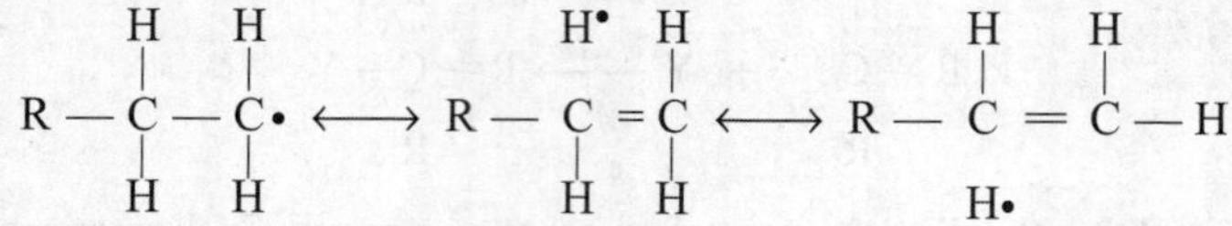

Greater the number of the resonating structures, greater is the stability of the free radical.

2. Allylic and benzylic type of free radicals are more stable and comparatively less reactive than simple alkyl radicals. The reason is the delocalisation of the unpaired electron over the π-orbital system.

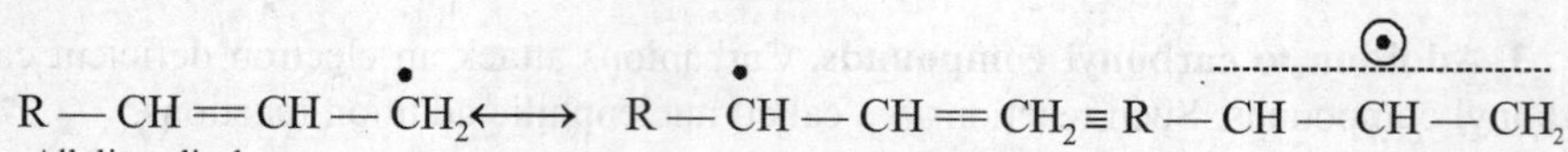

Allylic radical

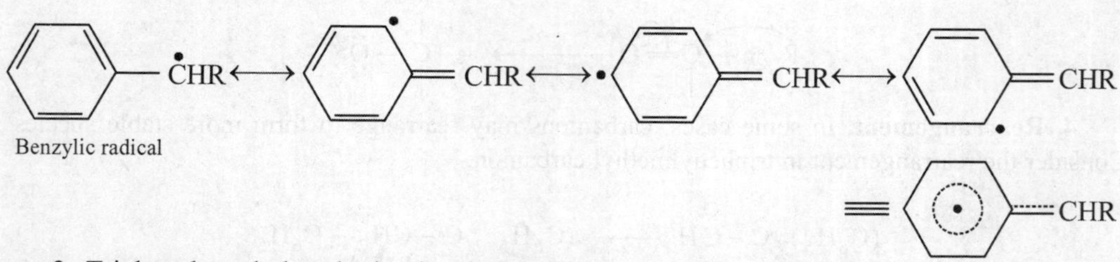

Benzylic radical

3. Triphenyl methyl and triarylmethyl radicals are much more stable in solution at room temperature. The stability of such radicals is due to resonance.

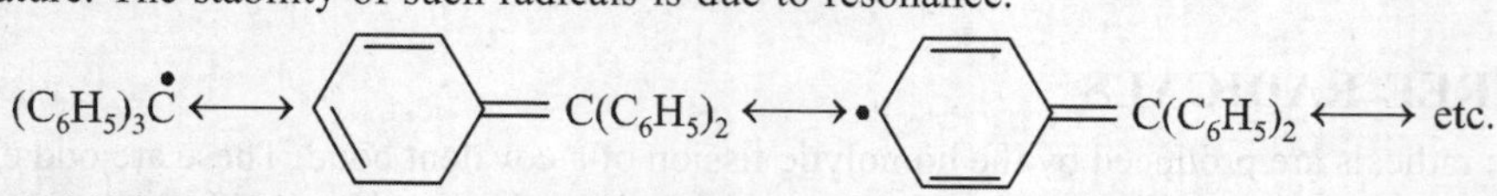

Steric hindrance to dimerisation is probably the major cause of their stability. If each aromatic nucleus in the radical has a bulky *p*-substitutent, then irrespective of any substitution at the *o*-positions, dimerisation will be greatly inhibited and hence radical stability increases.

Structure and reactions of free radicals

Structure. The state of hybridisation of carbon atom having the unpaired electron is not clearly established. However, it is believed to be either pyramidal (like ammonia) or planar. But experimental evidence suggests strongly that the alkyl radicals such as methyl radical are actually planar with sp^2 hybridisation. The three coplanar hybrid orbitals are used in the formation of three sigma bonds with other atoms. The unhybridised orbital which lies in a plane at right angles to the plane of the hybrid orbitals, carries the unpaired electron. ESR spectra of $\dot{C}H_3$ and other simple alkyl radicals indicate that these radicals have planar structures. This is also in accordance with the fact that optical activity is lost when a free radical is generated at an asymmetric carbon. As a general rule we can say that simple alkyl free radicals prefer a planar or near planar shape. However, the free radicals in which the carbon is connected to atoms of high electronegativity prefer a pyramidal shape. The increase in electronegativity causes the deviation from the planar geometry.

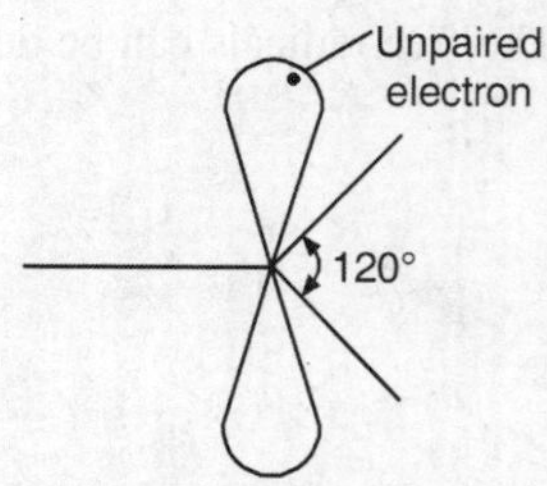

Structure of a free radical

Reactions

Some important reactions of free radicals are described below:

1. Halogenation of aliphatic hydrocarbons

$$Cl-Cl \xrightarrow{h\nu} 2Cl\bullet$$

$$CH_4 + Cl\bullet \longrightarrow CH_3\bullet + HCl$$

$$\dot{C}H_3 - Cl_2 \longrightarrow CH_3Cl + Cl\bullet$$

In the presence of sunlight, chlorination of methane gives chloromethane, dichloromethane, chloroform and carbon tetrachloride.

2. Addition. Halogen addition to alkenes takes place with free radical mechanism. The addition of chlorine to tetrachloroethene is photochemically catalysed. A molecule of chlorine undergoes homolytic fission giving two chlorine radicals. Each radical is capable of initiating a reaction chain.

$$Cl-Cl \underset{}{\overset{h\nu}{\rightleftharpoons}} C\bullet - Cl\bullet$$

$$Cl_2C = C.Cl_2 + Cl\bullet \longrightarrow Cl_2\dot{C} - CCl_3$$

$$\downarrow Cl-Cl$$

$$\dot{Cl} + Cl_3C - CCl_3$$

$\dot{Cl}$ radical further adds to tetrachloroethene. This continues till the whole of tetrachloroethene has been converted into hexachloroethane. Chain termination occurs through radical-radical collision. The radical reactions are inhibited by the presence of oxygen. The reason is that the molecule of oxygen has two unpaired electrons and behaves as a biradical. This biradical combines with highly reactive radical intermediate and converts it to less reactive peroxy radical which is unable to propagate the chain reaction.

3. Consider the addition of chlorine to benzne in the presence of light. The reaction proceeds by radical intermediates and gives benzene hexachloride. But the attack of chlorine radical on toluene results in preferential hydrogen abstraction giving substitution in CH_2 group. Clearly, it is because of the greater stability of benzyl radical due to delocalisation.

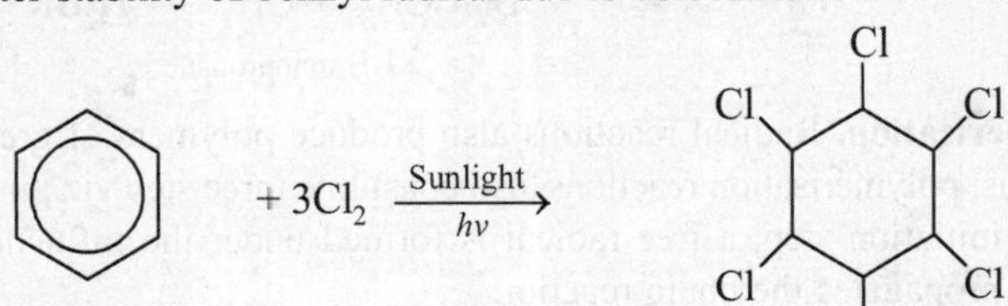

Benzene

Benzene hexachloride

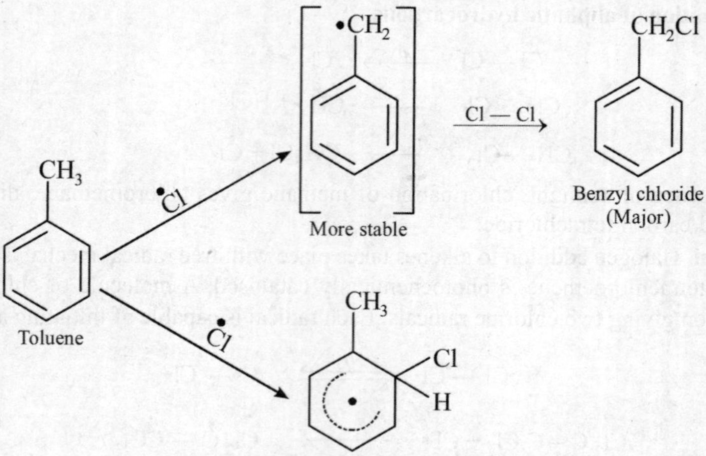

Less stable, no longer aromatic

4. The addition of HBr to propene in presence of peroxides yields 1-Bromopropane. The reaction involves the formation of radical intermediates giving anti-Markownikoff addition. In this reaction, $\dot{Br}$ radical attacks propene to form a more stable secondary free radical.

$$ROOR \longrightarrow 2R\dot{O}$$

$$R\dot{O} + H—Br \longrightarrow RO—H + \dot{Br}$$

$$CH_3CH{=}CH_2 + \dot{Br} \longrightarrow CH_3\dot{C}H—CH_2BR$$

$$\downarrow HBr$$

$$\dot{Br} + CH_3—CH_2—CH_2BR$$

1-Bromopropane

5. Vinyl polymerisation. Radical reactions also produce polymers of great importance. Like other radical reactions, polymerisation reactions also constitute three step viz.; initiation, propagation and termination. In initiation step, a free radical is formed under the influence of peroxide. The radical formed then propagates the chain reaction.

$$\dot{R} + CH_2 = CH \longrightarrow RCH_2 — \dot{C}H_2 \xrightarrow{CH_2 = CH_2} RCH_2 — CH_2 — CH_2 — \dot{C}H_2 \text{ and so on.}$$

In the final step, the collisions between the radicals terminate the reaction

$$R(CH_2)_n^{\bullet} + \dot{R} \longrightarrow R(CH_2)_n R$$

$$R(CH_2^{\bullet})_n + R(\dot{C}H_2)_n \longrightarrow R(CH_2)_{2n} R$$

Polythene, PVC, teflon etc. are the polymers which are formed from the respective monomers by radical pathways.

***Example 1.* Arrange the following free-radicals in increasing order of stability and explain your answer.**

$$\dot{C}_2H_5, (C_2H_5)_2\dot{C}(C_2H_5), (C_2H_5)_2\dot{C}H, \dot{C}H_3$$

Solution. The order of stability can be given considering the hyperconjugation phenomenon.

(*i*) Ethyl radical is a hybrid of four resonating structures.

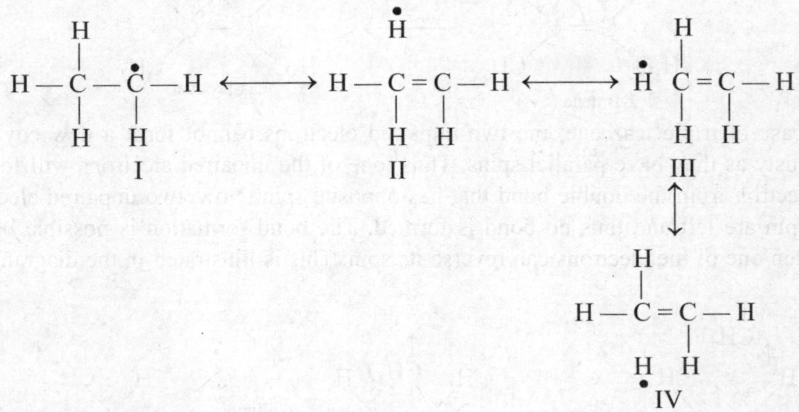

2.10 CARBENES

These are defined as the neutral organic species containing a divalent carbon atom having a sextet of electrons but no charge on it. For example,

(*a*) $:CH_2$ Methyl carbene (*ii*) $:CCl_2$ Dichlorocarbene

Formation of carbenes

They can be generated by the following methods:

(*b*) **By the action of UV light on diazomethane**

$$CH_2N_2 \longrightarrow \underset{\text{Methylene (Carbene)}}{:CH_2} + N_2$$

(*ii*) **By the action of sodium ethoxide on chloroform**

$$\underset{\text{Chloroform}}{CHCl_3} + C_2H_5ONa \longrightarrow \underset{\text{Dichloro carbene}}{:CCl_2} + C_2H_5OH + NaCl$$

These are highly reactive species.

Stability

1. The two unbonded electrons of a carbene may be either paired or unpaired. When the two electrons on the carbon atom are paired, then it is called *singlet* carbene. In the *triplet* state, it is called a biradical as the two electrons on carbon atom are unpaired.

$$\underset{\text{Singlet methylene}}{H-\ddot{C}-H} \quad \text{or} \uparrow\downarrow CH_2 \qquad \underset{\text{Triplet methylene}}{H-\dot{\underset{\cdot}{C}}-H} \quad \text{or} \uparrow\uparrow CH_2$$

In the triplet state, carbene is relatively more stable as it has lower energy content.

2. The singlet and the triplet species can be distinguished by the common addition reaction of carbene to double bond to form cyclopropane derivatives. The addition of singlet species to *cis*-Butene forms a *cis* product. The reason is that the movement of two pairs of electrons occur either simultaneously or with one rapidly succeeding the other.

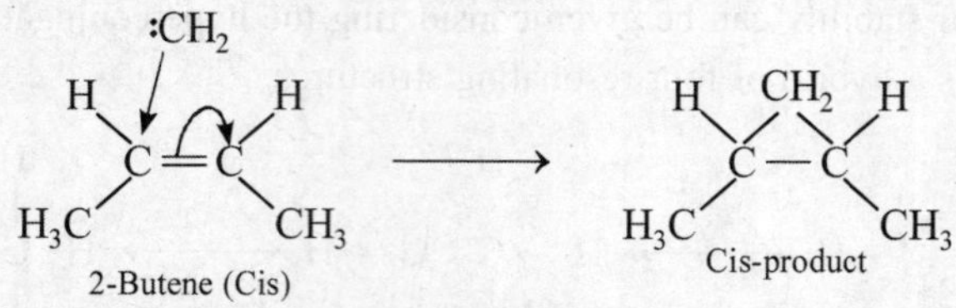

3. In case of triplet carbene, the two unpaired electrons cannot form a new covalent bond simultaneously as they have parallel spins. Thus, one of the unpaired electrons will form a bond with the electron from the double bond that has opposite spin. Now, two unpaired electrons with the same spin are left and thus no bond is formed. The bond formation is possible by collision process when one of the electrons can reverse its spin. This is illustrated in the diagram below:

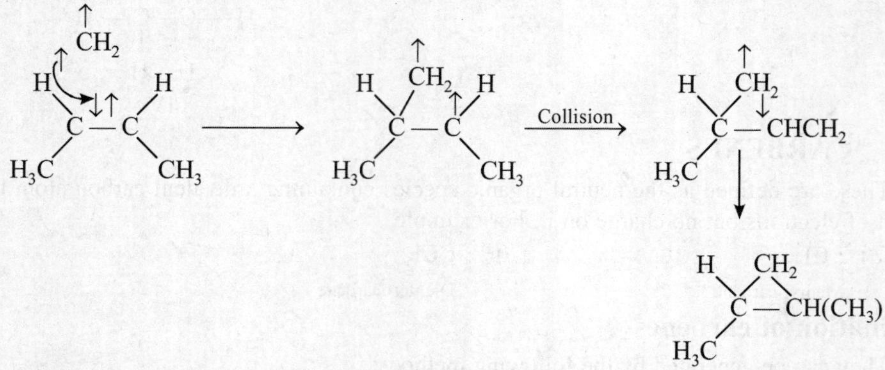

During this time of collision, there is a free rotation around C—C single bond and the product formed is a mixture of *cis* and *trans* isomers. It has been found that the difference in energy between the singlet and the triplet methylene is about 37–46 kJ/mole.

Structure and reactions of carbenes

Structure

The structure of two types of carbenes are different. : Singlet carbenes have their carbon in sp^2 hybridised state. Of the three sp^2 hybrid orbitals, two are used in forming two single bonds with monovalent atoms or groups attached to carbon. The unshared pair of electrons is present in the third sp^2 hybrid orbital and the unhybridised *p*-orbital is empty.

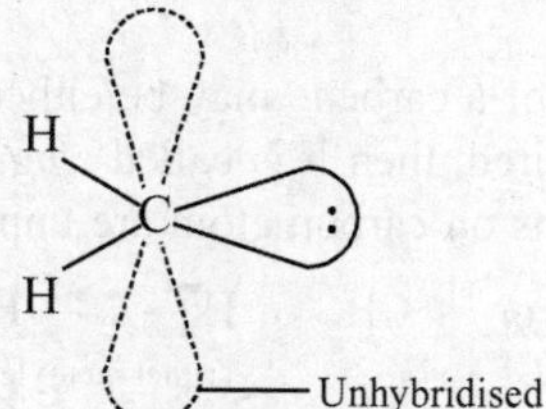

Singlet methylene

Triplet carbene has its carbon in *sp* hybridised state. The two *sp* hybrid orbitals form two bonds with monovalent atoms or groups attached to carbon. The two unhybridised p-orbitals contain one electron each.

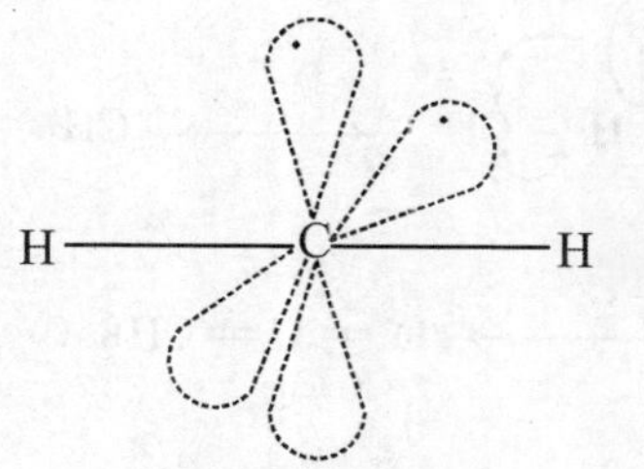

Triplet methylene

Triplet state aromaticity

The structure of triplet methylene is revealed by ESR measurements since these species are biradicals. Its geometry is found to be bent with bond angle of about 136°. The electronic spectra of singlet methylene formed in flash photolysis of diazomethane tell that it is also bent and the bond angle is about 103°.

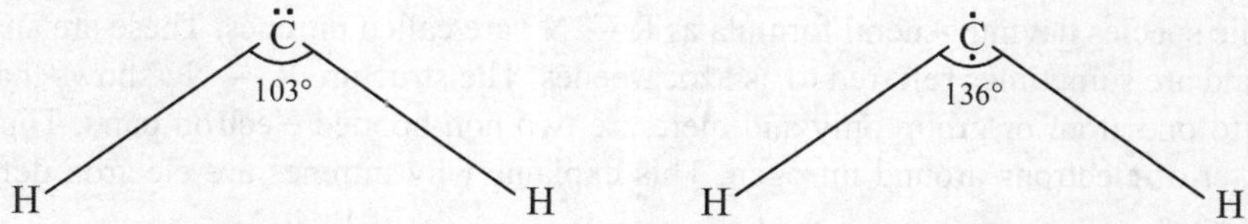

Singlet methylene Triplet methylene

The bond angles in CCl_2 and CBr_2 are 100°and 114° respectively.

Reactions

Some important types of reactions of carbenes are described below:

1. Addition to carbon-carbon double bonds. The reactions of singlet and triplet methylene to 2-Butene has been discussed above. Carbenes also add to aromatic systems and the intermediate products rearrange with ring enlargement reaction. Carbenes also show addition reactions to C = N bonds.

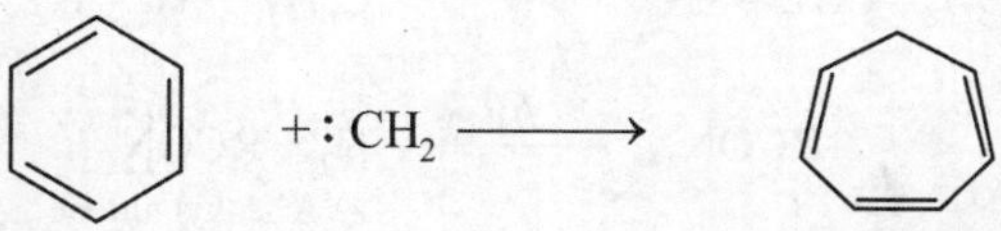

Benzene Cycloheptatriene

2. Addition reactions to alkanes. Methylene ($:CH_2$) reacts with ethane to form propane. Propane adds methylene to yield *n*-butane and isobutane. This shows the greater reactivity of carbene.

$$\underset{\text{Propane}}{CH_3-CH_2-CH_3} \xrightarrow{:CH_2} \underset{n\text{-Butane}}{CH_3-CH_2-CH_2-CH_3} + CH_3 + \underset{\text{Isobutane}}{CH_3-\underset{\displaystyle CH_3}{\underset{|}{CH}}-CH_3}$$

The reaction of carbenes with higher alkanes yield a number of possible products. It shows that the addition of carbenes is not all selective. Dichlorccarbenes do not give reactions involving insertion.

3. An important reaction of carbenes is dimerisation.

$$R_2\ddot{C} + R_2'\ddot{C} \rightleftharpoons R_2C = CR_2'$$

As the carbenes are highly reactive, the dimer formed carries so much energy that it dissociates again.

4. We also come across rearrangement reactions in carbenes with the migration of alkyl group or hydrogen atom. These rearrangments are much more rapid. Thus, the addition and insertion reactions of carbenes are seldom encountered with alkyl or dialkyl carbenes. Many rearrangements of carbenes directly give stable molecules. Some rearrangements involved in carbenes are:

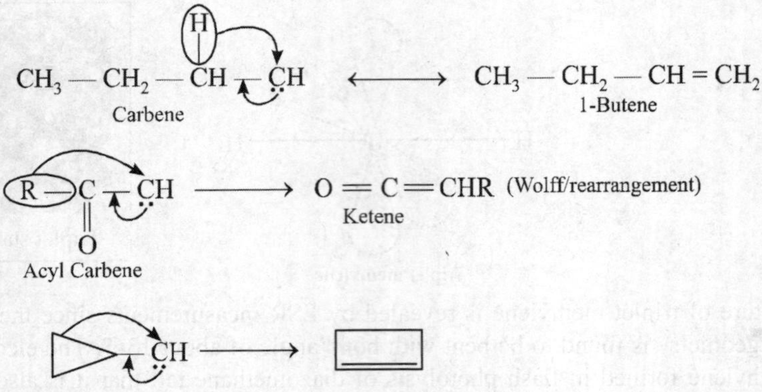

2.11 NITRENES

Organic species having general formula as R — N : are called nitrenes. These are analogous to carbenes and are sometimes referred to as azocarbenes. The structure R — N : shows that nitrogen is bonded to one atom or group only and there are two non-bonded electron pairs. Thus, there is only a sextet of electrons around nitrogen. This explains why nitrenes are electron deficient and reactive species.

The most common method of nitrene generation involves photolytic or thermolytic decomposition of azides, acyl azides and isocyanates

$$HN_3 \xrightarrow[\text{or } \Delta]{hv} H — \ddot{N}: + N_2 \uparrow$$

$$CH_3 — N_3 \xrightarrow{hv} CH_3 — \ddot{N}: + N_2 \uparrow$$

$$C_6H_5 — N_3 \xrightarrow{hv} C_6H_5 — \ddot{N}: + N_2 \uparrow$$

$$RCON_3 \xrightarrow{hv} N_2 + RCO\ddot{\underset{..}{N}}$$

Alkyl nitrene

$$R — N = C = O \xrightarrow{hv} CO + R — \ddot{N}$$

Stability of nitrenes

Nitrenes are highly reactive species and cannot be isolated under ordinary conditions. Alkyl nitrenes have been isolated by trapping in matrices at 4K while aryl nitrenes (less reactive) can be trapped at 77 K. Nitrenes can be generated both in the singlet and the triple states but its ground state is probably the triplet state.

$$\underset{\text{Singlet}}{R — \ddot{N}:} \qquad \underset{\text{Triplet}}{R – \dot{\underset{..}{N}}\cdot}$$

Reactions of Nitrenes

Nitrenes show the following reaction:

1. Addition to C = C bonds. Nitrenes add to alkenes to form cyclic product. Such reactions are common for acyl nitrenes.

Cycloaddition of nitrenes to alkenes produces acridines.

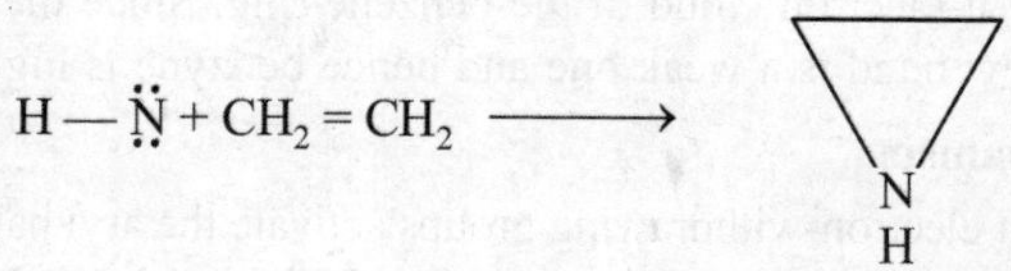

Ethylenimine

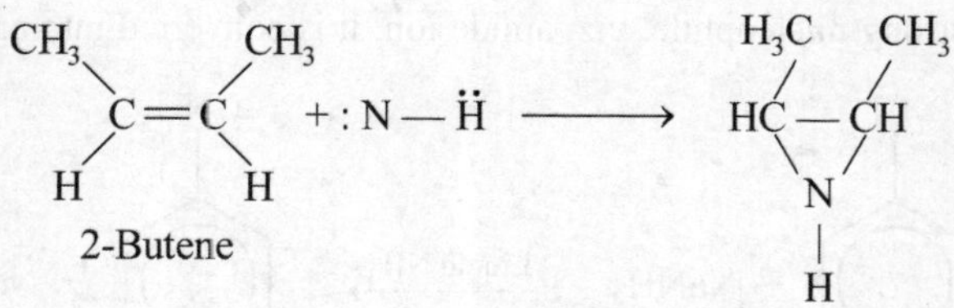

2-Butene

Acridine

2. Insertion. Nitrenes, especially acyl and sulphonyl nitrenes can insert into C — H and certain other bonds.

$$R'—\overset{O}{\overset{\|}{C}}—\ddot{N}\!: + R_3C—H \longrightarrow R'\overset{O}{\overset{\|}{C}}—\overset{H}{\overset{|}{N}}—CR_3$$

3. Abstraction. Nitrenes are also capable of extracting hydrogen atom form an alkane to form an alkyl radical.

$$R—\ddot{\underset{\cdot\cdot}{N}} + R—H \longrightarrow R\dot{\underset{\cdot\cdot}{N}}—H + \dot{R}$$

4. Rearrangements. Nitrenes undergo rearrangements also due to the migration of a group from adjacent atom to the electron deficient nitrogen. Insertion and addition to C = C bond reactions do not proceed in alkyl nitrenes as the rearrangement is much more faster.

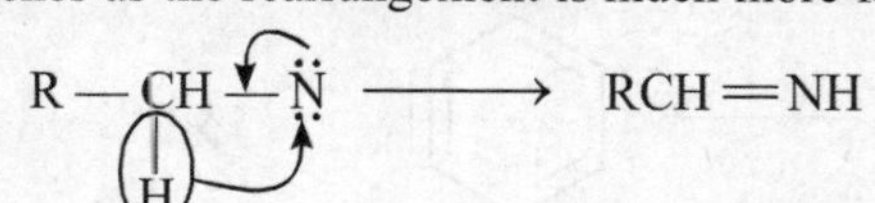

5. Dimerisation. Nitrenes also dimerise to form a di-imide. Aryl nitrenes yield azobenzenes.

$$2C_6H_5—\ddot{\underset{\cdot\cdot}{N}} \longrightarrow C_6H_5—N=N—C_6H_5$$

6. Ring enlargement. Azepines are prepared by the ring enlargement of phenyl azide.

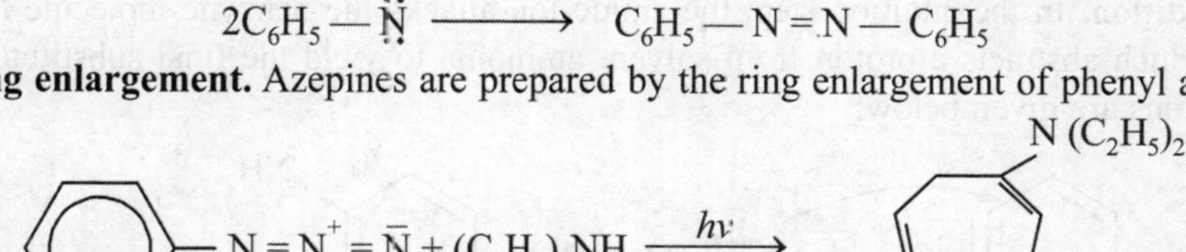

2.12 ARYNES

These are highly reactive species which contain an additional bond between two carbon atoms of the benzene rings. Taking benzyne as an example of aryne, the new bond of benzyne is formed by the sideways overlap of sp^2 orbitals belonging to two neighbouring carbon atoms. It is depicted as:

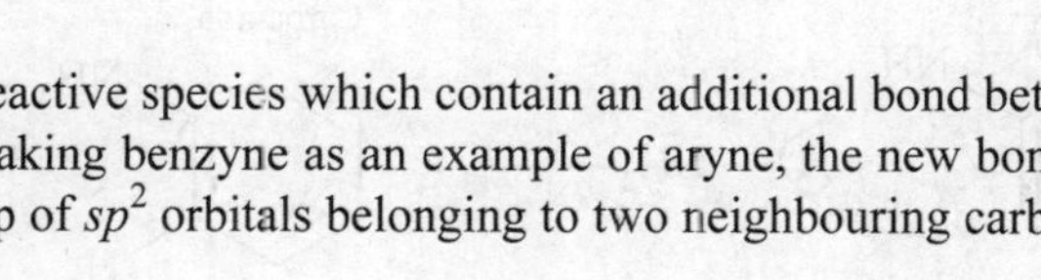

Representation of benzyne

It is evident that the new extra bond orbital lies along the side of the benzene ring and it has little interaction with the π-electron cloud of the benzene ring. Since the sideways overlapping is not very effective, the new bond is a weak one and hence benzyne is highly reactive molecule.

Aryne (Benzyne) Mechanism

It has been found that electron-withdrawing groups activate the aryl halides towards bimolecular nucleophilic substitution reactions. In the absence of such activation, aryl halides can also be made to undergo nucleophilic substitution in the presence of strong nucleophiles. Thus, when chlorobenzene is treated with very strong nucleophile, viz. amide ion, it is converted into aniline.

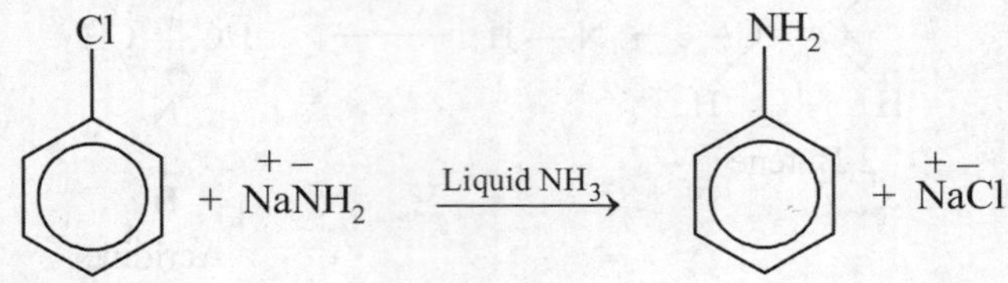

Chlorobenzene Aniline

The above type of nucleophilic substitution reaction occurs by benzyne mechanism. It involves both **elimination** and **addition** as represented below:

(I) Elimination. In the elimination step, the amide ion abstracts a proton from one of the ortho-positions with respect to the halogen. The resulting carbanion loses the halides ion to form the benzyne, as illustrated below:

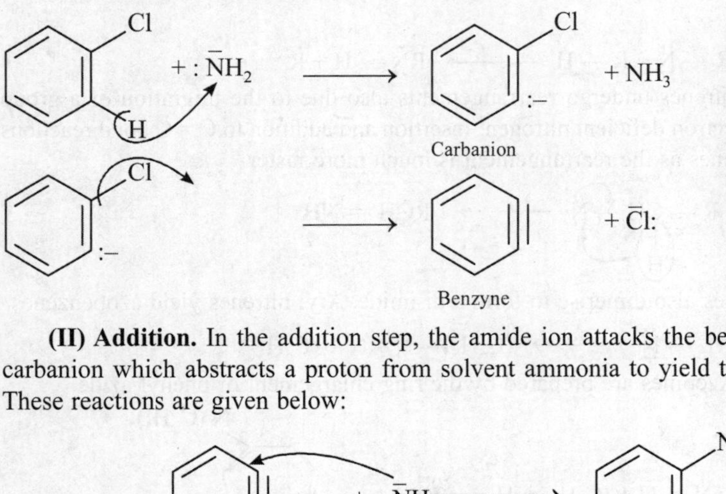

Carbanion

Benzyne

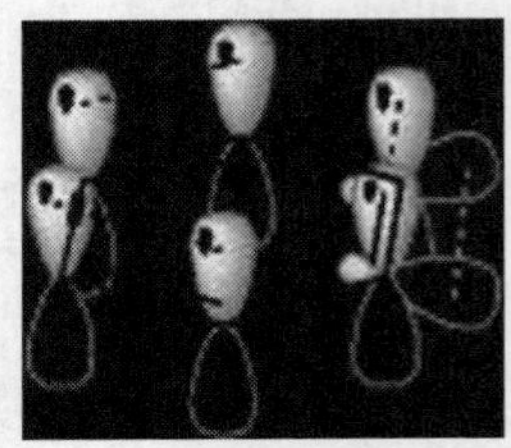

pi orbitals in arynes

(II) Addition. In the addition step, the amide ion attacks the benzyne molecule to form the carbanion which abstracts a proton from solvent ammonia to yield the final substituted product. These reactions are given below:

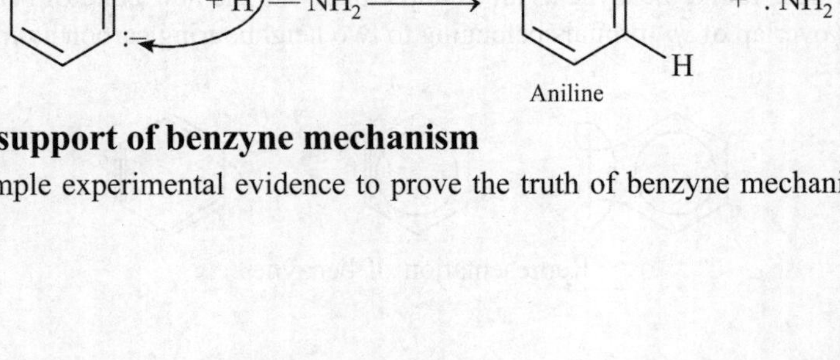

Carbanion

Aniline

Evidence in support of benzyne mechanism

There is ample experimental evidence to prove the truth of benzyne mechanism of aromatic substitution.

***(i)* Isotope effect.** The amination of *o*-deuteriobromobezene is slower than that of bromobenzene. This observation indicates that the cleavage of *ortho* hydrogen is involved in the rate determining step during the elimination stage. This can further be illustrated as follows:

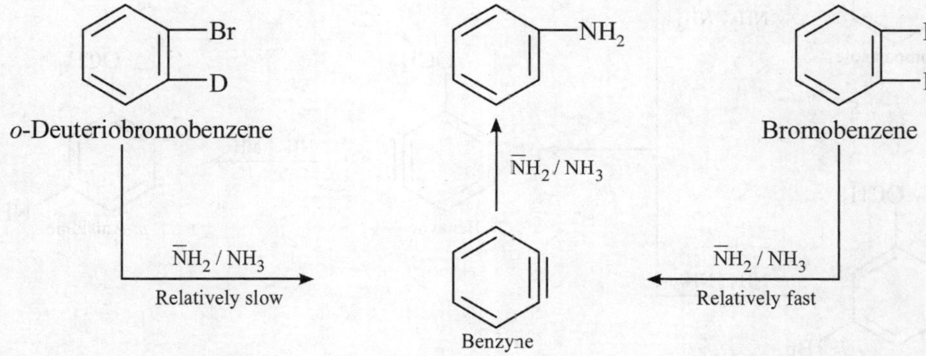

***(ii)* Absence of ortho hydrogen.** Aryl halides containing two groups ortho to hydrogen, like 2-bromo-3-methylanisol, do not react at all. This is because benzyne intermediate cannot be formed due to the absence of ortho hydrogens with respect to the halogen.

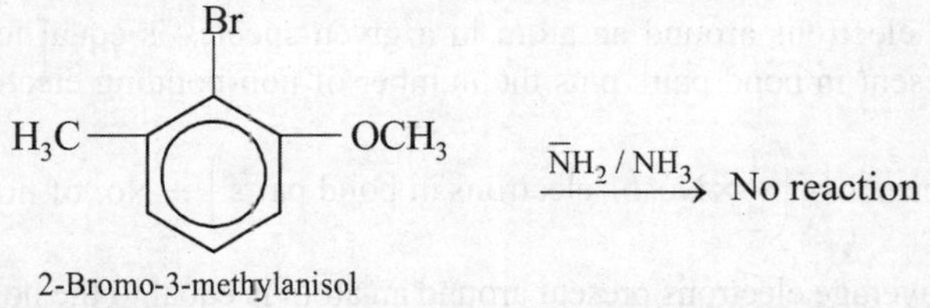

***(iii)* Tracer studies.** When labelled chlorobenzene in which chlorine is linked to C^{14} isotope (C^*) is allowed to react with amide ion, two types of aniline are formed. In one type, the —NH_2 group is bonded to C^{14} while in the other it is attached to carbon *ortho* to the labelled carbon. This can be explained as follows:

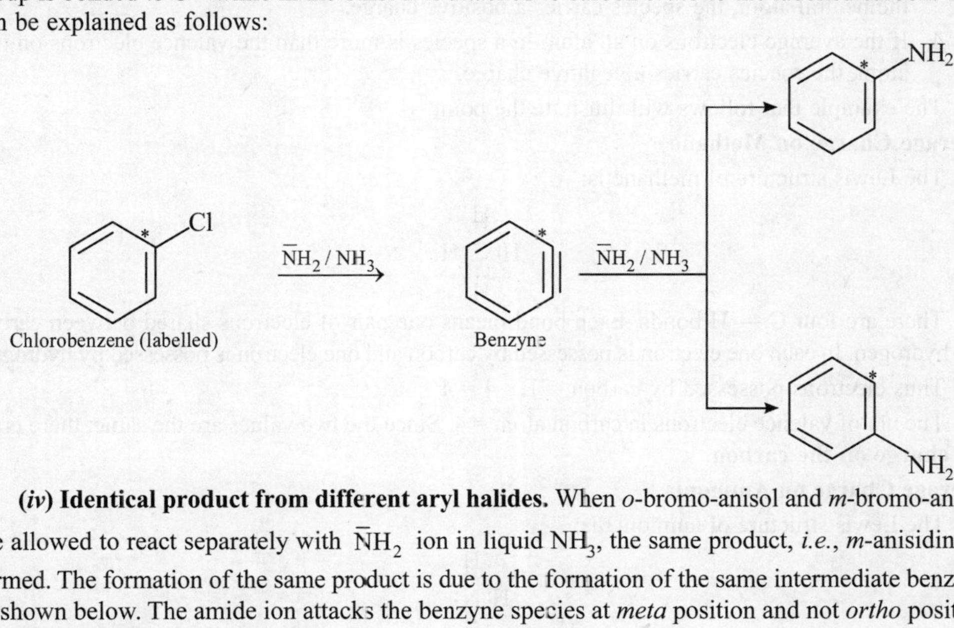

***(iv)* Identical product from different aryl halides.** When *o*-bromo-anisol and *m*-bromo-anisol are allowed to react separately with $\bar{N}H_2$ ion in liquid NH_3, the same product, *i.e.*, *m*-anisidine is formed. The formation of the same product is due to the formation of the same intermediate benzyne as shown below. The amide ion attacks the benzyne species at *meta* position and not *ortho* position probably because of steric factor.

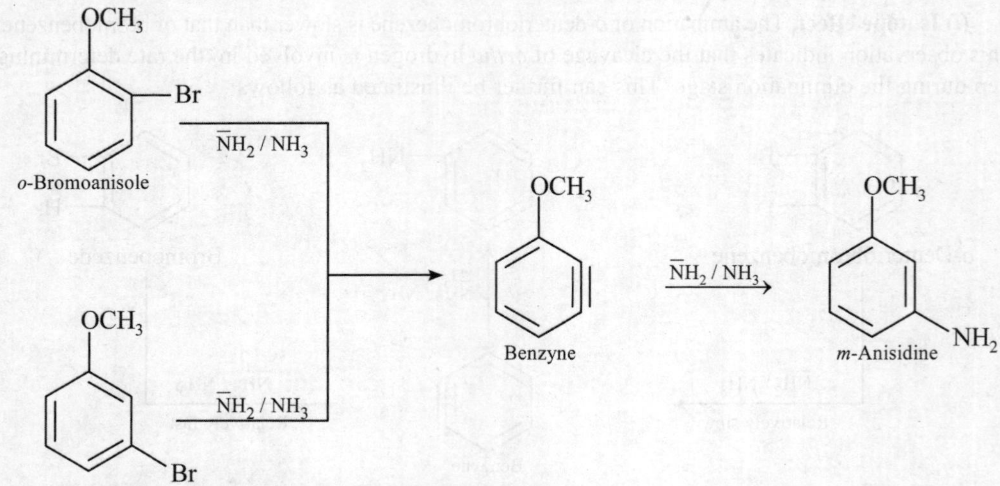

2.13 ASSIGNING FORMAL CHARGE

Formal charge on a intermediate or an ionic species is calculated from Lewis structure. The main features of this method are given below:

1. The average electrons around an atom in a given species is equal to half the number of electrons present in bond pairs plus the number of non-bonding electrons.

$$\text{Average electrons} = \left(\frac{1}{2} \times \text{No. of electrons in bond pairs}\right) + \text{No. of non-bonding electrons}$$

2. If the no. of average electrons present around an atom is equal to the no. of valence electrons in the neutral atom, there is no charge on the species.
3. If the average electrons present around an atom is less than the no. of valence electrons in the neutral atom, the species carries a positive charge.
4. If the average electrons on an atom in a species is more than the valence electrons on that atom, the species carries a negative charge.

The example that follows will illustrate the point.

Average Charge on Methane

The Lewis structure of methane is:

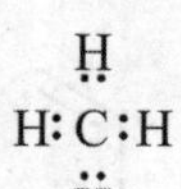

There are four C — H bonds. Each bond means one pair of electrons shared between carbon and hydrogen. In each one electron is possessed by carbon and one electron is possessed by hydrogen.

Thus electrons possessed by carbon = 1 × 4 = 4

The no. of valence electrons in carbon atom = 4. Since the two values are the same, there is **no net charge on the carbon.**

Average Charge on Ammonia

The Lewis structure of ammonia is:

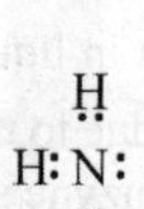

There are three bond pairs and one lone pair of electrons in ammonia. Bond pair electrons are shared between N and H whereas lone pair electrons are solely possessed by N.

$$\text{Average electrons on N} = \left(\frac{1}{2}\times 6\right) + 2 = 5.$$

This is equal to the number of valence electrons in neutral N. Hence there is not net charge on ammonia.

Average Charge on $\dot{C}H_3$ Radical

The structure of methyl radical is

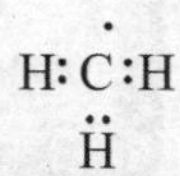

There are three bond pairs (or 6 elecrons) shared between C and H. Additionally one electron is solely possessed by C.

Therefore, Average electrons present on C = 3 + 1 = 4

Again there is no net charge on methyl radical.

Average Charge on Carbocation

Let us consider the charge on methyl carbocation, which has the structure

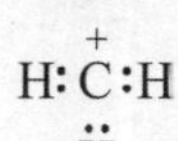

Here average charge on carbon = 3.

This is one less than the valence electrons in neutral carbon.

Therefore charge on CH_3 = 4 – 3 = 1

Average Charge on Carbanion

Let us consider the charge on methyl carbanion having structure

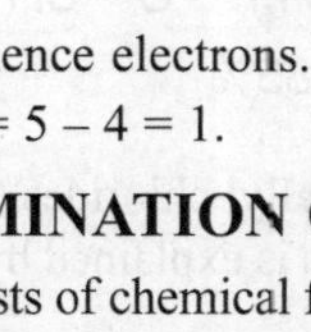

It carries extra electron which is responsible for its negative charge. No. of average electrons on C = 3 + 2 = 5.

This is one more than the no. of valence electrons.

Hence charge on methyl carbanion = 5 – 4 = 1.

2.14 METHODS OF DETERMINATION OF REACTION MECHANISM

The stoichiometry of a reaction consists of chemical formulas and relative molecular proportions of starting materials and products. Obviously, these have a bearing on the mechanism of the reaction, for the overall reaction course must proceed from the starting materials to the products. Stoichiometry of the reaction may be misleading, however, because the participants in the overall reaction may not be involved directly in the rate determining step.

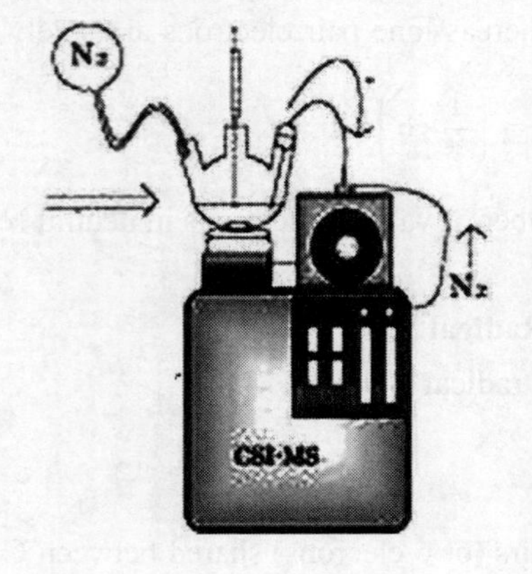

Analysis of organic reactions

Mechanism of a reaction can be determined by the following methods/techniques.

1. Product analysis
2. Intermediates
3. Isotope effects
4. Kinetic studies
5. Stereochemical studies
6. Crossover experiments

These are being described separately.

1. ***Product anlysis.*** In this method, we identify and determine the quantities of product and by products (if any). The following reactions are being discussed.

   (i) **Hydrolysis of isomeric allyl chlorides.** The isomeric allyl chlorides A and B as given below on hydrolysis give 85% of tertiary alcohol and 15% primary alcohol

$$\underset{(A)}{(CH_3)_2C=CH-CH_2Cl} \xrightarrow{H_2O} \underset{85\%}{(CH_3)_2-\overset{OH}{\overset{|}{C}}-CH=CH_2} + \underset{15\%}{(CH_3)_2C=CH-CH_2OH}$$

$$\underset{(B)}{(CH_3)_2-\overset{Cl}{\overset{|}{C}}-CH=CH_2} \xrightarrow{H_2O} \underset{85\%}{(CH_3)_2-\overset{OH}{\overset{|}{C}}-CH=CH_2} + \underset{15\%}{(CH_3)_2-C=CH-CH_2OH}$$

It is intriguing that two different allyl chlorides (A and B) on hyrolysis give the same two products in same percentage. This is explained by the following mechanism.

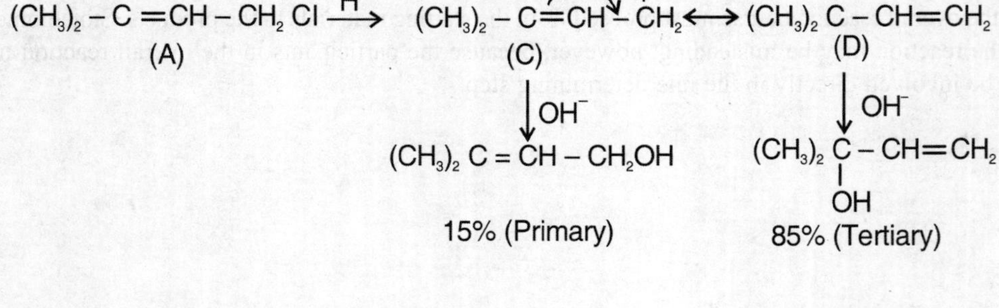

$$\underset{\text{(B)}}{(CH_3)_2\overset{OH}{\overset{|}{C}}-CH{=}CH_2} \xrightarrow{H^+} (CH_3)_2\overset{+}{C}-CH{=}CH_2 \longleftrightarrow (CH_3)_2C{=}CH-\overset{+}{C}H_2$$

$$\downarrow OH^- \qquad\qquad \downarrow OH^-$$

$$\underset{\text{85 \% (Tertiary)}}{(CH_3)_2\overset{OH}{\overset{|}{C}}-CH{=}CH_2} \qquad \underset{\text{15 \% (Primary)}}{(CH_3)_2C{=}CH-CH_2OH}$$

(ii) **Beckmann reaction.** It is observed that out of cis - 2 bromo - 5 nitroacetophenone (A) and trans - 2 - bromo - 5 - nitroacetophenone (B), only the former undergoes ring closure. The latter oxime (B) does not undergo ring closure. It rather gives Beckmann reaction forming substituted amide. This difference in behaviour of the two oximes explains the mechnism by which the reaction takes place with the two compounds cis-oxime (A) and trans-oxime (B). The latter compound does not undergo rign closure because of trans geometry.

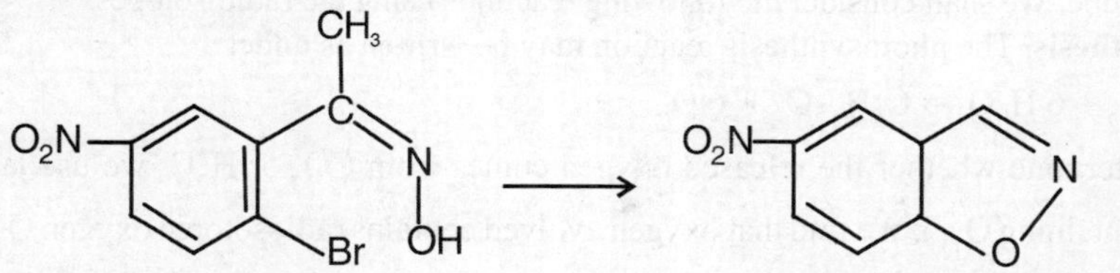

(A) (Ring closure)

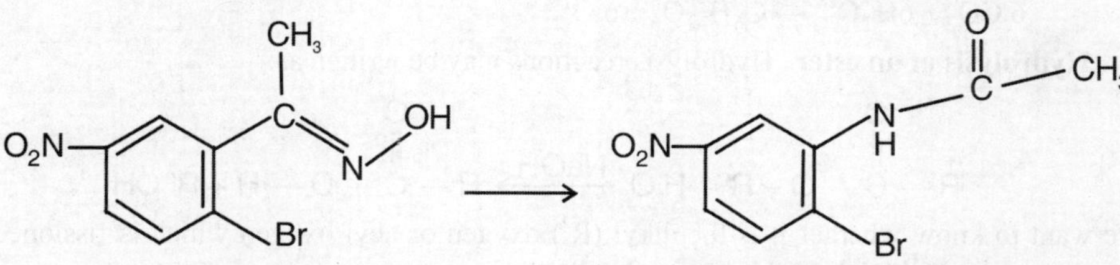

(B) Substituted amide

2. *Determination of presence of intermediates.* The detection and isolation of the intermediate can provide information regarding the mechanism of the reaction.

Stable intermediates can be isolated by interrupting the reaction before it is completed. Take for example, Hofmann bromamide reaction. In this reaction, three intermediates viz. $RCONHBr$, $RCONBr^-$ and $RNCO$ have been isolated. This reaction gives the conversion of an amide into an amine.

$$RCONH_2 \xrightarrow{Br_2|KOH} R-NH_2$$

Thus the mechanism which will take into account the formation of the above intermediate would be the valid mechanism for the reaction. A sure short proof for the existence fo an intermediate would be to trap it with an appropriate trapping agent. For example, in reactions with diazomethane, nitrogen gas is evolved and a carbene is produced. This carbene can be trapped as cyclopropane derivative using cyclohaxene.

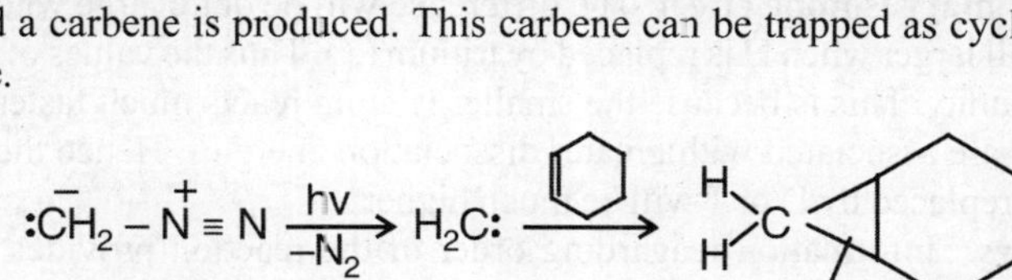

Cyclopropane

Similarly mechanisms involving benzyne can be confirmed by trapping benzyne with anthracene and furan.

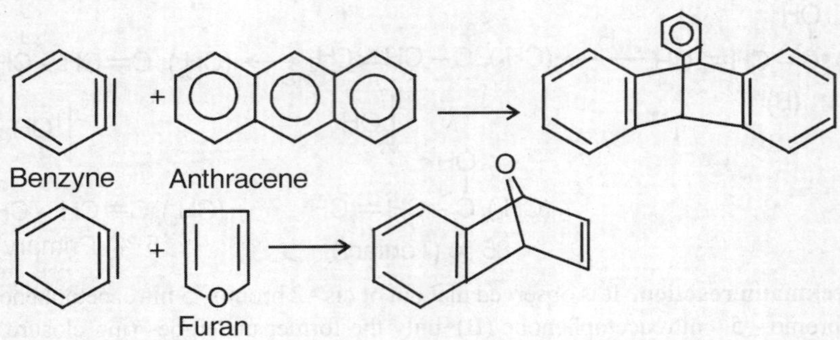

3. *Isotope effect.* Radioisotopes like deuterium ^2_1H, tritium ^3_1H and $^{14}_6C$ are extensively used in the determination of mechanism of organic reactions. The principle of the method is that one of the reactants contains an isotopic atom. We can establish the mechanism of the reaction by locating the position of that isotope. We shall consider the following reactions using the radioisotopes.

I. Photosynthesis. The photosynthesis reaction may be written as under

$$6\,CO_2 + 6\,H_2O \rightarrow C_6H_{12}O_6 + 6\,O_2$$

In order to determine whether the released oxygen comes from CO_2 or H_2O, we use labelled water (i.e. water containing O_{18}). We find that oxygen evolved contains radiosotopic oxygen O_{18} thus proving that oxygen released in photosynthesis of plants comes from water.

$$6\,CO_2 + 6H_2O^{18} \rightarrow C_6H_{12}O_6 + 6\,O_2^{18}$$

II. Hydrolysis of an ester. Hydrolysis reactions may be written as:

$$R-\overset{\overset{\large O}{\|}}{C}-O-R' + H_2O \xrightarrow{NaOH} R-\overset{\overset{\large O}{\|}}{C}-O-H + R'\,OH$$

We want to know whether it is the alkyl (R′) oxygen or acyl oxygen which is fissioned. We perform the reaction with labelled water i.e. H_2O^{18}.

$$R-\overset{\overset{\large O}{\|}}{C}-O-R' + H_2O^{18} \longrightarrow R-\overset{\overset{\large O}{\|}}{C}-O-H + R'-O^{18}-H$$

We obtain the carboxy acid which does not contain radioactive oxygen (O^{18}), but the alcohol obtained contains redioactive oxygen O^{18}. Thus, we can say that it is the fission of alkyl oxygen and not the acyl oxygen that takes place in the mechanism of the reaction.

The rate of a reaction changes when an atom is replaced with a radioisotope. Variation in the rate of a reaction due to the presence of an isotope in the molecule is called **isotope effect**. Normally the magnitude of isotope effect is small. However, when the cleavage of the bond involving the isotope is the rate determining stap, the magnitude of isotope effect may be quite significant.

This is called **primary isotope effect**. The difference will be quite large when H is replaced by deuterium (D) and still larger when H is replaced by tritium (T). Thus the values of K_H/K_D and K_H/K_T with be greater than unity. This is because the smaller H atom reacts much faster than D or T. Also C–D and C–T bonds are associated with greater dissociation energies. Hence the activation energy required in case H is replaced by D or T will be much higher.

4. *Kinetic studies.* Information reagarding order of the reaction provides us the clue to the mechanism of the reaction. For example the rate law for acid hydrolysis of $(CH_3)_3$ CBr is as follows.

$$\frac{-d\left[(CH_3)_3\,CBr\right]}{dt} = k\left[(CH_3)_3\,CBr\right]$$

This shows that the rate of reaction is not dependent on the concentration of H_2O or H^+ amd therefore the following mechanism justifies the kinetic studies

First Step $(CH_3)_3\,C\,Br \rightarrow (CH_3)_3\,\overset{+}{C} + B\overset{-}{r}$ (Slow and rate determining step)

Second Step $(CH_3)_3\,\overset{+}{C} + \overset{-}{OH} \rightarrow (CH_3)_3\,C\,OH$ (Fast)

The rate low obtained in the kinetic studies of hydrolysis of CH_3Br in the alkaline hydrolysis is as follows.

$$-\frac{d[CH_3Br]}{dt} = k[CH_3Br]\,[O\overset{-}{H}]$$

That means both the halide and hydroxl ions participate in the rate determining step. Hence, the following mechanism is formulated.

(i) $CH_3\ Br + O\overset{-}{H} \rightarrow [OH\ .....\ CH_3.....Br]^-$ (Slow and rate detemining step)

(ii) $[OH\ ....CH_3....Br]^- \rightarrow CH_3 - OH + Br^-$ (Fast)

5. ***Sterochemical studies*** Stereochemical evidance is of great importance in estabilishing the mechanism of reactions. Mechanism of nucleophilic substitution may be arrived at by performing experiments with optically active molecules. We check the optical activity of the products and then proceed to establish the mechanism. For example, if a racemic mixture is obtained after the nucleophilic substitution of an optically active compound, we can say that the reaction involves the formation of a planar intermediate as the first step.

In the second step, the nucleophide attaches itself to the planar intermediate. Since there is an equal opportunity of the nucleophile attaching from the two sides of the intermediate, we obtain a racemic mixture.

These steps indicate that the mechanism of the reaction in SN^1. However, if the product is optically active but the configuration of the product is opposite of the configuration of the reactant, it points to SN^2 mechanism. For example, we observe an inversion of configuration on alkaline hydrolysis of $C_6H_{13}(CH_3)$ CHBr. We therefore sketch the mechanism as follows.

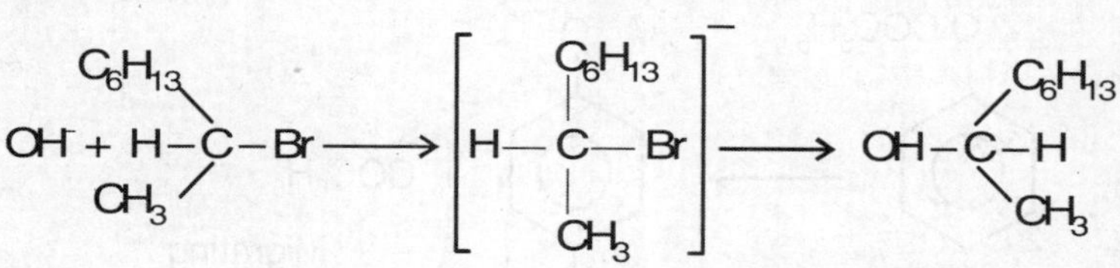

(With inversion of configuration)

The reaction is known as **Walden inversion**

6. ***Crossover experiment.*** This technique is employed if we want to know whether the rearrangment takes place in one step or two steps. The reaction is carried out by taking two similar but different reactants. Analysis of the products leads to the mechanism. If the reaction involes two steps, the migrating group becomes free. In such a case, the migrating group has the possibility to attach to the intermediates of both the reactants. As we have taken two reactants, two migrating group becoms free. In such a case, the migrating group has the possibility to attach to the intermediates of both the reactants. As we have taken two reactants, two migrating groups will be formed, each of which has the possibility of attaching to the two intermediates, thus yielding a total of four products. The mechanism of **Fries rearrangement** was established by this method.

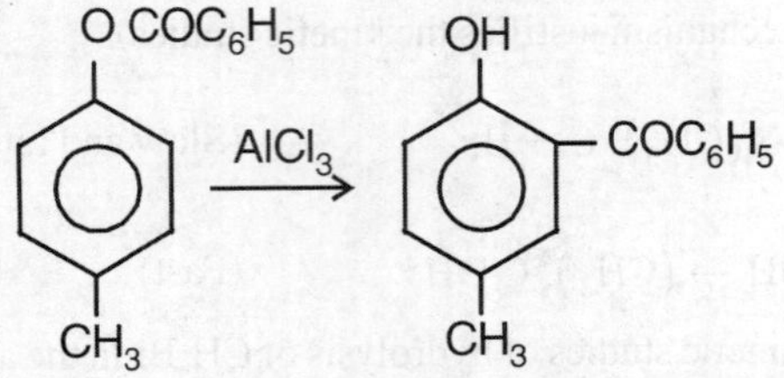

(Fries Rearragement)

The following mechanism has been suggested for this rearrangement

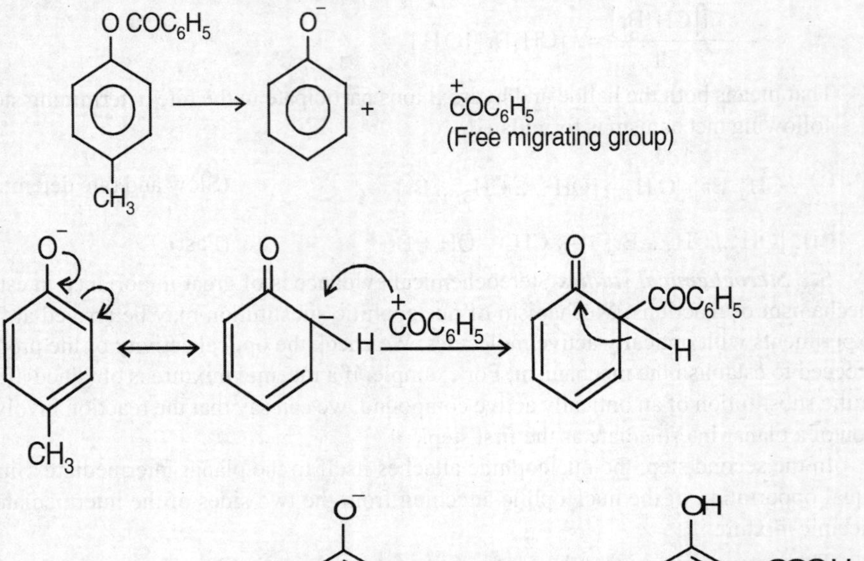

To confirm this mechanism, we take a mixture of the two compounds (A) and (B) as shown below.

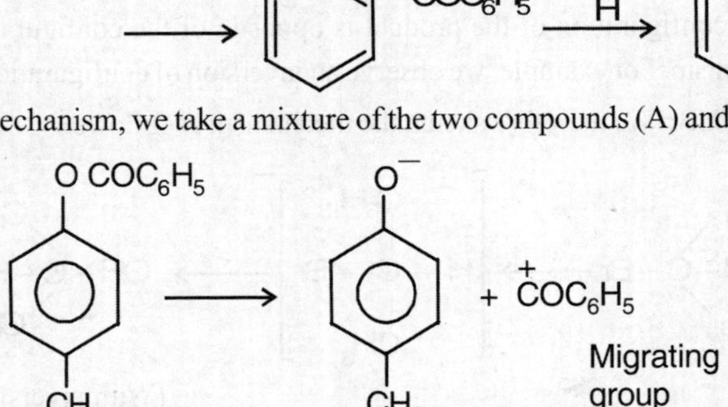

(A) Intermediate

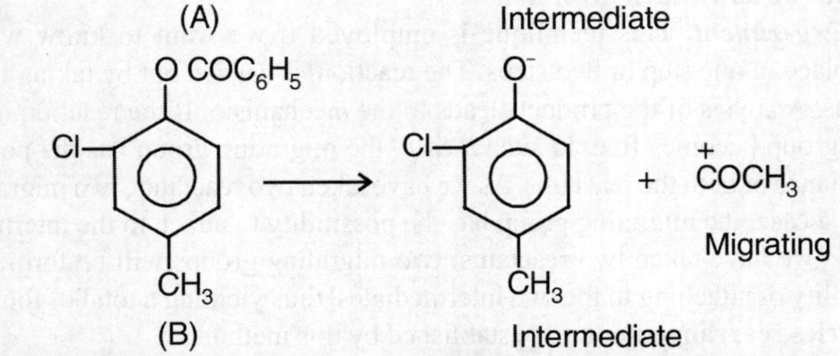

(B) Intermediate

Now the migrating group from the compound A has the possibilities of attaching to the intermediates of both (A) and (B). Similarly the migrating group from the compound (B) has the possibility of attaching to the intermediates of both (A) and (B). Thus, we expect to obtain four products if this really happens. The four products as listed below have actually been isolated

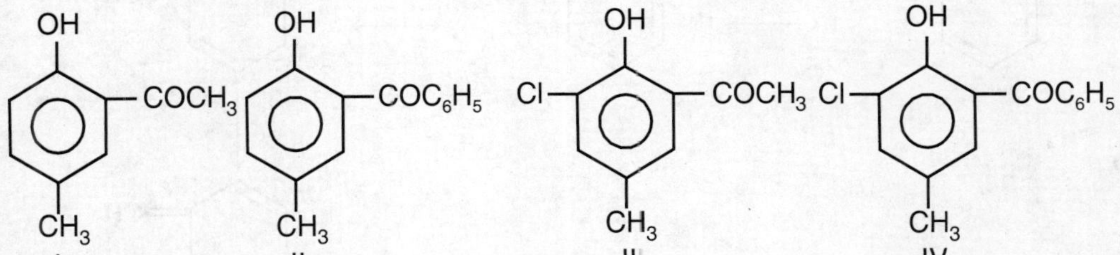

I II III IV

This lends support to the proposed mechanism

EXAMPLES

***Example 1.* Write down possible products in the following reactions:**

1. $CH_2(COOEt)_2 \xrightarrow[C_2H_5I]{NaOMe}$

2. $CH_3-CH_2-CH_2^+ \xrightarrow{OH^-}$

Solution.

(1) $CH_2(COOEt)_2 \xrightarrow[-MeOH]{NaOMe} \overset{-}{C}H(COOEt)_2$ (Carbanion)

$\xrightarrow{C_2H_5I} \underset{C_2H_5}{CH(COOEt)_2} \xrightarrow{Hydrolysis} C_2H_5CH_2COOH$ (Butanoic acid)

(2) $CH-CH_2-\overset{+}{C}H_2$
(*n*-Propyl carbocation)

$\downarrow$

$CH_3-\overset{+}{C}H-CH_3$
(more stable carbocation)

$\downarrow OH^-$

$CH_3-CH(OH)CH_3$
Propanol-2

***Example 2.* How do you explain that**

(*i*) Benzyl carbocation is much more stable than phenylethyl carbocation?

(*ii*) Triphenyl methyl carbocation is a very stable carbocation?

Solution.

(*i*)

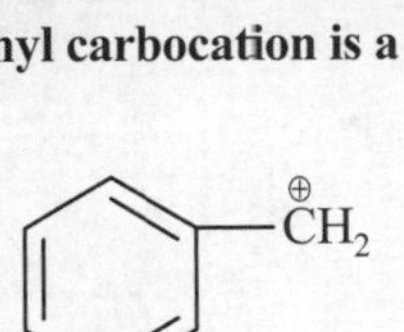

Benzyl carbocation

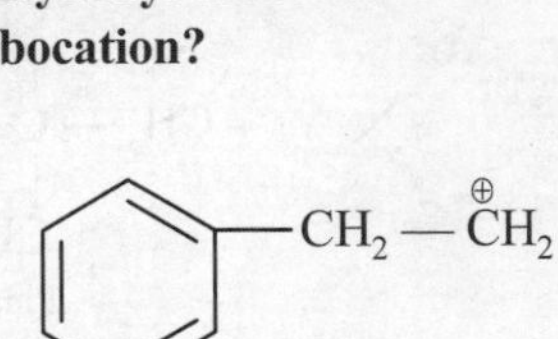

Phenylethyl carbocation

Benzyl carbocation exists in a number of canonical forms and hence is more stable than phenyl ethyl carbocation in which there is no possibility of resonance.

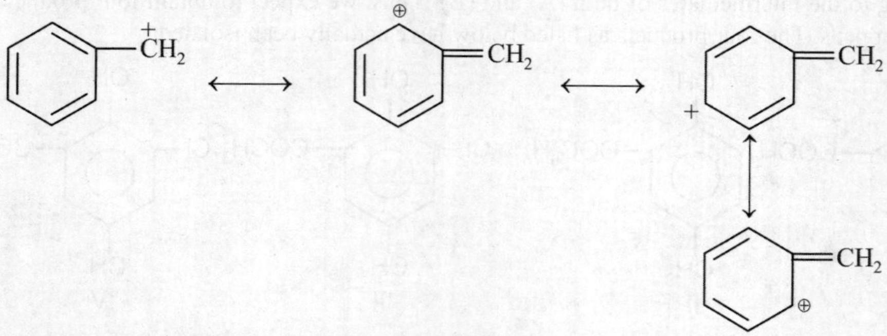

(*ii*) Triphenylmethyl carbocation is a very stable carbocation. See Art. 2.8.

***Example 3.* What happens when 2, 3-Dimethyl butan-2, 3-diol is treated with dil. H_2SO_4?**

Solution.

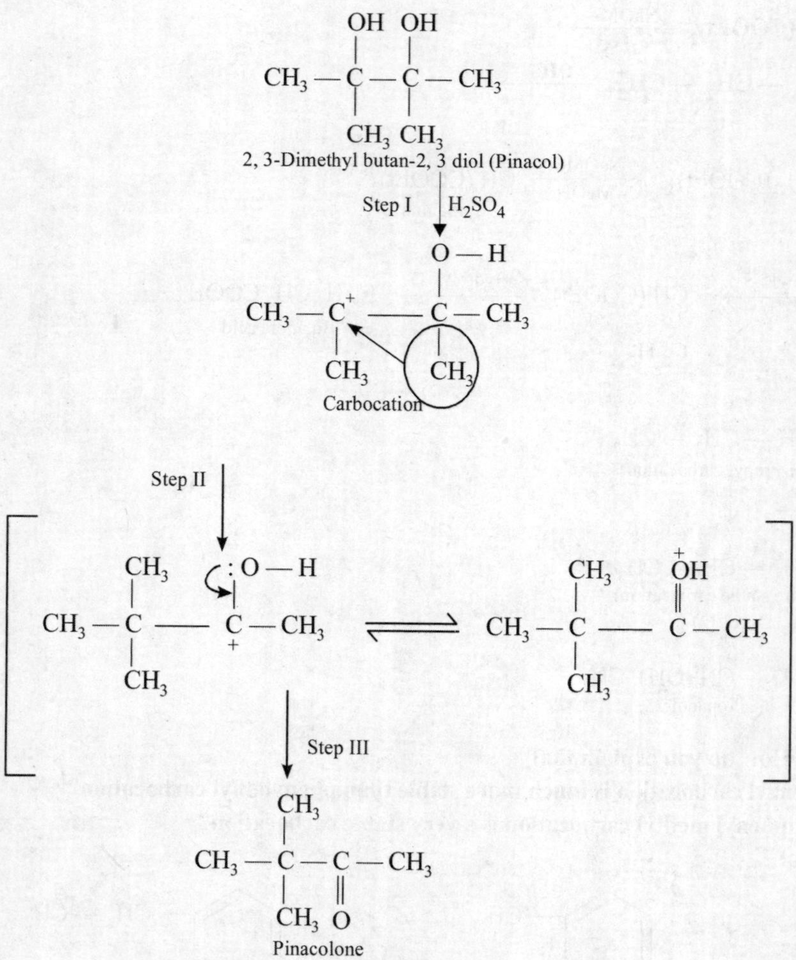

Migration of methyl group in step II is facilitated by the resonance possible in the product.

Example 4. **Arrange the following carbocations in the order of increasing stability:**

$$C_6H_5CH_2^+, (C_6H_5)_2\ CH^+, (CH_3)_3\ C^+, (CH_3)_3\ CH^+$$

Solution. $(C_6H_5CH_2^+ < (C_6H_5)_2\ CH^+ < (CH_3)_2\overset{+}{C}H < (CH_3)\overset{+}{C}$

Example 5. **Which of the two, single and triplet carbene is more stable and why?**

Solution. Triplet carbenes are more stable than singlet carbenes because two unhybridised p-orbitals contain one electron each while unshared pair of electrons is present in the third sp^2 hybrid orbital and the unhydridised *p*-orbital is empty.

Example 6. **Mention the methods by which you can infer that a particular reaction involves a free radical intermediate or not.**

Solution. Methods:

1. Formation of free radicals is favoured by the presence of u.v. light, heat and organic peroxides.
2. Free radicals are paramagnetic due to the presence of unpaired electrons.
3. Reactions involving free radicals widely occur in the gas phase.

Example 7. **Explain why benzyl carbocation is more stable than cyclohexyl carbocation.**

Solution. Benzyl carbocation is able to stabilise itself through resonance as shown below:

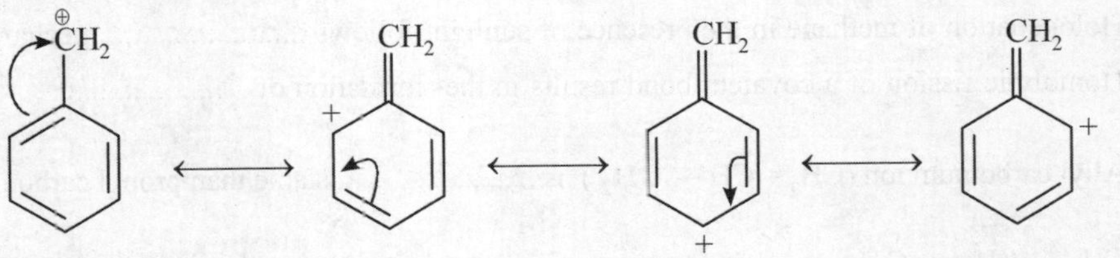

There is no such possibility in cyclohexyl carbocation

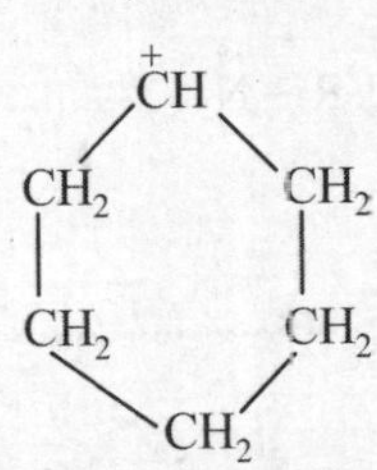

Cyclohexyl carbocation

Example 8. **Give the orbital structure of $\bar{C}H_3$. Mention its hybridisation and its shape.**

Solution. The central carbon atom in CH_3 with a pair of unshared electrons is sp^3 hybridised. The three sp^3 hybrid orbitals are used in forming three sigma bonds with hydrogen. The fourth sp^3 hybrid orbital is occupied by a lone pair of electrons. The structure is similar to that of ammonia in which one hybrid orbital of nitrogen is occupied by a lone pair of electrons. The structure may be depicted as:

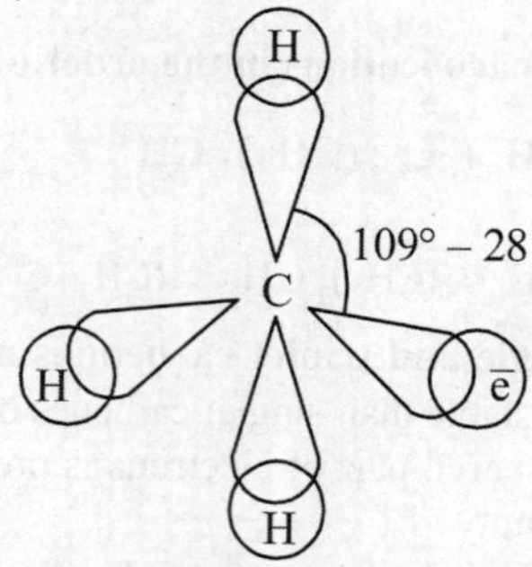

Thus it has a tetrahedral geometry. Three hydrogens are directed towards three corners of the tetrahedron. The fourth position is occupied by unshared pair of electrons.

EXERCISES
(INCLUDING QUESTIONS FROM DIFFERENT UNIVERSITY PAPERS)

Objective questions

Fill in the blanks

1. Hologenation of methane in the presence of sunlight follows .... mechanism
2. Homolytic fission of a covalent bond results in the formation of ....
3. Allyl carbonium ion ($CH_2 = CH— \overset{+}{C}H_2$) is .... stable than propyl carbonium ion $CH_3 — CH_2 — \overset{+}{C}H_2$
4. Out of $:NH_2$, $H - \ddot{N}:$, $R - \ddot{N}:$ and $R - \ddot{N}:$, ... is not nitrene
5. $H - \ddot{O} - H$, $R - \ddot{S} - H$, $R - \ddot{O} - H$ are .... mucleophiles due to the presence of .... pair on central atom
6. BF_3, $Fe\,Cl_3$ and $Al\,Cl_3$ are .... electrophiles because their central atom has .... elelctrons.
7. SN^2 reaction always involves .... of configuration
8. Triphenylmethyl free radical exists in .... resonating forms
9. The terms SN^1 and SN^2 were first introduced by .... in 1928
10. Heterolytic fission results in the formation of .... and ....

Multiple Choice Questions

Choose the correct option

1. The least stable carbanion is :

(*a*) $C_6H_5\overline{C}H_2$ (*b*) $(CH_3)_3\overline{C}$ (*c*) $\overline{C}Cl_3$ (*d*) $\overline{C}H_3$

2. Which of the following is an electrophile ?

(a) CH_3O^- (b) $CH_3\overset{+}{C}H_2$ (c) NH_3 (d) $CH_3\overset{-}{C}H_2$

3. Which of the following statements is correct regarding nucleophiles ?
(a) They have an over-all positive charge
(b) They have a lone-pair of electrons
(c) They have an unpaired electron
(d) They have empty orbitals

4. Which of the following is not a nucleophile ?

(a) NH_3 (b) HSO_3^- (c) $AlCl_3$ (d) HO^-

5. Which of the following is a Lewis acid?
(a) $AlCl_3$ (b) CH_3OH (c) NH_3 (d) CH_3OCH_3

6. Homolytic fission of C-C bond leads to the formation of:
(a) free radicals (b) carbonium ions (c) carbanions (d) none of these

7. Which of the following is not an electrophile ?
(a) NH_3 (b) Br^+ (c) $AlCl_3$ (d) NO_2^+

8. Which alkyl free radical is the most stable ?
(a) Methyl (b) Primary (c) Secondary (d) Tertiary

9. Which of the following is a nucleophile ? '

(a) $AlCl_3$ (b) H_3O^+ (c) BF_3 (d) CN^-

10. Which of the following is not a nucleophile?

(a) $\overline{C}N$ (b) BF_3 (c) $CH_3\overline{O}$ (d) NH_3

11. Which of the following ranges best represents the strength of a hydrogen bond ?
(a) 5 – 10 kcals (b) 60-80 kcals (c) 80-100 kcals (d) 100-120 kcals

12. Homolytic fission of covalent bond between carbon atoms will produce :
(a) two carbonium ions (b) two molecules
(c) free radicals (d) carbonium ion and carbanion

13. The average charge on carbon in $\overset{+}{C}H_3$ is
(a) 1 (b) 2 (c) 3 (d) 4

14. Which of the following free radicals is most stable ?

(a) $R_3\dot{C}$ (b) $R_2\dot{C}H$ (c) $R\dot{C}H_2$ (d) $\dot{C}H_3$

15. Which of the following is the least stable carbanion ?
(a) cyclopropyl (b) vinyl (c) isobutyl (d) ethyl

ANSWERS

Fill in the blanks

1. Free-radical	**2.** free-radicals	**3.** much more	**4.** $:NH_2$
5. neutral, lone	**6.** neutral, six	**7.** inversion	**8.** 36
9. Ingold	**10.** carbocation, carbanion		

Multiple choice questions

1. (*b*)	**2.** (*b*)	**3.** (*b*)	**4.** (*c*)	**5.** (*a*)
6. (*a*)	**7.** (*a*)	**8.** (*d*)	**9.** (*d*)	**10.** (*b*)
11. (*a*)	**12.** (*c*)	**13.** (*c*)	**14.** (*a*)	**15.** (*c*)

General Questions

1. What is meant by curved arrow notation for showing movement of electrons? Explain with an example.
2. Write notes on
   (*i*) half-headed arrows (*ii*) double headed arrows
3. Classify the following into electrophiles and nucleophiles and give reasons?
   (*i*) Water (*ii*) Silver cation (*iii*) Methylene carbene
4. In what way a covalent bond is fissioned? What are its results?
5. Which of the following will generate the most stable carbocation and why?
   CH_3CH_2OH, $CH_3CHOHCH_3$, $C_6H_5CHOHCH_3$
6. What do you understand by electrophilic and nucleophilic reagents. Which of the two will readily attack a carbocation and why?
7. In how many ways can fission of covalent bonds take place? Illustrate with examples.
8. Discuss the structure and shape of carbocations and carbanion.
9. Discuss the relative stabilities of carbocations and free radicals.
10. Define electrophile and nucleophile.
11. Explain the structure and stability of carbenes.
12. What are carbanions? How are they formed?
13. What is a free radical? Discuss its stability
14. What are arynes? How can you represent benzyne is different ways?
15. What are the different types of organic reactions? Explain any one of them
16. What are E_1 and E_2 reactions? Explain with the help of examples.
17. SO_3 (sulphonium ion) appears to the neutral, but still it acts as an electrophile, why?
18. What is meant by reactive intermediate? Make a list of all of them.
19. List the factors that determine the stability of a carbocation. Give an example where field effect controls the stability.
20. Draw the structure of carbocation. Give a reaction where the carbocation reacts with an alkene.
21. What are carbanions? Give a reaction where the rearrangement of a carbanion takes place to give another stable carbanion.
22. Explain vinyl polymerisation showing different steps.

23. What is meant by singlet and triplet carbenes? How are then represented?

24. How can we convert benzene into cycloheptatriene using a reaction intermediate?

25. Draw the structures of singlet and triplet carbene.

26. Give three methods of generation of nitrenes. Give any three properties of nitrenes.

27. How do we determine the formal charge on an intermediate and an ion?

28. What are the different methods for the determination of mechanism of a reaction.

29. Explain why ethyl carbocation ($CH_3-\overset{+}{C}H_2$) is more stable than *n*-propyl carbocation ($CH_3-CH_2-\overset{+}{C}H_2$)?

Hint: Greater the number of hyperconjugative structures of a carbocation, greater is the stability.

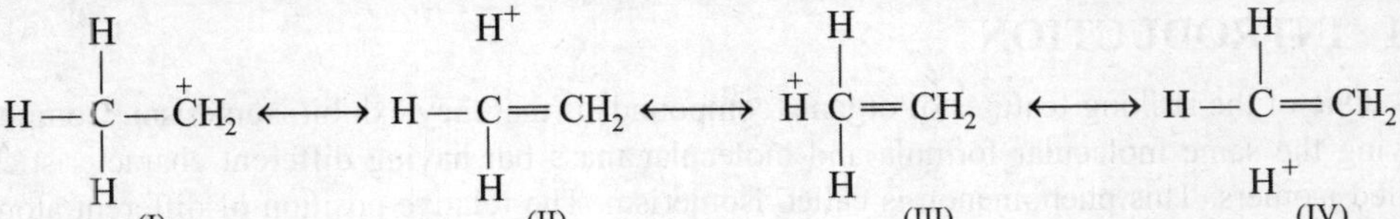

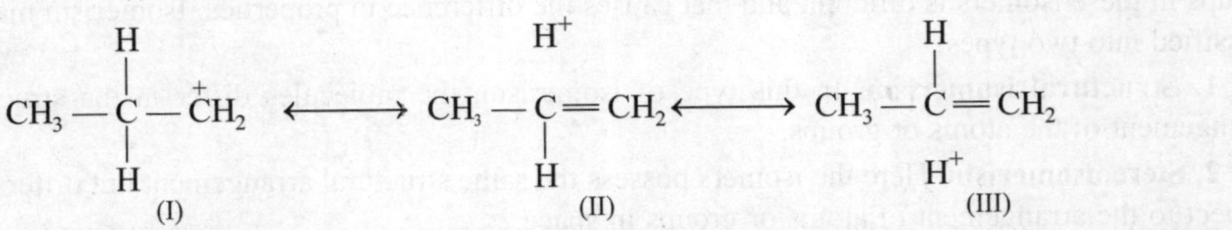

We find that ethyl carbocation has four hyperconjugation structures while *n*-propyl carbocation has three. Thus, ethyl carbocation is more stable.

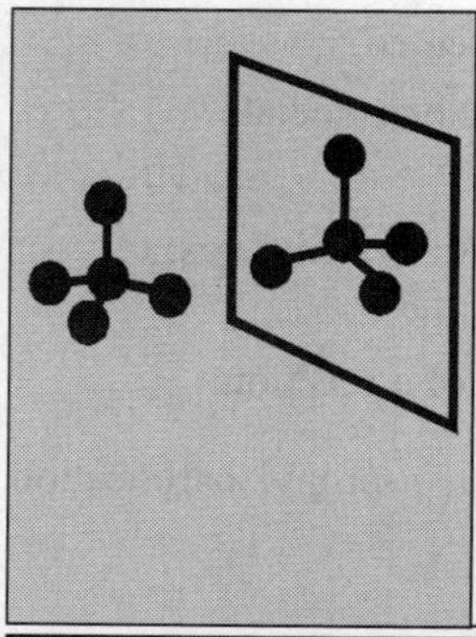

3

Stereochemistry of Organic Compounds

3.1 INTRODUCTION

One of the striking features of organic compounds is that they exhibit isomerism. Compounds having the same molecular formula and molecular mass but having different characteristics are called isomers. This phenomenon is called isomerism. The relative position of different atoms or groups in these isomers is different and that causes the difference in properties. Isomerism may be classified into two types:

1. Structural isomerism. In this type of isomerism, the molecules differ in the structural arrangement of the atoms or groups.

2. Stereoisomerism. Here the isomers possess the same structural arrangement but differ with respect to the arrangement of atoms or groups in space.

The two types of isomerisms are discussed in detail as under :

3.2 STRUCTURAL ISOMERISM

If two or more compounds differ in the relative arrangement of atoms in the molecule, they are said to be structural isomers and this phenomenon is known as **structural isomerism.** There are different kinds of structural isomerisms as under:

(*i*) Chain isomerism or Nuclear isomerism.
(*ii*) Position isomerism.
(*iii*) Functional isomerism.
(*iv*) Metamerism.
(*v*) Tautomerism.

Each one of these is discussed as under.

Chain Isomerism or Nuclear Isomerism

If different compounds of the same class of organic compounds, having the same molecular formula, differ in the structure of carbon chain, they are called chain isomers. Examples of this type of isomerism are:

(*a*) *n-Butane and Isobutane* (*Mol Formula* = C_4H_{10})

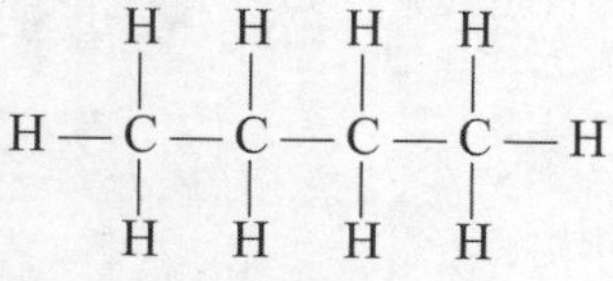

n-Butane

and

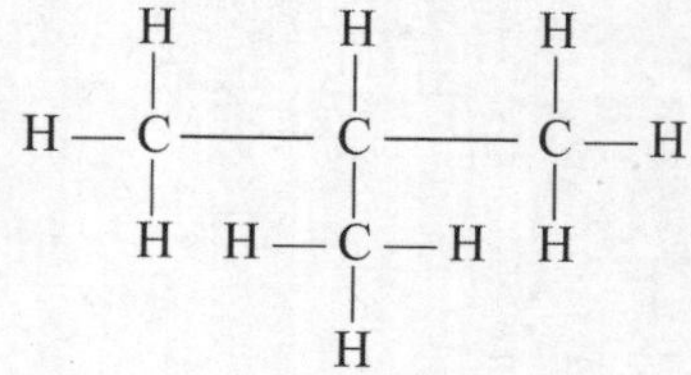

Isobutane

(*b*) *n-Pentane, isopentane and neo pentane* (*Mol. Formula* = C_5H_{12})

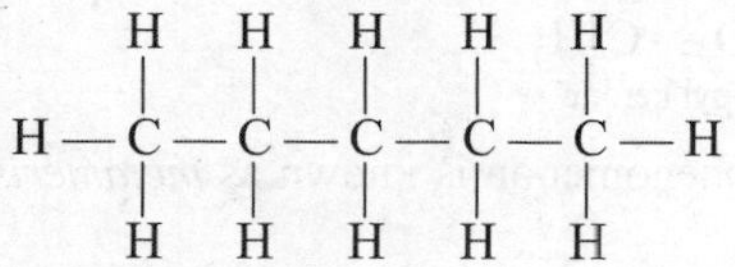

n-Pentane

;

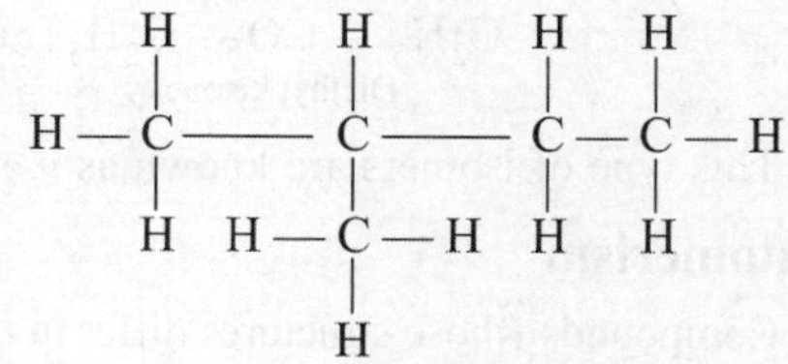

Isopentane

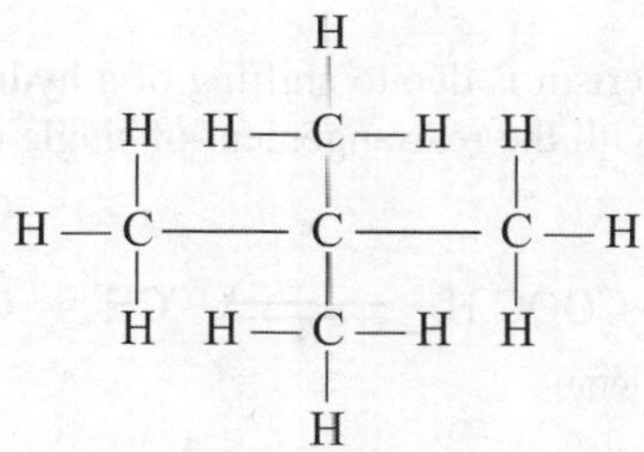

Neo-pentane

Position Isomerism

If different compounds, belonging to some homologous series, with same molecular formula have same carbon skeleton but differ in the position of substituent or functional group; these are known as *position isomers*. Examples of this type of isomers are:

(*a*) 1-Propanol and 2-propanol (Mol. Formula = C_3H_8O)

$$\underset{\text{1-Propanol}}{CH_3-CH_2-CH_2OH} \quad \text{and} \quad \underset{\text{2-Propanol}}{CH_3-\underset{\substack{| \\ OH}}{CH}-CH_3}$$

(*b*) 1-Butene and 2-Butene (Mol. Formula = C_4H_8)

$$\underset{\text{1–Butene}}{CH_2=CH_2-CH_2-CH_3} \quad \text{and} \quad \underset{\text{2–Butene}}{CH_3-CH=CH-CH_3}$$

Functional Isomerism

Different compounds, with same molecular formula but different functional groups are known as *functional isomers*. For example:

(*a*) Ethyl alcohol and Dimethyl ether (Mol. Formula = C_2H_6O)

$$\underset{\text{Ethyl alcohol}}{CH_3CH_2OH} \quad \text{and} \quad \underset{\text{Dimethyl ether}}{CH_3-O-CH_3}$$

(*b*) Propionaldehyde and acetone (Mol. Formula = C_3H_6O)

$$\underset{\text{Propionaldehyde}}{CH_3CH_2CHO} \quad \text{and} \quad \underset{\text{Acetone}}{CH_3-CO-CH_3}$$

Metamerism

This is a special kind of structural isomerism in which different compounds, with same molecular formula, belong to same homologous series but differ in the distribution of alkyl groups around a central atom. Examples are:

(*a*) Diethyl ether and methyl propyl ether (Mol. Formula = $C_4H_{10}O$)

$$\underset{\text{Diethyl ether}}{C_2H_5-O-C_2H_5} \quad \text{and} \quad \underset{\text{Methyl propyl ether}}{CH_3-O-C_3H_7}$$

(*b*) Diethyl ketone and methyl propyl ketone (Mol. Formula = $C_5H_{10}O$)

$$\underset{\text{Diethyl ketone}}{C_2H_5 - CO - C_2H_5} \quad \text{and} \quad \underset{\text{Methyl propyl ketone}}{CH_3 - CO - C_3H_7}$$

This type of isomers are known as *metamers* and the phenomenon is known as *metamerism*.

Tautomerism

Compounds whose structures differ in the arrangement of atoms but which exist simultaneously in dynamic equilibrium with each other, are called *tautomers*. This phenomenon is called *tautomerism*.

In most of the cases tautomerism is due to shifting of a hydrogen atoms from one carbon (or oxygen or nitrogen) to another with the rearrangement of single or double bonds. For example,

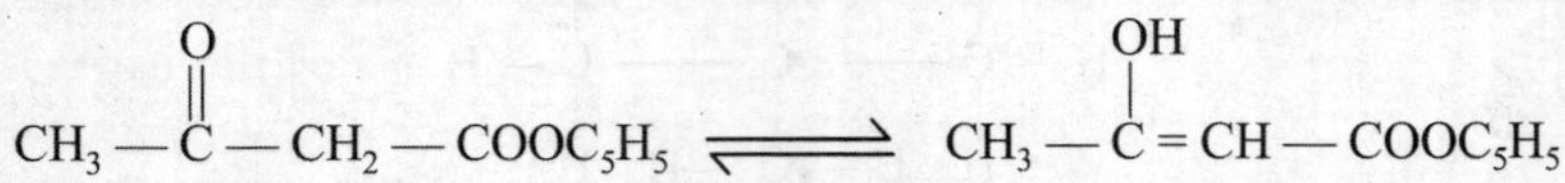

Ethyl acetoacetate (keto form) ⇌ Ethyl acetoacetate (Enolic form)

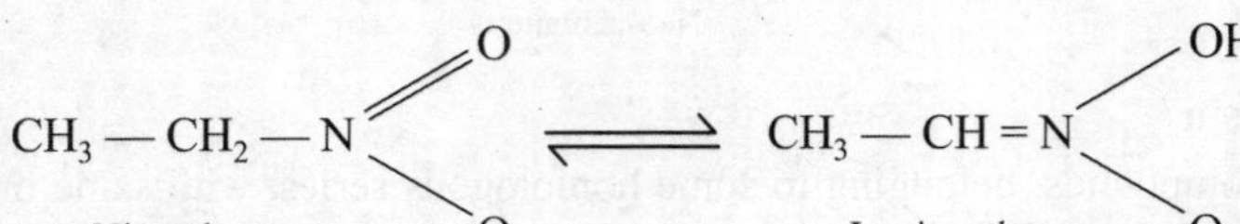

Nitroethane ⇌ Isonitroethane

Example. **Write the structures of all possible isomers of C_3H_8O giving their IUPAC names.**

Solution. **Position isomers**

$$\underset{\text{Propanol–1}}{CH_3CH_2CH_2OH} \qquad \underset{\text{Propanol–2}}{CH_3 - CHOHCH_3}$$

Functional Isomers

$$\underset{\text{Propanol–1}}{CH_3CH_2CH_2OH} \qquad \underset{\text{Methoxyethane}}{CH_3OCH_2CH_3}$$

Stereoisomerism

Compounds having different three-dimensional relative arrangement of atoms in space are called stereoisomers. This phenomenon is called *stereoisomerism*. These compounds are said to have different configurations. Stereoisomerism is of the following different kinds:

(*i*) Optical isomerism

(*ii*) Geometrical isomerism

(*iii*) Conformational isomerism.

3.3 OPTICAL ISOMERISM OR ENANTIOMERISM

Before taking up optical isomerism, let us understand the terms: plane polarised light, optical activity and specific rotation.

Plane polarised light

Ordinary light has vibrations taking place at right angles to the direction of propagation of light spread in all the possible planes. If we pass ordinary light through **nicol**. prism vibrations in all planes except one are cut off. Thus light coming out of nicol prism has vibrations only in one plane. Such a light is called *plane polarised light*.

Optical activity

Behaviour of certain substances is strange. When a plane polarised light is passed through the solution of such substances, the light coming out of the solution is found to be in a different plane.

The plane of polarised light is rotated. *Such substances, which rotate the plane of plane polarised light when placed in its path are known as optically active susbtances* and the phenomenon is known as *optical activity*. The angle of rotation (α) of plane polarised light is known as *Optical rotation.* The substances which rotate the plane of polarised light to the clockwise or right direction are known as dextrorotatory or having positive (+) rotation and those which rotate the plane polarised light to the anticlockwise or left direction are known as *laevorotatory* or having negative (–) rotation. Substances which do not rotate the plane of polarised light are said to be optically inactive.

The instrument used for measuring optical rotation is called polarimeter. It consists of a light source, two nicol prisms and in between a tube to hold the solution of organic substance. The schematic representation of a polarimeter is given in Fig. 3.1.

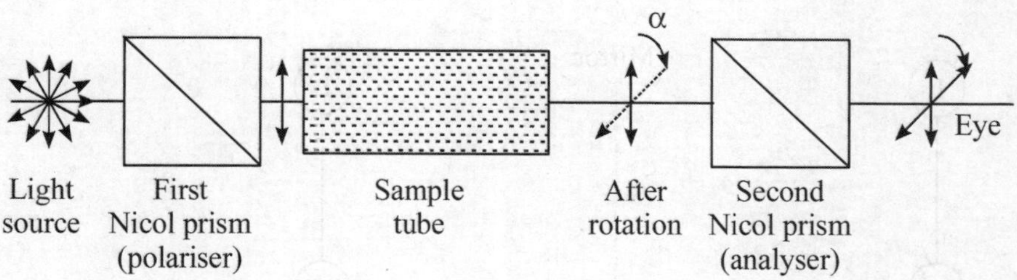

Specific rotation. The angle of rotation of plane-polarised light or optical rotation (α) of an organic substance depends not only on the kind of molecules but also varies considerably with the number of molecules that light encounters in its path which in turn depends on the concentration of the solution used and the length of polarimeter tube containing it. Besides this, it depends on temperature, wavelength of light and nature of solvent used.

The optical activity of a substance is expressed in terms of *specific rotation.* $[\alpha]_\lambda^t$ which is a constant quantity, characteristic of a particular substance,

$$[\alpha]_\lambda^t = \frac{\alpha}{l \times c}$$

where, α = observed rotation in degrees

ℓ = length of polarimeter tube in decimeter

c = concentration of substance in gm per ml of solution

t and λ signify the temperature and wavelength of light used.

When $l = 1$ and $c = 1$, $[\alpha]_\lambda^t = \alpha$

Specific rotation is thus defined as the optical rotation produced by a compound when plane polarised light passes through one decimeter length of the solution having concentration one gram per millilitre. Usually the monochromatic light used is D line of sodium (λ = 589 nm). Thus specific rotation of cane sugar can be expressed as

$$[\alpha]_D^{20°C} = +66.5° \text{ (water)}$$

In this expression D stands for D line of sodium, 20°C is temperature of measurement, + sign shows the dextrorotation and water is the solvent used.

Louis Pasteur, while studying the crystallography of salts of tartaric acid made a peculiar observation. He observed that optically inactive sodium ammonium tartarate existed as a mixture of two different types of crystals which were mirror images of each other. With the help of a hand lens and a pair of forceps, he carefully separated the mixture into two different types of crystals. These crystals were mirror images of each other and were called enantiomorphs and the phenomenon as *enantiomorphism*. Although the original mixture was optically inactive; each type of crystals when dissolved in water, were found to be optically active. Moreover the specific rotations of the two solutions were exactly equal but of opposite sign *i.e.* one solution rotated the plane polarised light to the right or clockwise while the other to the left or anticlockwise and to the same extent. Two types of crystals or solutions were identical in all other physical and chemical properties. *Isomers which are non-superimposable mirror images of each other are called enantiomers*.

According to La Bell and Van't Hoff the four valencies of a carbon atom are directed towards the four corners of a regular tetrahedron at the centre of which lies the carbon atom. Consider a

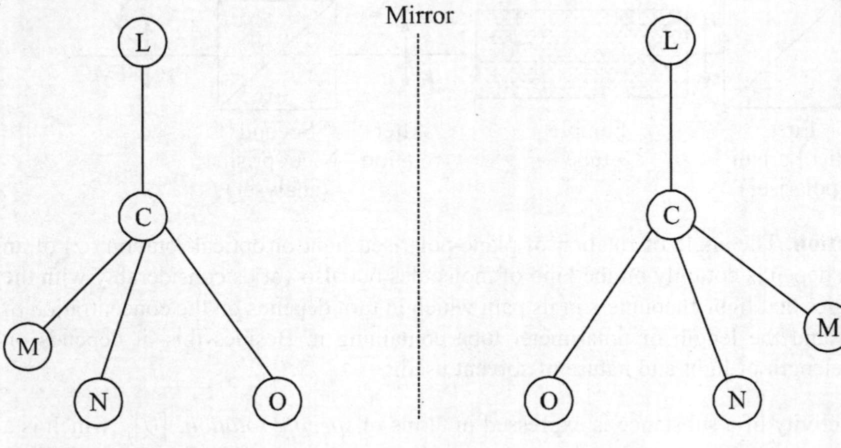

compound of formula C_{LMNO} having four different groups L, M, N and O attached to a carbon atom. This compound can be represented by two models which look like mirror images of each other.

Enantiomerism in nature

Van't Hoff

It is important to note here that these two molecules cannot be superimposed on each other *i.e.* they will not coincide in all their parts. We may turn them in as many ways as we like but we find that though two groups of each may coincide, the other two do not. Hence these must represent two isomers of formula C_{LMNO}. Lactic acid $CH_3CHOHCOOH$ and sec-Butyl chloride $C_2H_5CHClCH_3$, exist as two optically active isomers which are enantiomers *i.e.* mirror images of

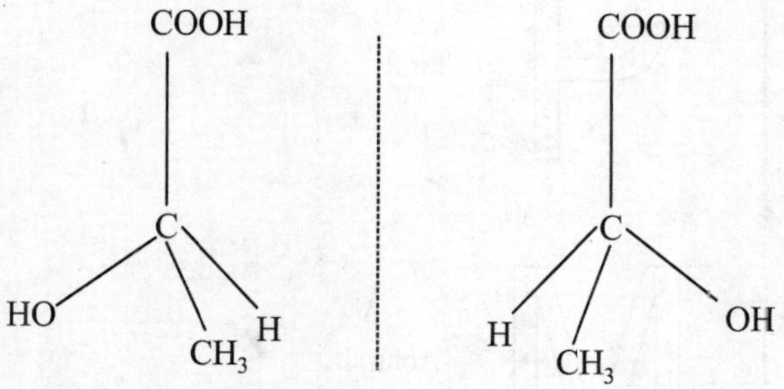

Enantiomers of Lactic Acid

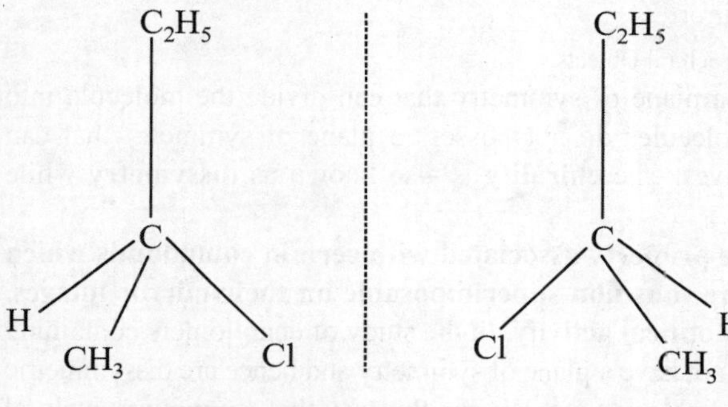

Enantiomers of Sec. Butyl Chloride

Enantiomers

each other. Mirror images of the two compounds are represented as above.

The carbon atom to which four different groups are attached, is known as *asymmetric* or *chiral carbon atom* or *stereogenic centre*.

If two of the groups attached to carbon are same, we shall observe that it is possible to superimpose the mirror images on each other. Such a compound will not show optical isomerism or *enantiomerism*.

Hence non-superimposability of the mirror images is responsible and essential for the type of stereoisomerism known as enantiomerism.

The term optical isomerism is used for the existence of stereoisomers which differ in their behaviour towards the plane polarised light. Thus enantiomeric molecules are always non-superimposable mirror images of each other. The non-superimposability of mirror images arises due to **chiral** or **asymmetric** nature of molecule. A molecule is said to be chiral if it has no plane of symmetry and is therefore non-superimposable on its mirror image.

It may be concluded with the remarks that chirality is the fundamental condition of enantiomerism or optical isomerism.

Chirality and dissymetry

An object which is non-superimposable on its mirror image is called **chiral** object. And an object which is superimposable on its mirror image is called **achiral** object. Examples of chiral objects including letters from English alphabets are: A pair of hands, shoes, gloves, letters P, F, J etc.

Examples of achiral objects and letters are: Ball, sphere, letters A, O, M, etc.

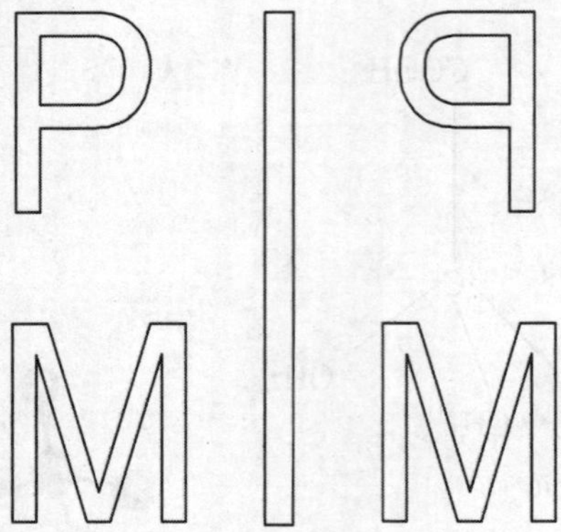

Chiral

Achiral

Mirror

Chiral and achiral Objects

Achiral objects or molecules possess a plane of symmetry that can divide the molecule into two identical halves. Chiral objects or molecules do not possess a plane of symmetry that can divide the molecule into two identical halves. The **chirality** is also known as **dissymetry** while **achirality** is also known as **symmetry**.

Thus chirality or dissymmetry is the property associated with certain compounds which do not have a plane of symmetry and are thus non-superimposable on their mirror images. Dissymmetry is an essential condition for optical activity. In the study of enantiomers containing one chiral carbon atom, we find that they do not have a plane of symmetry and hence are dissymmetric in nature, for example, lactic acid, sec-butyl chloride etc. We can thus say that compounds a chiral carbon atom are optically active in nature.

This is not true, however, for compounds with two or more chiral carbon atoms. Here, one or more isomers may be optically inactive in spite of the presence of chiral carbon atoms. Consider for example, the case of tartaric acid, CH(OH) COOHCH (OH) COOH. It has two chiral carbons (stereogenic centres) marked * and thus has four isomers as shown below:

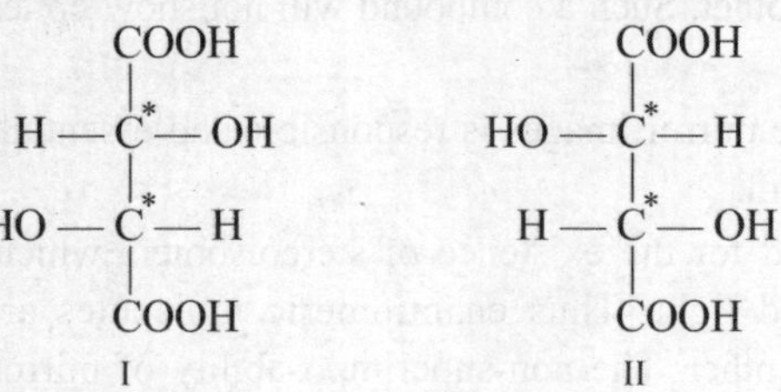

I II

I and II are optically active due to the absence of plane of symmetry

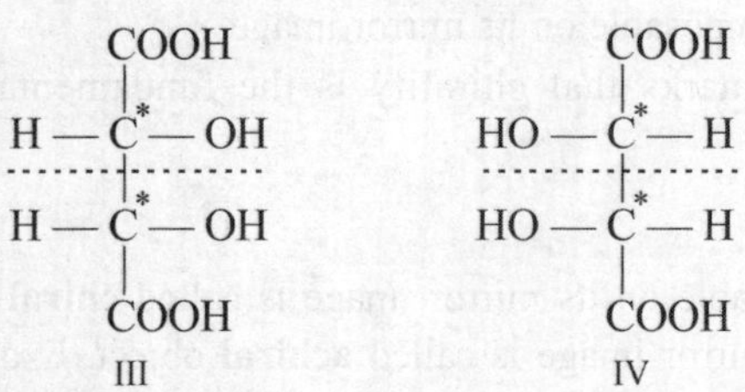

III IV

III and IV are optically inactive due to the presence of plane of symmetry (shown by dotted line)

Isomers I and II are non-superimposable mirror images of each other, they are dissymetric and optically active. But III and IV have a plane of symmetry. Therefore, they are symmetric and optically inactive in nature.

In other words, optical activity in organic compounds is due to the presence of dissymetry and not due to chiral carbon atoms alone.

No. of optical isomers for a compound is given by 2^n where n is the number of chiral carbon atoms.

It may be noted that some organic compounds are optically active even though they do not contain chiral atom. For example,

Substituted allenes. Dienes with double bonds in adjacent positions are called allenes. Substituted allenes may be represented as RCH = C = CHR. Pentane –2, 3-diene exists in two optically active forms. This is because the groups linked to one end carbon are in different plane than those linked to the other end carbon.

In allenes, the central carbon is *sp* hybridized, and the terminal carbons are sp^2 hybridized. The central carbon forms two sp-sp^2 σ bonds. The central carbon also has two *p* orbitals which are mutually perpendicular. These forms π bonds with the *p* orbitals on the other carbon atoms. As a result, the substituent at one end of the molecule are in a plane which is perpendicular to that of the substituents at the other end, so that the compound exists in two forms which are non-superimposable mirror images and are optically active.

Substituted biphenyl. Benzene rings in substituted biphenyl lie in different planes. Hence the compound exhibits optical activity.

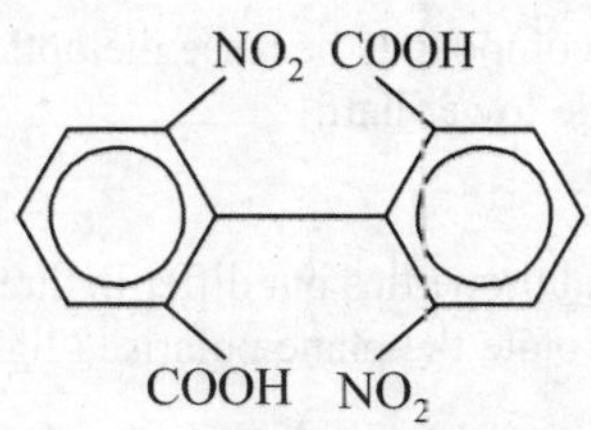

We can conclude from the above discussion that the most essential condition for a compound to show optical activity is the presence of dissymmetry and not chirality. However, compounds containing only one chiral carbon are always dissymmetric since they do not have a plane of symmetry.

The compounds, with two or more similar chiral carbon atoms which are optically inactive due to the presence of plane of symmetry, are called **meso** compounds.

Consider the following isomer of tartaric acid

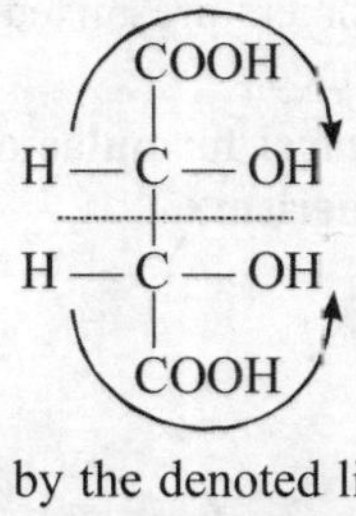

It has a plane of symmetry denoted by the denoted line. The optical rotation of upper half of the molecule is neutralized by that of the lower half as the two rotations are in opposite directions. The net result is that the molecule is optically inactive. Such compounds are known as meso compounds.

Similarly, we come across meso structures in 2, 3-Dichlorobutane. This compound also possesses two asymmetric carbon atoms (stereogenic centres) marked with *.

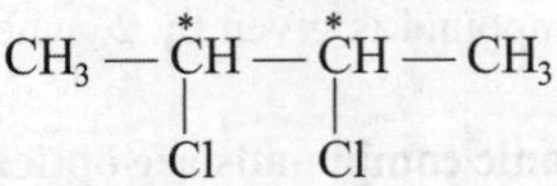

The different isomers can be written as

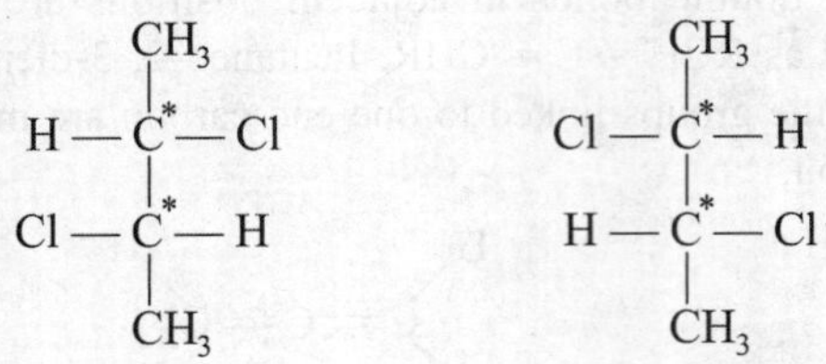

I II

I and II show optical activity due to absence of plane of symmetry (Dissymmetry)

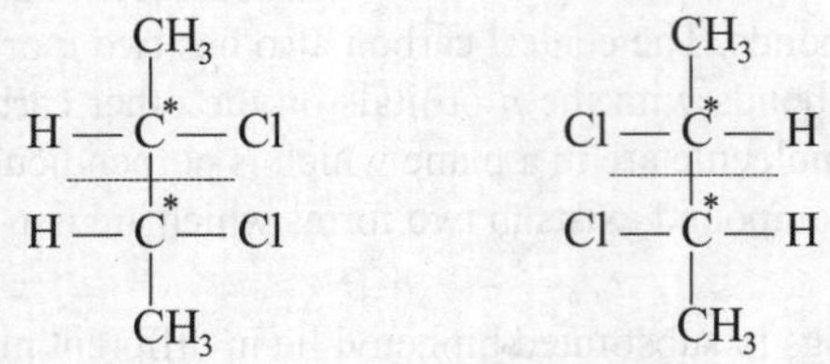

III IV

III and IV are optically inactive because of the presence of plane of symmetry

Structures III and IV are meso compounds because the optical activity of the upper half of the molecule is neutralised by that of the lower half.

Properties of enantiomers

(*i*) They have identical physical properties but differ in direction of rotation of plane polarised light. Though the two enantiomers rotate the plane polarised light in opposite direction, the extent of rotation is the same.

(*ii*) They have identical chemical properties except towards optically active reagents. The rates of reaction of optically active reagents with two enantiomers differ and sometimes one of the enantiomers does not react at all.

(*iii*) In biological system (–) or *l*-glucose is neither metabolised by animals nor fermented by yeast whereas (+) or *d*-glucose undergoes both these processes and plays an important role in animal metabolism and fermentation. Similarly mould *penicillium glaucum* consumes only *d*-tartaric acid when fed with a mixture of equal quantities of *d*- and *l*-tartaric acid.

(*iv*) When equal amounts of enantiomers are mixed together an optically inactive racemic modification denoted by (±) or *dl* is obtained.

***Example 1.* Given below are the structural formulae of all alkanes with molecular formula C_6H_{14}. Which of these exhibit enantiomerism?**

(*i*) $CH_3(CH_2)_4CH_3$

(*ii*) 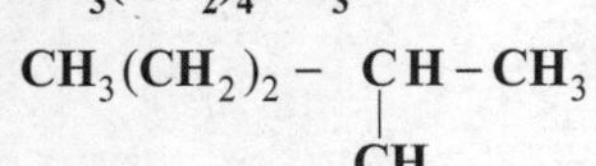

(*iii*) 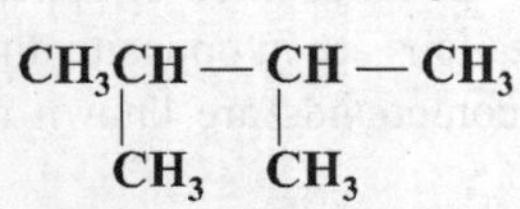

(*iv*) $(CH_3)_3C-CH_2CH_3$

Solution. There is no carbon in any of the compounds above which is chiral *i.e.* attached to four different groups.

Chirality is the necessary condition for a molecule to exhibit enantiomerism. Hence none of the compounds above shows enantiomerism.

Example 2. **Which of the following compounds exhibit enantiomerism?**

(*i*) $CH_2OH\ CHOH\ CHO$

(*ii*)
$$\begin{array}{l} \quad\quad\; CH_3 \\ \quad\quad\;\; | \\ CH_3 - CH - CHCl - CH_3 \end{array}$$

(*iii*)
$$\begin{array}{l} CH_3 - CH_2 - CH - CH_3 \\ \quad\quad\quad\quad\quad\;\; | \\ \quad\quad\quad\quad\quad CH_3 \end{array}$$

(*iv*) $CH_3CHOHCH_3$

(*v*) CH_2NH_2COOH

Solution. Compounds (*i*) and (*ii*) have chiral carbons marked with asterisks and hence these two compounds show enantiomerism.

$$CH_2OH\overset{*}{C}HOHCHO$$

$$\begin{array}{l} \quad\quad\; CH_3 \\ \quad\quad\;\; | \\ CH_3 - CH - \overset{*}{C}HCl - CH_3 \end{array}$$

Compounds (*iii*), (*iv*) and (*v*) have no chiral carbon atom in the molecule.

Hence they do not show enantiomerism.

Example 3. **Point out the optically active compounds out of the following:**

(*i*) $CH_2OH - CHOH - CHO$ **(*ii*) $CH_3 - CHOH - CH_2OH$**

(*iii*)
$$\begin{array}{c} CHOH \\ | \\ H - C - OH \\ | \\ HO - C - H \\ | \\ COOH \end{array}$$

Solution. Compounds (*i*) and (*ii*) have one chiral carbon (the middle one) and hence both these compounds are optically active compounds. (*iii*) has two chiral carbons (the middle ones) and therefore this compound is also optically active.

Fischer's projection formula for planar representation of three dimensional molecules.

Emil Fischer in 1891 introduced a simple method for representing three dimensional molecules in one plane. It is known as *Fischer projection formula.* Following points are to be observed for this purpose:

(*i*) The chiral molecule is imagined in such a way that two groups point towards the observer and two away from the observer. The groups pointing towards the observer are written along the horizontal line (shown by thick wedge-like bonds) and those pointing away are written along the vertical line. The central carbon is present at the crossing of the horizontal and vertical lines.

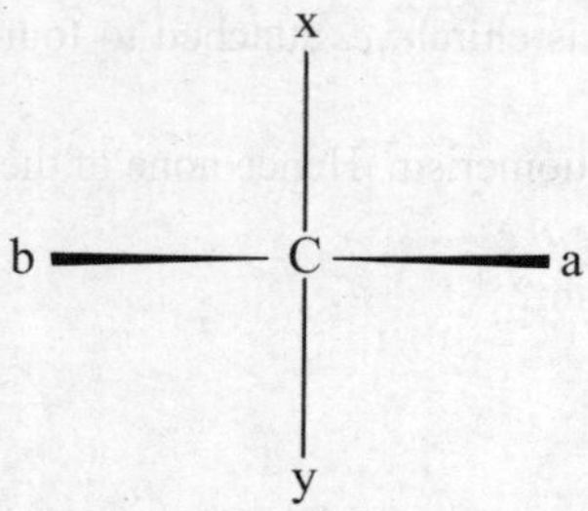

Flying wedge representation

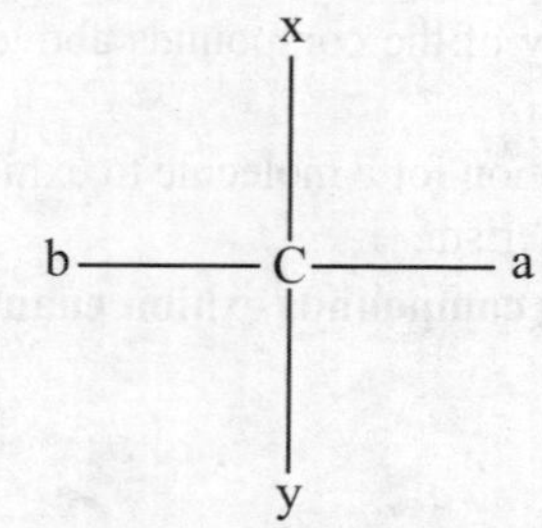

Normal representation

Emil Fischer

Thus if *a*, *b*, *x* and *y* are four groups attached to a carbon, the molecule will be represented by the projection formula as above. Here *a* and *b* groups point towards the observer (or above the plane) and groups *x* and *y* are away from the observer (or below the plane).

(*ii*) The longest chain of carbon atoms in the molecule should be represented along the vertical line. Lactic acid, therefore, according to the above conventions will be represented as

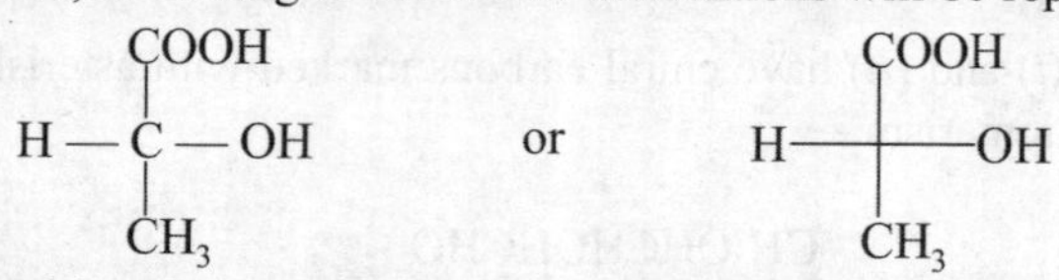

(*iii*) We can avoid writing carbon at the crossing of the vertical and horizontal lines. A crossing automatically means the presence of a carbon.

(*iv*) If necessary, planar formula may be imagined to be rotated from end to end without lifting it from the plane of the paper. Rotation by 180° in the plane of the paper does not create any change in the configuration of the molecule.

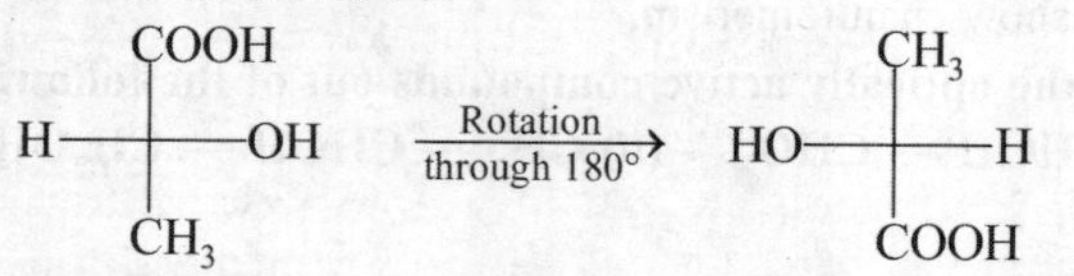

I (–) lactic acid II (–) lactic acid

In the above rotation by 180°, II has been obtained from I. There has been no change in configuration of the molecule. I and II are in fact the same thing.

(*v*) Rotation by 90° or 270° brings about a change in configuration of the molecule. Consider the following rotation.

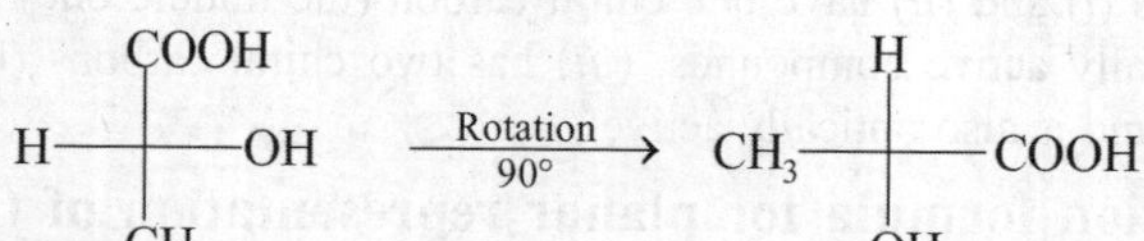

(–) lactic acid (+) lactic acid

(*vi*) If the positions of two groups across the chiral atom are interchanged, it leads to inversion of configuration. Two consecutive such changes neutralise the effect.

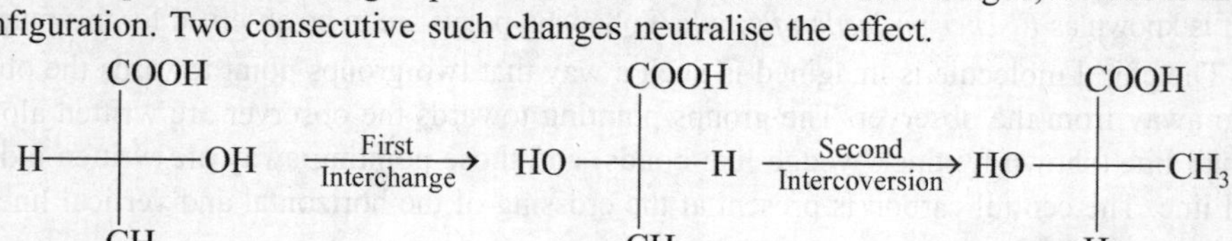

(–) lactic acid (+) lactic acid (–) lactic acid

I II III

Structure III and I are the same because as per rule (iv) above, III on rotation through 180° will give I.

Absolute configuration of optical isomers

In the earlier days, as the modern techniques of finding out configuration were not available, Fischer assigned the following configurations to the (+) and (–) enantiomers of glyceraldehyde arbitrarily and denoted them by capital letters D and L respectively. Small letters *d* and *l* represent sign of rotation, while capital letters D and L represent configuration.

MIRROR

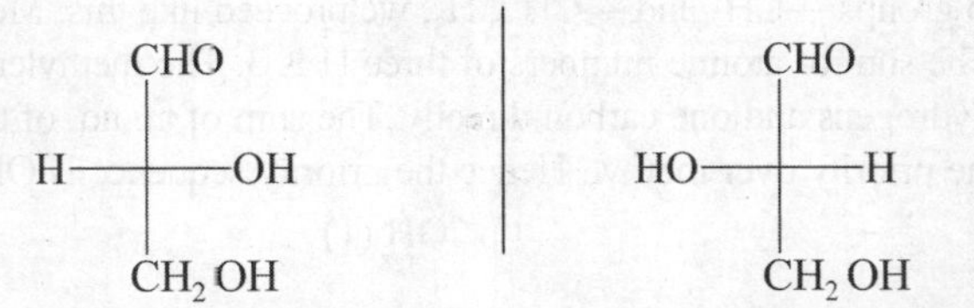

D (+) glyceraldehyde L (+) glyceraldehyde

The relative configurations of a number of other optically active compounds have been established by correlating them with D(+) or L(–) glyceraldehyde. All those optically active compounds, which are obtained from D(+) glyceraldehyde through a sequence of reactions *without breaking the bonds of asymmetric carbon atom*, are designated as D configuration irrespective of their sign of rotation and the other enantiomer as L configuration.

For example,

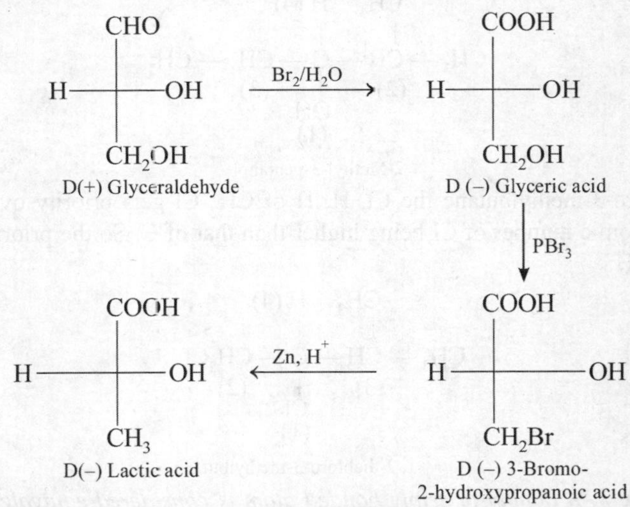

In all the D configurations, — OH attached to asymmetric carbon atom is written on the right hand side of Fischer projection formula. Similarly, in all the L-configurations, –OH attached to the lowest asymmetric carbon atom is written on the left hand side in the Fisher's projection formula.

R and S specification for the configuration of an optically active compound

Cahn, Ingold and Prelog developed a method which can be used to designate the configuration of all the molecules containing asymmetric carbon atom (chiral centre). This system is known as *Cahn-Ingold-Prelog system* or R and S system and involves two steps.

Step I. The four different atoms or groups of atoms attached to chiral carbon atom are assigned a sequence of priority according to the following set of sequence rules.

Sequence Rule 1. *If the four atoms, directly attached to asymmetric carbon atom, are all different, the priority depends on their atomic number. The atom of higher atomic number gets higher priority.* For example, in chloroiodomethane sulphonic acid the priority sequence is I, Cl, SO_3H, H

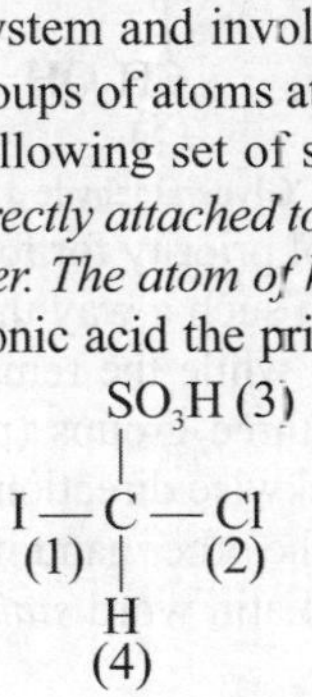

We consider the atom of the group which is directly linked to the central carbon.

Sequence Rule 2. *If Rule 1 fails to decide the relative priority of two groups it is determined by similar comparison of next atoms in the group and so on.* In other words, if two atoms directly attached to chiral centre are same, the next atoms attached to each of these atoms, are compared. For example in 2-butanol two of the atoms directly attached to chiral centre are carbon themselves. To decide the priority between the two groups —CH_3 and —CH_2CH_3, we proceed like this. Methyl carbon is further linked to H, H and H. The sum of atomic numbers of three H is 3. The methylene carbon of the ethyl group is linked to two hydrogens and one carbon directly. The sum of at. no. of two H and one C is 8. Thus ethyl group gets the priority over methyl. Hence the priority sequence is OH, C_2H_5, CH_3, H.

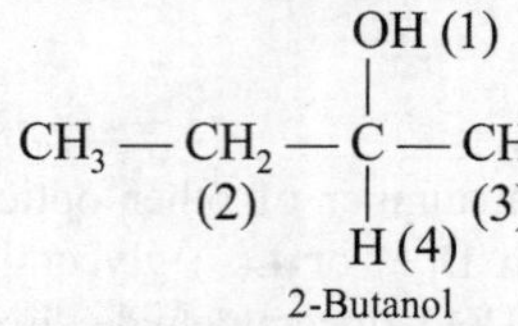

2-Butanol

In 2-methyl - 3-pentanol, the C, C, H of isopropyl gets priority over the C, H, H of ethyl, so the priority sequence is OH, isopropyl, ethyl, H.

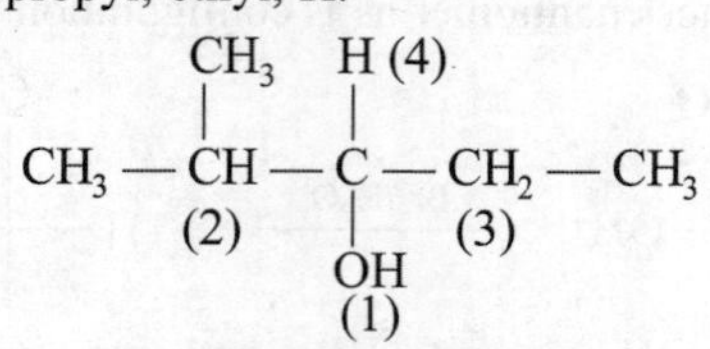

2-Methyl-3-pentanol

In 1, 2-dichloro-3-methylbutane the Cl, H, H of CH_2 Cl gets priority over the C, C, H of isopropyl due to atomic number of Cl being higher than that of C. So the priority sequence is Cl, CH_2Cl, isopropyl, H.

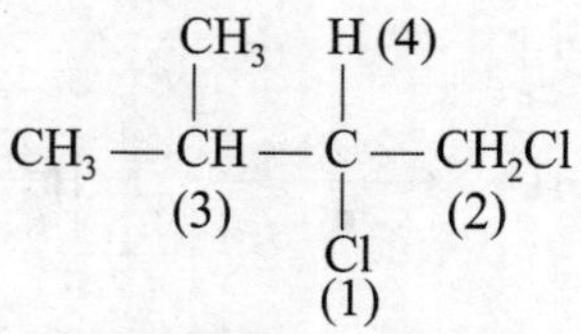

1, 2-dichloro-3-methylbutane

Sequence Rule 3. *A doubly or triply bonded atom is considered equivalent to two or three such atoms; but two or three atoms, if attached actually, get priority over doubly or triply bonded atom.* In glyceraldehyde, O, O, H of — CHO gets priority over the O, H, H of — CH_2OH; so the priority sequence is — OH, — CHO — CH_2OH — H.

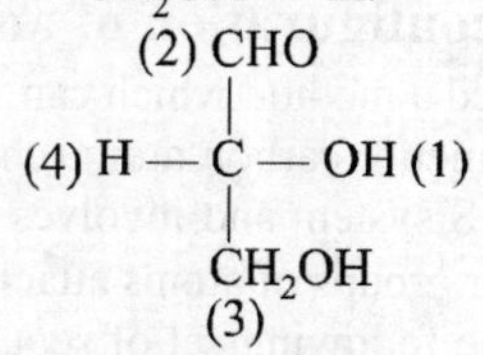

Glyceraldehyde

Step II. After deciding the sequence of priority for four atoms or groups attached to asymmetric carbon atom; the molecule is visualised in such a way that the atom or group of lowest or last (*i.e.* fourth) priority is directed away from us, while the remaining three atoms or groups are pointing towards us. Now if on looking at these three groups (pointing towards us) in the order of their decreasing priority, our eye moves in clockwise direction, the configuration is specified as R (from Latin word *rectus* meaning right) and on the other hand if our eye moves in anticlockwise direction the configuration is specified as S (from Latin word *sinister* meaning left).

The following examples illustrate the above method for specification of configuration as R and S to molecules of compounds containing an asymmetric or chiral carbon atom.

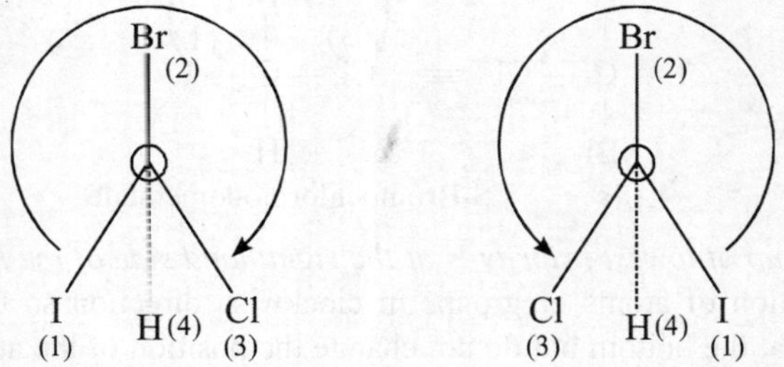

R S

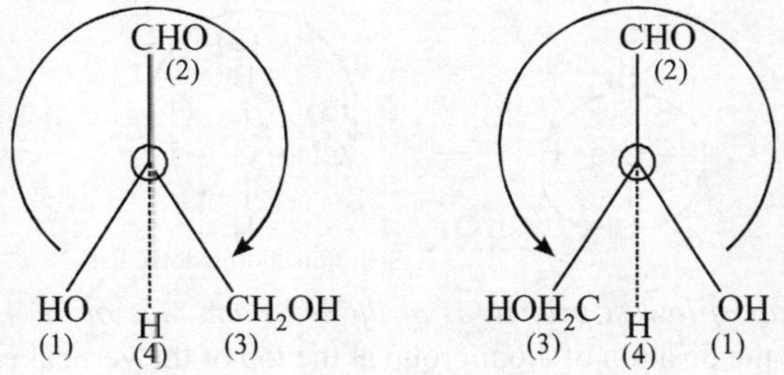

R S

Configuration on the basis of projection formula

When a compound is represented by the Fischer projection formula, the configuration can be easily determined without constructing the model. To determine whether the eye travels clockwise or anticlockwise, we have to place the group or atom of the lowest priority at the bottom of the Fischer projection formula. The following four situations arise:

(*i*) *The atom/group of lowest priority is at the bottom*. In such a case, simply rotate the eye in the order of decreasing priorities. The configuration is R if the eye travels in clockwise direction and S if the eye travels in *anticlockwise* direction.

For example, Glyceraldehyde, represented by the following projection formula has R configuration:

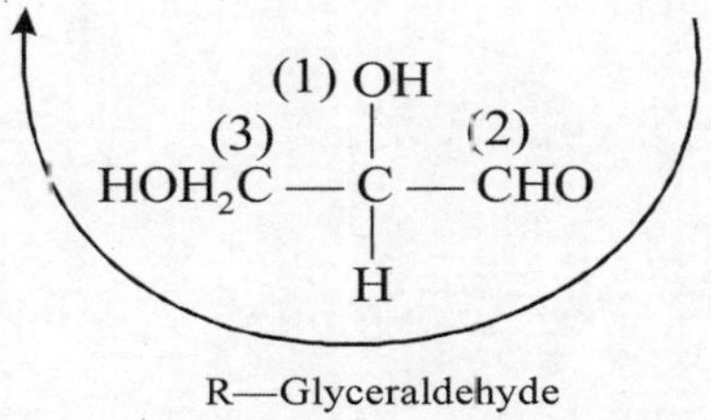

R—Glyceraldehyde

(*ii*) *The atom/group of lowest priority is at the top*. In such case, rotate the molecule by 180° so as to bring atom/group of lowest priority at the bottom. This can be done by reversing the position of all the atoms or groups. Then find the direction in the order of decreasing priorities. For example, the compound CHBrClI, represented by following projection formula, has S configuration:

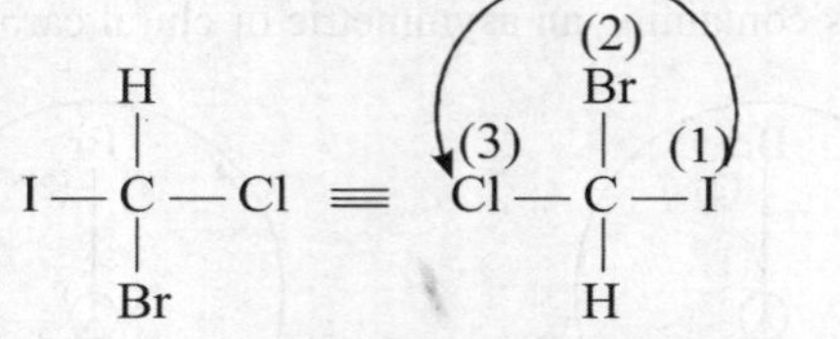

S-Bromochloroiodomethane

(*iii*) *The atom/group of lowest priority is at the right hand side of the horizontal line.* In such case, change the position of atoms or groups in clockwise direction so that atom/group of the lowest priority comes at the bottom but do not change the position of the atom/group at the top of the vertical end. Then find the direction in the order of decreasing priorities. For example, CHBrClI, represented by following projection formula, has S configuration.

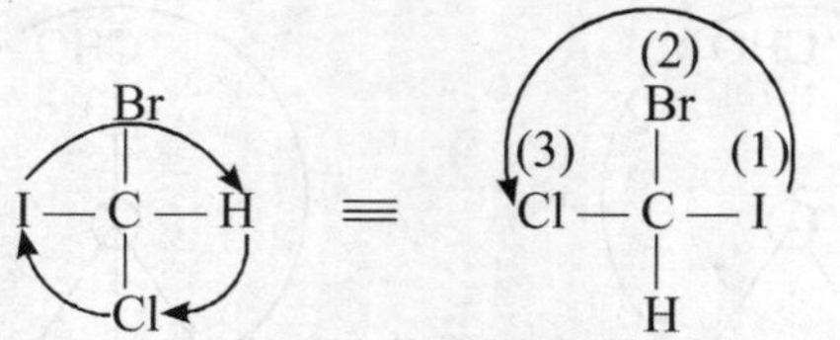

S-Bromochloroiodomethane

(*iv*) *The atom/group of lowest priority is at the left hand side of the horizontal line.* In such case, without changing the position of atom/group at the top of the vertical end, change the position of other atoms/groups in the anticlockwise direction so that atom/group of lowest priority comes at the bottom. Then find the direction in the order of decreasing priorities. For example, CHBrClI, represented by following projection formula, has R configuration.

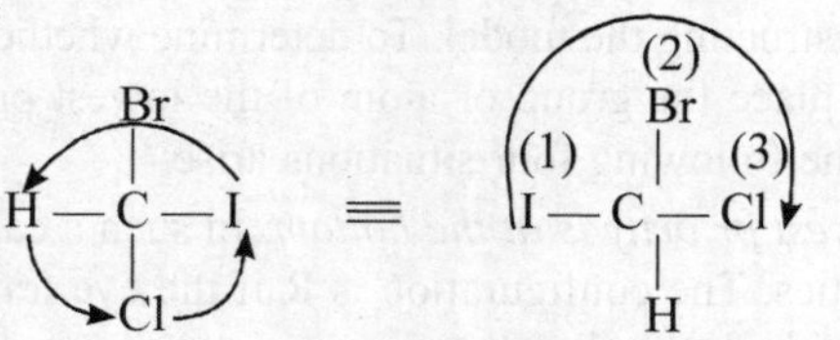

R-Bromochloroiodomethane

***Example 4.* Assign R and S configuration to the following:**

(*i*) 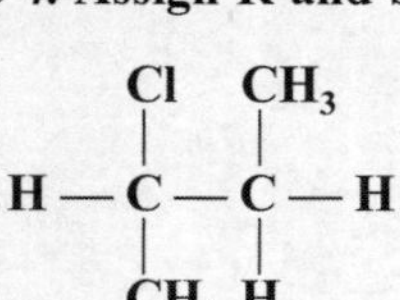

(*ii*) 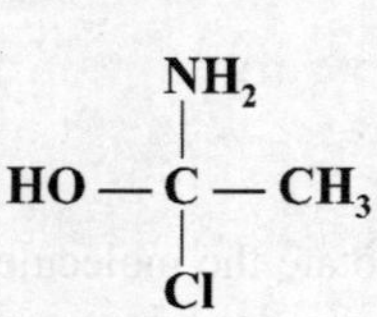

(*iii*) 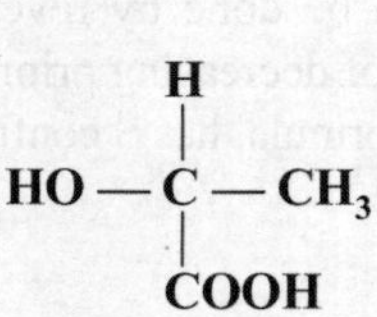

(*iv*) 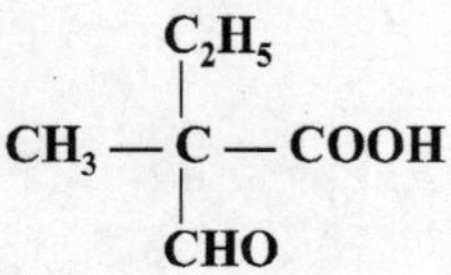

Solution. (*i*) 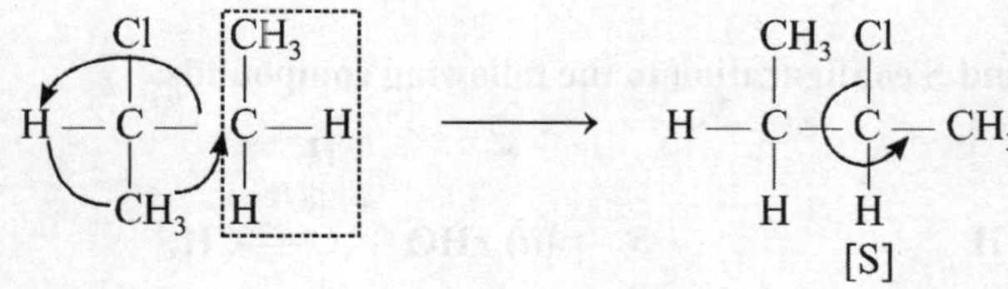

(*ii*) 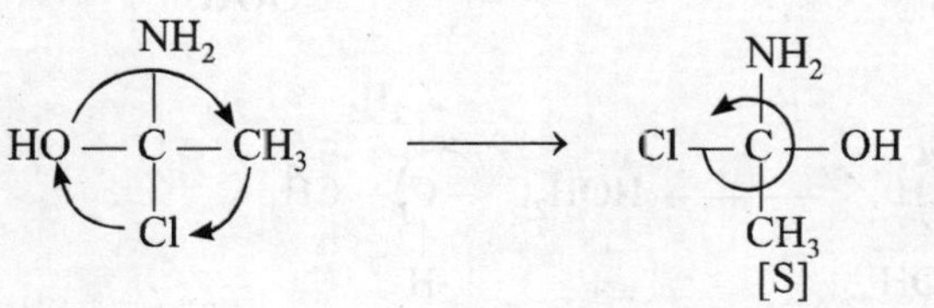

(*iii*) 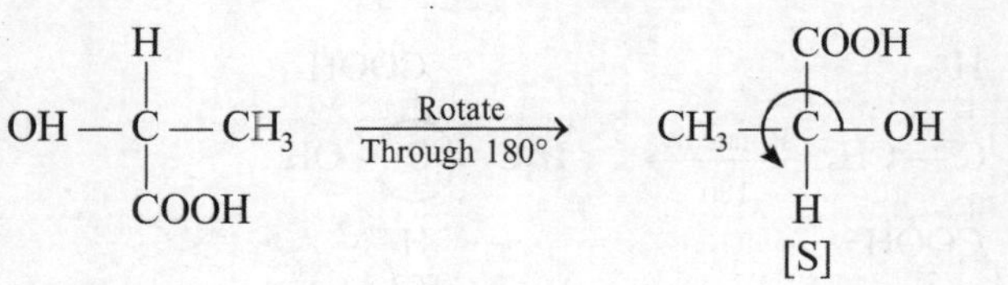

(*iv*) 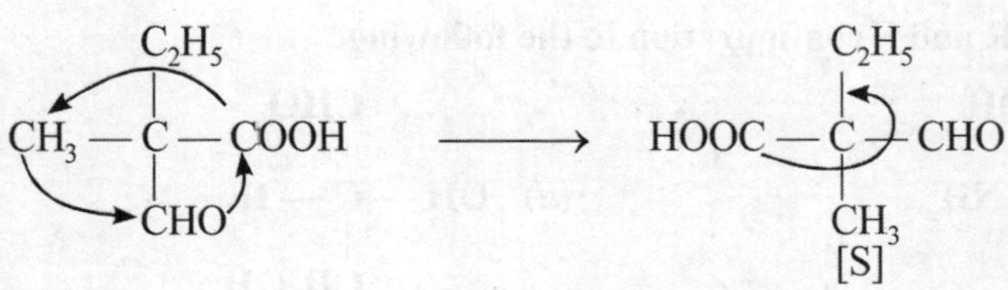

Example 5. **Assign R and S configuration to the following:**

(*i*) 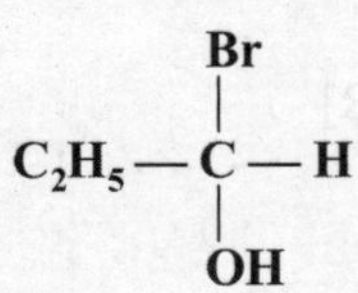 (*ii*) 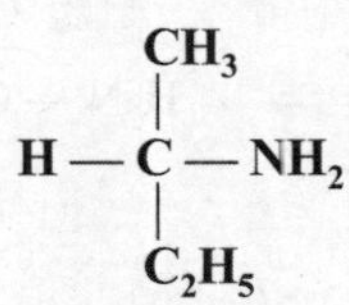 (*iii*) 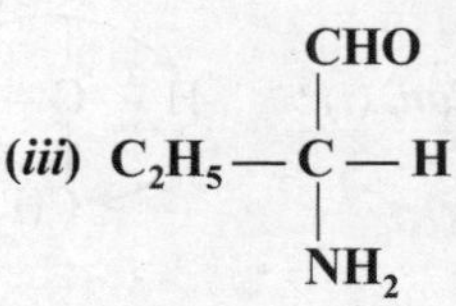

Solution. (*i*) 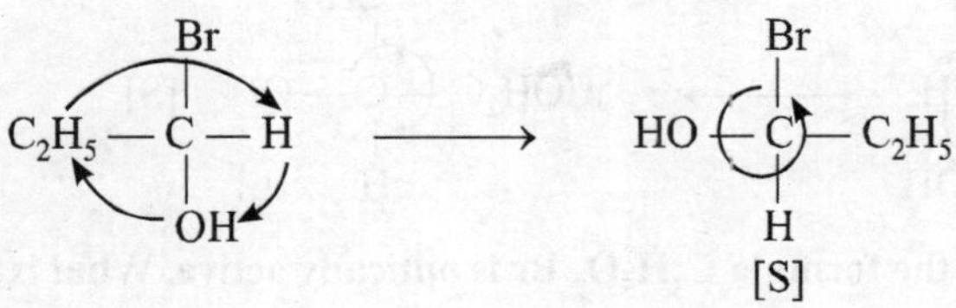

(*ii*) 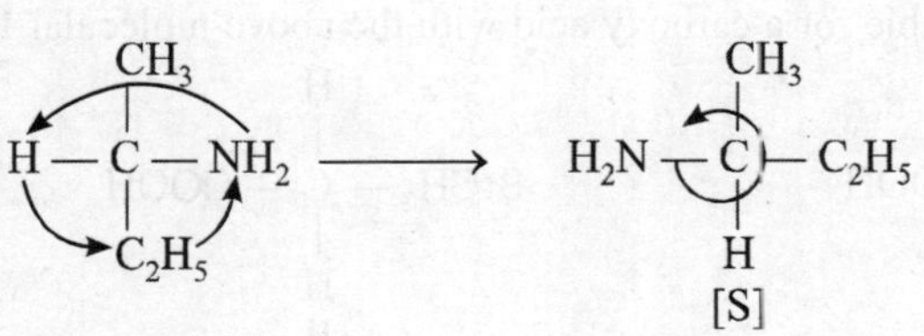

(*iii*) 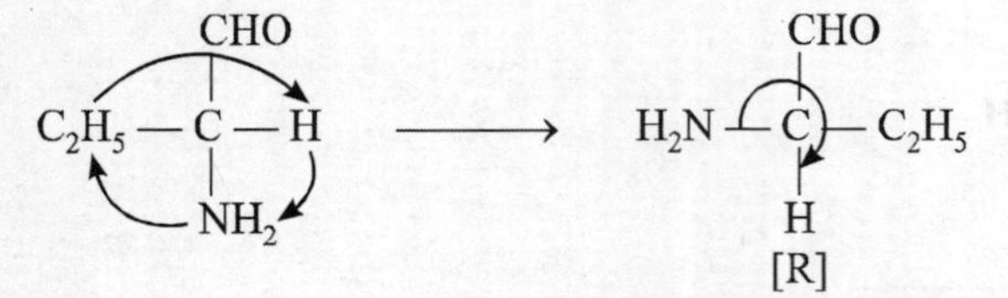

Example 6. **Assign R and S configuration to the following compounds:**

(*i*) 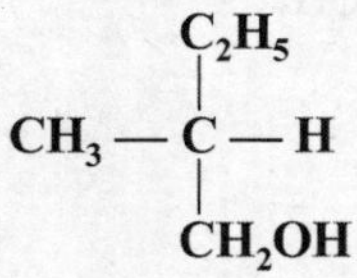 (*ii*) 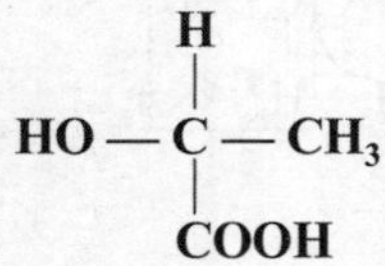

Solution. (*i*) 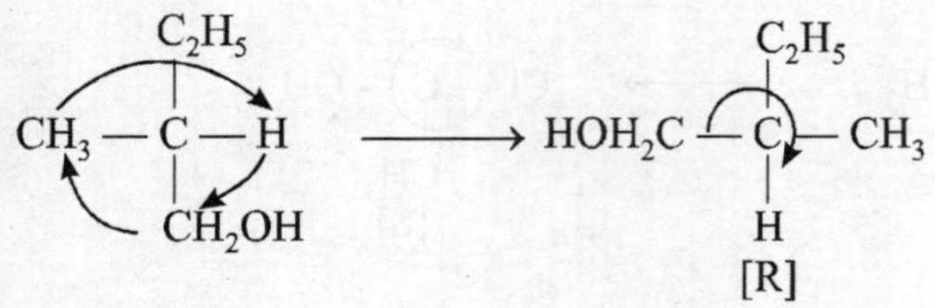

(*ii*) 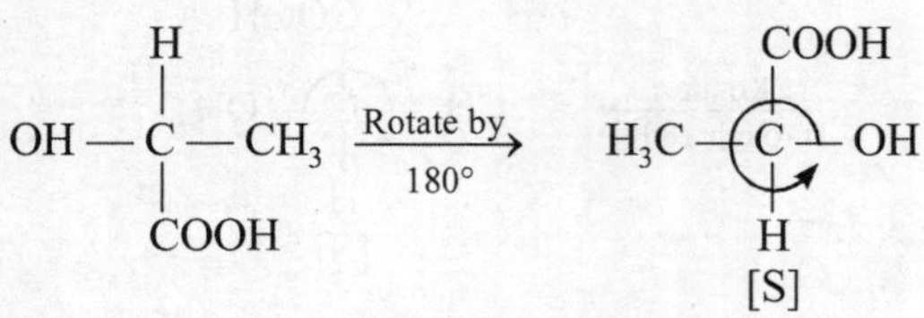

Example 7. **Assign R and S configuration to the following:**

(*i*) 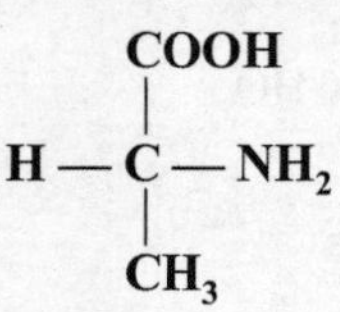 (*ii*) 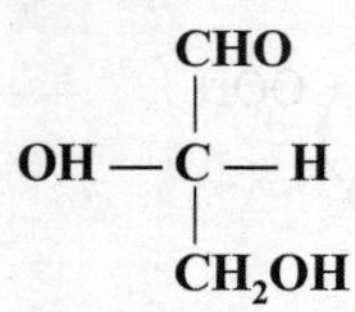

Solution. (*i*) 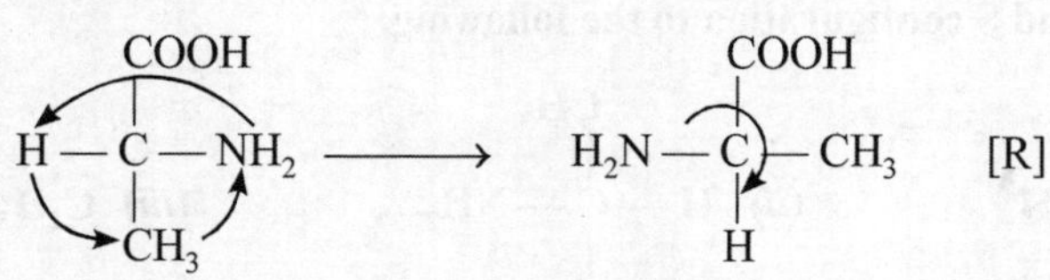

(*ii*) 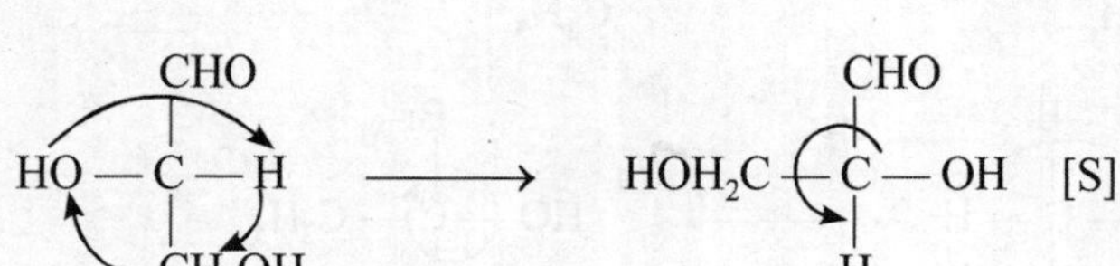

Example 8. **A carboxy acid of the formula $C_3H_5O_2$ Br is optically active. What is its structure?**

Solution. Two structures are possible for a carboxy acid with the above molecular formula

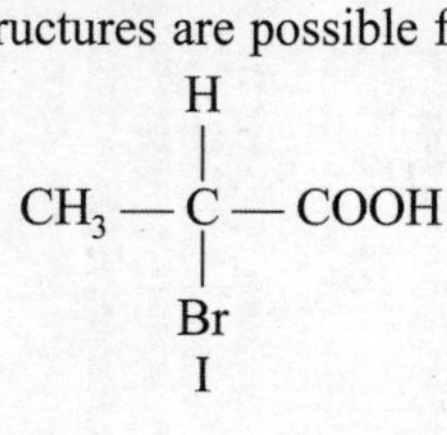

I

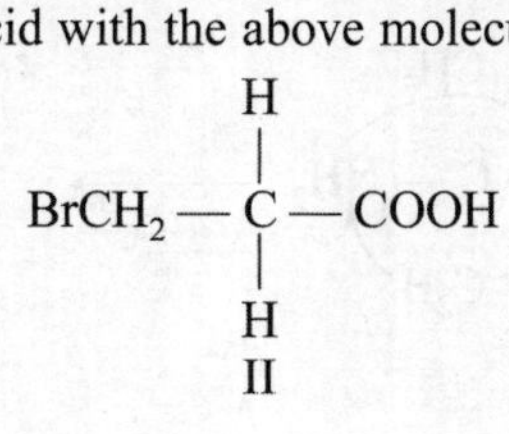

II

Of the two structures, structure I has a chiral (or asymmetric) carbon atom whereas structure II has none. Therefore the structure showing optical activity is I.

Configuration of compounds containing more than one chiral carbon

In such a case, firstly the configuration about each of the chiral carbon is specified and then with the help of numbers, the specification pertaining to the carbon atom of that number is written. Thus the configurations of isomers of 2, 3, 4-Trihydroxybutanal are:

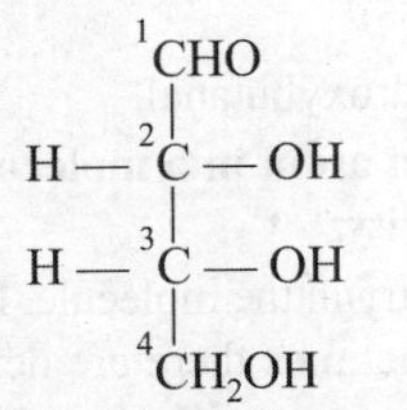

(2R, 3R) -2, 3, 4-Trihydroxy butanal (A)

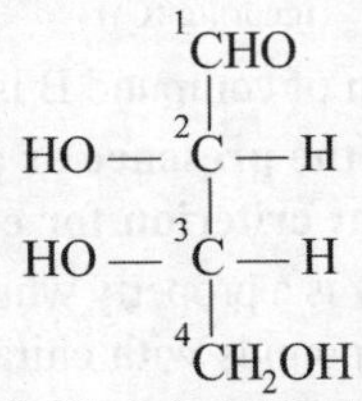

(2S, 3S) -2, 3, 4-Trihydroxy butanal (B)

The configurations are explained as follows:

In the compound A above, sequence of groups attached to C_2 is OH, CHO, CHOH — CH_2OH and H.

Now in order to fix specification, first we consider C_2 and ignore C_3

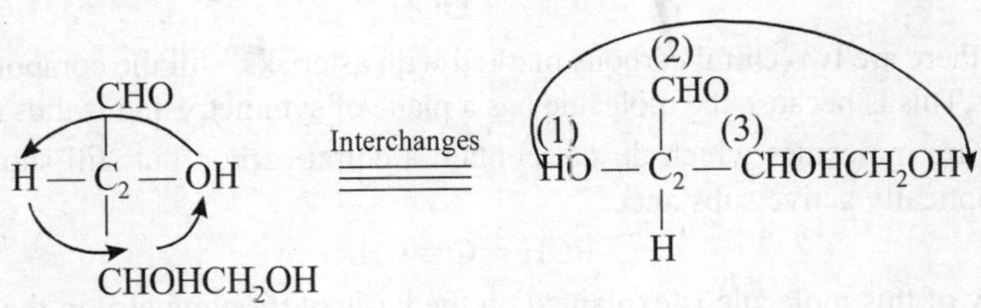

Ignoring (C_3) (2R)

Similarly we consider C_3 and we ignore C_2. Thus

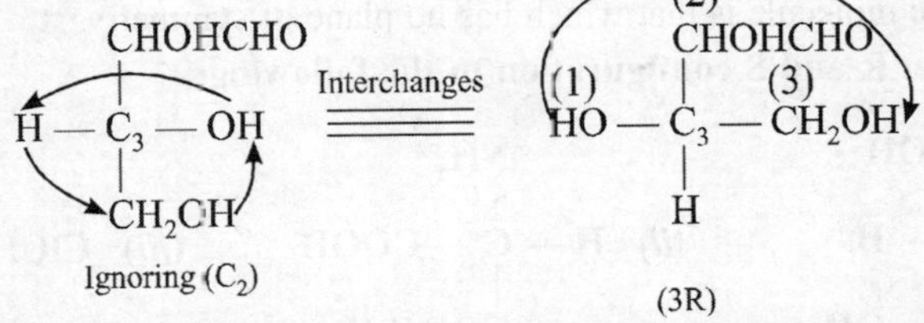

Ignoring (C_2) (3R)

(Here the sequence of groups attached to C_3 is OH, CHOHCHO, CH_2OH and H)

Hence the configuration of compound A is (2R, 3R) – 2, 3, 4-Trihydroxy butanal. Similarly the configuration of compund B can be derived as explained below:

For C_2:

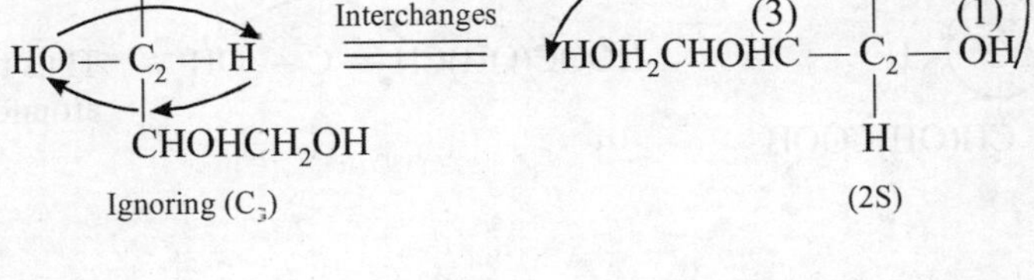

Ignoring (C_3) (2S)

For C_3:

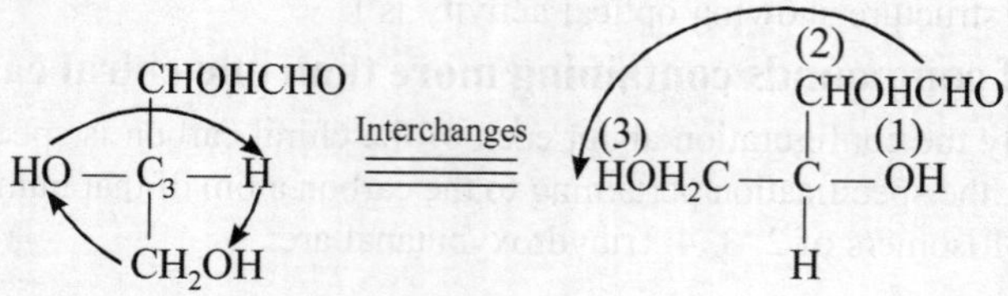

Ignoring (C_2) (3S)

Hence configuration of compund B is (2S, 3S) – 2, 3, 4-Trihydroxybutanal.

Example 9. **Prove that the presence or absence of chiral carbon atom in a molecule is not the necessary and sufficient criterion for existence of optical activity.**

Solution. Optical activity is a property which is related to dissymmetry in the molecule. Dissymmetry occurs normally in compounds with chiral carbon atoms. But sometimes there are deviations. For example,

(*i*) Consider the case of meso tartaric acid

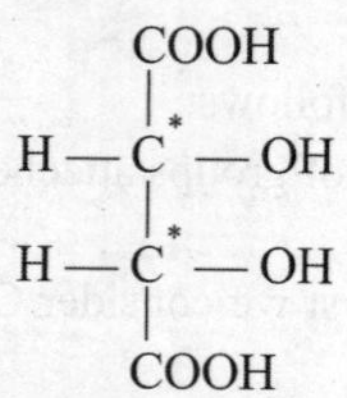

Although there are two chiral carbons marked with asterisks, still the compound does not show optical activity. This is because the molecule has a plane of symmetry and is thus non-dissymmetric.

(*ii*) There are molecules which do not contain a chiral carbon but still show optical activity. Consider the optically active substance

$$RCH = C = CHR$$

Optical activity of this molecule is explained on the basis of dissymmetry in the molecule. Thus it can be remarked that presence or absence of chiral C-atom is not the necessary and sufficient criterion from the existence of optical activity.

In fact it is the dissymmetry criterion which is responsible for the same.

A dissymmetric molecule is that which has no plane of symmetry.

Example 10. **Assign R and S configuration to the following:**

(*i*) 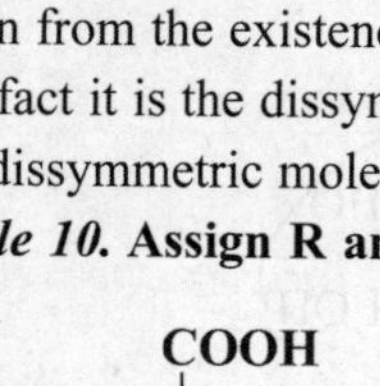 **(*ii*)** 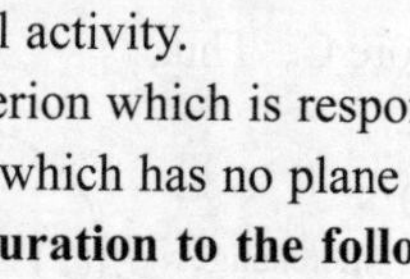 **(*iii*)** 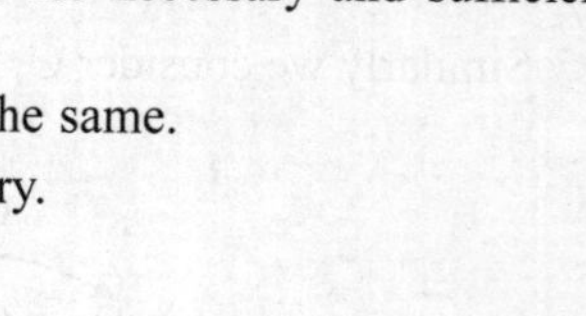

Solution. (*i*) **First consider C_2.** The groups in order of priority attached to C_2 are — OH, —COOH, —CH(OH) COOH, H

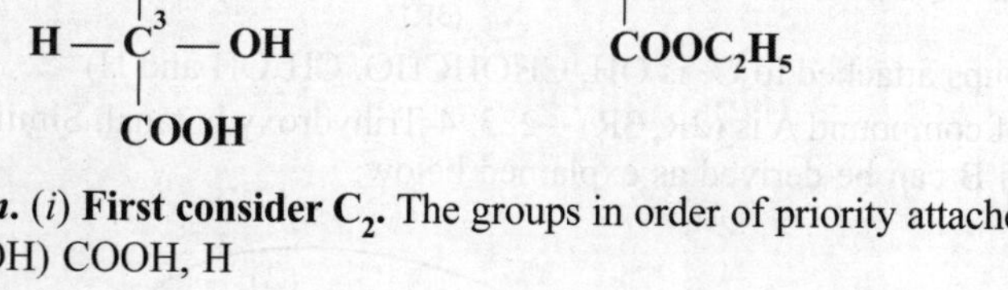

Thus the configuration around C_2 is S

Now consider C_3. The groups in order of priority attached to C_3 are —OH, —COOH, —CH(OH)COOH, H

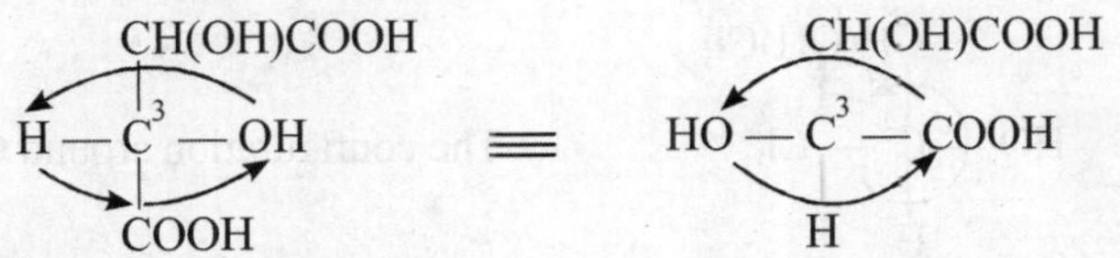

Thus the configuration around C_3 is S

Thus this structure has the configuration 2S, 3S

(*ii*)

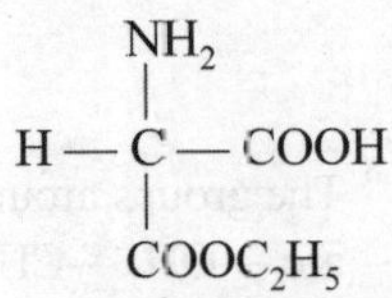

The groups attached to central carbon, in order of priority are $—NH_2$, $—COOC_2H_5$, —COOH and H

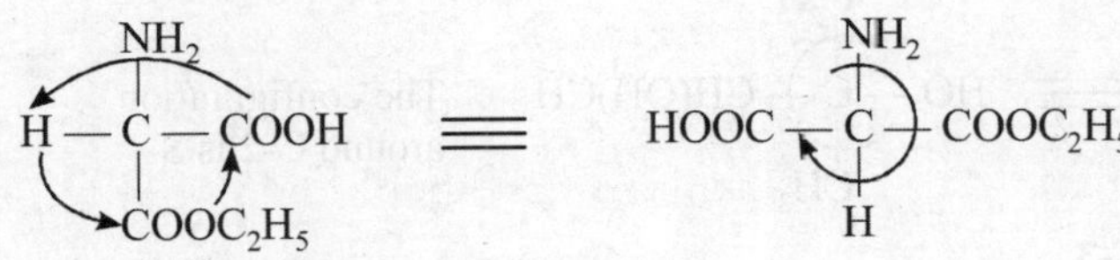

It has R configuration

(*iii*)

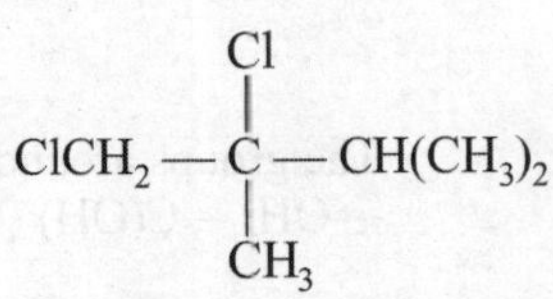

The groups attached, in order of priority, to the central carbon are —Cl, $—CH_2Cl$, $—CH(CH_3)_2$ and CH_3.

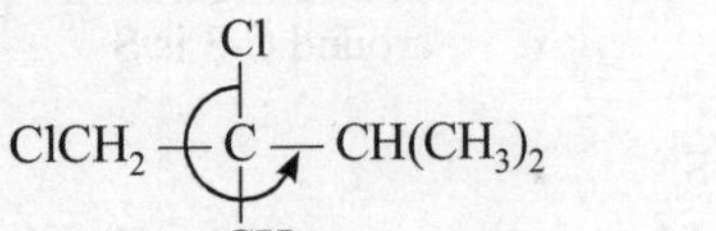

The compound has thus S configuration

***Example 11.* Assign R and S configuration to the following Fischer projection.**

(*i*)

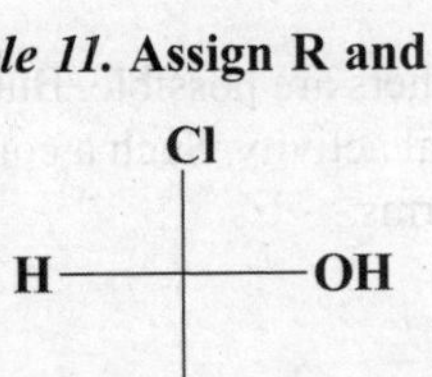

(*ii*)

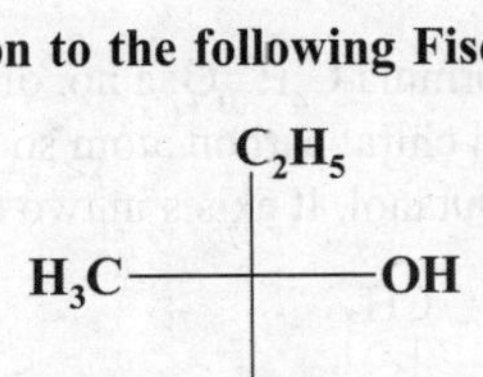

Solution. (*i*) The groups attached to C-2, in order of priority are —Cl, —OH, —CH(OH) Cl and —H

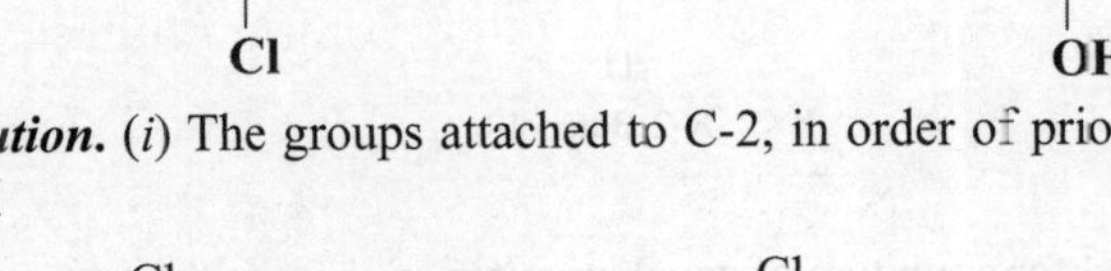

Configuration around C-2 is S

Now consider configuration around C-3.

The groups around C-3 in order of priority are —Cl, —OH, —CH(OH)Cl and H

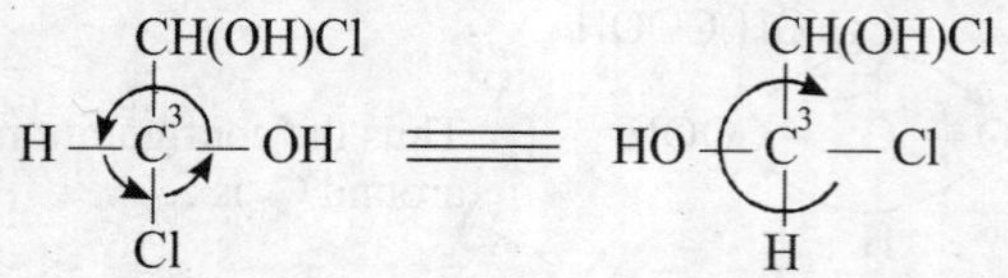

The configuration around C-2 is *R*.

Thus complete configuration of the compound is 2S, 3R

(*ii*)

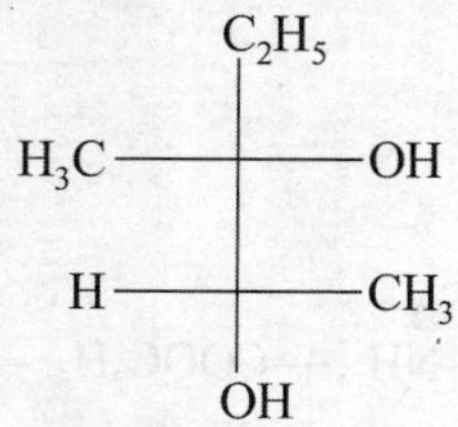

The groups around C_2, in order of priority are —OH, —CH(OH) CH_3, C_2H_5 and CH_3

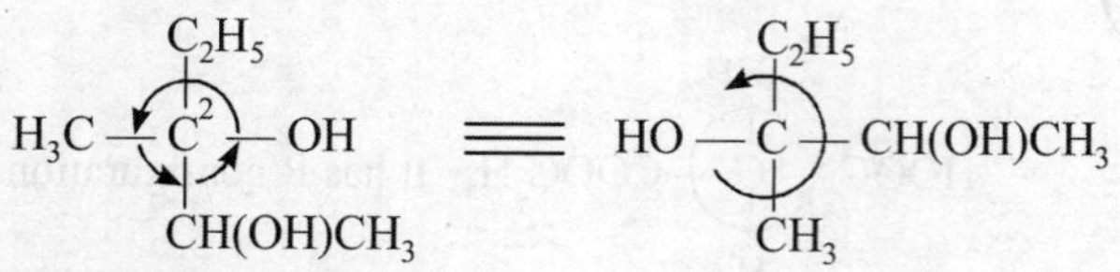

The configuration around C-2 is S

Configuration around C-3

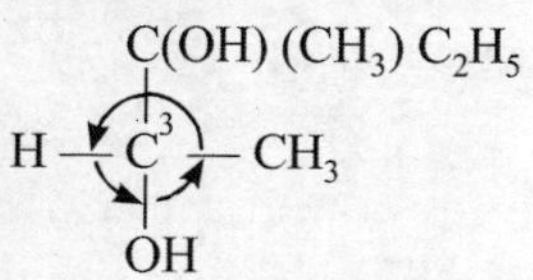

The groups around C-3, in order of priority are —OH, —C(OH) (CH_3) C_2H_5, —CH_3 and —H

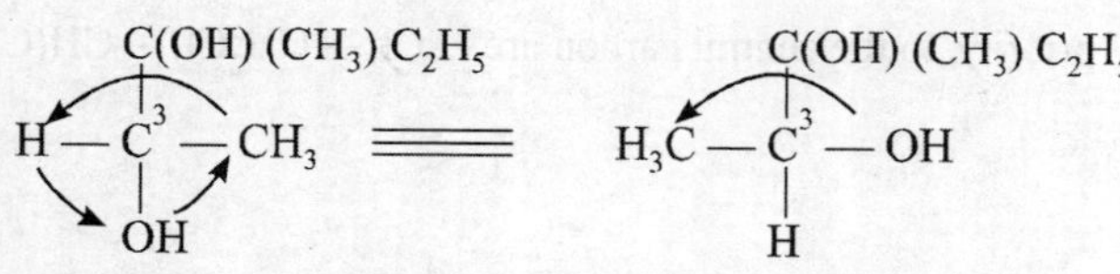

The configuration around C-3 is S

Thus complete configuration of the compound is 2S, 3S.

***Example 12.* A compound $C_4H_{10}O$ shows optical activity. Identify the compound and write the possible stereoisomers.**

Solution. With the molecular formula $C_4H_{10}O$, a no. of alcohols and ethers are possible. But we are interested in a compound with a chiral carbon atom so as to give optical activity. Such a compound with the formula $C_4H_{10}O$ is 2-Butanol. It exists in two enantiomeric forms.

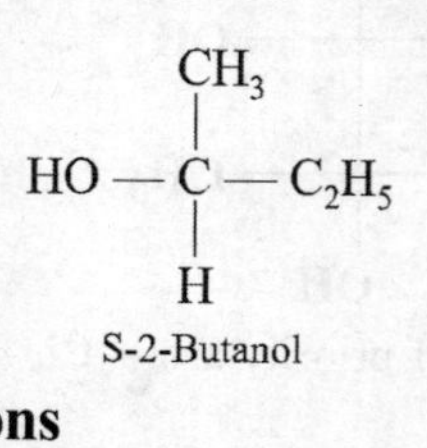

S-2-Butanol

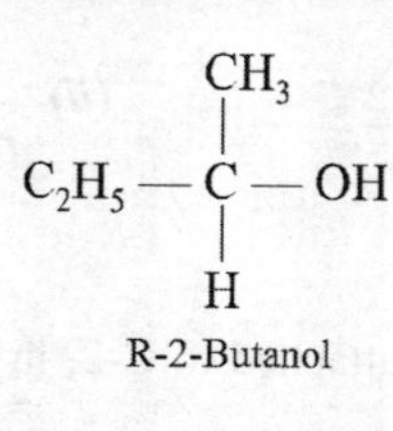

R-2-Butanol

Racemic modifications

Racemic modification is the term used for a mixture of equal amounts of enantiomers. A racemic mixture is optically inactive because of external compensation. The optical activity caused by one enantiomer is neutralised by the activity of the other enantiomer. The notation for a racemic modification or mixture is ± or *dl*. A racemic mixture may also be denoted by the letters R and S. For examples RS-sec. butyl chloride.

When a chiral compound is synthesised from an achiral reactant, a racemic variety of products is obtained. For example, when propionic acid is borminated, α-bromo propionic acid (a chiral product) is obtained. The two enantiomers (+) and (–) α-bromopropionic acids are formed in equal quantities and the product is a racemic mixture. It is optically inactive.

Optical isomerism (enantiomerism) in compounds having two dissimilar chiral carbon atoms

Two asymmetric carbon atoms are said to be dissimilar when atoms or groups attached to one asymmetric carbon atom are different from those attached to the other. Compounds of this type exist in 2^2 i.e., 4 stereoisomers. For example 3-chloro-2-butanol is a compound with two dissimilar chiral carbons marked 1 and 2 and exists in the following four forms:

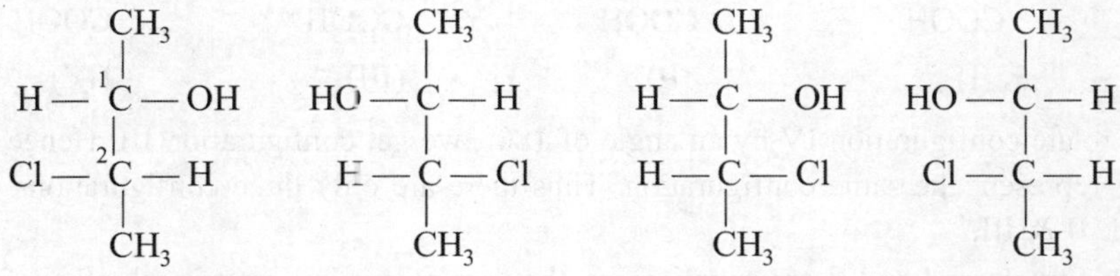

(I) (II) (III) (IV)

The structures I and II are non-superimposable mirror images, so these are a pair of enantiomers. Similarly structures III and IV are another pair of enantiomers. These two pairs of enantiomers give rise to two racemic modifications.

The structures I and III are neither enantiomers nor superimposable. Such type of stereoisomers are called *diastereomers*. Similarly I and IV, II and III, and II and IV, are the other pairs of diastereomers. *Diastereomers can be defined as those stereoisomers which are not mirror images of each other*.

Unlike enantiomers, diastereomers have different physical properties and may rotate the plane of polarised light in the same or different directions and to different extent.

Another example of this type is 2, 3-dichloropentane.

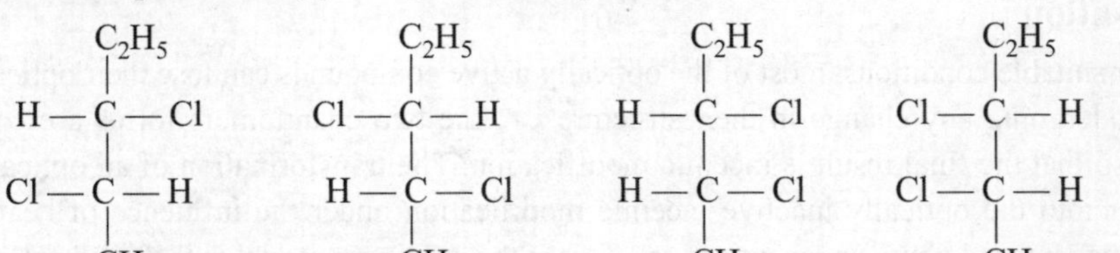

(I) (II) (III) (IV)

A pair of enantiomers — Second pair of enantiomers

Properties of Diastereomers

(1) They show similar, but not identical, chemical properties (as they contain the same functional groups). Rates of reactions of diastereomers with a given reagent are generally different.

(2) They have different physical properties like m.p. and b.p., densities, refractive indices, specific rotations, solubilities etc. in a given solvent.

(3) They can be separated by techniques like fractional crystallisation, fractional distillation and chromatography.

Example 13. **Write four configurations of tartaric acid and select the pairs forming (i) Enantiomers, (*ii*) Diastereomers. Which of them are optically active and which of them are not. Why?**

Solution. Tartaric acid is an example of compounds with two similar chiral carbon atoms. Here the four groups attached to both carbons (C-2 and C-3) are the same. Various configurations of tartaric acid are:

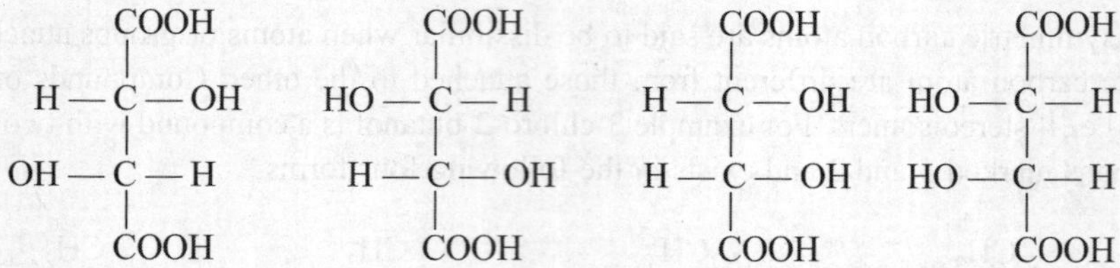

(I) (II) (III) (IV)

If we rotate configuration IV by an angle of 180°, we get configuration III. Hence structures III and IV represent the same configuration. Thus there are only three configurations of tartaric acid viz., I, II & III.

(*i*) Structures I and II are enantiomers, they are mirror images of each other. There is no plane of symmetry in either of them. Therefore both I and II are optically active but display the optical activity in opposite directions.

(*ii*) Structures I and III are not mirror images of each other. Hence they are not enantiomers. Such pairs of compounds having the identical molecular formula and identical groups are called diastereomers.

Similarly structures II and III also form a pair of diastereomers.

Compound III has two chiral carbons, but still it optically inactive. This is because there is a plane of symmetry as indicated by dotted line. There being no dissymmetry, the compound is not optically active. Such a compound is caused meso comp. Thus III and IV represent meso tartaric acid.

Racemisation

Under suitable conditions, most of the optically active compounds can lose their optical activity without undergoing any change in their structure *i.e.*, the two enantiomers forms are convertible into each so that the final result is racemic modification. The transformation of an optically active enantiomer into the optically inactive racemic modification under the influence of heat, light or chemical reagents is known as racemisation. Thus if the starting material is the (+) form, then after treatment, half will be converted into (–) form. If the starting material is (–) form, half will be converted into (+) form. For example, (+) or (–) lactic acid on warming with sodium hydroxide gets converted into (±) lactic acid.

Though different mechanisms have been presented for the racemisation of different types of compounds in most of the cases it occurs via the formation of an intermediate which is no longer dissymmetric or chiral. For example, a ketone in which chiral carbon atom is joined to a hydrogen atom can undergo racemisation via the formation of enolic form by tautomeric change.

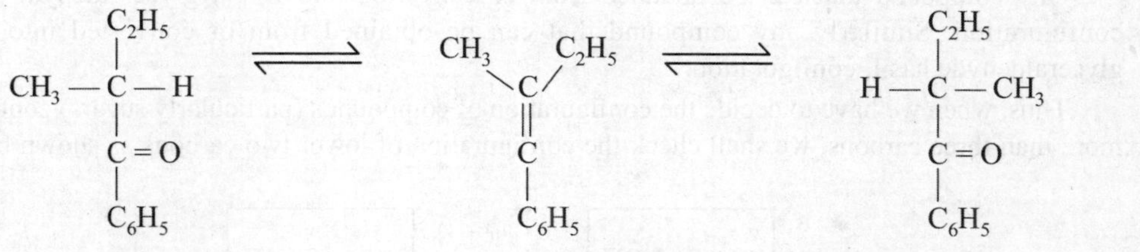

(+) form loss of chirality (–) form

The intermediate enolic form, which is no longer chiral when reverts to the stable keto form, it is equally likely to produce (+) or (–) forms and thus racemisation takes place.

Example 14. **Giving examples distinguish between the following: Meso and Racemic forms.**

Solution. Meso form of a compound is optically inactive form, in spite of the presence of asymmetric carbon atoms in it. This is because there is a plane of symmetry in the molecule. The activity of one part is neutralised by the activity of the other part. An example is meso tartaric acid.

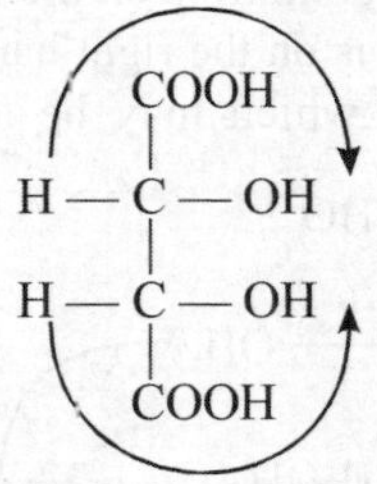

meso tartaric acid

Racemic form means a mixture of equal amounts of *d* and *l* forms of a compound, like mixture of equal amounts of *d* and *l* lactic acid or ± tartaric acid.

Such a mixture does not show any optical activity.

It may be mentioned that optical inactivity of the meso form is due to **internal compensation** as the activity of one part of the molecule is neutralised by that of the other part.

Optical inactivity of the racemic form is due to external compensation as the two forms *d*- and *l*-neutralise the optical activity of each other.

Threo and Erythro Notations

Diastereomers are optical isomers of a compound which are not mirror images of each other and hence are not enantiomers. For example *d*-tartaric acid is a diastereomer of mesotartaric acid. But the two are not enantiomers.

As for as the absolute configuration of a compound is concerned we derive it from the configuration of glyceraldehyde which has been taken as arbitrary standard.

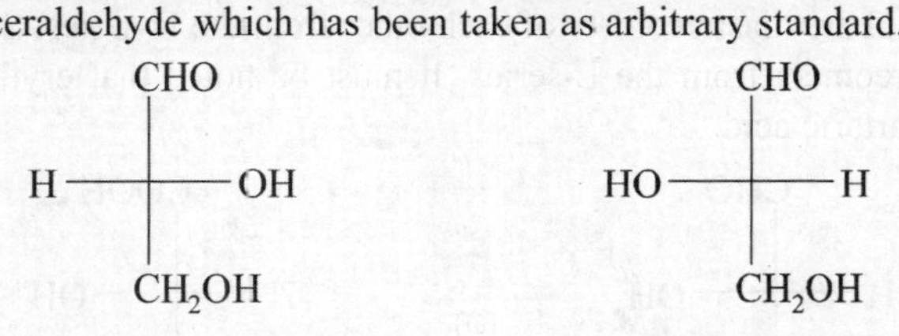

D (+) glyceraldehyde L (–) glyceraldehyde

D and L stand for the configuration while (+) and (–) signs denote the actual direction of rotation of plane polarised light. If H and OH are on the LHS and RHS of central carbon, it donates D configuration. If OH and H are on the LHS and RHS of central carbon it denotes L configuration.

Any compound that can be obtained from or converted into D (+) glyceraldehyde has D configuration. Similarly any compound that can be obtained from or converted into L (–) glyceraldehyde has L configuration.

Thus, when we have to decide the configuration of compounds (particularly sugars) containing more than three carbons, we shall check the configuration of lower two carbons as shown below.

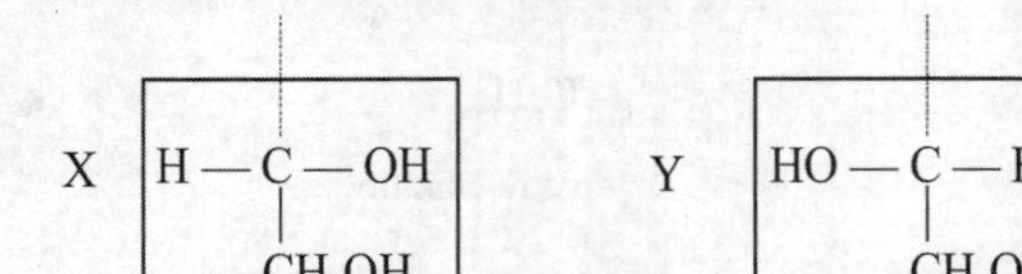

D-series L-series

If the arrangement corresponds to X, the compound has D configuration. If the arrangement corresponds to Y, the compound belongs to L configuration.

Let us take the example of CHO $\overset{*}{C}$H OH $\overset{*}{C}$HOH CH_2OH. This contains two asymmetric carbon atoms marked with asterisks. There are four optical isomers possible, all of which are known.

If the H on the third carbon atom (from the bottom) is on the left hand side, the compound is *erythro*, while if the H on the third carbon atom is on the right hand side, the compound is *threo*. The structures of D-erythrose and D-threose which may be assumed to be obtained from D-glyceraldehyde are given below

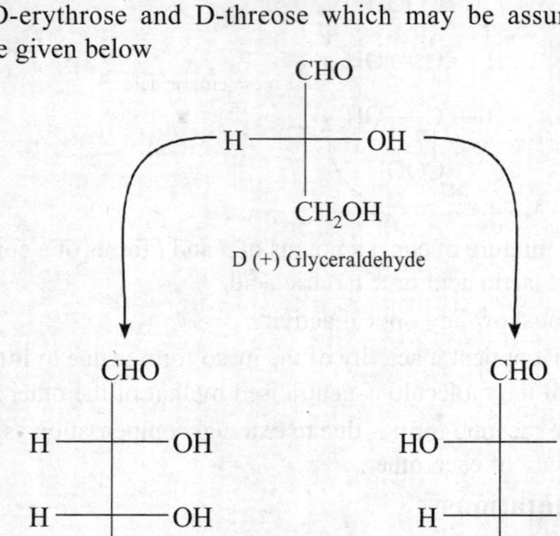

D (+) Glyceraldehyde

D (–) erythrose
I

D (–) Threose
II

Compounds I and II are diastereomers of each other from the D-series. Similarly there will be another pair of diastereomers from the L-series. It must be noted that erythro compound (I) on oxidation gives mesotartaric acid.

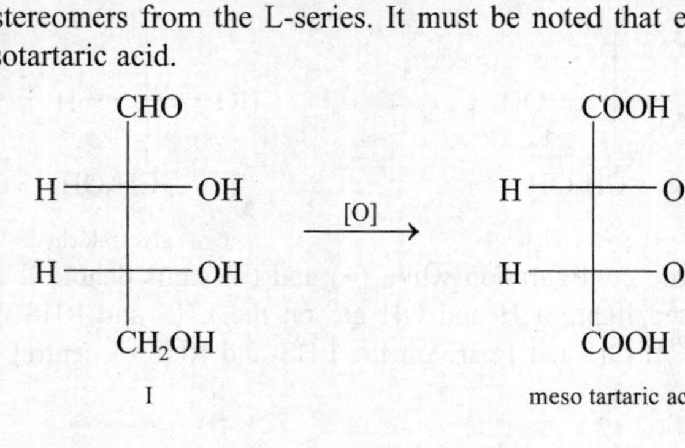

I

meso tartaric acid

Similarly the erythro compound from L-series on oxidation will also produce meso tartaric acid.

Example 15. **Predict whether 3-chlorohexane will be optically active or not? Give reasons for your answer.**

Solution.

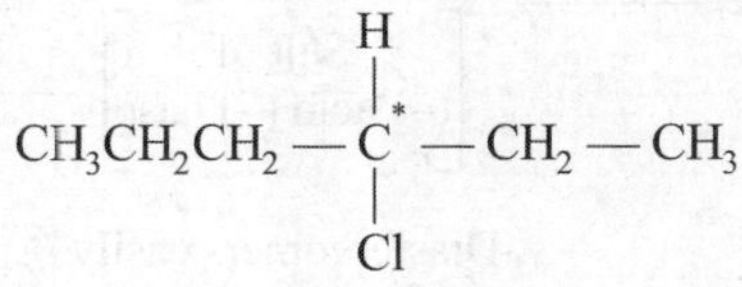

3-chlorohexane

The compound contains are asymmetric carbon atom (marked with *). The four different groups attached are

1. — H
2. — Cl
3. — CH_2 — CH_3
4. — CH_3 — CH_2 — CH_3

Hence the compund shows optical activity.

Methods for the resolution of racemic mixtures

The separation of racemic modification into enantiomers is called resolution. Since the two enantiomers in a racemic mixture have identical physical and chemical properties, these cannot be separated by usual methods of fractional distillation or fractional crystallisation. Special methods are adopted for their separation as given below:

(*i*) Mechanical separation. This method was first adopted by Pasteur for separating the enantiomers of ammonium tartarate. When racemic modification is crytallized from a solution, two types of crystals are obained. These are mirror images of each other consisting of (+) and (–) forms which can be separated by hand picking with the help of a pair of tweezers and a powerful lens. This is a very laborious method and can be applied only to those compounds which give well defined distinguishable crystals of enantiomers.

(*ii*) Biochemical method. Certain micro-organisms grow in a racemic mixture, consuming only one of the enantiomers while leaving the other unaffected. Thus *penicillium glaucum* when placed in (±) tartaric acid, consumes only (+) tartaric acid and leaves (–) tartaric acid unused. The major disadvantage of this method is that one of the enantiomers get destroyed.

(*iii*) Chemical method. This method is mostly used for the resolution of racemic modification. In this method the racemic modification is treated with an optically active reagent to get a pair of *diastereomers*. Since diastereomers differ in their physical properties, it is possible to separate them by physical methods such as fractional crystallisation, fractional distillation etc. The pure diastereomers are then decomposed, into a mixture of optically active reagent and corresponding enantiomer, which can be separated.

Suppose the racemic modification is an (±) acid. When it is treated with an optically acitve, say (–) base, it gives a mixture of two salts, one of (+) acid (–) base, the other of (–) acid (–) base.

These salts are neither superimposable nor mirror images; so these are diastereomers having different physical properties and can be separated by fractional crystallisation. After separation the optically active acids can be recovered in pure forms by adding a mineral acid.

$$\left.\begin{array}{l}(+)\text{ acid}\\ \\ (-)\text{ acid}\end{array}\right] + (-)\text{ base} \longrightarrow \begin{array}{l}\left[\begin{array}{c}\text{Salt of}\\ (+)\text{ acid }(-)\text{ base}\end{array}\right] \xrightarrow{H^+} (+)\text{ acid}\\ \\ \left[\begin{array}{c}\text{Salt of}\\ (-)\text{ acid }(-)\text{ base}\end{array}\right] \xrightarrow{H^+} (-)\text{ acid}\end{array}$$

Racemic modification — Diastereomers easily separable — Enantiomers in pure form

The commonly used optically active bases for the purpose are naturally occurring alkaloids such as (–) brucine, (–) quinine, (–) strychnine and (+) cinchonine.

Similarly, the resolution of racemic bases can be carried out using a naturally occurring optically active acid such as (–) malic acid. Alcohols can be resolved in a similar way by ester formation using an optically active acid.

Asymmetric Synthesis

The preparation of an optically active dissymmetric (chiral) compound from non-dissymmetric molecules under the influence of an optically active substance is known as asymmetric synthesis.

Whenever a dissymmetric product is synthesised from a non-dissymmetric reactants, it is always an optically inactive racemic modification (as described earlier). However, by the use of an optically active reagent, one of the enantiomers can be obtained in excess so that the resulting product is optically active.

For example pyruvic acid on direct reduction yields the optically inactive racemic lactic acid, but pyruvic acid, pre-esterified with an optically active alcohol say (–) menthol, on reduction and subsequent hydrolysis yields predominantly (–) lactic acid.

$$\underset{\text{Pyruvic acid}}{CH_3-\overset{\overset{\displaystyle O}{\|}}{C}-COOH} \xrightarrow[\text{reduction}]{2H} \underset{(\pm)\text{ Lactic acid}}{CH_3-CHOH-COOH}$$

$$\underset{\text{Pyruvic acid}}{CH_3-\overset{\overset{\displaystyle O}{\|}}{C}-COOH} + \underset{(-)\text{ Menthol}}{C_{10}H_{19}OH} \longrightarrow \underset{(-)\text{ Menthyl pyruvate}}{CH_3-\overset{\overset{\displaystyle O}{\|}}{C}-COOC_{10}H_{19}}$$

$$\Big\downarrow \text{reduction, } 2H$$

$$\underset{(-)\text{ Menthol}}{C_{10}H_{19}OH} + \underset{\substack{(-)\text{ Lactic acid}\\ \text{(predominantly)}}}{CH_3-CHOH-COOH} \xleftarrow{H_2O,\ H^+} \underset{(-)\text{ Menthyl lactate}}{CH_3-CHOH-COOC_{10}H_{19}}$$

Walden Inversion

In 1893, Walden was able to convert an optically active compound into its enantiomer by a series of replacement reactions. For example, conversion of (–) malic acid into (+) malic acid and vice-versa can be achieved as under :

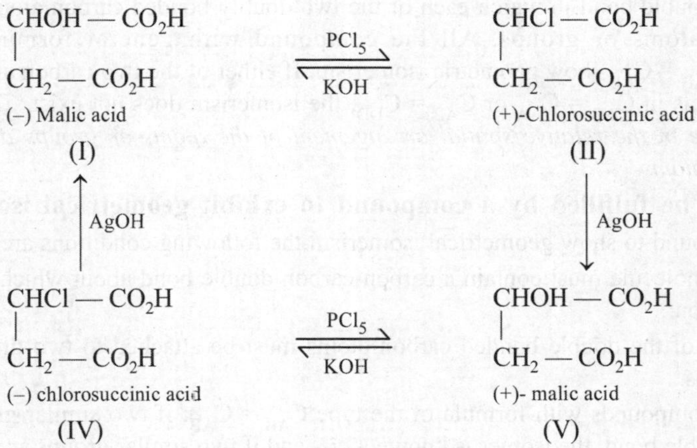

(–) Malic acid (I); (+)-Chlorosuccinic acid (II); (–) chlorosuccinic acid (IV); (+)- malic acid (V)

During the change of (–) malic acid (I) or (+) malic acid (III) there must be a change in configuration in one of the two steps. If the configuration of II and III is same, the change must have taken place between I and II. Any single reaction in which change of configuration takes place, is termed as Walden Inversion.

In other words if an atom or group of atoms directly attached to chiral carbon atoms is replaced by another atom or group of atoms the configuration of product is sometimes found to be different from that of starting compound. Such phenomenon which involves inversion of configuration during a reaction is called *Walden Inversion or Optical Inversion.*

For example, in SN^2 hydrolysis of 2-bromooctane inversion of configuration takes place.

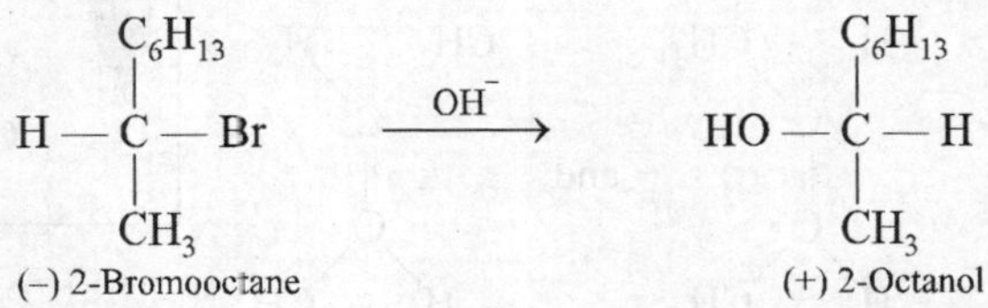

(–) 2-Bromooctane (+) 2-Octanol

It must be kept in mind here that inversion of configuration may or may not lead to a change in direction of rotation. The change, if any, is only a matter of chance as in the example given above.

3.4 GEOMETRICAL ISOMERISM

Two carbon atoms joined by a single bond (σ bond) are capable of free rotation around each other, but this rotation is hindered in case of compounds containing carbon-carbon double bond. According to molecular orbital theory, carbon atoms involved in double bond formation are sp^2 hybridised so that each carbon atom has three planar sp^2 hybridised orbitals and fourth p orbital having its lobes at right angles to the plane of sp^2 orbitals. The formation of π bond involves the overlapping of p orbitals. With the formation of a π bond between C — C along with a σ bond which is already existing, there remains no possibility of rotation along C — C axis. Neither of the two doubly bonded carbon atoms can be rotated about double bond without destroying π orbital which requires large amount of energy. Thus at ordinary temperature, the rotation about a carbon-carbon double bond is restricted or hindered and gives rise to a kind of stereoisomerism known as *Geometrical Isomerism.*

Geometrical Isomerism, also known as *cis-trans* isomerism takes place in compounds containing carbon-carbon double bond in which each of the two doubly bonded carbon atoms is attached to two different atoms or groups. All the compound with general formula of the type $C_{AB} = C_{DE}$ or $C_{AB} = C_{AB}$ show geometric isomerism. If either of the two carbon atoms carries two identical groups as in $C_{AB} = C_{AA}$ or $C_{AB} = C_{DD}$, the isomerism does not exist. *This isomerism is due to difference in the relative spatial arrangement of the atoms or groups about the doubly bonded carbon atoms.*

Conditions to be fulfilled by a compound to exhibit geometrical isomerism

For a compound to show geometrical isomerism the following conditions are necessary:

(*i*) The molecule must contain a carbon-carbon double bond about which there is no free rotation.

(*ii*) Each of the double bonded carbon atoms must be attached to two different atoms or groups.

In case of compounds with formula of the type $C_{AB} = C_{AB}$; if two similar groups are on the same side of double bond, the isomer is known a *cis-* and if two similar groups are on the opposite sides of the double bond the isomer is known as *trans-* such as:

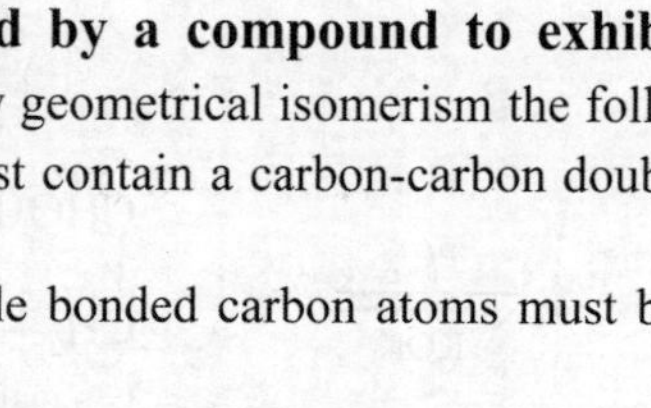

cis-isomer *trans*-isomer

Geometrical isomers of 2-butene

For example,

2-butene exists in two isomeric forms.

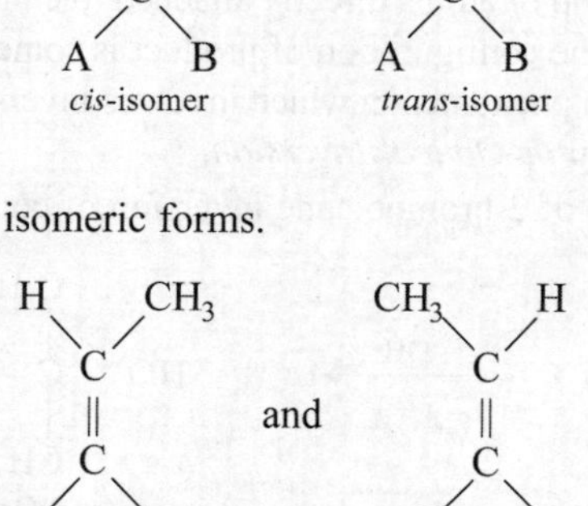

cis-2-butene *trans*-2-butene

Similarly butene-dioic acid exists in two isomeric forms; *cis-* form is called maleic acid and *trans-* form is called fumaric acid

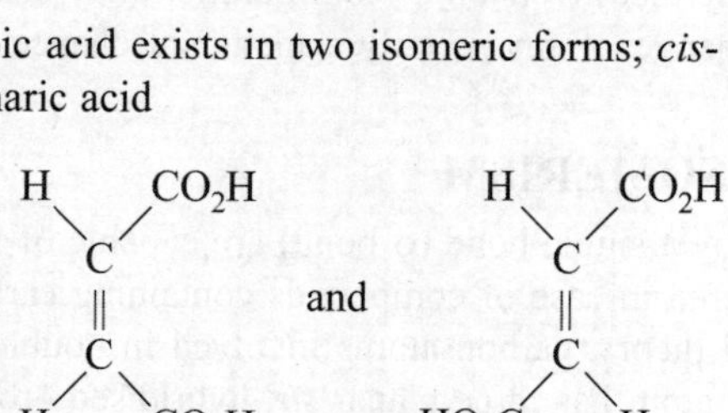

Maleic acid (*cis*-isomer) Fumaric acid (*trans*-isomer)

Determination of configuration of a geometrical isomer

Different methods available for determination of configuration of a geometrical isomers are described below:

(*i*) From Dipole moments. Generally *cis*-isomer has greater dipole moment as compared to *trans*-isomer. In case of cis- the similar groups being on the same side, the electronic effects are additive; while in case of *trans*-isomer, the similar groups being on opposite side, the electronic effects cancel each other.

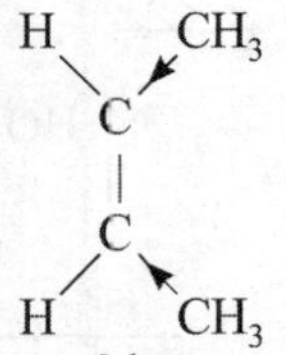

cis-2-butene
$\mu = 0.4$

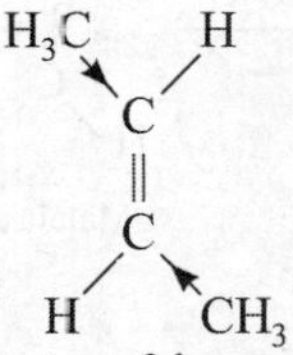

trans-2-butene
$\mu = 0$

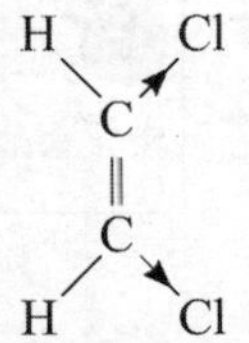

cis-2-Dichlorethene
$\mu = 1.85$

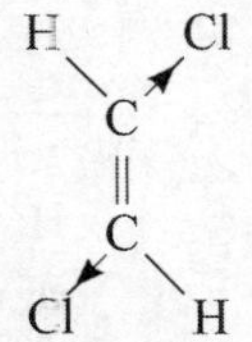

trans-1, 2-dichloroethene
$\mu = 0$

***(ii)* From boiling point.** Generally speaking, a *cis* isomer has a higher boiling point compared to the *trans* isomer. This is because of higher dipole moment and higher polarity in the molecule which acts as the binding force and is responsible for higher b.p. of the *cis* isomer. Boiling point of *cis*-2-butene is 277 K while the *trans*-2-butene boils at 274 K.

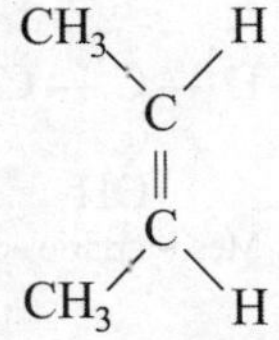

cis-2-butene
b.p. 277 K

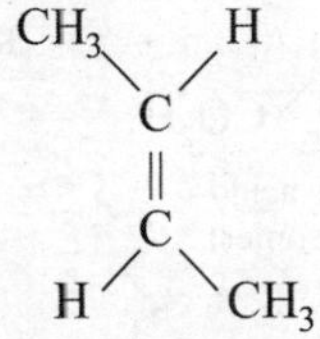

trans-2-butene
b.p. 274 K

***(iii)* From melting point.** The isomers show a reverse trand here. A *cis* isomer has a lower melting point compared to the *trans* isomer. Here the factor that is important is the size of the molecule. It can be realised that a *cis* isomer will occupy a smaller volume compared to the *trans* isomer as illustrated by dotted lines. Thus maleic acid melts at 403 K whereas fumaric acid melts at 575 K.

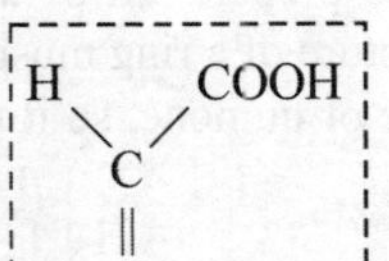

Maleic acid (*cis*)
m.p. 403 K

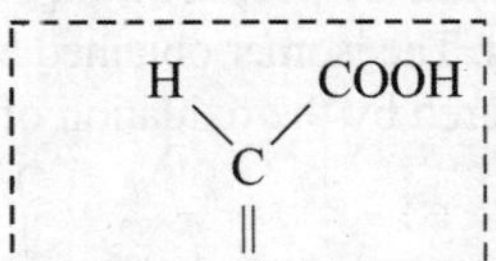

Fumaric acid (*trans*)
m.p. 575 K

***(iv)* From the formation of cyclic compounds.** Two geometric isomers (*cis* and *trans*) can be distinguished through reactions that lead to formation of ring. *Cis*-isomer undergoes ring closure more readily than the *trans*-isomer. For example maleic acid readily loses water when heated to about 423 K, to give an anhydride; while fumaric acid does not give anhydride at this temperature. Rather it must be heated to 573 K to get the same anhydride. Hydrolysis of anhydride yields only maleic acid.

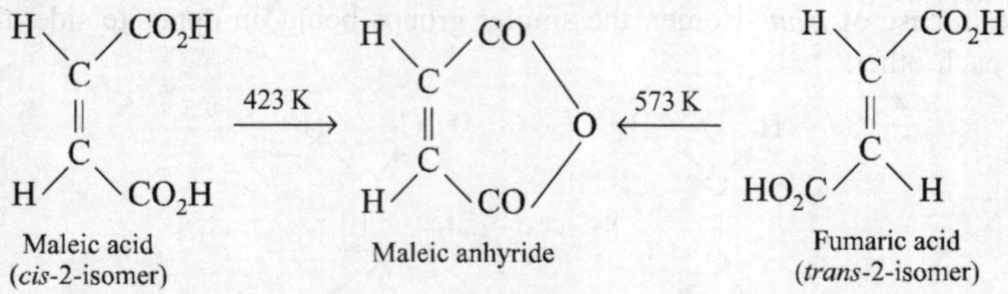

Maleic acid (*cis*-2-isomer) — Maleic anhyride — Fumaric acid (*trans*-2-isomer)

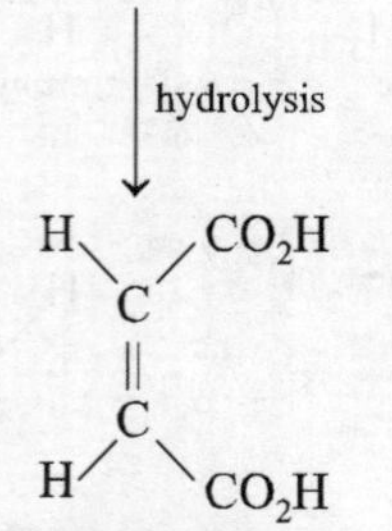

only maleic acid and not fumaric acid

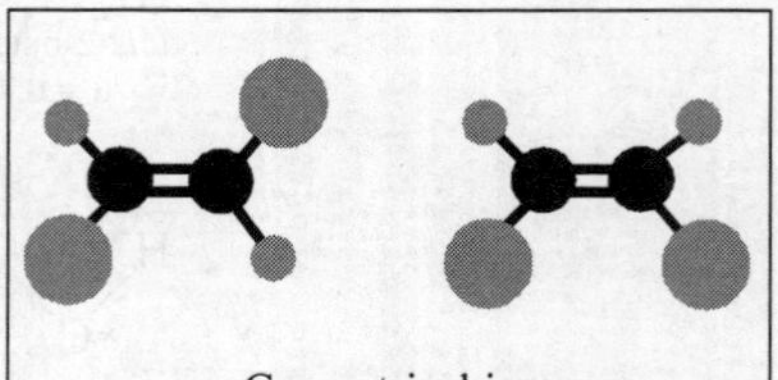

Geometrical isomers

(*v*) From the formation of the type of optical isomer. Maleic acid and fumaric acid, both on treatment with $KMnO_4$ or OsO_4 yield optically inactive variety of tartaric acid. Maleic acid yields meso tartaric acid while fumaric acid yields racemic (± or *dl*) tartaric acid.

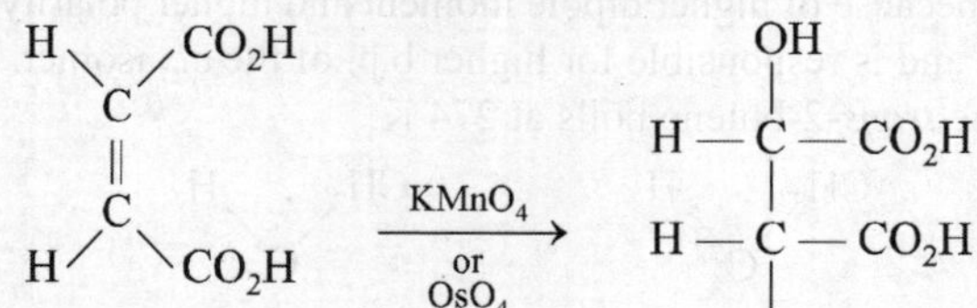

Maleic acid (*cis*-2-isomer) — Meso-tartaric acid

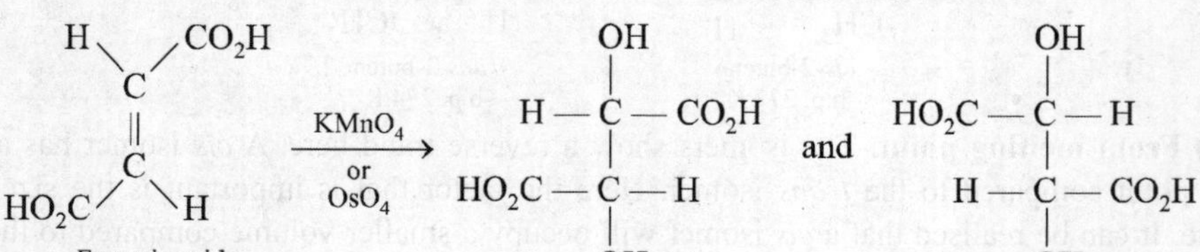

Fumaric acid (*trans*-isomer) — ± or *dl*-tartaric acid

(*vi*) From the method of preparation. Method of preparation of a compound sometimes leads to its configuration. The isomer obtained by the rapture of a ring must be the *cis*-isomer, *e.g.*, maleic acid can be prepared by the oxidation of benzene or quinone, so it must be a *cis*-isomer.

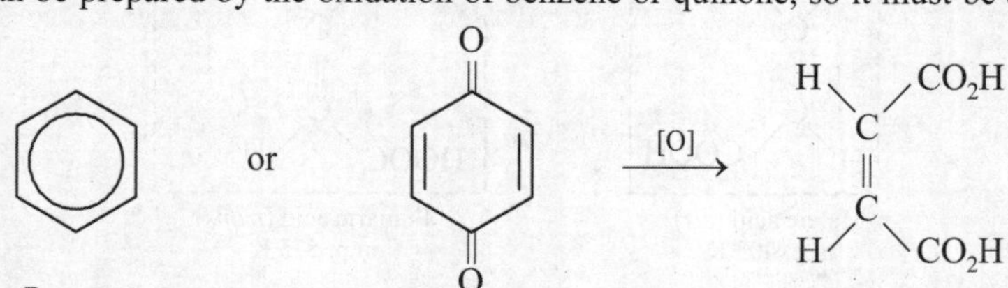

Benzene — Quinone — Maleic acid (*cis*-2-isomer)

E and Z designations of geometrical isomers

The *cis* and *trans*- designated can be used only for the compounds in which two doubly bonded carbon atoms are having similar atoms or groups *e.g.*, of the type $C_{AB} = C_{AB}$. But, when the two doubly bonded carbon atoms are having different atoms or groups attached to them *e.g.*, of the type $C_{AB} = C_{DE}$; it is not possible to assign them *cis* or *trans* configurations. To overcome this difficulty, a more general system for designating the configuration of geometric isomers has been adopted. This system developed by *Cahn*, Ingold and Prelog originally for the absolute configuration of optical isomers is known as *E* and *Z system* and is based on priority of attached groups. The atoms or groups attached to each carbon of the double bond, are assigned first and second priority. If the atoms or groups having higher priority attached to two carbons are on the same side of double bond the configuration is designated as Z (derived from German word *Zusammen* meaning together) and if the atoms or groups of higher priority are on the opposite side of the double bond, the configuration is designated as E (derived from German word *-entgegen* meaning across or opposite).

Priorities of atoms or groups are determined in the same way as for R & S configurations of optical isomers. At. weights or atomic numbers of atoms directly linked with ethylenic carbon atoms are taken into consideration.

Let us consider an example in which two doubly bonded atoms are attached to four different halogens such as $C_{BrF} = C_{ICl}$. Since Br is having higher priority over F and I is having priority over Cl (due to their higher atomic numbers). The isomer in which Br and I are on the same side of double bond will be called Z and the isomer in which Br and I are on the opposite sides of double bond will be called E.

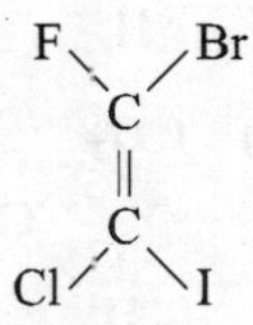
Z-isomer

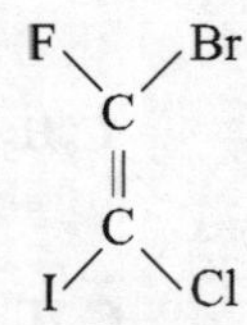
E-isomer

In the same way *cis* and *trans* isomers of 2-butene can be called Z and E-2-butenes respectively.

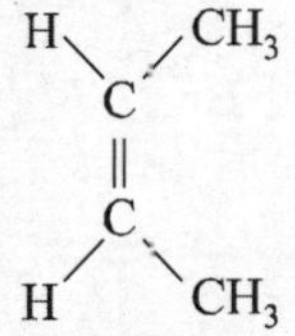

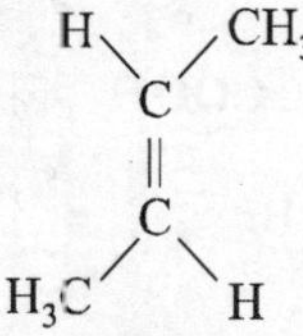

Similarly maleic acid can be specified as Z-isomer and fumaric acid as E-isomer.

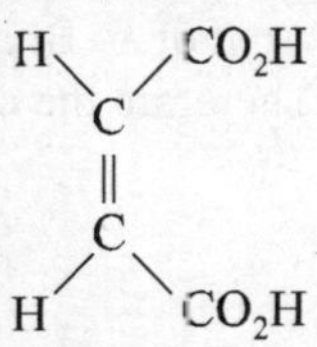
Maleic acid
(*Z*-isomer)

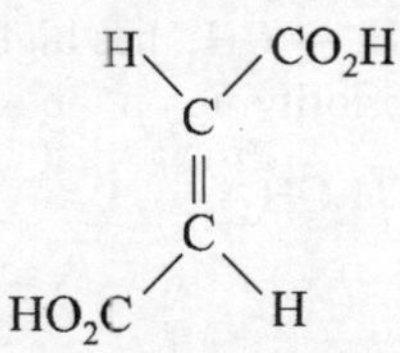
Fumaric acid
(*E*-isomer)

In determining the configuration, we have to select the group of higher priority on one carbon. Similarly we select the group of higher priority on the other carbon atom. If these two groups are on the same side of double bond, the configuration is Z, otherwise it is E.

Some more examples are:

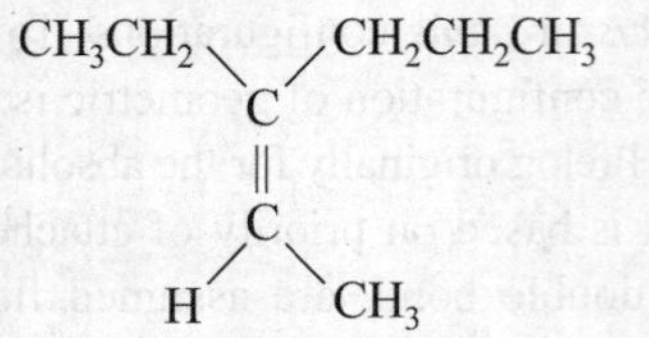

and

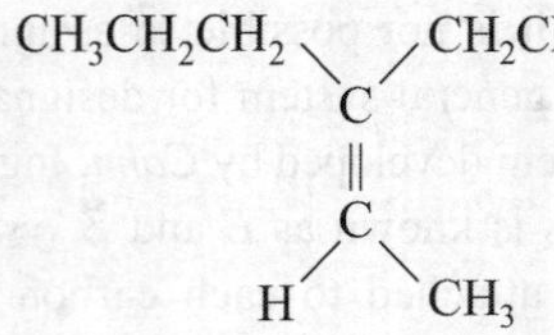

Z-3-Ethyl-2-hexene E-3-Ethyl-2hexene

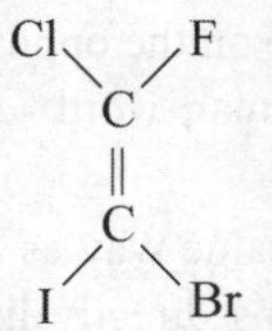

and

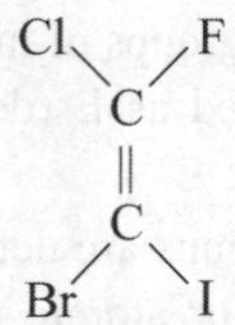

Z-1-Bromo-2-chloro 2-fluoro-1-iodoethene *E*-1-Bromo-2-chloro 2-fluoro-1-iodoethene

***Example 1.* Assign E and Z configurations to the following compounds.**

(*i*)

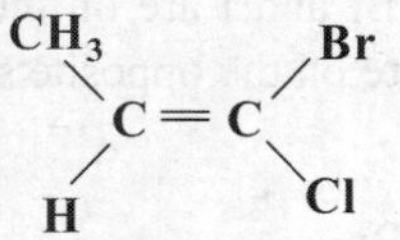

(*ii*)

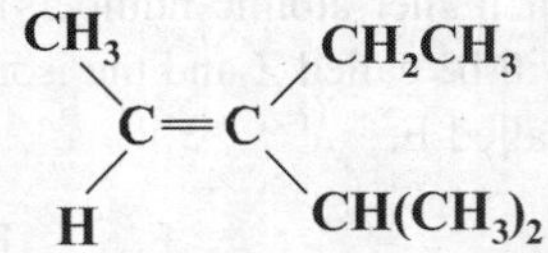

(*iii*)

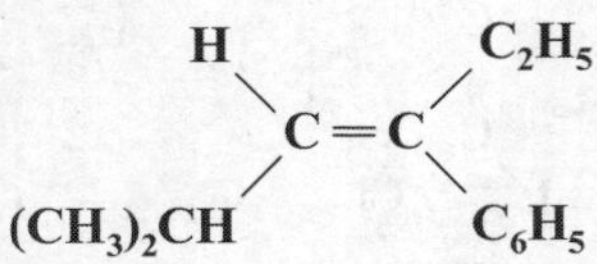

(*iv*)

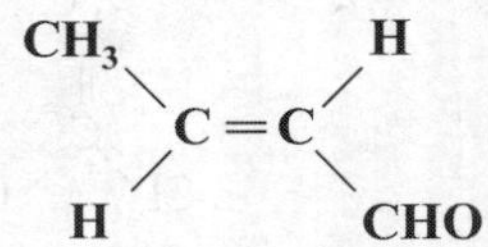

(*v*)

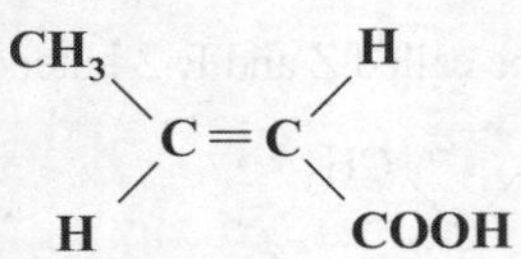

(*vi*)

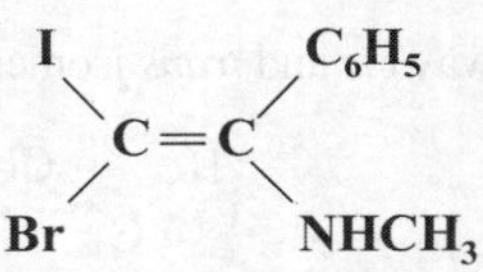

Solution. (*i*)

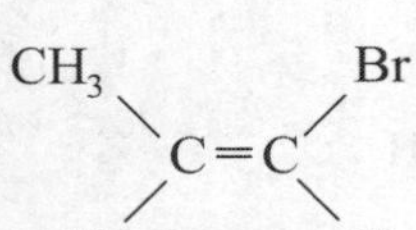

Out of — CH_3 and — H, — CH_3, has higher priority, out of — Br and — Cl, — Br has higher priority. Groups of higher priority lie on the same side. Therefore the configuration is Z.

(*ii*)

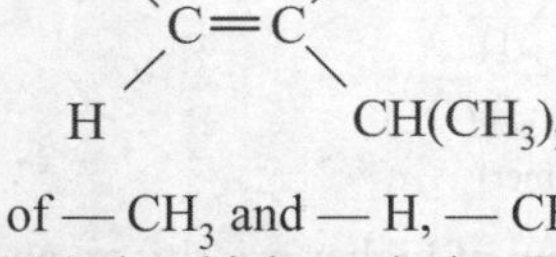

Out of — CH_3 and — H, — CH_3 has higher priority. Out of two groups on the second carbon, — $CH(CH_3)_2$ has higher priority. The groups of higher priority lie on opposite sides of the double bond. Therefore the configuration is E.

Following the same analogy, the students can see the configurations of compounds (*iii*), (*iv*), (*v*) and (*vi*) are Z, E, E and E respectively.

Example 2. **Assign E and Z configurations to the following :**

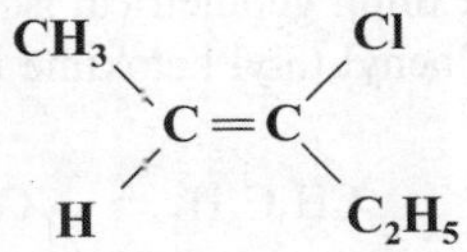

I

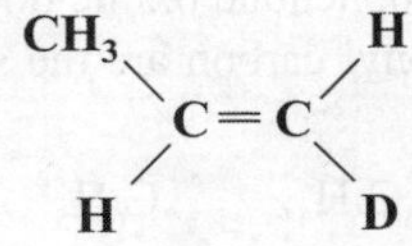

II

Solution. In compound I, the groups of higher priority on the two carbons linking double bond are CH_3— and Cl. Since these are on the same side of the double bond it has Z configuration.

In compound II, the groups of higher priority on the two carbons linking the double bond are CH_3— and D-(Deuterium) since these are on opposite sides of the double bond, it has E configuration.

Example 3. **Assign E and Z configuration to the following :**

(*i*) Maleic acid **(*ii*) Trans-But-2-enoic acid.**

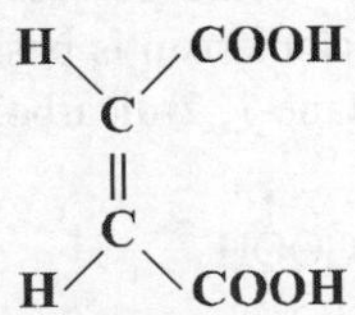

Maleic acid

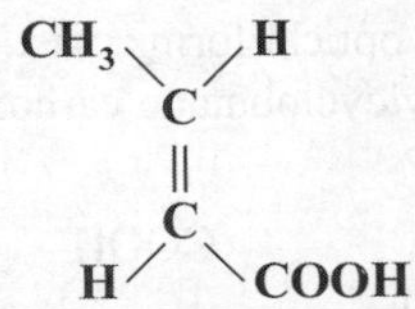

***Trans*-But-2-enoic acid**

Solution. In maleic acid, groups of higher priority on the two carbon atoms are — COOH and — COOH. These two groups are on the same side of the double bond. Hence configuration is Z.

In *trans*-but-2-enoic acid, the groups of higher priority on the two carbons are — CH_3 and — COOH, since these are on the opposite sides of the double bond, it has E configuration.

Geometrical isomerism in oximes

Oximes of aldehydes having the general formula R — CH = N — OH are capable of exhibiting geometrical isomerism as — H and — OH groups may be present on the same side or opposite sides of the double bond. The two stereoisomers thus obtained are named as *syn* (equivalent of *cis*) and *anti* (equivalent of *trans*). Two geometrical isomers of an aldoxime may be represented as :

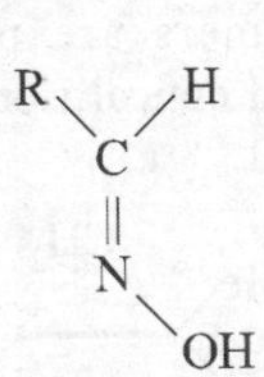

syn-aldoxime

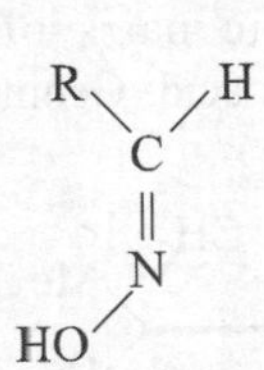

anti-aldoxime

Geometrical isomers of benzaldoxime have been actually isolated.

syn-banzaldoxime has m.p. 35°C while *anti*-benzaldoxime has m.p. 130°C. The two compounds may be shown as under :

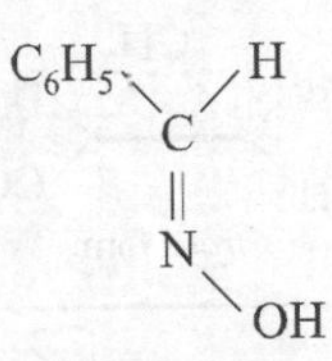

syn-benzaldoxime

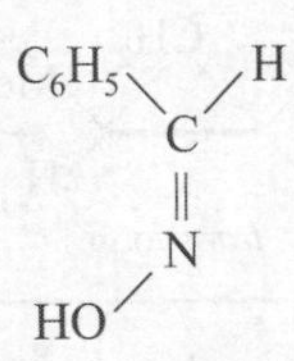

anti-benzaldoxime

Oximes of ketones (RR′C = NOH) can also show geometrical isomerism provide R and R′ are not the same. Thus benzophenone oxime does not exhibit geometrical isomerism because the two groups attached to carbonyl carbon are the same. Phenyl tolyl ketoxime is known to exist in two geometrical forms

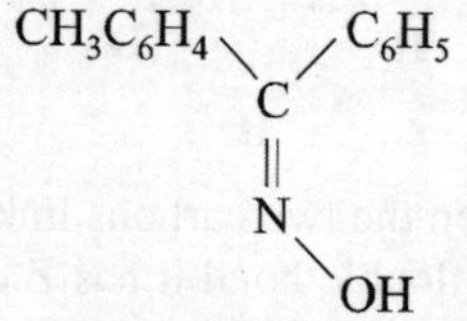

syn-phenyl tolyl oxime

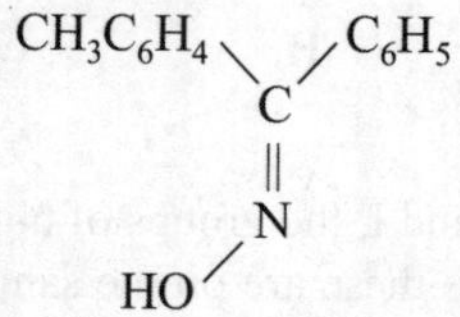

anti-phenyl tolyl oxime

Stereochemistry (geometrical and optical isomerism) of alicyclic compounds

The discussion can be made under two headings

Even-membered rings (containing 4, 6 carbon atoms)

If the substituents are across the ring from each other, the molecule possesses a plane of symmetry. Therefore, no optical forms exist. However geometrical isomerism is possible. *Cis* and *trans* isomers of 3-methylcyclobutane carboxylic acid and cyclobutane-1, 2-dicarboxylic acid are shown below

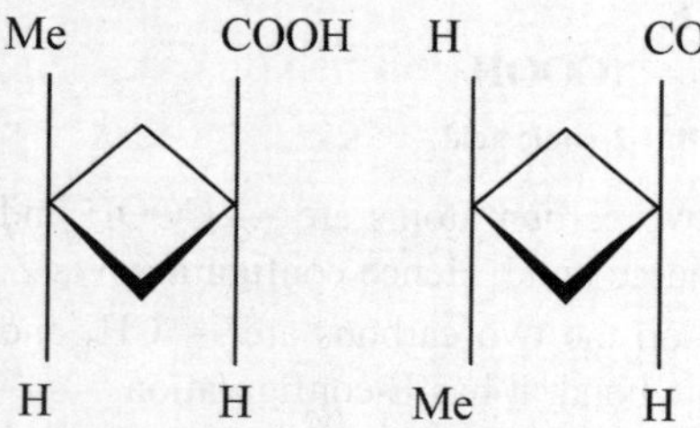

cis *trans*

3-methylcyclobutane carboxylic acids

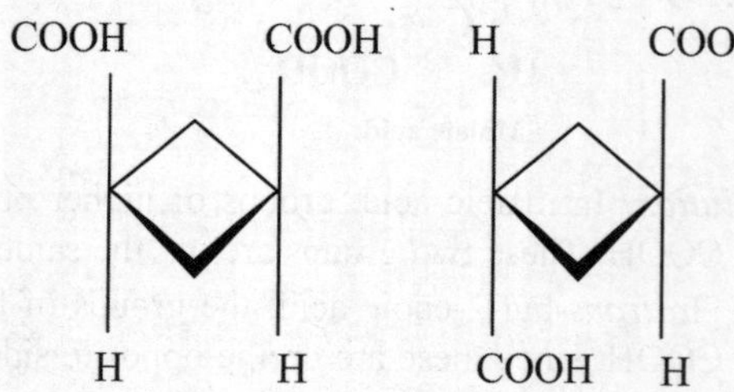

cis *trans*

Cyclobutane-1, 2-dicarboxylic acids

Odd membered rings (containing 3, 5 carbon atoms)

A disubstituted (on different carbons) odd membered ring has two asymmetric carbon atoms. Therefore two diastereomeric pairs of enantiomers are possible. Take the case of 2-methylcyclopropanecarboxylic acid. Geometrical and optical isomers have been represented as under

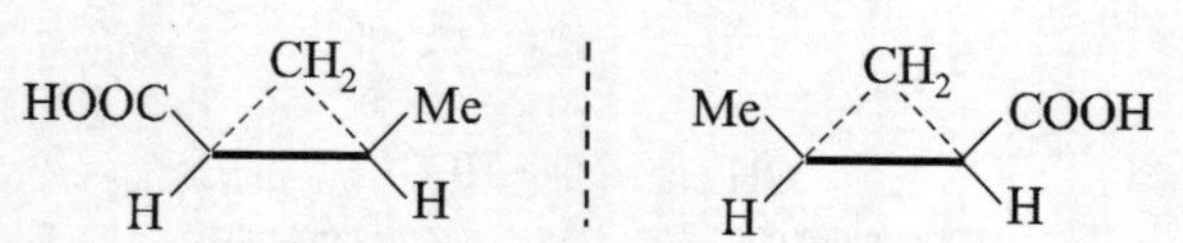

cis form Mirror *cis* form

One pair of enantiomers

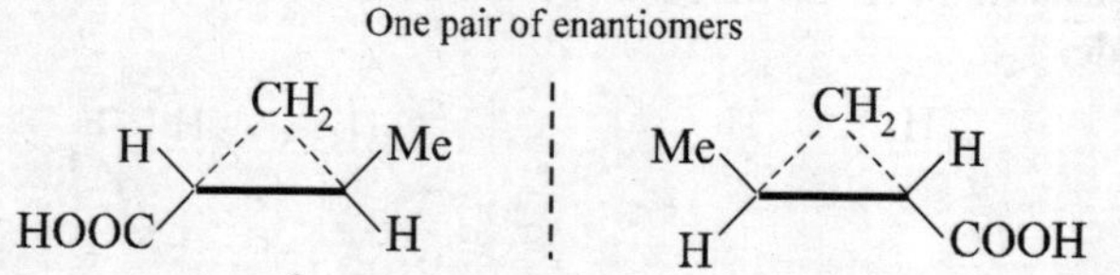

trans form Mirror *trans* form

Second pair of enantiomers

If however, the two substituents are the same, there will be one pair of enantiomers and a meso form. Take the case of cyclopropane-1, 2-dicarboxylic acid. *trans* form exists as a pair of enantiomers while the *cis* form as meso compound

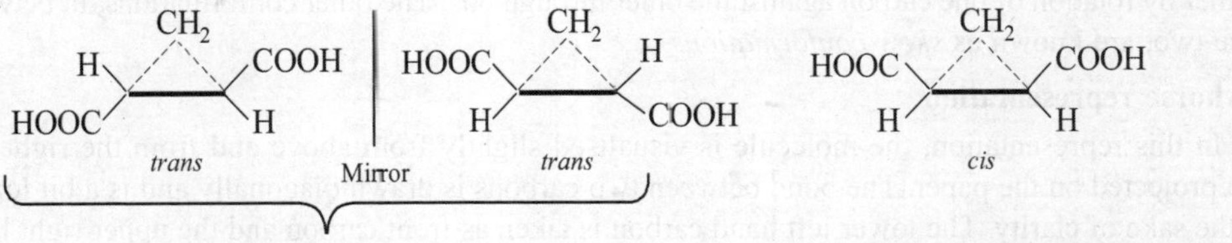

Pair of enantiomers   Meso cyclopropane-1, 2-dicarboxylic acid

3.5 CONFORMATIONAL ISOMERISM

Carbon-carbon single bond in alkanes is a sigma bond formed by the overlapping of sp^3 hybrid atomic orbitals along the inter-nuclear axis. The electron distribution in such a bond is symmetrical around internuclear axis; so that free rotation of one carbon against the other is possible without breaking sigma or single covalent bond. Consequently such compound can have different arrangements of atoms in space, which can be converted into one another simply by rotation around single bond, without breaking it. These different arrangements are known as *Conformational Isomers or Rotational Isomers or Conformers*. Since the potential energy barrier for their inter-coversion is very low, it is not possible to isolate them at room temperature. At least 60-85 kJ/mole must be the energy difference between two conformers to make them isolatable at room temperature.

Hence, conformations can be defined as different arrangements of the atoms which can be converted into one another by rotation around single bonds.

Newman and Sawhorse representations for the conformations of ethane

Alkanes can have an infinite number of conformations by rotation around carbon-carbon single bonds. In ethane two carbon atoms are linked by a single bond and each carbon atom is further linked with three hydrogen atoms. If one of the carbon atoms is allowed to rotate about carbon-carbon single bond keeping the other carbon stationary, an infinite number of arrangements of the hydrogens of one carbon, with respect to those of the other, are obtained. All these arrangements are called conformations (Bond angles and bond lengths remain the same).

Newman reppresentation

This can be easily understood with the help of Newman Projection formulae. The molecule is viewed from front to back in the direction of carbon-carbon single bond. The carbon nearer to the eye is represented by a point and three hydrogen attached to it are shown by three lines at an angle of 120° to one another. The carbon atom away from the eye is represented by a circle and three hydrogens attached to it are shown by shorter lines at an angle of 120° to each other.

Out of infinite number of conformations, Newman Projection formulae for two extreme cases are as shown below:

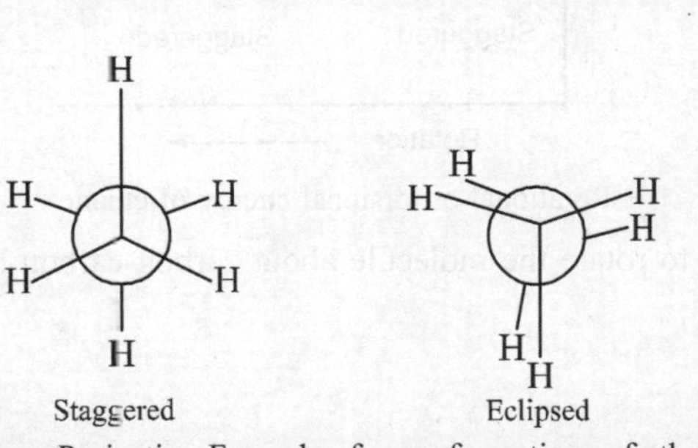

Newman Projection Formulae for conformations of ethane.

The conformation in which the H atoms of two carbons are as far apart as possible, is known as *Staggered conformation* and the conformation in which the H atoms of back carbon are just behind those of the front carbon is known a *Eclipsed conformation.* These are converted into one another by rotation of one carbon against the other through 60°. The other conformations, in between these two, are known as *skew conformations.*

Sawhorse representation

In this representation, the molecule is visualised slightly from above and from the right and then projected on the paper. The bond between two carbons is drawn diagonally and is a bit longer for the sake of clarity. The lower left hand carbon is taken as front carbon and the upper right hand carbon is taken as back carbon. The Sawhorse representation of staggered and eclipsed conformations of ethane are given below :

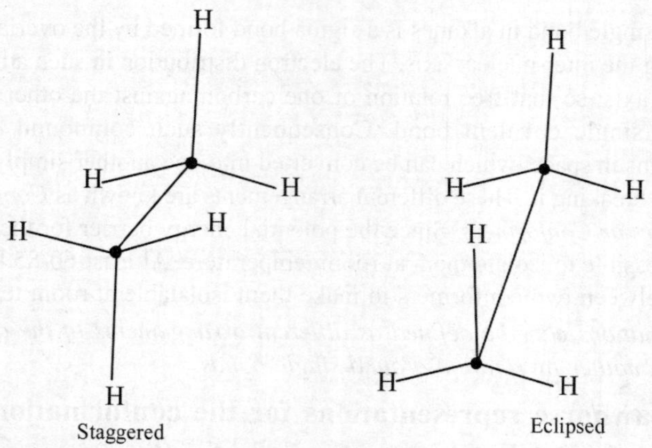

Sawhorse representation for conformations of ehtene

Pitzer in 1936 found that the rotation is not completely free. Rather there exists a potential energy barrier which restricts the free rotations. It means that the molecule spends most of its time in the most stable staggered conformation and it spends least time in the least stable eclipsed conformation; the energy difference being 12 kL/mole in the case of ethane.

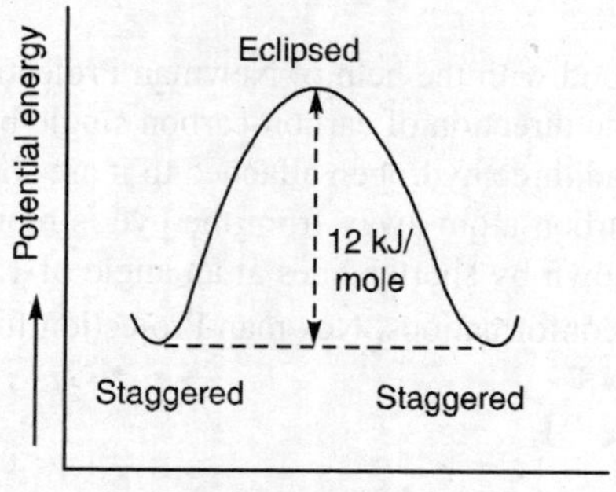

Rotational or torsional energy of ethane

The energy required to rotate the molecule about carbon-carbon bond is called rotational or torsional energy.

Difference between conformation and configuration

Conformation

Structures containing different arrangement of atoms of a molecule in space which can arise by rotation about a single bond are called *conformers*. For example ethane exists in different conformations called staggered, eclipsed and skew conformations. The energy difference between different conformers is rather small. This phenomenon is called conformation. The property of conformation is also exhibited by *n*-butane, cyclohexane, stilbene dichloride etc.

Configuration

Structures of a compound differencing in the arrangement of atoms or groups around a particular atom in space are called *configurations*. Enantiomers, distereomers and geometrical isomers come under this category. For example *d*-and *l*-lactic acids are configurations of lactic acid.

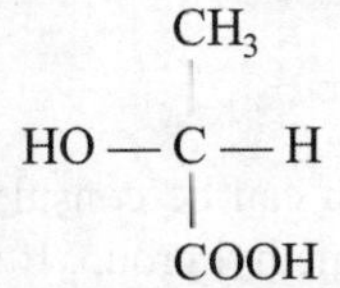

d-lactic acid

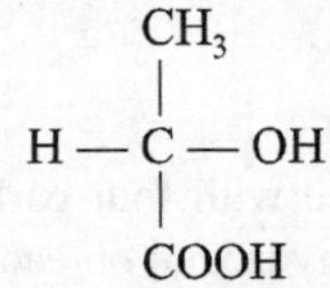

l-lactic acid

cis and *trans* butenes are configuration of butene

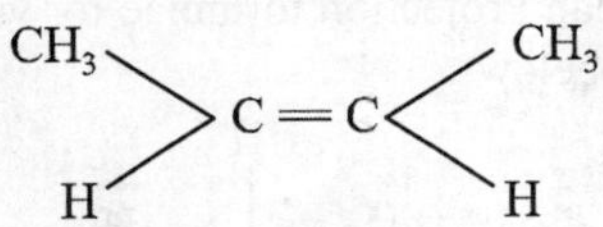

cis-2-butene

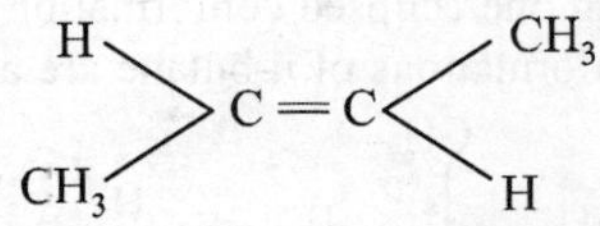

trans-2-butene

Factors affecting relative stability of conformations

The following factors play a vital role in the stability of conformations.

1. Angle strain. Every atom has the tendency to have the bond angles that match those of its bonding orbitals. If there is any deviation from this normal bond angle, the molecule suffers from *angle strain*. Conformations suffering from angle strain are found to be less stable.

2. Torsional strain. There is a tendency on the part of two carbons linked to each other to have their bonds staggered. That is why the staggered form of any molecule like ethane, *n*-butane is most stable. As the bonds of two connected carbons move towards eclipsed state, a *torsional strain* is set up in the molecule thus raising its energy. Thus the staggered conformations have the least and eclipsed the highest torsional strain. The energy required to rotate the molecule around the carbon-carbon bond is called *torsional energy.*

3. Steric strain (van der Waals' Strain). Groups attached to two linked carbons can interact in different ways depending upon their size and polarity. These interactions can be attractive or repulsive. If the distance between the groups or atoms is just equal to the sum of their van der Waals' radii, there will be attractive interactions between them. And if these atom or groups are brought closer than this distance, there will be repulsions leading to *van der Waals' strain or steric strain* in the molecule.

4. Dipole-dipole Interactions. Atoms or groups attached to bonded carbons orient or position themselves to have favourable dipole-dipole interactions. It will be their tendency to have maximum dipole-dipole attractions. Hydrogen bond is a particular case of powerful dipole-dipole attractions.

The stability of a conformer is determined by the net effect of all the above factors.

Conformations and change in dipole-moment of 1, 2-dibromoethane with temperature

The conformations of 1, 2 dibromoethane have been extensively studied by dipole-moment measurement. The conformations are depicted as under :

In the liquid state, the percentage of anti forms is 65 corresponding to conformational free energy of 3.5 kJ mol^{-1} in favour of *anti.*

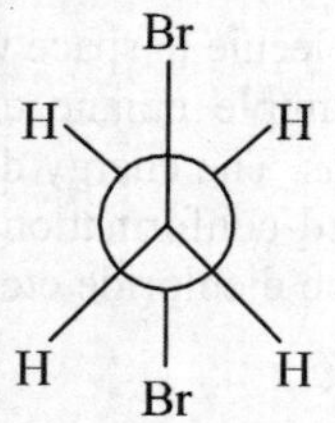

anti

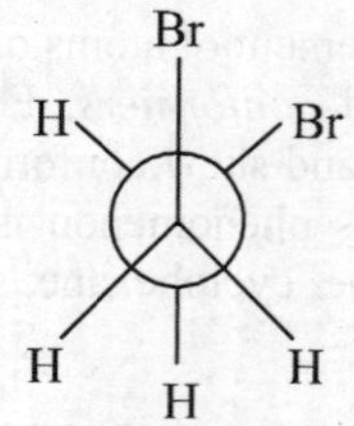

gauche

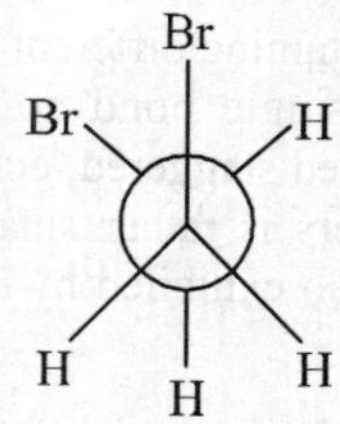

gauche

stability of *anti*-form is due to combined effects of steric factor and dipole-dipole interactions. It goes in favour of formation of *anti*- conformer. Dipole movement increases with increase of temperature.

Conformations of butane

n-butane is an alkane with four carbon atoms, which can be considered to be derived from ethane by replacing one hydrogen on each carbon with a methyl group. If we consider the rotation about the central carbon-carbon bond (C_2 — C_2), the situation is somewhat similar to ethane; but *n*-butane has more than one staggered and eclipsed conformations (unlike ethane which has only one staggered and one eclipsed conformation). Newman Projection formulae for various staggered and eclipsed conformations of *n*-butane are as given below :

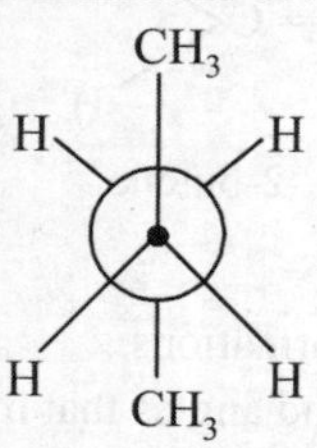

Completely staggered or anti form

I

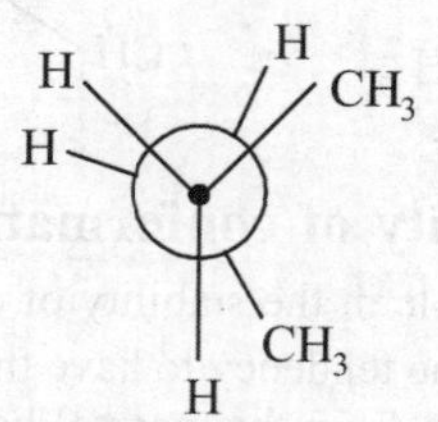

Partially eclipsed form

II

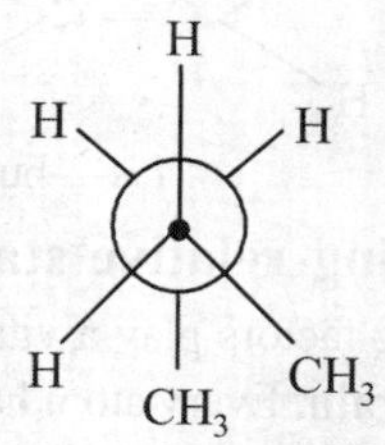

Partially staggered or gauche form or skew form

III

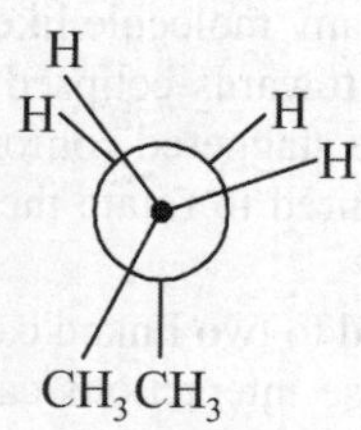

Completely staggered or antiform

IV

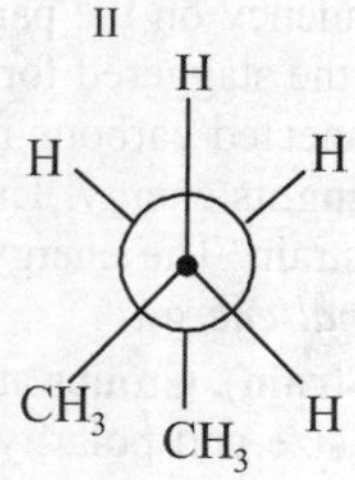

Partially staggered or gauche

V

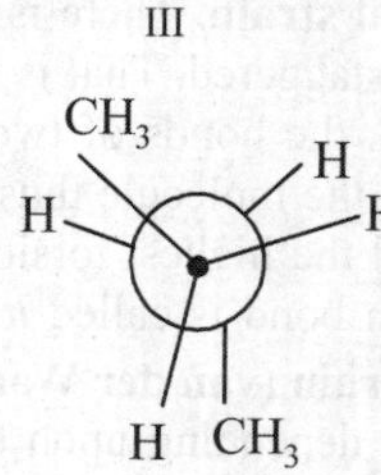

Partially eclipsed form

VI

The *completely staggered* conformation, (I) also known as *anti* form, is having the methyl goups as far apart at possible. Let us see how these forms have been obtained. Let us start from structure I. Holding the back carbon (represented by circle) fixed along with its groups, — H, — H and — CH_3, rotate the front carbon (shown by a dot) in clockwise direction by an angle of 60°. Groups attached to it will also move. *Partially eclipsed* form (II) is obtained. In this conformation (II), methyl group of one carbon is at the back of hydrogen of the other carbon. Further rotation of 60° leads of a *partially staggered* conformation (III), also known as *gauche* form, in which the two methyl groups are at an angle of 60°. Rotation by another 60° gives rise to a *fully eclipsed* form (IV) having two methyl groups at the back of each other. Further rotation of 60°, again leads to *partially staggered or gauche* form (V) having methyl groups at an angle of 60° (as in III). Still further rotation of 60° leads to *partially eclipsed* form (VI), having methyl group of one carbon at the back of hydrogen of the other (as in II). If a further rotation of 60° is operated (completing the rotation of 360°), again form I is obtained. Of course, there will be an infinite number of other conformations in between these six conformations (I to VI). (*Gauche* form is also known as skew form).

Out of these six conformations, the *completely staggered* or *anti conformation* (I) is most stable and *partially staggered* or *gauche* conformation (III or V) is slightly less stable: the energy difference being only 3.8 kJ/mole. On the other hand the *completely eclipsed* conformation (IV) is least stable and *partially eclipsed* conformation (II or VI) is slightly more stable, again the energy difference being 3.8 kJ/mole. (This is due to presence of steric strain between two methyl groups). The energy difference between most stable conformation (I) and least stable conformation (IV) is about 18.4 kJ/mole while that between I and II (or VI) is about 14.6 kJ/mole.

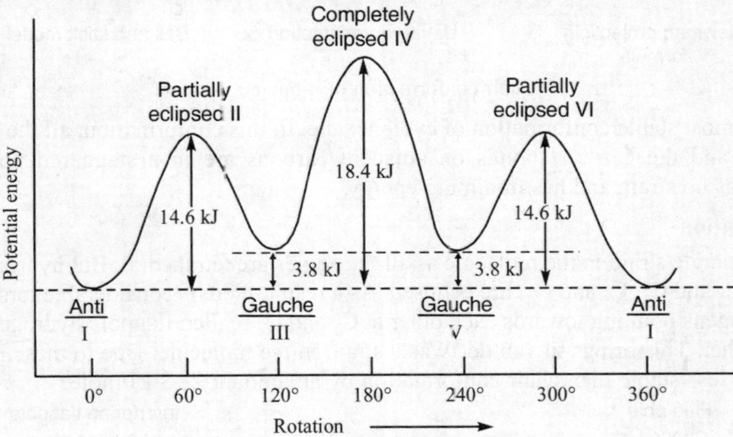

Rotational or torsional energy of *n*-butane.

Thus at ordinary temperature, *n*-butane molecule exists predominently in *anti* form with some *guache* forms.

Conformational enantiomers of 2, 3-Dimethylbutane and 2, 2, 3-Trimethylbutane

Conformational isomers of 2,3-Dimethylbutane

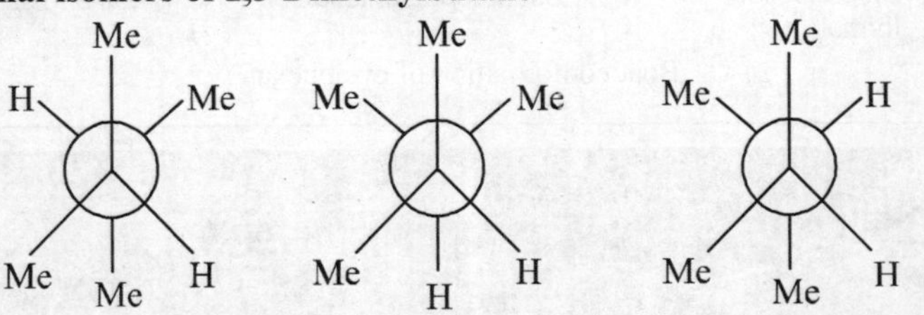

I II III

Structure I is supposed to have maximum stability out of these three isomers, II and III are conformational enantiomers (mirror-images) also. This can be visualized after rotation of III as a whole.

Conformational isomers of 2, 2, 3-Trimethylbutane

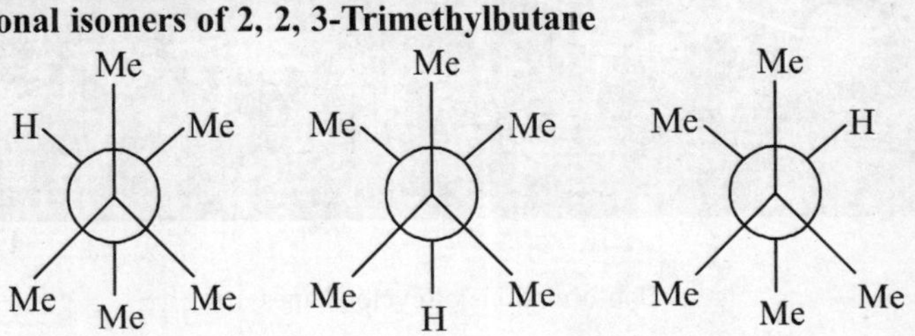

I II III

All these structures have the same stability and the compound may be assumed to occur in one conformation only.

Conformations of cyclohexane

Chair conformation

Chair conformation of cyclohexane is represented below :

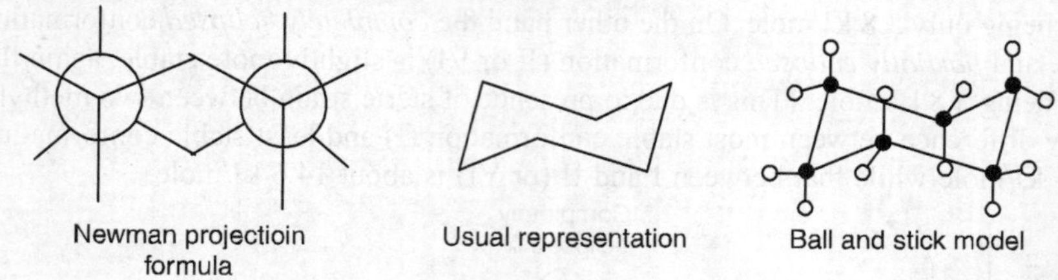

Chair conformation of *n*-butane.

This is the most stable conformation of cyclohexane. In this conformation, all the bond angles are tetrahedral and the C — H bonds on adjacent carbons are in a staggered position. This conformation has no strain and has minimum energy.

Boat Conformation

There is no angle strain in the molecule as all the angles are tetrahedral. But hydrogens on four carbon atoms (C_2 and C_3, C_5 and C_6) are eclipsed. As a result, there is considerable torsional strain. Also, two hydrogens pointing towards each other at C_1 and C_4 (called flagpole hydrogens) are very close to each other. This brings in van der Waals strain in the molecule. Due to these reasons boat conformation is less stable than chair conformation by an amount 28.8 kJ/mole.

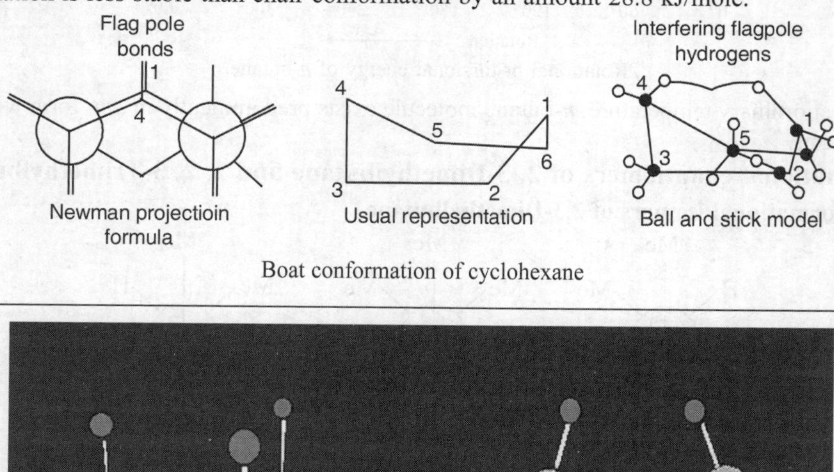

Boat conformation of cyclohexane

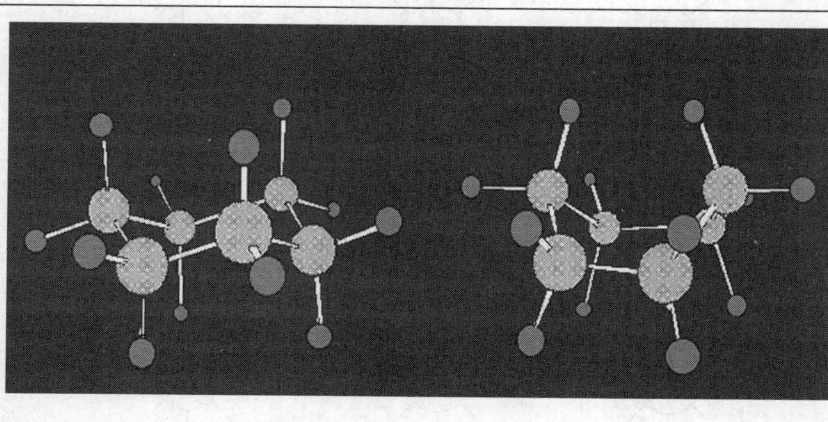

Conformations of cycloalkanes

Twist conformation of cyclohexane. Sequence of changes in going from chair form to boat form

Besides chair and boat conformation, cyclohexane can have several other possible conformations. Consider model of boat conformation of cyclohexane. Hold C_2 — C_3 bond in one hand and C_5 — C_6 in the other and twist the model so that C_2 and C_5 come upwards and C_3 and C_6 go downwards. We will get another conformation known as *twist form* or *skew boat* form.

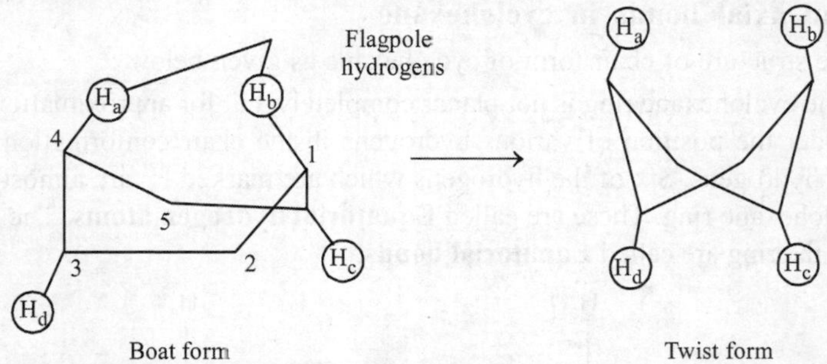

Boat form Twist form

During this twisting, the flagpole hydrogens (H_a and H_b) move apart while the hydrogens H_c and H_d move closer. If this motion is continued another boat form will be obtained in which H_c and H_d become the flagpole hydrogens.

In twist forms the distance between H_a and H_b is equal to that of H_c and H_d and the steric strain is minimum; also the torsional strain of C_2 — C_3 and C_2 — C_6 (due to their being eclipsed) is partly relieved. Thus the twist form is more stable than boat form by about 5.4 kJ/mole, but it is much less stable than chair form by 23.4 kJ/mole.

If we want to convert the chair form into boat form it will have to pass through a half chair form having considerable angle strain and torsional strain. The energy difference between chair form and half chair form being about 11 kcal/mole, half chair form is quite unstable.

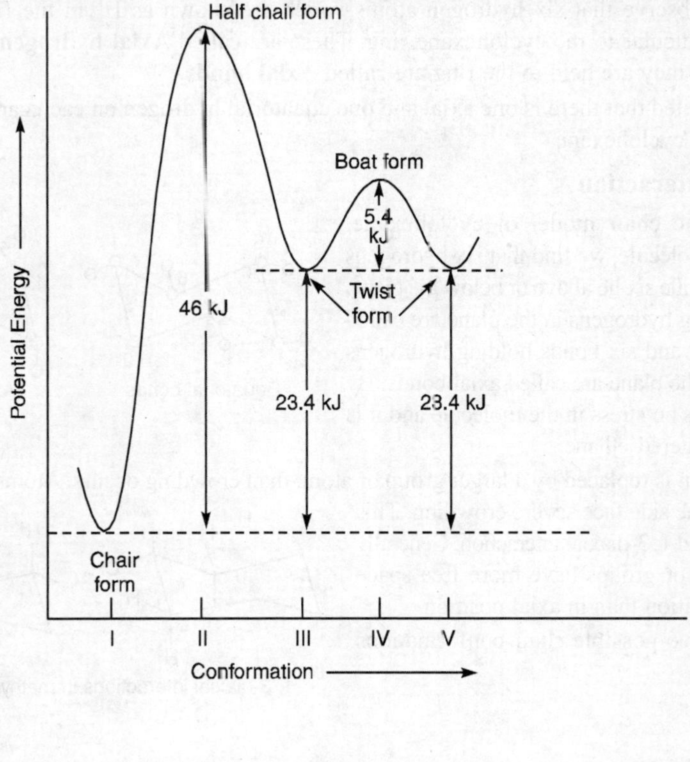

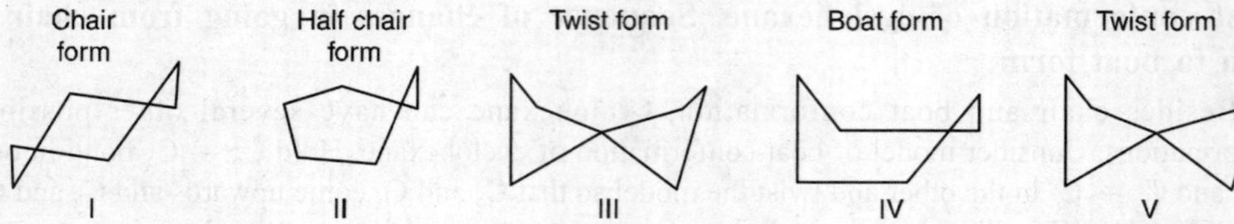

Equatorial and axial bonds in cyclohexane

Consider the structure of chair form of cyclohexane as given below :

Although, the cyclohexane ring is not planar completely, but for approximation, we can take it as planar. Consider the position of various hydrogens in the chair conformation. There are two distinct kinds of hydrogens, Six of the hydrogens which are marked H_e are almost oriented within the plane of cyclohexane ring. These are called **Equatorial hydrogen atoms.** The bonds by which these are held to the ring are called **Equatorial bonds.**

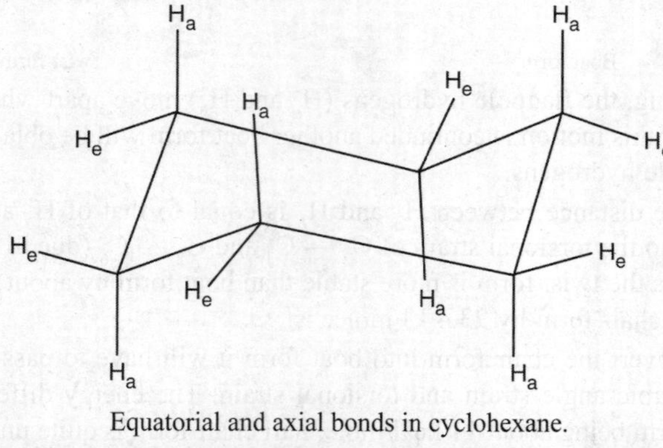

Equatorial and axial bonds in cyclohexane.

We again observe that six hydrogen atoms which are shown as H_a in the figure above are oriented perpendicular to the cyclohexane ring. These are called **Axial hydrogen atoms** and the bonds by which they are held to the ring are called **Axial bonds.**

It may be noted that there is one axial and one equatorial hydrogen on each carbon in the chair conformation of cyclohexane.

1, 3-diaxial interaction

Consider the chair model of cyclohexane. Looking at the molecule, we find that six hydrogens lie in the plane while six lie above or below the plane. Six bonds holding hydrogens in the plane are called *equatorial bonds* and six bonds holding hydrogens above or below the plane are called axial bonds. By and large, there is no stress in the molecule and it is as stable as staggered ethane.

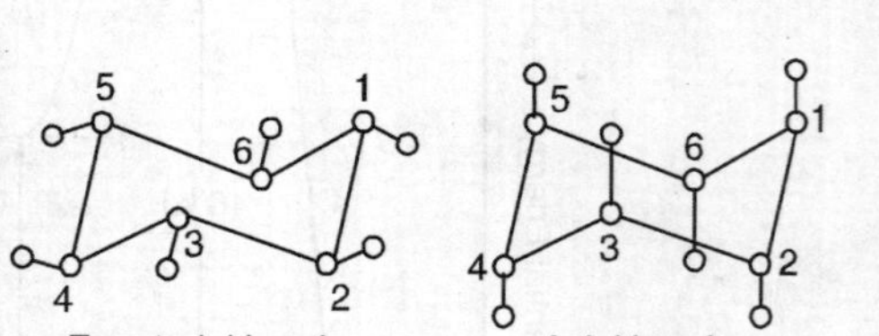

If a hydrogen is replaced by a larger group or atom, than crowding occurs. Atoms linked by axial bonds on the same side face severe crowding. This interaction is called 1, 3-diaxial interaction. Generally speaking, atoms or groups have more free space in equatorial position than in axial position.

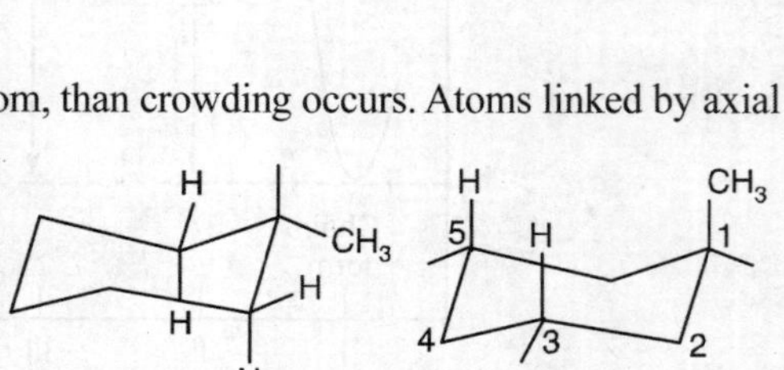

1, 3-diaxial interactions in methyl cyclohexane

There are two possible chair conformations

of methyl cyclohexane one with – CH_3 in equatorial position and the other with – CH_3 in axial position.

It is observed that $-CH_3$ in equatorial position faces less crowding by hydrogens compared to $-CH_3$ in the axial position. Methyl group in the axial position is approached more closely by axial hydrogens on C – 3 and C – 5. This is called *1, 3-diaxial interaction.*

Conformations of substituted cyclohexanes

A methyl group is bulkier than a hydrogen atom. When the methyl group in methylcyclohexane is in the axial position, the methyl group and the axial hydrogens of the ring repel each other. These interactions are called **Axial-Axial Interactions.** When the methyl group is in the equatorial position, the repulsions are minimum. Thus, the energy of the conformation with equatorial methyl group is lower. At room temperature, about 95% methylcyclohexane molecules are in the conformation in which the methyl group is equatorial.

The bulkier the group, the greater is the energy difference between equatorial and axial conformations. In other words, a cyclohexane ring with a bulky substituent (e.g., *t*-Butyl group) is more likely to have that group in the equatorial position.

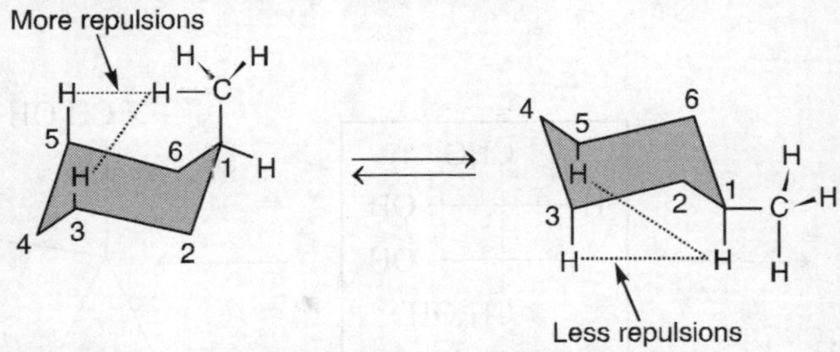

Conversion of Axial Bonds to Equatorial Bonds And Vice Versa

Each of the six carbon atoms of cyclohexane has one equatorial one axial hydrogen atom. Thus, there are six equatorial hydrogens and six axial hydrogens. In the flipping and reflipping between conformations, axial hydrogens become equatorial hydrogen while equatorial hydrogens become axial. This is represented in Fig. below :—

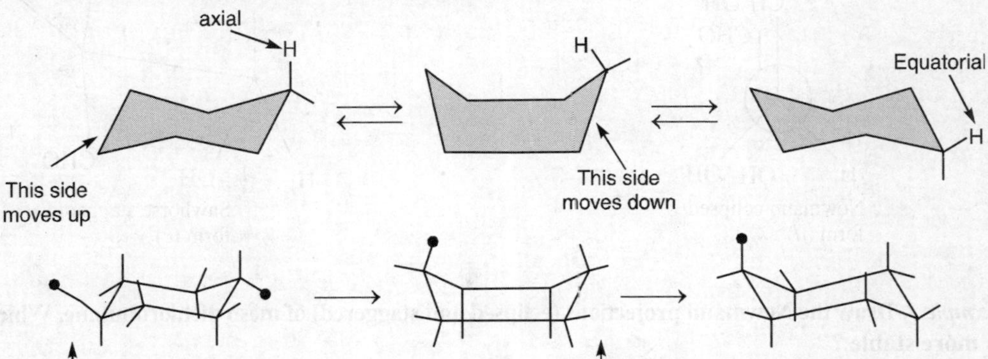

***Example 1.* Out of the following two conformational forms of 1, 2-dibromo ethane which will have higher dipole moment ?**

(*a*)

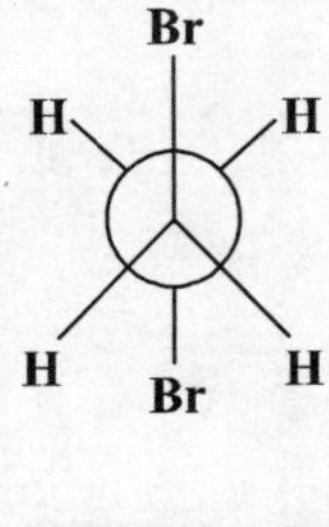

(*b*)

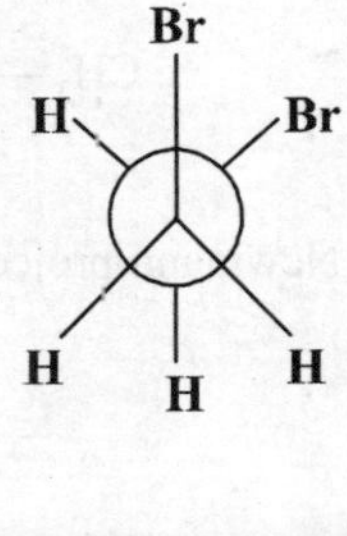

Solution. In structure (*b*) two Br groups are close to each other. This will result in repulsion between them leading to separation of charge and thus creation of higher dipole moment.

Example 2. **Convert the Fischer's projection of D (–) erythrose sugar (*a*) to the following :**

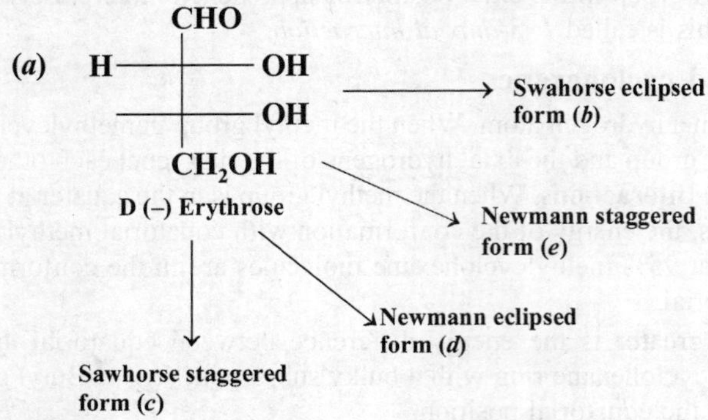

Solution.

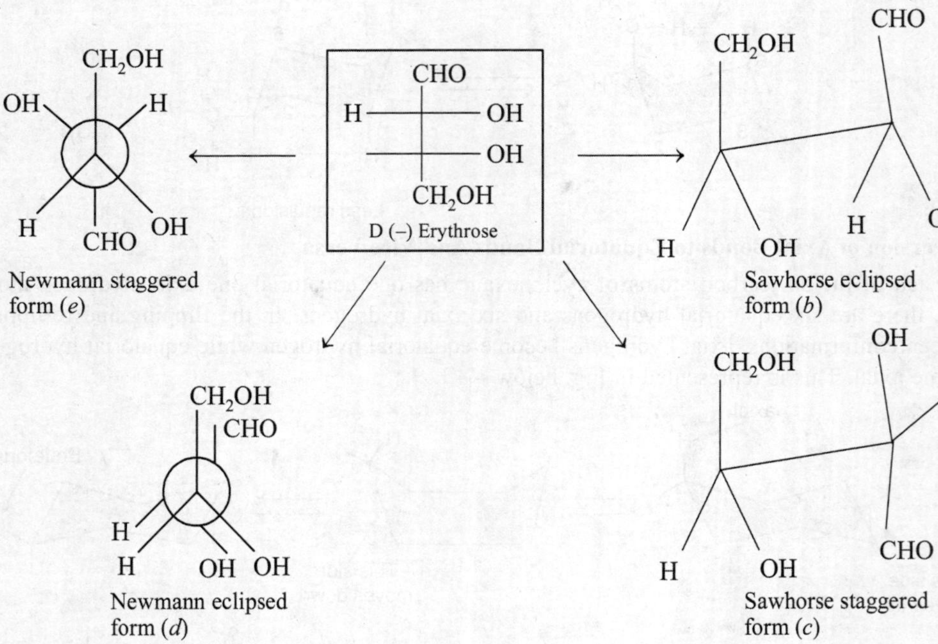

Exmple 3. **Draw the Newmann projections (eclipsed and staggered) of meso dichlorobutane. Which is more stable ?**

Solution. Meso dichlorobutane may be written as :

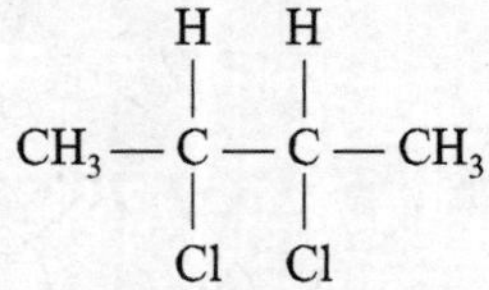

Eclipsed and staggered Newmann projections may be written as :

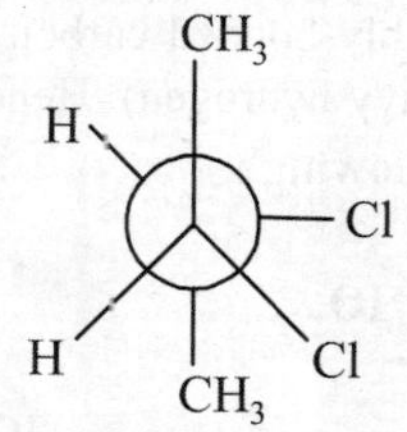

(Staggered)

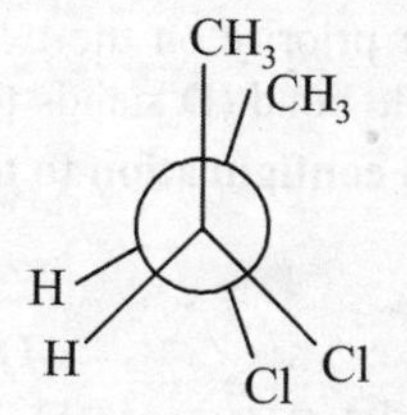

(Eclipsed)

MISCELLANEOUS EXAMPLES ON ISOMERISM

Example 1. **Assign E or Z configuration to each of the following compounds :**

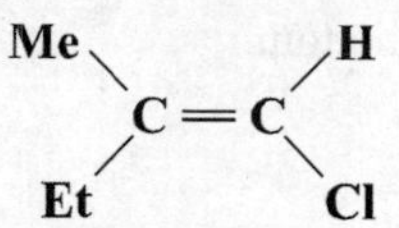

(A)

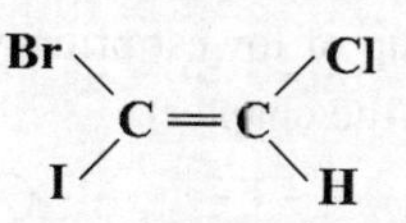

(B)

Solution. In compound (A), groups of higher priority on the doubly bonded carbon atoms are Et and Cl, which lie on the same side of the double bond. Hence, it is Z isomer

In compound (B), groups of higher priority on the doubly bonded carbon atoms are I and Cl, which lie on the opposite sides of the double bond. Hence, it is E isomer.

Example 2. **Indicate which of the following compounds has E or Z designation**

(*i*) 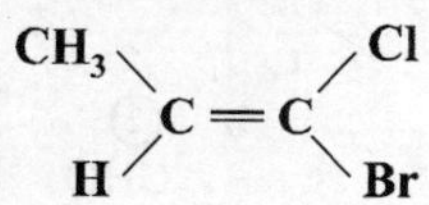

(*ii*) 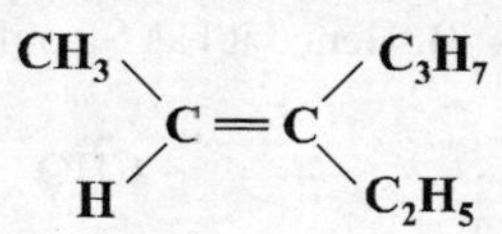

(*iii*) 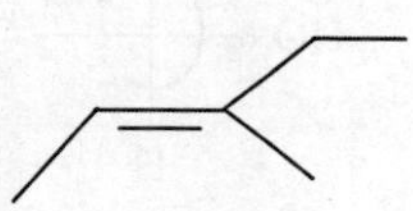

(*iv*) 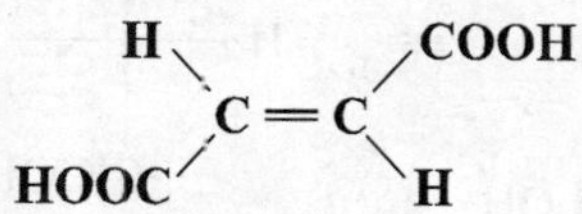

Solution.

(*i*) 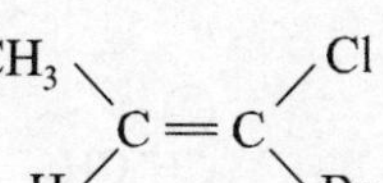

E-isomer

(*ii*) 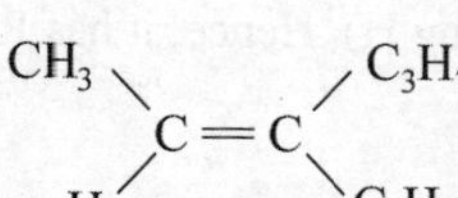

Z-isomer

(*iii*) 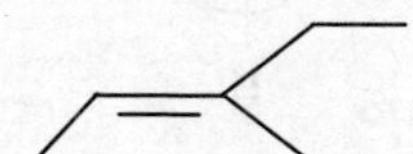

E-isomer

(*iv*) 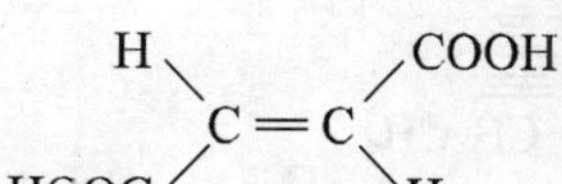

E-isomer

Example 3. **Assign E or Z configuration to the following :**

(*a*) 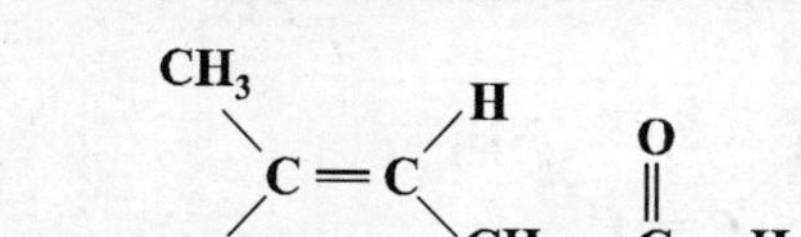

(*b*) 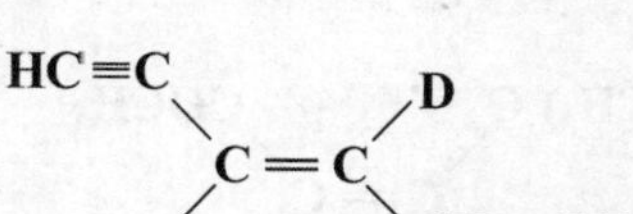

Solution.

(*a*) Groups of higher priority on the two doubly bonded carbon atoms are on the same side of the double bond. Hence, its configuration is Z.

(*b*) Groups of higher priority on the two doubly bonded carbon atoms are on the opposite sides of the double bond (D stands for heavy hydrogen). Hence, its configuration is E.

***Example 4.* Assign R or S configuration to the following :**

(*a*) 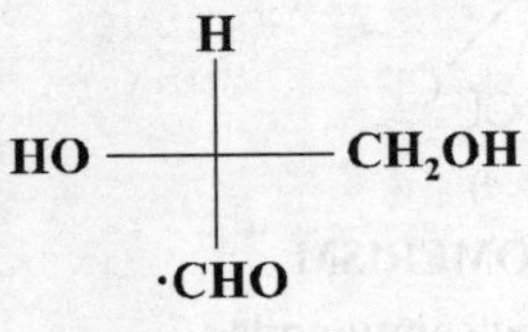 **(*b*)** 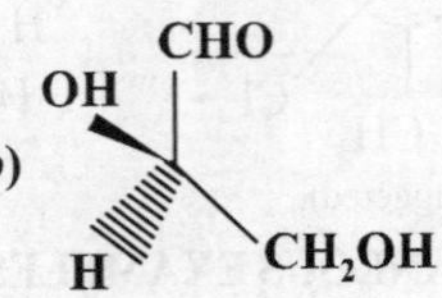 **(*c*)** 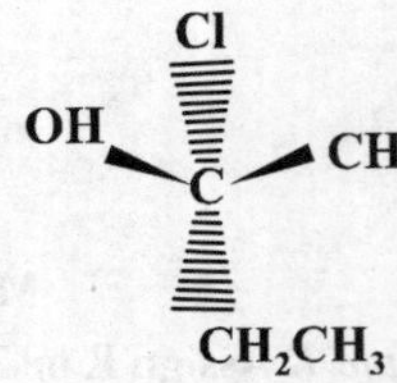

Solution. We need to bring the group of lowest priority *i.e.* H at the bottom. (Follow the rules given in the chapter)

(*a*) 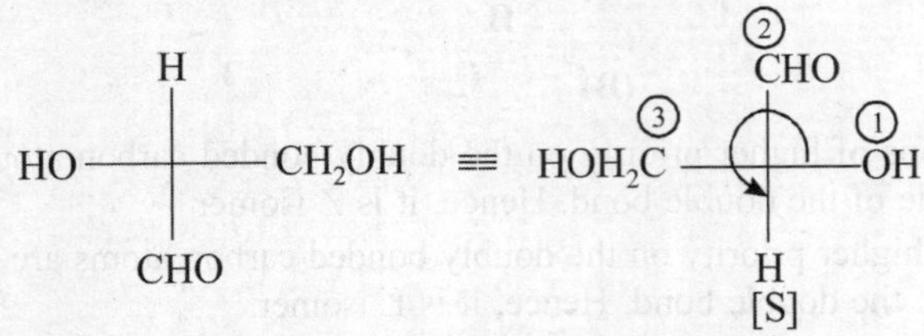

We move in the anticlockwise direction in moving from group of highest priority to the group of lowest priority ($1 \rightarrow 2 \rightarrow 3$). Hence it has S configuration.

(*b*) 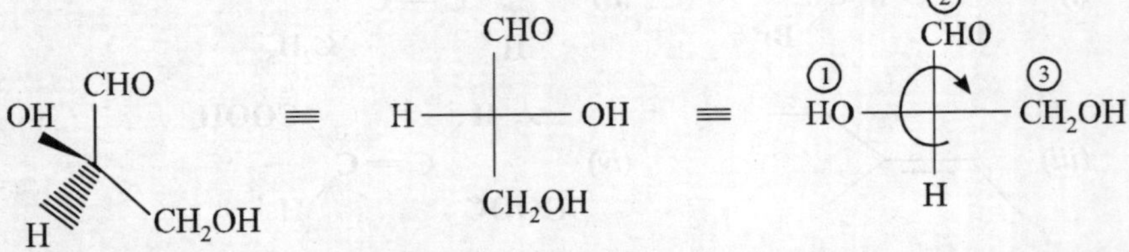

We move in the clockwise direction in moving from group of highest priority to the group of lowest priority, (leaving H). Hence, it has R-configuration.

(*c*) 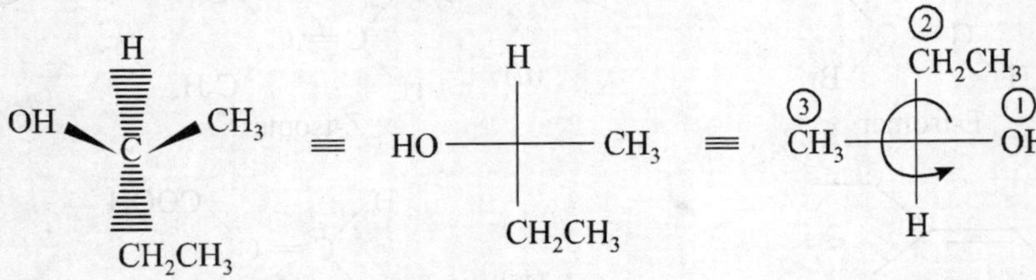

We follow anticlockwise direction in moving from group of highest priority to the group of lowest priority. Hence, it has S configuration.

***Example 5.* Assign A or Z rotations to the following compounds showing priorities for various groups :**

(*i*) 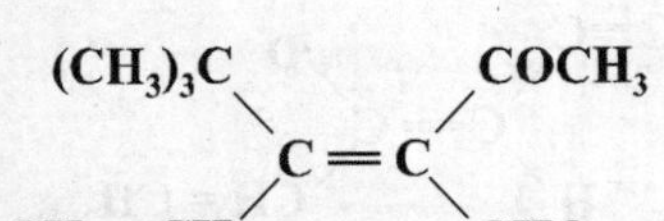 **(*ii*)** 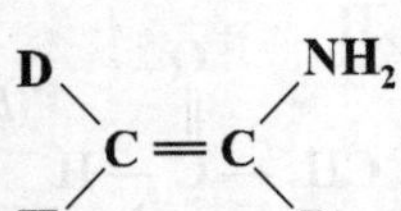

Solution. (*i*)

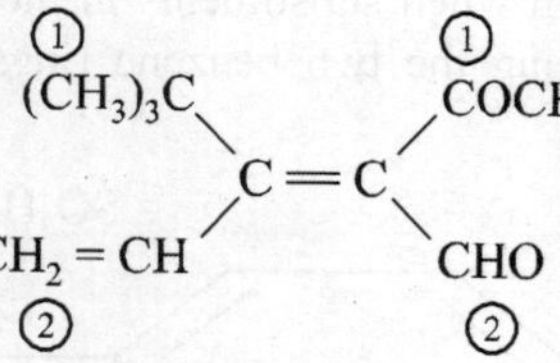

Groups of higher priority on the two doubly bonded carbon atoms are on the same side of the double bond. Hence it has Z configuration.

(*ii*)

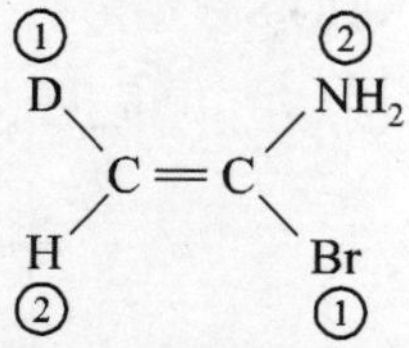

Here, D stands for deuterium *i.e.* heavy hydrogen. Groups of higher priorities on the two doubly bonded carbon atoms are on the opposite sides of the double bond. Hence, it is E configuration.

***Example 6.* Assign R or S configuration to the following compounds showing priorities for various groups.**

(*i*)

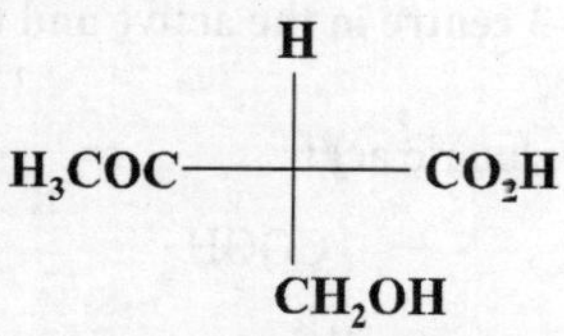

(*ii*)

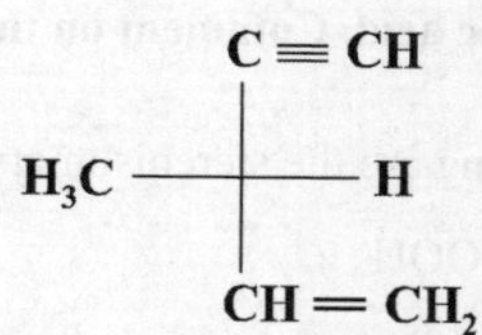

Solution. To assign the configuration to the compounds, we need to bring the group of lowest priority *i.e.* H at the bottom as per the rules explained in the chapter

(*i*)

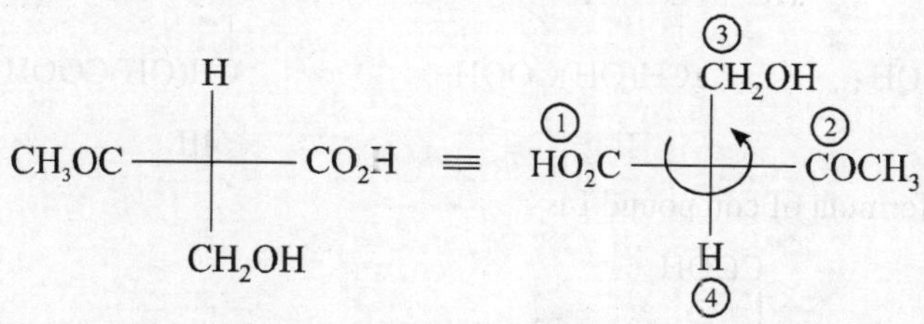

We have to trace anti-chlockwise direction in moving from the group of lowest priority to the group of highest priority. Hence, it has S configuration.

(*ii*)

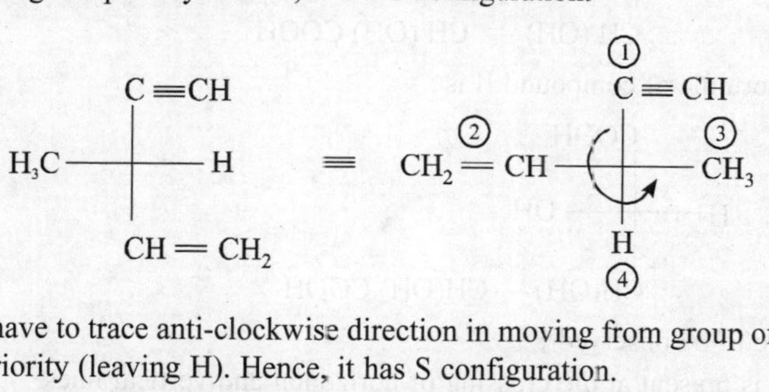

We have to trace anti-clockwise direction in moving from group of highest priority to group of lowest priority (leaving H). Hence, it has S configuration.

***Example 7.* Discuss the optical isomerism exhibited by diphenyl compounds.**

Solution. Substituted biphenyls show optical isomerism when substituents in the 2-positions are large enough to prevent rotation about the bond joining the two benzene rings. For example, biphenyl-2, 2′-disulphonic acid exists in two forms.

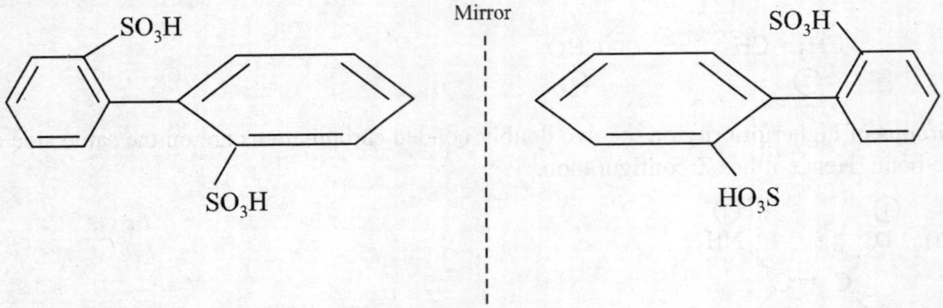

Biphenyl-2, 2′-disulphonic acid
(*a biphenyl*)

These two forms are non-superimposable mirror images. They do not interconvert at room temperature because the energy required to twist one ring through 180° relative to the other is too high. This is because, during the twisting process, the two-SO_3H group must come into very close proximity when the two benzene rings become coplaner and strong repulsive forces are introduced.

***Example 8.* Draw the Fischer projection formulas for all possible stereoisomers of 2, 3, 4-trihydroxyglutaric acid. Comment on the stereogenicity of C-3 centre in the active and the meso isomers.**

Solution. Following are the stereoisomers of 2, 3, 4-trihydroxyglutaric acid :

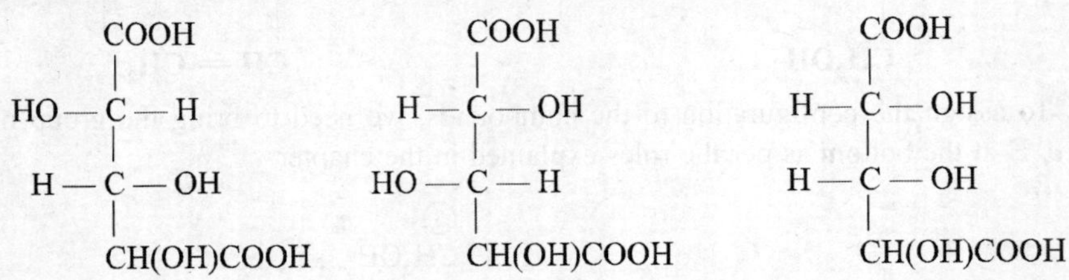

I II III

The Fischer projection formula of compound I is :

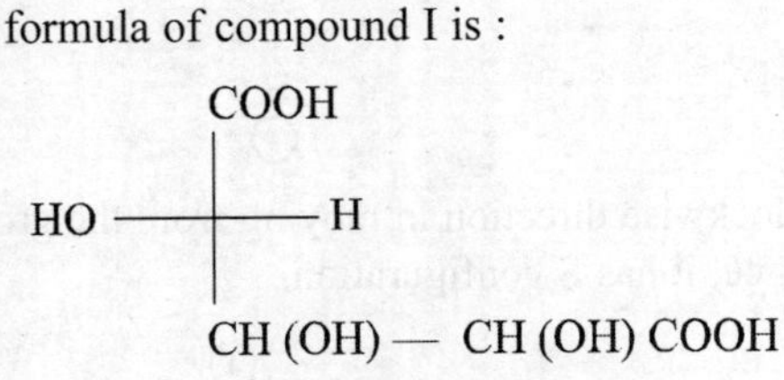

The Fischer projection formula of compound II is :

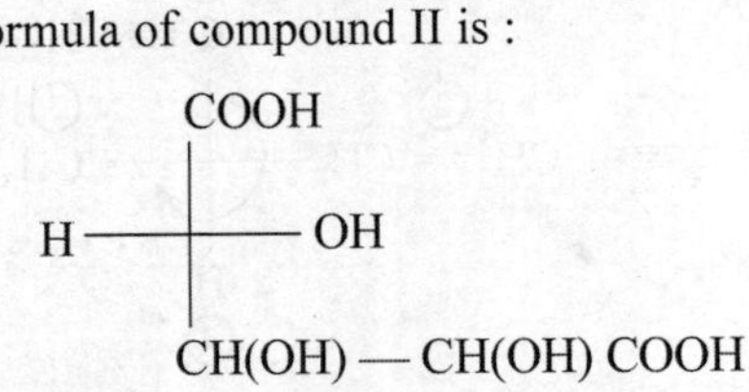

The following points are to be observed :

(*i*) The central carbon is present at the crossing of horizontal and vertical lines.

(*ii*) The longest chain of carbon atoms is present along the vertical line.

EXERCISES
(INCLUDING QUESTIONS FROM DIFFERENT UNIVERSITY PAPERS)

Objective Questions

Fill in the blanks

1. R and S-2 - butanols can be distinguished by the .... of polarized light
2. Of the compounds $X_2C = CY_2$, $X_2C = CXY$, $XYC = CXY$ and $Y_2C = CXY$, geometrical isomerism is shown by ....
3. The configuration of the compound 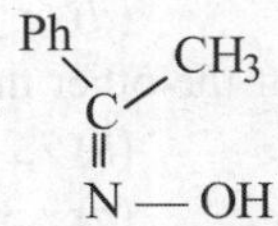 is ....
4. The following optically active molecules are called .... of each other

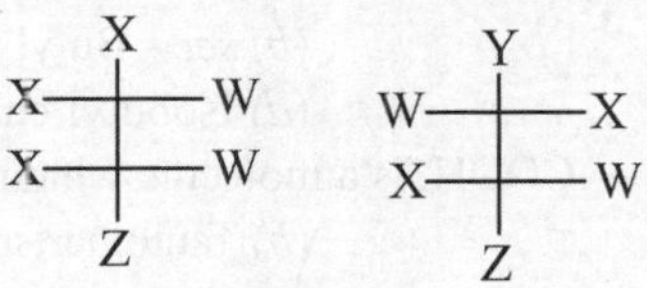

5. Lactic acid exhibits .... isomerism
6. Molecules containing *n* different asymmetric carbon atoms will have .... optically different forms
7. Out of butene –1 and butene – 2, ... shows geometrical isomerism
8. Mesotartaric acid has a .... of symmetry
9. Separation of racemic modification into its enantiomorphs is called ....
10. A molecule that is not superimposable on its mirror image is ....

Multiple Choice Questions

Choose the correct option

1. The isomers of a substance must have
   (*a*) same chemical properties (*b*) same molecular weight
   (*c*) same structural formula (*d*) same functional groups
2. Ethanol (CH_3CH_2OH) and dimethyl ether (CH_3OCH_3) are best considered as
   (*a*) structural isomers (*b*) stereoisomers
   (*c*) enantiomers (*d*) diastereomers
3. Which of the following statements is false about tautomers ?
   (*a*) Tautomers are structural isomers
   (*b*) Tautomers are structural isomers which exist in dynamic equilibrium
   (*c*) Tautomerism involves movement of atoms
   (*d*) Tautomers have independent existence
4. How many structural isomers are possible for C_4H_9Br ?
   (*a*) 2 (*b*) 3 (*c*) 4 (*d*) 5
5. Alkenes show geometrical isomerism due to :
   (*a*) asymmetry (*b*) rotation around a single bond
   (*c*) resonance (*d*) restricted rotation around a double bond
6. Two compounds have the same composition and also have the same atoms attached to the same atoms, although with different orientations in space. These compounds are
   (*a*) identical (*b*) position isomers (*c*) structural isomers (*d*) stereoisomers

7. Compounds with the same molecular formula but different structural formulas are called
(*a*) alkoxides (*b*) iso compounds (*c*) isomers (*d*) ortho compounds

8. How many isomeric aromatic hydrocarbons are possible for C_8H_{10} ?
(*a*) 3 (*b*) 4 (*c*) 5 (*d*) 6

9. How many isomers are possible for the compound with molecular formula C_4H_8 ?
(*a*) 2 (*b*) 4 (*c*) 6 (*d*) 8

10. The compounds $CH_3CH_2OCH_2CH_3$ and $CH_3OCH_2CH_2CH_3$ are
(*a*) enantiomers (*b*) conformational isomers
(*c*) metamers (*d*) optical isomers

11. Which compound is not an isomer of the other three ?
(*a*) *n* - Pentane (*b*) 2,2 - Dimethylpropane
(*c*) 2 - Methylbutane (*d*) 2,3 - Dimethylbutane

12. Optical isomerism is shown by
(*a*) *n* - Butyl chloride (*b*) *sec* - Butyl chloride
(*c*) *tert* - Butyl chloride (*d*) Isobutyl chloride

13. Lactic acid, CH_3 — CHOH — COOH, is a molecule which shows
(*a*) Geometrical isomerism (*b*) Tautomerism
(*c*) Optical isomerism (*d*) Metamerism

14. Consider (R)–and (S)–2–butanol. Which physical property distinguishes the two compounder ?
(*a*) Melting point
(*b*) Solubility in common solvents
(*c*) Rotation of plane-polarized light
(*d*) Infrared spectrum

15. How many optical isomers are possible for lactic acid ?
(*a*) 2 (*b*) 4 (*c*) 6 (*d*) 8

16. Which of the following compounds is an optically active compound ?
(*a*) $CH_3-CH(OH)-COOH$ (*b*) $CHCl_3$
(*c*) CH_3CH_2COOH (*d*) CH_3CH_2OH

17. Which of the following represents a racemic mixture ?
(*a*) 75% (R)–2–butanol, 25% (S)–2–butanol
(*b*) 25% (R)–2–butanol, 75% (S)–2–butanol
(*c*) 50% (R)–2–butanol, 50% (S)–2–butanol
(*d*) none of the above

18. Geometrical isomerism is shown by
(*a*) lactic acid (*b*) maleic acid
(*c*) 1 – butene (*d*) 1, 1– dichloroethylene

19. It is possible to distinguish between optical isomers by
(*a*) using chemical tests (*b*) mass spectrometry
(*c*) IR spectroscopy (*d*) polarimetry

20. Which of the following compounds will not show geometrical isomerism ?
(*a*) BrCH = CHBr (*b*) BrCH = CHCl
(*c*) $CH_3-C(Br)=CHBr$ (*d*) $CH_3-C(CH_3)=CHCH_3$

ANSWERS

Fill in the blanks

1. rotation	**2.** XYC = CXY	**3.** Z	**4.** diastereomers
5. optical	**6.** 2^n	**7.** butene – 2	**8.** plane
9. resolution	**10.** chiral		

Multiple choice questions

1. (*b*)	**2.** (*a*)	**3.** (*d*)	**4.** (*c*)	**5.** (*d*)
6. (*d*)	**7.** (*c*)	**8.** (*b*)	**9.** (*c*)	**10.** (*c*)
11. (*d*)	**12.** (*b*)	**13.** (*c*)	**14.** (*c*)	**15.** (*a*)
16. (*a*)	**17.** (*c*)	**18.** (*b*)	**19.** (*d*)	**20.** (*d*)

General Questions

1. Discuss the optical isomerism in 3, 4-dichlorohexane.

2. What are enantiomers? Write their characteristics?

3. What is chiral carbon atom?

4. What is the most essential condition of compound to be optically active in nature?

5. Assign R and S configurations to the following Fischer projection formulae :

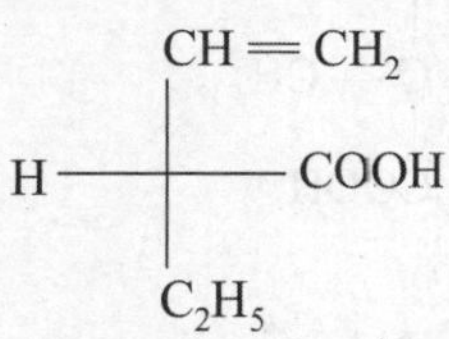

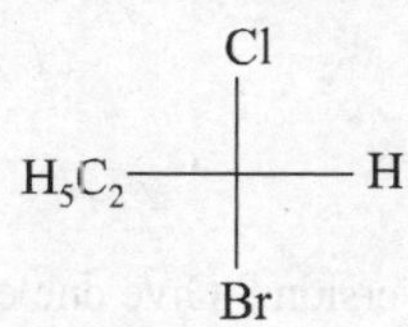

6. Assign R and S configurations to the following :

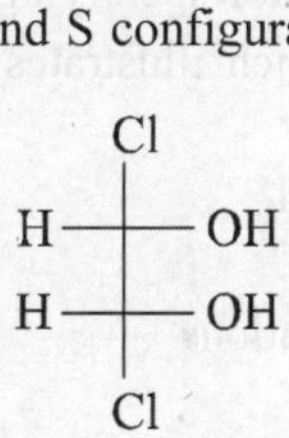

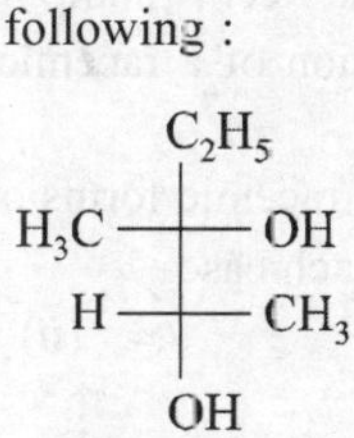

7. Which of the following can exhibit geometrical isomerism :

(*i*) 2-Hexene (*ii*) 2, 3-Dimethyl-2-butene

(*iii*) 4-Methyl-2-pentene (*iv*) 2, 3-Dimethyl-1-butene.

8. Draw the structures of each of the following compounds showing its stereochemistry :

(*i*) (2Z, 4Z)-2, 4-hexadiene (*ii*) (2R, 3S)-Tartaric acid.

9. Explain the racemisation of optical isomers with suitable examples.

10. Glucose contains four chiral carbon atoms in its molecule. What is the number of the optical isomers possible ?

11. What is resolution ? How will you resolve a racemic mixture by a chemical method?

12. Explain the difference between

(*i*) Diastereoisomers and (*ii*) Enantiomers.

13. What do you understand by the E and Z notations for representing geometrical isomers? Assign E and Z configurations to the following :

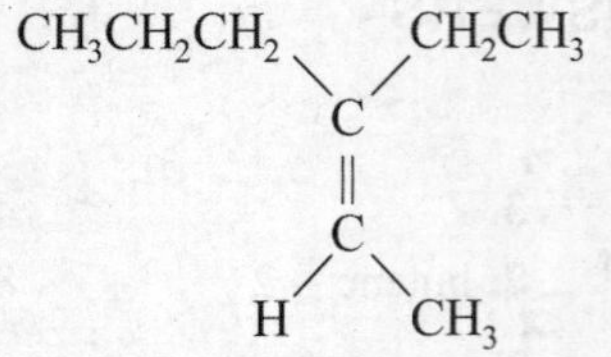
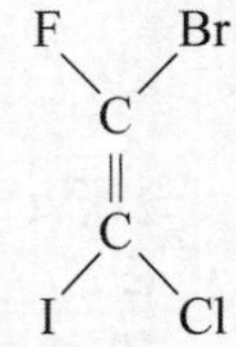

14. What are the essential differences between enantiomers and diastereoisomers?

15. Explain the term 'meso' form with suitable examples.

16. Assign R and S configurations to the following compounds :

(*i*) 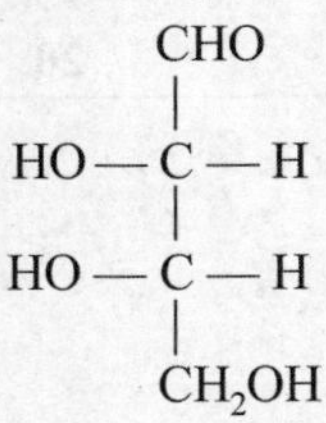
(*ii*) 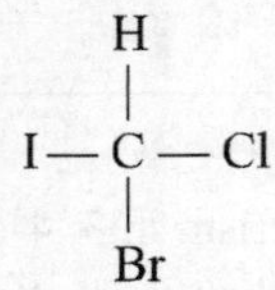

(*iii*) 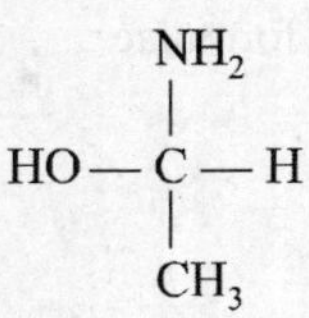
(*iv*) 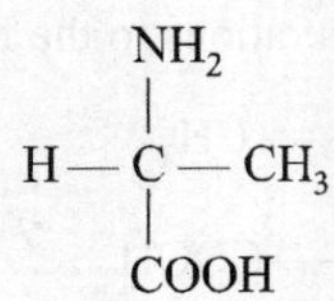

17. What is Walden Inversion ? Give one example.

18. What do you understand by the *Rectus* and *Sinister* system for specifying a configuration?

19. Discuss the method of resolution of a racemic mixture which illustrates the formation of diastereomers.

20. Distinguish between meso and racemic forms of tartaric acid.

21. Explain with one example in each case

(*i*) R and S configurations (*ii*) E and Z notations

(*iii*) Racemisation.

22. Assign R and S configurations to the following :

(*i*) 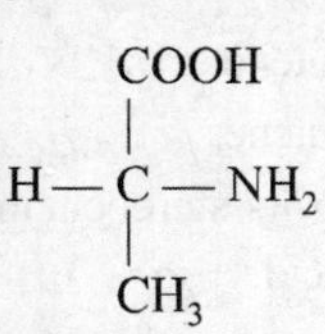
(*ii*) 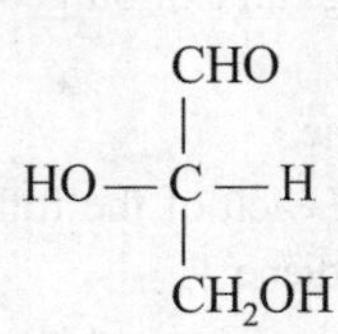

23. What is functional isomerism ? Illustrate with an example.

24. Write down the configuration of 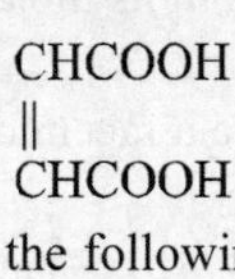 according to E and Z sequence rules.

25. Assign R and S configuration to the following :

(*i*) 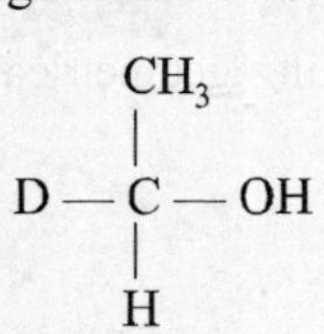
(*ii*) 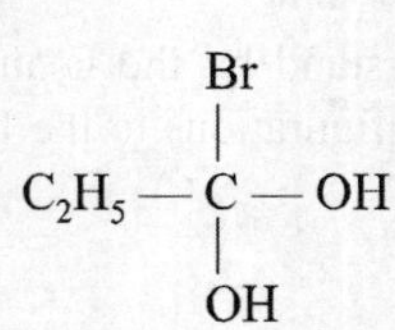

26. Explain biological method for the resolution of the racemic modification.

27. How will you prepare (–) Lactic acid from pyruvic acid?

28. Write a note on optical inversion.

29. Name the instrument to measure the angle of optical rotation.

30. Predict whether 3-chlorohexane will be optically active or not. Give reasons for your answer.

31. What are *cis-trans* isomers ? Give two examples of each.

32. What are diastereomers ? Write their main characteristics.

33. Give two methods of resolution of a racemic mixture.

34. Which of the following compounds exhibit geometrical isomerism ? Write *cis* and *trans* isomers :

(*i*) 2-Butene (*ii*) 2-Pentene

(*iii*) 1, 1-Dibromo-1-butene (*iv*) 1, 1-Dichloro-1-propene

(*v*) 1-Pentene (*vi*) 2-Methyl-2-butene.

35. What is geometrical isomerism ? How will you determine the configuration of geometrical isomers by

(*i*) Formation of cyclic compounds (*ii*) Dipole moment?

36. What do you understand by the E and Z notations for representing geometrical isomers. Assign E and Z configurations to the following :

(*i*) 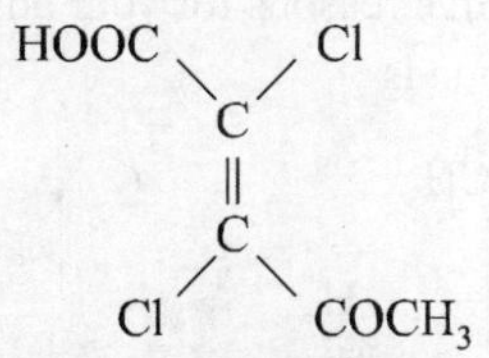

(*ii*) 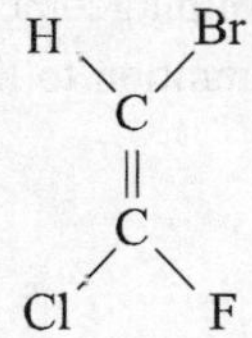

37. What are R and S configurations ? State sequence rules with examples.

38. Give a brief account of the following :

(*a*) Fischer's projection formula (*b*) Asymmetric synthesis.

39. Show that the presence or absence of chiral carbon atom in a molecule is not the necessary and sufficient condition for the existence of optical activity.

40. Optical activity is linked with the presence of asymmetric carbon atom. Justify the statement with a suitable example.

41. Define optical activity. Discuss the various methods for the resolution of a racemic mixture.

42. Explain the following terms :

(*i*) Specific rotation (*ii*) Chiral centre

(*iii*) Meso compounds.

43. Write a note on E and Z method of the configuration designation of a geometrical isomer.

44. Assign R and S configurations to the following compounds :

(*i*) 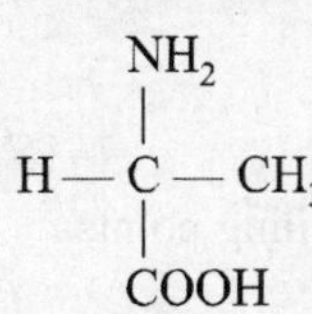

(*ii*) 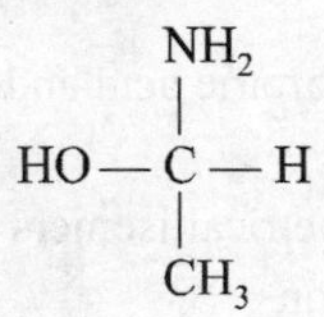

(*iii*) 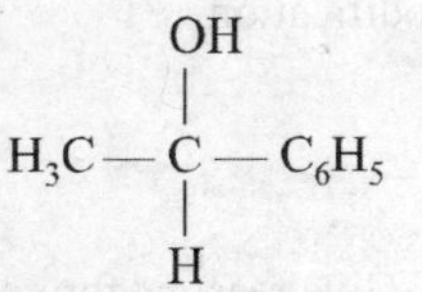   (*iv*) 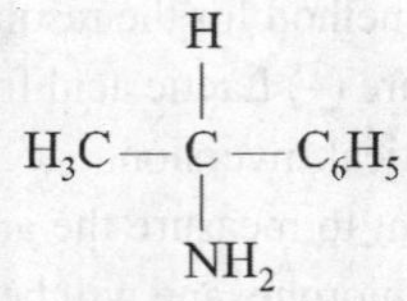

45. (*a*) What do the symbols E and Z stand for ? Illustrate briefly the E and Z system of naming a pair geometrical isomers. What are the advantages of E and Z system over conventional *cis-trans* system.

(*b*) Using E and Z notations write the IUPAC names of

(*i*) 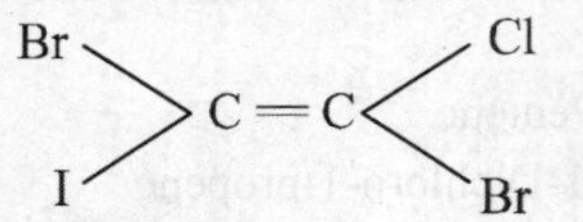   (*ii*) 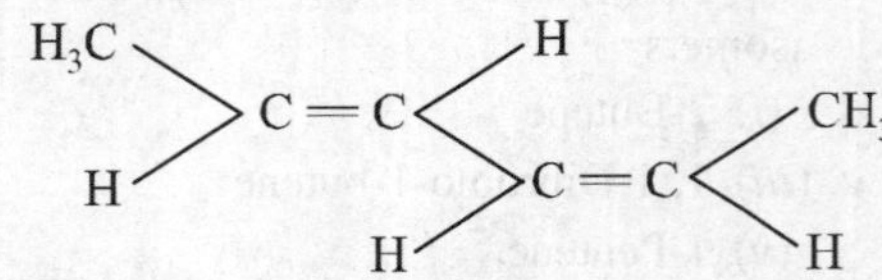

46. How will you resolve racemic lactic acid with optically active forms ? Give two methods for this purpose

47. What is resolution ? How will you resolve a racemic mixture by chemical method?

48. What makes a compound optically active ? How do you allot the configuration to a chiral centre. Distinguish between enantiomers and diastereoisomers.

49. (*a*) Will trichloroethane exhibit geometrical isomerism ? Give reasons for your answer.

(*b*) Assign R and S configuration to the following compounds :

(*i*) 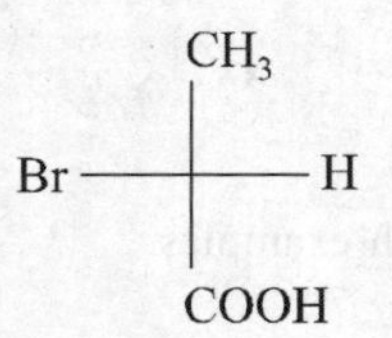   (*ii*) 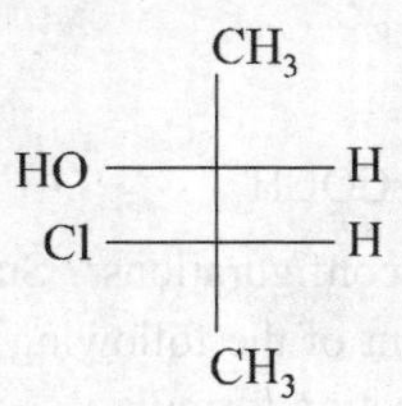

(*c*) Denote E and Z notations and write IUPAC names of the following compounds :

(*i*) 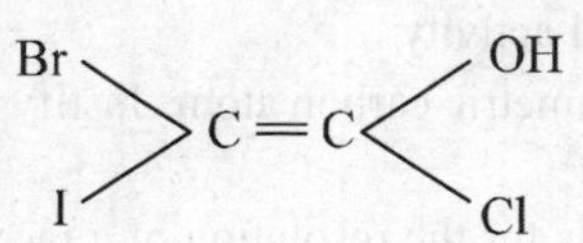   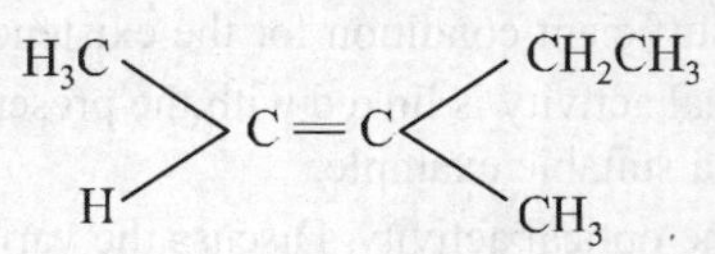

50. (*a*) What is sequence rule ? How is it used for assigning configuration ? Discuss with suitable example.

(*b*) When a compound is expected to be optically active and why?

51. What are the requirements for a compound to be optically active ? Comment on the observation that presence of a chiral carbon is not always essential for a compound to be optically active.

52. Write four isomers of tartaric acid and select the pairs forming :

(*i*) Enantiomers   (*ii*) Diastereoisomers.

53. (*a*) Discuss how geometrical isomers affect melting points and boiling points.

(*b*) Write short notes on :

(*i*) Sequence rule

(*ii*) Absolute configurations in terms of D and L notations.

(*c*) Assign R and S configurations to the following compounds.

(*i*) 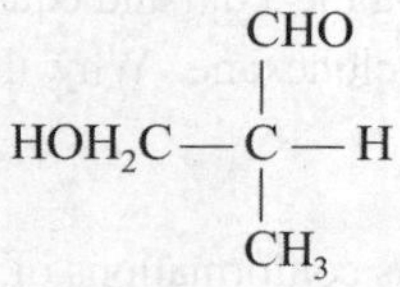 (*ii*) 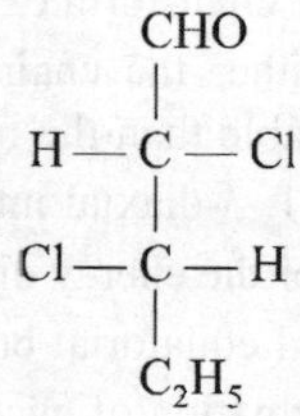

(*d*) Compare external compensation and Internal compensation.

54. Assign R and S to the following :

(*i*) 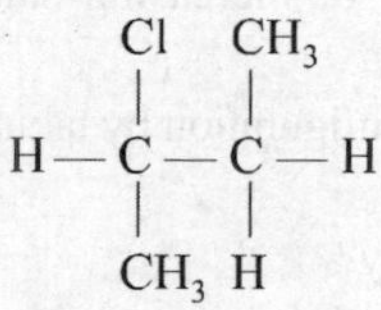 (*ii*) 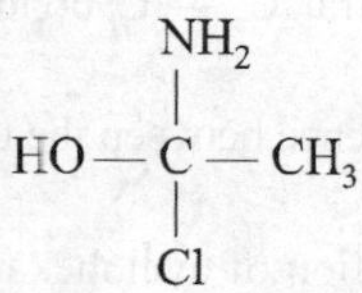

(*iii*) 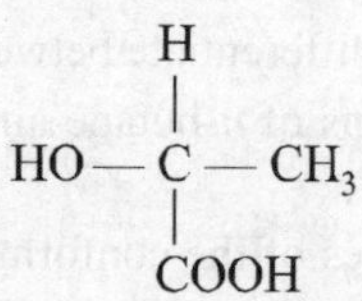 (*iv*) 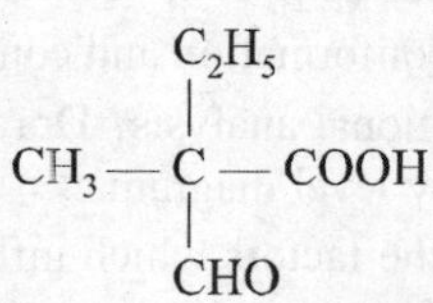

55. (*a*) Define optical isomerism. Mention the conditions and selecting lactic acid as example, explain how many isomers are possible in it.

(*b*) Giving examples, distinguish between the following :

(*i*) Enantiomers and Diastereoisomers

(*ii*) Meso and Racemic forms.

56. Represent the different conformations of ethane with the help of space models. Which out of them is more stable and why ?

57. Discuss the effect of temperature on the dipole moment of a conformation by taking a suitable example.

58. What do you understand by the following :

(*i*) Chair conformation (*ii*) Boat conformation

(*iii*) Twist boat conformation?

59. What do you understand by the term conformation ? How are the conformations formed?

60. How will you differentiate between the conformations and configurations?

61. The rotation about the C — C bond is almost free and not completely free. Justify.

62. Why is chair conformation of cyclohexane more stable than the boat conformation?

63. Which are two main conformations of cyclohexane ? How will you represent them?

64. How do angle strain and steric strains influence the stability of a conformation?

65. The eclipsed conformation of ethane is less stable than the staggered conformation. Explain.

66. Name the different factors which influence the relative stabilities of the conformation of a molecule.

67. Normally *anti*-conformation of a molecule is more stable as compared to the *gauche* conformation. But the same is not true in case of ethylene glycol. Explain.

68. With the help of suitable example discuss the following :

(*i*) Axial hydrogen atoms (*ii*) Flag pole hydrogen atoms.

69. Draw the structure of the chair form of cyclohexane depicting the axial and equatorial bonds.

70. There is no strain in either the chair or boat form of cyclohexane. Why then the chair conformation is more stable than the boat conformation?

71. Give a brief account of 1, 3-diaxial interactions.

72. How will you account for the energy difference in the various conformations of cyclohexane.

73. Write a note an axial and equatorial bonds in cyclohexane.

74. (*a*) Why is chair conformation of cyclohexane more stable than the boat conformation?

(*b*) Differentiate between conformation and configuration.

75. (*a*) Sketch the Newman's projection formulae of important conformations of *n*-butane with reference to central C — C bond. Arrange them in order of increasing stability and assign reason.

(*b*) Explain the difference between the terms conformation and configuration by taking suitable examles.

(*c*) Which conformation of cyclohexane is more stable and why?

(*d*) Explain the chair and boat conformations of cyclohexane.

76. Explain the term conformation and configuration. How will you differentiate between them?

77. What is conformational analysis? Draw the various conformations of *n*-butane and arrange them on the energy level diagram.

78. Discuss the brief the factors which influence the relative stabilities of the conformations.

79. Explain the following terms with suitable examples.

(*a*) Torsional strain (*b*) Steric strain (*c*) Dipole-dipole interactions.

80. Discuss the different conformations of *n*-butane by taking into account the rotation about C — C bond. Arrange them in increasing order of stability and assign suitable reason for the same.

81. Discuss the various factors which influence the relative stabilities of the conformation of a molecule.

4 Alkanes and Cycloalkanes

4.1 ALKANES

Alkanes which are the hydrocarbons, represent the parent family from which all other families have been derived by replacing one or more hydrogen atoms with suitable functional groups. These are also known as the saturated hydrocarbons or paraffins (*para*-little, *affin*-reactive). These have the general formula C_nH_{2n+2} and are open chain in nature. The bonds present in alkane molecules are single covalent bonds and are sigma (σ) bonds in nature. Moreover, the carbon atoms involved in these molecules have their normal tetrahedral bond angles. Alkanes have low reactivity because of their stable structure.

4.2 STRUCTURE AND NOMENCLATURE OF ALKANES

The first member of the alkane family is methane (CH_4). In this molecule, the carbon atom is located in the centre of a regular tetrahedron [Fig. 4.1(*a*)] and the four H atoms joined to it by C — H bonds, are directed towards its four corners. The H — C — H bond angle is 109° 28′. The spatial formula of methane can also be indicated by *Wedge projection* [Fig. 4.1(*b*)] where the thick line indicates the C — H bond that is directed towards the eye of observer. The normal line represents the bond that lies in the plane of the paper while the two C — H bonds indicated by dotted lines project behind the plane of the paper.

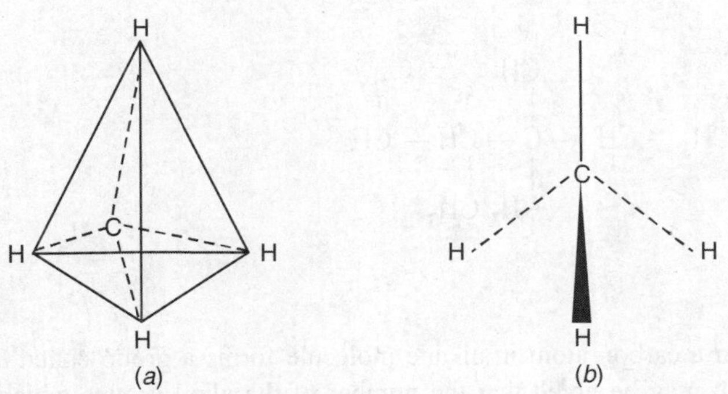

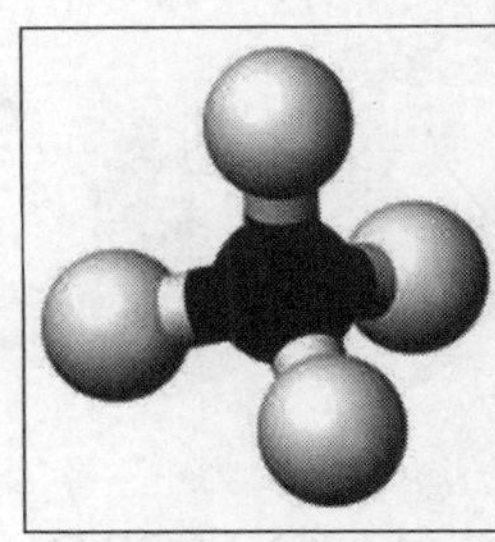

Model of methane

Fig. 4.1. Spatial and planar representation of methane.

The next member of the family is ethane (C_2H_6) and can be obtained by replacing on H atom in methane molecule with a methyl (CH_3) group.

$$\underset{\text{Methane}}{CH_4} \xrightarrow[+\,CH_3]{-\,H} \underset{\text{Ethane}}{CH_3CH_3}$$

Similarly, propane (C_3H_8), the third member of the family is formed in a similar way by replacing one H–atom in ethane with CH_3 group.

$$\underset{\text{Ethane}}{CH_3CH_3} \xrightarrow[+CH_3]{-H} \underset{\text{Propane}}{CH_3CH_2CH_3}$$

The IUPAC and common names of the first six members of the alkane family are given in the following table.

Value of *n*	Alkane	IUPAC name	Common name
1	CH_4	Methane	Methane
2	$CH_3 \cdot CH_3$	Ethane	Ethane
3	$CH_3 \cdot CH_2 \cdot CH_3$	Propane	Propane
4	$CH_3 \cdot CH_2 \cdot CH_2 \cdot CH_3$	Butane	*n*-Butane
5	$CH_3 \cdot (CH_2)_3 \cdot CH_3$	Pentane	*n*-Pentane
6	$CH_3 \cdot (CH_2)_4 \cdot CH_3$	Hexane	*n*-Hexane

The prefix *n*- is used in the common names of the compounds with four or more carbon atoms. This indicates a straight or normal chain of carbon atoms and no branching is involved.

4.3 CLASSIFICATION OF THE CARBON ATOMS IN ALKANES

The carbon atoms in an alkane molecule are classified as *primary* (1°), *secondary* (2°) *tertiary* (3°) and *quaternary* (4°).

(*i*) A primary carbon is linked to one or no carbon atom.

(*ii*) A secondary carbon is linked to two carbon atoms.

(*iii*) A tertiary carbon is linked to three carbon atoms.

(*iv*) A quaternary carbon is linked to four carbon atoms.

The H-atoms attached to these carbon atoms are also designated accordingly *i.e.*, as 1°, 2° and 3°. Quaternary carbon atom has no hydrogen atom attached to it. The following example shows different types of carbon atoms in the molecule.

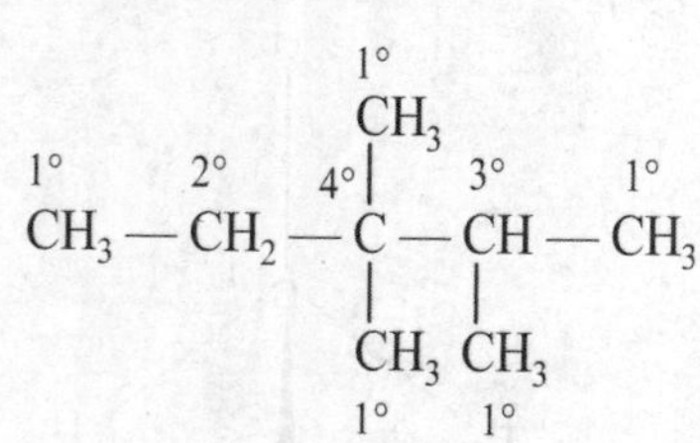

4.4 ALKYL GROUPS

Removal of the H-atom from a carbon atom in alkane molecule forms a group called *alkyl group* with formula $C_n H_{(2n+1)}$. It may be noted that the number of the alkyl groups which are formed from a particular alkane molecule depends upon the total number of different carbon atoms present. For example, methane and ethane molecules have only primary carbon atom and give only one alkyl group while propane, and *n*-butane with both primary and secondary carbon atoms yield two different alkyl groups as shown below :

$$\underset{\text{Ethane}}{\overset{1°}{CH_3} - \overset{1°}{CH_3}} \xrightarrow{-H(1°)} \underset{\text{Ethyl group}}{CH_3 - CH_2 -}$$

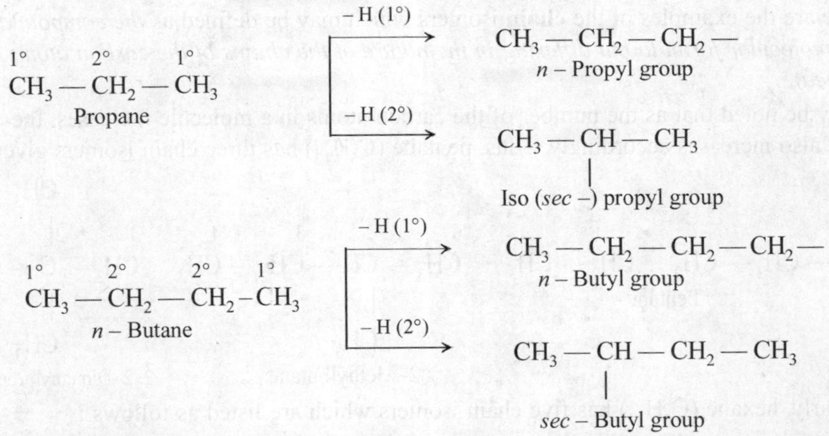

Similarly, isobutane having primary & tertiary carbon atoms also gives two alkyl groups, as under :

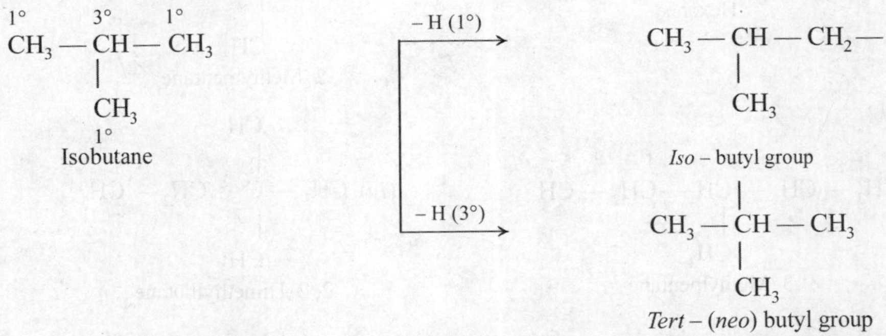

iso represents branching by two similar groups (*iso*-propyl and *iso*-butyl), *neo* indicates branching by *three* such groups (*neo*-butyl). A particular alkyl group will be secondary in case the H-atom attached to secondary carbon atom has been replaced. Thus, *iso* and *sec* prefixes for these groups should not be confused with each other.

4.5 STRUCTURAL ISOMERISM IN ALKANES

Alkanes exhibit only **chain isomerism** which is discussed as follows :

In the first three members of the alkane family, the carbon atoms are linked to each other to form straight chains (C — C or C — C — C). However, in the next member butane (C_4H_{10}), there is a scope for branching *i.e.* there may be a straight chain or branched chain of carbon atoms as shown below :

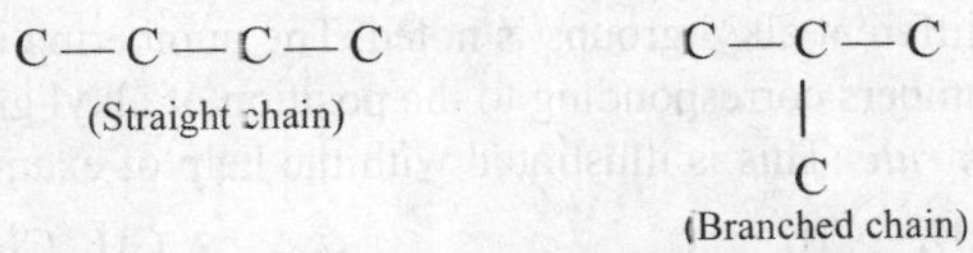

The alkanes corresponding to these structures are called butane (*n*-butane) and 2-Methylpropane (isobutane).

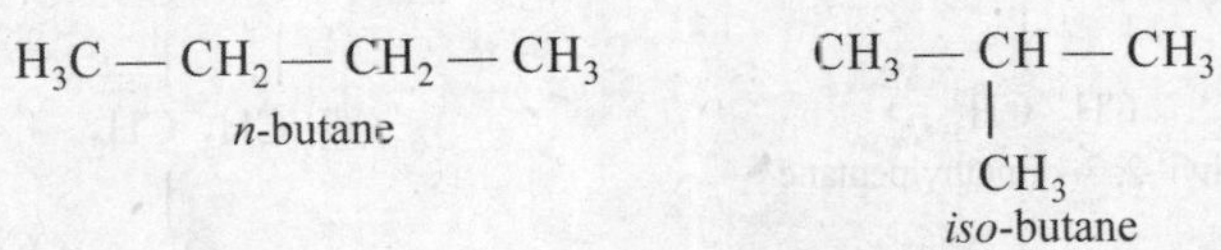

These are the examples of the **chain** isomers which may be defined as *the compounds having the same molecular formula but differing in the nature of the chains of the carbon atoms (straight or branched).*

It may be noted that as the number of the carbon atoms in a molecule increases, the extent of branching also increases accordingly. Thus, pentane (C_5H_{12}) has three chain isomers given below:

$CH_3—CH_2—CH_2—CH_2—CH_3$

Pentane

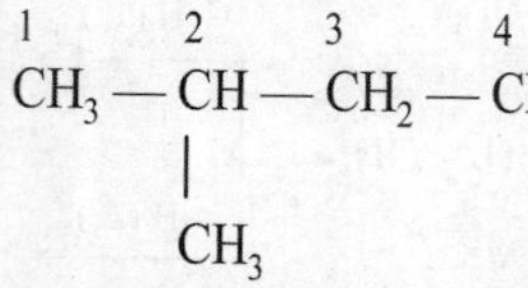

2–Methylbutane

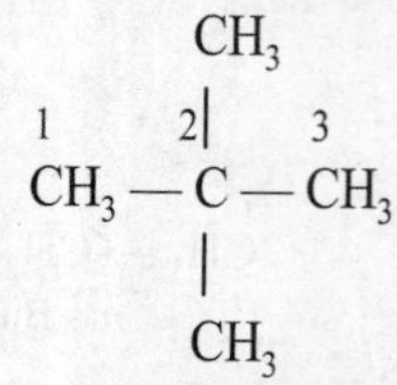

2, 2–Dimethylpropane

Similarly, hexane (C_6H_{14}) has five chain isomers which are listed as follows :

(*i*) $CH_3—CH_2—CH_2—CH_2—CH_2—CH_3$

Hexane

(*ii*) 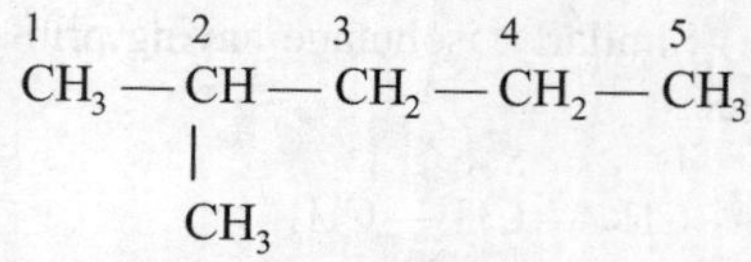

2–Methylpentane

(*iii*) 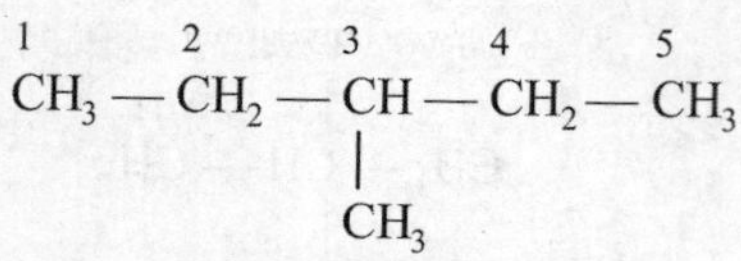

3–Methylpentane

(*iv*) 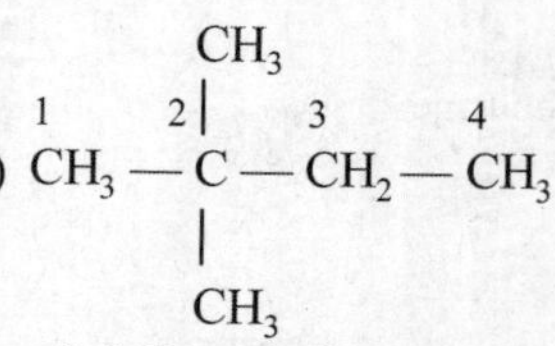

2, 2–Dimethylbutane

(*v*) 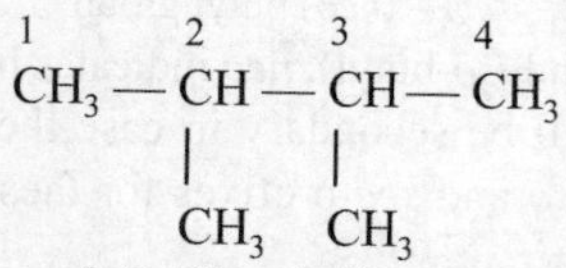

2, 3–Dimethylbutane

For stereoisomerism and conformational isomerism in alkanes refer to chapter 3.

IUPAC Nomenclature

(*i*) The longest carbon chain in the molecule is identified. The compound is named as a derivative of hydrocarbon corresponding to the longest chain. For example, if there are five carbon atoms in the longest chain, the hydrocarbon will be named as a derivative of pentane and so on.

(*ii*) The position of different alkyl groups is noted. The numbering is done in such a way that the sum of the numbers corresponding to the position of alkyl groups is minimum. This is called *lowest sum rule.* This is illustrated with the help of examples given below:

(*i*) 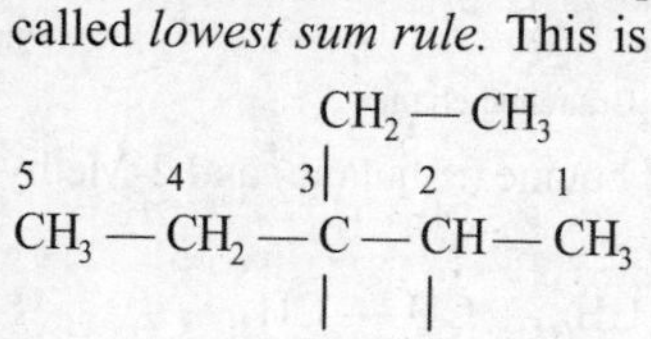

3–Ethyl–2, 3–dimethylpentane

(*ii*) 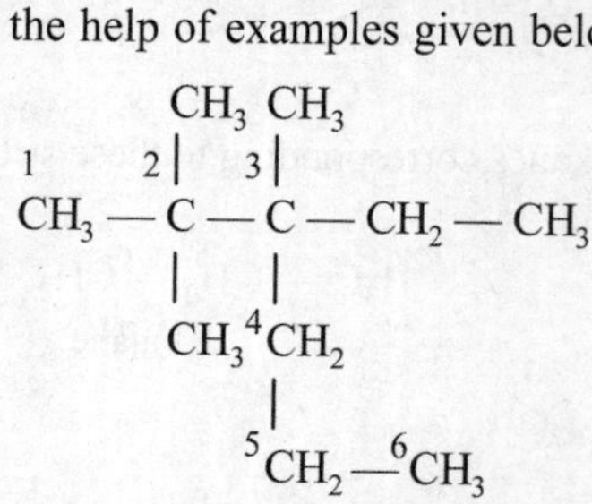

3–Ethyl–2, 2, 3–trimethylhexane

Example 1. **Give IUPAC names of the following compounds:**

(*i*) 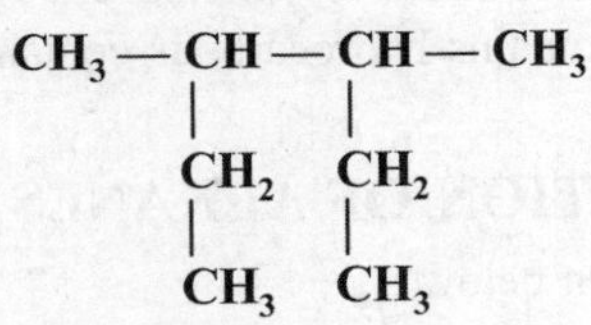

(*ii*) 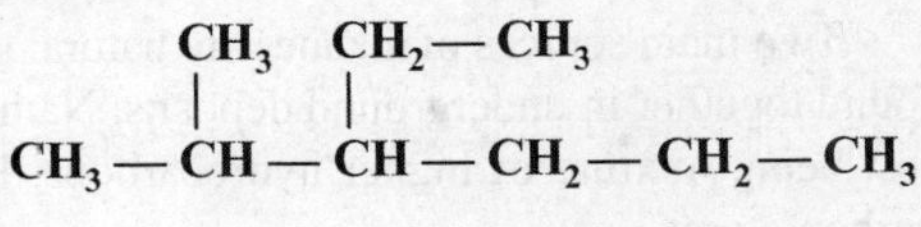

(*iii*) 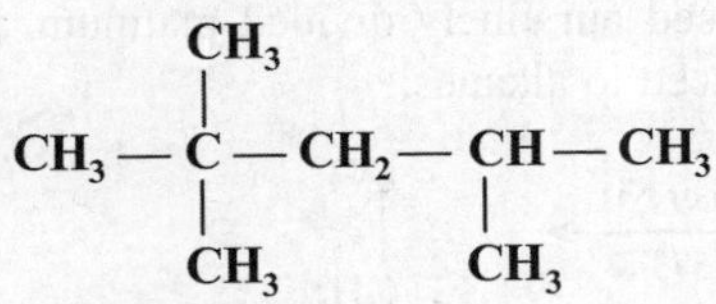

(*iv*) 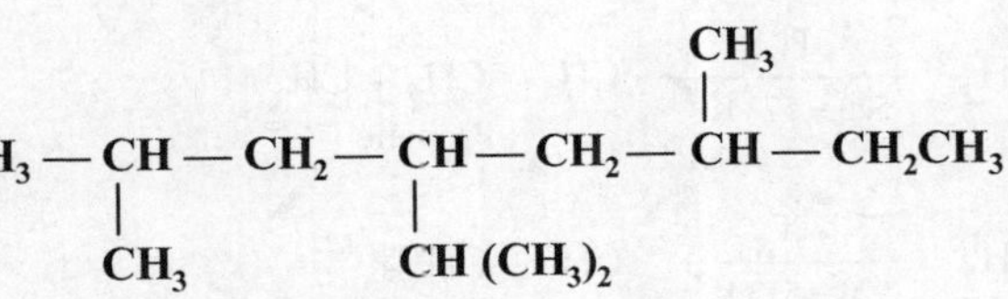

Solution.

(*i*) 3, 4 – Dimethylhexane.

(*ii*) 3-Ethyl-2 – methylhexane.

(*iii*) 2, 2, 4-Trimethylpentane.

(*iv*) 2, 6-Dimethyl-4 – isopropyloctane.

Example 2. **Write the structural formulae and IUPAC names of various compounds possible with the molecular formula C_6H_{14}.**

Solution. Various structures possible are given below:

(*i*) $CH_3CH_2CH_2CH_2CH_2CH_3$ — *n*-hexane

(*ii*) 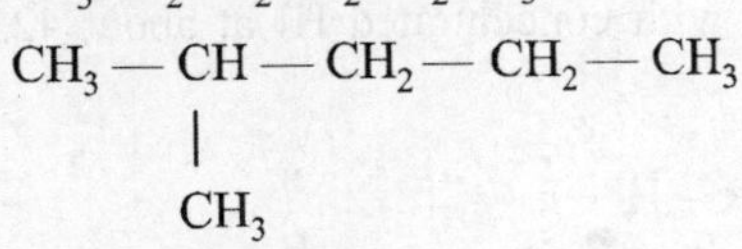 2-methylpentane

(*iii*) 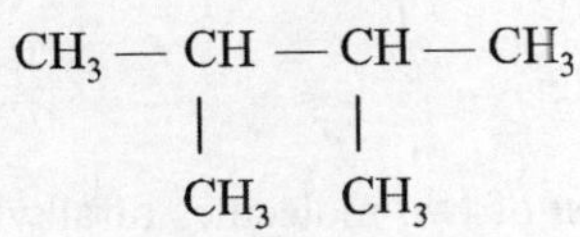 2, 3-dimethylbutane

(*iv*) 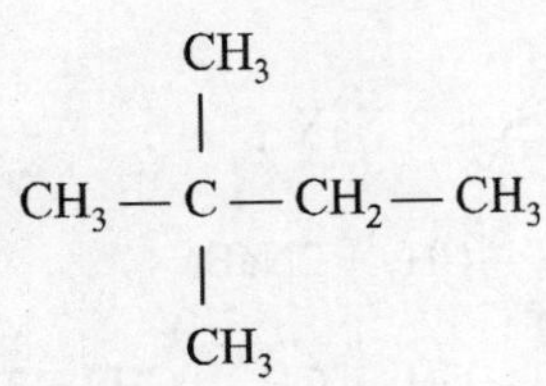 2, 2-dimethylbutane

4.6 SOURCES OF ALKANES

Two main sources of alkanes are natural gas and petroleum. Both these substances are generally found together in underground deposits. Natural gas contains about 80% methane, 10% ethane, the rest being mixture of higher hydrocarbons. Petroleum contains hydrocarbons containing upto 40 carbon atoms.

4.7 GENERAL METHODS OF PREPARATION OF ALKANES

Different methods of preparation of alkanes are given below :

(1) Hydrogenation of unsaturated hydrocarbons (Sabatier-Senderen's reaction). A mixture of unsaturated hydrocarbons and hydrogen is passed our finely divided platinum, palladium or nickel at 523-573 K. Alkenes and alkynes are reduced to alkanes.

$$\begin{matrix} CH \\ \| \\ CH_2 \end{matrix} + H_2 \xrightarrow[523-573\text{ K}]{\text{Pt, Pd or Ni}} \begin{matrix} CH_3 \\ | \\ CH_3 \end{matrix}$$

$$\underset{\text{Propene}}{CH_3 - CH = CH_2} + H_2 \xrightarrow[523-573\text{ K}]{\text{Pt}} \underset{\text{Propane}}{CH_3 - CH_2 - CH_3}$$

$$\underset{\text{Acetylene}}{CH \equiv CH} + 2H_2 \xrightarrow[523-573\text{ K}]{\text{Pt}} \underset{\text{Ethane}}{CH_3 - CH_3}$$

(2) Reduction of alkyl halides. This may be done in different ways:

(*i*) Chemical reducing agents like zinc and acetic acid or hydrochloric acid, zinc-copper couple and alcohol; magnesium-amalgam and water etc., convert alkyl halides into alkanes.

$$\underset{\text{Alkyl halide}}{R - X} + 2[H] \xrightarrow[\text{Zn / Acetic acid}]{\text{Zn – Cu / alcohol}} \underset{\text{Alkane}}{R - H} + H - X$$

$$C_2H_5 - Br + 2[H] \xrightarrow[\text{Zn / Acetic acid}]{\text{Zn – Cu / alcohol}} \underset{\text{Ethane}}{C_2H_6} + HBr$$

(*ii*) Alkyl iodides are easily reduced by heating with concentrated HI at about 423 K in a sealed tube.

$$R - I + HI \xrightarrow[423\text{ K}]{\text{P}} R - H + I_2$$

$$C_2H_5I + HI \xrightarrow[423\text{ K}]{\text{P}} C_2H_6 + I_2$$

(3) Wurtz reaction. This reaction involves the interaction of two molecules of alkyl halides and sodium metal in dry ether.

$$R - \boxed{X + 2Na + X} - R \xrightarrow[\text{ether}]{\text{Dry}} R - R + 2\ NaX$$

$$CH_3 - \boxed{Br + 2Na + Br} - CH_3 \xrightarrow{\text{Dry ether}} CH_3 - CH_3 + 2NaBr$$

$$C_2H_5 - \boxed{Br + 2Na + Br} - C_2H_5 \xrightarrow{\text{Dry ether}} \underset{n\text{-Butane}}{CH_3 - CH_2 - CH_2 - CH_3} + 2NaBr$$

When different alkyl halides are used, a mixture of three alkenes is obtained as shown below:

$$\underset{\text{Methyl bromide}}{CH_3 - Br} + 2Na + \underset{\text{Ethyl bromide}}{Br - C_2H_5} \longrightarrow \underset{\text{Propane}}{CH_3 - CH_2 - CH_3} + 2NaBr$$

$$CH_3 - Br + 2Na + Br - CH_3 \longrightarrow \underset{\text{Ethane}}{CH_3 - CH_3} + 2NaBr$$

$$C_2H_5 - Br + 2Na + Br - C_2H_5 \longrightarrow CH_3 - CH_2 - CH_2 - CH_3 + 2NaBr$$

Wurtz reaction is suitable for the preparation of symmetrical alkanes. With secondary and tertiary alkyl halides, many side products are formed.

Mechanism of Reaction

Two mechanisms have been suggested for the reaction.

Mechanism I

The reaction takes place in two steps as shown below:

$$RX + 2Na \longrightarrow R - Na + NaX$$

$$RNa + XR \longrightarrow R - R + NaX$$

Mechanism II

Again the reaction takes place in two steps as under:

$$RX + Na \longrightarrow \dot{R} + NaX$$

$$\dot{R} + \dot{R} \longrightarrow \underset{\text{Hydrocarbon}}{R - R}$$

(4) Corey-House Synthesis. It is a better method than Wurtz reaction. An alkyl halide and a lithium dialkyl copper are reacted to give a higher hydrocarbon.

$$\underset{\text{Alkyl halide}}{R' - X} + \underset{\text{Lithium dialkyl copper}}{R_2\,CuLi} \longrightarrow \underset{\text{Alkane}}{R - R'} + R - Cu + LiX$$

(R and R′ may be same or different)

For example,

$$\underset{n\text{–Propyl bromide}}{CH_3CH_2CH_2Br} + \underset{\text{Lithium dimethyl copper}}{(CH_3)_2CuLi} \longrightarrow \underset{n\text{–Butane}}{CH_3CH_2CH_2CH_3} + CH_3Cu + \underset{\text{Methyl copper}}{LiBr}$$

A notable feature of this reaction is that it can be used for preparing symmetrical, unsymmetrical, straight-chain or branched-chain alkanes.

For a better yield of the product, the alkyl halide used should be primary whereas lithium dialkyl copper may be primary, secondary or tertiary.

(5) Indirect reduction of alkyl halides. Alkyl halides react with magnesium in dry ether to form alkyl magnesium halides (Grignard reagent). Decomposition of the Grignard reagent with water yield alkanes.

$$\underset{\text{Alkyl halide}}{R - X + Mg} \xrightarrow{\text{Dry ether}} \underset{\text{Grignard reagent}}{R - MgX}$$

$$R - MgX + H - OH \longrightarrow \underset{\text{Alkane}}{R - H} + Mg\begin{matrix} OH \\ X \end{matrix}$$

$$\underset{\text{Methyl magnesium bromide}}{CH_3 - MgBr} + H - OH \longrightarrow CH_4 + Mg\begin{matrix} \diagup OH \\ \diagdown X \end{matrix}$$

(6) Decarboxylation of Carboxylic Acid Salts. Sodium salts of a fatty acid are heated with soda lime (NaOH + CaO) to form alkanes.

$$R - \boxed{COONa + NaO} - H \xrightarrow{CaO} R - H + Na_2CO_3$$

$$\underset{\text{Sodium acetate}}{CH_3 - COONa} + NaOH \xrightarrow{CaO} CH_4 + Na_2CO_3$$

(7) Kolbe's Electrolytic Process. By the electrolysis of conc. solution of sodium or potassium salts of fatty acids, alkanes are obtained.

$$2R - COONa \longrightarrow 2RCOO^- + 2Na^-$$

$$2R - COO^- \longrightarrow \begin{matrix} RCOO + 2e^- \\ | \\ RCOO \\ \text{(Unstable)} \\ \downarrow \\ R - R + 2CO_2 \end{matrix} \quad \text{[At anode]}$$

$$2Na^+ + 2e^- \longrightarrow \begin{matrix} 2Na \\ \downarrow 2H_2O \\ 2NaOH + H_2 \end{matrix} \quad \text{[At cathode]}$$

$$2CH_3COONa \xrightarrow{\text{Electrolysis}} \underbrace{CH_3 - CH_3 + 2CO_2}_{\text{[At anode]}} + \underbrace{2Na + 2H_2O}_{\text{[At cathode]}}$$

(8) By Reduction of Alcohols, Aldehydes, Ketones and Acids. The reduction is carried out with the help of HI and P.

(*i*)
$$\underset{\text{Alcohol}}{R - OH} + 2HI \xrightarrow[423\ K]{P} R - H + H_2O + I_2$$

$$\underset{\text{Ethyl alcohol}}{C_2H_5 - OH} + 2HI \xrightarrow[423\ K]{P} \underset{\text{Ethane}}{C_2H_6} + H_2O + I_2$$

(*ii*)
$$\underset{\text{Aldehyde}}{R - \overset{\overset{\displaystyle O}{\|}}{C} - H} + 4HI \xrightarrow[423\ K]{P} \underset{\text{Alkane}}{R - CH_3} + H_2O + 2I_2$$

$$\underset{\text{Acetaldehyde}}{CH_3 - \overset{\overset{\displaystyle O}{\|}}{C} - H} + 4HI \xrightarrow[423\ K]{P} \underset{\text{Ethane}}{CH_3 - CH_3} + H_2O + 2I_2$$

$$(iii)\quad \underset{\text{Ketone}}{R-\overset{\overset{\displaystyle O}{\|}}{C}-R+4HI} \xrightarrow[423\ K]{P} \underset{\text{Alkane}}{R-CH_2-R} + H_2O + 2I_2$$

$$\underset{\text{Acetone}}{CH_3-\overset{\overset{\displaystyle O}{\|}}{C}-CH_3+4HI} \xrightarrow[423\ K]{P} \underset{\text{Alkane}}{CH_3-CH_2-CH_3} + H_2O + 2I_2$$

$$(iv)\quad \underset{\text{Acid}}{R-\overset{\overset{\displaystyle O}{\|}}{C}-OH+6HI} \xrightarrow[473K]{P} \underset{\text{Alkane}}{R-CH_3} + 2H_2O + 3I_2$$

$$\underset{\text{Formic acid}}{H-COOH + 6HI} \xrightarrow[473K]{P} CH_4 + 2H_2O + 3I_2$$

$$CH_3-COOH + 6HI \xrightarrow[473K]{P} CH_3-CH_3 + 2H_2O + 3I_2$$

4.8 PHYSICAL PROPERTIES OF ALKANES

1. Solubility. Alkanes being non-polar in nature are insoluble in water but soluble in non polar solvents like ether, benzene, etc. Solubility decreases with increase in molecular size.

2. Boiling Points. As alkanes are non-polar molecules, they have only weak van der Waals' forces operating between them. Van der Waals' forces of attraction depend upon the molecular size and increase with increase in molecular size. Among the chain isomers of a compound, the branched chain isomer has lower boiling point compared to straight-chain isomer. For example, the boiling points of three isomeric pentanes are given below:

$$\underset{\substack{n\text{–Pentane}\\(\text{b.p}=309\text{ K})}}{CH_3CH_2CH_2CH_2CH_3} \qquad \underset{\substack{\text{Isopentane}\\(\text{b.p}=301\text{ K})}}{CH_3-\underset{\underset{\displaystyle CH_3}{|}}{CH}-CH_2-CH_3} \qquad \underset{\substack{\text{Neopentane}\\(\text{b.p}=281\text{ K})}}{CH_3-\overset{\overset{\displaystyle CH_3}{|}}{\underset{\underset{\displaystyle CH_3}{|}}{C}}-CH_3}$$

Explanation. The boiling point of a substance depends upon the force of attraction between the molecules. Force of attraction between the molecules depends upon the area of contact between them. Area of contact and force of attraction will be maximum for straight chain compound. As branching increases, the area of contact decreases. Consequently, the force of attraction and boiling point will decrease.

3. Melting Point. Alkanes show lower melting point because of absence of charge. Melting point is expected to increase with increase in molecular size. This is found to be true but the increase is not uniform.

Alkane	Melting Point (K)
C_3H_8	85.9
n-C_4H_{10}	138
n-C_5H_{12}	143.3
n-C_6H_{14}	179

We find that increase in melting point is relatively more in going from odd-carbon to even-carbon alkane compared to when we move from even-carbon to odd-carbon alkane.

Explanation. It may be realised that in even-carbon alkanes (*e.g.* C_6H_{14}) the packing of molecules can take place to a greater extent, resulting in greater attraction forces and hence greater melting points. This is not so with odd-carbon alkanes.

Representation of two chain of n-C_6H_{14} (greater attractive forces)

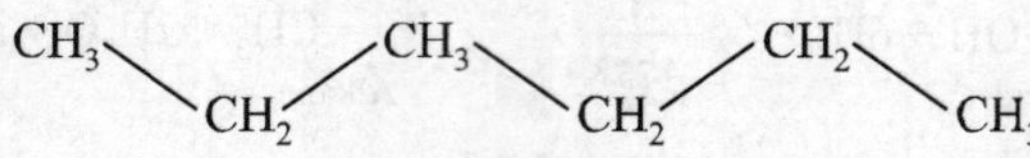

n-hexane

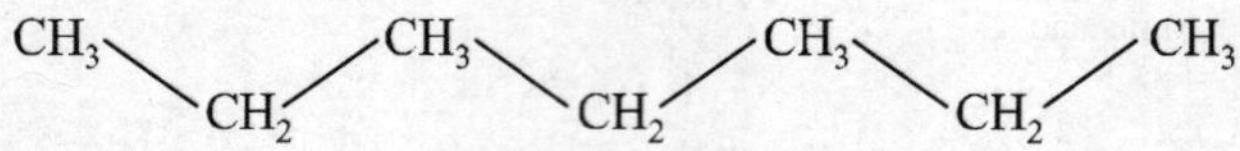

n-heptane

Example 3. **Write the structural formula of *n*-pentane, isopentane and neopentane. Which of these has highest boiling point and why?**

Solution.

Structural formulae of pentanes

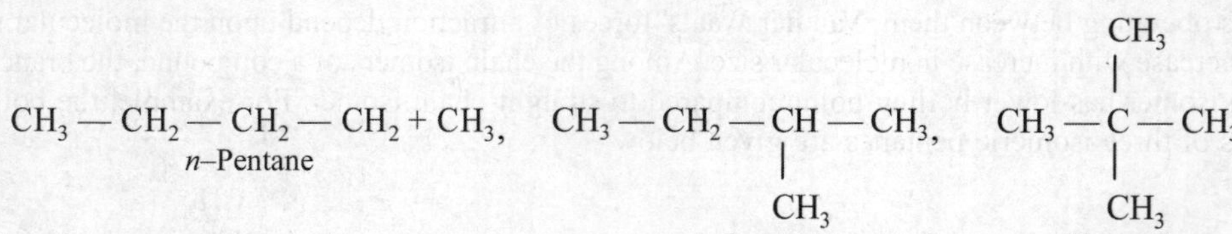

n–Pentane, Isopentane, Neopentane

Highest boiling point. Of the three pentanes, *n*-pentane has the highest boiling point.

Explanation. All the three pentanes are non-polar compounds having only weak intermolecular forces of attraction (van der Waals' forces). The molecular weights of the three isomers are naturally the same but they have different surface areas. As the surface area of *n*-pentane is larger than those of the other two, it has strongest intermolecular forces of attraction. Therefore it has highest boiling point followed by iso- and neopentanes.

Example 4. **As we move up from *n*-pentane to *n*-hexane, there is an increase of about 35° in the melting point but on moving from *n*-hexane to *n*-heptane the increase is only about 4°. How can you account for these observations?**

Solution.

Disproportionate variation in melting points of the three alkanes can be explained on the basis of their molecular weights and their ability to fit in the crystal structures as follows.

In alkanes, the carbon atoms form zig-zag chains as depicted below for *n*-pentane, *n*-hexane and *n*-heptane.

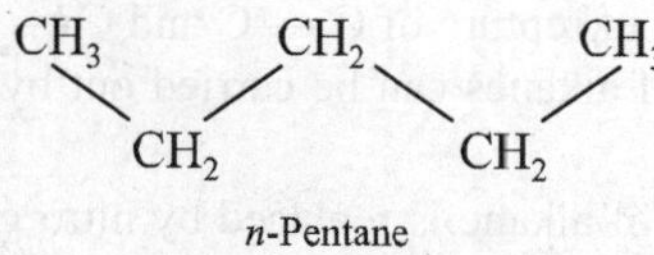

n-Pentane

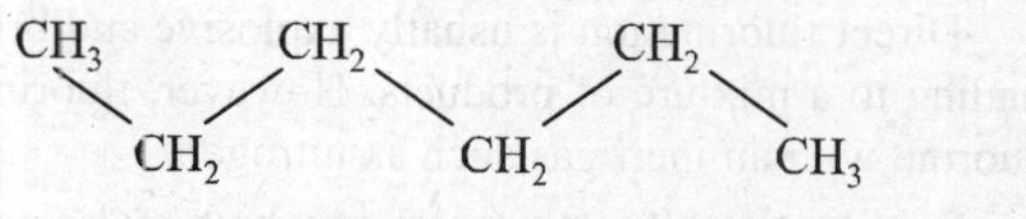

n-Hexane

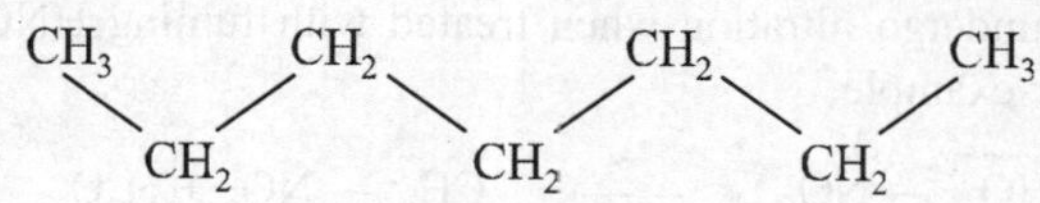

n-Heptane

It is apparent that in *n*-hexane, the two terminal methyl groups lie on opposite sides of the zig-zag chain which enables this chain to fit in closely in the crystal structure. But in *n*-pentane and *n*-heptane, the two terminal groups lie on the same side so that the crystal structure is not so closely packed.

As we move from *n*-pentane to *n*-hexane, the molecular weight increases, also the packing arrangement becomes more closed. Both these factors require greater energy to melt the solid. As such there is a large increase in melting point.

On the other hand, as we move from *n*-haxane to *n*-heptane, the molecular weight increases but the packing arrangement does not remain so close. While the melting point would tend to increase due to increase in molecular weight, the decrease in close packing would tend to have the opposite effect. The net result is that there is only a small increase in melting point in this case.

4.9 CHEMICAL PROPERTIES OF ALKANES

Chemical properties of alkanes are described as under:

1. Halogenation. Displacement or substitution of hydrogen atoms by halogens is known as *halogenation.* Methane on treatment with chlorine in the presence of diffused sunlight or by heating the reaction mixture to 523-673 K gives chloromethane.

$$CH_4 + Cl_2 \xrightarrow[\text{or } 523-673K]{\text{Sunlight}} \underset{\text{Chloro methane}}{CH_3Cl} + HCl$$

The reaction does not stop here. Chloromethane is further chlorinated to dichloromethane. Dichloromethane is chlorinated to trichloromethane and trichloromethane is finally chlorinated to tetrachloromethane. This sequence of reactions is given below:

$$CH_3Cl + Cl_2 \longrightarrow \underset{\text{Dichloromethane}}{CH_2Cl_2} + HCl$$

$$CH_2Cl_2 + Cl_2 \longrightarrow \underset{\text{Trichloromethane}}{CHCl_3} + HCl$$

$$CHCl_3 + Cl_2 \longrightarrow \underset{\text{Tetrachloromethane}}{CCl_4} + HCl$$

Bromination of alkanes takes place in similar manner but less readily. Direct iodination of alkanes is not possible as the reaction is reversible.

$$R—H + I_2 \rightleftharpoons R-I + HI$$

However, iodination can be carried out in the presence of oxidising agent such as iodic acid (HIO_3), which destroys the hydrogen iodide as soon as it is formed and so drives the reaction to the right.

$$5HI + 4HIO_3 \longrightarrow 3I_2 + 3H_2O$$

Direct fluorination is usually explosive and brings about repture of C — C and C — H bonds leading to a mixture of products. However, fluorination of alkanes can be carried out by diluting fluorine with an inert gas such as nitrogen.

2. Nitration. It is a process in which hydrogen atom of alkane is replaced by nitro group (— NO_2). Alkanes undergo nitration when treated with fuming HNO_3 in the vapour phase between 423–748 K. For example,

$$CH_3 — \boxed{H + HO} — NO_2 \longrightarrow \underset{\text{Nitro methane}}{CH_3 — NO_2} + H_2O$$

Higher alkanes on nitration give a mixture of all possible mono derivatives. For example,

$$CH_3 — CH_2 — CH_3 + HNO_3 \xrightarrow{673\ K} \underset{\text{1–Nitropropane}}{CH_3 — CH_2 — CH_2 — NO_2} +$$

$$\underset{\text{2–Nitropropane}}{CN_3 — \overset{\overset{NO_2}{|}}{CH} — CH_3} + \underset{\text{Nitroethane}}{CH_3 — CH_2 — NO_2} + \underset{\text{Nitromethane}}{CH_3 — NO_2}$$

3. Sulphonation. It is a process in which hydrogen atom of alkane is replaced by sulphonic acid group (— SO_3H). Sulphonation of hexane and higher members may be carried out by treating the alkane with oleum (fuming sulphuric acid). For example,

$$\underset{n\text{–Hexane}}{C_6H_{13} — H} + HO — SO_3H \xrightarrow{\Delta} \underset{\text{Hexane sulphonic acid}}{C_6H_{13} — SO_3H} + H_2O$$

The ease of replacement of hydrogen atom is

Tertiary hydrogen > Secondary hydrogen > Primary hydrogen

4. Oxidation. (*i*) *Combustion.* Alkanes burn readily in excess of air or oxygen to form CO_2 and H_2O. This process is called combustion and is accompanied by the evolution of a large amount of heat. For example,

$$C_n H_{2n+2} + O_2 \xrightarrow{\Delta} nCO_2 + (n + 1) H_2O + \text{Heat}$$

$$CH_4 + 2O_2 \xrightarrow{\Delta} CO_2 + 2H_2O + 890\ kJ$$

Alkanes are, therefore, used as fuels in the form of kerosene oil, gasoline, LPG.

(*ii*) *Catalytic oxidation.* When alkanes are heated in a limited supply of air or oxygen at high pressure and in the presence of suitable catalyst, they are oxidised to alcohols, aldehydes or fatty acids. For example,

$$2CH_4 + O_2 \xrightarrow[\text{Cu Tube}]{523\ K, 100\ Atm} 2CH_3OH$$

$$CH_4 + O_2 \xrightarrow[Mo_2O_3]{673\ K, 200\ Atm} HCHO + H_2O$$

(*iii*) Alkanes containing tertiary hydrogens can be oxidised by such reagents to form alcohols. For example,

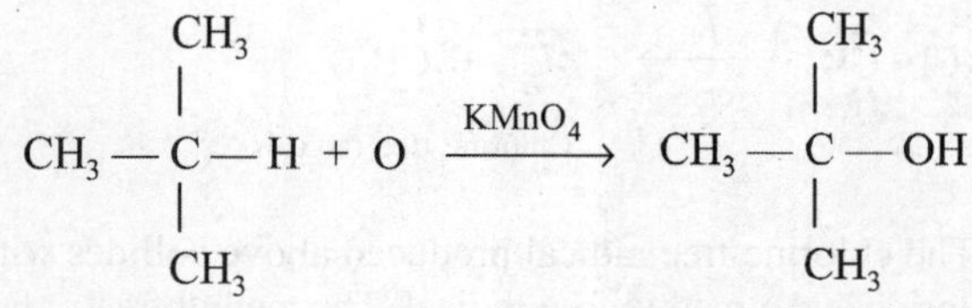

Iso-butane tert–Butyl alcohol

5. Aromatisation. When alkanes containing six or more carbon atoms are heated to high temperature (773 K) under high pressure and in the presence of a catalyst (oxides of chromium, vanadium, molybdenum supported on alumina), they are converted into aromatic hydrocarbons. The process involves cyclisation, isomerisation and dehydrogenation and is known as *aromatisation.* For example,

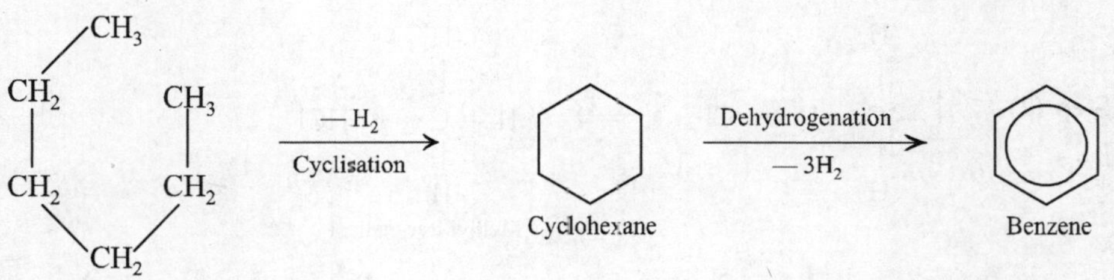

n–hexane Cyclohexane Benzene

Similarly

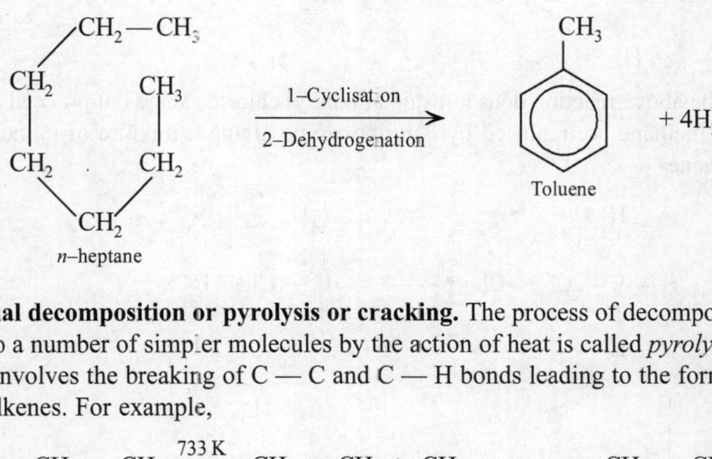

n–heptane Toluene

6. Thermal decomposition or pyrolysis or cracking. The process of decomposition of bigger molecules into a number of simpler molecules by the action of heat is called *pyrolysis or cracking.* This process involves the breaking of C — C and C — H bonds leading to the formation of lower alkanes and alkenes. For example,

$$CH_3 - CH_2 - CH_3 \xrightarrow{733\text{ K}} \underset{\text{Ethylene}}{CH_2 = CH_2} + \underset{\text{Methane}}{CH_4} \quad \text{or} \quad \underset{\text{Propene}}{CH_3 - CH = CH_2} + H_2$$

4.10 MECHANISM OF HALOGENATION OF ALKANE

The halogenation of alkanes occurs by a free radical mechanism. It involves three steps: (*a*) Chain Initiation, (*b*) Chain propagation, (*c*) Chain termination. Mechanism of halogenation is explained by considering the chlorination of methane.

(*a*) Chain Initiation. When a mixture of CH_4 and Cl_2 is heated or subjected to diffused sunlight, Cl_2 absorbs energy and undergoes homolytic fission producing chlorine free radicals. One molecule gives rise to two radicals.

Step (1) 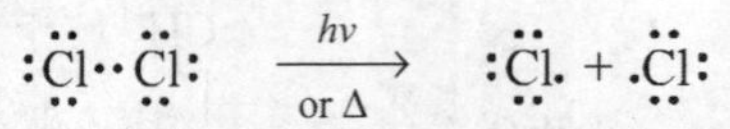

Chlorine free radicals

***(b)* Chain propagation.** The chlorine free radical produced above collides with a molecule of methane forming hydrogen chloride and a methyl free-radical. The methyl free-radical in turn reacts with a molecule of chlorine forming methyl chloride and chlorine free-radical. The newly formed chlorine radical can react with another molecule of methane as in step (2) generating methyl free-radical and hydrogen chloride. The methyl free-radical can again repeat step (3) and so on. Thus, the sequence of reactions in steps (2) and (3) is repeated over and over again and thus the chain is propagated. In other words, a single photon of light initially absorbed by chlorine can bring about the conversion of a large number of molecules of methane into methyl chloride.

Step (2) 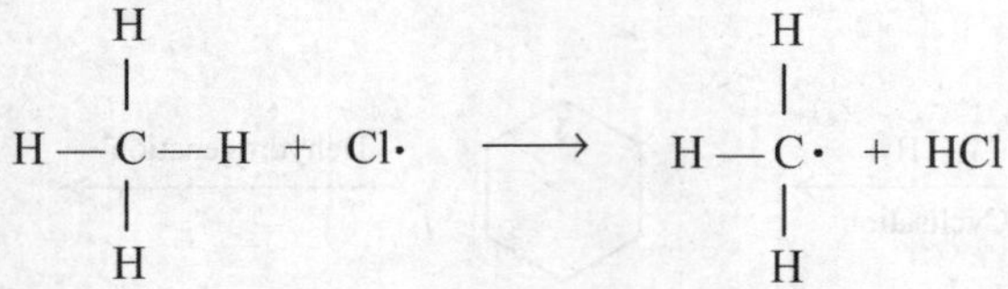

Methyl free-radical

Step (3) 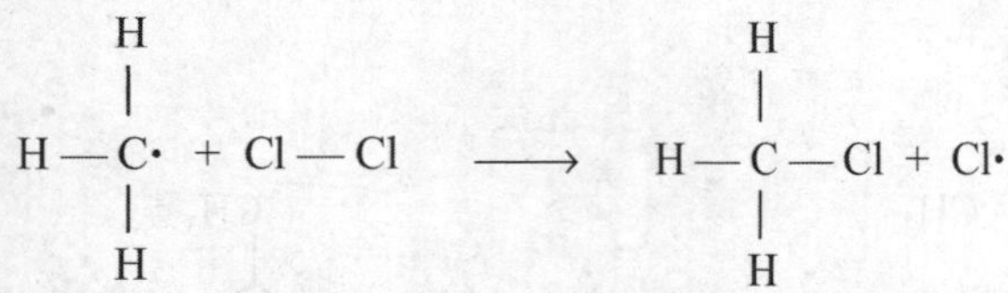

However, the above reaction does not stop at methyl chloride stage but proceeds further till all the H-atoms of methane are replaced by chlorine atoms giving a mixture of mono-, di-, tri- and tetra-chloromethane

Step (4) 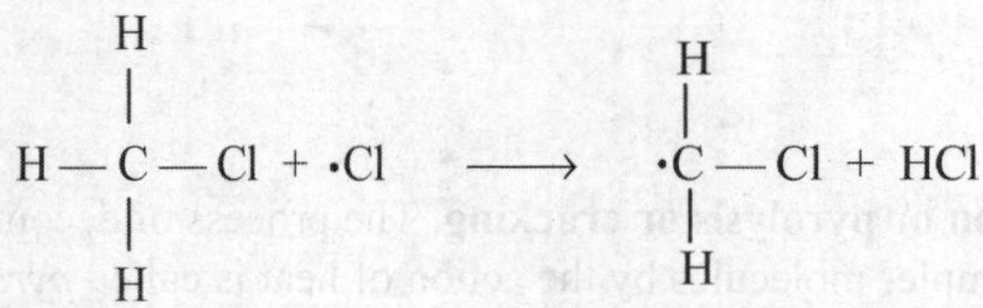

Step (5) 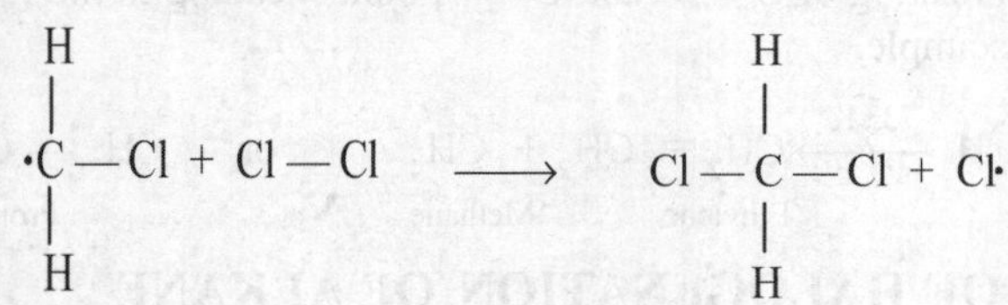

Methyledine chloride

Step (6) 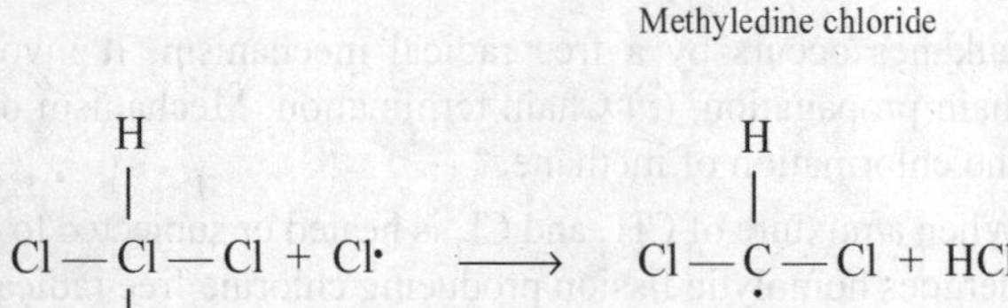

Step (7) 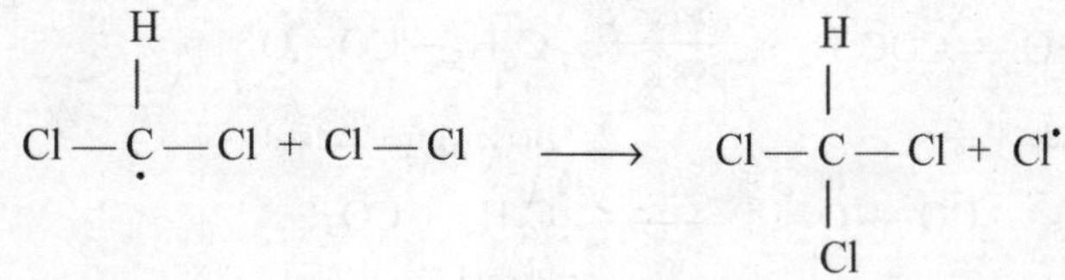

Chloroform

Step (8) 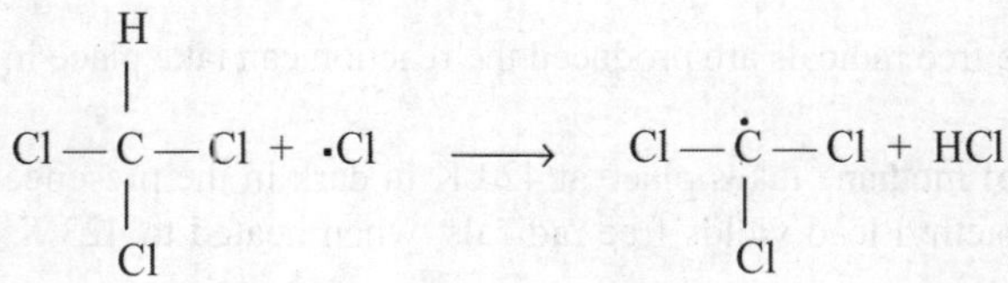

Step (9) 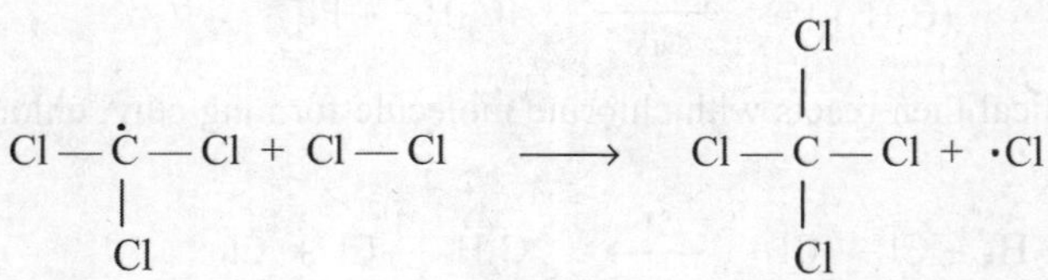

Carbon tetrachloride

Chain Termination

The chain reactions mentioned above, however, come to an end if the free radicals combine amongst themselves to form neutral molecules. Some of the chain terminating steps are:

$$\bullet Cl + \bullet Cl \longrightarrow Cl - Cl$$

$$\bullet CH_3 + \bullet Cl \longrightarrow CH_3 - Cl$$

$$\bullet CH_3 + \bullet CH_3 \longrightarrow CH_3 - CH_3$$

Evidence in support of free radical mechanism of halogenation of alkanes

The following points support the free radical mechanism:

(*i*) Reaction does not take place in dark at room temperature but requires energy in the form of heat or light. This is due to the fact that the chain initiation step (1) is endothermic and hence needs a large amount of energy to break the Cl — Cl bond into radicals.

(*ii*) The reaction has a high quantum yield *i.e.*, many thousand molecules of methyl chloride (or alkyl halide in general) are formed for each photon of light absorbed. This fact can be explained on the basis of chain propagation step (2) and (3).

(*iii*) Oxygen acts as an inhibitor. This is due to the fact oxygen combines with the alkyl free radical to form peroxyalkyl radical, (R — O — O•). This radical is much less reactive than alkyl free radical (R•) to continue the chain. As a result, the halogenation of alkyl in the presence of oxygen is slowed down or stopped. Thus the role of inhibitors like oxygen in this reaction gives support to the above mechanism.

(*iv*) If the above mechanism actually involves free radicals as reactive intermediate, then the addition of substances which are source of free radicals should initiate the reaction even in the dark at room temperature or much below 523 K. This has actually been found to be so. Thus chlorination of methane can be carried out in the dark at room temperature in the presence of small amount of benzoyl peroxide.

$$\underset{\text{Benzoyl peroxide}}{C_6H_5CO - O - O - COC_6H_5} \xrightarrow[298\text{ K}]{\text{dark}} \underset{\text{Benzoyl free radical}}{C_6H_5 - CO - \ddot{\underset{\cdot\cdot}{O}}\cdot}$$

$$C_6H_5 - CO - \ddot{\underset{\cdot\cdot}{O}}\cdot \longrightarrow \underset{\text{Phenyl free radical}}{C_6H_5^{\bullet} + CO_2}$$

$$C_6H_5^{\bullet} + Cl - Cl \longrightarrow C_6H_5 - Cl + Cl\cdot$$

Once, the chlorine free radicals are produced the reaction can take place in the manner explained above.

(*v*) Chlorination of methane takes place at 423 K in dark in the presence of a little $(C_2H_5)_4$ Pb (Tetraethyl lead). Tetraethyl lead yields free radicals, when heated to 423 K.

$$(C_2H_5)_4Pb \xrightarrow[\text{dark}]{423\text{ K}} 4C_2H_5^{\bullet} + Pb$$

The ethyl free radical then reacts with chlorine molecule forming ethyl chloride and chlorine free radical.

$$\cdot C_2H_5 + Cl - Cl \longrightarrow C_2H_5 - Cl + Cl\cdot$$

The chlorine free radical thus formed brings about chlorination of methane as explained above.

Chlorination and bromination of propane, butane and isobutane

Halogenation of an alkane containing more than one type (primary, secondary, tertiary) of hydrogens, gives a mixture of isomeric products. For example, chlorination of propane, butane and isobutane give the following products in the proportion given below each.

(*i*) $$CH_3 - CH_2 - CH_3 \xrightarrow[\text{light, 298K}]{Cl_2} \underset{\substack{n\text{–Propyl chloride}\\(45\%)}}{CH_3 - CH_2 - CH_2 - Cl} + \underset{\substack{\text{Isopropyl chloride}\\(55\%)}}{CH_3 - \overset{\overset{\displaystyle Cl}{|}}{CH} - CH_3}$$

(*ii*) $$CH_3 - CH_2 - CH_2 - CH_3 \xrightarrow[\text{light, 298K}]{Cl_2} \underset{\substack{n\text{-Butyl chloride}\\(28\%)}}{CH_3 - CH_2 - CH_2 - CH_2 - Cl} + \underset{\substack{sec\text{–Butyl chloride}\\(72\%)}}{CH_3 - CH_2 - \overset{\overset{\displaystyle Cl}{|}}{CH} - CH_3}$$

(*iii*) $$\underset{\text{Isobutane}}{CH_3 - \overset{\overset{\displaystyle CH_3}{|}}{CH} - CH_3} \xrightarrow{Cl_2} \underset{\substack{\text{Isobutyl chloride}\\(64\%)}}{CH_3 - \overset{\overset{\displaystyle CH_3}{|}}{CH} - CH_2 - Cl} + \underset{\substack{\text{tert–Butyl chloride}\\(36\%)}}{CH_3 - \underset{\underset{\displaystyle CH_3}{|}}{\overset{\overset{\displaystyle CH_3}{|}}{C}} - Cl}$$

Bromination gives the corresponding bromides but in different proportions:

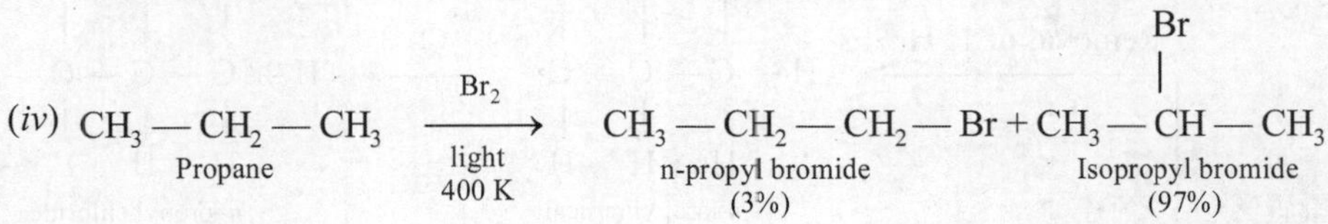

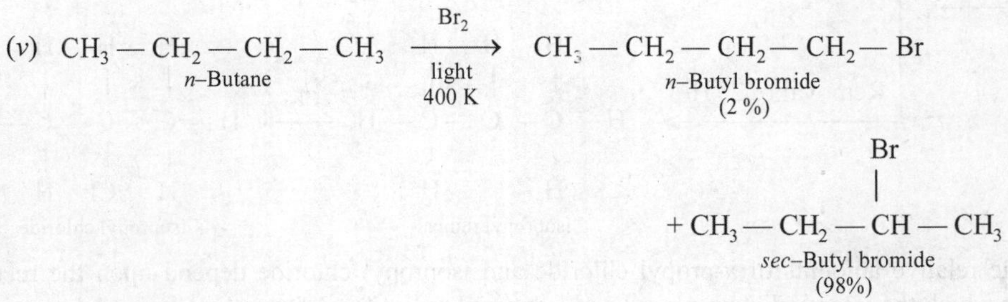

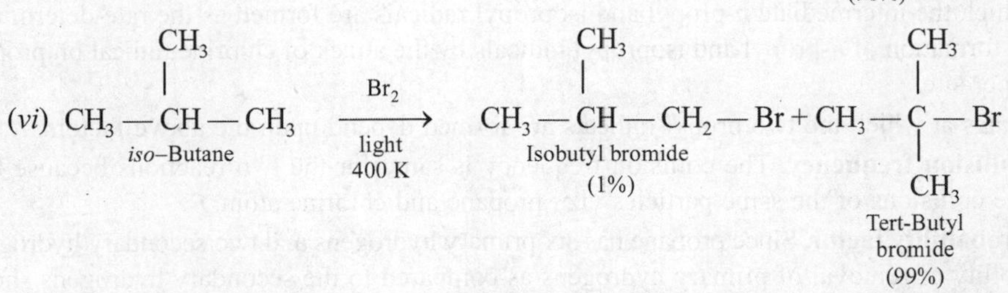

The result given above shows that the relative amounts of the different isomeric products differ largely depending upon the halogen used. It is also important to note that the bromination, in constrast to chlorination, leads to the formation of only one of the possible isomeric products. This is reflected in the percentages like 97%, 98% and 99% for one of the products in each reaction. Thus bromine atom is more selective in the site of attack than chlorine.

4.11 FACTORS THAT INFLUENCE THE ORIENTATION OF HALOGENATION OF ALKANES

Factors that influence the halogenation process are:

(*i*) Collision frequency

(*ii*) Probability factor

(*iii*) Energy factor

Let us take the example of chlorination of propane.

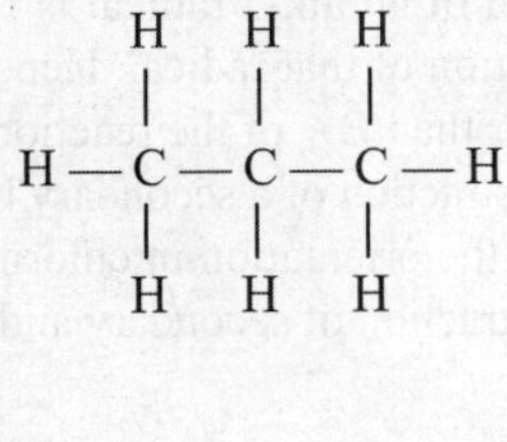

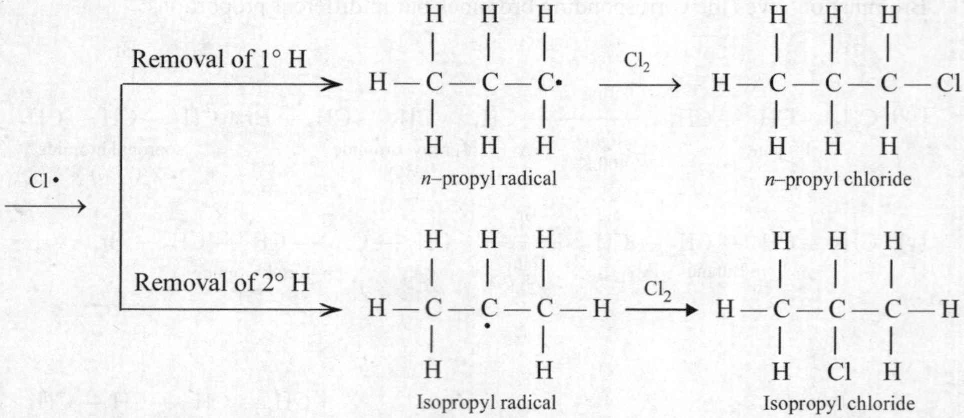

The relative amounts of *n*-propyl chloride and isopropyl chloride depend upon the relative rates at which the intermediate *n*-propyl and isopropyl radicals are formed as the rate-determining step is the formation of *n*-propyl and isopropyl radicals by the attack of chlorine radical on propane at the proper site.

The rates at which the two propyl radicals are formed depend upon the above factors.

(*i*) Collision frequency. The collision frequency is same for the two reactions because both involve the collisions of the same particles viz., propane and chlorine atom.

(*ii*) Probability factor. Since propane has six primary hydrogens and two secondary hydrogens, the probability of removal of primary hydrogens as compared to the secondary hydrogens should be in the ratio of 6 : 2 or 3 : 1.

If we consider the probability factor and collision frequency only, the chlorination of propane would be expected to yield *n*-propyl chloride and isopropyl chloride in the ratio of 3 : 1. But in actual practice, the two chloride are formed roughly in equal amounts *i.e.*, the ratio is 1 : 1 or 3 : 3 (see the experimental data).

(*iii*) Energy factor. The probability of formation of isopropyl chloride is about three times as large as that of formation of *n*-propyl chloride. This means that collisions with secondary hydrogens are about three times more effective than the collisions with primary hydrogens. This implies that isopropyl radicals are formed more easily than *n*-propyl radicals in the above reactions. We know that isopropyl radical is secondary radical while *n*-propyl radical is a primary radical and the decreasing order of stability is:

Tert-radical > sec-radical > primary radical

It is, therefore, clear that isopropyl chloride would be more stable (and hence easily formed) than *n*-propyl chloride. Now the stability or the ease of formation of an alkyl radical is related to the E_{act} (energy of activation) of the reaction leading to the formation of that radical. Hence E_{act} of the reaction leading to the formation of isopropyl radical is smaller than E_{act} of the reaction leading to the formation of *n*-propyl radical. In other words, E_{act} for the abstraction of a secondary hydrogen is less than E_{act} for the abstraction of a primary hydrogen. Hence the orientation in chlorination of propane is determined by the energy factor, *i.e.*, the E_{act} for the abstraction of secondary and primary hydrogens in propane.

Chlorination of isobutane presents a similar problem. Here there are nine primary hydrogens and one tertiary hydrogen. The abstraction of one of the nine primary hydrogens leads to the formation of isobutyl chloride while the abstraction of single tertiary hydrogen leads to the formation of tert-butyl chloride. We would, therefore, expect that the probability factor favours the formation of *iso*-butyl chloride (where primary hydrogen is replaced by chlorine) by the ratio of 9 : 1. But isobutyl chloride and tert-butyl chloride are formed roughly in the ratio of 2 : 1 or 9 : 4.5. Naturally, about 4.5 times as many collisions with tertiary hydrogens are effective as with primary hydrogens.

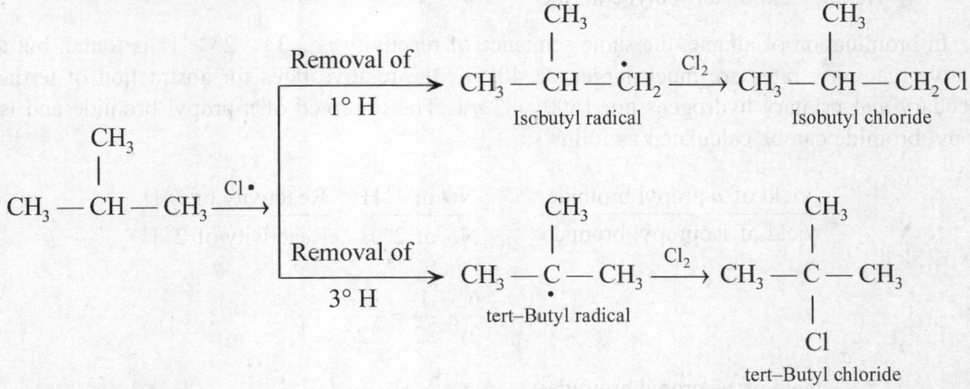

This means that E_{act} for the abstraction of a tertiary hydrogen (3° H) is less than E_{act} for the abstraction of a primary hydrogen (1° H) and, in fact, even less than for the abstraction of secondary hydrogen (2° H).

Study of chlorination of many alkanes have shown that at room temperature, the relative rates of abstraction of tertiary, secondary and primary hydrogens are in the ratio 5.0 : 3.8 : 1.0. With the help of these values, we can precict fairly well the ratio of isomeric chlorination products obtained from a given alkane. The expected ratio of *n*-propyl chloride and isopropyl chloride in the chlorination of propane can be calculated as follows:

$$\frac{\text{yield of } n\text{–propyl chloride}}{\text{yield of isopropyl chloride}} = \frac{\text{No. of } 1^\circ \text{ H}}{\text{No. of } 2^\circ \text{ H}} \times \frac{\text{Reactivity of } 1^\circ \text{ H}}{\text{Reactivity of } 2^\circ \text{ H}}$$

$$= \frac{6}{2} \times \frac{1}{3.8} = \frac{6}{7.6}$$

or $$\frac{\text{yield of isopropyl chloride}}{\text{yield of } n\text{–propyl chloride}} = \frac{7.6}{6} = 1.26$$

Thus yield is isopropyl chloride = 1.26 × yield of *n*-propyl chloride.

$$\frac{\text{Actual yield of isopropyl chloride}}{\text{Actual yield of } n\text{–propyl chloride}} = \frac{55}{45} = 1.22$$

Similarly, in case of isobutane, we get the following result:

$$\frac{\text{yield of isobutyl chloride}}{\text{yield of tert-butyl chloride}} = \frac{\text{No. of } 1^\circ \text{ H}}{\text{No. of } 3^\circ \text{ H}} \times \frac{\text{Reactivity of } 1^\circ \text{ H}}{\text{Reactivity of } 3^\circ \text{ H}}$$

$$= \frac{9}{1} \times \frac{1}{5} = \frac{9}{5}$$

$$\frac{\text{yield of isobutyl chloride}}{\text{yield of tert-butyl chloride}} = \frac{9}{5} = 1.8$$

Thus yield of isobutyl chloride = 1.80 × yield of tert-butyl chloride

$$\frac{\text{Actual yield of isobutyl chloride}}{\text{Actual yield of tert-butyl chloride}} = \frac{16}{9} = 1.78$$

In bromination of alkane, the same sequence of reactivity *i.e.*, 3° > 2° > 1° is found, but the relative reactivity rates are much larger. At 400 K, the relative rates for abstraction of tertiary. Secodary and primary hydrogens are 1600 : 82 : 1. The observed of *n*-propyl bromide and iso-propyl bromide can be calculated as follows:

$$\frac{\text{yield of } n\text{-propyl bromide}}{\text{yield of isopropyl bromide}} = \frac{\text{No. of 1° H}}{\text{No. of 2° H}} \times \frac{\text{Reactivity of 1° H}}{\text{Reactivity of 2° H}}$$

$$= \frac{6}{2} \times \frac{1}{82} = \frac{3}{82}$$

or
$$\frac{\text{yield of } n\text{-propyl bromide}}{\text{yield of iso-propyl bromide}} = \frac{3}{82}$$

Similarly,

$$\frac{\text{yield of iso-butyl bromide}}{\text{yield tert-butyl bromide}} = \frac{\text{No. of 1° H}}{\text{No. of 3° H}} \times \frac{\text{Reactivity of 1° H}}{\text{Reactivity of 3° H}}$$

$$= \frac{9}{1} \times \frac{1}{1600} = \frac{9}{1600}.$$

***Example 5.* An alkane with molecular weight 72 formed only one monochloro substitution product. Suggest a structure for the alkane.**

Solution.

Molecular weight 72 suggests that the alkane has the molecular formula C_5H_{12}. Out of three isomeric pentanes viz., *n*-pentane, isopentane and neopentane the last one is the answer of the question. This is because only neopentane has one type of hydrogen (primary) and hence would give one monochloro substitutional product.

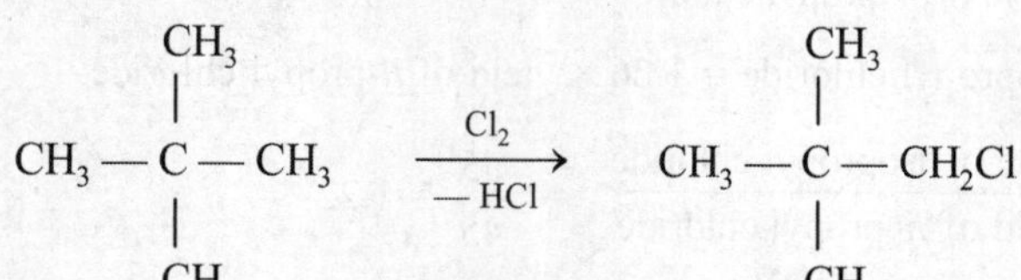

Neopentane (2, 2-Dimethylpropane) — Neopentyl chloride

***Example 6.* Give the structure of pentane that would be expected to produce the largest number of isomeric monochloro derivatives.**

Solution.

There are three isomers of pentanes.

$CH_3—CH_2—CH_2—CH_2—CH_3$
n–pentane

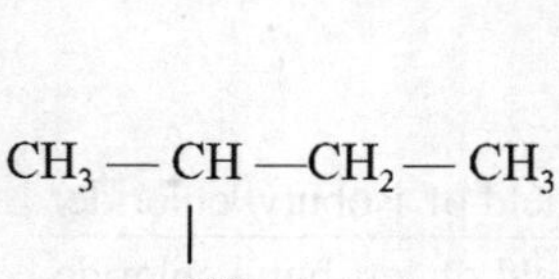

Isopentane

$$H_3C-\underset{\displaystyle CH_3}{\overset{\displaystyle CH_3}{\overset{|}{\underset{|}{C}}}}-CH_3 \quad \text{(neopentane)}$$

Out of the three isomers, isopentane would give largest number of isomeric derivatives as it contains three types of replaceable hydrogen (primary, secondary and tertiary).

***Example 7*. Calculate the percentage of expected isomers during monobromination of *n*-butane. The relative rates of substitution per 3°, 2° and 1° hydrogen are 1600 : 82 : 1.**

Solution.

$$\underset{n\text{-butane}}{CH_3-CH_2-CH_2-CH_3} \xrightarrow{Br_2}$$

$$\underset{n\text{-butyl bromide}}{CH_3-CH_2-CH_2-CH_2Br} + \underset{sec\text{-butyl biomide}}{CH_3\,CH_2\,CH\,Br\,CH_3}$$

$$\frac{\text{Yield of } n\text{-butyl bromide}}{\text{Yield of } sec\text{-butyl bromide}} = \frac{\text{No. of } 1^o \text{ H} \times \text{Reactivity of } 1^o \text{ H}}{\text{No. of } 2^o \text{ H} \times \text{Reactivity of } 2^o \text{ H}}$$

$$= \frac{6 \times 1}{4 \times 82} = \frac{6}{328}$$

$$\%\text{ of } n\text{-butyl bromide} = \frac{6}{334} \times 100 = 1.79\%$$

$$\%\text{ of sec-butyl bromide} = 100 - 1.79 = 98.21\%.$$

***Example 8*. Predict the proportions of isomeric products from chlorination at room temperature of (*i*) *n*-Butane (*ii*) 2, 3-dimethyl butane.**

***Solution.* (*i*) *n*-Butane**

$$CH_3-CH_2-CH_2-CH_3 \xrightarrow{Cl_2} \underset{n\text{–Butyl chloride}}{CH_3-CH_2-CH_2-CH_2Cl} + \underset{\text{sec–Butyl chloride}}{CH_3-CH_2-\overset{\displaystyle Cl}{\overset{|}{CH}}-CH_3}$$

$$\frac{\text{yield of } n\text{-Butyl chloride}}{\text{yield of sec-Butyl chloride}} = \frac{\text{No. of } 1^o \text{ H}}{\text{No. of } 2^o \text{ H}} \times \frac{\text{Reactivity of } 1^o \text{ H}}{\text{Reactivity of } 2^o \text{ H}}$$

$$= \frac{6}{4} \times \frac{1}{3.8} = \frac{3}{7.6} = \frac{28\%}{72\%}$$

(*ii*) 2, 3-dimethylbutane

$$H_3C-\overset{\displaystyle CH_3}{\overset{|}{HC}}-\overset{\displaystyle CH_3}{\overset{|}{HC}}-CH_3 + Cl_2 \longrightarrow \underset{\text{1-Chloro-2, 3-dimethylbutane}}{CH_3-\overset{\displaystyle CH_3}{\overset{|}{CH}}-\overset{\displaystyle CH_3}{\overset{|}{CH}}-CH_2Cl} \qquad \text{(I)}$$

$$+ CH_3 - \underset{}{CH}(CH_3) - CH(Cl) - CH_3 \qquad \text{(II)}$$

2-Chloro-3-methylbutane

$$\frac{\text{yield of I (Primary)}}{\text{yield of II (Sec.)}} = \frac{\text{No. of } 1^\circ \text{ H}}{\text{No. of } 3^\circ \text{ H}} \times \frac{\text{Reactivity of } 1^\circ \text{ H}}{\text{Reactivity of } 3^\circ \text{ H}}$$

$$= \frac{12}{2} \times \frac{1}{5} = \frac{6}{5} = \frac{55\%}{45\%}$$

Example 9. **Chlorination reaction of certain higher alkanes can be used for laboratory preparations. For example, we can prepare neopentyl chloride from neopentane and cyclopentyl chloride from cyclopentane. How do you account for this fact?**

Solution.

Chlorination of neopentane $\left[CH_3 - C(CH_3)_2 - CH_3 \right]$ and cyclopentane 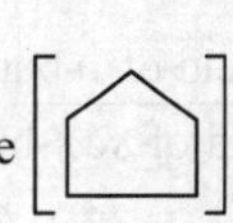 can be carried out because all the hydrogens of each compound are equivalent and replacement of any one yields the same product, *i.e.* neopentyl chloride $\left[CH_3 - C(CH_3)_2 - CH_2Cl \right]$ and cyclopentyl chloride 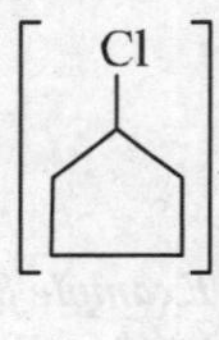 respectively.

Example 10. **Monochlorination of an equimolar mixture of methane and ethane does not yield a mixture of methyl chloride and ethyl chloride in equimolar proportions. How do you account for this observation?**

Solution.

Monochlorination of an equimolar mixture of methane and ethane yields a mixture of methyl chloride and ethyl chloride in which ethyl chloride is formed about 400 times as much as methyl chloride. This can be attributed to two factors as discussed below:

(1) Probability factor. Ethane contains six hydrogen atoms each of which can be replaced by chlorine while methane contains only four hydrogen atoms. Therefore the probabilities of formation of ethyl chloride and methyl chloride will be in the ratio 6 : 4 or 3 : 2.

(2) Relative reactivities of different classes of hydrogen atoms. This is the most important factor which determines the relative amount of products in the halogenation of methane and ethane (or other alkanes). This in turn depends upon the relative stabilities of alkyl radicals formed in the step which controls the overall process of halogenation. In case of methane and ethane, this step may be written as:

$$CH_4 + \dot{C}l \longrightarrow \dot{C}H_3 + HCl$$

$$CH_3 - CH_3 + Cl\cdot \longrightarrow CH_3 - \dot{C}H_2 + HCl$$

Now the stability of ethyl radical (relative to ethane) is more than the stability of methyl radical (relative to methane). Therefore the transition state involved in the formation of ethyl radical from ethane would be more stable than the transition state involved in the formation of methyl radical from methane. As such transition state of ethyl radical requires lesser E_{act} than transition state of methyl radical and therefore the former is formed more readily than the latter (Fig. 4.2). This means that carbon hydrogen bond in ethane breaks more readily than carbon hydrogen bond in methane or the primary hydrogen atoms in ethane are more reactive than hydrogen atoms in methane. As a consequence ethane undergoes chlorination much more readily than methane so that the mixture formed contains a very high proportion of ethyl chloride as compared to methyl chloride.

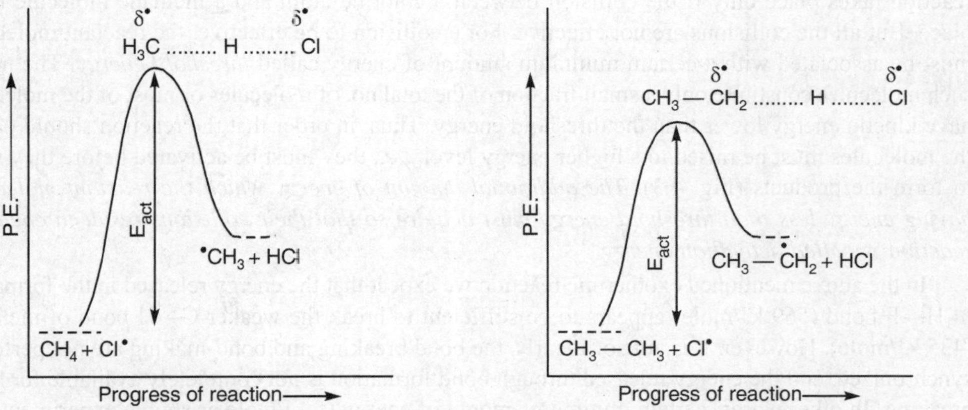

Greater E_{act} of transition state, lesser ease of formation of free radical, lesser reactivity of hydrogens.

Lesser E_{act} of transition state, greater ease of formation of free radical, greater reactivity of hydrogens.

Fig. 4.2. Illustration of relative reactivities of hydrogens of methane and primary hydrogens of ethane.

Example 11. **On chlorination, an equimolar mixture of neopentane and ethane yields neopentyl chloride and ethyl chloride in the ratio of 2.3 : 1. How does the reactivity of a primary hydrogen in neopentane compare with that of primary hydrogen in ethane?**

Solution. No of primary hydrogen atoms in ethane = 6

No. of primary hydrogen atoms in neopentane

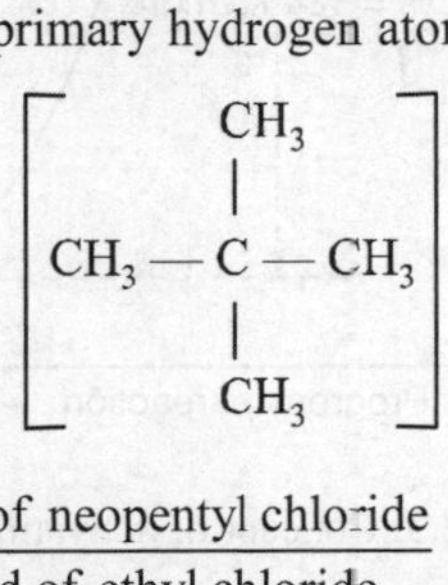

= 12

$$\frac{\text{yield of neopentyl chloride}}{\text{yield of ethyl chloride}} = \frac{\text{No. of } 1^{\circ}\text{ H} \times \text{Reactivity of } 1^{\circ}\text{ H in neopentane}}{\text{No. of } 1^{\circ}\text{ H} \times \text{Reactivity of } 1^{\circ}\text{ H in ethane}}$$

$$\frac{2.3}{1} = \frac{12 \times \text{Reactivity in neopentane}}{6 \times \text{Reactivity in ethane}}$$

or $\dfrac{\text{Reactivity in neopentane}}{\text{Reactivity in ethane}} = \dfrac{2.3\times 6}{12} = \dfrac{1.15}{1}$

or 1.15 : 1.

4.12 ACTIVATION ENERGY

Consider the reaction

$$F^{\bullet} + CH_3 - H \longrightarrow \bullet CH_3 + H - F \quad \Delta H = -134 \text{ kJ}$$
$$\Delta H^{\circ} = 435 \text{ kJ} \qquad 569 \text{ kJ} \qquad E_{act} = 5 \text{ kJ}$$

For a reaction to occur, collision between reactant molecules is a must. Thus the above mentioned reaction takes place only if the collision between a chlorine atom and a methane molecule takes place. But all the collisions are not effective. For a collision to be effective, the reactant molecules must be associated with a certain minimum amount of energy called *threshold energy*. The no. of such molecules constitute only a small fraction of the total no. of molecules or most of the molecules have kinetic energy lower than the threshold energy. Thus, in order that the reaction should occur, the molecules must be raised to a higher energy level, *i.e.*, they must be activated before they react to form the products (Fig. 4.3). *The additional amount of energy which the reactant molecules having energy less than threshold energy must acquire so that their collisions result in chemical reaction is called activation energy.*

In the above mentioned exothermic reaction we expect that the energy released in the formation of H—F bond (569 kJ/mole) appears to be sufficient to break the weaker C—H bond of methane (435 kJ/mole). However, it is not so. Clearly, the bond breaking and bond-making are not perfectly synchronized, and the energy released through bond formation is not completely available for bond breaking. In other words, certain minimum amount of energy (5 kJ/mole as shown experimentally) must be supplied to initiate the reaction. This is the *activation energy* for the reaction.

There is another aspect which must be kept in mind. In addition to the adequate energy (kinetic energy) which the colliding molecules possess at the time of collision, they must be properly oriented before the collision can be really effective in casuing the reaction. The number of properly oriented collisions in a reaction is quite small as compared to the total number of collisions.

Thus in the above mentioned example, the methane molecule must be oriented in such a way that it presents a hydrogen atom to the full force of the impact. In this case, only about one collision in eight is properly oriented.

In general, in collisions, sufficient energy (E_{act}) and proper orientation are essential for a reaction to take place.

As the reaction proceeds, changes in the potential energy takes place and these energy changes are shown in Fig. 4.3.

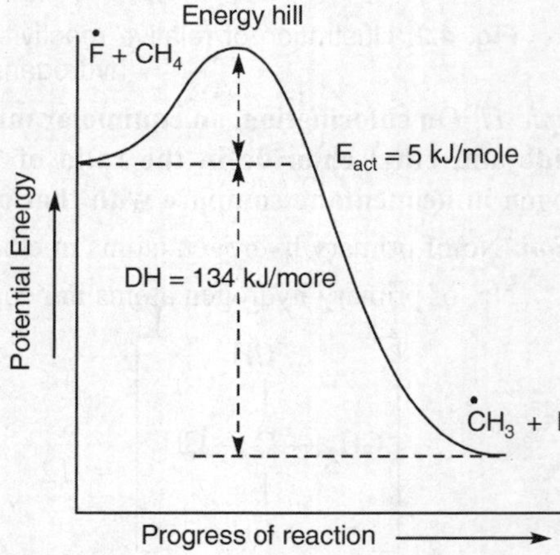

Fig. 4.3. Concept of activation energy

The potential energy of methane molecule and fluorine atom is shown on the left and that of the methyl free radical (CH_3) and hydrogen fluoride on the right (Fig. 4.3). As the reaction begins, the kinetic energy of the reactants (due to their motion) is changed into potential energy.

With the increase in potential energy, we move up the energy hill and move down the other side. During the descent, the potential is changed back into kinetic energy till we reach the level of the products ($CH_3\bullet$ and HF). As the potential energy of the product is less than that of the reactants; there must be a corresponding increase in kinetic energy. In other words, the molecules of the products ($\bullet CH_3$ and HF) move faster than the molecules of the reactants (CH_4 and $\bullet F$) so there would be a rise in temperature or heat will be evolved. The height of the hill top or energy hill (in kJ/mole) above the level of the reactant is the energy of activation (5 kJ/mole).

4.13 RELATIVE REACTIVITIES OF HALOGENS IN THE HALOGENATION OF ALKANES

The magnitude of E_{act} determines the rate of a reaction. Reactions with low E_{act} proceed at faster rate than the one with a high E_{act} at the same temperature. This relationship between the rate of a reaction and the magnitude of the energy of activation can be explained with the help of halogenation of methane as follows:

$$CH_4 + X_2 \xrightarrow[(X_2 = F_2, Cl_2, Br_2, I_2)]{} CH_3 - X + HX$$

The rate determining step here is the abstraction of hydrogen by halogen atom *i.e.*, how fast methyl radical is formed.

$$CH_4 + X\bullet \longrightarrow CH_3\bullet + HX$$

Formation of CH_3 radical in chain propagation step is difficult but once formed it gets readily converted into the alkyl halide.

Thus, the overall rate of halogenation of methane depends upon how fast the methyl radical is formed which, in turn, depends on the E_{act} for the formation of methyl radical. Thus,

$$\underset{435\text{ kJ}}{F\bullet + CH_3 - H} \longrightarrow \underset{569\text{ kJ}}{H - F + \bullet CH_3} \quad \Delta H = -134\text{ kJ}, \; E_{act} = 5\text{ kJ}$$

$$\underset{435\text{ kJ}}{Cl\bullet + CH_3 - H} \longrightarrow \underset{431\text{ kJ}}{H - Cl + \bullet CH_3} \quad \Delta H = +4\text{ kJ}, \; E_{act} = 17\text{ kJ}$$

$$\underset{435\text{ kJ}}{Br\bullet + CH_3 - H} \longrightarrow \underset{368\text{ kJ}}{H - Br + \bullet CH_3} \quad \Delta H = +67\text{ kJ}, \; E_{act} = 78\text{ kJ}$$

$$\underset{435\text{ kJ}}{I\bullet + CH_3 - H} \longrightarrow \underset{297\text{ kJ}}{H - I + \bullet CH_3} \quad \Delta H = +138\text{ kJ}, \; E_{act} = 138\text{ kJ}$$

Potential energy diagrams for the above mentioned reactions are given in Fig. 4.4

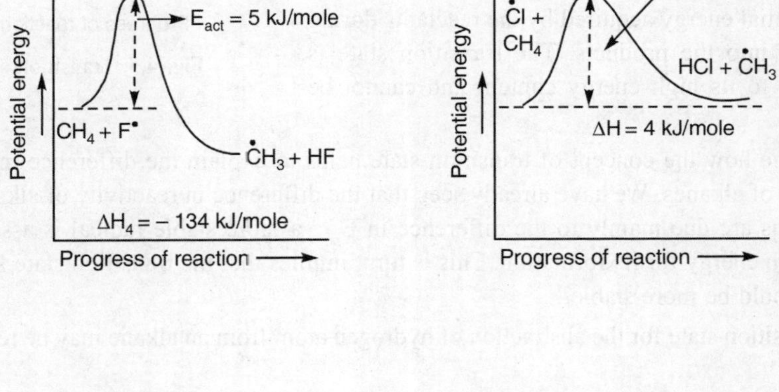

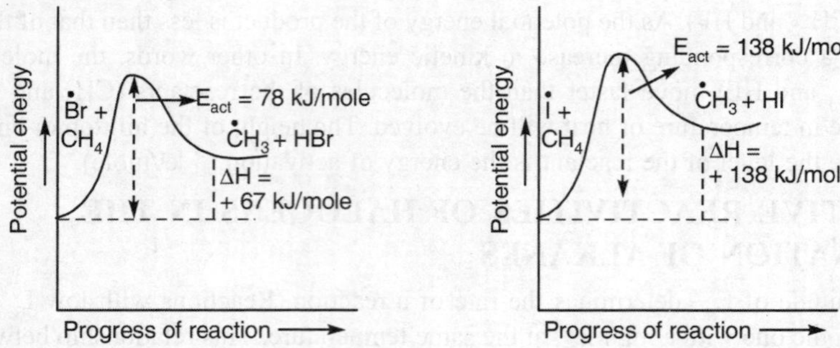

Fig. 4.4. Potential energy diagrams.

It is clear the decreasing order of E_{act} for chain propagation step is:

Iodination > Bromination > Chlorination > Fluorination

Hence order of halogenation should be:

Fluorination > Chlorination > Bromination > Iodination

This order is in conformity with the observed order of reactivity of F_2, Cl_2, Br_2 and I_2 in halogenation of methane in particular and alkanes in general.

4.14 EFFECT OF ACTIVATION ENERGY ON THE STRUCTURE OF TRANSITION STATE IN HALOGENATION OF ALKANES

For a collision to be effective collision, the reactant molecules must be associated with a certain amount of energy called threshold energy. Thus, they also have to cross an energy barrier (hill top) between the reactants and products before reaction can be accomplished. The top of the energy hill (energy barrier) represents an intermediate structure called the *transition state.* The transition state corresponds to a particular arrangement of atoms of reacting species as they are converted from reactants into product. It represents a configuration in which old bonds are partially broken and new bonds are partially formed. *The activation energy* (E_{act}) *is the difference in energy content between the transition state and reactants,* as shown in Fig. 4.5.

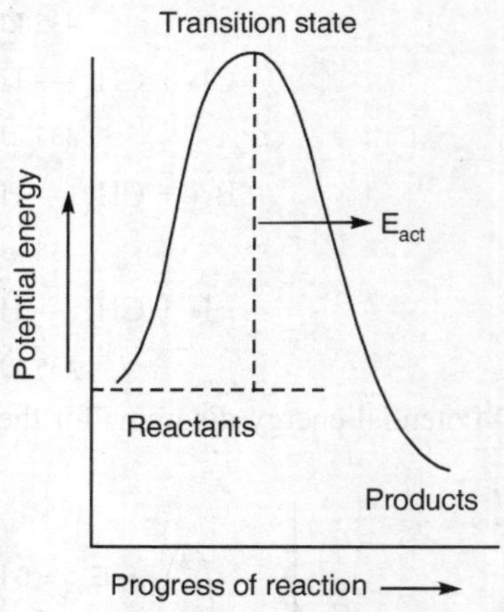

Fig. 4.5. Transition state.

Hence the transition state of a reaction is the state of highest potential energy acquired by the reactants during their change into the products. The transition state is unstable due to its high energy content and cannot be isolated.

Let us see how the concept of transition state helps to explain the difference in the rates of halogenation of alkanes. We have already seen that the difference in reactivity of alkanes towards halogen atoms are due mainly to the difference in E_{act}; a more stable radical is associated with low activation energy for its formation. This is turn, implies that the transition state leading to its formation would be more stable.

The transition state for the abstraction of hydrogen atom from an alkane may be represented as follows:

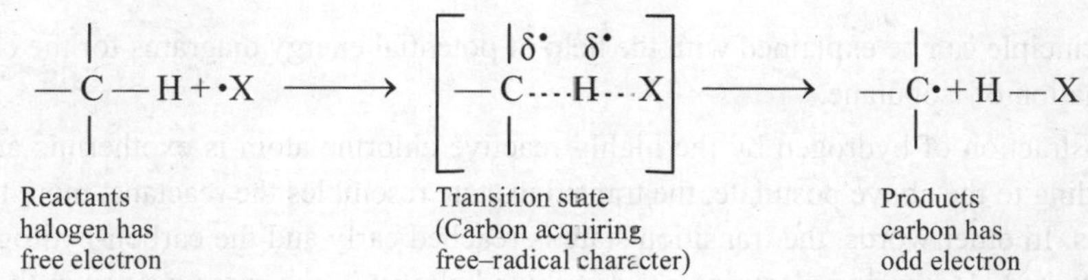

The C—H bond is partly broken and H—X bond is partly formed (Fig. 4.6). Depending upon the extent to which the C—H bond is broken, the alkyl group is associated with the character of free radical it will form. Factors that stabilise the resulting free radical (delocalisation of the odd electron) tend to stabilize the nascent free radical in the transition state.

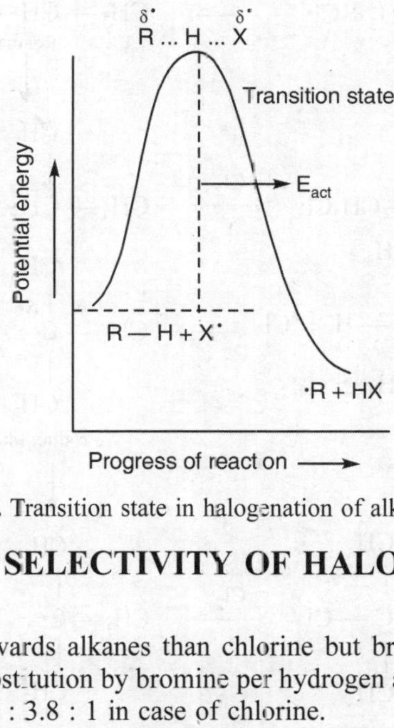

Fig. 4.6. Transition state in halogenation of alkane.

4.15 REACTIVITY AND SELECTIVITY OF HALOGENS TOWARDS ALKANES

Bromine is less reactive towards alkanes than chlorine but bromine is more selective than chlorine. The relative rates of substitution by bromine per hydrogen atom (3°, 2°, 1° hydrogen) are 1600 : 82 : 1 as compared to 5.0 : 3.8 : 1 in case of chlorine.

Thus the reaction of isobutane and bromine, for example, gives mainly tert-butyl bromide.

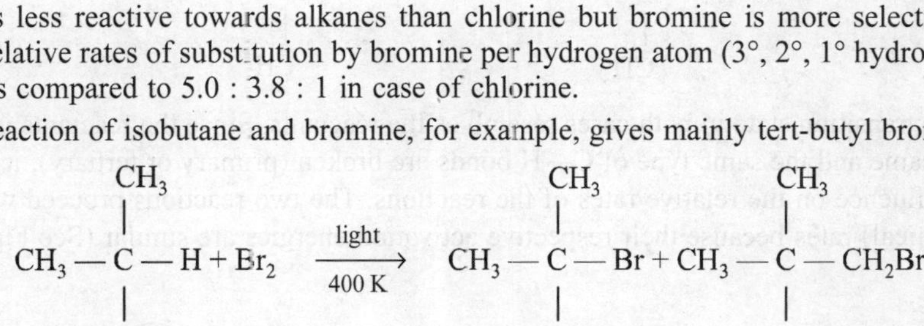

Further, bromine atom is less reactive than chlorine atom. Thus, selectivity is related to reactivity and it can be generalised that the less reactive the reagent the more selective it is in its attack under a set of similar reaction. The greater selectivity of bromine can be explained in terms of transition state theory. According to a postulate made by Professor G.S. Hammond **the structure of transition state of endothermic step (with high value of E_{act}) of a reaction resembles the products of that step more than it does the reactants. The structure of transition state of exothermic state**

with low value of E_{act} of a reaction resembles the reactant of that step more than it does the products.

This principle can be explained with the help of potential energy diagrams for the chlorination and bromination of isobutane.

The abstraction of hydrogen by the highly reactive chlorine atom is exothermic and has low E_{act}. According to the above postulate, the transition state resembles the reactants more than it does the products. In other words, the transition state is reached early and the carbon-hydrogen bond is slightly stretched. Atoms and electrons are distributed almost in the same way as in the reactants; carbon is still tetrahedral. The alkyl radical has developed very little free radical character. The transition state for the hydrogen (1° and 3°) abstraction steps may be shown as follows :

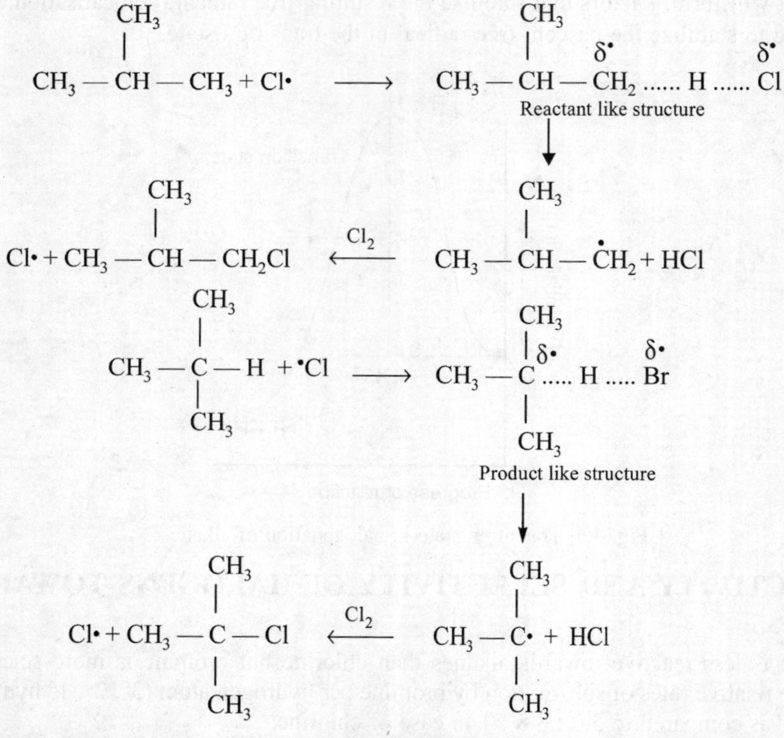

The transition state in both cases resembles the reactants. Since the reactants in both the cases are the same and the same type of C—H bonds are broken (primary or tertiary), it has a relatively small influence on the relative rates of the reactions. The two reactions proceed with similar (but not identical) rates because their respective activation energies are similar (See Fig. 4.7).

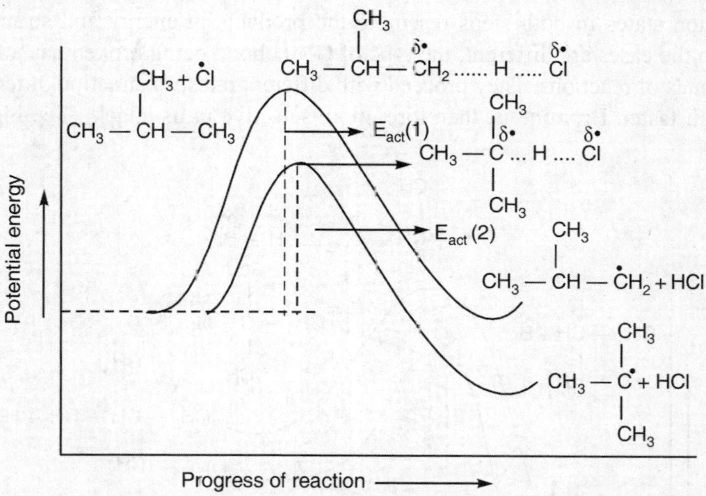

Fig. 4.7. Activation energies in exothermic reactions.

In the above diagram it is clear that E_{act} are similar, but because tertiary C — H bond is broken more easily than primary C—H bond, the reaction (2) has low E_{act} and proceeds comparatively at faster rate.

On the other hand, the abstraction of hydrogen by the less reactive bromine atom is endothermic and has a very high E_{act}. The transition state resembles the products more than it does the reactants. In other words, the transition state is reached late and the carbon-hydrogen bond is broken to a considerable extent. Atoms and electrons are distributed almost in the same way as in the products; the carbon is almost trigonal. The alkyl radical has developed considerable free radical character.

The transition states may be shown in the following ways:

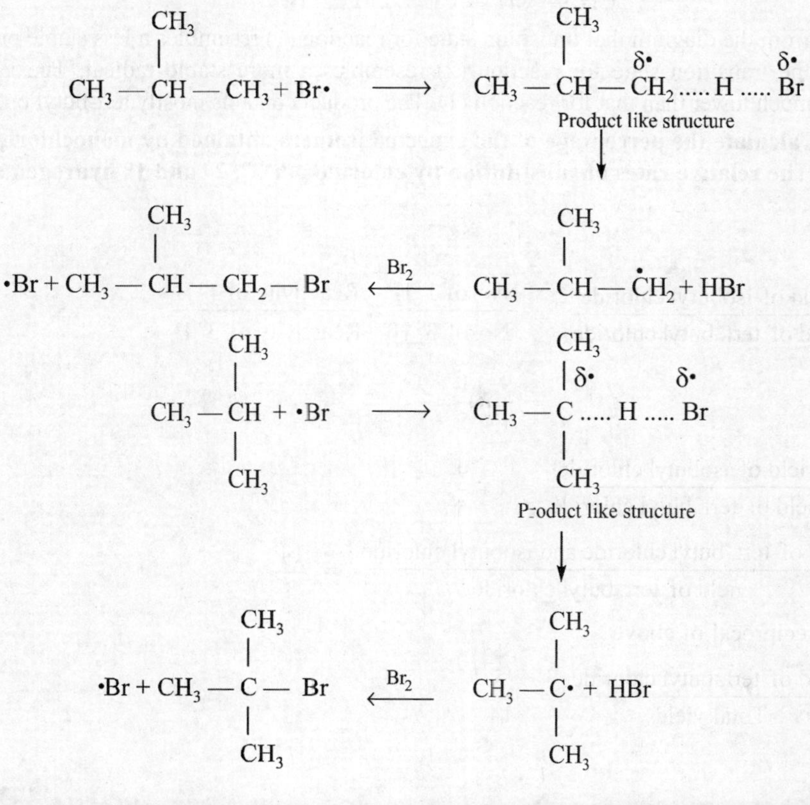

The transition states in both steps resemble the products in energy and structure. Since the products in both the cases are different, the type of C—H bond being broken has a lot of influence on the relative rates of reactions. They proceed with different rates. Abstraction of tertiary hydrogen takes place much faster. Bromine is, therefore, more selective in its attack. (See Fig. 4.8)

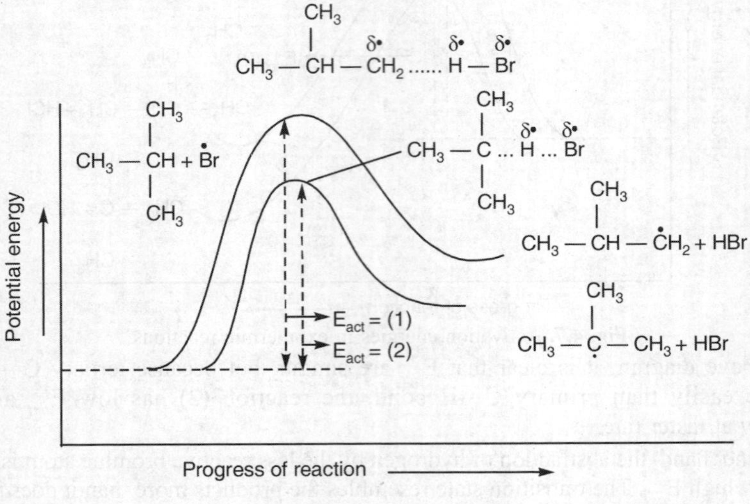

Fig. 4.8. Activation energies in endothermic reactions.

$$\begin{array}{c} CH_3 \\ | \\ CH_3 - CH - \overset{\delta\bullet}{CH_2} \cdots\cdots H \cdots\cdots \overset{\delta\bullet}{Br} \end{array}$$

It is clear from the diagram that transition state for reaction (1) resembles a less stable primary radical while the transition state for reaction (2) resembles a more stable radical. The E_{act} for reaction (2) is much lower than that for reaction (1). The product contains mostly tert-butyl bromide.

Example 12. **Calculate the percentage of the expected isomers obtained by monochlorination of isobutane. The relative rates of substitution by chlorine per 3°, 2° and 1° hydrogen atoms are 5 : 3.8 : 1.**

Solution.

$$\frac{\text{yield of isobutyl chloride}}{\text{yield of tert. butyl chloride}} = \frac{\text{No. of 1° H}}{\text{No. of 3° H}} \times \frac{\text{Reactivity of 1° H}}{\text{Reactivity of 3° H}}$$

$$= \frac{9}{1} \times \frac{1}{5} = \frac{9}{5}$$

$$1 + \frac{\text{yield of isobutyl chloride}}{\text{yield of tert. butyl chloride}} = 1 + \frac{9}{5}$$

$$\frac{\text{yield of tert. butyl chloride and isobutyl chloride}}{\text{yield of tert. butyl chloride}} = \frac{14}{5}$$

Take the reciprocal of above

$$\frac{\text{yield of tert. butyl chloride}}{\text{Total yield}} = \frac{5}{14}$$

$$\% \text{ of tert. butyl chloride} = \frac{5}{14} \times 100 = 35.7$$

$$\% \text{ or isobutyl chloride} = 64.3.$$

Example 13. **Calculate the percentage of expected isomers obtained by monobromination of propane. The relative rates of substitution by bromine per 3°, 2° and 1° hydrogen atoms are 1600 : 82 : 1.**

Solution.

$$\frac{\text{yield of } n\text{-propyl bromide}}{\text{yield of iso-propyl bromide}} = \frac{3}{82}$$

Add 1 to both sides

$$1 + \frac{\text{yield of } n\text{-propyl bromide}}{\text{yield of iso-propyl bromide}} = 1 + \frac{3}{82}$$

$$\frac{\text{Total yield}}{\text{yield of iso-propyl bromide}} = \frac{85}{82}$$

or
$$\frac{\text{yield of iso-propyl bromide}}{\text{Total yield}} = \frac{82}{85}$$

$$\% \text{ of iso-propyl bromide} = 96.5$$

$$\% \text{ of } n\text{-propyl bromide} = 3.5.$$

Example 14. **Fluorocarbon (C_5F_{12}) has lower boiling point than pentane (C_5H_{12}) even though it has a far higher molecular weight. Explain**

Solution.

Molecular weight is not the only criterion that affects the boiling point of a compound. There is a wide gap between the electronegativities of C and F. This results in separation of charges and creation of dipoles on the molecules of C_5F_{12}. This causes repulsions between the molecules of C_5F_{12}.

C_5H_{12} on the other hand does not exhibit this type of repulsive interactions between the molecules because of comparable electronegativities of C and H.

Example 15. **Which would have a higher boiling point and why?**

$CH_3CH_2CH_2CH_2CH_3$ or 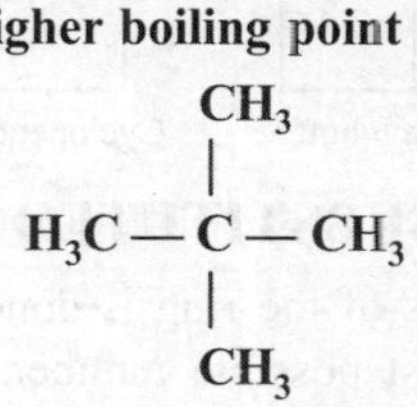

Solution.

Boiling point of an alkane depends upon the van der Waals force between the molecules which in turn depends upon the area of contact or surface area of the molecules. Out of the straight chain and branched chain alkanes, with the same number, of carbon atoms, the branched chain molecule has a smaller surface area because the molecule tends to be spherical.

Thus, straight chain molecules n-pentane which has a larger surface area will experience greater van der Waals forces and show a greater boiling point.

Thus, *n*-pentane has a higher boiling point than neopentane.

4.16 CYCLOALKANES

Cycloalkanes are cyclic compounds containing closed hydrocarbon chains. They are also called *alicyclic* compounds and resemble alkanes in many respects. Examples of cycloalkanes are cyclopropane, cyclobutane etc. They can expressed by a general formula $(CH_2)_n$ where n is 3, 4, 5...

Nomenclature. Common and IUPAC systems of nomenclature of cycloalkanes are given below:

Compound	Common name	IUPAC name
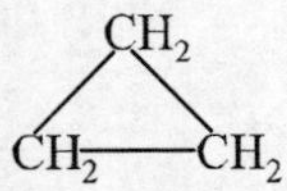	Trimethylene	Cyclopropane
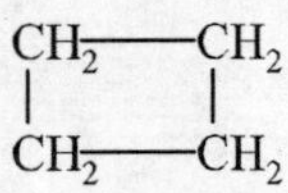	Tetramethylene	Cyclobutane
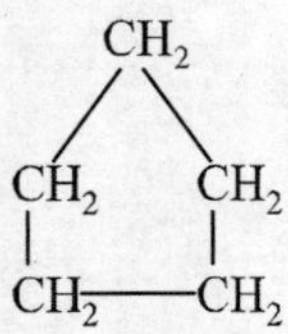	Pentamethylene	Cyclopentane
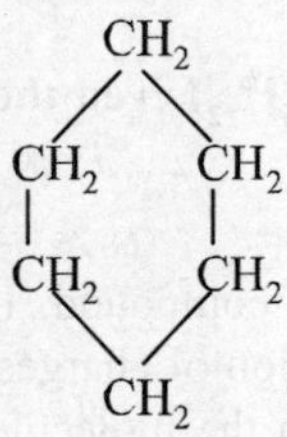	Hexamethylene	Cyclohexane

For the sake of simplicity, cyclopropane, cyclobutane, cyclopentane and cyclohexane are represented by the following geometrical figures.

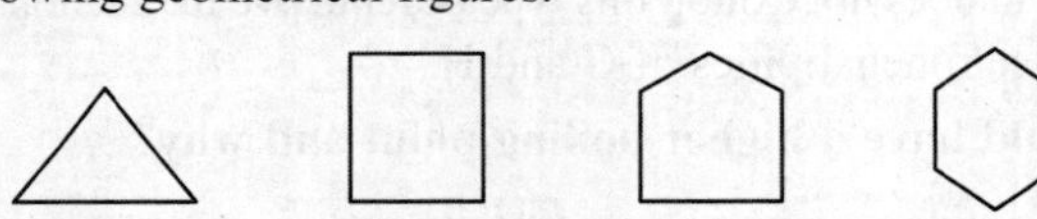

Cyclopropane Cyclobutane Cyclopentane Cyclohexane

4.17 NOMENCLATURE OF SUBSTITUTED CYCLOALKANES

Nomenclature of the carbon atoms of the ring is done in such a way that the carbon atom holding the substituent gets the lowest possible number. In order to write the name of such compounds, the name of the alkyl group is prefixed to the name of the parent cycloalkane molecule. The following examples will illustrate

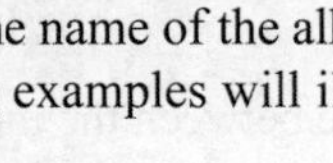

Propyl cyclopentane

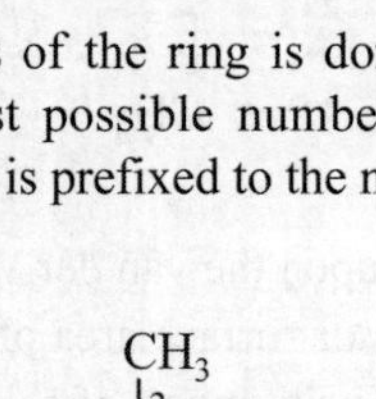

1-Ethyl-3-Methyl cyclopentane

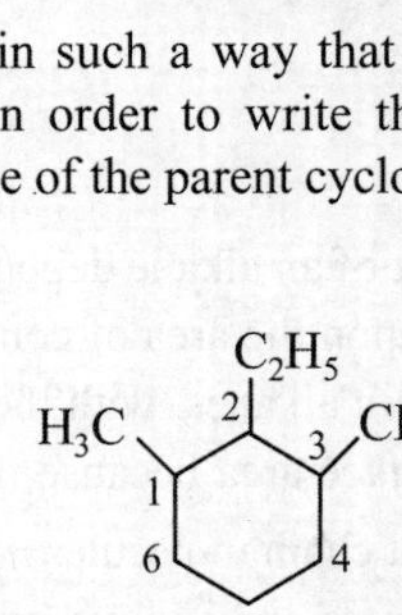

2-Ethyl-1, 3-Dimethyl cyclohexane

If the number of the carbon atoms in the ring is less the number of carbon atoms in the alkyl chain, then the ring structure is taken as the substituent and is named as cycloalkyl group. For example,

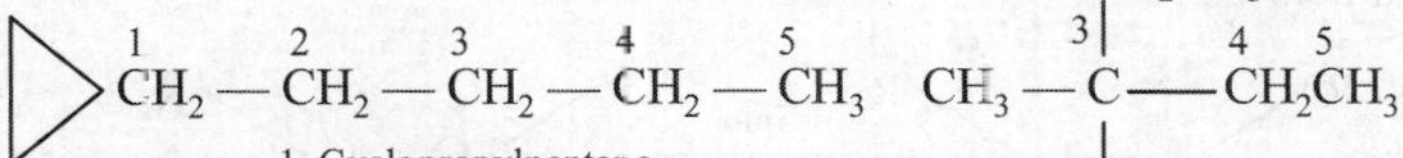

1–Cyclopropylpentane

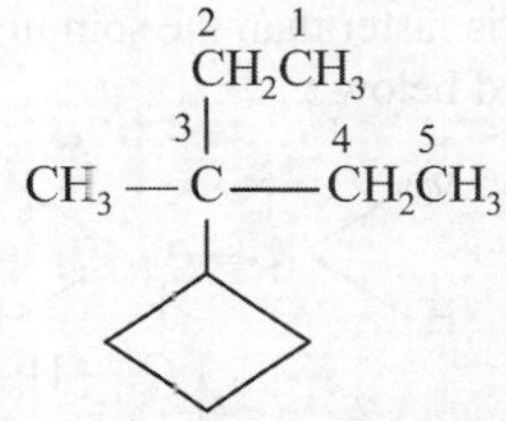

3–Methyl 3–cyclobutylpentane

4.18 METHODS OF PREPARATION OF CYCLOALKANES AND ITS DERIVATIVES

(1) By the addition of Carbenes to Olefins :

Carbenes are the natural divalent carbon intermediates in which carbon is covalently bonded to two atoms and has two non bonded orbitals containing two electrons between them. Thus, a carbene has a sextet of electrons without any charge on it. Carbenes are of two types, singlet carbenes (↑ ↓) and triplet carbenes (↑ ↑). Both of them react with alkenes to form cycloalkane derivatives and the reaction is called **cycloaddition reaction.**

(*a*) **Singlet carbenes.** They react with alkanes in a stereospecific manner in which stereochemistry of alkane is retained in the cycloaddition product formed. This means that a *cis* alkene forms *cis* addition product while *trans* alkene gives rise to *trans* addition product explained below :

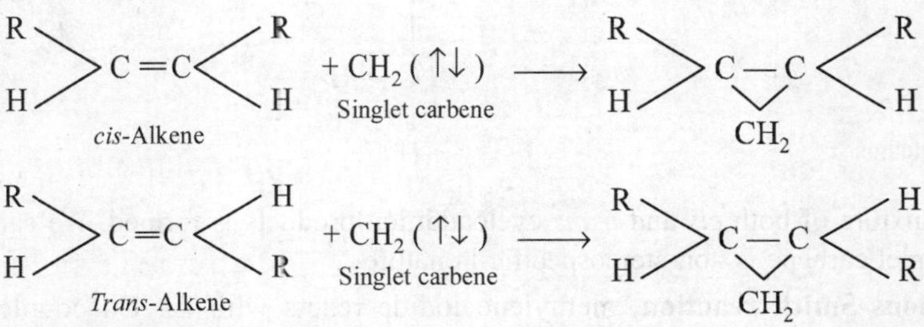

Mechanism. Addition of singlet carbene to an alkene is a **concerted** process and there is no spin restriction on the simultaneous formation of two new σ-bonds of the cyclopropane. Therefore, it is stereospecific in nature.

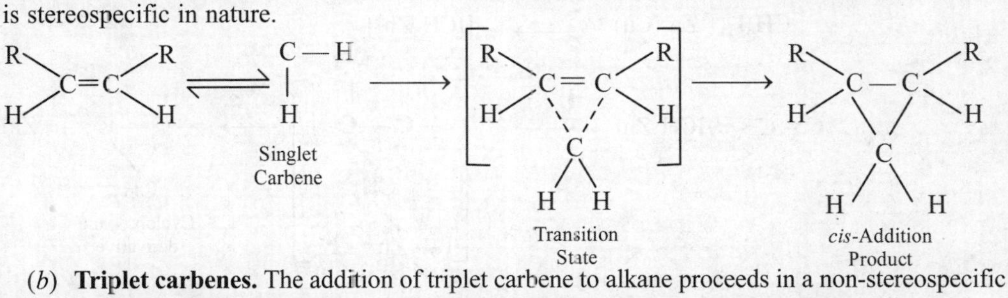

(*b*) **Triplet carbenes.** The addition of triplet carbene to alkane proceeds in a non-stereospecific manner, therefore, both *cis* and *trans* alkenes give mixture of cycloaddition products.

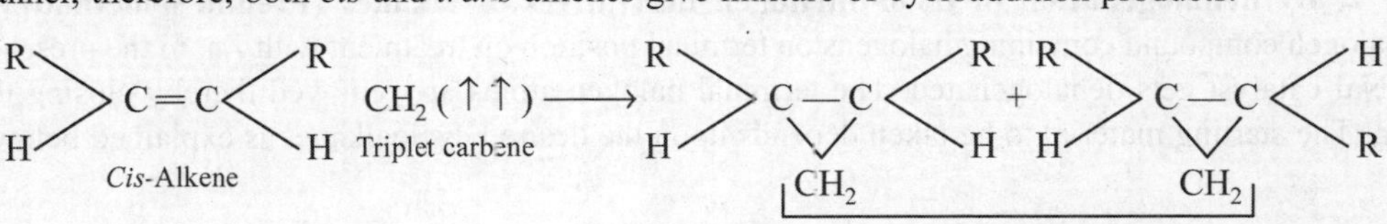

Mixture of products

Mechanism. The addition of triplet carbene to an alkene follows a different mechanism. The first step in the addition gives a biradical. It may undergo *spin inversion* before it cyclises into cyclopropane derivative. But at the same time, there is also possibility of free rotation about C — C bond. In case, this rotation is faster than the spin inversion then another cycloaddition product may also be formed as explained below :

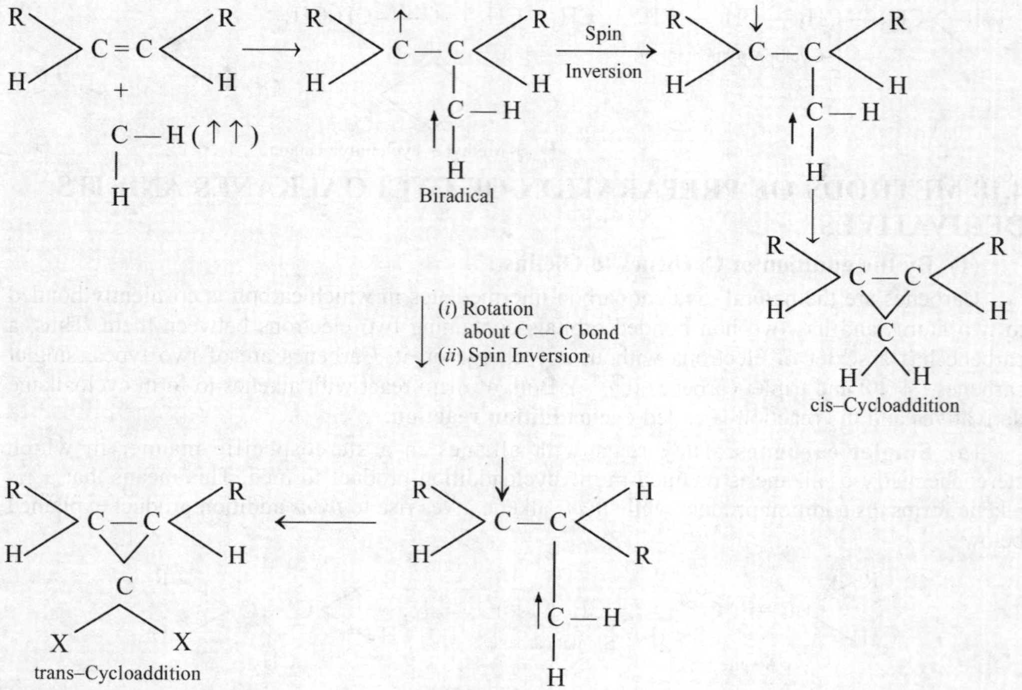

Thus, a mixture of both *cis* and *trans* cycloaddition products is formed. We can say that the addition of triplet carbene is non-stereospecific in nature.

In Simmons-Smith reaction, methylene iodide reacts with Zn–Cu couple to form an intermediate compound. This compound transfers a carbon to double bond to form a cyclopropane derivative as follows :

$$CH_2I_2 + Zn\ (Cu) \longrightarrow [ICH_2ZnI]$$

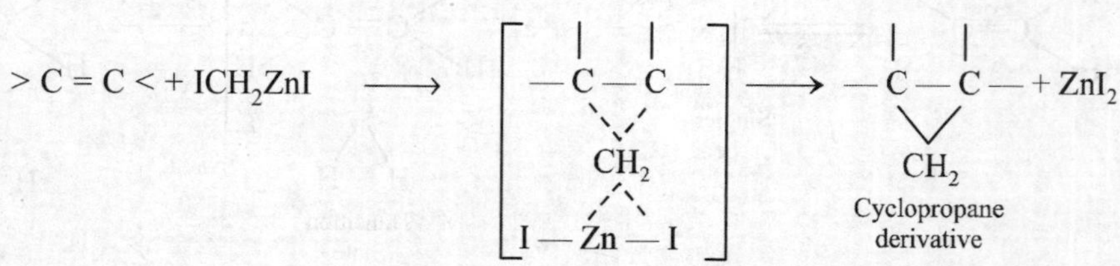

2. By dehalogenation of α, ω-dihalogen derivatives of alkanes (Freund's method). A dihalogen compound containing halogens on terminal position on treatment with Zn, in the presence of NaI catalyst gets dehalogenated. The terminal halogen atoms are removed thereby closing the ring. The starting material to be taken depends upon the desired cycloalkane, as explained below :

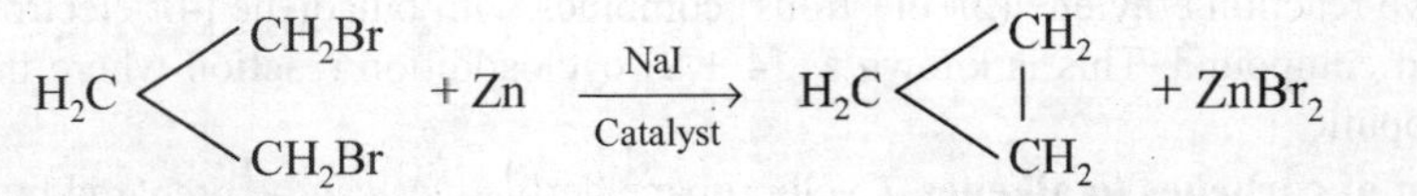

1, 3-dibromopropane Cyclopropane

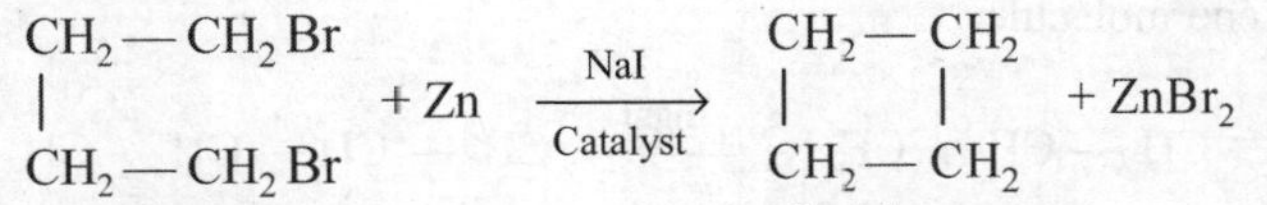

1, 4-dibromobutane Cyclobutane

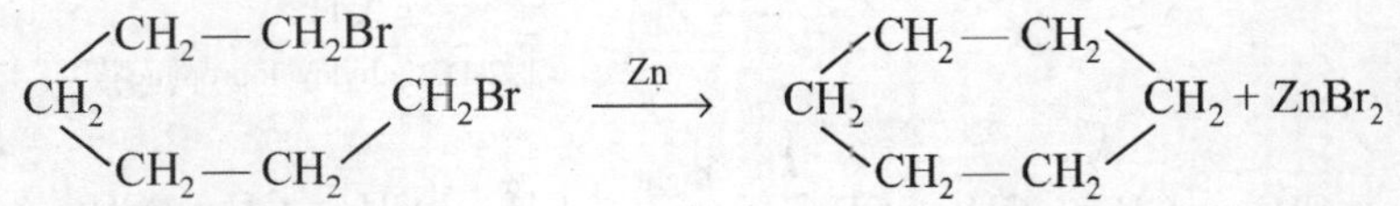

1, 6-Dibromohexane Cyclohexane

3. Distillation of calcium or barium salts of dicarboxylic acids followed by reduction (Wislicenus method). A cyclic ketone results when a calcium or barium salt of a dicarboxylic acid is distilled. Ketone obtained is subjected to Clemmenson reduction, giving the cycloalkane as illustrated by the following reaction.

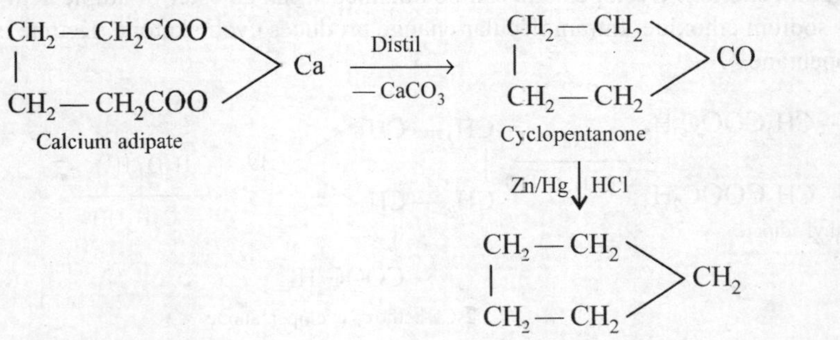

Calcium adipate Cyclopentanone Cyclopentane

4. Reduction of aromatic compounds. Cyclohexane and its derivatives can be prepared by passing hydrogen gas through benzene or its derivatives, in the presence of Ni at 423–473 K temperature.

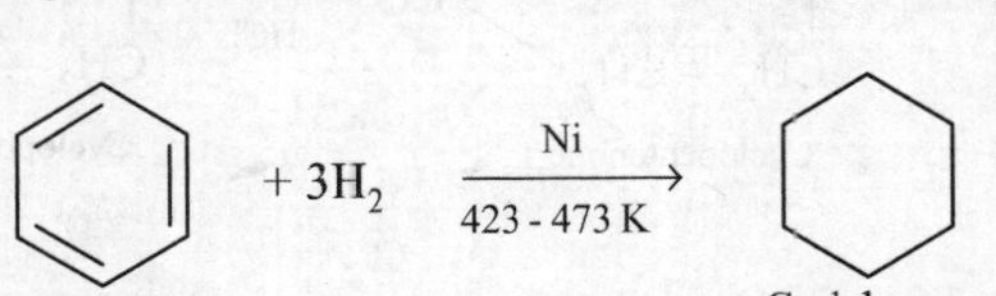

Benzene Cyclohexane

5. Diels-Alder addition. Two unsaturated compounds can be combined to produce a cycloalkane. When a mixture of 1, 3-butadiene and ethylene is heated to 475 K, cyclohexene is obtained. It can be reduced with hydrogen to yield cyclohexane.

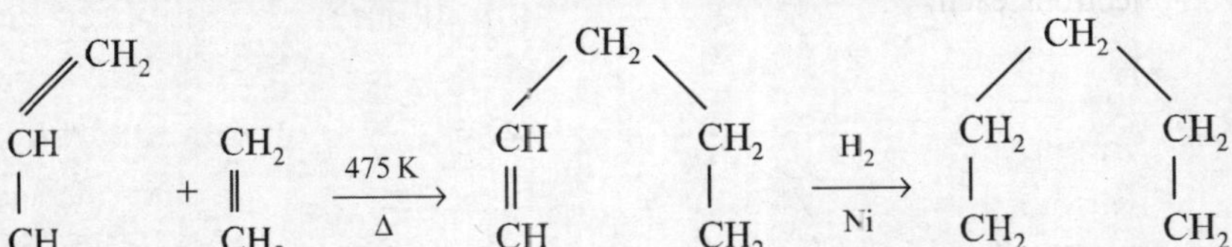

1, 3-butadiene Cyclohexene Cyclohexane

In the above reaction ethylene (2π electrons) combines with butadiene (4π electrons) to form a six membered compound. This is known as [4 + 2] cycloaddition reaction where the alkene is known as dienophile.

6. Addition of carbenes to alkenes. Cyclopropane derivatives can be prepared by treating an alkene with diazomethane. Methylene group in the form of carbene : CH_2 is attached across the double bond in the alkene molecule.

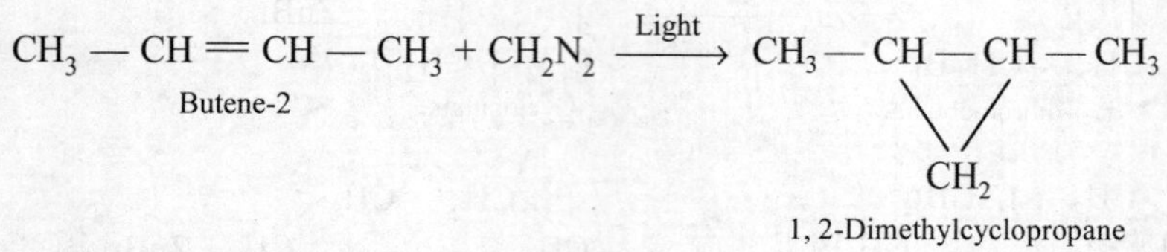

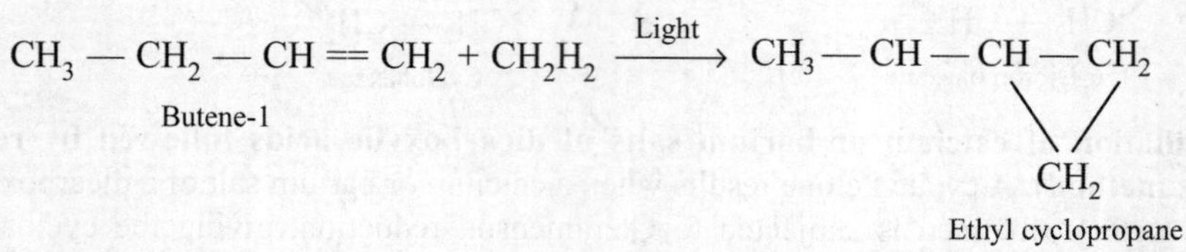

7. Dieckmann reaction. Cyclopentane can be obtained when an ester of adipic acid is treated with sodium or sodium ethoxide. Intramolecular change produces cyclic ketone. Ketone is reduced to obtain cyclopentane.

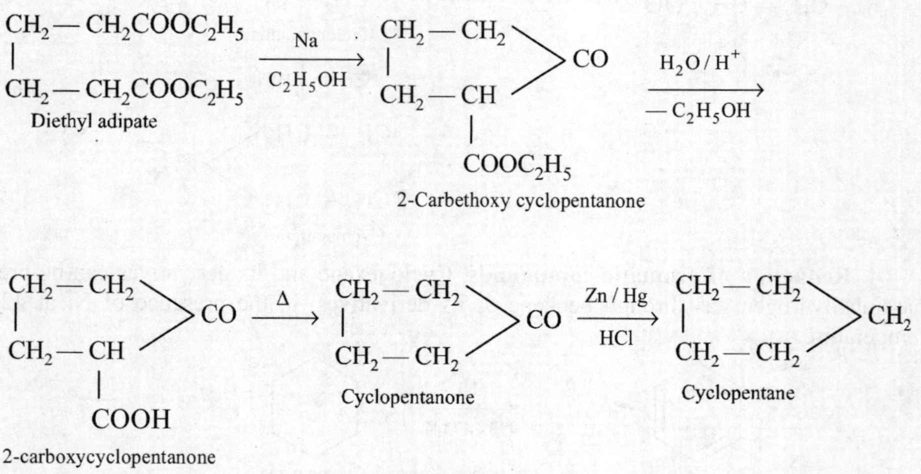

8. By [2 + 2] Photochemical Cycloaddition reaction. Cyclobutane derivatives can be obtained by addition of two alkenes under photochemical conditions. [2 + 2] means combination of two molecules using two π electrons each.

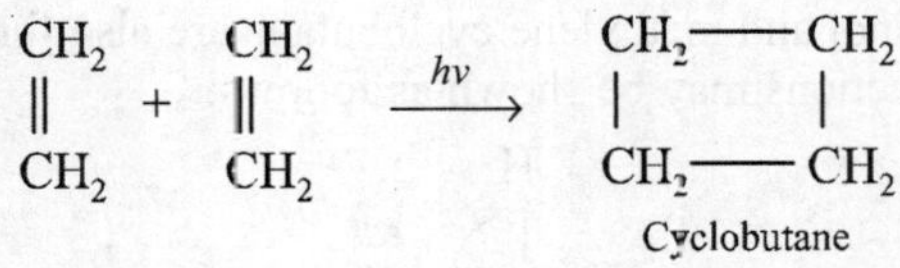

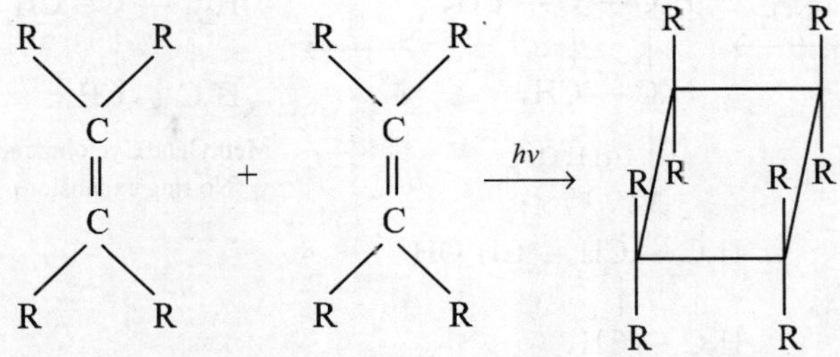

9. Tharpe-Ziegler reaction. A dinitrile is treated with LiN $(C_2H_5)_2$ in a large volume of benzene or toluene as solvent. A cyclic imino compound is obtained which on hydrolysis yields a cycloketone.

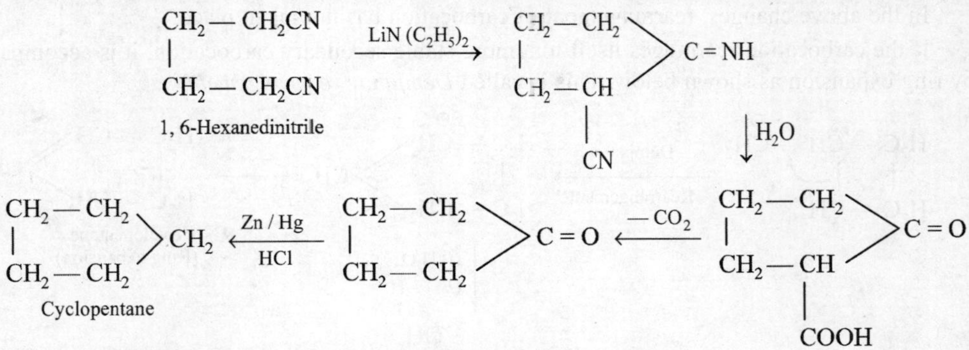

10. From Molonic ester and Acetoacetic ester

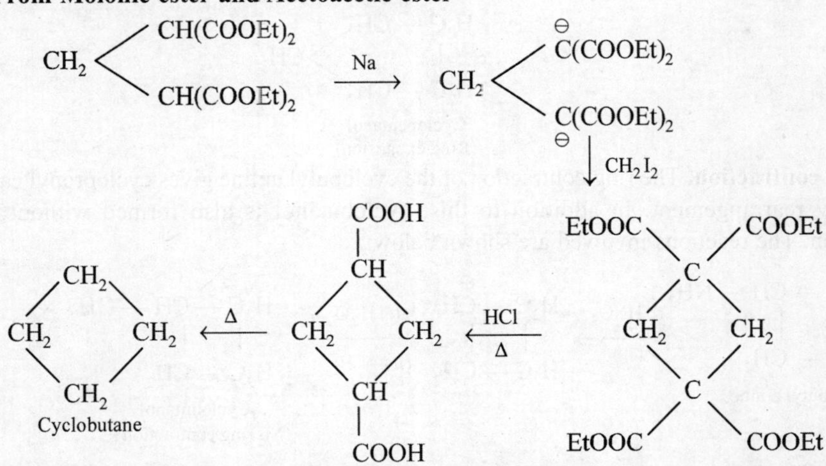

11. By Demjanov rearrangement. This is a convenient method of expanding and contracting alicylic ring systems through the formation of carbocation intermediates. This is achieved by deamination of alicyclic amines with nitrous acid. The carbocations may undergo reactions such as addition of nucleophiles, elimination of a proton and rearrangement to a more stable carbocation. The following examples will illustrate.

Ring expansion. When cyclobutyl methyl amine is treated with nitrous acid, the product is a mixture of cyclopentanol and cyclopentene. Both these products are formed by ring expansion. In addition to this, cyclobutyl carbinol and methylene cyclobutane are also formed where there is no ring expansion. The series of reactions may be shown as follows:

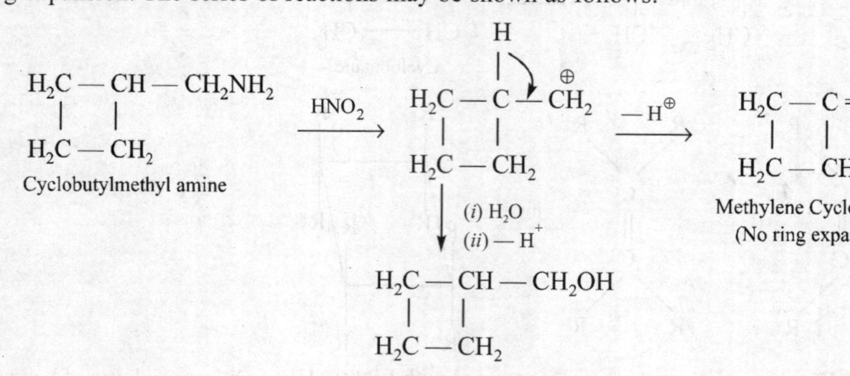

In the above changes, rearrangement of carbocation has not taken place.

If the carbocation rearranges itself to a more stable secondary carbocation, it is accompanied by ring expansion as shown below. This is called *Demjanov rearrangement.*

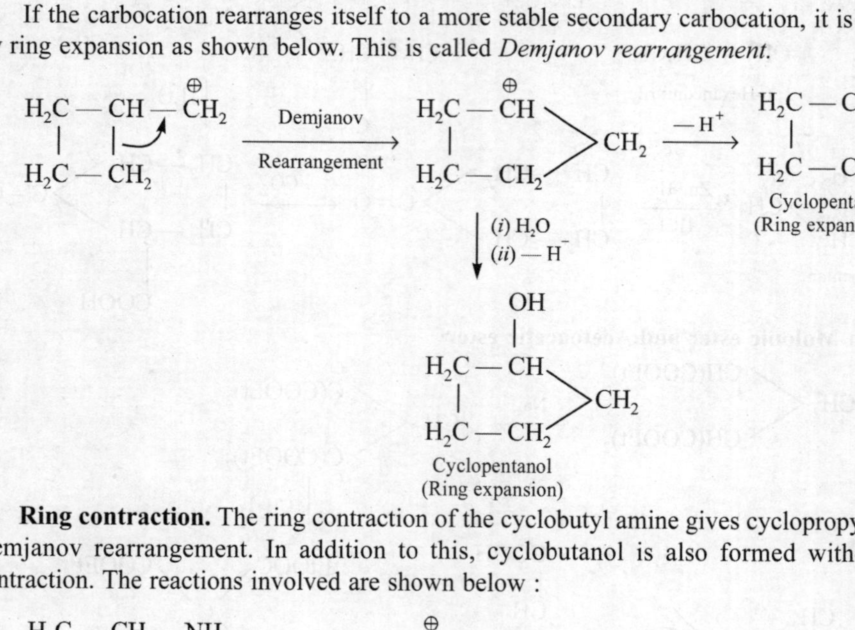

Ring contraction. The ring contraction of the cyclobutyl amine gives cyclopropyl carbinol by Demjanov rearrangement. In addition to this, cyclobutanol is also formed without any ring contraction. The reactions involved are shown below :

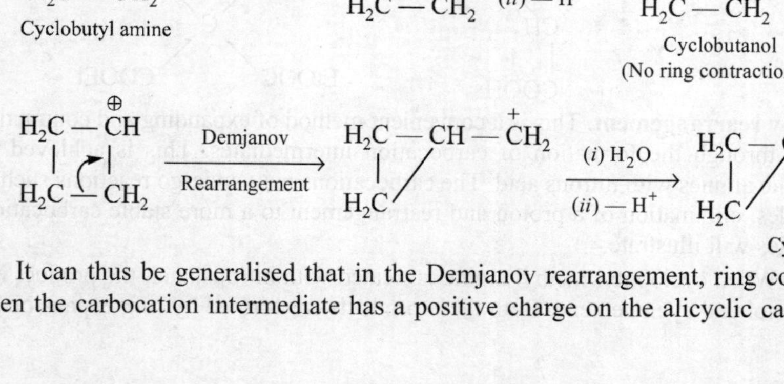

It can thus be generalised that in the Demjanov rearrangement, ring contraction takes place when the carbocation intermediate has a positive charge on the alicyclic carbon atom. Similarly, ring expansion takes place when the carbocation has a positive charge on the carbon atom α- to the alicyclic ring.

4.19 BLANC'S RULE

Blanc's rule states that :

Pyrolysis of 1, 4- and 1, 5-linear carboxylic acids yields cyclic anhydrides while 1, 6- and 1, 7-dicarboxylic acids give cyclic ketones. 1, 8- and higher dicarboxylic acids remain unaffected.

Thus

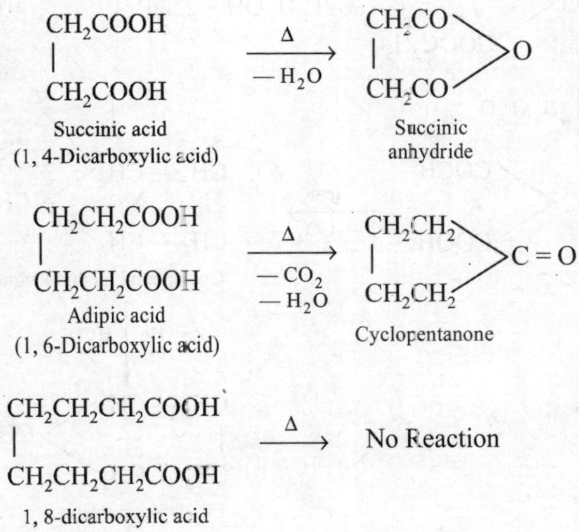

Utility of the blanc's rule

This rule finds use in the determination of size of the ring and chain length of the carboxylic acid. The number of carbons in the ring is reduced by one when the cyclization of the dicarboxylic compound takes place (Refer to cyclization of adipic acid into cyclopentanone).

In an unknown ring compound, whose size is desired to be determined, a double bond is created by a known method. The compound is then subjected to oxidation with $KMnO_4$ when the fission of double bond takes place and a dicarboxylic compound is obtained. This dicarboxylic compound is pyrolysed to obtain a cyclic compound and this can provide information about the size of the original ring.

Example 1. **How can you obtain (*i*) Ethyl cyclopentane (*ii*) Methyl cyclopentane?**

Solution. **(*i*) Ethyl cyclopentane**

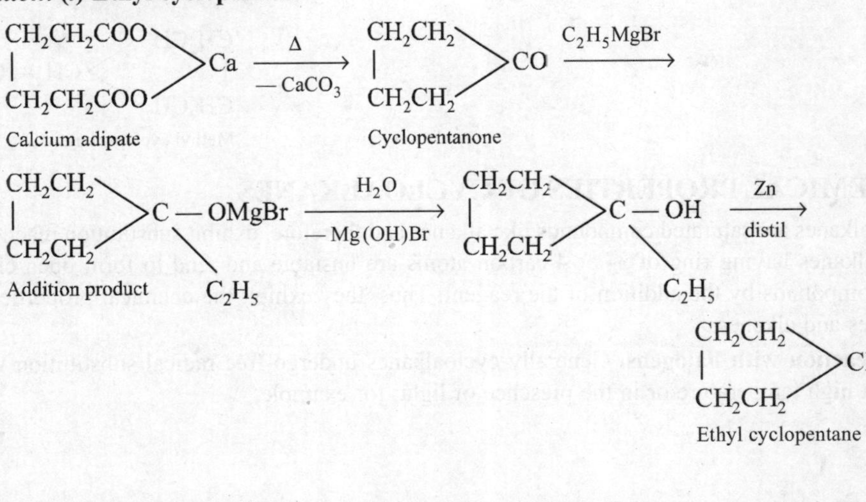

It can alternatively be prepared as under

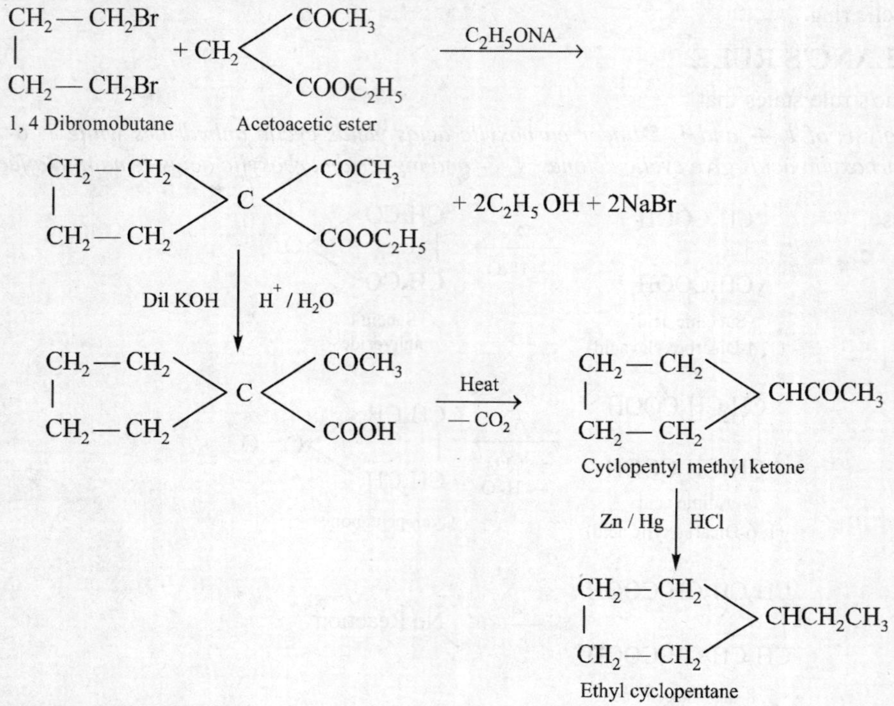

(ii) **Methyl cyclopentane**

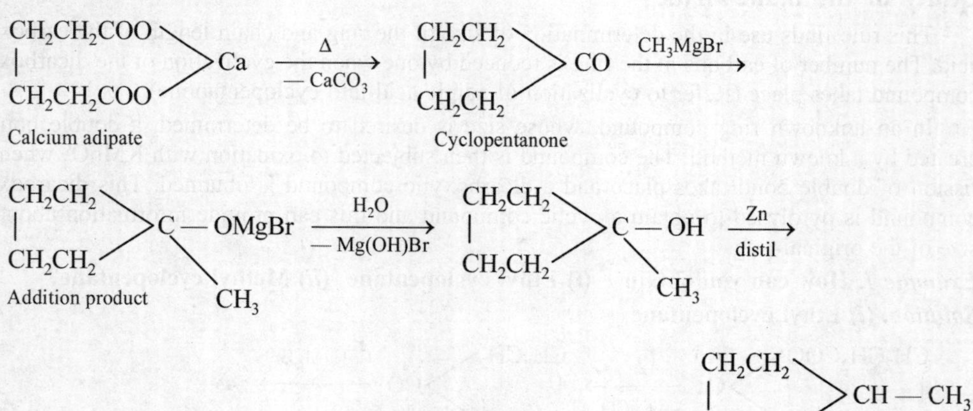

4.20 CHEMICAL PROPERTIES OF CYCLOALKANES

Cycloalkanes are saturated compounds like alkanes and therefore, exhibit substitution reactions. But cycloalkanes having ring of 3- or 4-carbon atoms are unstable and tend to form open chain aliphatic compounds by the addition of the reagent. Thus, they exhibit the chemical properties of both alkanes and alkenes.

(1) Reaction with halogens. Generally cycloalkanes undergo free radical substitution with halogens at high temperature or in the presence of light, for example,

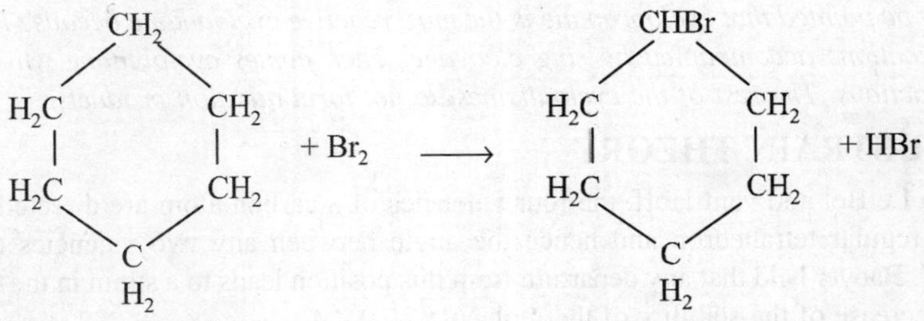

Cyclohexane Bromocyclohexane

In case of cyclopropane, bromine and chlorine in the presence of $FeCl_3$ break the ring system and give open chain addition products. For example,

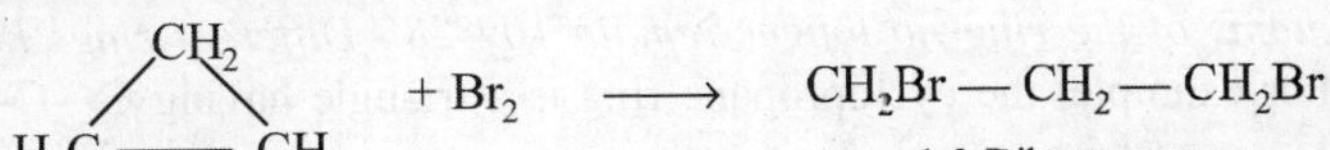

1-3-Dibromopropane

Cyclopropane

(2) Reaction with hydrogen. When heated with hydrogen in the presence of nickel, cyclopropane and cyclobutane give addition products whereas higher members do not give this reaction.

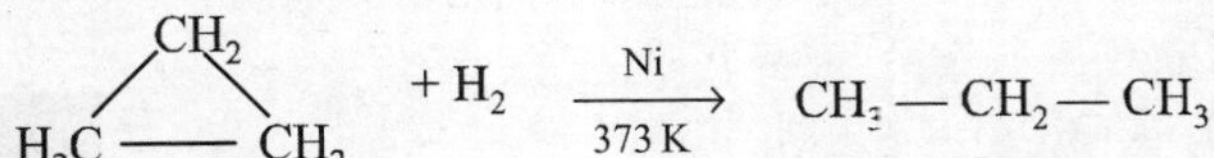

Cyclopropane *n*-Propane

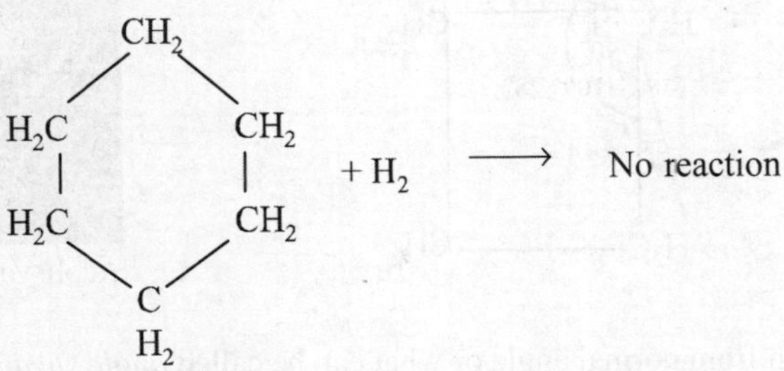

Cyclohexane

(3) Reaction with halogen acids. Generally cycloalkanes do not react with halogen acids. However, cyclopropane and cyclobutane add on halogen acids to give open chain alkyl halides.

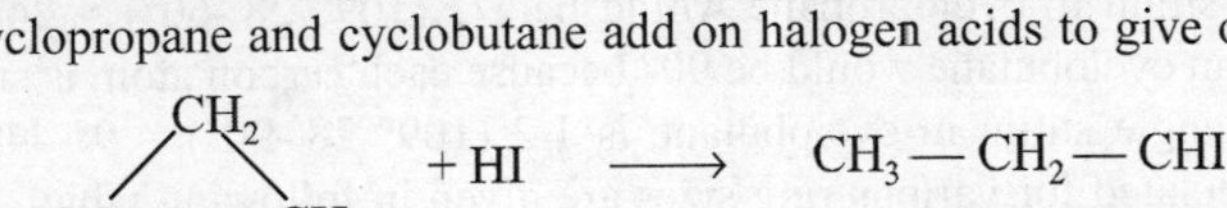

n-Propyl iodide

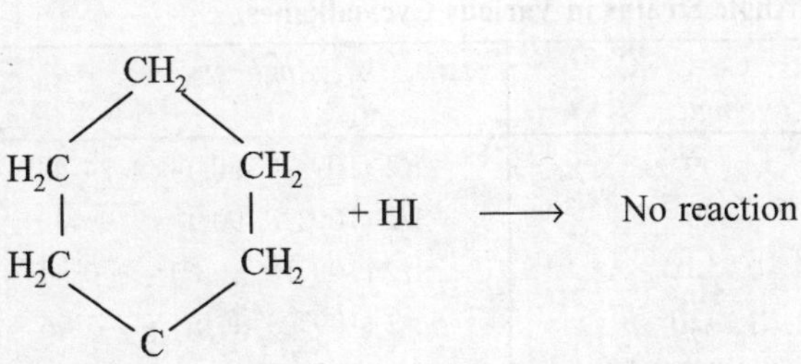

Cyclohexane

Hence it may be pointed that cyclopropane is the most reactive cycloalkane because it exhibits many addition reactions accompanied by ring cleavage. Then comes cyclobutane which shows some addition reactions. The rest of the cycloalkanes do not form addition products.

4.21 BAEYER STRAIN THEORY

According to Le Bel and vant Hoff, the four valencies of a carbon atom are directed towards the corners of a regular tetrahedron and hence the angle between any two valencies (any two bonds) is 109° 28′. Baeyer held that any departure from this position leads to a strain in the molecule resulting in the decrease of the stability of the molecule.

In the light of this concept Baeyer gave a theory which is known as Baeyer's Strain Theory. The main points of the theory are as given below :

(*i*) *The carbon atoms constituting the rings lie in the same plane. Hence bond angles between the adjacent carbon atoms of the ring no longer remain 109°28′. Different rings have different values of this angle.* For example, the cyclopropane ring is a triangle having C—C—C angle of 60° only.

(*ii*) *Any deviation, positive or negative from the normal tetrahedral bond angle of 109°28′ during the formation of a ring creates a strain in the molecule which makes the molecule unstable.*

(*iii*) *The larger (more) the deviation from the normal angle, the greater is the strain and thus lesser is the stability. However, it should be noted that the sign of deviation does not make any difference.*

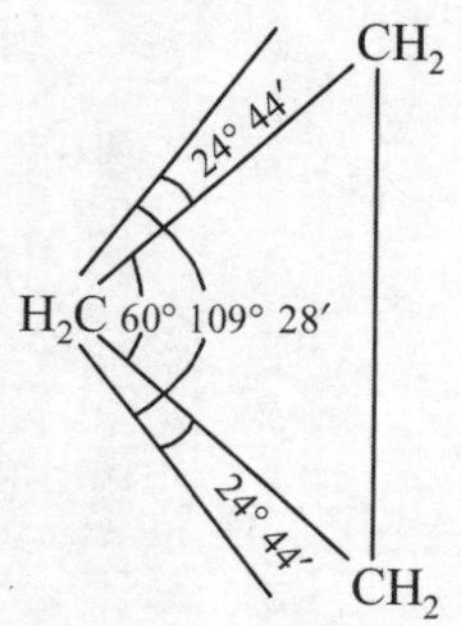

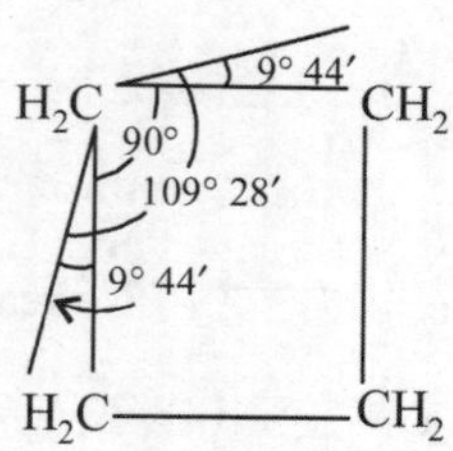

Adolf Von Baeyer

Baeyer worked out the deviation from normal angle or what can be called *angle strain* involved in the formation of cycloalkanes of various ring sizes. For instance the C—C—C angle of a cyclopropane ring is 60° because each carbon atom occupies a corner of an equilateral triangle. This shows that the angle strain in cyclopropane would be 1/2 (109° 28′-60°) = 24°-44′. In the same way C—C—C angle in cyclobutane would be 90° because each carbon atom is situated at the corner of the square. The angle strain in cyclobutane is 1/2 (109° 28′-90°) = 9°-44′. The angle strain for cycloalkanes calculated for various ring sizes are given in following table.

Table 4.2. Angle Strains in Various Cycloalkanes.

Compound	C—C—C *Angle*	*Angle strain*
Cyclopropane, C_3H_6	60°	1/2 (109°28′-60°) = + 24°44′
Cyclobutane, C_4H_8	90°	1/2 (109°28′-90°) = + 9°44′
Cyclopentane, C_5H_{10}	108°	1/2 (109°28′-108°) = + 0°44′
Cyclohexane, C_6H_{12}	120°	1/2 (109°28′-120°) = – 5°16′
Cycloheptane, C_7H_{14}	128°35′	1/2 (109°28′-128°34°) = – 9°33′
Cyclooctane, C_8H_{16}	134°	1/2 (109°28′-135°) = – 14°46′

It is clear from the above data that angle strain decreases as we move from cyclopropane to cyclopentane. This predicts ease of formation and stability of the compound from cyclopropane to cyclopentane. This is actually found to be so.

However the stability of cyclohexane and higher members cannot be explained by the angle strain values. According to angle strains, cyclohexane and higher members should be quite unstable as there is a lot of strain in the molecule. But this is not true. Actually cyclohexane and higher members are not planar molecules. Different carbon atoms in such molecules lie in different planes, thereby relieving the strain.

Example 2. **Giving examples of first three cycloalkanes, explain the relationship of relative stabilities with the heat of combustion.**

Solution.

Relationship of relative stability with heat of combustion.

Size of the ring	*Heat of combustion per — CH_2 group kJ mol^{-1}*	*Size of ring*	*Heat of combustion per — CH_2 group kJ mol^{-1}*
3	697	6	658.5
4	686	7	662.3
5	664	8	663.6

Heat of combustion per CH_2 group of open chain alkanes = 658.0 kJ mol^{-1}.

Heat of combustion may be defined as the quantity of heat evolved when one mole of the compound is burnt in excess of air.

Heat of combustion per CH_2 group gives a fair idea of the relative stabilities of cycloalkanes. If the heat of combustion is large, the compound contains more energy and is therefore less stable.

With this principle, let us examine the stabilities of cyclopropane, cyclobutane and cyclopentane. Heat of combustion of cyclopropane is 697 kJ mol^{-1}. It is 39 kJ more than that of open chain alkane. Thus it is less stable than the open chain alkane by this amount.

In the case of cyclobutane, the heat of combustion is more by an amount 28 kJ as compared to open-chain alkane. Thus cyclobutane is less stable than open chain alkanes but it is more stable than cyclopropane. Cyclopentane having heat of combustion 664 kJ is more stable than cyclobutane. The relative stability of first three members of cycloalkanes in decreasing order is given as under.

Cyclopentane > Cyclobutane > Cyclopropane

On having a look at the heat of combustion of cyclohexane and higher members, we observe that the values do not differ much. Thus higher members are expected to display almost uniform stability which is confirmed by their reactions.

4.22 SACHSE MOHR THEORY OF STRAINLESS RINGS AND STABILITY OF CYCLOHEXANE

To explain the greater stability of cyclohexane as compared with cyclopentane and stability of large rings, Sachse in 1890 advanced the Theory of Strainless Rings. *According to this theory large rings due to their bigger size get twisted in different planes with the result that the carbon atoms do not lie in the same plane as is assumed by Baeyer.* Thus, because of carbon atoms being in the different planes, their valency bonds are not distorted from their original angle and consequently no strain is set up in the molecules. In other words, the rings with six or more carbon atoms get *puckered* and thus become strainless.

Thus according to Sachse cyclohexane exists in the following strain-free forms called *boat* and *chair* forms.

Mohr however suggested that the two forms cyclohexane are continuously undergoing transformation from one form to another.

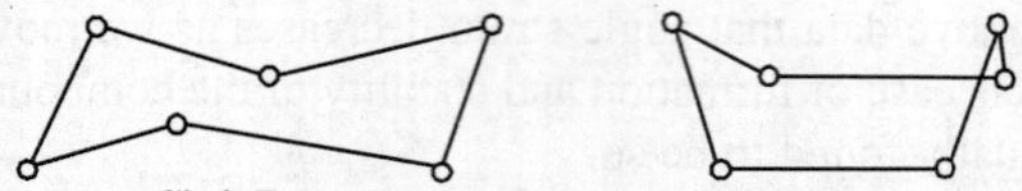

4.23 DIFFICULTIES IN THE SYNTHESIS OF LARGE-MEMBERED RINGS

It has been observed that heat of combustion per — CH_2 group in cycloalkanes from cyclopentane onwards remains nearly constant and is comparable to that of alkanes. Thus these cycloalkanes (5-membered and higher) should be as stable as corresponding alkanes. But still it is not easy to prepare such higher cyclohexane. To obtain a cycloalkane from a linear compound, it is necessary that the terminal groups of the linear compound should come sufficiently close to each other so that ring closure can take place.

In normal circumstances, the probability of end-groups coming close to provide cyclization are not very bright. This, then, results in intermolecular reaction leading to polymerisation. Baeyer advanced the views that because of tremendous strain, higher cycloalkanes are not formed easily. But this is not true. As postulated by Sachse-Mohr, such higher cycloalkanes are strain-free because the constituent atoms are in different planes and they retain the normal valency angle of 109°-28′. It is rather improbability of the end groups of the linear molecules coming closer that prevents the formation of ring compounds.

How is the difficulty overcome?

High dilution technique is resorted to in the preparation of higher cycloalkanes. The substance is taken in low concentration which minimises the collision between two molecules preventing polymerisation. Due to flopping, the two ends of the molecule come closer enough to react, giving rise to cyclization.

It may be remarked that during the synthesis of cyclic compounds, two processes are competing with each other. Process I is the ring closure by the coming together of the end groups and Process II is chain lengthening (polymerization) by the interaction of end group of one molecule with that of the other molecules as illustrated below. By reducing concentration of the reactant, the chances of Process II are minimised

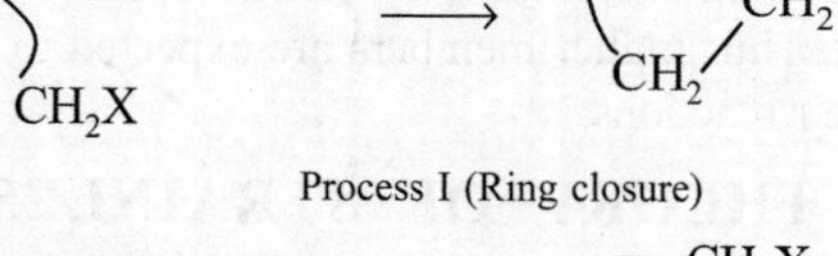

Process I (Ring closure)

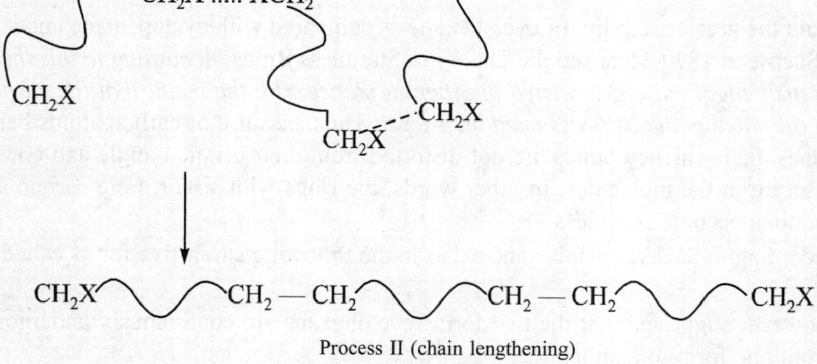

Process II (chain lengthening)

It is amazing that, once formed, these higher cyclic compounds are as stable as their straight chain counterparts.

Example 3. **Do the cycloalkanes exhibit isomerism with any class of open chain compounds? If so give examples.**

Solution. Cycloalkanes, which can be represented by the general formula $(CH_2)_n$ are isomeric with alkenes (open-chain compounds) having the general formula $C_n H_{2n}$.

Thus cyclopropane is isomeric with propene.

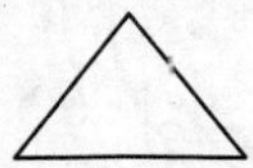

Cyclopropane

$CH_2 — CH = CH_2$

Propene

And cyclobutane is isomeric with butene-1 and butene-2

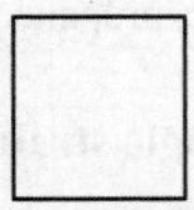

Cyclobutane

$CH_3 — CH = CH — CH_2$

Butene-2

$CH_3 — CH_2 — CH = CH_2$

Butene-1

Example 4. **Cyclopropane is the least stable member of cycloalkanes. How do you justify this is terms of orbital picture of 3-membered rings?**

OR

There are three banana bonds in cyclopropane. Explain.

Solution. In cyclopropane, the carbon atoms are sp^3 hybridised. The sp^3 hybrid orbitals are inclined at an angle of 109°28′ to one another. In order to form a single bond between two carbon atoms, the orbitals of the two carbon atoms must overlap to a maximum extent. Therefore when a carbon atom forms single bonds with two adjacent carbon atoms, the angle C—C—C should be 109°28′ as shown in the diagram below. This is not actually so in cyclopropane molecule. The angle C—C—C in this case is only 60°. This orientation will not allow perfect overlapping of the orbitals and therfore the bonds formed will be weaker.

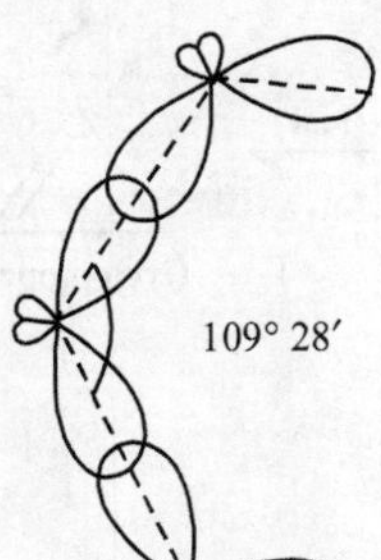

sp^3 orbitals overlaping along their axis; Maximum overlap, strong bond.

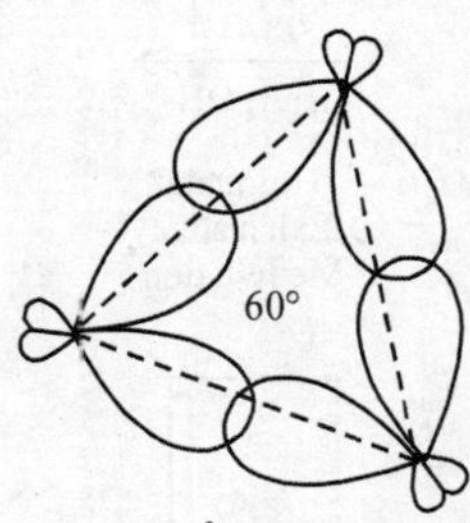

Overlap of sp^3 orbitals in cyclopropane; Poor overlap, weak bond.

This is the reason why cyclopropane molecule is unstable. Calculations show (on geometrical considerations) that the maximum deviation from the normal tetrahedral angle of 109°28′ is in the case of cyclopropane. Hence this is the least stable member of cycloalkanes.

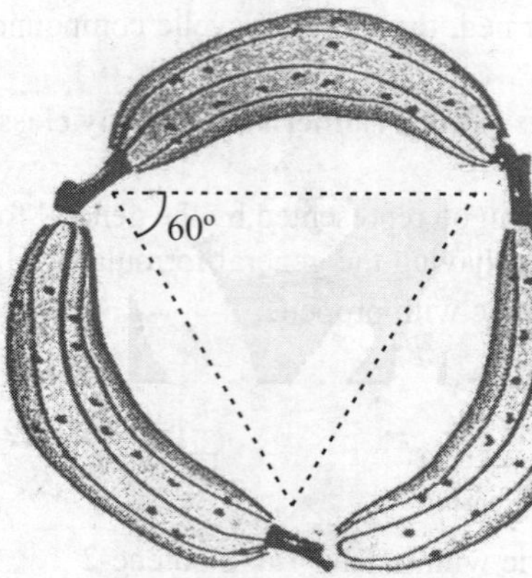

The C-C bonds in cyclopropane are weaker than the C-C bonds in propane.
They are called *Banana* bonds or *Bent* bonds.

Example 5. **Even though cycloheptane and higher members are free of angle strain, they cannot be synthesised easily. Explain.**

Solution. It is necessary for the formation of a cyclic compound that its terminal carbon atoms should come closer. In the case of a large hydrocarbon chain, the possibility that end carbons will come sufficiently closer and form the closed ring is very small. That is why cycloheptane and higher members are difficult to synthesis.

Example 6. **Give the methods of preparation of cyclopropene, cyclobutene cyclopentene and cyclohexene.**

Solution.

1. Cyclopropene, 

Cyclopene may be prepared by the following sequence of reactions

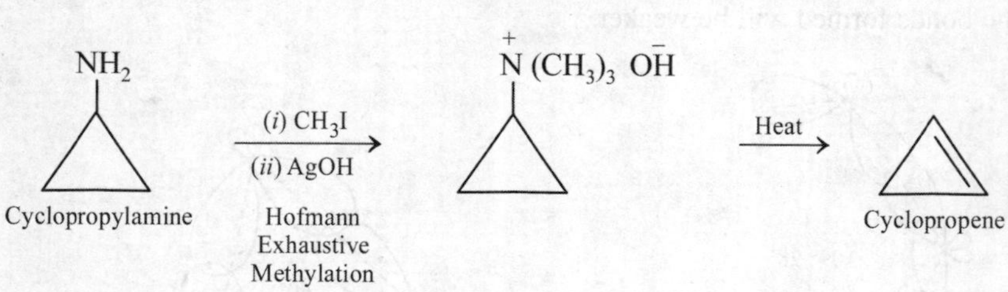

2. Cyclobutene, 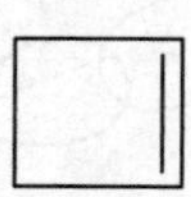

Cyclobutene may be prepared by Hofmann exhaustive methylation of cyclobutylamine in a similar process as in cyclopropene

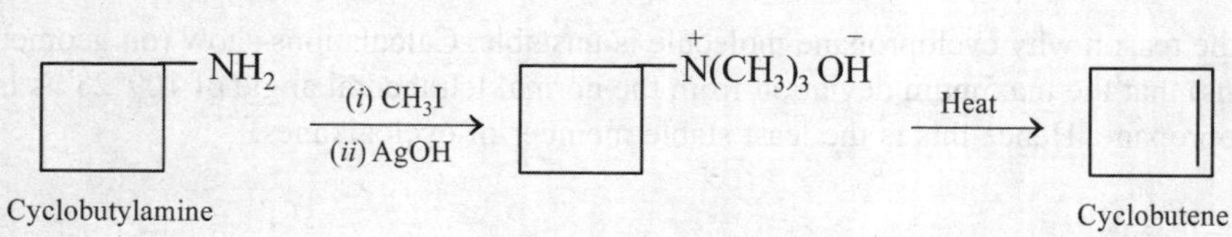

3. Cyclopentene, 

(*i*) It may be prepared by dahydrating cyclopentanol

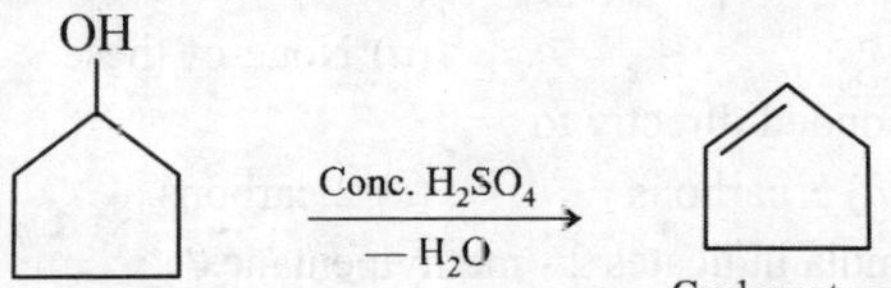

Cyclopentanol Cyclopentene

(*ii*) It may also be prepared by heating cyclopentyl bromide with ethanolic potassium hydroxide

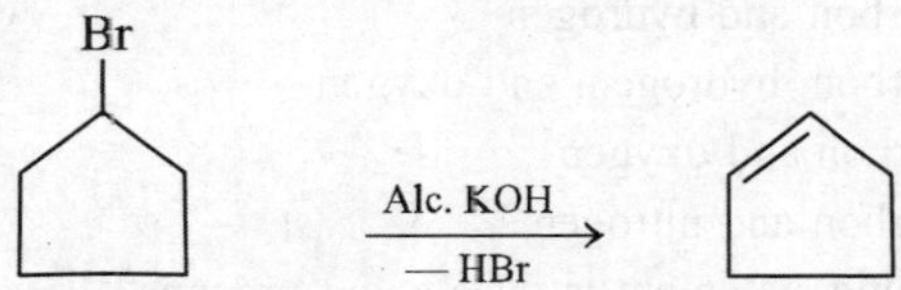

Cyclopentyl bromide Cyclopentene

4. Cyclohexene 

It may be prepared by dehydrating cyclohexanol with sulphuric acid

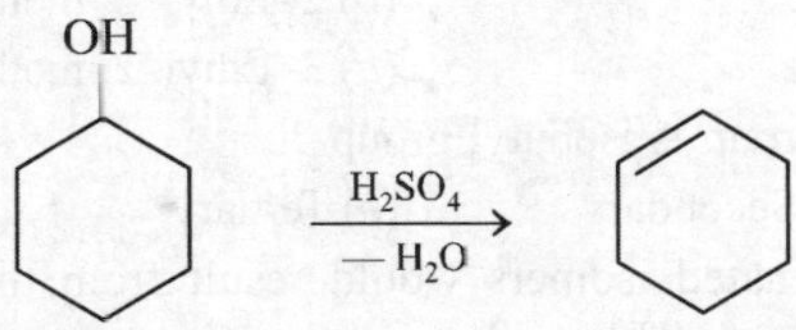

Cyclohexanol Cyclohexene

EXERCISES
(INCLUDING QUESTIONS FROM DIFFERENT UNIVERSITY PAPERS)

Objective Questions

Fill in the blanks

1. Complete: $R-H+Cl_2 \xrightarrow{\text{.........}} R-Cl+HCl$
2. Boiling point of alkanes .... on branching
3. Following is the puckered structure of ....

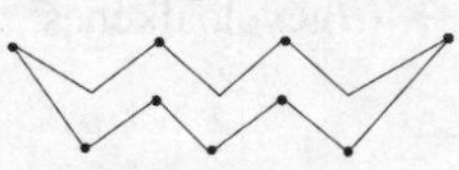

4. Halogenation of alkane is .... substitution
5. Strainless ring theory was proposed by .... and ....
6. Chair form of cyclohexane is .... stable than boat form
7. The evidence in support of Baeyer strain theory comes from heat of .... data
8. Alicyclic compounds are also known as ....
9. According to Baeyer strain theory, cyclopentane is .... stable than cyclohexane
10. The C – C – C angle in non – puckered cyclohexane is ....

Multiple Choice Question

Choose the correct option

1. Hydrocarbons are
   (*a*) insoluble in water (*b*) composed of carbon and hydrogen
   (*c*) both (*a*) and (*b*) (*d*) None of these
2. A tertiary carbon is bonded directly to :
   (*a*) 2 hydrogens (*b*) 3 carbons (*c*) 2 carbons (*d*) 4 carbons
3. Which molecular formula indicates 2 - methylpentane ?
   (*a*) C_5H_{12} (*b*) C_6H_{14} (*c*) C_5H_{10} (*d*) C_6H_{12}
4. Hydrocarbons are
   (*a*) Composed of carbon and hydrogen
   (*b*) Composed of carbon, hydrogen, and oxygen
   (*c*) Composed of carbon and oxygen
   (*d*) Composed of carbon and nitrogen
5. Which of the following statement is false about propane?
   (*a*) All bond angles are 109.5°
   (*b*) Each carbon is sp^3 hybridised
   (*c*) The compound is combustible
   (*d*) The compound undergoes polymerisation to give polypropylene.
6. Which of the following is a correct name according to the IUPAC rules ?
   (*a*) 2–methylcyclohexane (*b*) 2–Ethyl–2–methylpentane
   (*c*) 3, 4–Dimethylpentane (*d*) 3–Ethyl–2–methylpentane
7. What type of an alkyl group is isobutyl group ?
   (*a*) Primary (*b*) Secondary (*c*) Tertiary (*d*) None of these
8. How many monochlorinated isomers would result from the reaction of chlorine with *n*–butane in the presence of UV light ?
   (*a*) 2 (*b*) 4 (*c*) 6 (*d*) 8
9. A compound of formula C_3H_8 does not react with bromine in Cl_4 in the dark. The compound could be
   (*a*) alkane (*b*) cycloalkane (*c*) alkene (*d*) cycloalkene
10. Which halogen does not react appreciably with methane in a free–radical substitution reaction?
    (*a*) Chlorine (*b*) Bromine
    (*c*) Iodine (*d*) Fluorine
11. Cycloalkanes have the same molecular formula as
    (*a*) alkanes (*b*) alkenes (*c*) alkynes (*d*) cycloalkenes
12. Which of the cycloalkanes is not expected to have ring strain ?
    (*a*) Cyclopropane (*b*) Cyclobutane
    (*c*) Cycloheptane (*d*) None of these
13. Most stable conformation of cyclohexane is the
    (*a*) Haworth form (*b*) boat form (*c*) Newman form (*d*) Chair form
14. Which of the following compounds will give cyclopropane on treatment with sodium in dry ether ?
    (*a*) 1, 3–Dibromopropane (*b*) 1, 1–Dibromopropane

(*c*) 1, 2–Dibromopropane (*d*) 2, 2–Dibromopropane

15. Which of the following compounds will react most readily with concentrated sulphuric acid?
(*a*) Ethane (*b*) Cyclohexane (*c*) Propane (*d*) Cyclohexane

16. Cyclohexanol can be converted into cyclohexane by heating with
(*a*) Zn (Hg) and HCl (*b*) Conc.H_2SO_4
(*c*) $SOCl_2$ (*d*) H_2 and Ni

17. What percentage of cyclohexane molecule is estimated to be in the boat form at any given time ?
(*a*) Over 99% (*b*) Between 90% and 99%
(*c*) Approximately 50% (*d*) Less than 1%

18. Which of the following cycloalkanes is most reactive ?
(*a*) Cyclopropane (*b*) cyclohexane (*c*) Cyclobutane (*d*) Cycloheptane

19. Which of the following statements is wrong about cyclohexane ?
(*a*) It is a saturated cyclic hydrocarbon
(*b*) All C – C – C bond angle are 109° – 28′
(*c*) It is very unstable and strained compound
(*d*) It can exist in two conformations designated as boat and chair forms

20. A compound of formula C_6H_{12} does not react with concentrated sulphuric acid. The compound could be
(*a*) alkane (*b*) cycloalkane (*c*) alkene (*d*) cycloalkene

ANSWERS

Fill in the blanks

1. sunlight **2.** decreases **3.** docecane **4.** free – radical
5. Sache, Mohr **6.** more **7.** hydrogenation
8. Cyclopolymethylenes **9.** more **10.** 120°

Multiple choice questions

1. (*c*)	**2.** (*b*)	**3.** (*b*)	**4.** (*a*)	**5.** (*d*)
6. (*d*)	**7.** (*a*)	**8.** (*a*)	**9.** (*a*)	**10.** (*c*)
11. (*b*)	**12.** (*c*)	**13.** (*d*)	**14.** (*a*)	**15.** (*d*)
16. (*b*)	**17.** (*d*)	**18.** (*a*)	**19.** (*c*)	**20.** (*b*)

General Questions

1. Give the IUPAC and common names and also find the number of primary, secondary and tertiary carbon atoms in the compound :

$$CH_3 - CH_2 - \underset{\underset{CH_3}{|}}{CH} - CH_3$$

2. Complete the following equations

(*i*) $$CH_3 - \underset{\underset{CH_3}{|}}{CH} - COONa \xrightarrow[\text{heat}]{\text{Soda lime}}$$

(*ii*) $$CH_3COONa \xrightarrow{\text{Electrolysis}} (?) \xrightarrow{\text{At cathode}} (?) \xrightarrow{\text{At anode}} (?)$$

3. Write note on relative rate of halogenation of alkanes in terms of activation energy.
4. Out of the three isomeric pentanes, which one would be expected to produce the largest number of monochloro derivatives. Write the structures and give their IUPAC names.
5. The iodination of alkanes is done in the presence of HIO_3 but the chlorination and bromination is done as such. Explain.
6. Explain that in the chlorination and bromination of propane, the percentages of 2-haloderivatives are 56 and 97 respectively.
7. Arrange the isomeric pentanes in increasing order of their boiling points and give suitable explanation.
8. An alkane with molecular mass 72 forms only monochloro substituted product. Suggest a structure for the alkane.
9. Explain the predominance of 2-chlorobutane as one of the products when *n*-butane is treated with Cl_2 in the presence of U.V. light.
10. Bromine is less reactive but more selective whereas chlorine is more reactive and less selective in its attack on alkanes. Explain
11. How will you justify that the formation of H—X bond determines the rates of various halogenation processes in methane ?
12. Give an explanation for the variation in m.p. in alkanes.
13. Discuss the relative rates of halogenation of alkanes in terms of bond dissociation energy.
14. How does the stability of a free radical plays a role in the halogenation of alkanes.
15. Give mechanism of chlorination of methane.
16. Out of the isomeric pentanes, which has the maximum boiling point and why ?
17. Name the different conformations of butane and arrange them in increasing order of stability.
18. Calculate the percentage of expected isomers obtained by monobromination of propane. The relative rates of substitution by bromine per 3°, 2° and 1° hydrogen atoms are 1600, 82 and 1 respectively.
19. Write a note on the selectivity and reactivity in the halogenation reactions of alkanes.
20. Explain the following statements:
    (*i*) A branched chain hydrocarbon boils at lower temperature than the straight chain isomer.
    (*ii*) The smaller the activation energy of a reaction, greater is the rate of reaction.
    (*iii*) The stronger the bond of the halogen with hydrogen, the greater is the activity of halogenation of alkanes.
    (*iv*) Halogenation with tetraethyl lead proceeds at a lower temperature than when it is done in its absence.
21. (*a*) What are the products of chlorination of methane ? Describe the mechanism of the formation of each.
    (*b*) Distinguish between heat of reaction and activation energy. What are kinetic controlled and thermodynamic controlled reaction products. Give one example of each.
22. Draw the Newman projection formulae for the different conformations of *n*-butane. Which of them is the least stable and why ?
23. Discuss the orientation of halogenation of alkanes taking into account the probability factor and the relative ease with which the different hydrogen atoms are abstracted by the halogen atom.
24. Explain the terms activation energy and transition state. How does activation energy help in predicting the relative rates of similar reactions ?

25. In the chlorination of alkanes, the abstraction rates of the tertiary, secondary and primary hydrogens are 5, 3.8 and 1 respectively. Calculate the relative proportions in which the various chloroderivatives of iso-butane are formed.

26. Give mechanism of chlorination of methane and cite three evidence to support this mechanism.

27. (*a*) Chlorination of *n*-butane in the presence of ultra-violet light gives a mixture containing 72 per cent 2-chlorobutane and 28 per cent 1-chlorobutane. How will you account for this ?

(*b*) Explain the predominance of 2-bromobutane as one of the products when bromine is treated with butane.

28. (*a*) Why does the iodination of methane require an oxidising agent while no such reagent is needed in the chlorination and bromination of methane ?

(*b*) Out of isomeric pentanes, which one has the maximum boiling point and why ?

(*c*) Arrange butane, propane and pentane in decreasing order of melting point and assign reason.

29. Justify the statements :

(*a*) The halogenation of alkanes in the presence of tetraethyl lead proceeds at a lower temperature than when it is carried in its absence.

(*b*) The chlorination of alkanes is an exothermic reaction yet it requires ultra-violet light for its success.

(*c*) Fluorination of methane is faster as compared to its chlorination.

30. Write notes on :

(*i*) Relative reactivity of alkanes in halogenation. (*ii*) Oxidation of alkanes.

31. (*a*) What do you understand by conformation ? Draw the potential energy diagram for the conformations of *n*-butane and arrange them in increasing order of stability.

(*b*) In the halogenation of methane with Cl and Br free radicals, the values of ΔH under similar conditions are :

$$\dot{C}l + CH_4 \longrightarrow HCl + \dot{C}H_3 \quad \Delta H = 1 \text{ k cal/mole}$$
$$E_{act} = 4 \text{ k cal/mole}$$

$$\dot{B}r + CH_4 \longrightarrow HBr + \dot{C}H_3 \quad \Delta H = +16 \text{ k cal/mole}$$
$$E_{act} = \text{Not given}$$

Give the relative rates of the above reactions giving reasons for your answer.

32. What is the order of reactivity of the different halogens towards methane ? Give a suitable explanation in support of your answer.

33. Write down the various structural isomers of an alkane with molecular mass 72. Which of them has minimum boiling point ? Give reasons.

34. Calculate the percentage of isomers expected during the monochlorination of butane at room temperature. The relative reactivities for 3°, 2° and 1° hydrogens are 5.0, 3.8 and 1.0 respectively.

35. (*a*) How are alkanes prepared by :

(*i*) Corey-house reaction (*ii*) Wurtz reaction

(*b*) Give the structure of pentane which would be of expected to give largest number of isomeric monochloro derivative.

(*c*) Discuss the trends in mp and bp in alkanes.

(*d*) The relative rates of substitution of bromine per 2° and 1° hydrogen atoms are 82 : 1. Calculate the percentage of expected isomers obtained during monobromination of propane at 400 K.

36. (*a*) Write notes on :

(*i*) Wurtz reaction (*ii*) Corey-House synthesis

(*b*) Discuss the mechanism of halogenation of alkanes. Also give evidences in support of the above mechanism.

37. (*a*) Explain why chlorination of *n*-butane in the presence of light gives a mixture of 72% of 2-chloropropane and 28% of 1-chlorobutane.

(*b*) Explain why

(*i*) The boiling points of *n*-alkanes increase as the molecular mass increases.

(*ii*) The boiling points of isomeric alkanes decrease with increase in the branching of the chain.

(*c*) Give two methods for the preparation of alkanes.

38. (*a*) Discuss the mechanism of chlorination of methane giving potential energy diagram.

(*b*) Justify that the order of relative reactivity of halogenation of alkanes is

$$F_2 > Cl_2 > Br_2 > I_2$$

39. Write the structures of the compounds from their IUPAC names :

(*i*) 1-Ethyl-3-methylcyclopentane (*ii*) 1-cyclopropylbutane

(*iii*) 2-Ethyl-1, 3-dimethylcyclohexane (*iv*) 3-Methyl-3-cyclobutylpentane

40. Give two methods of preparation of cycloalkanes.

41. What is the cause of angle strain in cyclic compounds ? Discuss in terms of oriental concept.

42. Discuss the orbital structure of angle strain.

43. Give a brief account of Wislicenus method for the synthesis of cycloalkane molecules.

44. What are cycloalkanes ? Write their general formula.

45. How are cycloalkanes formed by (2 + 2) cycloaddition reactions. Can (4 + 2) cycloaddition reactions also take place under similar conditions ?

46. Give a brief account of Thorpe-Ziegler reaction.

47. Cycloalkane molecules do not have any double bond but they take part in the addition reactions. Explain.

48. What is Sachse-Mohr theory of strainless rings ?

49. Give a brief account of Simmon-Smith reaction for the synthesis of cyclopropane derivatives.

50. What is Balnc's rule ? Discuss its interpretation.

51. Give a brief account of addition of carbenes to alkenes.

52. Whereas a cyclopentane ring is strained, there is no such strain on cyclohexane ring. Discuss.

53. How will you prepare cyclobutane molecule by starting from propylene dibromide and malonic ester ?

54. What happens when 1, 3-dibromopropane is treated with zinc ? What is this reaction known as ?

55. What is meant by cycloaddition reaction ? Discuss one such reaction under photochemical conditions.

56. What happens when calcium adipate is heated ?

57. What are the difficulties encountered in the synthesis of large membered rings ?

58. Give a brief account of Dieckmann Cyclisation reaction.

59. What is Baeyer's strain theory ? What are its limitations ?

60. Give a brief account of Dieckmann cyclisation reaction. Discuss its mechanism.

61. (*a*) Explain the following :

(*i*) Thorpe-Ziegler reaction (*ii*) Demjanov Rearrangement

(*b*) Explain the validity of Sachse Mohr's theory using heat of combustion value of various cycloalkanes. Give orbital picture of angle strain.

62. (*a*) Explain Demjanov rearrangement pointing clearly the nature of the products formed.

(*b*) Give the structure and names of main organic products in the following reactions :

(*i*) Cyclopropane + $Br_2 \xrightarrow{(CCl_4)}$

(*ii*) Cyclopentane + $Br_2 \xrightarrow{(575\ K)}$

(*iii*) Cyclohexane + $Br_2 \longrightarrow$

63. Addition of singlet carbenes to alkenes is stereospecific in nature while that of triplet carbenes is non-stereospecific in nature. Explain.

64. What are cycloalkanes ? How will you account for their reactive nature ?

65. What type of compounds undergo ring expansion and what type of compounds undergo ring contraction in Demjanov rearrangement ? Give details of the reactions.

66. (*a*) Write the following reactions with suitable examples.

(*i*) Dieckmann Cyclisation

(*ii*) Thorpe Ziegler reaction.

(*b*) Explain Baeyer's strain theory. Why is it not applicable to cyclohexane ?

67. (*a*) How would you prepare cyclopentane from diethyl adipate ?

(*b*) Cyclobutyl amine on reaction with HNO_2 yields two products. Explain with mechanism.

(*c*) Choosing one suitable example differentiate between the terms "conformation" and "configuration".

68. (*a*) Discuss Demjanov rearrangement in detail with suitable examples.

(*b*) Write notes on the following :

(*i*) Schase Mohrs theory of strainless rings.

(*ii*) Orbital pircture of angle strain.

69. Give a brief account of Thorpe-Ziegler method and Dieckmann cyclisation reaction for the preparation of cycloalkanes.

70. How will you prepare cycloalkane derivatives by the following methods ?

(*i*) Simmons-Smith reaction

(*ii*) Diels-Alder reaction

(*ii*) Dehalogenation of dihalides.

71. What are α, ω dihalogen derivatives ? How will you prepare cycloalkanes from them ?

72. Give the main features of Baeyer's strain theory. Why does this theory fail to account for the stability of cyclohexane ?

73. How does heat of combustion value of methylene group account for the stability of cycloalkanes ?

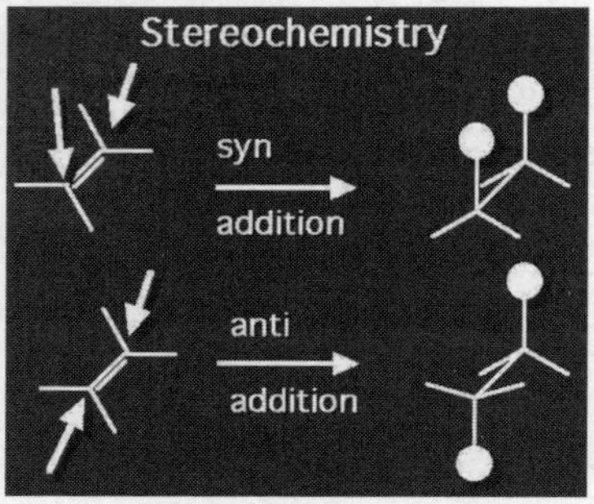

5 Alkenes, Cycloalkenes, Dienes And Alkynes

ALKENES

5.1 NOMENCLATURE OF ALKENES

Alkenes are unsaturated hydrocarbons having the general formula C_nH_{2n}. They contain two hydrogen atoms less than required to form alkanes with the same number of carbon atoms. This is made possible by introducing a double bond between two carbon atoms. This symbolises unsaturation in the compound. Example of alkenes are ethylene and propylene.

Nomenclature

Nomenclature of alkenes is explained in the form of table below:

S.No.	*Structure*	*Common name*	*IUPAC name*
1.	$CH_2 = CH_2$	Ethylene	Ethene
2.	$CH_3CH = CH_2$	Propylene	Propene
3.	$CH_3CH_2CH = CH_2$	β-butylene	But-1-ene
4.	$CH_3CH = CH - CH_3$	α-butylene	But-2-ene
5.	$CH_3 - C(CH_3) = CH_2$	Isobutylene	2-MethylPropene

Example 1. **Write the IUPAC names of isomers having the molecule formula C_5H_{10}.**

Solution. Names and structure of five isomers having the molecules formula C_5H_{10} are given below.

1. $CH_3 - CH_2 - CH_2 - CH = CH_2$ — pent-1-ene
2. $(CH_3 - CH_2)(H)C = C(CH_3)(H)$ — *cis*-pent-2-ene
3. $(CH_3 - CH_2)(H)C = C(H)(CH_3)$ — *trans*-pent-2-ene
4. $\overset{4}{C}H_3 - \overset{3}{C}H = \overset{2}{C}(CH_3) - \overset{1}{C}H_3$ — 2-Methylbut-2-ene

5.2 HINDERED ROTATION AROUND DOUBLE BOND

The lack of freedom of rotation of the bonded atoms around a carbon-carbon double bond is known as hindered rotation.

Explanation. A double bond is constituted of a σ bond and a π bond between two carbon atoms.

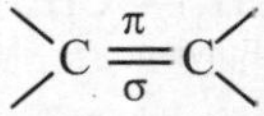

It is thus associated with greater bond dissociation energy compared to a single bond. Bond length is also shorter compared to a single bond. It is therefore much more difficult to break a double bond.

Rotation round a double bond would involve breaking the double bond which requires a large amount of energy. Hence rotation around the double bonds is difficult.

5.3 ORBITAL STRUCTURE OF ETHYLENE

In the molecule of ethylene (C_2H_4), the two carbon atoms are linked to each other through a double bond while each carbon atom is separately linked to two hydrogen atoms through single bonds.

Since hybridised orbitals lie in the same plane all the carbon and hydrogen atoms of ethylene also lie in the same plane.

The unhybridised 2π orbitals of the two carbon atoms overlap in a plane perpendicular to the plane of carbon and hydrogen atoms, to form carbons-carbon π bonds as shown below.

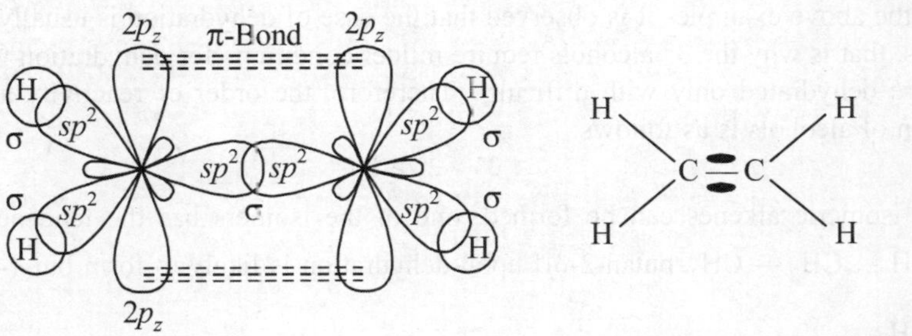

The carbon-carbon double bond length in ethylene is smaller than the carbon-carbon single bond length in ethane.

The carbon-carbon double bond length in ethylene is somewhat smaller than the carbon-carbon single bond length in ethane because of the following reasons:

(*i*) In ethylene, the two carbon atoms are sp^2 hybridised while in ethane the carbon atoms are sp^3 hybridised. The size of sp^2 hybrid orbitals is slightly smaller than that of sp^3 orbitals. Therefore, the length of the bond formed by the mutual overlapping of sp^2 orbitals would be smaller than bond formed by the mutual overlapping of sp^3 orbitals.

(*ii*) Carbon-carbon double bond also involves the additonal sidewise overlap of unhybridised $2p_z$ orbitals. As a result, the carbon atoms are brought still closer and the bond length decreases.

5.4 METHODS OF PREPARATION OF ALKENES

By Dehydration of Alcohols

An alcohol is converted into an alkene by dehydration. Dehydration is brought about by the use of an acid say conc. H_2SO_4 and application of heat. It involves heating the alcohol with H_2SO_4 or by passing the vapours of alcohol over alumina Al_2O_3 at 623-673 K.

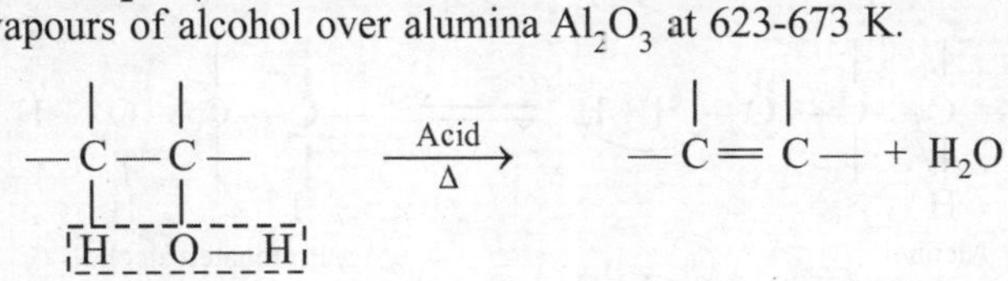

1. $CH_3CH_2OH \xrightarrow[433\text{-}443\ K]{H_2SO_4} CH_2{=}CH_2 + H_2O$
   Ethylene

2. $CH_3\,CH_2{-}CH_2{-}CH_2OH \xrightarrow[413\ K]{H_2SO_4} CH_3{-}CH{=}CH{-}CH_3$
   Butan-1-ol → But-2-ene [Major product]

3. 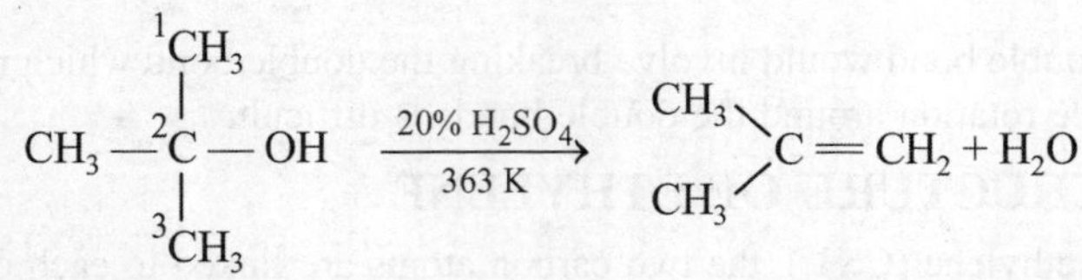

2-Methyl propan-2-ol → 2-Methyl prop-1-ene

4. 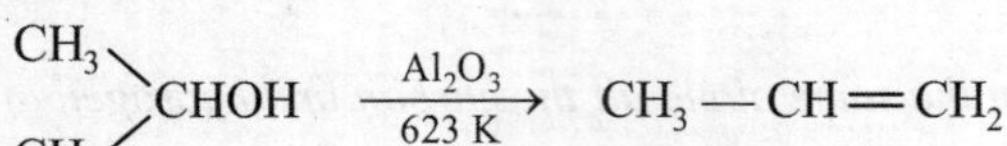

Isopropyl alcohol → Propylene

Reactivity and Orientation (Regioselectivity)

From the above examples it is observed that the ease of dehydration is usually the greatest in 3° alcohols, that is why the 3° alcohols require milder conditions for dehydration while simple 1° alcohols are dehydrated only with difficulty. Therefore, the order of reactivity and the ease of dehydration of alcohols is as follows:

$$3° > 2° > 1°$$

When isomeric alkenes can be formed, one of the isomers has the tendency to dominate [$CH_3{-}CH(OH){-}CH_2{-}CH_3$, butan-2-ol] upon dehydration is likely to form but-1-ene and but-2-ene, but actually yields almost exclusively but-2-ene. This is in accordance with Saytzeff rule.

Saytzeff rule states that in the dehydration of alcohols, hydrogen is preferentially eliminated from the carbon atom which contains fewer number of hydrogens attached to it This is because more highly alkylated olefins are more stable than others.

Since dehydration is an E_1 elimination the Saytzeff rule [Elimination yielding the most highly alkylated olefins] applies to those cases where more than one olefins can be formed.

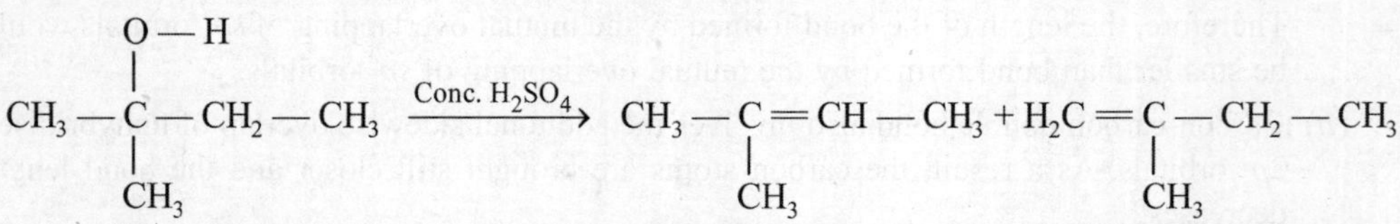

2-Methylbutan-2-ol → 2-Methylbut-2-ene (Major product) + 2-Methylbut-1-ene (Minor)

Mechanism of dehydration of alcohols. The mechanism of dehydration of alcohols is given schematically below:

1. Attachment of the proton to alcoholic oxygen

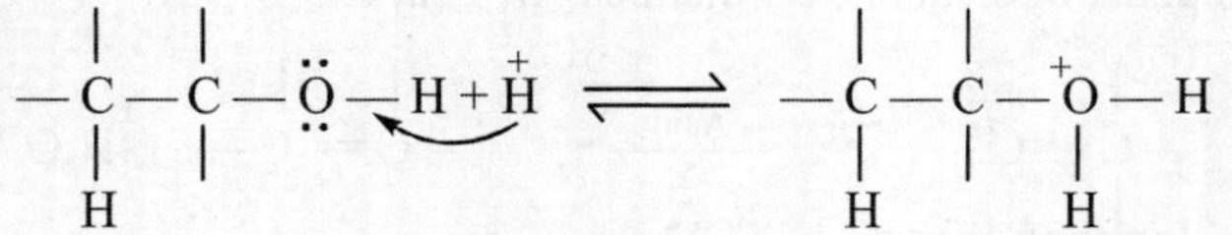

Alcohol ⇌ Protonated alcohol

2. Removal of water molecule

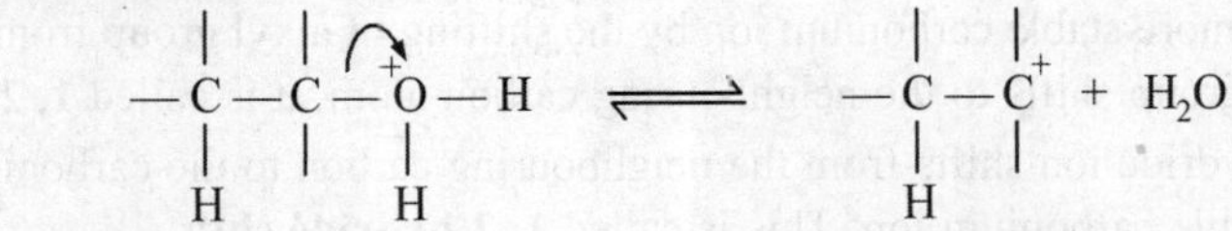

Carbonium ion

3. Removal of proton

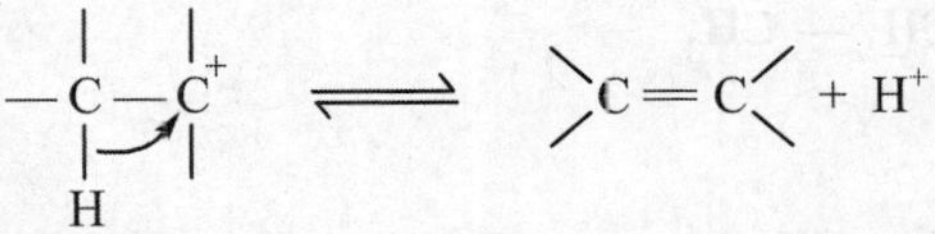

*For those alcohols that yield volatile olefins, the reaction can always be driven to completion by distilling off the olefin as it is formed.

Alkyl shift and hydrogen shift

Sometimes it is found that the alkene obtained by dehydration of alcohols does not fit in the mechanism.

For example,

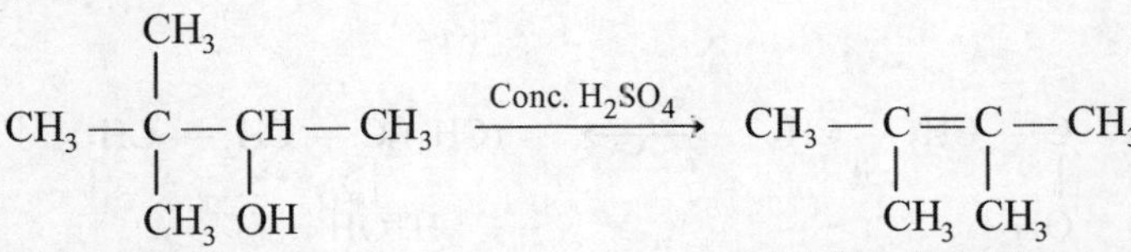

3, 3-Dimethylbutan-2-ol. 2, 3-Dimethylbut-2-ene (Major product)

This can be explained by considering a rearrangement of carbonium ions. A carbonium ion can arrange to form a more stable carbonium ion. The mechanism of rearrangement is given below:

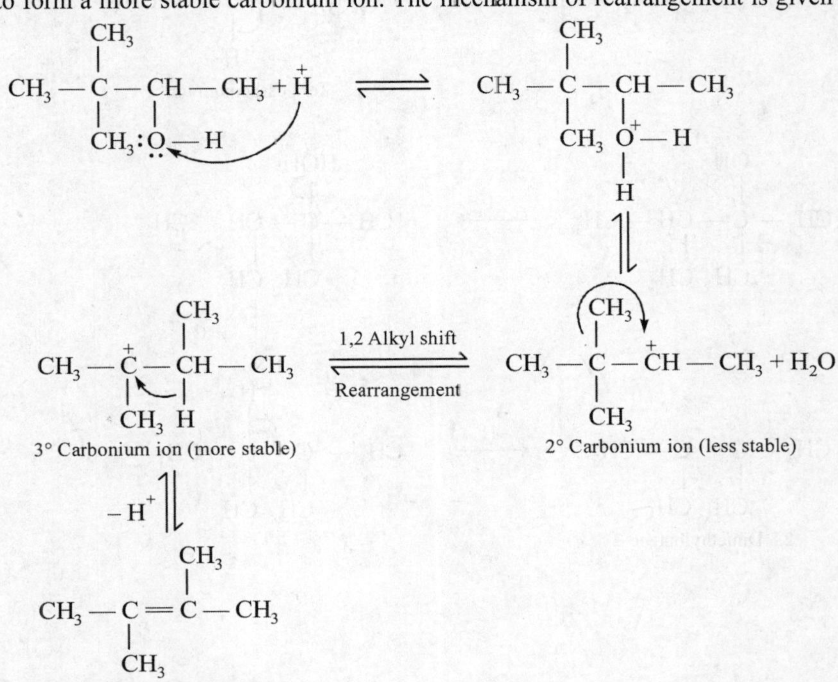

3° Carbonium ion (more stable)

2° Carbonium ion (less stable)

2, 3-Dimethylbut-2-ene (major product)

Thus, we see that the carbonium ion which is the intermediate product in the dehydration of alcohol changes to a more stable carbonium ion by the shifting of alkyl group from one position to another. If the alkyl group shifts to the neighbouring carbon atom, it is called **1, 2 alkyl shift**.

Somethings, a hydride ion shifts from the neighbouring carbon to the carbonium ion, thereby producing a more stable carbonium ion. This is called **1, 2 hydride shift**.

Example 2. **Predict the major product obtained on dehydration of the follwoing alcohols:**

(*i*) $(CH_3)_2\ C(OH)\ CH_2 - CH_3$

(*ii*) 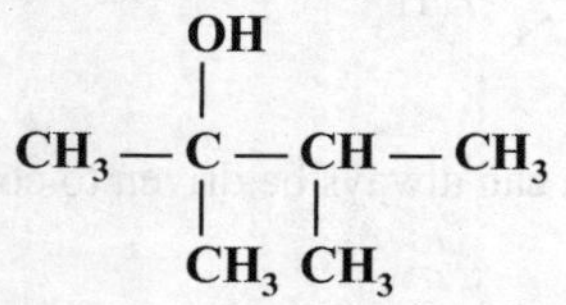

(*iii*) 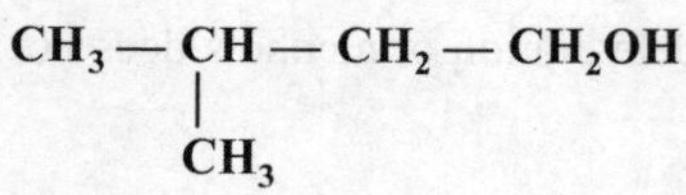

Solution.

(*i*) 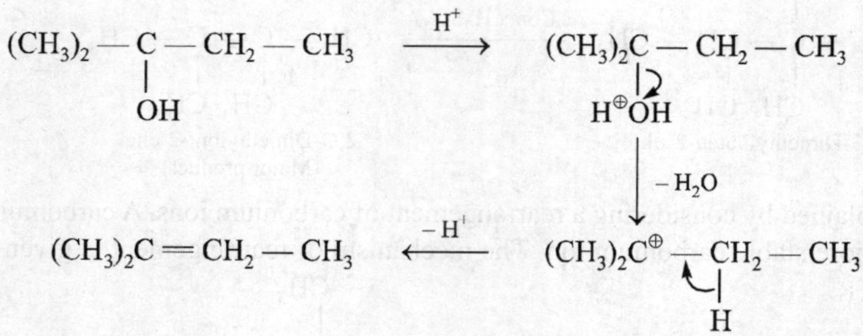

Tert. carbonium ion

(*ii*) 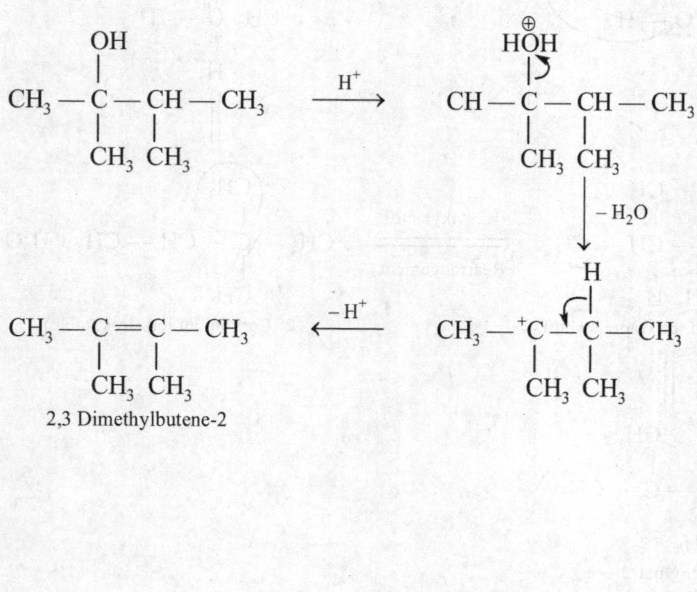

2,3 Dimethylbutene-2

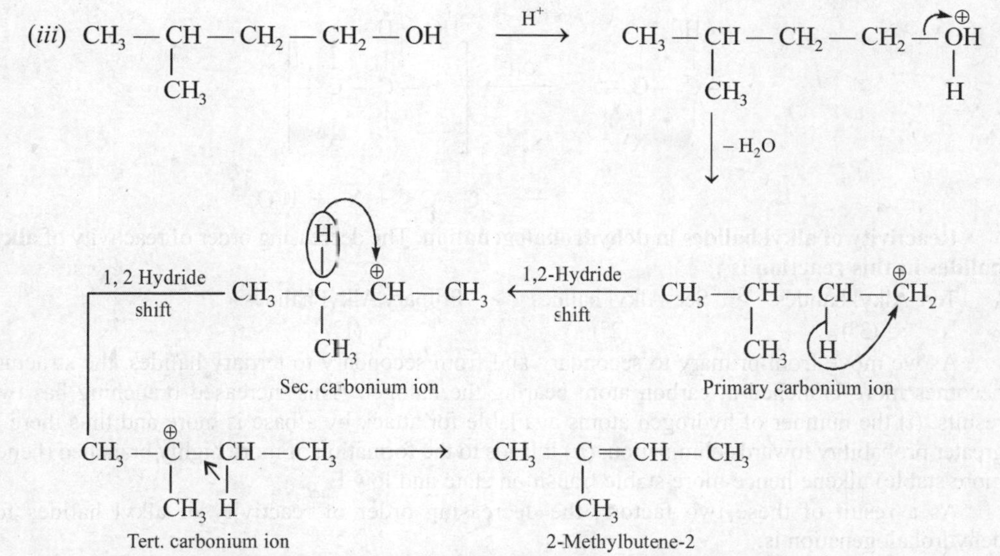

Preparation of alkenes by dehydrohalogenation of alkyl halides

Alkyl halides can form alkenes by the loss of molecule of hydrogen halide under the influence of a base catalyst (alcoholic potash). The reaction is known as *dehydrohalogenation of alkyl halides*. It involves the removal of the halogen atom together with the hydrogen atom from the adjacent carbon atom.

$$CH_3 - CH_2X \xrightarrow{OH^-} CH_2 = CH_2 + H_2O + X^-$$

Mechanism of reaction (E_2 mechanism)

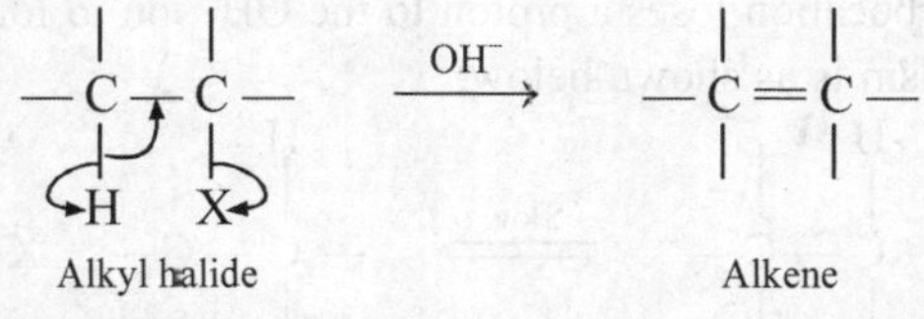

Orientation in dehydrohalogenation. In some cases, this reaction yields a single product (alkene) and in other cases yields a mixture of alkenes. For example, out of 1-chlorobutane and 2-chlorobutane the former yields only but-1-ene while the latter gives but-1-ene and but-2-ene. Thus, the preferred product is the alkene that has the greater number of alkyl groups attached to the doubly bonded carbon atoms. In other words, the ease of formation of alkene has the following order:

$$R_2C = CR_2 > R_2C = CHR > R_2C = CH_2 > RCH = CHR > RCH = CH_2 > CH_2 = CH_2$$

This order is based on heat of hydrogenation values.

Hence in dehydrohalogenation, the more stable the alkene more easily it is formed.

In the dehydrohalogenation of alkyl halide to form alkene, the preferred product is the alkene which is more highly alkylated at the doubly bound carbon atoms. This is another statement of Saytzeff Rule.

This is also clear from the transition state formed in the dehydrohalogenation of alkyl halides as discussed below. The double bond is partly formed and the transition state acquires alkene character. Factors that stabilise an alkene also stabilise an *incipient* alkene in the transition state.

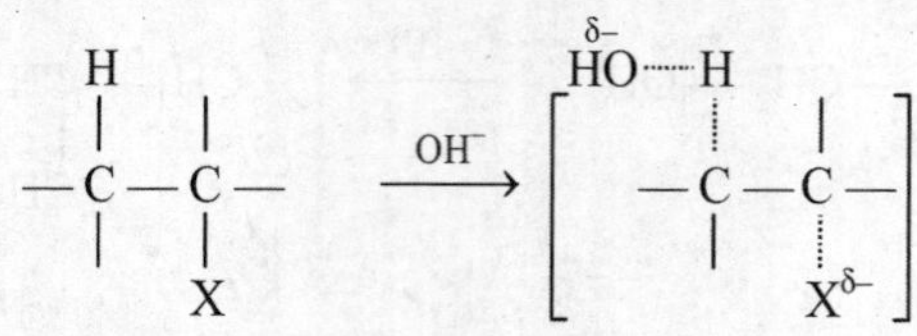

$$\longrightarrow\ >C=C<\ +\ X^- \ +\ H_2O$$

Reactivity of alkyl halides in dehydrohalogenation. The decreasing order of reactivity of **alkyl halides in this reaction is :**

Tert-Alkyl halides	>	Sec-Alkyl halides	>	Primary Alkyl halides
(3°)		(2°)		(1°)

As we move from primary to secondary and from secondary to tertiary halides, the structure becomes more branched at carbon atom bearing the halogen. This increased branching has two results. (*i*) the number of hydrogen atoms available for attack by a base is more and thus there is greater probability towards elimination, (*ii*) it leads to the formation of more highly branched (hence more stable) alkene hence more stable transition state and low E_{act}.

As a result of these two factors, the decreasing order of reactivity of alkyl halides for dehydrohalogenation is:

$$3° > 2° > 1°$$

The mechanism discussed above is called E_2 type of mechanism of dehydrohalogenation.

Mechanism of Dehydrohalogenation (E_1 mechanism)

Some secondary and tertiary alkyl halides undergo dehydrogenation in a solution of low base concentration by a different mechanism known as E_1 mechanism (E for elimination; 1 for unimolecular). *This mechanism operates through a two-step process in which the rate deteming step involves only one molecule.* It is believed that in the first step, which is a slow and rate-determining step, the alkyl halide dissociates into halide ion and carbocation. In the second step which is a fast one, the carbocation loses a proton to the OH^- ion to form the alkene.

The complete mechanism is as shown below:

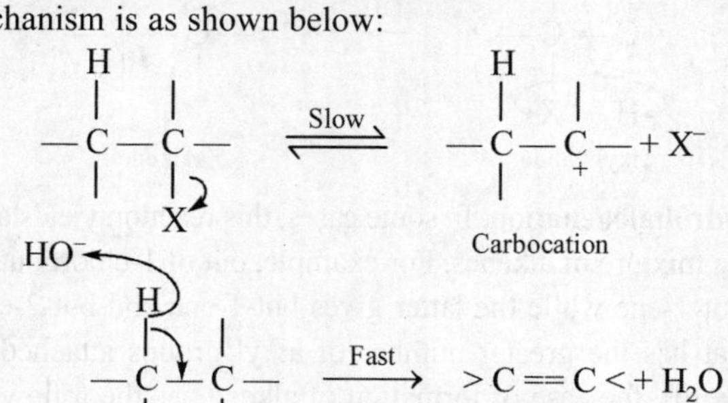

In certain alkyl halides (say 3°), there can be slight variation in the mechanistic approach as the carbon cation initially formed undergoes rearrangement to form more stable carbocation and thus yields highly branched alkene as the chief product of the reaction.

For example, action of alcoholic potash on 2-chloro-3, 3-dimethylbutane yields chiefly 2, 3-dimethylbut-2-ene as the carbocation initially formed undergoes rearrangement as shown below:

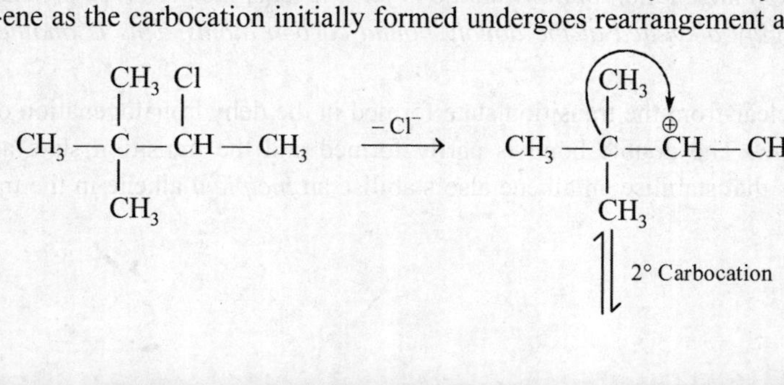

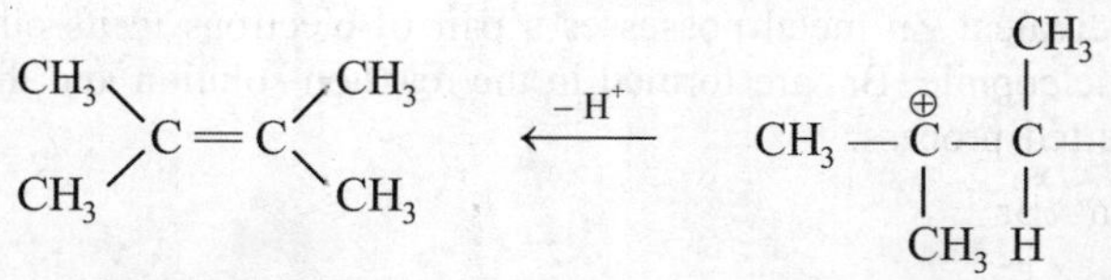

2,3 Dimethylbut-2-ene (Chief product) — 3° Carbocation (More stable)

This is an example of E_1 type of mechanism of dehydrohalogenation.

For a given alkyl group, the order of reactivity of different alkyl halides is RI > RBr > RCl > RF. For the same halogen atom, the order of reactivity is: Tertiary alkyl halide > sec. alkyl halide > Primary alkyl halide. We thus observe that the order of reactivity of alkyl halides in both E_1 and E_2 types of reactions is the same. It is tertiary > secondary > primary.

***Example 3.* For a given halogen, what is the order of reactivity of various alkyl halides? Explain.**

Solution. For a given halogen, the order of reactivity of alkyl halides is:

tertiary > secondary > primary

This is because in a tertiary alkyl halide, there is maximum number of hydrogens on the neighbouring carbon atom. So the chances of attack by hydroxyl ion (E_2 mechanism) are maximum leading to maximum reactivity. This is followed by secondary and primary alkyl halides.

Even if the reaction takes place by E_1 mechanism, tertiary alkyl halide has greatest reactivity because the carbonium ion formed in the first step has the maximum stability leading to completion of reaction. A tertiary carbonium ion (carbocation) has maximum stability followed by secondary and followed by primary carbonium ion.

***Example 4.* For a given alkyl group, what is the order of reactivity with different halogens?**

Solution. Order of reactivity with different halogens, with a given alkyl group is

I > Br > Cl > F

Alkyl iodide has the maximum reactivity because C — I bond can be broken most easily leading to the formation of products, C — I bond has the minimum bond dissociation energy. Reactivity of alkyl halide decreases as we go to bromide, chloride and fluoride.

Preparation of alkenes by dehalogenation of dihalides

Dehalogenation of 1, 2 dihaloalkanes (*vic*-dihalides) involves the treatment to the vic-dihalide with reactive metals like Zn in acetic acid and it yields the corresponding alkenes.

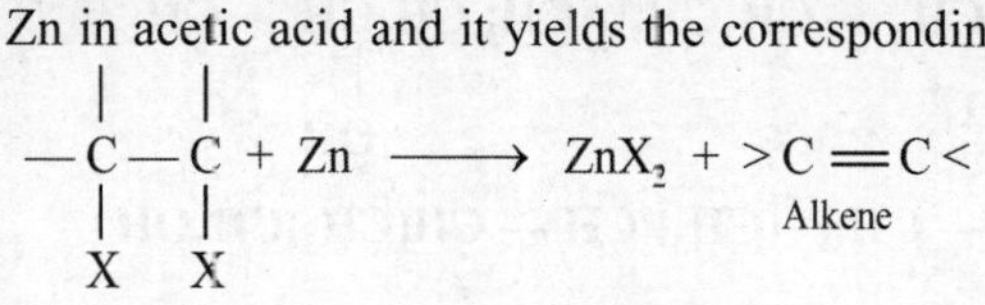

1,2 dihaloalkane

For example:

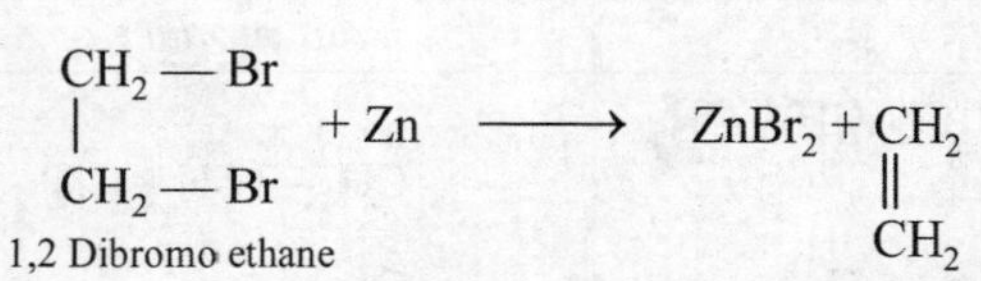

1,2 Dibromo ethane — Ethylene

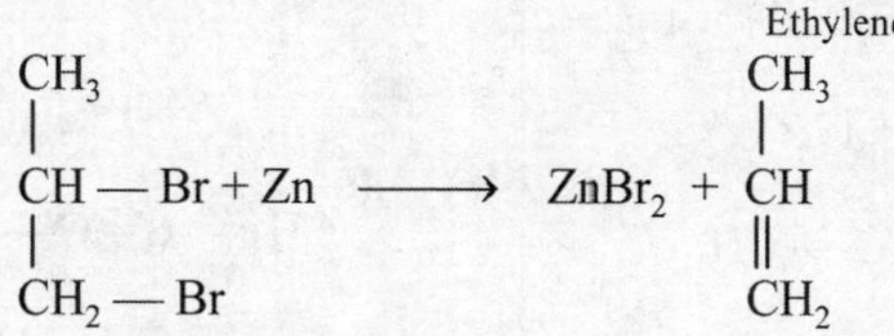

1,2 Dibromopropane — Propylene

Mechanism. The divalent Zn metal possesses a pair of electrons in its outermost shell and therefore it acts as a nucleophile, Br^- are formed in the reaction solution and the reaction can be represented by a concerted process.

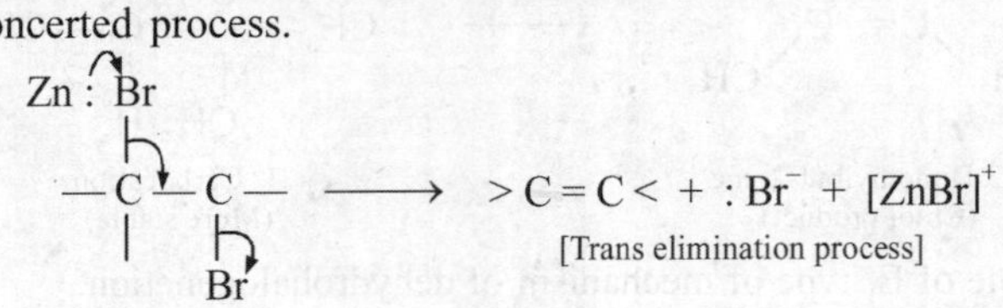

$$[Zn\,Br]^+ \longrightarrow Zn^{2+} + Br^-$$

Preparation of alkenes by (*i*) reduction of alkynes and (*ii*) dehydrogenation of alkanes

(*i*) Alkenes can be prepared by the partial reduction of alkynes using metallic *sodium* or *lithium* in liquid ammonia, or by using a calculated quantity of hydrogen over palladium catalyst.

Reduction with sodium in liquid ammonia yields *trans* alkene, whereas *cis* alkene results from H_2 in the presence of Pd.

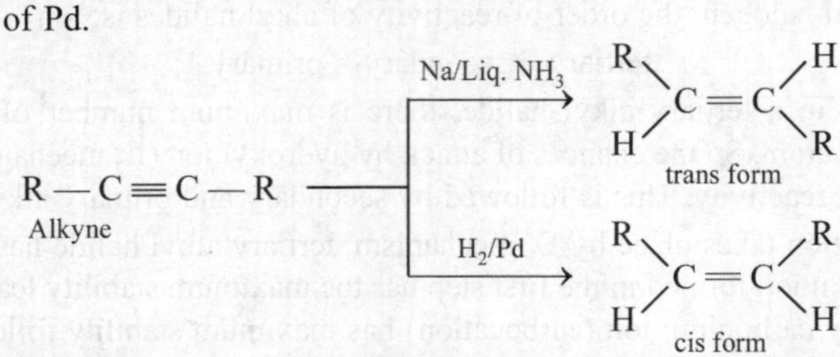

(*ii*) **Dehydrogenation of alkanes.** It involves the passing of the alkane in the vapour phase over a bed of catalysts (oxides of chromium and aluminium).

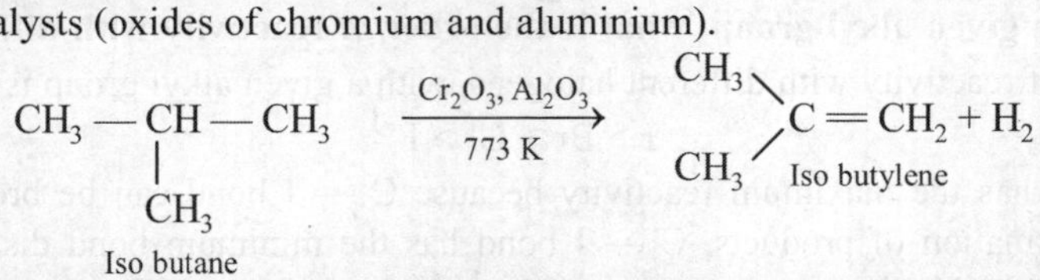

Example 5. **Which alcohol of each pair would you expect to be more easily dehydrated? Also name the main products of dehydration.**

1. $CH_3-CH_2-CH_2-CH_2OH$, $CH_3CH_2CH(OH)-CH_3$

2. $(CH_3)_2CH-C(CH_3)(OH)-CH_3$, $(CH_3)_2CH-CH(CH_3)CH_2OH$

Solution.

	Alcohol	*Main product of dehydration*
1.	$CH_3CH_2CH(OH)-CH_3$ Butan-2-ol	$CH_3-CH=CH-CH_3$ But-2-ene
2.	$CH_3-CH(CH_3)-C(CH_3)(OH)-CH_3$ 2, 3-Dimethylbutan-2-ol	$CH_3-C(CH_3)=C(CH_3)-CH_3$ 2, 3-dimethylbut-2-ene

Example 6. **When neopentyl alcohol $(CH_3)_3.C.CH_2OH$ is heated with acid, it is slowly converted into a mixture of two isomeric alkenes of the formula C_5H_{10}. Name these alkenes according to IUPAC system of nomenclature and also show how they are formed. Which one of them is the major product and why ?**

Solution.

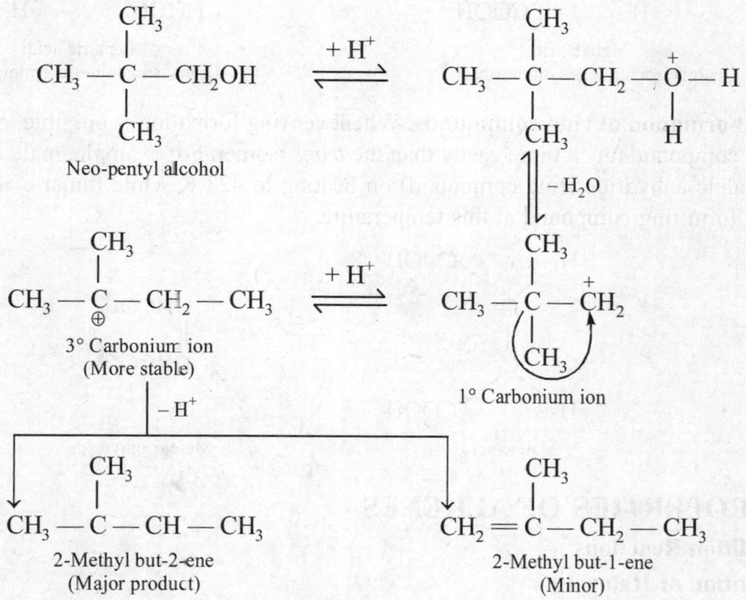

Distinguishing properties of geometrical isomers

The geometrical isomers generally differ from each other in the following poperties:

(*i*) Dipole moment. In case of compounds having polar bonds in their molecules, the *trans* isomer has generally smaller dipole moment than *cis* isomer. For example:

cis 1, 2 dibromoethene ($\mu = 1.35$) trans 1, 2 dibromoethene ($\mu = 0$)

This is due to the fact that in *trans* isomer, similar groups are on opposite sides of the molecule and their bond polarities tend to cancel each other. But in *cis* isomers the bond polarities are not cancelled out.

(*ii*) Boiling points. Generally speaking *cis* isomers have higher boiling points than *trans* isomers. This is due to the fact that *cis* isomers have greater intermolecular forces of attraction on account of higher dipole moments of their molecules. For example, boiling point of *cis*-2-butene is 277 K while that of *trans*-2-butene is 274 K.

(*iii*) Melting points. Out of *cis* and *trans* isomers, the *trans* compound has higher melting point. This is because *trans* compounds are more symmetrical than their *cis* isomers and are, therefore, more closely packed in the crystal lattice. For example, melting point of maleic acid (*cis* isomer) is 403 K while fumaric acid (*trans* isomer) is 503 K.

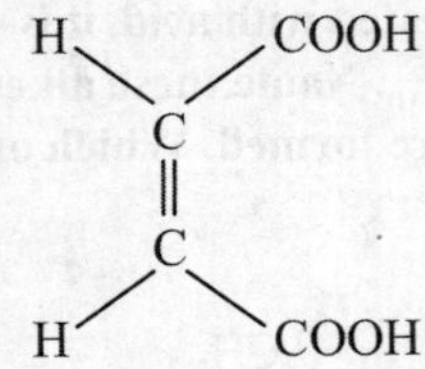

Maleic acid
(cis isomer, less symmetrical)

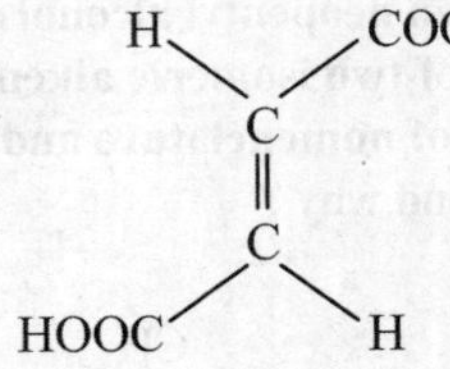

Fumaric acid
(trans isomer, more symmetrical)

***(iv)* Formation of ring compounds.** Whenever ring formation is possible, *cis* isomer changes into ring compound much more easily than the *trans* isomer. For example, maleic acid (*cis* isomer) forms maleic anhydride (ring compound) on heating to 423 K while fumaric acid (*trans* isomer) does not form ring compound at this temperature.

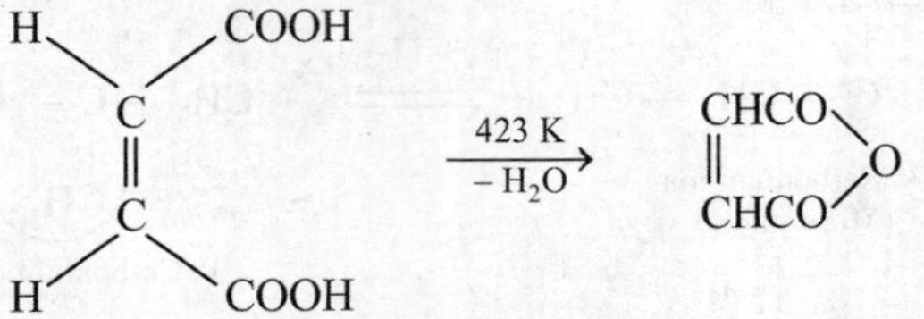

Maleic acid Maleic anhydride

5.6 PROPERTIES OF ALKENES

Addition Reactions

Addition of Halogens

Alkenes add a molecule of halogen to form alkyl halides.

$$CH_2=CH_2 + \underset{\text{Brown colour}}{Br_2} \longrightarrow \underset{\text{1, 2 Dibromoethane (colourless)}}{CH_2Br - CH_2Br}$$

On adding bromine water to a compound containing a double bond, brown colour due to bromine disappears.

Mechanism of halogenation

Halogenation of an alkene takes place by electrophilic addition mechanism. A halogen molecule (Cl_2, Br_2 or I_2) is non-polar in nature. But when this molecule approaches the alkene molecule, the π bonds have the effect of polarising the halogen molecule. The positive end of the polarised halogen molecule is attached to one ethylenic carbon atom with the simultaneous attachment of the negative end to the second ethylenic carbon atom forming a halonium ion, with the release of halide ion. The halide ion then attacks the halonium ion from the back side to avoid steric hindrance, thus giving rise to a dihalide. The steps of the mechanism are shown as under, taking the example of bromination.

$$Br - Br \xrightarrow[\text{cloud}]{\pi \text{ Electron}} \overset{\delta^+}{Br} - \overset{\delta^-}{Br}$$

Polarised bromine molecule

Step 1. 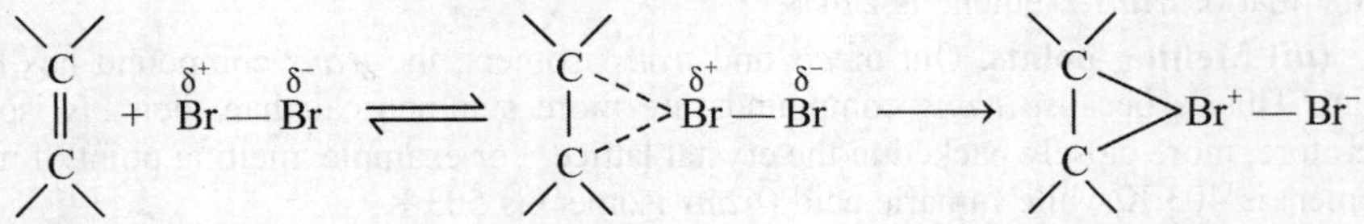

Alkene π-Complex Bromonium ion

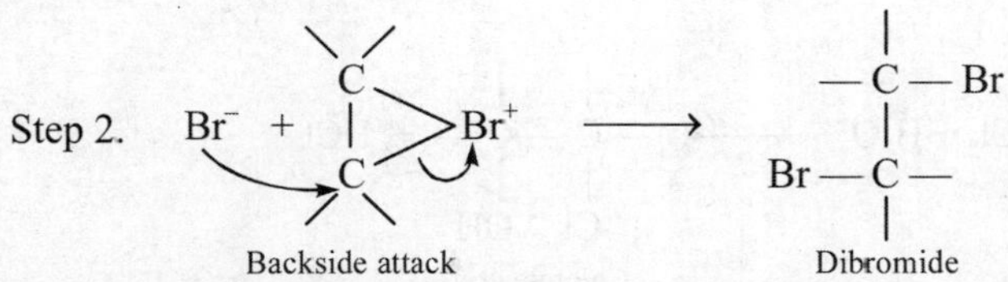

Evidence to prove that addition of halogen to an alkene proceeds through a halonium ion and not through carbonium ion (carbocation).

Addition of halogen to an alkene proceeds through the formation of a halonium ion and not through a carbonium ion. This can be proved like this:

It is observed that halogenation of alkenes gives rise to products that are optically active. This phenomenon can be explained by halonium ion mechanism and not by carbonium ion mechanism.

In halonium ion mechanism, the halide (in the second step) attacks from the back of halonium ion, giving rise to *trans* dihalide product.

In carbonium ion mechanism, the halide can attack from both sides of the carbonium ion (as it is flat in shape), giving rise to both *cis* and *trans* isomers.

Trans product obtained from halonium ion mechanism explains optical activity of the actual product obtained.

If we assume the carbonium ion mechanism, the mixture of *cis* and *trans* products will give us a racemic modification which will have no net optical activity. This is against the observations. Practically, we get an optically active compound.

This is schematically shown as under taking the example of bromination of cyclohexene.

Carbonium ion mechanism

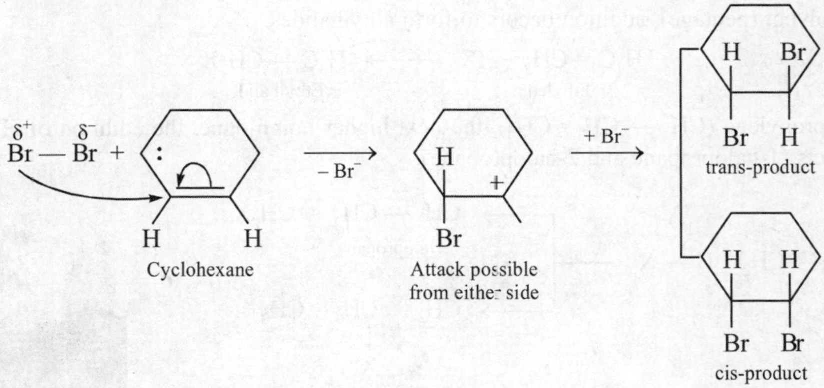

Halonium ion mechanism

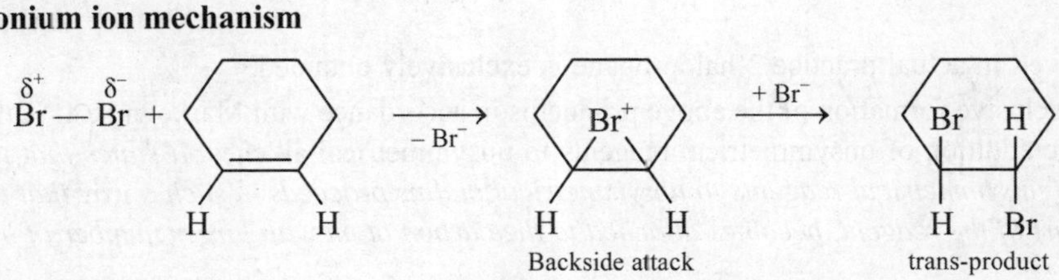

Hence halonium ion mechanism is confirmed.

Reaction of Alkenes with Halogens in the Presence of Water

Alkenes react with halogens in the presence of water to form halohydrins.

For example:

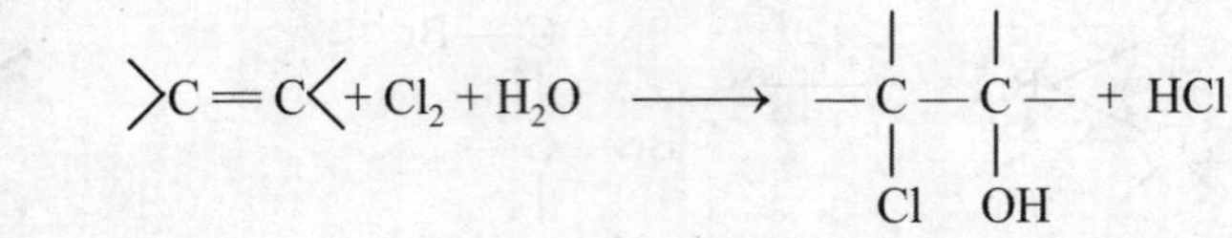

Chlorohydrin

Mechanism of the Reaction

The reaction takes place through the intermediate formation of a halonium ion as shown below:

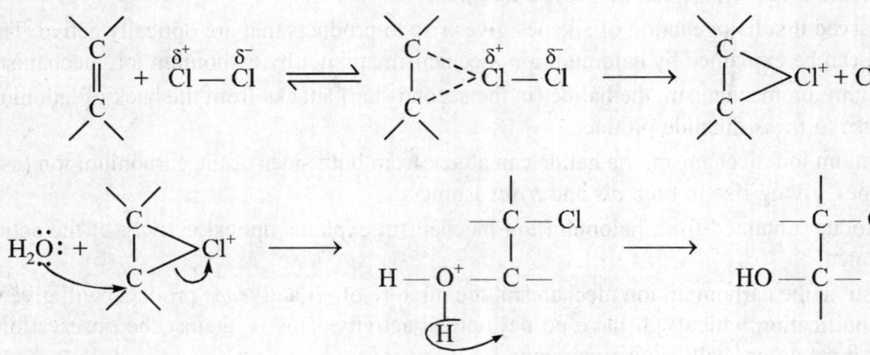

Addition of an unsymmetrical molecule over a double bond. Markownikoff's rule

When olefins are treated with hydrogen halides either in the gas phase or in an inert non-ionizing solvent (pentane), addition occurs to form alkylhalides.

$$\underset{\text{Ethylene}}{H_2C = CH_2} + HX \longrightarrow \underset{\text{Ethylhalide}}{H_3C - CH_2X}$$

With propylene ($CH_3 - CH = CH_2$), the next higher homologue, the addition of HX yields two products (1-halopropane and 2-halopropane).

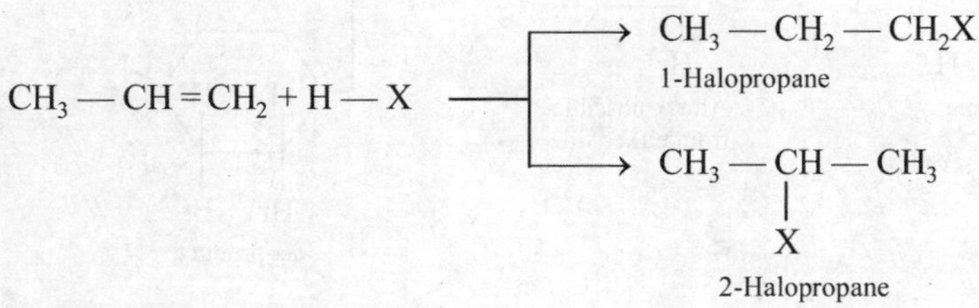

Vladimir Markownikoff

However, in actual practice 2-halopropane is exclusively obtained.

The exclusive formation of the above product is in accordance with Markownikoff Rule which governs the addition of unsymmetrical reagents to unsymmetrical alkenes. *It states that the ionic addition of unsymmetrical reagents to unsymmetrical olefins proceeds in such a way that the more positive part of the reagent, becomes attached to the carbon atom with larger number of hydrogen atoms.*

When the hydrogen halides or water is added, hydrogen atom constitutes the more positive part or in other words, the hydrogen atom becomes attached to the olefinic carbon atom which carries the larger number of hydrogen atoms, for example:

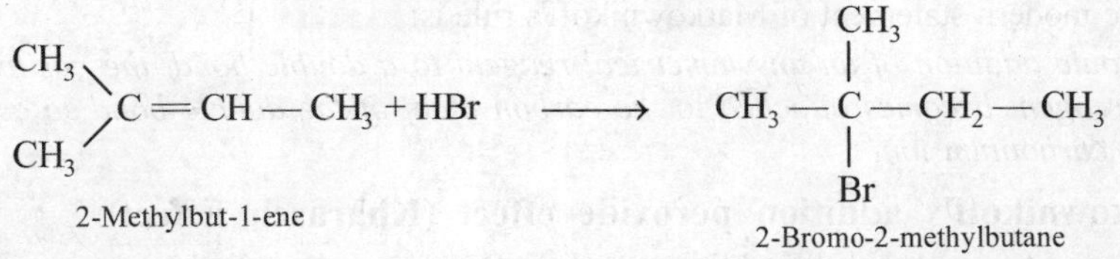

Theoretical explanation of the orientation of addition to olefins is related to the relative stability of carbonium ions. We know that relative stability of carbonium ions is: tertiary > secondary > primary > CH_3^+. If there is the probability of more then one carbonium ions being formed, the addition of electrophile yields the more stable one.

That is why **2 bromopropane** is exclusively formed by the addition of HBr to propylene (an unsymmetrical alkene) and not **1-bromopropane** because secondary isopropyl carbonium ion is more stable than primary *n*-propyl carbonium ion.

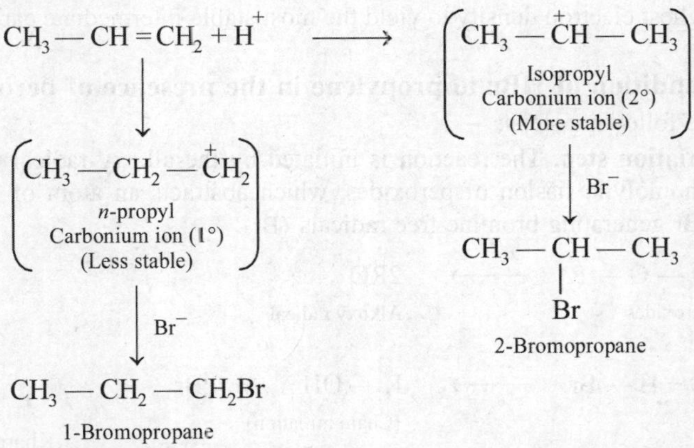

Considering the electronic effect of methyl group in propylene, the polarization of the double bond due to the (+I) electron-repelling inductive effect of the methyl group is depicted below.

$$CH_3 \rightarrow \overset{\delta^+}{CH}=\overset{\delta^-}{CH_2}$$

Thus, the +ve and –ve ends of the dipole (H — X) will add to the –ve and +ve ends of the double bonds, respectively yielding **2-halopropane.**

Mechanism

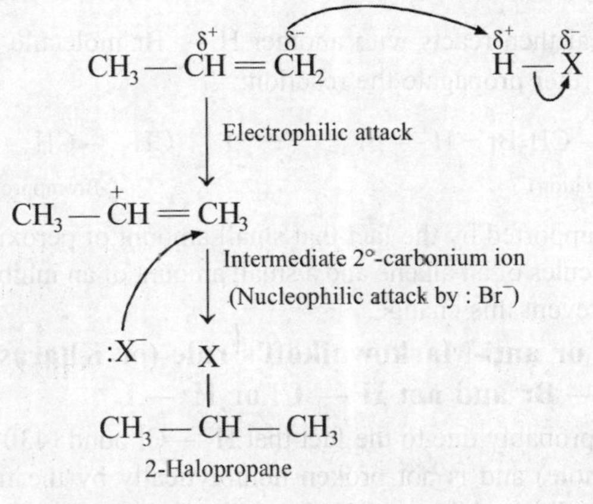

Thus the modern statement of Markownikoff's rule is:

In the ionic addition of an unsymmetrical reagent to a double bond, the positive portion of the adding reagent becomes attached to the carbon atom of the double bond so as to yield the more stable carbonium ion.

Anti-Markownikoff's addition peroxide effect (Kharasch Effect)

If has been observed that the addition of H — Br to propylene in the presence of peroxides, yields predominantly 1-bromopropane, that is the reagent adds onto the olefins under these conditions in a manner contrary to Markownikoff's rule thus suggesting a change in the mechanism.

This change in the orientation of addition due to the presence of peroxides is known as the *peroxide effect*.

Here the change in the mode of addition of the reagent is due to a change from an ionic *mechanism to a free-radical mechanism*. Markownikoff's addition requires the initiation by H^+. Anti-Markownikoff's addition requires initiation by ($\dot{B}r$). Each species attacks the olefin molecule at the centre of highest electron density to yield the most stable intermediate carbonium ion or free radical.

Mechanism of addition of HBr to propylene in the presence of peroxides

It involves the following steps:

***(i)* Chain initiation step.** The reaction is initiated by the alkoxy radical produced by the homolytic fission of peroxides, which abstracts an atom of hydrogen from HBr generating bromine free radicals ($\dot{B}r$).

Kharasch in his laboratory

$$\underset{\text{Peroxides}}{R-O-O-R} \longrightarrow \underset{\text{Alkoxy radical}}{2R\dot{O}}$$

$$R-\dot{O}+H-Br \longrightarrow \underset{\text{(Chain initiation)}}{R-OH} + \dot{B}r$$

***(ii)* Chain propagation step.** The $\dot{B}r$ then attacks the propylene molecule to give a more stable secondary radical addition species:

$$CH_3-CH=CH_2+\dot{B}r \longrightarrow \underset{\substack{2° \text{ free radical} \\ \text{(More stable)}}}{CH_3-\dot{C}H-CH_2Br} + \underset{\substack{\text{Br} \\ 1° \text{ free radical} \\ \text{(Less stable)}}}{CH_3-CH-\dot{C}H_2}$$

The secondary radical then reacts with another H — Br molecule to yield the product, and another $\dot{B}r$ which can further propagate the reaction:

$$\underset{\text{(Chain propagation)}}{CH_3-\dot{C}H-CH_2Br+H-Br} \longrightarrow \underset{\text{1-Bromopropane}}{CH_3-CH_2-CH_2Br}+\dot{B}r$$

This mechanism is supported by the fact that small amount of peroxide can influence addition to a large number of molecules of an alkene and a small amount of an inhibitor such as hydroquinone or diphenyl amine can prevent this change.

Peroxide oxide effect or anti-Markownikoff's rule (or Kharasch effect) is observed in the addition of H — Br and not H — Cl or H — I.

In case of HCl, it is probably due to the fact that H — Cl bond (430 kJ/mole) is stronger than H — Br bond (368 kJ/mole) and is not broken homolytically by the free radicals generated by peroxides. As such free radical addition of HCl to alkenes is not possible. In case of HI, H — I bond (297 kJ/mole) is no doubt weaker than H — Br bond and can be broken easily. But the iodine atoms (I•) thus formed readily combine amongst themselves to form iodine molecules rather than add to the olefins.

Addition of (*i*) Sulphuric acid (*ii*) water to an alkene

Sulphuric acid adds to an alkene to form alkyl hydrogen sulphate white water adds in the presence of an acid to form alcohol.

Mechanism of addition of (*i*) sulphuric acid and (*ii*) water to an alkene

(*i*) Mechanism of addition of sulphuric acid to alkene. It is a two step electrophilic addition reaction which takes place as follows:

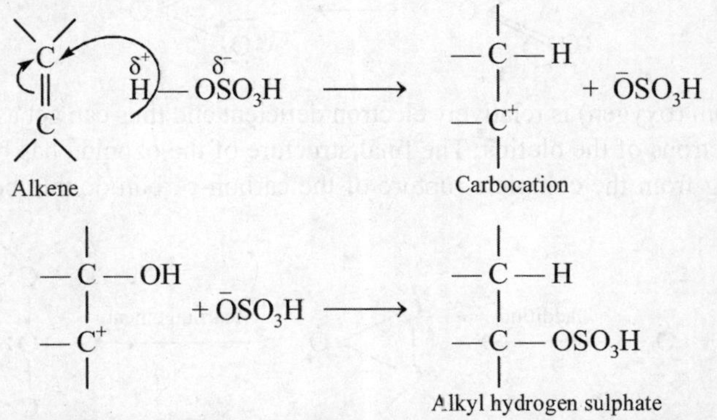

(*ii*) Mechanism of addition of water to an alkene. Alkenes react on water in the presence of an acid to form alcohols.

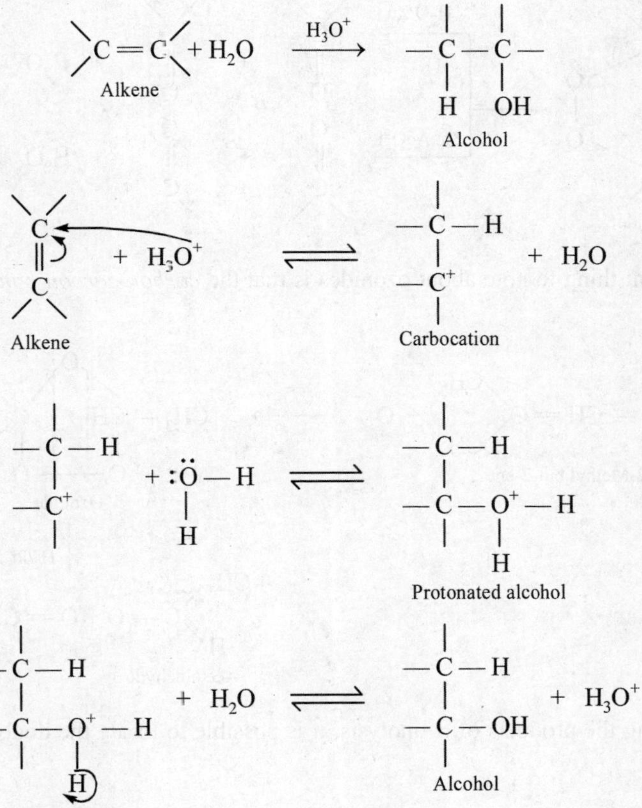

Ozonolysis

Ozone undergoes a reaction with olefins in an inert solvent at low temperatures to yield unstable addition compounds called 'ozonides'. These ozonides are not easily isolated and yield carbonyl compounds on further treatment either with boiling water and Zn or zinc and acetic acid.

This overall reaction is called as *Ozonolysis*.

Ozonolysis is a very useful reaction for illucidating the structures of olefinic compounds.

Ozone has a dipolar resonance structure with a bond angle of about 120°.

The central atom (oxygen) is relatively electron deficient and thus can act as an electrophile for attack on the π electrons of the olefins. The final structure of the ozonide has been shown to have a structure resulting from the complete rupture of the carbon-carbon double bond.

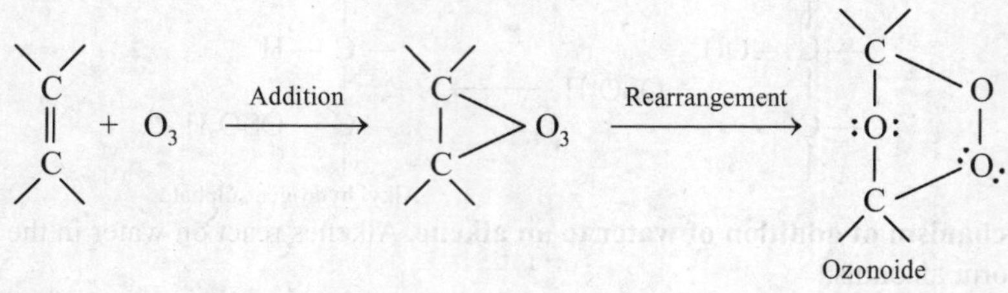

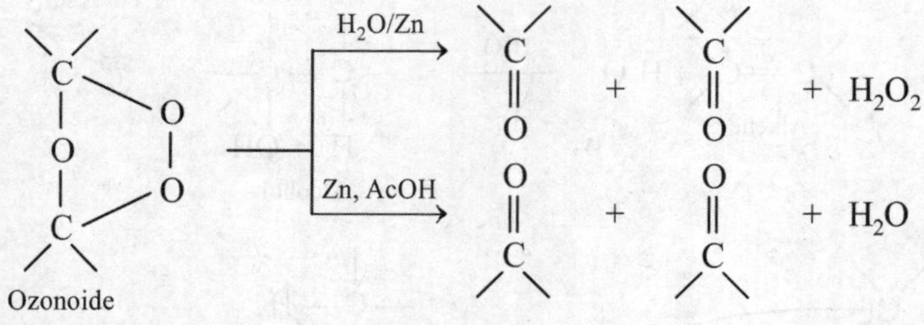

The important thing to note about ozonides is that the *carbon-carbon bond has been broken*. For example:

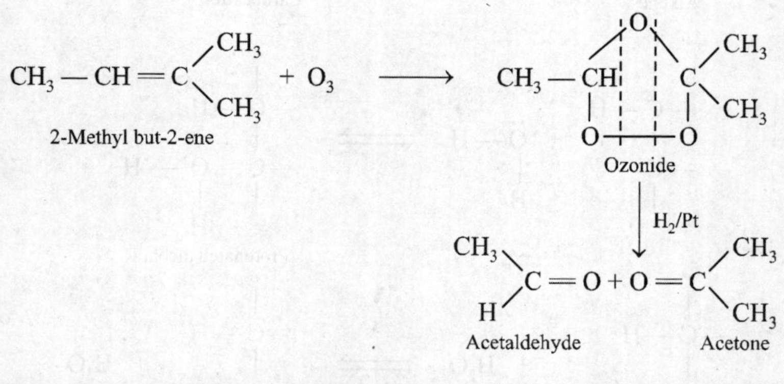

By identifying the products of ozonolysis, it is possible to locate the double bond in alkene.

***Example 7.* On reductive ozonolysis, an unsaturated hydrocarbon gave the following compounds:**

(*i*) Ethanediol (*ii*) Propanone (*iii*) Ethanol

Write the structural formula of the hydrocarbon and write its IUPAC name.

Solution. The hydrocarbon is

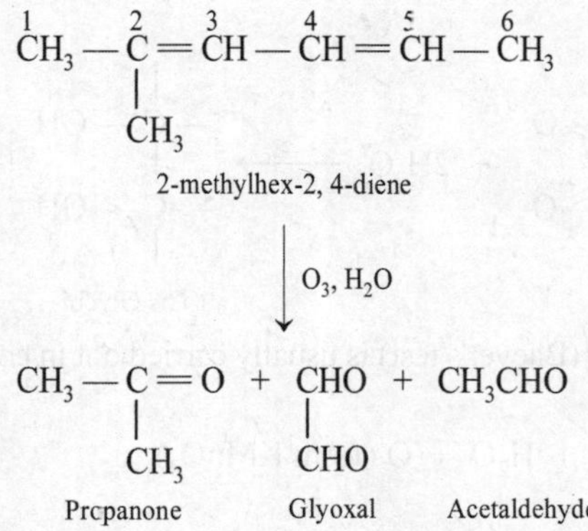

Glyoxal and acetaldehyde on reduction change to ethanediol and ethyl alcohol respectively.

***Example 8.* Complete the following reaction**

$$\mathbf{CH_3CH{=}CHCH_3 + O_3 \longrightarrow ? \xrightarrow{Zn/H_2O} ?}$$

Solution.

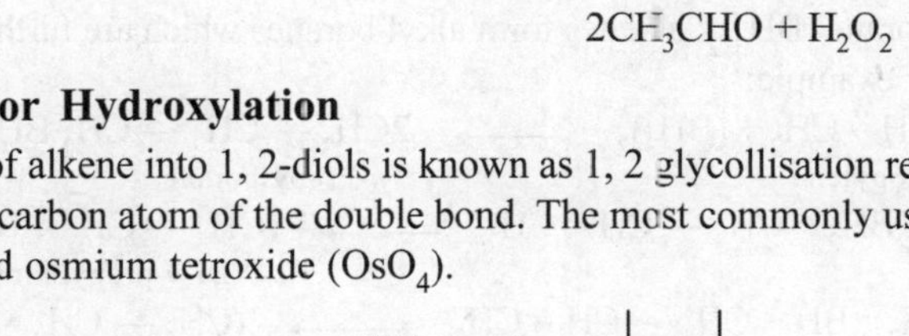

Glycollisation or Hydroxylation

Conversion of alkene into 1, 2-diols is known as 1, 2 glycollisation reaction; an — OH group is added on to each carbon atom of the double bond. The most commonly used reagents are potassium permanganate and osmium tetroxide (OsO_4).

$$>C{=}C< \xrightarrow[\text{or } OsO_4]{KMnO_4} -\underset{OH}{\overset{|}{C}}-\underset{OH}{\overset{|}{C}}- \quad \text{(1, 2 diol)}$$

The two reagents ($KMnO_4$ and OsO_4) are known to react by similar mechanism which involves cyclic intermediates of the following type which are subsequently hydrolyzed in aqueous solutions to glycol and the reduced form of the reagent.

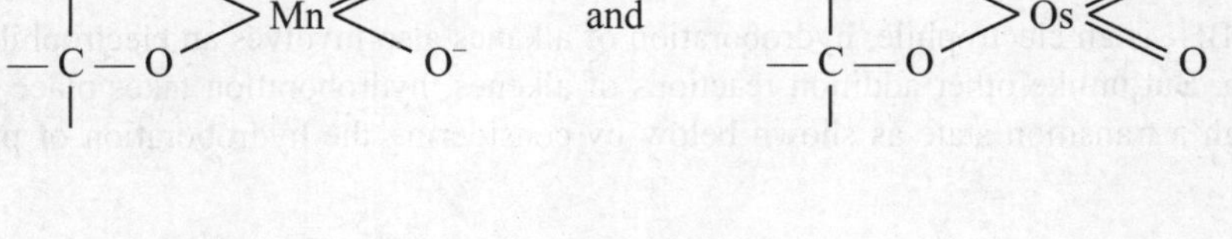

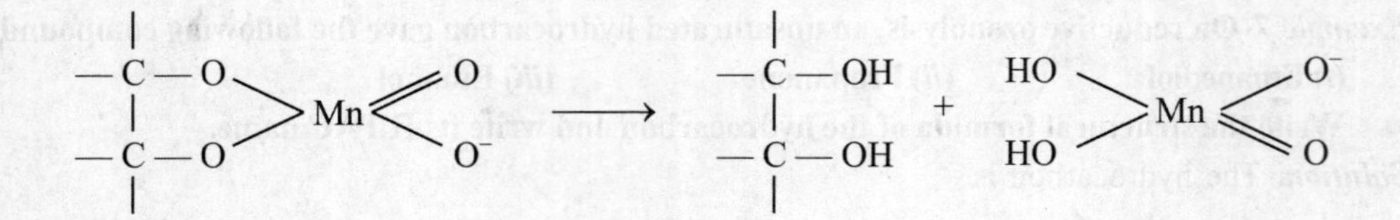

Cyclic manganese ester

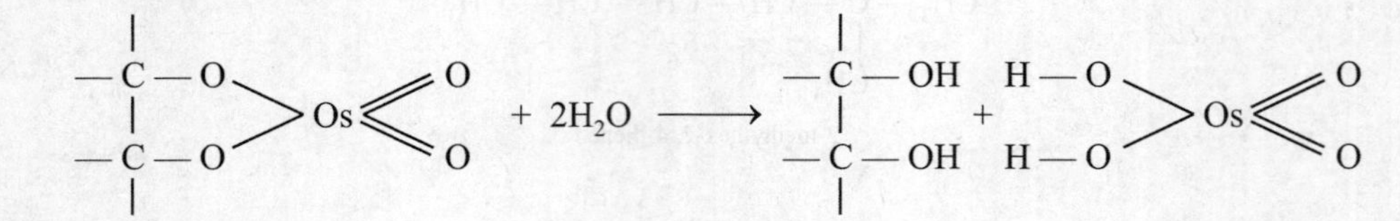

Cyclic osmium ester   Cis Glycol   Osmic acid

The permanganate reaction (Baeyer's test) is usually carried out in cold, aqueous alkaline solution.

$$\begin{array}{c} CH_2 \\ \| \\ CH_2 \end{array} + H_2O + O \text{ (from } KMnO_4) \longrightarrow \begin{array}{c} CH_2OH \\ | \\ CH_2OH \end{array}$$

Glycol

$$\underset{\text{Propylene}}{CH_3 - CH = CH_2} + H_2O + O \text{ (from } KMnO_4) \longrightarrow \underset{\text{(Propylene glycol)}}{CH_3 - \underset{OH}{\underset{|}{CH}} - \underset{OH}{\underset{|}{CH_2}}}$$

Hydroboration of alkenes

On treatment with diborane $(BH_3)_2$, alkenes form alkyl boranes which are further converted into di and tri alkylboranes. For example:

$$\underset{\text{Propylene}}{2CH_3 - CH = CH_2} + (BH_3)_2 \longrightarrow \underset{n\text{-Propylborane}}{2CH_3 - CH_2 - CH_2\,BH_2}$$

$$CH_3 - CH_2 - CH_2BH_2 + CH_3 - CH = CH_2 \longrightarrow \underset{\text{Di-}n\text{-propylborane}}{(CH_3 - CH_2 - CH_2)_2\,BH}$$

$$(CH_3 - CH_2 - CH_2)_2\,BH + CH_3 - CH = CH_2 \longrightarrow \underset{\text{Tri-}n\text{-propylborane}}{(CH_3 - CH_2 - CH_2)_3\,B}$$

Alkyboranes in turn are readily oxidised by alkaline solution of H_2O_2 to form alcohols. For example:

$$(CH_3CH_2CH_2)_3\,B + 3H_2O \xrightarrow{H_2O_2/OH^-} 3CH_3 - CH_2 - CH_2OH + H_3BO_3$$

This reactions for the preparation of alcohol from alkene is called **hydroboration.**

Mechanism. Diborane participates in the reaction in its monomeric form *i.e.*, BH_3. It behaves as an electrophile since the boron atom in the molecule is electron deficient as it has only six valence electrons.

Since BH_3 is an electrophile, hydroboration of alkanes also involves an electrophilic addition mechanism. But unlike other addition reactions of alkenes, hydroboration takes place in a single step through a transition state as shown below by considering the hydroboration of propylene.

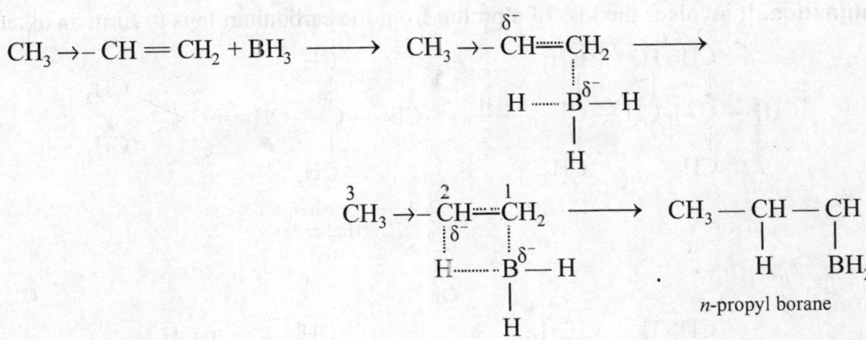

Alkylation of alkenes

Isobutylene may be converted into iso-octane by the addition of isobutane in the presence of H_2SO_4 as a catalyst. This process is known as '*alkylation*' and is of industrial importance as it is extensively used in oil refineries in the production of high grade motor fuels.

The probable route adopted may be shown taking place as:

Initiation

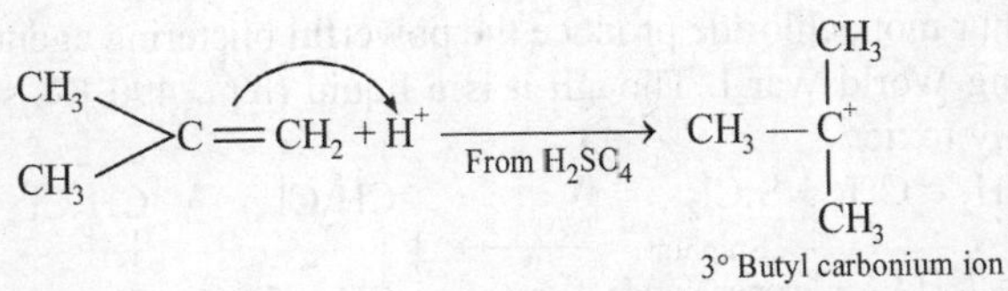

Propagation. There are two propagation steps. in the first step; addition of 3° butyl carbonium ion with isobutylene to form a new carbonium ion ($^+C_8H_{17}$) takes place.

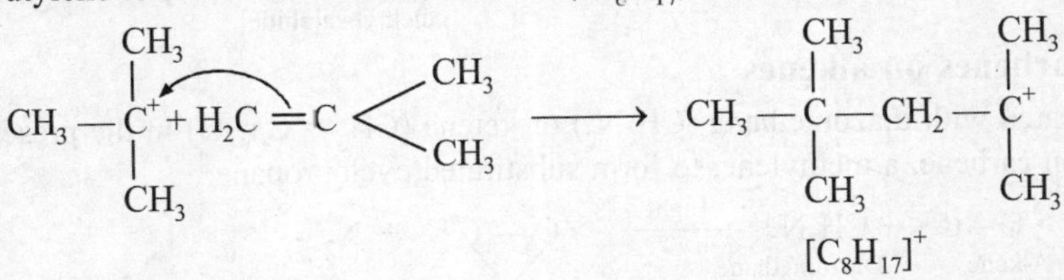

In the second step this new carbocation then abstracts a hydride ion from iso-butane to form isooctane and a new 3° butyl carbonium ion.

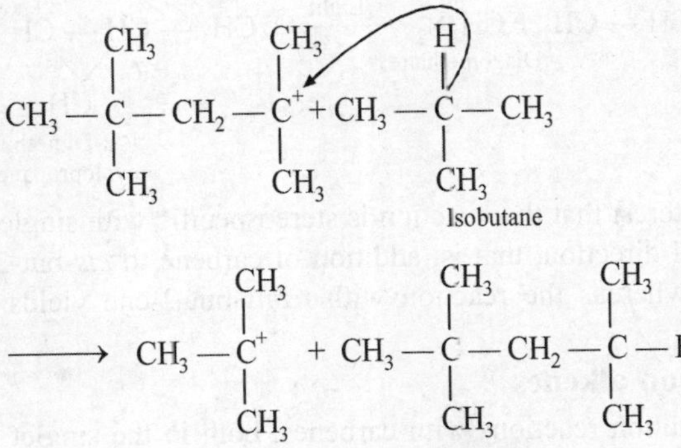

Termination. It involves the loss of a proton from the carbonium ions to form an alkene.

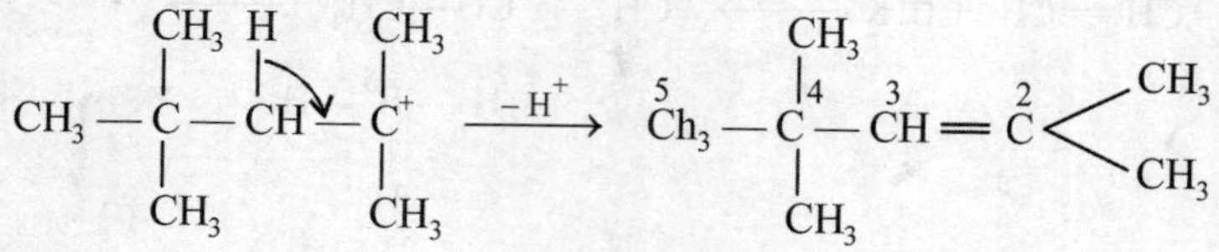

2, 4, 4-Tri methylpent-2-ene
(Higher %)
(A)

Or

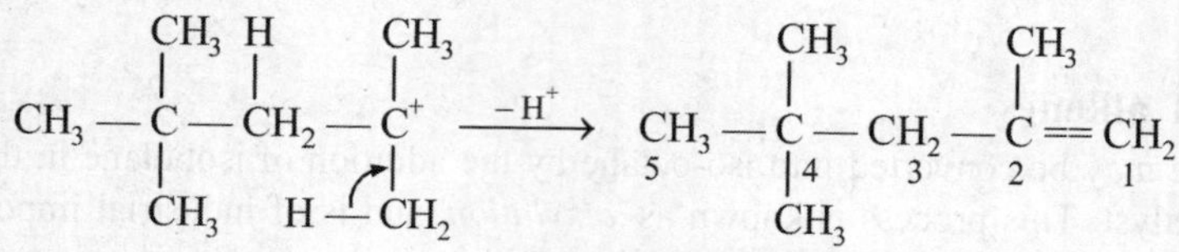

2, 4, 4-Tri methyl pent-1-ene
(B)

Reaction of alkenes with sulphur monochloride

Ethylene and sulphur monochloride produce the powerful blistering agent (Vesicant) known as mustard gas used during World War I. Though it is a liquid (b.pt. 490 K), small quantities of its vapours in air are highly toxic.

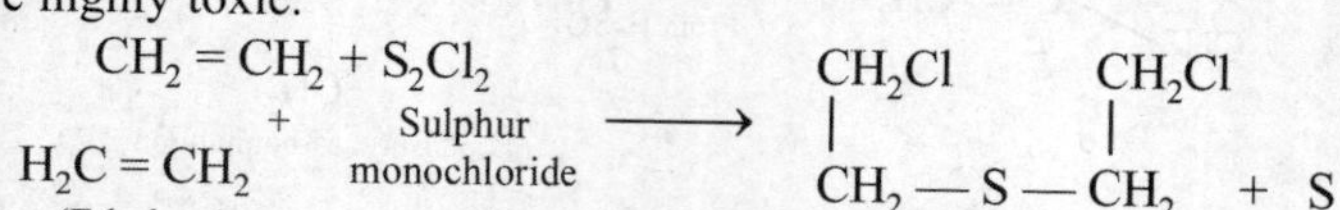

Mustard gas
[β, β′-Dichloro-diethyl sulphide]

Action of carbenes on alkenes

When treated with diazomethane (CH_2N_2) or ketene ($CH_2 = C = O$) in the presence of light, alkenes add on carbene, a methylene, to form substituted cyclopropane.

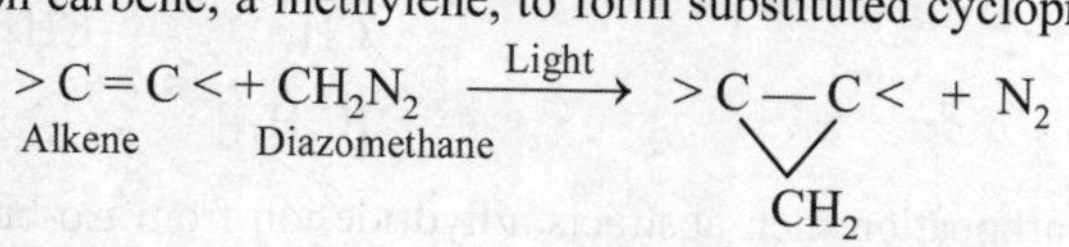

Substituted cyclopropane

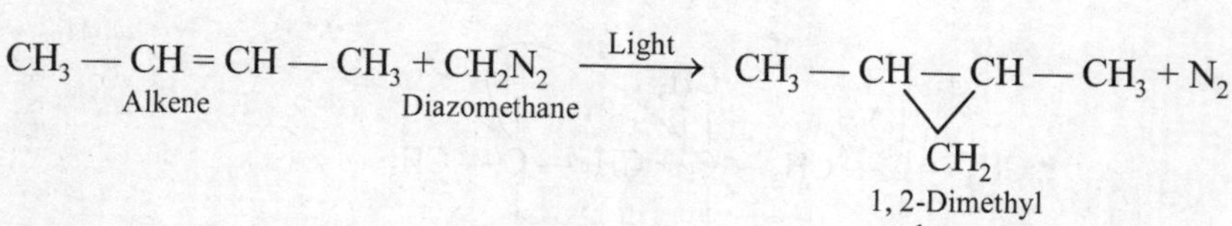

1, 2-Dimethyl cyclopropane

It is of theoretical interest that this reaction is stereospecific with singlet methylene occurring exclusively in one spatial direction, that is, addition of carbene to *cis*-but-2-ene yields *cis* –1, 2-dimethyl cyclopropane whereas the reaction with *trans*-but-2-ene yields *trans* –1, 2-dimethyl cyclopropane.

Addition of carbenes on alkenes

Alkenes undergo addition reactions with carbenes, both in the singlet and triplet forms*, to form cyclopropanes. Since a ring is generated as a result of such an addition reaction, the reaction

*For difference between singlet an triplet forms of carbenes, see chapter 2, under carbenes.

is known as *cycloaddition.* For example, when treated with diazomethane ketone, (CH_2N_2) or ($CH_2 = C = O$) which produce carbene in the presence of light, alkenes add on methylene to from cyclopropanes.

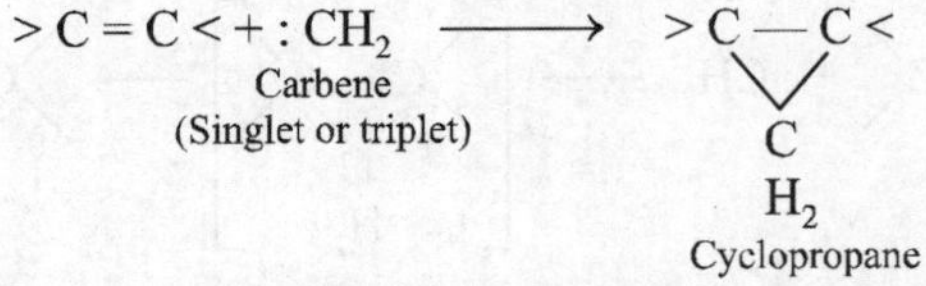

Addition of singlet methylene takes place in a *sterospecific* manner to yield product having the same stereochemistry as the alkene. For example, singlet methylene reacts with *cis*-2 butene to form *cis*-1, 2-dimethylcyclopropane. Similarly, it reacts with *trans*-2 butene to form *trans*-1, 2-dimethylcyclopropane.

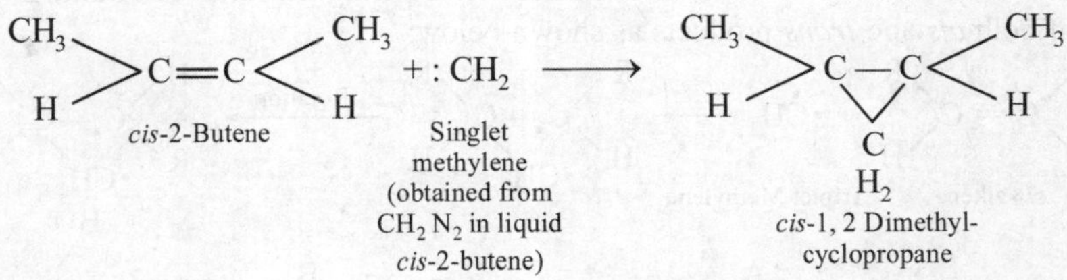

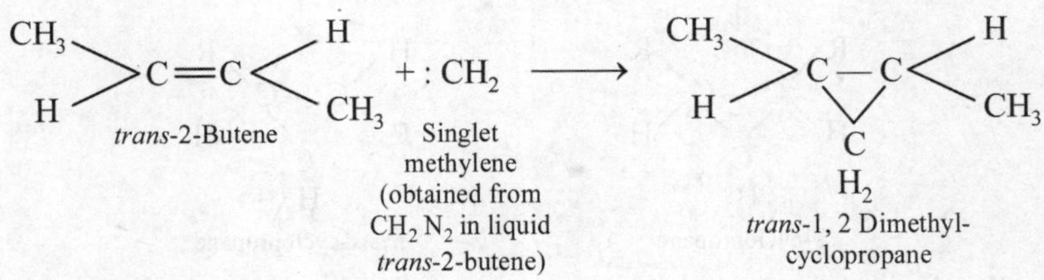

Addition of **triplet methylene** is *non-stereospecific* and produces a mixture of products. For example, addition of triplet methylene to *either cis*-2-butene or *trans*-2 butene produces a *mixture of cis and trans*-1, 2-dimethylcyclopropanes.

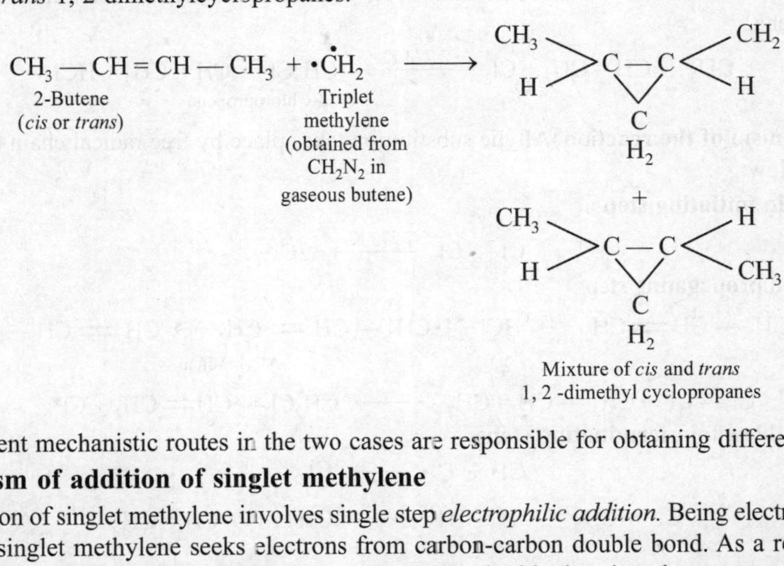

Different mechanistic routes in the two cases are responsible for obtaining different results.

Mechanism of addition of singlet methylene

Addition of singlet methylene involves single step *electrophilic addition.* Being electron deficient in nature, singlet methylene seeks electrons from carbon-carbon double bond. As a result singlet methylene gets simultaneously attached to both the doubly bond carbon atoms. Therefore, stereochemistry of the reactant remains preserved in the products, *cis*-alkenes yield *cis*-dialkyl cyclopropanes while *trans*-alkenes yield *trans*-dialkylcyclopropanes. The complete mechanism may be depicted as:

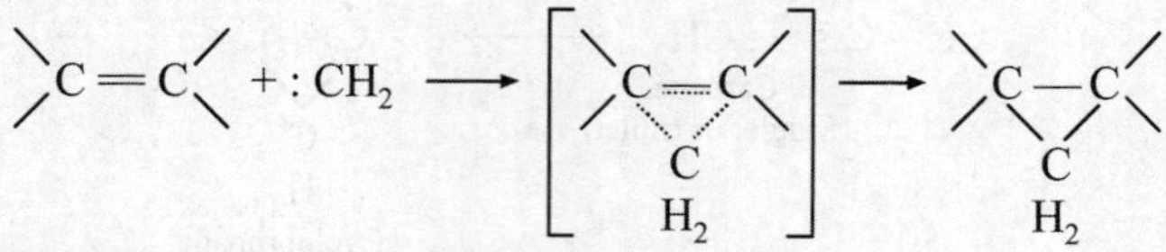

Mechanism of addition of triplet methylene

Triplet methylene reacts with alkenes in a two step free radical mechanism. Triplet methylene is a diradical and it attacks the alkene to form an intermediate diradical having the conformation A. This diradical has a life time long enough for rotation of groups to take place around central carbon-carbon bond. This gives rise to conformation B. Ring closures of A and B lead to the formation of both *cis* and *trans* products as shown below:

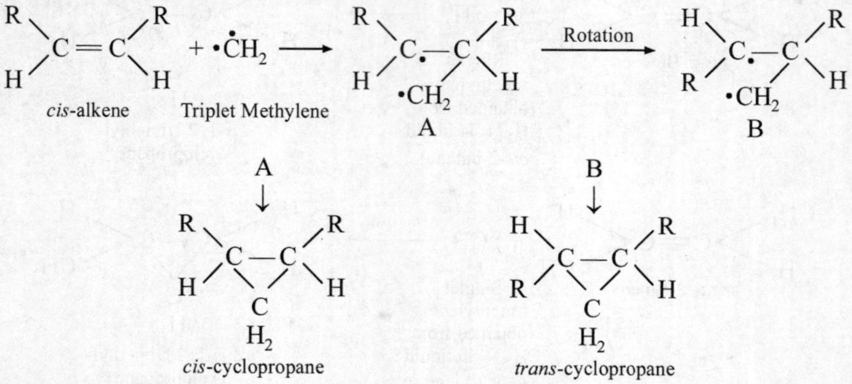

Allylic substitution

It is a substitution reaction which takes place at the carbon atom next to the double bond. Higher homologues of ethylene undergo allylic substitution under specific conditions. For example, when propene is treated with chlorine at 773 K, it undergoes allylic substitution to yield 3-chloropropene.

$$CH_3-CH=CH_2+Cl_2 \xrightarrow{773\text{ K}} \underset{\text{3-Chloropropene}}{CH_2Cl-CH=CH_2}+HCl$$

Mechanism of the reaction. Allylic substitution takes place by free radical chain mechanism as given below:

1. Chain initiating step

$$Cl-Cl \xrightarrow{773\text{ K}} 2Cl^{\bullet}$$

2. Chain propagating step

$$Cl^{\bullet}+CH_3-CH=CH_2 \longrightarrow HCl+\underset{\text{Allyl radical}}{[{}^{\bullet}CH_2-CH=CH_2 \longleftrightarrow CH_2=CH-\dot{C}H_2]}$$

$$Cl-Cl+{}^{\bullet}CH_2-CH=CH_2 \longrightarrow CH_2Cl-CH=CH_2+Cl^{\bullet}$$

3. Possible chain terminating steps

$$Cl^{\bullet}+Cl^{\bullet} \longrightarrow Cl_2$$

$$Cl^{\bullet}+{}^{\bullet}CH_2-CH=CH_2 \longrightarrow CH_2Cl-CH=CH_2$$

More conveniently, allylic bromination is done with the help of N-Bromosuccinimide (abbreviated as NBS). The reaction is carried in the presence of light

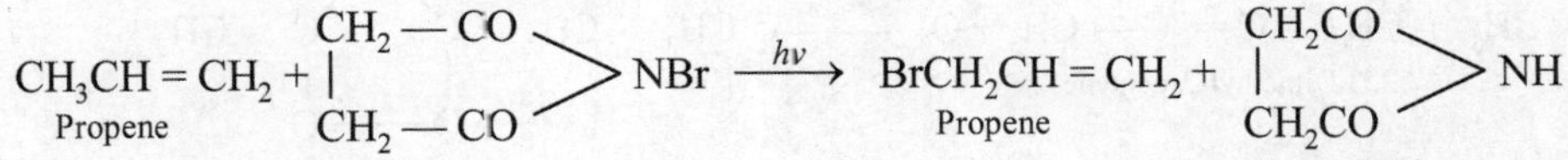

At high temperature Alkenes undergo free radical substitution and not free radical addition of halogens

When an alkene is treated with a halogen at high temperature, what happens is that the halogen molecule readily decomposes due to the high temperature to form halogen free radicals or atoms. The halogen atom in turn attacks the alkene molecule at the allylic carbon to form an allyl free radical. The allyl free radical then abstracts a halogen from another halogen molecule to form the substitution product (refer to mechanism above)

The question is why the halogen atom does not attack the alkene at the double bond to form the free radical and then add another halogen atom to form an addition product. The answer is that halogen atom does attack the alkene even at the double bond to form a free radical. But due to very high temperature, this free radical readily eliminates the halogen atom before another halogen atom could be added to form the addition product.

That is:

$$Cl\bullet + CH_3-CH=CH_2 \underset{\text{Elimination of Cl}}{\overset{\text{Addition of Cl}^\bullet}{\rightleftharpoons}} CH_3-\dot{C}H-CH_2Cl$$

Thus the formation of addition product is prevented.

Epoxidation reaction

When an alkene is treated with peroxybenzoic acid, an epoxide is formed. This reaction is called **epoxidation** reaction

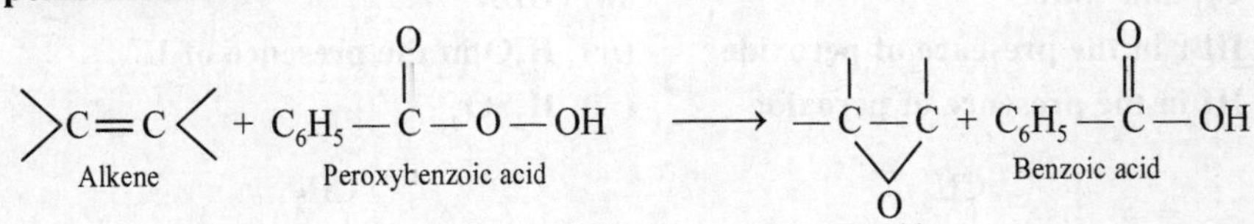

Mechanism of the reaction

$$C_6H_5\overset{\overset{O}{\|}}{C}-O-O-H \underset{}{\overset{\text{Slow}}{\rightleftharpoons}} R-\overset{\overset{O}{\|}}{C}-O^- + \overset{+}{O}H$$

Electrophile

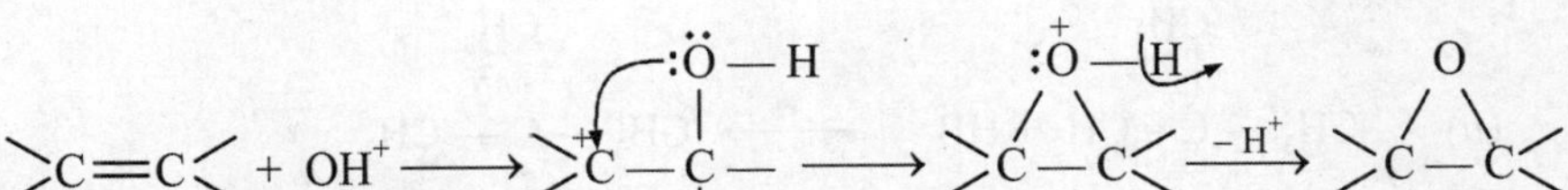

Example 9. **What happens when:**

(*i*) 2, 3-Dimethyl-2-pentene is subjected to ozonolysis.

(*ii*) when propylene is treated with sodium borohydride and boron trifluoride followed by the treatment of product thus formed with alkaline hydrogen peroxide.

Solution. (*i*) When 2, 3-dimethyl-2-pentene is subjected to ozonolysis, a mixture of acetone and ethyl methyl ketone is obtained.

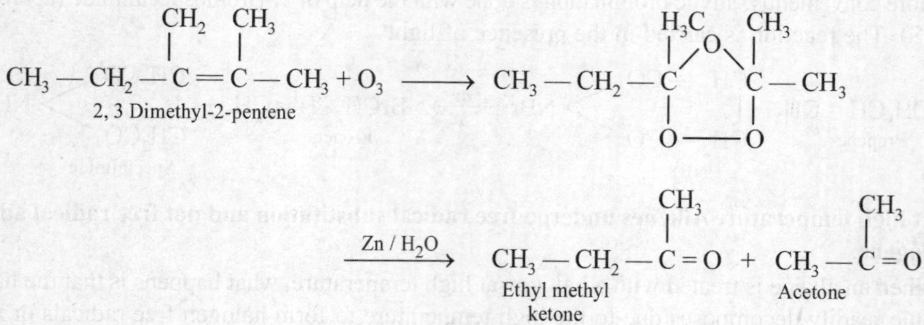

(*ii*) On treatment with sodium borohyride and boron tri-fluoride propylene yields tripropyl borane which on treatment with alkaline hydrogen peroxide yields *n*-propyl alcohol.

$$3NaBH_4 + 4BF_3 \xrightarrow{273\ K} 2(BH_3)_2 + 3Na\ BF_4$$

$$2CH_3CH = CH_2 + (BH_3)_2 \longrightarrow 2CH_3 - CH_2 - CH_2BH_2$$

$$CH_3 - CH_2 - CH_2BH_2 \xrightarrow{CH_3CH=CH_2} (CH_3 - CH_2 - CH_2)_2\ BH$$

$$\xrightarrow{CH_3 - CH = CH_2} (CH_3 - CH_2 - CH_2)_3\ B$$

Tri *n*-propyl borane

$$(CH_3 - CH_2 - CH_2)_3\ B + 3H_2O_2 \xrightarrow{NaOH} 3CH_3CH_2OH + H_3BO_3$$

n-propyl alcohol

Example 10. **Write equations for the reactions and name the product formed when isobutylene reacts with:**

(*i*) Cl_2 and water **(*ii*) HBr**

(*iii*) HBr in the presence of peroxide **(*iv*) H_2O in the presence of H^+**

(*v*) HI in the presence of peroxide **(*vi*) H_2SO_4**

Solution. (*i*)

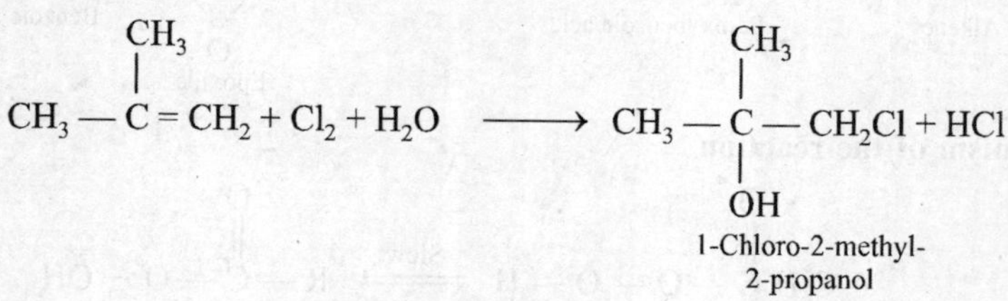

(*ii*)

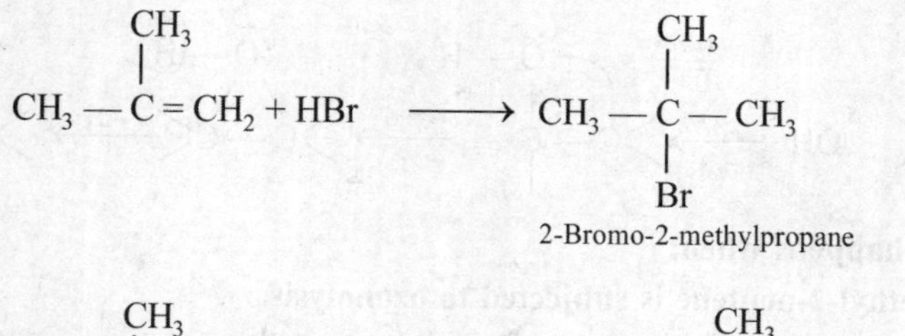

(*iii*)

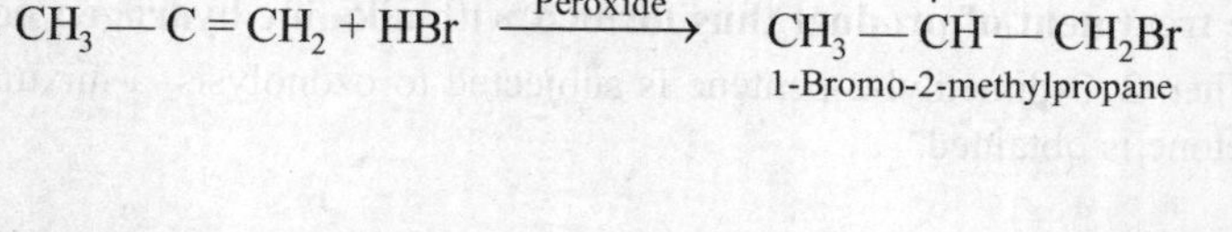

(*iv*)
$$CH_3-\overset{\overset{\displaystyle CH_3}{|}}{C}=CH_2+H_2O \xrightarrow{H^+} CH_3-\underset{\underset{\displaystyle OH}{|}}{\overset{\overset{\displaystyle CH_3}{|}}{C}}-CH_3$$
2-methyl-2- propanol

(*v*)
$$CH_3-\overset{\overset{\displaystyle CH_3}{|}}{C}=CH_2+HI \xrightarrow{\text{Peroxide}} CH_3-\underset{\underset{\displaystyle I}{|}}{\overset{\overset{\displaystyle CH_3}{|}}{C}}-CH_3$$
2-Iodo-2-methylpropane

(*vi*)
$$CH_3-\overset{\overset{\displaystyle CH_3}{|}}{C}=CH_2+H_2SO_4 \qquad CH_3-\underset{\underset{\displaystyle OSO_3H}{|}}{\overset{\overset{\displaystyle CH_3}{|}}{C}}-CH_3$$
Tert. butyl hydrogen sulphate

5.7 POLYMERISATION

It is the process of formation of a large molecule from small molecules. For example, when heated at a high temperature and pressure in the presence of traces of oxygen, ethylene undergoes polymerisation to form polythene.

$$nCH_2=CH_2 \xrightarrow[\text{Traces of } O_2]{\text{High temp. and pressure}} (-CH_2-CH_2-)_n$$

Free radical chain mechanism of addition polymerisation

Many alkenes and substituted alkenes undergo polymerisation by free radical chain mechanism. Such reactions usually occur at high temperature under pressure or in the presence of catalysts such as peroxides and salts of peracids. The catalyst provides the free radical to initiate the chain process which is carried on by chain propagating steps. The mechanism is illustrated below with the help of polymerisation of ethylene in the presence of peroxide.

1. Chain initiating state

$$\underset{\text{Peroxide}}{R-O-O-R} \xrightarrow{hv} \underset{\text{Free radical}}{2\dot{R}O}$$

2. Chain propagating stage

$$\dot{R}O+CH_2=CH_2 \longrightarrow RO-CH_2-\dot{C}H_2$$

$$RO-CH_2-\dot{C}H_2+CH_2=CH_2 \longrightarrow RO-CH_2-CH_2-CH_2-\dot{C}H_2$$

The free radical formed in each step adds to a fresh molecule of alkene and so the chain progresses till a large molecule of the polymer is obtained and the chain is terminated.

3. Possible chain terminating stage

(*i*) **Coupling** of free radicals to form a deactivated molecule.

For example:

$$RO-(CH_2-CH_2)_n-CH_2-\dot{C}H_2+\dot{C}H_2-(CH_2-CH_2)_n-OR$$
$$\longrightarrow RO-(CH_2-CH_2)_n-CH_2-CH_2-CH_2-(CH_2-CH_2)_n-OR$$

(*ii*) Disproportionation of free radicals in which one free radical acquires a hydrogen from another and both get deactivated.

$$RO-(CH_2-CH_2)_n-CH_2-\dot{C}H_2+\dot{C}H_2-CH_2-(CH_2CH_2)_n-OR$$
$$\longrightarrow RO-(CH_2-CH_2)_n-CH=CH_2+CH_3-CH_2-(CH_2-CH_2)_n-OR$$

Cationic and anionic mechanism of polymerisation

(*a*) Cationic mechanism. Alkene molecules having electron releasing groups (propylene, isobutylene etc.) in the acidic medium may polymerise by cationic mechanism. In fact, the electrophile attack of proton of acid on the alkene molecule leads to a carbocation. Such cations may further take part in the chemical combination leading to the formation of polyalkene. The cationic polymerisation is illustrated with the example of isobutyne.

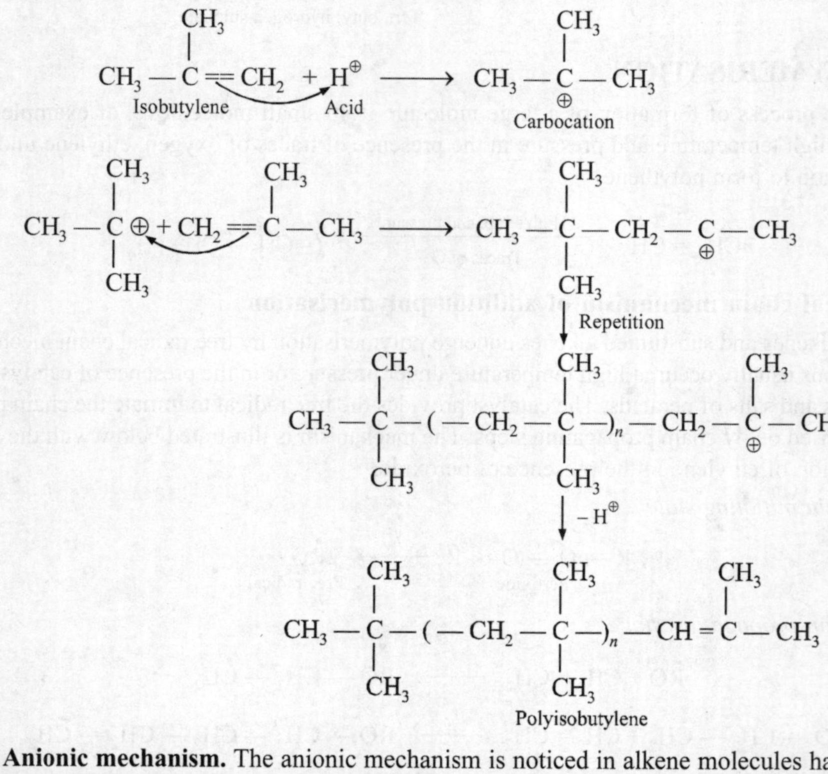

Polyisobutylene

(*b*) Anionic mechanism. The anionic mechanism is noticed in alkene molecules having some electron withdrawing groups present in them *e.g.* vinyl chloride, etc. The anionic polymerisation is carried in the presence of a suitable base such as sodamide ($NaNH_2$), *n*-butyl lithium (*n*-C_4H_9Li) etc.

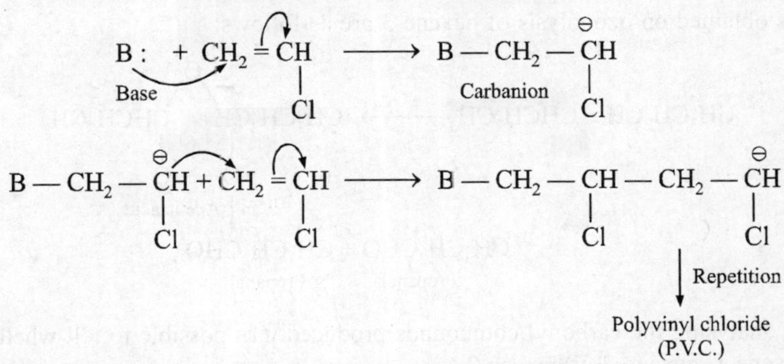

Polymarisation in alkenes may take place by either by free radical or ionic mechanism. The free radical polymerisation takes place in the presence of an organic peroxide. Cationic polymerisation takes place in the presence of an acid while the anionic polymerisation proceeds in the presence of a base acting as a catalyst

5.8 INDUSTRIAL APPLICATIONS OF ETHYLENE AND PROPENE

Ethylene

It is used in the manufacture of many industrial polymers *viz.* PVC, EPR, Sarans and polyethylene. These polymers are used in the manufacture of floor tiles, shoe soles, synthetic fibres, raincoats, purses, saran wraps and pipes

Propene

Propene is used in the manufacture of ethylene-propylene rubbers (EPR), acetone, isopropyl alcohol, *n*-propyl alcohol and allyl bromide

MISCELLANEOUS EXAMPLES

Example 1. **How is it possible to know whether the given bexene is hexene-1, hexene-2 or hexene-3?**

Solution. It is possible to distinguish between the three compounds by carrying out ozonolysis of the substance. An alkene forms an ozonide with ozone. The ozonide is subjected to hydrolysis when we obtain a mixture of carbonyl compounds. By analysing the carbonyl compounds, we can tell about the position of double bond in the molecule. The products obtained on ozonolysis of hexene-1 are as given below:

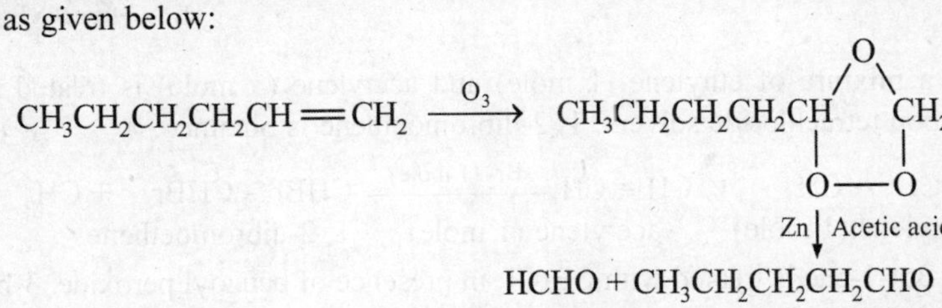

Products obtained on ozonolysis of hexene-2 are as follows:

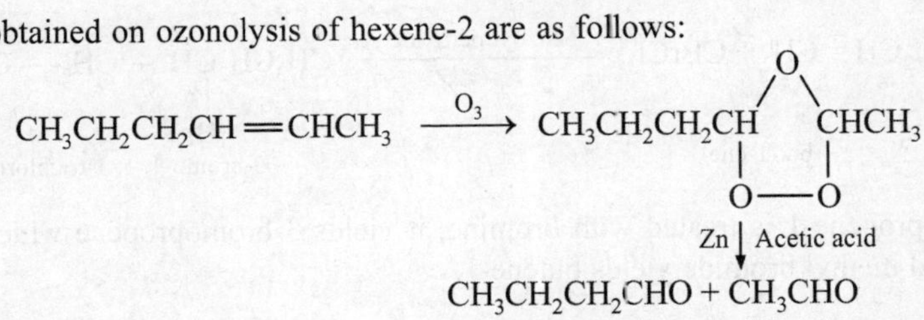

Products obtained on ozonolysis of hexene-3 are as follows:

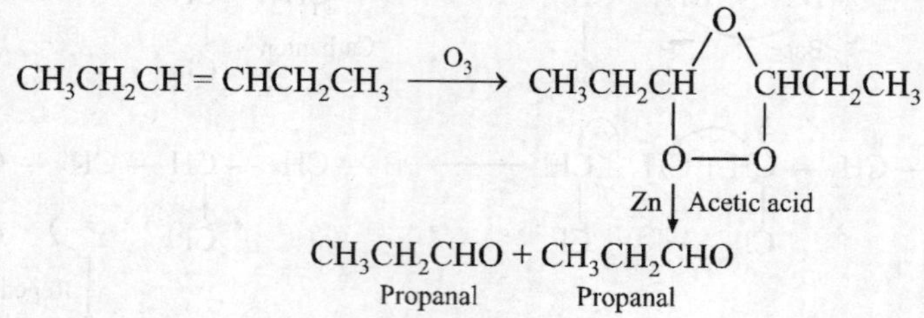

Thus by analysing the carbonyl compounds produced it is possible to tell whether the given hexene is hexene-1, hexene-2 or hexene-3.

Example 2. **What happens when :**

(*i*) 2, 2, 2-trimethyl -1-bromoethane reacts with alcoholic potassium hydroxide.

(*ii*) A mixture of ethylene (1 mole) and acetylene (1 mole) is treated with bromine (1 mole) in carbon tetrachloride solvent.

(*iii*) But-1-ene is treated with $CBrCl_3$ in presence of catalytic amount of benzoyl peroxide.

(*iv*) Propene-1 is heated with bromine followed by the treatment of product thus formed with sodium and methyl bromide.

Solution. (*i*) When 2, 2, 2-trimethyl -1-bromoethane reacts with alc. potassium hydroxide, 1, 2, 2-trimethyl ethene is obtained.

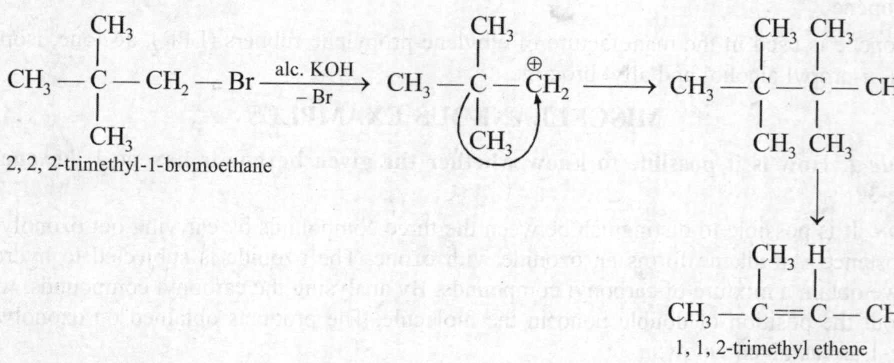

(*ii*) When a mixture of ethylene (1 mole) and acetylene (1 mole) is treated with bromine (1 mole) in carbon tetrachloride solvent, 1, 2-dibromoethene is obtained.

$$\underset{\text{ethylene (1 mole)}}{CH_2=CH_2} + \underset{\text{acetylene (1 mole)}}{CH\equiv CH} \xrightarrow{Br_2\text{ (1 mole)}} \underset{\text{1, 2-dibromoethene}}{CHBr=CHBr} + CH_2=CH_2$$

(*iii*) When But-1-ene is treated with $CBrCl_3$ in presence of benzoyl peroxide, 3-bromo-1, 1, 1-trichloropentane is formed.

$$\underset{\text{but-1-ene}}{CH_3CH_2CH=CH_2} + CBrCl_3 \xrightarrow{\text{benzoyl peroxide}} \underset{\text{3-bromo-1, 1, 1-trichloropentane}}{CH_3CH_2\underset{\substack{|\\Br}}{CH}-CH_2-CCl_3}$$

(*iv*) When propene-1 is treated with bromine, it yields 3-bromopropene which on treatment with sodium and methyl bromide yields butene-1.

$$CH_3CH=CH_2 \xrightarrow{Br_2} BrCH_2CH=CH_2 \xrightarrow[CH_3Br]{2Na} \underset{\text{butene-1}}{CH_3CH_2CH_2=CH_2}$$

Example 3. **How will you differentiate between 1-pentene and 2-pentene?**

Solution. The products obtained on ozonolysis of pentene-1 are as given below:

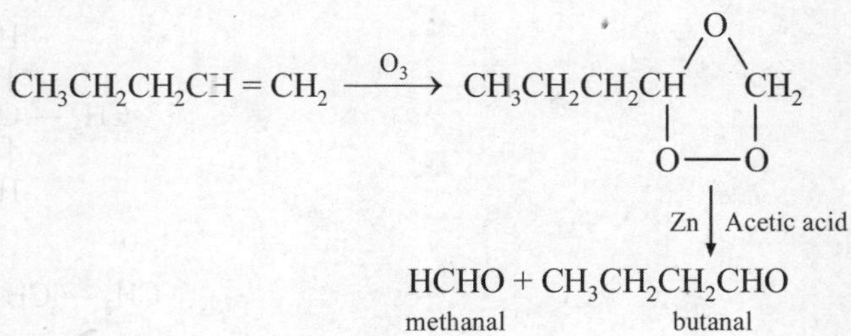

Products obtained on ozonalysis of pentene-2 are as follows:

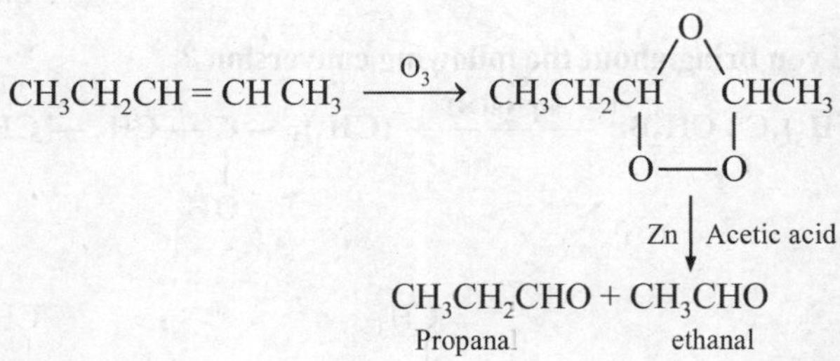

By identifying the reaction products, 1-pentane and 2-pentane can be differentiated.

Example 4. **Compare with proper justification the stabilities of isobutene, Z-2-butene and E-2-butene.**

Solution. In isobutene, two methyl groups are attached to same carbon atom due to which it experiences steric hinderance. Hence, it is least stable. In Z-2-butene, two methyl groups are attached to two different carbon atoms but on the same side of the double bond due to which it experiences steric hinderance but less than in isobutene. Hence, it is more stable than isobutene.

In E-2-butene, the two methyl groups are attached to two different carbon atoms on the opposite side of the double bond. Hence, it experiences least steric hinderance. So, it is stable to the maximum extent.

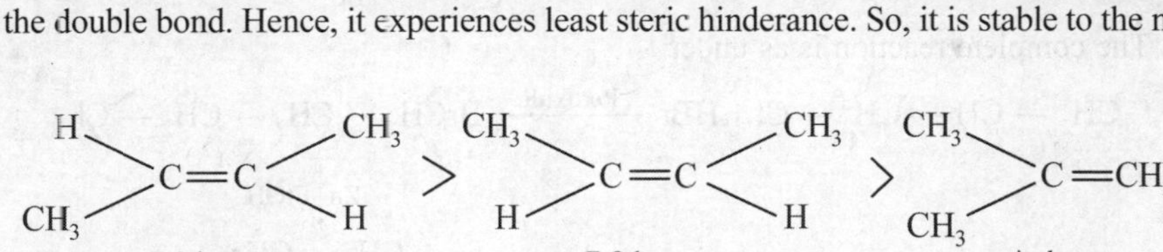

Z-2-butene E-2-butene isobutene

Example 5. **Write the major product (products) in the following reaction :**

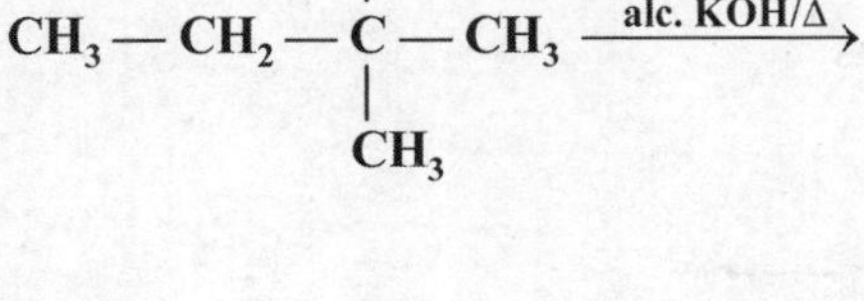

Solution. 2-methyl-butene-2 is the major product which is formed in the reaction as under:

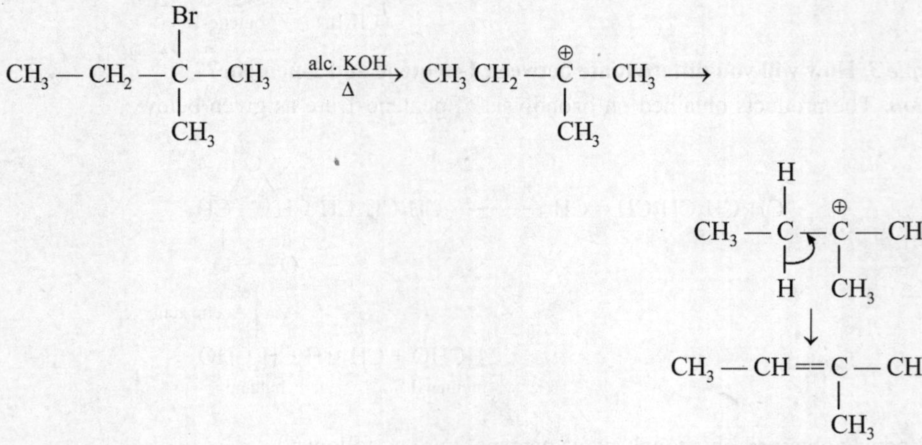

2-methyl-butene-2
(major product)

Example 6. **How will you bring about the following conversion ?**

$$(CH_3)_3C \cdot CH_2Br \xrightarrow{\text{aq.NaOH}} (CH_3)_3 - \underset{\underset{OH}{|}}{C} - CH_2 - CH_3$$

Solution.

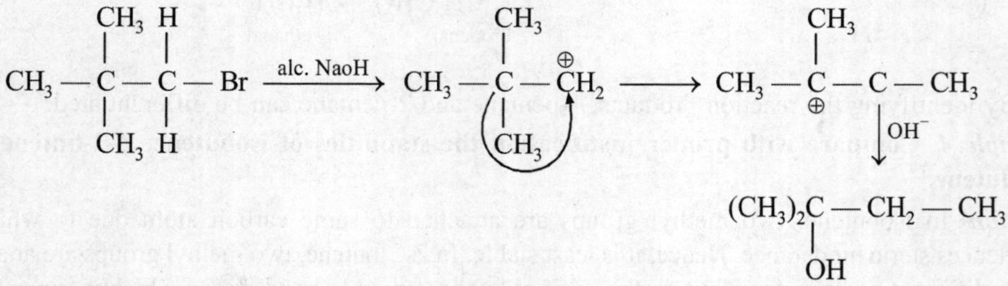

Example 7. **Identify A and B in the following reaction :**

$$CH_2 = CH - CH_2 - Cl + HBr \xrightarrow{\text{Peroxide}} A \xrightarrow[\text{Et}]{\text{Zn}} B + ZnBrCl$$

Solution. The complete reaction is as under :

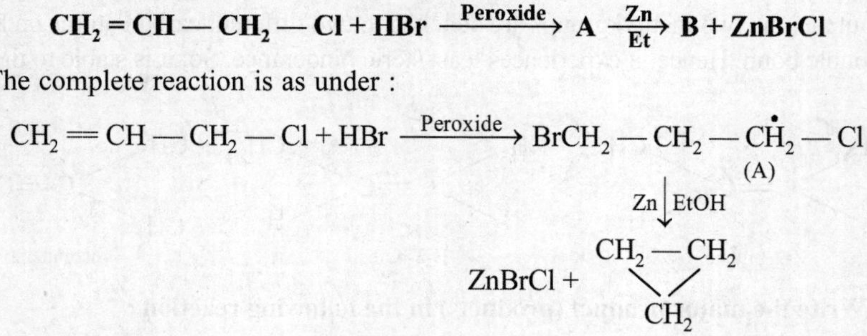

Example 8. **Draw geometrical isomers of the following:**

(*i*) Monobromopropenes

(*ii*) Hept-2-en-5 yne

(*iii*) Pentene-2

Solution. (*i*) *Geometrical isomers of monobromopropene :*

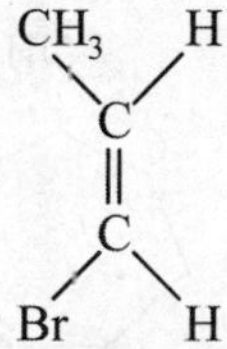

cis-form

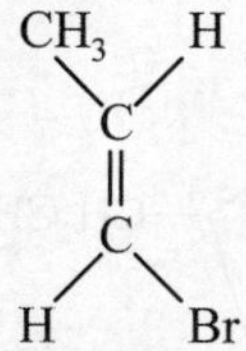

trans-form

(*ii*) *Geometrical isomers of hept-2-ene-5-yne :*

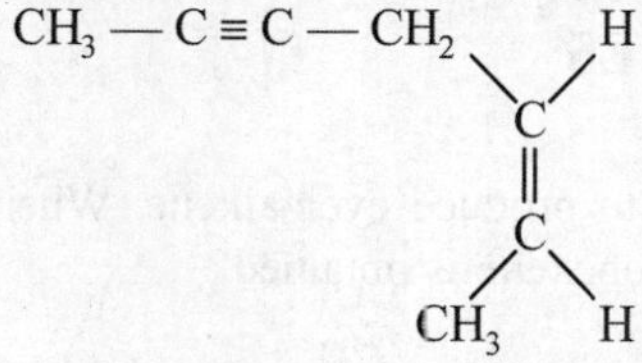

cis-form

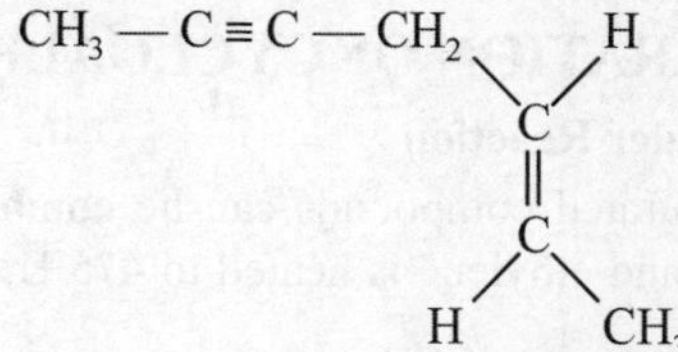

trans-form

(*iii*) *Geometrical isomers of pentene-2*

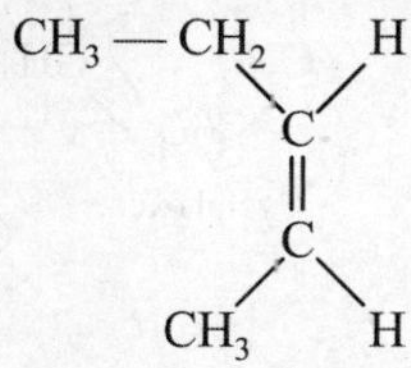

cis-form

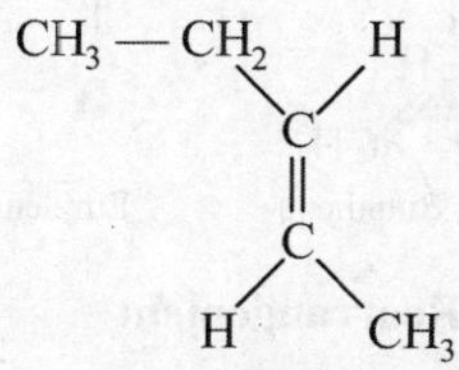

trans-form

Example 9. **How will you distinguish between 1-hexene and *n*-hexane ?**

Solution. This can be done with the help of following tests.

1. Baeyer's test. (reaction with cold dil. $KMnO_4$)

n-hexane does not react. 1-hexene reacts with Baeyer's reagent. Purple colour of $KMnO_4$ is discharged.

$$CH_2CH_2CH_2CH_2CH = CH_2 \xrightarrow[H_2O + O]{\text{dil } KMnO_4} CH_3CH_2CH_2CH_2\underset{|}{\underset{OH}{CH}} - \underset{|}{\underset{OH}{CH_2}}$$

2. Reaction with Br_2 in CCl_4.

1-hexene will discharge the yellow colour of bromine

$$CH_3CH_2CH_2CH_2CH = CH_2 \xrightarrow[CCl_4]{Br_2 \text{ in}} \underset{\text{(Colourless)}}{CH_3CH_2CH_2CH_2CH\,BrCH_2Br}$$

n-hexene will react only on heating to a high temperature.

CYCLOALKENES

5.9 INTRODUCTION

Cycloalkenes are ring compounds with one or more double bonds in the molecule.

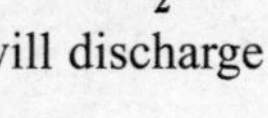

Cyclopropene

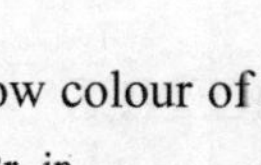

Cyclobutene

Cyclohexadiene

Nomenclature of Cycloalkenes

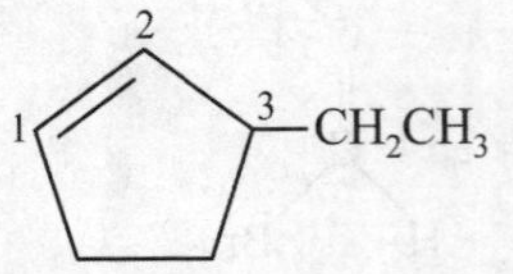

3-Ethylcyclopentene

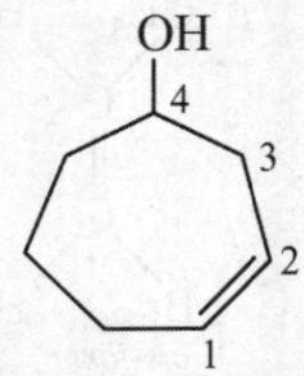

4-Hydroxycycloheptene

5.10 PREPARATION OF CYCLOALKENES

1. Diels-Alder Reaction

Two unsaturated compounds can be combined to produce cycloalkene. When a mixtrue of 1, 3-butadiene and ethylene is heated to 475 K, cyclohexene is obtained

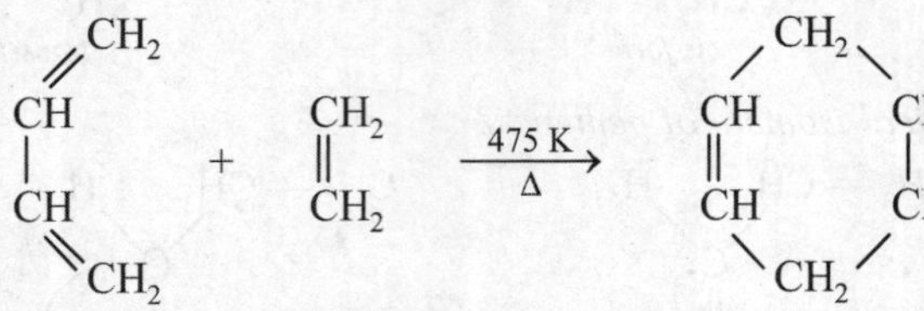

Butadiene Ethylene Cyclohexene

2. By Demjanov Rearrangement

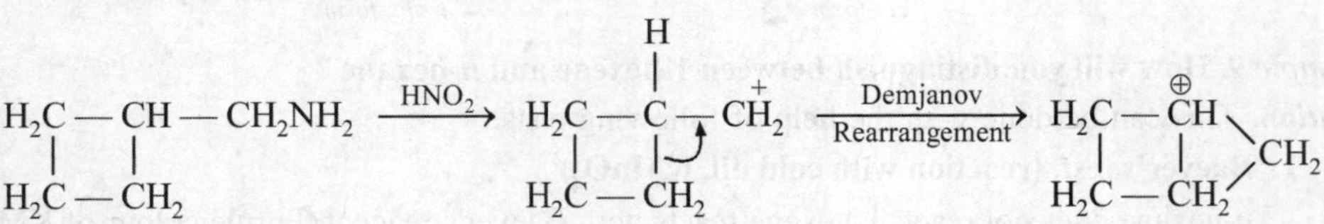

Cyclobutylmethyl amine

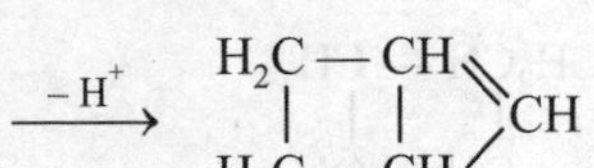

5.11 REACTION OF CYCLOALKENES

1. Addition of Halogens. Addition of halogen takes place

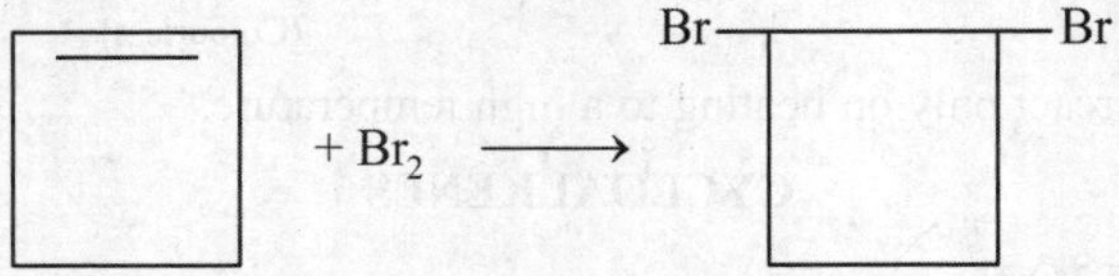

Cyclobutene 1, 2 Dibromocyclobutane

2. Addition of Hydrogen

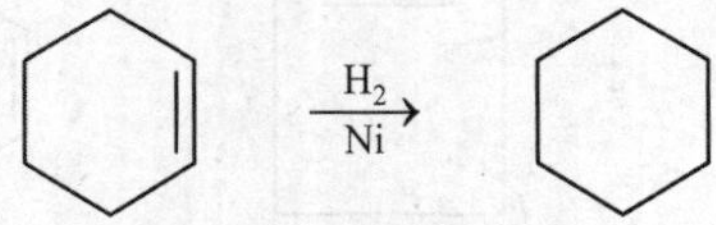

Cyclohexene Cyclohexane

3. Addition of halogen acids

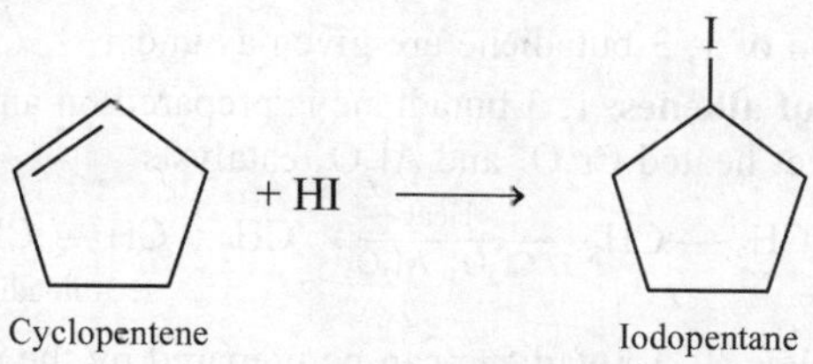

Conformations of cycloalkanes have been dealt with in chapter 3

DIENES

5.12 INTRODUCTION

Hydrocarbons containing two double bonds are called dienes. These are also called alkadienes. General formula for such compounds is $C_n H_{2n-2}$. These compounds are isomeric with alkynes which also have the same general formula. But there are striking differences between the two classes of compounds.

Dienes are classified into three types depending upon the location of double bonds.

(*a*) Isolated dienes (*b*) Cumulated dienes

(*c*) Conjugated dienes

(*a*) Isolated Dienes. The dienes in which the double bonds are separated by more than one single bond are isolated dienes.

For example:

$$\underset{5}{H_2C} = \underset{4}{CH} - \underset{3}{CH_2} - \underset{2}{CH} = \underset{1}{CH_2} \qquad \text{1, 4-pentadiene}$$

$$\underset{6}{H_2C} = \underset{5}{CH} - \underset{4}{CH_2} - \underset{3}{CH_2} - \underset{2}{CH} = \underset{1}{CH_2} \qquad \text{1, 5-pentadiene}$$

(*b*) Cumulated dienes. These dienes contain the grouping $>C=C=C<$ and are characterized by the presence of at least one carbon atom joined to both the neighbouring C-atoms by double bonds.

For example,

$$H_2C = C = CH_2 \quad \text{allene} \quad \text{or} \quad \text{1, 2-propadiene}$$

$$CH_3 - CH = C = CH_2 \qquad \text{1, 2-butadiene}$$

(*c*) Conjugated dienes. In these dienes, a single bond intervenes two double bonds or in other words, they have an arrangement of alternate single and double bonds.

For example,

$$CH_2 = CH - CH = CH_2 \qquad \text{1, 3-butadiene}$$

β^- carotene which is found in carrots and green leaves, contains a system of eleven conjugate double bonds.

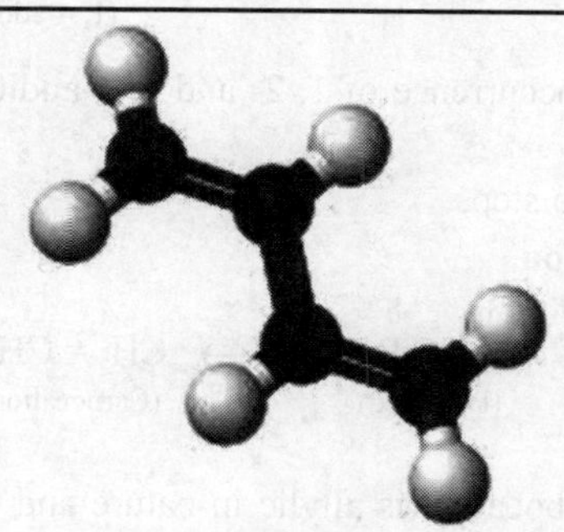

trans-butadiene –1,3

5.13 METHODS OF FORMATION OF 1, 3-BUTADIENE

Methods of preparation of 1, 3-butadiene are given as under.

***(i)* Dehydrogenation of alkenes.** 1, 3-butadiene is prepared on an industrial scale by passing the vapours of n-butane over heated Cr_2O_3 and Al_2O_3 catalysts.

$$\underset{n\text{-butane}}{CH_3-CH_2-CH_2-CH_3} \xrightarrow[Cr_2O_3,\ Al_2O_3]{\text{Heat}} \underset{1,\ 3\text{-butadiene}}{CH_2=CH-CH=CH_2} + 2H_2O$$

***(ii)* Dehydration of diols.** 1, 3-butadiene can be prepared by the dehydration of butane 1, 4-diol (or butane-1, 3-diol) in the presence of H_2SO_4 or phosphoric acid.

$$\underset{OH}{\underset{|}{CH_2}}-CH_2-CH_2-\underset{OH}{\underset{|}{CH_2}} \xrightarrow[\Delta]{\text{Conc. } H_2SO_4} \underset{1,\ 3\text{-butadiene}}{CH_2=CH-CH=CH_2} + 2H_2O$$

***(iii)* Pyrolysis of Cycloalkenes.** When the vapours of cyclohexene are passed over heated Nichrome (Ni – Cr – Fe) alloy, 1, 3-butadiene results.

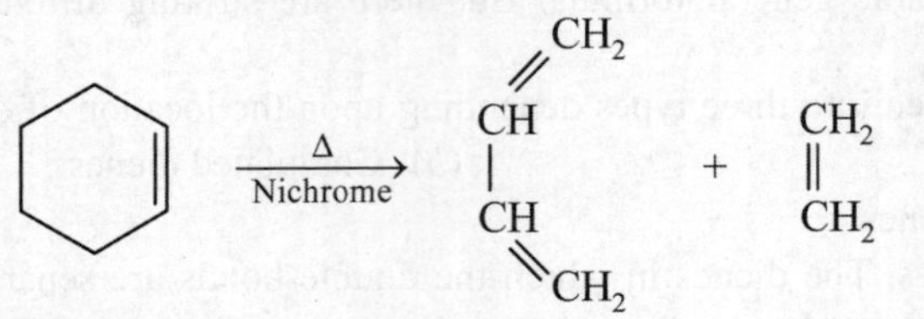

Cyclohexadiene 1, 3-Butadiene Ethylene

5.14 PROPERTIES OF 1, 3-BUTADIENE

Addition of Br_2 to 1, 3-Butadiene

When one mole of the addendum such as Br_2 is added to a conjugated diene, the resulting product obtained is a mixture of 1, 2 and 1, 4-addition products. Such an addition is known as 1, 2- and 1, 4-addition to conjugated dienes. Consider the addition of 1 mole of Br_2 to 1, 3-butadiene.

$$\underset{1,\ 3\text{-Butadiene}}{CH_2=CH-CH=CH_2} \xrightarrow{+Br_2} \begin{cases} CH_2=CH-\underset{Br}{\underset{|}{CH}}-\underset{Br}{\underset{|}{CH_2}} & \text{3, 4-Dibromo-1-butene (1, 2-addition product)} \\ \underset{Br}{\underset{|}{CH_2}}-CH=CH-\underset{Br}{\underset{|}{CH_2}} & \text{1, 4-Dibromo-2-butene (1, 4-addition product)} \end{cases}$$

In order to understand the occurrence of 1, 2- and 1, 4-addition reactions, the mechanism of the above reaction is discussed.

Mechanism. It involves two steps:

***(i)* Formation of carbocation.**

$$CH_2=CH-CH=CH_2 + \underset{(\text{Polarised})}{\overset{\delta^+}{Br}-\overset{\delta^-}{Br}} \longrightarrow \underset{(\text{Carbocation})}{CH_2=CH-\overset{+}{CH}-\underset{Br}{\underset{|}{CH_2}}} + Br^-$$

The above intermediate carbocation is allylic in nature and hence can be considered to be a resonance hybrid of two equivalent resonating structures I and II.

$$\left[\underset{\text{I}}{CH_2=CH-\overset{\oplus}{CH}-\underset{\underset{Br}{|}}{CH_2}} \longleftrightarrow \underset{\text{II}}{\overset{\oplus}{CH_2}-CH=CH-\underset{\underset{Br}{|}}{CH_2}}\right] \equiv \overset{+}{\overline{CH_2 \cdots CH \cdots CH}}-\underset{\underset{Br}{|}}{CH_2}$$

Resonance stabilised carbocation

***(ii)* Combination of bromide with carbocation.**

$$\underset{\text{I}}{CH_2=CH-\overset{\oplus}{CH}-\underset{\underset{Br}{|}}{CH_2}} + Br^- \longrightarrow \underset{(1,\,2\text{-addition})}{CH_2=CH-\underset{\underset{Br}{|}}{CH}-\underset{\underset{Br}{|}}{CH_2}}$$

Bromide ion combines with carbocation I and II to from 1, 2 and 1, 4 addition products respectively.

$$\underset{\text{II}}{\overset{\oplus}{CH_2}-CH=CH-\underset{\underset{Br}{|}}{CH_2}} + Br^- \longrightarrow \underset{(1,\,4\text{-addition})}{\underset{\underset{Br}{|}}{CH_2}-CH=CH-\underset{\underset{Br}{|}}{CH_2}}$$

Addition of HBr to 1, 3-butadiene

Addition of HBr to 1, 3-butadiene gives a mixture of 1, 2 and 1, 4 addition products as given below:

$$\underset{\text{1, 3-Butadiene}}{CH_2=CH-CH=CH_2} \xrightarrow{HBr} \begin{cases} \underset{\text{3-Dibromo-1-butene (1, 2-addition)}}{CH_2=CH-\underset{\underset{Br}{|}}{CH}-CH_3} \\ \underset{\text{1-Bromo-2-butene (1, 4-addition)}}{\underset{\underset{Br}{|}}{CH_2}-CH=CH-CH_3} \end{cases}$$

This can be explained in terms of the mechanism which assumes the formation of carbocations I and II.

Mechanism

(*i*) $$\underset{\text{1, 3-butadiene}}{CH_2=CH-CH=CH_2} \;\overset{\delta^+}{H}-\overset{\delta^-}{Br} \longrightarrow CH_2=CH-\overset{+}{CH}-CH_3 + Br^-$$

The above carbocation is a resonance hybrid of the following two equivalent structure (I and II).

$$\underset{\text{I}}{CH_2=CH-\overset{\oplus}{CH}-CH_3} \longleftrightarrow \underset{\text{II}}{\overset{\oplus}{CH_2}-CH=CH-CH_3} \equiv \underset{\text{Resonance stabilised carbocation}}{\overset{\delta^+}{CH_2}\cdots CH\cdots\overset{\delta^+}{CH}-CH_3}$$

(*ii*) Carbocations I and II react with bromide ion to give 1, 2 and 1, 4 products respectively.

$$\underset{\text{I}}{CH_2=CH-\overset{+}{C}H-CH_3} + Br^- \longrightarrow \underset{\text{(1, 2-addition)}}{CH_2=CH-\underset{|\atop Br}{CH}-CH_3}$$

$$\underset{\text{II}}{\overset{+}{C}H_2-CH=CH-CH_3} + Br^- \longrightarrow \underset{\text{(1, 4-addition)}}{\underset{|\atop Br}{CH_2}-CH=CH-CH_3}$$

Effect of temperature on 1, 2-and 1, 4-addition to conjugated dienes

It has been observed that temperature has a marked effect on the proportions in which 1, 2- and 1, 4-products are obtained as a result of electrophilic additions to conjugated dienes. For example:

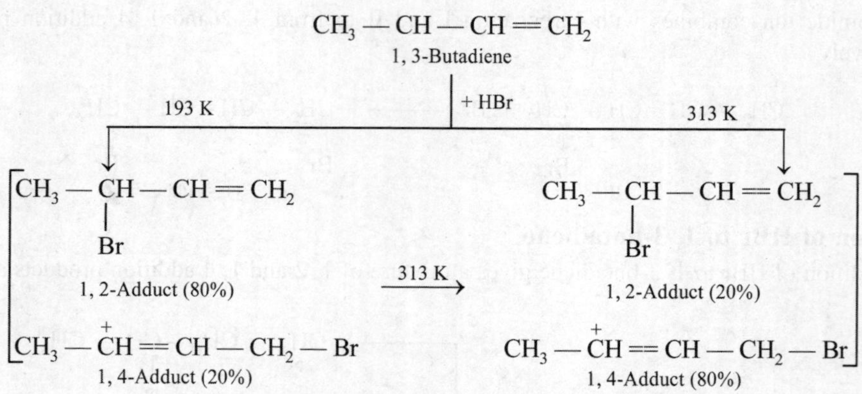

The observations can be summed up as follows:

(*a*) At 193 K, a mixture containing 80% of the 1, 2-product and 20% of the 1, 4-product is obtained.

(*b*) Reaction at 313 K yields a mixture containing 80% of the 1, 4-product and 20% of the 1, 2-product.

(*c*) At intermediate temperatures, mixtures of intermediate compositions are obtained.

(*d*) Although each isomer is quite stable at low temperatures, prolonged heating of either 1,2-product or 1, 4-product yields the same mixture.

Interpretation of the above observations

The fact that 1, 2-product dominates at 193 K, indicates that 1, 2-product is formed faster than 1, 4-product. Thus E_{act} leading to 1, 2-product is less than that leading to 1, 4-product.

At 313 K, 1, 4-product dominates over 1, 2-product. this shows that 1, 4-product is more stable.

Both the above isomers undergo ionisation forming a common carbocation (I). Thus, an equilibrium between the isomers and the carbocation (I) is established as indicated below:

$$\underset{\underset{(1,\,2\text{-product})}{\overset{|}{Br}}}{HC_3-CH}-CH=CH_2 \rightleftharpoons \underset{\underset{+Br^-}{(I)}}{CH_2-\overset{\delta^+}{CH}-CH-\overset{\delta^+}{CH_2}} \rightleftharpoons CH_3-CH=CH-\underset{\underset{(1,\,4\text{-product})}{\overset{|}{Br}}}{CH_2}$$

Potential energy changes during the progress of the reaction have been shown in Fig. 5.1.

It is clear that E_{act} for the ionisation of 1, 2-product to carbocation (I) is less than for 1, 4-product. As a result, 1, 2-product will ionise much more readily than 1, 4-product. Equilibrium is reached when the rates of these two opposing reactions become equal.

When temperature is low, ionisation of the product is quite low and thus 1, 2-product which is formed rapidly is the main product. At high temperature, less stable 1, 2-product readily ionises while the more stable, 1, 4-product ionises very slowly. Thus, at high temperature, 1, 4-product tends to stay and hence constitutes the major product of the reaction.

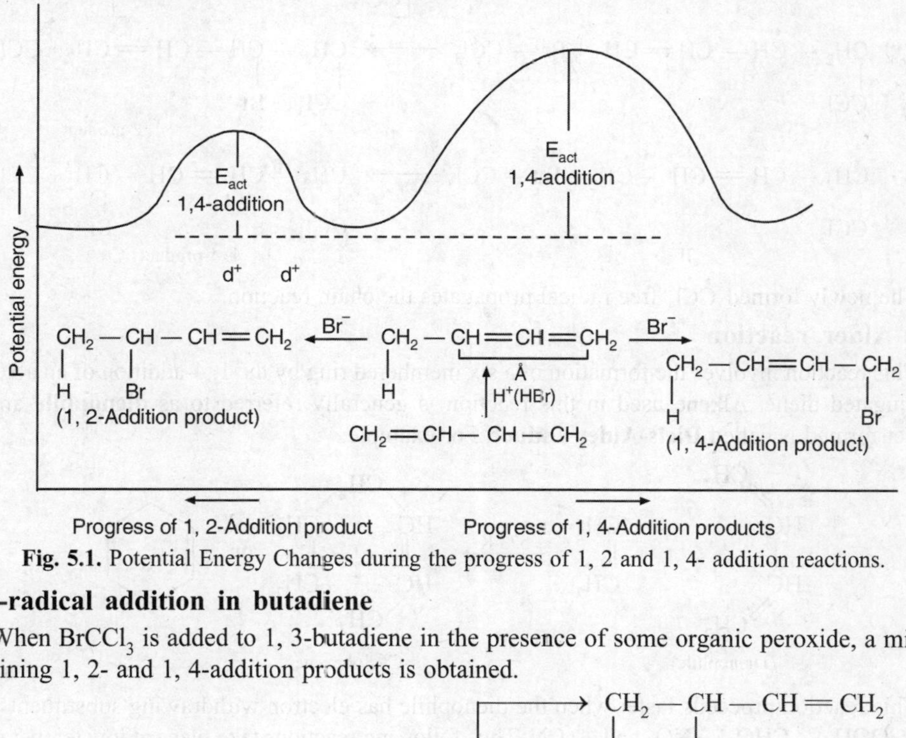

Fig. 5.1. Potential Energy Changes during the progress of 1, 2 and 1, 4- addition reactions.

Free-radical addition in butadiene

When $BrCCl_3$ is added to 1, 3-butadiene in the presence of some organic peroxide, a mixture containing 1, 2- and 1, 4-addition products is obtained.

$$\underset{\text{1, 3-Butadiene}}{CH_2=CH-CH=CH_2} + \underset{\text{Bromotrichloromethane}}{BrCCl_3} \xrightarrow{\text{Peroxide}} \underset{\underset{\text{3-bromo-4-trichloromethyl-1-butene}}{\text{(1, 2-addition)}}}{\underset{CCl_3}{\overset{|}{CH_2}}-\underset{Br}{\overset{|}{CH}}-CH=CH_2}$$

$$\xrightarrow{\text{Peroxide}} \underset{\underset{\text{1-Bromo-4-trichloromethyl-2-butene}}{\text{(1, 4-addition)}}}{\underset{CCl_3}{\overset{|}{CH_2}}-CH=CH-\underset{Br}{\overset{|}{CH_2}}}$$

Mechanism. The above reaction is assumed to occur by a free radical mechanism as illustrated below:

$$(i)\quad C_6H_5-\overset{\overset{O}{\|}}{C}-O-O-\overset{\overset{O}{\|}}{C}-C_6H_5 \xrightarrow{\Delta} 2C_6H_5-\overset{\overset{O}{\|}}{C}-\dot{O} \longrightarrow 2C_6\dot{H}_5 + 2CO_2$$

Benzoyl peroxide — Phenyl free radical

$$(ii)\quad C_6\dot{H}_5 + Br-CCl_3 \longrightarrow C_6H_5Br + \dot{C}Cl_3$$

Trichloromethyl free radical

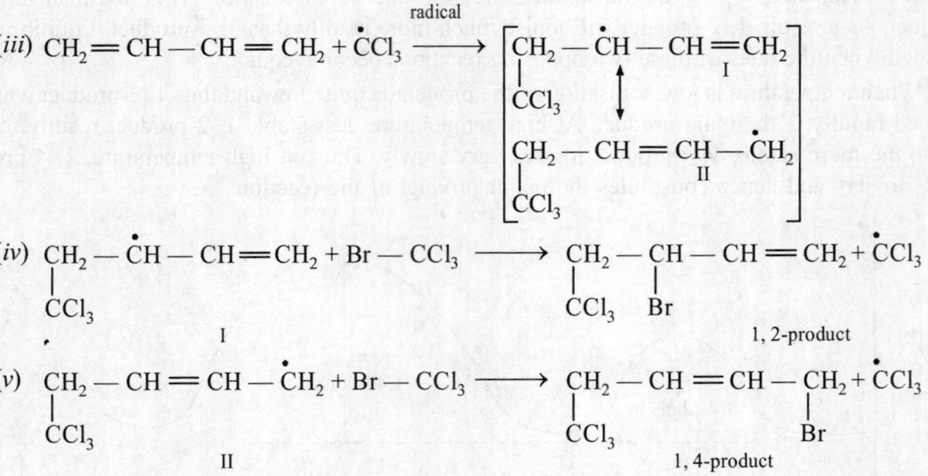

The newly formed $\dot{C}Cl_3$ free radical propagates the chain reaction.

Diels-Alder reaction

This reaction involves the formation of a six membered ring by the 1, 4-addition of an alkene to a conjugated diene. Alkene used in this reaction is generally referred to as **dienophile** and the product formed is called **Diels-Alder adduct**. For example:

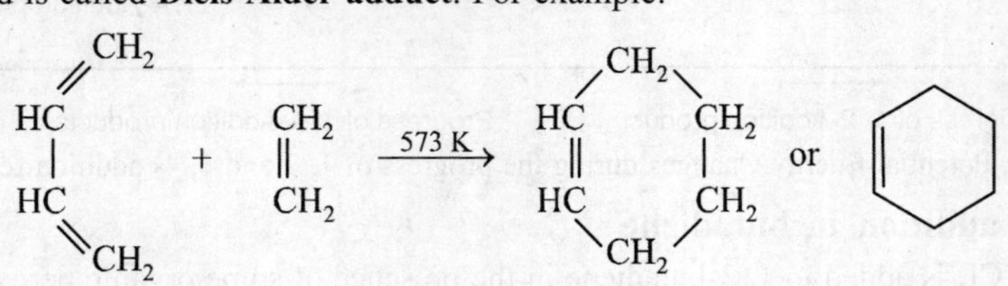

(Dienophile) Cyclohexene

This reaction proceeds faster when the dienophile has electron withdrawing substituents such as — COOH, — CHO, — NO_2 and — CN. Thus following reactions take place at low temperatures.

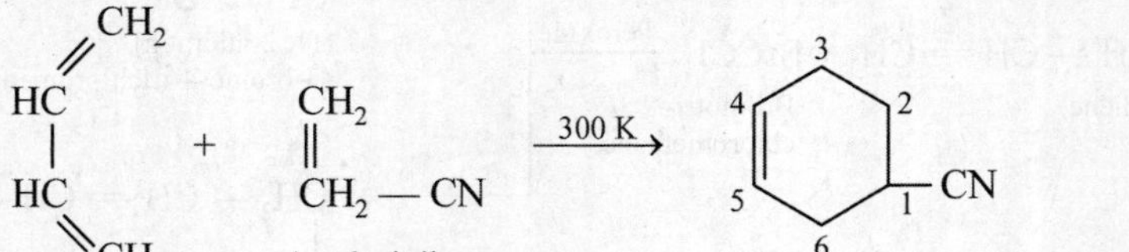

1, 3-Butadiene 1, 2, 3, 6-Tetrahydrobenzonitrile

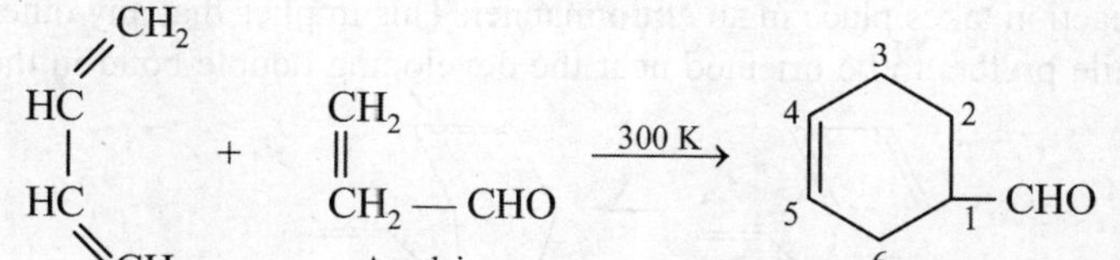

Acrolein

1, 2, 3, 6-Tetrahydrobenzaldehyde

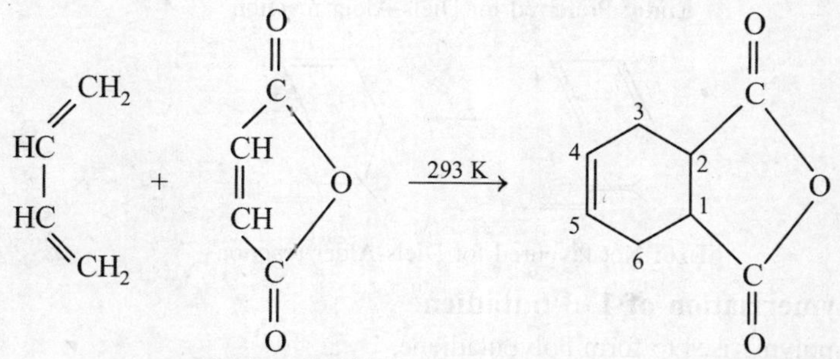

Maleic anyhydride

cis-1, 2, 3, 6-Tetrahydrophthalic anhydride

Stereochemistry of Diels-Alder reaction

1. 1, 3-butadiene exists in two planar conformations called *s*-cis and *s*-trans as shown below:

s-cis-1, 3-Butadiene *s*-trans-1, 3-Butadiene

2. The reaction takes place by *syn*-addition of the diene to the dienophile and the configuration of the dienophile is retained in the product.

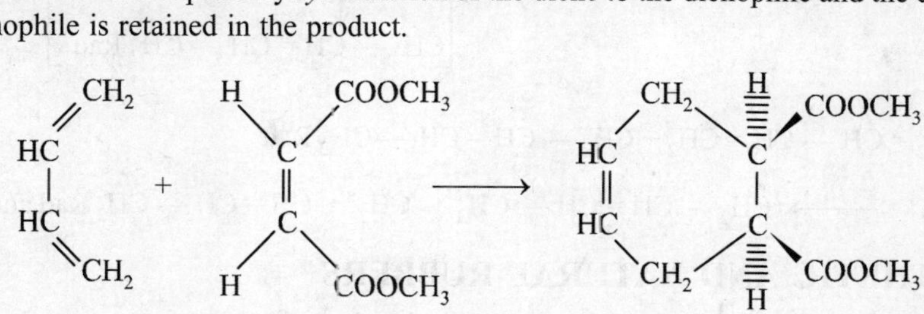

1, 3-Butadiene Dimethylmaleate (*cis*-configuration) Dimethyl *cis*-4-cyclohexane-1,2-dicaryboxylate

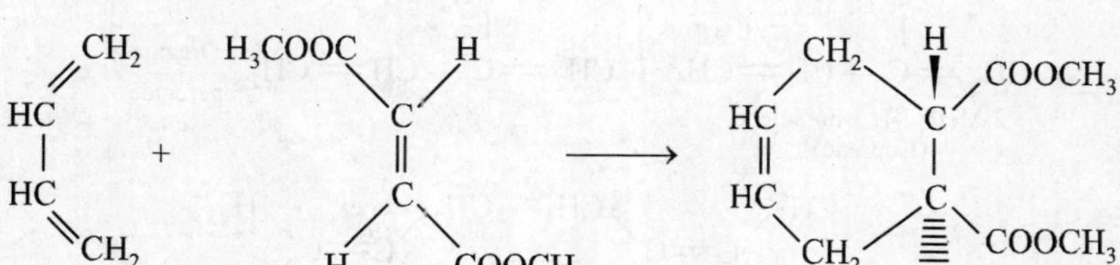

Dimethyl fumarate (*trans*-configuration) Dimethyl *trans*-4-cyclohexane-1, 2-dicarboxylate

3. Diels-Alder reaction takes place in an **endo** manner. This implies that any other unsaturated group in the dienophile prefers to be oriented near the developing double bond in the product.

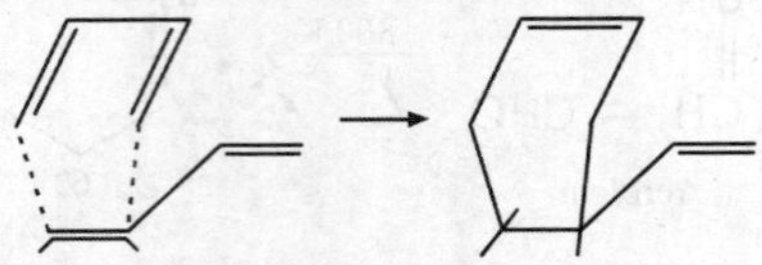

Endo: Preferred for Diels-Alder reaction

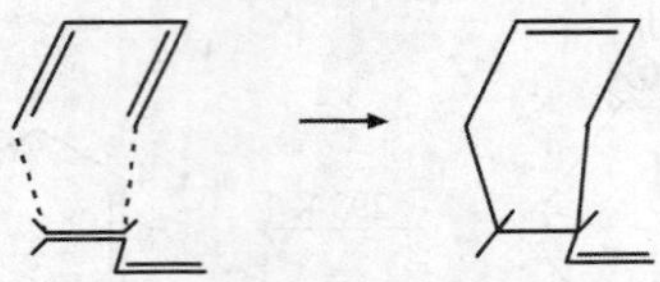

Exo: Not favoured for Diels-Alder reaction

Free radical polymerisation of 1, 3-butadiene

1,3-Butadiene polymerises to form polybutadiene.

$$n\text{CH}_2=\text{CH}-\text{CH}-\text{CH}_2 \xrightarrow{\quad} (-\text{CH}_2-\text{CH}=\text{CH}-\text{CH}_2-)_n$$

1, 3-Butadiene — Polybutadiene

The formation of polybutadiene involves mainly 1,4-addition of butadiene units to each other so that the polymer obtained still contains a double bond for each butadiene unit.

Mechanism. In the presence of some organic peroxides, the polymerisation of 1, 3-butadiene proceeds through the formation of free radicals as illustrated below:

(*i*) Organic peroxide $\longrightarrow$ Rad•

(*ii*) $\text{CH}_2=\text{CH}-\text{CH}=\text{CH}_2+\text{Rad}\bullet \longrightarrow \left[\begin{array}{c}\text{CH}_2=\text{CH}-\dot{\text{C}}\text{H}-\text{CH}_2\,\text{Rad}\\ \downarrow\\ \dot{\text{C}}\text{H}_2-\text{CH}=\text{CH}-\text{CH}_2\,\text{Rad}\end{array}\right]$

(*iii*) $\text{CH}_2=\text{CH}-\text{CH}=\text{CH}_2+\dot{\text{C}}\text{H}_2-\text{CH}=\text{CH}-\text{CH}_2\,\text{Rad}$

$\longrightarrow \dot{\text{C}}\text{H}_2-\text{CH}=\text{CH}-\text{CH}_2-\text{CH}_2-\text{CH}=\text{CH}-\text{CH}_2\,\text{Rad}$ and so on.

5.15 SYNTHETIC AND NATURAL RUBBERS

Synthetic rubber. It is obtained by free radical polymerisation of isoprene. It has *trans*-configuration at almost every double bond.

$$\text{CH}_2=\underset{\text{CH}_3}{\overset{|}{\text{C}}}-\text{CH}=\text{CH}_2+\text{CH}_2=\underset{\text{CH}_3}{\overset{|}{\text{C}}}-\text{CH}=\text{CH}_2 \xrightarrow[\text{perocide}]{\text{Organic}}$$

(CH$_3$ groups drawn above the C atoms.)

2-Methyl-1, 3-butadiene
(Isoprene)

$$\begin{array}{ccccc}\text{CH}_3 & & \text{CH}_2-\text{CH}_2 & & \text{H}\\ & \text{C}=\text{C} & & \text{C}=\text{C} & \\ -\text{CH}_2 & & \text{H} \quad \text{CH}_3 & & \text{CH}_2-\end{array}$$

trans-Polyisoprene or Gutta-percha
(Synthetic rubber)

Natural rubber. It is also a polymer of isoprene in which all the double bonds have mainly *cis*-stereochemistry as indicated below:

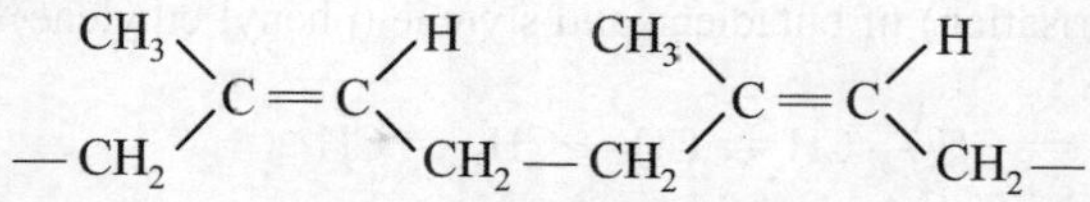

cis-Polyisoprene
(Natural rubber)

It is obtained from *latex* which is a milky white liquid.

Latex contains 33-40% rubber which is coagulated by adding salt and acetic acid to it.

Vulcanization of rubber

The products obtained from raw natural and synthetic rubbers become sticky in hot weather and stiff in cold weather. This drawback can be removed by heating rubber with sulphur. The process of heating rubber with 5—10% of sulphur is called *vulcanization*. During vulcanization, cross links are produced between polymer chains through reactive allylic positions as indicated below:

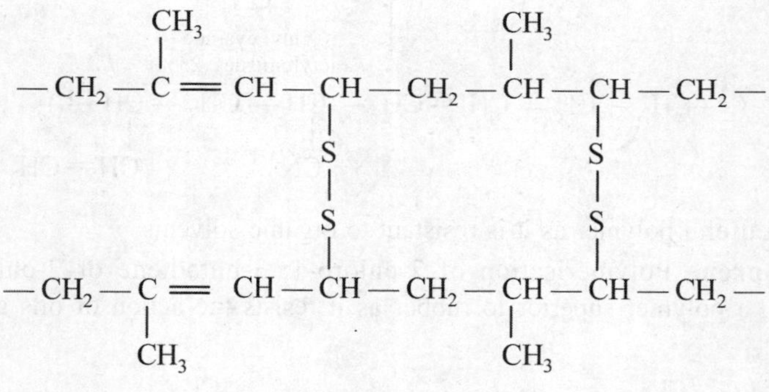

Vulcanized Rubber

Vulcanisation removes the drawbacks of natural rubber and makes it tougher and stronger.

Buna, Buna S, Buna N And Polychloroprene

Buna. 1, 4 addition polymerisation of 1, 3 butadiene yields two products as given below.

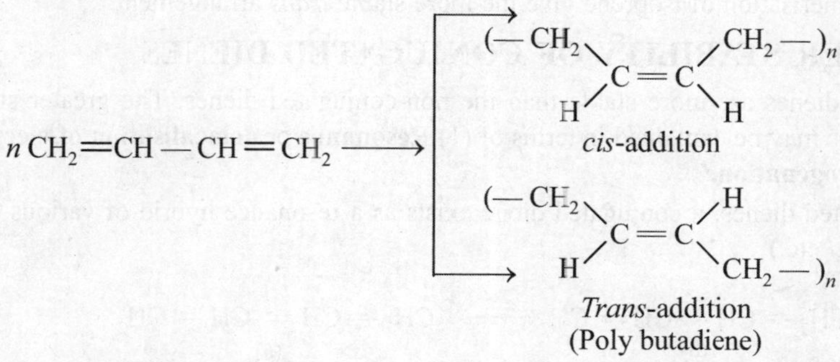

Trans-addition
(Poly butadiene)

$$\underset{\text{Butadiene}}{n\,CH_2{=}CH-CH{=}CH_2} \xrightarrow[\text{or Peroxide}]{\text{Na, }\Delta} \underset{\text{Buna}}{(-CH_2-CH{=}CH-CH_2-)_n}$$

Buna (Bu-Butadiene, Na-Natrium) is a rubber like polymer.

Buna S Synthetic rubber (Buna S) is formed by the copolymerisation (two or more different monomers for polymerisation) of butadiene and styrene (phenyl ethylene)

Butadiene rubber

$$\underset{\text{Butadiene (75\%)}}{CH_2=CH-CH=CH_2} + \underset{\text{Styrene (25\%)}}{CH_2=\underset{|\\C_6H_5}{CH}} \xrightarrow{Na} (-CH_2-CH=CH-CH_2-\underset{|\\C_6H_5}{CH}-CH_2-\underset{|\\CH=CH_2}{CH}-)_n$$

Buna S (S for styrene)

Buna N. [Copolymer of butadiene and acrylonitrile or vinyl cyanide)

$$\underset{\text{Butadiene (75\%)}}{CH_2=CH-CH=CH_2} + \underset{\text{Vinyl cyanide (acrylonitrile) (25\%)}}{CH_2=\underset{|\\CN}{CH}} \longleftrightarrow (-CH_2-CH=CH-CH_2-\underset{|\\CN}{CH}-CH_2-\underset{|\\CH=CH_2}{CH}-)_n$$

Buna N is a useful polymer as it is resistant to organic solvents.

Polychloroprene Polymerisation of 2-chloro-1, 3-butadiene or 2-chloroprene yields polychloroprene, a polymer superior to rubber as it resists the action of oils and other organic solvents.

$$\underset{\text{2-Chloro 1, 3-butadiene or Chloroprene}}{n\,CH_2=\overset{Cl\\|}{C}-CH=CH_2} \longrightarrow \underset{\text{polychloroprene}}{(-CH_2-\overset{Cl\\|}{C}=CH-CH_2-)_n}$$

The double bonds in natural rubber are known to have the *cis* configuration whereas various methods of polymerisation of isoprene give the more stable *trans* arrangement.

5.16 GREATER STABILITY OF CONJUGATED DIENES

Conjugated dienes are more stable than the non-conjugated dienes. The greater stability of conjugated dienes may be explained in terms of (1) **Resonance** or **delocalisation of electrons** and (2) **heat of hydrogenation.**

Unlike isolated dienes, a conjugated diene exists as a resonance hybrid of various canonical structures ($a \rightarrow d$ etc.)

$$\underset{(a)}{CH_2=CH-CH=CH_2} \longleftrightarrow \underset{(b)}{\overset{-}{C}H_2-CH=CH-\overset{+}{C}H_2}$$

$$\underset{(c)}{\overset{+}{C}H_2-CH=CH-\overset{-}{C}H_2} \longleftrightarrow \underset{(d)}{\dot{C}H_2-CH=CH-\dot{C}H_2} \longrightarrow \text{etc.}$$

In canonical structures (*b*, *c* and *d*) C_2 — C_3 bonds has some double bond character and C_1 — C_2 bond and C_3 — C_4 bonds some single bond character as confirmed by bond length measurements; C_2 — C_3 (1.46 Å) as against 1.54 Å for C — C single bond; C_1 — C_2 and C_3 — C_4 (1.35 Å). In structure ***d*** it has got only a formal bond between C_1 and C_4 due to the tendency of the odd electron on each of these carbons to get paired up.

$$[\dot{C}H_2 - CH{=}CH - \dot{C}H_2]$$

The formation of an effective covalent bond between C_1 and C_4 is not possible due to the large distance between them.

The canonical structures *b* and *c* involve charge separation and structure *a* which does not involve any charge separation and has more number of effective covalent bonds (π-bonds) is most stable and therefore, the most important of all the contributing structures.

Thus, the actual structure of 1, 3-butadiene is a resonance hybrid which lies somewhere in between the canonical structures ($a \rightarrow d$).

Greater the number of resonating structures of a compound, greater is the stability. Since non-conjugated dienes do not exist in more than one form, they are less stable than conjugated dienes.

Heat of hydrogenation

Heats of hydrogenation of conjugated dienes when compared with the heats of hydrogenation of non-congjugated dienes, reveal that in case of conjugated dienes, the observed values are less than the anticipated values.

Heats of hydrogenation of monosubstituted alkenes ($RCH = CH_2$), disubstituted alkenes ($R_2C = CH_2$ or $RCH = CHR$) and trisubstituted alkenes ($R_2C = CHR$) have been found to be 125.5, 117 and 113 kJ mol^{-1} respectively. For a diene, the total heat of hydrogenation is expected to be sum of heats of hydrogenation of individual double bonds. This is found to be nearly so for non-conjugated bonds. For example, heat of hydrogenation of 1, 4-pentadiene ($CH_2 = CH - CH_2 - CH = CH_2$) is 254.4 kJ mol^{-1} which is almost equal to the calculated value ($2 \times 125.5 = 251$).

But the situation is different in case of conjugated dienes. Consider the case of 1, 3-butadiene. It has observed that heat of hydrogenation is equal to 238.0 kJ mol^{-1}. This is 13.0 kJ less than the calculated value. Similarly 1, 3-pentadiene has heat ofhydrogenation 16.7 kJ less than expected value.

$$\underset{\text{1, 4 Pentadiene}}{CH_2{=}CH - CH_2 - CH{=}CH_2} \xrightarrow{H_2} \underset{n\text{-Pentane}}{CH_2CH_2CH_2CH_2CH_3}; \Delta H = -254.4 \text{ kJ}$$

$$\underset{\text{1, 3 Pentadiene}}{CH_3 - CH{=}CH - CH = CH_2} \xrightarrow{H_2} \underset{n\text{-Pentane}}{CH_3CH_2CH_2CH_2CH_3}; \Delta H = -234.3 \text{ kJ}$$

Smaller heat of hydrogenation means greater stability of the diene. Hence 1, 3-pentadiene is more stable than 1,4-pentadiene. In general, a conjugated diene is more stable than non-conjugated diene.

5.17 STABILITY OF CONJUGATED DIENES ON THE BASIS OF MOLECULAR ORBITAL THEORY

Each carbon atom in 1, 3-butadiene involve sp^2 hybridization thus leaving on each carbon atom an unhybridized *p*-orbital which overlaps sideways, with the *p*-orbitals of the two adjacent carbon atoms forming π bond(s). (Figs. B and C)

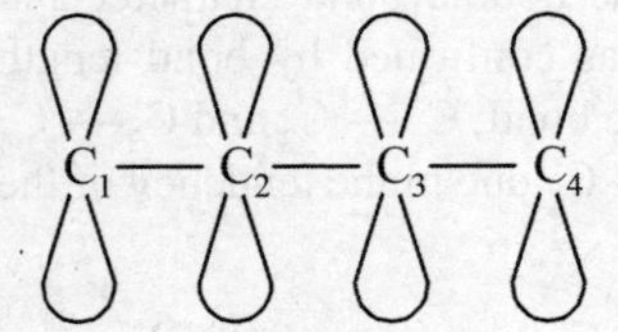

A

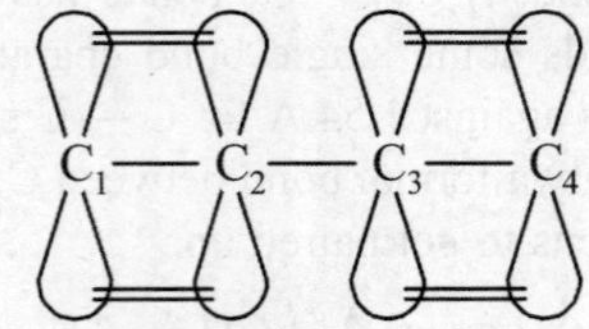

B

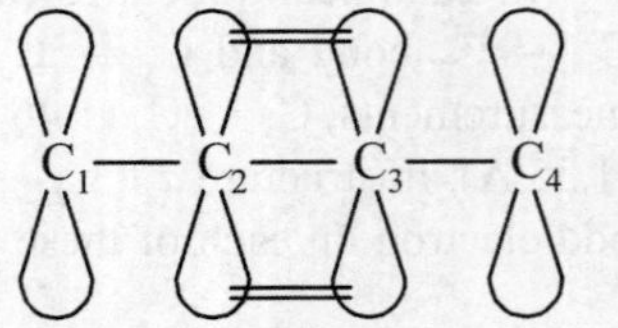

C

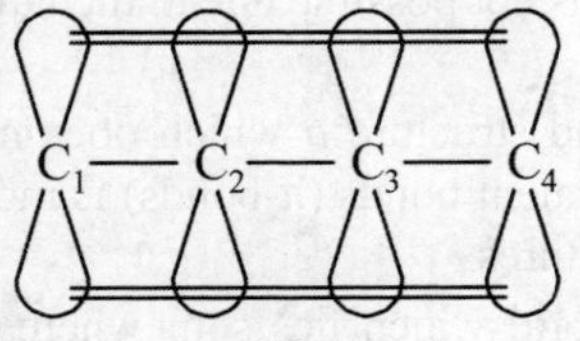

D

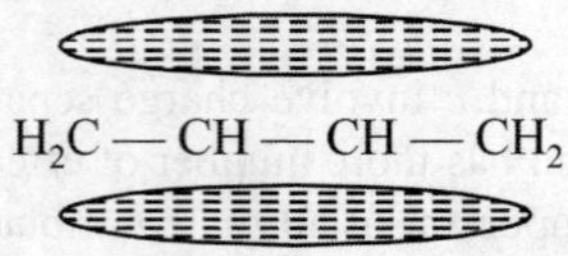

E

Since the four carbons and the hydrogens attached to them lie in the same plane, the *p*-orbital of C_2 and C_3 can also overlap.

Thus, considering the various methods of overlapping of *p*-orbitals, a π molecular orbital will be formed which covers all the four carbon atoms (Fig. D/E) and the formation of delocalized molecular orbitals (M.O. spreading over the nuclei of more than two C-atoms) result in *delocalized electrons* and delocalized electrons account for the extra stability of conjugated dienes. As a result of delocalisation, all the bonds acquire intermediate character between C — C and C = C.

Thus, the extra stability imparted to the molecule (1, 3-butadiene) is due to the presence of delocalized electrons as each pair of delocalized electrons attracts and is attracted by four, and not two nuclei.

Recently, it has been pointed out that resonance stabilization in case of 1, 3-butadiene is negligible and stability to the extent of 15 kJ $mole^{-1}$ is due to the hybridization of $C_2 — C_3$ bond from $sp^3 - sp^3$ to $sp^2 - sp^2$ hybridization. This also accounts for the shortening of $C_2 — C_3$ bond length.

***Example 1.* Complete the following reactions;**

(*i*) $\underset{\displaystyle OH}{CH_2} — CH_2 — CH_2 — \underset{\displaystyle OH}{CH_2} \xrightarrow{H_2SO_4}$

(*ii*) $CH_2{=}CH — CH{=}CH — CH_3 \xrightarrow{O_3,\ H_2O,\ Zn}$

Solution.

(*i*) $\underset{\displaystyle OH}{CH_2} — CH_2 — CH_2 — \underset{\displaystyle OH}{CH_2} \xrightarrow[-2H_2O]{H_2SO_4} \underset{\text{1, 3-Butadiene}}{CH_2{=}CH — CH{=}CH_2}$

Butan-1, 4-diol

(*ii*)

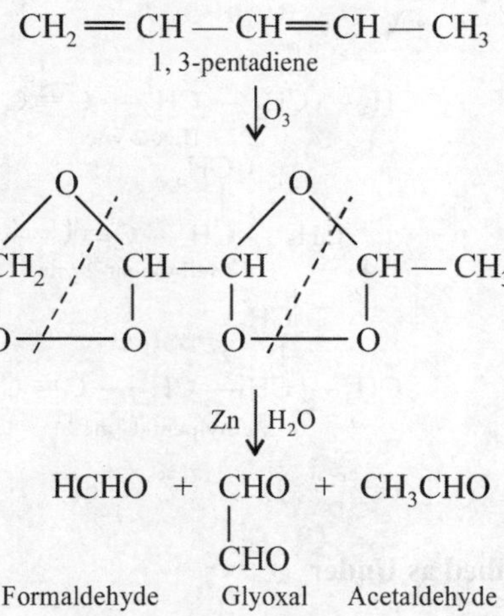

Example 2. **Predict the major Products of the following reaction.**

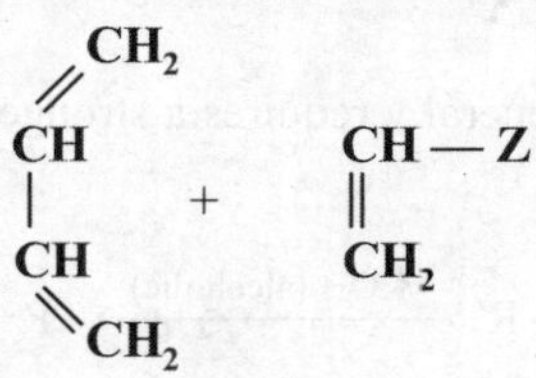

Solution.

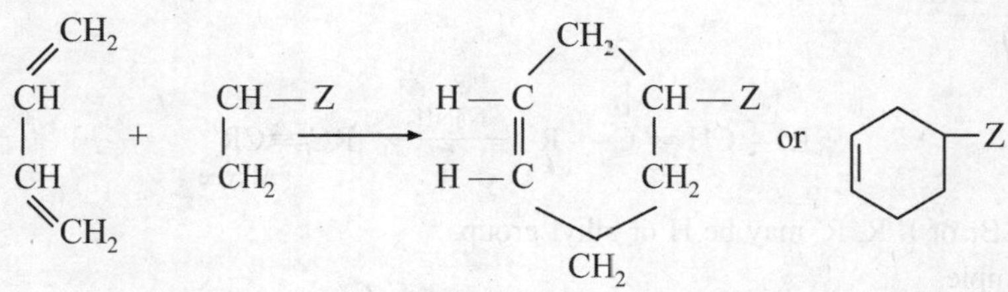

ALKYNES

5.18 NOMENCLATURE OF ALKYNES

Alkynes are unsaturated hydrocarbons having the general formula C_nH_{2n-2}. Such compounds are more unsaturated than alkenes. This is because alkynes contain a triple bond in a molecule. One triple bond is equivalent to two double bonds. First memeber of this series is acetylene C_2H_2 with two carbon atoms.

Nomenclature of alkynes, according to common system and IUPAC system is given below:

Compound	*Common Name*	*IUPAC Name*
$CH \equiv CH$	Acetylene	Ethyne
$CH_3C \equiv CH$	Methyl acetylene	Propyne
$CH_3CH_2C \equiv CH$	Ethyl acetylene	But-1-yne
$CH_3C \equiv CCH_3$	Dimethyl acetylene	But-2-yne
$CH_3CH_2C \equiv CCH_3$	Ethyl methyl acetylene	Pent-2-yne

Example. **Draw the structures of all isomeric alkynes having the formula C_6H_{10} and give their IUPAC names.**

Solution. $CH_3—CH_2—CH_2—CH_2—C{\equiv}CH$
Hex-1-yne

$CH_3—CH_2—CH_2—C{\equiv}C—CH_3$
Hex-2-yne

$CH_3—CH_2—C{\equiv}C—CH_2—CH_3$
Hex-3-yne

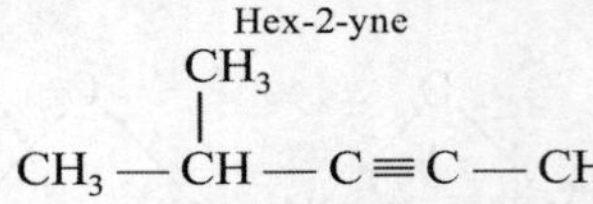
4-Methylpent-2-yne

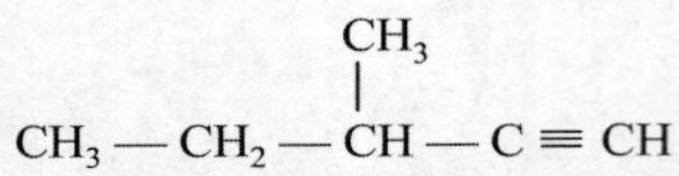
3-Methylpent-1-yne

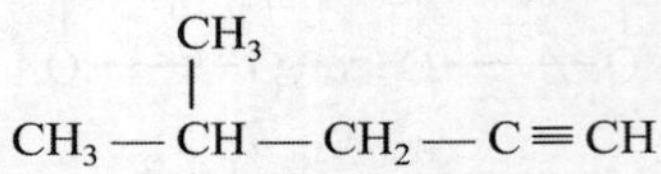
4-Methylpent-1-yne

5.19 PREPARATION OF ALKYNES

Methods of Preparation of Alkynes are Described as Under

(*i*) Dehydrohalogenation of vicinal dihalides.

When 1, 2-dihaloalkane is heated with alcoholic potash, it undergoes dehydrohalogenation, yielding an alkyne. The reaction takes place in two stages involving the intermediate formation of vinyl halide (haloalkene).

The second stage of reaction generally requires a stronger base (sodium amide).

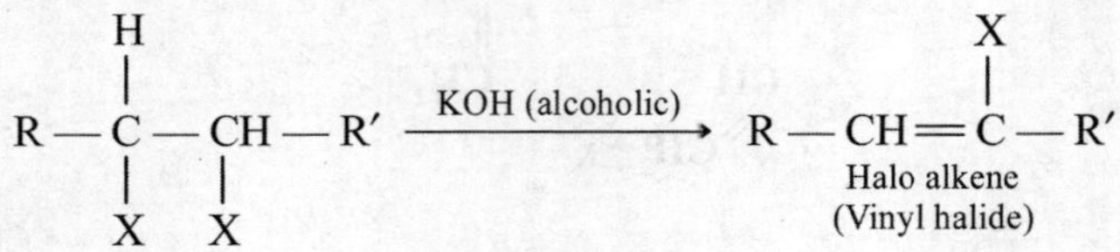
Halo alkene (Vinyl halide)

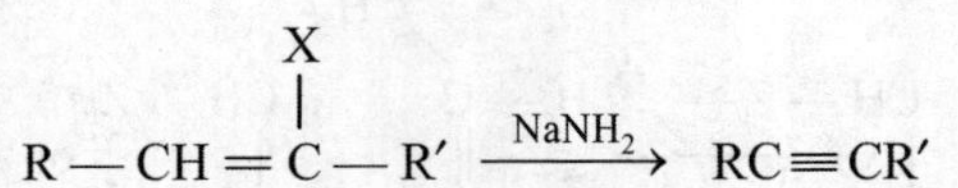
Alkyne

X = Cl, Br or I. R, R′ may be H or alkyl group.

For example,

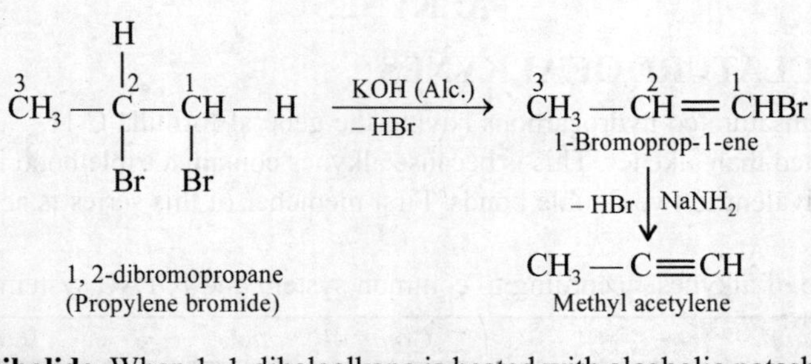
1, 2-dibromopropane (Propylene bromide) → 1-Bromoprop-1-ene → Methyl acetylene

From Gem dihalide. When 1, 1-dihaloalkane is heated with alcoholic potash or sodium amide, it undergoes dehydrohalogenation yielding an alkyne. The reaction proceeds in two stages involving the intermediate formation of haloalkene.

The sequence of the reactions is similar to that in dehydrohalogenation of *vic.*, dihalides.

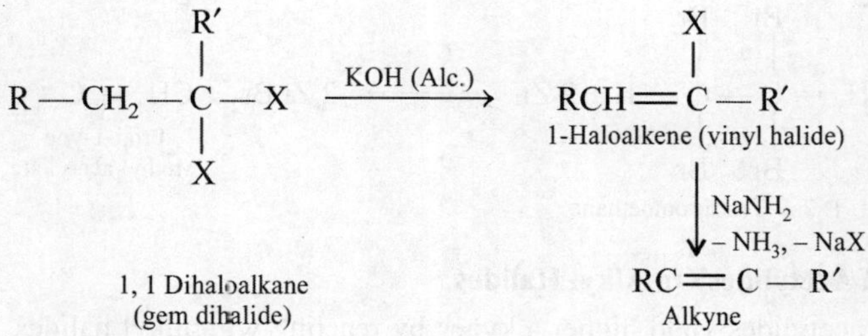

For example,

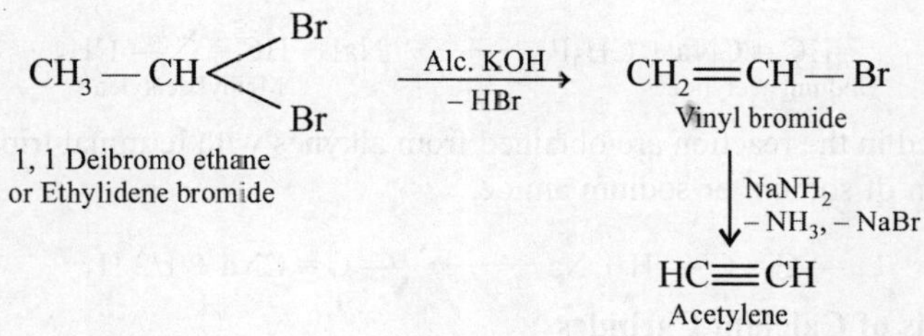

(*b*)

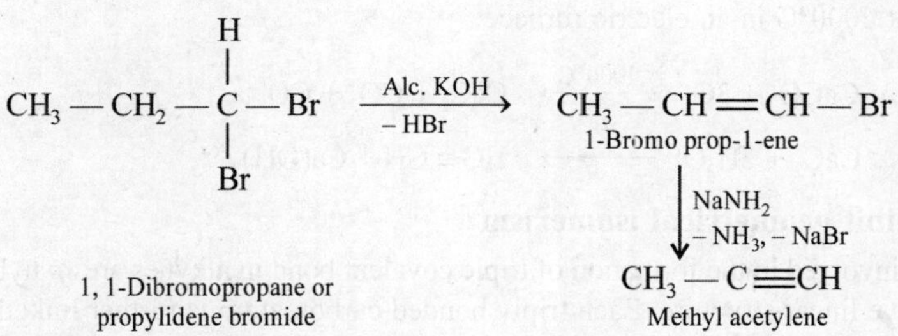

(*ii*) Dehalogenation of Tetrahalides

Dehalogenation of tetra halides (1, 1, 2, 2-tetra haloalkanes) is carried out by passing their vapours over heated zinc and it results in the formation of alkyne.

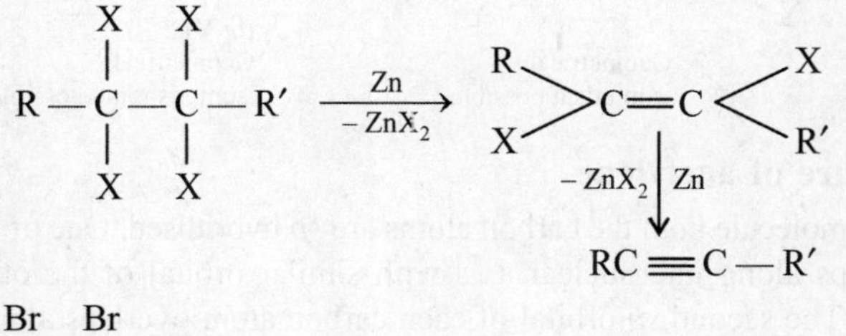

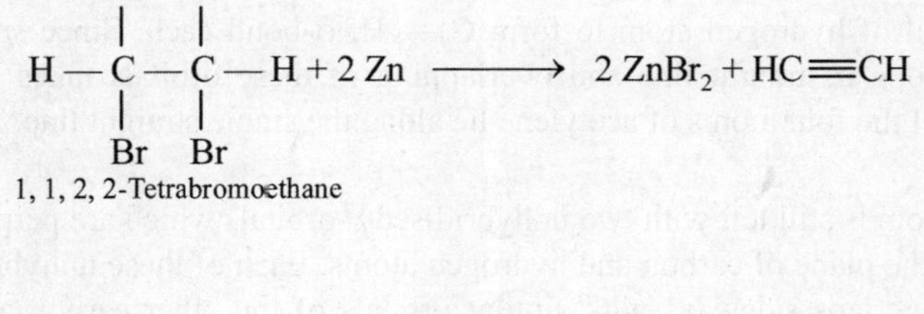

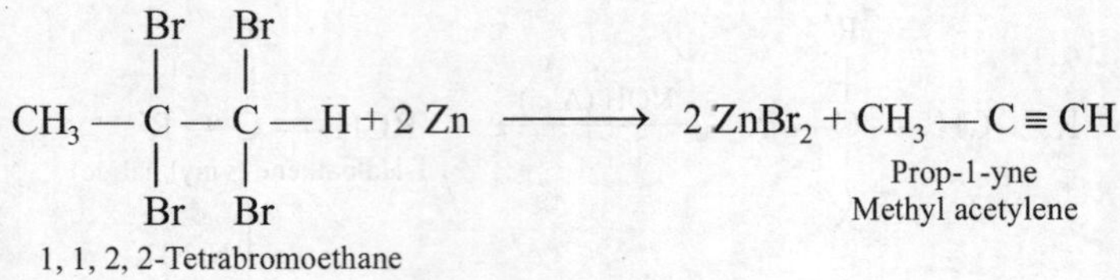

(*iii*) Action of Acetylides On Alkyl Halides

The metallic acetylides yield higher alkynes by reacting with alkyl halides. It is a very good method of converting lower alkynes into higher alkynes.

For example,

$$\underset{\text{Sodium acetylide}}{HC \equiv \overset{-}{C}\overset{+}{Na}} + CH_3I \longrightarrow NaI + \underset{\text{Methyl acetylene}}{HC \equiv C - CH_3}$$

Acetylides used in the reaction are obtained from alkynes with terminal triple bond (— C ≡ C — H) by the action of sodium or sodium amide.

$$-C \equiv C - H + Na \longrightarrow -C \equiv \overset{-}{C}\overset{+}{Na} + 1/2\ H_2$$

(*iv*) Hydrolysis of Calcium Carbides

Acetylene is prepared by the hydrolysis of calcium carbide. The latter is obtained by heating lime stone with coke at 2000°C in an electric furnace.

$$CaCO_3 + 3C \xrightarrow{2000°C} CaC_2 + CO_2 + CO$$

$$CaC_2 + 2H_2O \longrightarrow HC \equiv CH + Ca(OH)_2$$

Alkynes do not exhibit geometrical isomerism

The carbon atoms involved in the formation of triple covalent bond in alkynes are *sp* hybridised. As a result, alkynes have linear structure. Each triply bonded carbon atom is further linked to only one atom or group of atoms. Hence inspite of hindered rotations, the possibility of different relative spatial arrangement of the groups does not arise. Thus, alkynes do not exhibit geometrical isomerism. This type of isomerism is possible when there are at least two groups attached to one carbon.

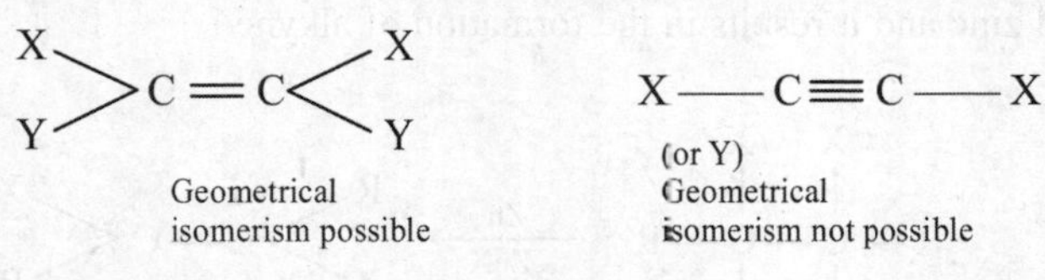

Orbital structure of acetylene

In acetylene molecule both the carbon atoms are *sp* hybridised. One of the *sp* hybridised orbitals of carbon overlaps along internuclear axis with similar orbital of the othe carbon atom to form C — C, σ-bond. The second *sp* orbital of each carbon atom overlaps along internuclear axis with half filled 1*s* orbital of hydrogen atom to form C — H, σ-bond each. Since *sp* orbitals of each carbon atom lie along a straight line and overlapping of these orbitals takes place along their internuclear axis, all the four atoms of acetylene lie along the single straight line. Thus, acetylene is a linear molecule.

Each carbon atom is still left with two unhybridised *p*-orbitals which are perpendicular to each other as well as to the plane of carbon and hydrogen atoms. Each of these unhybridised orbitals of one carbon atom overlaps sideways with similar orbitals of the other carbon atom to form two π-bonds. There is overlapping between these two π electron clouds. As a result, the four lobes of two π-bonds merge to form a single electron cloud which is cylindrically symmetrical about the internuclear axis.

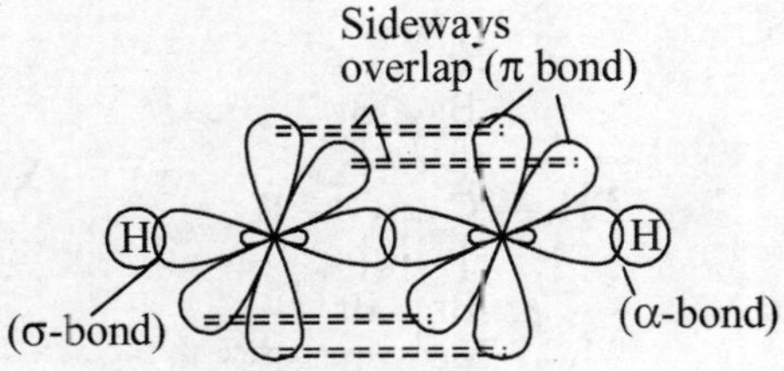

Orbital picture
of acetylene

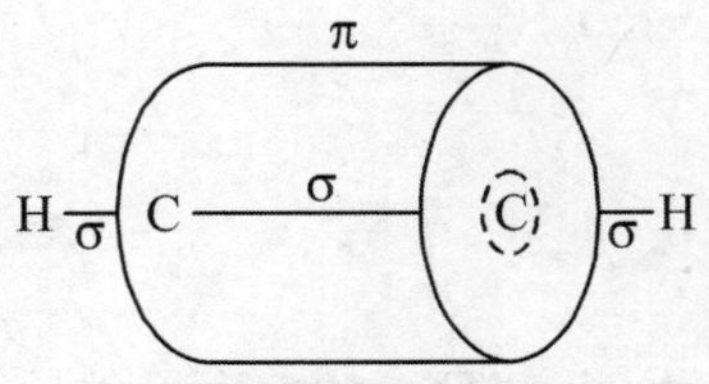

Cylindrical π
electron cloud

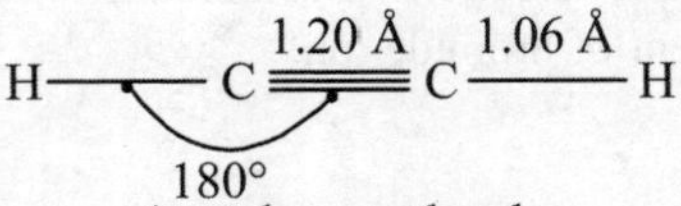

Acetylene molecule
(shape and size)

The carbon-carbon triple bond is thus made up of one strong σ-bond and two weak π-bonds.

Bond parameters. The various bond lengths and energies for the molecule of acetylene are given below:

(*i*) C ≡ C	bond length = 1.20 Å
(*ii*) C — H	bond length = 1.06 Å
(*iii*) H — C — C	bond angle = 180°
(*iv*) C ≡ C	bond dissociation energy = 830 kJ mol^{-1}
(*v*) C — H	bond dissociation energy = 522 kJ mol^{-1}

5.20 CHEMICAL REACTIONS OF ALKYNES

Addition Reactions

Alkynes are less reactive than alkenes towards electrophilic addition reactions. Following factors are responsible for this:

(*i*) Cylindrical π-electron cloud. In alkynes, the four lobes of two π-bonds merge to form a single electron cloud which is cylindrically symmetrical about the internuclear axis and occupies a big volume. Thus, electron density per unit volume becomes low. Due to decrease in electron density, π-electrons are less available to an electrophile. Hence alkynes are less reactive than alkenes towards electrophilic addition reactions.

(*ii*) *sp*-Hybridisation of carbon. The carbon atoms in alkynes are *sp*-hybridised while in alkenes are sp^2-hybridised. Greater the *s* character of an orbital, the more closely the electrons in that orbital are held by the nucleus. Thus π-electrons in alkynes are more strongly held by the carbon atoms than in case of alkenes and are less easily available for reactions with electrophiles. Thus makes alkynes less reactive than alkenes in electrophilic reactions

Bromine readily adds to propyne first forming *trans*-1, 2-dibromopropane and then 1, 1, 2, 2-tetrabromopropane.

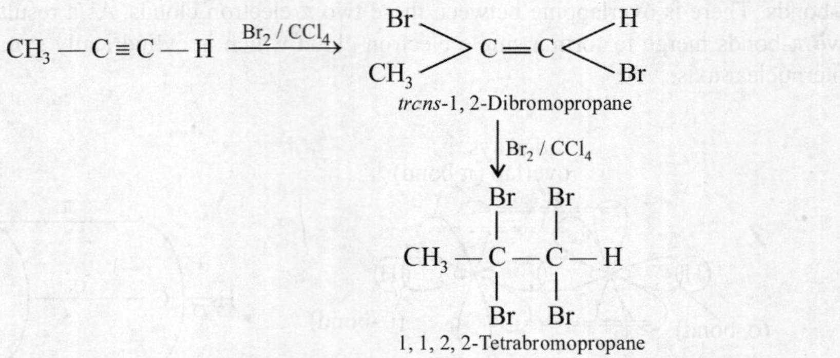

Mechanism. The mechanism of the reaction involves electrophilic addition. It takes place in two steps. This is known as halonium ion mechanism of addition. Bromine (or any halogen) gets polarised under the influence of π-electrons. Bromonium ion (Br^+) adds first forming a bridge bond, followed by the attachment of bromide ion.

First step

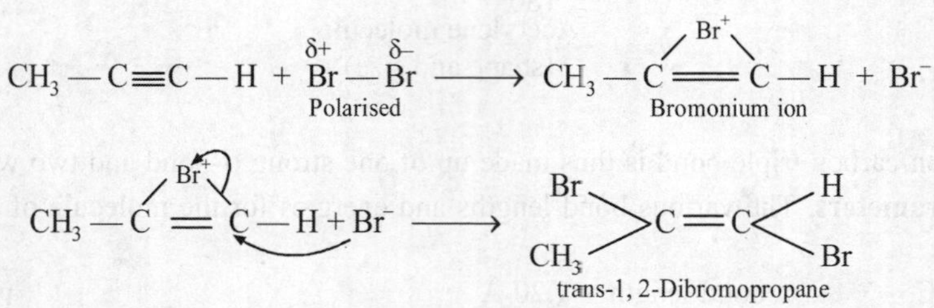

This sequence is repeated in the addition of another bromine molecule.

Second step

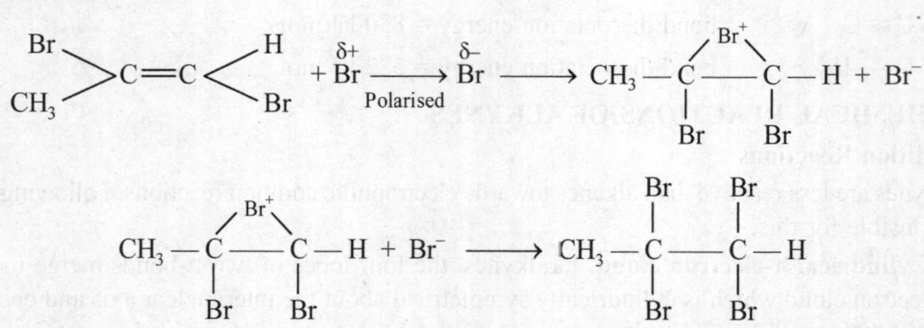

Acidic nature of acetylenic protons

In acetylene and terminal alkynes, the hydrogen atom is attached to a *sp* hybridised carbon which is more electronegative because of increase in *s*-character. Due to greater electronegativity of the *sp* hybridised carbon, the electron pair of C — H bond gets displaced more towards carbon and this helps in the release of proton by strong bases. Consequently acetylene and terminal alkynes behave as acids.

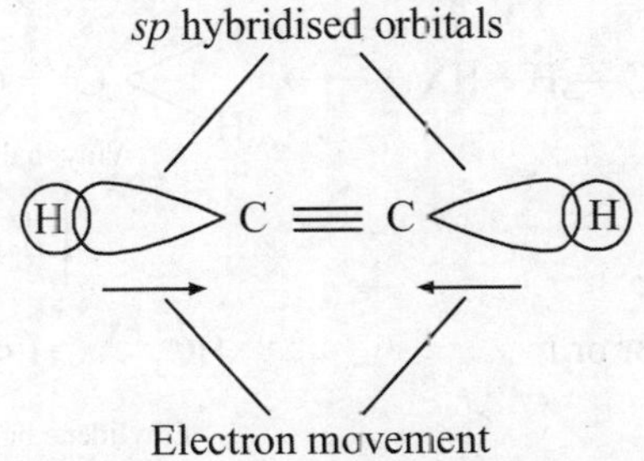

Following reactions illustrate the acidic nature of 1-alkynes:

$$\underset{\text{Acetylene}}{2HC{\equiv}C-H} + 2Na \longrightarrow \underset{\text{Sodium acetylide}}{2HC{\equiv}C\overset{-}{}\overset{+}{Na}} + H_2$$

$$\underset{\text{Propyne}}{CH_3-C{\equiv}C-H} + NaNH_2 \longrightarrow \underset{\text{Sodium methyl acetylide}}{CH_3-C{\equiv}\overset{-}{C}\overset{+}{Na}} + NH_3$$

Comperison of acidic strength of acetylene, ethylene and ethane

Strength of an acid depends upon the ease with which it can lose a proton. This will depend upon the type of bonding between carbon and hydrogen.

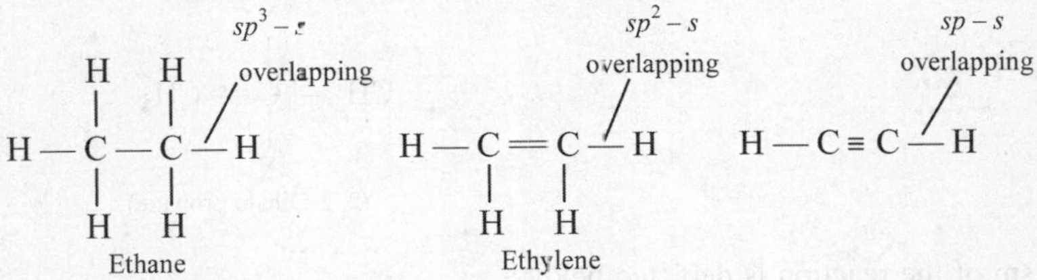

(*i*) In the case of ethane, C — H bond involves overlapping of sp^3 hybrid orbital of carbon and *s*-orbital of H.

(*ii*) In case of ethylene, overlapping is between sp^2 hybrid orbital of carbon and *s*-orbital of H.

(*iii*) In case of acetylene, overlapping in C — H bond is between *sp* hybrid orbital of carbon and *s*-orbital of hydrogen, *s*-orbital has greater electron density compared to a *p*-orbital, which can be explained in terms of their shape. Out of sp^3, sp^2 and *sp* hybrid orbitals, the *sp* orbital has the greatest *s*-character followed by sp^2 and sp^3. Thus carbon atom possessing *sp* hybrid orbital will have maximum electronegativity followed by carbon having sp^2 hybrid orbitals and then followed by carbon having sp^3 hybrid orbital.

In view of this, the hydrogen atom in acetylene will be replaceable with maximum ease followed by that in ethylene and ethane. Hence acidic strength decreases in the order.

acetylene > ethylene > ethane

Addition of halogen acids on alkynes

Treatment of alkynes with halogen acid initially yields vinyl halides and finally alkylidene halides. Combination with second molecule takes place in accordance with Markownikoff's rule.

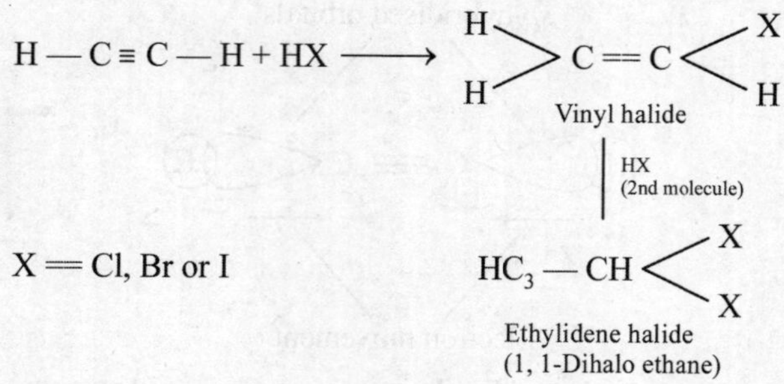

In case of propyne (unsymmetrical alkyne), the first molecule of HX adds according to Markownikoff's rule.

For example,

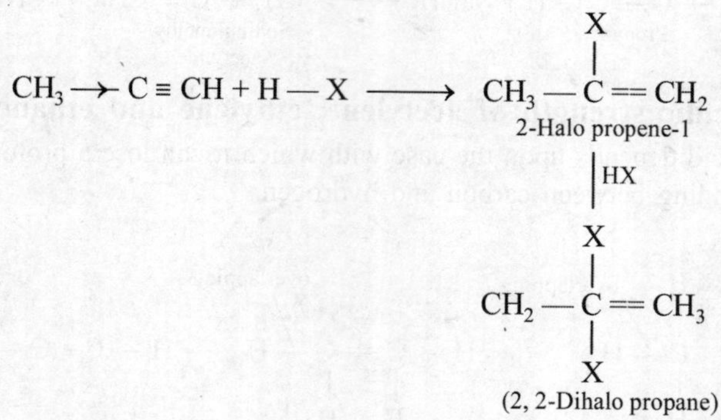

Mechanism of the reaction is depicted below:

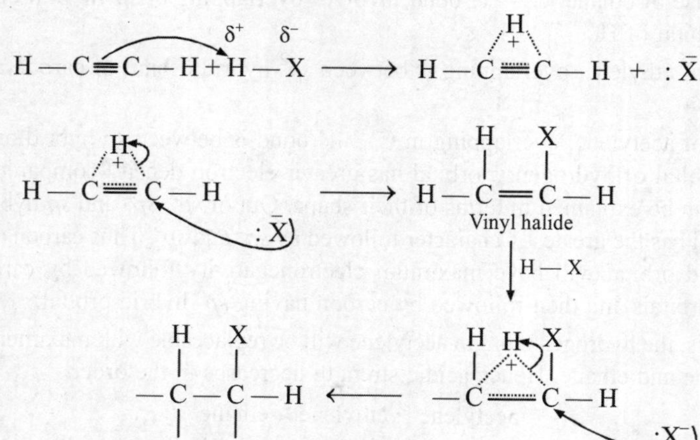

Hydrogenation of alkynes

Catalytic hydrogenation of alkynes yields alkanes through the intermediate formation of an *alkene*.

The reaction can be stopped at the alkene stage with the proper choice of reagent. Predominantly, *trans* alkene is obtained if the reduction of alkyne is carried out in the presence of sodium or lithium in liquid ammonia. Almost entirely *cis* alkene is obtained if the hydrogenation of alkyne is carried out with Lindlar's catalyst (Palladium supoorted over $BaSO_4$ partially poisoned with quinoline) or Nickel Boride catalyst. Thus, the reduction of alkynes is a stereoselective reaction.

For example,

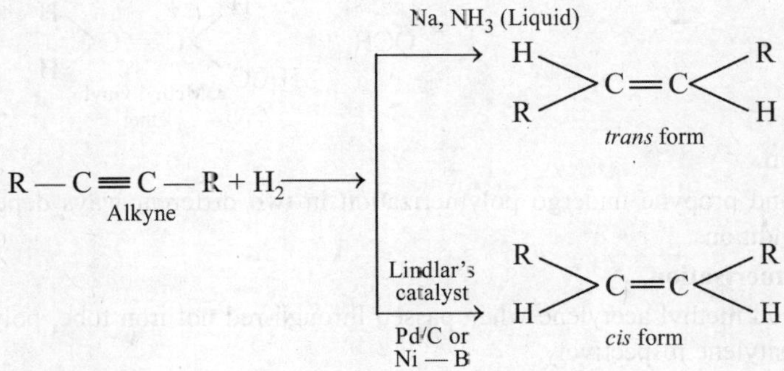

Addition of HCN

Acetylene reacts with hydrogen cyanide in the presence of barium cyanide yielding vinyl cyanide (acrylonitrile).

$$H-C\equiv C-H + HCN \xrightarrow{Ba^{++}} \underset{\text{Vinyl cyanide (acrylonitrile)}}{H_2C=CH-CN}$$

Acrylonitrile is used in the manufacture of Buna N.

The probable mechanism of the reaction is depicted below :

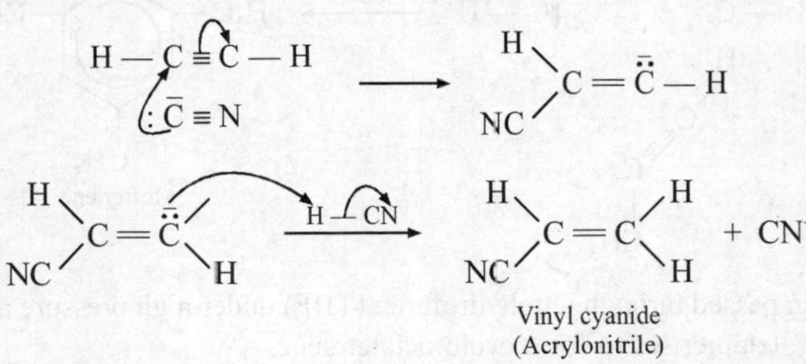

Addition of Methanol

Acetylene adds methanol in the presence of sodium methoxide yielding vinyl ether.

$$H-C\equiv C-H + CH_3OH \xrightarrow{CH_3ONa} \underset{\text{Methyl vinyl ether}}{H_2C=CH.OCH_3}$$

Methyl vinyl ether is used in the manufacture of polyvinyl ether plastics.

The probable mechanism of the reaction is shown below:

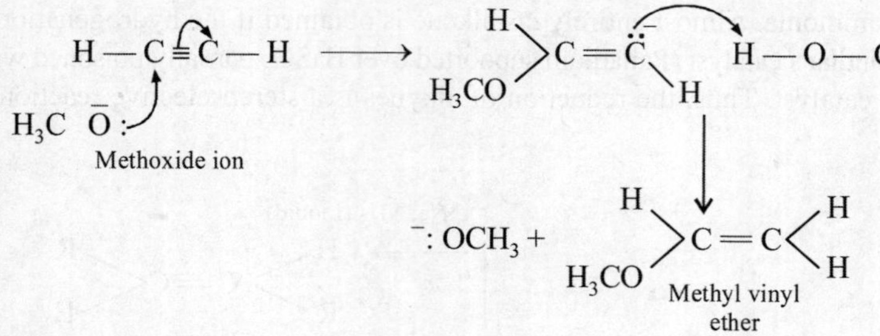

Polymerisation.

Acteylene and propyne undergo polymerization in two different ways depending upon the experimental conditions

Cyclic polymerisation.

Acetylene and methyl acetylene when passed through red hot iron tube, polymerize to form benzene and mesitylene respectively.

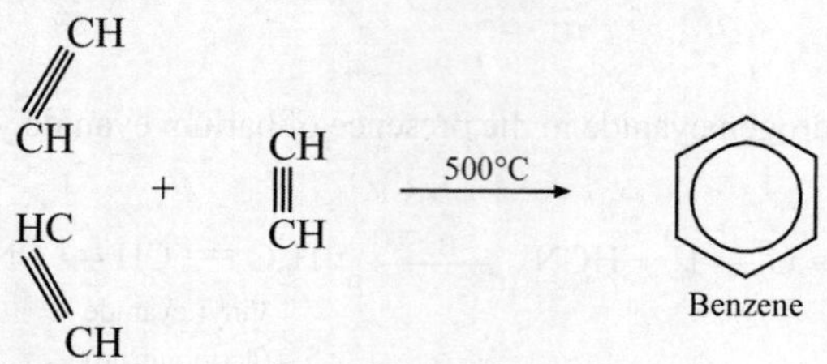

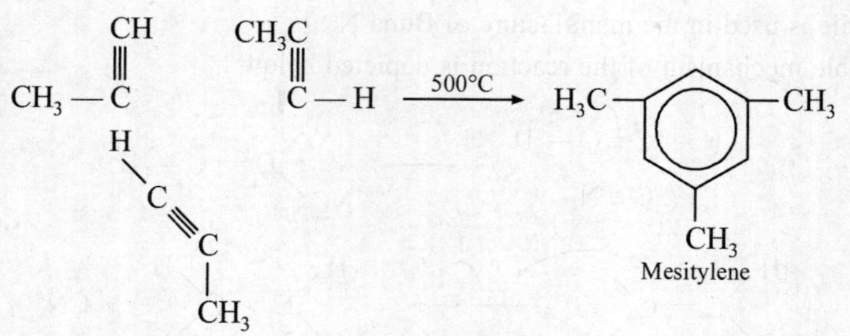

Acetylene when passed through tetrahydrofuran (THF) under high pressure and in the presence of catalyst $Ni(CN)_2$, tetamerises to form cyclo octatetraene.

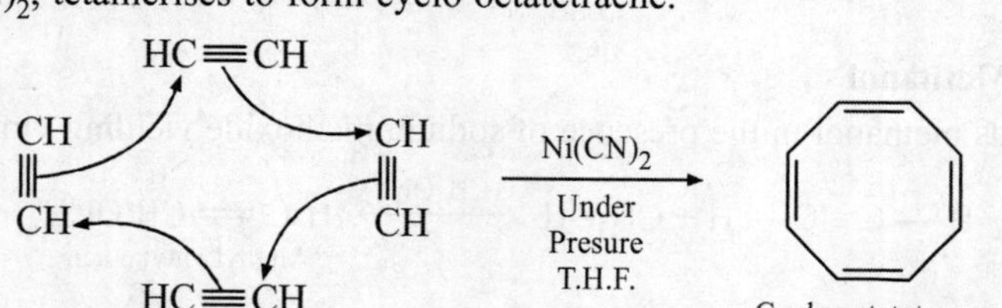

Linear polymerization. Acetylene when passed through a solution of cuprous chloride and ammonium chloride at 343 K, dimerises to form *vinyl acetylene* and finally trimerises to form *divinyl acetylene.*

$$HC \equiv CH + HC \equiv CH \xrightarrow{Cu_2Cl_2,\ NH_4Cl} \underset{\text{Vinyl acetylene}}{H_2C = CH - C \equiv CH}$$

$$HC \equiv CH + HC \equiv CH + HC \equiv CH \xrightarrow{Cu_2Cl_2,\ NH_4Cl} \underset{\text{Divinyl acetylene}}{H_2C = CH - C \equiv C - CH = CH_2}$$

Ozonolysis

Alkynes add on ozone to form ozonides. The ozonides are hydrolysed by water to form dicarbonyl compounds (1, 2-diketones) which undergo oxidative cleavage by H_2O_2 to form *acids*. The identification of the acids formed helps to locate the position to triple bond in the original alkyne. Thus, ozonolysis followed by oxidative cleavage can be used as an unambiguous method for locating the position of triple bond in the original alkyne.

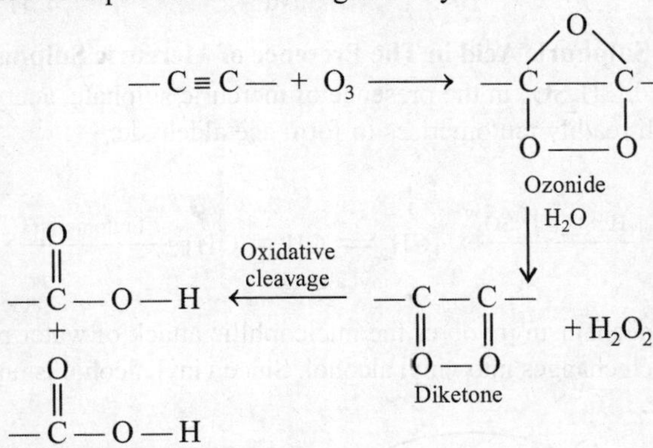

Nucleophilic Addition Reactions

Alkynes undergo nucleophilic addition reactions because of the greater stability of the resulting carbanion formed by the attack of nucleophile. Thus, consider the possible formation of carbanions formed by the attack of nucleophile, $\ddot{N}u^-$.

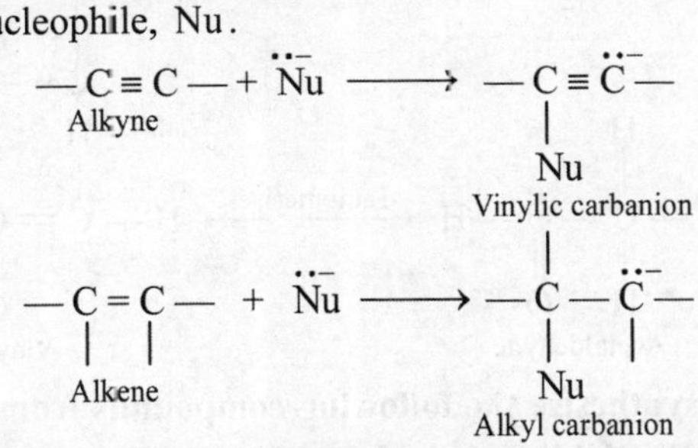

In vinylic carbanion, the negative charge can be dissipated more easily by the electronegative sp^2 hybridised carbon. In other words, vinylic carbanion is more stable than the alkyl carbanion. Since the rate determining step involves the formation of carbanion, it is clear that alkynes are more susceptible to nucleophilic attack than alkenes.

Mechanism of Nucleophilic Addition to Alkynes

It is observed that such reactions are generally catalysed by the presence of heavy metal salts such as those of mercury and barium. This suggests that metal ions form some sort of complex with alkyne by co-ordinating with the electrons of the pie (π) bonds as shown below:

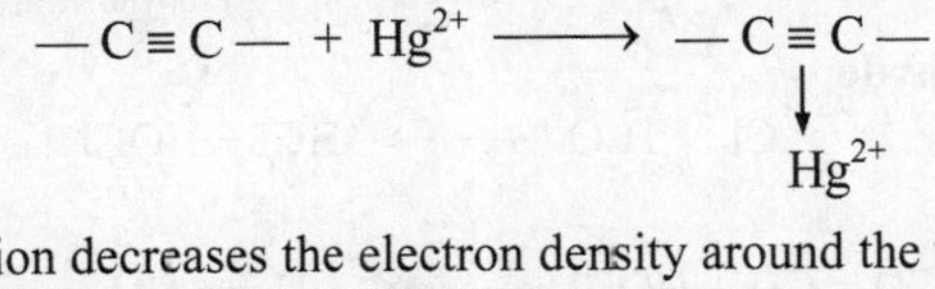

This complex formation decreases the electron density around the triply bonded carbon atoms and thus helps in the attack by the nucleophilic reagent.

Like electrophilic addition reactions, nucleophilic addition also takes place in two steps.

In the first step, the nucleophile attacks the alkyne to form a vinyl carbanion. In the second step, the carbonion takes up a proton to form the final product. The complete mechanism may be depicted as follows:

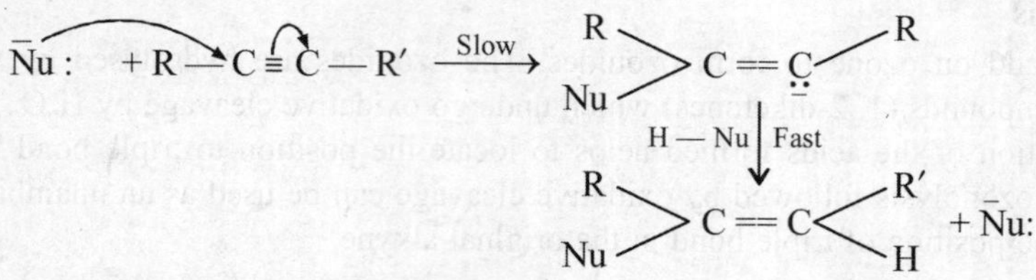

Reaction with dil. Sulphuric Acid in The Presence of Mercuric Sulphate

When treated with dil. H_2SO_4 in the presence of mercuric sulphate, acetylene gets hydrated to form first an enol which readily tautomerises to form acetaldehyde.

$$CH \equiv CH + HOH \xrightarrow{HgSO_4 / H_2SO_4} \underset{\text{Vinyl alcohol}}{[CH_2 = CH - OH]} \xrightarrow{\text{Tautomerises}} \underset{\text{Acetaldehyde}}{CH_3 - \overset{\overset{O}{\|}}{C} - H}$$

Mechanism. The mechanism involves the nucleophilic attack of water molecule on acetylene to form a carbanion which changes into vinyl alcohol. Since vinyl alcohol is unstable, it tautomerises to form stable aldehyde.

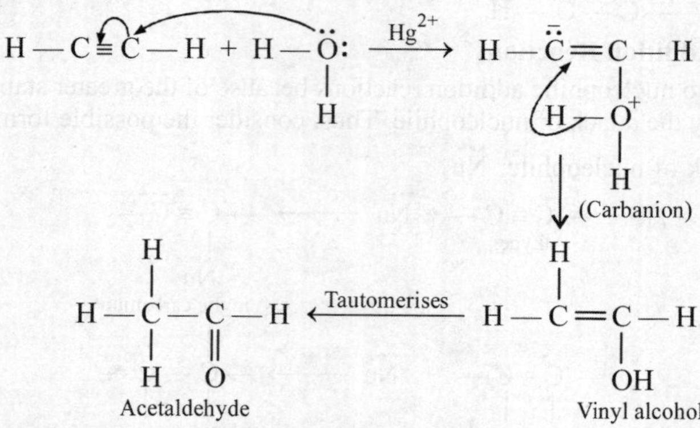

Example 1. **How will you synthesise the following compounds from acetylene?**

(I) Propyne (II) Dichloroacetaldehyde (III) Benzene

(IV) Acrylonitrile (V) Ethylidene chloride.

Solution. Synthesis of the given compounds from acetylene:

(I) Propyne

$$CH \equiv CH + Na\,NH_2 \longrightarrow \underset{\text{Sodium acetylide}}{CH \equiv \overset{-}{C}\overset{+}{Na}} + NH_3$$

$$CH_3 - Cl + \overset{+}{Na}\overset{-}{C} \equiv CH \longrightarrow \underset{\text{Propyne}}{CH_3 - C \equiv CH} + NaCl$$

(II) Dichloroacetaldehyde

$$Cl_2 + H_2O \longrightarrow HCl + HOCl$$

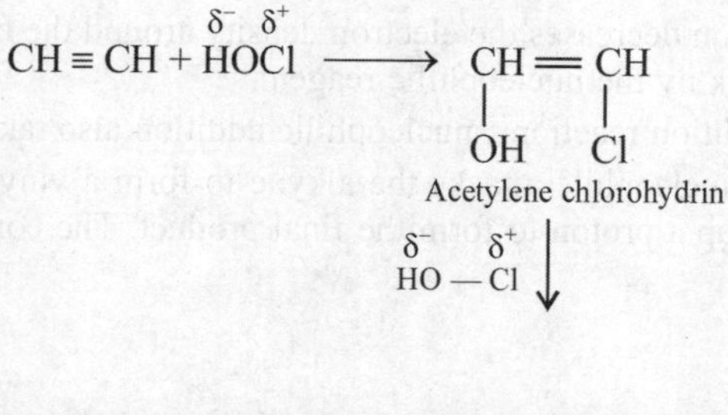

$$\underset{\text{Dichloroacetaldehyde}}{O{=}\underset{|}{C}(H)-CHCl_2} \xleftarrow{-H_2O} \left[\underset{\text{Unstable}}{HO-\underset{|}{CH}(OH)-CHCl_2}\right]$$

(III) Benzene

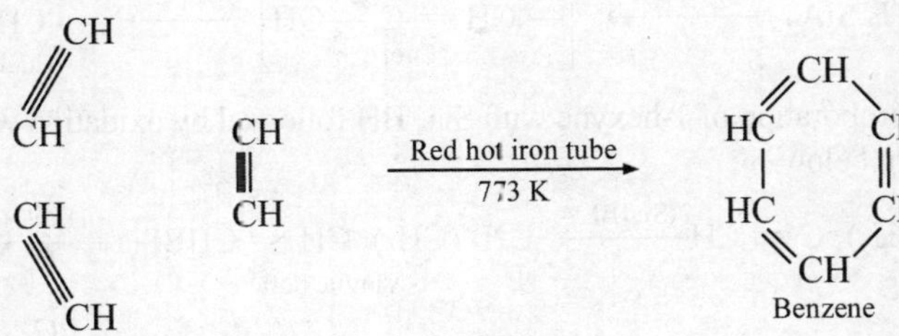

(IV) Acrylonitrile

$$CH \equiv CH + HCN \xrightarrow{Ba(CN)_2} \underset{\substack{\text{Acrylonitrile}\\ \text{(Vinyl cyanide)}}}{CH_2 = CH - CN}$$

(V) Ethylidene chloride

$$CH \equiv CH \xrightarrow{HCl} \underset{\substack{\text{Vinyl chloride}\\ \text{(Chloroethene)}}}{CH_2 = CH\,Cl} \xrightarrow{HCl} \underset{\substack{\text{Ethylidene chloride}\\ \text{(1, 1-Dichloroethane)}}}{CH_3 - CHCl_2}$$

Markownikoff rule is followed in the second step.

Hydroboration-Oxidation of Alkynes

When treated with diborane, *terminal alkynes* (*i.e.*, alkynes which contain a — C ≡ C — H group) undergo addition in two stages, forming firstly a *vinylborane* and finally a *gem-dibora derivative. It is not possible to* **stop** *the reaction at the stage of vinyl borane.*

$$\underset{\text{Terminal alkyne}}{2R - C \equiv C - H} + (BH_3)_2 \longrightarrow \left[\underset{\text{Vinyl borane}}{2\,RCH = \underset{|}{CH}(BH_2)}\right] \xrightarrow{(BH_3)_2} \underset{\substack{gem\text{-dibora}\\ \text{derivative}}}{2RCH_2 - \underset{|}{CH}(BH_2) - BH_2}$$

In case of *internal alkynes* (*i e.*, alkynes in which triply bonded carbons do not carry any hydrogen), the reaction can be stopped at the vinylic stage by using suitable amounts of the reagent.

If in place of diborane, a bulky sterically hindered borane such as bis (1, 2-dimethylpropyl) borane $\left[CH_3 - \underset{|}{CH}(CH_3) - \underset{|}{CH}(CH_3) - \right]_2$ BH commonly known as *disiamylborane* (abbreviated as Sia_2 BH) is used, reaction can be stopped at the vinyl borane stage for all alkynes whether terminal or internal.

That is:

$$\underset{\substack{\text{Terminal}\\ \text{alkyne}}}{R - C \equiv C - H} - Sia_2\,BH \longrightarrow R - CH = CH - BSia_2$$

$$\underset{\substack{\text{Internal}\\\text{alkyne}}}{R-C\equiv C-R} + Sia_2BH \longrightarrow R-CH=\underset{\substack{|\\BSia_2}}{C}-R$$

The vinylic boranes can then be oxidised with H_2O_2 in basic solution to form aldehydes or ketones via enolic intermediates.

$$-CH=\overset{|}{C}-B\,SiA_2 \xrightarrow{H_2O_2,\,OH^-} \left[\underset{\text{Enol}}{-CH=\overset{|}{C}-OH}\right] \longrightarrow \underset{\text{Aldehyde or ketone}}{-CH_2-\overset{|}{C}=O}$$

For example, hydroboration of 1-hexyne with Sia_2 BH followed by oxidation with H_2O_2 / OH^- to yield hexanal is shown below.

$$\underset{\text{1-Hexyne}}{CH_3(CH_2)_3C\equiv CH} \xrightarrow{Sia_2BH} \underset{\text{Vinylic borane}}{CH_3(CH_2)_3CH=CHBSia_2} \xrightarrow{H_2O_2,\,OH^-}$$

$$\underset{\text{Enol}}{[CH_3(CH_2)_3CH=CH-OH]} \rightleftharpoons \underset{\text{Hexanal}}{CH_3(CH_2)_3CH_2-\overset{\overset{O}{\|}}{C}-H}$$

Similarly hydroboration-oxidation of 4-octyne gives rise to 4-octanone.

$$\underset{\text{4-Octyne}}{C_3H_7-C\equiv C-C_3H_7} \xrightarrow{Sia_2BH} \underset{\text{Vinylic borane}}{C_3H_7-\underset{\substack{|\\BSia_2}}{CH}=C-C_3H_7} \xrightarrow{H_2O_2,\,OH^-}$$

$$\underset{\text{Enol}}{\left[C_3H_7-\underset{\substack{|\\OH}}{CH}=C-C_3H_7\right]} \rightleftharpoons \underset{\text{4-Octanone}}{C_3H_7-CH_2-\underset{\substack{\|\\O}}{C}-C_3H_7}$$

It may be noted that hydroboration oxidation leads to **anti-Markownikoff** *addition of H — and — OH. It may also be observed that hydroboration oxidation of* **terminal alkynes** *furnishes aldehydes while that of internal alkynes gives rise to ketones.*

Addition of boranes to carbon-carbon triple bond can also produce cis-alkenes

$$R-C\equiv C-R \xrightarrow{B_2H_6} \begin{matrix} R-C-H \\ \| \\ R'-C-H \end{matrix} \xrightarrow{CH_3COOH} \underset{\textit{cis}\text{-alkene}}{\begin{matrix} R & & R' \\ & C=C & \\ H & & H \end{matrix}}$$

Acetylene generator

Example 2. **What happens when:**

(I) 2-butyne is treated with Pd/C (Lindlar calalyst and hydrogen)

(II) Propyne is treated with chlorine water

(III) Acetylene is treated with Br_2 dissolved in CCl_4

(IV) Acetylene is treated with ammoniacal cuprous chloride solution

(V) Acetylene is treated with ammoniacal $AgNO_3$

(VI) 2-Butyne is treated with sodium dissolved in liquid ammonia

(VII) Acetylene is subjected to ozonolysis.

Solution.

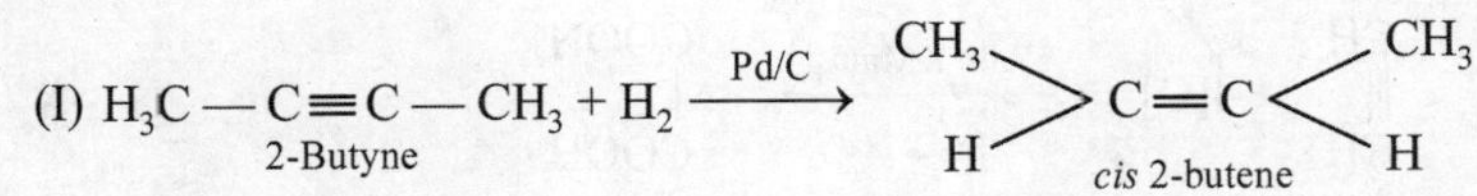

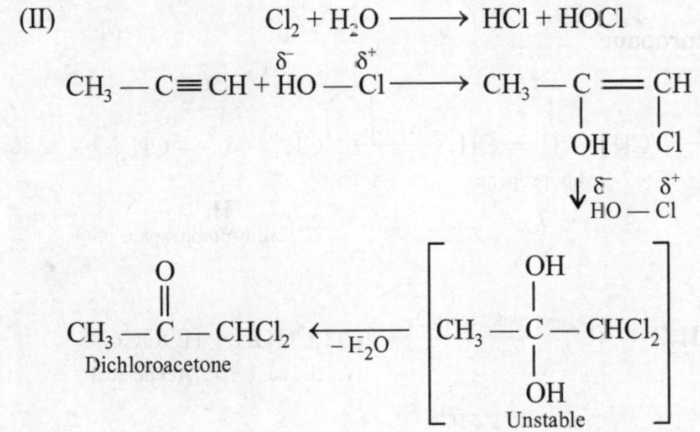

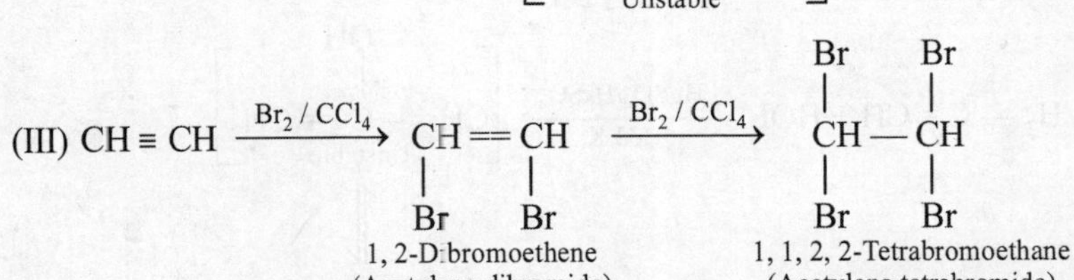

(IV) $CH\equiv CH + Cu_2Cl_2 + 2NH_4OH \longrightarrow CuC\equiv CCu + 2NH_4Cl + 2H_2O$

Copper acetylide (Red ppt.)

(V) $CH\equiv CH + 2NH_4OH + 2AgNO_3 \longrightarrow AgC\equiv CAg + 2NH_4NO_3 + 2H_2O$

Silver acetylide

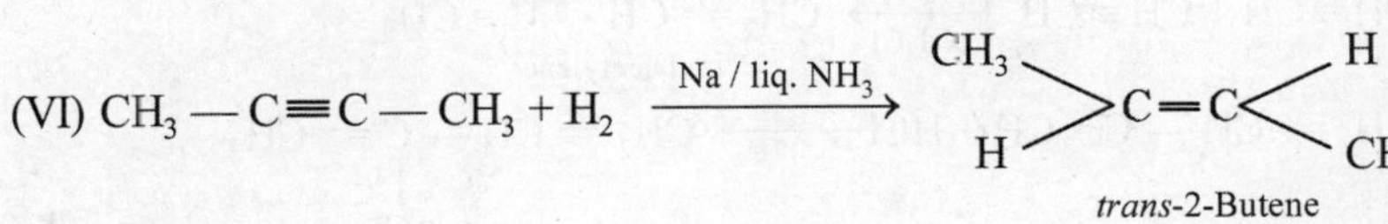

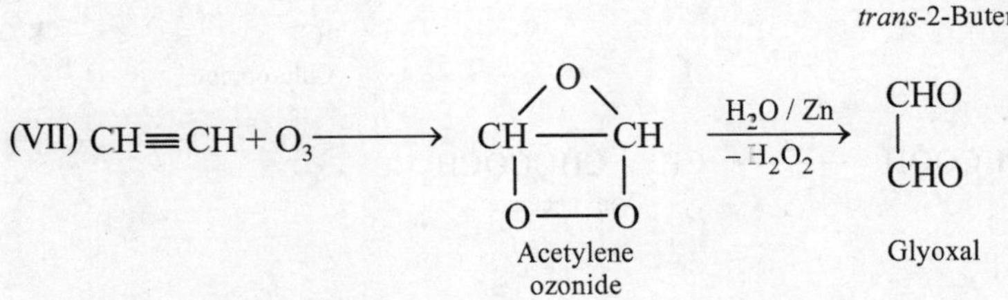

Example 3. **How will you bring about the following conversions?**

(I) Acetylene into oxalic acid

(II) Propyne into 2, 2-dibromopropane

(III) 2-Butyne into acetic acid

(IV) Propyne into propanone

(V) Acetylene into chloroprene

(VI) Acetylene into vinyl acetate.

Solution.

(I) Acetylene into oxalic acid

$$\underset{\text{Acetylene}}{\begin{matrix} CH \\ ||| \\ CH \end{matrix}} + 4[O] \xrightarrow{\text{Alk. } KMnO_4} \underset{\text{Oxalic acid}}{\begin{matrix} COOH \\ | \\ COOH \end{matrix}}$$

(II) Propyne into 2, 2-dibromopropane

$$\underset{\text{Propyne}}{CH_3 - C \equiv CH} \xrightarrow{HBr} \underset{\text{2-Bromopropene}}{CH_3 - \overset{\overset{Br}{|}}{C} = CH_2} \xrightarrow{HBr} \underset{\text{2, 2-dibromopropane}}{CH_3 - \overset{\overset{Br}{|}}{\underset{\underset{Br}{|}}{C}} - CH_3}$$

(III) 2-butyne into acetic acid

$$\underset{\text{2-Butyne}}{CH_3 - C \equiv C - CH_3} + H_2O + 3[O] \xrightarrow{\text{Alk. } KMnO_4} \underset{\text{Acetic acid}}{CH_3COOH} + \underset{\text{Acetic acid}}{HOOCCH_3}$$

(IV) Propyne into propanone

$$CH_3 - C \equiv CH + HOH \xrightarrow[333\ K]{HgSO_4/H_2SO_4} \left[\underset{\text{Unstable}}{CH_3 - \overset{\overset{OH}{|}}{C} = CH_2} \right] \rightleftharpoons \underset{\text{Propanone}}{CH_3 - \overset{\overset{O}{||}}{C} - CH_3}$$

(V) Acetylene into chloroprene

$$CH \equiv CH + CH \equiv CH \xrightarrow[NH_4Cl]{CuCl} \underset{\text{Vinyl acetylene}}{CH_2 = CH - C \equiv CH}$$

$$CH_2 = CH - C \equiv CH + HCl \longrightarrow \underset{\text{Chloroprene}}{CH_2 = CH - \underset{\underset{Cl}{|}}{C} = CH_2}$$

(VI) $CH \equiv CH + CH_3COOH \xrightarrow{Hg^{2+}} \underset{\text{Vinyl acetate}}{CH_2 = CHOCOCH_3}$

Example 4. **Complete the following reactions**

$$(i)\ HC \equiv CH + H_2O \xrightarrow[H_2SO_4]{HgSO_4}$$

$$(ii)\ HC \equiv CH + HBr \longrightarrow$$

Solution.

$$(i)\ \underset{\text{Acetylene}}{HC \equiv CH} + H_2O \xrightarrow{HgSO_4} \underset{\text{Vinyl alcohol}}{CH_2 = CHOH} \xrightarrow{\text{Rearranges}} \underset{\text{Acetaldehyde}}{CH_3CHO}$$

$$(ii)\ HC \equiv CH + HBr \xrightarrow{HgSO_4} CH_2 = CHBr \xrightarrow{HBr} \underset{\text{Ethylidene bromide}}{CH_3CHBr_2}$$

Example 5. **Justify with example the following statement: Acetylene undergoes electrophilic as well as nucleophilic addition reactions.**

Solution. Acetylene undergoes nucleophilic reaction like addition of water in the presence of dil. H_2SO_4 and $HgSO_4$ as follows:

$$CH \equiv CH + H_2O \xrightarrow[HgSO_4]{HgSO_4} CH_2CHO$$

Nucleophilic reaction with acetylene takes place because Hg^{2+} forms a complex with the π electrons of carbon-carbon triple bond thereby reducing election density. This consequently encourages attack by nucleophiles. Moreover the intermediate allylic carbanion in the nucleophilic attack gets stabilized.

Acetylene undergoes electrophilic reaction as given by alkenes because of π electrons of the carbon-carbon triple bond.

$$CH \equiv CH + Br_2 \longrightarrow CHBr = CHBr \xrightarrow{Br_2} CHBr_2 - CHBr_2$$

Example 6. **How will you bring about the following conversion ?**

$$CH_3C \equiv CH \longrightarrow CH_3C \equiv CCH_3$$

Solution. $CH_3C \equiv CH + Na \longrightarrow CH_3C \equiv CNa \xrightarrow{CH_3Br} CH_3C \equiv CCH_3$

Example 7. **How will you differentiate propyne-1 from butyne-2?**

Solution. The compounds will be treated with ammoniacal cuprous chloride and ammoniacal silver nitrate separately. Propyne-1 gives a red precipitate with ammoniacal cuprous chloride and gives a white precipitate with ammoniacal silver nitrate. While butyne-2 does not give any precipitate with these reagents.

$$2CH_3C \equiv CH + Cu_2Cl_2 + 2NH_4OH \longrightarrow \underset{\text{(red ppt.)}}{2CH_3C \equiv C.Cu} + 2NH_4Cl + 2H_2O$$

$$CH_3C \equiv CH + 2NH_4OH + 2AgNO_3 \longrightarrow \underset{\text{(white ppt.)}}{2CH_3C \equiv CAg} + 2NH_4Cl + 2H_2O$$

$$CH_3.C \equiv C.CH_3 + Cu_2Cl_2 + 2NH_4OH \longrightarrow \text{No reaction}$$

$$CH_3.C \equiv C.CH_3 + 2NH_4OH + 2AgNO_3 \longrightarrow \text{No reaction}$$

***Example 8.* How will you synthesise acetone from propene?**

Solution. Following steps are involved

$$\underset{\text{Propene}}{CH_3CH = CH_2} \xrightarrow{Br_2} CH_3\overset{\overset{Br}{|}}{C}H - \overset{\overset{Br}{|}}{C}H_2 \xrightarrow{NaNH_2} \underset{\text{Propyne}}{CH_3C \equiv CH} \xrightarrow[H_2SO_4/HgSO_4]{H_2O} \underset{\text{Acetone}}{CH_3 - \overset{\overset{O}{||}}{C} - CH_3}$$

***Example 9.* How will you differentiate between but-1-yne and but-2-yne ?**

Solution. The compounds will be treated with ammoniacal cuprous chloride and ammoniacal silver nitrate. Butyne-1-gives red ppt and silver mirror respectively. Butyne-2 reacts with neither. For reactions, see Example 7.

EXERCISES
(INCLUDING QUESTIONS FROM DIFFERENT UNIVERSITY PAPERS)

Objective Questions

Fill in the blanks

1. IUPAC name of $CH_2 = \underset{\underset{CH_3}{|}}{C} - CH = CH_2$ is ....
2. Complete the reaction: 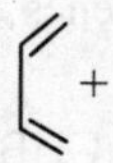 $+ \begin{matrix} CH_2 \\ \| \\ CH_2 \end{matrix} \xrightarrow[\Delta]{200°C}$ ....
3. Reaction of alkene with diborane is called ....
4. $CH_3 - CH_2 - CH\,Br_2 + Zn \xrightarrow{CH_3COOH}$ ....
5. Cummulated diene is called ....
6. Products of cyclic polymerisation of acetylene are .... and ....
7. Complete: $CH_3 - CH = CH_2 + HBr \xrightarrow{\text{Peroxide}}$ ....
8. In Diels Alder reaction, diene reacts with ....
9. $\begin{matrix} CH\,COO\,Na \\ \| \\ CH\,COO\,Na \end{matrix} \xrightarrow[\text{electrolysis}]{H_2O}$ ....
10. $CH_3 - \underset{\underset{CH_3}{|}}{C} = CH_2 + H_2O \xrightarrow{\text{Acid}}$ ....

Multiple Choice Questions

Choose the correct option

1. Markovnikov's addition of HBr is not applicable to

(*a*) Propene (*b*) 1-butene

(*c*) I-pentene (*d*) 2-butene

2. Which of following compounds will react most readily with bromine in CCl_4?

(*a*) $CH_3CH_2CH_3$ (*b*) $(CH_3)_3CH$ (*c*) $CH_3CH = CH_2$ (*d*) $(CH_3)_4C$

3. Ethylene reacts with HI to give

(*a*) Iodoethane (*b*) 2,2-Diiodoethane (*c*) 2,2-Diiodoethane (*d*) None of these

4. Which of the following compounds will have zero dipole moment?

(*a*) *cis*-1,2-dibromoethylene

(*b*) 1,1-dibromoethylene

(*c*) *trans*-1,2-dibromoethylene

(*d*) all of these

5. 2- Butene reacts with HBr to give

(*a*) 1-Bromobutane

(*b*) 2,3 Dibromobutane

(*c*) 2-Bromobutane

(*d*) 2,3-Dibromobutene

6. 2-Methylpropene reacts with HBr to give

(*a*) *tert*-Butyl bromide (*b*) Isobutane

(*c*) *n*-Butyl bromide (*d*) None of these

7. Which of the following alkenes reacts with HBr in the presence of a peroxide to give anti Markovnikov's product?

(*a*) 1-Butene (*b*) 2,3-Dimethyl-2-butene

(*c*) 2-Butene (*d*) 3-Hexene

8. In the addition of HX to a double bond, the hydrogen goes to the carbon that already has more hydrogens, is a statement of

(*a*) Hund's rule (*b*) Markovnikov's rule

(*c*) Huckel rule (*d*) Saytzeff rule

9. In the reaction of $CH_3CH_2CH = CH_2$ with HCl, the H of the HCl will become attached to which carbon ?

(*a*) C-1 (*b*) C-2 (*c*) C-3 (*d*) C-4

10. The disappearance of the purple colour of $KMnO_4$ in its reaction with alkene is known as

(*a*) Markovnikov test (*b*) Grignard test

(*c*) Baeyer test (*d*) Wurtz test

11. Which of the following reagents will react with propene?

(*a*) Hot alkaline $KMnO_4$ (*b*) Sodium metal

(*c*) Cold dilute HNO_3 (*d*) $LiAlH_4$

12. Which of the following will give a negative test when treated with bromine in carbon tetrachloride?

(*a*) Butane (*b*) 2-Butene (*c*) 1,3-Butadiene (*d*) 3-Butyne

13. Which of the following compounds on hydrolysis gives acetylene ?

(*a*) CaC_2 (*b*) Mg_2C_3 (*c*) Al_4C_3 (*d*) Cu_2Cl_2

14. Which alkyne yields propanoic acid as the only product upon treatment with ozone followed by hydrolysis?

(*a*) 1–Butyne (*b*) 2– Hexyne (*c*) 1–Pentyne (*d*) 3–Hexyne

15. Lindlar's Catalyst is

(*a*) Li Al H_4 (*b*) Pd/$BaSO_4$ in quinoline

(*c*) NH_2 NH_2 (*d*) HCl / Zn Cl_2

16. A triple bond consists of

(*a*) 2 sigma bonds and 1 pi bond (*b*) 2 sigma bonds

(*c*) 1 sigma bond and 2 pi bonds (*d*) 3 pi bonds

17. When acetylene is passed through hot iron tube at 400°C, it gives

(*a*) benzene (*b*) toluene (*c*) o–xylene (*d*) mesitylene

18. Which of the following compounds does not react with ammoniacal Ag NO_3 solution?

(*a*) Acetylene (*b*) 1–Butyne (*c*) Propyne (*d*) 2–Butyne

19. Propyne reacts with aqueous H_2SO_4 in the presence of Hg SO_4 to form

(*a*) acetone (*b*) proponol–1 (*c*) acetaldehyde (*d*) propanol–2

20. 1–Butyne can be distinguished from 2–butyne by using

(*a*) potassium permanganate (*b*) bromine in C Cl_4

(*c*) Tollen's reagent (*d*) chlorine in C Cl_4

ANSWERS

Fill in the blanks

1. 2-Methyl–1, 3– butadiene **2.**  **3.** Hydroboration

4. CH_3 C ≡ CH **5.** allene

6. benzene and cyclooctatetraene **7.** CH_3– CH_2 – CH_2 Br **8.** dienophile

9. CH ≡ CH **10.**

OH
|
H_3C– C – CH_3
|
CH_3

Multiple choice questions

1. (*d*)	**2.** (*c*)	**3.** (*a*)	**4.** (*c*)	**5.** (*c*)
6. (*a*)	**7.** (*a*)	**8.** (*b*)	**9.** (*a*)	**10.** (*c*)
11. (*a*)	**12.** (*a*)	**13.** (*a*)	**14.** (*d*)	**15.** (*b*)
16. (*c*)	**17.** (*a*)	**18.** (*d*)	**19.** (*a*)	**20.** (*c*)

General Questions

Alkenes

1. How will you bring about the following conversions:

(*i*) $CH_3 - CH = CH_2 \longrightarrow CH_3CH_2CH_2Br$

(*ii*) $CH_3 - CH = CH_2 \longrightarrow CH_3CH(Br)CH_3$

(*iii*) $CH_3 - C(CH_3) = CH_2 \longrightarrow (CH_3)_3COH$

Give mechanism in each case.

2. What is Markownikoff's rule ? Illustrate with suitable examples.

3. What are alkenes ? Why do we call them olefins ?

4. How would you detect the presence of unsaturation in an organic molecule ?

5. Write all the isomeric alkenes with the formula C_5H_{10} and give their IUPAC names.

6. Ozonolysis of an alkene [X] followed by decomposition with water and a reducing agent gave a mixture of two isomers of the formula C_3H_6O. Write the structure of X and also of the two isomers.

7. Why do alkenes take part in the electrophilic addition reactions?

8. Explain why addition of chlorine to propene at ordinary temperature gives 1, 2-dichloropropane but at 773K, the product is 3-chloropropene.

9. What is meant by ionic polymerisation ? Explain giving mechanism of the reaction.

10. Explain why Kharasch effect is shown by HBr only and not by HCl or HI.

11. Complete the following reaction and discuss its mechanism.

$$CH_3 - CH_2 - OH \xrightarrow{H_2SO_4}$$

12. On reductive ozonolysis, an unsaturated hydrocarbon gave the following products :

(*i*) Ethane diol (*ii*) Propanone (*iii*) Ethanol

Write the structural formula of the hydrocarbon and write its IUPAC name.

13. Fill in the blanks :

(*i*) NBS is used for the bromination of alkenes at the .... position.

(*ii*) In the dehydrohalogenation of alkyl halides, the .... alkene is obtained as the preferred product.

14. Complete the following reactions and name the products

(*i*) $CH_3 - CH = CH_2 + Cl_2 \xrightarrow{CCl_4}$ (*ii*) $CH_3 - CH = CH_2 + Cl_2 \xrightarrow{773\text{ K}}$

(*iii*) $CH_2 = CH_2 + O_3 \xrightarrow{CCl_4}$

15. Complete the following reaction

$$CH_3CH = CHCH_3 + O_3 \longrightarrow ? \xrightarrow{Zn + H_2O} ?$$

16. Why do *trans* alkenes possess higher melting points but lower boiling points as compared to *cis* alkenes ?

17. Predict the structure of alkene which would give acetaldehyde and formaldehyde upon ozonolysis. Write equations for the reaction.

18. What is polymerisation ? Describe the mechanism of anionic polymerisation of alkenes.
19. Write the reaction for the hydroxylation of alkene with osmium tetroxide.
20. Acidic hydration of propene yields 2-propanol while the hydroboration oxidation gives 1-propanol. Explain.
21. How will you justify that the addition of halogens to alkenes is of ionic nature ?
22. Why peroxide effect is shown only by HBr and not by HCl or HI ?
23. Name a polymer which can be obtained by anionic addition mechanism. Discuss this mechanism in general.
24. What happens when :
(*i*) 2-Methyl-2-butene is heated with HCl ?
(*ii*) 2-Bromobutane is treated with alcoholic KOH ?
25. Out of $CH_2 = CH_2$ and $CH_3CH = CH_2$, which will respond more to the electrophilic addition ?
26. How does ozonolysis help in locating the position of double bond in an alkene ?
27. What is Baeyer's test ? Give the reaction.
28. Explain Markownikoff's rule.
29. How will you convert propene to :
(*i*) 2-Propanol (*ii*) 1-Bromopropane (*iii*) 1-Propanol
30. Predict the structures of alkenes form the following sets of ozonolysis products :
(*i*) Only CH_3CHO (*ii*) CH_3COCH_3 and HCHO
(*iii*) Only CH_3COCH_3 (*iv*) CH_3CH_2CHO and CH_3CHO.
31. Write the structure of alkene that on ozonolysis gives :
(*i*) Acetone and formaldehyde
(*ii*) CH_3CHO and $(CH_3)_2$ CHCHO.
32. Give mechanism of the following reactions :
(*i*) Dimerisation of isobutylene
(*ii*) Anti-Markownikoff's addition in alkenes.
33. Describe the mechanism of addition of HBr to propene in the presence of peroxide
34. Define addition polymerisation. Describe with the help of at least one example the free radical mechanism of addition polymerisation.
35. (*a*) Discuss the mechanism of the following reactions :
(*i*) Dimerisation of isobutylene (*ii*) Addition of HI to propylene
(*b*) Deduce IUPAC name of an alkene which gives a mixture of propanal and methanal on ozonolysis.
36. How each of the following substances can be converted to propylene :
(*i*) *n*-propyl alcohol (*ii*) 1, 2-dibromopropane
(*iii*) Isopropyl bromide ?
37. (*a*) Discuss the mechanism of the polymerisation in alkenes under suitable conditions.
38. (*a*) Discuss the mechanism of the following reactions :
(*i*) Formation of halohydrin (*ii*) Addition of methylene to alkenes
(*iii*) hydroboration oxidation of alkenes (*iv*) Alkylation of alkenes.
39. (*a*) State and explain Markownikoff's rule. Justify it theoretically with an example.
(*b*) Give structures and names of products (if any) expected form the reactions of propylene with
(*i*) H_2SO_4 (*ii*) Br_2 (*iii*) H_2O/H^+ (*iv*) HBr/peroxide.

40. Describe very briefly the following giving examples :

(*i*) Markownikoff's rule (*ii*) Carbenes.

41. (*a*) Explain why chlorine forms a substituted product and not addition product upon heating with propene as 773 K.

(*b*) Give the mechanism of free radical polymerisation of alkenes.

(*c*) Arrange the following in increasing order of their stability : 1-butene, *trans*-2-butene, *cis*-2-butene and isobutylene.

42. What is allylic substitution ? Explain.

43. More highly alkylated alkenes are more stable. Explain why ?

44. Explain the mechanism of hydroboration oxidation of propylene.

45. Discuss the mechanism of the following reactions :

(*i*) $CH_3-CH_2-CH=CH_2+HCl \longrightarrow$

(*ii*) $\underset{H}{\overset{R}{>}}C=C\underset{H}{\overset{R}{<}} + CH_2 \longrightarrow$

46. (*a*) Write short notes on Ozonolysis and hydroxylation of olefins.

(*b*) Predict the main product and discuss the reaction mechanism.

(*i*) $CH_3-\underset{CH_3}{\overset{CH_3}{\underset{|}{\overset{|}{C}}}}=CH_2-\underset{CH_3}{\underset{|}{CH_2}} \xrightarrow[\text{Heat}]{H_2SO_4}$ (*ii*) $CH_3-CH=CH_2 \xrightarrow{\text{Peroxide}}$

(*iii*) Discuss the mechanism of addition of carbenes to alkenes.

47. (*a*) Give evidence in support of ionic mechanism of addition of halogens to alkenes. Illustrate by the addition of bromine to ethylene

(*b*) What is peroxide effect ? Give its mechanism. Why is it only applicable to the addition of HBr and not to HCl and HI to unsymmetrical double bond ?

(*c*) Discuss the mechanism of free radical polymerisation of alkenes.

(*d*) Explain why allyl free radical is more stable than *n*-propyl free radical

49. Discuss the E_1 mechanism of dehydrohalogenation of alkyl halides.

50. Give a brief account of :

(*i*) Markownikoff's rule (*ii*) Polymerisation

51. Discuss the mechanism of the following reactions :

(*i*) $CH_3CH_2OH \xrightarrow[443\ K]{H_2SO_4}$ (*ii*) $CH_2=CH_2 \xrightarrow{Br_2+H_2O}$

(*iii*) $CH_3-\overset{CH_3}{\overset{|}{C}}=CH_2+CH_3-\underset{CH_3}{\overset{CH_3}{\underset{|}{\overset{|}{C}}}}-H \xrightarrow{H_2SO_4}$

52. (*a*) How would you detect the presence of unsaturation in an organic compound.

(*b*) Give the mechanism of the following reactions.

(*i*) Hydroboration oxidation of alkenes (*ii*) Peroxide addition to alkenes

53. (*a*) Give the mechanism of dehydration of alcohols.

(*b*) Explain that the addition of bromine to ethene is a two step process.

(*c*) What is Markownikoff's rule? Discuss its mechanism?

54. Discuss the mechanism of hydroboration oxidation of alkenes.

55. (*a*) Discuss the mechanism of the following reactions :

(*i*) Hydration of propene in acidic medium

(*ii*) Electrophilic addition to alkenes

(*iii*) Formation of halohydrins

Dienes

1. Which out of the following compounds is more stable and why ?

(*i*) $CH_2 = CH — CH = CH — CH_3$

(*ii*) $CH_2 = CH — CH_2 — CH = CH_2$

2. Discuss the mechanism of free radical addition to conjugated dienes.

3. Discuss the effect of temperature on 1, 2 and 1, 4 addition to conjugated dienes.

4. Explain the structure of conjugated dienes.

5. How will you justify the extra stability of butadiene ?

6. A conjugated diene is always more stable than a non-conjugated diene. Explain why ?

7. Complete the following and name the product:

1, 3-Butadiene + Maleic anhydride $\longrightarrow$

8. What are dienes ? How do conjugated and commulated dienes differ in their structures ?

9. Discuss the vulcanisation of natural rubber.

10. What happens when 1, 3-butadiene is treated with $BrCCl_3$ in the presence of peroxide ?

11. Explain why does 1, 3-butadiene undergo both electrophilic and nucleophilic addition reactions.

12. A conjugated diene is more stable than non-conjugated diene. Discuss.

13. Give the mechanism of 1, 2 and 1, 4 addition to 1, 3-butadiene.

14. Write a short note on Diel-Alders reaction.

15. Predict the product of the following reaction :

1, 3-Butadiene $\xrightarrow[\text{Peroxide}]{Br\,C\,Cl_3}$

16. Why does butadiene undergo both 1, 2 and 1, 4-addition reactions ?

17. Give the mechanism of conversion of 1, 3-butadiene into polybutadiene.

18. Give the possible products and give their IUPAC names when 1, 3-butadiene is treated with 1 and 2 moles of HBr ?

19. Conjugated dienes are more stable than the isolated dienes. Explain.

20. What do you understand by (4 + 2)-cycloaddition reactions ?

21. 1, 3-Pentadiene is more stable than 1, 4-Pentadiene. Explain.

22. Give the mechanism of electrophilic addition to conjugated dienes.

23. Why are conjugated dienes more stable than the isolated dienes ?

24. Complete the following and name the products :

(i) 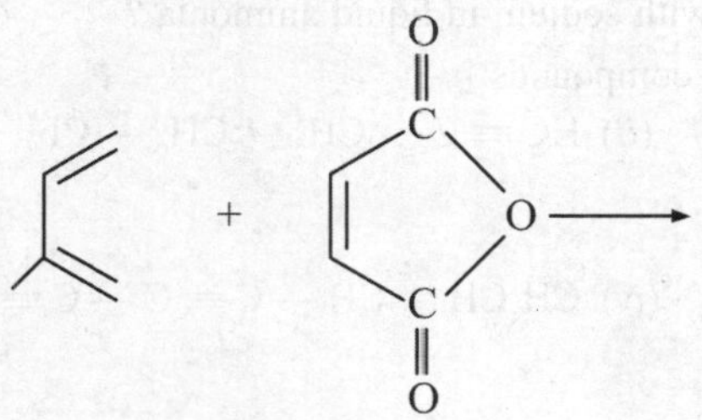

(ii) 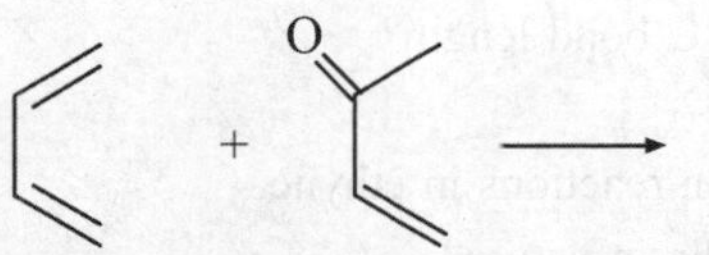

25. On the basis of molecular orbital picture, explain why conjugated dienes are more stable than non-conjugated dienes.

26. (*a*) Discuss the effect of temperature on 1, 2 and 1, 4 addition reactions in conjugated dienes.
(*b*) Why is Diel-Alder's reaction referred to as (4 + 2) cyclo addition reaction.

27. Give the mechanism of 1, 2 and 1, 4 additions to 1, 3-butadiene. What is the effect of temperature to the addition of HBr to 1, 3-butadiene ?

28. What are dienes ? How are they classified ? Give examples. On the basis of the molecular orbital theory explain the stability of conjugated dienes over the isolated dienes.

29. What are the various classes of alkadienes ? Why are conjugated dienes more stable than others ?

30. Give the mechanism and applications of the following :
(*i*) Diel Alders Reaction (*ii*) Polymersiation Reaction

31. Complete the following and name the products :
(*i*) 1, 3-Butadiene + Maleic anhydride $\longrightarrow$
(*ii*) 1, 3-Pentadiene + $BrCCl_3$ $\xrightarrow{\text{Peroxide}}$
(*iii*) 1, 3-Butadiene $\xrightarrow{\text{Peroxide}}$

32. (*a*) Write a short note on Diel Alders reaction.
(*b*) Discuss the relative stabilities of the conjugated dienes over isolated dienes.

33. What happens when bromine is added to 1, 3-Butadiene. Give the mechanism of the reaction.

34. What are the various classes of dienes ? Illustrate with structures. Out of these which class is more stable and why?

35. Name the products formed by the addition of bromine to 1, 3-butadiene. Give mechanism to account for the formation of these products.

36. What are conjugated dienes? Why are they more stable than the isolated dienes?

37. Which is more stable ; 1, 4-pentadiene or 1, 3-butadiene? Explain your answer.

38. How do natural and synthetic rubber differ in their structures ? Explain the process of vulcanisation of rubber.

Alkynes

1. Discuss the orbital structure of acetylene.

2. How will you explain the acidic nature of C — H bond in acetylene ?

3. Why are alkynes linear in nature ?

4. What happens when 2-butyne is treated with sodium in liquid ammonia ?

5. Write the IUPAC names of the following compounds.

(*i*) $CH\equiv C-CH_2-CH=CH_2$ (*ii*) $HC\equiv C-CH_2-CH_2-Cl$

(*iii*) $CH_3-\underset{\underset{H}{|}}{\overset{\overset{CH_3}{|}}{C}}-C\equiv C-\underset{\underset{CH_3}{|}}{\overset{\overset{CH_3}{|}}{CH}}$ (*iv*) $CH_3CH=CH-C\equiv C-C\equiv CH$

6. Why $C\equiv C$ bond length shorter than the $C=C$ bond length?

7. What is the product of ozonolysis of 2-butyne?

8. Discuss the mechanism of nucleophilic addition reactions in ethyne.

9. Why are the hydrogen atoms in acetylene acidic in nature?

10. Give a brief account of nucleophilic addition reaction in alkynes.

11. What happens when :

(*i*) Acetylene vapours are passed through HCl solution?

(*ii*) Acetylene vapours are passed through red hot copper tubes at 773 K?

12. Why do alkynes participate in nucleophilic addition reactions while alkenes do not?

13. How will you convert acetylene into dichloracetaldehyde?

14. How does the orbital picture of triple bond show the absence of geometrical isomerism?

15. How will you synthesise the following from acetylene :

(*i*) Vinyl chloride (*ii*) Benzene.

16. Explain the mechanism of the following reaction.

$$HC\equiv CH \xrightarrow[HgSO_4]{\text{dilute } H_2SO_4}$$

17. Discuss the orbital structure of acetylene.

18. Do all alkynes show acidic nature? Explain.

19. What is the product of ozonolysis of acetylene?

20. Out of ethyne and ethene, which is more reactive towards the electrophilic addition?

21. What is the role of Hg^{2+} ion in the nucleophilic addition reactions of alkynes?

22. Explain why alkynes do not exhibit geometrical isomerism.

23. Predict the products of the following reactions:

(*i*) $CH\equiv CH + C_2H_5OH \xrightarrow{KOH}$ (*ii*) $CH_3-C\equiv C-H + H_2O \xrightarrow[H_2SO_4]{HgSO_4}$

24. How will you test the presence of unsaturation in an organic compound ?

25. What is the reason for the low reactivity of alkynes towards electrophilic addition ?

26. Starting from acetylene, how will you prepare :

(*i*) Ethylidene chloride (*ii*) Dichloroaetaldehyde

(*iii*) Vinyl acetate (*iv*) Ethylene

27. (*a*) How will you synthesise the following from acetylene :

(*i*) Vinyl chloride (*ii*) Benzene

(*b*) Discuss briefly the acidic character of alkynes.

28. What happens when acetylene vapours are passed through dilute sulphuric acid in the presence of mercuric sulphate ? Give the mechanism of the reaction.

29. Discuss keto-enol tautomerism with reference to the products of reaction of acetylene with water in the presence of $HgSO_4$ and H_2SO_4.

30. How will you convert:

(*i*) Acetylene into acetic acid. (*ii*) Acetylene into chloroprene

(*iii*) Acetylene into acetaldehyde (*iv*) Acetylene into Vinyl chloride

31. Write short notes on :

(*i*) Acidity of treminal alkynes (*ii*) Nucleophilic additions of alkynes.

32. Justify with examples, the following statement :

Acetylene undergoes electrophilic as well as nucleophilic addition.

33. (*a*) How will you account for acidic character of acetylene H-atoms ?

(*b*) An organic compound (A) C_4H_6 takes up two molecules of hydrogen. it reacts with ammoniacal $AgNO_3$ to form a white precipitate. On treatment with dilute H_2SO_4 in the presence of $HgSO_4$, (A) forms a ketone (B) C_4H_8O. Write the reactions involved and propose the structures of (A) and (B).

34. What are alkynes ? Why are these unsaturated in nature ? Illustrate with the help of orbital structure.

35. Discuss the mechanism of the addition of the following reactions in case of acetylene :

(*i*) Hydrochloric acid (*ii*) Hypobromous acid.

36. How will you account for the fact that the alkynes take part in the nucleophilic addition reactions while alkenes do not ?

37. What happens when acetylene is passed through dilute sulphuric acid containing mercuric sulphate at 333 K ? Write the mechanism of this reaction.

38. Complete the following reactions :

(*i*) $CH \equiv CH \xrightarrow{\text{Iron tubes}}$ (*ii*) $CH \equiv CH \xrightarrow[NH_4Cl]{CuCl}$

(*iii*) $CH_3C \equiv CH \xrightarrow[HgSO_4]{H_2SO_4}$ (*iv*) $CH \equiv CH + O_3 \xrightarrow{Zn/CH_3COOH}$

39. Predict the products of the following reactions :

(*i*) $CH \equiv CH + Cl_2 \longrightarrow$ (*ii*) $CH \equiv CH + CH_3OH \longrightarrow$

(*iii*) $CH \equiv C — CH = CH_2 - HCl \longrightarrow$

40. Alkynes participate in both electrophilic and nucleophilic addition reactions. Discuss.

41. Explain why alkynes are less reactive towards electrophilic addition than alkenes.

42. Write notes on :

(*i*) Acidity of terminal alkynes. (*ii*) Nucleophilic addition reactions in alkynes.

6

Arenes And Aromaticity

6.1 INTRODUCTION

Some sweet smelling compounds were obtained from natural sources during the study of organic chemistry in earlier times. These compounds had contrasting properties compared to aliphatic compounds. The sweet smelling compounds were called **aromatic** (Greek **aroma** = pleasant smell). Further studies on these compounds revealed that they contained benzene rings involving six carbon atoms in a ring. Later a large number of aromatic compounds were discovered which lacked pleasant smell. Some of them had even foul smell. Therefore, the word 'aroma' lost its significance in relation to aromatic compounds.

Aromatic compounds are now regarded as a class of compounds which contain at least one benzene ring. They are also called **benzenoids.** It may be mentioned that there are some compounds which do not contain benzene ring but they still behave as aromatic compounds. Such compounds are called **non-benzenoids.**

6.2 ARENES

Arenes are mixed aromatic aliphatic compounds. An alkyl, alkenyl or alkynyl group attached to a benzene ring constitutes an arene. Thus, arenes can be classified into the following types:

Alkyl benzenes

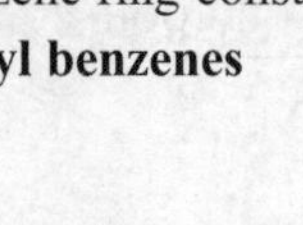

Toluene

Ethyl benzene

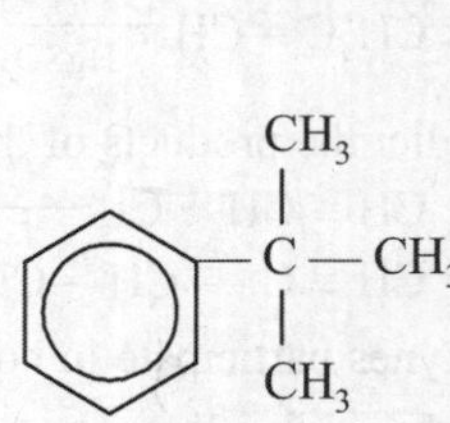

tert-Butyl benzene

Alkenyl benzenes

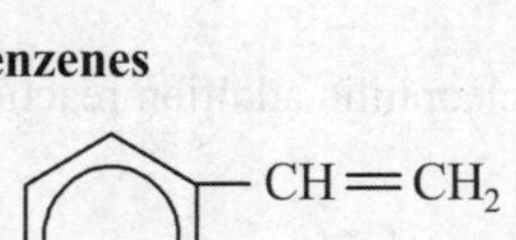

Styrene

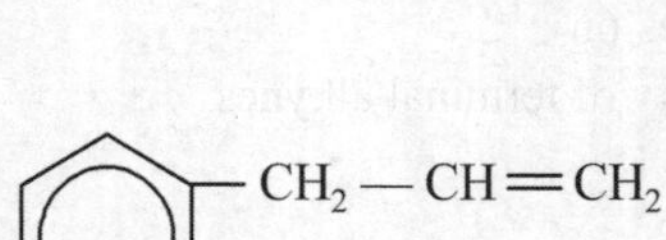

3-Phenyl propene-1

Alkynyl benzenes

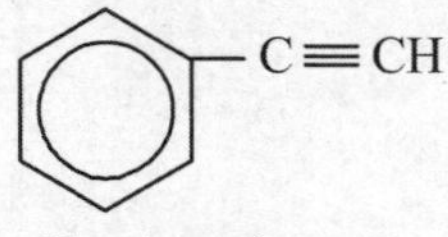

Phenyl acetylene

6.3 METHODS OF OBTAINING BENZENE ON INDUSTRIAL SCALE

***(i)* From Coal Tar (Industrial method).** The light oil fraction (boiling range 353 K to 443 K) of coal tar which contains mainly benzene, toluene and xylenes, is treated with conc. H_2SO_4 to remove impurities of pyridine and thiophene. To remove acidic impurities such as phenol, the organic layer is washed successively with water and aq. KOH and finally after washing with water it is subjected to fractional distillation to obtain three main fractions as given below:

(*a*) *90% Benzol* (boiling range 353 K – 383 K) consists of 70% benzene, 24% toluene and rest xylenes. Its careful fractionation gives benzene, toluene and xylene.

(*b*) *90% Toluol* (boiling range 383 K – 413 K) consists mainly of xylenes and small amounts of benzene and toluene.

(*c*) *Solvent naphtha or Benzine* (boiling range 413 K — 443 K) mainly consists of xylenes.

***(ii)* From Petroleum (Industrial method).** Aromatic hydrocarbons are synthesized by catalytic reforming of alkanes obtained from petroleum which involves cyclization and aromatization. For example, hexane can be converted to cyclohexane and then to benzene by passing its vapours over Cr_2O_3, V_2O_5 or MoO_3 at 723 K — 823 K. Similarly, *n*-heptane gives toluene.

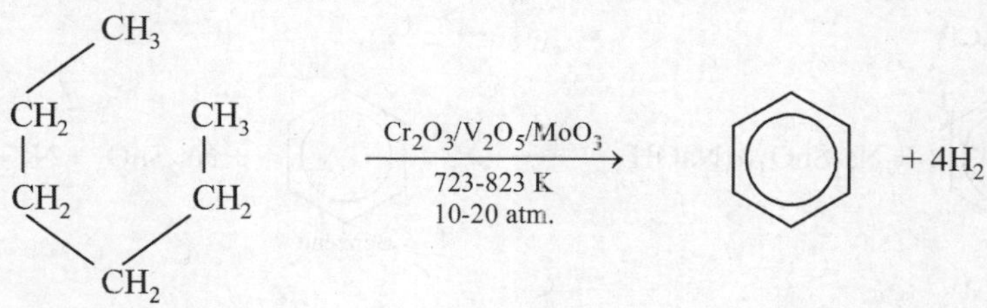

n-Hexane

***(iii)* From acetylene.** Acetylene on passing through red hot tube trimerizes to benzene

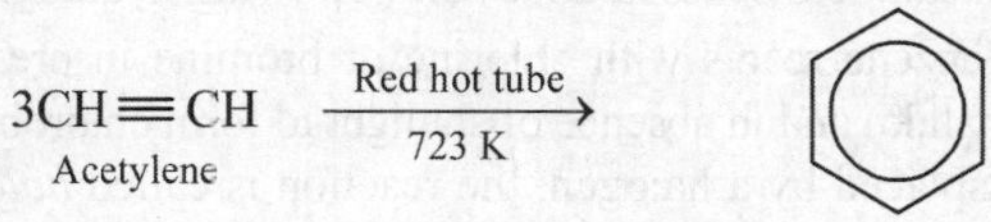

6.4 LABORATORY METHODS OF PREPARING BENZENE

Laboratory methods of preparing benzene are given as under :

(*a*) Decarboxylation of benzoic acid/phthalic acid with soda lime.

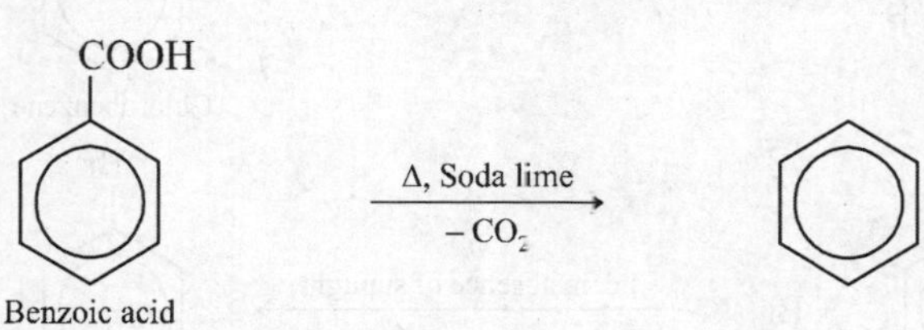

Benzoic acid

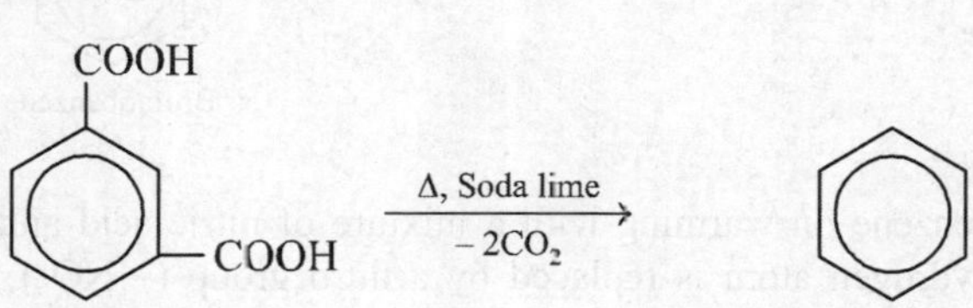

o-, *m*- or *p*-
phathalic acid

(*b*) Distillation of phenol with zinc dust (reduction).

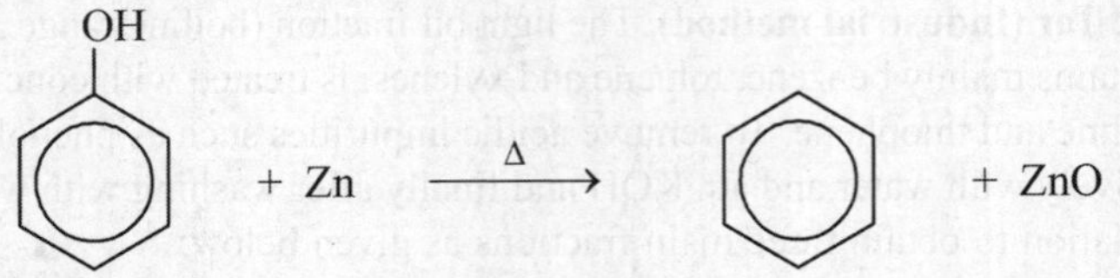

(*c*) By hydrolysis of benzene sulphonic acid with super-heated steam or boiling with HCl under pressure.

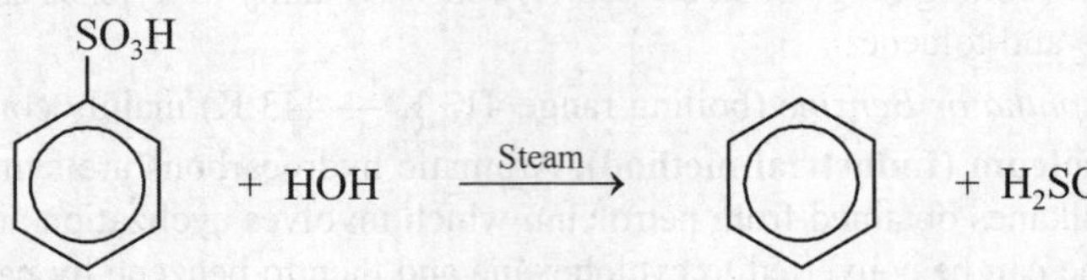

(*d*) By reduction of benzene diazonium chloride with alkaline sodium stannate.

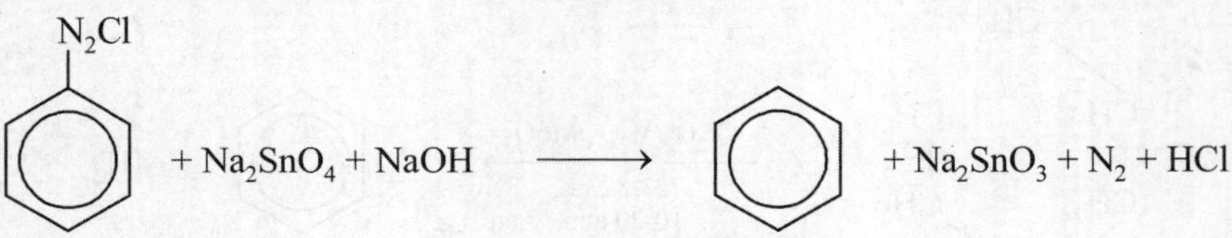

Benzene

6.5 PROPERTIES OF BENZENE

Substitution Reactions of Benzene

Substitution reactions can be discussed under the following headings:

(*a*) Halogenation. Benzene reacts with chlorine or bromine in presence of a catalyst (iron, iron halide or aluminium halide) and in absence of sunlight to form chlorobenzene or bromobenzene. As a hydrogen atom is displaced by a halogen, the reaction is called *halogenation*.

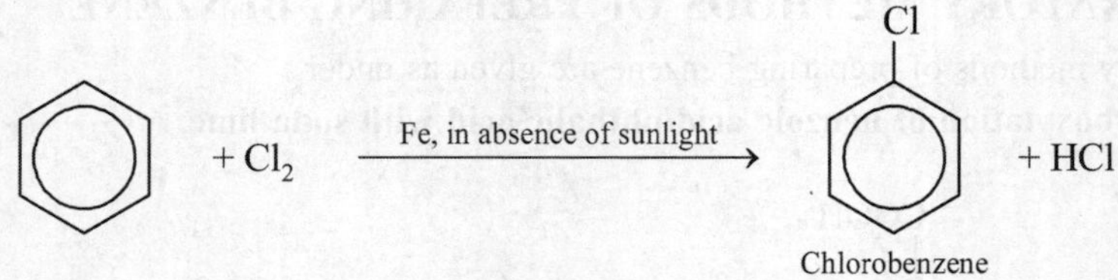

Chlorobenzene

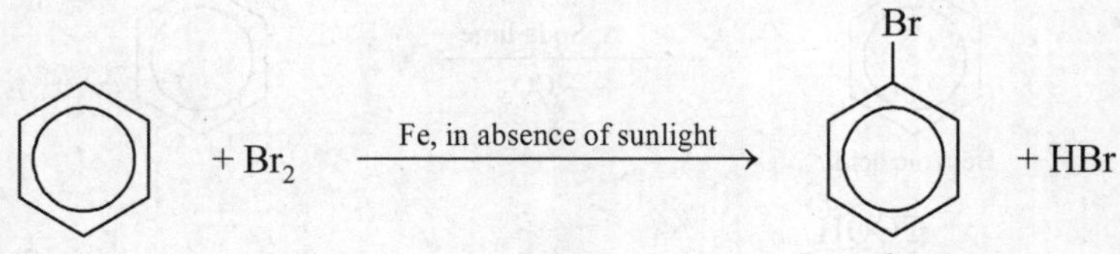

Bromobenzene

(*b*) Nitration. Benzene on warming with a mixture of nitric acid and sulphuric acid forms *nitrobenzene*. As a hydrogen atom is replaced by a nitro group ($-NO_2$), the reaction is called nitration.

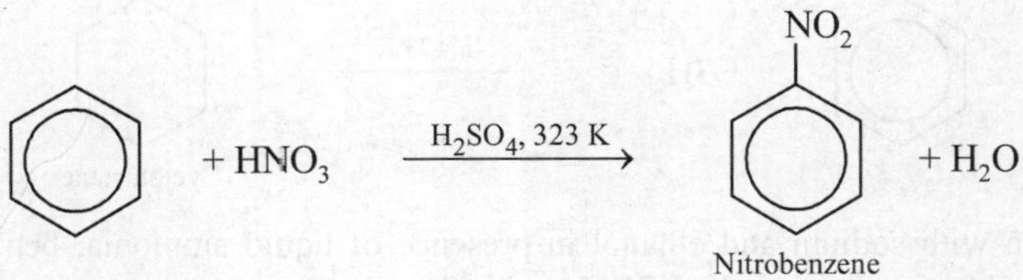

(*c*) **Friedel-Crafts alkylation.** This reaction is used for introducing an alkyl group in benzene ring.

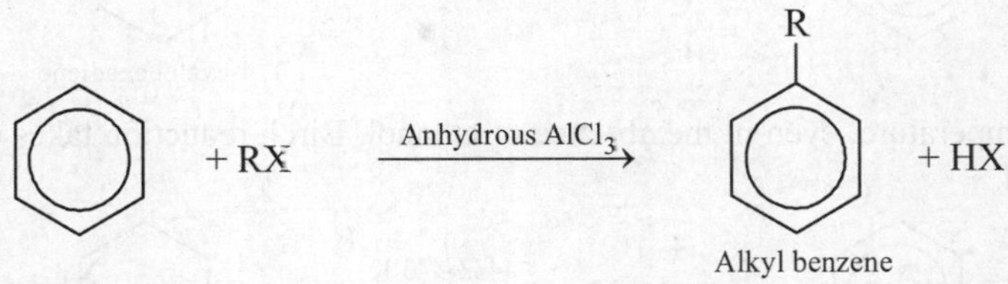

(*d*) **Friedel-Crafts acylation.** This reaction is used for the preparation of aromatic aldehydes and ketones. Benzene is treated with acid chloride in the presence of anhydrous aluminium chloride.

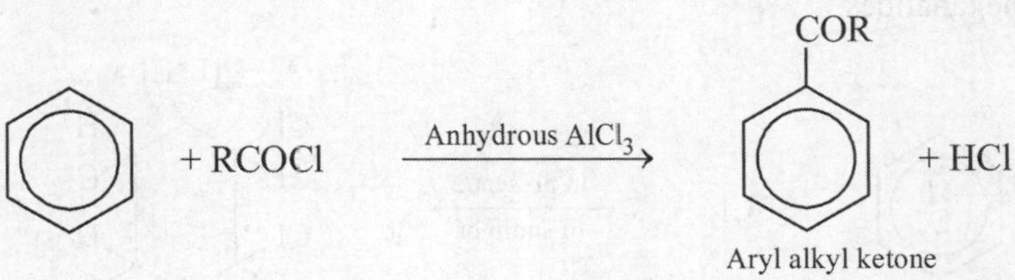

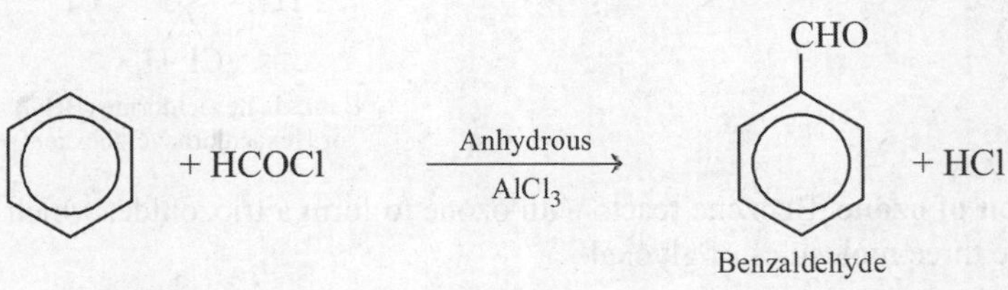

(*e*) **Sulphonation.** Benzene on warming with conc. sulphuric acid forms benzene sulphonic acid. As a hydrogen atom is replaced by a sulphonic acid group (— SO_3 H), the reaction is called *sulphonation.*

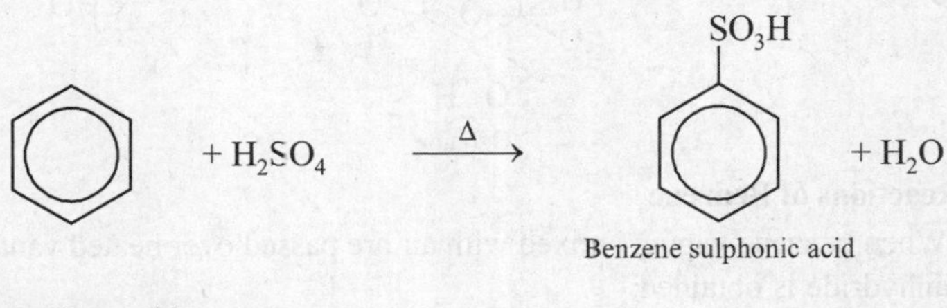

Addition reactions of benzene.

(*a*) **Addition of hydrogen.** When benzene vapours mixed with hydrogen are passed over nickel catalyst at 473 K, cyclohexane is formed.

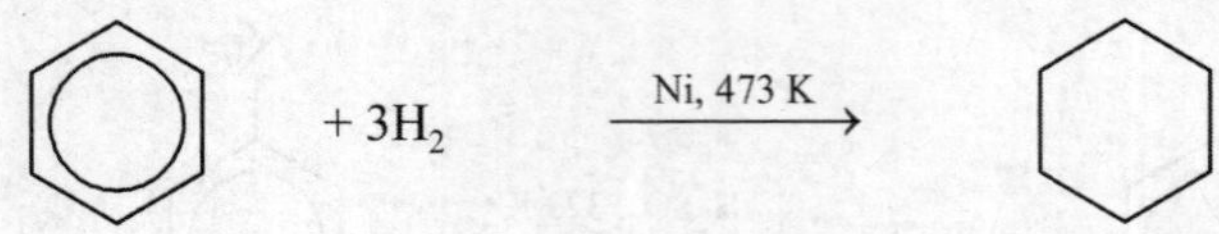

Cyclohexane

On reduction with sodium and ethanol in presence of liquid ammonia, benzene gives 1, 4-cyclohexadiene. This reaction is called **Birch reduction.**

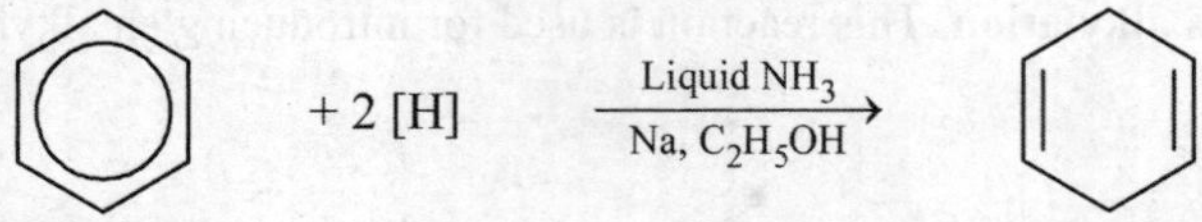

1, 4-cyclohexadiene

At higher temperature, even in the absence of ethanol, Birch reduction takes place and cyclohexene is formed.

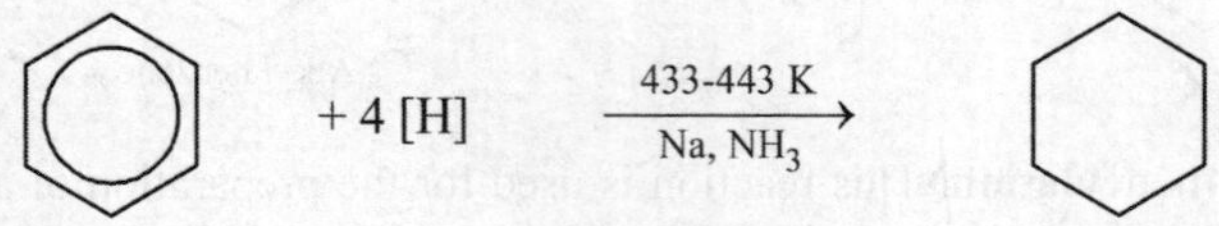

Cyclohexene

***(b)* Addition of halogens.** Benzene on treatment with halogens in the presence of sunlight gives benzene hexahalides.

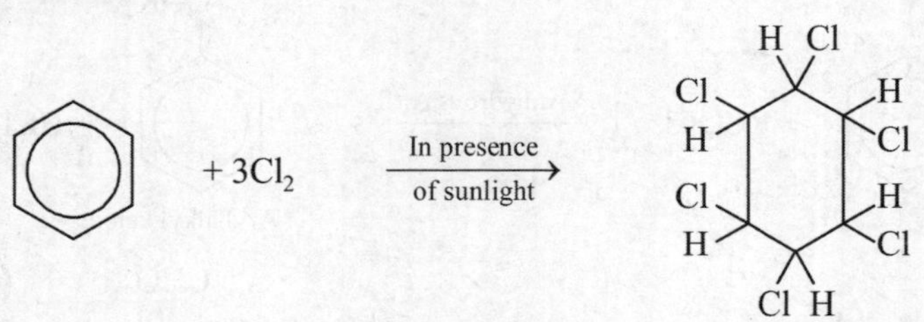

Benzene hexachloride (BHC)
or Hexachlorocyclohexane

(*c*) Addition of ozone. Benzene reacts with ozone to form a triozonide, which is decomposed by water to give three molecules of glyoxal.

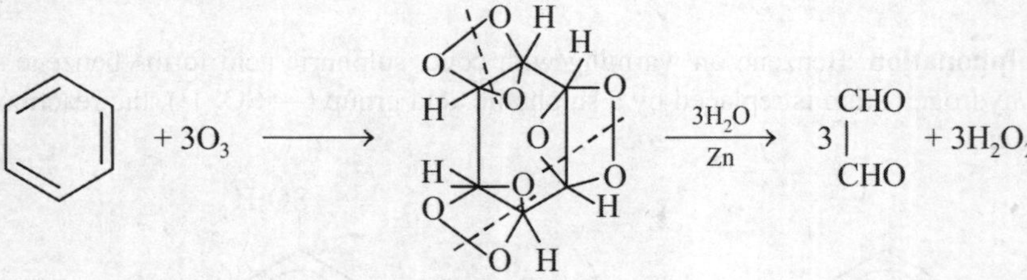

Ozonide

Oxidation Reactions of Benzene

Oxidation. When benzene vapours mixed with air are passed over heated vanadium pentoxide catalyst, maleic anhydride is obtained.

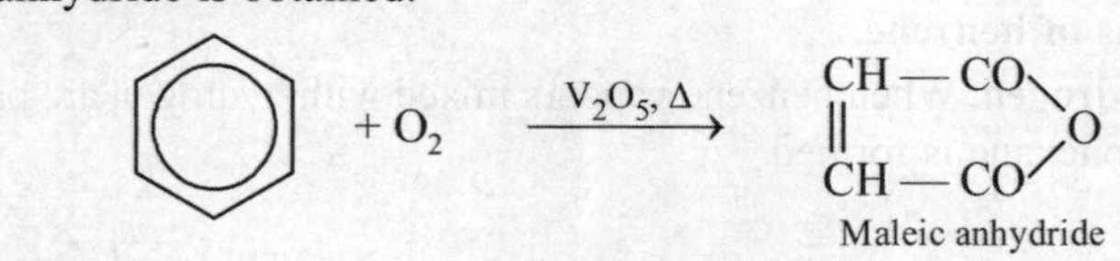

Maleic anhydride

Combustion. When burnt in air or oxygen, carbon dioxide and water are formed

$$2C_6H_6 + 15O_2 \longrightarrow 12CO_2 + 6H_2O$$

6.6 STRUCTURE OF BENZENE

Kekule Structure of Benzene

According to Kekule, six carbon atoms of benzene are linked to each other by alternate single and double covalent bonds to form a hexagonal ring as shown in the following figure:

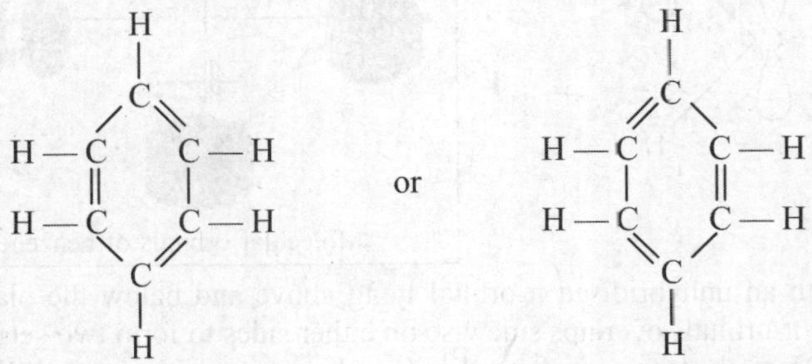

Kekule structure of benzene

August Kekule

Each carbon atom is linked to one hydrogen atom thus conforming to its molecular formula.

Drawbacks of Kekule structure. Kekule structure does not explain the following observations:

***(i)* Chemical reactions.** Benzene does not give addition reactions and fails to decolorise Baeyer's reagent. It readily undergoes electrophilic substitution reactions in which the benzene ring is retained.

***(ii)* Heat of combustion.** On the basis of Kekule structure, the heat of combustion of benzene is expected to be 3449.0 kJ mol^{-1}. But the experimental value is 3298.5 kJ mol^{-1}. Thus, benzene has 150.5 kJ mol^{-1} of energy less than Kekule structure. Benzene is, therefore, more stable than the structure proposed by Kekule.

***(iii)* Heat of hydrogenation.** On the basis of Kekule structure, the heat of hydrogenation of benzene is expected to be 358.0 kJ mol^{-1}. But the experimental value is 208.5 kJ mol^{-1}. This again shows that benzene is more stable than expected from Kekule structure.

***(iv) o*-Disubstituted product.** Benzene forms only one ortho disubstituted product whereas Kekule structure predicts two *o*-disubstituted products as shown below:

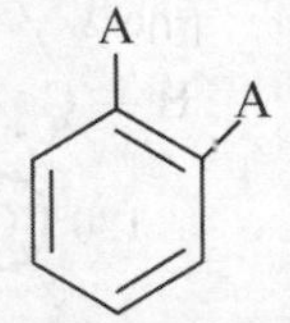

Presence of double bond between the substituents

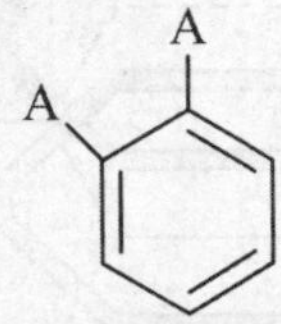

Presence of single bond between the substituents

***(v)* Carbon-carbon bonds lengths.** X-ray diffraction studies show that all the carbon-carbon bond lengths are identical, *viz.*, 1.397 Å and lie in between that of single and double bonds. This is not in accordance with Kekule structure which contains two kinds of carbon-carbon bonds.

Molecular orbital structure of benzene

Structure of benzene can be best described by using the orbital concept. Each carbon atom in benzene is sp^2 hybridised and thus forms three σ bonds, two with adjacent carbon atoms and one with hydrogen. Thus, all the six carbons and six hydrogen atoms lie in the same plane and the angle between two adjacent σ bonds is 120° [Fig. (*a*)].

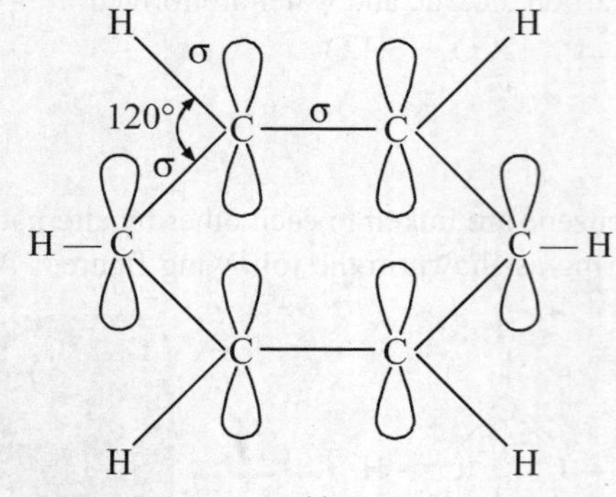

(a)

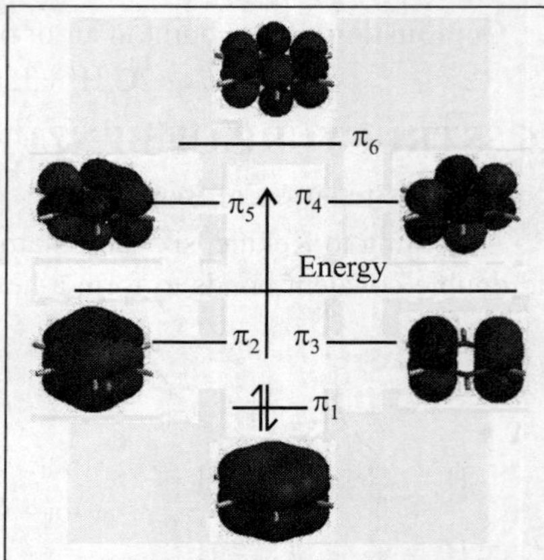

Molecular orbitals of benzene

Each carbon is still left with an unhybridised π-orbital lying above and below the plane of benzene ring . Each one of these π-orbitals overlaps sidewise on either sides to form two sets of π-electron clouds [Fig. (*b*) and (*c*)]. π electrons are delocalised as these can move over all the six carbon atoms [Fig. (*d*)]. As a result of this delocalisation two continuous ring like electron clouds one above and the other below the plane of carbon atoms [Fig. (*e*)] are formed. Bond angles and bond lengths in the molecule of benzene are given in [Fig. (*f*)].

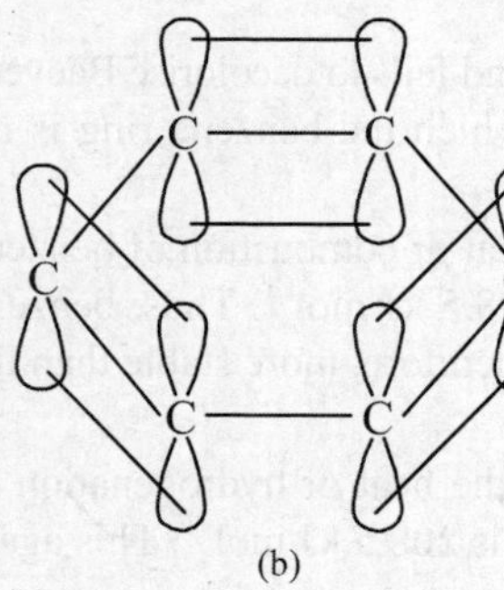

(b)

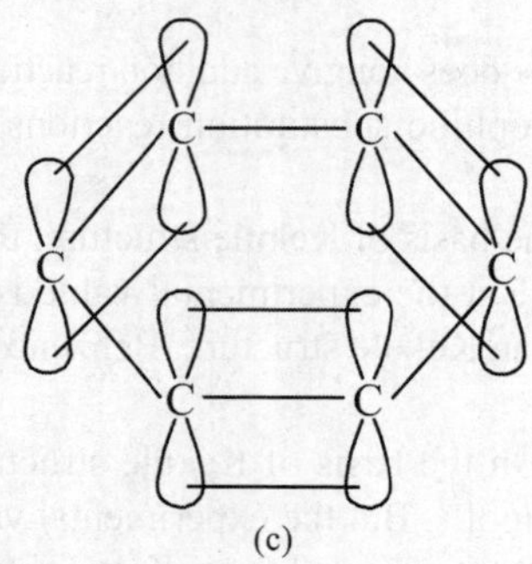

(c)

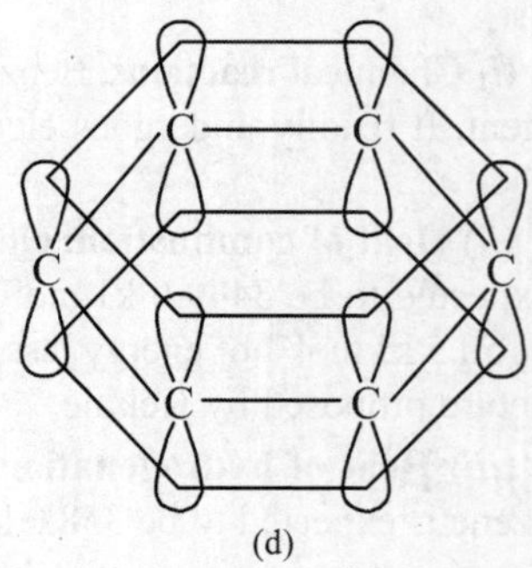

(d)

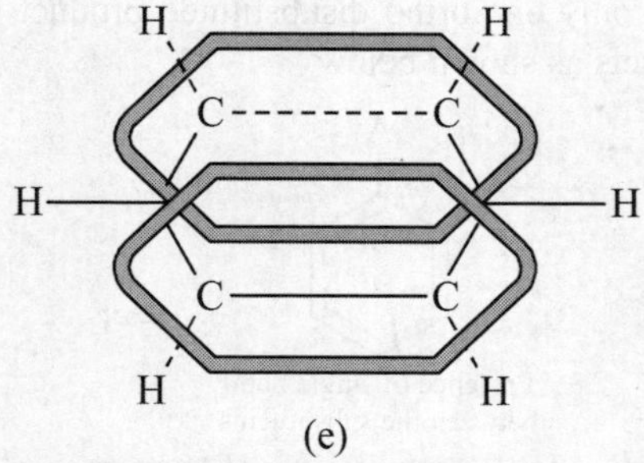

(e)

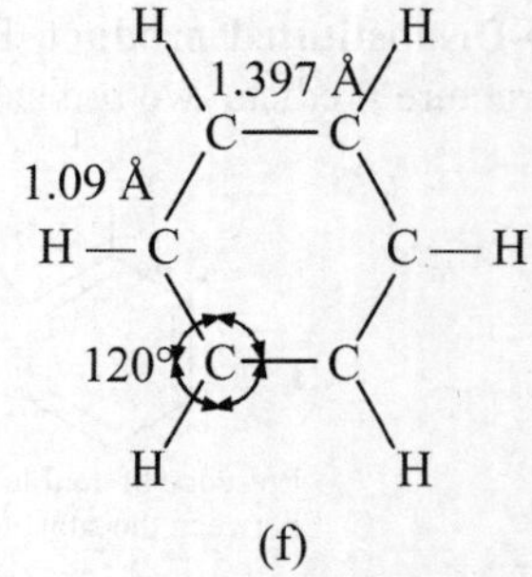

(f)

Evidence in support of orbital structure of benzene

(*i*) Unusual stability. Benzene molecule exhibits unusual stability and resists the formation of addition products. This can easily be understood in terms of delocalisation of π-electrons which is responsible for aromaticity.

(*ii*) Isomer number. According to orbital concept all the six carbons in benzene are completely equivalent. Similarly, all the six hydrogen atoms also occupy identical positions. Thus, benzene should form only one monosubstituted and three disubstituted products. This has been found to be so in actual practice.

***(iii)* Electrophilic substitution reactions.** There are two continuous ring like π-electron clouds one above and the other below the plane of carbon atoms. The π-electrons are easily attacked by electrophiles. Hence benzene undergoes electrophilic substitution reactions.

6.7 AROMATICITY

Organic compounds which resemble benzene in their chemical behaviour are called aromatic compounds. They exhibit certain characteristic properties which are quite different from those of aliphatic and alicyclic compounds. These characteristic properties are collectively referred to as **aromaticity or aromatic character.**

Characteristics.

(*i*) Although their molecular formulae suggest a high degree of unsaturation, yet they do not respond to tests characteristic of unsaturated compounds. Thus, aromatic compounds fail to decolorise an aqueous solution of potassium permanganate (Baeyer's test).

(*ii*) They undergo readily certain electrophilic substitution reactions such as nitration, halogenation, sulphonation, Friedel-Crafts alkylation and acylation, etc.

(*iii*) Their molecules are flat or nearly flat as shown by physical methods such as X-ray and electron diffraction methods.

(*iv*) They are associated with high thermodynamic stability as is indicated by their low heats of combustion and hydrogenation.

6.8 HUCKEL'S RULE AND AROMATICITY

This rule is based upon molecular orbital treatment and is employed for predicting aromaticity in organic compounds. The main theoretical requirements for a substance to possess aromaticity are:

(*i*) The molecule or ion must be flat or nearly flat.

(*ii*) It must have cyclic clouds of delocalised π-electrons above and below the plane of the molecule.

(*iii*) The π-clouds in the molecule or ion must contain a total of $(4n + 2)$ π electrons where $n = 0, 1, 2, 3, ...,$ etc.

The above requirements are collectively known as *Huckel rule* or $(4n + 2)$ *rule*. This rule can be applied successfully to cyclic polyenes, polycyclic compounds and non-benzenoid compounds to predict aromaticity in them. Thus, molecules or ions having 2, 6, 10, 14 ... π electrons will be aromatic and others will be non-aromatic.

Aromatic molecules

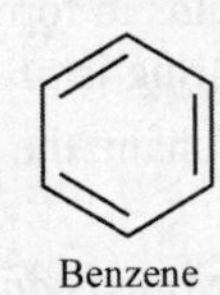

Benzene
(6π electrons)

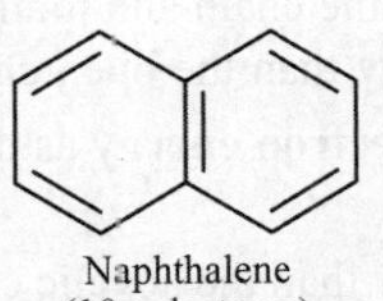

Naphthalene
(10π electrons)

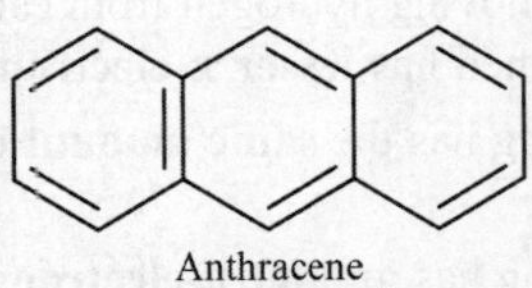

Anthracene
(14π electrons)

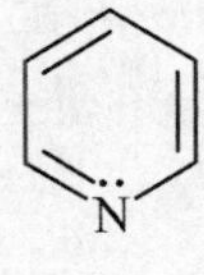

Pyridine
(6π electrons)

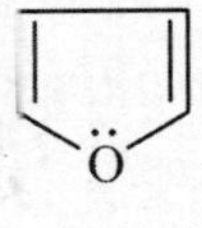

Furan
(6π electrons)

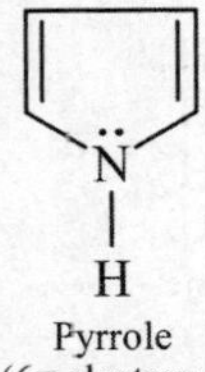

Pyrrole
(6π electrons)

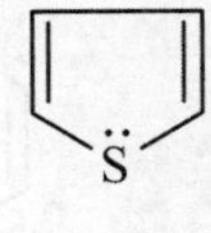

Thiophene
(6π electrons)

Aromatic ions

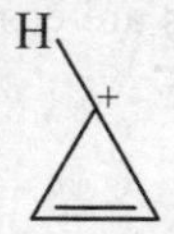

Cyclopropenyl anion
(2π-electrons)

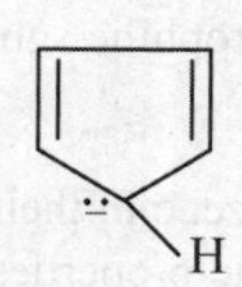

Cyclopentadienyl anion
(6π-electrons)

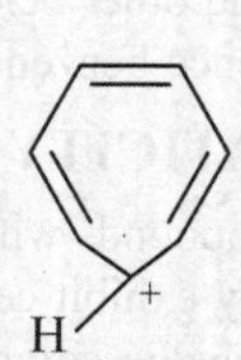

Cycloheptatrienyl anion
(6π-electrons)

Non-Aromatic species

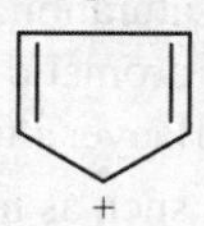

Cyclopentadenyl cation
(4π electrons)

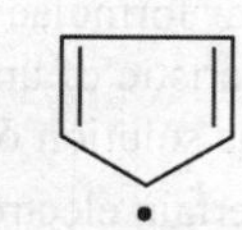

Cyclopentadenyl radical
(5π electrons)

6.9 ANNULENES

Annulenes are monocyclic compounds having a system of single and double bonds in alternate positions. They may be called cyclic polyenes and can be represented by a general formula $(-CH = CH-)_n$ where n is a positive integer. They are named by writing the ring size followed by the word "annulene". Some examples of annulenes are given below:

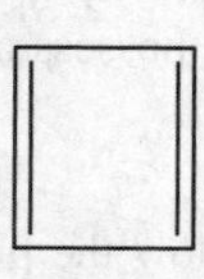

[4] Annulene
or Cyclobutadiene

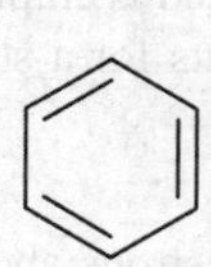

[6] Annulene
or Benzene

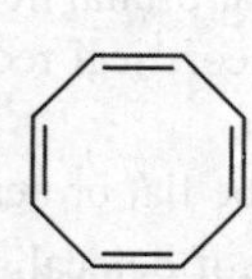

[8] Annulene
or Cyclo-octadiene

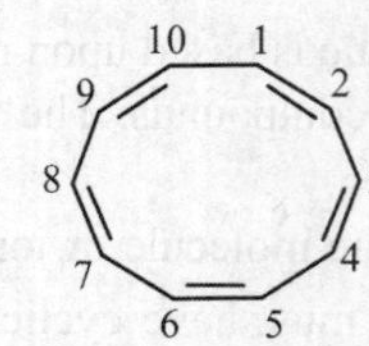

[10] Annulene

According to Huckel's rule, Annulenes containing $(4n + 2)$ π-electrons and having a coplaner ring should be aromatic in nature. The synthesis of a number of annulenes has confirmed this. Thus [6] annulene or benzene is aromatic. [4] annulene and [8] annulene are not aromatic because they do not conform to Huckel's rule. They are rather anti-aromatic. [10] annulene although expected to be aromatic as per Huckel's rule is not so because the ring is not coplaner.

Aromatic, Anti-aromatic and Non-aromatic compounds

(*i*) If on removing hydrogen from each end the chain and joining the ends to form a ring, the ring formed has **lesser π-electron energy** than the open chain, the ring is *aromatic*.

(*ii*) If the ring has the **same amount of π-electron energy** as the open chain, the ring is *non-aromatic*.

(*iii*) If the ring has **greater π-electron energy** than the acyclic chain, the ring is *anti-aromatic*.

The example given below illustrate the point.

(1) Aromatic system

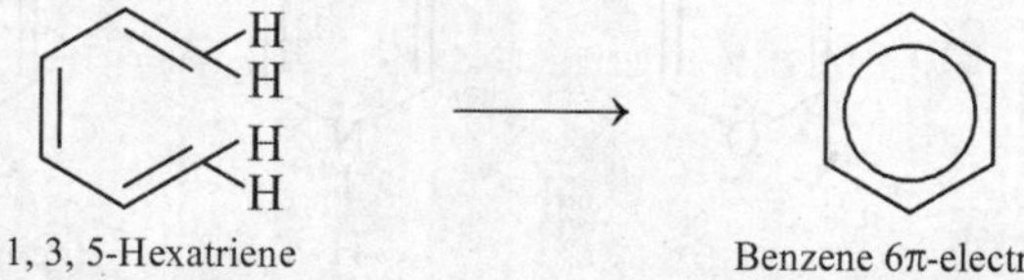

1, 3, 5-Hexatriene
6π-electrons

Benzene 6π-electrons
(Lesser π electron energy)

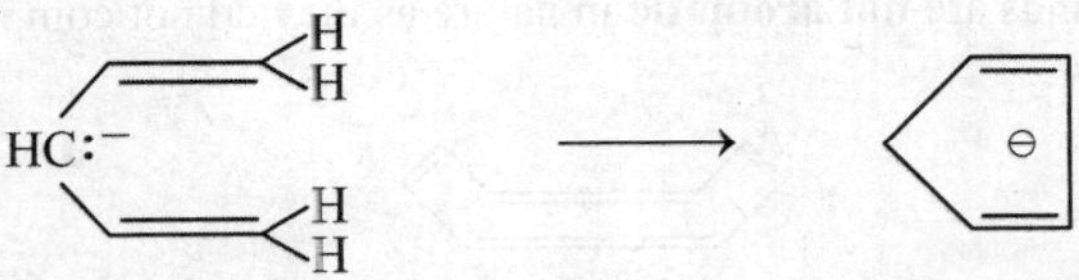

Pentadienyl anion 6π-electrons — Cyclopentadienyl anion 6π-electrons (Lesser π electron energy)

(2) Anti-aromatic system

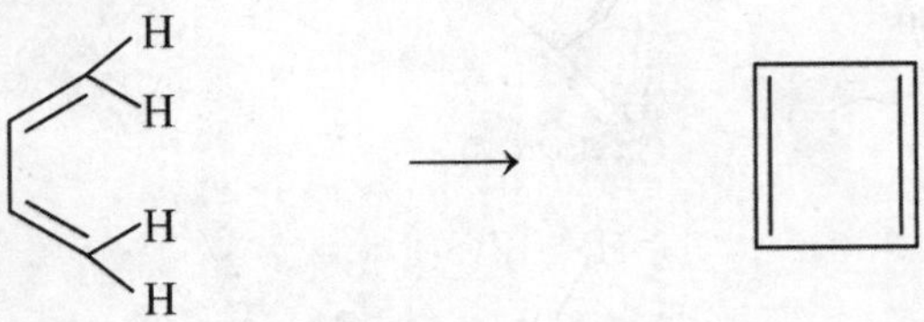

1, 3, Butadiene 4π-electrons — Cyclobutadiene 4π-electrons (Greater π electron energy)

Similarly other systems having $4n\pi$ electrons are anti-aromatic.

(3) Non-aromatic

$$CH_2{=}CH-(CH{=}CH)_{15}-CH{=}CH_2 \longrightarrow [30]-\text{ANNULENE}$$

30π-electrons — 30π-electrons (π-electron energy same as in acyclic system)

It may be pointed out that in general as the value of n increases, the difference between the cyclic and corresponding acyclic system decreases both in system having $(4n)$ π-electrons and $(4n + 2)$ π-electrons. In other words, such cyclic systems show non-aromatic behaviour.

***Example 1.* Which of the following species will exhibit aromatic character? Write their structural formulae.**

Cyclooctatetraene, tropylium ion, cycloheptatrienyl anion, furan, cyclopropenyl anion, cyclopentadienyl cation, cyclobutadiene, anthracene, phenanthrene, pyridine.

Solution. (*a*) The compounds mentioned below are aromatic in nature in terms of Huckel's rule. No. of π-electrons in these compounds is $(4n + 2)$ where $n = 0, 1, 2$ etc.

I Tropylium ion (cycloheptatrienyl cation)

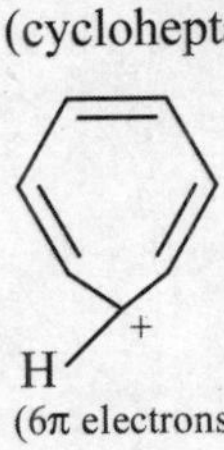

(6π electrons)

II Furan

(6π electrons)

III Anthracene

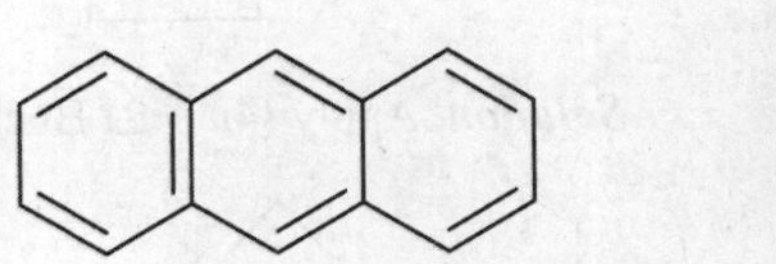

(14π electrons)

IV Phenanthrene

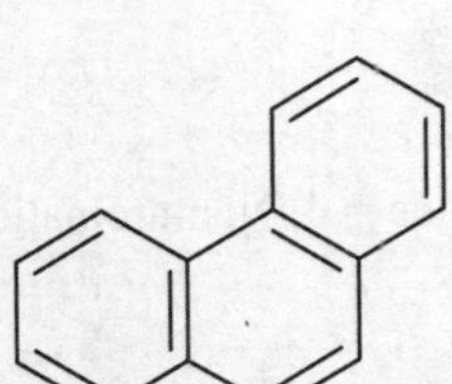

(14π electrons)

V Pyridine

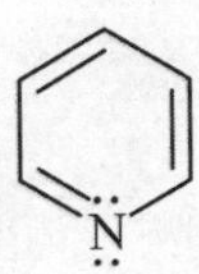

(10π electrons)

(*b*) Following compounds are **not aromatic** in nature as they do not comply with the $(4n + 2)$ rule.

Cyclooctatetraene

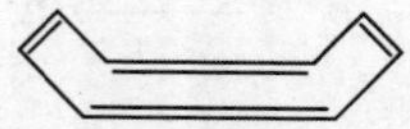

(8π electrons)

Cycloheptatrienyl anion

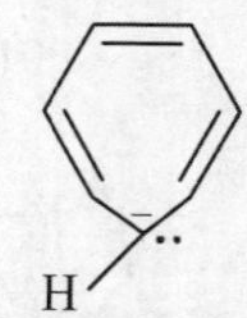

(8π electrons)

Cyclopropenyl anion

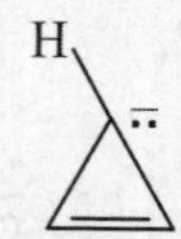

(4π electrons)

Cyclopentadienyl cation

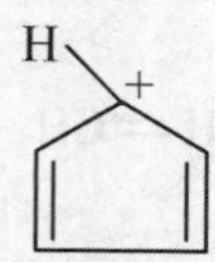

(4π electrons)

Cyclobutadiene

(4π electrons)

Example 2. **Which of the following have aromatic character and why?**

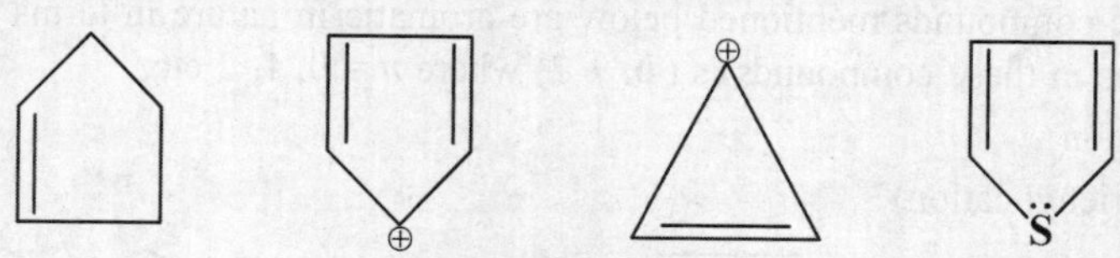

Solution. Apply $(4n + 2)$ Huckel's Rule

I

Cyclopentene

No. of π electrons = 2, Aromatic

II

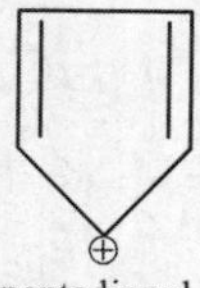

Cyclopentadienyl cation

No. of π electrons = 4, Non-aromatic

III 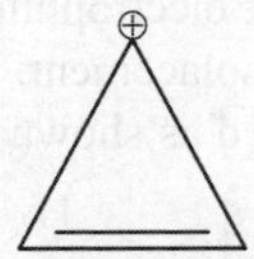 No. of π electrons = 2, Aromatic

Cyclopropenyl cation

IV 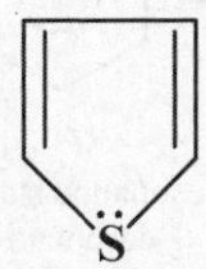 No. of π electrons = 6, Aromatic

Thiophene

Example 3. **Which of the following compounds possess aromaticity? Explain briefly.**

(*a*) 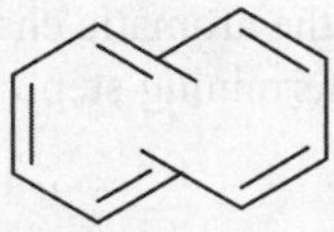

[10] - Annulene

(*b*) 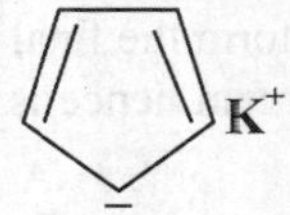

Solution. **(*a*) [10]-Annulene**

It satisfies the Huckel's rule $(4n + 2)$ π electrons. There are a total of 10π electrons. Thus $n = 2$. Theoretically, it is an aromatic compound. But, because of interaction of two hydrogen atoms inside the planar ring, it becomes non-planar. Thus, it might behave as non-aromatic.

(*b*) Potassium cyclopentadienide

This compound follows $(4n + 2)$ Huckel's rule with $n = 1$

There are a total of six π electrons.

Example 4. **Comment on the aromatic character of 1, 3-cyclopentadiene and 1, 3-cyclopentadienyl anion.**

Solution.

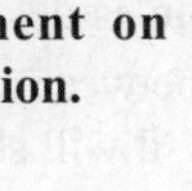

1, 3-cyclopentadiene
(4π electrons)

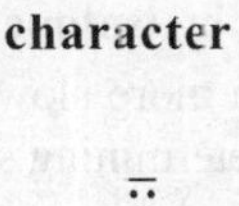

1, 3-cyclopentadienyl anion
(6π electrons)

1, 3-cyclopentadiene does not conform to $(4n + 2)$ Huckel's rule.

Thus, it is not aromatic.

1, 3-cyclopentadienyl anion conforms to $(4n + 2)$ Huckel's rule with $n = 1$. Thus, it is an aromatic species.

6.10 GENERAL MECHANISM OF ELECTROPHILIC SUBSTITUTION IN BENZENE

Electrophilic substitution reactions are initiated by substances which are either electrophilic themselves or which generate some electrophilic species. General mechanism involves the following sequence of steps:

(I) Generation of electrophile. To start with, there is preliminary reaction which generates an electrophile.

$$A—B \longrightarrow \underset{\text{Electrophile}}{A^{\oplus}} + B^{-}$$

(II) Formation of intermediate carbocation (carbonium ion). The electrophile attacks the π electron cloud of the benzene ring and thus brings about an electronic displacement. This results in the formation of **intermediate carbocation** which is resonance stabilized as shown below:

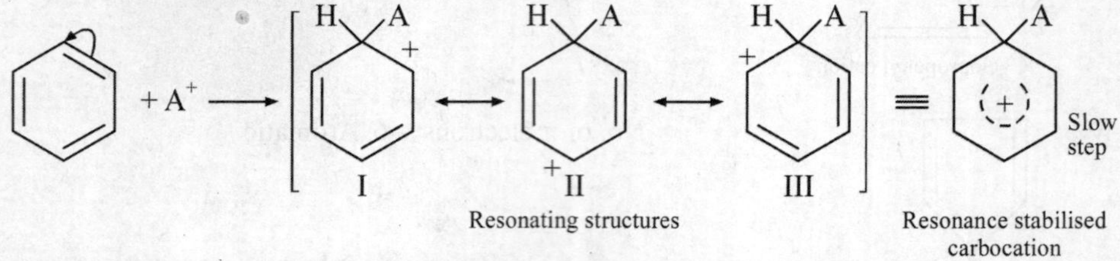

During the formation of the carbocation, the aromaticity of the benzene ring is destroyed. Consequently, the formation of carbocation is slow and hence is the **rate determining step.**

(III) Abstraction of a proton from the carbocation. The base present in the reaction mixture abstracts the proton from the carbocation to form the final product. Since the aromatic character of the benzene ring is restored, this step is **fast** and hence is not the rate determining step.

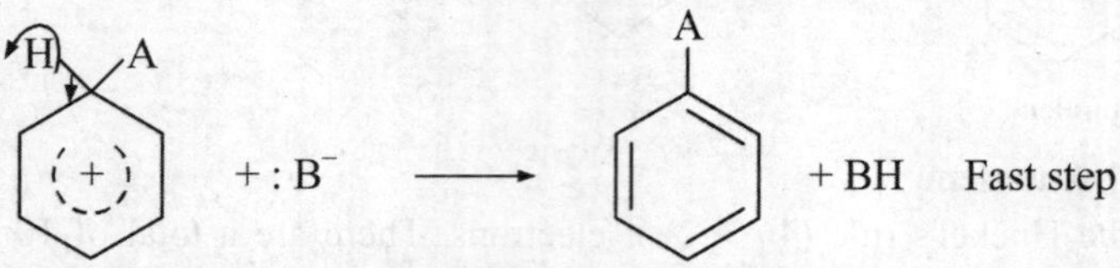

Evidence in support of the mechanism. Electrophilic aromatic substitution involves two steps viz. formation of carbocation and abstraction of proton from carbocation, after the generation of electrophile. The first step involving the formation of intermediate carbocation is slow and hence is the rate determining step of the reaction. The second step which involves the abstraction of a proton from the carbocation is fast.

The above mechanism is supported by **isotope tracer technique** as described below.

A carbon-deuterium bond is broken more slowly than a bond between carbon and hydrogen. Thus, if a hydgrogen is lost in the rate determining step of a reaction, it will show an **isotope effect** on replacing hydrogen by deuterium. In other words, a deuterated compound should undergo substitution more slowly than a non-deuterated compound. But this isotope effect has not been observed in aromatic electrophilic substitution reactions. For example, the rate of nitration of deuterobenzene is the same as the rate for benzene. Thus, proton elimination is not the rate determining step of the reaction. The step leading to the formation of carbocation does not involve the cleavage of carbon-hydrogen bond and hence it is the rate determining step of the reaction.

Electrophilic substitution

Nitration of Benzene

Benzene has a tendency to give electrophilic substitution reactions because of the π electron cloud above and below benzene ring. In these reactions, a hydrogen atom of benzene is substituted by $-NO_2$, – X (halogen), $-SO_3H$, – R (alkyl) or – COR group, under different conditions and by taking a suitable electrophile. Such reactions are called electrophilic substitution reactions.

Nitration. This reaction involves the treatment of an aromatic compound with a mixture of nitric acid and sulphuric acid. For example :

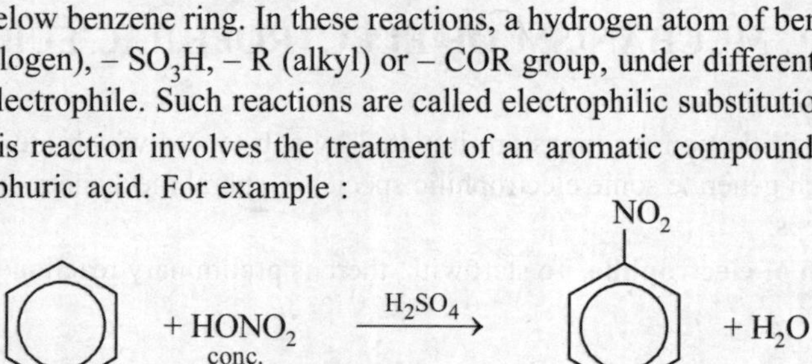

Mechanism. The various steps in the nitration of benzene can be outlined as below:

(I) Generation of electrophile

$$HONO_2 + 2H_2SO_4 \rightleftharpoons \underset{\text{Nitronium ion}}{NO_2^+} + H_3O^+ + 2HSO_4^-$$

(II) Formation of carbocation

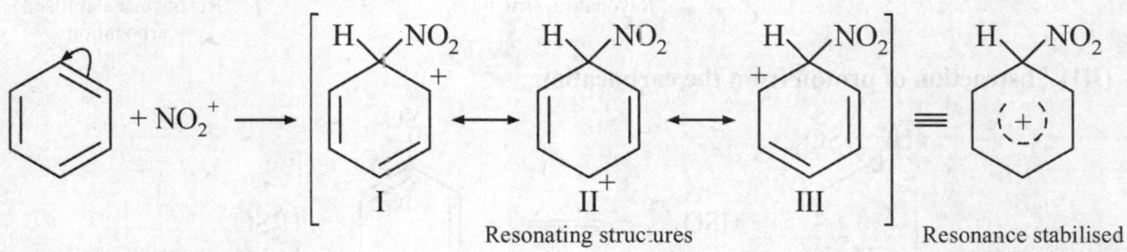

Resonating structures

Resonance stabilised carbocation

(III) Abstraction of proton from the carbocation

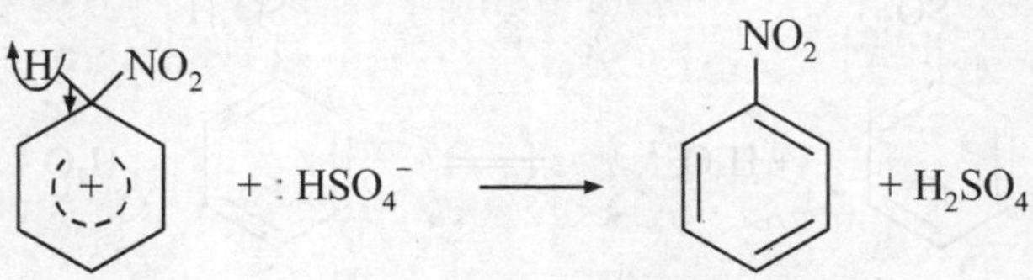

Nitrobenzene

Sulphonation of Benzene

The sulphonation of an aromatic compound can be brought about by the action of conc. H_2SO_4 or oleum. For example:

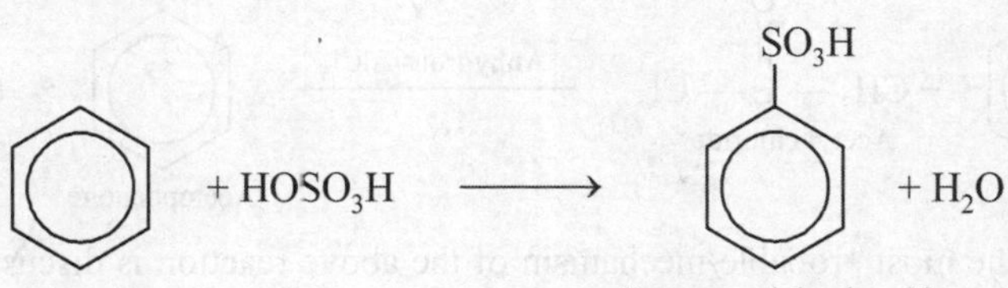

Benzene sulphonic acid

Mechanism. The attacking electrophilic reagent in this reaction is believed to be sulphr trioxide which is present as such in oleum or may be formed by the dissociation of sulphuric acid. The electrophilic nature of SO_3 molecule is due to the presence of electron deficient sulphur atom.

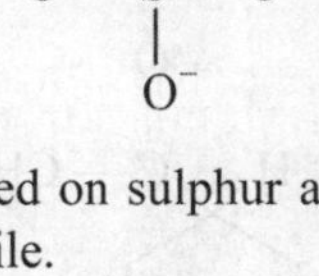

As the positive charge is concentrated on sulphur and negative charges are scattered on two oxygens, SO_3 molecule acts as electrcphile.

The complete mechanism for this reaction may be outlined as follows:

(I) Generation of electrophile

$$2H_2SO_4 \rightleftharpoons \underset{\text{Electrophile}}{SO_3} + HSO_4^- + H_3O^+$$

(II) Formation of carbocation

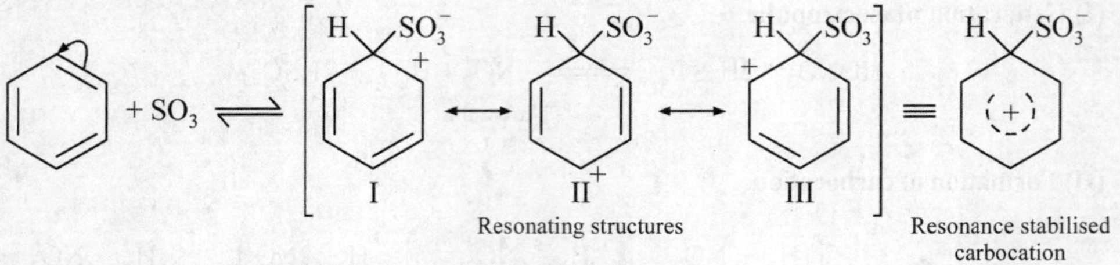

(III) Abstraction of proton from the carbocation

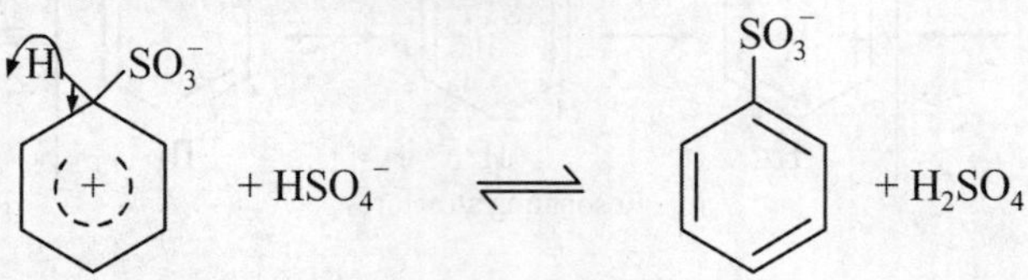

(IV) Formation of the final product

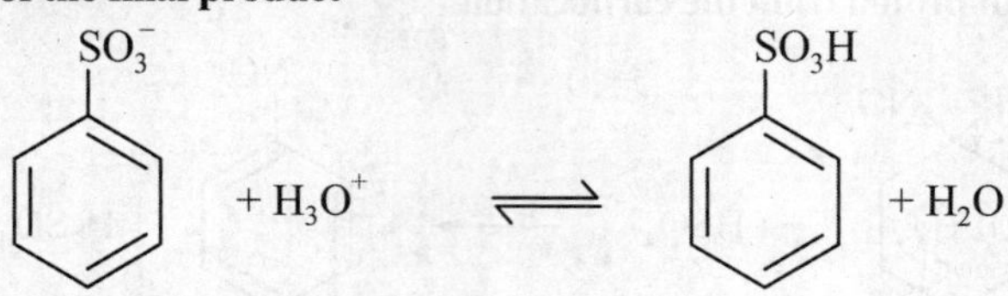

Friedel-Crafts Acylation of Benzene

When benzene is treated with acetyl chloride in the presence of anhydrous aluminium chloride, the formation of acetophenone takes place. This reaction is known as *Friedel-Crafts acylation.*

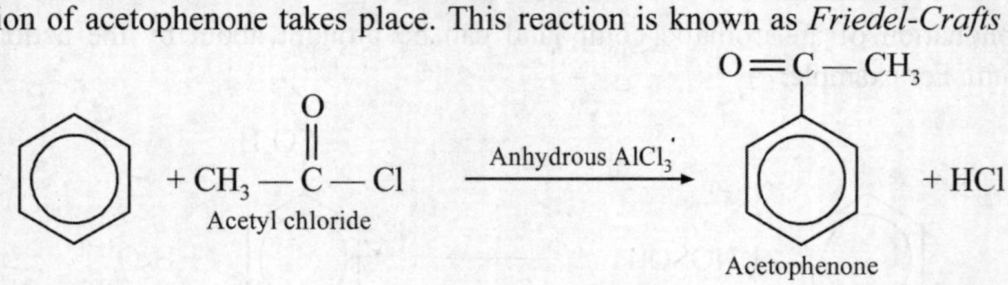

Mechanism. The most probable mechanism of the above reaction is discussed as under:

(I) Generation of electrophile

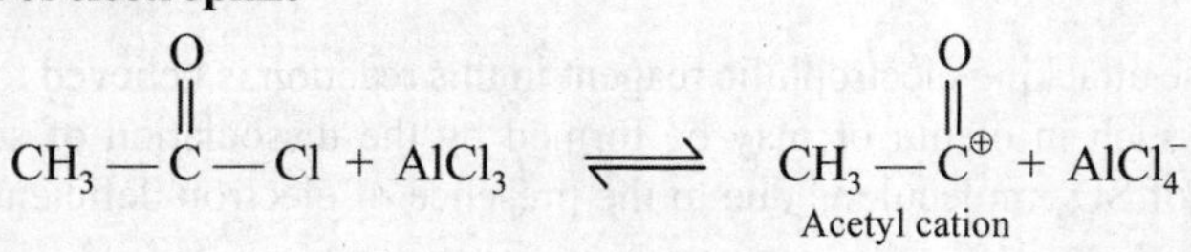

(II) Formation of carbocation

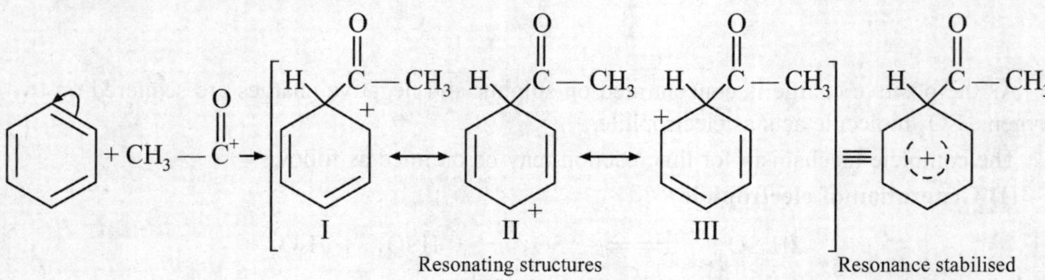

(III) Abstraction of proton from the carbocation

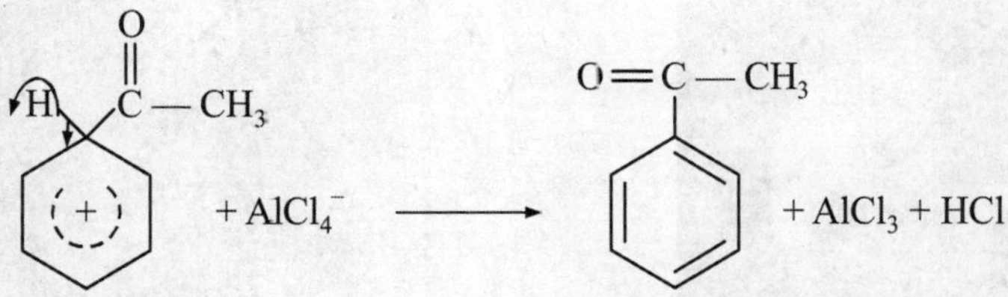

Acetophenone

Friedel-Crafts Alkylation of Benzene.

The reaction consists in treating benzene or substituted benzenes with an alkyl halide in the presence of small amounts of Lewis acids ($AlCl_3$, BF_3, $FeCl_3$ etc.). It leads to the direct introduction of an alkyl group into the benzene ring.

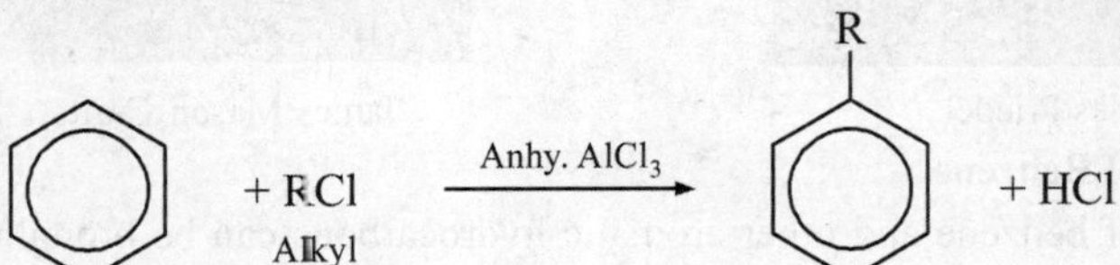

Alkyl chloride — Alkyl benzene

Mechanism. The reaction is believed to take place through the following steps:

(I) Generation of electrophile

$$R-Cl + AlCl_3 \rightleftharpoons \underset{\text{Electrophile}}{\overset{+}{R}} + AlCl_4^-$$

(II) Formation of carbocation

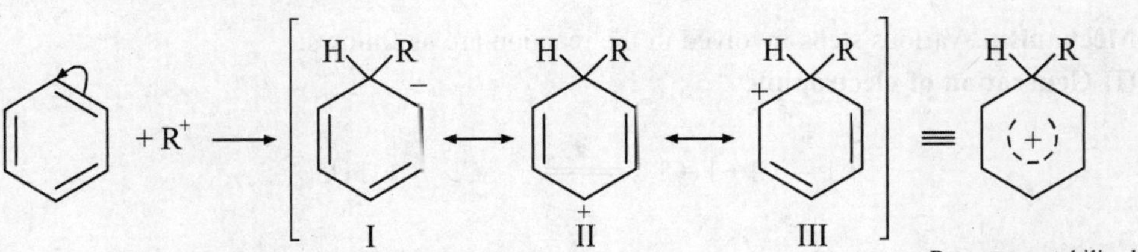

I II III

Resonating structures

Resonance stabilised carbocation

(III) Abstraction of proton from the carbocation

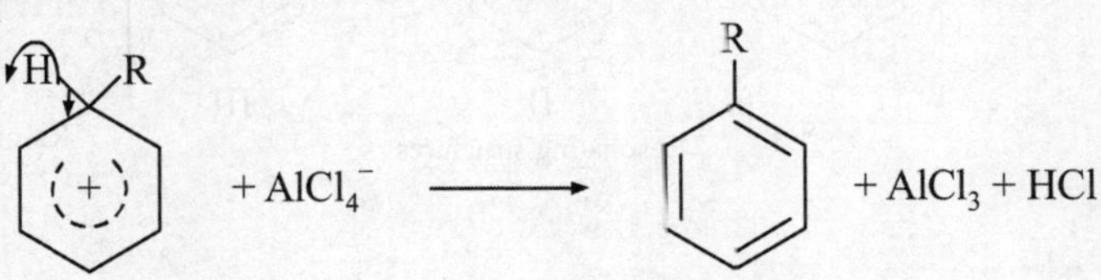

Alkyl benzene

Charles Friedel

James Mason Crafts

Halogenation of Benzene

Halogenation of benzene and other aromatic hydrocarbons can be brought about by treating with halogens in the presence of Lewis acids such as ferric halides, anhydrous aluminium chloride. Chlorination of benzene is considered below:

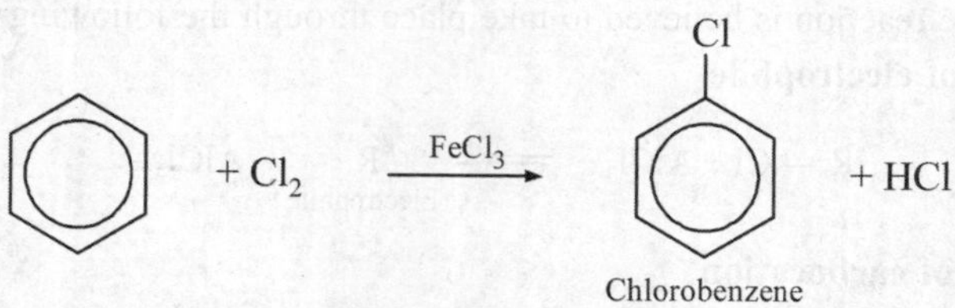

Mechanism. Various steps involved in the reaction are as follows:

(I) Generation of electrophile

$$Cl—Cl + FeCl_3 \rightleftharpoons \overset{+}{Cl} + FeCl_4^-$$

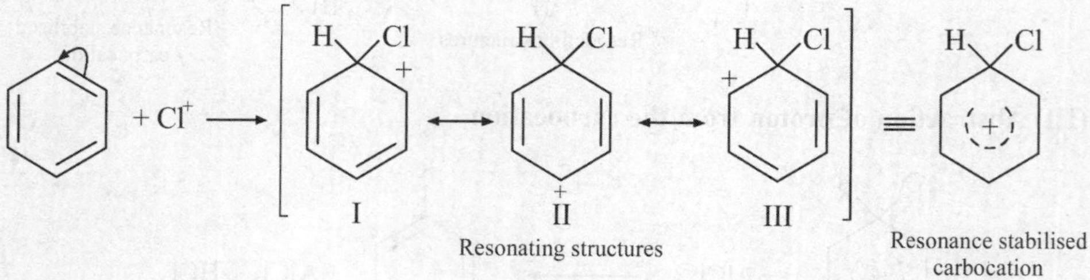

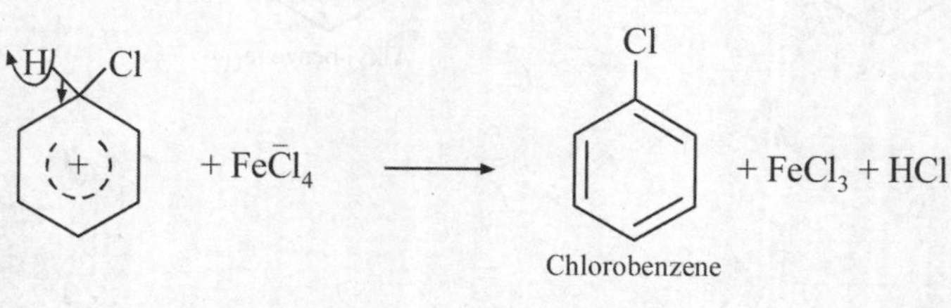

6.11 EFFECT OF SUBSTITUENTS ON ORIENTATION AND REACTIVITY OF BENZENE RING

It has been observed that substituents already attached to the benzene ring not only govern the orientation of further substitution but also affect the reactivity of the benzene ring. It is discussed in brief as below:

(I) Effect of substituents on orientation. As stated above, the nature of the group already attached to benzene ring determines the position of the incoming group. In general, groups have been classified into two categories:

(*a*) Ortho-para directing groups. The groups which direct the incoming group towards ortho and para positions are called ortho-para directing groups. Groups such as – R, – C_6H_5, – OH, – SH, – OR, – NH_2, – NHR, – NR_2, – Cl, – Br, – I, etc. are all ortho-para directing groups.

(*b*) Meta directing groups. The groups which direct the incoming groups towards meta position are called meta directing groups. Groups such as – COOH, – CHO, – CN, – NO_2, – COR, – SO_3H, etc., are all meta directing groups.

It may be mentioned that groups which contain double or triple bond are usually meta directing while those which do not contain multiple bonds are ortho-para directing. However, there are certain exceptions to this rule.

(II) Effect of substituents on reactivity. Reactivity of the benzene ring in electrophilic substitution reactions depends upon the tendency of the substituent group already present in the benzene ring to release or withdraw electrons. A group that releases electrons activates benzene ring while the one which draws electrons deactivates the benzene ring. It is found that except halogens all ortho-para directing groups activate the ring and all meta directing groups deactivate the ring towards further electrophilic substitution. Thus, nitration of toluene can be carried out at room temperature, while that of nitrobenzene requires more drastic conditions. This is illustrated as below:

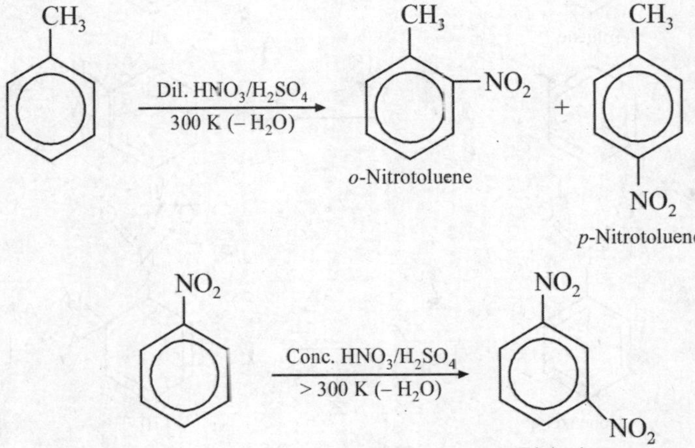

o-Nitrotoluene

p-Nitrotoluene

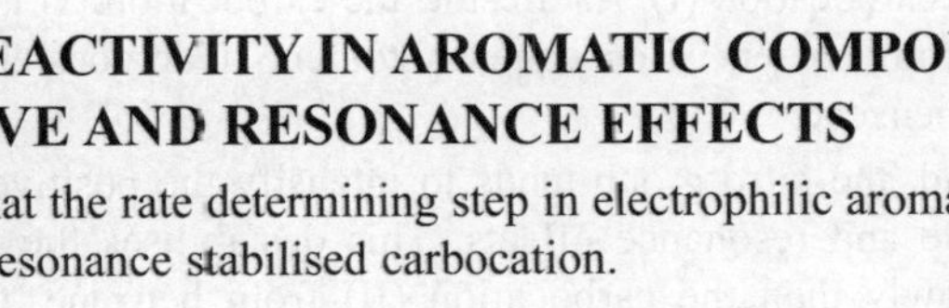

m-Dinitrobenzene

6.12 THEORY OF REACTIVITY IN AROMATIC COMPOUNDS ON THE BASIS OF INDUCTIVE AND RESONANCE EFFECTS

It has been observed that the rate determining step in electrophilic aromatic substitution is the formation of intermediate resonance stabilised carbocation.

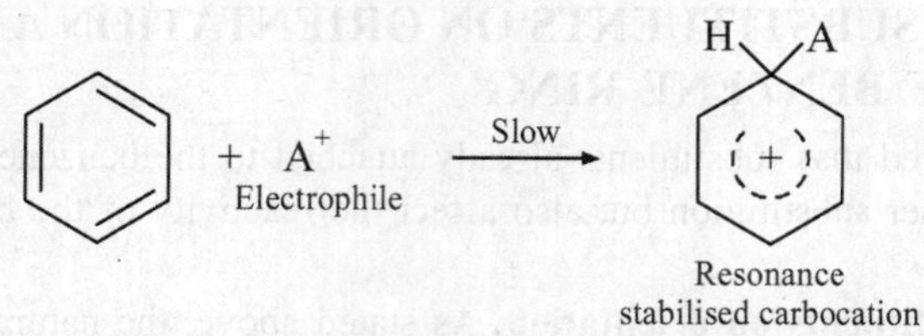

It is thus clear that any factor which stabilises the intermediate carbocation will also stabilise the transition state leading to its formation. Consequently, the carbocation will be formed more quickly and the rate of the overall reaction increases. On the other hand, factors which destabilise the carbocation will decrease the ease of its formation and hence decrease the rate of the reaction.

It is a familiar fact that dispersion of charge leads to the stability of the system. Now electron releasing groups tend to decrease the positive charge on the carbocation and thus stabilise the ion. As a result, the rate of the reaction increases. On the other hand, electron withdrawing groups destabilise the carbocation by intensifying the positive charge on it. Consequently the rate of further electrophilic substitution reaction decreases. It may be mentioned that release or withdrawal of electrons may occur due to inductive effect alone or through the net result of inductive and resonance effects.

The effect of substituents on reactivity can be illustrated by considering electrophilic substitution in benzene, toluene and nitrobenzene. The relative rates in these three substitution reactions will depend upon the relative stabilities of their corresponding intermediate carbocations formed as shown below :

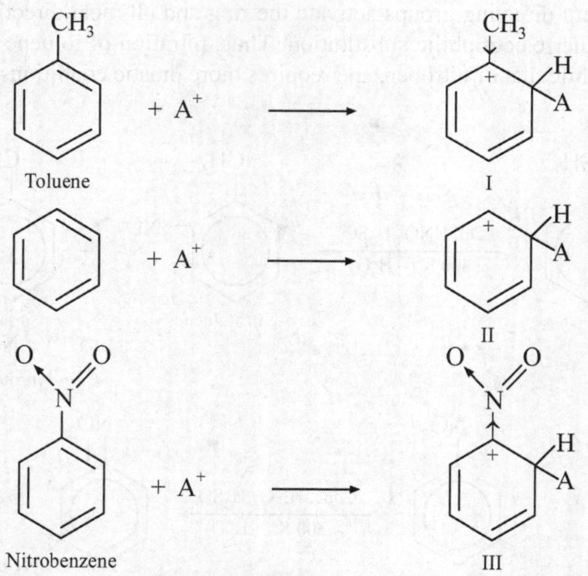

Due to its electron releasing inductive effect, methyl group tends to disperse the positive charge and thus stabilise the carbocation (I). As aresult, the carbocation (I) is formed more quickly than the carbocation (II) from benzene. Consequently toluene undergoes electrophilic substitution at a faster rate than does benzene.

On the other hand, the nitro group tends to intensify the positive charge due to its electron withdrawing inductive and resonance effects. This destabilises the carbocation (III) which is, therefore, formed slowly than the carbocation (II) from benzene. Consequently nitrobenzene undergoes electrophilic substitution at a slower rate as compared to benzene.

Electronic Interpretation of the Ortho-para Directing Influence of Amino Group

Amino group exerts electron withdrawing (– I) and electron releasing (+M) effects. Of the two opposing effects, the resonance effect dominates and thus overall behaviour of – NH_2 group is electron releasing, *i.e.*, it acts as an activator.

The ortho and para directing influence of – NH_2 group can be explained by assuming that nitrogen can share more than a pair of electrons with the benzene ring and can accommodate a positive charge. Thus, consider the case of further electrophilic substitution in aniline. The various resonating structures of the carbocations formed by ortho, para and meta attack are given below:

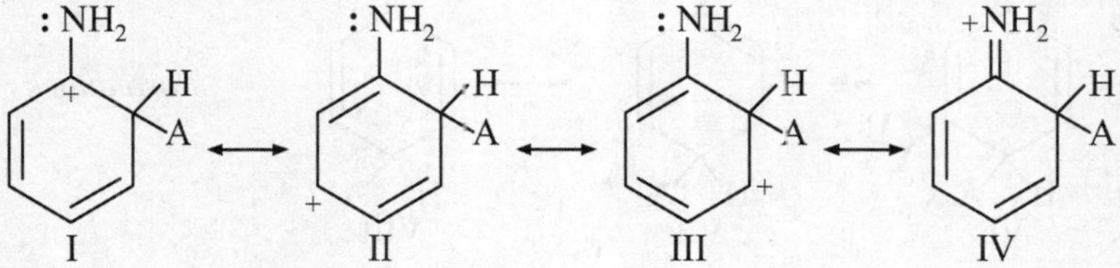

Ortho attack

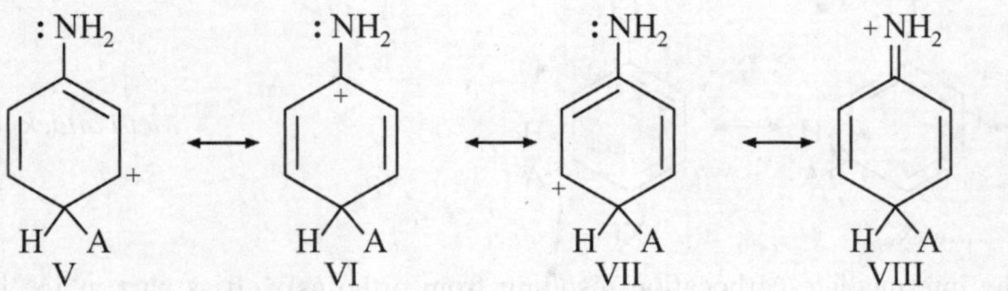

Para attack

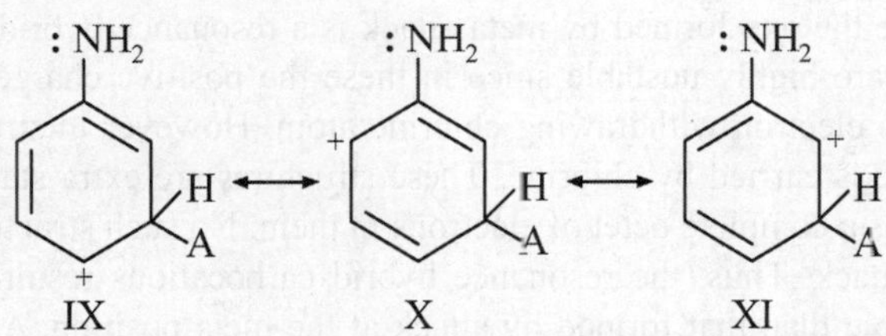

Meta attack

The intermediate carbocation resulting from *ortho* as well as *para* attack is a resonance hybrid of four structures while the one formed by *meta* attack is a resonance hybrid of three structures. Further, in structures IV and VIII, the positive charge is carried by nitrogen. These structures are more stable since in every atom (except hydrogen) has a complete octet of electrons in them. No such structure is, however, possible in case of *meta* attack. It is, therefore, clear that the resonance hybrid carbocations resulting from *ortho* and *para* attack are more stable than the carbocation formed by attack at the *meta* position. Consequently further electrophilic substitution in aniline occurs faster at the *ortho* and *para* positions than at the *meta* position. In other words – NH_2 group is an *ortho* and *para* directing group.

Ortho-Para Directing Influence of Halogens

Halogens exert electron withdrawing (–I) and electron releasing (+M) effects. Due to high electronegativities of halogens, the inductive effect predominates over the mesomeric effect and thus the overall behaviour of halogens is electron withdrawing. In other words, halogens act as deactivators for further substitution.

In order to account for their ortho and para directing nature, it has been assumed that halogens can share more than one pair of electrons with the benzene ring and can accommodate positive charge. Thus, consider the case of further electrophilic substitution in chlorobenzene. The various resonating structures of the carbocation formed by ortho, para and meta attack are given below:

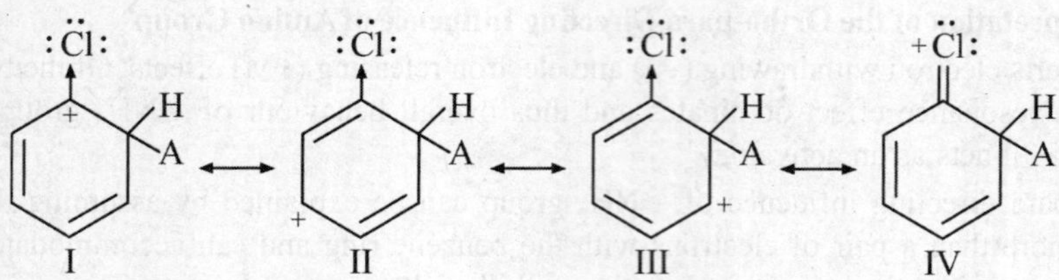

Ortho attack

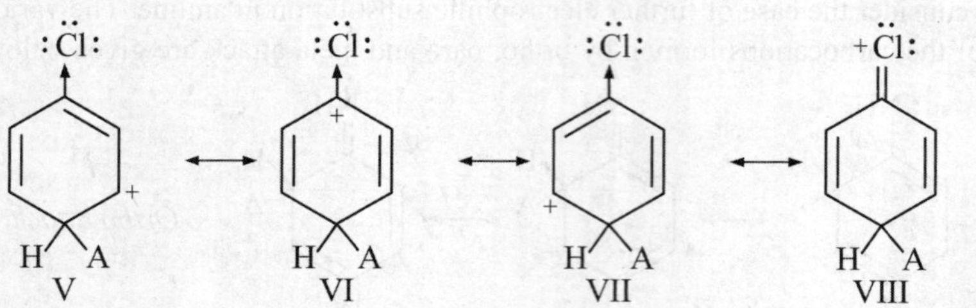

Para attack

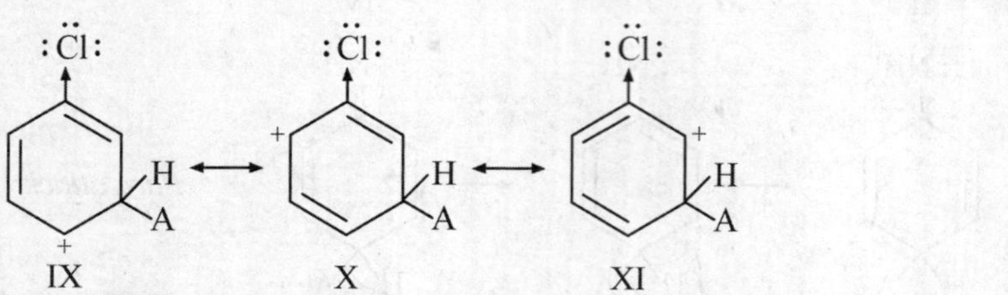

Meta attack

It is clear that the intermediate carbocation resulting from ortho as well as para attack is a resonance hybrid of four structures while the one formed by meta attack is a resonance hybrid of three structures. Structures (I) and (VI) are highly unstable since in these the positive charge is carried by that carbon which is linked to electron withdrawing chlorine atom. However in structures (IV) and (VIII), the positive charge is carried by chlorine. These structures are extra stable since in every atom (except hydrogen) has a complete octet of electrons in them. No such structure is, however, possible in case of meta attack. Thus, the resonance hybrid carbocations resulting from ortho and para attack are more stable than that formed by attack at the meta position. As a result, chlorine is ortho and para directing. The same is true for other halogens. It may thus be concluded that in case of halogens, the reactivity is controlled by the stronger inductive effect and the orientation is determined by mesomeric effect.

Ortho and Para Directing Influence of Alkyl Groups

Alkyl groups exert electron releasing inductive effect (+I effect). Consider the case of further electrophilic substitution in toluene which contains an electron releasing methyl group. The various resonating structures of the carbocations formed by ortho, para and meta attack are given below:

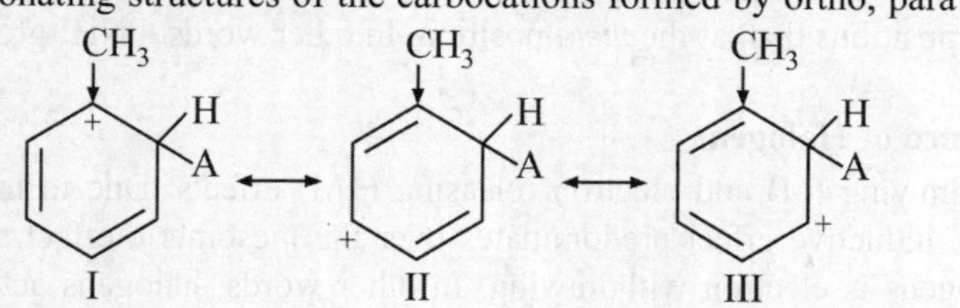

Ortho attack

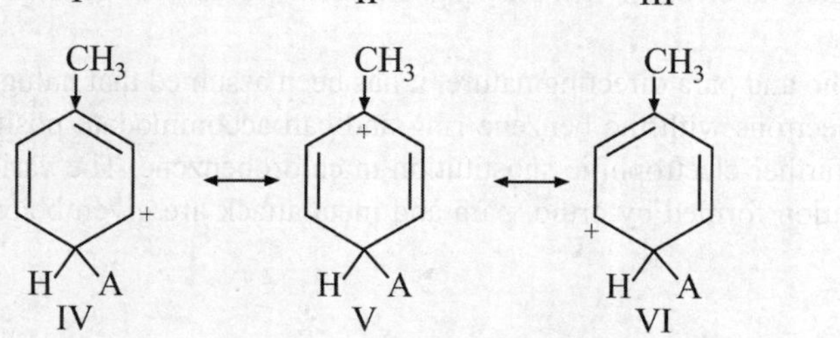

Para attack

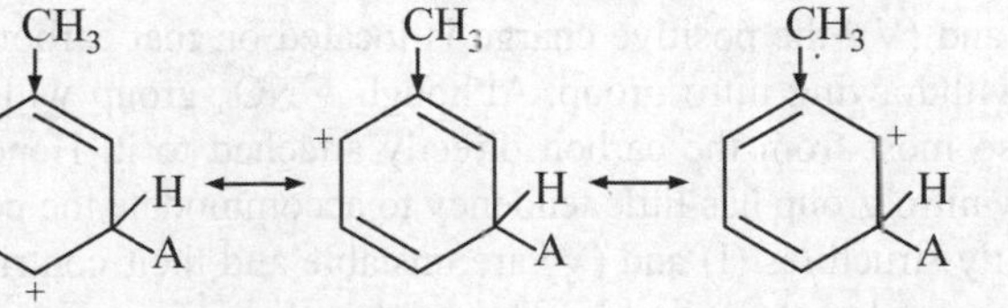

VII VIII IX

Meta attack

We find that in each case, the intermediate carbocation is a resonance hybrid of three structures. In structures (I) and (V), the positive charge is located on the carbon atom to which electrons releasing methyl group is attached. Therefore, the positive charge on such a carbon is highly dispersed and thus the corresponding structures (I) and (V) are more stable than all other structures. No such structure is, however, possible in case of meta attack. Hence the resonance hybrid carbocations resulting from ortho and para attack are more stable than the one formed by attack at the meta position. Therefore further electrophilic substitution in toluene occurs faster at the ortho and para position than at the meta position. In other words, methyl group is an ortho and para directing group. Other alkyl groups behave similarly.

Deactivating and Meta Directing Nature of Nitro Group Towards Electrophilic Aromatic Substitution.

Nitro group is electron withdrawing in nature. In this case, the electron withdrawal occurs through electron withdrawing effect (– I effect) as well as electron withdrawing resonance effect (– M effect). Due to its electron withdrawing character, nitro group deactivates the benzene ring towards further electrophilic substitution.

Let us examine the directing influence of $-NO_2$ group by considering electrophilic substitution in nitrobenzene. The various resonating structures of the carbocations formed by ortho, para and meta attack are given below:

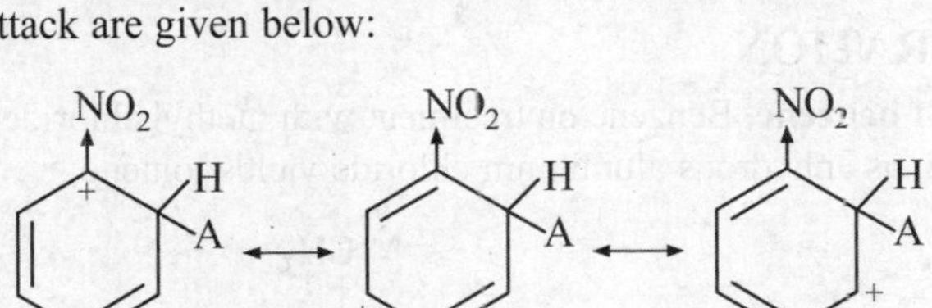

I II III

Specially unstable

Ortho attack

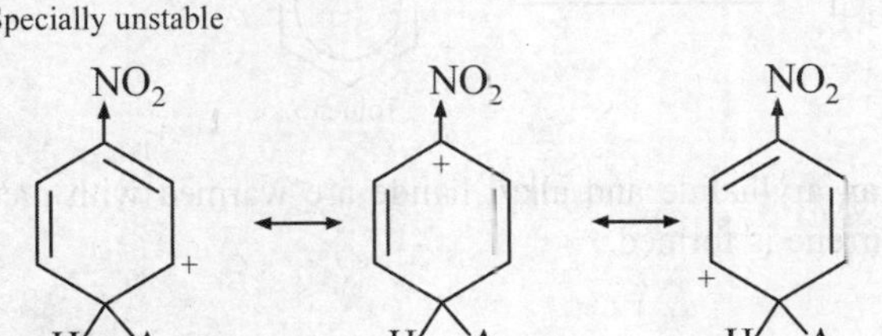

IV V VI

Specially unstable

Para attack

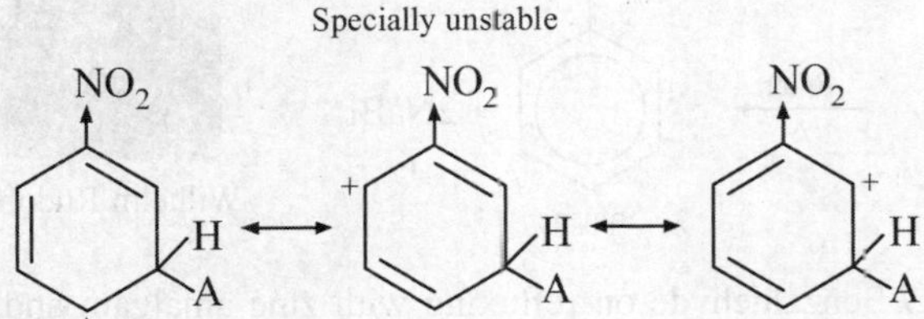

VII VIII IX

Meta attack

In the contributing structures (I) and (V), the positive charge is located on that carbon atom which is directly linked to electron withdrawing nitro group. Although – NO_2 group withdraws electrons from all positions, it does so most from the carbon directly attached to it. Hence this carbon atom, already made positive by nitro group has little tendency to accommodate the positive charge of the carbocation. Consequently structures (I) and (V) are unstable and their contribution towards stabilization of the carbocation is almost negligible. Thus carbocations formed by ortho and para attack are virtually resonance hybrids of only two structures while the one formed by meta attack is a resonance hybrid of three structures. Therefore the resonance hybrid carbocation resulting from meta attack is more stable than the carbocations resulting from ortho and para attack. Consequently, further electrophilic substitution takes place at the meta position. Thus, nitro group is meta directing.

Ortho-Para Directing Influence of Alkoxy Group

Alkoxy group is *o-p* directing because the lone-pair of electrons on oxygen takes part in resonance as illustrated below:

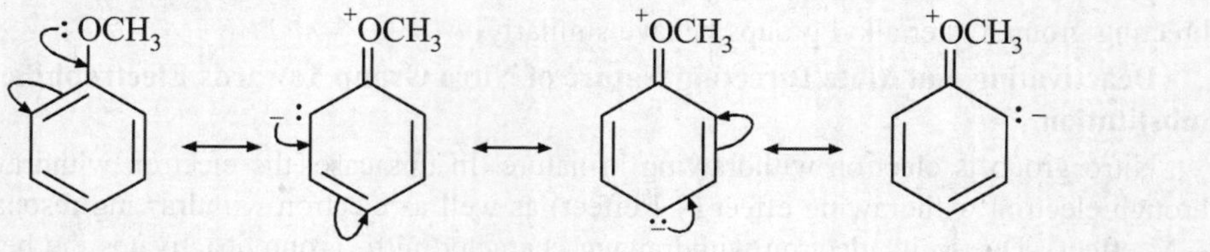

Ortho and para positions relative to methoxy group are negative. Hence in further substitution the electrophile will preferentially attach at these negative points.

TOLUENE

6.13 METHODS OF PREPARATION

***(i)* Friedel-Crafts methylation of benzene.** Benzene on treatment with methyl chloride in the presence of a Lewis acid catalyst such as anhydrous aluminium chloride yields toluene.

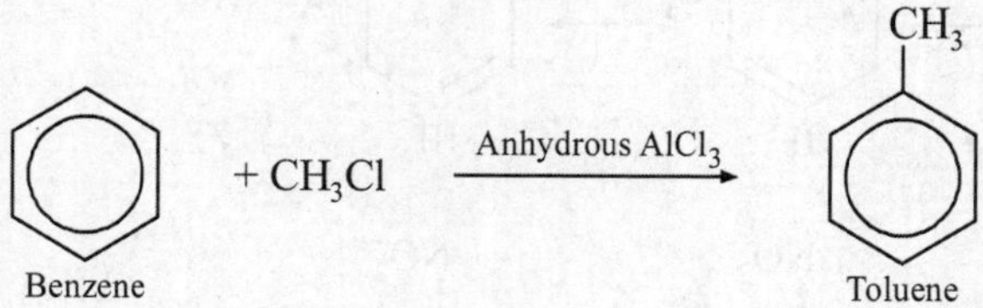

***(ii)* Wurtz-Fittig reaction.** When an arylhalide and alkyl halide are warmed with metallic sodium in ethereal solution, an alkyl benzene is formed.

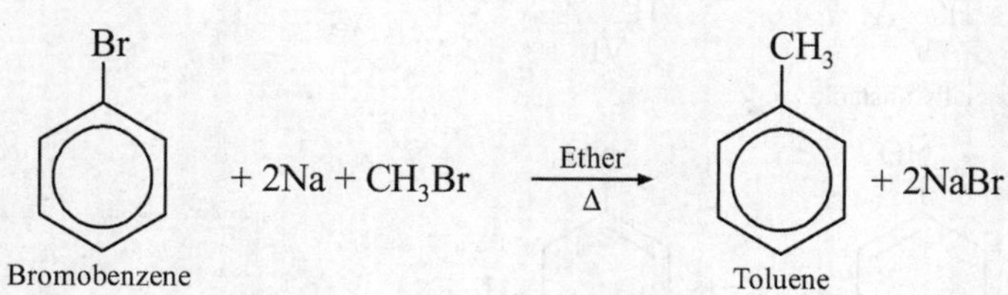

Wilhelm Rudolf Fittig

***(iii)* Reduction of benzaldehyde.** Benzaldehyde on refluxing with zinc amalgam and conc. hydrochloric acid (*Clemmensen reduction*) yields toluene.

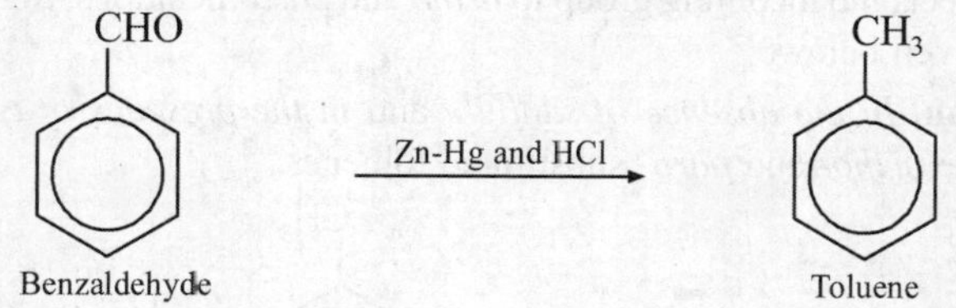

The reduction can also be carried out by heating benzaldehyde with hydrazine and a strong base like potassium hydroxide or potassium *tert*- butoxide (*Wolff-Kishner reduction*).

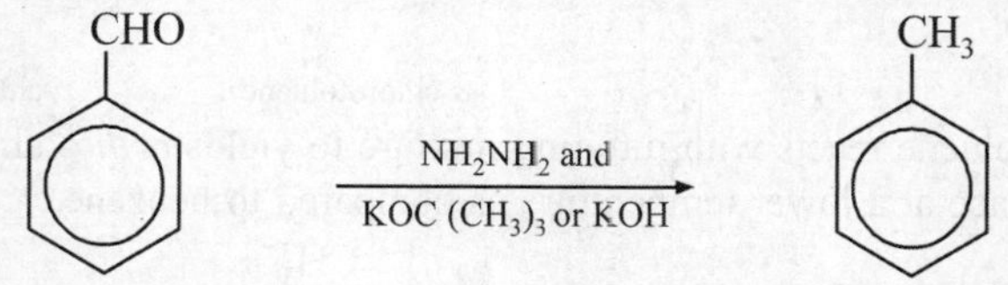

***(iv)* Grignard reaction.** Phenyl magnesium bromide on treatment with methyl halide yields toluene.

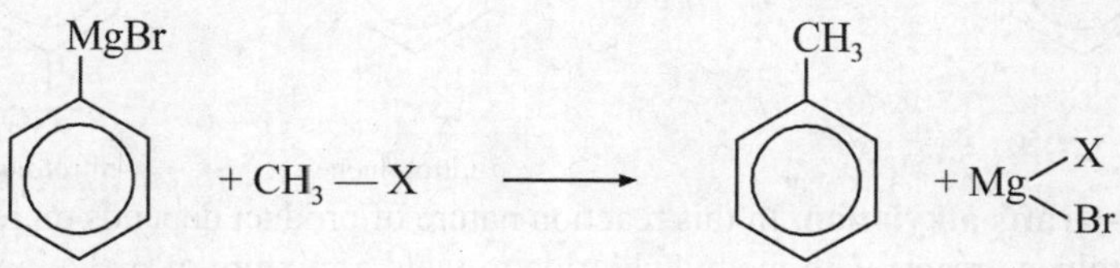

***(v)* Decarboxylation of toluic acids.** Toluene can also be prepared by heating *ortho-, meta-* or *para*-toluic acid with soda lime.

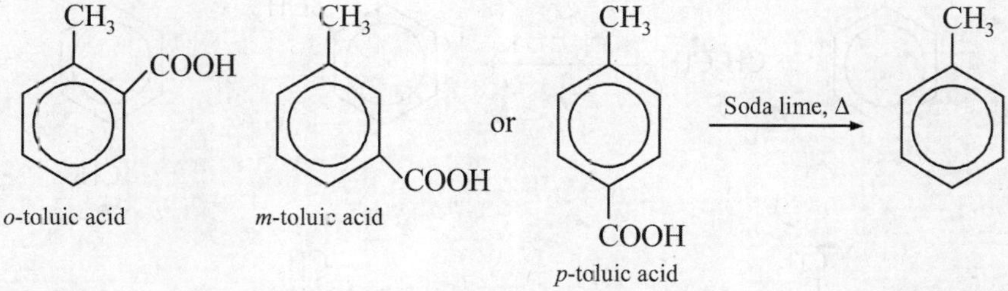

***(vi)* Reduction of cresols with zinc dust.** *ortho, meta* or *para*-cresol on heating with zinc dust yields toluene.

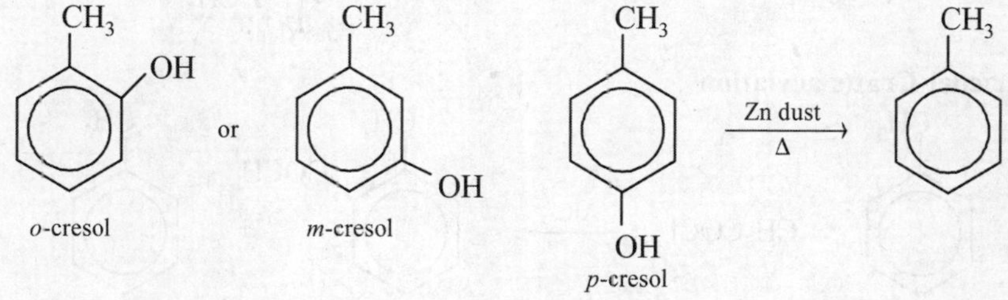

6.14 CHEMICAL PROPERTIES OF TOLUENE

***(i)* Electrophilic aromatic substitution reactions.** Methyl group in toluene activates the aromatic ring and directs the incoming group to *ortho* and *para-* positions. Some important reactions of this type are as given below:

(*a*) Halogenation. *In the absence of sunlight* and *in the presence of catalyst*, toluene reacts with halogens to give *ortho-* and *para-* substituted toluenes.

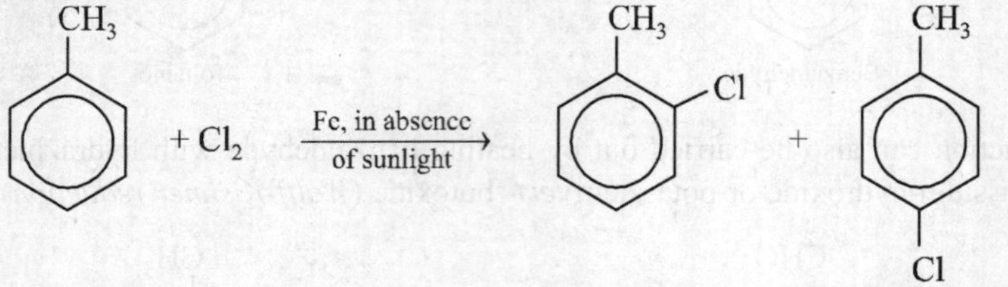

o-chlorotoluene *p*-chlorotoluene

(*b*) Nitration. Toluene reacts with nitrating mixture to yields *ortho-* and *para-* nitrotoluenes. The reaction takes place at a lower temperature as compared to benzene.

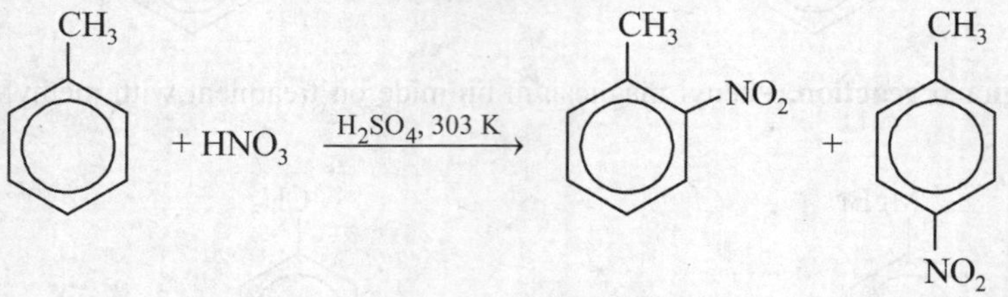

o-nitrotoluene *p*-nitrotoluene

(*c*) Friedel-Crafts alkylation. In this reaction nature of product depends on reaction temperature. At 273 K toluene reacts with methyl chloride to yield a mixture of *ortho-* and *para-* xylenes while at 353 K, it gives *meta*-xylene.

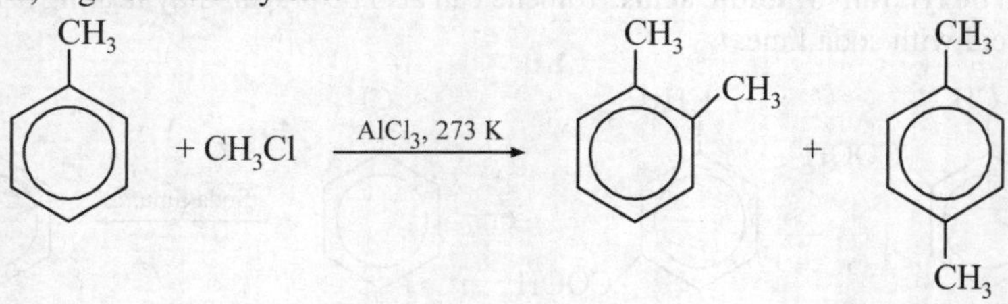

o-xylene *p*-xylene

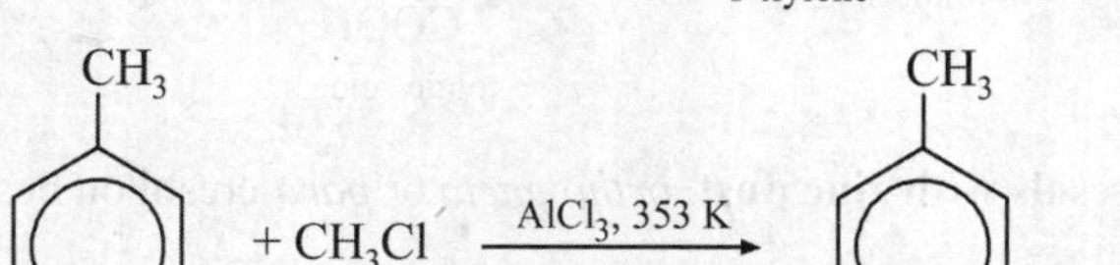

m-xylene

(*d*) Friedel-Crafts acylation

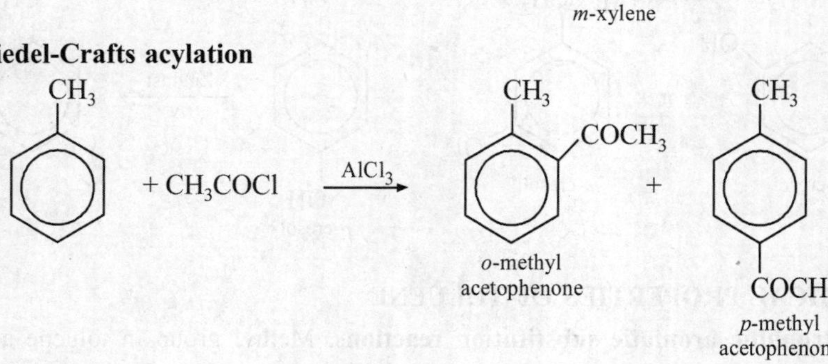

o-methyl acetophenone *p*-methyl acetophenone

(e) Sulphonation

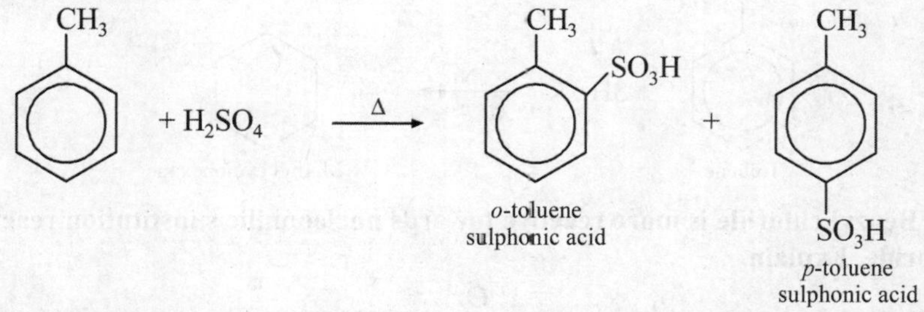

(ii) Substitution at the side chain (methyl group). *Halogenation.* In presence of sunlight and in absence of a catalyst or at high temperature in presence of free radical initiators, halogens displace the hydrogen atoms of the side chain *i.e.*, methyl group.

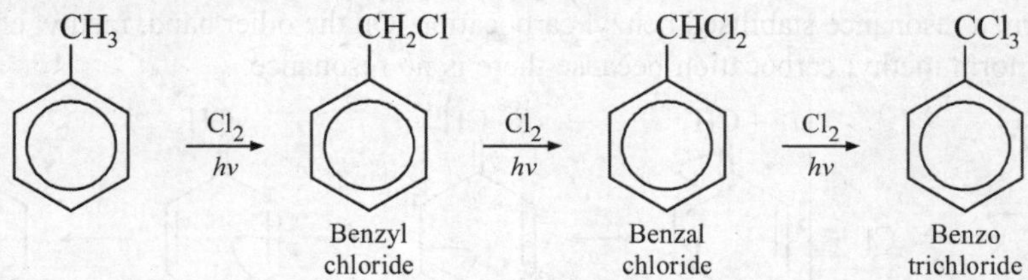

This reaction takes place via free radical chain mechanism as in the case of halogenation of alkanes.

(iii) Oxidation of the side chain. Though benzene and alkanes are quite stable towards oxidation, toluene or any monoalkyl substituted benzene on oxidation with potassium permanganate, potassium dichromate or even with dilute nitric acid, yields benzoic acid.

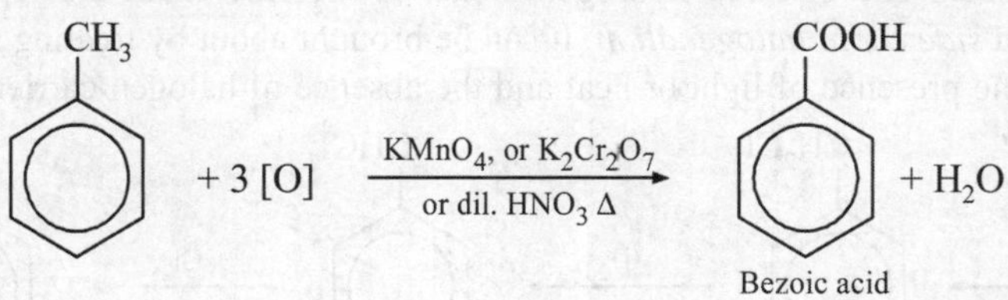

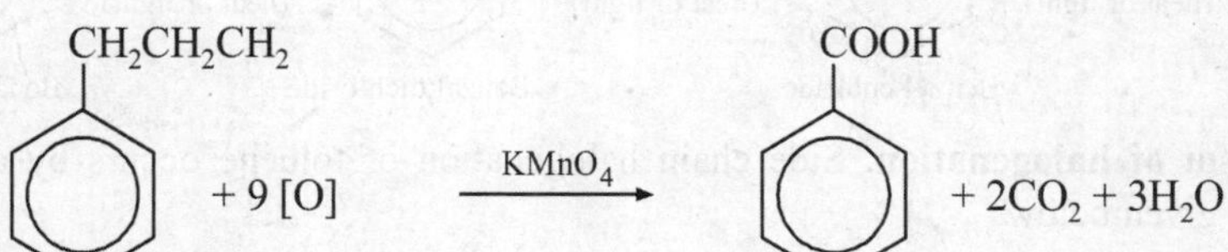

On treatment with a milder oxidising agent such as chromyl chloride in carbon tetrachloride solution, toluene yields benzaldehyde.

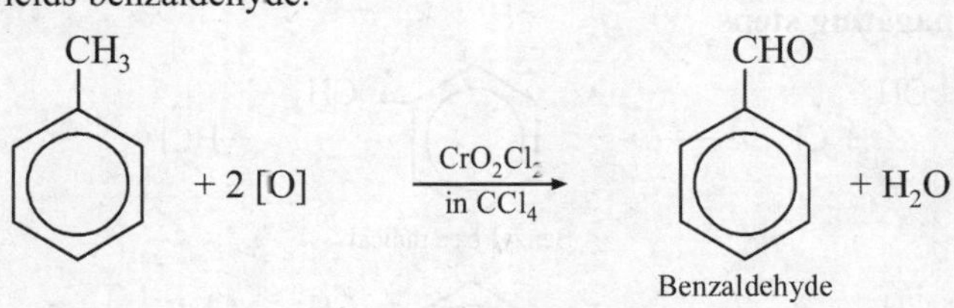

(iv) Addition of hydrogen. Like benzene, toluene also add on hydrogen in the presence of catalyst to form methyl cyclohexane.

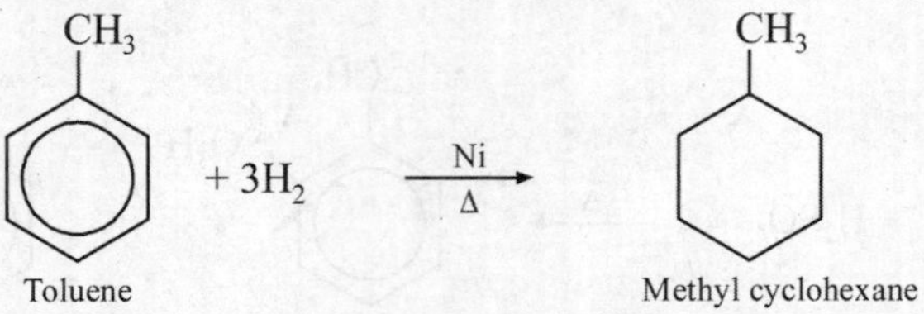

Example 5. **Benzyl chloride is more reactive towards nucleophilic substitution reactions than methyl chloride. Explain.**

Or

Both benzyl and methyl carbocations are primary but benzyl cation is much more stable than methyl cation.

Solution. The greater reactivity of benzyl chloride is mainly due to the fact that it readily ionises to form a resonance stabilised benzyl carbocation. On the other hand, methyl chloride does not ionise to form methyl carbocation because there is no resonance.

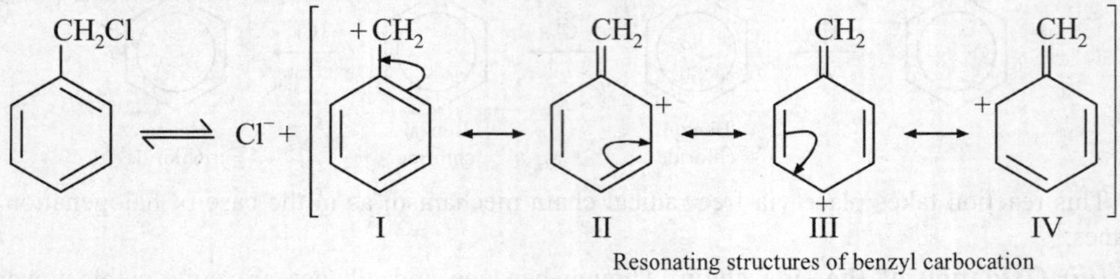

Resonating structures of benzyl carbocation

6.15 SIDE CHAIN HALOGENATION

The reaction in which one or more hydrogen atoms of the side chain are replaced by the halogen atoms is called *side chain halogenation.* It can be brought about by treating an arene with a suitable halogen in the presence of light or heat and the absence of halogen carriers.

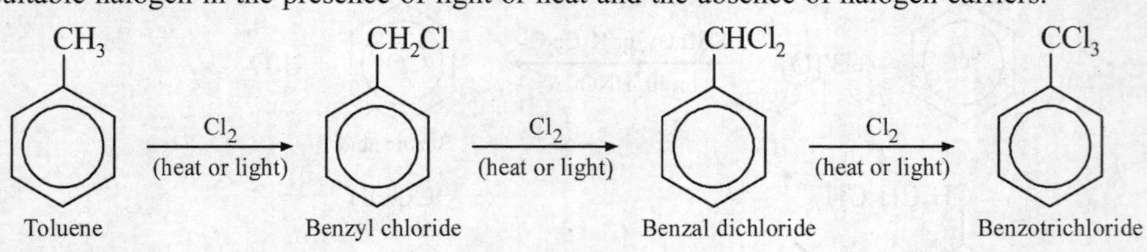

Mechanism of halogenation. Side chain halogenation of toluene occurs by a free radical mechanism as given below:

(*a*) Chain initiation step

$$Cl—Cl \xrightarrow[\text{(homolytic fission)}]{\text{Heat or light}} 2\,Cl\cdot$$

(*b*) Chain propagating steps

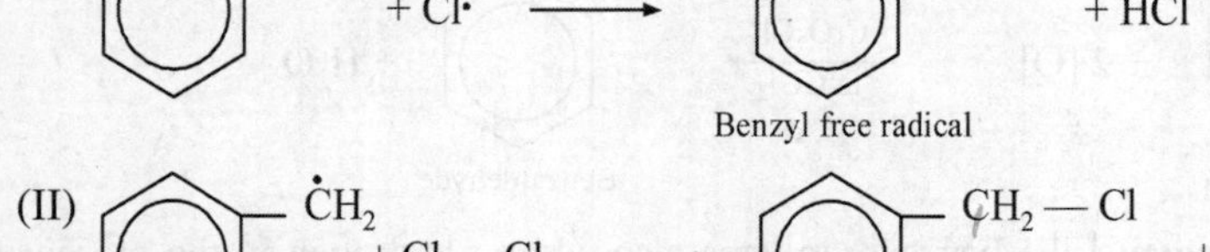

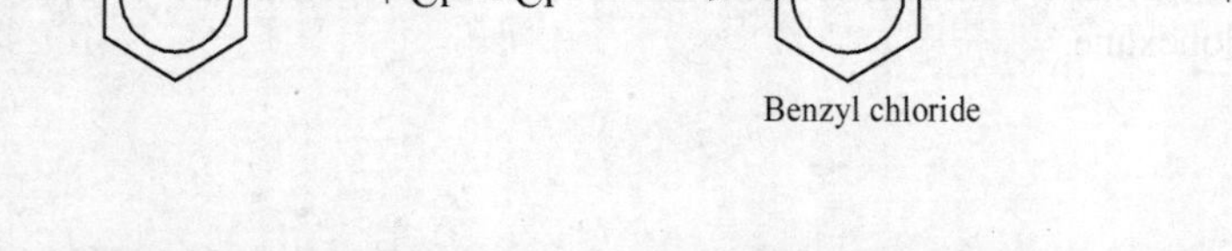

The above steps (I) and (II) are repeated again and again.

***(c)* Chain terminating steps.** When the free radicals combine with each other, the reaction comes to a stop and thus the chain of reaction gets terminated.

$$Cl\cdot + Cl\cdot \longrightarrow Cl-Cl$$

When excess of chlorine is used, other hydrogen atoms are also replaced in the same manner. Thus,

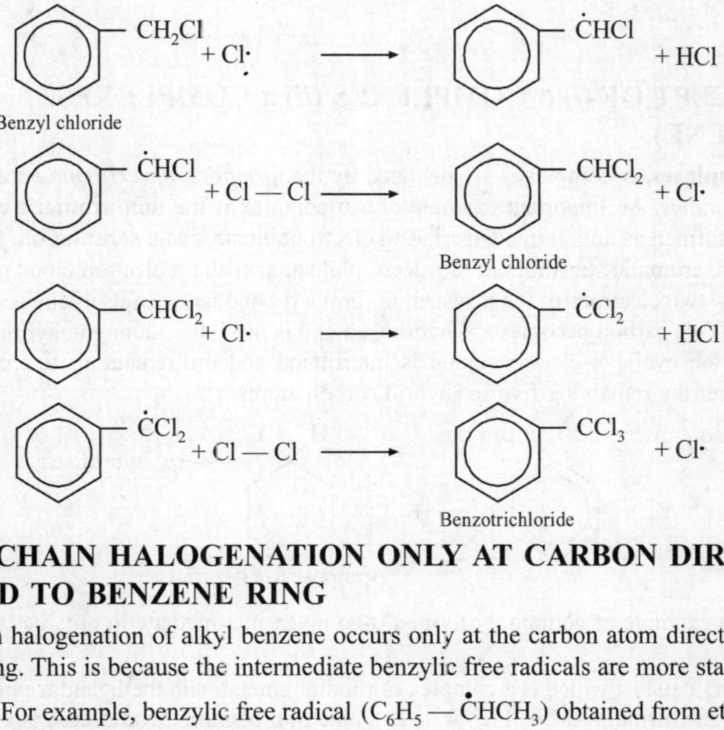

6.16 SIDE CHAIN HALOGENATION ONLY AT CARBON DIRECTLY ATTACHED TO BENZENE RING

Side chain halogenation of alkyl benzene occurs only at the carbon atom directly attached to the benzene ring. This is because the intermediate benzylic free radicals are more stable and hence easily formed. For example, benzylic free radical ($C_6H_5 - \dot{C}HCH_3$) obtained from ethylbenzene is a resonance hybrid of the following structures:

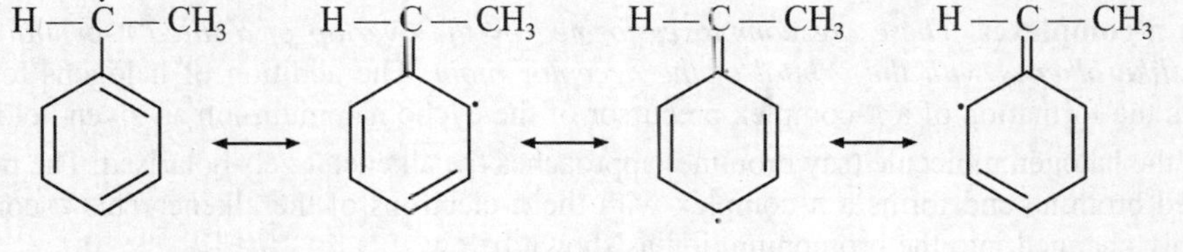

On the other hand, the second free radical ($C_6H_5 - CH_2 - \dot{C}H_2$) that can be obtained from ethylbenzene is a resonance hybrid of only two structures.

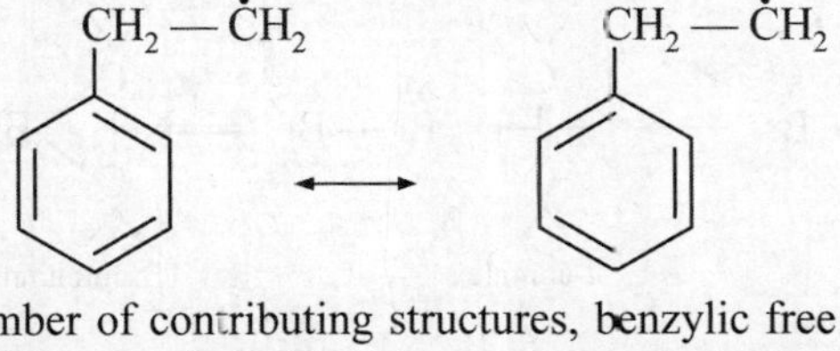

Due to its greater number of contributing structures, benzylic free radical is more resonance stabilised than non-benzylic free radical. Thus, bromination of ethylbenzene occurs exclusively at the **benzylic position.**

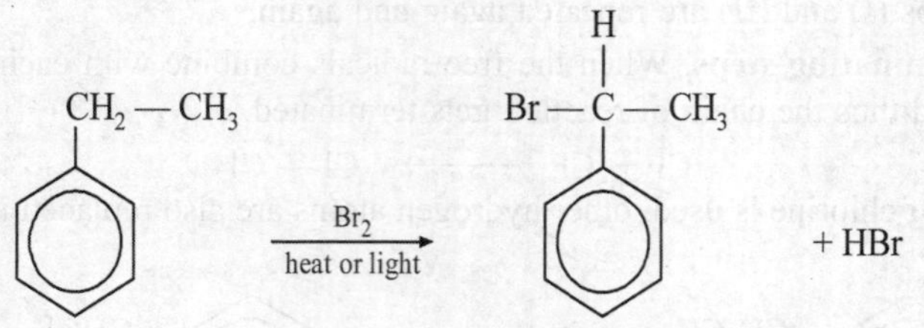

1-Bromo-1-phenylethane

6.17 CONCEPT OF (*i*) σ COMPLEXES (*ii*) π COMPLEXES. (FERROCENE)

***(i)* σ-complexes.** σ-complexes are obtained by the *formation of a σ-bond between the donor and acceptor atoms*. An important example of a σ-complex is the **non-aromatic carbocation** or **arenium ion** formed as an intermediate during electrophilic aromatic substitution. In the first step of electrophilic aromatic substitution, the electrophile attacks the π electron cloud of benzene ring and takes away two electrons of the π system to form a σ bond between itself and one carbon of the benzene ring. This carbon becomes sp^3 hybridised and is not left with any unhybridised π-orbital. As a results, the cyclic π electron cloud is interrupted and the remaining four π-electrons are delocalised over the remaining five sp^2 hybrid carbon atoms.

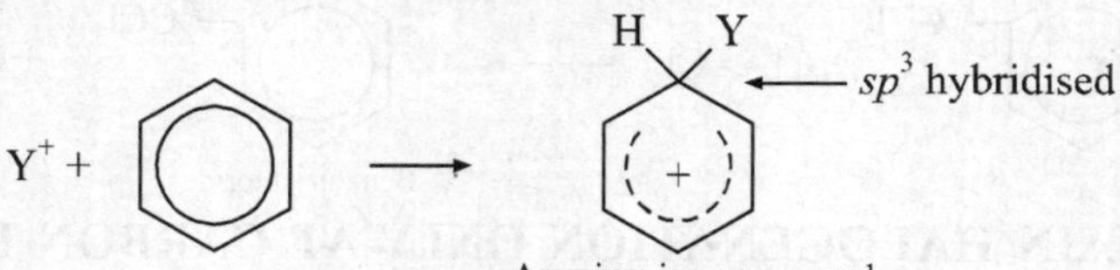

Aremium ion or σ-complex

This is an example of commonly formed σ-complex intermediate in organic reaction mechanism.

Wilkinson's catalyst which is a complex of rhodium metal with the ligand triphenyl phosphine (PPh_3) having the formula $RhCl(PPh_3)$ is an example of transition metal σ complex. It is used as a catalyst in the hydrogenation of alkenes by forming a π-complex with the alkene by means of π electron cloud of alkene.

***(ii)* π-complexes.** *These are complexes formed by the overlap of a filled π-orbital of compounds like alkenes with the orbital of the acceptor atom.* The addition of halogens to alkenes involves the formation of a π-complex precursor of the cyclic halonium ion as given below.

As the halogen molecule (say bromine) approaches the alkene, it gets polarised. The positively polarised bromine end forms a π-complex with the π-electrons of the alkene. This π-complex is reversibly changed into the bromonium ion as shown below.

It may be noted that in the formation of above π-complex, bromine gets bonded, not only to one of the alkene carbons, but to both by π-electrons of alkene.

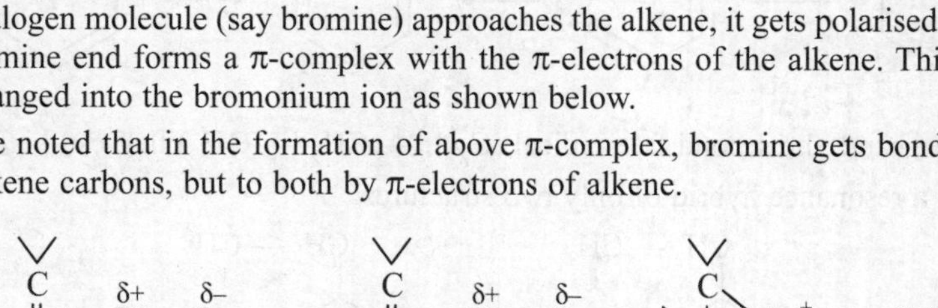

π-complex   Bromonium ion

Dicyclopentadienyl iron $[(C_5H_5^-)]_2$ Fe, commonly known as ferrocene is an example of transition metal π complex. It is a stable molecule in which iron is "sandwiched" between two flat cyclodienyl rings. The carbon iron bond results by overlap between π-cloud of cyclopentadienyl anions and 3*d* orbitals of iron.

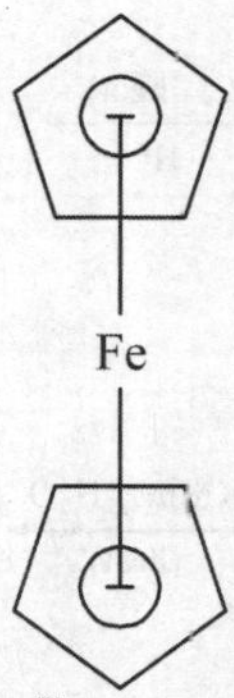

Ferrocene

Example 6. **What happens when:**

(*i*) Benzyl chloride is subjected to oxidation with lead nitrate.

(*ii*) Toluene is treated with methyl chloride in the presence of anhydrous aluminium chloride.

(*iii*) Benzyl chloride is treated with silver acetate.

(*iv*) Chlorine is passed through boiling toluene.

(*v*) *o*-Xylene is treated with hot aqueous $KMnO_4$ solution.

Solution.

(*i*)

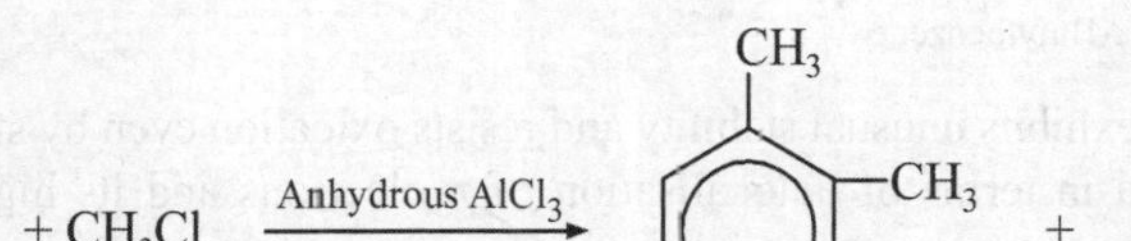

Benzaldehyde

(*ii*) Benzyl chloride

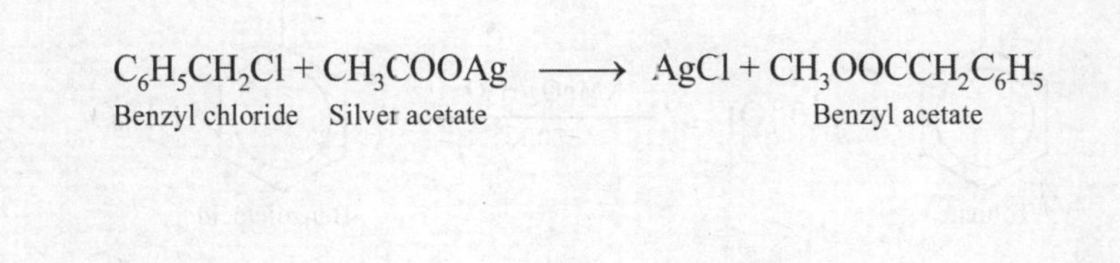

(*iii*) $\underset{\text{Benzyl chloride}}{C_6H_5CH_2Cl} + \underset{\text{Silver acetate}}{CH_3COOAg} \longrightarrow AgCl + \underset{\text{Benzyl acetate}}{CH_3OOCCH_2C_6H_5}$

(*iv*)

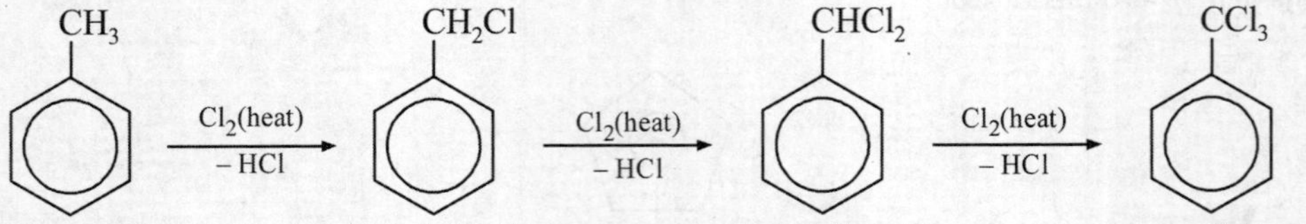

Toluene Benzyl chloride Benzal dichloride Benzotrichloride

(*v*)

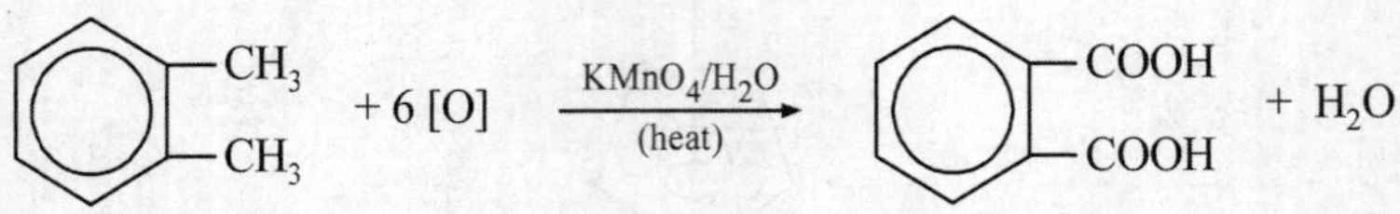

o-Xylene Phthalic acid

***Example 7.* Explain the following:**

(*i*) *n*-Butylbenzene on oxidation gives benzoic acid but *t*-butylbenzene does not.

(*ii*) Benzene is stable towards oxidising agents but toluene can be oxidised easily to benzoic acid.

Solution.

(*i*) It is believed that oxidation involves the attack on benzylic hydrogen. *t*- Butylbenzene does not possess benzylic hydrogen and hence it can't undergo oxidation easily. On the other hand, *n*-butylbenzene is easily oxidised due to the presence of benzylic hydrogens. Thus,

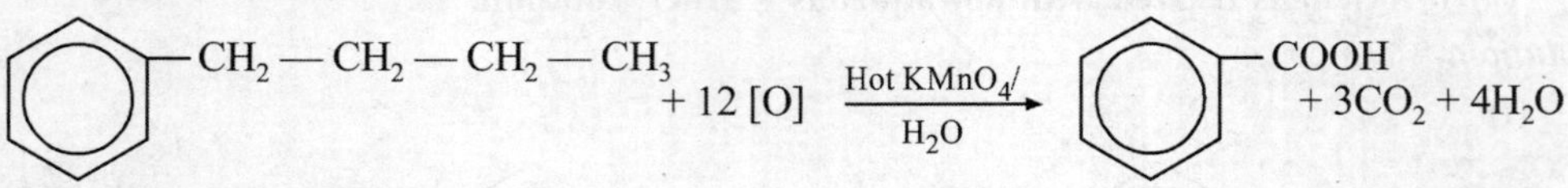

n-Butylbenzene

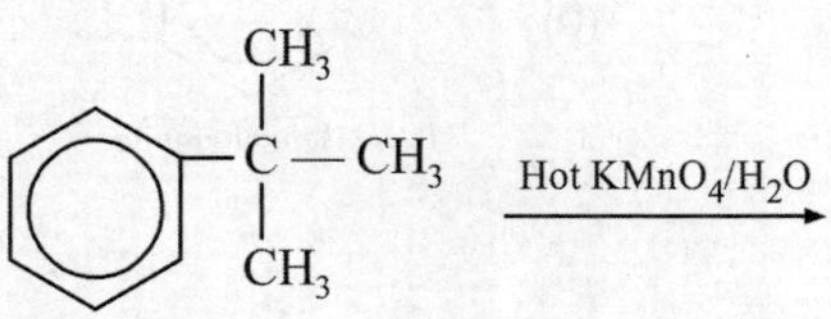

t-Butylbenzene

(*ii*) Benzene ring exhibits unusual stability and resists oxidation even by strong oxidising agents. This can be explained in terms of delocalisation of π electrons and its high value of resonance energy.

Toluene can easily be oxidised due to the presence of benzylic hydrogens *i.e.*, hydrogens attached to carbon linked directly to the benzene ring which are highly reactive. Thus

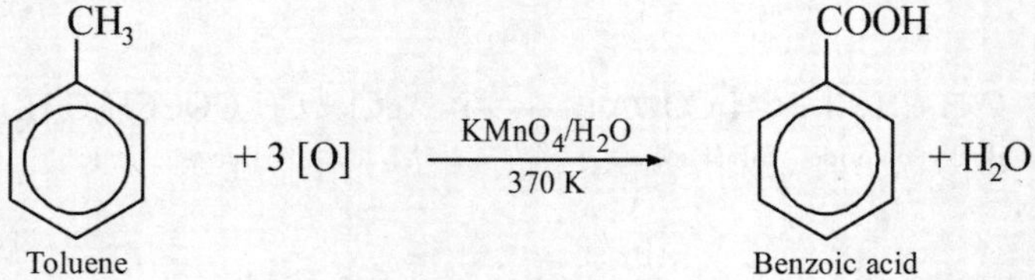

Toluene Benzoic acid

***Example 8.* Fill in the blanks, giving names of products.**

(*i*)

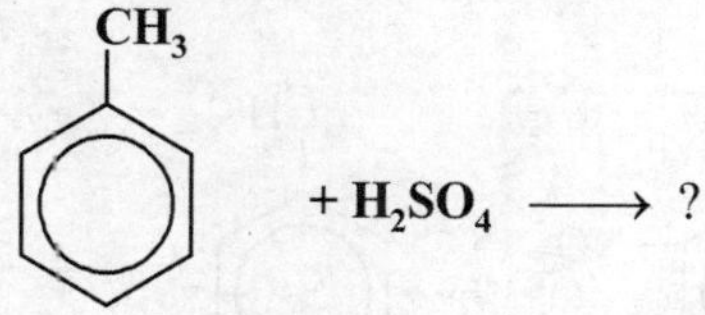

(*ii*)

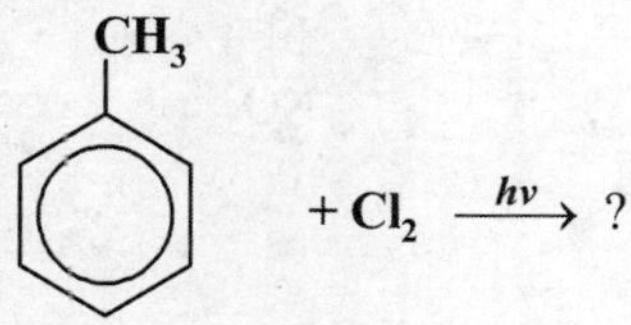

Solution.

(*i*)

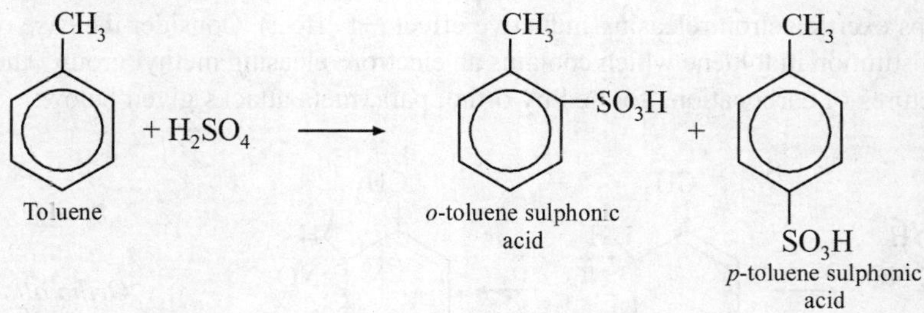

(*ii*)

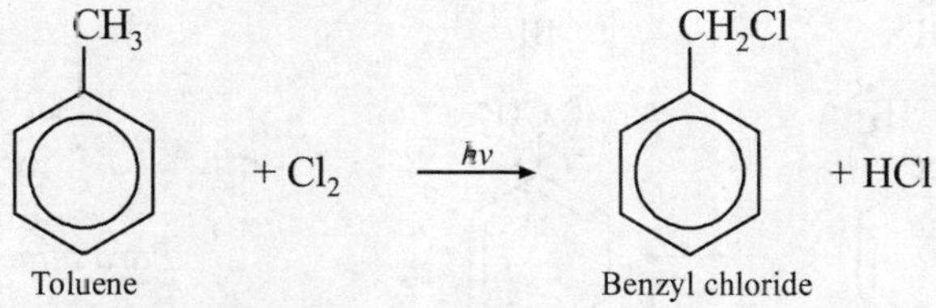

***Example 9.* Indentify the products A and B.**

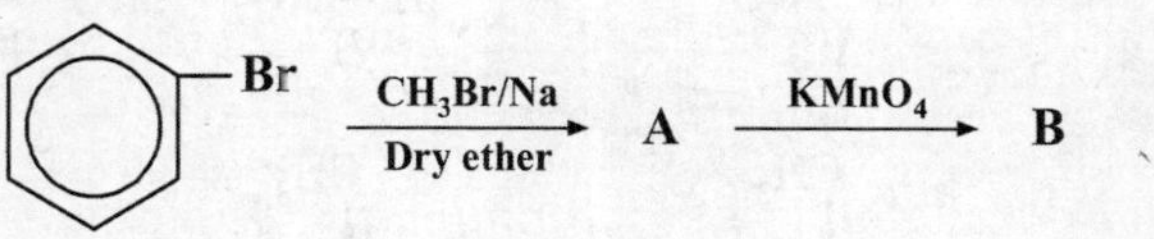

Solution.

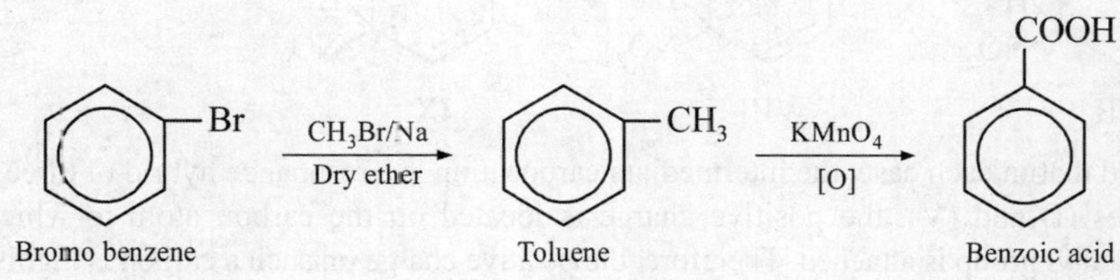

***Example 10.* Give the mechanism of nitration of toluene.**

Solution.

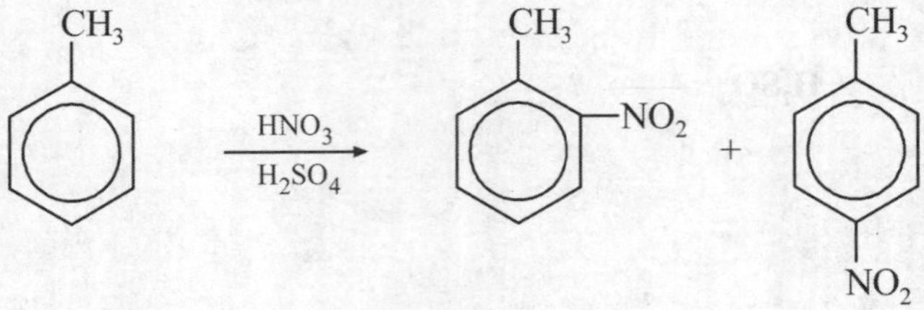

Mechanism. (*i*) Generation of an electrophile

$$HONO_2 + 2H_2SO_4 \longrightarrow \overset{+}{N}O_2 + H_3\overset{+}{O} + 2HSO_4^-$$

Formation of carbocation

Alkyl groups exert electron releasing inductive effect (+I effect). Consider the case of further electrophilic substitution in toluene which contains an electron releasing methyl group. The various resonating structures of carbocations formed by ortho, para, meta attacks given below:

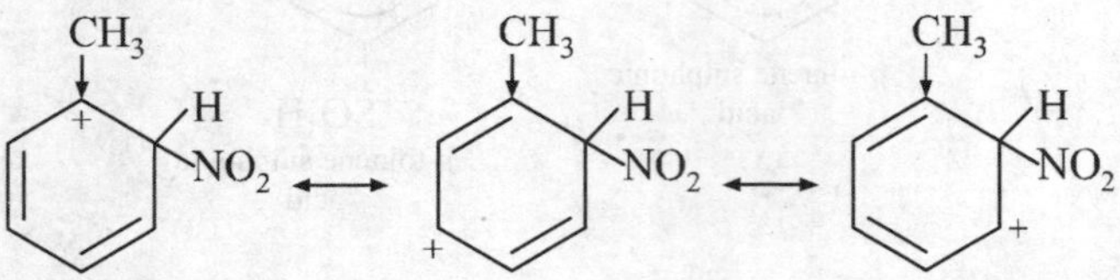

I II III *Ortho attack*

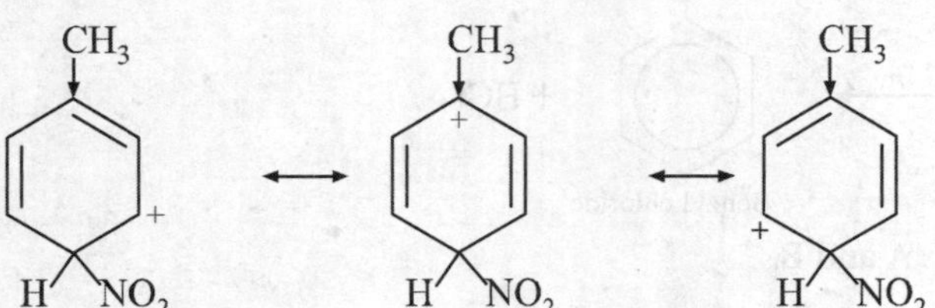

IV V VI *Para attack*

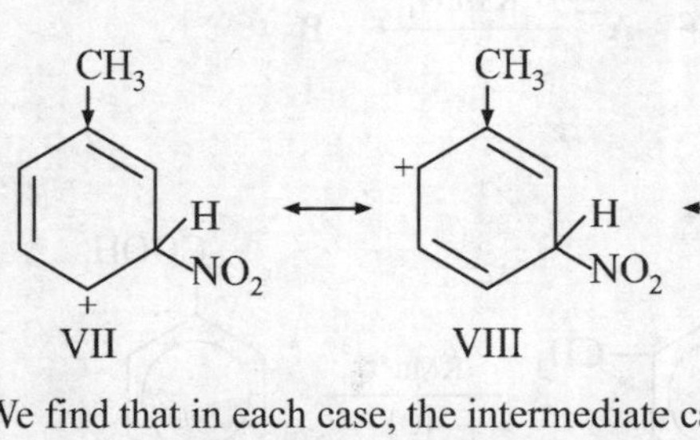

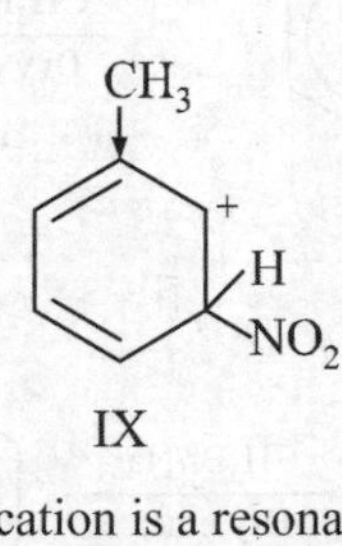

VII VIII IX *Meta attack*

We find that in each case, the intermediate carbocation is a resonance hybrid of three structures. In structures (I) and (V), the positive charge is located on the carbon atom to which electron releasing methyl group is attached. Therefore, the positive charge on such a carbon is highly dispersed and thus the corresponding structures (I) and (V) are more stable than all other structures. No such structure is, however, possible in case of meta attck. Hence the resonance hybrid carbocations resulting from ortho and para attack are more stable than the ones formed by attack at the meta position. Therefore, further electrophilic substitution in toluene occurs faster at ortho and para position than at meta position.

ALKENYL BENZENE AND ALKYNYL BENZENE

6.18 PREPARATION AND PROPERTIES OF STYRENE

(Phenylethylene, Vinyl benzene), $C_6H_5—CH=CH_2$

Methods of Preparation

1. By heating phenyl methyl methonal with sulphuric acid

Phenylmethyl methanol is obtained by the reaction between phenyl magnesium bromide and ethanal and subsequent hydrolysis.

$$C_6H_5MgBr + CH_3CHO \longrightarrow C_6H_5\underset{\displaystyle CH_3}{\underset{|}{C}}HOMgBr \longrightarrow \underset{\text{Phenylmethyl methanol}}{C_6H_5CHOHCH_3}$$

$$C_6H_5CHOHCH_3 \xrightarrow[\Delta,\ -H_2O]{H_2SO_4} \underset{\text{styrene}}{C_6H_5CH=CH_2}$$

2. By heating 2-phenylethanol with alkali

$$\underset{\text{2-Phenylethanol}}{C_6H_5CH_2CH_2OH} \xrightarrow{NaOH} C_6H_5CH = CH_2 + H_2O$$

3. By heating cinnamic acid in the presence of a small quantity of quinol

$$\underset{\text{Cinnamic acid}}{C_6H_5CH=CHCOOH} \xrightarrow{\Delta} C_6H_5CH = CH_2 + CO_2$$

Quinol prevents the polymerisation of styrene

4. Styrene is obtained on industrial scale by dehydrogenating ethylbenzene in the presence of ZnO

$$\underset{\text{Ethyl benzene}}{C_6H_5CH_2CH_3} \xrightarrow[600°C]{ZnO} C_6H_5CH = CH_2 + H_2$$

Properties

1. It adds a molecule of bromine to from dibromide

$$C_6H_5CH = CH_2 + Br_2 \longrightarrow \underset{\text{styrene dibromide}}{C_6H_5CHBrCH_2Br}$$

2. It polymerises when exposed to sunlight or in the presence of sodium

$$n\,C_6H_5CH = CH_2 \xrightarrow{\text{Sunlight}} (C_8H_8)_n$$

3. Substitution in the side chain gives two possibilities

$$C_6H_5CH = CH_2 \xrightarrow{Cl_2} \underset{\beta\text{-chlorostyrene}}{C_6H_5CH = CHCl}$$

$$C_6H_5CH = CH_2 \xrightarrow{Cl_2} \underset{\alpha\text{-chlorostyrene}}{C_5H_5CCl = CH_2}$$

6.19 METHODS OF PREPARATION AND PROPERTIES OF PHENYLACETYLENE, $C_6H_5C \equiv CH$

Methods of Preparation

1. By decarboxylation of phenyl propiolic acid

$$\underset{\text{Propiolic acid}}{C_6H_5C \equiv C\,COOH} \xrightarrow{\Delta} \underset{\text{Phenyl acetylene}}{C_6H_5C \equiv CH} + CO_2$$

2. *By heating ω-bromostyrene with ethanolic potassium hydroxide*

$$\underset{\omega\text{-bromostyrene}}{C_6H_5CH = CH\,Br} \xrightarrow{\text{Alc. KOH}} C_6H_5\,C \equiv CH + HBr$$

Properties

1. It forms metallic compounds with alkali metals

$$C_6H_5C \equiv CH + Na \longrightarrow \underset{\text{Sodium phenyl acetylide}}{C_6H_5\,C \equiv C\,Na} + \frac{1}{2}H_2$$

2. It is reduced to styrene in the presence of zinc and acetic acid

$$C_6H_5C \equiv CH + 2[H] \xrightarrow{\text{Zn / Acetic acid}} \underset{\text{Styrene}}{C_6H_5\,CH = CH_2}$$

3. It adds a water molecule in the presence of sulphuric acid to form acetone

$$C_6H_5C \equiv CH + H_2O \xrightarrow{H_2SO_4} \underset{\text{unstable}}{C_6H_5\overset{OH}{\overset{|}{C}} = CH_2} \longrightarrow \underset{\text{acetophenone}}{C_6H_5\,COCH_3}$$

MISCELLANEOUS EXAMPLES

Example 11. **Identify the compounds A, B and C.**

$$\textbf{Toluene} \xrightarrow{KMnO_4} \textbf{A} \xrightarrow[H^+]{MeOH} \textbf{B} \xrightarrow[H_2SO_4]{HNO_3} \textbf{C}$$

Solution.

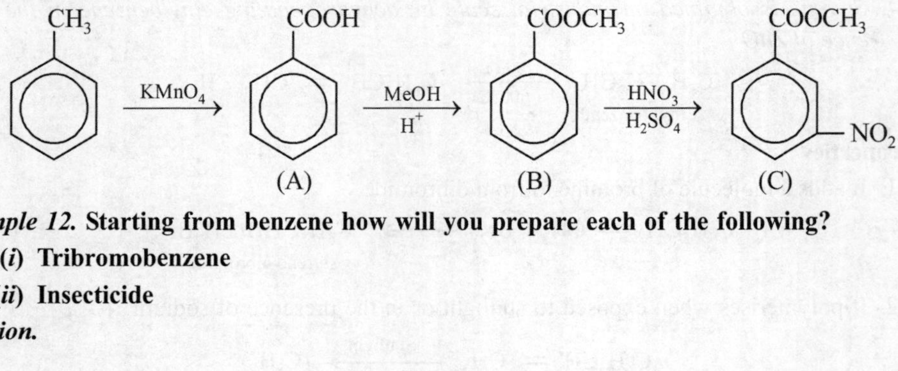

(A) (B) (C)

Example 12. **Starting from benzene how will you prepare each of the following?**

(*i*) Tribromobenzene

(*ii*) Insecticide

Solution.

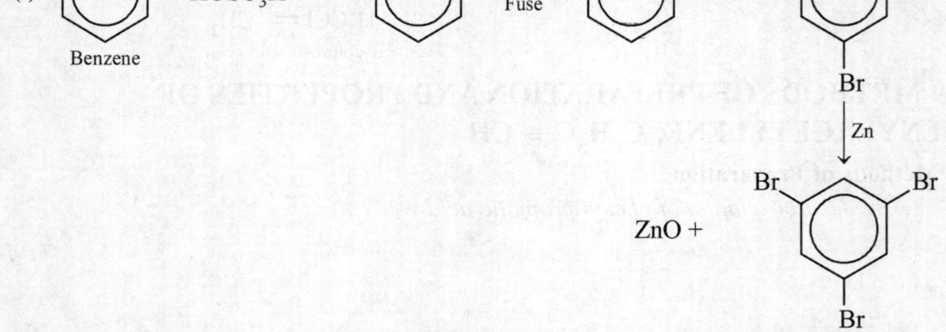

Tribromobenzene

(ii) 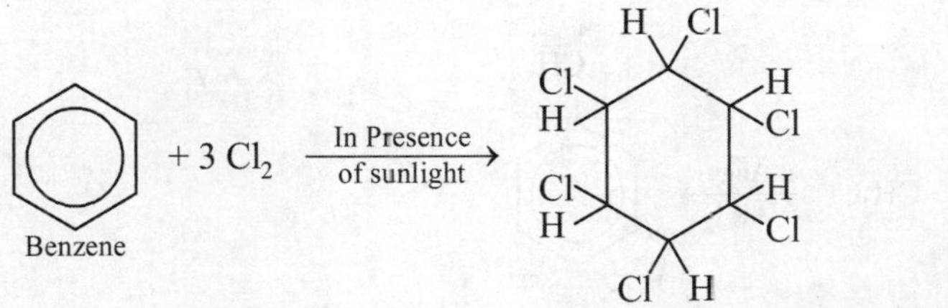

benzene hexachloride (insecticide)

Example 13. **Describe mechanism involved in the follwoing reaction :**

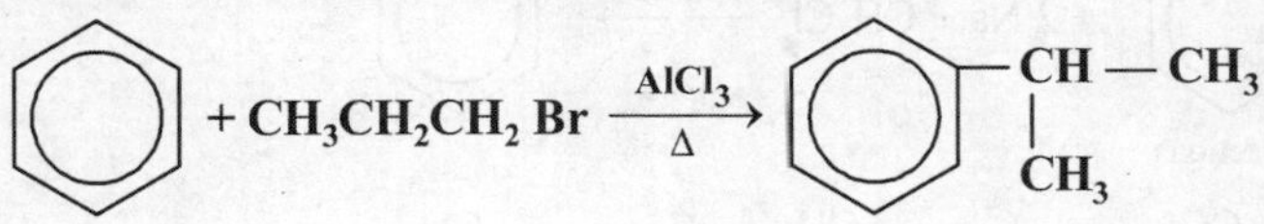

Solution.

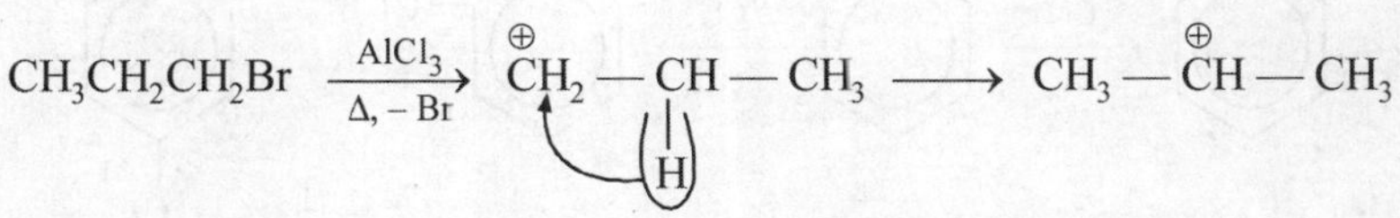

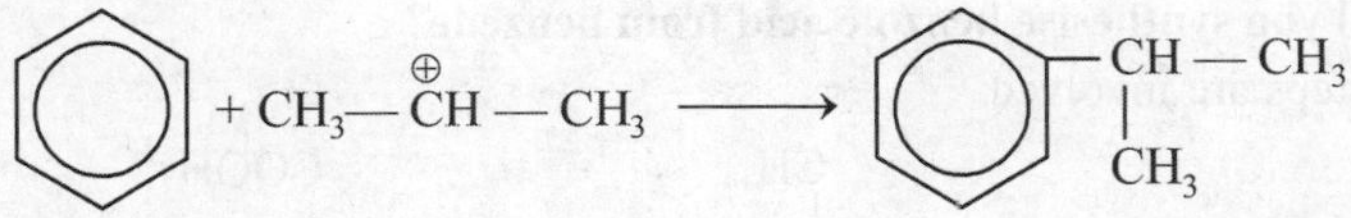

Example 14. **When chlorine is passed into boiling toluene three different chlorinated products are obtained. What products will be obtained if each of these chlorinated products is separately treated with aqueous sodium hydroxide solution?**

Solution. When chlorine is passed in to boiling toluene chlorine displaces the hydrogen atoms of the side chain *i.e.*, methyl group, one by one.

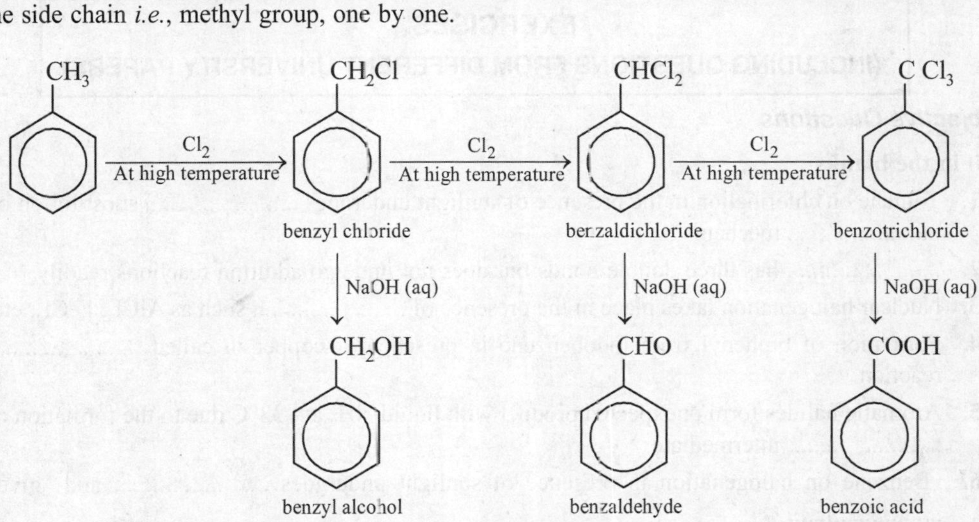

Example 15. **Give the preparation of toluene from (*i*) benzene (*ii*) chlorobenzene. How does toluene react with (*i*) Cl_2 in presence of $FeCl_3$ (*ii*) Cl_2 in sunlight?**

Solution

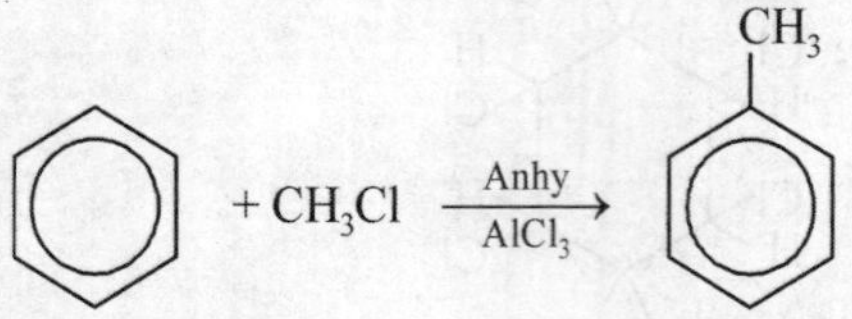

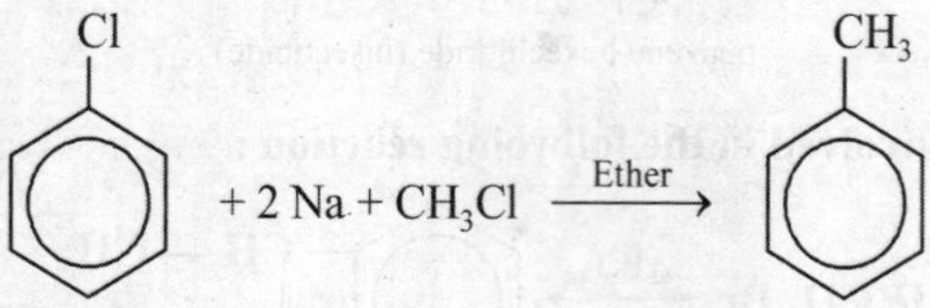

Reactions

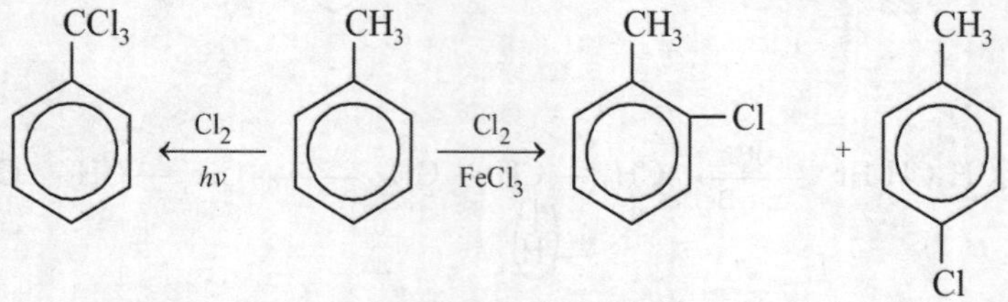

Example 16. **How will you synthesise benzoic acid from benzene?**

Solution. Following steps are involved

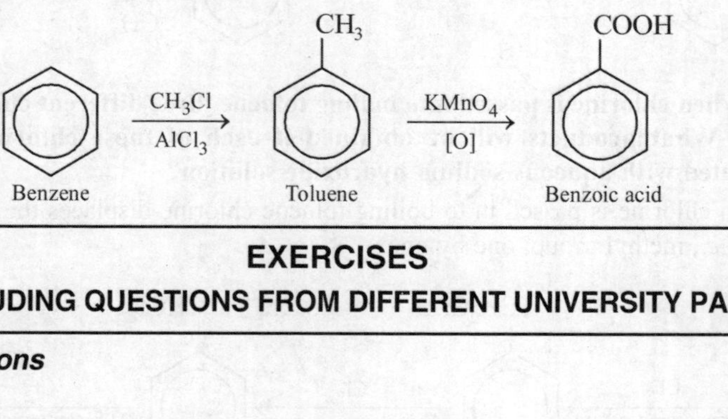

Benzene Toluene Benzoic acid

EXERCISES
(INCLUDING QUESTIONS FROM DIFFERENT UNIVERSITY PAPERS)

Objective Questions

Fill in the blanks

1. Toluene on chlorination in the presence of sunlight undergoes .... substitution by .... mechanism.

2. .... has three double bonds but does not undergo addition reactions readily.

3. Nuclear halogenation takes place in the presence of .... such as $AlCl_3$, $FeCl_3$, etc.

4. Formation of biphenyl from iodobenzene in presence of copper in called .... reaction

5. Aromatic halides form unexpected product with liquid NH_3 at –33°C due to the formation of .... intermediate

6. Benzene on halogenation in presence of sunlight undergoes .... and gives ....

7. Side chain substitution takes place through ....

8. – Cl, – Br, – I, – F are $o - p$ directing but .... groups

9. In Friedel Crafts alkylation reaction, the electrophile is a .... ion
10. An aromatic molecule must obey .... rule and must be ....

Multiple Choice Questions

Choose the correct option

1. In sulphonation of benzene, the attacking species is
(*a*) H^+ (*b*) SO_2 (*c*) SO_3 (*d*) HSO^-_4
2. Benzene reacts with H_2 at 150° C at 30 atm in the presence of Ni catalyst to give
(*a*) Cyclohexene (*b*) Cyclohexane (*c*) *n*-Hexane (*d*) No reaction occurs
3. Benzene reacts with benzoyl chloride in the presence of anhydrous aluminium chloride to form
(*a*) benzyl chloride (*b*) benzaldehyde (*c*) benzal chloride (*d*) benzophenone
4. Benzene reacts with propene in the presence of H_2SO_4 catalyst to give
(*a*) *n*-Propylbenzene (*b*) benzophenone (*c*) cumene (*d*) nothing happens
5. Gammexane is
(*a*) hexachloroethane (*b*) DDT
(*c*) hexachlorocyclohexane (*d*) TNT
6. Benzene undergoes Friedel-Crafts reaction with isopropyl alcohol in the presence of H_2SO_4 catalyst to give
(*a*) *n*-Propylbenezene (*b*) benzophenone
(*c*) isopropylbenzene (*d*) nothing happens
7. Benzene reacts with chlorine in the presence of $FeCl_3$ catalyst to form
(*a*) hexachlorobenzene (*b*) chlorobenzene
(*c*) hexachlorocyclohexane (*d*) benzyl chloride
8. Consider the following reaction:

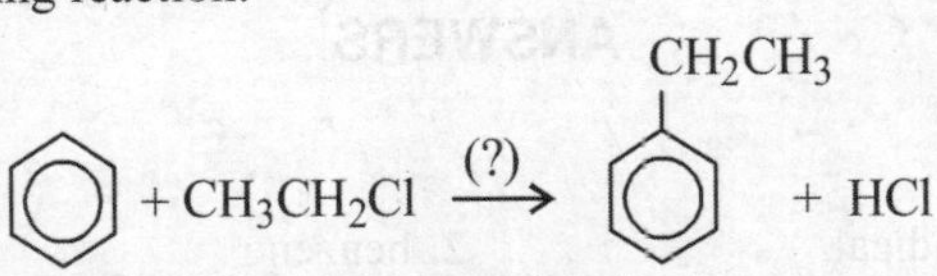

The catalyst used to complete the above reaction is
(*a*) $LiAlH_4$ (*b*) $AlCl_3$ (*c*) Na (*d*) KOH
9. Ozonolysis of benzene gives
(*a*) formic acid (*b*) glyoxal (*c*) formaldehyde (*d*) glycine
10. Benzene reacts with acetic anhydride in the presence of $AlCl_3$ to form
(*a*) acetophenone (*b*) benzophenone (*c*) phenylacetic acid (*d*) phenyl acetate
11. Which of the following compounds uses only sp^2 hybridised carbon atoms for bond formation?
(*a*) 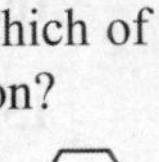 (*b*) 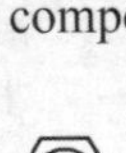 (*c*) 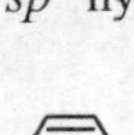 (*d*) 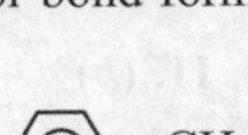

12. Which of the following compounds is aromatic?
(*a*) 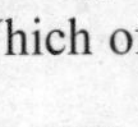 (*b*) 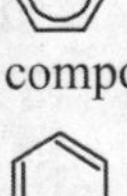

(*c*) 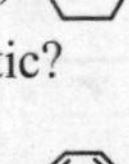 (*d*) 
13. Benzene undergoes substitation reaction more readily then addition reaction because

(*a*) it has a cyclic structure (*b*) it has three double bonds
(*c*) it has six hydrogen atoms (*d*) there is delocalisation of electrons

14. Which of the following is used in order to make benzene react with concentrated nitric acid to give nitrobenzene?
(*a*) Concentrated H_2SO_4 (*b*) $FeCl_3$, Catalyst
(*c*) Lindlar's catalyst (*d*) Ultraviolet light

15. Oxidation of toluene with chromyl chloride gives benzaldehyde. This reaction is known as
(*a*) Perkin's reaction (*b*) Benzoin condensation
(*c*) Etard's reaction (*d*) Ozonolysis

16. Which of the following compounds undergoes nitration most easily?
(*a*) Benzene (*b*) Toluene (*c*) Nitrobenzene (*d*) Benzoic acid

17. Which compounds undergoes substitution slower than benzene?

(*a*) 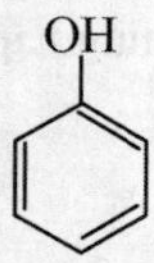

(*b*) 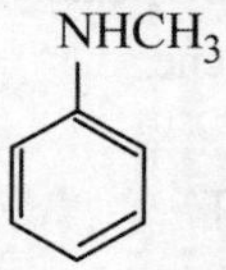

(*c*) 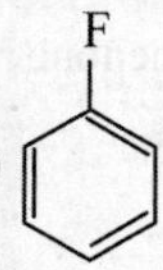

(*d*) 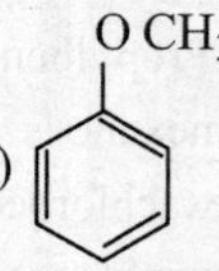

18. Which of the following will undergo substitution in the ***ortho*** and ***para*** positions rather than in the ***meta***, position?
(*a*) Nitrobenzene (*b*) Benzoic acid (*c*) Acetanilide (*d*) Benzaldehyde

19. Compared to benzene, nitration of toluene takes place at
(*a*) same rate (*b*) faster rate (*c*) slower rate (*d*) unpredictable

20. Toluene reacts with bromine in the presence of UV light to give.
(*a*) *m*-bromotoluene (*b*) benzyl bromide (*c*) *o*-bromotoluene (*d*) benzoyl bromide

ANSWERS

Fill in the blanks

1. Side chain, free radical **2.** benzene **3.** Lewis acid
4. Ullman reaction **5.** benzyne
6. Free radical addition, gammaxene **7.** free radical **8.** deactivating
9. carbonium **10.** $(4n + 2)\ \pi$, planar

Multiple Choice Questions

1. (*c*)	**2.** (*a*)	**3.** (*d*)	**4.** (*c*)	**5.** (*c*)
6. (*c*)	**7.** (*b*)	**8.** (*b*)	**9.** (*b*)	**10.** (*a*)
11. (*b*)	**12.** (*b*)	**13.** (*d*)	**14.** (*a*)	**15.** (*c*)
16. (*b*)	**17.** (*c*)	**18.** (*c*)	**19.** (*b*)	**20.** (*b*)

General Questions

1. Discuss the orbital structure of benzene.

2. Define Huckel's rule. What do you understand by the term aromaticity? Give examples.

3. Which of the following have aromatic character and why?

(i) 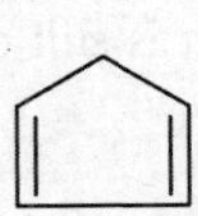 (ii) 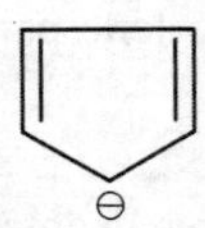 (iii) 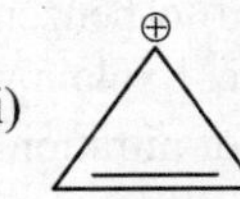 (iv) 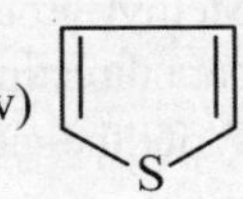

4. State Huckel's rule. Illustrate with suitable examples.
5. What is the role of catalyst in the electrophilic substitution reactions in benzene?
6. State Huckel's rule and explain the stability of :
   (*i*) tropyllium cation (*ii*) Cyclopentadienyl anion.
7. Give the mechanism of Friedel Crafts acylation of benzene.
8. What is Etard's reaction?
9. Justify that the carboxyl group attached to the benzene ring is meta directing for electrophilic substitution.
10. What happens when :
    (*i*) Toluene is sulphonated?
    (*ii*) *o*-xylene is treated with alkaline $KMnO_4$?
11. Describe the Friedel-Crafts acylation. Give its mechanism.
12. Complete the following :

(*i*) 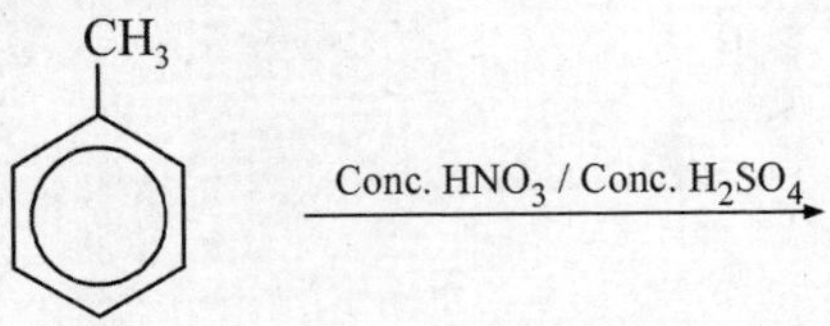

(*ii*) 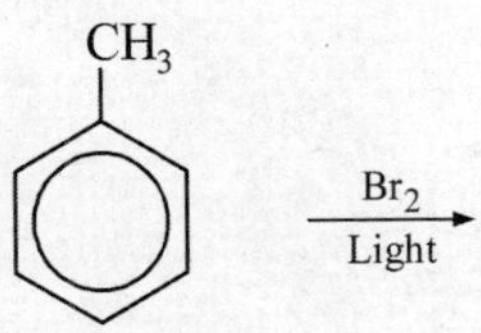

13. Define the terms: Aromatic, Non-Aromatic and Anti-Aromatic compounds and classify the following into these:
    (*i*) Cyclopropenyl anion (*ii*) Naphthalene (*iii*) Cyclooctatetraene.
14. Although benzene contains three double bonds, it resists addition reactions. Explain.
15. Write a short note on annulenes.
16. Discuss the mechanism of the efollowing reaction:

$$C_6H_6 + (CH_3CO)_2O \xrightarrow[AlCl_3]{Anhydrous} C_6H_5COCH_3 - CH_3COOH$$

17. Explain meta directing nature of nitro group.
18. Discuss the mechanism of electrophilic substitution in benzene.
19. Halogens are electron withdrawing in nature but they are ortho and para directing in benzene. Explain.
20. Give the mechanism of sulphonation of benzene.
21. How will you explain that the unsaturated character of benzene is somewhat different from olefins?
22. What happens when benzene is heated with V_2O_5 in presence of air?

23. Methyl group attached to benzene ring has no lone pair of electrons but is still ortho and para directing in nature. Explain why.

24. Write the mechanism of nitration of benzene.

25. What are non-benzenoid compounds? Give three examples along with their structures.

26. Chlorine deactivates the benzene ring yet it is ortho and para directing in nature. Explain.

27. Arrange the following compounds in decreasing order of reactivity towards nitration :
(*i*) Benzene (*ii*) Toluene (*iii*) Nitrobenzene.

28. What is aromaticity ?

29. What is the cause of aromatic character of benzene in terms of Huckel's rule?

30. Explain why methyl group in toluene is ortho and para directing in nature.

31. Give the mechanism of nitration of benzene.

32. What happens when (Give chemical equations)
(*i*) Chlorine is passed through boiling toluene in the presence of sunlight.
(*ii*) Chlorine is passed through toluene in the presence of $FeCl_3$ and absence of sunlight.

33. Explain why – NH_2 group when attached to benzene ring is ortho and para directing in nature.

34. (*a*) Complete the following reactions :

(*i*)
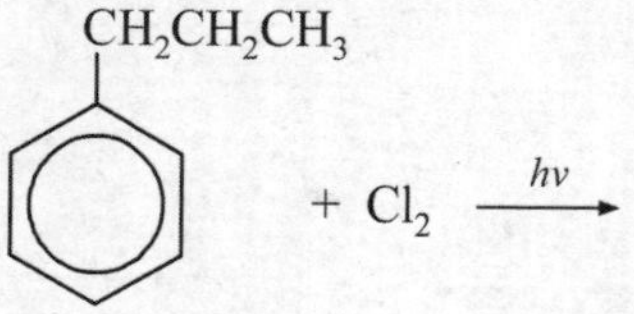

(*ii*)
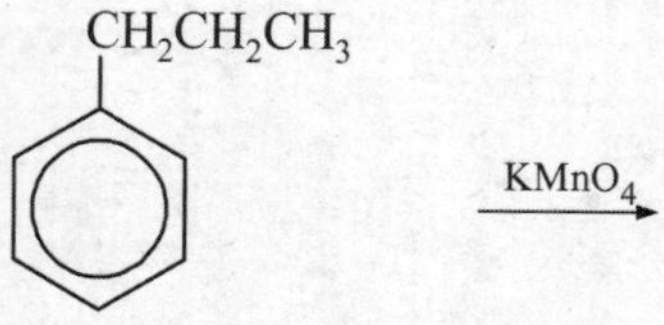

(*iii*)
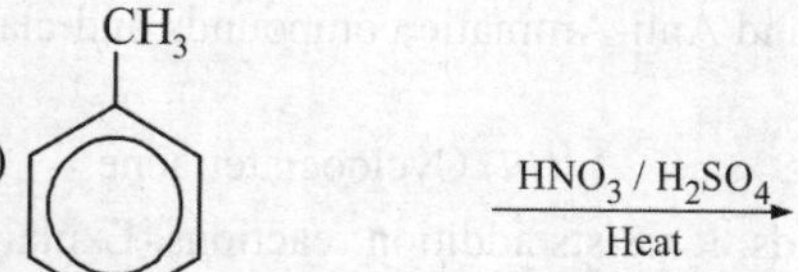

35. (*a*) How will you account for the fact that an – OH group in the benzene ring is ortho and para directing while – CHO group is meta directing in nature?
(*b*) Out of toluene and nitrobenzene, which will be nitrated more easily?

36. (*a*) Chlorine atom is electron withdrawing but still ortho and para directing. Why?
(*b*) Why is it easier to introduce a nitro group at ortho and para positions in phenol than in toluene?
(*c*) What happens when :
(*i*) o-Xylene is heated with hot $KMnO_4$ solution?
(*ii*) Chlorine is passed in boiling toluene?
(*iii*) Toluene is subjected to oxidation by chromyl chloride?

37. (*a*) Explain the stability of benzene on the basis of orbital structure.
(b) Predict the products of the following reactions :

(*i*) Nitration of chlorobenzene (*ii*) Methylation of toluene
(*iii*) Chlorination of toluene (*iv*) Chlorination of nitrobenzene.

(*c*) Halogens are electron withdrawing and yet they are ortho and para directing. Explain

38. Assign suitable reasons for the following :

(*i*) Halogens are ortho and para directing though they are deactivating in nature.

(*ii*) Methyl group has no lone pair of electrons, yet it is ortho and para directing in nature.

(*iii*) Nitration of toluence can take place more easily as compared to benzene.

(*iv*) In case of ortho and para disubstitution, the para isomer generally dominates.

39. What do you understand by the term aromaticity? How is it related with Huckel's rule? Give examples.

40. Comment on the statement
"In case of halogens, the reactivity is determined by stronger inductive effect while the orientation is largely controlled by resonance effect".

41. (*a*) How will you prepare the following compounds from benzene?

(*i*) Acetophenone (*ii*) Toluene
(*iii*) Chlorobenzene (*iv*) Benzene hexachloride.

(*b*) Explain the mechanism of nitration of benzene.

(*c*) Explain why nitration of chlorobenzene gives ortho and para chloronitrobenzene but the chlorination of nitrobenzene gives meta chloronitrobenzene.

42. Predict the major products in the following reactions :

(*i*)
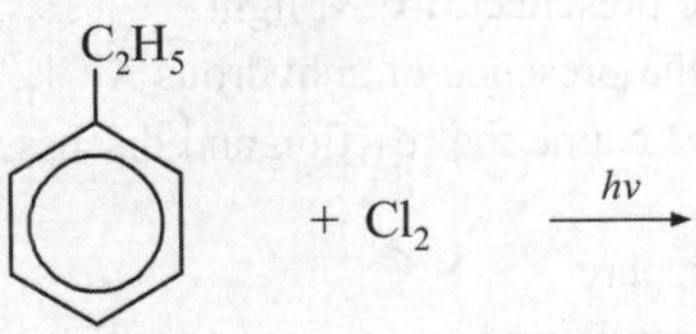

(*ii*)
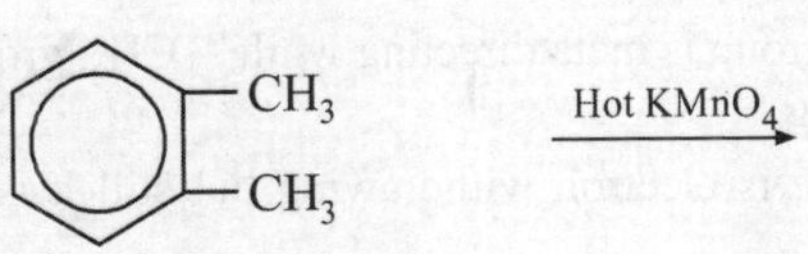

(*iii*)
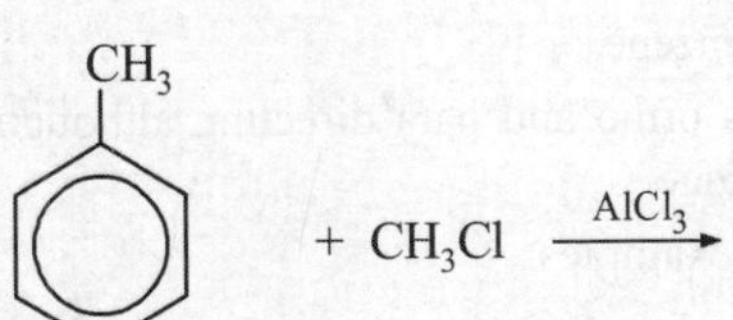

43. Predict the products of the following reactions :

(*i*)
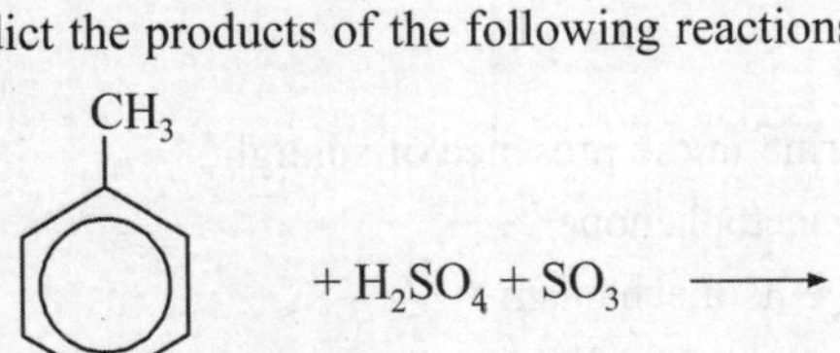

(*ii*) 

(*iii*) 

44. Discuss the orientation of electrophilic substitution in nitrobenzene.

45. What is Huckel's rule of aromaticity ? Which of the following exhibit aromatic character?
(*i*) Cyclooctatetraene (*ii*) Cyclopentadiene (*iii*) Tropyllium cation.

46. Discuss the mechanism of the electrophilic substitution for the following reactions in benzene :
(*i*) Halogenation (*ii*) Sulphonation

47. Give the mechanism for ortho and para directing influence of groups in benzenoids.

48. (*a*) Define aromaticity and state Huckel's rule. Will Cyclooctatetraene show aromatic character?
(*b*) Explain why nitration of toluene is easier than that of benzene.
(*c*) What happens when :
(*i*) o-xylene is treated with hot $KMnO_4$ solution.
(*ii*) Ethyl benzene is treated with chlorine in the presence of U.V. light.
(*iii*) Toluene is treated with methyl chloride in the presence of anhydrous $AlCl_3$.

49. (*a*) How is benzene converted to acetophenone ? Name the reaction and discuss the mechanism explaining each step.
(*b*) Starting from nitrobenzene, how will you prepare
(*i*) *p*-nitrotoluene (*ii*) Isopropylbenzene

50. (*a*) Explain with mechanism that – SO_3H group is meta directing while – CH_3, group is ortho and para directing when present in the ring.
(*b*) Due to high electronegativity, chlorine is electron withdrawing but still it is ortho and para directing. Explain.

51. (*a*) Give the mechanism of sulphonation of benzene.
(*b*) Explain why chlorine in chlorobenzene is ortho and para directing although it has – I effect during electrophilic substitution in benzene.
(*c*) What is Huckel's rule? Illustrate with two examples.

52. What is Friedel-Crafts reaction ? Discuss the mechanism of acylation of benzene. Give evidence in support of the mechanism.

53. (*a*) Write a note on Huckel's rule.
(*b*) Give a reaction of toluene with chlorine in the presence of sunlight.
(*c*) How can benzene be converted into acetophenone ?

54. (*a*) Discuss Friedel-Crafts acylation. Give its mechanism.
(*b*) Discuss the stability of benzene with the help of resonance theory.
(*c*) Discuss how benzene is converted into benzene sulphonic acid. Explain its mechanism.
(*d*) Write a short note on annulenes.

55. (*a*) What are electrophilic aromatic substitution reactions? Discuss the mechanism of chlorination of benzene. What is the role of catalyst used ?

(*b*) Why does side halogenation in alkyl benzenes take place preferentially at position alpha to the aromatic ring.

56. (*a*) Give the mechanism of reaction of benzene with $CH_3-\overset{\overset{\displaystyle O}{\|}}{C}-Cl$ in the presence of $AlCl_3$.

(*b*) State Huckel's rule. Explain the aromatic character in benzene, [18] annulene and furan.

(*c*) Halogens are ortho and para directing though they are deactivating in nature. Comment.

(*d*) Nitration of toluene can take place more readily as compared to benzene. Comment.

57. (*a*) Give the mechanism and the synthetic applications of Friedel-Crafts reaction.

(*b*) What is Huckel's rule? Briefly discuss aromatic and anti-aromatic compounds. How do they differ from benzenoids?

58. How is benzene converted to nitrobenzene? Give mechanism.

59. (*a*) Give the mechanism of sulphonation of benzene.

(*b*) Discuss the mechanism of Friedel-Crafts reaction.

60. What happens when :

(*i*) *o*-xylene is heated with $KMnO_4$ solution.

(*ii*) Chlorine is passed through boiling toluene.

(*iii*) Toluene is subjected to oxidation with chromyl chloride.

61. What are electrophilic substitution reactions? Discuss the mechanism of sulphonation of benzene and give evidence in support of the mechanism.

62. (*a*) Define annulenes. Comment upon the aromatic nature of the following annulenes :
[4] Annulene, [10] Annulene, [14] Annulene

(*b*) State and explain Huckel's rule.

(*c*) Predict the main product and discuss the mechanism :

(*i*)  $+ H_2SO_4 \xrightarrow{\text{Heat}}$

(*ii*)  $+ CH_3CH=CH_2 \xrightarrow{H_2SO_4}$

63. (*a*) What is aromaticity?

(*b*) Explain the mechanism of nitration of toluene.

(*c*) Complete the following reactions :

 $+ CH_3COCl \xrightarrow{\text{Anhyd. } AlCl_3}$

 $+ HNO_3 \xrightarrow{H_2SO_4}$

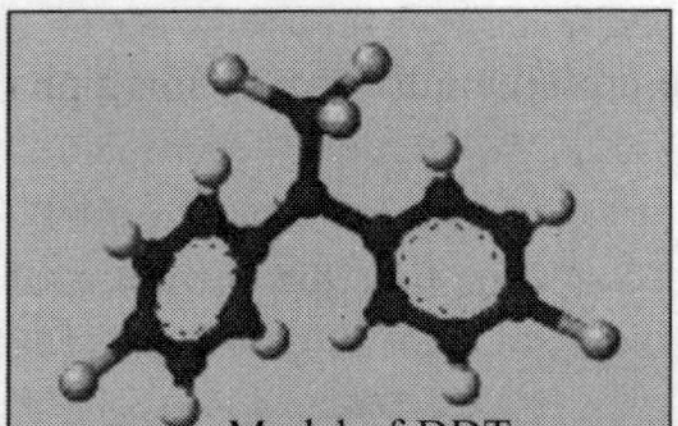

Model of DDT

7

Alkyl and Aryl Halides

7.1 CLASSIFICATION OF ALIPHATIC HALOGEN COMPOUNDS

Aliphatic halogen compounds can be classified as follows:

(*a*) Monohalogen compounds. Compounds such as methyl chloride, which contain one halogen in the molecule are called monohalogen compounds.

(*b*) Dihalogen compounds. Compounds such as ethylene dichloride which contain two halogen atoms in the molecule are called dihalogen compounds. These can be further classified into two types:

(*i*) Vicinal dihalides. Compounds containing two halogens on neighbouring carbon atoms are called *vicinal* dihalides.

(*ii*) Geminal dihalides. Compounds containing two halogens on the same carbon in the molecule are called *geminal* dihalides. Vicinal (or vic.) and geminal (or gem.) dihalides are discussed in detail later in this chapter.

(*c*) Trihalogen compounds. Compounds such as chloroform or iodoform which contain three halogens in the molecule are called trihalogen compounds.

(*d*) Tetrahalogen compounds. Compounds such as carbon tetrachloride, which contain four halogen atoms in the molecule are called tetrahalogen compounds.

Monohalogen compounds are further classified into primary, secondary and tertiary alkyl halides.

Primary alkyl halides are those compounds in which the halogen is linked to a carbon which is further linked to no carbon or one carbon atom. For example, $CH_3 — CH_2 — Cl$.

Secondary alkyl halides are those compounds in which the halogen is linked to a carbon which is further linked to two carbon atoms. For example,

$$\begin{array}{c} CH_3 — \underset{\displaystyle Cl}{\underset{|}{CH}} — CH_3 \end{array}$$

Tertiary alkyl halides are those compounds in which the halogen is linked to a carbon which is further linked to three carbon atoms. For example,

$$\begin{array}{c} CH_3 — \overset{\displaystyle CH_3}{\overset{|}{\underset{\displaystyle Cl}{\underset{|}{C}}}} — CH_3 \end{array}$$

7.2 NOMENCLATURE OF MONOHALOGEN ALIPHATIC COMPOUNDS

Nomenclature of monohalogen compounds is explained as under by taking a few compounds.

Structural formula	Common name	IUPAC name
CH_3Cl	Methyl chloride	Chloromethene
CH_3-CH_2Br	Ethyl bromide	Bromoethane
$CH_3-CH_2-CH_2I$	*n*-Propyl iodide	1-Iodopropane
$CH_3-CH(Cl)-CH_3$	*Iso*-propyl chloride	2-Chloropropane
$CH_3-CH_2-CH_2-CH_2-Br$	*n*-Butyl bromide	1-Bromobutane
$CH_3-CH_2-CH(Br)-CH_3$	*Sec*-butyl bromide	2-Bromobutane
$CH_3-CH(CH_3)-CH_2-Br$	*Iso*-butyl bromide	1-Bromo-2-methylpropane

Example 1. **Write all the possible structures for a compounds having molecular formula $C_5H_{11}Cl$. Name these according to IUPAC system. How many are optically active?**

Solution.

Structural formula	IUPAC Name
(*a*) $CH_3-CH_2-CH_2-CH_2-CH_2-Cl$	1-Chloropentane
(*b*) $CH_3-CH_2-CH_2-CH(Cl)-CH_3$	2-Chloropentane
(*c*) $CH_3-CH_2-CH(Cl)-CH_2-CH_3$	3-Chloropentane
(*d*) $CH_3-CH_2-CH(CH_3)-CH_2-Cl$	1-Chloro-2-methylbutane
(*e*) $CH_3-CH_2-C(CH_3)(Cl)-CH_3$	2-Chloro-2-methylbutane
(*f*) $CH_3-CH(Cl)-CH(CH_3)-CH_3$	2-Chloro-3-methylbutane
(*g*) $CH_3-CH(CH_3)-CH_2-CH_2-Cl$	1-Chloro-3-methylbutane
(*h*) $CH_3-C(CH_3)_2-CH_2-Cl$	1-Chloro-2, 2-dimethylpropane

Optically active forms. The isomers (*b*), (*d*) and (*f*) are optically active due to the presence of **chirality** in their structures.

7.3 METHODS OF PREPARATION OF ALKYL HALIDES

Alkyl halides can be prepared by the following methods:

1. Halogenation of Alkanes Alkanes react with halogens in the presence of UV light or by heating at high temperature (400°C) to give alkyl halide along with other polyhalogen compounds

$$CH_4 \xrightarrow[\text{UV light}]{Cl_2} CH_3Cl + CH_2Cl_2 + CHCl_3 + CCl_4$$

2. Addition of Halogen Acids to Alkenes. Halogen acids add to alkenes to produce alkyl halides. Addition takes place according to Markownikof rule. However, in the presence of peroxide, HBr adds contrary to the rule.

$$\underset{\text{Ethylene}}{CH_2{=}CH_2} + HI \longrightarrow \underset{\text{Ethyl codide}}{CH_3{-}CH_2{-}I}$$

$$CH_3{-}CH{=}CH_2 + HBr \longrightarrow \underset{\text{2-Bromopropane}}{CH_3CHBrCH_3}$$

$$CH_3{-}CH{=}CH_2 + HBr \xrightarrow{\text{Peroxide}} \underset{\text{1-Bromopropane}}{CH_3CH_2CH_2Br}$$

3. Action of Halogen Acids on Alcohols. Alcohols react with HBr or HI to produce alkyl bromide or alkyl iodide. Alkyl chlorides are obtained by treating alkenes with dry HCl in presence of zinc chloride.

$$CH_3CH_2OH + HCl \xrightarrow{ZnCl_2} \underset{\text{Ethyl Chloride}}{CH_3CH_2Cl} + H_2O$$

$$CH_3CH_2OH + HBr \longrightarrow \underset{\text{Ethyl bromide}}{CH_3CH_2Br} + H_2O$$

4. Action of phosphorus Halides on Alcohols. Alcohols react with PX_5 or PX_3 to give alkyl halides.

$$2CH_3CH_2OH + PCl_5 \longrightarrow 2CH_3CH_2Cl + POCl_3 + H_2O$$

$$3CH_3CH_2OH + PBr_3 \longrightarrow 3CH_3CH_2Br + H_3PO_3$$

PBr_3 or PI_3 is produced by the action of red phosphorus on Br_2 or I_2

5. Action of Thionyl Chloride on Alcohols. Alcohols react with thionyl chloride $SOCl_2$ in presence of pyridine to form alkyl halides. The function of pyridine is to absorb HCl as it is formed.

$$CH_3CH_2OH + SOCl_2 \xrightarrow{\text{Pyridine}} CH_3CH_2Cl + SO_2 + HCl$$

6. Halogen Exchange Reaction. Alkyl chloride or bromide on heating with a solution of sodium iodide in acetone gives alkyl iodide

$$\underset{\text{Ethyl Chloride}}{CH_3Cl} + NaI \xrightarrow[\Delta]{\text{Acetone}} \underset{\text{Ethyl Iodide}}{CH_3CH_2I} + NaCl$$

7.4 PHYSICAL PROPERTIES OF ALKYL HALIDES

1. Alkyl halides containing one carbon atom are gases at room temperature. Other alkyl halides upto C_{18} are colourless liquid while those beyond C_{18} are colourless solids.
2. Alkyl halides are soluble in organic solvents but insoluble in water. There is no possibility of hydrogen-bonding in alkyl halide.

3. Alkyl bromides and iodides are heavier than water while chloride and fluorides are lighter than water.
4. For a given halogen atom, the boiling point increases with increase in the size of the alkyl group.
5. For a given alkyl group, the boiling point increases in the order.
   RI > RBr > RCl > RF
6. In the case of isomeric halides, the boiling point decreases with branching. This is because surface area decreases with branching and the intermolecular forces of attraction decrease with surface area.

REACTIONS OF ALKYL HALIDES

7.5 NUCLEOPHILIC SUBSTITUTION REACTIONS

The reaction involving the replacement of an atom cr a group of atoms by a suitable nucleophile is called nucleophilic substitution reaction. The carbon-halogen bond in alkyl halides is polar due to greater electronegativity of the halogen as compared to carbon. It is depicted as below:

$$-\overset{|}{\underset{|}{C}}^{\delta^+}\text{—}X^{\delta^-}$$

The presence of partial positive charge on carbon atom makes it susceptible to nucleophilic attack. Thus, when a stronger nucleophile approaches the positively charged carbon of the alkyl halide, the halogen is readily displaced as halide ion as shown below:

$$-\overset{|}{\underset{|}{C}}^{\delta^+}\text{—}X^{\delta^-} + \underset{\text{Attacking nucleophile}}{:Nu^-} \longrightarrow -\overset{|}{\underset{|}{C}}\text{—}Nu + \underset{\text{Halide ion}}{X^-:}$$

Nucleophilic Substitutions of Alkyl halides

A few important nucleophilic substitution reactions are discussed below:

(*i*) Replacement by hydroxyl group (Formation of alcohols). Alkyl halides react with aqueous KOH or moist silver oxide, *i.e.*, AgOH to form alcohols.

$$RX + :OH^- \longrightarrow ROH + :X^-$$

$$\underset{\text{Ethyl bromide}}{C_2H_5Br} + KOH\,(aq) \longrightarrow \underset{\text{Ethyl alcohol}}{C_2H_5OH} + KBr$$

(*ii*) Replacement by cyano group (Formation of cyanides). Alkyl halides react with alcoholic potassium cyanide to form cyanides or nitriles.

$$RX + CN^-: \longrightarrow RCN + X^-:$$

$$C_2H_5Br + KCN \longrightarrow \underset{\substack{\text{Ethyl cyanide} \\ \text{(Propanenitrile)}}}{C_2H_5CN} + KBr$$

(*iii*) Replacement by alkoxy group (Formation of ethers). When treated with sodium alkoxide, alkyl halides yield ethers.

$$RX + :OR^- \longrightarrow R-O-R + :X^-$$

$$C_2H_5Br + \underset{\substack{\text{Sodium} \\ \text{ethoxide}}}{C_2H_5ONa} \longrightarrow \underset{\text{Diethyl ether}}{C_2H_5-O-C_2H_5} + NaBr$$

***(iv)* Replacement by alkynyl group (Formation of alkynes).** When an alkyl halide is treated with sodium salt of an alkyne, higher alkynes are formed.

$$RX + :\bar{C} \equiv CR' \longrightarrow R - C \equiv C - R' + \bar{X}:$$

$$\underset{\text{Sodium methyl acetylide}}{CH_3Br + NaC \equiv C - CH_3} \longrightarrow \underset{\text{2-Butyne}}{CH_3 - C \equiv C - CH_3} + NaBr$$

$$\underset{\text{Sodium acetylide}}{CH_3Br + NaC \equiv CH} \longrightarrow \underset{\text{Propyne}}{CH_3 - C \equiv CH} + NaBr$$

***(v)* Replacement by carboxylate group (Formation of a esters).** When heated with silver salts of carboxylic acids, alkyl halides give esters.

$$RX + R'CO\bar{O}: \longrightarrow R'COOR + \bar{X}:$$

$$C_2H_5Br + \underset{\text{Silver acetate}}{CH_3\,COO\,Ag} \longrightarrow \underset{\substack{\text{Ethyl acetate}\\ \text{(Ethyl ethanoate)}}}{CH_3COOC_2H_5} + AgBr$$

7.6 ELIMINATION REACTIONS

Reaction with Alcoholic KOH. Alkyl halides react with alcoholic potassium hydroxide to form alkenes

$$\underset{\text{1-Bromopropane}}{CH_3 - \underset{|}{\overset{H}{C}}H - \underset{|}{\overset{Br}{C}}H_2} + KOH \xrightarrow[\Delta]{\text{Ethanol}} \underset{\text{Propene}}{CH_3 - CH = CH_2} + KBr + H_2O$$

For mechanism of reaction, refer to chapter on alkenes.

If an alkyl halide can eliminate hydrogen halide in two different ways, alkene formed will be the one which is most highly substituted (Saytzeff Rule). For example,

$$\underset{\substack{\text{1-Butene}\\ \text{Less alkylated (20\%)}}}{CH_3 - CH_2 - CH = CH_2} \xleftarrow[\text{alc.}]{\text{KOH}} \underset{\text{2-Bromobutane}}{CH_3 - CH_2 - \overset{Br}{\overset{|}{C}}H - CH_3} \xrightarrow[\text{alc.}]{\text{KOH}} \underset{\substack{\text{2-Butene}\\ \text{More alkylated (80\%)}}}{CH_3 - CH = CH - CH_3}$$

It may be noted that aqueous solution of KOH causes substitution in alkyl halides while alcoholic solution causes elimination. In aqueous solution, KOH ionises to give OH^- ions which have a tendency to get hydrated. Hydrated ion being large in size is not in a position to abstract the β-hydrogen from alkyl halide and the elimination is checked. In alcoholic medium, the base gives OH^- ions to affect elimination to form an alkene.

7.7 ACTION WITH ACTIVE METALS

1. Action with Magnesium. Magnesium metal reacts with alkyl halide in the presence of dry ether to form organomagnesium halide known as Grignard reagents

$$C_2H_5Cl + Mg \xrightarrow{\text{Ether}} \underset{\text{Ethyl magnesium Chloride}}{C_2H_5\,Mg\,Cl}$$

2. Action with Sodium (Wurtz Reaction). Alkyl halides react with sodium metal in the presence of anhydrous ether to form higher alkanes.

$$C_2H_5\,Br + 2\,Na + Br\,C_2H_5 \xrightarrow[\text{Ether}]{\text{Dry}} \underset{\text{Butane}}{C_2H_5C_2H_5} + 2\,NaBr$$

3. Action with lead-sodium alloy. When heated with an alloy of lead and sodium, alkyl halides form tetraalkyl lead

$$4C_2H_5Br + 4\,Na/Pb \longrightarrow \underset{\text{Tetraethyl lead}}{(C_2H_5)_4Pb} + 4\,NaBr + 3\,Pb$$

Tetraethyl lead (T.E.L.) is used as anti-knocking agent in gasoline

Reduction Reactions of Alkyl Halides

1. With Zn/Cu couple and alcohol

$$C_2H_5Cl + 2[H] \xrightarrow[\text{Alcohol}]{\text{Zn / Cu}} \underset{\text{Ethane}}{C_2H_6} + HCl$$

2. With Zn and dil HCl

$$C_2H_5Cl + 2[H] \xrightarrow{\text{Zn / HCl}} \underset{\text{Ethane}}{C_2H_6} + HCl$$

Example 1. **Halogenation of alkanes in the presence of tetraethyl lead proceeds at low temperature than in its absence. Explain.**

Solution. Halogenation of alkanes is a free-radical reaction in which the reaction is started by halogen free-radical (say Cl•). Tetraethyl lead decomposes into ethyl redicals on slight heating as follows:

$$(C_2H_5)_4Pb \xrightarrow{\Delta} 4C_2H_5\bullet + Pb$$

Ethyl radicals combine with chlorine molecule to produce chlorine radicals

$$C_2H_5\bullet + Cl_2 \longrightarrow C_2H_5Cl + Cl\bullet$$

Chlorine radicals so produced, carry out halogenation of alkanes.

7.8 S_N1 AND S_N2 REACTIONS OF ALKYL HALIDES

S_N1 reaction means nucleophilic substitution reaction of first order.

In such reactions, there is only one species in the rate determining step. An example of S_N1 reaction is:

$$\underset{\text{Tert. butyl chloride}}{(CH_3)_3C - Cl} \xrightarrow{OH^-} \underset{\text{Tert. butyl alcohol}}{(CH_3)_3C - OH} + Cl^-$$

Here the rate-determining step of the reaction is:

$$(CH_3)_3C - Cl \longrightarrow (CH_3)_3\overset{+}{C} + Cl^-$$

It involves only one substance on the reactant side.

S_N2 reaction means nucleophilic substitution reaction of second order. In such reactions there are two species in the rate-determining step.

An example of $S_N{}^2$ reaction is:

$$CH_3Br + \bar{O}H \longrightarrow CH_3OH + Br^-$$

In this reaction the rate determining step of the reaction is:

$$CH_3Br + \bar{O}H \longrightarrow [HO —— CH_3 —— Br]^-$$

As it involves two species on the reactant's side, it is a bimolecular or S_N2 reaction.

7.9 S_N2 REACTION MECHANISM OF SUBSTITUTION IN ALKYL HALIDES

Hydrolysis of methyl bromide proceeds by S_N2, (meaning substitution nucleophilic reaction of the second order), mechanism. S_N2 reaction may also be termed as bimolecular substitution reaction. The term bimolecular implies that there are two reacting species in the rate determining step of the reaction.

A simple and typical example of S_N2 substitution is the hydrolysis of methyl bromide with aqueous sodium hydroxide.

$$CH_3Br + OH^- \longrightarrow CH_3OH + Br^-$$

The kinetic data reveal that the rate of the reaction is dependent upon the concentration of both these reactants

$$\text{Rate} = k[CH_3\,Br]\,[OH]^-$$

The reaction is of second order and thus it proceeds by a direct diplacement mechanism in which both the reactants are present in the rate determining step. It is assumed that the nucleophile attacks the side of the carbon atom opposite to that of bromine. This is referred to as **backside attack.** As a result of this attack, a transition state is formed in which carbon atom is partially bonded to both — OH and — Br group. In the transition state, the central carbon is sp^2 hybridised and the three hydrogens attached to it lie in the same plane with mutual bond angles of 120°. The reaction may be depicted as:

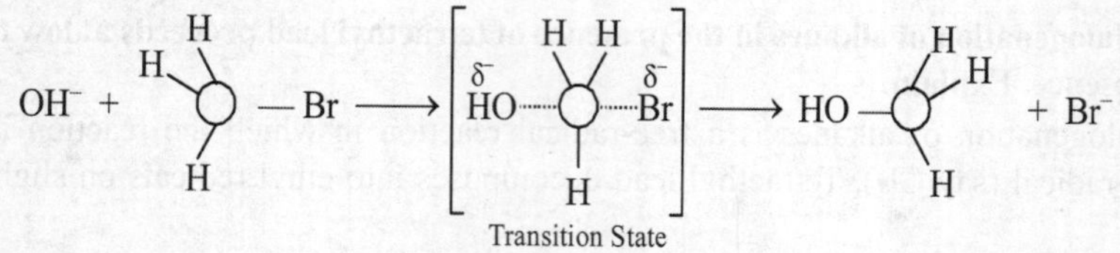

It may be noted that in the transition state, the hydroxide ion has diminished negative charge since it has started sharing its electrons with carbon. Similarly bromine develops a partial negative charge as it is tending to depart with the bonding electrons. When carbon-oxygen bond is completely formed, the carbon-bromine bond is altogether broken. The energy required for breaking a bond is compensated by the formation of a new bond. The overall reaction is thus a **concerted process** ocurring in one step through the intervention of a single transition state. In a single step, C—Br bond is broken and C — OH bond is formed resulting in the formation of alcohol molecule.

The potential energy diagram (energy profile diagram) for the above reaction is shown in Fig. 7.1.

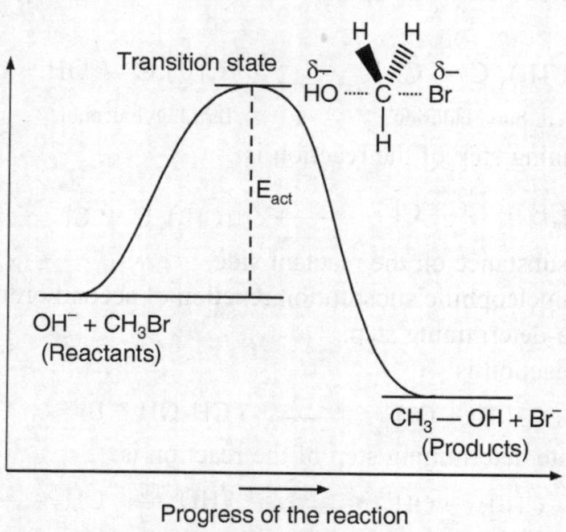

Fig. 7.1. Progress of S_N2 reaction

Stereochemistry of S_N2 reactions. In S_N2 reactions, it is assumed that the nucleophile (OH^- in the present case) attacks the side of the carbon opposite to that of the leaving group. As a result, the configuration of the resulting product is inverted. In other words S_N2 reaction proceeds with **stereochemical inversion.** This has been found to be so in actual practice. Thus when (–) 2-bromooctane is hydrolysed under S_N2 conditions, it gives exclusively (+) -2-octanol. The reaction may thus be represented as:

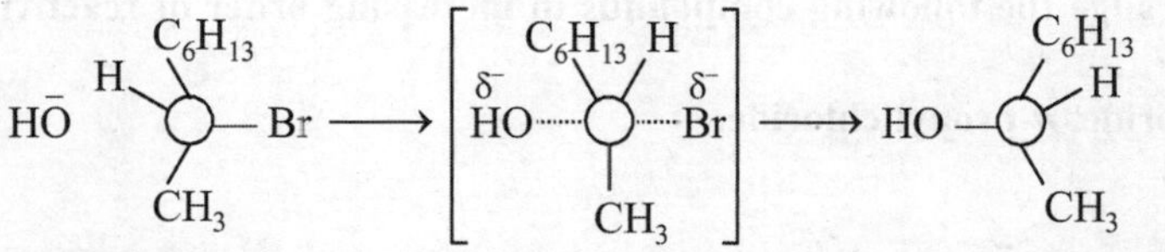

(–)-2-Bromooctane Transition State (Inversion in SN^2 reaction) (+)-2-Octanol

We have cited the example of (–)-2-bromooctane because it contains a chiral carbon atom (a carbon linked to four different groups). The configurations of the reactant and the products are different in such S_N2 reaction. We normally check the configuration by measuring the direction of rotation of plane polarised light although it is not necessary always that inversion of configuration should be accompanied by change in direction of rotation of plane-polarised light.

Relative Reactivities of Alkyl Halides in S_N2 Substitution

In S_N2 reactions, the nucleophile attacks the side of the carbon opposite to that of the leaving group because both groups are to be accommodated in the transition state. Thus consider the alkaline hydrolysis of methyl bromide which follows S_N2 mechanism. It involves the backside attack of the nucleophile (OH^-) on the carbon bearing the bromine atom. Evidently, when the hydrogen atoms of methyl group are successively replaced by bulkier alkyl groups such as methyl, ethyl, etc., there is an increased crowding around the central carbon both in the reactant and transition state. Due to steric hindrance of the alkyl groups, the nucleophile finds it more and more difficult to attack the carbon carrying the halogen. Thus, in the reaction of hydrolysis of an alkyl bromide, simultaneous presence of – OH and – Br in the transition state becomes difficult because of overcrowding due to bulky groups. Greater the size of the alkyl group attached to halogen bearing carbon atom, greater is the difficulty in the transition state formation and thus slower will be the rate of such S_N2 reaction. The overcrowding of alkyl groups also leads to strong non-bonded interactions in the transition state in which carbon is bonded to five atoms.

In general, the order of relative reactivities of alkyl halides in S_N2 reactions follows the sequence:

Methyl halides > Primary halides > Secondary halides > Tertiary halides

Reactions of Alkyl Halides with Acetylide Ion

Reactions of alkyl halide with acetylide ion is nucleophilic substitution reaction (S_N2). It is successful only with primary halides. With secondary and tertiary halides, the halogen carrying carbon is heavily crowded. The nucleophile (acetylide ion) finds it difficult to attach to this carbon because of steric hindrance.

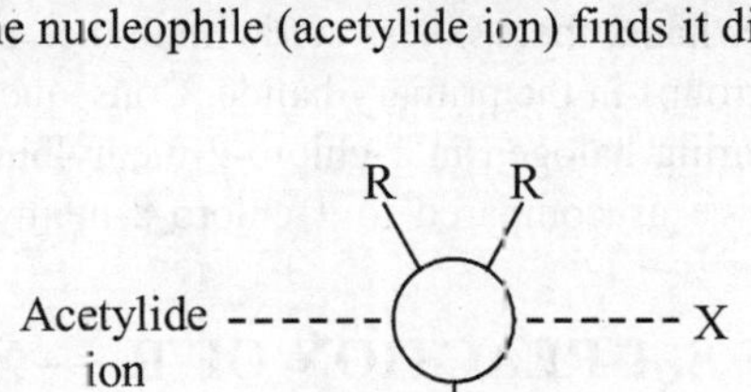

Hence, reaction with secondary and tertiary halides does not take place easily. With primary halides, there is comparatively less crowding around the halogen carrying carbon. Here R-groups are replaced by H-groups. Hence, the reaction is comparatively easier.

Example 2. **Arrange the following compounds in increasing order of reactivity towards S_N2 reaction.**

***t*-butyl chloride, *n*-propyl chloride.**

Solution.

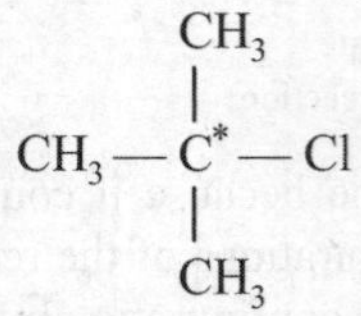

t-Butyl chloride (I)

$CH_3CH_2\overset{*}{C}H_2\,Cl$

n-propyl chloride (II)

In S_N2 reaction, the nucleophile attacks the halide from the side opposite to that where halogen is attached. Attachment of the nucleophile becomes easier if the carbon carrying the halogen is not heavily crowded *i.e.*, if minimum number of alkyl groups are attached to this carbon. There is no crowding with hydrogen atoms. In the two compounds above, the carbon atom marked with an asterisk, where the nucleophile will attach itself, contains three and one alkyl gorups respectively in compounds I and II. Hence S_N2 reaction proceeds faster in compounds II.

Example 3. **Why are S_N2 displacements more difficult with 2-chloro-2-methylbutane than with 1-chloro-2-methylbutane?**

Solution. 2-chloro-2-methylbutane and 1-chloro-2-methylbutane are tertiary and primary halides respectively as indicated below:

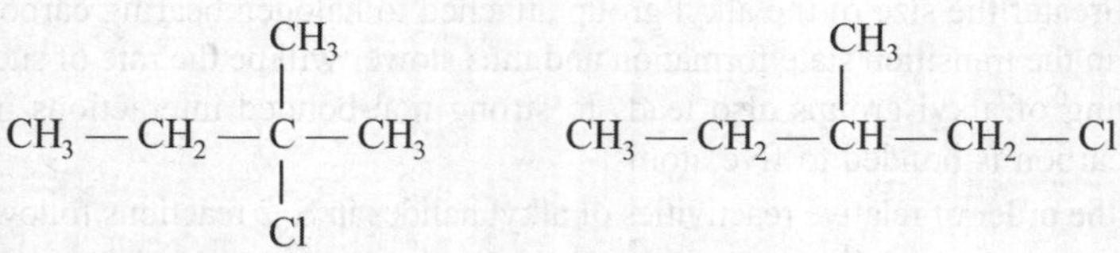

2-Chloro-2-methylbutane (Tertiary alkyl halide) 1-Chloro-2-methylbutane (Primary alkyl halide)

In S_N2 displacements, the nucleophile attacks the side of the carbon opposite to that of the leaving group. In tertiary halides, there is steric hindrance because of the presence of bulky (two methyl and one ethyl) groups and hence the nucleophile finds it difficult to attack the carbon carrying chlorine. On the other hand, there is less steric hindrance because of the presence of two hydrogen and one sec. butyl groups in the primary halide. Consequently the backside attack of the nucleophile on the carbon bearing halogen in 1-chloro-2-methylbutane is easier. Hence 2-chloro -2-methylbutane is less reactive as compared to 1-chloro-2-methylbutane in S_N2 displacement reactions.

7.10 MECHANISM OF S_N1 REACTION OF R — X

An example of unimolecular substitution reaction (S_N1) is the hydrolysis of tert. butyl bromide

$$(CH_3)_3\,C — Br + OH^- \longrightarrow (CH_3)_3\,C — OH + \bar{Br}$$

Chemical kinetic studies reveal that the rate of this reaction depends only on the concentration of tert-butyl bromide. Thus,

$$\text{Rate} = k[(CH_3)_3\,C — Br]$$

The reaction is of first order and thus it is believed to occur in two steps.

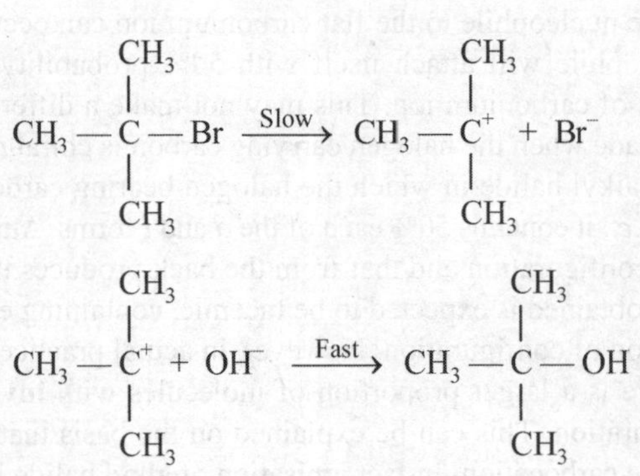

It is thus clear that the rate determining step of the reaction is the slow ionisation of *tert*-butyl bromide to form tert-butyl carbocation. The energy required for ionisation is supplied by the formation of many ion-dipole bonds between the ions produced and the polar solvent molecules. The second step involving the combination of carbocation with hydroxide ion to form alcohol is fast.

The potential energy diagram for the reaction is as shown in the following diagram. (Fig. 7.2)

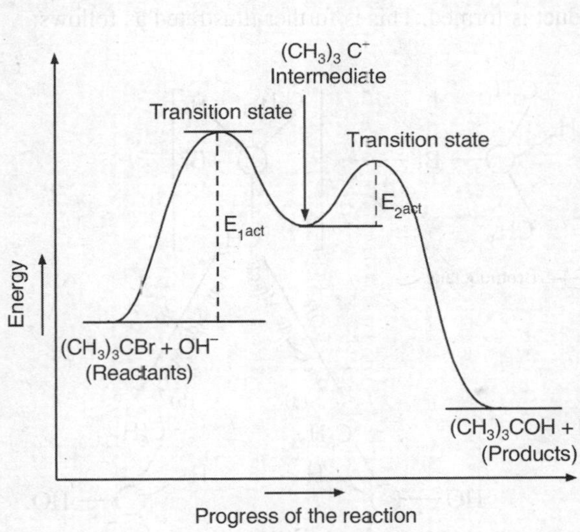

Fig. 7.2. Progress of S_N1 reaction.

From the above diagram, it is evident that there are two transition states. The first transition state is formed when the reactants combine to form the intermediate. The second transition state is formed when the intermediate compound reacts with OH^- ions to form the product. As the activation energy ($E_{1\ act}$) involved in the first transition state is greater than that in the formation of second transition state ($E_{2\ act}$), the first step in the reaction (formation of carbocation) is a slow step while the second step, in which OH^- ions react with the carbocation, is a fast step.

Stereochemistry of S_N1 reactions. In unimolecular nucleophilic substitution, the rate determinings step involves the formation of a carbocation. The carbon atom in the carbocation is in the sp^2 hybridised state. Thus the carbocation has a flat structure in which all the three substituents attached to carbon lie in a plane, making angles of 120° between them. The empty *p*-orbital lies perpendicular to the plane.

The attachment of the nucleophile to the flat carbonium ion can occur probably from the back of carbonium ion. Nucleophile will attach itself with 50% probability from the front and 50% probability from the back of carbonium ion. This may not make a difference in a simple case, but the difference is clearly made when the halogen carrying carbon is chiral *i.e.*, linked to four different groups. Hydrolysis of an alkyl halide in which the halogen bearing carbon is chiral will produce a product which is racemic *i.e.*, it contains 50% each of the *d* and *l* forms. Attachment of the nucleophile from front produces one configuration and that from the back produces the other configuration. As a result, the final product obtained is expected to be racemic, containing equal number of molecules with retention and inversion of configuration. However, in actual practice, the product as a whole is not racemic. Usually there is a larger proportion of molecules with **inverted configuration** than those of the same configuration. This can be explained on the basis that initial ionisation of alkyl halide does not form a free carbocation. In fact ionisation of alkyl halide leads first to the formation of an **ion pair** in which the departing halide ion is still in close proximity to the carbocation. Consequently, the attack on the front side of the carbocation which leads to a product with retention of configuration is slightly hindered. On the other hand, the attack on the backside leading to a product with inversion of configuration is somewhat preferred. Thus, the actual product formed is **partially recemised** and the enantiomer with inverted configuration predominates. For example, when (–)-2-bromooctane is hydrolysed under SN^1 conditions (low concentration of OH^- ions), a partially racemised product is formed. This is further illustrated as follows:

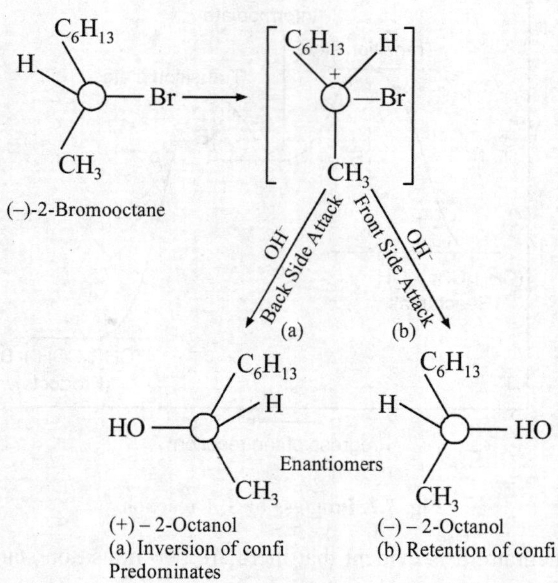

(Partial racemisation in SN^1 reaction)

***Example 1.* Halogenation of alkanes in the presence of tetraethyl lead proceeds at low temp. than in its absence.**

Evidence for the Formation of Carbocation Intermediates in S_N1 Reactions

It is well known that carbocations undergo rearrangement to form more stable carbocations wherever possible. Since the carbocations are the intermediates in S_N1 reactions, it is expected that these intermediates will undergo rearrangement if the structure permits. As a result, the rearranged substitution product will be obtained. This is actually found to be so. For example, when 2-bromo-3-methylbutane is hydrolysed under S_N1 conditions, the formation of rearranged product, 2-methyl-2-butanol and not 3-methyl butanol-2 takes place as shown below:

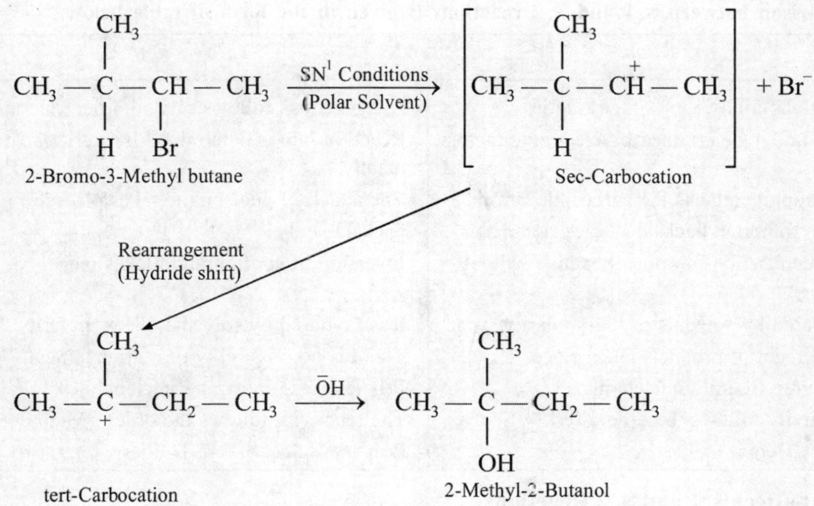

Relative Reactivities of Alkyl Halides in Unimolecular Nucleophilic Substitution (S_N1) Reactions

Formation of the carbocation is the rate determining step in unimolecular nucleophilic substitution (S_N1). More the stability, more rapidly is the carbocation formed and hence faster is the reaction. Now the relative stability of the carbocations follows the sequence:

Benzyl, allyl > Tertiary > Secondary > Primary > Methyl carbocations. Therefore, the reactivity of alkyl halides is S_N1 reactions follows the same order *i.e.*,

Benzyl, allyl > Tertiary > Secondary > Primary > Methyl halides

Effect of Polarity of Solvent on the Rate of S_N1 and S_N2 Reactions

Effect of Solvent in S_N1 Reactions. The rate determining step in these reactions involves the ionisation of alkyl halide to form carbocation and halide ion. The energy required for the ionisation is provided by the solvation of these ions. Now polar solvent such as water, alcohols, *etc.*, have a greater capacity as compared to non-polar solvents to solvate the ions and thus liberate considerable amount of energy. The energy released facilitates the ionisation and hence increases the rate of the reaction. It is thus obvious that S_N1 reactions are favoured by polar solvents. For example,

$$R - X \xrightarrow{\text{Ionization}} R^+ + X^- \quad \text{slow}$$

$$R^+ + \bar{O}H \xrightarrow{\text{Polar solvent}} ROH \quad \text{fast}$$

Effect of solvent in S_N2 reactions. S_N2 reaction is a concerted process occurring in one step through the intervention of a transition state.

$$\bar{N}u + R - X \longrightarrow \underset{\text{Transition state}}{\left[\overset{\delta^-}{Nu} \cdots\cdots R \cdots\cdots \overset{\delta^-}{X}\right]} \longrightarrow Nu - R + \underset{\text{Halide ion}}{\bar{X}:}$$

It is clear that in the transition state, the negative charge is dispersed over nucleophile and halogen while in the reactants, it is concentrated only on the nucleophile. Reactants are thus more polar than the transition state. Evidently polar solvent will solvate the reactants more strongly than the transition state and hence slow down the rate of S_N2 reaction. In other words S_N2 reactions are favoured by solvents of low polarity.

Comparison between S_N1 and S_N2 Reactions

Comparison between S_N1 and S_N2 reactions is given in the form of table below:

S.No.	S_N1 reactions	S_N2 reactions
1.	S_N1 reactions follow first order kinetics.	S_N2 reactions follows second order kinetics.
2.	Reaction rate is determined by electronic factors mainly.	Reaction rate is determined by steric factors mainly.
3.	The nucleophile attacks the carbocation from both sides although backside attack dominates.	The attack of nucleophile takes place from the backside only.
4.	Partial racemisation of optically active halides takes place.	Inversion of configuration takes place.
5.	It is favoured by solvents of high polarity.	It is favoured by solvents of low polarity.
6.	Rearrangement of products takes place.	There is no possibility of rearrangement.
7.	It is favoured by mild nucleophiles	It is favoured by strong nucleophiles.
8.	The reactivity follows the order: Tert. > Sec. > Primary > Methyl halide.	The reactivity follows the order : Methyl > Primary > Sec. > Tert. halides.

Factors that affect S_N1 and S_N2 Reactions

The factors that affect S_N1 and S_N2 reactions are described hereunder.

(i) Nature of alkyl groups. *S_N1 reactions.* The relative reactivity of alkyl halides is the same as the ease of formation and stability of carbonium ion. The stability of carbonium ion is in the order.

Tert. > Sec. > Primary > Methyl carbonium ion

Therefore, the reactivity of alkyl halides is also in the same order.

S_N2 reactions. Here the steric factors are to be taken into consideration. Lesser the crowding around halogen carrying carbon, greater the reactivity. For such reactions, the reactivity follows the order:

Methyl carbonium ion > Primary > Sec. > Tert.

(ii) Nature of halogen Atom. *$S_N{}^1$ reactions.* The ease of elimination of the leaving group is I^- > Br^- > Cl^- > F^-, therefore, the order of reactivity of alkyl halidies is RI > RBr > RCl > RF.

S_N2 reactions. Same as in S_N1.

(iii) Nature of the nucleophile. *S_N1 reaction.* As the nucleophile is not involved in the rate determining step on SN^1 reaction, its nature does not affect the reaction.

S_N2 reaction. Stronger the nucleophile, greater will be the rate of reaction. Increasing strength of a few nucleophiles is given below:

$$H_2O < OH^- < RO^- < CN^-$$

(iv) Nature of the solvent. *S_N1 reaction.* Polarity of the solvent is important in the ease of formation of carbocation. Greater the polarity of the solvent, greater is the rate of reaction.

$S_N{}^2$ reaction. Polarity of the solvent has a role to play in this reaction also but in a reverse manner. Higher Polarity of the solvent is likely to destroy less polar intermediate product. Hence, less polar the solvent, greater is the reactivity.

Example 4. **Explain giving reason why electronic factors have no significant effect on S_N2 reaction.**

Solution. S_N2 reaction is a reaction which depends upon the steric environment around halogen carrying carbon. Greater the crowding of alkyl groups around this carbon, smaller the reactivity.

By electronic effects, we mean the electron withdrawing or repelling effect of various substituents. Such electron withdrawing or repelling groups have no effect on the stability of the intermediate complex, therefore, they do not affect S_N2 reactions.

***Example 5.* Hydrolysis of ethyl chloride is S_N2 but hydrolysis of tert. butyl chloride is S_N1. Explain with mechanism.**

***Solution.* Hydrolysis of Ethyl Chloride**

Hydrolysis of ethyl chloride is a bimolecular reaction (S_N2). Here there is no steric hindrance to the attachment of hydroxide nucleophile to form the transition state. For complete mechanism, see relevant reaction in this chapter

Hydrolysis of Tert. Butyl Chloride

The nucleophile OH^- faces steric hindrance in attaching itself to the carbon bearing halogen. This is because of the steric hindrance of three bulkyl methyl groups. Hence it follows S_N1 mechanism. For complete mechanism, refer to relevant section in this chapter.

***Example 6.* Which alkyl halide to do you expect to react more rapidly by S_N2 mechanism ?**

(*a*) 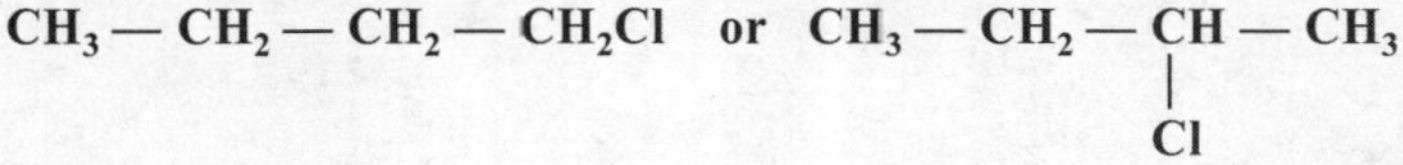

(*b*) 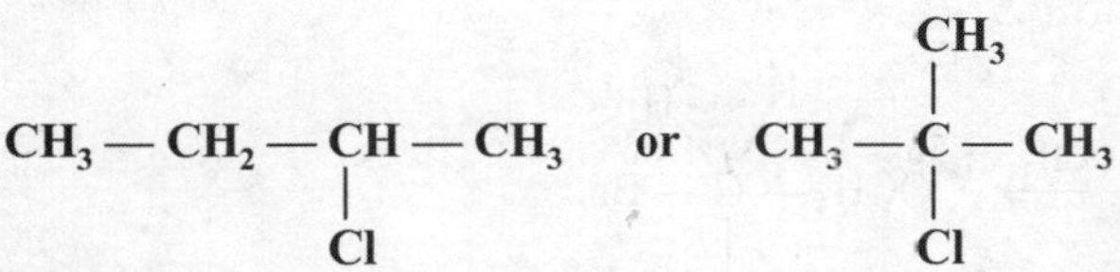

Solution. In S_N2 reaction, the nucleophile and the halogen are both attached to the halogen bearing carbon in the transition state as given below:

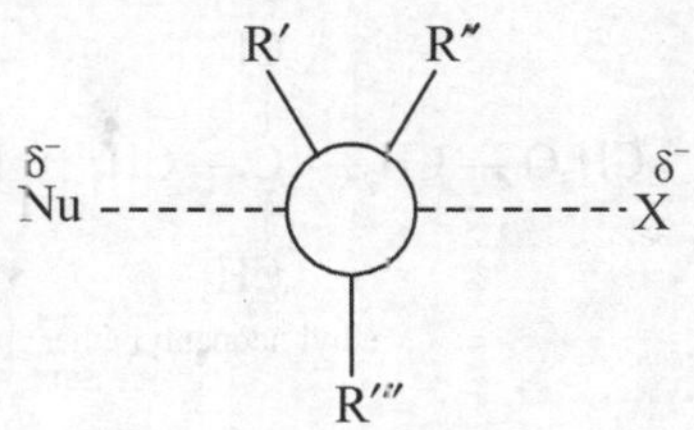

where R′, R″ and R‴ are alkyl or hydrogen groups. Obviously, greater the size of three alkyl groups, greater will be the hindrance faced by the nucleophile in forming the transition state.

Therefore out of 1-chloro and 2-chlorobutane, the former will offer less hindrance to the nucleophile and therefore will react more rapidly.

Similarly, out of 2-chlorobutane and 2-chloro-2-methyl propane, the former will offer less hindrance to the nucleophile in forming transition state and therefore will react more rapidly.

***Example 7.* Predict the products in the following reactions:**

(*i*) 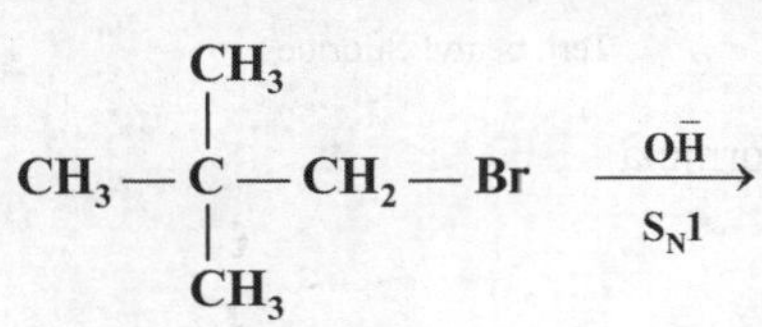

(*ii*) 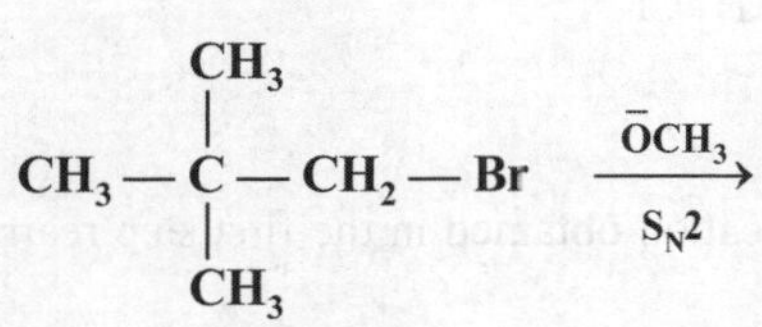

Solution.

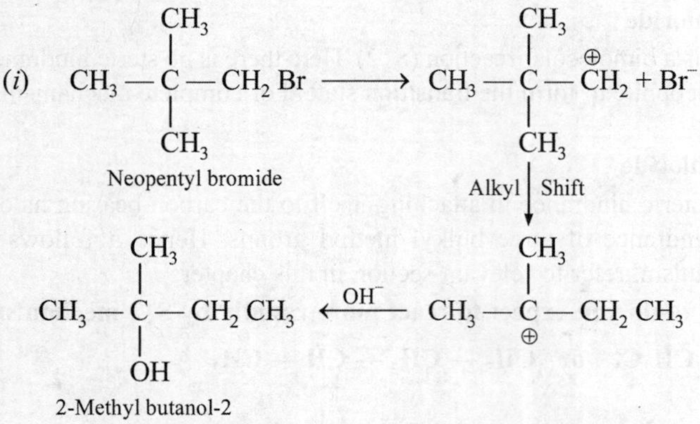

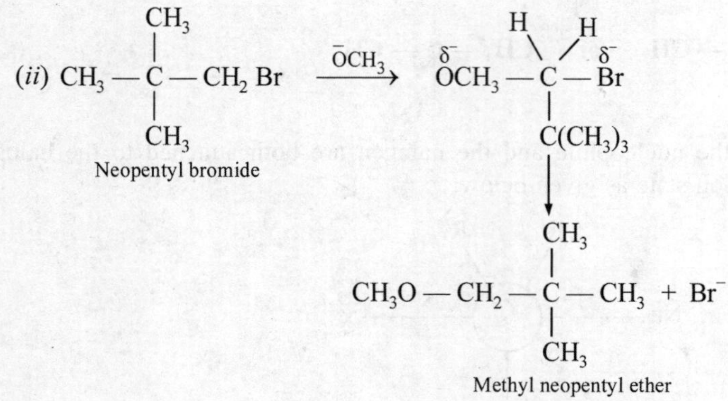

***Example 8.* Neopentyl alcohol on reaction with HCl forms tert. pentyl chloride and not neopentyl chloride.**

Solution.

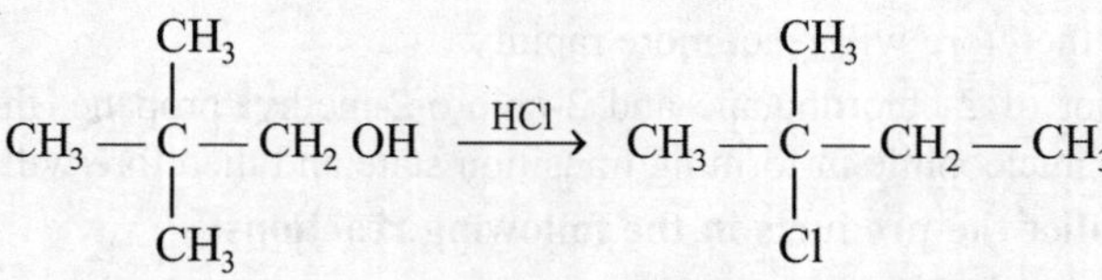

It does not form neopentyl chloride having the formula

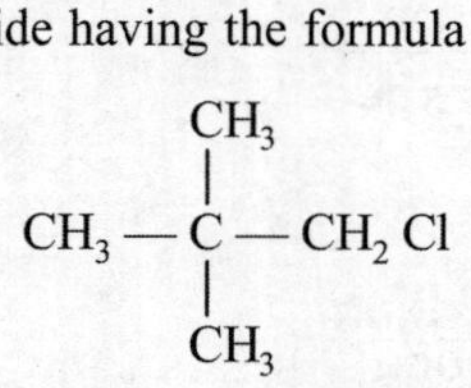

This reaction is S_N1 reaction in which, the carbocation obtained in the first step rearranges to more stable carbocation.

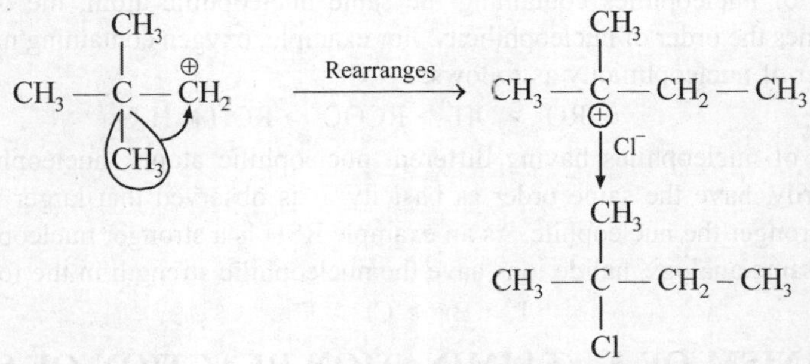

Tert. pentyl chloride

***Example 9. n*-butyl chloride and tert-butyl chloride behave differently when treated with aqueous sodium hydroxide. Discuss reaction path in both cases.**

Solution.

$CH_3 — CH_2 — CH_2 — CH_2Cl$ (I) (*n*-Butyl chloride)

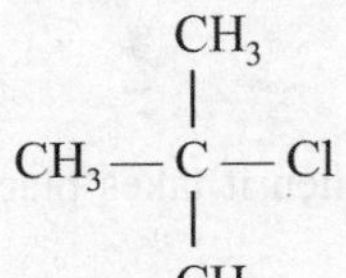

(II) (tert-Butyl chloride)

Compound (I) when treated with aqueous NaOH gives *n*-butyl alcohol by S_N2 mechanism because the nucleophile attaches itself to the carbon carrying chlorine, as there is not much crowding on this carbon.

Compound (II) when treated with aqueous NaOH gives tert-butyl alcohol by S_N1 mechanism. Three alkyl groups attached to central carbon exert their inductive effect and polarise the molecule as:

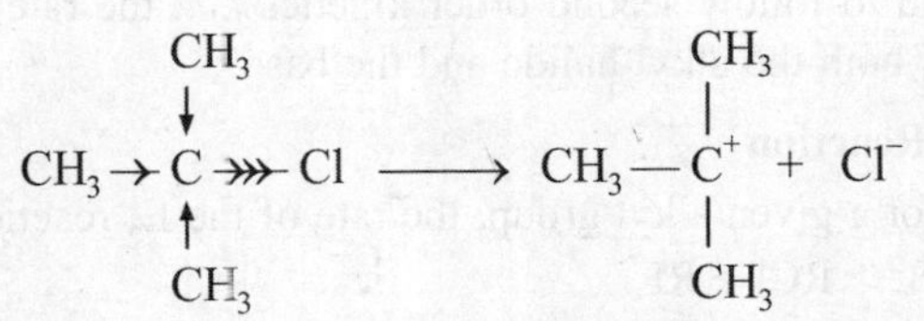

tert-Butyl carbocation

For reaction path, see relevant sections in the chapter.

7.11 NUCLEOPHILICITY OR NUCLEOPHILIC STRENGTH

Nucleophilicity of reagent represents its ability to donate a pair of electrons to the electron deficient carbon. Greater the nucleophilicity, greater will be its tendency to participate in S_N2 reaction.

Comparison of nucleophilic strength of reagents can be made with the help of following general rules.

(*i*) A negatively charged nucleophile is stronger than its conjugate acid. For example OH^- is a stronger nucleophile than its conjugate acid H_2O and RO^- is a stronger nucleophile than its conjugate acid ROH.

(*ii*) In case of nucleophiles containing the same nucleophilic atom, the order of basicity determines the order of nucleophilicity. For example, oxygen containing nucleophiles have the order of nucleophilicity as follows:

$$RO^- > OH^- > RCOO^- > ROH > H_2O$$

(*iii*) In case of nucleophiles having different nucleophilic atoms, nucleophilicity does not necessarily, have the same order as basicity. It is observed that larger the nucleophilic atom, stronger the nucleophile. As an example RSH is a stronger nucleophilic than ROH. On the same analogy, halide ions have the nucleophilic strength in the following order.

$$I^- > Br^- > Cl^- > F^-$$

7.12 MECHANISM OF E_2 (ELIMINATION REACTION OF SECOND ORDER. REACTIONS OF ALKYHALIDES

An alkyl halide on treatment with alcoholic KOH gives an alkene. This is an elimination reaction in which a molecule of hydrogen halide is eliminated.

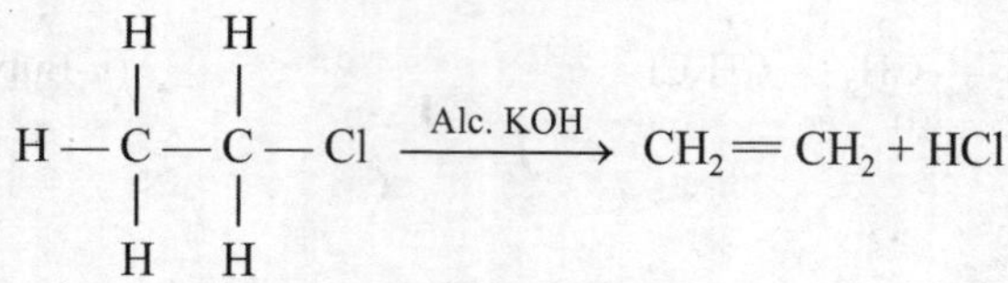

This reaction could take place either by E_2 or E_1 mechanism. When it takes place by E_2, it follows the following steps.

Mechanism of E_2 Reaction

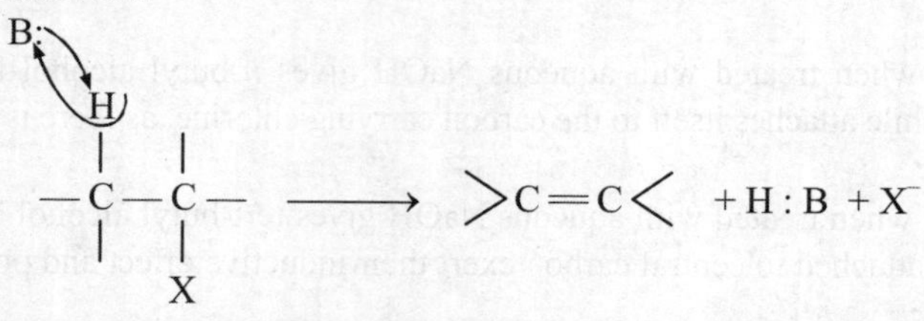

This reaction is found to follow second order kinetics *i.e.*, the rate of reaction is dependent upon the concentration of both the alkyl halide and the base.

Factors Influencing E_2 Reaction

(*i*) It is found that for a given alkyl group, the rate of the E_2 reaction varies in the order.

RI > RBr > RCl > RF

(*ii*) For a given halogen, the rate of reaction follows the order.

tert. halide > sec. halide > prim. halide

(*iii*) Stronger the base, greater will be the rate of reaction. A notable feature of such reactions is that the alkene which has maximum possible branching (Saytzeff's Rule) is obtained.

7.13 MECHANISM OF E_1 (ELIMINATION REACTION OF FIRST ORDER)

REACTIONS OF ALKYL HALIDES

Some alkyl halides particularly secondary and tertiary alkyl halides lose a molecule of hydrogen halide by E_1 mechanism.

In such reactions, it is believed that in the first step, which is rate detemining step, the molecule of alkyl halide undergoes ionisation to give a halide ion and a carbonium ion.

In the second step, which is a fast step, the carbonium ion (or carbocation) loses a proton from the neighbouring carbon to yield the alkene.

The mechanism of the reaction is illustrated as under:

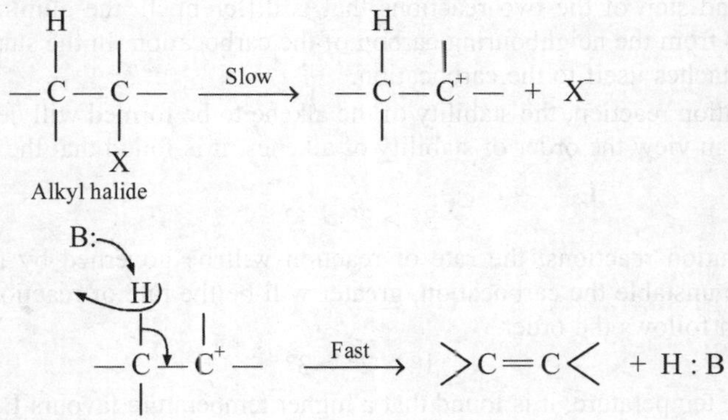

Factors Influencing E_1 Mechanism

(*i*) For a given alkyl group, it is found that rate of reaction varies in the order.

$$RI > RBr > RCl > RF$$

(*ii*) For a given halogen, the order of reactivity of alkyl halides for E_2 reaction follows the order.

Tert. halides > sec. halides > prim. halides

It is noteworthy that unlike in S_N1 and S_N2 reactions, the order of reactivity in E_2 and E_1 is the same.

7.14 S_N^2 VERSUS E_2 REACTIONS OF ALKYL HALIDES

It is observed that when we carry out an S_N2 reaction, some amount of E_2 products are also obtained. Similarly when it is intended to carry out E_2 reaction, some amounts of S_N2 products are also obtained. Thus the two reactions proceed concurrently although with different speeds.

It is possible to direct the reaction in one or the other way by regulating some parameters as discussed hereunder.

(*i*) The order of reactivity for S_N2 reaction is

1° halide > 2° halide > 3° halide

The order of reactivity for E_2 reaction is

3° halide > 2° halide > 1° halide

Thus E_2 reaction is increasingly favoured as we move from primary to secondary to tertiary alkyl halide.

(*ii*) The size of the alkyl group and branching at β-carbon influences the relative rates of S_N2 and E_2 reactions due to steric reasons. Formation of intermediate product in S_N^2 reactions becomes difficult with bulky groups. Thus rate of $S_N2$reaction decreases when we move from CH_3CH_2Br to $CH_3CH_2CH_2Br$ to $(CH_3)_2$ $CHCH_2Br$.

It is observed that nature of halogen does not affect the relative ratio of $S_N2$and E_2 products.

It is observed that by decreasing the polarity of the solvent, we can shift more towards elimination (E_2).

By increasing the temperature also, it is possible to shift towards E_2 reaction.

As far as the solvent is concerned, S_N2 reactions are favoured by strong nucleophiles and E_2 reactions are favoured by strong bases. And in general, strong nucleophiles are strong bases as well.

7.15 S_N1 VERSUS E_1 REACTIONS OF ALKYL HALIDES

Whether it is S_N1 reaction or E_1 reaction, the first step is the slow rate-determining step of dissociation of alkyl halide into carbocation and the halide ion.

It is the second step of the two reactions that is different. In the elimination reaction, the proton is removed from the neighbouring carbon of the carbocation. In the substitution reactions, the nucleophile attaches itself to the carbocation.

In the elimination reaction, the stability of the alkene to be formed will determine the rate of reaction. Keeping in view the order of stability of alkenes, it is found that the rate of elimination follows the order.

$$3° > 2° > 1°$$

In the substitution reactions, the rate of reaction will be governed by the stability of the carbocation. More unstable the carbocation, greater will be the rate of reaction. Accordingly the rate of substitution follows the order.

$$1° > 2° > 3°$$

Coming to the temperature, it is found that a higher temperature favours E_1 reaction over S_N1 reaction.

CHLOROFORM

7.16 METHODS OF PREPARATION

Chloroform is prepared by the following methods :

***(i)* From Chloral hydrate.** Pure Chloroform can be obtained by distilling chloral or chloral hydrate with concentrated aqueous NaOH solution.

$$NaOH + \underset{\text{Chloral}}{CCl_3CHO} \longrightarrow \underset{\text{Sod. formate}}{HCOONa} + CHCl_3$$

$$NaOH + \underset{\text{Chloral hydrate}}{CCl_3CH(OH)_2} \longrightarrow HCOONa + \underset{\text{Chloroform}}{CHCl_3} + H_2O$$

***(ii)* From Methane.** Chloroform is manufactured by chlorination of methane in the presence of light or catalysts when a mixture of methyl chloride, methylene dichloride, chloroform and carbon tetrachloride is obtained.

$$CH_4 + Cl_2 \longrightarrow CH_3Cl + HCl$$
$$CH_3Cl + Cl_2 \longrightarrow CH_2Cl_2 + HCl$$
$$CH_2Cl_2 + Cl_2 \longrightarrow CHCl_3 + HCl$$
$$CHCl_3 + Cl_2 \longrightarrow CCl_4 + HCl$$

The mixture of CH_3Cl, CH_2Cl_2, $CHCl_3$ and CCl_4 can be separated by fractional distillation.

***(iii)* Laboratory Method.** Ethanol or acetone is treated with a paste of bleaching powder and water.

In this reaction, bleaching powder serves as a source of chlorine which first oxidises ethanol to acetaldehyde, which is then further chlorinated to chloral. Chloral is hydrolysed with $Ca(OH)_2$ given by $CaOCl_2$ to give chloroform and calcium formate.

(a) In the case of ethanol, the following reactions are involved :

$$CaOCl_2 + H_2O \longrightarrow Ca(OH)_2 + Cl_2$$

$$\underset{\text{Ethanol}}{CH_3CH_2OH} + Cl_2 \xrightarrow{\text{Oxidation}} \underset{\text{Ethanal}}{CH_3CHO} + 2HCl$$

$$CH_3CHO + 3Cl_2 \xrightarrow{\text{Chlorination}} \underset{\text{Chloral}}{CCl_3CHO} + 3HCl$$

$$Ca(OH)_2 + 2CCl_3CHO \xrightarrow{\text{Hydrolysis}} \underset{\text{Chloroform}}{2CHCl_3} + \underset{\text{Calcium formate}}{(HCOO)_2Ca}$$

(*b*) In the case of acetone, the following reactions are involved :

$$CaOCl_2 + H_2O \longrightarrow Ca(OH)_2 + Cl_2$$

$$\underset{\text{Acetone}}{CH_3COCH_3} + 3Cl_2 \xrightarrow{\text{Chlorination}} \underset{\text{Trichloroacetone}}{CCl_3COCH_3}$$

$$Ca(OH)_2 + CCl_3COCH_3 \xrightarrow{\text{Hydrolysis}} \underset{\text{Chloroform}}{CHCl_3} + \underset{\text{Calcium acetate}}{(CH_3COO)_2Ca}$$

(*iv*) From carbon tetrachloride. By partial reduction of carbon tetrachloride with iron filings and water.

$$CCl_4 + 2[H] \xrightarrow[\text{heat}]{Fe/H_2O} CHCl_3 + HCl$$

7.17 PHYSICAL PROPERTIES

1. Chloroform is a colourless, heavy liquid which has sweetish sickly odour and burning taste.
2. It is heavier than water.
3. It boils at 334 K and is slightly soluble in water.
4. Inhaling of the vapours of chloroform induces unconsciousness.

Model of chloroform

7.18 CHEMICAL PROPERTIES

1. Action of sunlight and air. In the presence of sunlight, chloroform gets oxidized by air to produce highly poisonous compound phosgene, $COCl_2$.

$$2CHCl_3 + O_2 \xrightarrow{\text{Sunlight}} 2COCl_2 + 2HCl$$

As chloroform is used for anaesthetic purposes, therefore, high level of purity is desirable. To prevent oxidation of chloroform it is stored in dark bottles, completely filled upto brim. *The use of dark bottles cuts radiations and filling upto brim keeps out air.* A small amount of enthanol (1%) is usually added to bottle of chloroform. *Addition of a little ethanol converts the toxic $COCl_2$ to non-poisonous diethyl carbonate.*

$$COCl_2 + 2C_2H_5OH \longrightarrow \underset{\text{Diethyl carbonate}}{O=C\begin{cases} OC_2H_5 \\ OC_2H_5 \end{cases}} + 2HCl$$

2. Reduction. Chloroform can be reduced to methylene chloride with zinc and HCl.

$$Zn + 2HCl \longrightarrow ZnCl_2 + 2[H]$$

$$CHCl_3 + 2[H] \longrightarrow \underset{\substack{\text{Dichloromethane} \\ (\textit{Methylene chloride})}}{CH_2Cl_2} + HCl$$

Reduction of chloroform with zinc and water (neutral medium) produces methane.

$$CHCl_3 \xrightarrow{Zn + H_2O} CH_4 + 3HCl$$

3. Hydrolysis. When boiled with aqueous KOH, chloroform is hydrolysed to potassium formate.

$$H-C\begin{cases} Cl \\ Cl \\ Cl \end{cases} + \begin{matrix} K\ OH \\ K\ OH \\ K\ OH \end{matrix} \xrightarrow{-3KCl} \underset{\text{Unstable}}{H-C\begin{cases} OH \\ OH \\ OH \end{cases}} \xrightarrow{-H_2O} H-\overset{\overset{\large O}{\|}}{C}-OH$$

$$\xrightarrow[-H_2O]{+KOH} \underset{\substack{\text{Potassium} \\ \text{formate}}}{HCOOK}$$

4. Reaction with acetone. Chloroform reacts with acetone in the presence of a base such as KOH and forms the addition product, *chloretone.*

$$\underset{\text{Acetone}}{(CH_3)_2C=O} + CHCl_3 \xrightarrow{KOH} \underset{\text{Chloretone}}{CH_3-C(CH_3)(OH)-CCl_3}$$

Chloretone is used as sleep inducing drug.

5. Reaction with primary amines (*Carbylamine reaction*). On warming chloroform with a primary amine (aliphatic or aromatic) in the presence of alc. KOH, an offensive smell of *isocyanide* or *carbylamine* is obtained.

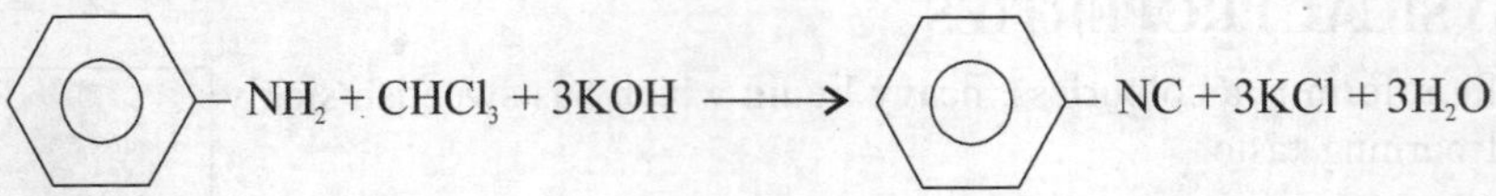

Phenyl isocyanide

6. Reaction with nitric acid. Chloroform reacts with conc. HNO_3 on heating.

$$CHCl_3 + HONO_2 \longrightarrow \underset{\text{Chloropicrin}}{CCl_3.NO_2} + H_2O$$

Chloropicrin is used as an insecticide and war-gas.

7. Chlorination. Further chlorination of chloroform gives CCl_4.

$$CHCl_3 + Cl_2 \xrightarrow{hv} CCl_4 + HCl$$

8. Reimer-Tiemann reaction. Chloroform reacts with phenol and alc. KOH to form salicylaldehyde which is used to prepare a number of drugs.

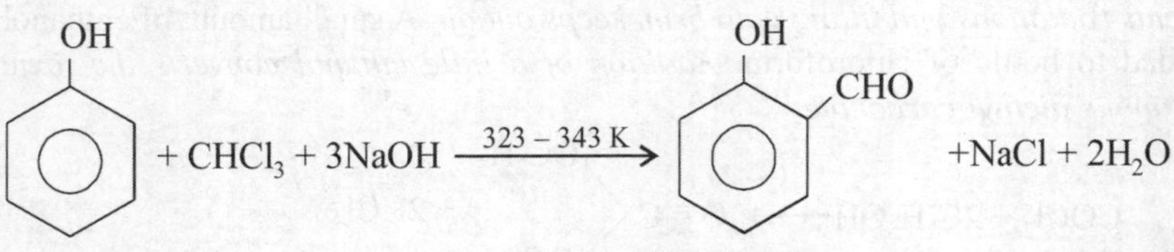

Phenol

Salicylaldehyde (2-Hydroxybenzaldehyde)

Daniel Tiemann

9. Reaction with silver powder. On heating with silver powder, chloroform forms acetylene.

$$2CHCl_3 + 6Ag \xrightarrow{heat} \underset{\text{Acetylene}}{CH \equiv CH} + 6AgCl$$

USES OF CHLOROFORM

1. As laboratory reagent.
2. As solvent in oils and varnishes.
3. As an anaesthetic.
4. As preservative for anatomical specimens.
5. In medicines such as cough syrups.

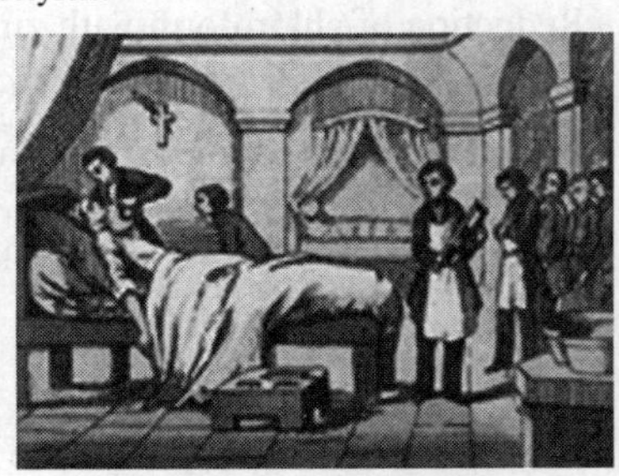

Chloroform was used as anaesthetic earlier

TETRACHLOROMETHANE (Carbon Tetrachloride)

7.19 METHODS OF PREPARATION

Carbon tetrachloride is prepared by the following methods:

***(i)* From Methane.** By chlorination of methane in the presence of sunlight. Chlorine gas is taken in excess in the reaction.

$$CH_4 + 4Cl_2 \xrightarrow{h\nu} CCl_4 + 4HCl$$

***(ii)* From Carbon disulphide.** By reaction of chlorine with carbon disulphide in the presence of aluminium chloride as catalyst.

$$CS_2 + 3Cl_2 \xrightarrow{AlCl_3} CCl_4 + \underset{\text{Sulphur-monochloride}}{S_2Cl_2}$$

CCl_4 separated from S_2Cl_2 by fractional distillation.

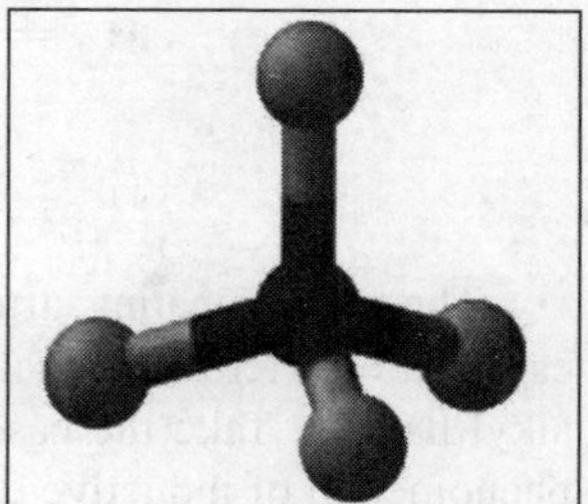

Model of carbon tetrachloride

7.20 PHYSICAL PROPERTIES

1. Carbon tetrachloride is a colorless, heavy liquid with characteristic sickly smell.
2. It is heavier than water.
3. It is insoluble in water but is soluble in organic solvents such as ether.

7.21 CHEMICAL PROPERTIES

***(i)* Stability.** CCl_4 is stable to red heat but when the vapours come in contact with water, phosgene is liberated.

$$CCl_4 + H_2O \longrightarrow COCl_2 + 2HCl$$

It is because of thermal stability that CCl_4 is used as a fire extinguisher under the name pyrene.

***(ii)* Hydrolysis.** On heating with alc. KOH, it undergoes hydrolysis and gives CO_2 which dissolves in KOH to yield potassium carbonate.

$$CCl_4 + 4KOH \xrightarrow[-4KCl]{\Delta} \underset{\text{Unstable}}{[C(OH)_4]} \longrightarrow CO_2 + 2H_2O$$

$$2KOH + CO_2 \longrightarrow K_2CO_3 + H_2O$$

***(iii)* Reduction.** On reduction with moist iron fillings, CCl_4 gives chloroform.

$$CCl_4 + 2H \xrightarrow{Fe/H_2O} CHCl_3 + HCl$$

***(iv)* Reimer-Tiemann reaction.** On heating with phenol and sodium hydroxide, salicyclic acid is obtained.

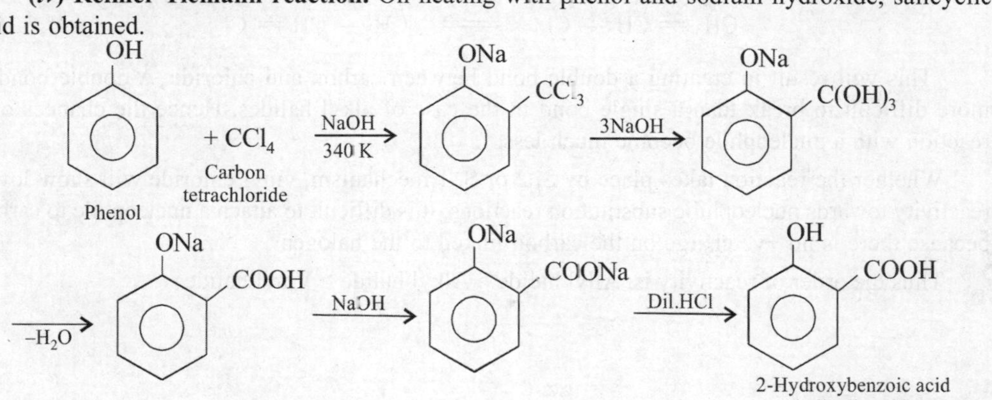

2-Hydroxybenzoic acid
(salicylic acid)

7.22 USES OF CCl_4

1. As a laboratory reagent.
2. As a fire extinguisher.
3. As an industrial solvent.
4. In dry cleaning.
5. For the manufacture of chloroform.

7.23 COMPARISON OF REACTIVITIES OF ALLYL HALIDES, VINYL HALIDE AND ALKYL HALIDE

Allyl chloride (in fact any halide) is more reactive than alkyl halide because it can give more stable intermediate carbocation as illustrated below.

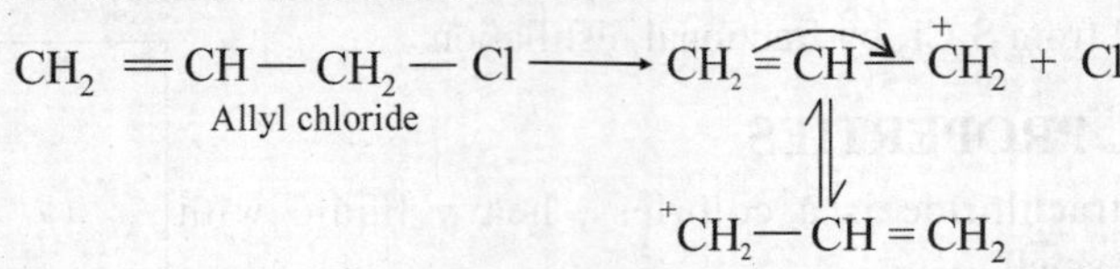

The two resonating structures are of equal energies thereby making it most stable intermediate carbocation. Here stabilization is by means of resonance which is more effective than in case of alkyl halides. Take the case of *t*-butyl chloride. Here the carbocation is stabilized by a weaker phenomenon of inductive effect.

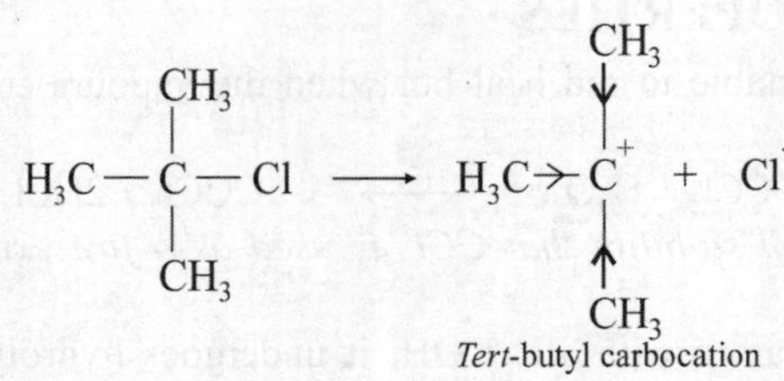

Tert-butyl carbocation

A substance that can furnish more stable carbocation is naturally more reactive.

Therefore allyl chloride is more reactive than alkyl halide.

Even if the reaction takes place by $S_N{}^2$ mechanism, allyl chloride will show greater reactivity because being linear in shape, will not hinder the attachment of a nucleophile to form the intermediate product.

Vinyl chloride is less reactive than alkyl halide because of the following resonance phenomenon.

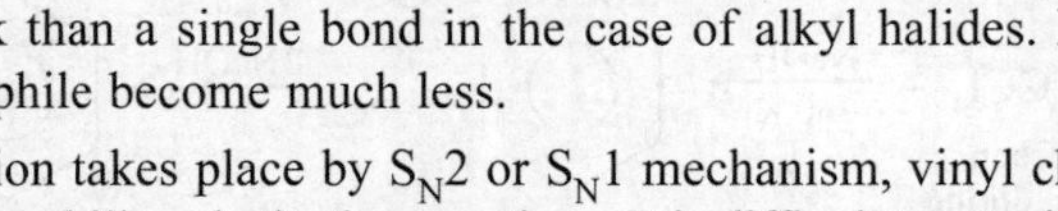

This will result in creating a double bond between carbon and chloride. A double bond is more difficult to break than a single bond in the case of alkyl halides. Hence the chances of a reaction with a nucleophile become much less.

Whether the reaction takes place by S_N2 or S_N1 mechanism, vinyl chloride will show lower reactivity towards nucleophilic substitution reactions. It is difficult to attach a nucleophile to carbon because there is no +ve charge on the carbon linked to the halogen.

Thus the order of reactivity is: Allyl halide > Alkyl halide > Vinyl halide

MISCELLANEOUS EXAMPLES

Example 1. **How will you convert ethyl iodide to propionic acid ?**

Solution. $\underset{\text{Ethyl Iodide}}{C_2H_5I} \xrightarrow{KCN} CH_3CH_2CN \xrightarrow{H_2O} \underset{\text{Propionic acid}}{CH_3CH_2COOH}$

Example 2. **How will you bring about the following conversions :**

(*i*) *n*–propyl bromide → *n*–propane

(*ii*) $CH_3\,CH\,Br\,CH_3 \rightarrow CH_3\,CH_2\,CH_2\,OH$

Solution. (*i*) $\underset{n\text{-propyl bromide}}{CH_3-CH_2-CH_2Br} \xrightarrow[\text{Zn/Acetic acid}]{\text{Zn-Cu/alcohol}} \underset{n\text{-propane}}{CH_3\,CH_2-CH_3} + HBr$

(*ii*) $CH_3-CH_2-BrCH_3 \xrightarrow{\text{alc.KOH}} CH_3\,CH=CH_2 \xrightarrow[\text{HBr}]{\text{Peroxide}} CH_3CH_2-CH_2\,Br \xrightarrow{H_2O}$
$CH_3-CH_2-CH_2OH$

Example 3. **Identify the compounds P, Q and R in the following reaction :**

$$CH_3-CH_2-CH_2I \xrightarrow{\text{alc.KOH}} P \xrightarrow{H_2O/H^+} Q \xrightarrow{SOCl_2} R$$

Solution. $CH_3-CH_2-CH_2I \xrightarrow{\text{alc.KOH}}$

$$\underset{\substack{(P)\\ \text{Propene-2}}}{CH_3-CH=CH_2} \xrightarrow{H_2O/H^+} \underset{\substack{(Q)\\ \text{Propan-2-ol}}}{CH_3-\underset{OH}{\underset{|}{CH}}-CH_3} \xrightarrow{SOCl_2} \underset{\substack{(R)\\ \text{2-chloropropane}}}{CH_3-\underset{Cl}{\underset{|}{CH}}-CH_3}$$

Example 4. **Complete the following reaction.**

$$\mathbf{CH_3-CH_2-\overset{Br}{\overset{|}{\underset{CH_3}{\underset{|}{C}}}}-CH_3 \xrightarrow{\text{alc.KOH}}}$$

Solution. $CH_3-CH_2-\overset{Br}{\overset{|}{\underset{CH_3}{\underset{|}{C}}}}-CH_3 \xrightarrow{\text{alc.KOH}} \underset{\text{2-methylbutene-2}}{CH_3-CH_2=\underset{CH_3}{\underset{|}{C}}-CH_3}$

ARYL HALIDES

7.24 CLASSIFICATION AND NOMENCLATURE OF AROMATIC HALOGEN COMPOUNDS

Aromatic halogen compounds are those compounds which contain at least one benzene ring and one halogen atom in the molecule. They are classified as under:

Nuclear Halogen Compounds

These are the compounds in which the halogen is directly attached to the benzene ring. For example,

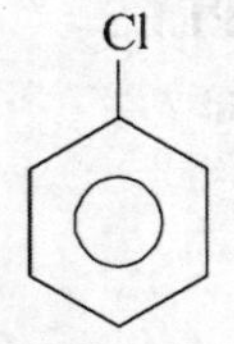

Chlorobenzene

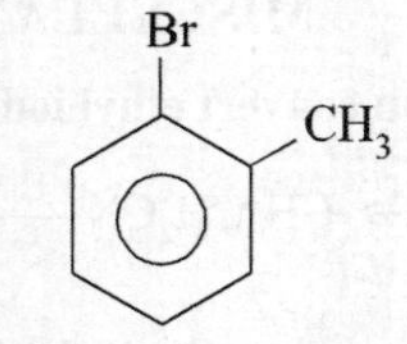

o-Bromotoluene

Side-chain Halogen Compound

These are the compounds in which the halogen is attached to side chain of benzene ring. For example,

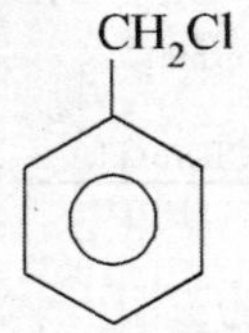

Benzyl chloride

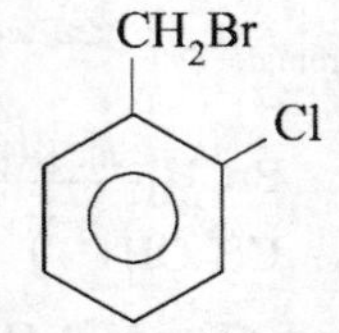

o-Chloro benzyl bromide

More than one halogen could be present in the same side-chain in a compound.

Nomenclature

Following rules are generally followed in the nomenclature of aromatic halogen compounds. Only one system of nomenclature is prevalent. There are no separate common and IUPAC name.

(*i*) In monohalogen compounds, the name of the compound is written by mentioning the halogen followed by the word "benzene". For example,

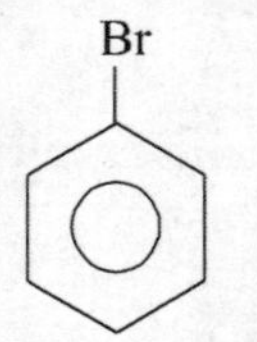

Bromobenzene

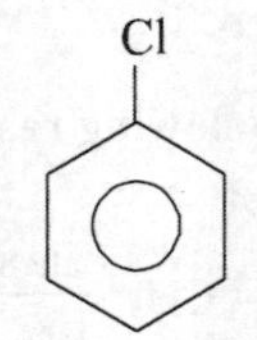

Chlorobenzene

All the six positions of benzene are equivalent. Hence, monohalogen compounds could be represented by attaching the halogen to any of the six positions.

(*ii*) In nuclear dihalogen compounds, there are two halogen groups attached. Now two groups can be attached in different relative positions. If the two groups are located in neighbouring positions, we say that the two groups are *ortho* to each other. If there is a gap of one position, we say that the two groups are *meta* to each other. If there is a gap of two positions we say that the two groups are *para* to each other. The names of a few compounds are given below. For *ortho, meta* and *para,* we use abbreviations *o*-, *m*- and *p*-respectively.

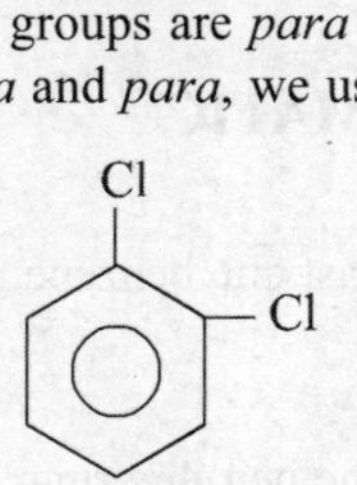

o-Dichlorobenzene
or
1, 2 Dichlorobenzene

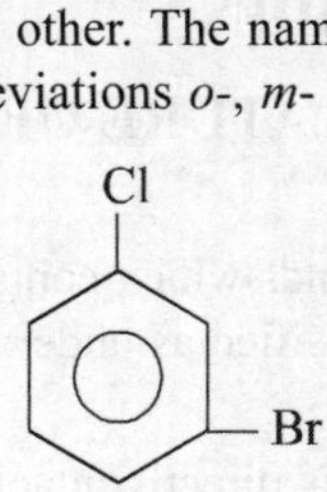

m-Bromo
chlorobenzene

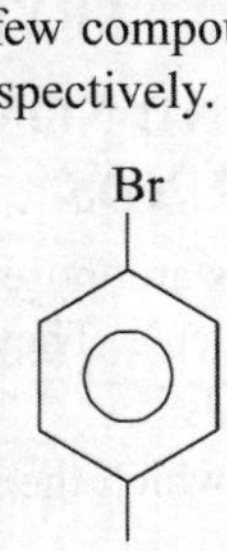

p-Iodo
bromobenzene

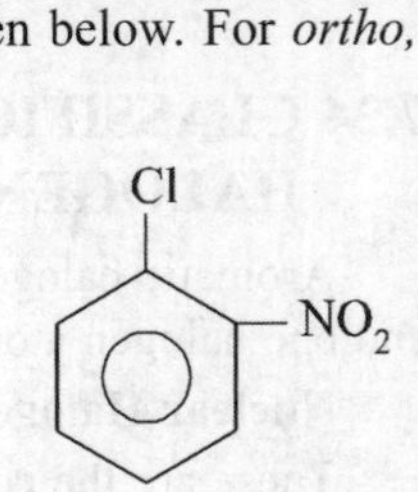

o-Nitro
chlorobenzene

(*iii*) For naming trihalogen compounds, we use the number system. Number 1 is given to the position of one of the groups and then the other groups are numbered accordingly. Clockwise direction is followed for this purpose.

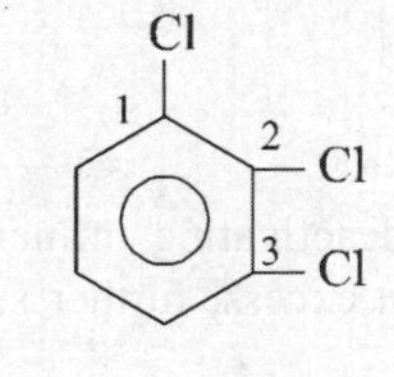
1, 2, 3 Trichlorobenzene

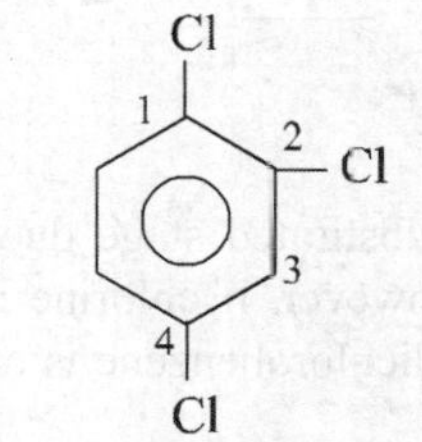
1, 2, 4 Trichlorobenzene

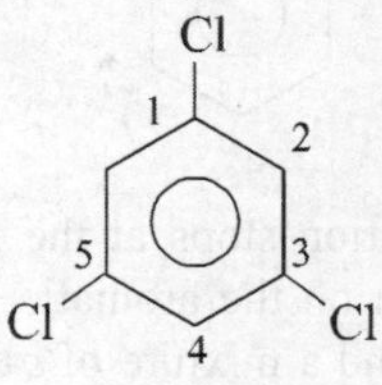
1, 3, 5 Trichlorobenzene

(*iv*) Side-chain halogen compounds are named as follows:

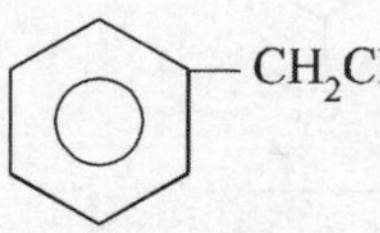
Benzyl chloride

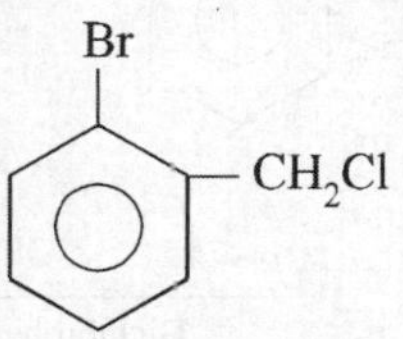
o-Bromo benzyl chloride

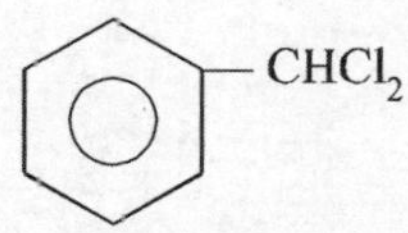
Benzal chloride

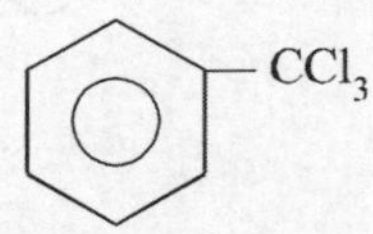
Benzo Trichloride

Example 1. **Write the structures and names of all possible isomers of an aromatic compound with molecular formula C_7H_7Cl.**

Solution. The structures and names of possible isomers of the compound with molecular formula C_7H_7Cl are given below:

(*i*)
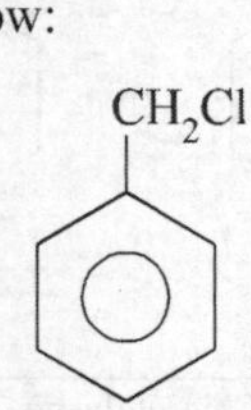
Benzyl chloride

(*ii*)
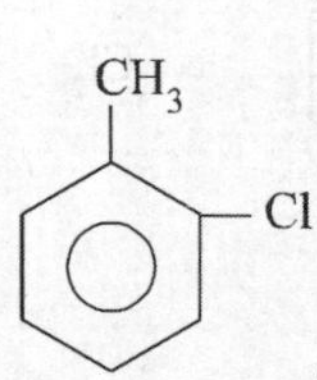
o-Chlorotoluene (or *o*-Methyl chlorobenzene)

(*iii*)
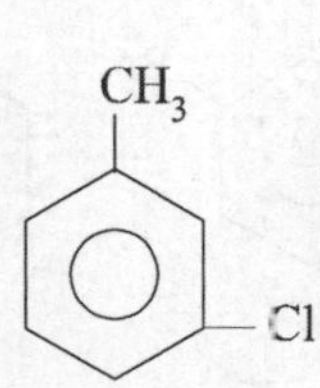
m-Chlorotoluene

(*iv*)
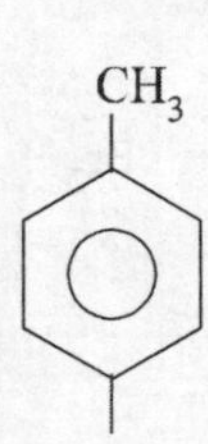
p-Chlorotoluene

7.25 METHODS OF PREPARATION OF ARYL HALIDES

Aryl halides are prepared by the following methods:

(1) Direct halogenation. It consists in treating the aromatic compound with halogen (chlorine or bromine only) at low temperature and in the absence of sunlight and in the presence of halogen carriers like ferric halides or aluminium halides. Iron is mostly used which is converted into the corresponding ferric halide. For example:

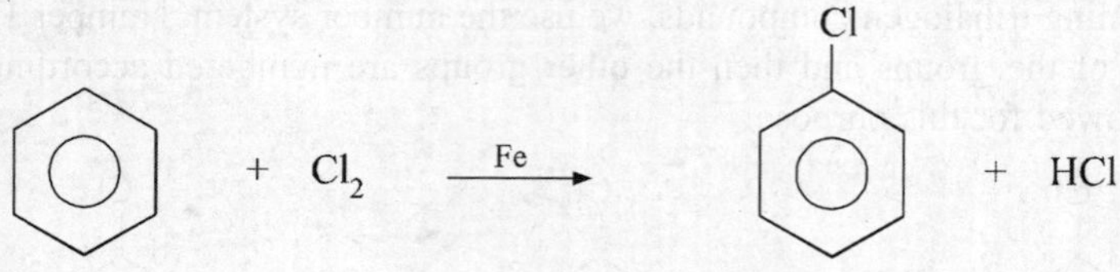

Chlorobenzene

The reaction stops at the mono substituted stage due to the deactivating influence of the halogen atom on the aromatic ring. However, if chlorine is used in excess, further substitution takes place and a mixture of *o*- and *p*-dichlorobenzene is obtained.

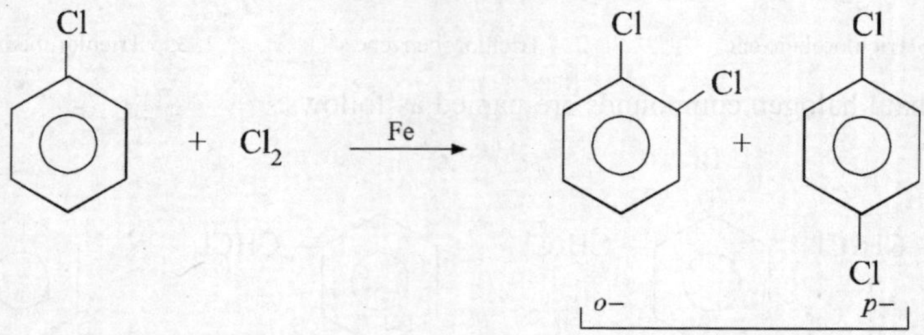

Dichlorobenzene

Similarly, when toluene is chlorinated or brominated (using 1 molecule of chlorine or bromine), a mixture *o*-, *p*-chlorotoluene or bromotoluene is obtained.

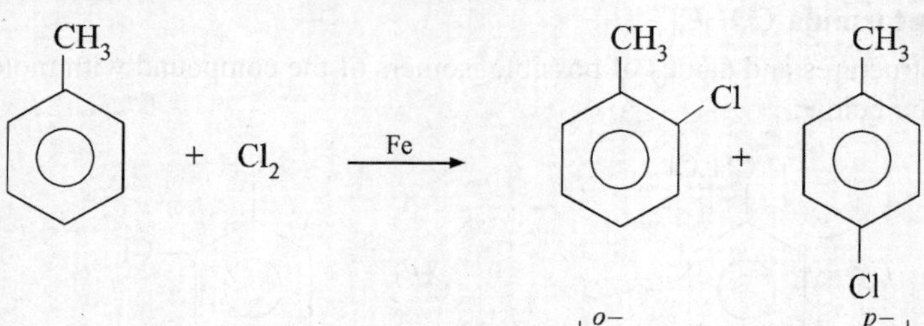

Chlorotoluene

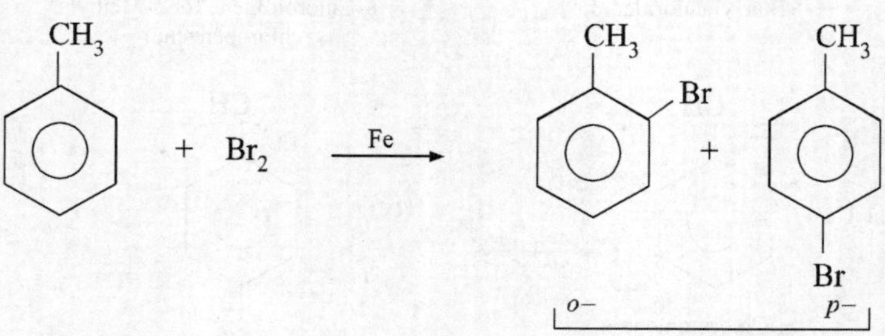

Bromotoluene

When aromatic compounds contain strongly deactivating groups such as $-NO_2$, $-COOH$, etc., high temperature is required and the halogen enters the meta position.

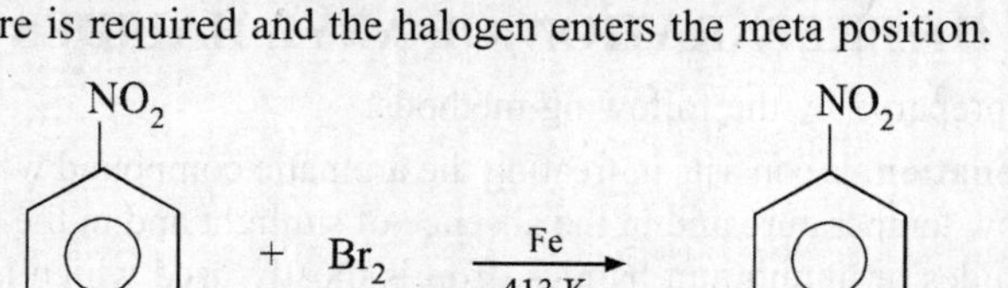

Nitrobenzene   *m*-Bromo nitrobenzene

(2) From diazonium salts (Sandmeyer's reaction). This is an important method for the preparation of aryl halides, specially those which cannot be prepared by the direct halogenation. Aryl chlorides and bromides are obtained by treating the diazonium salt solution with cuprous chloride or bromide in the corresponding halogen acid.

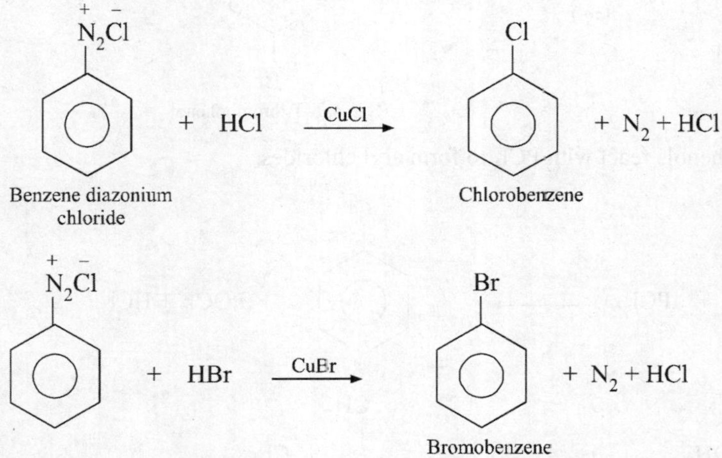

Traugott Sandmeyer

Modified form of above mentioned reaction, called **Gattermann** reaction, involves the use of copper powder and halogen acid instead of cuprous chloride or bromide.

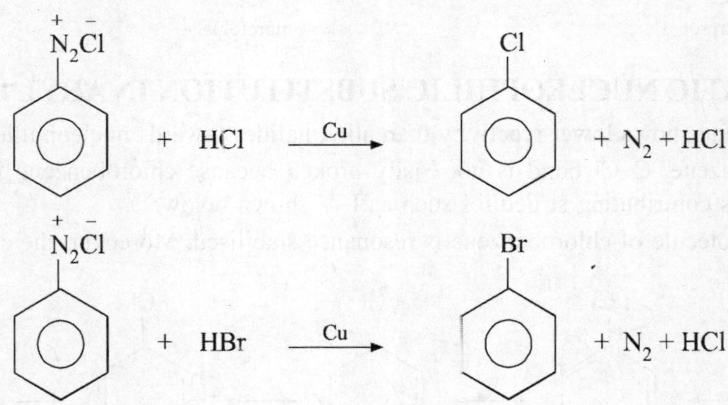

Aryl iodides are obtained by warming the diazonium salt solution with potassium iodide solution.

When aromatic compounds contain highly activating groups such as –OH, $-NH_2$ etc. they undergo halogenation in the absence of halogen carries and halogen enters the *ortho* and *para* positions.

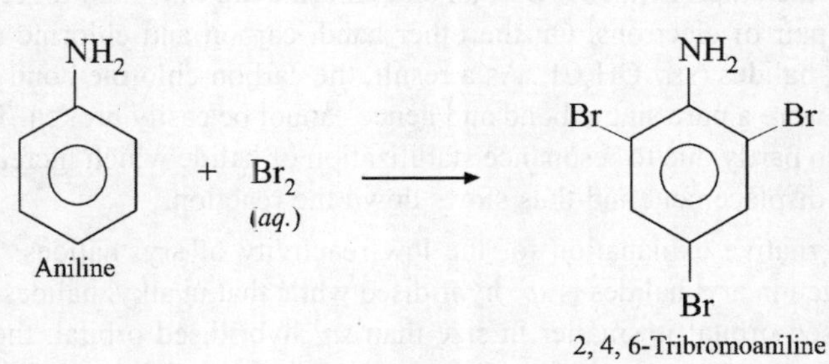

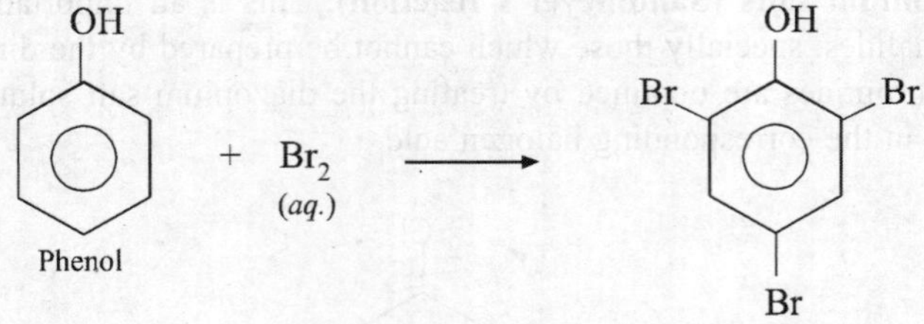

Phenol 2, 4, 6-Tribromophenol

(3) From Phenols. Phenols react with PCl_5 to form aryl chlorides.

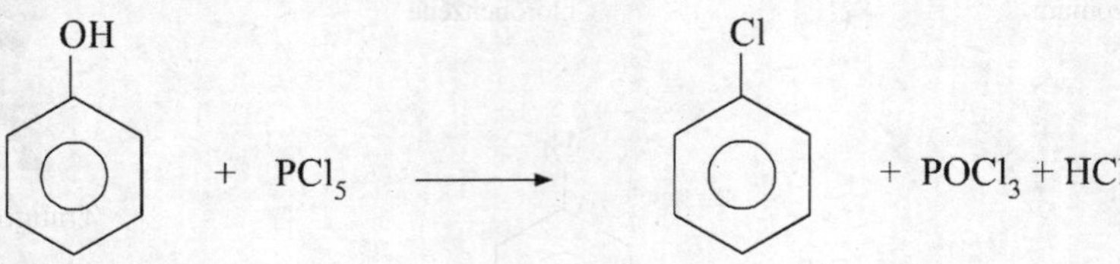

Phenol

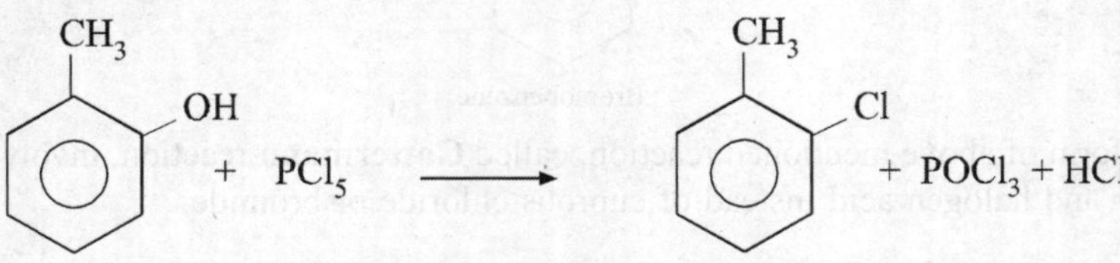

o-Cresol *o*-Chlorotoluene

7.26 AROMATIC NUCLEOPHILIC SUBSTITUTION IN ARYL HALIDES

Chlorobenzene shows lower reactivity than alkyl halides towards nucleophilic substitution.

In chlorobenzene, C–Cl bond is not easily broken because chlorobenzene is a resonance hybrid of various contributing structures such as I–V shown below:

Thus, the molecule of chlorobenzene is resonance stabilised. Moreover, the contribution of

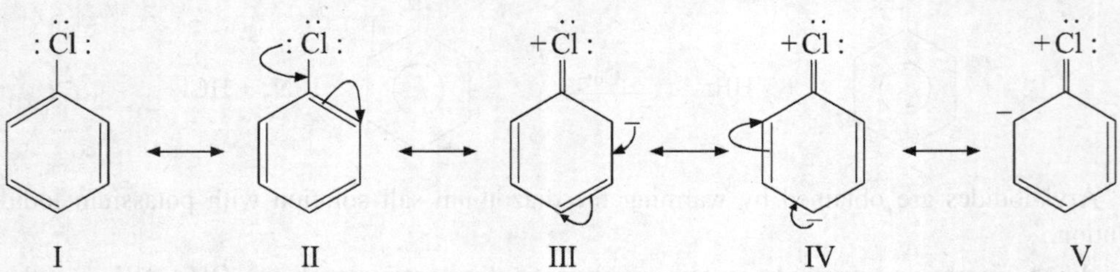

I II III IV V

structures like III to V imparts a partial double bond character to the carbon halogen bond in the resonance hybrid of the chlorobenzene. Carbon and chlorine are thus held together by little more than a single bond pair of electrons. On the other hand, carbon and chlorine are attached by a single bond in alkyl halides (say CH_3Cl). As a result, the carbon-chlorine bond in chlorobenzene is stronger than if it were a pure single bond and hence cannot be easily broken. The low reactivity of aryl halides is also partly due to resonance stabilization of halide which increases the energy of activation (E_{act}) for displacement and thus slows down the reaction.

There is an alternative explanation for the low reactivity of aryl halides. The carbon atom attached to the halogen in aryl halides is sp^2 hybridised while that in alkyl halides is sp^3 hybridised. Since a sp^2 hybridised orbital is smaller in size than sp^3 hybridised orbital, therefore, the C–Cl bond in chlorobenzene is shorter and hence stronger than in methyl chloride.

This has been confirmed by the X-ray analysis which shows that the C–Cl bond in chlorobenzene is 1.69 Å while in CH_3Cl it is 1.77 Å.

7.27 PROPERTIES OF ARYL HALIDES

Important reactions of aryl halides are discussed below:

(1) Nucleophilic substitution reactions.

(*i*) **Reaction with NaOH**. When heated with aqueous solution of NaOH at 573 K and under a pressure of 200 atmosphere, aryl halides form phenol.

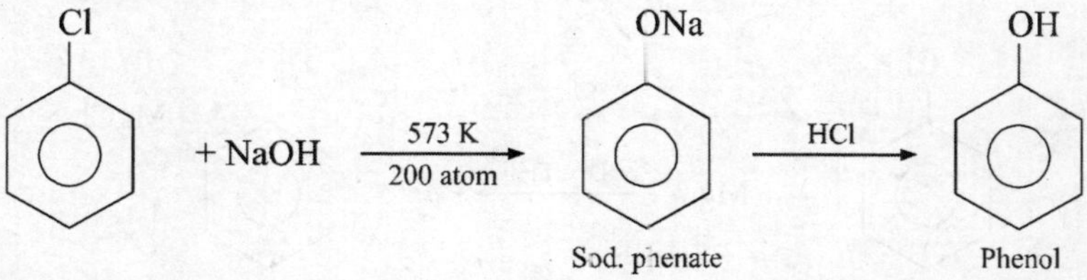

(*ii*) **Reaction with NH_3**. When aryl halides are heated with aqueous NH_3 at 473 K in the presence of Cu_2O as catalyst under a pressure of 60 atmosphere, amino compounds are obtained.

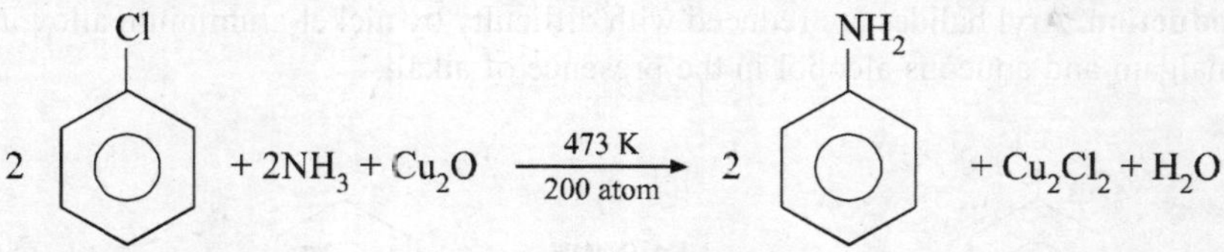

(*iii*) **Reaction with cuprous cyanide.** When aryl halides are heated with cuprous cyanide at 473 K in the presence of pyridine, the halogen atom is replaced by –CN group.

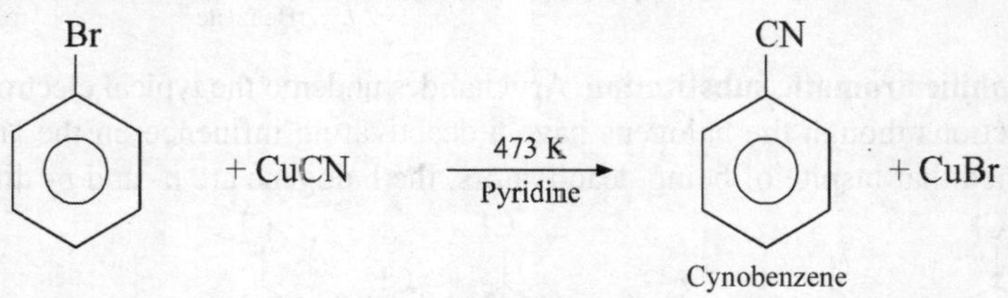

(2) Wurtz-Fittig reaction. When aryl halide is treated with ethereal solution of alkyl halide, in the presence of sodium, an alkyl benzene is formed.

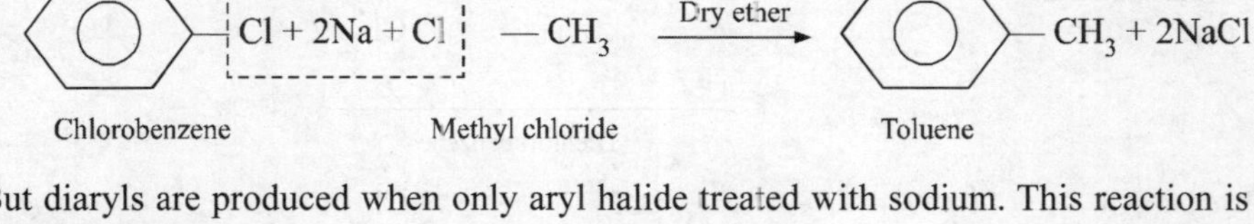

But diaryls are produced when only aryl halide treated with sodium. This reaction is called **Fittig's reaction.**

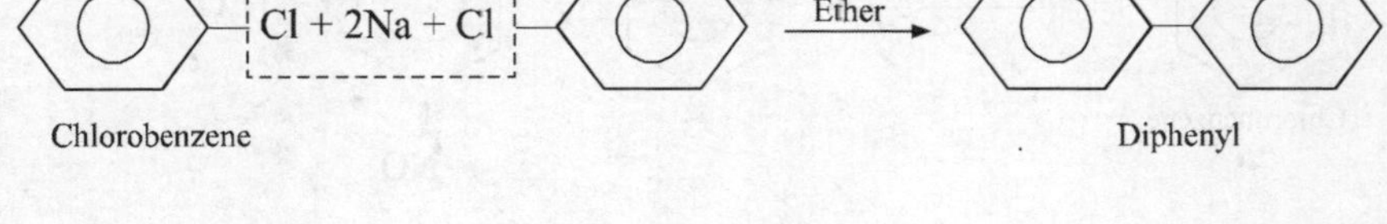

(3) Formation of Grignard reagent. Aryl bromides and iodides form Grignard reagent when they are treated with magnesium turnings in dry ether. Aryl chlorides form Grignard reagents only when the reaction is carried out in dry tetrahydrofuran (THF) as solvent

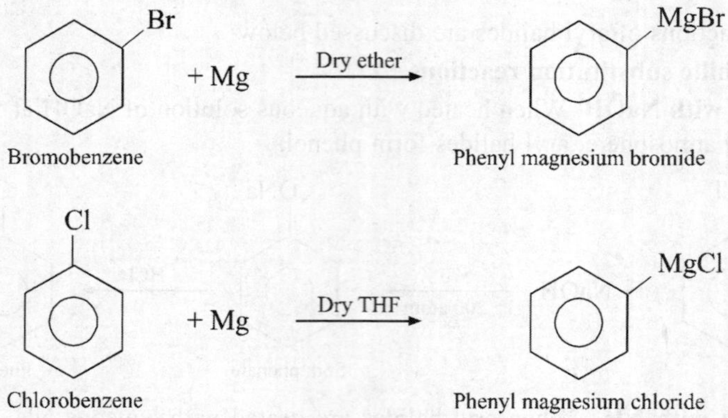

THF stands for tetrahydrofuran.

(4) Reduction. Aryl halides are reduced with difficulty by nickel-aluminium alloy in alkali or sodium amalgam and aqueous alcohol in the presence of alkali.

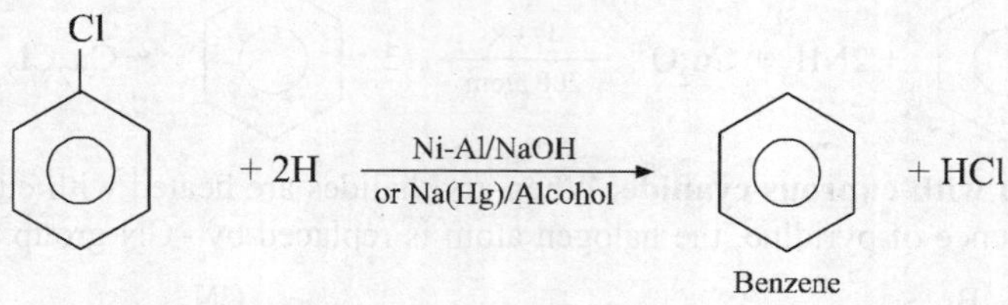

(5) Electrophilic aromatic substitution. Aryl halides undergo the typical electrophilic aromatic substitution reactions though the halogens have a deactivating influence on the aromatic ring. It may be mentioned that inspite of being deactivators, the halogens are *o*- and *p*- directing groups.

(*i*)

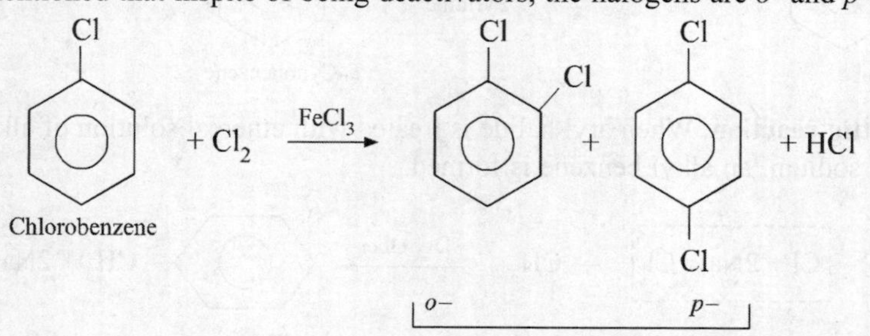

(*ii*)

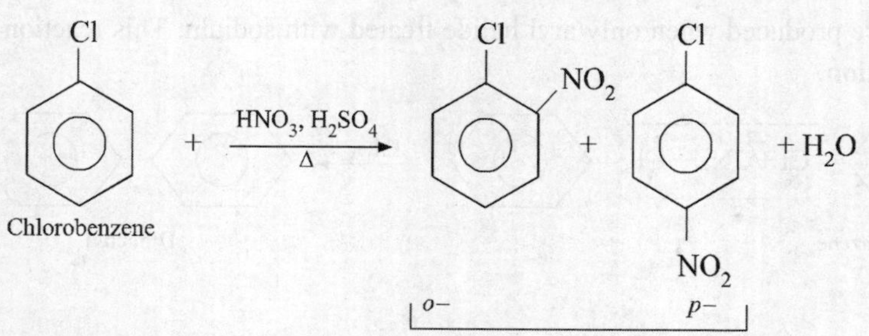

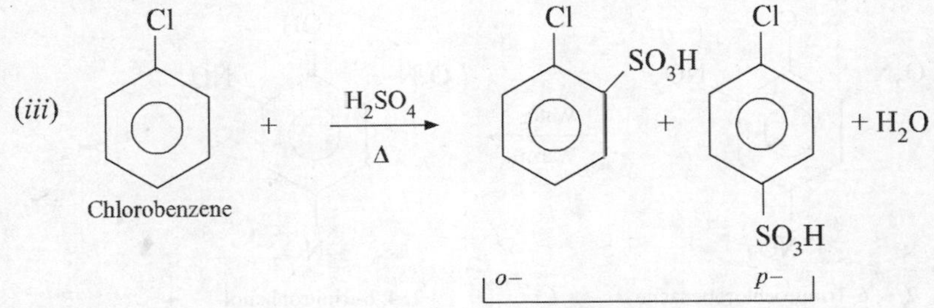

Chlorobenzene; o– / p– Chlorobenzene sulphonic acid

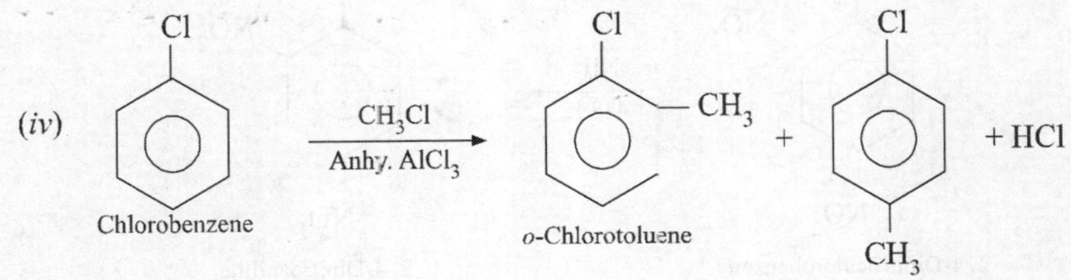

Chlorobenzene; o-Chlorotoluene; p-Chlorotoluene

This particular reaction is Friedel Crafts reaction.

(6) Ullmann reaction

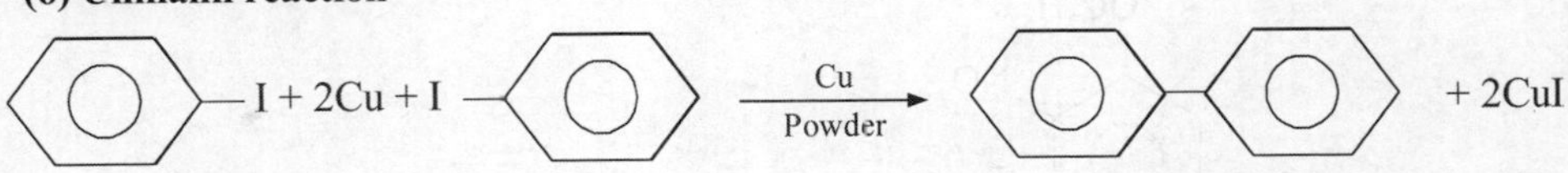

Diphenyl

7.28 NUCLEOPHILIC SUBSTITUTION IN ARYL HALIDES IN THE PRESENCE OF ACTIVATING AND DEACTIVATING GROUPS

As already discussed, aryl halides undergo nucleophilic substitution under drastic conditions. On the other hand, when powerful electron-withdrawing groups such as —NO_2, > C = O, —CHO, —COOH and —SO_3H etc. are present in *o*- and/or *p*-position with respect to the halogen atom, replacement of the latter by nucleophilic reagents take place under moderate or even ordinary conditions. For example,

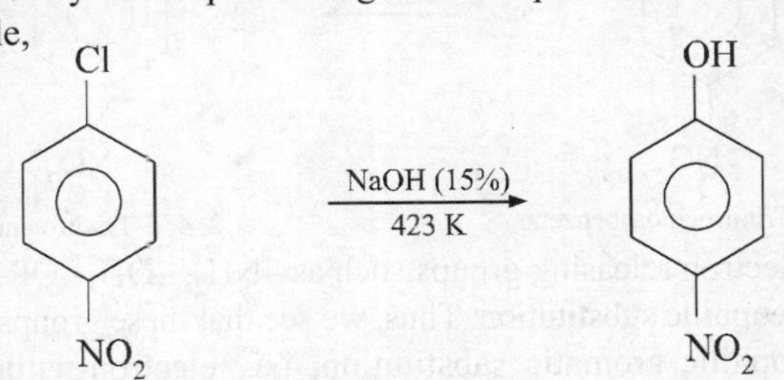

p-Chloronitrobenzene; p-Nitrophenol

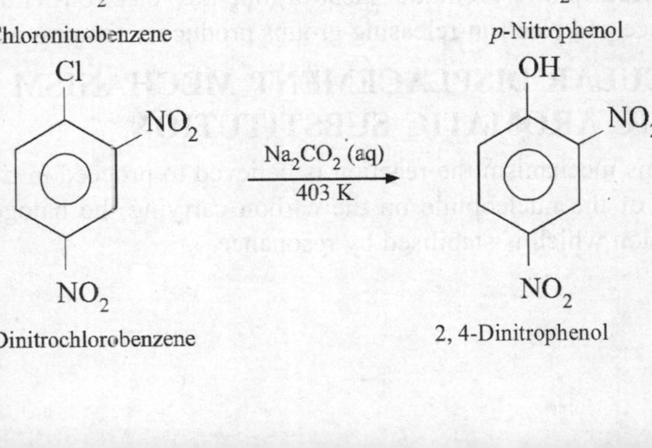

2, 4-Dinitrochlorobenzene; 2, 4-Dinitrophenol

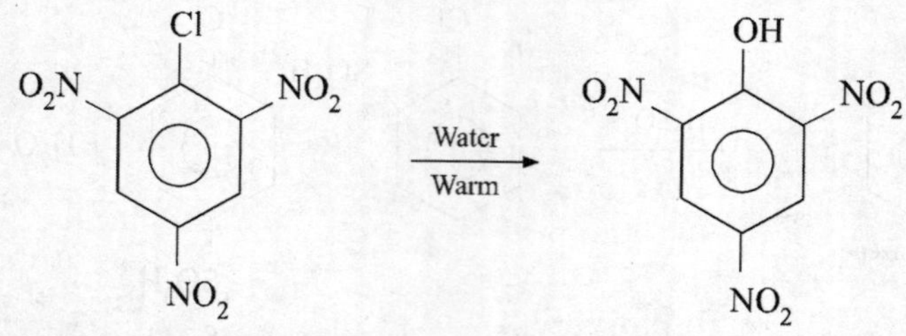

2, 4, 6-Trinitrochlorobenzene 2, 4, 6-trinitrophenol

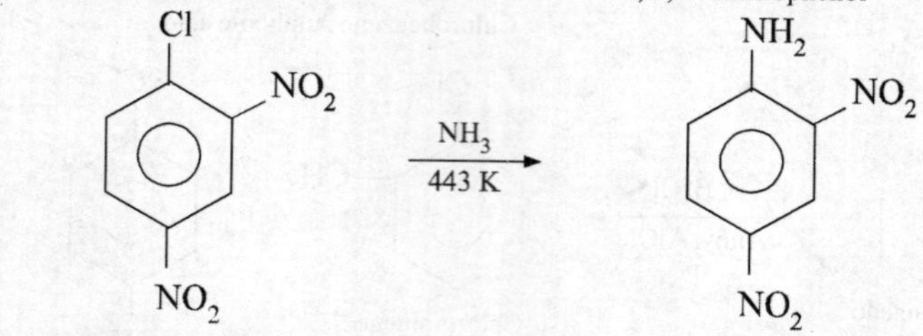

2, 4-Dinitrochlorobenzene 2, 4-Dinitroaniline

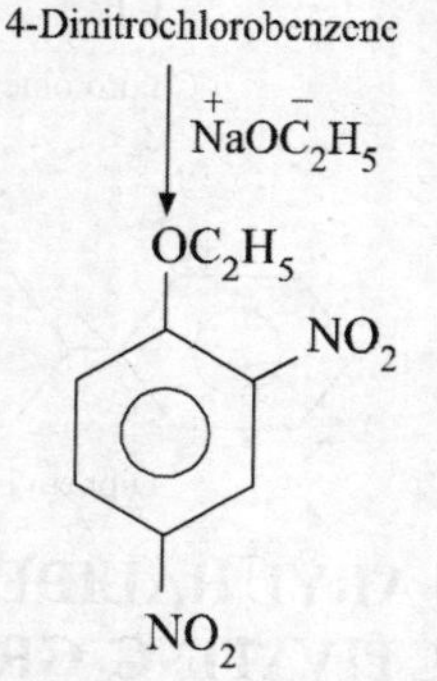

2, 4-Dinitrophenetole

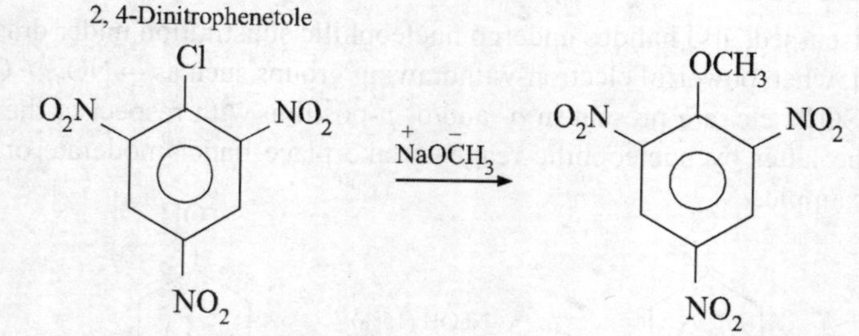

2, 4, 6-Trinitrochlorobenzene 2, 4, 6-Trinitro anisole

On the other hand electron releasing groups such as $–NH_2$, –OH, –OR, –R etc. deactivate the aryl halides towards nucleophilic substitution. Thus, we see that these groups show just the opposite influence on the electrophilic aromatic substitution, i.e., electron-withdrawing groups show deactivating influence and electron-releasing groups produce activating influence.

7.29 BIMOLECULAR DISPLACEMENT MECHANISM FOR NUCLEOPHILIC AROMATIC SUBSTITUTION

According to this mechanism the reaction is believed to proceed in two steps. The first step involves the attack of the nucleophile on the carbon carrying the halogen. This results in the formation of carbanion which is stabilised by resonance.

During the formation of intermediate carbanion, there is a change of hybridisation of the carbon involved from sp^2 to sp^3. As a result the aromatic character of the benzene ring is destroyed. Consequently, this step is slow and hence is the rate determining step of the reaction.

In the second step, the carbon loses the halide ion to form the product in which aromaticity is regenerated. General bimolecular displacement mechanism is represented as:

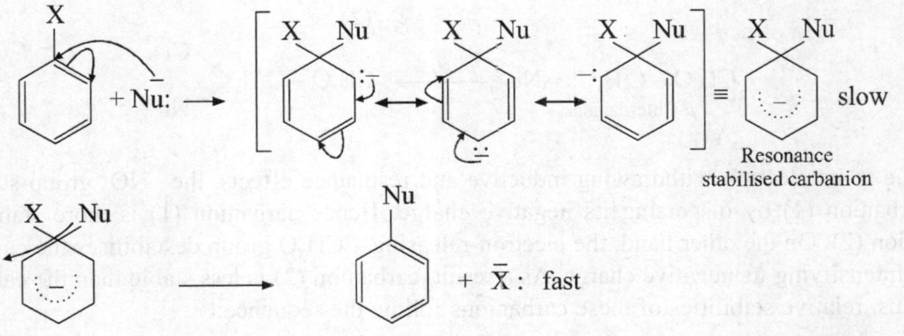

As two molecular species participate in the slow and rate determining step, it is a bimolecular reaction.

It may be mentioned that the above bimolecular mechanism is applicable mainly to activated aryl halides containing electron-withdrawing groups such as $-NO_2$, $-C \equiv N$, $-COOH$, etc. at the *ortho* and *para* positions.

Evidence in support of this mechanism. The rate determining step in bimolecular displacement reactions involves the formation of carbanion in which there is no cleavage of carbon-halogen bond. In other words, the rate of the reaction is independent of the strength of carbon-halogen bond. This has actually been verified by the absence of **element effect.** For example, there is only a little difference in the reactivity of aryl iodides, bromides, chlorides and fluorides in nucleophilic aromatic substitution reactions. This lends supports to the proposed mechanism.

7.30 EFFECT OF SUBSTITUENTS ON THE REACTIVITY OF ARYL HALIDES IN NUCLEOPHILIC SUBSTITUTION REACTIONS

In terms of bimolecular displacement mechanism, the nucleophile forms a carbanion with the aryl halide.

$$\overline{Nu}: + Ar-X \longrightarrow \underset{\text{Carbanion}}{Nu-\bar{Ar}-X}$$

Greater the stability of the carbanion, more rapidly the carbanion is formed and hence faster is the reaction. Thus, factors which stabilize the carbanion will increase the rate of the reaction, while factors which destabilise the carbanion decrease the rate of the reaction.

Consider the nucleophilic substitution reaction of *p*-nitrochlorobenzene, chlorobenzene and *p*-chloroanisol.

The relative rates of these substitution reactions depend upon the relative stabilities of the intermediate carbanions.

$$\underset{p\text{-Nitrochlorobenzene}}{O_2N-C_6H_4Cl} + \bar{N}u: \longrightarrow \underset{(1)}{O_2N-C_6\bar{H}_4\begin{matrix}\diagup Cl\\ \diagdown Nu\end{matrix}}$$

$$\underset{\text{Chlorobenzene}}{C_6H_5Cl} + \bar{N}u: \longrightarrow \underset{(2)}{C_6\bar{H}_5\begin{matrix}\diagup Cl\\ \diagdown Nu\end{matrix}}$$

$$\underset{p\text{-Chloroanisol}}{CH_3O-C_6H_4Cl} + N\bar{u}: \longrightarrow \underset{(3)}{CH_3O-\bar{C}_6H_4\begin{matrix}\diagup Cl\\ \diagdown Nu\end{matrix}}$$

Due to its electron-withdrawing inductive and resonance effects, the $-NO_2$ group stabilises the carbanion (1) by dispersing its negative charge. Hence carbanion (1) is more stable than carbanion (2). On the other hand, the electron-releasing $-CH_3O$ group destabilises the carbanion (3) by intensifying its negative charge. As a result, carbanion (3) is less stable than the carbanion (2). Thus, relative stabilities of these carbanions follow the sequence:

$$(1) > (2) > (3)$$

Therefore the order of reactivity of the above aryl halides in nucleophilic aromatic substitution reaction is:

p-Nitrochlorobenzene > Chlorobenzene > *p*-Chloroanisol.

It may, thus, be concluded that electron-withdrawing groups ($-NO_2$, $-CN$, $-COOH$ etc.) increase the reactivity while electron-releasing groups, ($-CH_3$, $-OH$, $-OCH_3$ etc.) decrease the reactivity of aryl halides towards nucleophilic substitution reactions.

Electron withdrawing substituents affect the reactivity of aryl halides towards nucleophilic substitution reactions only when present at the ortho and para positions with respect to halogen.

In order to explain the effect of substituents on orientation, a comparison of the carbanions formed during the attack of *ortho*, *para* and *meta*-nitrochlorobenzenes by a nucleophilic is made.

It is evident that the carbanions, resulting from *ortho* and *para*-nitrochlorobenzenes are resonance hybrid of four structures and that resulting from *meta*-nitrobenzene is a resonance hybrid of three structures. It is also clear that contributing structures (III) and (VI) are particularly stable because the negative charge in these is located on the carbon directly attached to the electron-withdrawing $-NO_2$ group. It is further noted that in the structures (IV) and (VIII), the negative charge is carried by highly electronegative oxygen and hence such structures are especially stable. No such structures are possible for the carbanion resulting from *m*-nitrochlorobenzene. Consequently, *o*-nitrochlorobenzene and *p*-nitrochlorobenzene undergo nucleophilic substitution much faster than *m*-nitrochlorobenzene.

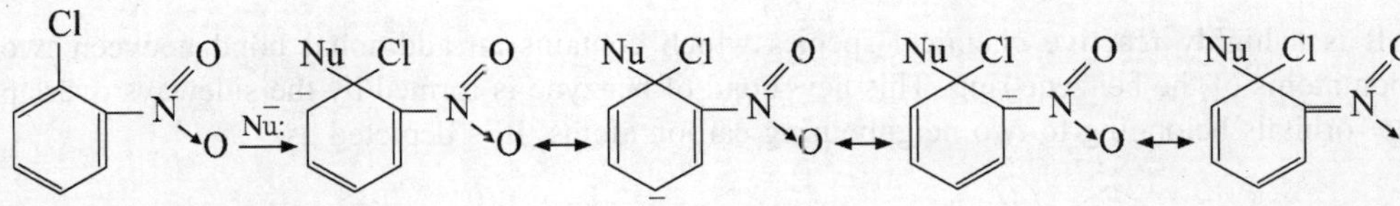

o-Nitrochlorobenzene I II III IV

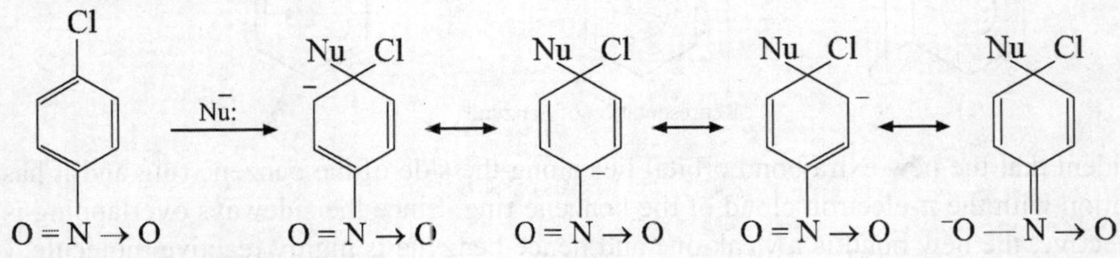

p-Nitrochlorobenzene V VI VII VIII

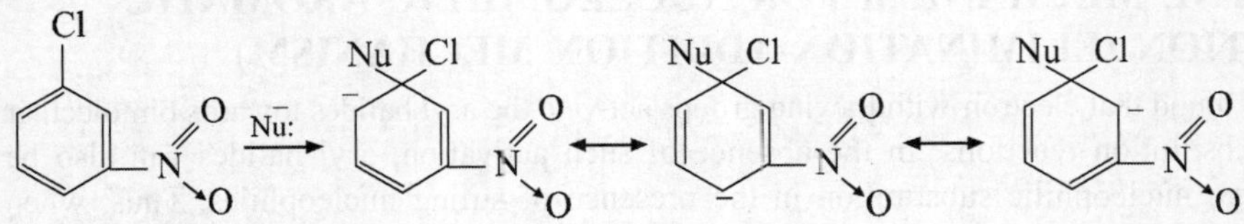

o-Nitrochlorobenzene IX X XI

It can be said, therefore, that electron-withdrawing groups present at *ortho* and *para* positions with respect to the halogen activate the aryl halide towards nucleophilic substitution reaction.

Example 2. **2, 4, 6-Trinitrochlorobenzene is easily hydrolysed with water but chlorobenzene is not hydrolysed. Explain**

Solution. It has been observed that the presence of electron-withdrawing groups at the *o*- and *p*-positions with respect to the halogen, always activates the aryl halide towards nucleophilic aromatic substitution reactions. Greater the number of such groups at *o*- and *p*-positions, more reactive is the corresponding aryl halide. Since there are three electron-withdrawing nitro groups present at the *ortho*- and *para*-positions, trinitrochlorobenzene is highly active and hence can undergo nucleophilic substitution reaction easily. There is no such group present at the *o*- and *p*-positions in chlorobenzene. Hence it requires drastic conditions to change into phenol.

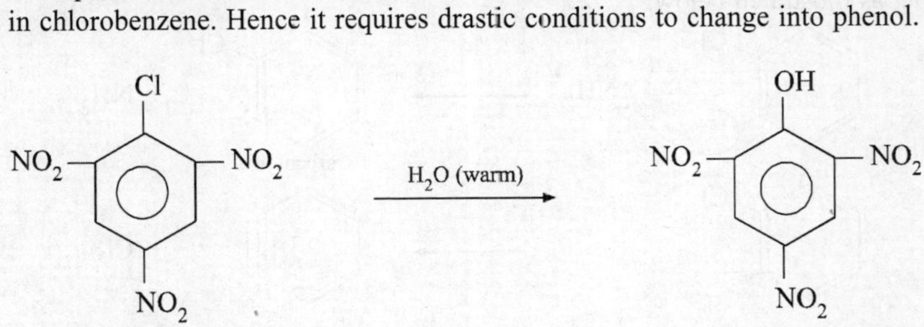

2, 4, 6-Trinitrochlorobenzene (Picryl chloride) → 2, 4, 6-Trinitrophenol (Picric acid)

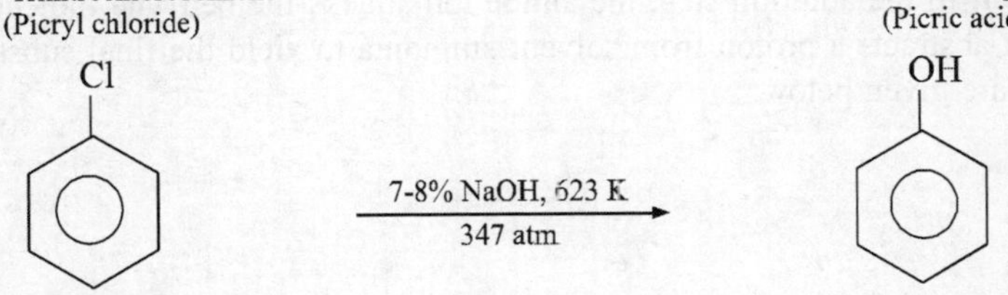

Chlorobenzene → Phenol

7.31 BENZYNE

It is a highly reactive chemical species which contains an additional bond between two carbon atoms of the benzene ring. This new bond of benzyne is formed by the sideways overlap of sp^2 orbitals belonging to two neighbouring carbon atoms. It is depicted as:

Representation of benzyne

It is evident that the new extra bond orbital lies along the side of the benzene ring and it has little interaction with the π-electron cloud of the benzene ring. Since the sideways overlapping is not very effective, the new bond is a weak one and hence benzyne is highly reactive molecule.

7.32 BENZYNE MECHANISM FOR NUCLEOPHILIC AROMATIC SUBSTITUTION (ELIMINATION-ADDITION MECHANISM)

It has been found that electron-withdrawing groups activate the aryl halides towards bimolecular nucleophilic substitution reactions. In the absence of such activation, aryl halides can also be made to undergo nucleophilic substitution in the presence of strong nucleophiles. Thus, when chlorobenzene is treated with very strong nucleophile, *viz.* amide ion, it is converted into aniline.

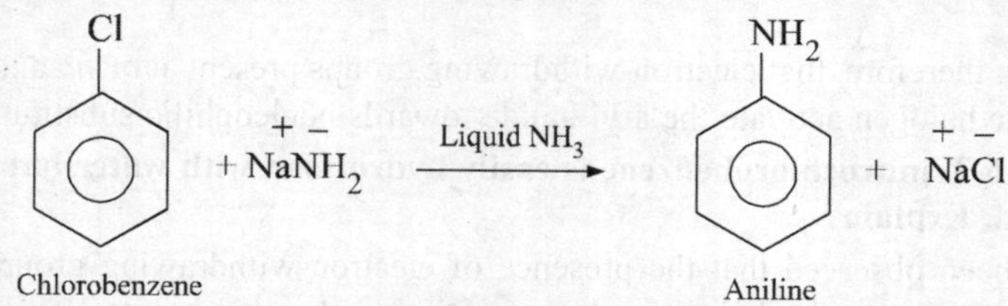

The above type of nucleophilic substitution reaction occurs by benzyne mechanism. It involves both **elimination** and **addition** as represented below:

(I) Elimination. In the elimination step, the amide ion abstracts a proton from one of the *ortho*-positions with respect to the halogen. The resulting carbanion loses the halide ion to form the benzyne, as illustrated below:

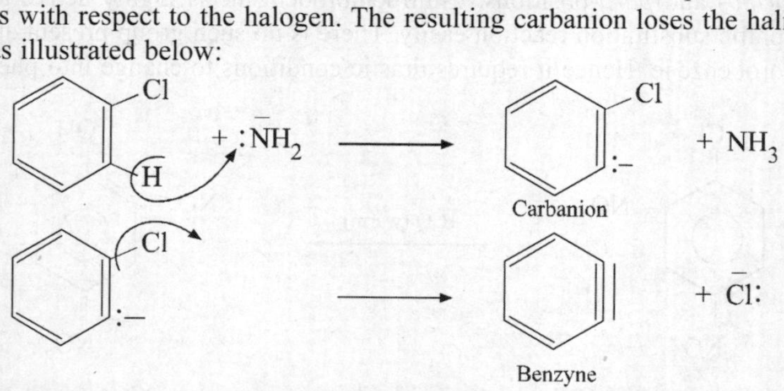

(II) Addition. In the addition step, the amide ion attacks the benzyne molecule to form the carbanion which abstracts a proton from solvent ammonia to yield the final substituted product. These reactions are given below:

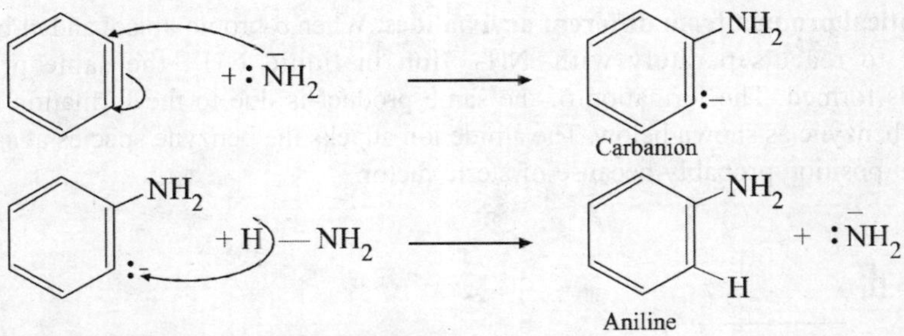

Evidence in support of benzyne mechanism

There is ample experimental evidence to prove the truth of benzyne mechanism of aromatic substitution.

(*i*) **Isotope effect.** The amination of *o*-deuteriobromobenzene is slower than that of bromobenzene. This observation indicates that the cleavage of *ortho* hydrogen is involved in the rate determining step during the elimination stage. This can further be illustrated as follows:

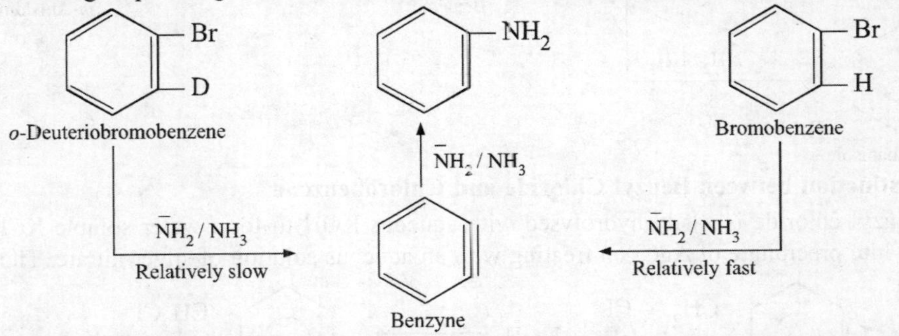

(*ii*) **Absence of ortho hydrogen.** Aryl halides containing two groups ortho to halogen, like 2-bromo-3-methylanisol, do not react at all. This is because benzyne intermediate cannot be formed due to the absence of ortho hydrogens with respect to the halogen.

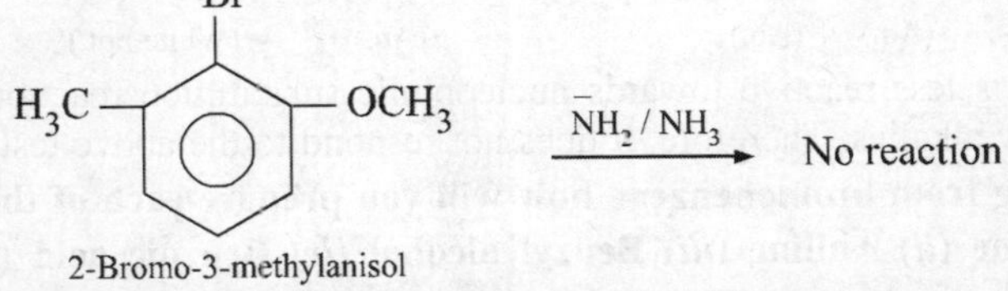

(*iii*) **Tracer studies.** When labelled chlorobenzene in which chlorine is linked to C^{14} isotope (C^*) is allowed to react with amide ion, two types of aniline are formed. In one type, the $-NH_2$ group is bonded to C^{14} while in the other it is attached to carbon *ortho* to the labelled carbon. This can be explained as follows:

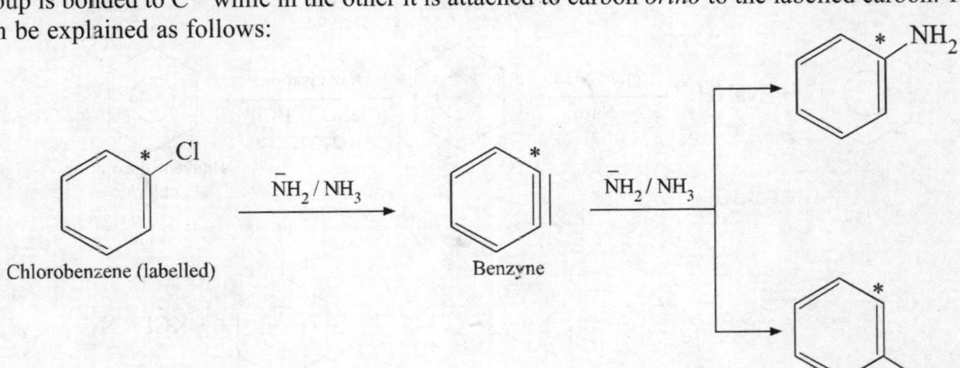

(*iv*) **Identical product from different aryl halides.** When *o*-bromo-anisol and *m*-bromo-anisol are allowed to react separately with NH_2^- ion in liquid NH_3, the same product, *i.e.*, *m*-anisidine is formed. The formation of the same product is due to the formation of the same intermediate benzyne as shown below. The amide ion attacks the benzyne species at *meta* position and not *ortho* position probably because of steric factor.

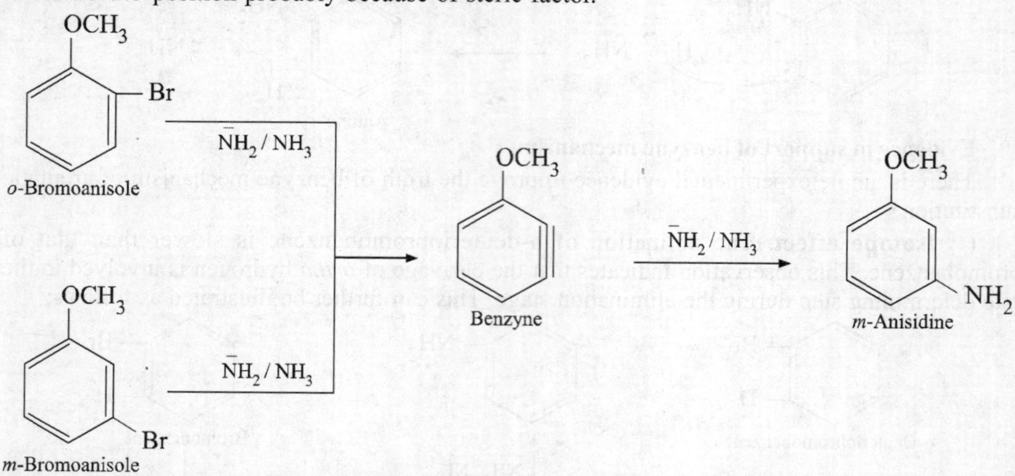

Distinction between Benzyl Chloride and Chlorobenzene

Benzyl chloride is easily hydrolysed with aqueous KOH to form water soluble KCl which gives white precipitate of AgCl on treating with an aqueous solution of silver nitrate. Thus,

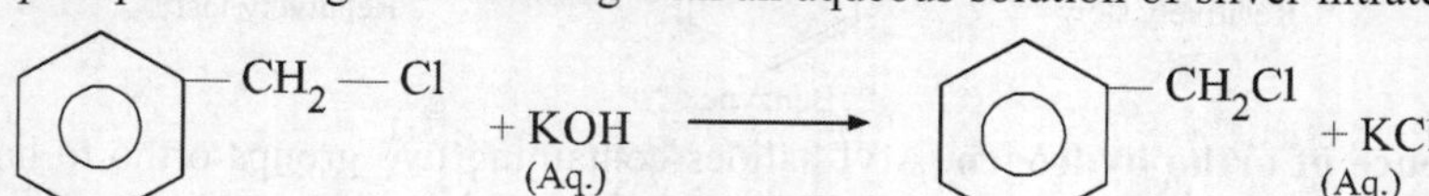

$$\underset{(Aq.)}{KCl} + \underset{(Aq.)}{AgNO_3} \longrightarrow \underset{(Aq.)}{KNO_3} + \underset{(White\ ppt.)}{AgCl}$$

Chlorobenzene is less reactive towards nucleophilic substitution reactions. Hence it is not easily hydrolysed by alkalies. Therefore, it does not respond to the above test.

Example 3. **Starting from bromobenzene how will you prepare each of the following:**

(*i*) Iodobenzene (*ii*) Aniline (*iii*) Benzyl alcohol (*iv*) Benzoic acid (*v*) Fluorobenzene (*vi*) Benzene

Solution. **(*i*) Iodobenzene**

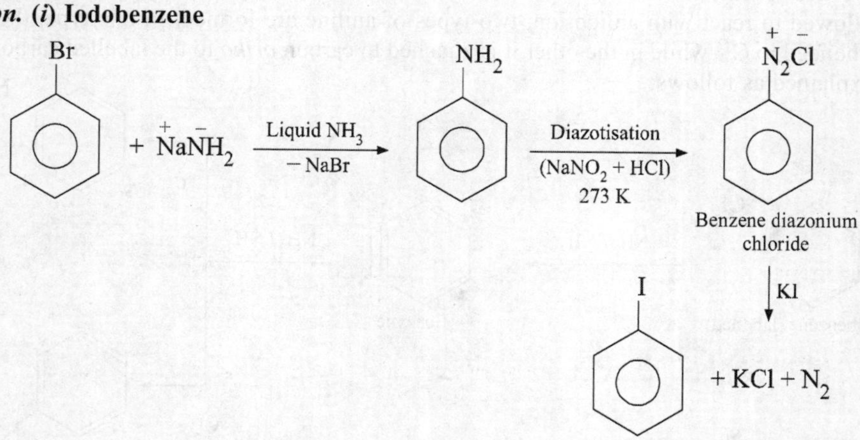

(*ii*) Aniline

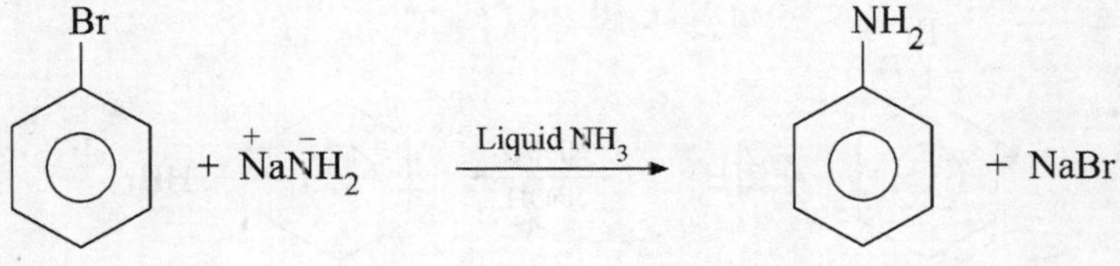

Aniline

(*iii*) Benzoic acid

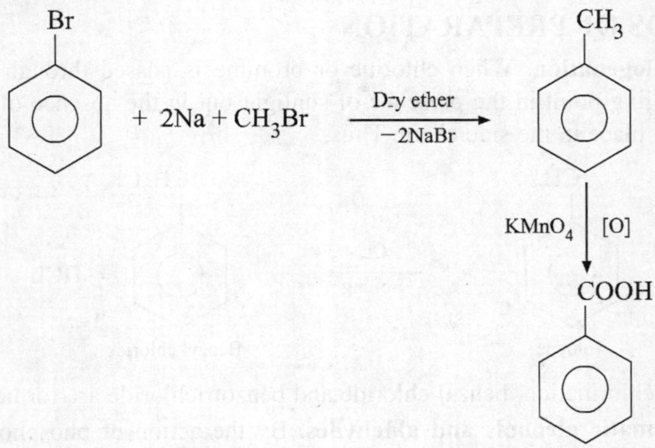

Benzoic acid

(*iv*) Benzyl alcohol

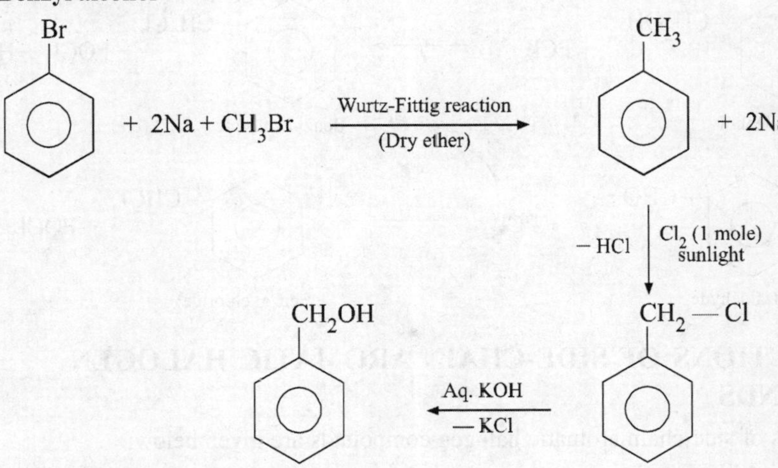

Benzyl alcohol

(*v*) Fluorobenzene

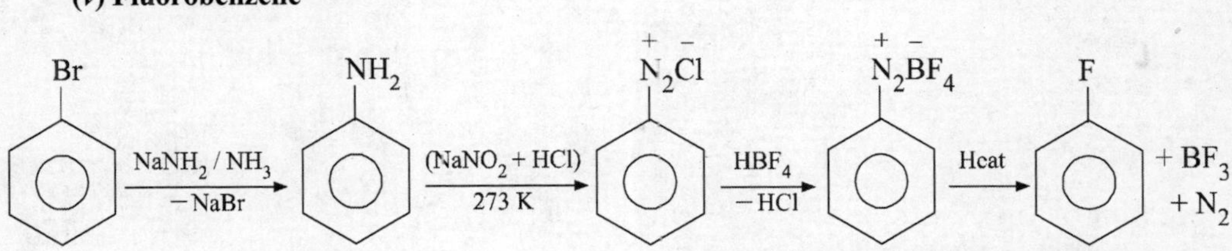

Benzene diazonium fluoborate

Fluorobenzene

(*vi*) Benzene

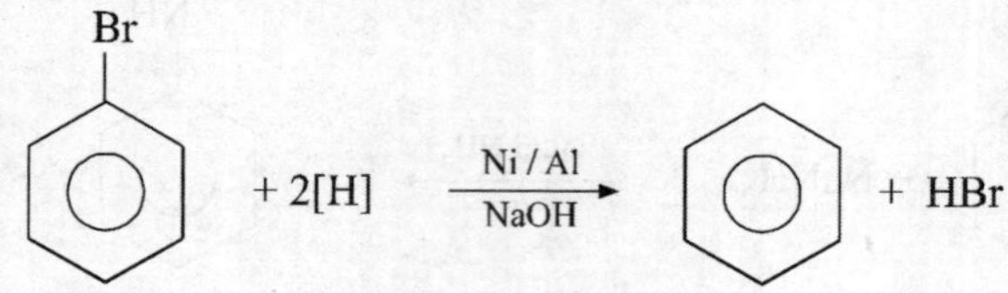

SIDE CHAIN HALOGEN COMPOUNDS

7.33 METHODS OF PREPARATION

(1) Direct halogenation. When chlorine or bromine is passed through toluene or higher homologues at boiling point in the presence of sunlight but in the absence of a halogen carrier, halogenation takes place in the side chain. Thus,

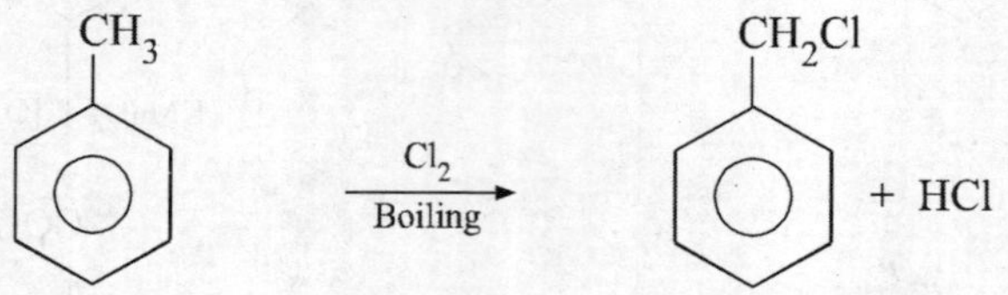

Toluene Benzyl chloride

On continued chlorination, benzal chloride and benzotrichloride are formed.

(2) From aromatic alcohols and aldehydes. By the action of phosphorus pentachloride. Thus,

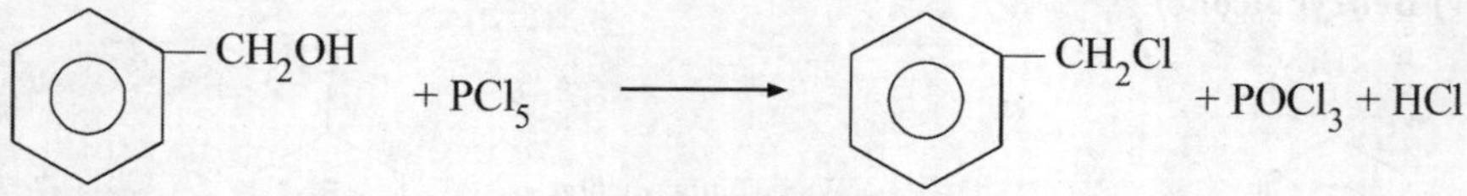

Benzyl alcohol Benzyl chloride

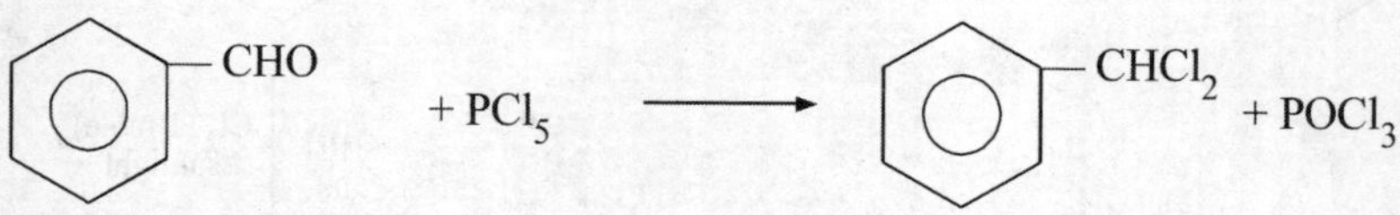

Benzaldehyde Benzal chloride

7.34 REACTIONS OF SIDE-CHAIN AROMATIC HALOGEN COMPOUNDS

Reactions of side-chain aromatic halogen compounds are given below:

1. Nucleophilic reactions

(*i*) Replacement by hydroxyl group

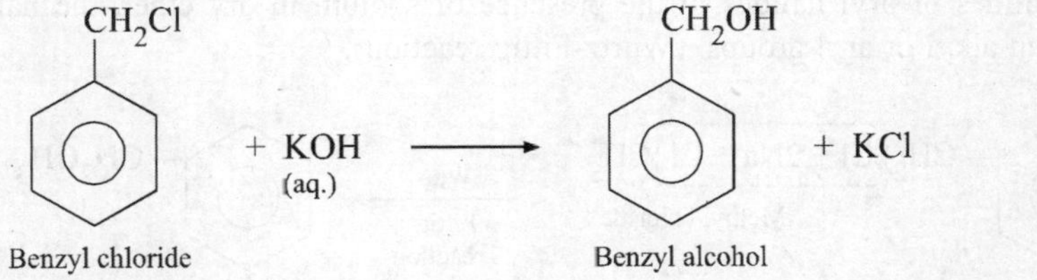

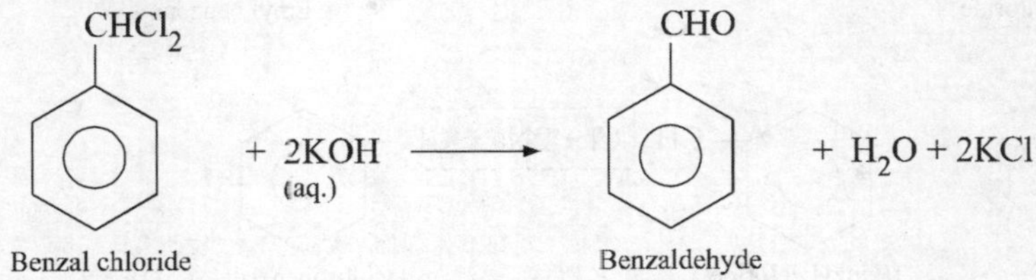

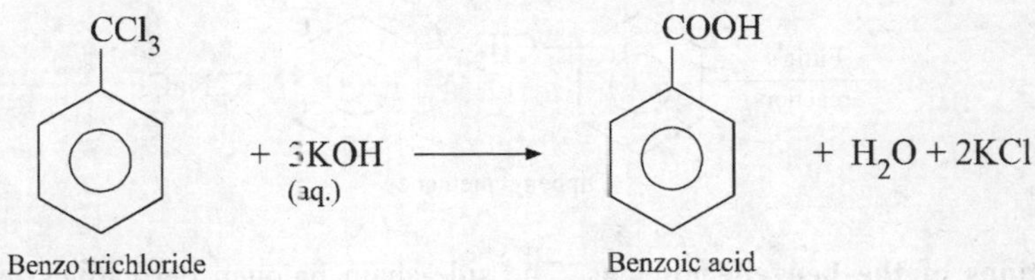

(*ii*) Replacement by amino group

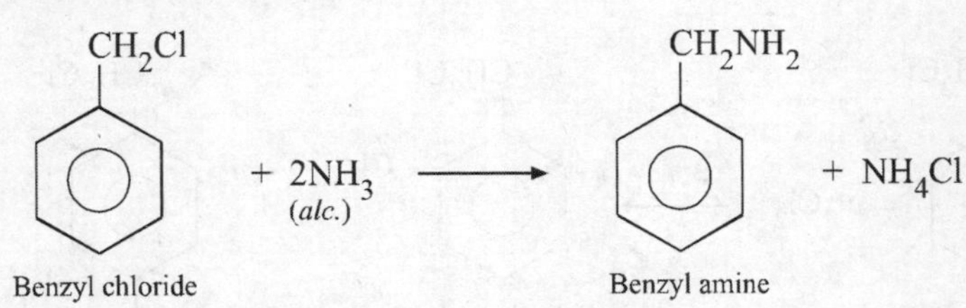

(*iii*) Replacement by cyano group

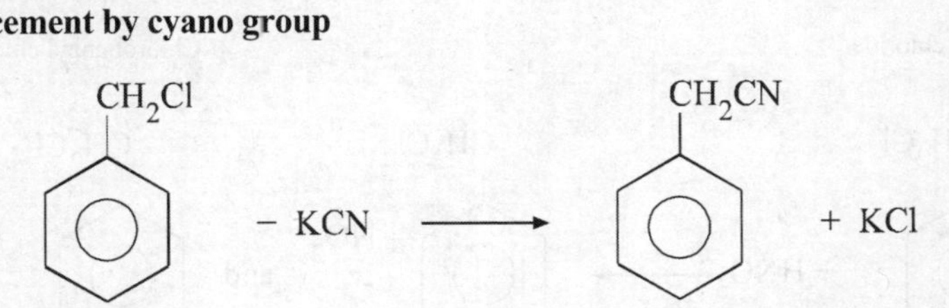

(*iv*) Replacement by hydrogen (Reduction). Reduction can be brought about by zinc-copper couple and alcohol to give parent hydrocarbon.

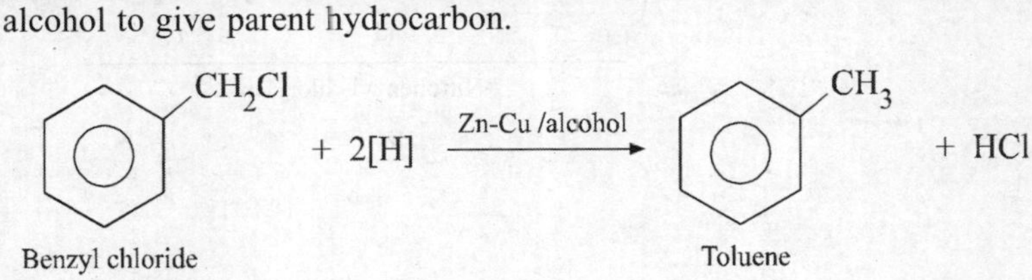

(v) Replacement by an alkyl or an aryl radical. On treating side-chain halogen derivative with alkyl halides or aryl halides in the presence of sodium in dry ether, the halogen atom is replaced by an alkyl or aryl groups. (Wurtz-Fittig reaction).

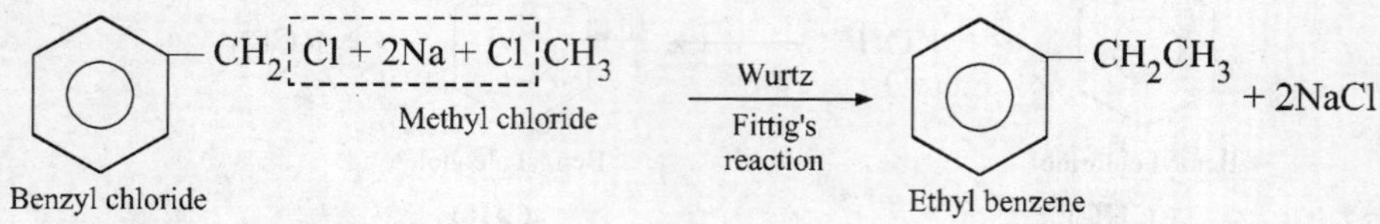

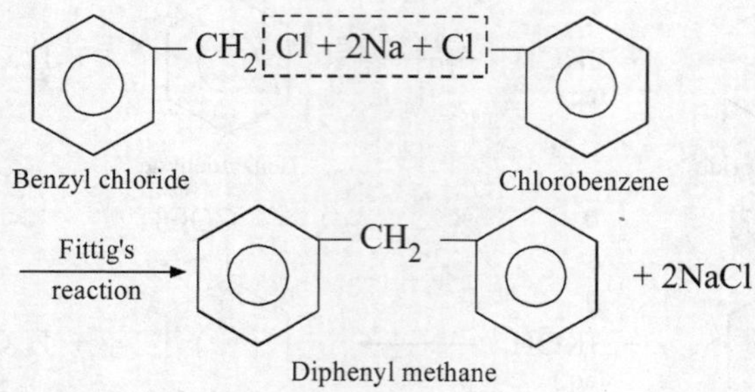

2. Reactions of the benzene nucleus. The side-chain halogen derivatives give the usual substitution reactions of benzene nucleus, *viz*., they can be nitrated, sulphonated and halogenated, the new entrant occupying the *ortho* and *para* positions.

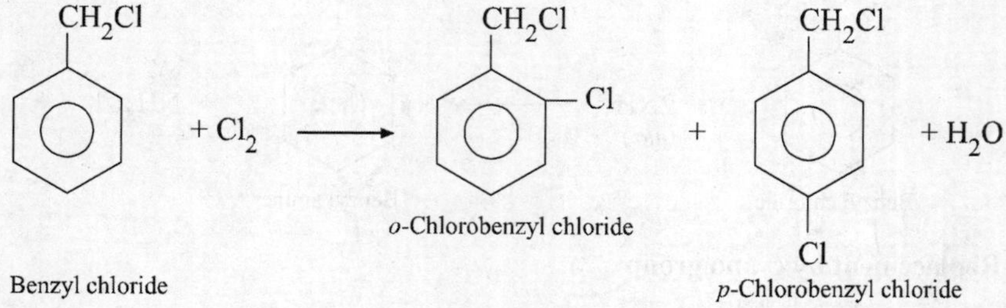

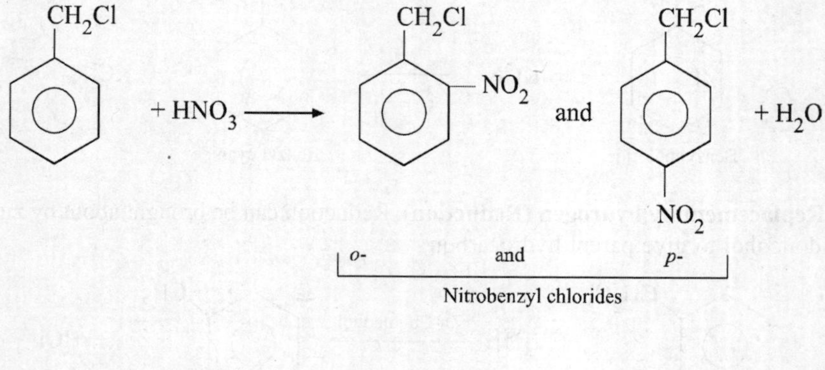

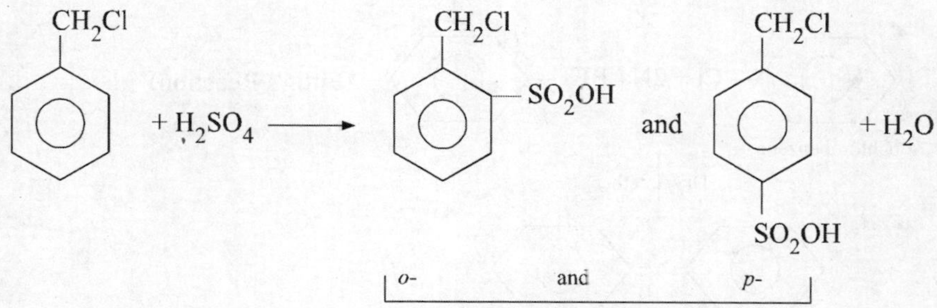

Benzyl chloride sulphonic acids

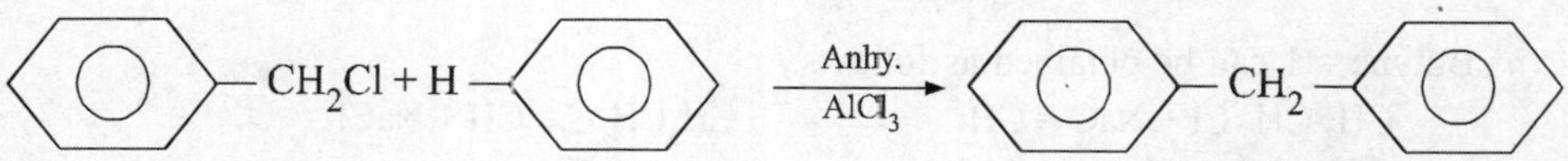

Diphenyl methane

This is Friedel Crafts type of reaction.

(3) Other reactions. (*i*) Formation of Grignard Reagents. Like nuclear halogen compounds, side chain derivatives form Grignard reagents on treatment with dry magnesium powder in dry ether.

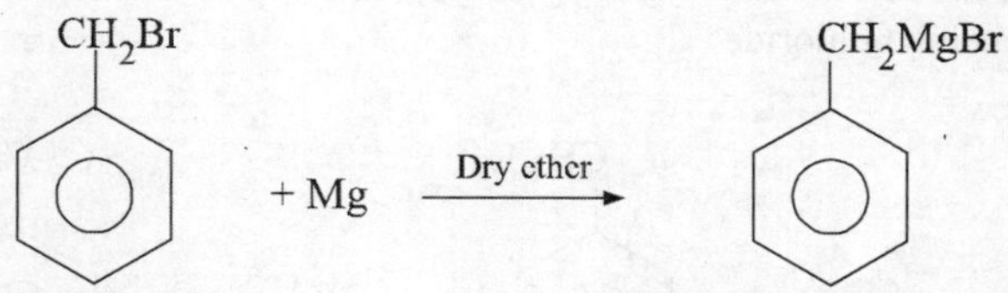

Benzyl bromide — Benzyl magnesium bromide

(*ii*) Oxidation. Action of oxidising agents results in the oxidation of the side chain including the halogen atom present in the chain. Mild oxidising agents like lead nitrate convert the side chain into –CHO group while reagents like nitric acid change it into –COOH group.

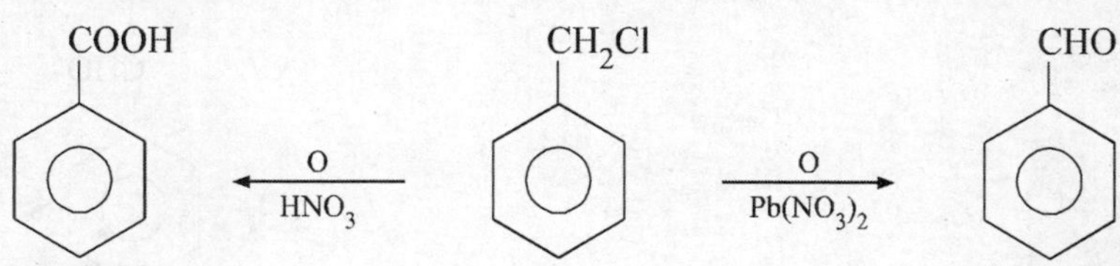

Benzoic acid — Benzaldehyde

(*iii*) Halogenation of side chain. On treatment with halogens at higher temperature, the presence of sunlight and absence of a halogen carrier, the side chain gets further halogenated.

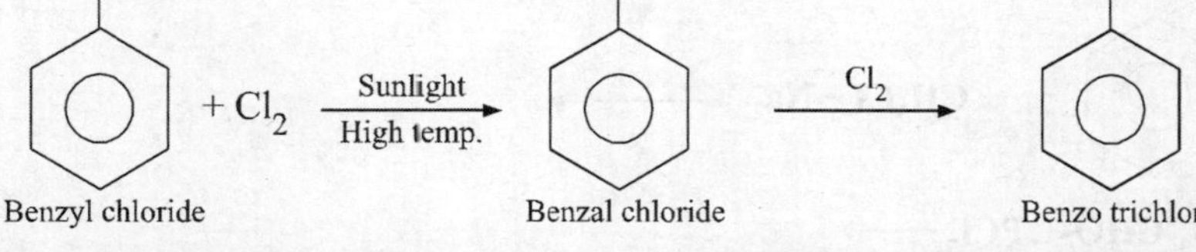

Benzyl chloride — Benzal chloride — Benzo trichloride

Example 4. **Starting with a suitable alkyl or aryl halide, how will you obtain (*i*) Diphenyl (*ii*) Butyne (*iii*) Benzaldehyde.**

Solution.

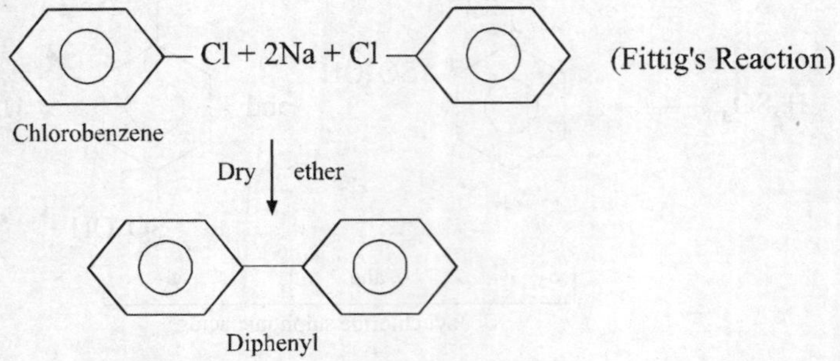

(*ii*) Butyne – 1 can be obtained as follows:

$$\underset{\text{Chloroethane}}{CH_3\,CH_2\,Cl} + \underset{\text{Sod.acetylide}}{NaC \equiv CH} \longrightarrow \underset{\text{Butyne-1}}{CH_3CH_2C \equiv CH} + NaCl$$

Butyne-2 can be obtained as follows:

$$\underset{\text{Chloro methane}}{CH_3 - Cl} + \underset{\text{Disodium acetylide}}{NaC \equiv CNa} + Cl - CH_3 \longrightarrow \underset{\text{Butyne-2}}{CH_3 - C \equiv C - CH_3} + 2NaCl$$

(*iii*) Benzaldehyde can be obtained by the hydrolysis of benzal chloride, which can be obtained by the chlorination of benzyl chloride

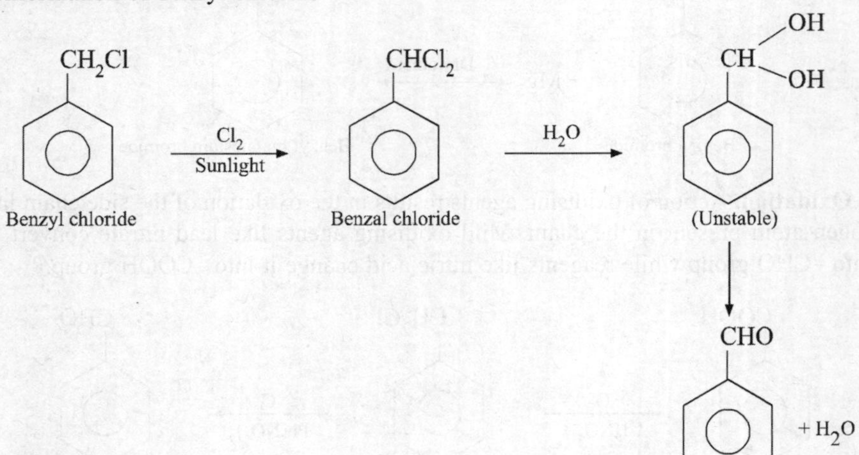

Example 5. **Complete the following reactions, giving the names of products.**

(*i*) 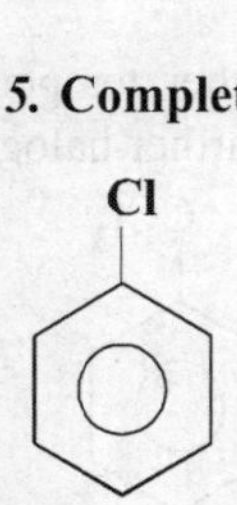

$+ CH_3Cl + Na \longrightarrow$

(*ii*) $CH_3\,CHO + PCl_5 \longrightarrow$

***Solution.* (*i*)**

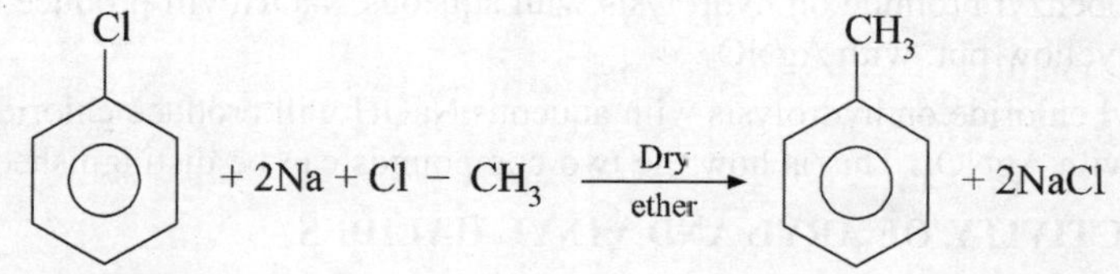

Chlorobenzene Toluene

(*ii*) $CH_3CHO + PCl_5 \longrightarrow CH_3CHCl_2 + POCl_3$

Acetaldehyde Benzal chloride Phosphorus oxy trichloride

***Example 6.* How would you distinguish between *o*-chlorobenzyl bromide and *o*-bromobenzyl chloride?**

Solution.

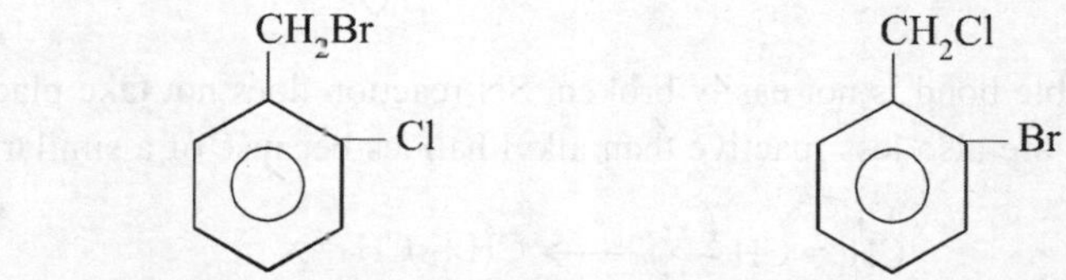

o-Chlorobenzyl bromide *o*-Bromobenzyl chloride

The two compounds can be distinguished by performing the hydrolysis reaction of the two compounds with aqueous NaOH under ordinary conditions. Under ordinary reaction conditions *i.e.* at room temp. and atmospheric pressure, nuclear chloro and bromo groups are not replaced. Only the halogen in side chain is substituted as given below:

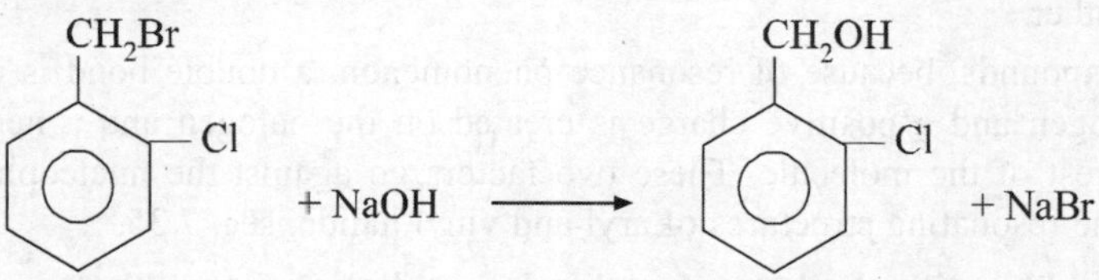

I *o*-Chlorobenzyl alcohol

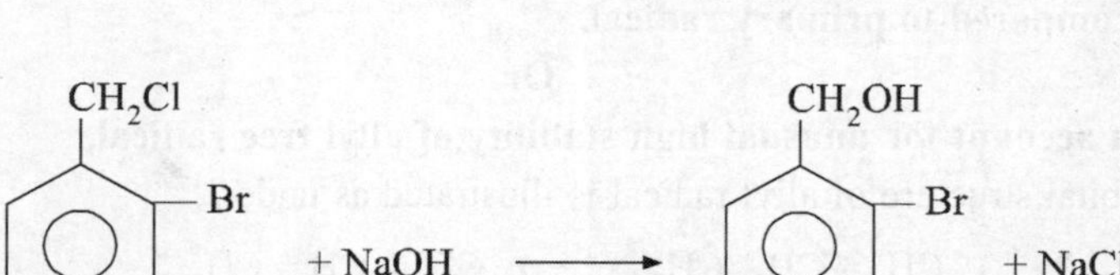

II *o*-Bromobenzyl alcohol

Thus compounds I and II give sodium bromide and sodium chloride respectively in the solution. These are ionic compounds and furnish bromide and chloride ions in solutions. These ions can be analysed by reaction with silver nitrate solution as given below:

$$Br^- + AgNO_3 \longrightarrow \underset{\text{Pale yellow ppt}}{AgBr} + NO_3^-$$

$$Cl^- + AgNO_3 \longrightarrow \underset{\text{White ppt}}{AgCl} + NO_3^-$$

Thus, *o*-chlorobenzyl bromide on hydrolysis with aqueous NaOH will produce bromide ions which give a pale-yellow ppt. with $AgNO_3$.

o-bromobenzyl chloride on hydrolysis with aqueous NaOH will produce chloride ions which give a white ppt. with $AgNO_3$. This is how the two compounds can be distinguished.

7.35 LOW REACTIVITY OF ARYL AND VINYL HALIDES

(*i*) Aryl halides are less reactive than alkyl halides towards nucleophilic substitution reactions because of the resonance stabilization of aryl halide.

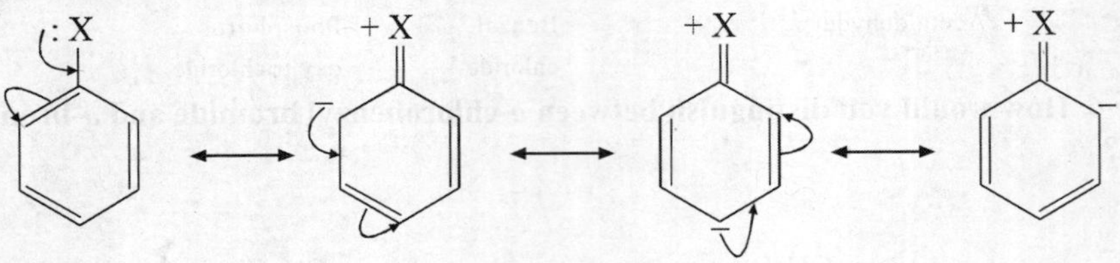

Since C – X double bond is not easily broken, SN reaction does not take place easily.

(*ii*) Vinyl halides are also less reactive than alkyl halides because of a similar reason.

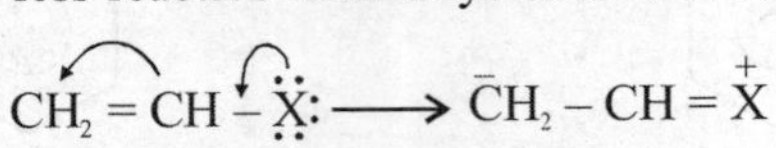

Double bond between C and X and presence of a positive charge on X prevent the SN reaction from taking place.

***Example 7.* Vinyl chloride is an aliphatic halide but it resembles chlorobenzene in its properties. Explain.**

Solution. Because of low reactivity of both towards nucleophilic substitution reactions, they resemble each other.

In both compounds, because of resonance phenomenon, a double bond is created between carbon and halogen and a positive charge is created on the halogen and a negative charge is created on the rest of the molecule. These two factors go against the nucleophilic substitution reactions. For the resonating structures of aryl and vinyl halide, sec. 7.35.

***Example 8.* Describe with the help of molecular orbital structure that an allyl radical is more stable as compared to primary radical.**

Or

How do you account for unusual high stability of allyl free radical.

Solution. The orbital structure of allyl radical is illustrated as under.

$$CH_2 = CH - \dot{C}H_2 \longleftrightarrow \dot{C}H_2 - CH = CH_2$$

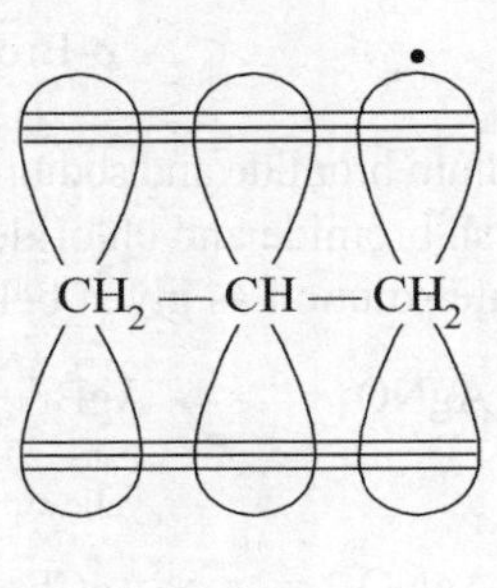

The delocalisation of π electron cloud lends stability to the allyl radical. It is believed that *p* orbitals of the doubly bonded carbon atoms and the single electron of the third carbon are in the same plane and overlap. This type of delocalisation is not possible in alkyl radicals

$$CH_3 = CH_2 - \dot{C}H_2$$

Example 9. **Bromobenzene does not give precipitate even on prolonged heating with alcoholic silver nitrate. Explain.**

Solution. 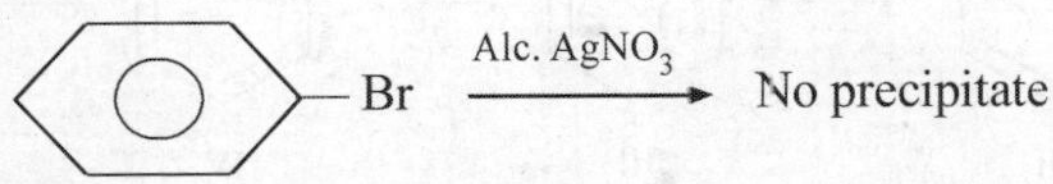

Aromatic halides, do not give nucleophilic substitution reactions easily.

As a result, bromine is not displaced as bromide ion in the solution, consequently a precipitate is not obtained.

Example 10. **An alcoholic solution of butyl bromide when heated with silver nitrate gives a white precipitate but pentyl bromide does not give precipitate of silver bromide even on prolonged heating with $AgNO_3$. Explain why.**

Solution. Butyl bromide and pentyl bromide are expected to give butyl and pentyl alcohol respectively on treatment with alcoholic silver nitrate by S_N2 mechanism. The bromide ions produced in the reaction react with $AgNO_3$ to give a precipitate of silver bromide.

The success of an S_N2 reaction depends upon the ease with which a nucleophile, can attach itself to the halogen carrying carbon.

In the case of *n*-butyl chloride, the groups attached to this carbon are H, H and C_3H_7.

$$C_3H_7 - \underset{\underset{H}{|}}{\overset{\overset{H}{|}}{C}} - Cl \qquad\qquad C_4H_9 - \underset{\underset{H}{|}}{\overset{\overset{H}{|}}{C}} - Cl$$

Butyl chloride   Pentyl chloride

In case of pentyl chloride, the groups attached are H, H and C_4H_9. Thus there is greater crowding and smaller chance of S_N2 reaction to take place in the case of pentyl chloride. Therefore, no precipitate is obtained.

Example 11. **Whereas Chlorobenzene reacts with sodamide in the presence of liquid ammonia, 2-bromo-3-methyl anisole fails to react. Explain.**

Solution. Chlorobenzene reacts with sodamide dissolved in liquid ammonia by benzyne mechanism to form aniline. In 2-bromo-3-methyl anisole both the ortho positions adjacent to the bromine atom are occupied in the ring. As a result, the ortho hydrogen atoms are not available to be eliminated.

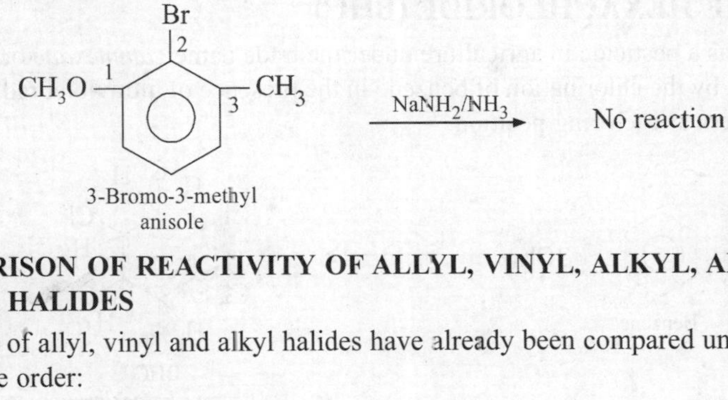

3-Bromo-3-methyl anisole

7.36 COMPARISON OF REACTIVITY OF ALLYL, VINYL, ALKYL, ARYL AND BENZYL HALIDES

Reactivities of allyl, vinyl and alkyl halides have already been compared under Alkyl Halides which follow the order:

Allyl halide > Alkyl halide > Vinyl halide

Benzyl halide is most reactive because the benzyl carbocation that is obtained as intermediate can stabilize itself to the maximum because of a number of resonating structures (I to V)

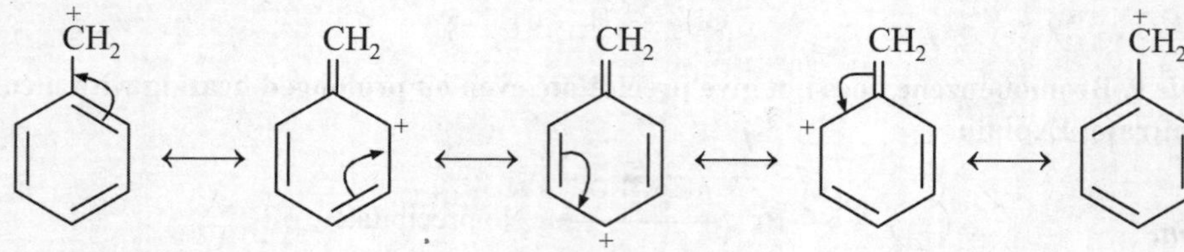

I II III IV V

Resonating structures of benzyl carbocation

Aryl halide is the least reactive because the molecule as a whole stabilizes by means of a number of resonating structures as opposed to only two structures in vinyl chloride.

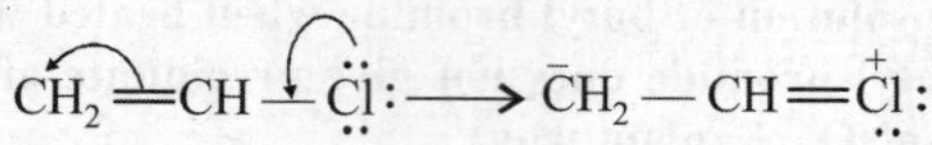

The double bond between chlorine and carbon in aryl chloride is thus, most difficult to break in aryl halide. The order of reactivity of halides is therefore as:

Benzyl > Allyl > Alkyl > Vinyl > Aryl.

7.37 1, 1-BIS (4-CHLOROPHENYL)-2, 2, 2-TRICHLOROETHANE, (DDT)

It is prepared by heating a mixture of chloral and chlorobenzene in the presence of concentrated sulphuric acid. It is used as an insecticide. DDT is not biodegradable. Its residues accumulate in environment and prove to be toxic to living beings.

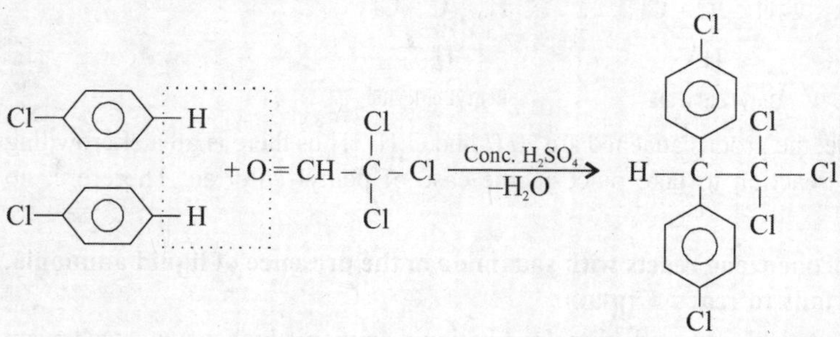

DDT (p, p- Dichlorodiphenyl trichloromethane)

DDT is used for pest control.

7.38 BENZENE HEXACHLORIDE (BHC)

BHC is used as a pesticide in agriculture under the trade name *gammexane* or *Lindane* or *666*. It is manufactured by the chlorination of benzene in the presence of ultraviolet radiation where one chlorine atom enters at each ring position.

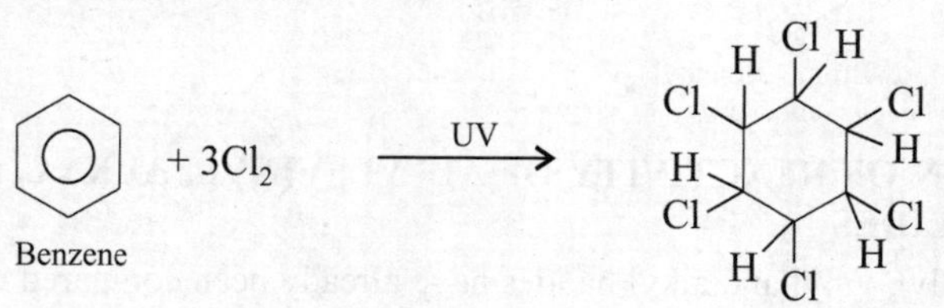

Benzene

BHC

(*Benzene hexachloride*)

EXERCISE
(INCLUDING QUESTIONS FROM DIFFERENT UNIVERSITY PAPERS)

Objective Questions

Fill in the blanks

1. In halogenation of benzene, the function of Al Cl_3 is to facilitate the formation of ....
2. Addition of chlorine to benzene in the presence of UV light proceeds through .... mechanism.
3. Benzene diazonium chloride on reaction with Cu_2 Cl_2/HCl gives ....
4. C—Cl bond of chlorobenzene has a .... bond character and is .... than a C—Cl bond when Cl is not linked to benzene ring.
5. Benzyl chloride behaves like ethyl chloride because the former is aromatic equivalent of ....
6. Nucleophilic aromatic substitution is understood by .... mechanism.
7. Structure of DDT is ....
8. Nuclear halogenation of benzene needs presence of ....
9. SN^2 mechanism is a ....step mechanism.
10. Benzyne mechanism is also called ....mechanism.

Multiple Choice Questions

Choose the correct option

1. Alkyl halides undergo
(*a*) Electrophilic substitution reactions (*b*) Electrophilic addition reactions
(*c*) Nucleophilic substitution reactions (*d*) Nucleophilic addition reactions
2. 1-Bromobutane reacts with alcoholic KOH to mainly give
(*a*) 1-Butene (*b*) 2-Butene (*c*) 1-Butanol (*d*) 2-Butanol
3. 2,2-Dichloropropane reacts with aqueous KOH to give
(*a*) 2,2-propanediol (*b*) propanal (*c*) acetone (*d*) propene
4. When chloroform is heated with aqueous NaOH, it gives
(*a*) formic acid (*b*) sodium formate (*c*) acetic acid (*d*) sodium acetate
5. 1,1-Dichlororopane reacts with aqueous KOH to give
(*a*) 1,1-propanediol (*b*) acetaldehyde (*c*) propanone (*d*) propyne
6. *n*-Propyl iodide reacts with sodium ethoxide to give:
(*a*) $CH_3CH_2OCH_2CH_3$ (*b*) $CH_3CH_2OCH_2CH_2CH_3$
(*c*) $CH_3CH_2OCH_3$ (*d*) $CH_3OCH_2CH_2CH_3$
7. Isopropyl bromide reacts with alcoholic KOH to give
(*a*) Propene (*b*) Isopropyl alcohol (*c*) Propane (*d*) *n*-Propyl alcohol
8. Which of the following compounds will give a white precipitate with alcoholic $AgNO_3$?
(*a*) Vinylbenzene (*b*) Chlorobenzene (*c*) Vinyl chloride (*d*) Allyl chloride
9. 2-Bromobutane reacts alcoholic KOH to mainly give
(*a*) 1-Butene (*b*) 2-Butene (*c*) 1-Butanol (*d*) 2-Butanol
10. Which of the following compounds will not give a white precipitate with alcoholic $AgNO_3$?
(*a*) Bromobenzene (*b*) 1-Bromopropane (*c*) Ethyl bromide (*d*) 2-Bromopropane
11. Carbylamine test involves heating a mixture of
(*a*) alcoholic KOH, methyl iodide, and sodium metal

(*b*) alcoholic KOH, methyl iodide, and primary amine
(*c*) alcoholic KOH, chloroform, and primary amine
(*d*) alcoholic KOH, methyl alcohol, and primary amine

12. Aryl halides are less reactive towards nucleophilic substitution as compared to alkyl halides due to
(*a*) inductive effect (*b*) resonance stabilization
(*c*) tautomerism (*d*) stereoisomerism

13. Hydrolysis of benzal chloride gives
(*a*) phenol (*b*) benzaldehyde (*c*) benzyl alcohol (*d*) benzoyl chloride

14. Which of the following reagents can be used to distinguish between chlorobenzene and benzyl chloride?
(*a*) alcoholic $AgNO_3$ (*b*) Br_2 in CCl_4
(c) KCN (*d*) Br_2 in H_2O

15. The following compound is called?

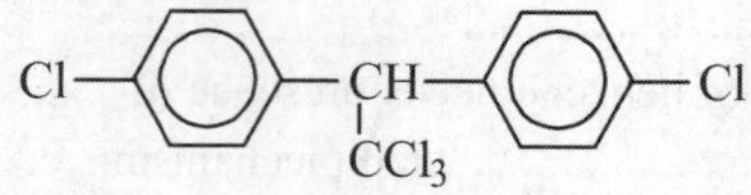

(*a*) Chloral (*b*) DDT (*c*) Lindane (*d*) BHC

16. Hydrolysis of benzotrichloride gives
(*a*) benzophenone (*b*) benzoic acid (*c*) benzyl alcohol (*d*) phenol

17. Which of the following compounds would be unaffected by boiling aqueous NaOH?
(*a*) Chloroethane (*b*) Acetaldehyde (*c*) Chlorobenzene (*d*) Ethyl acetate

18. DDT is prepared by the reaction of chlorobenzene with (in the presence of conc. H_2SO_4)
(*a*) chloral (*b*) chlorine (*c*) chloroform (*d*) carbon tetrachloride

19. Which reagent can be used to carry out the following conversion?

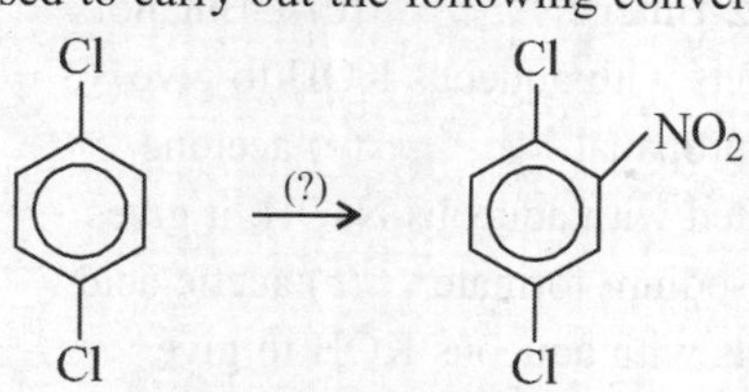

(*a*) $HNO_3 + H_2SO_4$ (*b*) $NaNO_2$
(*c*) $NaNO_2$ – HCl at 5°C (*d*) NO_2

20. Which of the following reagents can be used to distinguish between chlorobenzene and allyl chloride?
(*a*) H_2/Ni (*b*) Br_2 in CCl_4 (*c*) Zn/HCl (*d*) NH_2NH_2

ANSWERS

Fill in the blanks

1. electrophile **2.** free-radical **3.** chlorobenzene **4.** double, shorter

5. aliphatic halide **6.** benzyne **7.** $(Cl-Ph)_2CH-CCl_3$

8. halogen carrier **9.** two **10.** elimination addition

Multiple choice questions

1. (*c*)	**2.** (*b*)	**3.** (*c*)	**4.** (*b*)	**5.** (*b*)
6. (*b*)	**7.** (*a*)	**8.** (*d*)	**9.** (*b*)	**10.** (*a*)
11. (*c*)	**12.** (*b*)	**13.** (*b*)	**14.** (*a*)	**15.** (*b*)
16. (*b*)	**17.** (*c*)	**18.** (*a*)	**19.** (*a*)	**20.** (*b*)

General Questions

ALKYL HALIDES

1. What happens when 2-bromobutane is treated with alcoholic KOH ?

2. Enlist the factors which influence S_N2 reactions.

3. What happens when 2-bromobutane is heated with alcoholic KOH ?

4. Draw the structure of isobutyl chloride and secondary butyl chloride and write their IUPAC name.

5. How will you show that S_N2 reactions in alkyl halides are accompanied by racemisation.

6. Justify that all S_N 2 reactions proceed by inversion of configuration.

7. How will you convert ethyl chloride into :

(*i*) *n*-Propyl chloride (*ii*) 1-Butyne (*iii*) Ethylene

8. Aqueous KOH causes mainly substitution in alkyl halides while alcoholic KOH results in elimination. Explain.

9. Compare the important features of S_N1 and S_N2 reactions.

10. Why do primary alkyl halides take part mainly in S_N2 reactions while the tertiary alkyl halides participate in S_N 1 reactions.

11. Give the mechanism of S_N1 reactions in alkyl halides.

12. Allyl chloride is hydrolysed more easily than ethyl chloride. Explain.

13. Haloalkanes are polar in nature but are still insoluble in water. Explain.

14. How will you convert :

(*i*) Ethylene to 2-butyne (*ii*) Ethane to butane (*iii*) Ethyl bromide to ethane.

15. Identify X and Y in the following sequence of reactions :

(*i*) $CH_3CHCl_2 \xrightarrow{KCN} X \xrightarrow[Heat]{H_2O/H^+} y.$

(*ii*) $\underset{\displaystyle |\atop \displaystyle I}{CH_3-CH}-\underset{\displaystyle |\atop \displaystyle I}{CH_2} \xrightarrow{\Delta} X \xrightarrow{HBr} y$

16. Explain the rearrangement of the carbocation during S_N1 mechanism.

17. What are nucleophilic substitution reactions ? Discuss the mechanism, stereochemistry and kinetics of S_N2 reactions.

18. Discuss the relative reactivity of alkyl chloride and vinyl chloride towards nucleopiles.

19. (*a*) *n*-butyl chloride and tertiary butyl chloride behave differently when treated with aqueous sodium hydroxide. Discuss the reaction pathway in both the cases.

(*b*) Discuss the mechanism of E_2 reaction.

20. Explain S_N2 mechanism. Arrange the following compounds in increasing order of their reactivity towards S_N2 reactions:

(*i*) *t*-butyl chloride (*ii*) *n*-propyl chloride

21. What is the order of reactivity of various alkyl halides towards nucleophilic displacement reactions ? How is this order justified ?
22. (*a*) Discuss the stereochemistry of S_N1 and S_N2 reactions.
(*b*) Discuss the structure of aryl halides and explain their low reactivity.
23. (*a*) What are the factors which affect S_N1 and S_N2 reactions ?
(*b*) How are the following synthesised from ethyl chloride ?
(*i*) Propionic acid (*ii*) *n*-Propyl amine (*iii*) Ethyl acetate
24. (*a*) Explain with the help of suitable examples that S_N1 reactions of alkyl halides are accompanied by partial racemisation.
(*b*) Give the factors which influence S_N1 reactions.
25. (*a*) Describe the benzyne mechanism for nucleophilic aromatic substitution
(*b*) Enlist in a tablular form the distinguishing points of S_N1 and S_N2 mechanism.
26. (*a*) Boiling points of alkyl halides are higher as compared to those of corresponding alkanes, why ?
(*b*) Starting from methyl bromide, how will you prepare the following ?
(*i*) Acetic acid (*ii*) Ethylene
(*iii*) Propyne (*iv*) Ethoxyethane
27. Discuss the relative reactivities of alkyl halides in S_N1 and S_N2 reactions.
28. What do you understand by S_N1 and S_N2 reactions ? Discuss the stereochemistry of S_N1 reactions.
29. Discuss the stereochemistry of S_N2 reactions.
30. Discuss the mechanism of dehydrohalogenation of ethyl chloride.

ARYL HALIDES

1. How does resonance theory explain the low reactivity of aryl halides ?
2. Hydrolysis of $C_2H_5 - Cl$ is easier than that of $C_6H_5 - Cl$. Explain with mechanism.
3. Complete the following reactions and give mechanism:

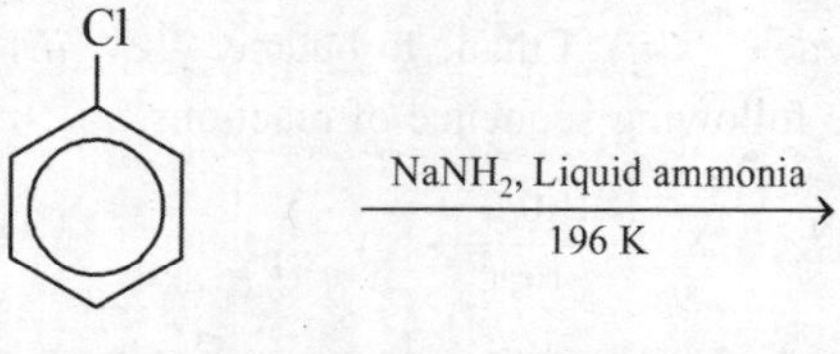

4. How do chemical properties of chlorobenzene and benzyl chloride differ from each other.
5. Give the evidence in support of elimination addition mechanism of nucleophilic aromatic substitution.
6. Starting from a suitable aryl halide how will you prepare
(*i*) Diphenyl (*ii*) Benzoic acid (*iii*) Iodobenzene.
7. Write the structures and names of all the possible isomers of an aromatic compound with formula C_7H_7Cl.
8. Due to high electronegativity, chlorine is electron withdrawing but still ortho and para directing. Explain.
9. How do chemical properties of chlorobenzene and benzyl chloride differ from each other.
10. Explain why aryl halides have low reactivity during the nucleophilic substitution.
11. Discuss the mechanism of nucleophilic substitution reactions of aryl halides.

12. What happens when :
    (*i*) Chlorobenzene reacts with ammonia.
    (*ii*) Benzyl chloride reacts with potassium cyanide.
13. Chlorobenzene does not give a precipitate even on prolonged heating with $AgNO_3$. Why ?
14. Describe the benzyne mechanism of nucleophilic aromatic substitution.
15. What happens when chlorobenzene is treated with sodamide in the presence of liquid ammonia? Discuss the mechanism of the reaction. Give evidence in support of benzyne intermediate formed during the reaction.
16. (*a*) Describe the benzyne mechanism for nucleophilic aromatic substitution.
    (*b*) Aryl halides are less reactive than alkyl halides. Explain.
17. (*a*) What happens when chloro benzene reacts with sodamide in the presence of liquid ammonia? Discuss mechanism of reaction.
    (*b*) "Benzyl chloride is more reactive than chlorobenzene towards nucleophiles." Explain.
    (*c*) Discuss the mechanism of bimolecular nucleophilic aromatic substitution. How does the reactivity of reaction affected by the presence of substituents at ortho/para positions ?
18. Discuss nucleophilic susbtitution in aryl halides through addition-elimination mechanism.
19. Discuss the Benzyne mechanism for nucleophilic aromatic substitution reactions. Give evidence in support of the mechanism.
20. Explain the bimolecular displacement mechanism of nucleophilic substitution of aryl halides.
21. What happens when chlorobenzene is treated with sodamide in the presence of liquid ammonia? Discuss the mechanism of this reaction. Give evidence in support of the benzyne intermediate formed during the reaction.
22. Give the elimination addition mechanism for nucleophilic aromatic substitution. Give evidences in favour of this mechanism.
23. Give evidence in support of elimination addition mechanism of nucleophilic aromatic substitution.
24. Give mechanism of Elimination–Addition reactions in aryl halides and also give evidence in support of it.
25. Discuss the structure of aryl halides and explain their low reactivity.
26. Name the products formed when chlorobenzene reacts with the following :
    (*i*) Conc. HNO_3 / Conc. H_2SO_4 (*ii*) NH_3 / Cu_2O
    (*iii*) Mg (*iv*) Ni / Al and NaOH
27. Discuss the benzyne mechanism for nucleophilic aromatic substitution.
28. (*a*) Arrange the following in order of their increasing reactivity towards nucleophiles:
    $CH_2 = CHCl$, C_6H_5Cl, CH_3Cl, $C_6H_5CH_2Cl$, $CH_2 = CHCH_2Cl$
    (*b*) Explain why 2, 4, 6-trinitrochloobenzene readily reacts with NH_3 whereas chlorobenzene fails to do so.
29. Explain why chlorobenze is less reactive than ethyl chloride during nucleophilic substitution.

PART – III
PHYSICAL CHEMISTRY

1

Mathematical Concepts and Computers

(A) MATHEMATICAL CONCEPTS

1.1 LOGARITHMIC RELATIONS

Definition of Logarithm. Logarithm of a number with respect to a given base is the power to which base must be raised to represent the number.

For example
$$1000 = 10^3$$
$$\log_{10} 1000 = 3$$
In general $N = a^x$
then $\log_a N = x$

As $a^0 = 1$ It follows that $\log_a 1 = 0$

$a^1 = a$ It follows that $\log_a a = 1$.

Four standard formulae of logarithm. These four formulae are commonly used. We are mentioning them without proof.

$$\log_a mn = \log_a m + \log_a n \quad \text{(Product formula)}$$
$$\log_a \frac{m}{n} = \log_a m - \log_a n \quad \text{(Quotient formula)}$$
$$\log_a m^n = n \log_a m \quad \text{(Power formula)}$$
$$\log_a m = \log_b m \times \log_a b \quad \text{(Base changing formula)}$$

Example 1. **Express $\log_a \dfrac{x^3 \sqrt[5]{y^2}}{\sqrt{z^3}}$ in terms of $\log_a x$, $\log_a y$ and $\log_a z$.**

Solution.

$$\log_a \frac{x^3 \sqrt[5]{y^2}}{\sqrt{z^3}} = \log_a \frac{x^3 y^{2/5}}{z^{3/2}}$$
$$= \log_a \left(x^3 y^{2/5}\right) - \log_a z^{3/2} \quad \text{(Quotient formula)}$$
$$= \log_a x^3 + \log_a y^{2/5} - \log_a z^{3/2} \quad \text{(Product formula)}$$
$$= 3 \log_a x + \frac{2}{5} \log_a y - \frac{3}{2} \log_a z \quad \text{(Power formula)}$$
which is the result.

Example 2. **If *a, b, c, d* are four positive numbers, prove that $\log_b a \times \log_c b \times \log_d c = \log_d a$**

Solution.

$$\begin{aligned}
\text{L.H.S.} &= \log_b a \times \log_c b \times \log_d c \\
&= \frac{\log_m a}{\log_m b} \times \frac{\log_m b}{\log_m c} \times \frac{\log_m c}{\log_m d} \quad \text{(by Base changing formula)} \\
&= \frac{\log_m a}{\log_m d} \\
&= \log_d a \\
&= \text{R.H.S.}
\end{aligned}$$

Hence the result.

EXERCISES

1. Express in terms of log *x*, log *y* and log *z*.

   (*i*) $\log_a \dfrac{x^3}{y^2 z^4}$ (*ii*) $\log_a \dfrac{\sqrt[n]{x^m}\, y^r}{3\sqrt{z}}$

2. Prove that $\log\dfrac{a^2}{bc} + \log\dfrac{b^2}{ca} + \log\dfrac{c^2}{ab} = 0$.

3. Prove that $\log_a mnp = \log_a m + \log_a n + \log_a p$.

4. $\log\dfrac{42}{55} = \log 2 + \log 3 + \log 7 - \log 11 - \log 5$. Prove it.

ANSWERS

1. (*i*) $3 \log_a x - 2 \log_a y - 4 \log_a z$

   (*ii*) $\dfrac{m}{n}\log_a x + r\log_a y - \dfrac{1}{3}\log_a z$

1.2 TWO SYSTEMS OF LOGARITHM

The two systems of logarithm in common use are:

(*i*) Neperian log or Natural log. Here the base is *e*, where $e = 1 + \dfrac{1}{1!} + \dfrac{1}{2!} + \dfrac{1}{3!} + ...$ In all theoretical calculations, we use natural logarithm.

(*ii*) Common logarithm. Here, the base of the log is 10. In all numerical calculations, we use common log. We can convert natural log into common log as follows.

$$\begin{aligned}
\log_e m &= \log_{10} m \times \log_e 10 \\
\text{and} \quad \log_e 10 &= 2.3026 \times \log_{10} 10 \\
&= 2.3026 \times 1 = 2.3026 \qquad |\because \log_{10} 10 = 1 \\
\therefore \quad \log_e m &= 2.3026 \log_{10} m
\end{aligned}$$

Log tables. Log tables are the standard common log tables which are available to us for calculations. There are usually four digit tables and are given at the end of the book. The logarithm of any given number can be read from these tables. The logarithm of a number consists of two parts :

(*i*) Integral part called *Characteristic* (*ii*) Decimal part called *Mantissa.*

***(i)* Characteristic.** It can be zero, positive or negative, we use the following two rules for finding the characteristic.

Rule I : If the given number > 1, its characteristic is +ve. Count the number of digits to the left of the decimal point. Subtract one from this number. This is the characteristic of the log of given number.

For example,

Characteristic of log of 1798.3 is $4 - 1 = 3$

Characteristic of log of 28.78 is $2 - 1 = 1$

Characteristic of log of 1.923 is $1 - 1 = 0$

Rule II : If the given number is less than one, its characteristic is negative. It is put under the bar. Count the number of zeros immediately after the decimal point. Add 1 to this number. It gives us the characteristic of the log of the given number.

For example,

Characteristic of log 0.5828 $= -(0 + 1) = \bar{1}$

Characteristic of log 0.0628 $= -(1 + 1) = \bar{2}$

Characteristic of log 0.0008 $= -(3 + 1) = \bar{4}$

***(ii)* Mantissa.** As mentioned above, it is the decimal part of log of a given number. It is always positive, and can be read from the log tables as follows.

(*i*) In the given number, omit the decimal point and zeros in the beginning and at the end.

For example in the log of 0.001175, we have to look up the mantissa of 1175.

(*ii*) Take first two digits *i.e.*, 11 and locate it in the first vertical column of the four figure log table.

(*iii*) Go through the horizontal row beginning with 11 and look up the value under the column headed by the third digit *i.e.* 7. From the log tables, we find this number as 0682.

(*iv*) Continue moving in the same horizontal line and note the number in the small differences column headed by the fourth digit, i.e., 5. This number is 18.

(*v*) Add 18 to 0682, we got 0700. $\therefore$ The mantissa is 0.0700.

Note: If the number consists of a single digit, put zeros at the end of the number to complete four digits. Read mantissa as explained above.

For example log 7 = log 7.000 = 0.8451

Mantissa for 7, 70, 700, 0.7 etc. is the same.

Anti-logarithm: Antilogarithm is the reverse process of logarithm. Thus antilogarithm to logarithm is as division to multiplication.

For example, if log 7 = 0.8451 then Antilog 0.8451 = 7.

Standard Antilog tables are also given at the end of the book for calculations. We follow the rules given below for reading antilog tables.

(*i*) Read the antilog of mantissa only (i.e. of decimal part). Locate the four digits the way we did in reading mantissa from log tables.

(*ii*) If the characteristic is positive say *n*, the decimal is placed after $(n + 1)$ digits in the value read.

(*iii*) If the characteristic is –ve, say $\bar{n}$, then $(n - 1)$ zeros are placed before the left side of the number read; then decimal point is placed.

For example antilog of $\bar{2}.3478 = 0.02227$

antilog of 1.7192 = 52.38

antilog of 0 = 1.000

Use of logarithm. Logarithm is practically very useful in simplifying the complicated calculations. Logarithm means a rule to shorten arithmetic. It was invented by John Napier in 1614.

Following examples will enable the students to use the log tables for that purpose.

***Example 1.* Find the fifth root of 0.0076.**

Solution.

John Napier

Let $x = (0.0076)^{1/5}$

Taking log on both sides, we get

$$\log x = \frac{1}{5} \log (0.0076)$$

$$= \frac{1}{5}(\bar{3}.8808) \quad \text{(From log tables)}$$

$$= \frac{1}{5}(\bar{3}.8808+2-2)$$

$$= \frac{1}{5}(\bar{5}.8808+2)$$

$$= \frac{1}{5}(\bar{5}+2.8808)$$

$$\log x = \bar{1}.5762$$

Taking antilog on both sides, we get

$$x = 0.3769$$

***Example 2.* Simplify $\mathbf{\dfrac{1.792 \times 77.4}{(129.7)^{2/3}}}$.**

Solution.

$$\text{Let} \quad x = \frac{1.792 \times 77.4}{(129.7)^{2/3}}$$

Taking log on both sides, we get

$$\log \quad x = \log 1.792 + \log 77.4 - \frac{2}{3} \log (129.7)$$

$$= 0.2534 + 1.8887 - \frac{2}{3} (2.1130)$$

$$= 0.2534 + 1.887 - 1.4087$$

$$= 2.1421 - 1.4087$$

$$= 0.7334$$

$$x = \text{Antilog } 0.7334$$

$$= 1.5413 \quad \text{which is the result.}$$

EXERCISES

1. Simplify the following

(*i*) $(0.0036)^{1/5}$

(*ii*) $(0.056)^{2/3}$

(*iii*) $10^{-1/5}$

2. Simplify the following

(*i*) $(0.05246)^{1/8} - 2.6055$

(*ii*) $\dfrac{3.142\times(80.2)^{1/2}}{(9.8)^{1/2}}$

3. Find the antilog of

(*i*) –2.6674 [**Hint.** $(-2.6674 = \bar{3}.3326)$]

(*ii*) –3.01

ANSWERS

1. (*i*) 0.3245

(*ii*) 0.1464

(*iii*) 0.6310

2. (*i*) –1.9137

(*ii*) 8.989

3. (*i*) 0.002151

(*ii*) 0.0009772

1.3 CURVE SKETCHING

When a variable point $P(x, y)$ moves under a given set of conditions, then the path traced by P is called graph or locus of P. The equation which relates the variables x and y is called the equation of the graph of P.

Every point which is on the graph satisfies the given set of conditions and any point satisfying the given set of conditions lies on the graph.

To draw the graph or to sketch the curve we take y as the dependent variable and x is the independent variable. To draw the graph the following points are followed:

(*i*) Write the given equation equal to y.

(*ii*) Now give three, four values of x and find the corresponding values of y.

(*iii*) Plot these points in the graph by taking two perpendicular lines. Horizontal line as x-axis and vertical line as y-axis, we get the required curve.

Example 1. **Sketch the curve $y = x^2 + 2x - 3$.**

Solution.

First of all, we get table of values for $y = x^2 + 2x - 3$ as under.

x	–2	–1	0	2	–4
y	–3	–4	–3	5	5

Plot these points in the graph and join it with free hand as illustrated below.

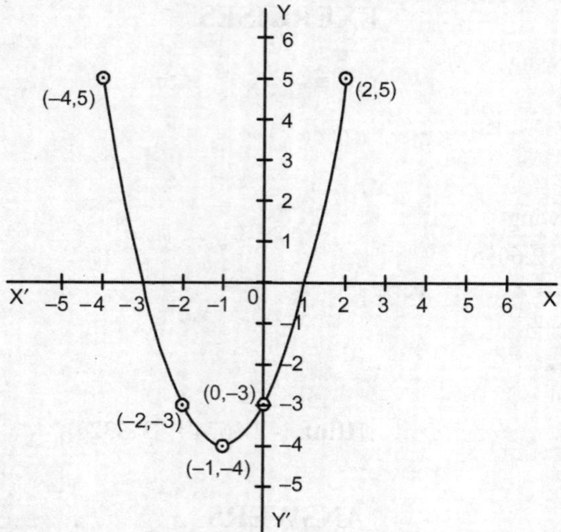

***Example 2.* Sketch the curve $y = x^2 - 9$.**

Solution.

First of all we get table of values for $y = x^2 - 9$ by taking x as independent and y as dependent variable. We get the table of values as under.

x	–3	–2	–1	0	1	2	3
y	0	–5	–8	9	–8	–5	0

Plot these points in the graph and join it with free hand as shown below.

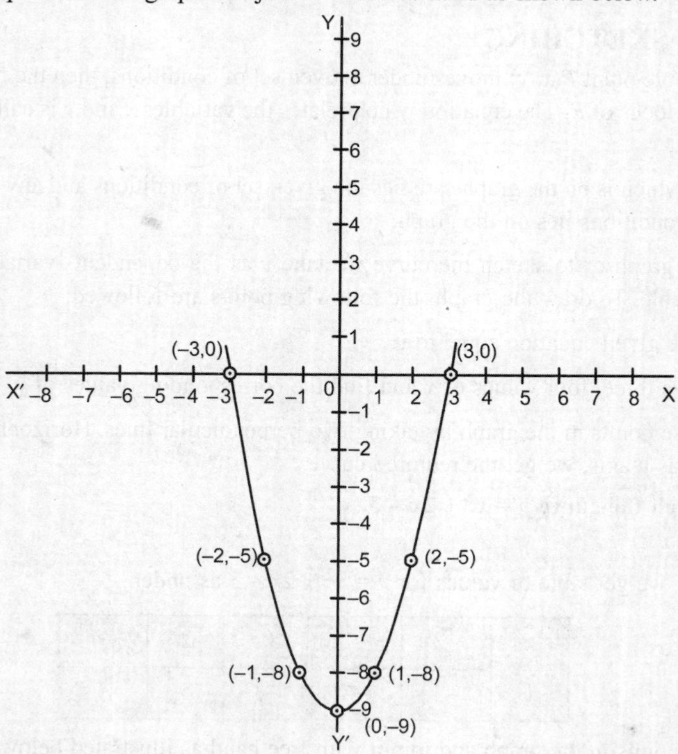

Example 3. **Sketch the curve $y = x^3 + 1$.**

Solution:

First of all we get table of values for $y = x^3 + 1$ by taking x as independent variable and y as dependent variable, we get the table of values as under.

x	−2	−1	0	1	2
y	−7	0	1	2	9

Plot these points in the graph and join it with the free hand as shown below.

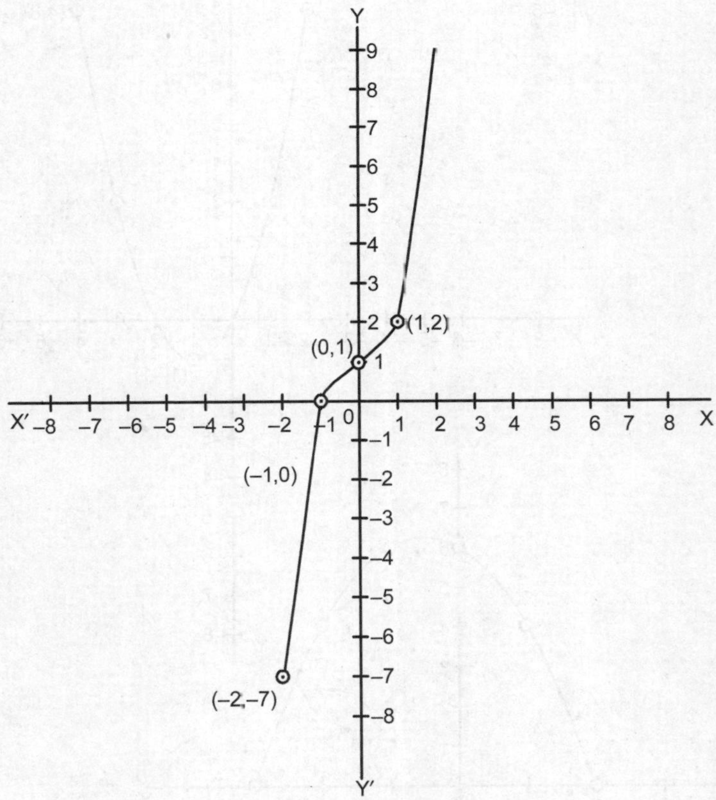

EXERCISES

1. Sketch the graph $y = x^4 - 1$.
2. Sketch the curve $y = x^2 - 2x$.
3. Sketch the curve $y = x^2 + x + 6$.

ANSWERS

1.

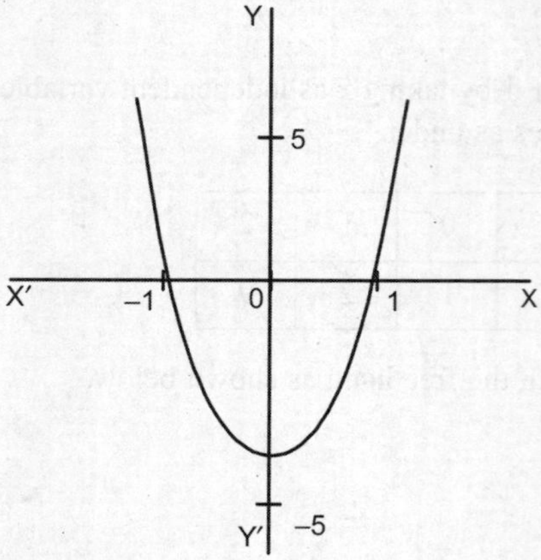

Y
5
X′
–1
0
1
X
Y′
–5

2.

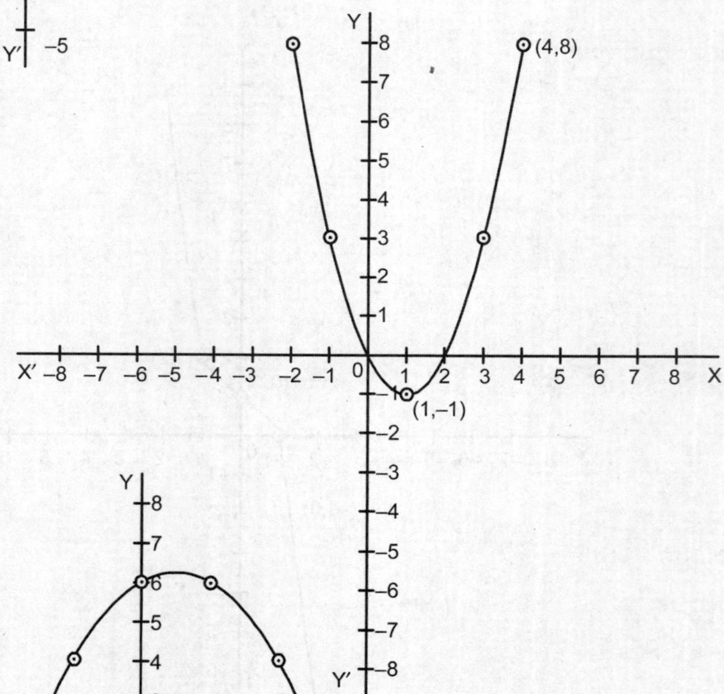

Y
8
7
6
5
4
3
2
1
(4,8)
X′ –8 –7 –6 –5 –4 –3 –2 –1 0 1 2 3 4 5 6 7 8 X
(1,–1)
–1
–2
–3
–4
–5
–6
–7
–8
Y′

3.

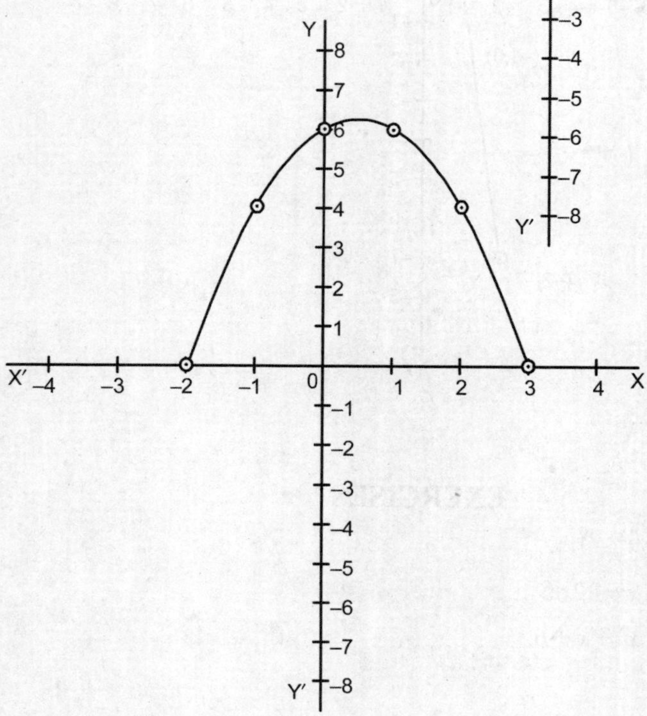

Y
8
7
6
5
4
3
2
1
X′ –4 –3 –2 –1 0 1 2 3 4 X
–1
–2
–3
–4
–5
–6
–7
–8
Y′

1.4 LINEAR GRAPHS AND CALCULATIONS OF SLOPES

Linear graph. The graph of a linear equation is called the linear graph.

Linear Equation. An equation of the form $ax + by + c = 0$, where a, b, c are constants and x and y are two variables, is called a linear equation of two variables. Every linear equation in x and y represents a straight line. The various cases are discussed below one by one.

(*i*) The equation $x = a$ represents a straight line parallel to y-axis and at a distance a from it.

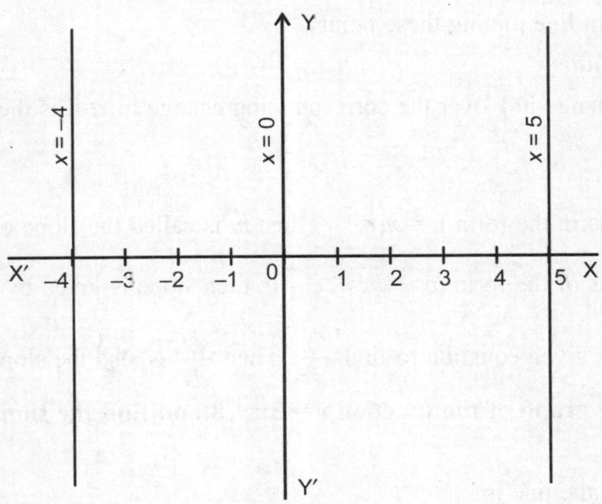

If the line is to the right of y-axis, a is +ve.

If the line is to the left of y-axis, a is –ve.

If the line coincides with y-axis, $a = 0$, the equation of the y-axis is $a = 0$.

The graph of $x = 5$, $x = 0$ and $x = -4$ is shown in the above figure.

(*ii*) The equation $y = b$ represents a straight line parallel to x-axis and at a distance b from it.

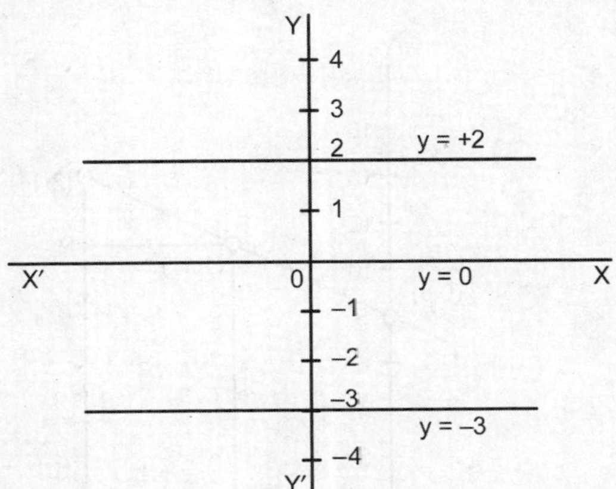

If the line is above x-axis, b is +ve.

If the line is below x-axis, b is –ve.

If the line coincides with x-axis, $b = 0$.

The equation of x-axis is $y = 0$.

The graphs of $y = 2$, $y = 0$ and $y = -3$ are shown in the figure above.

General method to draw the graph of linear equation $ax + bx + c = 0$ and to find the slope of the line.

1. Write the given equation in the form showing one variable in terms of the other variable (generally in the form $y = mx + c$)
2. Now give three, four values to x and find the corresponding values of y.
3. Plot these points named by ordered pairs.
4. Draw the straight line joining these points.

To find slope of line

1. Calculate the change in y over the corresponding change in x. It is the slope of the line *i.e.*, $\frac{\text{Change in } y}{\text{Change in } x}$.
2. If the equation is in the form $y = mx + c$, then m is called the slope of the line.
3. If the equation is of the form $ax + bx + c = 0$, then slope is given by $-\frac{\text{Coeff. of } x}{\text{Coeff. of } y} = -\frac{a}{b}$.
4. Differentiate the given equation to find $\frac{dy}{dx}$. Then $\frac{dy}{dx}$ is also the slope of the line.

***Example 1.* Draw the graph of the function $y = 3x + 8$ and find the slope of the line.**

Solution.

The equation of the line is

$$y = 3x + 8.$$

Putting $x = 0$ we get $y = 8$

Putting $x = 1$ we get $y = 11$

Putting $x = 2$ we get $y = 14$

$A(0, 8)$, $B(1, 11)$ and $C(2, 14)$ are three points on the line. Plot these points and join them by a foot ruler.

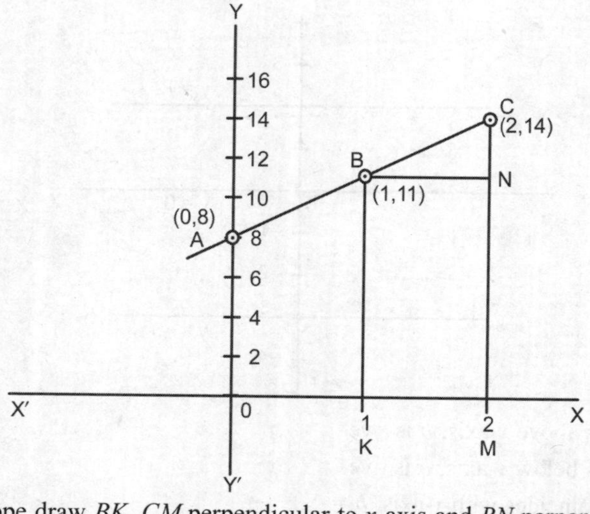

To find the slope draw BK, CM perpendicular to x-axis and BN perpendicular to CM.

The slope is given by $\frac{CN}{BN} = \frac{\text{Change in } y}{\text{Change in } x}$

Now

$$CN = CM - NM = 14 - 11 = 3$$
$$BN = KM = OM - OK = 2 - 1 = 1$$

$$\text{Slope} = \frac{3}{1} = 3. \text{ Hence the result.}$$

EXERCISES

Draw the graphs and find the slopes.

(*i*) $x + y = 0$ (*ii*) $2x - y = 3$

(*iii*) $x - y = 2$ (*iv*) $4x + y = 3$

ANSWERS

(*i*)

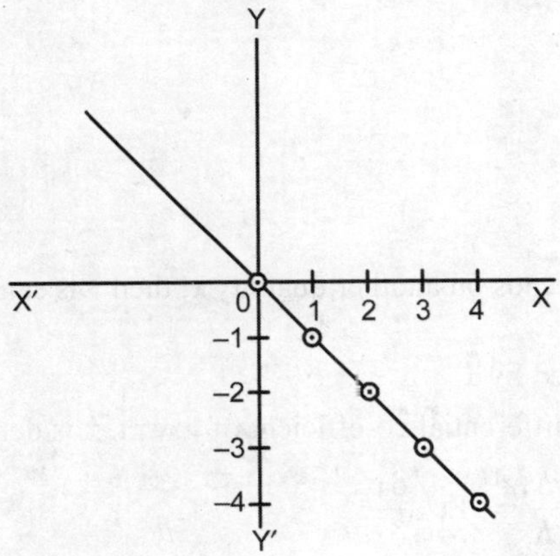

Slope = –1

(*ii*)

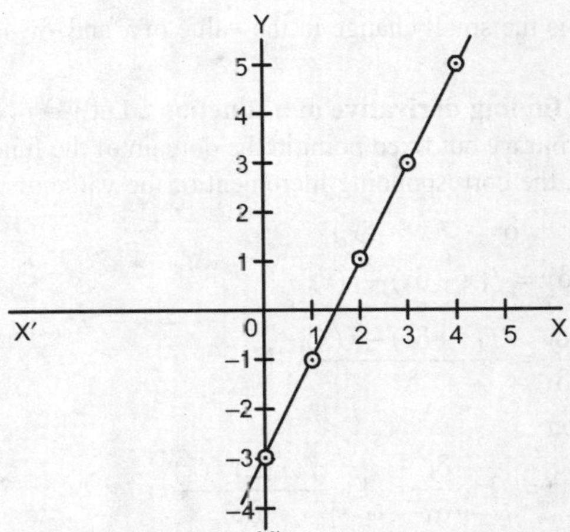

Slope = 2

(*iii*)

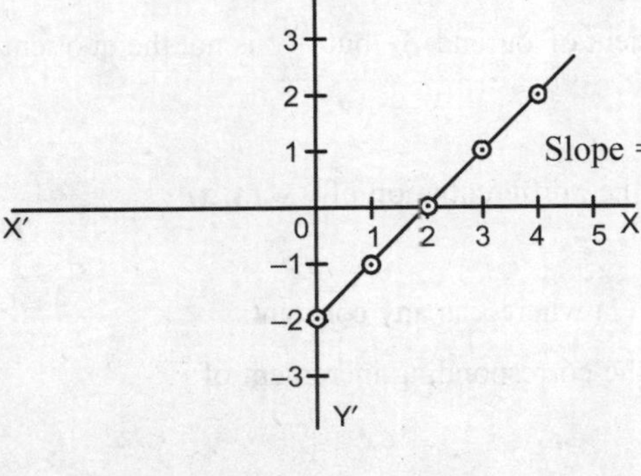

Slope = +1

(*iv*)

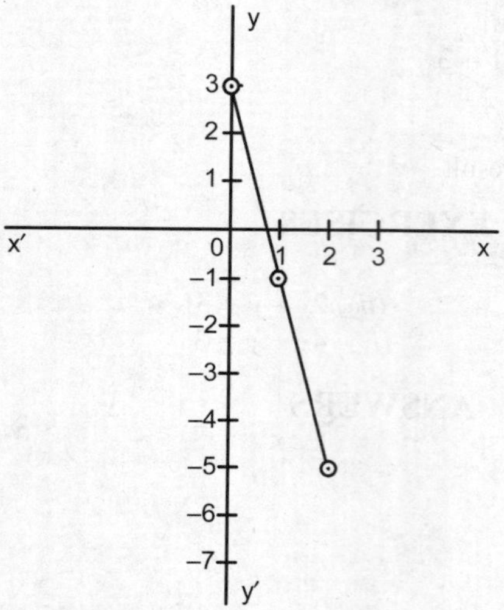

Slope = –4

1.5 DIFFERENTIATION

Function. When the value of a quantity y depends on another quantity x, then y is said to be a function of x. We represent it as $y = f(x)$.

Note that $f(x)$ is a symbol. It is not f multiplied by x.

Differential co-efficient : The derivative or differential co-efficient of y w.r.t. x is denoted by

$$\frac{d}{dx}(y) = \frac{dy}{dx} = \underset{\delta x \to 0}{\text{Lt}} \frac{\delta y}{\delta x}$$

Where δx is the small change in the value of x and δy is the corresponding small change in the value of y.

Method of finding derivative of a function : Let $y = f(x)$ be a derivable function of x. Let x represent any arbitrary but fixed point in the domain of the function. Let δx be the small increment of x and let δy, the corresponding increment of the value of y.

$$\therefore \quad y + \delta y = f(x + \delta x)$$

$$\Rightarrow \quad \delta y = f(x + \delta x) - f(x)$$

$$\Rightarrow \quad \frac{\delta y}{\delta x} = \frac{f(x + \delta x) - f(x)}{\delta x}$$

By definition

$$f'(x) = \underset{\delta x \to 0}{\text{Lt}} \frac{\delta y}{\delta x} = \underset{\delta x \to 0}{\text{Lt}} \frac{f(x \delta x) f(x)}{\delta x}$$

It is worthwhile to note that $\frac{\delta y}{\delta x}$ is the quotient of δy and δx but $\frac{dy}{dx}$ is not the quotient of dy and dx.

Instead, $\frac{dy}{dx}$ is to be interpreted as $\frac{d}{dx}(y)$ i.e., differentiation of y w.r.t. x.

1. **Derivative of kx w.r.t. x**

Let $\qquad y = kx \qquad$ .... (1) where k is any constant.

Let δx be an increment of x and δy, the corresponding increment of y.

$\therefore \quad y + \delta y = k(x + \delta x)$ .... (2)

$$(2) - (1) \Rightarrow \quad y + \delta y - y = k(x + \delta x) - kx$$

$$\delta y = k\,\delta x$$

$$\Rightarrow \quad \delta y = k\,\delta x$$

$$\Rightarrow \quad \frac{\delta y}{\delta x} = k$$

Taking limits $\delta x \to 0$

$$\underset{\delta x \to 0}{\text{Lt}} \frac{\delta y}{\delta x} = \underset{\delta x \to 0}{\text{Lt}} k$$

$$\frac{dy}{dx} = k$$

$$\frac{d}{dx}(kx) = k$$

2. Derivative of e^x w.r.t. x

Let $\quad y = e^x$ .... (1)

Let δx be an increment of x and δy by the corresponding increment of y.

$$y + \delta y = e^{x+\delta x}$$

$$(2) - (1) \Rightarrow \delta y = e^{x+\delta x} - e^x$$

$$\Rightarrow \delta y = e^x . e^{\delta x} - e^x$$

$$\Rightarrow \frac{\delta y}{\delta x} = \frac{e^x \cdot e^{\delta x} - e^x}{\delta x}$$

$$= \frac{e^x (e^{\delta x} - 1)}{\delta x}$$

Taking limits $\delta x \to 0$

$$\underset{\delta x \to 0}{\text{Lt}} \frac{\delta y}{\delta x} = \underset{\delta x \to 0}{\text{Lt}} \frac{e^x (e^{\delta x} - 1)}{\delta x}$$

$$\frac{dy}{dx} = e^x \log e \qquad \left|\because \underset{\delta x \to 0}{\text{Lt}} \frac{e^{\delta x} - 1}{\delta x} = \log e\right.$$

$$\frac{dy}{dx} = e^x \qquad |\because \log e = 1$$

$$\frac{dy}{dx}(e^x) = e^x$$

3. Derivative of x^n w.r.t. x

Let $\quad y = x^n$ .... (1) where n is any natural number.

Let δx be an increment of x and δy, the corresponding increment of y.

$$y + \delta y = (x + \delta x)^n. \quad ... (2)$$

$$(2) - (1) \Rightarrow \quad \delta y = (x + \delta x)^n - x^n$$

$$\Rightarrow \quad \delta y = x^n + n_{C_1} x^{n-1}\delta x + n_{C_2} x^{n-2}(\delta x)^2 + ... + n_{C_n}(\delta x)^n - x^n$$

$$\Rightarrow \quad \delta y = n x^{n-1}\delta x + \frac{n(n-1)}{\underline{|2}} x^{n-2}(\delta x)^2 + ... + (\delta x)^n$$

$$\Rightarrow \quad \delta y = \delta x\left[nx^{n-1} + \frac{n(n-1)}{\underline{|2}} x^{n-2}\delta x + ... + (\delta x)^{n-1}\right]$$

Dividing both sides by δx

$$\Rightarrow \quad \frac{\delta y}{\delta x} = \frac{\delta x\left[nx^{n-1} + \frac{n(n-1)}{\underline{|2}} x^{n-2}\delta x + ... + (\delta x)^{n-1}\right]}{\delta x}$$

$$\Rightarrow \quad \frac{\delta y}{\delta x} = nx^{n-1} + \frac{n(n-1)}{\underline{|2}} x^{n-2}\delta x + ... + (\delta x)^{n-1}$$

$$\Rightarrow \quad \underset{\delta x \to 0}{\text{Lt}} \frac{\delta y}{\delta x} = \underset{\delta x \to 0}{\text{Lt}}\ nx^{n-1} + \frac{n(n-1)}{\underline{|2}} x^{n-2}\delta x + ... + (\delta x)^{n-1}$$

$$\frac{\delta y}{\delta x}(x^n) = nx^{n-1}$$

$$\therefore \quad \frac{d}{dx}(x^n) = nx^{n-1}$$

4. Derivative of sin *x* w.r.t. *x*

Let $\quad y = \sin x \quad .... (1)$

Let δx be an increment of x and δy, the corresponding increment of y.

$$\therefore \quad y + \delta y = \sin(x + \delta x) \quad .... (2)$$

(2) – (1) $\quad \Rightarrow \delta y = \sin(x + \delta x) - \sin x$

$$= 2\cos\left(x + \frac{1}{2}\delta x\right)\sin\frac{\delta x}{2}$$

$$\frac{\delta y}{\delta x} = \frac{2\cos\left(x + \frac{\delta x}{2}\right)\sin\frac{\delta x}{2}}{\delta x} \qquad \left|\ \because \sin C - \sin D = 2\cos\frac{C+D}{2}\sin\frac{C-D}{2}\right.$$

$$\Rightarrow \quad \frac{\delta y}{\delta x} = \frac{\cos\left(x + \frac{\delta x}{2}\right)\sin\frac{\delta x}{2}}{\frac{\delta x}{2}}$$

Taking limits

$$\underset{\delta x \to 0}{\text{Lt}} \frac{\delta y}{\delta x} = \underset{\delta x \to 0}{\text{Lt}} \cos\left(x + \frac{\delta x}{2}\right) \underset{\delta x \to 0}{\text{Lt}} \frac{\sin\frac{\delta x}{2}}{\frac{\delta x}{2}}$$

$$= \cos x \cdot 1 \qquad \left|\ \because \underset{x \to 0}{\text{Lt}} \frac{\sin x}{x} = 1\right.$$

$$\therefore \quad \frac{dy}{dx} = \cos x$$

i.e., $\frac{d}{dx}(\sin x) = \cos x$

On the lines of sin x we can find the following derivatives:

$$\frac{d}{dx}(\cos x) = -\sin x \quad \frac{d}{dx}(\tan x) = \sec^2 x$$

$$\frac{d}{dx}(\cot x) = -\operatorname{cosec}^2 x \quad \frac{d}{dx}(\sec x) = \sec x \tan x$$

$$\frac{d}{dx}(\operatorname{cosec}) = -\operatorname{cosec} x \cot x \quad \frac{d}{dx}(a^x) = a^x \log a \qquad \text{(on the lines of } e^x\text{)}$$

5. Derivative of log x w.r.t. x

Let $y = \log x$ .... (1)

Let δx be the small increment in x, then the corresponding increment in y be δy.

Then $y + \delta y = \log (x + \delta x)$ .... (2)

(2) – (1) $\Rightarrow \delta y = \log (x + \delta x) - \log x$

$$\Rightarrow \delta y = \log \frac{(x+\delta x)}{x}$$

$$\Rightarrow \frac{\delta y}{\delta x} = \frac{1}{\delta x}\log\left(1+\frac{\delta x}{x}\right)$$

$$\Rightarrow \frac{\delta y}{\delta x} = \frac{1}{x}\cdot\frac{x}{\delta x}\log\left(1+\frac{\delta x}{x}\right)$$

$$\Rightarrow \frac{\delta y}{\delta x} = \frac{1}{x}\cdot\log\left(1+\frac{\delta x}{x}\right)^{\frac{x}{\delta x}}$$

Taking limits

$$\underset{\delta x \to 0}{\text{Lt}} \frac{\delta y}{\delta x} = \underset{\delta x \to 0}{\text{Lt}} \frac{1}{x}\log\left(1+\frac{\delta x}{x}\right)^{\frac{x}{\delta x}}$$

| when $\delta x \to 0$ then $\frac{\delta x}{x} \to 0$

$$\frac{dy}{dx} = \frac{1}{x}\cdot \underset{\delta x \to 0}{\text{Lt}} \log\left(1+\frac{\delta x}{x}\right)^{\frac{x}{\delta x}}$$

| $\Rightarrow \underset{\frac{\delta x}{x} \to 0}{\text{Lt}} \left(1+\frac{\delta x}{x}\right)^{\frac{x}{\delta x}} = e$

$$\Rightarrow \frac{dy}{dx} = \frac{1}{x}\log e$$

$$= \frac{1}{x} \qquad |\because \log e = 1$$

$$\Rightarrow \frac{d}{dx}(\log x) = \frac{1}{x}.$$

Example 1.

(*i*) $\frac{d}{dx}(x^7) = 7x^6$ $\qquad \left|\because \frac{d}{x}(x^n) = nx^{n-1}\right.$

(ii) $\frac{d}{dx}\left(\sqrt{x}\right) = \frac{d}{dx}\left(x^{1/2}\right) = \frac{1}{2}x^{1/2-1} = \frac{1}{2}x^{-\frac{1}{2}} = \frac{1}{2}\sqrt{x}$

(iii) $\frac{d}{dx}(x^2+3) = \frac{d}{dx}(x^2) + \frac{d}{dx}(3) = 2x + 0 = 2x$ $\quad\Big|\because \frac{d}{dx}(C) = 0$

(iv) $\frac{d}{dx}(C \cdot v) = C \cdot \frac{d}{dx}(v) = C \cdot \frac{dv}{dx} = C$ $\quad|\because v$ is the function of x

(v) $\frac{d}{dx}(5x^2) = 5 \cdot \frac{d}{dx}x^2 = 5 \cdot 2x = 10x$

(vi) $\frac{d}{dx}(\sin 7x) = \cos 7x \frac{d}{dx}(7x) = 7\cos 7x$

(vii) $\frac{d}{dx}\left[(2x+3)^5\right] = 5(2x+3)^4 \frac{d}{dx}(2x+3) = 5(2x+3)^4 \cdot 2 = 10(2x+3)^4$

(viii) $\frac{d}{dx}(\log x^2) = \frac{d}{dx}(2\log x) = 2\frac{d}{dx}\log x = \frac{2}{x}$

(ix) $\frac{d}{dx}(e^{5x}) = e^{5x}\frac{d}{dx}5x = 5e^{5x}$

EXERCISES

Find the derivative of following functions.

1. (*i*) $x^2 + 2$ (*ii*) $\frac{1}{x^2}$ (*iii*) $x + \frac{1}{x}$

2. (*i*) $x^{3/2}$ (*ii*) $x^{-2} - \frac{x^3}{x^4}$ (*iii*) $(ax + b)^n$

3. (*i*) $a^{\sin x}$ (*ii*) e^{x^2+5} (*iii*) $5x^{2/3}$

4. (*i*) $\sin 5x$ (*ii*) $\tan 5x$ (*iii*) $\cos 5x$

5. If $x = a(\theta + \sin\theta)$, $y = a(1 - \cos\theta)$

find $\frac{dy}{dx}$.

(**Hint.** $\frac{dx}{d\theta} = a(1 + \cos\theta)$

$\frac{dy}{d\theta} = a\sin\theta \qquad \frac{dy}{dx} = \frac{\frac{dy}{d\theta}}{\frac{dx}{d\theta}}$)

ANSWERS

1. (*i*) $2x$ (*ii*) $-\frac{2}{x^3}$ (*iii*) $1 - \frac{1}{x^2}$

2. (*i*) $\frac{3}{2}x^{1/2}$ (*ii*) $-\frac{2}{x^3}+\frac{1}{x^2}$ (*iii*) $an(ax + b)^{n-1}$

3. (*i*) $a^{\sin x} \log a \cos x$ (*ii*) $2x\,e^{x^2+5}$ (*iii*) $5 \times \frac{2}{3} x^{-1/3}$

4. (*i*) $5 \cos 5x$ (*ii*) $5 \sec^2 5x$ (*iii*) $-5 \sin 5x$

5. $\tan\frac{\theta}{2}$

Some important results of Trigonometry which are used in differentiation and integration

1. $\sin^2\theta + \cos^2\theta = 1$
2. $\sec^2\theta = 1 + \tan^2\theta$
3. $\text{cosec}^2\theta = 1 + \cot^2\theta$
4. $\sin(A \pm B) = \sin A \cos B \pm \cos A \sin B$ and $\sin 2A = 2 \sin A \cos A$
5. $\cos(A \pm B) = \cos A \cos B \mp \sin A \sin B$ and $\cos 2A = \cos^2 A - \sin^2 A$ or $1 - 2\sin^2 A$ or $2\cos^2 A - 1$
6. $\sin C + \sin D = 2\sin\frac{C+D}{2}\cos\frac{C-D}{2}$
7. $\sin C - \sin D = 2\cos\frac{C+D}{2}\sin\frac{C-D}{2}$
8. $\cos C + \cos D = 2\cos\frac{C+D}{2}\cos\frac{C-D}{2}$
9. $\cos C - \cos D = 2\sin\frac{C+D}{2}\sin\frac{D-C}{2}$
10. $1 + \cos A = 2\cos^2\frac{A}{2}$
11. $1 - \cos A = 2\sin^2\frac{A}{2}$
12. $\tan(A + B) = \frac{\tan A + \tan B}{1 - \tan A \tan B}$ and $\tan 2A = \frac{2\tan A}{1 - \tan^2 A}$

θ	0	$\frac{\pi}{6}$	$\frac{\pi}{4}$	$\frac{\pi}{3}$	$\frac{\pi}{2}$
$\sin\theta$	$\sqrt{\frac{0}{4}}$	$\sqrt{\frac{1}{4}} = \frac{1}{2}$	$\sqrt{\frac{2}{4}} = \frac{1}{\sqrt{2}}$	$\sqrt{\frac{3}{4}} = \frac{\sqrt{3}}{2}$	$\sqrt{\frac{4}{4}} = 1$
$\cos\theta$	$\sqrt{\frac{4}{4}} = 1$	$\sqrt{\frac{3}{4}} = \frac{\sqrt{3}}{2}$	$\sqrt{\frac{2}{4}} = \frac{1}{\sqrt{2}}$	$\sqrt{\frac{1}{4}} = \frac{1}{2}$	$\sqrt{\frac{0}{4}} = 0$

$\sin(-\theta) = -\sin\theta$	$\cos(-\theta) = \cos\theta$	$\tan(-\theta) = -\tan\theta$
$\sin(90^\circ+\theta) = \cos\theta$	$\cos(90^\circ+\theta) = -\sin\theta$	$\tan(90^\circ+\theta) = -\cot\theta$
$\sin(90^\circ-\theta) = \cos\theta$	$\cos(90^\circ-\theta) = \sin\theta$	$\tan(90^\circ-\theta) = \cot\theta$
$\sin(180^\circ+\theta) = -\sin\theta$	$\cos(180^\circ+\theta) = -\cos\theta$	$\tan(180^\circ+\theta) = \tan\theta$
$\sin(180^\circ-\theta) = \sin\theta$	$\cos(180^\circ-\theta) = -\cos\theta$	$\tan(180^\circ-\theta) = -\tan\theta$
$\sin(270^\circ+\theta) = -\cos\theta$	$\cos(270^\circ+\theta) = \sin\theta$	$\tan(270^\circ+\theta) = -\cot\theta$
$\sin(270^\circ-\theta) = -\cos\theta$	$\cos(270^\circ-\theta) = -\sin\theta$	$\tan(270^\circ-\theta) = \cot\theta$
$\sin(360^\circ+\theta) = \sin\theta$	$\cos(360^\circ+\theta) = \cos\theta$	$\tan(360^\circ+\theta) = \tan\theta$
$\sin(360^\circ-\theta) = -\sin\theta$	$\cos(360^\circ-\theta) = \cos\theta$	$\tan(360^\circ-\theta) = -\tan\theta$

1.6 MAXIMA AND MINIMA

Neighbourhood of a point : If a is any real number, then any open interval containing a is called a neighbourhood of a. For example (–2, 2), (–2, 1), (–0.6, 0.7) are all neighbourhoods of real number $\frac{1}{2}$.

Local maximum value. A function $f(x)$ is said to have a local maximum value at $x = a$ if $f(a)$ is greatest of all its values of x lying in same neighbourhood of a. A local maximum value is also called a relative maximum or simply a maximum value.

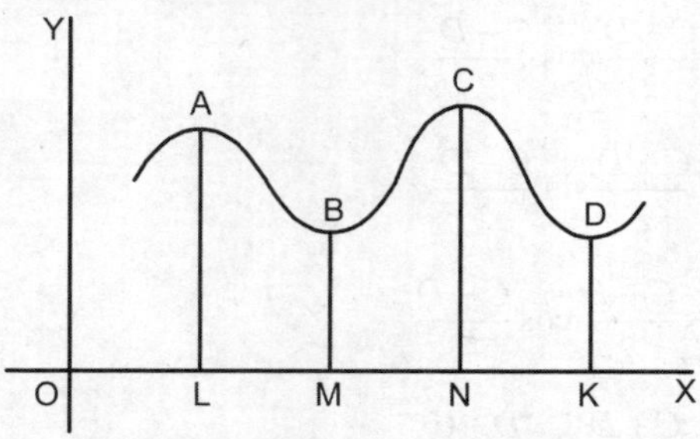

The above figure shows the graph of same function $f(x)$. It is seen that at A and C. The ordinate ceases to increase and if at A, $x = a$, we find that $AL = f(a)$ and it is greater than all other ordinates i.e., values of $f(x)$ in the neighbourhood of A i.e., $x = a$.

Thus from the above definition $f(x)$ is said to have a maximum value of A, C etc.

Local minimum value : A function $f(x)$ is said to have a local minimum value at $x = a$ if $f(a)$ is the smallest of all its values for values of x lying in same neighbourhood of a. A local minimum value is also called a relative minimum value or simply minimum value.

In the above figure B, D are points where the ordinates are less than those in the small neighbourhood of these points and as such we find that $f(x)$ has minimum values at B, D etc.

Local extremum value : A local maximum or a local minimum of a function is called a local extremum value.

Working rule for solving problems

Step I : Differentiate the given function and put $f'(x) = 0$ to get the critical values of $f(x)$.

Step II : Arrange the critical values in the ascending order of magnitude.

Step III : For a particular critical values at $x = a$

(*i*) If $f''(a) < 0 \Rightarrow f(a)$ is a local maximum value.

(*ii*) If $f''(a) > 0 \Rightarrow f(a)$ is a local minimum value.

(*iii*) If $f''(a) = 0 \Rightarrow f'''(a)$ must be studied.

If $f'''(a) \neq 0$ then $(a, f(a))$ is called a point of inflexion. If $f'''(a) = 0$ then $f^{iv}(a)$ is studied at $x = a$.

Example 1. **Show that the following functions do not have local maxima or minima.**

(*i*) $2x + 7$ (*ii*) $\log x$

Solution.

(*i*) Let $f(x) = 2x + 7$

$\therefore$ $f'(x) = 2$

$f''(x) = 0 \Rightarrow 2 = 0$ which is impossible.

$\therefore$ Given function does not have any critical values.

$\therefore$ Given function cannot have any local maximum or minimum value.

(*ii*) Let $f(x) = \log x$

$\therefore$ $f'(x) = \dfrac{1}{x}$

Put $f'(x) = 0 \Rightarrow \dfrac{1}{x} = 0$ which is impossible.

$\therefore$ Given function does not have any critical value.

$\therefore$ Given function cannot have any local maximum or local minimum value.

Example 2. **Find the local maximum and local minimum values of the following functions.**

(*i*) $2x^2 - 3x + 5$ (*ii*) $-5x^2 + 20x + 9$

Solution:

(*i*) Let $f(x) = 2x^2 - 3x + 5$

$\therefore$ $f'(x) = 4x - 3$ and $f''(x) = 4$

Put $f'(x) = 0 \Rightarrow 4x - 3 = 0 \Rightarrow x = \dfrac{3}{4}$

The critical value for $f(x)$ is $x = \dfrac{3}{4}$

$$f''\left(\frac{3}{4}\right) = 4 \text{ which is} > 0$$

$\therefore x = \dfrac{3}{4}$ is a local minimum and the corresponding local minimum value is

$$f\left(\frac{3}{4}\right) = 2\left(\frac{3}{4}\right)^2 - 3\left(\frac{3}{4}\right) + 5$$

$$= 2\left(\frac{9}{16}\right) - \frac{9}{4} + 5$$

$$= \frac{9}{8} - \frac{9}{4} + 5$$

$$= \frac{31}{8}$$

Local minimum value at $x = \dfrac{3}{4}$ is $\dfrac{31}{8}$.

(ii) Let $f(x) = -5x^2 + 20x + 9$

$f'(x) = -10x + 20 \qquad f''(x) = -10$

Put $f'(x) = 0 \Rightarrow -10x + 20 = 0 \Rightarrow x = 2$

The critical value for $f(x)$ is $x = 2$.

Now $f''(x) = -10$ which is < 0

$x = 2$ is a local maximum and the corresponding local maximum value is

$$f(2) = -5(2)^2 + 20 \times 2 + 9 = 29$$

$\therefore$ Local maximum value = 29 at $x = 2$.

Example 3. **Find the local maxima and local minima, if any, of the function $f(x) = x^3 - 6x^2 + 9x + 15$. Find also the local maximum and local minimum values.**

Solution.

We have $f(x) = x^3 - 6x^2 + 9x + 15$

$$f'(x) = 3x^2 - 12x + 9 = 3(x^2 - 4x + 3)$$
$$= 3(x-1)(x-3)$$

Put $f'(x) = 0 \Rightarrow 3(x-1)(x-3) = 0$

$\Rightarrow x = 1, 3$

The critical values for $f(x)$ in the ascending order are $x = 1, 3$.

Also $f''(x) = 6x - 12$

Critical value at $x = 1$

$$f''(1) = 6(1) - 12 = -6 \text{ which is } < 0$$

$x = 1$ is a local maximum and the corresponding local maximum value is

$$f(1) = (1)^3 - 6(1)^3 + 9(1) + 15 = 19$$

Critical value at $x = 3$

$$f''(3) = 6(3) - 12 = 6 \text{ which is } > 0$$

$x = 3$ is a local minimum and the corresponding local minimum value is

$$f(3) = (3)^3 - 6(3)^2 + 9(3) + 15 = 15$$

Hence local maximum value = 19 at $x = 1$

and local minimum value = 15 at $x = 3$.

EXERCISES

1. Find the critical values for the following functions.

(*i*) $x^2 - 2x + 4$ (*ii*) $x^3 - 1$

2. Show that the following functions do not have local maxima or minima.

(*i*) $x + 5$ (*ii*) $\log(x + 4)$

3. Find the points of local maxima and local minima and also find the local maximum and local minimum values of the following functions.

(*i*) x^2 (*ii*) $2x^3 - 21x^2 + 36x - 20$

ANSWERS

1. (*i*) $x = 1$ (*ii*) $x = 0$

3. (*i*) Local minimum value = 0 at $x = 0$

(*ii*) Local maximum value = –3 at $x = 1$

Local minimum value = –128 at $x = 6$

1.7 PARTIAL DIFFERENTIATION

Partial Derivatives of first order : Let $z = f(x, y)$ be a real valued function of two independent variables x and y, then $\underset{\delta x \to 0}{\text{Lt}} \dfrac{f(x+\delta x, y) - f(x, y)}{\delta x}$, if it exists, is called partial derivative of z (or f) w.r.t. x and it is generally denoted by $\dfrac{\partial f}{\partial x}$ or $\dfrac{\partial z}{\partial x}$ or $f\,x(x, y)$ or fx or zx. $\dfrac{\partial f}{\partial x}$ is read it as curly f by curly x.

Similarly $\underset{\delta x \to 0}{\text{Lt}} \dfrac{f(x, y+\delta y) - f(x, y)}{\delta y}$ if it exists is called partial derivative of z(or f) w.r.t. y and it is generally denoted by $\dfrac{\partial f}{\partial y}$ or $\dfrac{\partial z}{\partial y}$ or $f\,y(x, y)$ or $f\,y$ or zy.

From the above definition we conclude that if $\mu = f(x, y)$ is a function of two variables x and y, then the partial derivative of μ w.r.t. x is the differentiation of μ w.r.t. x taking y as constant.

***Example 1.* Find fx and fy for the function $f(x, y) = y^x$.**

Solution.

$$f(x, y) = y^x$$

Differentiating partially w.r.t. x, we get

$$fx = \frac{\partial f}{\partial x} = y^x \log y \qquad \left|\ \because \frac{d}{dx}(a^x) = a^x \log a\right.$$

Now differentiating partially w.r.t. y we get

$$fy = \frac{\partial f}{\partial y} = xy^{x-1} \qquad \left|\ \because \frac{d}{dx}(x^n) = nx^{n-1}\right.$$

***Example 2.* If $f(x, y) = \log(x^2 + y^2)$, find fx and fy.**

Solution.

$$f(x, y) = \log(x^2 + y^2)$$

Differentiating partially w.r.t. x we get

$$fx = \frac{\partial f}{\partial x} = \frac{1}{x^2 + y^2} 2x$$

$$= \frac{2x}{x^2 + y^2}$$

Differentiating partially w.r.t. y we get

$$\frac{\partial f}{\partial y} = \frac{1}{x^2 + y^2} \times 2y = \frac{2y}{x^2 + y^2}$$

EXERCISES

1. Find the first order derivatives of the following functions.

(*i*) $\dfrac{1}{\sqrt{x^2 + y^2}}$ (*ii*) $ax^2 + 2hxy + by^2$

2. Find $\dfrac{\partial \mu}{\partial x}, \dfrac{\partial \mu}{\partial y}, \dfrac{\partial \mu}{\partial z}$ if it is given that $\mu = x^2y + y^2z + z^2x$.

ANSWERS

1. (*i*) $\dfrac{-x}{(x^2+y^2)^{3/2}}$; $\dfrac{-y}{(x^2+y^2)^{3/2}}$ (*ii*) $2(ax + hy)$; $2(hx + by)$

2. $\dfrac{\partial \mu}{\partial x} = 2xy + z^2$ $\dfrac{\partial \mu}{\partial y} = x^2 + 2yz$ $\dfrac{\partial \mu}{\partial z} = y^2 + 2zx$

1.8 PARTIAL DERIVATIVES OF HIGHER ORDER

Since each of the partial derivatives $\dfrac{\partial f}{\partial x}, \dfrac{\partial f}{\partial y}$, in general, is a function of x and y, so each of them possesses partial derivatives w.r.t. the independent variables x and y.

(*i*) The partial derivative of $\dfrac{\partial f}{\partial x}$ w.r.t. x is given by $\dfrac{\partial}{\partial x}\left(\dfrac{\partial f}{\partial x}\right) = \dfrac{\partial^2 f}{\partial x^2}$ or fxx

(*ii*) The partial derivative of $\dfrac{\partial f}{\partial y}$ w.r.t. y is given by $\dfrac{\partial}{\partial y}\left(\dfrac{\partial f}{\partial y}\right) = \dfrac{\partial^2 f}{\partial y^2} = fyy$

(*iii*) The partial derivative of $\dfrac{\partial f}{\partial x}$ w.r.t. y is given by $\dfrac{\partial}{\partial y}\left(\dfrac{\partial f}{\partial x}\right) = \dfrac{\partial^2 f}{\partial y \partial x}$ or fyx

(*iv*) The partial derivative of $\dfrac{\partial f}{\partial y}$ w.r.t. x is given by $\dfrac{\partial}{\partial x}\left(\dfrac{\partial f}{\partial y}\right) = \dfrac{\partial^2 f}{\partial x \partial y}$ or fxy

$\therefore \dfrac{\partial^2 f}{\partial x^2}, \dfrac{\partial^2 f}{\partial y^2}, \dfrac{\partial^2 f}{\partial x \partial y}, \dfrac{\partial^2 f}{\partial y \partial x}$ are all called second order partial derivatives.

Example 1. **Verify that** $\dfrac{\partial^2 f}{\partial x\, \partial y} = \dfrac{\partial^2 f}{\partial y\, \partial x}$ **for the function** $\log \dfrac{x^2 + y^2}{xy}$.

Solution.

Let
$$f(x, y) = \log \frac{x^2+y^2}{xy} = \log(x^2 + y^2) - \log xy$$

$$\frac{\partial f}{\partial x} = \frac{1}{x^2+y^2} \times 2x - \frac{1}{xy} \times y = \frac{2x}{x^2+y^2} - \frac{1}{x}$$

$$\frac{\partial^2 f}{\partial y \partial x} = 2x \cdot \frac{-2y}{(x^2+y^2)^2} - 0 = \frac{-4xy}{(x^2+y^2)^2} \quad \text{.... }(i)$$

Now
$$\frac{\partial f}{\partial y} = \frac{1}{x^2+y^2} \times 2y - \frac{1}{xy} \times x = \frac{2y}{x^2+y^2} - \frac{1}{y}$$

$$\frac{\partial^2 f}{\partial x \partial y} = 2y\left(\frac{-2x}{(x^2+y^2)^2}\right) - 0 = \frac{-4xy}{(x^2+y^2)^2} \quad \text{.... }(ii)$$

From (*i*) and (*ii*), we see that

$$\frac{\partial^2 f}{\partial x \partial y} = \frac{\partial^2 f}{\partial y \partial x}.$$ Hence the result.

Example 2. $f = \tan^{-1}\left(\dfrac{y}{x}\right)$ **verify that** $\dfrac{\partial^2 f}{\partial y\, \partial x} = \dfrac{\partial^2 f}{\partial x\, \partial y}$.

Solution.

Given
$$f = \tan^{-1}\left(\frac{y}{x}\right) \quad \text{.... } (i)$$

Differentiating (*i*) partially w.r.t. *x*, we get

$$\frac{\partial f}{\partial x} = \frac{1}{1+\left(\frac{y}{x}\right)^2} \times \left(\frac{-y}{x^2}\right) = \frac{-y}{x^2+y^2} \quad \text{.... } (ii)$$

Differentiating (*i*) partially w.r.t. *y*, we get

$$\frac{\partial f}{\partial y} = \frac{1}{1+\left(\frac{y}{x}\right)^2} \cdot \frac{1}{x} = \frac{x}{x^2+y^2} \quad \text{.... } (iii)$$

Differentiating (*ii*) partially w.r.t. *y*, we get

$$\frac{\partial^2 f}{\partial y \partial x} = \frac{\partial}{\partial y}\left(\frac{-y}{x^2+y^2}\right)$$

$$= \frac{(x^2+y^2)(-1)-(-y)(2y)}{(x^2+y^2)^2}$$

$$= \frac{y^2-x^2}{(x^2+y^2)^2} \quad \text{.... } (iv)$$

Differentiating (*iii*) partially w.r.t. *x*, we get

$$\frac{\partial^2 f}{\partial x \partial y} = \frac{\partial}{\partial x}\left(\frac{x}{x^2+y^2}\right) = \frac{(x^2+y^2)(1)-x(2x)}{(x^2+y^2)^2}$$

$$= \frac{y^2-x^2}{(x^2+y^2)^2} \quad \text{.... } (v)$$

From (*iv*) and (*v*), we get $\dfrac{\partial^2 f}{\partial y \partial x} = \dfrac{\partial^2 f}{\partial x \partial y}$. Hence the result.

EXERCISES

1. Verify that $\dfrac{\partial^2 z}{\partial y \partial x} = \dfrac{\partial^2 z}{\partial x \partial y}$ when $z = \log \dfrac{x^2+y^2}{xy}$.

2. Verify that $\dfrac{\partial^2 z}{\partial y \partial x} = \dfrac{\partial^2 z}{\partial x \partial y}$ when $z = ax^2 + 2hxy + by^2$.

1.9 INTEGRATION

Integration is just the reverse process of differentiation.

The purpose of integration is to find out the function whose differential co-efficient is given.

The symbol $\int dx$ represents the operation of finding integral of the function w.r.t. x. Thus

$$\int \cos x \, dx = \sin x + c \qquad \therefore \frac{d}{dx}(\sin x) = \cos x$$

where c is called the constant of integration.

Important integration formulae

1. $\int x^n dx = \frac{x^{n+1}}{n+1} + C$ provided $n \neq -1$ as $\frac{d}{dx}\left(\frac{x^{n+1}}{n+1}\right) = x^n$

2. $\int 1 dx = x + C$ as $\frac{d}{dx}(x) = 1$

3. $\int (\mu + v) dx = \int \mu dx + \int v dx$

4. $\int \frac{1}{x} dx = \log x + C$ as $\frac{d}{dx}(\log x) = \frac{1}{x}$

5. $\int a^x dx = \frac{a^x}{\log_e a} + C$ as $\frac{d}{dx}(a^x) = a^x \log a$

6. $\int e^x dx = e^x + C$ as $\frac{d}{dx}(e^x) = e^x$

7. $\int \cos x \, dx = \sin x + C$ as $\frac{d}{dx}(\sin x) = \cos x$

8. $\int \sin x \, dx = -\cos x + C$ as $\frac{d}{dx}(\cos x) = -\sin x$

9. $\int \sec^2 x \, dx = \tan x + C$ as $\frac{d}{dx}(\tan x) = \sec^2 x$

10. $\int \text{cosec}^2 x \, dx = -\cot x + C$ as $\frac{d}{dx}(\cot x) = -\text{cosec}^2 x$

11. $\int \sec x \tan x \, dx = \sec x + C$ as $\frac{d}{dx}(\sec x) = \sec x \tan x$

12. $\int \text{cosec}\, x \cot x \, dx = -\text{cosec}\, x + C$ as $\frac{d}{dx}(\text{cosec}\, x) = -\text{cosec}\, x \cot x$

1.10 DEFINITE INTEGRAL

A definite integral is one whose lower and upper limits are given.

It is denoted by $\int_a^b f(x) dx$, and is read as integral of $f(x)$ w.r.t. x between $x = a$ and $x = b$.

If $\int f(x) dx = g(x)$

then, $\int_a^b f(x) dx = [g(x)]_a^b = g(b) - g(a)$.

Example 1. **Evaluate** $\int_1^2 x^2 dx$

Solution.

$$\int_1^2 x^2 dx = \left[\frac{x^3}{3}\right] = \frac{2^3}{3} - \frac{(1)^3}{3} = \frac{8}{3} - \frac{1}{3} = \frac{7}{3}$$

Example 2. $\int_0^{\pi/4} \sin x \cos x \, dx$

Solution.

$$\int_0^{\pi/4} \sin x \cos x dx = \int_0^{\frac{\pi}{4}} \frac{1}{2}(2\sin x \cos x)dx = \frac{1}{2}\int_0^{\frac{\pi}{4}} \sin 2x \, dx = \frac{1}{2}\left[\frac{-\cos 2x}{2}\right]_0^{\frac{\pi}{4}}$$

$$= -\frac{1}{4}[\cos 2x]_0^{\frac{\pi}{4}}$$

$$= -\frac{1}{4}\left[\cos 2\frac{\pi}{4} - \cos 0\right]$$

$$= -\frac{1}{4}\left[\cos\frac{\pi}{2} - \cos 0\right]$$

$$= -\frac{1}{4}[0-1]$$

$$= \frac{1}{4}.$$

EXERCISES

1. Integrate the following functions w.r.t. x and find within the limits $x = 2$ to $x = 4$.

(*i*) x^3 (*ii*) $\sqrt{x}$

2. Evaluate the following integrals

(*i*) $\int_0^{30°} \sin 7x \, dx$ (*ii*) $\int_{15°}^{30°} \cos(4x-3) \, dx$

3. Evaluate the following integrals

(*i*) $\int_0^5 \frac{dx}{5^x}$ (*ii*) $\int \sqrt{1+x} dx$

4. Evaluate $\int (1+\cos x)^{1/2} dx$ Hint. $\int (1+\cos x) = 2\cos^2 \frac{x}{2}$

ANSWERS

1. (*i*) 60 (*ii*) 3.448

2. (*i*) 0.2666 (*ii*) $\frac{1}{4}$ [sin 117° – sin 57°]

3. (*i*) $\frac{1}{5}\log\frac{5}{2}$ (*ii*) $\frac{2}{3}(1+x)^{3/2}+C$

4. $2\sqrt{2}\cos\frac{x}{2}$

1.11 FACTORIALS

Factorial Notation : The continued product of first n natural numbers i.e., product of all natural numbers beginning with 1 and ending with n is written as $\underline{|n}$ or $n!$ and read as factorial n. Thus

$$n\,! = 1 \cdot 2 \cdot 3 \cdot .... n$$

or writing the factors in reverse order

$$n\,! = n(n-1)\,(n-2)\, ...\, 3 \cdot 2 \cdot 1$$

For example $4\,! = 4 \cdot 3 \cdot 2 \cdot 1 = 24$

$3\,! = 3 \cdot 2 \cdot 1 = 6$

and $0\,!$ is always 1 (i.e., $0! = 1$)

A property of factorials :

$$\begin{aligned} 6\,! &= 6 \cdot 5 \cdot 4 \cdot 3 \cdot 2 \cdot 1 = 6 \cdot 5\,! \\ &= 6 \cdot 5 \cdot 4\,! = 6 \cdot 5 \cdot 4 \cdot 3\,! \\ &= 6 \cdot 5 \cdot 4 \cdot 3 \cdot 2\,! \end{aligned}$$

Also we know that $n! = n(n-1)\,(n-2)\, ...\, 3 \cdot 2 \cdot 1$

which can be written as $n = n(n-1)\,!$

or $n = n(n-1)\,(n-2)\,!$

or $n = n(n-1)\,(n-2)\,(n-3)!$ and so on.

Example 1. **Find the value of** $\frac{11!}{7!\;4!}$.

Solution.

$$\frac{11\,!}{7!\;4!} = \frac{11 \cdot 10 \cdot 9 \cdot 8 \cdot 7!}{7!\;4!}$$

$$= \frac{11 \cdot 10 \cdot 9 \cdot 8}{4 \cdot 3 \cdot 2 \cdot 1} = 330$$

Example 2. **Prove that** $(2n)\,! = 2^n\, n\,!\,[1 \cdot 3 \cdot 5 \cdot ......\,(2n-1)]$.

Solution.

$$\begin{aligned} (2n)! &= 1 \cdot 2 \cdot 3 \cdot 4 \cdot ....\,(2n-1)\,2n \\ &= (2 \cdot 4 \cdot 6 \cdot ....\,2n)\,[1 \cdot 3 \cdot 5 \cdot ....\,(2n-1)] \\ &= 2^n(1 \cdot 2 \cdot 3 \cdot ....\,n)\,[1 \cdot 3 \cdot 5 \cdot ....\,(2n-1)] \\ &= 2^n\;\; n\,!\;\;[1 \cdot 3 \cdot 5 \cdot ....\,(2n-1)] \end{aligned}$$

Hence the result.

EXERCISES

Find the value of

1. (a) $\frac{8!}{6!}$ (b) $\frac{10! - 9!}{9!}$

2. Prove that $12\,! = 2^6 \cdot 6!(1 \cdot 3 \cdot 5 \cdot 7 \cdot 9 \cdot 11)$

3. $\dfrac{(n+2)!}{n!} = n^2 + 3n + 2$. Prove it.

4. $(2n)!\,(n-1)! = 2(n!)\,(2n-1)!$. Prove it.

ANSWERS

1. (*a*) 56 (*b*) 9

1.12 PERMUTATIONS AND COMBINATIONS

Permutations. The different arrangements that can be made with a given number of things taking some or all of them at a time are called permutations.

The symbol ^nP_r is used to denote the number of permutations of n things taken r at a time.

To find the value of ^nP_r

The number of permutations of n things taken r at a time is the same as the number of ways in which r places in a row can be filled with n different things.

Suppose that we have n things. The first place can be filled up by any one of these n things. Thus there are n ways of filling up the first place.

When the first place has been filled up by any one of the n ways, we are left with $(n-1)$ things and any one of them can be put in the second place. Therefore the second place can be filled up in $(n-1)$ ways. So by the principle of Association, the number of ways of filling up the first two places is $n(n-1)$.

When the first two places have been filled up in any one of the $n(n-1)$ ways, then we are left with $(n-2)$ things. So the third place can be filled up in $(n-2)$ ways.

By the principle of Association, the number of ways of filling up the first three places is $n(n-1)(n-2)$.

Proceeding in this way, the number of ways in which r places can be filled up is

$$\begin{aligned} ^nP_r &= n(n-1)(n-2) \text{ .... upto } r \text{ factors} \\ &= n(n-1)(n-2) \text{ .... } (n-(r-1)) \\ &= n(n-1)(n-2) \text{ .... } (n-r+1) \end{aligned}$$

Therefore $\quad ^nP_r = n(n-1) \text{ .... } (n-r+1)$

For example $\quad ^7P_4 = 7 \times 6 \times 5 \times 4 = 840$

$$^5P_4 = 5 \times 4 \times 3 \times 2 = 120$$

When $r = n$, $\quad ^nP_r = {^nP_n} = n(n-1)(n-2) \text{ .... to } n \text{ factors}$

$$\begin{aligned} &= n(n-1)(n-2) \text{ .... } (n-(n-1)) \\ &= n(n-1)(n-2) \text{ .... } 1 \\ &= n! \end{aligned}$$

So $\quad ^nP_n = n!$

Now to prove that $\quad ^nP_r = n \times {^{n-1}P_{r-1}} = \dfrac{n!}{(n-r)!}$

$$\begin{aligned} \text{L.H.S.} = {^nP_r} &= n(n-1)(n-2) \text{ ....} (n-r+1) \\ &= \frac{\{n(n-1)(n-2)\text{.......}(n-r+1)\}\{(n-r)(n-r-1)\text{.....}3\cdot2\cdot1\}}{(n-r)(n-r-1)\text{.....}3\cdot2\cdot1} \\ &= \frac{n!}{(n-r)!} \end{aligned}$$

$$\text{R.H.S.} = n \times {^{n-1}P_{r-1}}$$

$$= n \times \frac{(n-1)!}{((n-r)\cdot(r-1))!} \qquad \left| \begin{array}{l} \because \text{We have proved above} \\ {}^nP_r = \dfrac{n!}{(n-r)!} \end{array} \right.$$

$$= \frac{n.(n-1)!}{(n-r)!} = \frac{n!}{(n-r)!}$$

$$\therefore \qquad {}^nP_r = n \times {}^{n-1}P_{r-1} = \frac{n!}{(n-r)!}$$

Example 1. **Find the value of $^{12}P_4$.**

Solution.

$$^{12}P_4 = 12 \times 11 \times 10 \times 9$$
$$= 11880$$

Example 2. If $^{n-1}P_3 : {}^{n+1}P_3 = 5 : 12$, find the value of n.

Solution.

$$\frac{^{n-1}P_3}{^{n+1}P_3} = \frac{5}{12}$$

or $$\frac{(n-1)(n-2)(n-3)}{(n+1)\, n(n-1)} = \frac{5}{12}$$

Simplifying $$\frac{(n-2)(n-3)}{(n+1)n} = \frac{5}{12}$$

Cross-multiplying , $12(n^2 - 5n + 6) = 5n\,(n+1)$

$\Rightarrow \qquad 12n^2 - 60n + 72 = 5n^2 + 5n$

$\Rightarrow \qquad 7n^2 - 65n + 72 = 0$

$$\Rightarrow \qquad n = \frac{65 \pm \sqrt{(-65)^2 - 4(7)(72)}}{2 \times 7}$$

$$= \frac{65 \pm 47}{14} = 8 \quad \text{or} \quad \frac{9}{7}$$

Rejecting $\frac{9}{7}$ as n is a +ve integer.

Hence $n = 8$.

Example 3. **In how many ways can the letters of 'Lahore' be arranged ?**

Solution.

Total number of letters in the word Lahore = 6

These are all different letters.

$\therefore$ Number of arrangements of these letters taken all at a time

$^6P_6 = 6! = 720$.

Example 4. **How many words can be formed from the letters of the word DAUGHTER so that the vowels are never together ?**

Solution.

Total number of letters in the word DAUGHTER is 8. These 8 letters are all different.

First of all, let us find those arrangements in which vowels are always together.

When the vowels A, E, U are always together, they can be supposed to be put in a bracket and treated as one letter (A, E, U). So the number of letters becomes 6.

i.e., (8 – 3 + 1) (D, G, H, T, R (A, E, U))

These 6 letters can be arranged in ${}^6P_6$ ways = 6! ways

Now the three letters in the group (A, E, U) can be arranged among themselves in ${}^3P_3 = 3!$ ways.

∴ Required numbers of such words = 6! 3! = 720 × 6 = 4320.

Now total numbers of arrangements of letters of the word DAUGHTER is 8!

∴ Number of words in which vowels are never together

= Total number of words – Numbers of words in which vowels are always together

= 8! – 4320

= 40320 – 4320 = 36000.

EXERCISES

1. How many different words can be formed with the letters of the word 'LUCKNOW'?

2. Prove that ${}^{2n}P_n = 1 \cdot 3 \cdot 5 \cdot \ldots\ldots (2n-1) \cdot 2^n$.

3. Find n if ${}^{n+3}P_6 : {}^{n+2}P_4 = 14 : 1$.

4. In how many ways 5 books on English and 4 books on Hindi can be placed on a shelf so that the books on the same subject always remain together?

5. How many words can be formed out of the letters of the word 'ORIENTAL' so that no two of the vowels are together?

ANSWERS

1. 7! = 5040

3. 4

4. 5! 4! 2!

5. $4!\ {}^5P_4 = 2880$.

Combinations : The different selections or groups that can be made out of a given number of things taking some or all of them at a time are called combinations.

The symbol ${}^nC_r$ is used to denote the number of combinations of n things taken r at a time.

To find the value of ${}^nC_r$

The number of combinations of n things taken r at a time is denoted by ${}^nC_r$. Let these combinations be x. Because each combination contains r things and these r things in any one of the combinations can be arranged among themselves in $r!$ ways (as we know by permutations). Hence one combination will give rise to $r!$ permutations. So x combinations will give rise to $x \cdot r!$ permutations. But the number of permutation of n things taken r at a time is ${}^nP_r$.

$${}^nP_r = x \cdot r! = {}^nC_r \cdot r!$$

$${}^nC_r = \frac{{}^nP_r}{r!} = \frac{n!}{(n-r)!\,r!}$$

$$\therefore \quad {}^nC_r = \frac{n!}{r!(n-r)!}$$

To prove that ${}^nC_r = {}^nC_{n-r}$.

We know that ${}^nC_r = \dfrac{n!}{r!(n-r)!}$

Changing r to $n - r$ we get

$$^nC_{n-r} = \frac{n!}{(n-r)!(n-(n-r))}.$$

$$= \frac{n!}{(n-r)!\,r!}$$

$\therefore \quad {}^nC_r = {}^nC_{n-r}$. Hence the result.

Example 1. **Evaluate $^{50}C_{47}$.**

Solution.

$$^{50}C_{47} = {}^{50}C_{50-47} = {}^{50}C_3 = \frac{\lfloor 50}{\lfloor 3 \lfloor 47} = \frac{50{\cdot}49{\cdot}48\lfloor 47}{\lfloor 3 \lfloor 47}$$

$$= \frac{50 \cdot 49 \cdot 48}{3 \cdot 2 \cdot 1} = 19600.$$

Example 2. **If $^nC_{10} = {}^nC_{15}$, find the value of $^{27}C_n$.**

Solution:

Given $\quad {}^nC_{10} = {}^nC_{15}$

Either $10 = 15$ or $10 + 15 = n$ $\quad \left|\because {}^nC_r = {}^nC_{n-r}\right.$

But $10 = 15$ is impossible

So $n = 10 + 15$

$\therefore \quad n = 25$

Now we have to calculate $^{27}C_n$ i.e., $^{27}C_{25}$ $\quad \left|\because n = 25\right.$

$$^{27}C_{25} = \frac{27!}{25!\,2!} = \frac{27 \cdot 26 \cdot 25!}{25!\ 2!}$$

$$= \frac{27 \cdot 26}{2 \cdot 1} = 351$$

So $\quad {}^{27}C_{25} = 351.$

Example 3. **If $^{n+1}C_{r+1} : {}^nC_r = 11 : 6$ and $^nC_r : {}^{n-1}C_{r-1} = 6 : 3$, find the values of n and r.**

Solution.

$$\frac{^{n+1}C_{r+1}}{^nC_r} = \frac{11}{6} \text{ and } \frac{^nC_r}{^{n-1}C_{r-1}} = \frac{6}{3}$$

$$\Rightarrow \frac{\dfrac{(n+1)!}{(r+1)!\{(n+1)-(r+1)\}!}}{\dfrac{n!}{r!(n-r)!}} = \frac{11}{6} \text{ and } \frac{\dfrac{\lfloor n}{\lfloor r \lfloor n-r}}{\dfrac{(n-1)!}{(r-1)!\{(n-1)-(r-1)\}!}} = \frac{6}{3} = 2$$

$$\Rightarrow \frac{(n+1)!}{(r+1)!(n-r)!} \times \frac{r!\,(n-r)!}{n!} = \frac{11}{6} \text{ and } \frac{\lfloor n}{\lfloor r \lfloor n-r} \times \frac{(r-1)!\,(n-r)!}{(n-1)!} = 2$$

$$\Rightarrow \frac{(n+1)!\ r!}{(r+1)!\,n!} = \frac{11}{6} \text{ and } \frac{n!\,(r-1)!}{r!\,(n+1)!} = 2$$

$$\Rightarrow \frac{(n+1)n!\ r!}{(r+1)\,r!\,n!} = \frac{11}{6} \text{ and } \frac{n(n-1)!\,(r-1)!}{r(r-1)!\ (n-1)!} = 2$$

$$\Rightarrow \frac{n+1}{r+1} = \frac{11}{6} \text{ and } \frac{n}{r} = 2$$

$\Rightarrow \quad 6n + 6 = 11r + 11$ and $\quad n = 2r$

$\Rightarrow \quad 6(2r) + 6 = 11r + 11 \qquad n = 2 \times 5$

$\Rightarrow \quad 12r + 6 = 11r - 11 \qquad = 10$

$r = 5.$

Hence $\quad n = 10, \; r = 5.$

EXERCISES

1. Evaluate (*i*) $^{51}C_{49}$ (*ii*) $^{100}C_{96}$

2. If $^nP_r = {}^nP_{r+1}$ and $^nC_r = {}^nC_{r-1}$, find n and r.

3. Prove that $^{2n}C_n = \dfrac{2^n \; 1\cdot3\cdot5\cdot......(2n-1)}{n!}$.

4. If $^{18}C_r = {}^{18}C_{r+2}$, evaluate $^rC_5$ and $^{11}C_r$.

ANSWERS

1. (*i*) 1275 (*ii*) 3921225

4. 56, 165

1.13 PROBABILITY

Probability means chance or possibility occurrence of some event.

Experiment. An experiment is defined as a process of well-defined outcomes.

Random experiment. A random experiment is defined as an experiment in which all possible outcomes are known in advance.

For example, If we toss a coin we will either get head or tail.

or, in throwing of a dice there are 6 possibilities 1 or 2 or 3 or 4 or 5 or 6.

Sample space. The sample space of a random experiment is the set of all possible outcomes.

For example,

	Random Experiment	***Sample Experiment***
1.	Throwing of a fair dice	$S = \{1, 2, 3, 4, 5, 6\}$
2.	Tossing of a coin	$S = \{H, T\}$
3.	Tossing of two coins	$S = \{HH, HT, TH, TT\}$
4.	A family of two children	$S = \{BB, BG, GB, GG]$

Event. An event is a subset of sample space. An event is called simple event, if it contains only one sample point. In the experiment of throwing a die, the event A of getting 3 is a simple event.

Equal likely outcomes. The outcomes of a random experiment are called equally likely, if all of these have equal preferences. In the experiment of tossing of a coin, the outcomes head and tail are equally likely.

Exhaustive outcomes. The outcomes of random experiment are called exhaustive if they cover all the possible outcomes of the experiment. In throwing of a dice, the outcomes 1, 2, 3, 4, 5, 6 are exhaustive.

Probability. Suppose in a random experiment there are n equally likely, exhaustive outcomes. Let A be an event and there are m outcomes favourable to the happening of it.

Then the probability $P(A)$ of the happening of the event A is defined as

$$P(A) = \frac{m}{n}$$

or $$P(A) = \frac{\text{Total no. of favourable cases in the happening of } A}{\text{Total no. of equally likely exhaustive cases}}$$

It is clear from the definition that

$$0 \le m \le n \quad \Rightarrow \quad 0 \le \frac{m}{n} \le 1$$

$$\Rightarrow \quad 0 \le P(A) \le 1$$

The number of cases favourable to the non-happening of the event A is $n - m$.

$$\therefore \quad P(\text{not } A) = \frac{n-m}{n}$$

$$= \frac{n}{n} - \frac{m}{n} = 1 - \frac{m}{n}$$

$$= 1 - P(A)$$

$$P(\bar{A}) = 1 - P(A) \qquad (\bar{A} = \text{not } A)$$

$$\Rightarrow P(\bar{A}) + P(A) = 1$$

***Example 1.* What is the probability of getting an even number in the throw of an unbiased dice?**
Solution.

In this experiment, there are 6 equally likely possible outcomes, *i.e.*, {1, 2, 3, 4, 5, 6} and let A be the event of getting even number.

$$A = \{2, 4, 6\}$$

Favourable cases = 3

Total no. of cases = 6

$$P(A) = \frac{3}{6} = \frac{1}{2}$$

***Example 2.* Find the probability of getting the sum 10 in a single throw of two dice.**
Solution.

Here $S = \{(1, 1), (1, 2), (1, 3) \ldots\ldots.. (6, 5), (6, 6)\}$

No. of possible outcomes are $6 \times 6 = 36$

Let A be the event of getting sum 10

$$A = \{(4, 6), (5, 5), (6, 4)\}$$

Favourable cases = 3

$$P(A) = \frac{3}{36} = \frac{1}{12}.$$

***Example 3.* Find the probability of getting a 'King' or 'Queen' in a single draw from a well shuffled pack of playing cards.**
Solution.

Let A be the event of getting a King or a Queen in the draw.

No. of favourable cases for happening of the event A is $4 + 4 = 8$

Total number of cases = 52

$$P(A) = \frac{8}{52} = \frac{2}{13}.$$

EXERCISES

1. Find the probability of getting an odd number in a single throw of a fair dice.
2. Find the probability of getting a number less than 2 in a single throw of a fair dice.

3. In a simultaneous toss of two coins, find the probability of getting exactly two heads.
4. Find the probability of getting no head in a single toss of three coins.
5. A ball is drawn at random from a box containing 4 white, 7 red and 12 black balls. Determine the probability that the ball drawn is red.

ANSWERS

1. $\frac{1}{2}$ **2.** $\frac{1}{6}$ **3.** $\frac{1}{4}$ **4.** $\frac{1}{8}$ **5.** $\frac{7}{23}$

1.14 MUTUALLY EXCLUSIVE EVENTS

Two events are said to be mutually exclusive events if both cannot occur together with same trial. In experiment of rolling a die, the events $A = \{1, 3\}$ and $B = \{2, 4, 6\}$ are mutually exclusive events. In the same experiment, the events $A = \{1, 3\}$ and $C = \{1, 3, 5, 6\}$ are not mutually exclusive events because if 3 appears on the die it is favourable to both A and C.

Addition Theorem (For mutually exclusive events). If A and B are two mutually exclusive events associated with a random experiment, then

$$P(A \text{ or } B) = P(A) + P(B).$$

Proof: Let n be the total number of exhaustive, equally likely cases of the experiment.

Let m_1 and m_2 be the number of cases favourable to the happening of the events A and B respectively.

$$P(A) = \frac{m_1}{n}$$

and

$$P(B) = \frac{m_2}{n}$$

As the events are given to be mutually exclusive, so there will be no sample point common to both events A and B.

$\therefore$ The event A or B can happen in exactly $m_1 + m_2$ ways

$$\therefore \quad P(A \text{ or } B) = \frac{m_1 + m_2}{n} = \frac{m_1}{n} + \frac{m_2}{n} = P(A) + P(B)$$

Hence $P(A \text{ or } B) = P(A) + P(B)$

Addition Theorem (For non-mutually events). If A and B are two non-mutually exclusive events associated with a random experiment, then

$$P(A \text{ or } B) = P(A) + P(B) - P(AB)$$

Proof. Let n be the total number of exhaustive and equally likely cases of the experiment.

Let m_1 and m_2 be the number of cases favourable to the happening of the events A and B respectively.

$$\therefore \quad P(A) = \frac{m_1}{n}$$

$$P(B) = \frac{m_2}{n}$$

As the events are given to be non-mutually exclusive, there will be some sample points common to both events A and B.

Let m_3 be the number of common sample points

$$P(AB) = \frac{m_3}{n}$$

These m_3 sample points are also included in the events A and B separately.

Total number of sample points in A or B

$$= m_1 + m_2 - m_3$$

$$P(A \text{ or } B) = \frac{m_1 + m_2 - m_3}{n} = \frac{m_1}{n} + \frac{m_2}{n} - \frac{m_3}{n} = P(A) + P(B) - P(AB)$$

Example 1. **The probability that a company executive will travel by bus is $\frac{2}{3}$ and that he will travel by train is $\frac{1}{5}$. Find the probability of his travelling by bus or train.**

Solution.

Let A = the event that the company executive travels by bus

$$\therefore \quad P(A) = \frac{2}{3}$$

Let B = the event that the company executive travels by train

$$P(B) = \frac{1}{5}$$

Events in this question are mutually exclusive because either he can travel by bus or by train.

$\therefore$ The probability of his travelling by bus or by train

$$P(A \text{ or } B) = P(A) + P(B) = \frac{2}{3} + \frac{1}{5} = \frac{13}{15}.$$

Example 2. **One number is drawn from numbers 1 to 150. Find the probability that it is either divisible by 3 or 5.**

Solution.

Here $\quad S = \{1, 2, 3, ..., 149, 150\}$

Let $\quad A$ = the event that number is divisible by 3

$A = \{3, 6, 9, ..., 147, 150\}$

$$P(A) = \frac{50}{150}$$

B = The event that the number is divisible by 5

$B = \{5, 10, 15, ..., 145, 150\}$

$$P(B) = \frac{30}{150}$$

The events are not mutually exclusive because some points are common to both A and B. The common points are 15, 30, 45, ..., 150.

So $\quad AB = \{15, 30, 45, ..., 135, 150\}$

$$P(AB) = \frac{10}{150}$$

So by the Addition theorem

$$P(A \text{ or } B) = P(A) + P(B) - P(AB)$$

$$= \frac{50}{150} + \frac{30}{150} - \frac{10}{150} = \frac{70}{150} = \frac{7}{15}$$

Hence the result.

EXERCISES

1. A bag contains 30 balls numbered from 1 to 30. One ball is drawn at random. Find the probability that number of the ball is a multiple of 5 or 6.

2. Find the probability that a card drawn from a pack of playing cards is either King or Spade.

3. A box contains 4 red balls, 4 green balls and 7 white balls. What is the probability that a ball drawn is either red or white?

ANSWERS

1. $\frac{1}{3}$ **2.** $\frac{4}{52}+\frac{13}{52}-\frac{1}{52}$ **3.** $\frac{11}{15}$

(B) COMPUTERS

1.15 INTRODUCTION

Definition: Computer is an electronic device which is able to process unmeaningful raw data into meaningful information.

This raw data is called **Input** and the generated information is called **Output** in computer terminology.

Analog Vs Digital

An **Analog** device deals with the continuous measurement, recording or processing of information.

A speedometer is an analog device which measures the speed of the vehicle continuously. Similarly a slide-rule is an analog computing device.

A **Digital** device deals with discrete samples of information.

Both types of information can be converted to another with sufficient accuracy using either ADC (Analog to Digital converter) or DAC (Digital to Analog converter).

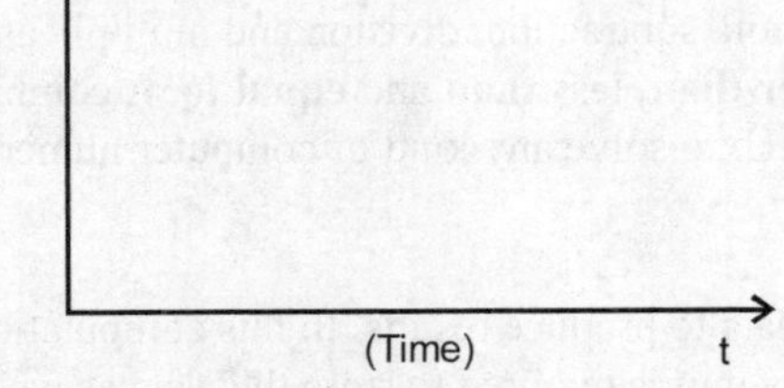

Analog Measurement

Digital Measurement

All modern computers are Digital in nature. They require Digital Output and can only generate Digital Output. Any interfacing with an Analog Input device requires use of ADC circuits. Similarly to produce an Analog output, DAC circuits are required.

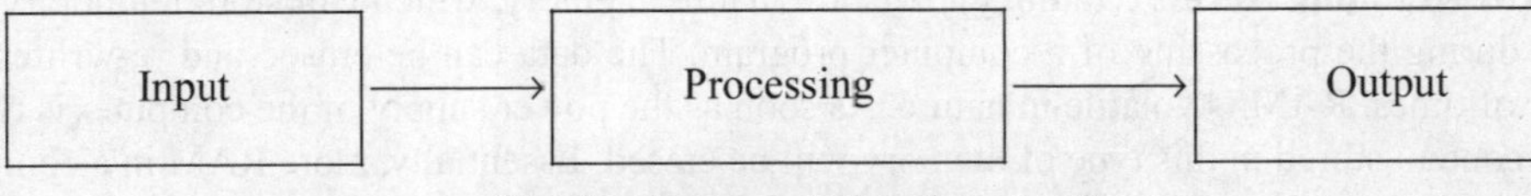

(The basic computational process)

1.16 COMPONENTS OF A COMPUTER

A computer system is made up of three basic components.

1. Input Devices
2. Output Devices
3. Central Processing Unit (CPU)

1. Input Devices provide data for the computer to operate upon. Some commonly used input devices are

(*a*) Keyboard (to enter data manually)

(*b*) Mouse (to re-position the cursor on the screen)

The **Keyboard** allows the entry of textual information. It has keys for characters (letters, numbers and punctuation marks) and special commands. Pressing the keys tells the computer what to do or what to write. The **Mouse** has a special ball that allows the selection of a point on the screen by moving a screen cursor to the point and pressing a mouse button. Alternative devices to the mouse are tracker balls, light pens and touch sensitive screens. Other types of input devices allow you to put images into the computer like **Scanner** and **Digital Camera**.

Besides these, there could be other specialised input devices, providing information about Industrial Automation System, Control Systems to the Computer, etc.

2. Output Devices are used to view or store the generated information. Some commonly used output devices are:

(*a*) Monitor (a T.V. screen to view results)

(*b*) Printer (to print the results)

(*c*) Speakers (outputs the sound)

The **Monitor** is a Cathode Ray Tube device like a T.V. which displays text and graphics. It allows you to 'see' what you and the computer are doing together. A **Printer** is another common part of a computer system. It takes computer data and generates a hard copy of that data. The most commonly used printers are inkjet printer and Laser printer. **Speakers** are output devices that allow you to hear sound from your computer. They are just like stereo speakers.

3. Central Processing Unit (C.P.U.)

CPU is the **brain** of the computer. It is responsible for converting the raw data into meaningful information based on certain instructions. These instructions are called **Programs**. All processing and calculations take place here. This is further divided into 3 parts.

(*a*) Arithmetic Logical Unit (A.L.U.):

It performs all the mathematical functions of addition, subtraction, division and multiplication. Besides this, it performs logical operations like **greater than, less than** and **equal to**. A combination of these basic functions makes it possible for A.L.U. to solve any kind of computer numerical problems.

(*b*) Memory:

Computer requires a set of instructions and input data to produce results. In this computational process, it may also generate temporary data. A memory unit is required to store this data as well as the generated result until it can be passed on to the output device.

Memory is further classified into 2 types.

Primary Memory. The memory which resides inside the computer system (CPU) is known as primary memory. It is further classified into 2 types.

RAM (Random Access Memory). It is a dynamic memory, which can store temporary data created during the processing of a computer program. The data can be erased and re-written any number of times. RAM is volatile in nature. As soon as the power supply of the computer is turned off, the content stored in this type of memory will be erased. Essentially, more RAM in a computer system means faster processing.

ROM (Read Only Memory) – This type of memory is used to store data, which is permanent in nature. This includes configuration data used to start a computer (BIOS—Basic Input Output System). Such information is related to hardware present in the system and does not change. Data stored in ROM cannot be erased or re-written.

Secondary Memory – Secondary memory resides outside the computer system (CPU). It is used to permanently store large amounts of data. Large amount of input data and the programs/instruction required to process this data into information, both reside on the secondary memory and can be accessed any time later. Data on secondary memory has to be transferred to primary memory (RAM), for the computer to utilize it.

The various types of secondary memory are: — Floppy Disk, Hard Disk, Compact Disk (CD). Some of the devices which were earlier in use were — punched cards, magnetic tape and magnetic drums.

(*c*) Control Unit. It controls the functioning of the CPU. The flow of data between the ALU and Memory is controlled by the Control Unit.

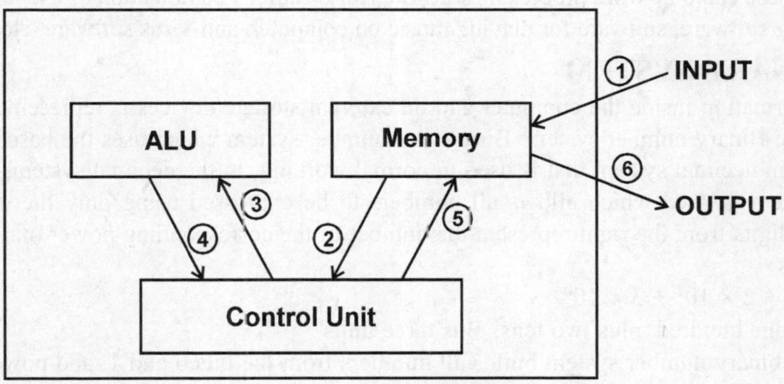

Central Processing Unit

It passes the input data present in the memory to ALU for processing. Once the data is processed, it sends the generated information to the memory so that it can be passed on to the output devices.

1.17 HARDWARE AND SOFTWARE

Hardware is a collective term used for all the physical parts of a computer which we can see and touch. Hardware consists of the monitor, keyboard, mouse, internal parts of the computer like Processor Chips, Memory Chips and other Integrated Circuits (I.Cs).

Software is a set of programs/instructions required by the hardware of the computer to perform the desired data-processing function. Software can be classified further into:

1. System Software
2. Application Software
3. Utility Programs

1. System Software consists of programs that are required for the functioning of the computer system. The following can be the examples of system software.

(*a*) Operating System. This software controls the loading of instructions stored in the Disk (Secondary memory) into the primary memory. The other components of O.S. help in controlling the overall flow of data to various parts of the computer. It also helps in the proper functioning of various peripheral devices like Printers, Modems, etc. Most popular operating systems are: Windows, Unix, Linux and MS-DOS.

(*b*) Instruction Interpreters (Languages): These sets of programs understand instructions written in English like language and convert them into Machine Language (Binary language) which the computer understands.

If they convert programs written in English like languages (called High Level Languages) into machine language, line by line, they are called **Interpreters**. If they convert the whole program together and then execute it, they are called **Compilers**. Some of the Interpreted Languages are Basic, Java. Some of the Compiled Languages are C, C^{++}, Pascal.

Internally, all the programs are converted to an intermediate Language called **Assembly Language** before getting converted to machine language. The program that converts Assembly Language programs to machine language are called **Assembler**.

2. Application Software. These are a set of programs written to solve a specific problem. These programs are specially written for such Applications like Railway Reservation, Library Management, Weather forecasting, etc.

3. Utility Programs are general purpose programs written to solve commonly faced problems/ functions. These could be word processing software (for creating a combination of text and graphics), photo-editing software, software for playing music on computer, anti-virus software, etc.

1.18 BINARY SYSTEM

All information inside the computer and on external storage devices is represented in a form related to the Binary number system. Binary is a number system which uses the base 2 instead of the base 10 in decimal system that is used in normal working. In the decimal system, a positional number system is used which allows all numbers to be expressed using only the digits 0 – 9. Successive digits from the right represent the number of the corresponding power of 10. Thus 123 in decimal is

$$1 \times 10^2 + 2 \times 10^1 + 3 \times 10^0$$

that is, one hundred, plus two tens, plus three units.

But the binary number system builds all numbers from the bits 0 and 1, and power of 2.

101 in the binary number system represents the number

$$1 \times 2^2 + 0 \times 2^1 + 1 \times 2^0$$

having value equal to 5 in the decimal system. Binary system allows all computation inside the computer to take place using cheap two-state electronic devices.

Modern computers organise information into small units called **bytes** which hold eight **bits**, each bit representing a 0 or 1. Each byte may hold a single character of text or a small integer. Larger numbers, computer instructions and character strings occupy several bytes.

Binary Arithmetic

All digital computers represent data as a collection of bits. A bit is the smallest possible unit of information. It can be in one of two states—**off** or **on, 0** or **1.**

1 Bit [ ]

can be [0] or [1]

If one bit has two different states, how many states do two bits have? The answer is four. Likewise, three bits have eight states as represented below.

2 Bits = 4 States

[0 | 0] [0 | 1] [1 | 0] [1 | 1]

3 Bits = 8 States

0	0	0

0	0	1

0	1	0

0	1	1

1	0	0

1	0	1

1	1	0

1	1	1

A common and convenient grouping of bits is the **byte** or **Octet**, composed of eight bits. If two bits have four combinations, and three bits have eight combinations, how many combinations do eight bits have? If you don't want to write out all possible byte patterns, just multiply eight twos together – one two for each bit, *i.e.*,

$$2^8 = 256$$

So, a byte has 256 possible states.

1 Byte = 8 Bits 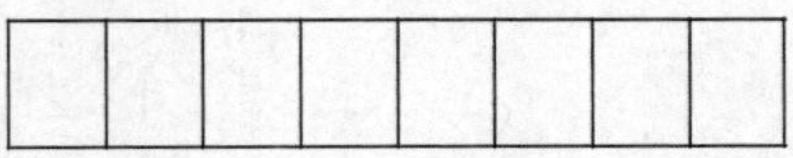

If a byte has 256 possible states, its exact state can be represented by a number from 1 to 256. However, since zero is a very important number, a byte is more typically represented by a number from 0 to 255. This is very common with bit pattern 00000000 representing zero, and bit pattern 11111111 representing 255. The numbers matching these two patterns, and everything in between, can be computed by assigning a weight to each bit, multiplying each bit's value (0 or 1) by its weight, and then adding the totals. For example, here's how 217 is represented as 11011001 in binary:

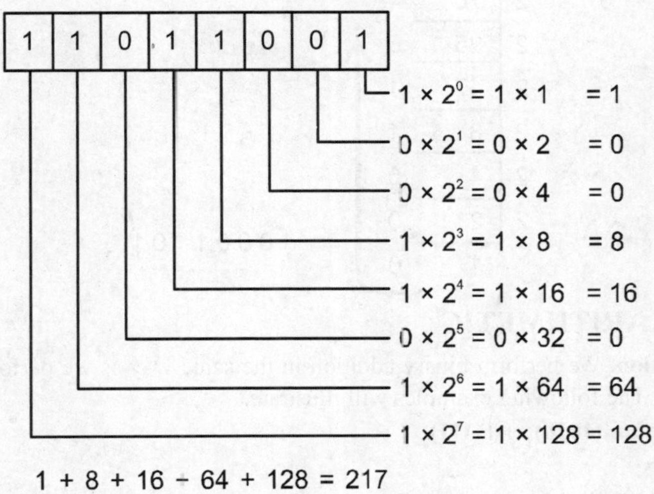

To convert a number from decimal to binary, begin at left most bit position (128). If the number is larger than or equal to the bit's weight, write a 1 in the bit position, subtract the bit's weight from the number, and continue with the difference. If the number is less than the bit's weight, write a 0 in the bit position and continue without any subtraction. Here's an illustration of converting 141 to binary:

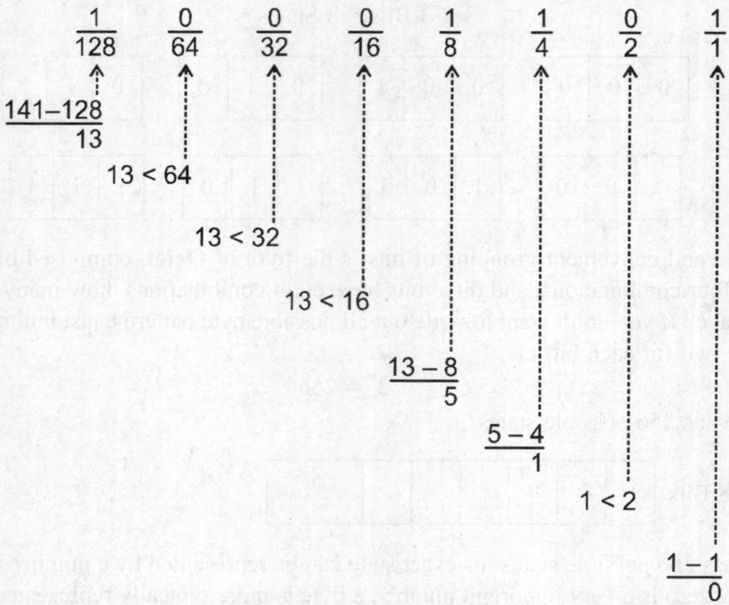

Another way of doing conversion from decimal to binary is to repeatedly divide the decimal number by 2, keeping only the whole part and noting down the remainder at each step, until we reach zero. At the end of this, we have the binary representation of the integer by reading the remainders in reverse order.

2	141	
2	70	1
2	35	0
2	17	1
2	8	1
2	4	0
2	2	0
	1	0

10001101

1.19 BINARY ARITHMETIC

Binary Addition. We perform binary addition in the same way as we perform addition with Decimal numbers. The following examples will illustrate.

***Example 1.* Compute $(1010)_2 + (1111)_2$**

Solution.

$$(1010)_2 + (1111)_2 = (?)_2$$

		1	1	1		→ carry
Augend			1	0	1	0
Addend	+		1	1	1	1
		1	1	0	0	1

We start the addition from the RHS as we do in Decimal system. As soon as the sum becomes two or more, we carry 1 to the next position on the left and continue addition.

Binary Subtraction. This is also performed in the same way as we do in Decimal system. While subtracting, a borrow is required when subtracting 1 from 0. In such a case, when 1 is borrowed from the higher column on LHS, the difference is also 1. The following example will illustrate.

Example 2. **Compute $(1100)_2 - (11)_2$**

Solution.

$$(1100)_2 - (11)_2 = (?)_2$$

$$\begin{array}{rrrr} 1 & 1 & 0 & 0 \\ -\;0 & 0 & 1 & 1 \\ \hline 1 & 0 & 0 & 1 \end{array}$$

As in the decimal system, we start subtraction from the RHS. We need to borrow from the second column from RHS. But the upper number in the second column is already 0. Thus, it needs to borrow from the third column from RHS. It may be assumed that second column borrows from third column. The number in the third column becomes 0. In the seond column, it becomes 2, out of which 1 is borrowed by the number in the first column. The difference in the first column will be (2 – 1), *i.e.*, 1. Now 1 is left in the second column. The difference in the second column will be (1 – 1), *i.e.*, 0. The upper number in the third column is now 0. Therefore the difference in third column will be (0 – 0), *i.e.*, 0.

Binary Multiplication

Binary multiplication is also similar to multiplication in decimal system. The following example will illustrate.

Example 3. **Compute $(1101)_2 \times (101)_2$**

Solution.

$$\begin{array}{rrrrrrr} & & & 1 & 1 & 0 & 1 \\ & & & \times & 1 & 0 & 1 \\ \hline & & & 1 & 1 & 0 & 1 \\ & & 0 & 0 & 0 & 0 & \times \\ & 1 & 1 & 0 & 1 & \times & \times \\ \hline 1 & 0 & 0 & 0 & 0 & 0 & 1 \end{array}$$

Binary Division. Binary division is performed by successive subtraction of the divisor from the dividend. The quotient bits are thus framed. The following rules are obeyed :

$$0 \div 1 = 0$$

$$1 \div 1 = 1$$

Division by 0 has no meaning.

The following example will illustrate.

Example 4. **Compute $(110)_2 \div (10)_2$**

Solution.

$$\begin{array}{r} 11 \quad \leftarrow \text{Quotient} \\ \text{Divisor} \rightarrow 10 \,\big)\, 110 \quad \leftarrow \text{Dividend} \\ \underline{10\;\;} \\ 10 \\ \underline{10} \\ \underline{00} \quad \text{Remainder} \end{array}$$

Thus, $(110)_2 \div (10)_2 = (11)_2$

1.20 OPERATING SYSTEM

An operating system is a layer of software which takes care of technical aspects of a computer's operation. It shields the user of the machine from the low-level details of the machines operation and provides frequently needed facilities. There is no universal definition of what an operating system is. You can think of it as being the software which is already installed on a machine, before you add anything of your own. Normally the operating system has a number of key elements:-

(*i*) A technical layer of software for driving the hardware of the computer, like disk drives, the keyboard and the screen.

(*ii*) A file system which provides a way of organising files logically.

(*iii*) A simple command language which enables users to run their own programs and to manipulate their files in a simple way.

(*iv*) Some operating systems also provide text editors, compilers, debuggers and a variety of other tools. Since the Operating System (OS) is in charge of a computer, all requests to use its resources and devices need to go through the OS. An OS therefore provides legal entry points into its code for performing basic operations like writing to devices.

Operating systems may be classified by how many tasks they can perform 'simultaneously' and by how many users can be using the system 'simultaneously'. That is: Single-user or multi-user and single-tasking or multi-tasking.

Single-user operating system: We are all familiar with the concept of sitting down at a computer system and writing documents or performing some task. Operating systems such as Windows 95, Windows NT Workstation and Windows 2000 Professional are essentially single user operating systems. They provide you the capability to perform tasks on the computer system such as writing programs and documents, printing and accessing files. In essence, a single-user operating system provides access to the computer system by a single user at a time. If another user needs access to the computer system, he must wait till the current user finishes what he is doing and leaves.

Students in Computer Labs at Colleges or University often experience this. You might also have experienced this at home, where you want to use the computer but some one else is currently using it. You have to wait for them to finish before you can use the computer system.

Multi-user operating system: A multi-user operating system lets more than one user access the computer system at one time. Access to the computer system is normally provided *via* a network, so that users access the computer remotely using a terminal or other computer.

In the early days of large multi-user computers, multiple terminals (keyboards and associated monitors) were provided. These terminals send their commands to the main multi-user computer for processing, and the results were then displayed on the associated terminal monitor screen. Terminals were hand-wired directly to the multi-user computer system.

Today, these terminals are generally personal computers and use a network to send and receive information to the multi-user computer system. Examples of multi-user operating systems are Unix, Linux (A Unix clone) and main frames such as the IBM A5400.

The Operating System for a large multi-user computer system with many terminals is much more complex than a single-user operating system. It must manage and run all user requests, ensuring they do not interfere with each other.

Advantages and Disadvantages of Multi-user Operating Systems. The advantage of having a multi-user operating system is that it lets a number of users share the very expensive hardware resource. This means the cost is divided amongst the users. It also makes better use of the resources. Since the resources are shared, they are more likely to be in use than sitting idle and being unproductive.

One problem with multi-user computer systems is that as more people access it, the performance becomes slower and slower. Another disadvantage is the increased cost of memory devices, as a multi-user operating system requires a lot of disk space and memory. In addition, the actual software for multi-user operating systems tend to cost more than single-user operating systems.

Multi-tasking Operating System. A multi-tasking operating system provides the ability to run more than one Program at a time. For example, a user could be running a word processing package, printing a document, copying files to the floppy disk and backing up selected files to a tape unit. Each of these tasks the user is doing appears to be running at the same time.

A multi-tasking operating system has the advantage of letting the user do more than one task at a time, so this leads to increased productivity. The disadvantage is that more the programs that are run by the user, more the memory that is required.

1.21 COMPUTER PROGRAMMING

Writing software or computer programs, is a lot like writing down the steps it takes to do something. But, if what you need to do is not obvious, programming will challenge you like when you are solving puzzles.

A computer is very dumb. It does exactly what you tell it to do, which is not necessarily what you wanted. Programming will help you to learn the importance of clarity of expression.

These days, most people don't need to know how a computer works. Most people can simply turn on a computer and point the mouse at some little graphical image on the display, double-click the left mouse button, and a computer program starts up (like a Web browser); and, this is all they need to know about their computer.

To learn how to write computer programs, you need to know a little bit about how a computer works. Your job will be to instruct the computer to do things. These lists of instructions that you will write are computer programs, and the stuff that these instructions manipulate are different types of objects.

Basically, computer perform operations on objects. A microprocessor, which is the heart of a computer, is really primitive but very fast. It takes group of binary numbers representing parts of objects and moves them around, adds pairs together, subtracts one from another, compares a pair, etc.

Computers manipulate numbers, symbolic information (think characters), visual things (images), sound (MP3), and sets of instruction (the computer's native language). As an example, computer displays consist of a bunch of coloured points called pixels. A pixel is an object. It has a position which consists of the row and column it is in. Its colour is specified as three numbers—called RGB (Red, Green, Blue) values. So, like pixels are used to compose graphics and images, there are also standard representations for the other things: numbers, characters, sound, and the computer's instructions. At the lowest level, everything is a bunch of ones and zeros (binary number).

Programming Languages (The Microprocessor's Language)

Just like we have many different languages in use throughout the world for us humans to communicate with, there are different languages that we use to communicate with a computer. Computer languages are used to tell the computer what to do, you instruct it.

If you instructed a computer in its native language (machine language), you would have to write instructions in the form of binary numbers. This is very, very hard to do. Although the pioneers of computers did this, no one does this these days.

One step above machine language is assembler language. In assembler, the operations that the microprocessor knows how to do are given names. Address in memory can also be given meaningful names. This is a big step over binary, but still very tedious to do any large software program with. It still has its place for little pieces of software that need to interact directly with the microprocessor and/or those that are executed many times. Here is an example of DEC system–10 assembler language that:

1. adds two numbers together,
2. adds in a third number if it is greater than zero, and
3. stores the results in a third location.

Op. Code	Accumulator	Operand
Move	T_1	Number 1
Add	T_1	Number 2
Skiple		Number 3
Add	T_1	Number 3
Move	T_1	Result

Programming Languages (High-Level Languages)

Most software written today is in high-level languages. There are many and some are quite old. COBOL, FORTRAN, and Lisp were written in the 1950s. Higher-level languages make it easier to describe the pieces of the Program you are creating. They abstract away the specifics of the microprocessor in your computer. Most come with large sets of common stuff you need to do.

High Level Language Programming doesn't require an intimate knowledge of the internal structure of the computer and can be ported over different computer architectures. High level language programs are processed by a program, known as a **Compiler**, which translates the high level language program into an equivalent machine language program. Thus, if a compiler was available for each machine architecture, then a program could be run on any computer without change.

At the moment there are several internationally standardised languages as given in the table below which are popular.

Language	Application Area
COBOL	Business Applications
FORTRAN	Engineering & Scientific Calculations
C	General Purpose Language
Prolog	Artificial Intelligence
Lisp	Artificial Intelligence
C++	Object Oriented version of C
Java	Object Oriented Language widely used on Web

Most popular high-level languages have been standardised by one or the other international body. Thus if Programs are written to conform strictly to the standard then they should execute correctly on any architecture for which there is a standard conforming compiler.

EXERCISES
(INCLUDING QUESTIONS FROM DIFFERENT UNIVERSITY PAPERS)

Objective Questions

Fill in the blanks

1. If N = a^x, then $\log_a$ N is equal to ....
2. Decimal part of logarithm of a given number is called ....
3. Derivative of e^x *w.r.t.* x is ....
4. What is cos (180 + θ) equal to ?
5. $\int_1^2 x^2\, dx$ is equal to ....
6. How many different words can be formed with the letters of the word 'LUCKNOW'?
7. A device which deals with the continuous measurement, recording or processing of information is called .... device
8. CPU in a computer is responsible for converting raw data into .... based on certain instructions
9. Number system uses the base 2 instead of the base 10 in decimal system that is used in normal working
10. High level language programs are processed by a program known as .... which translates the high level language program into machine language program

Multiple Choice Questions

Choose the correct option

1. $\log_a 1$ is equal to
   (*a*) 0 (*b*) 1
   (*c*) ∞ (*d*) 10
2. $\log_a m$ is equal to
   (*a*) $\log_a m \times \log_b a$ (*b*) $\log_b m \times \log_a b$
   (*c*) $\log_m b \times \log_b a$ (*d*) $\log_m b \times \log_a b$
3. Characteristic of log 0.5828 is
   (*a*) –2 (*b*) 1
   (*c*) $\bar{2}$ (*d*) $\bar{1}$
4. For the curve $y = x^3 + 1$, the values of y corresponding to values of x as –2 and 2 are
   (*a*) 7, –9 (*b*) – 3, 6
   (*c*) – 7, 9 (*d*) 6, – 3
5. Slope of the line $2x - y = 3$ is
   (*a*) – 2 (*b*) $-\frac{1}{2}$
   (*c*) $\frac{1}{2}$ (*d*) 2

6. Derivative of $x + \frac{1}{x}$ *w. r. t. x* is

(*a*) $1 - \frac{1}{x^2}$ (*b*) $1 + \frac{1}{x^2}$

(*c*) $-1 + \frac{1}{x^2}$ (*d*) $-1 - \frac{1}{x^2}$

7. Experssion for sin C – sin D is

(*a*) $2 \cos \frac{C-D}{2} \sin \frac{C+D}{2}$ (*b*) $2 \cos \frac{C+D}{2} \sin \frac{C-D}{2}$

(*c*) $2 \sin \frac{C+D}{2} \sin \frac{C-D}{2}$ (*d*) $2 \cos \frac{C+D}{2} \cos \frac{C-D}{2}$

8. Integration of x^3 w. r. t. x within the limits x = 2 to x = 4 has the value

(*a*) 60 (*b*) 40

(*c*) 20 (*d*) 10

9. The value of $\frac{8!}{6!}$ is

(*a*) 224 (*b*) 28

(*c*) 112 (*d*) 56

10. The number of ways 5 books on English and 4 books on Hindi can be placed on a shelf so that the books on the same subject always remain together, is

(*a*) 5 ! 4 ! 3 ! (*b*) 5 ! 4 ! 2 !

(*c*) 4 ! 3 ! 2 ! (*d*) 5 ! 3 ! 2 !

11. The decimal equivalent of $(11001)_2$ is

(*a*) 25 (*b*) 50

(*c*) 35 (*d*) 15

12. General purpose computer language is

(*a*) Prolog (*b*) COBOL

(*c*) C (*d*) FORTRAN

13. Binary representation of the decimal number 141 is

(*a*) 10110001 (*b*) 11110001

(*c*) 10001111 (*d*) 10001101

14. General purpose programs written to solve commonly faced problems are called

(*a*) system software *b*) application software

(*c*) utility program (*d*) none

15. The unit that performs the mathematical functions is

(*a*) ALU (*b*) CPU

(*c*) RAM (*d*) ROM

ANSWERS

Fill in the blanks

1. x	**2.** mantissa	**3.** e^x	**4.** $-\cos\theta$
5. $\frac{7}{3}$	**6.** 5040	**7.** analcgue	**8.** meaningful information
9. binary	**10.** compiler		

Multiple choice questions

1. (*a*)	**2.** (*b*)	**3.** (*d*)	**4.** (*c*)	**5.** (*d*)
6. (*a*)	**7.** (*b*)	**8.** (*a*)	**9.** (*d*)	**10.** (*b*)
11. (*a*)	**12.** (*c*)	**13.** (*d*)	**14.** (*c*)	**15.** (*a*)

General Questions

1. Define a computer. Give the terms used for raw data and the information generated by the computer.

2. Differentiate clearly between analog and digital divices. Give one example of each.

3. What are the components of a computer. Classify the following into input device and output device :

(*i*) keyboard (*ii*) printer

(*iii*) mouse (*iv*) monitor

4. (*a*) What do you mean by 'program'?

(*b*) Explain the function of arithmetical logical urit.

5. Give the classification and sub-classification of memory.

6. Give a schematic diagram for the flow of data between the ALU and Memory.

7. Explain the following terms :

(*i*) Hardware (*ii*) Software

Give examples of three operating systems.

8. What are utility programs ? Explain giving examples.

9. What is binary system used in computers ? Differentiate between byte and bit.

10. What are the advantages and disadvantages of multi-user operating system.

11. Give the application areas of the following programming languages :

(*i*) COBOL (*ii*) LISP

(*iii*) C^{++} (*iv*) FORTRAN

12. Find the binary equivalent of $(72)_{10}$ **[Ans.** $(1001000)_2$**]**

13. Find the decimal equivalent of $(11001)_2$ **[Ans.** 25**]**

14. Compute $(1010)_2 + (1101)_2$ **[Ans.** $(10111)_2$**]**

15. Compute $(1111)_2 - (1010)_2$ **[Ans.** $(0101)_2$**]**

16. Compute $(1111)_2 \times (101)_2$ **[Ans.** $(1001011)_2$**]**

17. Compute $(10000111)_2 \div (1001)_2$ **[Ans.** $[01111)_2$**]**

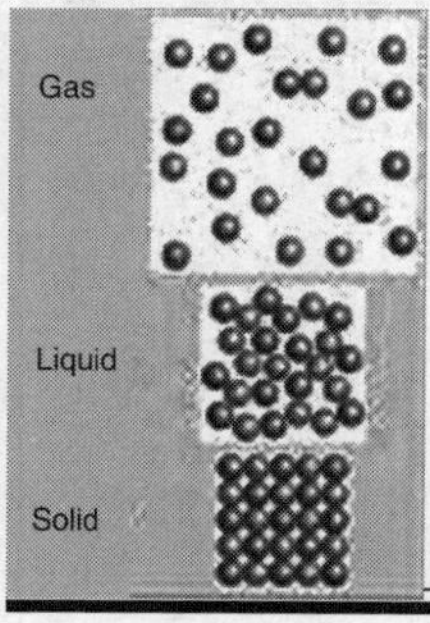

2

Gaseous State

2.1 INTRODUCTION

Matter exists in three states, gas, liquid and solid. A gas consist of molecules separated by wide distances. The molecules are free to move about in the space in which it is contained.

Intermolecular space in the liquids is much less but the molecules are free to move throughout the liquid. Molecules in a solid occupy fixed positions. They cannot move about, however, they can vibrate slightly from the fixed positions.

Some important characteristics of gases are given below:

(*i*) They fill up the whole vessel in which they are placed.

(*ii*) They exert pressure on the walls of the vessel in which they are taken.

(*iii*) They can be compressed by applying the pressure.

(*iv*) Cooling and compression of the gas brings about its liquefaction.

(*v*) They exhibit the property of intermixing with one another—a property known as *diffusion.*

(*vi*) At ordinary conditions of temperature and pressure, their densities are low.

2.2 KINETIC THEORY OF GASES

Maxwell and Boltzmann developed a mathematical theory to explain the behaviour of gases. This theory makes the following assumptions :

(*i*) Every gas is made up of a large number of extremely small particles called molecules. All the molecules of a particular gas are identical in mass and size and differ from gas to gas.

(*ii*) The molecules of a gas are separated from each other by large distances so that the actual volume of the molecules is negligible as compared to the total volume of the gas.

(*iii*) The distances of separation between the molecules are so large that the forces of attraction or repulsion between them are negligible.

Ludwig Boltzmann

(*iv*) The molecules are supposed to be moving continuously in different directions with different velocities. Hence, they keep on colliding with one another (called molecular collisions) as well as on the walls of the containing vessel.

(*v*) The pressure exerted on the walls of the containing vessel is due to the bombardment of the molecules on the walls of the containing vessel.

(*vi*) Since the molecules are moving with different velocities, they possess different kinetic energies. However, the average kinetic energy of the molecules of a gas is directly proportional to the absolute temperature of the gas.

(*vii*) Molecules are supposed to be perfectly elastic hard spheres so that no energy is lost when the molecules collide with one another or with the walls of the vessel.

(*viii*) The force of gravitation on the molecules is also supposed to be negligible.

2.2.1 Derivation of kinetic gas equation, $PV = \frac{1}{3} mnu^2$

Consider a cubical vessel containing the gas (Fig. 2.1). Suppose,

Total number of molecules present in the gas = n,

Mass of each molecule of the gas = m,

Length of each side of the cube = l cm.

According to the kinetic theory of gases, the molecules of the gas are moving in different directions with different velocities. Consider any one molecule, moving with a velocity ϑ_1 cm/sec in any direction. The velocity can be resolved into three mutually perpendicular components ϑ_x, ϑ_y and ϑ_z parallel to X-axis, Y-axis, and Z-axis respectively as shown in Fig. 2.1. Then mathematically,

$$\vartheta_1^2 = \vartheta_x^2 + \vartheta_y^2 + \vartheta_z^2 \quad ...(i)$$

Now, consider the motion of the molecule due to component ϑ_x only. Due to this component, the molecule will strike the opposite faces A and B which are at right angles to the X-axis.

Momentum of molecule due to component ϑ_x just before it strikes the face A

$$= m\vartheta_x$$

As the molecules are supposed to be perfectly elastic, after colliding the face A, the molecule will return with the same velocity. As the direction has changed, the velocity will now be taken as $-\vartheta_x$. Hence

Momentum of the molecule due to component ϑ_x just after it strikes the face A

$$= m \times (-\vartheta_x) = -m\vartheta_x$$

∴ Change of the momentum of the molecule in *one collision* on the face A due to component ϑ_x

$$= m\vartheta_x - (-m\vartheta_x) = 2\,m\vartheta_x$$

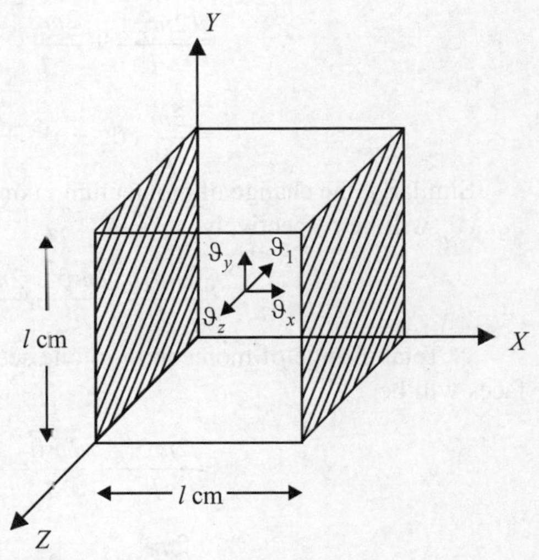

Fig. 2.1. A cubical vessel containing n molecules of the gas.

Now let us calculate the number of collisions which take place on the face A in one second due to component ϑ_x.

For next collision on the face A, the molecule has to travel from face A to B and then back from B to A *i.e.,* a total distance of

$$l + l = 2l \text{ cm}$$

As the component being considered is ϑ_x cm/sec, this means that the distance travelled by the molecule in one second is ϑ_x cm.

∴ Number of collisions which take place on the face A in one second due to component ϑ_x

$$= \frac{\vartheta_x}{2l}$$

As the change of momentum in one collision is $2m\vartheta_x$

∴ Change of momentum of the molecule in one second due to component ϑ_x for collisions on the face A

$$= 2m\vartheta_x \times \frac{\vartheta_x}{2l} = \frac{m\vartheta_x^2}{l}$$

Similarly, change of momentum of the molecule in one second due to component ϑ_x for collisions on the face B

$$= \frac{m\vartheta_x^2}{l}$$

∴ Change of momentum of the molecule in one second due to component ϑ_x for collisions on the opposite faces perpendicular to X-axis.

$$= \frac{m\vartheta_x^2}{l} + \frac{m\vartheta_x^2}{l} = \frac{2m\vartheta_x^2}{l}$$

Similarly, change of momentum of the molecule in one second due to component ϑ_y, for collisions on the opposite faces perpendicular to Y-axis and due to component ϑ_z for collisions on the opposite faces perpendicular to Z-axis will be respectively.

$$\frac{2m\vartheta_y^2}{l} \text{ and } \frac{2m\vartheta_z^2}{l}$$

∴ Change of momentum of the molecule in one second due to all the components for collisions on all the six faces

$$= \frac{2m\vartheta_x^2}{l} + \frac{2m\vartheta_y^2}{l} + \frac{2m\vartheta_z^2}{l}$$

$$= \frac{2m}{l}(\vartheta_x^2 + \vartheta_y^2 + \vartheta_z^2) = \frac{2m\vartheta_1^2}{l} \quad \text{from eqn. } (i)$$

Similarly, the change of momentum in one second, for molecules having velocities $\vartheta_2, \vartheta_3, \vartheta_4$, ... ϑ_n will be respectively.

$$\frac{2m\vartheta_2^2}{l}, \frac{2m\vartheta_3^2}{l}, \frac{2m\vartheta_4^2}{l} ..... \frac{2m\vartheta_n^2}{l}$$

∴ Total change of momentum in one second for all n molecules for collisions on all the six faces will be

$$= \frac{2m\vartheta_1^2}{l} + \frac{2m\vartheta_2^2}{l} + \frac{2m\vartheta_3^2}{l} + ... + \frac{2m\vartheta_n^2}{l}$$

$$= \frac{2m}{l}(\vartheta_1^2 + \vartheta_2^2 + \vartheta_3^2 + ... + \vartheta_n^2)$$

$$= \frac{2m}{l} \times n \times \frac{\vartheta_1^2 + \vartheta_2^2 + \vartheta_3^2 + ... + \vartheta_n^2}{n} = \frac{2mn}{l} \times u^2$$

(where $u = \sqrt{\frac{\vartheta_1^2 + \vartheta_2^2 + \vartheta_3^2 + ... + \vartheta_n^2}{n}}$ is called root mean square velocity.)

But the change of momentum in one second according to Newton's second law is equal to the force. Hence, the force (F) exerted by all the 'n' molecules on all the six faces is

$$F = \frac{2mnu^2}{l}$$

As $\quad$ Pressure $= \dfrac{\text{Force}}{\text{Area}}$

and $\quad$ area of all the faces $= 6\, l^2$

$\therefore$ Pressure (P) exerted on the walls of the cubical vessel will be

$$P = \frac{2mnu^2}{l} \times \frac{l}{6l^2} = \frac{1}{3}\frac{mnu^2}{l^3}$$

But l^3 is the volume of the cubical vessel and hence is equal to the volume (V) of the gas

$$\therefore \quad P = \frac{1}{3}\frac{mnu^2}{V}$$

or
$$PV = \frac{1}{3}\, mnu^2$$

This is called *kinetic gas equation.*

2.2.2 Derivation of (*i*) Boyle's law (*ii*) Charles law (*iii*) Avogadro's law from kinetic gas equation

(*i*) **Boyle's law**

According to kinetic gas equation,

$$PV = \frac{1}{3}\, mnu^2 = \frac{2}{3}\cdot\frac{1}{2}\, mnu^2 = \frac{2}{3}\cdot\frac{1}{2}\, Mu^2$$

(where $m \times n = M$ is the total mass of the gas)

But $\quad \frac{1}{2} Mu^2$ = Kinetic Energy of the gas

$$\therefore \quad PV = \frac{2}{3}\ \text{K.E.} \qquad \text{... } (i)$$

Further, according to one of the postulates of the kinetic theory of gases,

K.E. $\propto$ Absolute Temperature (T)

or $\quad$ K.E. $= k\, T$

where k is a constant of proportionality.

Putting this value in equation (*i*), we get

$$PV = \frac{2}{3} kT \qquad \text{... } (ii)$$

As $\frac{2}{3}$ is a constant quantity, k *is* also a constant, therefore, if T *is* kept constant, $\frac{2}{3} kT$ *wi*ll be constant. Hence

PV = constant, if T is kept constant.

Boyle's law is thus deduced.

(*ii*) **Charles Law**

Consider eqn. (*ii*) deduced under Boyle's law

$$PV = \frac{2}{3}\, kT$$

This may be rewritten as

$$\frac{V}{T} = \frac{2}{3}\frac{k}{P}$$

Now $\frac{2}{3}$ is a constant quantity, k is also constant, therefore, if P is kept constant, the whole quantity on the right hand side becomes constant. Hence

$$\frac{V}{T} = \text{constant, if } P \text{ is kept constant}$$

Charles law is thus deduced.

(*iii*) **Avogadro's Law**

Consider any two gases. According to kinetic gas equation,

For 1st gas $$P_1 V_1 = \frac{1}{3} m_1 n_1 u_1^2 \qquad ...(i)$$

For 2nd gas $$P_2 V_2 = \frac{1}{3} m_2 n_2 u_2^2 \qquad ...(ii)$$

where P_1 = Pressure of the 1st gas
V_1 = Volume of the 1st gas
m_1 = Mass of each molecule of the 1st gas
n_1 = Total number of molecules of the 1st gas
u_1 = Root mean square velocity of the molecules of 1st gas.

P_2, V_2, m_2 and u_2 are the corresponding values for the 2nd gas.

If conditions of pressure are similar for both the gases, then

$$P_1 = P_2 \qquad ...(iii)$$

If equal volumes of both the gases are taken, then

$$V_1 = V_2 \qquad ...(iv)$$

Combining equations (*iii*) and (*iv*), we get

$$P_1 V_1 = P_2 V_2 \qquad ...(v)$$

Thus left hand sides of equations (*i*) and (*ii*) are equal. Hence right hand sides must also be equal, *i.e.*,

$$\frac{1}{3} m_1 n_1 u_1^2 = \frac{1}{3} m_2 n_2 u_2^2$$

or $$m_1 n_1 u_1^2 = m_2 n_2 u_2^2 \qquad ...(vi)$$

Further, according to one of the postulates of the kinetic theory of gases,

Average K.E. of a gas ∝ absolute temperature.

Since conditions of temperature are similar in both the cases,

∴ Average K.E. of 1st gas = Average K.E. of 2nd gas

i.e., $$\frac{1}{2} m_1 u_1^2 = \frac{1}{2} m_2 u_2^2$$

or $$m_1 u_1^2 = m_2 u_2^2 \qquad ...(vii)$$

Dividing equation (*vi*) by equation (*vii*), we get

$$n_1 = n_2$$

Thus Avogadro's law is proved.

2.2.3 Derivation of (*i*) Ideal gas equation (*ii*) Graham's law of diffusion (*iii*) Dalton's law of partial pressure from kinetic gas equation

(*i*) Ideal gas equation

According to kinetic gas equation

$$PV = \frac{1}{3}mnu^2 = \frac{2}{3}\cdot\frac{1}{2}mnu^2$$

$$= \frac{2}{3}\text{K.E.}\left(\because \frac{1}{2}mnu^2 = \text{K.E. of the gas}\right)$$

But K.E. $\propto T$ (one of the postulates of Kinetic theory)

or $\quad \text{K.E.} = kT$

where k is constant of proportionality.

$$\therefore \quad PV = \frac{2}{3}kT$$

$\frac{2}{3}$ is a constant quantity, k is also constant

$\therefore \quad \frac{2}{3}k = R$, another constant, called gas constant

Hence $\quad PV = RT$

which is the equation of state for 1 mol of the ideal gas.

(*ii*) Graham's Law of Diffusion

According to kinetic gas equation

$$PV = \frac{1}{3}mnu^2 = \frac{1}{3}Mu^2 \qquad (\because mn = M, \text{ total mass of the gas})$$

or $\quad u^2 = \frac{3PV}{M} = \frac{3P}{M/V} = \frac{3P}{d} \qquad \left(\because \frac{M}{V} = d \text{ is the density of the gas}\right)$

or $\quad u = \sqrt{\frac{3P}{d}}$ ...(*i*)

But greater the velocity (u) of the molecules, greater is the rate of diffusion (r) of the gas

i.e., $\quad r \propto u$ ...(*ii*)

Combining equations (*i*) and (*ii*), we get

$$r \propto \sqrt{\frac{3P}{d}}$$

Hence, if pressure (P) is kept constant, then

$$r \propto \sqrt{\frac{1}{d}}$$

Thus, Graham's law follows from the kinetic gas eqation.

***(iii)* Dalton's Law of Partial Pressures**

Let us consider two gases. According to kinetic gas equation,

$$PV = \frac{1}{3} mnu^2$$

or

$$P = \frac{1}{3} \frac{mnu^2}{V}$$

Now, if only the first gas is enclosed in the vessel of volume *V*, the pressure exerted would be

$$p_1 = \frac{1}{3} \frac{m_1 n_1 u_1^2}{V}$$

Again,if only the second gas is enclosed in the same vessel (so that *V is* onstant), then the pressure exerted would be

$$p_2 = \frac{1}{3} \frac{m_2 n_2 u_2^2}{V}$$

Lastly, if both the gases are enclosed together in the same vessel then since the gases do not react with each other, their molecules behave independent of each other. Hence, the total pressure exerted would be

$$P = \frac{1}{3} \frac{m_1 n_1 u_1^2}{V} + \frac{1}{3} \frac{m_2 n_2 u_2^2}{V}$$

$$= P_1 + P_2$$

Similarly, if more than two gases are present, then it can be proved that

$$P = p_1 + p_2 + p_3 + ...$$

2.2.4 Derivation of relations between molecular velocity, pressure, temperature, volume and density of the gas

According to kinetic gas equation $PV = \frac{1}{3} mnu^2$

$$u^2 = \frac{3PV}{mn} = \frac{3PV}{M} \qquad (\because mn = M \text{ is the mass of the gas})$$

If 1 mol of the gas is taken, *M* will be equal to the molecular weight of the gas,

$$u = \sqrt{\frac{3PV}{M}} \qquad ...(i)$$

Further, for 1 mole of the gas

$$PV = RT$$

$$\therefore \qquad u = \sqrt{\frac{3RT}{M}} \qquad ...(ii)$$

Equation (*i*) can be rewritten as

$$u = \sqrt{\frac{3P}{M/V}}$$

But $\frac{M}{V} = d$ is the density of the gas

$$u = \sqrt{\frac{3P}{d}}$$

2.2.5 Derivation of relation between kinetic energy and temperature

For 1 mol of the gas containing *n* molecules, each of the mass *m*, the kinetic energy will be

$$\text{K.E.} = \frac{1}{2} mnu^2$$

According to kinetic gas equation

$$PV = \frac{1}{3} mnu^2$$

or $$mnu^2 = 3\ PV$$

$\therefore$ $$\frac{1}{2} mnu^2 = 3/2\ PV$$

$$\text{K.E.} = 3/2\ PV$$

Further for 1 mol of the ideal gas,

$$PV = RT$$

$\therefore$ $$\boxed{\text{K.E.} = 3/2\ RT}$$

2.3 MAXWELL DISTRIBUTION OF VELOCITIES

James Clerk Maxwell

Molecules of a gas are moving continuously in different directions with different velocities. Hence they keep on colliding with one another. However, the collisions are supposed to be perfectly elastic and there is no net loss of kinetic energy. In other words, whereas some molecules are speeded up, the others are slowed down after collision. Maxwell suggested that at a particular temperature, the fractions of molecules possessing particular velocities remain almost constant. Such a state is called a *steady state.* On the basis of the laws of probability, he calculated that the fraction of molecules having velocities between *c* and *c* + *dc* is given by

$$\frac{dn_c}{n} = 4\pi \left(\frac{M}{2\pi RT}\right)^{3/2} e^{-Mc^2/2RT} c^2 dc \qquad ...(i)$$

where dn_c is the number of molecules having velocities between *c* and *c* + *dc,* out of a total of *n* molecules of the gas and *M* and *T* are the molecular weight and absolute temperature of the gas respectively.

The above equation may be rewritten as

$$\frac{1}{n}.\frac{dn_c}{dc} = 4\pi \left(\frac{M}{2\pi RT}\right)^{3/2} e^{-Mc^2/2RT} c^2 \qquad ...(ii)$$

The quantity $\frac{1}{n}.\frac{dn_c}{dc}$ is called the probability (*p*) of finding molecules with velocity *c*.

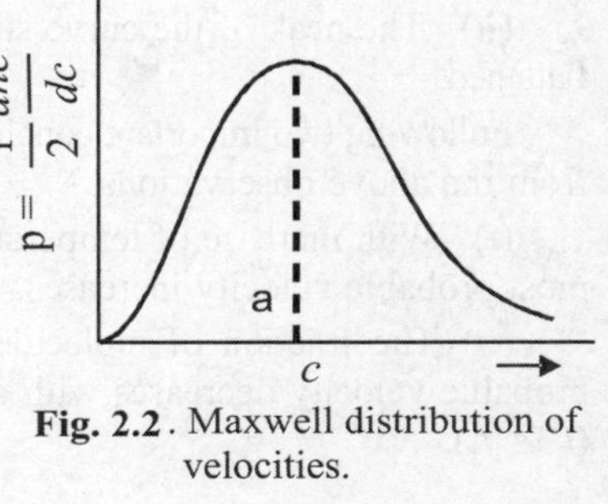

Fig. 2.2. Maxwell distribution of velocities.

The significance of the above equation may be seen by plotting the probability $\left(p = \frac{1}{n}\frac{dn_c}{dc}\right)$ versus the velocity *c* at a particular temperature *T.* Such a plot is known as Maxwell's distribution of velocities and is shown in Fig. 2.2. α represents the most probable velocity.

The area between two ordinates separated by dc is equal to dn_c/n *i.e.,* the fraction of the total number of molecules possessing velocities between c and $c + dc$ (Fig. 2.3) Hence the area under the whole curve is equal to the total number of molecules and the height of the ordinate corresponding to any velocity is a measure of the fraction of the molecules which have that velocity. From the Maxwell's distribution of velocities as given in Fig. 2.2, it may be observed that the fraction of molecules having zero velocity *i.e.,* of molecules which are motionless is zero and the fractions of molecules having velocities greater than zero increase with increase of velocities, reach a maximum value corresponding to a particular velocity and then begin to decrease. Thus, it may be concluded that

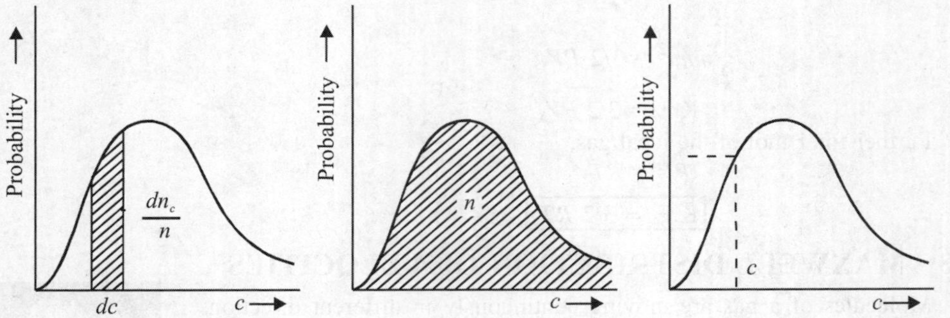

Fig. 2.3. Interpretation of Maxwell's distribution of velocities.

(*i*) The fractions of molecules having very low velocities or very high velocities are very small.

(*ii*) There is a velocity (α) corresponding to which the fraction of molecules is maximum.

In other words, α is the velocity possessed by the maximum fraction of the molecules. *This velocity* is called the *most probable velocity.* This *most probable velocity may be defined as the velocity possessed by the maximum fraction of the molecules of the gas at a particular temperature.*

Effect of Temperature on Maxwell's Distribution of Velocities. With increase of temperature, the curve shifts as shown in Fig. 2.4. Two important changes may be noticed:

(*i*) The peak of the curve shifts forward.

(*ii*) The peak of the curve shifts downward and is flattened.

Following two important conclusions may be drawn from the above observations.

(*i*) With increase of temperature, the value of the most probable velocity increases.

(*ii*) The fraction of molecules possessing the most probable velocity decreases with increase of temperature ($f_2 < f_1$).

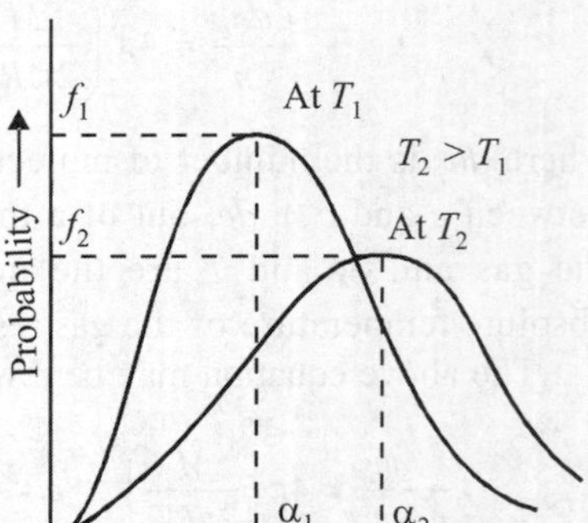

Fig. 2.4. Distribution of velocities at two different temperatures.

2.3.1 Mathematical representation of Maxwell's distribution of energies

Eqn. (*i*) and (*ii*) in The above section representing the distribution of velocities can be converted to represent the distribution of energies as follows:

$\frac{1}{2}Mc^2$ can be replaced by E, the kinetic energy of the gas per mole

$$\frac{1}{2}Mc^2 = E \quad ...(iii)$$

To change $c^2\,dc$ into the energy terms, Eqn. (*iii*) can be rewritten as

$$Mc^2 = 2E \quad ...(iv)$$

Differentiating both sides, we get

$$2Mc\,dc = 2dE$$

or
$$Mc\,dc = dE \quad ...(v)$$

Taking square roots of **both** sides of Eqn. (*iv*), we get

$$M^{1/2}c = (2E)^{1/2} \quad ...(vi)$$

Multiplying Eqns. (*v*) and (*vi*), we get

$$M^{3/2}c^2\,dc = (2E)^{1/2}\,dE$$

$$c^2\,dc = \frac{(2E)^{1/2}\,dE}{M^{3/2}} \quad ...(vii)$$

Substituting the value of $\frac{1}{2}Mc^2$ from Eqn. (*iii*) and that of $c^2\,dc$ from Eqn. (*vii*) in Eqn. (*i*), we get

$$\frac{dn_c}{n} = 4\pi\left(\frac{M}{2\pi RT}\right)^{3/2} e^{-E/RT}\,\frac{(2E)^{1/2}\,dE}{M^{3/2}}$$

or
$$\frac{dn_c}{n} = \frac{2\pi}{(\pi RT)^{3/2}}\,e^{-E/RT}E^{1/2}dE \quad ...(viii)$$

where dn_c/n now represents the fraction of the total number of molecules having kinetic energies between E and dE.

Eqn. (*viii*) can be rewritten as

$$p' = \frac{1}{n}\frac{dn_c}{dE} = \frac{2\pi}{(\pi RT)^{3/2}}\,e^{-E/RT}E^{1/2} \quad ...(ix)$$

where p' represents the probability of molecules having energy E.

The plots of p' (probability) vs E obtained are similar to those for the distribution of velocities.

2.4 (*i*) MOST PROBABLE VELOCITY (α) (*ii*) AVERAGE VELOCITY (ϑ) (*iii*) ROOT MEAN SQUARE (RMS) VELOCITY (U)

(*i*) Most probable Velocity (α). Most probable velocity of a gas at a particular temperature is the velocity possessed by the maximum fraction of the total number of molecules at that temperature. It is usually represented by 'α'.

(*ii*) Average Velocity (ϑ). It is defined as the arithmetic mean of the velocities possessed by the different molecules of a gas at a particular temperature. It is usually represented by ϑ. It is expressed in two different ways.

(*a*) If all the *n* molecules of the gas are supposed to have different velocities, say

$\vartheta_1, \vartheta_2, \vartheta_3, ..., \vartheta_n$ cm /sec, then

$$\vartheta = \frac{\vartheta_1, \vartheta_2, \vartheta_3......\vartheta_n}{n}$$

(*b*) If n_1 molecules of the gas have velocity ϑ_1 cm/sec, n_2 have velocity ϑ_2 cm/sec, n_3 have velocity ϑ_3 cm/sec and so on, then

$$\vartheta = \frac{n_1\vartheta_1 + n_2\vartheta_2 + n_3\vartheta_3 + ...}{n_1 + n_2 + n_3 + ...(= n)}$$

***(iii)* Root Mean Square Velocity (*u*).** It is defined as the square root of the mean of the squares of the different velocities. It is usually represented by '*u*'. Like the average velocity, it can be expressed in two different ways.

(*a*) Assuming all the *n* molecules to have different velocities *viz.* $\vartheta_1, \vartheta_2, ...\vartheta_n$ cm/sec, we have

$$u = \sqrt{\frac{\vartheta_1^2 + \vartheta_2^2 + \vartheta_3^2 + ... + \vartheta_n^2}{n}}$$

(*b*) Assuming n_1 molecules have velocity ϑ_1, n_2 molecules have velocity ϑ_2, n_3, have ϑ_3 and so on, we have

$$u = \sqrt{\frac{n_1\vartheta_1^2 + n_2\vartheta_2^2 + n_3\vartheta_3^2 + ...}{n_1 + n_2 + n_3.....(= n)}}$$

2.4.1 Derivation of expression for (*a*) most probable velocity (α) (*b*) average velocity (ϑ) (*c*) root mean square velocity (*u*) from maxwell distribution of velocities

(*a*) **Most Probable Velocity (α).** Maxwell's distribution law equation can be treated mathematically to get the condition for the maximum in the probability distribution curve. The condition obtained is

$$c = \sqrt{\frac{2RT}{M}}$$

Representing the most probable velocity by α in place of *c*, we have

$$\alpha = \sqrt{\frac{2RT}{M}} \qquad ...(i)$$

This result is obtained by differentiating right hand side of Eqn. (*ii*) in section 2.3 w.r.t. *c* and equating the result to zero *i.e.*, we get

$$\left(1 - \frac{Mc^2}{2RT}\right) ce^{-Mc^2/2RT} = 0$$

Taking $1 - \frac{Mc^2}{2RT} = 0$ we get $\frac{Mc^2}{2RT} = 1$

$$c = \sqrt{\frac{2RT}{M}}$$

(*b*) **Average Velocity (ϑ).** By using Maxwell's distribution law equation, it can be derived mathematically that

$$\vartheta = \sqrt{\frac{8RT}{\pi M}} \qquad ...(ii)$$

[This result is obtained by making use of the fact that

$$\vartheta = \frac{\int_0^\infty c\,dn_c}{n}$$

and then substituting the value of dn_c from Eqn. (*i*) in section 2.3 and solving it].

(*c*) **Root Mean Square Velocity (*u*).** From the kinetic gas equation,

$$PV = \frac{1}{3} mNu^2 = \frac{1}{3} Mu^2 \qquad (mN = M \text{ gram molecular mass})$$

or $$u^2 = \frac{3PV}{M} = \frac{3RT}{M} \qquad (\because PV=RT \text{ for 1 mol of an ideal gas})$$

$\therefore$ $$u = \sqrt{\frac{3RT}{M}} \qquad ...(iii)$$

2.4.2 Relationships between (*i*) Most probable velocity (*ii*) Average velocity (*iii*) Root mean square velocity

Expressions for α, ϑ and u are

$$(i)\ \alpha = \sqrt{\frac{2RT}{M}}, \qquad (ii)\ \vartheta = \sqrt{\frac{8RT}{\pi M}}, \quad (iii)\ u = \sqrt{\frac{3RT}{M}}$$

Combining (*i*) with (*iii*), we get $\alpha = \sqrt{\frac{2}{3}}u$

i.e., $$\alpha = 0.816\,u \qquad ...(i)$$

Combining (*ii*) with (*iii*), we get

$$\vartheta = \sqrt{\frac{8}{3\pi}}u$$

i.e., $$\vartheta = 0.921\,u \qquad ...(ii)$$

Alternatively, the relationship between α, ϑ and u will be

$$\alpha : \vartheta : u = \sqrt{\frac{2RT}{M}} : \sqrt{\frac{8RT}{\pi M}} : \sqrt{\frac{3RT}{M}}$$

$$= \sqrt{2} : \sqrt{\frac{8}{\pi}} : \sqrt{3}$$

$$= 1{\cdot}414 : 1{\cdot}595 : 1{\cdot}732$$

or $$\alpha : \vartheta : u = 1 : 1{\cdot}128 : 1{\cdot}224 \qquad ...(iii)$$

Example 1. **Calculate the temperature at which the hydrogen molecules will have an average speed of 176400 cm s^{-1}.**

Solution. We are given that, $\vartheta = 176400$ cm s^{-1}

Applying the formula $$\vartheta = \sqrt{\frac{8RT}{\pi M}}$$

Putting $R = 8{\cdot}314 \times 10^7$ ergs/degree/mol

$M = 2$ (for H_2)

$$176400 = \sqrt{\frac{8 \times 8.314 \times 10^7 \times T}{22/7 \times 2}}$$

or $$T = \frac{176400^2 \times 22 \times 2}{7 \times 8 \times 8.314 \times 10^7} = 300 \text{ K or } 27°\text{C}$$

Example 2. **Calculate the root mean square velocity, average velocity and most probable velocity of sulphur dioxide molecules at 427°C.**

Solution. (*a*) **Calculation of root mean square velocity**

$$u = \sqrt{\frac{3RT}{M}}$$

Here, $T = 427°C = 427 + 273\ K = 700\ K$

M = Mol. wt. of $SO_2 = 32 + 32 = 64$

$R = 8.314 \times 10^7$ ergs/degree/mol (C.G.S. Units)

Putting the values in the above equation, we get

$$u = \sqrt{\frac{3 \times 8{\cdot}314 \times 10^7 \times 700}{64}}$$

$$= \sqrt{\frac{3 \times 8{\cdot}314 \times 10^8 \times 70}{64}}$$

$$= \sqrt{\frac{3 \times 8{\cdot}314 \times 70}{64}} \times 10^4 = 5.224 \times 10^4\ \text{cm s}^{-1}$$

$$= \mathbf{52240\ cm\ s^{-1}}$$

(*b*) *Calculation of average velocity* (ϑ)

$$\vartheta = 0{\cdot}9213 \times u = 0{\cdot}9213 \times 5.2240 = \mathbf{48128\ cm\ s^{-1}}$$

(*c*) **Calculation of most probable velocity (α)**

$$= 0{\cdot}816 \times u = 0.816 \times 52240\ \text{cm s}^{-1} = \mathbf{42628\ cm\ s^{-1}}$$

Example 3. **At what temperature the root mean square velocity of chlorine gas will be equal to that of SO_2 at N.T.P. ?**

Solution. The root mean square velocity is given by the equation

$$u = \sqrt{\frac{3RT}{M}}$$

For SO_2 $\quad T = 273\ K$

$M = 64$

$$\therefore \quad u_{SO_2} = \sqrt{\frac{3R \times 273}{64}}$$

For Cl_2 $\quad T = ?$

$M = 71$

$$\therefore \quad u_{Cl_2} = \sqrt{\frac{3R \times T}{71}}$$

But $\quad u_{SO_2} = u_{Cl_2}$

$$\therefore \quad \sqrt{\frac{3R \times T}{71}} = \sqrt{\frac{3R \times 273}{64}}$$

or $$\frac{3R \times T}{71} = \frac{3R \times 273}{64}$$

or $$T = \frac{273 \times 71}{64} = 302{\cdot}9\ K$$

$$= 302{\cdot}9 - 273 = \mathbf{29{\cdot}0^\circ\ C}$$

Example 4. **Oxygen has a density of 1.429 g per litre at N.T.P. Calculate the R.M.S. and average velocity of oxygen molecules.**

Solution. (*a*) *Calculation of R.M.S. velocity*

R.M.S. velocity (u) is given by the formula

$$u = \sqrt{\frac{3P}{d}}$$

Here $P = 1$ atmosphere $= 76 \times 13{\cdot}6 \times 981$ dynes/sq cm

$$d = 1{\cdot}429\ gL^{-1} = \frac{1{\cdot}429}{1000}\ \text{g cm}^{-3} \quad \text{(C.G.S. Units)}$$

$$= 0.001429\ \text{g /cm}^{-3}$$

Putting these values in the above formula, we get

$$u = \sqrt{\frac{3 \times 76 \times 13.6 \times 981}{0.001429}}$$

$$= \sqrt{\frac{30.42}{1{\cdot}429} \times 10^8}$$

$$= 4{\cdot}61 \times 10^4\ \text{cm s}^{-1} = \mathbf{46100\ cm\ s^{-1}}$$

(*b*) *Calculation of Average Velocity* (ϑ)

$$\nu = 0{\cdot}9213 \times u = 0{\cdot}9213 \times 46100 = \mathbf{42461\ cm\ s^{-1}}$$

Example 5. **Calculate the root mean square velocity of chlorine molecules at 17°C and 800 mm pressure.**

Solution. This numerical problem involves two steps as follows:

Step 1. *To calculate the volume under given conditions* (for one mole of gas) one mole of a gas at N.T.P. occupies 22400 cm^3.

Applying the equation

$$\frac{P_1V_1}{T_1} = \frac{P_2V_2}{T_2}$$

$$\frac{760 \times 22400}{273} = \frac{800 \times V_2}{290}$$

$$\therefore\ V_2 = \frac{760 \times 22400 \times 290}{273 \times 800} = 22600\ \text{cm}^3$$

Step 2. *To calculate the R.M.S. velocity using the equation*

$$u = \sqrt{\frac{3PV}{M}}$$

Here, P = 800 mm = 80 cm = 80 × 13·6 × 981 dynes per sq cm (C.G.S. Units)

V = 22600 cm^3

M = 71 (for Cl_2)

$$u = \sqrt{\frac{3 \times 80 \times 13{\cdot}6 \times 981 \times 22600}{71}}$$

$$= \sqrt{10{\cdot}18 \times 10^8} = \mathbf{3{\cdot}19 \times 10^4\ cm\ s^{-1}}$$

Example 6. **Calculate translational K.E. of 2 moles of a gas at 27°C**

Solution. For 1 mol of a gas, K.E. $= \frac{3}{2} RT$

$\therefore$ For 2 moles of the gas,

$$\text{K.E.} = \left(\frac{3}{2} RT\right) \times 2$$

Putting $R = 8{\cdot}314 \times 10^7$ ergs per degree per mol

and T = 27°C = (27 + 273) K = 300 K

we get K.E. = 3 × 8·314 × 10^7 × 300 ergs (C.G.S. Units)

$= 7482{\cdot}5 \times 10^7$ ergs = **7482·6 joules.**

Example 7. **Calculate the root mean square velocity of nitrogen at N.T.P. ($R = 8{\cdot}314 \times 10^7$ ergs/degree/mol) .**

Solution. For N_2 at N.T.P.

$$P = 760 \text{ mm} = 76 \text{ cm} = 76 \times 13{\cdot}6 \times 981 \text{ dynes cm}^{-2} \text{ (C.G.S Units)}$$
$$V = 22400 \text{ cm}^3$$
$$M = 28$$

Putting these values in the equation

$$u = \sqrt{\frac{3PV}{M}}$$

we get
$$u = \sqrt{\frac{3 \times 76 \times 13.6 \times 981 \times 22400}{28}}$$
$$= \mathbf{4{\cdot}92 \times 10^4 \text{ cm s}^{-1}}$$

2.5 DEGREE OF FREEDOM

The number of coordinates required to specify the position of all the mass points *i.e.*, atoms in a molecule is called the number of **degrees of freedom.**

To represent one atom, we need three coordinates.

Thus a molecule made up of N atoms has 3 N degrees of freedom.

According to kinetic theory of matter, when thermal energy is absorbed by a molecule, it is stored within the molecule in the form of:

(*i*) translational motion of the molecule

(*ii*) internal movement of the atoms within the molecule, *i.e.*, rotational motion and vibrational motion.

The translational motion of the molecule means the motion of the centre of mass of the molecule as a whole. The centre of mass of the molecule can be represented by three coordinates. Hence there are *three translational degrees of freedom.* The remaining ($3N$– 3) coordinates represent the *internal degree of freedom.*

However, as already mentioned, the internal degrees of freedom may be subdivided into

(*i*) rotational degrees of freedom, and

(*ii*) vibrational degrees of freedom.

For a rotational motion, there are two degrees of freedom for a linear molecule and three for a non-linear molecule as shown in Fig. 2.5.

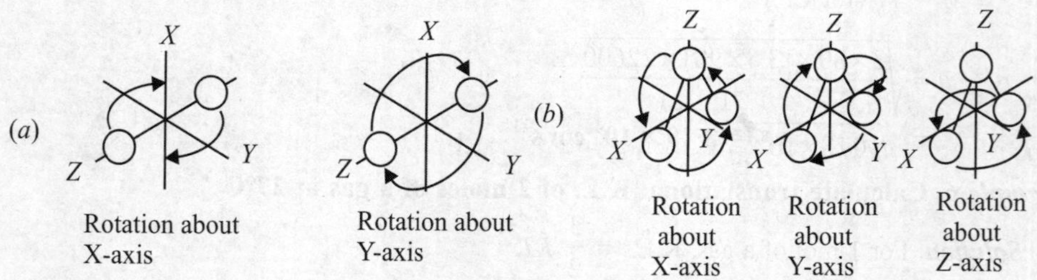

Fig. 2.5. (*a*) Rotations of a linear molecule about two mutually perpendicular axes (*b*) Rotations of a non-linear molecule about three mutually perpendicular axes.

Remaining degrees of freedom for the linear molecules are equal to ($3N$– 5) and for the non-linear molecules they are equal to ($3N$– 6). These describe the motion of the nuclei with respect to one another and thus represent the number of *vibrational degrees of freedom.* Hence,

Vibrational degrees of freedom of a linear molecule containing N atoms = $3N - 5$.

Vibrational degrees of freedom of a non-linear molecule containing N atoms = $3N - 6$.

The following examples illustrate calculation of vibrational degree of freedom of molecules:

(*i*) As a diatomic molecule is always linear, its vibrational degrees of freedom $= 3N - 5 = 3 \times 2 - 5 = 1$.

(*ii*) For the linear polyatomic molecule, HC≡CH, ($N = 4$), the vibrational degrees of freedom $= 3N - 5 = 3 \times 4 - 5 = 7$.

(*iii*) For the non-linear polyatomic molecule, HCHO, ($N = 4$) the vibrational degrees of freedom $= 3N - 6 = 3 \times 4 - 6 = 6$.

It may be summarised that for a molecule consisting of N atoms

	Translational	*Rotational*	*Vibrational*
For linear molecule :	3	2	$3N - 5$
For non-linear molecule:	3	3	$3N - 6$

Thus the total degrees of freedom in each case $= 3N$

2.6 PRINCIPLE OF EQUIPARTITION OF ENERGY

According to kinetic theory of gases, the molecules of a gas are moving in different directions with different velocities. Consider any one gas molecule moving in any direction with velocity ϑ. The velocity ϑ can be resolved into three components ϑ_x, ϑ_y and ϑ_z along the three mutually perpendicular axes (Fig. 2.6) such that

$$\vartheta^2 = \vartheta_x^2 + \vartheta_y^2 + \vartheta_z^2 \quad ...(i)$$

Multiplying both sides of the equation by $\frac{1}{2}m$, we get

$$\frac{1}{2}m\vartheta^2 = \frac{1}{2}m\vartheta_x^2 + \frac{1}{2}m\vartheta_y^2 + \frac{1}{2}m\vartheta_z^2 \quad ...(ii)$$

i.e., K.E. of the gas molecule $= (\text{K.E.})_x + (\text{K.E.})_y + (\text{K.E})_z$ ...(*iii*)

Considering all the molecules of the gas and taking their average value, we can say,

$$\overline{\text{K.E.}} = (\overline{\text{K.E.}})_x + (\overline{\text{K.E.}})_y + (\overline{\text{K.E.}})_z \quad ...(iv)$$

Since the molecules of the gas move in a random manner and do not have any preference for a particular direction, therefore, it follows that

$$\vartheta_x^2 = \vartheta_y^2 = \vartheta_z^2 \quad ...(v)$$

The mean value of any velocity component such as $\overline{\vartheta_x}$ is zero because all directions are equally possible so that there are equal probabilities for the components in opposite direction, *i.e.*, ϑ_x and $-\vartheta_x$ are equally probable. But $\overline{\vartheta_x^2}$ is not zero because the squares are always positive.

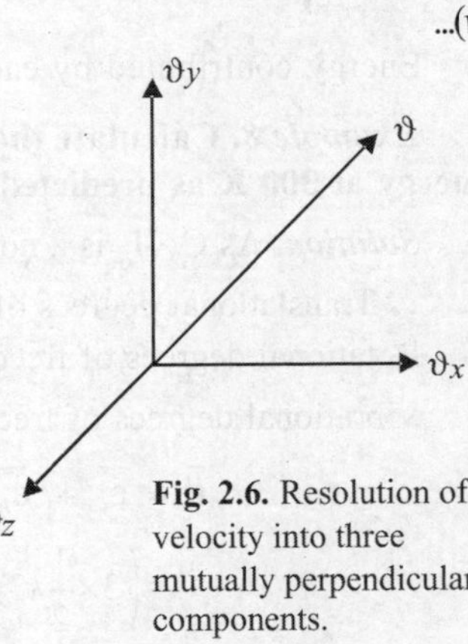

Fig. 2.6. Resolution of a velocity into three mutually perpendicular components.

Multiplying throughout by $\frac{1}{2}m$, we get

$$\frac{1}{2}m\overline{\vartheta_x^2} = \frac{1}{2}m\overline{\vartheta_y^2} = \frac{1}{2}m\overline{\vartheta_x^2} \quad ...(vi)$$

i.e $(\text{K.E.})_x = (\text{K.E})_y = (\text{K.E.})_z$ ...(*vii*)

Combining eqn. (*v*) and (*vii*), it becomes clear that the average translational kinetic energy of a molecule is the sum of their components along the three mutually perpendicular axes and these three components are equal to one another. From kinetic theory of gases, as the total is translational kinetic energy of a molecule is $\frac{3}{2}kT$, therefore, each component will be $\frac{1}{2}kT$ where k is Boltzmann constant and T is the absolute temperature.

Now let us consider the rotational and vibrational energies associated with the gas molecules. Just as translational kinetic energy of a molecule is the sum of the three squared terms, each squared term contributes $\frac{1}{2}$ kT to the total energy of the molecules, the rotational energy is kinetic in nature and is represented by one squared term ($1/2I\omega^2$ where, I is the moment of inertia and ω is the angular velocity of rotation). Each type of rotation (*i.e.*, each squared term) contributes $\frac{1}{2}kT$ to the energy of the molecule. However, vibrational energy is partly kinetic and partly potential and is represented by the sum of two squared terms. Each squared term contributes $\frac{1}{2}kT$ to the energy of the molecule so that each mode of vibration contributes kT to the energy of the molecule.

The above discussion leads to the following generalization known as the "principle of equipartition of energy".

The total energy of a molecule is divided equally among the different degrees of freedom, the amount of energy associated with each degree of freedom being equal to $\frac{1}{2}$ *kT where the term 'degree of freedom' refers to the number of independent squared terms (i.e., terms involving the square of a coordinate or momentum/velocity) required to express the total energy of the molecule.*

It may be summarised as :

Energy contributed by each translational degree of freedom = $\frac{1}{2}$ kT.

Energy contributed by each rotational degree of freedom = $\frac{1}{2}$ kT.

Energy contributed by each vibrational degree of freedom (or vibrational mode) = kT.

Example 8. **Calculate the average energy of the molecule of C_2H_6 as well as the molar energy at 300 K as predicted by the law of equipartition of energy.**

Solution. As C_2H_6 is a non-linear molecule, and $N = 8$

$\therefore$ Translational degrees of freedom = 3

Rotational degrees of freedom = 3

Vibrational degrees of freedom = $(3N - 6) = 18$

$$\therefore \quad \bar{\varepsilon} = \overline{\varepsilon_i} + \overline{\varepsilon_r} + \overline{\varepsilon_\vartheta}$$

$$= \left(3\times\frac{1}{2}kT\right)+\left(3\times\frac{1}{2}kT\right)+(18\times kT)$$

$$= \frac{3}{2}kT + \frac{3}{2}kT + 18\,kT = 21\,kT$$

$$= 21 \times \frac{8{\cdot}314}{6{\cdot}022'10^{23}} \times 300$$
$$= 8.70 \times 10^{-20} \text{ Joules}$$

Average energy per mole will be given by

$$\overline{E} = \overline{\varepsilon} \times N$$
$$= 8.70 \times 10^{-20} \times 6.022 \times 10^{23} \text{ Joules}$$
$$= 52.39 \text{ kJ}$$

Example 9. **Calculate the average internal energy of a diatomic molecule at 300 K using law of equipartition of energy.**

Solution. For a diatomic molecule, as it is linear, and $N = 2$.

Average translational energy

$$(\overline{\varepsilon}_t) = 3 \times \frac{1}{2} kT = \frac{3}{2} kT$$

Average rotational energy

$$(\overline{\varepsilon}_r) = 2 \times \frac{1}{2} kT = kT$$

Average vibrational energy

$$(\overline{\varepsilon}_\vartheta) = (3N - 5) \times kT = (3 \times 2 - 5)\, kT = kT$$

Hence the average internal energy of the molecule is

$$\overline{\varepsilon} = \overline{\varepsilon}_t + \overline{\varepsilon}_r + \overline{\varepsilon}_\vartheta$$
$$= \frac{3}{2} kT + kT + kT = \frac{7}{2} kT$$
$$= \frac{7}{2} \times \frac{8.314}{6.022 \times 10^{23}} \times 300 \text{ Joules}$$
$$\mathbf{= 1.45 \times 10^{-20} \text{ Joules}}$$

Example 10. **At what temperature would ethane molecules have the same r.m.s. velocity as methane molecules at 27°C ?**

Solution. r.m.s. velocity $= \sqrt{\dfrac{3RT}{M}}$

For methane, *r.m.s.* velocity at 27°C (300 K) $= \sqrt{\dfrac{3 \times R \times 300}{16}}$

Let ethane possess the same velocity at TK. Then,

$$\sqrt{\frac{3RT}{30}} = \sqrt{\frac{3 \times R \times 300}{16}} \quad \text{or} \quad \frac{T}{30} = \frac{300}{16} \quad \text{or} \quad T = \frac{30 \times 300}{16}$$

or $T = 562.5$ K or 289.5°C

2.7 (*i*) COLLISION DIAMETER (*ii*) COLLISION NUMBER (*iii*) COLLISION FREQUENCY (*iv*) MEAN FREE-PATH

(*i*) **Collision diameter**

Collision is an event in which the centres of two identical molecules come within a distance σ from one another as shown in the figure below.

Sigma (σ) which is the distance between centres of the molecules at the point of closest approach is called the **collision diameter.**

(*ii*) **Collision number**

The number of molecules with which a single molecule will collide per unit time per unit volume of the gas is called *collision number.*

It is given by

$$Z_1 = \sqrt{2}\,\pi\,\sigma^2\,\bar{c}\,\rho \qquad ...(i)$$

where Z_1 = Collision number

σ = Diameter of molecule

$\bar{c}$ = Average velocity

ρ = No. of molecules per unit volume of the gas

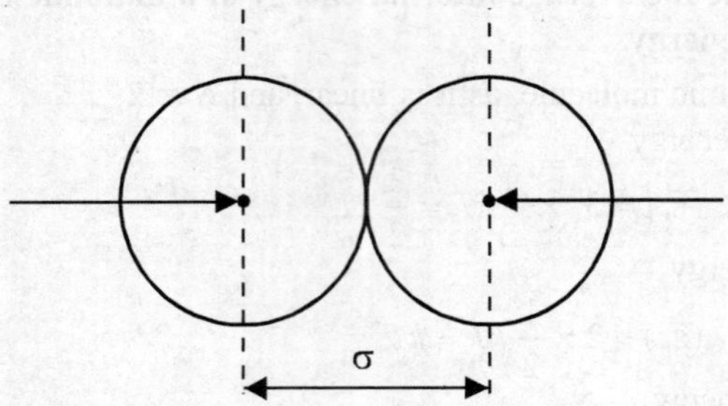

(*iii*) **Collision frequency**

It is the number of molecular collisions occurring per unit time per unit volume of the gas.

Total number of molecules colliding per unit volume per unit time is obtained by multiplying collision number by number density ρ. Thus total number of colliding molecules $= \sqrt{2}\,\pi\,\sigma^2\,\bar{c}\,\rho^2$. Since each collision involves two molecules of the same type, number of collisions is given by one half of this

$$Z_{11} = \frac{1}{2}\cdot\sqrt{2}\,\pi\sigma^2\,\bar{c}\,\rho^2$$

$$= \frac{1}{\sqrt{2}}\pi\,\sigma^2\,\bar{c}\,\rho^2 \qquad ...(ii)$$

It readily follows that number of collisions of molecules of type 1 with those of type 2 would be given by

$$Z_{12} = \frac{1}{\sqrt{2}}\pi\,\sigma^2\,\bar{c}\,\rho_1\,\rho_2 \qquad ...(iii)$$

where ρ_1 and ρ_2 are densities of molecules of type 1 and 2 respectively.

(*iv*) **Mean Free Path**

Mean free path (λ) is defined as the mean distance travelled by a gas molecule between two successive collisions. Thus

$$\lambda = \frac{\bar{c}}{Z_1} = \frac{\bar{c}}{\sqrt{2}\pi\sigma^2\,\bar{c}\,\rho^2} \qquad \text{[Substituting from (}i\text{)]} \qquad ...(iv)$$

Also $PV = nRT = nN_A\ kT$

where A_A = Avogadro's number

and k = Boltzmann constant

[Gas constant (R) = Avogadro's no. (N_A) × Boltzmann constant] or $P = \frac{nN_A kT}{V} = \frac{NkT}{V}$ where $N (= nN_A)$ is total no. of molecules in n moles of the gas

Since ρ is the number of molecules per unit volume, $\rho = \frac{N}{V}$

Thus $P = \rho\, kT$ or $\rho = \frac{P}{kT}$ ...(*v*)

Substituting the value of ρ of in (*iv*)

$$\lambda = \frac{\bar{c}}{\sqrt{2}\,\pi\sigma^2\,\bar{c}\,P/kt} = \frac{kT}{\sqrt{2}\,\pi\sigma^2 P}$$

Mean free path is also related to the viscosity of the gas. It is given by the equation

$$\lambda = \eta\sqrt{\frac{3}{Pd}}$$

where P = Pressure of the gas

d = Density of the gas

η = coefficient of viscosity of the gas

2.7.1 Effect of temperature and pressure on (*i*) mean free path (*ii*) collision frequency

Let us first take up *mean free path*. It is given by the equation

$$\lambda = \frac{kT}{\sqrt{2}\,\pi\sigma^2 P}$$ Refer to previous section.

Clearly, the variables on the R.H.S. of the equation are P and T. Other quantities are constants. T appears in the numerator and P in the denominator. We can say that

$$\lambda \propto \frac{T}{P}$$

or λ varies directly as the temperature and varies inversely as the pressure.

Taking the case of *collision frequency:*

Collision frequency $$Z_{11} = \frac{1}{\sqrt{2}}\pi\,\sigma^2\,\bar{c}\,\rho^2$$ ...(*i*)

Effect of temperature

From the above equation $Z_{11} \propto \bar{c}$

Also $\bar{c} \propto \sqrt{T}$ (R.M.S. velocity and average velocity vary directly as the square root of absolute temperature)

$\therefore$ $Z_{11} \propto \sqrt{T}$

Thus collision frequency varies directly as the square root of absolute temperature

Effect of pressure

From equation (*i*) above

$$Z_{11} \propto \rho^2$$

But $P \propto \rho$ (Pressure of a gas is directly proportional to P *i.e.,* number of molecules present in per unit volume of the gas)

$$\therefore \quad Z_{11} \propto P^2$$

or collision frequency varies directly as the pressure of the gas.

Example 11. **RMS velocity of hydrogen at STP is 1·85 × 10⁵ cm sec⁻¹ and its mean free path is 1·75 × 10⁻⁵ cm. Calculate the collision number.**

Solution. RMS velocity, $\mu = 1{\cdot}85 \times 10^5$ cm sec^{-1}

$$\text{Average velocity } \bar{c} = 0{\cdot}9213 \times \text{RHS Velocity}$$

$$= 0{\cdot}9213 \times 1{\cdot}85 \times 10^5 \text{ cm sec}^{-1}$$

$$\text{Mean free path} = \frac{\text{Average velocity}}{\text{Collision number}}$$

$$1{\cdot}75 \times 10^{-5} = \frac{0{\cdot}9213 \times 1{\cdot}85 \times 10^5}{\text{Collision number}}$$

$$\text{or} \quad \text{Collision number} = \frac{0{\cdot}9213 \times 1{\cdot}85 \times 10^5}{1{\cdot}75 \times 10^{-5}} = \mathbf{9{\cdot}734 \times 10^9\ sec^{-1}.}$$

Example 12. **At 0°C and 1 atmosphere pressure, molecular diameter of a gas is 4 × 10⁻⁸ cm. Calculate the mean free path of its molecule.**

Solution. Mean free path is given by the equation

$$\lambda = \frac{1}{\sqrt{2}\pi\sigma^2\rho}$$

where σ is molecular diameter.

22400 ml of a gas at 0°C and 1 atm. pressure contains $6{\cdot}02 \times 10^{23}$ molecules.

Hence, number of molecules per c.c,

$$\rho = \frac{6{\cdot}02 \times 10^{23}}{22400} = 2{\cdot}689 \times 10^{19}.$$

Substituting the values in the above equation

$$\lambda = \frac{1}{1{\cdot}414 \times 3{\cdot}14 (4 \times 10^{-8})^2 \times 2.689 \times 10^{19}} = \mathbf{0{\cdot}524 \times 10^{-5}\ cm.}$$

2.8 AN IDEAL GAS AND A REAL GAS

Ideal gas

A gas which obeys Boyle's law, Charles law etc., under all conditions of temperature and pressure is known as an ideal gas or a perfect gas.

An ideal gas may also be defined as:

A gas which obeys the gas equation (PV = nRT) under all conditions of temperature and pressure is called an ideal gas.

It is observed that there is no gas which obeys the gas laws or the gas equation under all conditions of temperature and pressure. The gases are found to obey the gas laws fairly well if the pressure is low or the temperature is high.

Real Gas

A real gas is one which obeys the gas laws fairly well under low pressure or high temperature.

All gases are real gases. They show more and more deviations from the gas laws as the pressure is increased or the temperature is decreased.

2.8.1 Deviation of real gases from ideal behaviour

According to Boyle's law, $PV=$ constant at constant temperature. This means that the product PV should remain constant at all pressures provided temperature is kept constant. Thus a plot of PV *vs* P should be a horizontal straight line parallel to X-axis. However, in actual practice, it is found that no gas gives a straight line plot. Two types of curves are generally obtained.

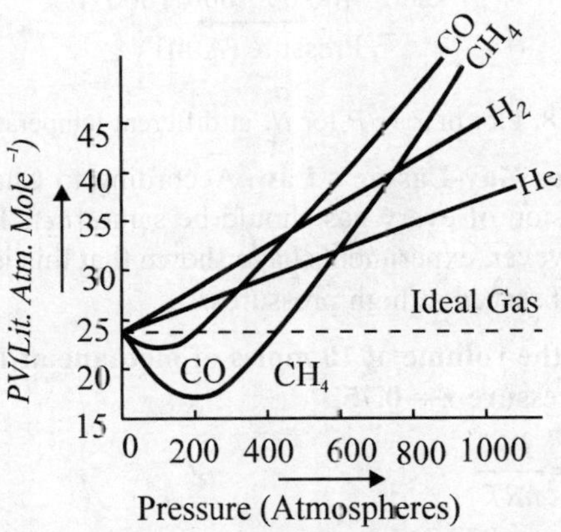

Fig. 2.7. Plots of PV *vs* P for different gases at 0°C (273 K).

(*i*) For gases like hydrogen and helium, the product PV continuously increases with increase of pressure (see Fig. 2.7).

(*ii*) For gases like CO and CH_4, the product PV first decreases with increase of pressure, reaches a minimum value and then begins to increase (see Fig. 2.7).

2.8.2 Deviation of real gases from Boyle's and Charles's law in terms of compressibility factor, z

The effect of temperature and pressure on the behaviour of a gas may be studied in terms of quantity 'z' called **compressibility factor,** which is defined as

$$z = \frac{PV}{nRT}$$

For an ideal gas, $z = 1$, at all temperatures and pressures. In case of real gases, *the factor z varies from values less than 1 to values greater than 1, with changes of temperature and pressure.* For example, the plots of compressibility factor z *vs.* pressure P at different constant temperatures for nitrogen gas are shown in Fig. 2.8. From these plots it is observed that as the temperature increases, the minimum in the curves shifts upwards. Ultimately, a temperature is reached at which the value of z remains close to 1 over an appreciable range of pressure. For example, in case of nitrogen gas, at 50°C, the value of z remains close to 1 upto nearly 100 atmospheres. *This temperature at which a real gas behaves like an ideal gas over an appreciable pressure range is called* **Boyle temperature** *or* **Boyle point** *because at this temperature Boyle's law is obeyed over a range of pressures.* Obviously, above the Boyle temperature, a gas shows positive deviations only.

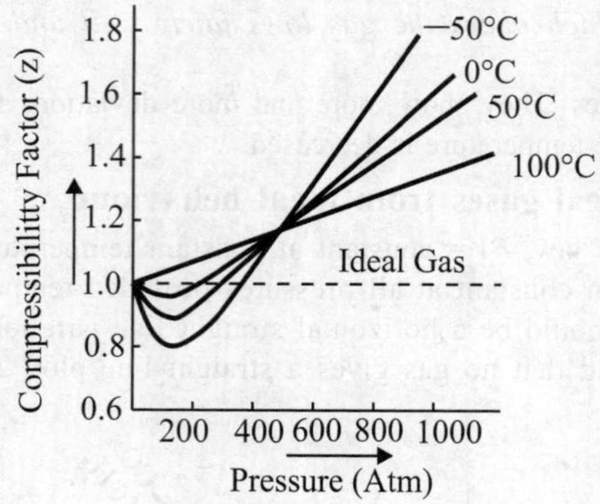

Fig. 2·8. Plot of Z *vs* P for N_2 at different temperatures.

Deviations from Charles Gay-Lussac's Law. According to Charles-Gay Lussac's law, the coefficient of volume expansion of every gas should be same [*i.e.*, 1/273 of its volume at 0°C] independent of pressure. However, experiments have shown that this is true only at low pressures. Appreciable deviations are observed at high pressures.

***Example 13*. Calculate the volume of 10 moles of methane at 100 atm pressure and 0°C. At this temperature and pressure z = 0.75.**

Solution.

$$z = \frac{PV}{nRT}$$

$$\therefore \quad V = \frac{znRT}{P}$$

$$= \frac{0.75 \times 10 \times 0.0821 \times 273}{100}$$

$$= 1.68 \text{ litre}$$

2.8.3 Causes of deviation of real gas from ideal behaviour

The derivation of the gas laws is based upon the Kinetic Theory of Gases which in turn is based upon certain assumptions. Thus there must be something wrong with certain assumptions. A careful study shows that at high pressure or low temperature, the following two assumptions of Kinetic Theory of Gases become invalid.

(*i*) *the volume occupied by the gas molecules is negligible as compared to the total volume of the gas.*

(*ii*) *The forces of attraction or repulsion between the gas molecules are negligible.*

The above two assumptions are true only if the pressure is low or the temperature is high so that the distance between the molecules is large. However, if the pressure is high or the temperature is low, the gas molecules come close together. Under these conditions:

(*i*) the forces of attraction or replusion between the molecules are not negligible.

(*ii*) the volume of the gas is so small that the volume occupied by the gas molecules cannot be neglected.

2.8.4 Derivation of van der Waals' equation for real gases

van der Waal obtained the equation for real gases by applying the corrections for volume and pressure as follows:

1. Correction for the volume. Suppose ϑ is the actual volume occupied by one mole of the gas molecules. Then since the gas molecules are in motion, it has been found that the *effective volume* occupied by the gas molecules is four times the actual volume *i.e.*, equal to 4ϑ. Let it be represented by *b*. The volume *b* is also called *co-volume* or *excluded volume.* Thus the free space available for the movement of the gas molecules is $(V - b)$. Hence in the real gas equation *V* should be replaced by $(V - b)$.

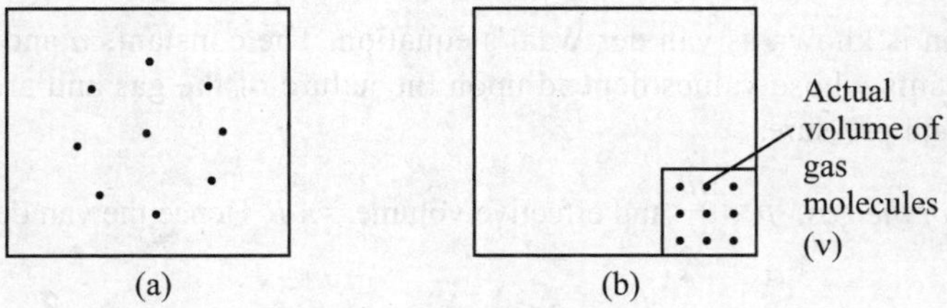

Fig. 2.9. (a) Gas molecules are dispersed in the whole space of container
(b) Actual volume of the gas molecules

2. Correction for the forces of attraction *i.e.*, pressure correction. A gas molecule lying in the interior of the gas such as molecule *A* in Fig. 2.10 is attracted by all other gas molecules surrounding it. Hence the net force of attraction exerted on such a molecule by the other molecules is zero. However, for a molecule lying near the wall of the container, such as molecule *B* in Fig. 2.10, the molecules lying inside the bulk of the gas exert some net inward pull. Thus the effect of such an inward pull is 'dragging back' of the molecule. Consequently, the pressure with which the gas molecule strikes the wall of the vessel is less than the pressure that would have been exerted if there were no such inward pull.

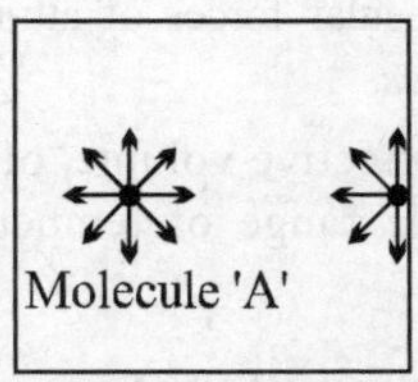

Fig. 2.10. Intermolecular forces of attraction.

Thus the ideal pressure would be greater than the observed pressure by a factor '*p*'. Hence

Corrected pressure $= P + p$

where *p* is the 'correction factor' due to the inward pull.

This inward pull on the gas molecules lying near the wall depends upon:

(*i*) Number of molecules surrounding a molecule,

(*ii*) Total number of molecules.

Each of these factors in turn depends upon the density (ρ) of the gas. Hence

correction factor, $p \propto \rho^2$

But for a given mass of the gas

$$\rho \propto \frac{1}{V}$$

where *V* is the volume of the gas

$$\therefore \qquad p \propto \frac{1}{V^2}$$

or

$$p = \frac{a}{V^2}$$

where *a* is a constant, depending upon the nature of the gas. Thus

Corrected pressure = $P + \frac{a}{V^2}$

Putting the corrected values of volume and pressure in the ideal gas equation, $PV = RT$, for 1mole of the gas, the equation is modified to

$$\left(p + \frac{a}{V^2}\right)(V - b) = RT$$

This equation is known as van der Waal's equation. The constants a and b are called van der Waal's constants whose values depend upon the nature of the gas and are independent of the temperature and pressure.

For n moles of the gas, $p \propto \frac{n}{V}$ and effective volume = $n\,b$. Hence the van der Waal's equation takes the form

$$\left(P + \frac{an^2}{V^2}\right)(V - nb) = nRT$$

2.8.5 Significance of van der Waals constants *a* and *b*

It is observed that the values of a for the easily liquefiable gases are greater than those for the so called permanent gases like H_2 and He. Moreover, the value of a increases with the ease of liquefaction of the gas. We know that a gas which is more easily liquefiable has greater intermolecular forces of attraction. Hence *'a' is a measure of the intermolecular forces of attraction in a gas.*

'b' is the 'effective volume' of the gas molecules. The constancy in the value of b for any gas over a wide range of temperature and pressure confirms that the *gas molecules are incompressible.*

2.8.6 Behaviour of real gases using van der Waal's equation (*i*) At low pressure, (*ii*) At moderate pressure, (*iii*) At high pressure, (*iv*) At high temperature

(*i*) **At low pressure.** At extremely low pressure, V is very large. Hence the correction term $\frac{a}{V^2}$ is very small. Similarly, the correction term b is also very small as compared to V. Thus both the correction terms can be neglected so that the van der Waal's equation reduces to $PV = RT$. This is why at extremely low pressures, the gases obey the ideal gas equation.

(*ii*) **At moderate pressure.** As the pressure is increased, the volume decreases and hence the factor $\frac{a}{V^2}$ increases. Thus the factor $\frac{a}{V^2}$ can no longer be neglected. However, if the pressure is not too high, the volume V is still sufficiently large so that b can be neglected in comparison with V. Thus the van der Waal's equation reduces to

$$\left(P + \frac{a}{V^2}\right)V = RT$$

or

$$PV + \frac{a}{V} = RT$$

or

$$PV = RT - \frac{a}{V}$$

Thus PV is less than RT by a factor $\frac{a}{V}$. As the pressure increases, V decreases, so that the factor $\frac{a}{V}$ increases. Thus PV decreases as the pressure is increased. This explains why a dip the plots of PV vs P of real gases is obtained (see Fig. 2.7).

(*iii*) **At high pressure.** As the pressure is increased further so that it is fairly high, V is so small that b can no longer be neglected in comparison with V. Although under these conditions the factor $\frac{a}{V^2}$ is quite large but since P is very high so $\frac{a}{V^2}$ can be neglected in comparison with P. Thus the van der Waal's equation reduces to

$$P(V-b) = RT$$

or
$$PV - Pb = RT$$

or
$$PV = RT + Pb$$

Thus PV is greater than RT by a factor Pb. Now as the pressure is increased, the factor Pb increases more and more. This explains why after the minima in the curves, the product PV increases continuously as the pressure is increased more and more (Fig. 2.7).

(*iv*) **At high temperature.** At any given pressure, if the temperature is sufficiently high, V is very large so that as in case (*i*), the van der Waal's equation reduces to $PV = RT$. Hence at high temperature, real gas behaves like the ideal gas.

2.8.7 Exceptional behaviour of hydrogen and helium

In case of hydrogen and helium, their molecules have very small masses. Hence in the case of these gases, intermolecular forces a of attraction are extremely small even at high pressures. In other words, the factor $\frac{a}{V^2}$ is negligible at all pressures. Hence the van der Waal's equation is applicable in the following form at *all pressures and ordinary temperatures.*

$$P(V - b) = RT \text{ or } PV = RT + Pb$$

This explains why hydrogen and helium show positive deviation only which increases with increase in the value of P.

Example 14. **Calculate from the van der Waal's equation the temperature at which 3 moles of SO_2 would occupy a volume of 0.01 m^3 at 1519875 Nm^{-2} pressure (a = 0.679 Nm^4 mol^{-2}, b = 5.64 × 10^{-5} m^3 mol^{-1}, R = 8.314 JK^{-1} mol^{-1}).**

Solution. van der Waal's equation can be written as :

$$T = \frac{1}{nR}\left(P + \frac{an^2}{V^2}\right)(V - nb)$$

Substituting the given values, we get,

$$\therefore \quad T = \frac{1}{3 \times 8.314}\left(1519875 + \frac{0.679 \times 3^2}{(0.01)^2}\right)(0.01 - 3 \times 564 \times 10^{-5})$$

$$= \frac{1}{24.942} \times 1580985 \times 0.0098308$$

$$= 623.1 \text{ K.}$$

Example 15. **Two moles of ammonia gas are enclosed in a vessel of 5-litre capacity at 27°C. Calculate the pressure exerted by the gas, assuming that**

(*i*) the gas behaves like an ideal gas (using ideal gas equation);

(*ii*) the gas behaves like a real gas (using van der Waal's equation).

Given that for ammonia a = 4.17 atm litre^2 mol^{-2} (or 4.17 atm dm^6 mol^{-2}) and b = 0.037 litre mol^{-1} (or 0.037 dm^3 mol^{-1})

Solution. We are given that

$$n = 2 \text{ moles}$$

$$V = 5 \text{ litres or dm}^3$$

$$T = 27°C = (27 + 273)\text{ K} = 300\text{ K}$$
$$a = 4.17\text{ atm litre}^2\text{ mol}^{-2}\text{ or atm dm}^6\text{ mole}^{-2}$$
$$b = 0.037\text{ litre m or }^{1}\text{ or dm}^3\text{ mol}^{-1}$$

We know that

$R = 0.0821$ litre atm K^{-1} mol^{-1} atm or dm^3 atm K^{-1} mol^{-1}

(*i*) If the gas behaves like an ideal gas, we have

$$PV = nRT$$

$$\therefore \quad P = \frac{nRT}{V}$$
$$= \frac{2 \times 0.0821 \times 300}{5} = \mathbf{9.85\,atm}$$

(*ii*) If the gas behaves like a real gas, we apply van der Waal's equation *i.e.*,

$$\left(P + \frac{an^2}{V^2}\right)(V - nb) = nRT$$

or $$P = \frac{nRT}{V - nb} - \frac{an^2}{V^2}$$
$$= \frac{2 \times 0.0821 \times 300}{5 - 2 \times 0.037} = \frac{4.17 \times (2)^2}{(5)^2} = \mathbf{9.33\ atm}$$

2.8.8 Kammerlingh-Onnes (virial) equation of state for real gases

It is an *empirical equation* which expresses PV as a power series of pressure P at any given temperature. It is written as

$$PV = A + BP + CP^2 + DP^3 + ...$$

where P is the pressure in atmospheres and V is the *molar* volume in litres. The coefficients A, B, C etc., are known respectively as the first, second, third etc., *virial coefficients.* The following points may be noted about the above equation:

(*i*) The virial coefficients are different for different gases.

(*ii*) At very low pressure, only the first virial coefficient is significant and it is equal to RT *i.e.*,

$$A = RT$$

(*iii*) At higher pressures, the other virial coefficients also become important and must also be considered.

(*iv*) For any particular gas, the values of $A, B, C,$ etc., are constant at constant temperature. Their values change with change of temperature. The first virial coefficient A is always positive and increases with temperature. On the other hand, the second virial coefficient B is negative at low temperatures. With increase of temperature it increases to zero and then becomes more and more positive. The temperature at which $B = 0$ is called *Boyle temperature.* At this temperature, Boyle's law is valid over a fairly wide pressure range.

The other form in which Onnes equation is often written is by expressing the compressibility factor (z) as a power series in $1/V$ *i.e.*,

$$Z = \frac{PV}{RT} = 1 + \frac{B}{V} + \frac{C}{V^2} + \cdots$$

or $$PV = RT\left[1 + \frac{B}{V} + \frac{C}{V^2} + \cdots\right]$$

where the coefficients B, C etc., (called second, third virial coefficients etc.) for a particular gas depend only on the temperature and not on the pressure.

Example 16. **What is the molar volume of N_2 (*g*) at 500 K and 600 atm according to (*a*) the perfect gas law and (*b*) the virial equation? The virial coefficient B of N_2 (*g*) at 500 K is 0.0169 litre mol^{-1}.**

Solution. (*a*) According to perfect gas law

$$V = \frac{RT}{P} = \frac{0.0821 \times 500}{600} = \mathbf{0.0684\ litre.}$$

(*b*) $$PV = RT\left(1 + \frac{B}{V}\right) = RT + \frac{RT}{V}B$$

or $$PV = RT + PB$$

or $$V = \frac{RT + PB}{P}$$

or $$V = \frac{RT}{P} + B = \left(\frac{0.0821 \times 500}{600}\right) + 0.0169$$

$$= 0.0684 + 0.0169 = \mathbf{0.0853\ litre\ mol^{-1}}$$

2.8.9 Boyle's temperature

It is the temperature at which a gas behaves like an ideal gas. This temperature is given by $T_B = \frac{a}{Rb}$.

It may be derived from the van der Waal's equation as follows :

$$\left(p + \frac{a}{V^2}\right)(V - b) = RT$$

It may be written in the form

$$PV = RT - \frac{a}{V} + bP + \frac{ab}{V^2} \quad ...(i)$$

As both *a* and *b* are small, and if the pressure is not too high so that *V* is not so small, $\frac{ab}{V^2}$ can be neglected.

Further *V* in the correction term *a* /*V* may be replaced by *RT*/*P*.

Then equation (*i*) reduces to

$$PV = RT - \frac{aP}{RT} + bP$$

$$= RT + P\left(b - \frac{a}{RT}\right)$$

Since the gas behaves ideally at Boyle's temperature, $PV = RT$

Hence the second term on R.H.S. should be zero.

Since *P* has a finite value

$$b - \frac{a}{RT} = 0 \quad \text{or} \quad T = \frac{a}{Rb}$$

or Boyles temp., $$T_b = \frac{a}{Rb} \quad ...(ii)$$

2.8.10 Second virial coefficient using van der Waal's equation

The general equation of state as given by Onnes is

$$PV = RT\left[1+\frac{B}{V}+\frac{C}{V^2}+...\right] \qquad ...(i)$$

The second virial coefficient B of this equation can be obtained by comparing it with van der Waal's equation

$$\left(P+\frac{a}{V^2}\right)(V-b) = RT$$

which may be rewritten in the form

$$PV = RT + bP + \frac{ab}{V^2} - \frac{a}{V}$$

or

$$PV = RT + bP - \frac{a}{V}+\frac{ab}{V^2}$$

As a and b are very small and if the pressure is not too high, $\frac{ab}{V^2}$ can be neglected. Further P may be replaced by $\frac{RT}{V}$. Hence we get

$$PV = RT + b\frac{RT}{V}-\frac{a}{V}$$

or

$$PV = RT\left[1+\frac{1}{V}\left(b-\frac{a}{RT}\right)\right] \qquad ...(ii)$$

Comparing eqns. (*i*) and (*ii*), we get

$$\boxed{B = b-\frac{a}{RT}}$$

Thus second virial coefficient can be calculated from a knowledge of the van der Waal's constants a and b of the gas at temperature T.

Example 17. **The van der Waal's constant b for CO_2 is 4.28×10^{-2} dm^3 mol^{-1}. Assuming CO_2 molecule to be spherical, calculate its molecular diameter.**

Solution. For 1 mole of the gas

$$b = 4\,N\vartheta_m$$

where N is Avogadro's number and ϑ_m is the actual volume of the gas molecule.

$$\vartheta_m = \frac{n}{4N} = \frac{0.0428}{4\times6.022\times10^{23}} \text{ litres}$$

$$= 1{\cdot}777 \times 10^{-26} \text{ litres} = 1{\cdot}777 \times 10^{-23} \text{ cm}^3$$

But

$$\vartheta_m = \frac{4}{3}\pi r^3 = \frac{4}{3}\pi\left(\frac{\sigma}{2}\right)^3$$

$\therefore$

$$\sigma^3 = \frac{6\vartheta_m}{\pi} = \frac{6\times1.777\times10^{-23}}{3.143} = 33.92\times10^{-24}$$

which gives $\sigma = 3.24 \times 10^{-8}$ cm = **3.24 Å.**

2.9 CRITICAL PHENOMENON

Andrews studied the pressure-temperature conditions for liquefaction of several gases. He concluded that for each gas there is a temperature above which the gas cannot be liquefied, however high pressure we may apply. This temperature is called **critical temperature** which is different for different gases.

At critical temperature and pressure, the gas becomes indentical with its liquid and is said to be in **critical state**. The smooth merging of the gas with its liquid state is referred to as the **critical phenomenon**.

2.9.1 Critical temperature, critical pressure and critical volume

Critical temperature (T_0) *of a gas may be defined as that temperature above which the gas cannot be liquefied, howsoever high pressure we may apply on the gas. The minimum pressure required to liquefy the gas at the critical temperature is called the* **critical pressure (P_c)**. *The volume occupied by one mol of the gas at the critical temperature and pressure is called the* **critical volume (V_c)**.

The critical temperature, critical pressure and critical volume of a gas are collectively called as the **critical constants** of the gas. For example, the critical constants of carbon dioxide (CO_2) are

$$T_c = 31.1°C \quad \text{or} \quad 304.2 \text{ K}$$
$$P_c = 72.9 \text{ atmospheres}$$
$$V_c = 95.65 \text{ cm}^3 \text{ mol}^{-1}$$

2.9.2 Measurement of critical constants (T_c, P_c and V_c)

Measurement of T_c and P_c. The measurement of critical temperature (T_c) is based upon the observation that when a liquid *e.g.* ether or carbon disulphide, is heated in a closed space the surface of separation between the liquid and the vapour disappears at a definite temperature and on cooling the surface of separation reappears at the same temperature. The temperature at which this occurs is the **critical temperature.**

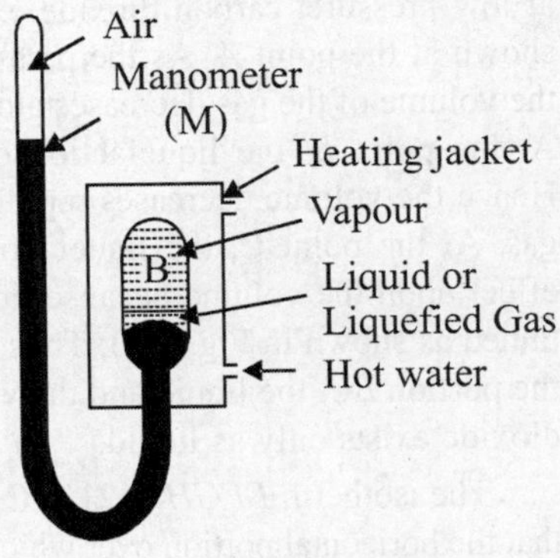

Fig. 2.11. Measurement of critical temperature and pressure.

The apparatus used for the measurement of the critical temperature is shown in Fig. 2.11. It helps to measure critical pressure simultaneously. It consists of a glass bulb *B* containing only the liquid and its vapour sealed over mercury. The bulb is connected to a manometer *M* containing air above the mercury.

Further, the bulb *B* is surrounded by a heating jacket as shown in the Fig. 2.11. The bulb *B* is heated gently by circulating hot water in the heating jacket and the temperature is noted at which the meniscus disappears. The bulb is then allowed to cool till the meniscus reappears or the cloudiness appears on the surface of the bulb due to condensation of vapour. The mean of the two temperatures gives the critical temperature of the liquid/gas enclosed in the bulb *B*.

Corresponding pressure from the manometer at the two stages is noted and taking their mean, gives the critical pressure (P_c).

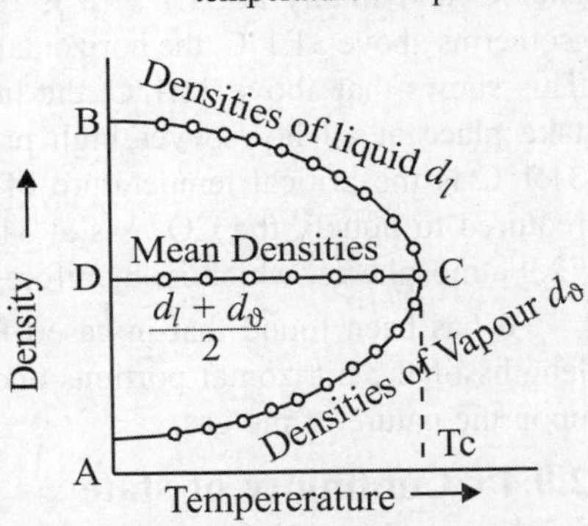

Fig. 2.12. Determination of critical density

Measurement of V_c. The measurement of critical volume (V_c) is based upon a rule which states that '*the mean of the densities of any substance in the state of liquid and saturated vapour at the same temperature is a linear function of the temperature.*'

Thus to determine the critical volume of any substance, the densities of the substance in the liquid state and those of the vapour in equilibrium with it are determined at a number of temperatures. If these densities are plotted against temperature, the curve obtained is as shown in Fig. 2.12. The densities of the vapour lie along the curve *AC* while those of the liquid lie along the curve *BC*. The two curves meet at the point *C* which corresponds to the critical temperature. If the mean values of the densities $\left(\frac{d_l + d_\vartheta}{2}\right)$ are plotted against temperature, a straight line *CD* is obtained. The line on extrapolation meets the density axis at *D* which corresponds to the critical density. Dividing the molecular weight of the substance by the critical density, *critical volume* is obtained.

2.9.3 Andrews experiments on critical phenomenon (Isotherms of carbon dioxide)

Andrews, in 1861, was the first to study the effect of temperature and pressure on the volume of carbon dioxide. Each time, keeping the temperature constant at a particular value, he measured the volume of carbon dioxide at different pressures. He then plotted the volumes against pressures, at the constant temperature. Such a plot of *P* vs *V* at constant *T* is called an *isotherm*. A number of such isotherms were obtained at different temperatures as shown in Fig. 2.13.

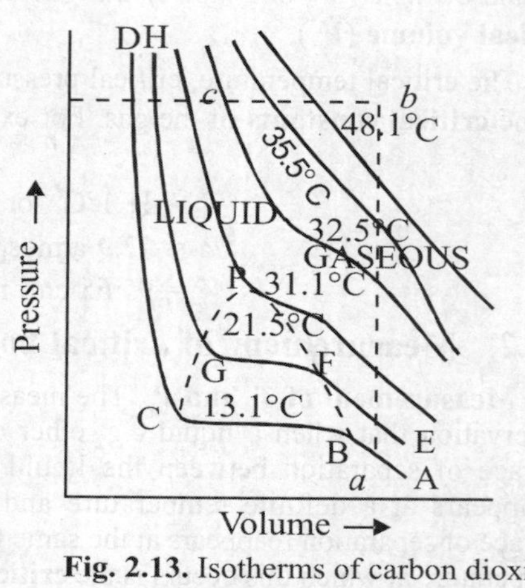

Fig. 2.13. Isotherms of carbon dioxide.

At the lowest temperature *i.e.*, 13.1°C (286.2 K), at low pressure, carbon dioxide exists as a gas, as shown at the point *A*. As the pressure is increased, the volume of the gas decreases along the curve *AB*. At the point *B*, the liquefaction of the gas starts. Hence the volume decreases rapidly along *BC* because the liquid has much less volume than the gas. At the point C, the liquefaction is complete. Now the increase of pressure has very little effect upon the volume because liquids are hardly compressible. Hence a steep curve *CD* is obtained as shown in Fig. 2.13. Thus along the portion *AB*, carbon dioxide exists only as gas, along the portion *BC*, the liquid and the vapour are in equilibrium with each other and along *CD*, carbon dioxide exists only as liquid.

The isotherm *EFGH* at 21.5°C is similar to that at 13.1°C except that the horizontal portion over which the liquefaction occurs is shorter. In fact, as the temperature is raised, this portion becomes smaller and smaller and finally at 31.1°C it is reduced only to a point. In the isotherms above 31.1°C, the horizontal portion completely disappears. This shows that above 31.1°C, the liquefaction of the gas does not take place at all howsoever high pressure may be applied. Thus, 31.1°C is the critical temperature of carbon dioxide. The pressure required to liquefy the CO_2 gas at 31.1°C (304.2 K) is found to be 72.9 atmospheres, which is, therefore, the critical pressure for CO_2.

Critical point apparatus

It has been found that in case of all other gases, similar isotherms are obtained. However, lengths of the horizontal portions and the peak of the dotted parabola are different, depending upon the nature of the gas.

2.9.4 Continuity of state

If the ends of the horizontal portions of the different isotherms are joined together, a parabolic curve (shown by dotted lines in Fig. 2.13) is obtained. The peak of this parabola represents the critical point. Within the parabola, the liquid and the vapour exist together in equilibrium. Outside the parabola, the substance exists only as gaseous under the conditions represented on the right side and only as liquid under the conditions represented on the left side. A careful thinking reveals that below the critical temperature it is possible to change a substance from gaseous state to liquid state without having both the phases present together at any stage. This is explained as under :

Suppose the substance under consideration is CO_2 so that Fig. 2.13 may be referred again. At 13.1°C, if pressure and volume correspond to the point *a,* the substance is gaseous. Keeping the volume constant, if the gaseous CO_2 is heated till it has pressure corresponding to the point *b,* the path *ab* is followed. Now keeping the pressure constant, suppose the gaseous CO_2 is cooled the volume decreases along the path *bc.* At *c,* CO_2 will exist as liquid. Thus nowhere along the path *abc,* there has been more than one phase present. This is expressed by saying that the change of state from gaseous to liquid is *continuous* or it is referred to as the *continuity of state.*

2.9.5 Expression for critical constants in terms of van der Waal's constants *a* and *b*

For 1mol of a gas, the van der Waal's equation is

$$\left(P+\frac{a}{V^2}\right)(V-b)=RT$$

On simplification, it gives

$$PV+\frac{a}{V}-Pb-\frac{ab}{V^2}=RT$$

Multiplying throughout by V^2, we get

$$PV^3+aV-PbV^2-ab=RTV^2$$

Dividing throughout by P, we get

$$V^3+\frac{aV}{P}-bV^2-\frac{ab}{P}=\frac{RT}{P}V^2$$

Rearranging in the descending powers of V, we get

$$V^3-\left(b+\frac{RT}{P}\right)V^2-\frac{a}{P}V-\frac{ab}{P}=0$$

Using the appropriate values of the van der Waal's constants *a* and *b* and the gas constant *R,* for any constant value of *T* (used by Andrews in his experiments), the values of *V* can be obtained for different values of *P.* The *theoretical isotherms* thus obtained are shown in Fig. 2.14. These curves are similar to those obtained experimentally (Fig. 2.13) except that in the curves below the critical temperature, the horizontal portions are replaced by curved portions. For example, in the isotherms at 13.1°C the horizontal portion *BC* has been replaced by the curved portion *BXYZC.* This is expected because the van der Waal's equation is cubic in *V,* as written above. Hence for every value of pressure *P* it has three roots (*i.e.* three values of *V*), all of which may be real or one may be real and the other two may be imaginary. *Below the*

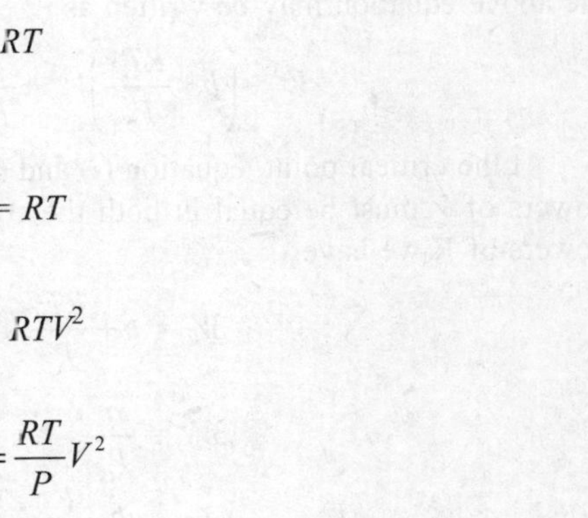

Fig. 2.14. Theoretical isothems of CO_2.

critical temperature, all the three roots are real in a certain pressure range, and these values come closer as the isotherm moves upwards. At the *critical temperature*, all the roots become equal for a particular pressure. *Above the critical temperature*, for any value of P, there is only one real root *i.e.*, only one value of V.

It has been observed that the portions BX, XYZ and ZC represent metastable conditions and can be actually obtained in practice by careful experiments.

As stated above, at the critical temperature the three values of V become identical for a particular value of pressure P. If we represent critical volume by V_c, then

$$V = V_c \text{ or } (V - V_c)^3 = 0$$

Expanding this, we get

$$V^3 - 3V_c V^2 + 3V_c^2 V - V_c^3 = 0 \qquad ...(i)$$

Further, as already explained, the van der Waal's equation may be written in the form

$$V^3 - \left(b + \frac{RT}{P}\right)V^2 + \frac{a}{P}V - \frac{ab}{P} = 0$$

At the critical point, the terms P and T may be replaced by the critical values P_c and T_c. Hence the above equation may be written as

$$V^3 - \left(b + \frac{RT_c}{P_c}\right)V^2 + \frac{a}{P_c}V - \frac{ab}{P_c} = 0 \qquad ...(ii)$$

At the critical point, equation (*i*) and (*ii*) must be identical. Hence the coefficients of similar powers of V must be equal in both the equations. Equating the coefficients of terms of similar powers of V, we have

$$3V_c = b + \frac{RT_c}{P_c} \qquad ...(iii)$$

$$3V_c^2 = \frac{a}{P_c} \qquad ...(iv)$$

and

$$V_c^3 = \frac{ab}{P_c} \qquad ...(v)$$

These equations can be solved to calculate the critical constants V_c, P_c and T_c in terms of van der Waal's constants a and b. Dividing equation (*v*) by (*iv*), we get

$$\frac{V_c^2}{3V_c^2} = \frac{ab}{P_c} \times \frac{P_c}{a}$$

or

$$\boldsymbol{V_c = 3b} \qquad ...(vi)$$

Putting this value in equation (*iv*), we get

$$3 \times (3b)^2 = \frac{a}{P_c}$$

or

$$\boldsymbol{P_c = \frac{a}{27b^2}} \qquad ...(vii)$$

Putting the values of V_c and P_c in equation (*iii*), we get

$$3 \times 3b = b + \frac{RT}{a/27b^2}$$

or

$$9b = b + \frac{RT_c \times 27b^2}{a}$$

or $$\frac{RT_c \times 27b^2}{a} = 8\,b$$

or $$\mathbf{T_c = \frac{8a}{27Rb}} \qquad ...(viii)$$

2.9.6 Derivation of relation between the critical constants P_c, V_c and T_c

Writing again the equations between critical constants and van der Waal's constants

$$V_c = 3\,b \qquad ...(i)$$

$$P_c = \frac{a}{27b^2} \qquad ...(ii)$$

$$T_c = \frac{8a}{27Rb} \qquad ...(iii)$$

To obtain a relationship between P_c, V_c and T_c we have to eliminate a and b from equations (*i*), (*ii*) and (*iii*). This may be done as follows :

From eqn. (*i*)

$$b = \frac{V_c}{3} \qquad ...(iv)$$

Substituting this value of b in eqn. (*ii*), we get

$$P_c = \frac{a}{27(V_c/3)^2} = \frac{a}{3V_c^2}$$

or $$a = 3\,P_c V_c^2 \qquad ...(v)$$

Substituting the values of a and b from eqns. (*iv*) and (*v*) in eqn. (*iii*), we get

$$T_c = \frac{8 \times 3P_c V_c^2}{27R \times V_c/3}$$

or $$T_c = \frac{8}{3}\frac{P_c V_c}{R}$$

or $$\mathbf{P_c V_c = \frac{3}{8} RT_c} \qquad ...(vi)$$

2.9.7 van der Waal's constants *a* and b in terms of critical constants

Expressions for a and b in terms of P_c and V_c are

$$b = \frac{V_c}{3} \text{ and } a = 3\,P_c^3\,V_c^3$$

Expressions in terms of P_c and T_c are

$$a = \frac{27R^2T_c^2}{64P_c} \quad \text{and} \quad b = \frac{RT_c}{8P_c}$$

These equations can be obtained by eliminating a and b separately using the equations

$$P_c = \frac{a}{27b^2} \text{ and } T_c = \frac{8a}{27Rb^2}$$

Example 18. **The critical temperature of carbon dioxide gas is 31.1°C and its critical density is 0.455 g/cm^3. Find out the values of *a, b* and P_c for the gas (*R*=0082 litre/atm/degree/mol).**

Solution. T_c = 31.1°C = 304.1 K

$$V_c = \frac{\text{mol. wt.}}{\text{critical density}} = \frac{M}{d} = \frac{44}{0.455} = 96.70\,\text{cm}^3 = 0.0967\,\text{litre}$$

$$b = \frac{V_c}{3} = \mathbf{0.0322\ litre\ mol^{-1}} \qquad ...(1)$$

$$P_c = \frac{a}{27b^2} \qquad ...(2)$$

$$T_c = \frac{8a}{27Rb^2} \qquad ...(3)$$

Dividing (2) by (3), we get

$$\frac{P_c}{T_c} = \frac{R}{8b}$$

or $$P_c = \frac{RT_c}{8b} = \frac{0.082 \times 304.1}{8 \times 0.0322} = \mathbf{96.80\ atm.}$$

From $$P_c = \frac{a}{27b^2}$$

$$\boldsymbol{a = 27\, P_c\, b^2}$$

Substitute the value of b from (1) in the above equation

$$a = 27 \times 96{\cdot}8 \times 0.0322^2$$
$$= 2.71 \text{ atm litre}^2\text{mol}^{-2}$$

***Example 19.* The critical temperature and critical pressure of chlorine are 146°C and 93.5 atmosphere respectively. Calculate the values of the van der Waal's constants *a* and *b*.**

Solution. Here we are given that

$$T_c = (146 + 273)\text{K} = 419 \text{ K}$$

and $$P_c = 93{\cdot}5 \text{ atmospheres}$$

The relations between critical constants and van der Waal's constant are

$$P_c = \frac{a}{27b^2} \qquad ...(i)$$

and $$T_c = \frac{8a}{27Rb} \qquad ...(ii)$$

Dividing equation (*ii*) by equation (*i*), we get

$$\frac{T_c}{P_c} = \frac{8a}{27Rb} \times \frac{27b^2}{a} = \frac{8b}{R}$$

or $$b = \frac{RT_c}{8P_c} \qquad ...(iii)$$

From equation (*i*), $$a = 27\, P_c\, b^2$$

$$= 27P_c\left(\frac{RT_c}{8P_c}\right)^2$$

$$= \frac{27R^2T_c^2}{64P_c} \qquad ...(iv)$$

Taking $R = 0.0821$ litre atm degree^{-1} and mol^{-1} and putting the given values of T_c and P_c in equations (*iii*) and (*iv*), we get

$$a = \frac{27R^2T_c^2}{64P_c} = \frac{27\times(0.0821)^2\times(419)^2}{64\times93.5}$$

$$= \mathbf{5.34\ atm\ litre^2\ mol^{-2}}$$

and $$b = \frac{RT}{8P_c} = \frac{0.0821\times419}{8\times93.5}$$

$$= \mathbf{0.046\ litre\ mol^{-1}}$$

Solution in terms of SI units:

$T_c = 419$ K

$P_c = 93.5$ atm $= 93.5 \times 101325$ Nm^{-2} ($\because$ 1 atm = 101325 Nm^{-2})

$R = 8.314$ JK^{-1} mol^{-1}

$$\therefore \quad a = \frac{27R^2T_c^2}{64P_c}$$

$$= \frac{27\times(8.314)^2\times(419)^2}{64\times93.5\times101325}\ \text{Nm}^4\ \text{mol}^{-2}$$

$$= \mathbf{0.54\ Nm^4\ mol^{-2}}$$

$$b = \frac{RT_c}{8P_c} = \frac{8.314\times419}{8\times93.5\times101325}\ \text{m}^3\ \text{mol}^{-1}$$

$$= \mathbf{4.6 \times 10^{-5}\ m^3\ mol^{-1}}$$

4.9.8 Reduced equation of state and the law of corresponding states

If the pressure, volume and temperature of a gas are expressed in terms of the critical pressure, volume and temperature respectively, then we may write

$$P = \pi P_c,\ V = \phi\ V_c,\ T = \theta\ T_c \qquad ...(i)$$

Thus the quantities π, ϕ and θ may be defined as

$$\pi = \frac{P}{P_c},\ \phi = \frac{V}{V_c} \text{ and } \theta = \frac{T}{T_c}$$

The quantities π, ϕ and θ are called reduced pressure, reduced volume and reduced temperature respectively.

Substituting the values of *P*, *V* and *T* given by equation (*i*), in van der Waal's equation

$$\left(P - \frac{a}{V^2}\right)(V - b) = RT$$

we get $$\left(\pi P_c + \frac{a}{\phi^2 V_c^2}\right)(\phi V_c - b) = R\theta T_c.$$

Further, substituting the values of P_c, T_c and V_c in terms of van der Waal's constant *viz.*

$$V_c = 3\ b$$

$$P_c = \frac{a}{27b^2}$$

$$T_c = \frac{8a}{27Rb}$$

we get $$\left(\pi\frac{a}{27b^2}+\frac{a}{\phi^2.9b^2}\right)(\phi.3b-b) = R\theta\frac{8a}{27Rb}$$

or $$\left(\pi a+\frac{3a}{\phi^2}\right)(3b\phi-b) = 8\theta ab$$

or $$\boxed{\left(\pi+\frac{3}{\phi^2}\right)(3\phi-1) = 8\theta}$$

This equation which relates the reduced pressure, volume and temperature (*i.e.*, π, φ and θ) is called the **reduced equation of state**. The most important point about this equation is that this does not involve the constants *a, b* and *R* and hence is applicable to all substances in the liquid and the gaseous state. Another important conclusion drawn from the above equation is as follows :

If two substances have the same reduced pressure and the same reduced temperature, they must have the same reduced volume.

The above statement is called the **law of corresponding states** and the substances having the same reduced pressure and same reduced temperature and hence the same reduced volume are said to be in the corresponding states.

Example 20. **Calculate the pressure exerted by one mol of CO_2 at 40°C when confined to a volume of 0.17 dm^3 by making use of the law of corresponding states. The values of the critical pressure (P_c), critical volume (V_c) and critical temperature (T_c) are 73 atmospheres, 0.0957 dm^3 and 30.42°C respectively.**

Solution. $T = 40 + 273 = 313$ K, $V = 0{\cdot}7$ dm^3

$P_c = 73$ atm, $V_c = 0.0957$ dm^3, $T_c = (30.42 + 273)$ K or 303.42 K

$$\theta = \frac{T}{T_c} = \frac{313}{303.42} = 1.032$$

and $$\phi = \frac{V}{V_c} = \frac{0.7}{0.0957} = 7.315$$

Substitute in the reduced equation of state and calculate π.

$$\left[\pi+\frac{3}{\phi^2}\right](3\phi-1) = 8\theta$$

$$\left[\pi+\frac{3}{7.315^2}\right](3\times7.315-1) = 8\times1.032$$

$$(\pi + 0.056) \times 20.945 = 8.256$$

$$\pi = \frac{8.256}{20.945} - 0.056 = 0.338$$

∴ $P = \pi P_c = 0.338 \times 73 =$ **24.674 atm**

Example 21. **The reduced volume and temperature of a gas are 10.2 and 0.7. What will be its pressure if its critical pressure is 42 atm ?**

Solution. It is given that

Reduced volume, $\phi = 10.2$, Reduced temperature, $\theta = 0.7$

Critical pressure, $P_c = 42$ atm

According to the reduced equation of state

$$\left(\pi+\frac{3}{\phi^2}\right)(3\phi-1)=8\theta$$

Substituting the value of ϕ and θ, we get

$$\left(\pi+\frac{3}{(10.2)^2}\right)(3\times10.2-1)=8\times0.7$$

or $$\left(\pi+\frac{3}{104.4}\right)(30.6-1)=5.6$$

or $$(\pi + 0.0288)\ (29.6) = 5.6$$

or $$\pi + 0.0288 = \frac{5.6}{29.6} = 0.1892$$

or $$\pi = 0.1604$$

i.e. $$\pi = \frac{P}{P_c} = 0.1604$$

or $$P = 0.1604 \times P_c\ (P_c = 42 \text{ atm})$$
$$= 0.164 \times 42 = 6.7368 \text{ atm.}$$

2.10 LIQUEFACTION OF GASES

On lowering the temperature and increasing the pressure, the molecules of a gas come close to one another. A stage reaches when the gas liquefies and appears as a liquid. Certain gases like carbon dioxide can be liquefied by applying pressure even at room temperature. There are others which require slightly higher pressure than the atmospheric pressure and the temperature of freezing mixtures like sodium chloride and ice is enough to achieve liquefaction. Faraday could liquefy a number of gases by a technique as shown in Fig. 2.15 below.

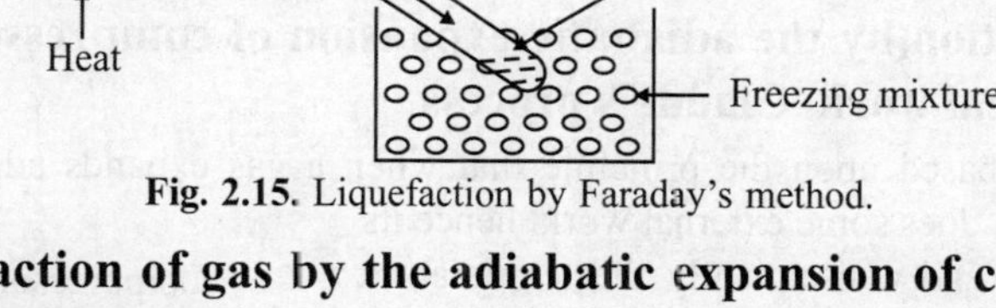

Fig. 2.15. Liquefaction by Faraday's method.

2.10.1 Liquefaction of gas by the adiabatic expansion of compressed gas—Linde's Process

This process is based upon Joule-Thomson effect which states as follows :

When a gas under high pressure is allowed to expand adiabatically through a fine orifice (hole) into a region of low pressure, it is accompanied by cooling.

An obvious reason for the cooling is that when the gas expands from a region of high pressure to the region of low pressure, the intermolecular distances increase. Some work is done in separating the molecules apart. This is done at the cost of kinetic energy of the gas. Thus the kinetic energy of the gas decreases and therefore, the temperature falls.

It may be mentioned here that in case of hydrogen and helium, instead of cooling, these gases are warmed up if allowed to expand as above, under ordinary temperature. However, if these gases are first cooled to some particular temperature and then allowed to expand, they behave like other gases. The temperature to which a gas should thus be cooled to give normal Joule-Thomson effect is called the *inversion temperature.* The inversion temperatures for hydrogen and helium are – 80°C and – 240°C respectively.

The process based upon the above principle is known as **Linde's process.** The apparatus used is shown in Fig. 2.16.

The above apparatus was used by Linde for the liquefaction of air. The procedure adopted is as follows:

The air is first compressed to about 200 atmospheres in the compressor and then passed through the condensing coils. As a result, the air is cooled and the water vapours present condense to form the liquid which is removed. The dry air is then passed through the copper spiral which terminates into a jet. The air expands through the jet into the chamber where the pressure of the air falls down to about 50 atmospheres. Thus as a result of Joule-Thomson effect, the air is cooled.

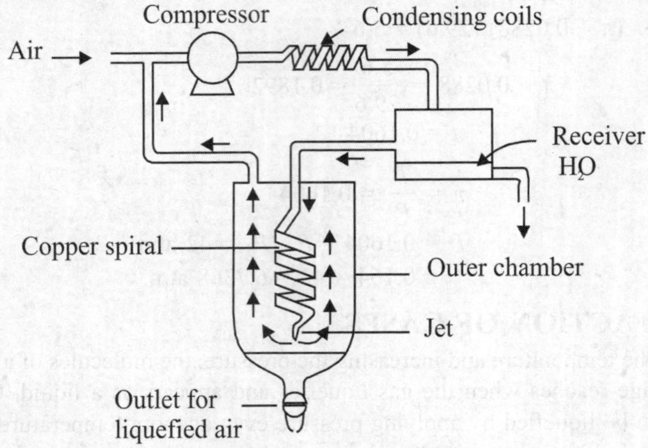

Fig. 2.16. Linde's process for liquefaction of air.

As this cooled air moves up, it further cools the incoming air. The cooled air is sent to the compressor again and the process is repeated a number of times till ultimately the air is cooled to such an extent that it liquefies. The liquefied air collects at the bottom of the chamber and can be drawn off. Any uncondensed air is recirculated.

2.10.2 Liquefaction by the adiabatic expansion of compressed gas involving mechanical work–Claude's process

This process is based upon the principle that when a gas expands adiabatically against a piston in an engine, it does some external work, hence its internal energy falls and consequently the temperature of the gas falls. This principle, combined with the Joule-Thomson effect, has been applied in the Claude's process for the liquefaction of air. The apparatus used is shown in Fig. 2.17.

The purified air *i.e.*, air freed from carbon dioxide and water vapour is first compressed to about 200 atmospheres. The compressed air then passes through the tube *T* which bifurcates at the point *X*. A part of the air goes into the cylinder fitted with an airtight piston and the rest of the air passes through the coil *C* having a jet *J* at its end. The air that goes into the cylinder pushes the piston and thus does some external work. As a result, the internal energy of the air falls and hence the temperature falls. The cooled air then enters the

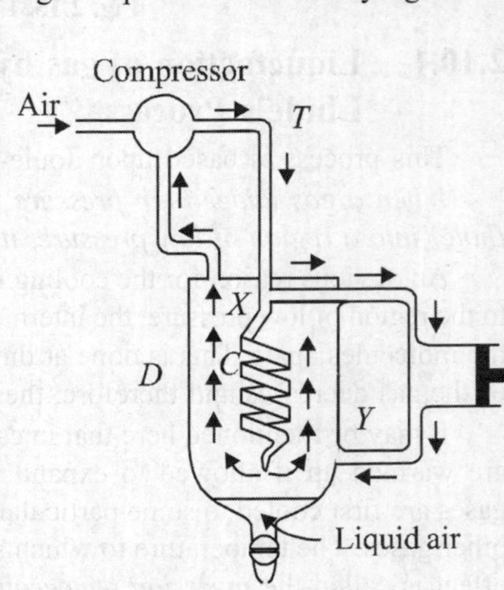

Fig. 2.17. Claude's process.

chamber D at the point Y. The air then passes through the coil C is cooled by Joule-Thomson effect as it comes out of the jet J into the chamber D. The cooled air going upwards in the chamber D further cools the incoming air in the coil C. The cooled air is led to the compressor and the process is repeated a number of times till the air is cooled to such an extent that it liquefies. The liquefied air collects in the chamber and is withdrawn.

2.10.3 Liquefaction by adiabatic demagnetisation

The process is based upon the principle that when a magnetised body is demagnetised adiabatically (*i.e.*, in such a way that heat can neither enter nor leave the system) the temperature of the body must fall. When a suitable solid is magnetised, the order is increased, since various magnetic entities are aligned. When demagnetisation is allowed to occur, the disorder returns and, therefore, the entropy increases. Hence heat must be absorbed and the cooling of the substance and its surroundings results. Now-a-days, the lower temperatures are commonly attained by this methods.

MISCELLANEOUS SOLVED EXAMPLES

***Example 22.* The values of van der Waal constant '*a*' for the two gases are *x* and *y* respectively (*x* >> *y*). Which of these can be liquefied easily ?**

Solution. Force of attraction between the molecules of a gas is one factor that governs the liquefactioin process. The quantity '*a*' in van der Waal equation for real gases is related to force of attraction between the molecules. A greater value of '*a*' means greater force of attraction, which in turn means greater chance of liquefaction thus the gas having '*x*' as the value of '*a*' can be liquafied more easily.

***Example 23.* An ideal gas can never be liquefied. Justify.**

Solution. According to kinetic theory of gases which is applicable to ideal gases, there are no forces of attraction between the molecules of a gas. In the absence of force of attraction, it is not possible to bring them together and obtain them in the liquified state.

***Example 24.* Can a gas of the type $P(V - b) = RT$ liquefied ? Explain.**

Solution. In the above equation, the quantity '*a*' which refers to the force of attraction between gaseous molecules, does not occur. In the absence of force of attraction, it is not possible to liquefy the gas.

***Example 25.* Prove that van der Waal's equation reduces to ideal gas equation at Boyle's temperature.**

Solution. At Boyle's temperature, the gas behaves in an ideal manner, therefore, the van der Waals' equation reduces to ideal gas equation.

NUMERICAL PROBLEMS FOR PRACTICE

1. The critical temperature and pressure of chlorine 4.19 K and 9.474×10^6 Nm^{-2} respectively. Calculate van der Waal's constants '*a*' and '*b*'. Given = 8.314 J.K^{-1}.

   (**Ans.** $a = 0.54$ Nm^4 mol^{-2}, $b = 4.59 \times 10^{-5}$ m^3 mol^{-1})

2. If van der Waal's equation of a gas is given by

   $$\left(P + \frac{0.00786}{V^2}\right)(V - 0.0024) = 0.0041\,(273 + t)$$

   Calculate T_c and P_c, if $R = 0.082$ Lt. atm. K^{-1} mol^{-}

   (**Ans.** $P_c = 50.54$ atm, $T_c = 236.39$ K)

3. Calculate the compressibility factor of a van der Waal's gas at the critical point.

   [**Hint:** Compressibility factor Z of gas is given by

   $$Z = \frac{PV}{RT}$$

At critical point $Z_c = \dfrac{P_c V_c}{RT_c}$

Substitute the value of critical constant

$$Z_c = \frac{\left(\dfrac{a}{27b^2}\right)(3b)}{R \cdot \dfrac{8a}{27Rb}}$$

$$= \frac{a}{9b} \times \frac{27b}{8a} = \frac{3}{8}$$

$$Z_c = \frac{3}{8} = 0.375.$$

4. A certain gas has the following critical constants: P_c = 45.6 atm., V_c = 0.0987 L mol^{-1} and T_c = 190.5 K. Calculate van der Waal's constants for the gas. Estimate the radius of the molecule, assuming it to be spheres.

(**Ans.** a = 1.333 L^2 atm mol^{-2} b = 0.0329 L mol^{-1}, r = 2.35 Å)

5. 1 Mole of the gas occupies a volume 10.7 litres at 313 K. Calculate the pressure using

(*i*) ideal gas equation

(*ii*) van der Waal's equation

$$V_c = 0.957 \text{ L mol}^{-1}$$
$$P_c = 73 \text{ atm.}$$
$$R = 0.0821 \text{ L atm. K}^{-1} \text{ mol}^{-1}$$

(**Ans.** 2.40 atm., 2.891 atm.)

6. What is the mean free path of argon at 2988 K and a pressure of 1 atm. Given σ = 1.86 A°.

(**Ans.** 1115 A°)

7. The compressibility factor $Z = \dfrac{PV}{nRT}$ for N_2 at 223 K at 800 atm. pressure is 1.95 and at 373 K and 200 atm., it is 1.10. A certain mass of N_2 occupied a volume of 1 L at 223 K and 800 atm. Calculate the volume occupied by the quantity of N_2 at 373 K and 800 diameter.

Sol. $Z_1 = \dfrac{P_1V_1}{nRT_1}$ at temperature T_1, volume V_1 and pressure P_1.

Similarly,

$Z_2 = \dfrac{P_2V_2}{nRT_2}$ at temperature T_2 volume V_2 and pressure P_2.

$$n = \frac{P_1V_1}{RT_1Z_1} = \frac{P_2V_2}{RT_2Z_2}$$

or
$$\frac{P_1V_1}{T_1Z_1} = \frac{P_2V_2}{T_2Z_2}$$

∴
$$V_2 = \frac{P_1V_1}{T_1Z_1} \times \frac{Z_2T_2}{P_2}$$

$$= \frac{800 \times 1 \times 1.10 \times 373}{200 \times 1.95 \times 223}$$

$$= 3.77 \text{ litres.}$$

8. van der Waal's constant for CO_2

$$a = 3.6 \text{ L}^2 \text{ atm mol}^{-2}$$

and $$b = 4.28 \times 10^{-2} \text{ L mol}^{-1}$$

Calculate the critical temperature and critical volume, $R = 0.0821$ L atm K^{-1} mol^{-1} (**Ans.** 31°C; 12.34 × 10^{-2} L mol^{-2})

9. Find the temperature at which the r.m.s. velocity of CO_2 gas is the same as that of Cl_2 at 293 K. (**Ans.** 181.57 K)

10. van der Waal's constant '*b*' for O_2 gas is 0.0318 L mol^{-1}. Calculate the diameter of O_2 molecule. (**Ans.** 2.934 × 10^8 cm.)

11. Calculate the r.m.s. velocity of N_2 at 298 K. The molecular weight of N_2 is 28.02. (**Ans.** μ = 5.14 × 10^4 cm sec^{-1})

12. Calculate the pressure exerted by 10 g of N_2 enclosed in a 1 L flask at 298 K. Assuming ideal behaviour. (**Ans.** 8.73 atm.)

13. Calculate the root mean square velocity and average velocity of CO_2 at 291 K and 400 mm pressure. (**Ans.** μ = 48390 cm. sec^{-1} ; C = 40920 cm sec^{-1})

14. 1 mole of CO_2 has volume 1.5 L at 300 K. If R = 0.0821 L atm K^{-1} mol^{-1}. Then the values of '*a*' and '*b*' are 1.04 and 0.04 respectively. Find the pressure of the gas using van der Waal's equation. (**Ans.** 16.39 atm.)

15. The density of H_2 at 273 K and 760 mm pressure is 0.00009 g mol^{-1}. Calculate r.m.s. velocity of H_2 molecule. (**Ans.** μ = 182900 cm/sec)

EXERCISES
(INCLUDING QUESTIONS FROM DIFFERENT UNIVERSITY PAPERS)

Objective Questions

Fill in the blanks

1. Moistured air diffuses .... through a small hole in comparison with dry air.
2. There is no effect of .... on the velocity of gas molecules
3. Pressure volume curves drawn at constant temperature are called ....
4. For an ideal gas C_p - C_v is equal to ....
5. The temperature at which real gases behave like ideal gases is called ....
6. Average velocity = ... × RMS
7. Heat capacity of a diatomic gas is .... than that of monoatomic gas
8. The distance covered by an atom of a gas between the consecutive collisions is called ....
9. Deviation from ideal behaviour of a gas is more prominent at ....
10. The vlaue of R is 0.0821. The unit of this is ....

Multiple Choice Questions

Choose the correct option

1. Which is not true in case of an ideal gas?

(*a*) It cannot be converted into a liquid.

(*b*) There is no interaction between the molecules.

(*c*) All molecules of the gas move with same speed.

(*d*) At a given temperature PV is proportional to the amount of the gas.

2. A real gas most closely approaches the behaviour of an ideal gas at:
(*a*) 15 atmospheres and 220 K. (*b*) 1 atm and 273 K
(*c*) 0-5 atm and 500 K (*d*) 15 atm and 500K

3. A gas is said to behave like an ideal gas when the relation PV/T = constant. When do you expect a real gas to behave like an ideal gas ?
(*a*) When the temperature is low
(*b*) When both the temperature and pressure are low
(*c*) When both the temperature and pressure are high
(*d*) When the temperature is high and pressure is low

4. An ideal gas cannot be liquefied because :
(*a*) its critical temperature is always above 0°C
(*b*) its molecules are relatively smaller in size
(*c*) it solidifies before becoming a liquid
(*d*) forces operative between its molecules are negligible

5. The gas which can be liquefied under high pressure and 40°C is :
(*a*) nitrogen (*b*) hydrogen (*c*) oxygen (*d*) ammonia

6. Real gas will approach the behaviour of ideal gas at:
(*a*) low temperature and low pressure (*b*) low temperature and high pressure
(*c*) high temperature and high pressure (*d*) high temperature and low pressure

7. Which set of conditions represent easiest way to liquefy a gas?
(*a*) Low temperature and high pressure (*b*) High temperature and low pressure
(c) Low temperature and low pressure (*d*) High temperature and high pressure

8. The gas that is heated up during Joule – Thomson effect at ordinary temperature is :
(*a*) O_2 (*b*) CO_2 (*c*) H_2 (*d*) SO_2

9. Gases deviate from the ideal gas behaviour because their molecules :
(*a*) possess negligible volume
(*b*) have forces of attraction between them
(*c*) are polyatomic
(*d*) are not attracted to one another

10. The temperature at which the second virial coefficient of a real gas is zero is called:
(*a*) critical temperature (*b*) eutectic point
(*c*) boiling point (*d*) Boyle's temperature

11. An ideal gas, obeying kinetic theory of gases, can be liquefied :
(*a*) it cannot be liquefied at any P and T
(*b*) its pressure is more than P at a temperature less than T
(*c*) its temperature is more than critical temperature T
(*d*) its pressure is more than critical pressure P.

12. The van der Waals equation explains the behaviour of :
(*a*) ideal gases (*b*) real gases (*c*) vapours (*d*) non-real gases

13. Formula for inversion temperature is :

(*a*) $T_i = \frac{2a}{nb}$ (*b*) $T_i = \frac{a}{Rb}$ (*c*) $T_i = \frac{2a}{Rb}$ (*d*) $T_i = \frac{a}{b}$

14. Which one of the following gases has the maximum value of root mean square velocity ?
(*a*) CH_4 (*b*) CO_2 (*c*) H_2 (*d*) CO_2

15. The correct representation of van der Waals equation is :
(*a*) $PV = RT$ (*b*) $\left(P + \frac{a}{V^2}\right)(V - b) = RT$
(*c*) $\left(P - \frac{a}{V^2}\right)(V + b) = RT$ (*d*) $PV = \frac{1}{2}mnu^2$

16. Degrees of freedom in water molecule are :
(*a*) 1 (*b*) 2 (*c*) 9 (*d*) 4

17. Critical temperature means above this temperature at any pressure, any gas can be:
(*a*) liquefied (*b*) not liquefied
(*c*) both statements are correct (*d*) solified

18. Expansion of a gas above inversion temperature creates :
(*a*) cooling (*b*) heating (*c*) constant temperature (*d*) none

19. According to the van der Waals correction to get the pressure of ideal gas in observed pressure a definite fraction for correction is:
(*a*) subtracted (*b*) added (*c*) divided (*d*) Unchanged

20. The value of C_p / C_v for inert gases is
(*a*) 1.33 (*b*) 1.66 (*c*) 2.13 (*d*) 1.99

ANSWERS

Fill in the blanks

1. faster **2.** gravitational **3.** isotherms **4.** R
5. Boyle's temperature **6.** 0.9213 **7.** greater
8. mean free path **9.** High temp. and low pressure **10.** L atm K^{-1} mol^{-1}

Multiple choice questions

1. (*c*)	**2.** (*c*)	**3.** (*d*)	**4.** (*d*)	**5.** (*d*)
6. (*d*)	**7.** (*a*)	**8.** (*c*)	**9.** (*b*)	**10.** (*d*)
11. (*b*)	**12.** (*b*)	**13.** (*c*)	**14.** (*c*)	**15.** (*b*)
16. (*c*)	**17.** (*b*)	**18.** (*b*)	**19.** (*b*)	**20.** (*b*)

General Questions

1. (*a*) Outline the basic assumptions of the kinetic theory of gases. Show that the total pressure (p) exerted by an ideal gas containing N molecules (each of mass m) and occupying volume V is given by

$$p = \frac{m N u^2}{3V}$$

where u^2 has the standard meaning of mean square velocity for molecules.

(*b*) Draw the curve showing Maxwell distribution of molecular velocities at a definite temperature in a gas and indicate on the curve the positions of the average $(\bar{v})$, most probable (v_{mp}) and the root mean square ($v_{r.m.s.}$) velocities. Also describe/illustrate how this distribution changes with an increase in temperature.

(*c*) How many vibrational degrees of freedom are there in the non-linear O_3 molecule ?

2. (*a*) State law of corresponding states.

(*b*) Derive the equation

$$\left[\pi + \frac{3}{\phi^2}\right][3\phi - 1] = 8\theta$$

(*c*) What are the units of van der Waal's constants '*a*' and '*b*' ?

3. Write down the assumptions of the kinetic theory of gases and point out those assumptions which were later on found valid for real gases. Also derive the kinetic gas equation.

4. (*a*) What do you understand by principle of corresponding states ? Deduce the expression of corresponding states from van der Waal's equation.

(*b*) What do you understand by continuity of state ? Explain with the help of isotherm of carbon dioxide.

5. (*a*) Isotherms of CO_2 at different temperatures have helped in establishing a relationship between critical constants and van der Waal's constants. Justify.

(*b*) What is an ideal gas ? What are the causes for deviation of real gases from ideal behaviour ?

(*c*) The reduced volume and temperature of the gas are 10.2 and 0.7 what will be its pressure if its critical pressure is 42 atmosphere ? (**Ans.** 6.7 atm.)

6. (*a*) What is the physical meaning of the constant '*a*' and '*b*' in the van der Waal's equation of state for real gases :

$$\left(P + \frac{n^2 a}{V^2}\right)(V - nb) = n\,RT$$

Show by deriving a relationship between the parameters of this equation and the critical constant that

$$\frac{P_c \cdot V_c}{R \cdot T_c} = \frac{3}{8}$$

(*b*) The critical volume for methane (CH_4) is 99 c.c. mol^{-1}. Assuming the molecules to be spherical, estimate the diameter of methane molecule.

(*c*) Draw isotherms of CO_2 below and above its critical temperature. Write briefly how these isotherms are used to understand the principle of liquefaction of gases.

7. What are the limitations of kinetic gas equation ?

(*a*) Give an account of the improvement suggested by van der Waal.

(*b*) Calculate, the pressure exerted by one mole of CO_2 gas at 40°C confined to a volume of 0.107 dm^3 using the law of corresponding states. The critical constants of the gas are $V_c = 0.0957$ dm^3, $T_c = 304$ K and $P_c = 73$ atm. (**Ans.** 74.46.*atm.*)

8. (*a*) Using the principle of equipartition of energy and taking various degrees of freedom of motion into consideration estimate the value of C_v for SO_2 (a bent molecule).

(*b*) What is the equation for the fraction of gas molecule with speed in a certain range according to Maxwell distribution law of speeds. Also illustrate the effect of temperature on :

(*i*) the distribution and

(*ii*) the most probable speed with the help of curves.

(*c*) Write van der Waal's equation for *n* moles of a real gas. Also calculate the van der Waal's constants of CO_2 for which the critical constants are :

$$T_c = 304.2 \text{ K}$$
$$P_c = 72.8 \text{ atm.}$$
$$V_c = 128.4 \text{ cm}^3 \text{ mol}^{-1}$$

(**Ans.** $b = 4.28 \times 10^{-2}$ *Lt* mole^{-1} ; $a = 3.612$ atm. mol^{-2})

9. (*a*) What is the principle of equipartition of energy ?

(*b*) What is *Z* called in the following equation :

$$Z = \frac{PV}{nRT} \ ?$$

Also write the value of *Z* for an ideal gas.

10. Explain, what is meant by critical constants of a gas. Derive the relations you will use to calculate these from van der Waal's constants.

11. What is virial equation of state :–

(*a*) Express the van der Waal's equation of state in the form of virial equation of state. Calculate Boyle's temperature.

(*b*) Calculate the volume occupied by 1 mole of a slightly imperfect gas at 4.75 atm. and 17°C. The second virial coefficient being 0.028 lt. mol^{-1}.

[**Hint:** Due to slight imperfect nature of the gas, the gas behaves more ideal. The pressure correction, *i.e.*,

$$\frac{a}{V^2} = 0, a = 0$$

The equation of state of the gas becomes

$$P(V - b) = RT$$

Second virial Co-eff.

$$B = \left(b - \frac{a}{RT}\right) = 0.028 \ Lt \text{ mol}^{-1}$$

As $a = 0$ $B = b = 0.028$

Now $$V = \frac{RT}{P} + b = \frac{0.0821 \times 290}{4.75} + 0.028$$

$$= 5.0404 \ Lt.]$$

12. (*a*) Derive the following :

$$P_c V_c = \frac{3}{8} RT_c$$

where the symbols have their usual meanings.

(*b*) When temperature of a gas is increased, what happens with the following :

(*i*) Most probable velocity.

(*ii*) Number of molecules with higher velocity

(*c*) Distinguish between an ideal and a real gas.

(*d*) If van der Waal's equation of a gas is given by

$$\left(P + \frac{0.00786}{V^2}\right)(V - 0.0024) = 0.0041\,(273 + t)$$

Calculate T_c and P_c ($R = 0.082$ L atm K^{-1} mol^{-1}

13. (*a*) What are the limitations of the equation $PV = RT$? What improvements have been suggested by van der Waal?

(*b*) Define compressibility factor and Boyle's point. What is the value of compressibility factor for an ideal gas ? Under what conditions of temperature and pressure, nitrogen gas behaves almost as an ideal gas ?

(*c*) Write virial equation in terms of

(*i*) Pressure

(*ii*) Volume

(*d*) One mole of ammonia gas is enclosed in a vessel of 2.5 dm^3 at 300 K. What will be the pressure of a real gas when

$$a = 4 \text{ dm}^6 \text{ mol}^{-2}$$
$$b = 0.04 \text{ dm}^3 \text{ mol}^{-1}$$
$$(1 \text{ Bar} = 10^5\ Pa)$$

14. (*a*) State Maxwell's law for distribution of velocities among the molecules of a gas. How does a change in the temperature of a gas influence this distribution ?

(*b*) Explain the following with reasons :

(*i*) The molar volumes of various liquids are studied at their boiling points.

(*ii*) An ideal gas cannot be liquefied.

(*iii*) Real gases show ideal behaviour at very low pressure.

(*c*) The reduced volume and temperature of a gas are 10.2 and 0.7. What will be the pressure if its critical pressure in 40 atmosphere.

15. (*a*) State the law of corresponding state and deduce the equation

$$\left(\pi + \frac{3}{\phi^2}\right)(3\phi - 1) = 8\theta$$

Boiling points are approximately the corresponding temperature. Justify this statement.

(*b*) Calculate the root mean square velocity for the following set of molecules: 10 molecules moving 5×10^2 ms^{-1}, 20 molecules moving 10×10^2 ms^{-1} and 5 molecules moving 15×10^2 ms^{-1} **(Ans.** $\mu = 2.8 \times 10^2$ ms^{-1})

(*c*) Explain the following :

(*i*) The kinetic energy of a gas increases with temperature.

(*ii*) Real gases approach ideal behaviour at low pressure and high temperature.

(*d*) Explain the term Degrees of freedom of motion. For a molecule made up of 3 atoms, what is the energy associated with

(*i*) translational degrees of freedom.

(*ii*) vibrational degrees of freedom for a non-linear molecule.

16. (*a*) Write van der Waal's equation of state for *n* moles of real gas and write the significance of van der Waal's constant '*a*' and '*b*'.

(*b*) A gas cannot be liquefied above 304 K but the same can be liquefied by applying a pressure of 7396725 Nm^2 at 304 K. Calculate van der Waal's constant ($R = 8.314 JK^{-1} mol^{-1}$)

(*c*) State the law of corresponding states and deduce the equation.

$$\left(\pi + \frac{3}{\phi}\right)(3\phi - 1) = 8\theta$$

What is the limit of its applicability.

17. (*a*) What is an ideal gas ? What are the causes for deviation of real gases from ideal behaviour ?

(*b*) Show that van der Waal's equation for the mole of gas can be written in the form

$$\frac{PV}{RT} = 1 + \frac{bP}{RT} + \frac{a}{RTV} + \frac{ab}{RTV^2}$$

(*c*) Derive the kinetic gas equation.

(*d*) van der Waal constant '*b*' for NO gas is 0.0278 $dm^3 mol^{-1}$. Calculate the diameter of NO molecule.

18. (*a*) State and explain the principles used in the liquefaction of gases by

(*i*) Claude's Process

(*ii*) Linde's Process

(*b*) What is the limitation of Law of Equipartition of Energy ?

(*c*) A 10 dm^3 flask containing 64 g of oxygen at 300 K. Calculate the pressure of the gas in kPa :

(*i*) if oxygen is an ideal gas and

(*ii*) if oxygen is a real gas

$a = 47$ kPa $dm^6 mol^{-2}$, $b = 0.037$ $dm^3 mol^{-1}$, $R = 8.314$ k Pa $dm^3 K^{-1} mol^{-1}$

19. (*a*) Derive the equation of state for a gas in critical state.

(*b*) The critical temperature and critical pressure of Cl_2 gas are 146° C and 93.5 atmosphere respectively. Calculate the value of van der Waal's Constants '*a*' and '*b*'. (1 Atm. = 101325 Nm^{-2})

(*c*) Explain the Linde's process for the liquefaction of gases.

(*d*) Define the principle of equipartition of energy.

20. (*a*) Briefly describe Maxwell's distribution of velocities of gases.

(*b*) Explain the following with reasons :

(*i*) Why H_2 gas cannot be liquefied at room temperature ?

(*ii*) Gases cannot be liquefied merely by the application of pressure above their critical temperature.

(*iii*) An ideal gas cannot be liquefied.

(*c*) Derive the kinetic gas equation.

21. (*a*) Starting from van der Waal's equation obtain an expression for critical constants in terms of van der Waal's Constant *a* and *b*.

(*b*) The van der Waal's Constants for gases HCl are $a = 0.367$ $Nm^4 mol^{-2}$ and $b = 0.0408 \times 10^{-8}$ $m^3 mol^{-1}$. Calculate T_c, V_c and P_c of the gas.

(*c*) What do you understand by Andrew's Critical phenomenon ? Explain.

3

Liquid State

3.1 STRUCTURAL DIFFERENCES BETWEEN SOLIDS, LIQUIDS AND GASES

We shall describe the structures of solids, liquids and gases in terms of kinetic molecular theory.

Molecules in a *gas* are in a state of constant random motion. The space between the molecules is large and therefore, there is a negligible force of attraction between the molecules.

In *liquids,* the force of attraction between the molecules are sufficiently strong to hold them in contact with one another. But, the molecules can move past one another through the vacant spaces or 'holes' here and there.

A liquid may be defined as **a collection of molecules held close to one another but showing random motion through the intervening empty spaces.**

In *solids*, there are strong intermolecular forces, thus fixing the position of each molecule. The molecules in a solid are not able to leave their position. However, small oscillation around the mean position is possible. Structures of gases, liquids and solids may be represented with the help of Fig. 3.1.

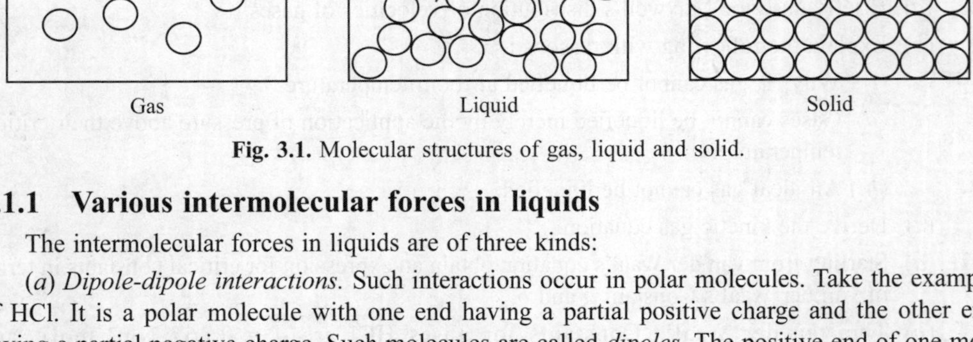

Fig. 3.1. Molecular structures of gas, liquid and solid.

3.1.1 Various intermolecular forces in liquids

The intermolecular forces in liquids are of three kinds:

(*a*) *Dipole-dipole interactions.* Such interactions occur in polar molecules. Take the example of HCl. It is a polar molecule with one end having a partial positive charge and the other end having a partial negative charge. Such molecules are called *dipoles*. The positive end of one molecule attracts the negative end of another molecule. There is a net attractive force between the polar molecules. Such attractive forces are called dipole-dipole attraction. The strength of such forces is generally 1% of that of a covalent bond. It may be observed that attractive forces between the opposite poles are stronger than the repulsive forces between like poles. Thus, there is a net force of attraction. It is illustrated in the Fig. 3.2

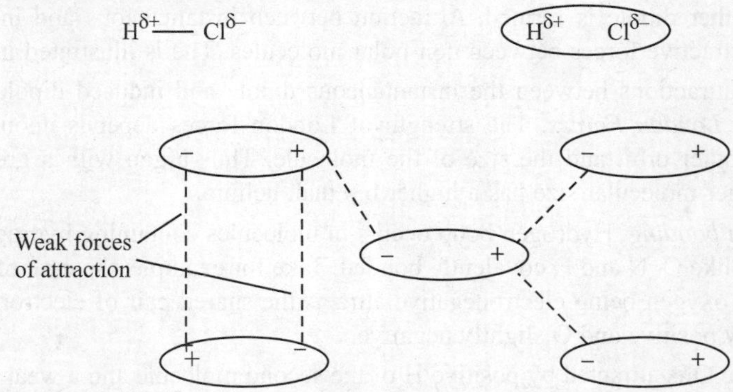

Fig. 3.2. Weak forces of attraction between polar HCl molecules.

(*b*) *London Forces*. Fritz London in 1930 explained the force of attraction between non-polar molecules. A non-polar molecule is formed between the atoms of similar electro-negativity by sharing of electrons. The shared electron pairs lie exactly in the middle.

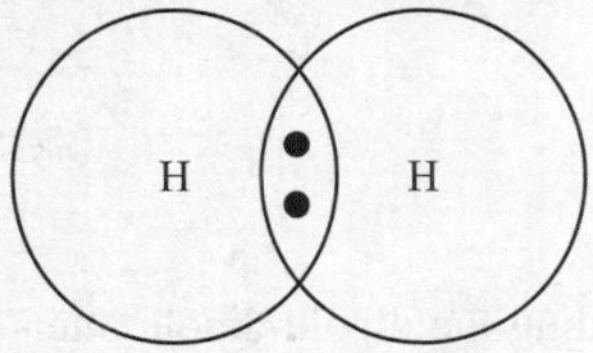

Hydrogen molecule

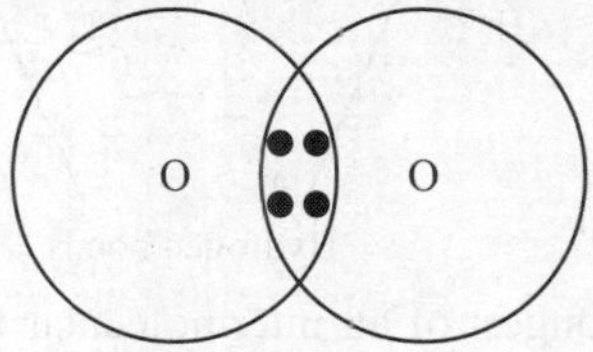

Oxygen molecule

Fritz London

Even these molecules exhibit weak attractive forces. London explained this by saying that according to principle of probability, at a particular instant, the electrons may shift to one side of the molecule making that side negative and the other side positive, thus creating a dipole.

Such a dipole is called *instantaneous dipole.* The negative pole of such a molecule repels the electrons of neighbouring non polar molecule making that end positive and the other end negative.

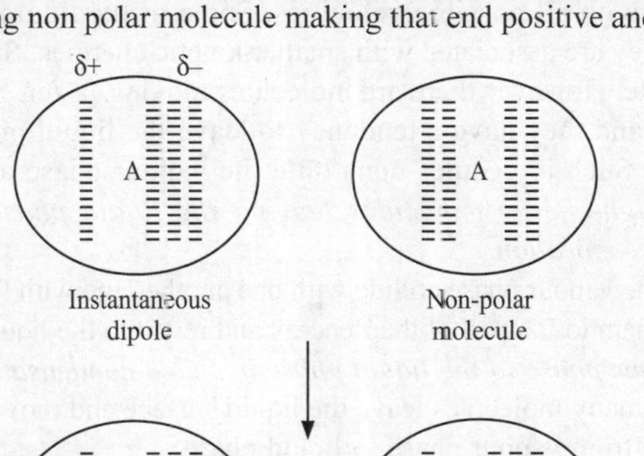

Fig. 3.3. Representation of London force.

Thus, an *induced dipole* is created. The instantaneous dipole and induced dipole will now attract each other. Because of the mobile nature of electrons, instantaneous dipole may disappear the next moment and another dipole is formed. Attraction between instantaneous and induced dipole is responsible for attractive forces between non-polar molecules. The is illustrated in Fig. 3.3.

Momentary attractions between the instantaneous dipole and induced dipole in liquid molecules are called *London Forces*. The strength of London forces depends upon the number of electrons in the outer orbit and the size of the molecule. Thus argon with a greater number of electrons and larger molecular size has a higher b.p than helium.

(*c*) *Hydrogen bonding*. Hydrogen bond occurs in molecules containing hydrogen and an electronegative atom like O, N and F, covalently bonded. Take for example, the case of H_2O molecule. In this molecule, oxygen being electronegative attracts the shared pair of electrons towards itself making H slightly positive and O slightly negative.

Now negative O is attracted by positive H of the second molecule and a weak bond is created between the two, which is called *hydrogen bond.*

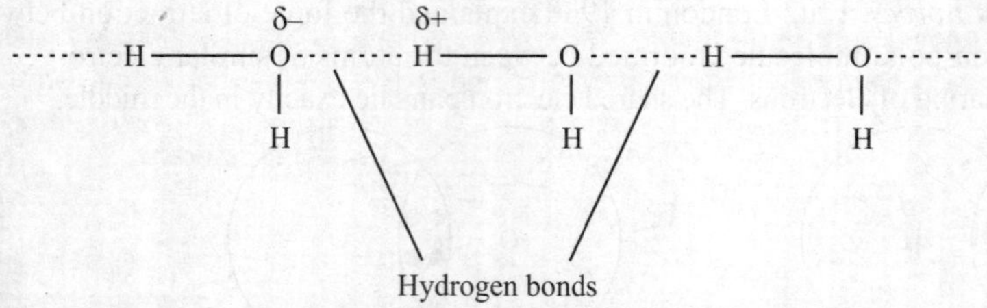

Hydrogen bond is the strongest of all intermolecular forces including the dipole-dipole interactions and London forces.

It may be mentioned that all the three intermolecular forces mentioned above are collectively called van der walls' forces.

3.1.2 Evaporation and condensation

According to kinetic theory of gases and liquids, the molecules are constantly moving in all possible directions and with all possible velocities. Some of the liquid molecules are moving with small velocities. They are associated with smaller kinetic energies. Such molecules continue moving in the liquid state. However there are molecules moving faster. Such molecules possess higher kinetic energies and they have a tendency to leave the liquid phase and move into the space above the liquid. Such molecules constitute the vapour phase above the liquid phase. *This phenomenon of high-energy molecules leaving the liquid phase and moving into the vapour phase is called evaporation.*

The molecules in the vapour phase collide with one another and with the walls of the container. In this process, some of them lose a part of their energy and revert to the liquid state. *Transference of molecules from the vapour phase to the liquid phase is called condensation.*

At equilibrium, as many molecules leave the liquid surface and move to the vapour phase as the molecules that move from vapour phase to liquid phase.

3.2 VAPOUR PRESSURE OF A LIQUID

Vapour pressure of a liquid at any temperature may be defined as the pressure exerted by the vapour present above the liquid in equilibrium with the liquid at that temperature. Vapour pressure exerted by the vapours obtained from the liquid may be illustrated as shown in Fig. 3.4.

Cooling caused by evaporation

When a liquid evaporates, the more energetic molecules leave the liquid. As a result, the average kinetic energy of the remaining liquid decreases and hence the temperature falls.

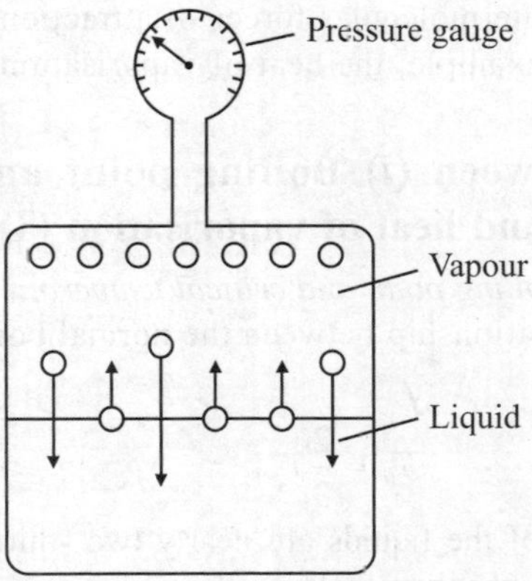

Fig. 3.4. Illustration of Vapour pressure.

3.2.1 Important factors that affect the vapour pressure

Two important factors on which the vapour pressure of a liquid depends are:

(*a*) *Nature of the liquid:* If the intermolecular forces of attraction in the liquid are weak, the molecules can easily leave the liquid and come into the vapour phase and hence the vapour pressure is higher. For example, the vapour pressures of ether, acetone, benzene etc. are higher than that of water at the same temperature. This is illustrated in Fig. 3.5. Acetic acid gives lower vapour pressure than water as the former has greater intermolecular forces caused by hydrogen bonding.

(*b*) *Effect of temperature :* As the temperature of a liquid is increased, the number of molecules with higher kinetic energy increases. Hence the vapour pressure of the liquid increases. This is again evident from each curve in Fig. 3.5

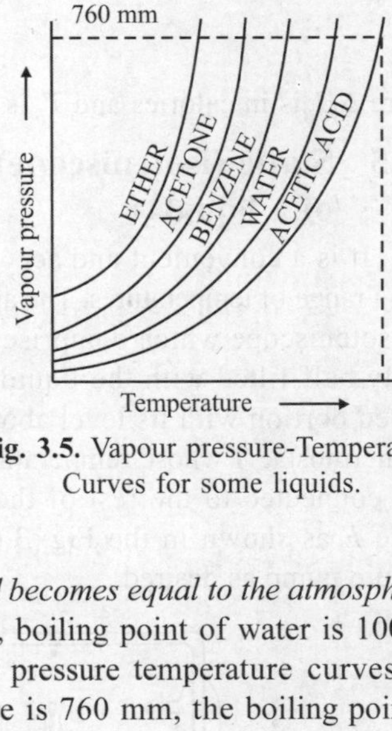

Fig. 3.5. Vapour pressure-Temperature Curves for some liquids.

3.2.2 Boiling point of a liquid

The temperature at which the vapour pressure of the liquid becomes equal to the atmospheric pressure is called the boiling point of the liquid. For example, boiling point of water is 100°C. The boiling point for any liquid is obtained from the vapour pressure temperature curves, as shown for acetic acid in Fig. 3.5. If the atmospheric pressure is 760 mm, the boiling point is called normal boiling point.

Obviously if the external pressure is higher, more heat will be required to make the vapour pressure equal to the external pressure and hence higher will be the boiling point. Similarly, if the external pressure is decreased, the boiling point is lowered. This is the reason that a liquid boils at a lower temperature on the top of a mountain (where pressure is low) than on the seashore.

3.2.3 Heat of vaporisation (ΔH_v) of a liquid

After the liquid starts boiling, the extra heat supplied to the liquid is used up not in increasing the temperature of the liquid but to overcome the intermolecular forces of attraction existing in the liquid and thus changing the liquid into vapour. Hence the temperature remains constant till whole of the liquid changes into vapour.

The amount of heat required to change 1 mole of the liquid into its vapour at the boiling point is called the heat of vaporisation of the liquid.

Obviously, greater the intermolecular forces of attraction present in a liquid, greater is the heat of vaporisation. For example, the heat of vaporisation of water is more than that of ether, acetone or benzene.

3.2.4 Relationship between (*i*) Boiling point and critical temperature (*ii*) Boiling point and heat of vaporisation (Trouton's rule)

(*i*) *Relationship between boiling point and critical temperature* (*Guldberg's Rule*): Guldberg gave the following general relationship between the normal boiling point (T_b) and the critical temperature (T_c) of the liquids :

$$T_b \approx \frac{2}{3} T_c$$

i.e., the normal boiling points of the liquids are nearly two -third of their critical temperatures when both are expressed on the absolute scale.

(*ii*) *Relationship between boiling point and heat of a vaporisation* (*Trouton's Rule*): Trouton observed the following generalization about liquids which are non-associated and which do not have too high boiling points.

$$\frac{\Delta H_\vartheta}{T_b} \approx 21$$

where ΔH_ϑ is in calories and T_b is in degrees absolute. This is known as Trouton's Rule.

3.2.5 Static (isoteniscope) method for the determination of vapour pressure of a liquid

It is a convenient and accurate method which can be used to measure the vapour pressure over a range of temperatures. The apparatus is shown in Fig. 3.6. The main part of the apparatus is the isoteniscope which comprises a bulb *B* and U-shaped portion of the tube *C*. The bulb *B* is nearly half-filled with the liquid under investigation and the same liquid is placed in the U-shaped portion with its level about 2-3 cm below that in the bulb. The isoteniscope is placed in the thermostat *A* whose temperature can be noted from the thermometer *T*. The isoteniscope is then connected to the rest of the apparatus which comprises a barometric tube *D* and a big bottle *E* as shown in the Fig. 3.6. The bottle *E* can be connected to the atmosphere or to the vacuum pump as desired.

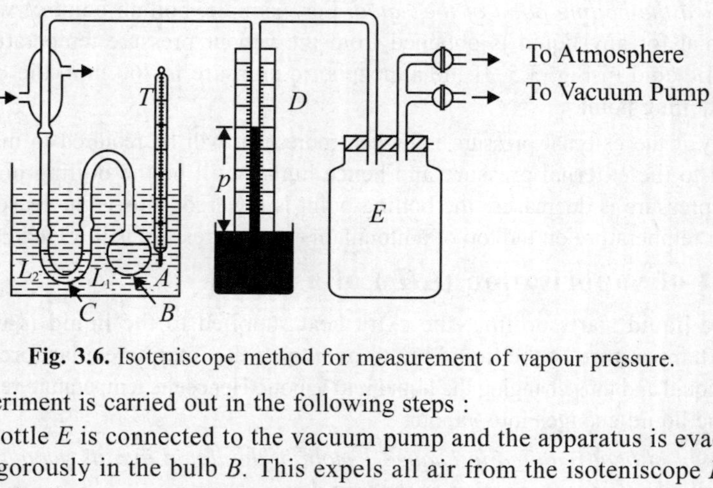

Fig. 3.6. Isoteniscope method for measurement of vapour pressure.

The experiment is carried out in the following steps :

(*i*) The bottle *E* is connected to the vacuum pump and the apparatus is evacuated till the liquid boils vigorously in the bulb *B*. This expels all air from the isoteniscope *BC*. The level of the liquid in the limb L_2 stands higher than in the limb L_1.

(*ii*) The temperature of the thermostat is now adjusted to the desired value and air is admitted slowly into the apparatus by connecting the bottle *E* to the atmosphere till the levels of the liquid in both the limbs L_1 and L_2 are equal. Under these conditions, the pressure in both the limbs L_1 and L_2 will be equal. Pressure in the limb L_1 is the vapour pressure whereas pressure in the limb L_2 is equal to the pressure in the rest of the apparatus and is equal to atmospheric pressure minus the height of the mercury column in the tube *D*. Hence, this difference is the vapour pressure of the liquid at the temperature of the thermostat.

3.2.6 Dynamic method for the determination of vapour pressure

The dynamic method is based upon the principle that a liquid boils at the temperature at which the vapour pressure becomes equal to the external pressure over the liquid. Thus by fixing the external pressure over the liquid at different values, the boiling point of the liquid is determined at each external pressure. Suppose the boiling points are t_1, t_2, t_3 etc. at the external pressure p_1, p_2, p_3 etc. respectively. Then obviously, the pressures p_1, p_2, p_3 etc. will be the vapour pressures at the temperatures, t_1, t_2, t_3 etc. respectively. By plotting the pressures against the corresponding temperatures for any particular liquid, a curve of the type shown in Fig. 3.5 is obtained. From this curve, the vapour pressure at any temperature can be found.

3.3 SURFACE TENSION OF A LIQUID

Molecules of the liquid at the surface are in a different situation than those in the interior of the liquid. For example, a molecule lying inside the liquid is surrounded by other molecules and so is attracted equally in all directions (Fig. 3.7). Hence the net force of attraction acting on the molecule is zero. But, a molecule lying at the surface is attracted downwards by the molecules lying in the bulk of the liquid. Thus a molecule lying at the surface experiences a net inward attraction. As a result of this inward pull on all molecules lying at the surface, the surface behaves as if it were under tension. This property of liquids is called surface tension.

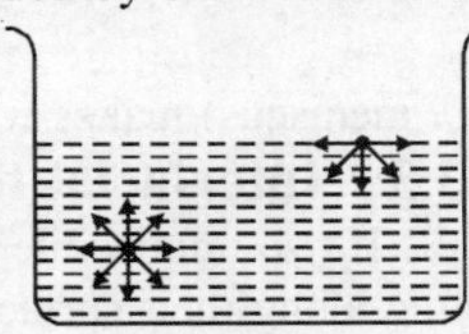

Fig. 3.7. Surface tension

The surface tension of a liquid is defined as the force acting at right angles to the surface along one centimetre length of the surface. Thus the units of surface tension are dynes per cm (or Newtons per metre *i.e.*, Nm^{-1} in the S.I. system).

Why is the drop of a liquid spherical ?

As a result of the inward pull on the molecules at the surface, the surface of the liquid tends to contract to the smallest possible area for a given volume of the liquid. It is for this reason that the drops of a liquid are spherical because for a given volume, a sphere has minimum surface area.

3.3.1 Surface energy

To increase the area of surface, some work has to be done against the inward pull. Consider a soap solution film contained in rectangular wire frame *ABCD* in which the side *CD* is movable (Fig. 3.8). In order to extend the surface area of the film, the movable wire has to be pulled from position *CD* to position *EF*. Thus some work has to be done against surface tension.

The work in ergs required to be done to increase or extend the surface area by 1 sq. cm is called surface energy. The *units* of surface energy are, therefore, ergs per sq. cm (or joules per sq. metre in S.I. system).

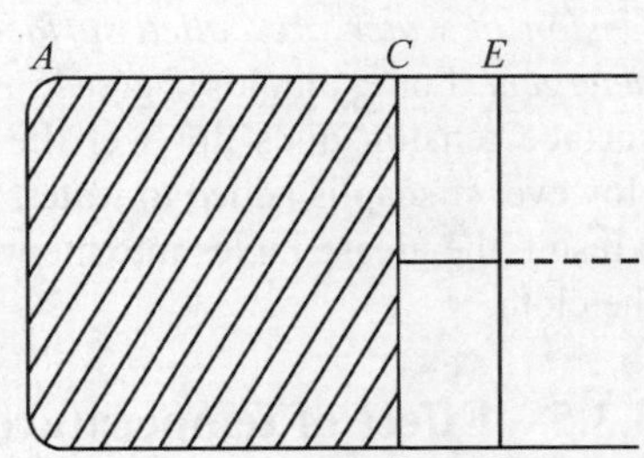

Fig. 3.8. Concept of surface energy.

In terms of dimensions,

Surface energy = work per sq. cm. = (Force × length) per sq. cm.

$$= \frac{\text{dynes} \times \text{cm}}{\text{cm}^2} = \text{dynes cm}^{-1}$$

These units are the same as those of surface tension. Thus the surface energy is same thing as surface tension.

3.3.2 Liquid rises in a capillary tube

When one end of a capillary tube is put into a liquid that *wets glass,* the liquid rises into the capillary tube to a certain height. This rise is obviously due to the inward pull of surface which pushes the liquid into the capillary tube (Fig. 3.9).

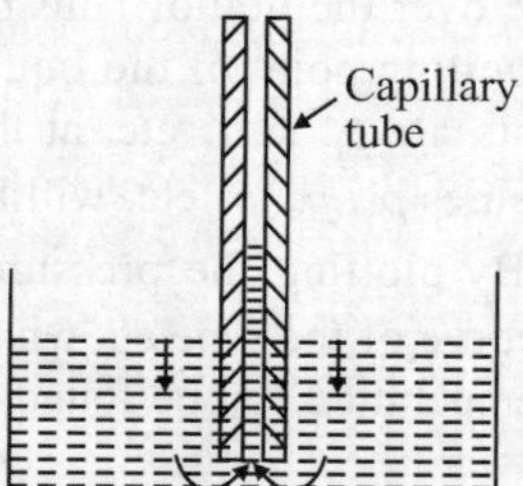

Fig. 3.9. Rise of a liquid in a Capillary tube

In case of liquids which do not wet glass *e.g.* mercury, the level inside the capillary falls below the level outside. Further whereas the upper surface of a liquid that wets glass is concave, that of mercury is convex. Such a surface is called **meniscus** (Greek : a little moon). The angle which the curved surface (*i.e.* meniscus) makes with the wall of the tube is called **contact angle**, Fig. 3.10.

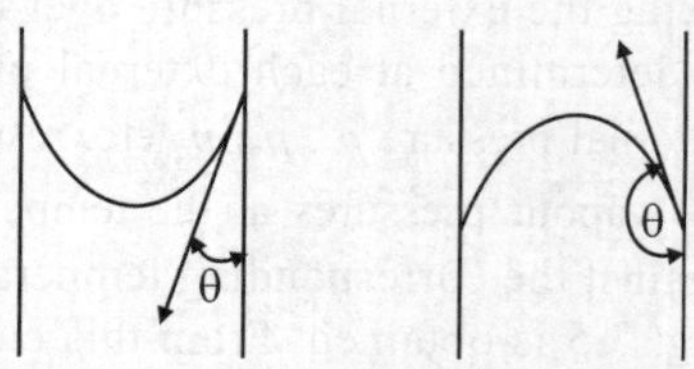

Fig. 3.10. Contact angle for a liquid
(a) that wets glass
(b) that does not wet glass.

3.3.3 Interfacial tension

If two immiscible or partially miscible liquids are taken in a vessel, then the surface tension acting along their surface of separation *i.e.*, along the interface is called the **interfacial tension.** Its value is generally intermediate between the surface tension of the two liquids but sometimes it is less than both. This is due to the fact that at the interface, the molecules of one liquid are attracted by the molecules of the other.

In fact, the surface tension of a liquid is also an interfacial tension, the two phases involved being the liquid and the vapour above it.

3.3.4 Surface active agents

There are certain substances like soaps, sulphonic acids and organic compounds like methyl alcohol, ethyl alcohol, acetone which when added to water even in small amount decrease the surface tension of water to a considerable extent. *Such substances which lower the surface tension of water are called surface active agents.* It is for this reason that the soap acts as a *detergent.* For example, if grease is sticking on the surface of a cloth, water because of high surface tension does not wet the grease and hence grease is not removed by water alone. However if soap is added to water, it lowers the interfacial tension between water and grease. As a result, the grease mixes into the water (soap solution) and hence is removed from the surface of the cloth.

3.3.5 Effect of temperature on surface tension

The surface tension of liquids generally decreases with increase of temperature and becomes zero at the critical temperature. The decrease in surface tension with increase of temperature is obviously due to the fact that the kinetic energy of the molecules increases and, therefore, the intermolecular attraction decreases.

An important relationship between temperature and surface tension is as follows:

$$\gamma\left(\frac{M}{D}\right)^{2/3} = k(t_c - t)$$

where M = molecular mass of the liquid

D = density of the liquid at temperature t°C

t_c = critical temperature of the liquid

γ = surface tension (or surface energy) of the liquid at temperature t°C

k = a constant

As the surface tension almost vanishes at 6° above the critical temperature instead of critical temperature, Ramsay and Shields gave the following equation for the temperature dependance of surface tension

$$r = (M/d)^{2/3} = k(t_c - t - 6)$$

Katayama suggested a more correct relation *viz.*

$$4\gamma\left(\frac{M}{D-d}\right)^{2/3} = k'(t_c - t)$$

where d is density of the vapour at temperature t. In this equation $\gamma = 0$ when $t = t_c$.

DETERMINATION OF SURFACE TENSION

A number of methods are available for the determination of surface tension of a liquid. Some of them are described hereunder.

3.3.6 Capillary rise method for determining surface tension of a liquid

The principle of the method is as follows :

Suppose one end of a capillary tube is put into a liquid which *wets* glass. Then the force of surface tension pushes the liquid into the capillary to a certain height (as shown in Fig. 3.11) till the force of surface tension acting upwards balances the weight of the liquid column acting downwards.

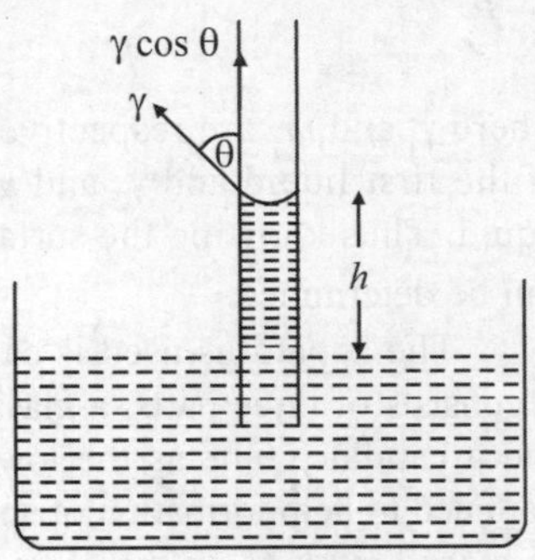

Fig. 3.11. Capillary rise method for measurement of surface tension.

Suppose,

height to which liquid rises in the capillary = h cm,

density of the liquid =d g/ml,

inner radius of the capillary = r cm,

surface tension of the liquid = γ dynes/cm and contact angle = θ

Then component of surface tension acting in the vertical direction = $\gamma \cos \theta$.

Inner circumference of the capillary = $2\pi r$

$\therefore$ Force of surface tension acting upwards all along the circumference = $2\pi r.\gamma \cos \theta$

Weight of the liquid acting downwards = $\pi r^2 hdg$ where g is acceleration due to gravity, and $\pi r^2 h$ is the volume of the liquid column.

Hence at equilibrium

$$2\pi r.\gamma\cos\theta = \pi r^2 hdg$$

or

$$\gamma = \frac{rhdg}{2\cos\theta}$$

For most of the liquids which wet glass,

$$\theta = 0$$

so that $\cos \theta = 1$

Hence $$\boxed{\gamma = \frac{rhdg}{2}}$$

Thus the only two parameters that need to be measured are the inner radius (r) of the capillary tube and the height (h) to which the liquid rises.

***Example 1.* At 20°C, toluene rises 1·95 cm in a capillary tube of radius 0·3412 mm. Calculate the surface tension of toluene. The density of toluene at 20°C is 0·866 g per c.c.**

Solution. Surface tension is given by the relation

$$\gamma = \frac{rhdg}{2}$$

$r = 0{\cdot}3412$ mm $= 0{\cdot}03412$ cm

$h = 1{\cdot}95$ cm

$d = 0{\cdot}866$ g cm^{-3}

$g = 981$ cm s^{-2}

Substituting the values in the equation above

$$\gamma = \frac{0{\cdot}03412 \times 1{\cdot}95 \times 0{\cdot}866 \times 981}{2} = 28{\cdot}26 \text{ dynes cm}^{-1}$$

3.3.7 Drop weight method for the determination of surface tension of a liquid using stalgmometer

This method is based upon the *principle* that the weight of drop falling slowly out of a capillary held vertically is directly proportional to its surface tension. Thus for two liquids, the drops of which fall almost at the same rate out of the same capillary,

$$\frac{\gamma_1}{\gamma_2} = \frac{m_1}{m_2}$$

where γ_1 and m_1 are respectively the surface tension and weight of each drop of the first liquid and γ_2 and m_2 are the corresponding values for the second liquid. Thus knowing the surface tension of one of the liquids, that of the other can be determined.

The apparatus used is called a stalgmometer and is shown in Fig. 3.12. It consists of three parts, a glass tube as the upper portion, a bulb in the middle and a capillary tube as the lower portion. The lowest point of the capillary is flattened to help formation of spherical drops and smooth flow of liquid. There is an etched mark M on the glass tube portion.

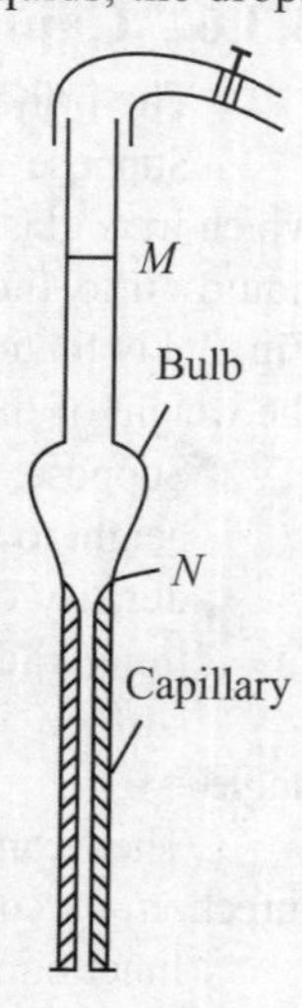

Fig. 3.12. Stalgmoemter.

The procedure consists of the following steps :

(*i*) Stalgmometer is clamped vertically in a clamp stand and a rubber tube with a pinch cock is attached to the upper portion.

(*ii*) Distilled water is sucked into the apparatus and then the pressure adjusted with the help of pinch cock so that the rate of fall of drops is about 15-20 drops per minute.

(*iii*) 10 drops of water are allowed to fall into a clean weighing bottle which has already been weighed when empty. The weight of the weighing bottle alongwith the drops is taken.

(*iv*) The apparatus is now rinsed with acetone and then allowed to dry.

(*v*) Now, the experimental liquid is sucked into the stalgmometer, the rate of fall of drops adjusted as before and 10 drops of the liquid taken into the empty weighing bottle and the bottle weighed.

Calculations :

Suppose weight of empty bottle $= w_1$ g
Weight of bottle + 10 drops of water $= w_2$ g
Weight of bottle + 10 drops of liquid $= w_3$ g
$\therefore$ Weight of 10 drops of water $= (w_2 - w_1)$ g
Weight of 10 drops of liquid $= (w_3 - w_1)$ g

If γ_w is surface tension of water, then the surface tension of the experimental liquid, γ_l is given by

$$\frac{\gamma_l}{\gamma_w} = \frac{w_3 - w_1}{w_2 - w_1} \quad \text{or} \quad \gamma_l = \frac{w_3 - w_1}{w_2 - w_1} \times \gamma_w$$

Knowing γ_w, γ_l can be calculated.

3.3.8 Drop number method for the determination of surface tension

As it is more convenient to count the number of drops formed from a certain volume of the liquid than to take the weight of a drop of a liquid, therefore the above method may be carried out in an alternative manner as follows :

Suppose the number of drops formed from v ml of a liquid $= n$.

Then weight of one drop of the liquid

$$= \frac{v \times d}{n}$$

where d is the density of the liquid. Thus if from the same volume (say v ml) of the two liquids, the number of drops formed are n_1, n_2 and d_1, d_2 are their respective densities, then

$$\frac{\gamma_l}{\gamma_w} = \frac{m_1}{m_2} = \frac{\dfrac{v \times d_1}{n_1}}{\dfrac{v \times d_2}{n_2}} = \frac{n_2 \times d_1}{n_1 \times d_2}$$

Hence knowing the number of drops (n_1 and n_2) formed from the same volume of two liquids, knowing their densities (d_1 and d_2) and knowing the surface tension of one of them, that of the other can be calculated.

In the procedure, the apparatus is cleaned and clamped vertically. Distilled water is first sucked into the apparatus upto mark *M*. The number of drops, while water flows from the mark *M* to the mark *N*, is noted. The apparatus is rinsed with acetone and dried. The experimental liquid is now sucked in upto the mark *M* and the number of drops formed, while the liquid flows from *M* to *N*, is counted. The result is then calculated using the formula derived above.

3.3.9 Nouy torsion balance method for the measurement of surface tension

The method is based upon the principle that *the force required to take out a horizontal platinum ring dipping just below the surface of a liquid is directly proportional to the surface tension of the liquid.* The platinum ring used is usually of about 4 cm diameter and the liquid is usually taken in a watch glass. The force required is measured either by attaching the platinum ring to one end of the beam of a sensitive balance and placing weights on the other side or by using a special balance called *torsion balance*. In this balance, the platinum ring is

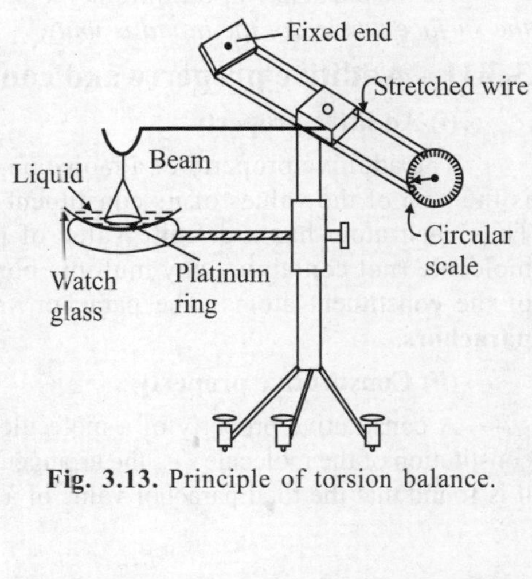

Fig. 3.13. Principle of torsion balance.

attached to one end of the beam and the other end of the beam is fixed to a wire which can be given a torsion by suitable device. The amount of torsion just sufficient to take out the ring can be recorded on a circular scale attached to one end of the wire as shown in Fig. 3.13. The experiment is first performed with water and then with the experimental liquid. Suppose the torsion angles are θ_1 and θ_2 and their surface tensions are γ_1 and γ_2 respectively.

Then $$\frac{\gamma_1}{\gamma_2} = \frac{\theta_1}{\theta_2}$$

Thus knowing the surface tension of water, that of the experimental liquid can be found.

3.3.10 Parachor

The following relationship between surface tension and density exists

$$\frac{\gamma^{1/4}}{D-d} = C \qquad ...(i)$$

where γ = surface tension of the liquid

D = density of the liquid

d = density of the vapour at the same temperature

and C = a constant, independent of temperature

Multiplying both sides of equation (*i*) with the molecular weight (M) of the liquid, we get

$$\frac{M\gamma^{1/4}}{D-d} = MC = [P] \qquad ...(ii)$$

The constant $[P]$ thus obtained is called **Parachor.**

At ordinary temperature, d is very small in comparison to D and hence can be neglected in the denominator. Hence equation (*ii*) becomes

$$\frac{M\gamma^{1/4}}{D} = [P] \qquad ...(iii)$$

Putting $\frac{M}{D} = V$, the molar volume of the liquid, equation (*iii*) become

$$V.\gamma^{1/4} = [P]$$

In this equation if $\gamma = 1$, then

$$[P] = V$$

Hence *parachor of a liquid is the molar volume of the liquid at a temperature at which the surface tension of the liquid is unity.*

3.3.11 Additive property and constitutive property

(*i*) Additive property

The additive property of a molecule is defined as that property, the total value of which is the sum of the values of its constituent atoms. Parachor is an example of such a property. Thus each atom has a definite value of parachor and the total parachor value of a simple molecule (not containing any multiple bond or a ring etc.) is the sum of the parachor values of the constituent atoms. The parachor values associated with the atoms are called **atomic parachors.**

(*ii*) Constitutive property

A constitutive property of a molecule is defined as that property which depends upon the constitution of the molecule *i.e.*, the arrangement of atoms within the molecule. In case of parachor, it is found that the total parachor value of a molecule containing some multiple bond or a closed ring etc. depends not only upon the atomic parachors but also upon the multiple bonds or rings present. The parachor values associated with structure of the molecule *i.e.*, the multiple bonds, closed rings etc. are called **structural parachors.**

Values of some atomic and structural parachors are given below in Table 3.1

Table 3.1. Some atomic and structural parachors.

Atomic Parachors		*Structural Parachors*	
Atom	*Parachor*	*Structure*	*Parachor*
C	4.8	Single bond	0.0
H	17.1	Double bond	23.2
O	20.0	Triple bond	46.6
Cl	53.8	3-membered ring	16.7
Br	68.0	4-membered ring	11.6
I	90.0	5-membered ring	8.5
N	12.5	6-membered ring	6.1
P	37.7	Co-ordinate bond	– 1.6
S	48.5	O_2 in esters	60.0

***Example 2.* Establish the structure of quinone based upon parachor values.**

Solution. The following two structures for quinone are possible, the molecular formula for quinone being $C_6 H_4 O_2$.

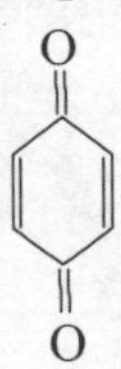

I

And

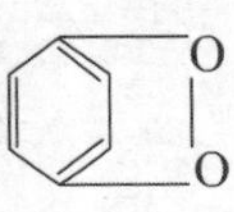

II

Parachor of quinone according to structure I (using values from Table 3.1)

$= 6\,[P]_c + 4\,[P]_H + 2\,[P]_O + 4\,[P]_{\text{Double bond}} + [P]_{6\text{-ring}}$

$= (6 \times 4.8) + (4 \times 17.1) + (2 \times 20.0) + (4 \times 23.2) + 6.1$

$=$ **236.1**

Parachor of quinone according to structure II (using values from Table 3.1)

$= 6\,[P]_c + 4\,[P]_H + 2\,[P]_O + 3[P\,]_{\text{Double bond}} + 2[P]_{6\text{-ring}}$

($\because$ there are two 6-membered rings in II)

$= (6 \times 4.8) + (4 \times 17.1) + (2 \times 20.0) + (3 \times 23.2) + (2 \times 6.1) =$ **219.0**

Observed value of parachor for quinone (From experiment) = **236.8**

This value is close to the value calculated for structure I.

Hence I must be the correct structure for quinone.

***Example 3.* Establish the structure of benzene based on parachor values.**

Solution. The structure of benzene is confirmed by the parachor values as follows :

Kekule's structure for benzene ($C_6\,H_6$) is 

Calculated value of parachor for benzene is (using values from Table 3.1)

$= 6\,[P]_c + 6\,[P]_H + 3\,[P]_{\text{Double bond}} + [P]_{6\text{–ring}}$

$= (6 \times 4.8) + (4 \times 17.1) + (3 \times 23.2) + (6.1)$

$=$ **207.1**

Density of benzene at 20°C, (D)

$= 0.878$ g/ml

Surface tension of benzene at 20°C, (γ)

$= 29.3$ dynes/cm

Molecular weight of benzene (C_6H_6)

$= 78$

∴ Experimental value of parachor, (P)

$$= \frac{M\gamma^{1/4}}{D}$$

$$= \frac{78 \times (29.3)^{1/4}}{0.878} = \mathbf{206.7}$$

Thus the experimental value is in close agreement with the calculated value. Hence Kekule's structure is confirmed.

Example 4. **The parachors of ethane and propane are 110.5 and 150.8 respectively. What values of parachor do you expect for hexane?**

Solution. $[P]\ C_2H_6 = 110.5$

$[P]\ C_3H_8 = 150.8$

$[P]\ CH_2 = [P]\ C_3H_8 - [P]\ C_2H_6$

$= 150.8 - 110.5 = 40.3$

$[P]\ C_6H_{14} = [P]\ C_3H_8 + 3[P]CH_2$

$= 150.8 + 3 \times 40.3 = 271.7$

Example 5. **How high will a liquid rise in a capillary tube of 0.1 mm diameter, if its density be 1.3 g cm^{-3} and surface tension 65 dynes cm^{-1} ? (g= 980 cm s^{-2})**

Solution. Diameter of the capillary $(2r)$

$= 0.1$ mm $= 0.01$ cm

∴ Radius, $r = 0.005$ cm

Density of the liquid $(d) = 1.3$ g cm^{-3}

Surface tension of the liquid (γ)

$= 65$ dynes cm^{-1}

The formula for the rise of liquid in the capillary is

$$\gamma = \frac{rhdg}{2}$$

$$h = \frac{2\gamma}{rdg}$$

$$= \frac{2\times65}{0.005\times1.3\times980}$$

$$= \mathbf{20.41\ cm}$$

***Example 6*. The density of acetone at 20°C is 0.7910 g/ml. Calculate the surface tension of acetone, given that the parachor values of C, H, O and double bond are 7.2, 16.2, 20.0 and 23.2 respectively.**

***Solution*.** Density of acetone at 20°C (D)

$$= 0.791 \text{ g/ml}$$

Parachor values : $[P]_c = 7.2$

$[P]_H = 16.2$

$[P]_O = 20.2$

$[P]_{\text{Double bond}} = 23.2$

$\therefore$ Parachor for acetone $(CH_3 - \overset{\overset{\displaystyle O}{\|}}{C} - CH_3)$

$$= 3[P]_c + 6[P]_H + [P]_o + [P]_{\text{Double bond}}$$

$$= 3 \times 7.2 + 6 \times 16.2 + 20.0 + 23.2 = 162$$

Molecular weight of acetone (M) = 58

Surface tension is related to parachor according to the formula

$$\frac{M\gamma^{1/4}}{D} = [P]$$

$$\therefore \qquad \gamma^{1/4} = \frac{[P]D}{M}$$

$$\text{or} \qquad \gamma = \left[\frac{[P]D}{M}\right]^4$$

Substituting the values for $[P]$, D and M

$$\gamma = \left[\frac{162\times0.791}{58}\right]^4$$

$$= \mathbf{23.83\ dynes/cm}$$

3.4 VISCOSITY

Consider a liquid flowing through a narrow tube, Fig. 3.14. All parts of the liquid do not move through the tube with the same velocity. Imagine the liquid to be made up of a large number of thin cylindrical coaxial layers. The layer which is in contact with the walls of the tube is almost stationary.

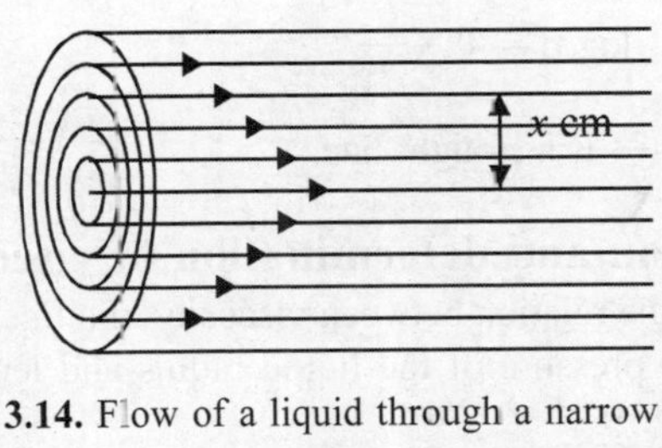

Fig. 3.14. Flow of a liquid through a narrow tube.

As we move from the walls towards the centre of the tube, the velocity of the cylindrical layers goes on increasing. It is maximum at the centre. Conversely, we may say that as we move from the centre towards the walls, the velocity of the layers keeps on decreasing. In other words, every layer offers some resistance or friction to the layer immediately below it. *This force of friction which one part of the liquid offers to another part of the liquid is called viscosity.*

It has been found that the *force of friction* (f) between two cylindrical layers each having area A sq. cm, separated by a distance x cm, and having a velocity difference of v cm/sec., is given by

$$f \propto Av/x$$

or

$$f = \eta \frac{Av}{x} \quad \text{or} \quad \eta = \frac{f\,x}{Av}$$

where η is a constant known as coefficient of viscosity.

If $x = 1$ cm, $A = 1$ sq. cm and $v = 1$ cm/sec, then $\eta = f$

Hence *coefficient of viscosity may be defined as the force of friction (in dynes) required to maintain a velocity difference of 1 cm/sec between two parallel layers, 1 cm apart and each having an area of 1 sq. cm.*

The unit of viscosity is dynes cm^{-2} s or poise.

In SI units

$$\eta = \frac{\text{Newton}}{\text{metre}^2} \times \frac{\text{metre}}{\text{metre/sec.}}$$

$$= \text{Newton sec. metre}^{-2}$$

$$\eta = \text{Ns m}^{-2} = \text{kg m}^{-1}\,\text{s}^{-1}$$

The unit of viscosity in SI system are Ns m^{-2} or kg m^{-1} s^{-1}.

It can be shown that 1 poise = $\frac{1}{10}$ kg m^{-1} s^{-1}.

FLUIDITY

Reciprocal of viscosity is fluidity represented by ϕ. Thus $\phi = \frac{1}{\eta}$ Unit of fluidity is poise^{-1}.

3.4.1 Viscosity change of a liquid with change in temperature

With increase of temperature, the kinetic energy of the molecules of the liquid increases. Hence the liquid starts flowing faster. In other words, *the viscosity of a liquid decreases with increase in temperature.* It has been found that the decrease is about 2% per degree rise of temperature.

Arrhenius gave the following relationship between viscosity and temperature:

$$\eta = Ae^{B/RT}$$

where A and B are the constants for the liquid and T is the absolute temperature. The above equation may be written as

$$\log \eta = \log A + \frac{B}{RT}$$

or

$$\log \eta = A' + \frac{B'}{T}$$

Thus the plot of log η vs $\frac{1}{T}$ is a straight line.

3.4.2 Poiseulle's equation and determination of viscosity of a liquid

Poiseulle's equation gives the relation between viscosity of a liquid flowing through a capillary tube and volume of the liquid, pressure of the liquid radius and length of the tube and the time taken for the flow

$$n = \frac{\pi P r^4 t}{8 l v} \qquad \text{(Poiseulle's equation)}$$

where v = volume of the liquid flowing thrcugh the capillary in time t

P = Pressure-head

r = Radius of the capillary tube

l = Length of the tube through which the liquid flows.

It is quite cumbersome to determine experimentally the values of P, r, l and v. Hence, it may not be practical to use Poiseulle's equation for the determination of viscosity of a liquid.

Viscosity is generally determined w.r.t. water. This value is called *relative viscosity.* We note down the time for flow of water from the point M to the point M' on the viscometer (see Fig. 3.15). We repeat the experiment with the unknown liquid. Let t_l and t_{H_2O} be the times for the flow of liquid and water in the same viscometer. As the viscometer is the same, the value of r, l and v will remain the same. We can write the Poisecuille's equation for the liquid and water as under:

$$n_l = \frac{\pi P_l r^4 t_l}{8 l v} \quad ...(i)$$

$$n_{H_2O} = \frac{\pi P_{H_2O} r^4 t_{H_2O}}{8 l v} \quad ...(ii)$$

Divide eq. (*i*) by eq. (*ii*)

$$\frac{n_l}{n_{H_2O}} = \frac{P_l t_l}{P_{H_2O} t_{H_2O}} \quad ...(iii)$$

Pressure is proportional to the density. Thus $P_l \propto d_l$ and $P_{H_2O} \propto d_{H_2O}$.

Therefore, equation (*iii*) is reduced to

$$\frac{n_l}{n_{H_2O}} = \frac{d_l t_l}{d_{H_2O} t_{H_2O}}$$

t_l and t_{H_2O} are determined from the experiment.

Substituting the values of d_l, d_{H_2O} and n_{H_2O}, we can find out the n_l, the viscosity of the liquid.

3.4.3 Ostwald's viscometer method for the determination of viscosity of a liquid

The apparatus used for the measurement of relative viscosities is called a *viscometer.* The simplest is the one given by Ostwald and is called Ostwald's viscometer. It is shown in Fig.3.15. It consists of a bulb B, to which is sealed a tube above and a capillary below. There is a mark M on the tube and mark M' on the capillary as shown. The lower end of the capillary is sealed to a bigger bulb C and a wider tube CD.

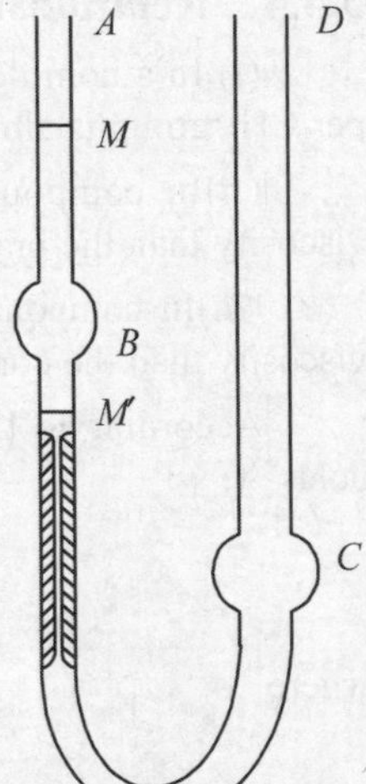

Fig. 3.15. Ostwald's viscometer.

The procedure for viscosity determination consists of the following steps :

(*i*) The apparatus is first washed with chromic acid and then thoroughly with water. It is then clamped vertically and a rubber tube is attached to the end A.

(*ii*) A definite volume of water is introduced into the bulb C through the end D.

(*iii*) Through the rubber tube attached to the end A, distilled water is sucked up into the bulb B so that the level of water is a little above the mark M. The water is then allowed to flow back and the time taken for the water to flow from the mark M to M' is noted.

(*iv*) The apparatus is then dried and the experiment is repeated as above taking the same volume of the experimental liquid. The result is then calculated using the formula as given below:

$$\frac{\eta_l}{\eta_{H_2O}} = \frac{d_l t_l}{d_{H_2O} t_{H_2O}}$$

where η_l = coefficient of viscosity of liquid

η_{H_2O} = Coefficient of viscosity of water

d_l = Density of liquid

d_{H_2O} = Density of water

t_l = Time of flow with the liquid

t_{H_2O} = Time of flow with water

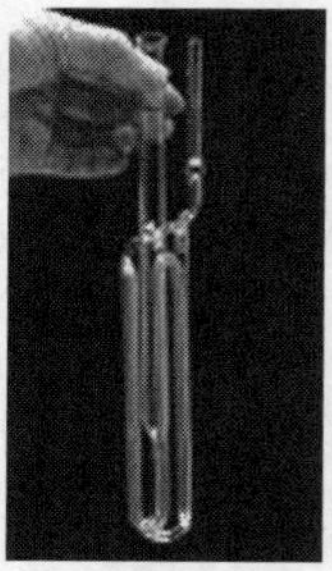

Extended Ostwald viscometer

knowing the values of all other terms, η_l can be calculated.

Example 7. **Why do we use the same viscometer for the liquid and water during the experimental determination of viscosity by ostwald viscometer? Describe the experiment.**

Solution. For determination of viscosity of the liquid by ostwald viscometer, we use the equation

$$\frac{n_l}{n_{H_2O}} = \frac{d_l\, t_l}{d_{H_2O}\, t_{H_2O}}$$

This equation is derived using Poisecuille's equation

$$n = \frac{\pi P r^4 t}{8 l v}$$

We have assumed that the values of *r*, *l* and *v* remain the same. This can happen when the oswald viscometer remains the same for measurement with the liquid and water.

For description of the experiment, refer to section 3.4.3

3.4.4 Relationship between viscosity and chemical constitution

(*i*) In a homologous series, the viscosity increases as we go up a series and the increase per CH_2 group is almost constant.

(*ii*)In compounds involving chain isomerism, the *n*-isomers have generally greater viscosity than the branched chain isomers.

(*iii*)In compounds involving geometrical isomerism, *trans*-isomers have higher value of viscosity than the corresponding *cis*-isomers.

According to Dunstan's rule, for non-associated or normal liquids, the following equation holds

$$\frac{d}{M}.\eta\times10^6 \approx 40 \text{ to } 80$$

where d = density of the liquid

η = coefficient of viscosity

M = Molecular mass

For associated liquids, the value is much greater than 80.

3.4.5 Rheochor

The following relation exists between rheochor [R], mol wt, density and coefficient of viscosity of the liquid.

$$[R] = \frac{M}{D} = (\eta)^{1/8}$$

where *M, D* and η are the molecular mass, density and coefficient of viscosity of the liquid respectively.

Thus if $\eta = 1,$

then $[R] = \frac{M}{D} = V$, molar volume

Hence, *rheochor may be defined as the molar volume of the liquid at a temperature at which its coefficient of viscosity is unity.*

3.4.6 Molecular weight of polymers using coefficient of viscosity

If the viscosity of a solution is η at concentration *c*, and the viscosity of pure solvent is η_O, then the specific viscosity of a solution is defined by the relationship

$$\frac{\eta - \eta_0}{\eta_0} = \eta_{sp}$$

The plot of $\frac{\eta_{sp}}{c}$ vs *c* is found to be linear. The value of $\frac{\eta_{sp}}{c}$ obtained by extrapolating the plot to zero concentration gives what is called **intrinsic viscosity**, [η]. It is found to be related to the average molecular weight of the polymer, $\overline{M}$,by the equation

$$[\eta] = k\,\overline{M^a}$$

where *k* and *a* are constants for a given type of polymer and a solvent. The values of these constants are found by first performing the experiments with two samples of the given polymer whose molecular masses have already been found by some other method.

Example 8. **At 20°C, the density of water is 0.9983 g/ml and its viscosity is 0.010087 poise. Explain how these figures indicate that water is an associated liquid.**

Solution. Here we are given that

$$d = 0.9983 \text{ g/ml}$$
$$\eta = 0.010087 \text{ poise}$$

For H_2O

$$M = 18.$$

According to Dunstan's rule, $\frac{d}{M}\eta \times 10^6$ for normal liquids is from 40 to 60.

In this case $\frac{d}{M}\eta \times 10^6 = \frac{0.9983}{18} \times 0.010087 \times 10^6 = 559.4$

which is much greater than 80. Hence water is an associated liquid.

Examle 9. **The water flow time for an Ostwald viscometer is 59.2 sec at 25°C. If 46.2 seconds are required for the same volume of ethyl benzene (density = 0.867 g cm^{-3}) to flow through the capillary, calculate its absolute viscosity at 25°C, that of water being 0.00895 poise at the same temperature.**

Solution. *For water :* $t_1 = 59.2$ sec

$$\eta_1 = 0.00895 \text{ poise}$$
$$d_1 = 1.0 \text{ g/cc}$$

For ethyl benzene : $t_2 = 46.2$ sec

$$d_2 = 0.867 \text{ g /cc}$$

$$\eta_2 = ?$$

Substituting the values in the formula

$$\frac{\eta_1}{\eta_2} = \frac{d_1 t_1}{d_2 t_2}$$

$$\eta_2 = \frac{\eta_1 \times d_2 \times t_2}{d_1 t_1}$$

$$= \frac{0.00895 \times 0.867 \times 46.2}{1 \times 59.2} = \mathbf{0.00606 \text{ poise}}$$

3.5 REFRACTIVE INDEX

When a ray of light travels from one medium to another, it undergoes a change in direction, as shown in Fig. 3.16. This phenomenon is called *refraction.* If a ray of light travels from air or vacuum to a more dense medium (*i.e.*, any liquid or solid), then the ratio of the sine of the angle of incidence, *i,* to the sine of the angle of refraction, *r,* is constant depending upon the nature of the medium. This constant is called the **refractive index** of the medium. It is usually represented by μ. Thus

$$\mu = \frac{\sin i}{\sin r}$$

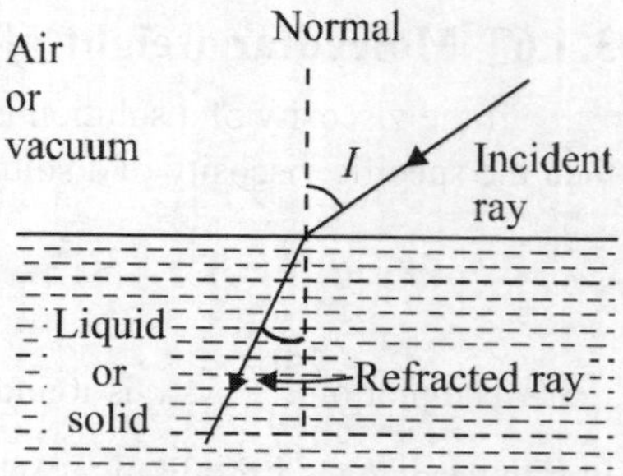

Fig. 3.16. Refraction of light from air to a more dense medium.

This is called **Snell's law.**

The value of refractive index is also found to depend upon the wavelength of the light used and the temperature. These are usually specified as subscripts and superscripts respectively. Thus $\mu_D^{20^0}$ indicates a refractive index taken at 20°C with light corresponding to yellow *D*-line of the sodium spectrum obtained from sodium vapour lamp.

Alternatively, *the refractive index of a medium is also defined as the ratio of the velocity of light in vacuum to the velocity of light in the medium, i.e.,*

$$\mu = \frac{\text{Velocity of light in vacuum}}{\text{Velocity of light in the medium}}$$

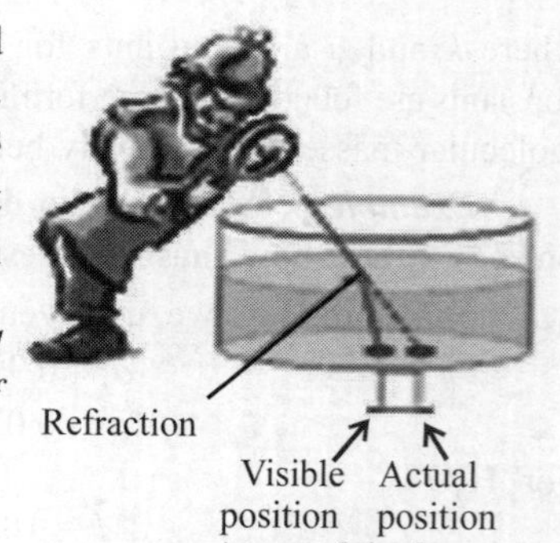

A gap of the image (by refraction of light) Refractive index effect

3.5.1 Refractometer and its types

The instrument used for the measurement of refractive index is called a *refractometer.* The two most common refractometers are :

(*i*) Pulfrich refractometer (*ii*) Abbe's refractometer.

The principle of the Abbe's refractometer is explained in Fig. 3.17. The actual shape of the instrument is shown in Fig. 3.18. It consists of two Thin layer of right angled prisms *P* and *Q* with their hypotenuse liquid faces together. These can be clamped together with a clamp *X* in a metal box hinged at *Y*, as shown in Fig. 3.18. The box is attached to an arm *R* with the help of which the prism system can be rotated and the value of refractive index can be read directly on the circular scale fixed at the top of the arm *R*. The hypotenuse face of the lower prism (*P*) is ground whereas that of the upper prism (*Q*) is polished. A drop of the liquid is placed on the hypotenuse face of the prism *P* which spreads between the two faces when the two prisms are fixed together. The light from a suitable source is then reflected by the mirror M on to the liquid layer. On striking the ground surface of the prism P, the light gets diffused in all directions. Thus the rays are incident to the polished face of the prism Q at angles less than 90° (in the liquid medium) and at angles greater than 90° (in the prism Q). The rays incident at angles greater than 90° undergo total internal reflection within the prism Q and do not come out of the prism Q. The rays incident at angles less than 90° undergo refraction and come out of the prism Q in the form of a bright band of light. To observe the edge of the bright band corresponding to the 'grazing incidence' a telescope is fixed above the prism system and the box containing the prism system is rotated till the edge of the bright band coincides with the cross-wire of the telescope. This setting corresponds to the refracted ray, refracted at the critical angle, e. However, instead of measuring the angles, the circular scale is graduated directly for reading the refractive indices.

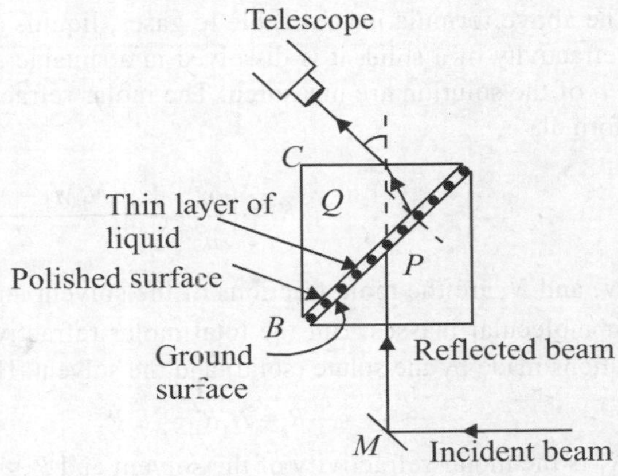

Fig. 3.17. Principle of Abbe's Refractometer.

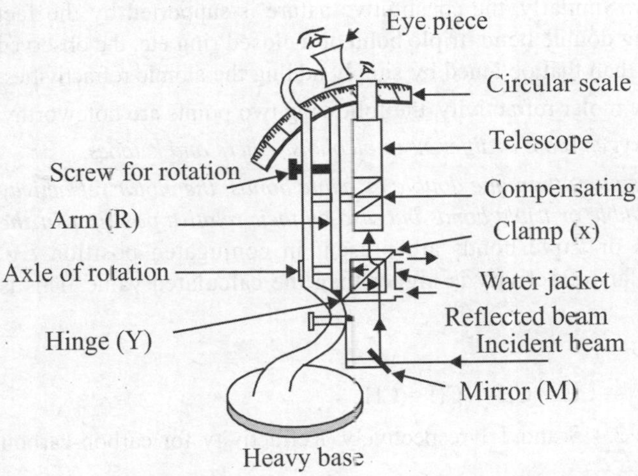

Fig. 3.18. Design of the instrument (Abbe's refractrometer)

Ernst Abbe

3.5.2 Specific refractivity and Molar refractivity

(i) Specific Refractivity: It has been found that for any substance having refractive index n and density d at the same temperature, the following equation holds good for the particular substance and is independent of temperature.

$$r = \left(\frac{n^2 - 1}{n^2 + 2}\right)\frac{1}{d}$$

The quantity r is called the specific refractivity or specific refraction of the substance.

(ii) Molar Refractivity: Multiplying the specific refractivity by the molecular mass M of the substance, we get what is known as molar refractivity or molar refraction R. Thus

$$R_m = r \times M = \left(\frac{n^2 - 1}{n^2 + 2}\right)\frac{M}{d}$$

The above formula is applicable to gases, liquids as well as to solids. To determine the molar refractivity of a solid, it is dissolved in a suitable solvent and the refractive index n and density d of the solution are measured. The molar refractivity R_s of the solution is then given by the formula

$$R_s = \left(\frac{n^2 - 1}{n^2 + 2}\right)\left(\frac{N_1M_1 + N_2M_2}{d}\right)$$

where N_1 and N_2 are the mole fractions of the solvent and solute respectively and M_1 and M_2 are their molecular masses. But the total molar refractivity of the solution is the sum of the contributions made by the solute (solid) and the solvent. Hence

$$R_s = N_1R_1 + N_2R_2$$

where R_1 is the molar refractivity of the solvent and R_2 that of the solid. Thus knowing R_1, R_2 can be calculated.

3.5.3 Molar refractivity in the determination of structure of compounds

The molar refractivity has been found to be additive as well as constitutive. The additive nature is supported again by the facts that (*i*) the difference in the molar refractivity of the successive members of a homologous series is nearly constant and (*ii*) the isomeric compounds with similar structures (*e.g.* iso-propyl alcohol and *n*-propyl alcohol) are found to have nearly same value of the molar refractivity. Similarly, the constitutive nature is supported by the fact that in case of compounds containing double bond, triple bond or a closed ring etc, the observed value of molar refractivity is higher than that obtained by simply adding the atomic refractivities.

However, while calculating the molar refractivity, the following two points are noteworthy.

(*i*) *The atomic refractivity of oxygen atom is different in alcohols, ethers and ketones.*

(*ii*) *When a compound contains more than one double or triple bonds, the molar refractivity depends not only on the number of double or triple bonds but also on their relative position in the molecule.* When the double bonds or triple bonds are present in conjugated position *e.g.* —CH = CH—CH = CH—, the molar refractivity is higher than the calculated value. This is known as **optical exaltation.**

Thus for hexatriene,

$$CH_2 = CH — CH = CH— CH = CH_2$$

Refractivities for C and H are 2.418 and 1.1 respectively. Refractivity for carbon-carbon double bond is 1.733.

Calculated value of $[R] = 28.52$

Experimental value of $[R] = 30.58$

Optical exaltation $= 30.58 - 28{\cdot}52 = 2.06$

***Examle 10*. Density of acetic acid at 25°C is 1.046 g/ml. Calculate the refractive index of acetic acid. Values of R_m (Molecular refractivity) for C = 2.42, H = 1.10, O in OH = 1.50, O in CO = 2.21.**

Solution. 1*st step* : To calculate the value of R_m for CH_3COOH

Here we are given

$$R_C = 2.42,\ R_H = 1.10,$$

$$R_O \text{ in} - OH = 1.50,\ R_O \text{ in} > CO = 2.21$$

$$R_m \text{ for } CH_3-\overset{\overset{\displaystyle O}{\|}}{C}-OH = 2\,R_c + 4R_H + R_O \text{ in—OH} + R_O \text{ in} > CO$$

$$= 2 \times 2.42 + 4 \times 1.10 + 1.50 + 2.21$$

$$= 12.95 \text{ cm}^3 \text{ mol}^{-1}.$$

2nd step : To calculate the value of *n*

Also we are given, $d = 1.046$ g/m1

Mol. mass of CH_3COOH (M) = 60

Substituting these values in the formula

$$R_m = \left(\frac{n^2-1}{n^2+2}\right)\frac{M}{d}$$

we get
$$12.95 = \left(\frac{n^2-1}{n^2+2}\right)\frac{60}{1.046}$$

or
$$\frac{n^2-1}{n^2+2} = 0.2258$$

$$n^2 - 1 = 0.2258\,n^2 + 0.4516$$

or
$$0.7742\,n^2 = 1.4516$$

$$n^2 = \frac{1.4516}{0.7742} = 1.8753$$

$$\mathbf{n = 1.369}$$

***Examle 11*. Calculate the molar refraction of acetic acid (CH_3COOH) at temperature at which its density is 1.046 g/cm^3. The experimentally observed value of refractive index at this temperature is 1.3715.**

Solution. Here we are given

$$d = 1.046 \text{ g /cm}^3$$

$$n = 1.3715$$

Molecular mass of CH_3COOH is $M = 60$

Substituting these value in the formula

$$R_m = \left(\frac{n^2-1}{n^2+2}\right)\frac{M}{d}$$

$$R_m = \frac{(1.3715)^2-1}{(1.3715)^2+2}\cdot\frac{60}{1.046}$$

$$= \mathbf{13.021 \text{ cm}^{-3} \text{ mol}^{-1}}$$

Example 12. **A substance having the formula C_3H_6O might be either acetone $\left[\begin{matrix}CH_3\\CH_3\end{matrix}\!\!>C=O\right]$ or allyl alchol ($CH_2 = CH—CH_2—OH$). Determine which substance it actually is if its molar refractivity is found to be 16.974 by experiment. Given that refractivities for C =2.418,H = 1.100, O (carbonyl) = 2.211, O(hydroxyl) = 1.525, C = C double bond = 1.733.**

Solution. Theoretical molar refractivities of the two compounds are calculated below :

Acetone		*Ally alcohol*	
		3 Carbons	= 7.254
3 Carbons	= 3 × 2.418 = 7.2	6 Hydrogens	= 6.600
6 Hydrogens	= 6 × 1.100 = 6.600	1 Double bond(C = C)	= 1.733
1 Carbonyl oxygen	= 2.211	1 Hydroxyl oxygen	= 1.525
Total	16.065	Total	= 17.112

As the experimental value is close to that calculated value for allyl alcohol, the compound must be allyl alcohol.

3.6 LIQUID CRYSTALS

A large number of organic compounds with long chain molecules such as :

Cholesteryl acetate ($CH_3COOC_{27}H_{45}$), Cholesteryl benzoate ($C_6H_5COOC_{27}H_{45}$), and ammonium oleate ($C_{17}H_{33}COONH_4$) on heating melt sharply to give a milky (turbid) liquid which on further heating becomes clear again at a definite temperature.

The liquid crystals have a structure between those of liquid and solid states.

Molecules have random arrangement and are able to move post each other in a liquid and they have an ordered arrangement and are in fixed positions in the solid state. Thus, the liquid crystals have the fluidity of a liquid and optical properties of solid crystals.

On cooling the clear liquid (melt), the reverse changes take place exactly at the same temperatures. The turbid liquids thus obtained are found to be *anisotropic,* a property possessed by crystalline substances. Hence these turbid liquids are called **liquid crystals.**

Liquid state → Liquid crystal → Solid state

However, it is found that the molecules of a liquid crystal do not have the arrangement like that of a crystal lattice. Moreover, they behave like true liquids, *e.g.*, they possess surface tension and have the property to flow. Hence in place of 'liquid crystals' certain other names have been proposed such as *anisotropic liquids or para crystals.* The best name for the substances in this state is **mesomorphic state.** However, in spite of all this the name 'liquid crystals' is still frequently used.

Uses of Liquid Crystals

(*i*) Liquid crystals are used for digital display of numbers and letters in watches, calculators and other electronic devices.

(*ii*) Thin films of cholesteric liquid crystals are used to determine temperature and in the rings to study a person's mood.

3.6.1 Different kinds of liquid crystals

Liquid crystals are of the following three types:

(*i*) **Smectic Liquid Crystals:** The smectic liquid crystals have limited mobility. They flow in layers as if different sheets are gliding over one another. The distribution of velocity in different layers is different from that found in a true liquid. The concept of viscosity is, therefore, not applicable to the liquid crystals of this type. When spread over a clean glass surfaces, they form a series of *terraces or strata,* as shown in Fig. 3.19. The edges of these terraces often show a series of fine lines even in ordinary light, though more so in polarised light.

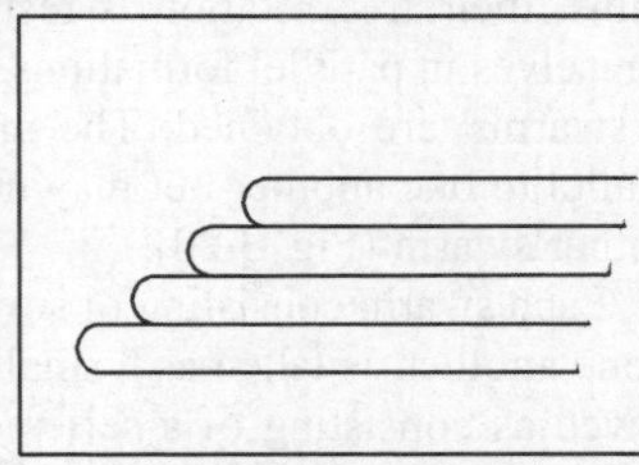

Fig. 3.19. Layer structure of smectic liquid crystals.

Smectic liquid crystals also give X-ray diffraction patterns like solid crystals but it is in one direction only. Thus, smectic phase has a complex internal structure.

When viewed in polarized light, smectic phases appear to have fan-like appearance. They are always uniaxial and are not affected by a magnetic field.

Ethyl-*p*-azoxy benzoate, ethyl-*p*-azoxy cinnamate are some examples of this class of liquid crystals.

(*ii*) **Nematic Liquid Crystals:** Nematic type of liquid crystals show near normal flow behaviour of liquids. They flow more readily than the smectic liquid crystals. Their flow is also Newtonian and the concept of viscosity is applicable to their flow. Their viscosity, however, is rather low as compared to that of liquids. Thus, nematic phases have liquid-like character and yet being *turbid and anisotropic,* they are unlike true liquids. *This type is, therefore, regarded less truly anisotropic than the smectic type.*

In polarised light, substances in nematic phase appear to have *thread-like structures*. They are uniaxial like smectic substances, but unlike the latter, they are affected by a strong magnetic field. Thus, when viewed in the direction of the lines of force of a magnetic field, the turbid liquid appears quite clear. If the magnetic field is cut off, the liquid appears turbid again. *p*-Azoxy anisole and *p*-Azoxy phenetole are some examples of this type.

p-Azoxyanisole :

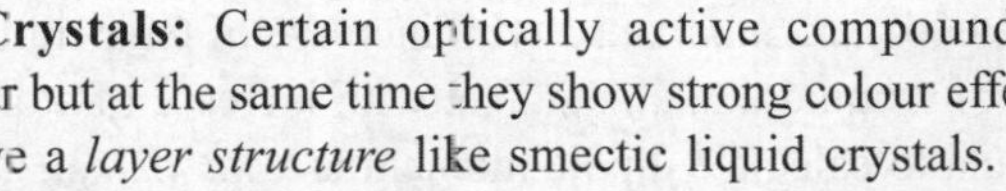

p-Azoxyphenetole :

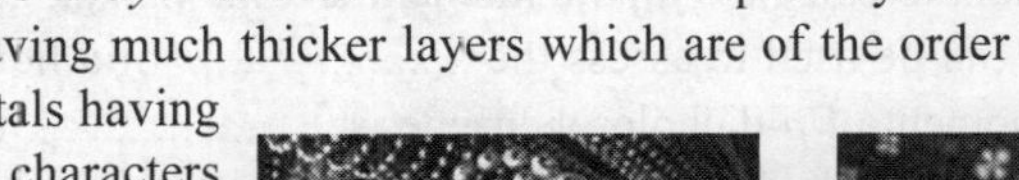

(*iii*) **Cholesteric Liquid Crystals:** Certain optically active compounds show many characteristics of nematic behaviour but at the same time they show strong colour effects in polarised light. This indicates that they have a *layer structure* like smectic liquid crystals. But they differ from smectic liquid crystals in having much thicker layers which are of the order of 500 to 5000 molecules thick. Such liquid crystals having some nematic and some smectic characters are called **cholesteric liquid crystals.** The first substance in which this type of mesomorphism was detected was cholesteryl benzoate. Its transition point is 145° and melting point 178°C. *l*-cholesteryl formate is another example of this type of liquid crystals.

Liquid crystals

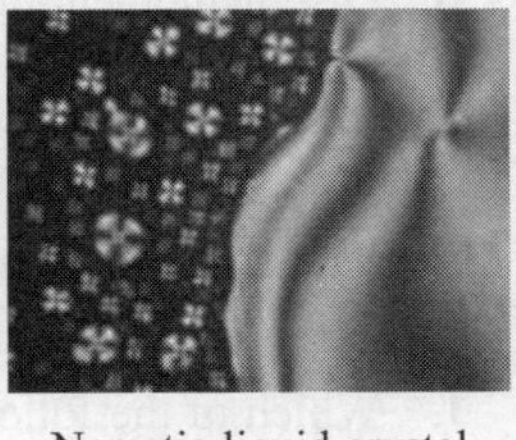

Nematic liquid crystal

3.6.2 Swarm theory of liquid crystals

The best explanation for turbidity is furnished on the basis of the 'Swarm Theory' proposed by Bose. According to this theory, since mesomorphic behaviour is shown by substances having long chains, their free rotation is restricted and they tend to arrange themselves in parallel formations. In this way, a number of groups or 'swarms' are obtained. The molecules in any one swarm are parallel to one another but may not be parallel to those present in another swarm (Fig. 3.20).

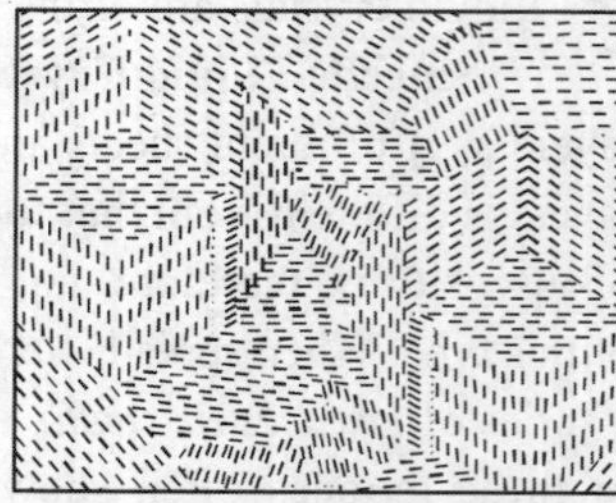

Fig. 3.20. Arrangement of molecules in anisotropic liquids.

Each swarm consisting of several molecules, arranged parallel to one another, is taken as a small crystal. Thus, a turbid liquid is viewed as consisting of a series of small `crystals' distributed at random in the body of the liquid. Lehmann actually found the each particular swarm appears as a *transparent liquid crystal* when viewed under a microscope. The turbidity of the liquid crystals is simply due to the *scattering of light at the boundaries between the various swarms*. The effect is similar to that observed when a transparent substance like glass is finely powdered. Each piece is transparent but when huddled together the mass appears somewhat opaque.

As the temperature of the liquid crystals is raised, the molecular motion increases and an interchange between the swarms occurs. The swarms become smaller and smaller with rise in temperature and ultimately at a certain temperature, the swarms become smaller than half the wavelength of light. They are then too small to scatter light. The turbidity now disappears. The liquid becomes clear. The double refraction and other anisotropic properties due to orientation of molecules within the swarms disappear. The liquid becomes singly refracting and isotropic. This temperature is the melting point of the substance. The arrangement of molecules in an isotropic liquid is depicted in Fig. 3.21.

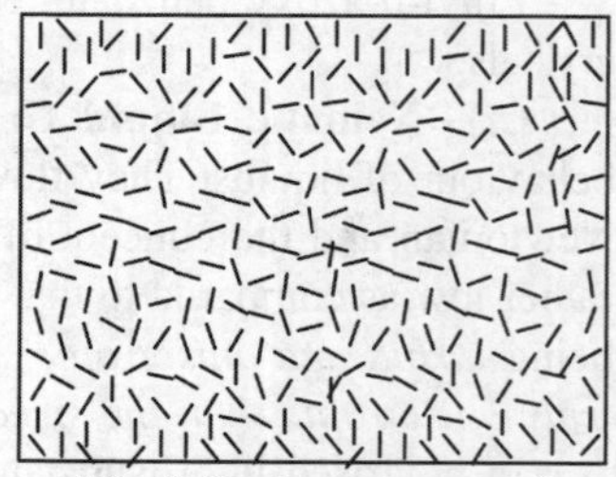

Fig. 3.21. Arrangement of molecules in isotropic liquids.

EXERCISES
(INCLUDING QUESTIONS FROM DIFFERENT UNIVERSITY PAPERS)

Objective Questions

Fill in the blanks

1. The equation to calculate parachor is ....
2. The unit of viscosity is ....
3. In a *trans* isomer, the dipole moment is ....
4. Surface tension .... with increase of temperature.
5. Viscosity of a liquid .... with increase of atomic mass.
6. The formula to calculate dipole moment is ....
7. Parachor can be used to assess the .... of a liquid compound.
8. Dispolemoment of *p*-dicholorobenzene is .... than that of *m* – dichlorobenzene
9. The formula to calculate rheochor is ....
10. The temperature at which, the vapour of a liquid is equal to .... pressure, is called its boiling point.

Multiple Choice Questions

Choose the correct option.

1. Which of the following statements is wrong ?

(*a*) Evaporation is a spontaneous process
(*b*) Evaporation is a surface phenomenon
(*c*) Vapour pressure decreases with increase of temperature
(*d*) The vapour pressure of a solution is always less than the vapour pressure of the pure solvent

2. Water boils at lower temperature on high altitudes because:
(*a*) atmospheric pressure is low there
(*b*) atmosphere pressure is high there
(*c*) water is weakly hydrogen bonded there
(*d*) water in pure form is found there

3. Mark the statement which is correct:
(*a*) Surface tension of a liquid increases with temperature
(*b*) Addition of chemicals reduces the surface tension of a liquid
(*c*) Stalgmometer is used for measuring viscosity of the liquid
(*d*) Viscosity of a liquid does not depend on intermolecular forces

4. The viscosity of which liquid is highest?
(*a*) Water (*b*) Glycol (*c*) Acetone (*d*) Ethanol

5. If η_1 and η_2 are coefficients of viscosity of two liquids, d_1 and d_2 their densities and t_1 and t_2 are their flow times in an ostwald's viscometer, then :
(*a*) $\frac{\eta_1}{\eta_2}=\frac{d_1t_2}{d_2t_1}$ (*b*) $\frac{\eta_1}{\eta_2}=\frac{d_2t_2}{d_1t_1}$ (*c*) $\frac{\eta_1}{\eta_2}=\frac{d_1t_1}{d_2t_2}$ (*d*) $\frac{\eta_1}{\eta_2}=\frac{d_2t_1}{d_1t_2}$

6. Association of molecules in water is due to :
(*a*) covalent bonding (*b*) hycrogen bonding
(*c*) ionic bonding (*d*) vander Waals forces

7. Normal boiling point of a liquid is that temperature at which vapour pressure of the liquid is equal to :
(*a*) zero (*b*) 380 mm of Hg (*c*) 760 mm of Hg (*d*) 100 mm Hg

8. When a student was given a viscometer, the liquid was sucked with difficultly, the liquid may be :
(*a*) benzene (*b*) toluene (*c*) water (*d*) glycerine

9. With increase in temperature, the fluidity of liquids :
(*a*) increases (*b*) decreases
(*c*) remains constant (*d*) may increase or decrease

10. The rise of a liquid in a capillary tube is due to :
(*a*) viscosity (*b*) osmosis (*c*) diffusion (*d*) surface tension

11. With the increasing molecular mass of a liquid, the viscosity :
(*a*) decreases (*b*) increases (*c*) not affected (*d*) none of above

12. Which of the following expressions regarding the unit of coefficient of viscosity is not true:
(*a*) dyne cm^{-2} sec (*b*) dyne cm^2 sec^{-1}
(*c*) Nm^{-2} sec (*d*) 1 poise= 10^{-1} Nm^{-2} sec

13. Which is correct for parachor ?
(*a*) $\frac{M}{D}\cdot\gamma^{1/8}$ (*b*) $\frac{M}{D}\cdot\gamma^{1/2}$ (*c*) $\frac{M}{D}\cdot\gamma^{1/6}$ (*d*) $\frac{M}{D}\cdot\gamma^{1/4}$

14. Unit of surface tension is :
(*a*) dyne cm (*b*) dyne cm^{-1} (*c*) dyne cm^{-2} (*d*) none

15. The molecular viscosity of a liquid is given by :
(*a*) viscosity × mol. wt. (*b*) viscosity × molar volume
(*c*) viscosity × molar surface (*d*) none

16. The boiling points of water, ethanol and diethylether are 100°C, 78.5°C and 34.6°C, respectively. The intermolecular forces will be in the order:
(*a*) water > ethanol > diethyl ether (*b*) ethanol > water > diethyl ether
(*c*) diethyl ether > ethanol > water (*d*) diethyl ether > water > ethanol

17. Parachors of two liquids can be compared by :
(*a*) their molar volumes (*b*) their molecular masses
(*c*) their atomic numbers (*d*) their vapour pressures

18. For any liquid the number of drops formed per unit volume depends on
(*a*) viscosity (*b*) surface tension (*c*) pressure (*d*) none

19. The correct expression for molar refraction is :
(*a*) $\frac{n^2-1}{n^2+2}\cdot\frac{1}{d}$ (*b*) $\frac{n^2-1}{n^2+2}\cdot\frac{M}{d}$ (*c*) $\frac{M}{D}\cdot\gamma^{1/4}$ (*d*) $\frac{M}{a}\cdot\eta^{1/8}$

20. Which of the following is a colligative property ?
(*a*) molecular weight (*b*) osmotic pressure
(*c*) parachor (*d*) surface tension

21. Molar refraction is a property which is :
(*a*) additive (*b*) colligative
(*c*) constitutive (*d*) additive and constitutive

22. Parachor is a property which is :
(*a*) additive-cum-constitutive (*b*) colligative
(*c*) additive (*d*) none

ANSWERS

Fill in the blanks

1. $[P]=\frac{M}{D}\gamma^{\frac{1}{4}}$ **2.** poise **3.** zero **4.** decreases **5.** increases
6. $u = e \times d$ **7.** stucture **8.** smaller **9.** $R_h=\frac{M}{d}[\eta]^{\frac{1}{8}}$ **10.** atmospheric

Multiple choice questions

1. (*c*) **2.** (*a*) **3.** (*b*) **4.** (*b*) **5.** (*c*)
6. (*b*) **7.** (*c*) **8.** (*d*) **9.** (*a*) **10.** (*d*)
11. (*b*) **12.** (*b*) **13.** (*d*) **14.** (*b*) **15.** (*c*)
16. (*a*) **17.** (*a*) **18.** (*b*) **19.** (*b*) **20.** (*b*)
21. (*d*) **22.** (*a*)

General Questions

1. Give in brief, the concept of liquid structure.
2. Describe isoteniscope method for the measurement of vapour pressure of the liquid.
3. What is meant by vapour pressure of a liquid ? How will you account for it ?
4. Describe the drop weight method for the determination of surface tension of a given liquid.
5. How does surface tension vary with temperature ? Describe Eotvos equation.

6. Describe the gas saturation method for the determination of vapour pressure of a liquid.
7. Explain the term "Parachor". Discuss its utility in the elucidation of structure of molecules.
8. (*a*) What are additive and constitutive properties ?
(*b*) How does the molar refraction determines the chemical constitution of an organic compound ?
9. Define Surface Tension. Describe one method for determining the surface tension of a liquid.
10. (*a*) Define surface tension. What is the surface tension of a liquid at its critical temperature ?
(*b*) Calculate the molar refraction of acetic acid at temperature at which its density is 1.046 g/cc. The experimentally observed value of refractive index at this temperature is 1.3715.
(*c*) Explain the term viscosity and viscosity coefficient ? What are the units of viscosity coefficient ?
(*d*) Give the relationship between boiling point and heat of vapourisation of a liquid.
11. Define the term refraction, specific refraction, molar refraction. How is refractive index measured experimentally ?
12. (*a*) How is boiling point of a liquid related to its vapour pressure ?
(*b*) The normal boiling point of benzene is 353.3 K. Assuming Trouton's Rule, estimate the boiling point of benzene at a pressure of 100 mm.
13. (*a*) Define the term coefficient of viscosity of a liquid. Discuss Ostwald's viscometer for measuring the viscosity of a liquid.
(*b*) Explain the following terms :
(*i*) Vapour pressure of a liquid
(*ii*) Fluidity
(*iii*) Surface tension
(*iv*) Parachor
(*v*) Specific and molar refraction.
(*c*) The coefficient of viscosity of two liquids at 300 K are 1.4×10^{-3} kg m^{-1} s^{-1} and 1.6×10^{-3} kg m^{-1} s^{-1} and their densities at the same temperature are 8×10^2 and 10×10^2 kg m^{-3} respectively. If the time of flow in the Ostwald's viscometer for the first liquid is 100 second, calculate the time of flow for the second liquid.
14. (*a*) What do you understand by viscosity, coefficient of viscosity and fluidity ? How fluidity is related to viscosity ? What are the units of viscosity ?
(*b*) What is the effect of temperature on viscosity ?
(*c*) Describe one method for the determination of viscosity of a given liquid.
(*d*) At 293 K, the density of acetone is 0.291 g/ce and its viscosity is 3.29 millipoise. Explain these figures which indicate that acetone is an associated liquid.
15. Define coefficient of viscosity. How does it vary with temperature ?
16. What is "Rheochor" ? How is it related with the viscosity of liquid ?
17. What is meant by surface tension of liquid ? How is it measured experimentally ?
18. Differentiate between surface tension and surface energy ? How are they related to each other ?
19. Describe a method to determine the viscosity of a liquid. In what units the viscosity of a liquid is measured ?
20. In what way a knowledge of surface tension help in deciding chemical constitution of the liquid ?
21. What do you know by the term refractive index ? How is it determined experimentally ?
22. Name different types of liquid crystals and briefly explain each type and give their uses.
23. Justify : The term mesomorphic state is also used to describe liquid crystals.
24. What are cholestic liquid crystals ? What are their uses ?
25. Explain briefly Swarm theory of liquid crystals.
26. Differentiate clearly between smectic liquid crystals and nematic liquid crystals.

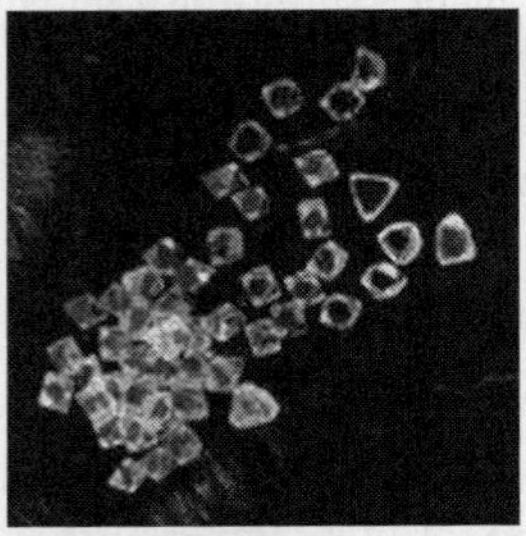

4 Solid State

4.1 TYPES OF SOLIDS

There two types of solids as described here under :

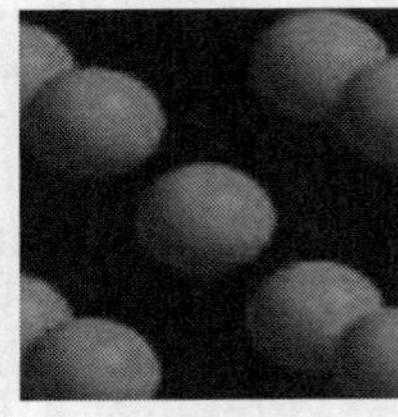

Solid state

(*i*) **Crystalline solids:** If various ions, atoms and molecules constituting the solid are arranged in definite geometric pattern within the solid, it is called a crystalline solid. All solid elements and compounds exist in this form.

(*ii*) **Amorphous solids:** A solid is known as amorphous solid if the constituent particles of the substance are not arranged in any regular fashion. Glass, pitch and polymers of high molecular mass appear in this category. Amorphous solids are not accepted as *true solids.* They are regarded as supercooled liquids of high viscosity.

4.1.1 Points of difference between crystalline and amorphous solids

Points of difference between the two solids are given as under:

(*i*) **Arrangement of particles:** The constituent atoms or ions are arranged in a definite pattern in a crystalline solids whereas there is no regular pattern of arrangement of particles in the amorphous solids.

(*ii*) **Melting Point:** Crystalline solids melt sharply at their melting points whereas amorphous solids melt gradually over a range of temperature.

(*iii*) **Isotropy and anisotropy:** Properties like refractive index and electrical conductivity are identical in all directions in the case of amorphous substances. This property also shown by gases and liquids, is called *isotropy* and substances showing this property are called *isotropic.* On the other hand, in the case of crystalline substan- ces, the properties mentioned above have different values in different directions. This type of behaviour is called *anisotropy* and the substances exhibiting this type of behaviour are called *anisotropic.* Thus whereas amorphous substances are isotropic in nature, the crystalline substances are anisotropic. The anisotropy exhibited by crystalline substances is obviously due to the reason that we come across, different particles in different directions (Fig. 4.1.).

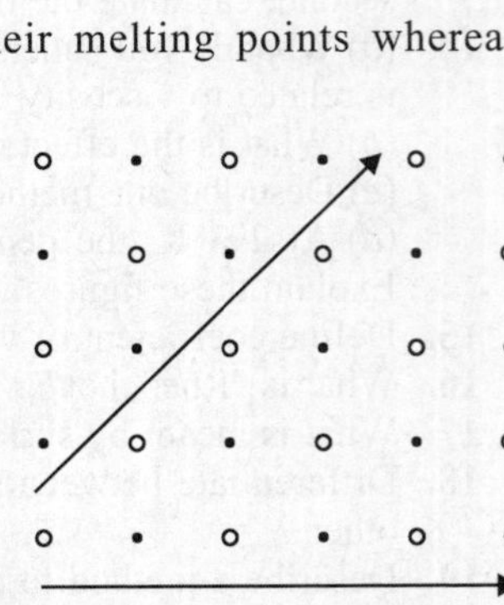

Fig. 4.1. Anisotropy exhibited by crystalline substances.

(*iv*) **Regular and irregular cut:** A crystalline solid on being cut with a sharp-edged knife gives a clean cleavage, whereas an amorphous solid breaks irregularly as shown in Fig. 4.2.

(*a*) A crystalline solid gives a clean cut. (*b*) An amorphous solid gives an irregular cut.

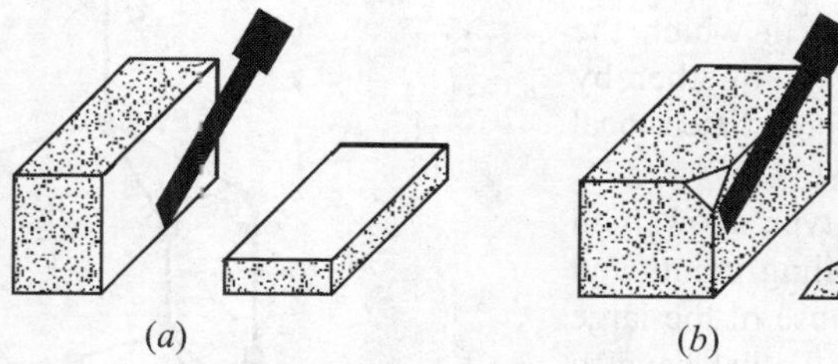

Fig. 4.2. Cutting of solids.

4.2 TYPES OF CRYSTALS

(*i*) **Ionic Crystals:** In these crystals the constituent particles occupying the lattice points are the *positive and negative ions.* These ions are held together by the *strong electrostatic forces of attraction.* Some of the important characteristics of these crystals are as follows:

(*i*) Because of strong electrostatic forces of attraction among the ions, they have high melting and boiling points.

(*ii*) They are good conductors of electricity in the molten state and in the solution.

(*iii*) They are soluble in polar solvents but insoluble in non-polar solvents.

(*iv*) Because of strong electrostatic forces of attraction, the ions are closely packed and hence the ionic crystals are hard. However, they are brittle because on applying force, their arrangement pattern is disturbed. Sodium chloride crystal is a common example of this type of solid.

(*ii*) **Molecular Crystals:** The constituent particles occupying the lattice points, in this case are the *molecules.* The intermolecular forces holding the molecules together in the crystal lattice are the *weak van der Waals forces.* In the case of polar molecules, (*e.g.* HCl, H_2O, NH_3 etc.) these van der Waals forces are the *dipole-dipole attractions* as shown in Fig. 4.3 (*a*). In case of non-polar molecules (*e.g.*, H_2, Cl_2, CH_4 etc.) the van der Waals forces are the *London dispersion forces.*

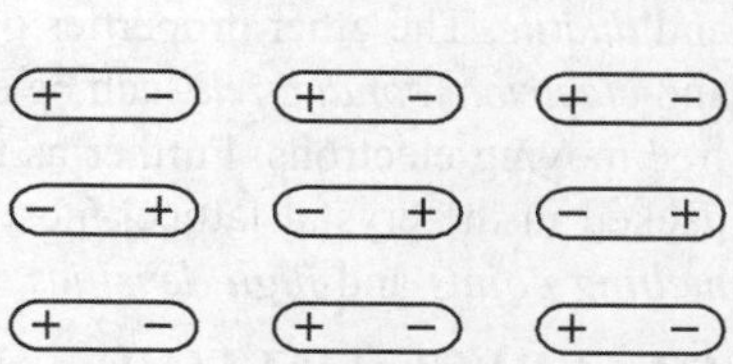

Fig. 4.3. (*a*) Dipole-dipole attraction in molecular crystals.

These forces are supposed to arise due to momentary dipole produced as a result of distortion of the electron cloud of one molecule which produces an induced dipole in the other molecule as shown in Fig. 4.3 (*b*).

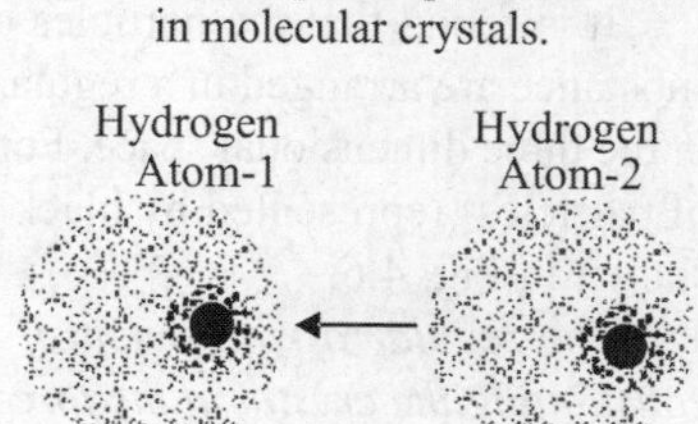

Fig. 4.3. (*b*) London dispersion forces in the molecular of hydrogen.

The main characteristics of the molecular crystals are as follows:

(*a*) As the van der Waal's forces are much weaker than the electrostatic forces existing in ionic crystals, the molecular crystals are soft and have low melting and boiling points.

(*b*) They do not conduct electricity in the solid or liquid state or in solution as there are no ions present. And electricity is conducted with the help of ions.

(*c*) They are less soluble in water and more soluble in non-polar solvents.

(*d*) As dipole-dipole attractions are stronger than London forces, therefore, polar molecular crystals generally have higher melting points and boiling points than crystals of non-polar molecules of *comparable molecular size and shape.*

(*iii*) **Network or Covalent Crystals:** In these crystals, the lattice points are occupied by atoms which are linked together by a network of covalent bonds to form a giant molecule. A common example of the crystals of this type is that of diamond in which the carbon atoms are linked together by covalent bonds to give a three dimensional structure as shown in Fig. 4.4.

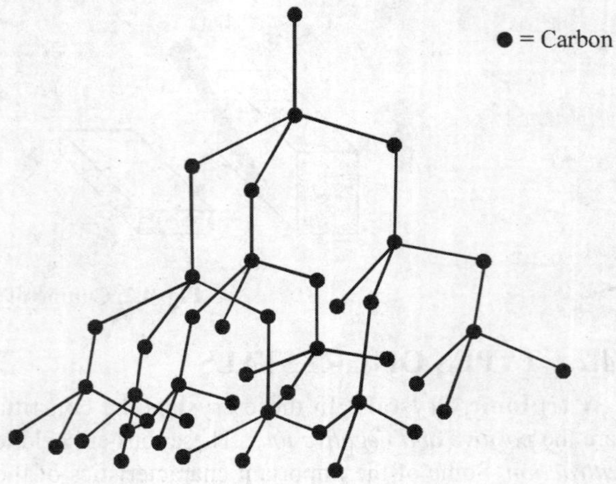

Fig. 4.4. Structure of diamond.

Substances of this type show high melting points, high boiling points and are extremely hard because of the large number of covalent bonds that have to be broken to destroy the crystal structure. Further, they are non-conductors of electricity.

(*iv*) **Metallic crystals:** These crystals consist of positively charged metals ions (kernels) occupying the lattice points which are held together by the metallic bond. The metallic bond arises due to the presence of *mobile electrons.* These mobile electrons undergo simultaneous attraction by a number of positive ions and hence the ions are held together (Fig. 4.5).

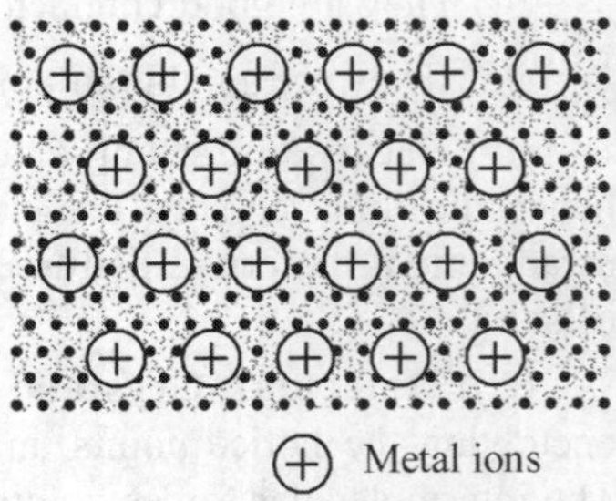

Fig. 4.5. Metallic crystal.

The positions of the positive ions can be changed without destroying the crystal because of the uniform charge distribution provided by the freely moving electrons. Thus metallic crystals can be easily deformed. That is why the metals are *malleable* and *ductile.* The other properties of metals like *lustre, thermal* and *electrical conductivity* can be explained on the basis of the free moving electrons. Further as the positive ions are closely packed in the crystal lattice, most of the metals possess *high melting points* and *high densities.*

4.3 SPACE LATTICE AND UNIT CELL OF A CRYSTAL

It is found that the particles constituting the crystalline substance are arranged in a regular fashion within the crystal in the three dimensional space. For example, the arrangement of particles (represented by black dots) for a cubic crystal is shown in Fig. 4.6.

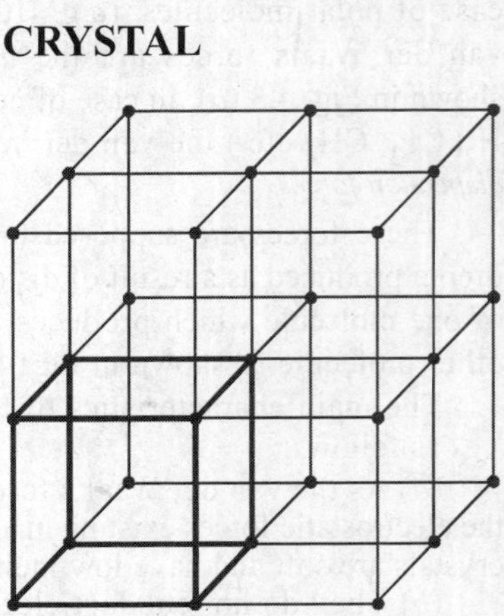

Fig. 4.6. Space lattice and unit cell.

The regular arrangement of ions, atoms, or molecules constituting the crystal in the three dimensional space within the crystal is called the **space lattice.**

If a big-sized crystal is broken more and more, ultimately a stage is reached when we get the smallest possible crystal. If it is broken further, it will break to give the constituent particles, i.e., ions, atoms or molecules. For example, if a big cubic crystal is broken, the smallest crystal obtained will be a small cube, as represented by thick lines in Fig. 4.6. This small cubic crystal has all the elements of symmetry as possessed by the big cubic crystal. Moreover it is evident that the complete lattice has been obtained by the repetition of this smallest unit in different directions.

The smallest portion of the complete space lattice, which when repeated over and again in different directions produces the complete space lattice, is called the **unit cell.**

4.3.1 Different types of unit cells (crystal systems)

Unit cells can be divided into four types.

(*i*) *Simple Unit Cell.* This type of unit cell is produced when the particles are present only at the corners of the unit cells. The simple unit cell is also called **primitive unit cell.**

(*ii*) *Face-Centred Unit Cell.* If the particles are located at the centre of each face in addition to the corners, **face-centred unit cell** is produced.

(*iii*) *End-Face Centred Unit Cell.* If the particles are located at the centre of end faces, in addition to the corners, **end face centred unit cell** is produced.

(*iv*) *Body-Centred Unit Cell.* If the particles are present at the centres of the cell in addition to the corners, **body-centred unit cell** is produced.

Different unit cells are shown in Fig. 4.7.

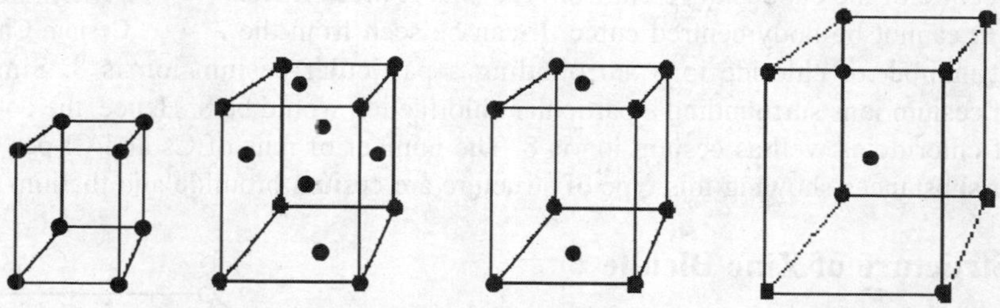

Simple | Face-centred | End Face-centred | Body-centred

Fig. 4.7. Types of unit cells.

4.3.2 Co-ordination Number

In solid state chemistry, we represent constituent particles by spheres.

The number of spheres which are in direct contact with a particular sphere is called co-ordination number of that sphere or (constituent particle).

In case of an ionic solid, the co-ordination number of a particular ion is the number of oppositely charged ion surrounding that particular ion.

4.3.3 Structure of Sodium Chloride (NaCl)

NaC1 has the structure as shown in Fig. 4.8.

(*i*) It has face-centred cubic (FCC) lattice. Na^+ are present at the corners and face-centres while Cl^- are present at body-centre and edge centres. It can also be vice-versa.

(*ii*) Every sodium ion is surrounded by six Cl^- ions, one above, one below, one on the left, one on the right, one in front and one at the back.

Similarly, every chloride is surrounded by six sodium ions. We express it by saying that sodium ion as well as chloride ion have co-ordination number equal to 6. Co-ordination number is the no. of atoms or ions immediately around a particular atom or ion. This structure has co-ordination number 6 : 6.

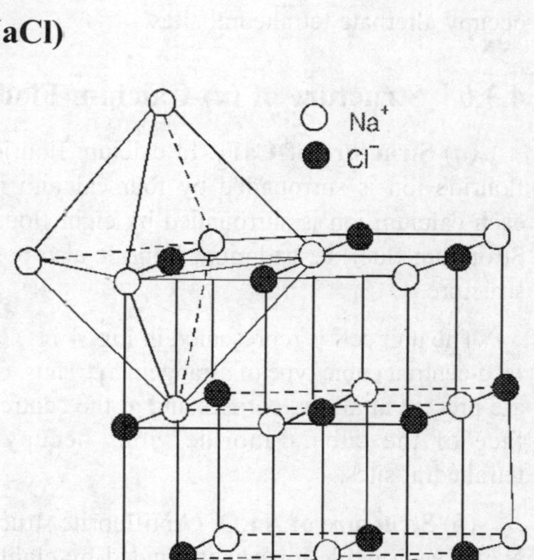

Fig. 4.8. Structure of Sodium Chloride. (Fluorite structure)

(*iii*) If we look carefully at the structure of sodium chloride, we shall find that there are 4 units of sodium and chloride ions each in a unit cell. Sodium bromide, sodium iodide, silver chloride and magnesium oxide also possess sodium chloride type structure, *i.e.*, FCC lattice.

4.3.4 Structure of cesium chloride

Cesium chloride has the structure as shown in Fig. 4.9.

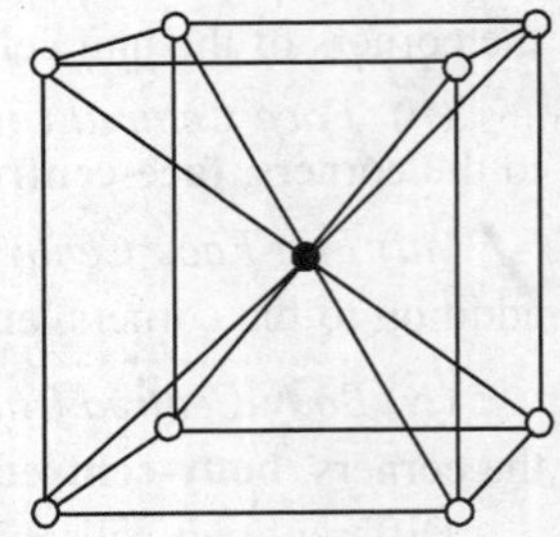

Fig. 4.9. Structure of Cesium Chloride.

This structure is simple cubic structure although it can be mistaken as body-centred cubic structure. The eight corners of the cube are occupied by Cl^-. Cesium occupies the centre of body of the cube. Similarly, we shall find that cesium occupies the corners of a cube and chloride is at the centre of that cube. We must note that the ion in the centre of the cube is different from the ions at the corners. Therefore it cannot be body centred cubic. It can be seen from the figure that number of chloride ions surrounding a particular cesium ion is 8. Similarly, the number of cesium ions surrounding a particular chloride ion would be 8. Hence, the co-ordination number of chloride as well as cesium ion is 8. The number of ions of Cs and Cl per unit cell is one. Other substances showing this type of structure are cesium bromide and thalium chloride.

4.3.5 Structure of Zinc Blende

Zinc Blende (ZnS). Zinc sulphide also exists in FCC (face-centred cubic) structure. Here the co-ordination number of Zn^{2+} as well as S^{2-} ions is 4. Thus, every Zn^{2+} is surrounded by 4 S^{2-} ions and every S^{2-} ion is surrounded by 4 Zn^{2+} ions. Copper chloride and mercuric sulphide also exist in this structure.

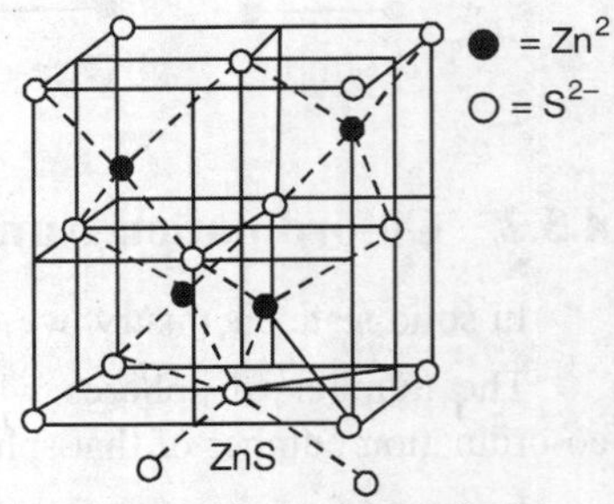

Fig. 4.10. Structure of Zinc Blende.

A unit cell representation of zinc blende type structure is shown in Fig. 4.10. The sulphide ions are present at the corners and at the centre of each face of the cube. Zn^{2+} ions occupy alternate tetrahedral sites.

4.3.6 Structure of (*a*) Calcium Flouride (CaF_2) (*b*) Na_2O

(*a*) **Structure of CaF_2.** In calcium flouride, each flouride ion is surrounded by four calcium ions and each calcium ion is surrounded by eight flouride ion. Strontium fluoride, cadmium flouride also posses this structure.

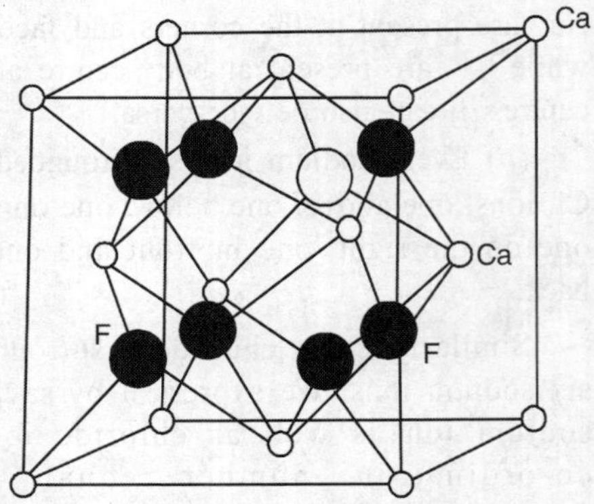

Fig. 4.11. Structure of CaF_2. (Fluorite structure)

The unit cell is represented in Fig. 4.11. Ca^{2+} form face-centred cubic type of arrangement. Thus, Ca^{2+} ions are present at all the corners and at the centre of each face of the cube. Fluoride ions occupy all the tetrahedral sites.

(*b*) **Structure of Na_2O.** (Antifluorite structure). In Na_2O, each oxide ion is surrounded by eight sodium ions and each sodium ion by four oxide ions. It has 4 : 8 co-ordination.

Table 4.1. Characteristics of different type of structures.

Structure type	*Characteristic with examples*	*Co-ordination number*
1. NaCl	It is FCC structure. The two ions inter-penetrate each other in sodium halides : AgCl, MgO.	$Na^+ = 6$ $Cl^- = 6$
2. CsCl	Cs^+ at the body centre of a cube of Cl^- ions and vice-versa : CsCl, CsBr.	$Cs^+ = 8$ $Cl^- = 8$
3. ZnS	FCC arrangement of both ions ZnS, CuCl, HgS.	$Zn^{2+} = 4$ $S^{2-} = 4$

4.3.7 Effect of temperature and pressure on crystal lattice

It is noticed that substances undergo changes in their structure when temperature and pressure are changed. As a general rule, high pressure favours greater co-ordination number and high temperature favours lower co-ordination number. For example, when sodium chloride is subjected to high pressure, it transforms into body-centred cubic structure of cesium chloride. Thus, the co-ordination number is increased from 6 : 6 to 8 : 8. On the other hand, if we heat CsCl to a high temperature of 760 K, it converts to a structure of sodium chloride.

$$\underset{(6:6 \text{ co-ordination})}{\textbf{NaCl structure}} \underset{\text{Heat}}{\overset{\text{Pressure}}{\rightleftharpoons}} \underset{(8:8 \text{co-ordination})}{\textbf{CsCl structure}}$$

4.4 PACKING OF PARTICLES IN A CRYSTAL

Considering the particles as spheres of equal size, we can visualize the formation of a solid as shown in Fig. 4.12.

One-Dimensional Packing. We can arrange the spheres in row, see Fig. 4.12 (A), such that they touch one another. This is to utilize the space economically and to account for the high density of the solids.

Two-Dimensional Packing. Then we can place another row of spheres by the side of the first row. This can be done in two ways. One way is depicted in Fig. 4.12 (B). Second method is shown in Fig. 4.12 (C). The first type [Fig. 4.12 (B)] is called **square close packing** and second type [Fig. 4.12 (C)] is called **Hexagonal close packing.**

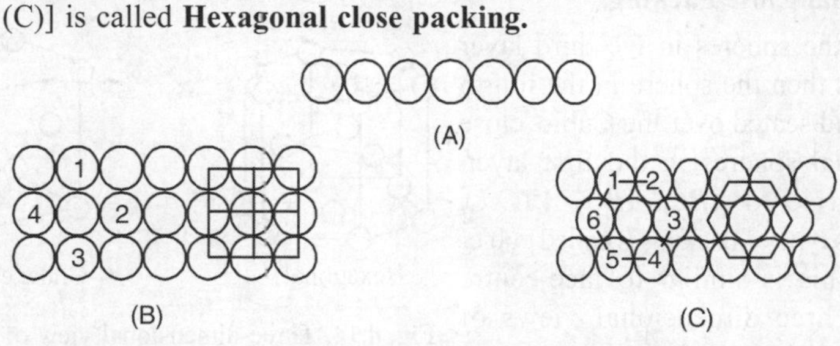

Fig. 4.12. Packing of spheres for efficient use of space. (A) Edge formation (B) Square close packing (C) Hexagonal close packing.

In square close packing, a sphere is surrounded by four spheres in the plane whereas in hexagonal close packing, a sphere is surrounded by six spheres in the plane. Thus, the empty space in square close packing is more as compared to the hexagonal close packing. Or in other words, hexagonal close packing is more dense than the square close packing.

Three-Dimensional Packing. Now let us form a second layer of spheres on the first layer in hexagonal close packing. (See Fig. 4.13)

Spheres in the first layer are named *A*.There are voids in between them. These voids are *C* voids and voids are indicated in Fig. 4.13 (A). We could place the spheres of the second layer on

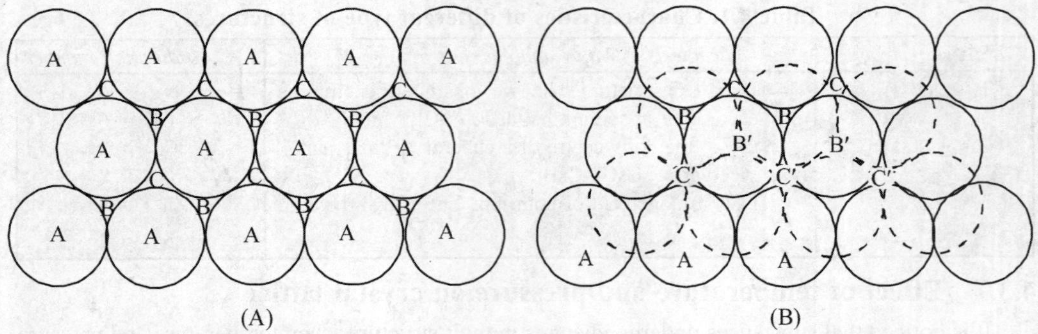

Fig. 4.13. (A) Hexagonal closed packed layer (B) Formation of the second layer.

the first layer in such a way that every sphere in the second layer lies vertically over the sphere in the first layer. But this would not to be an efficient packing as this would leave lots of gaps between them.

So let us place the spheres in the second layer over the voids *B* in the first layer as shown in Fig. 4.13 (B). It will then not be possible to place spheres over the voids *C* because of space limitation. Now in the second layer of sphere there are *B*′ voids and *C*′ voids. **In fact, *C*′ voids of the second layer lie over the C voids of the first layer.**

Let us now place a third layer of spheres over the second layer. There are two possibilities. Either we place spheres on the *B*′ voids or on *C*′ voids. If we place the spheres on the *B*′ voids in third layer, these spheres in fact would be seated over the spheres of the first layer. This arrangement would be *ABABAB* ... arrangement. This type of packing is called **Hexagonal Close Packing.**

If we place the spheres in the third layer over the *C*′ voids then the sphere in the fourth layer will be found seated over the Cubic close packing hexagonal spheres in the first layer. This will give rise to *ABC, ABC, ABC* ... arrangement. This type of packing is called cubic-close packing. This is similar to face-centre cubic system. Three dimensional views of hexagonal and cubic close packed lattices are shown in Fig. 4.14.

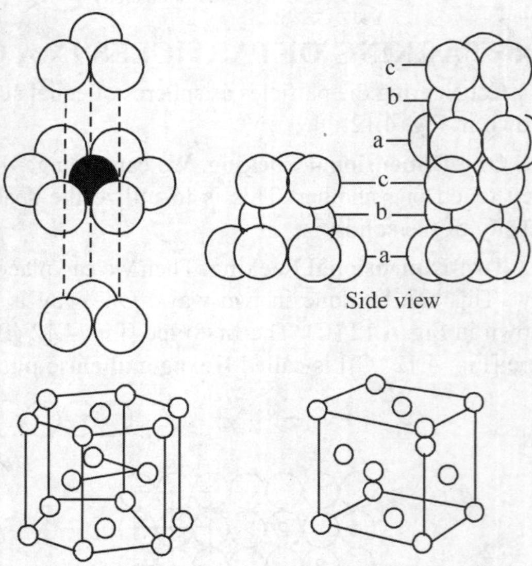

Fig. 4.14. Three-dimensional view of hexagonal and cubic close packed lattices.

4.4.1 Body-centred cubic (BCC) arrangement

This is deviation from the close packing. If the spheres of the first layer are loosened up, we obtain body-centred cubic structure. Each sphere in the second layer is in contact with four spheres of the first layer (see Fig. 4.15).

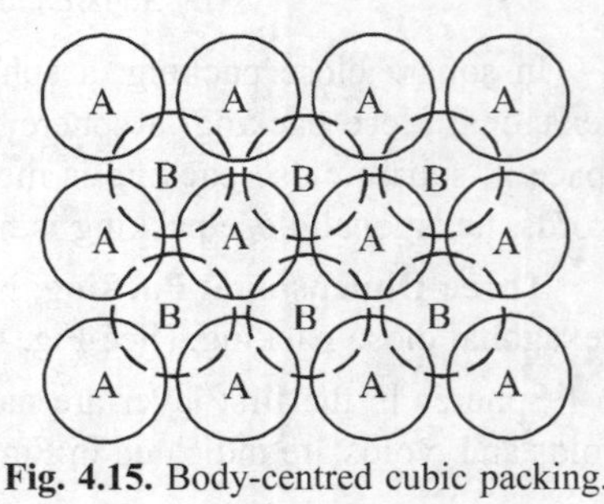

Fig. 4.15. Body-centred cubic packing.

The spheres in the second layer have the same packing as those in the first layer. The spheres in the third layer, having the same packing will be laying exactly over the corresponding spheres in the first layer. Thus, every sphere will be in contact with 8 spheres, four of the layer above and four of the layer below it. The co-ordination number is thus 8.

Alkali metals like Li, Na, K, etc. crystallise in BCC structure. The structure is less closely packed than the CCP and HCP structures and the empty space is about 32%.

4.4.2 Tetrahedral and octahedral voids

A void or empty space in a lattice which is surrounded by four spheres is called a tetrahedral void. For example, the void *B* in Fig. 4.13 are tetrahedral voids because they are surrounded by four spheres three in first layer and the fourth in the second layer.

A void which is surrounded by six spheres, is called the octaheral void. The voids *C* in Fig. 4.13 are octaheral voids because, they are surrounded by six spheres, three in the first layer and three in the second layer. These voids are shown in Fig. 4.16.

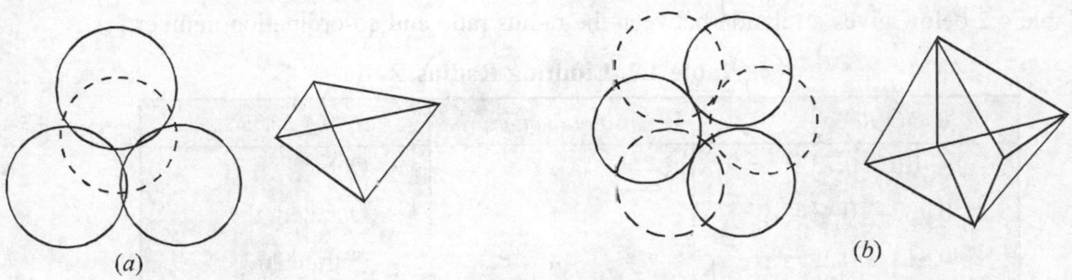

Fig. 4.16. Two types of voids (*a*) Tetrahedral (*b*) Octaheral.

4.4.3 Co-ordination number in HCP and CCP

Every shpere in either HCP or CCP has 12 spheres directly touching it, 6 in its own layer, 3 in layer below it and 3 in the layer above it. Thus, the co-ordination number of a sphere in HCP or CCP is 12. In other types of crystal, co-ordination number of 4, 6, 8 is quite common.

4.4.4 Some general observations that have been made in the study of crystals

Some observations are :

(*i*) Noble gases and most of metals crystallise in closed packed structures.

(*ii*) A large number of ionic compounds crystallise in structures which are related to close packed structures. Here, one of the ions forms the close packed structures and the other ion goes into voids.

(*iii*) If in a compound of the general formula *AB*, the *B* ions form the close packed structures, then *A* ions will occupy all the octahedral voids or half the tetrahedral voids.

(*iv*) In a compound of the type A_2B, all the tetrahedral sites will be occupied by *A* ions.

(*v*) In NaC1, Cl^- forms the CCP structure and Na^+ occupy the octahedral voids. In ZnS, S^{2-} forms the CCP structures with Zn^{2+} occupying alternate tetrahedral voids. In CaF_2, it is the Ca^{2+} which form the CCP lattice and F^- occupy all the tetrahedral voids.

(*vi*) In the packing of oxides of ion, *viz.*, FeO and Fe_2O_3, FeO has the NaC1 structure. Thus, Fe^{2+} should occupy the octahedral voids. But this oxide has stoichiometry with a formula $Fe_{0.95}O$. In converting two thirds of Fe_2^+ in FeO into Fe^{3-}, we get FeO. Fe_2O_3 or Fe_3O_4. In Fe_2O_4 all the Fe^{2+} are in octahedral voids and Fe^{3+} are equally distributed between octahedral and tetrahedral voids. $MgFe_2O_4$ has also a similar structure.

(*vii*) $MgAl_2O_4$ and $ZnFe_2O_4$ (ferrites) form spinal structures, Mg^{2+} in the former and Zn^{2+} in the latter (divalent ions) are in tetrahedral voids and Al^{3+} and Fe^{3+} are in the octahedral voids. These spinal type compounds are important magnetic materials used in memory loops in computers.

4.5 RADIUS RATIO

In a solid, the co-ordination number of an ion, *i.e.*, the number of ions surrounding a particular ion depend upon the radius ratio. In ionic solids, radius ratio is defined *as the ratio of the radius of the cation to the radius of anion.*

$$\text{Radius ratio} = \frac{\text{Radius of cation}\left(r^{+}\right)}{\text{Radius of anion}\left(r^{-}\right)}$$

It is observed that greater the radius ratio, greater is the co- ordination number of the ions. This can be understood as, the ions have a tendency of get surrounded by maximum number of ions of opposite charge. Thus, in NaC1, Na^+ ion has co- ordination number 6, while in CsCl, Cs^+ ion which is bigger than Na^+ ion has a co-ordination number 8. A bigger cation has a greater surface area and thus will have a tendency to get surrounded by large number of anions. The Table 4.2 below gives a relation between the radius ratio and co-ordination number.

Table 4.2. Limiting Radius Ratio.

Radius Ratio	*Co-ordination No.*	*Arrangement*
1.0 – 0.732	8	Cubic
0.732 – 0.414	6	Octahedral
0.414 – 0.225	4	Tetrahedral
0.225 – 0.155	3	Triangular

We may visualize the above results as follows :

If the anions are touching one another and also the cation together, it leads to stable arrangement. The stable arrangement for triangular structure will be as shown in Fig. 4.17 (*a*). If the cation becomes bigger in size, the anion will no longer be touching one another, leading to instability [Fig. 4.17 (*b*)]. To counter this they will transform into tetrahedral arrangement [Fig. 4.17 (*c*)], so that anions continue to touch one-another and also the cation in contact with all the three anions. Let us start from the octahedral arrangement. If the cation becomes smaller (radius ratio decrease), the arrangement changes to tetrahedral in order to maintain close contact between the anions and between the anions and the cations.

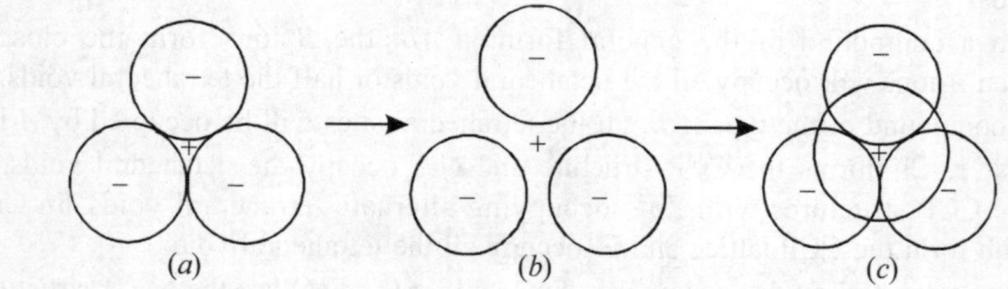

Fig. 4.17. The stable planar triangular arrangement (*a*) changing to tetrahedral arrangement (*c*) as $\overset{+}{r}/\overset{-}{r}$ increases.

On the other hand if the cation becomes bigger (radius ratio increases), the structure changes to BCC (co-ordination no. 8) again to maintain close contact between the anions and between anion and cation. This is illustrated in Fig. 4.18.

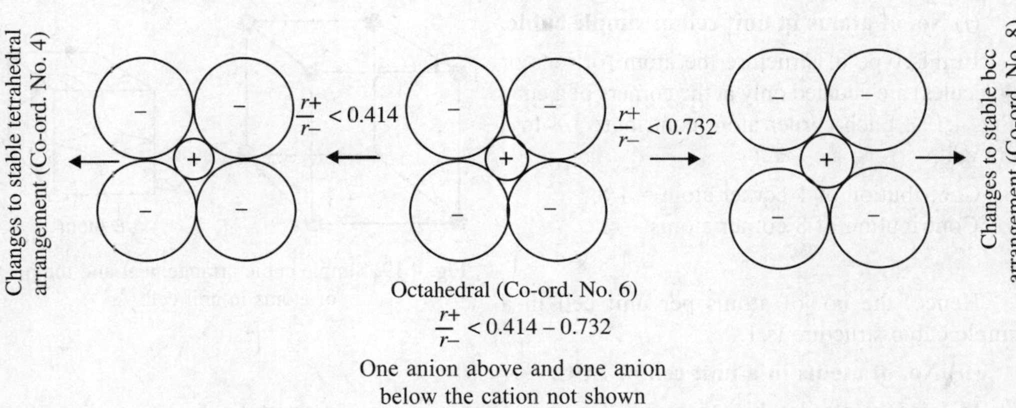

Fig. 4.18. Effect of radius ratio on co-ordination number.

***Example 1.* The radius of the cation in a solid measures 33 pm and the anion measures 70 pm. Predict the co-ordination no. of cation and structure of solid.**

Solution. Radius of cation (r^+) = 30 pm.

Radius of anion (r^-) = 70 pm.

$$\text{Radius ratio } (r^+/r^-) = \frac{30}{70} = 0.471 \text{ pm}$$

As the value lies in the range 0.732 – 0.414, the co- ordination no. of the cation is 6 and octahedral arrangement is predicted.

***Example 2.* A solid AB has the NaCl structure. If the radius of cation A is 100 pm, what is the radius of the anion B ?**

Solution. Since the solid has NaCl structure the co-ordination number of *A* and *B* each is 6. This is possible when the radius ratio is in the range 0.414 – 0.732 thus,

$$\frac{\text{Radius of cation}}{\text{Radius of anion}} = 0.414 \text{ to } 0.732$$

or

$$\text{Radius of anion} = \frac{\text{Radius of cation}}{0.414 \text{ to } 0.732}$$

$$= 241.6 \text{ to } 133.0 \text{ pm}$$

$$= 133.6 \text{ to } 241.6 \text{ pm}$$

4.6 NUMBER OF ATOMS IN A UNIT CELL FOR SIMPLE CUBIC, FCC AND BCC STRUCTURES

The following points are kept in mind while calculating the number of atoms in unit cell.

(*a*) Each corner atom in a cube is shared between eight unit cells. Therefore, a corner atom contributes 1/8 to each unit cell.

(*b*) An atom on the face of unit cell is shared by the two unit cells. Thus, a face atom contributes 1/2 to each unit cell.

(*c*) An atom on the edge of a cube is shared by four unit cells. Hence, an edge atom contribute 1/4 to each unit cell.

(*d*) An atom inside a cube belongs to that cube only. Hence, that atom contributes fully to that unit cell.

We shall calculate separately number of atoms in simple, face-centred and body-centred cubic structures.

(*i*) **No. of atoms in unit cell of simple cubic.**

In this type of structure the atom (or ions or molecules) are situated only at the corners of a cube (Fig. 4.19). Each corner atom contribute 1/8 to a unit cell.

Contribution of 1 corner atom = 1/8

Contribution of 8 corner atoms

$= 1/8 \times 8 = 1$

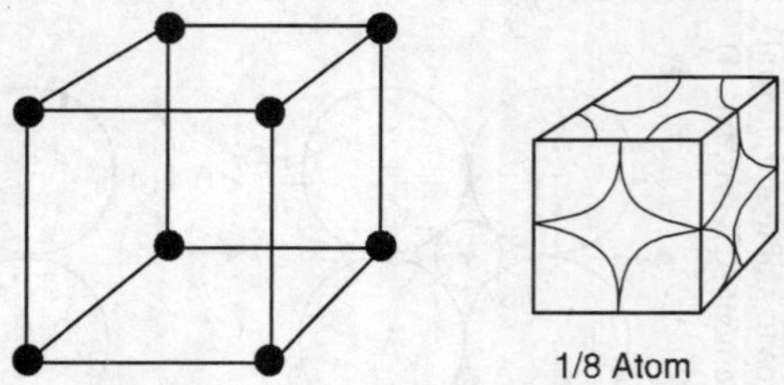

Fig. 4.19. Simple cubic arrangement and the no. of atoms in unit cell.

Hence, the no. of atoms per unit cell in a simple cubic structure is 1.

(*ii*) **No. of atoms in a unit cell of FCC.**

In a face-centred cubic, there are 8 atoms at the 8 corners of the cubic and 6 atoms at the centre of 6 faces of the cube. (Fig. 4.20)

Contribution of 1 corner atom = 1/8

Contribution of 8 corner atoms

$= 1/8 \times 8 = 1$

Contribution of 1 face atom = 1/2

Contribution of 6 face atoms $= 6 \times 1/2 = 3$

∴ Total no. of atoms in unit cell of FCC

$= 1 + 3 = 4$

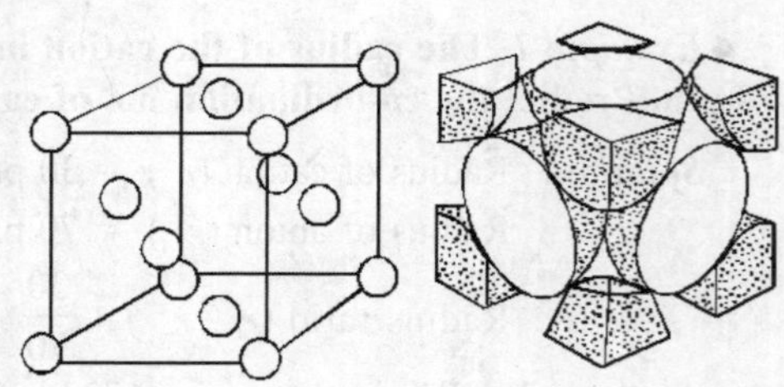

Fig. 4.20. Face-centred cubic arrangement.

(*iii*) **No. of atoms in a unit cell of BCC.**

In Body centred cubic structure, there are 8 atoms at the centre of the body of the cube (Fig. 4.21).

Contribution of 1 corner atom = 1/8

Contribution of 8 corner atoms $= 1/8 \times 8 = 1$

Contribution of 1 central atom = 1

Thus, total no. of atoms in a unit cell of body centred cubic $= 1 + 1 = 2$.

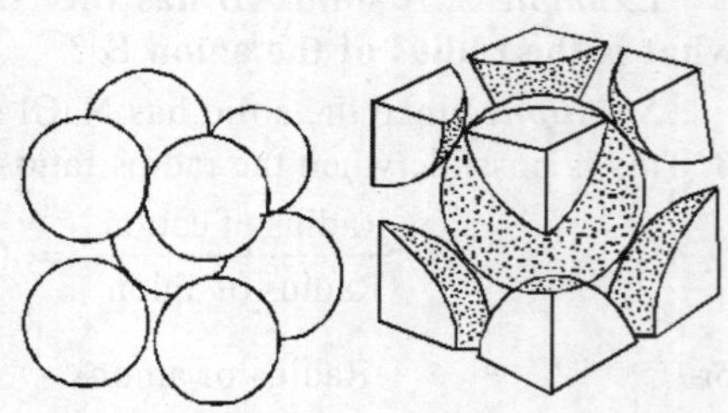

Fig. 4.21. Body-centred cubic arrangement.

Example 3. **A compound formed by elements *X* and *Y* crystallizes in the cubic structures. *X* atoms are situated at the corners and *Y* atoms are the centre of faces. What is the formula of compound ?**

Solution. There are 8 atoms of *X* at 8 corners and 6 atoms of Y at the centre of six faces of the face-centred cubic.

Contribution of 1 corner atom = 1/8

Contribution of 8 corner atoms $= 1/8 \times 8 = 1$

Contribution of 1 face-centred atom = 1/2

Contribution of 6 face-centred atoms $= 6 \times 1/2 = 3$

Thus, there is 1 atom of *X* and 3 atoms of *Y*/unit cell.

Hence, the formula of the compound is XY_3.

Ex*ample 4.* A cubic solid is made up of two elements *P* and *Q*. Atoms *Q* are present at the corner of cube and atoms *P* at the body centre. What is the formula of the compound ? What are the co-ordination number of *P* and *Q* ?

Solution. No. of atoms of Q (at the corner) in unit cell = $1/8 \times 8 = 1$

No. of atoms of P (at body-centre) in unit cell = 1

Hence, the formula of the compound is PQ.

Co-ordination of no. of P and Q each = 8.

4.6.1 Relation between atomic radius (*r*) and the edge length (*a*) in crystals

The following relations exist between d and a for various structures.

Structure	*Relation*
Simple	$d = a$
Body-centred	$d = \frac{\sqrt{3}}{2} a = 0.866\ a$
Face-centred	$d = \frac{a}{\sqrt{2}} = 0.707\ a$

Relation between atomic radius (r) which is equal to $d/2$ for pure elements and the edge (a). The following relations exist between the two quantities for various structures.

Structure	*Relation*
Simple	$r = \frac{a}{2}$
Body-centred	$r = \frac{\sqrt{3}}{4} a = 0.433\ a$
Face-centred	$r = \frac{a}{2\sqrt{2}}\ 0.3535a$

***Example 5.* CsCl has bcc arrangement and its unit cell edge length is 400 pm. Calculate the interionic distance in CsCl.**

Solution. Fig. 4.22, shows the arrangement of CsCl. Open circles represent Cl^- and solid circle Cs^+. Half of the distance CE is the interionic distance

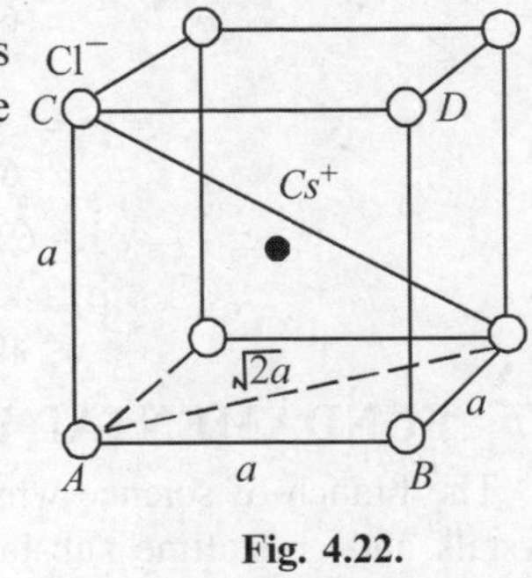

Fig. 4.22.

$$AE = \sqrt{a^2 + a^2} = \sqrt{2}\,a$$

$$CE = \sqrt{2a^2 + a^2} = \sqrt{3}\,a$$

$$= \sqrt{3} \times 400$$

$$\text{Interionic distance} = \frac{1}{2} \times \sqrt{3} \times 400 = 346.4 \text{ pm.}$$

4.6.2 Relation between the density and edge length of a cube

If we know the structure of the solid, then it is possible to calculate the density provided we know the cell parameters, *i.e.*, length of edge of the cube.

Let the length of edge of the cube = a pm.

$= 9 \times 10^{-10}$cm.

The volume of the cube = $(\text{Edge})^3$

$$= (a \times 10^{-10})^3 = a^3 \times 10^{-30}\ \text{cm}^3$$

$$\text{Density of the unit cell} = \frac{\text{Mass of the unit cell}}{\text{Volume of the unit cell}} \quad ...(1)$$

But mass of unit cell = No. of atoms in the unit cell × Mass of each atom

If we represent the number of atoms in the unit cell by Z and mass of each atom m, then

$$\text{Mass of unit cell} = Z \times m \quad ...(2)$$

Also,
$$\text{Mass of each atom, } m = \frac{\text{Atomic Mass}}{\text{Avogadro's no.}} \quad ...(3)$$

Substituting the value of m from eqn. (3), in eqn. (2)

$$\text{Mass of unit cell} = Z \times \frac{\text{Atomic Mass}}{6.023 \times 10^{23}} \quad ...(4)$$

Substituting the value of mass of unit cell from eqn. (4) in eqn. (1)

$$\text{Density of the unit cell of the solid} = \frac{Z \times \text{Atomic Mass}}{6.023 \times 10^{23} \times a^3 \times 10^{-30}}\ \text{g/cm}^3$$

Example 6. **An element (atomic mass = 60 a.m.u.) having face-centred cubic unit cell has a density of 6. 23 gcm^{-3} Calculate the edge length of the unit cell.**

Solution. Use the relation between density and edge length (a) :

$$\text{Density} = \frac{Z \times \text{atomic mass}}{6.023 \times 10^{23} \times a^3 \times 10^{-3}}\ \text{g/cm}^3 \quad (a \text{ has the units of picometre pm})$$

or
$$a^3 = \frac{Z \times \text{atomic mass}}{6.023 \times 10^{23} \times 10^{-30} \times \text{density}}\ \text{cm}^3$$

For a face-centred cubic, $Z = 4$ Substitute the values in the above equation

$$a^3 = \frac{4 \times 60}{6.023 \times 10^{-7} \times \text{density}}$$

or
$$a^3 = \frac{4 \times 60}{6.023 \times 10^{-7} \times 6.23}$$

or $a^3 = 63.96 \times 10^6$

or $a = 4 \times 10^2$ pm

$= 400$ pm or 400×10^{-10} cm or 4×10^{-8} cm or 4 Å

4.7 FUNDAMENTAL LAWS OF CRYSTALLOGRAPHY

The branch of science which deals with the study of geometry, properties and structure of crystals and crystalline substances is known as *Crystallography*. Three fundamental laws of crystallography are:

(1) The law of constancy of interfacial angles
(2) The law of rationality of indices
(3) The law of symmetry.

4.7.1 Some terms used in crystallography

Face: The plane surfaces of a crystal are called the faces of the crystal.

Edge: The line along which the two adjacent faces of a crystal intersect is called the edge.

Form: A set of similar faces of a crystal is called a form.

Zone: Faces are found to occur in sets, which meet in parallel edges or would meet in parallel edges if the planes of the faces were extended. Such a set of faces is known as zone.

Crystal habit: The external shape of the crystal is called crystal habit.

Interfacial angle: The angle between the adjacent faces of a crystal is called interfacial angle.

If we represent number of faces by F, number of interfacial angles by C and number of edges by E, then the following relation holds : F + C = E + 2

4.7.2 Law of constancy of interfacial angles

The crystals of a substance are obtained by cooling the liquid (or the melt) or the solution of that substance. The size of the crystals depends upon the rate of cooling. If cooling is carried out slowly, crystals of large size are obtained because the particles (*viz.,* ions, atoms or molecules) get sufficient time to arrange themselves in proper positions. Similarly, it is found that the shape of the crystals of a substance also depends upon the conditions under which the crystallization takes place *i.e.*, the absence or presence of any other substance *etc*. For example, shapes of the crystals of a particular substance are shown in Fig. 4.23.

However, inspite of the different sizes and shapes of the crystal of the same substance, it is observed that the *angles between the corresponding faces called the interfacial angles of the crystals of a particular substance are always the same.* This is called the **law of constancy of interfacial angles.** The instrument used for the measurement of interfacial angles is called a **goniometer.**

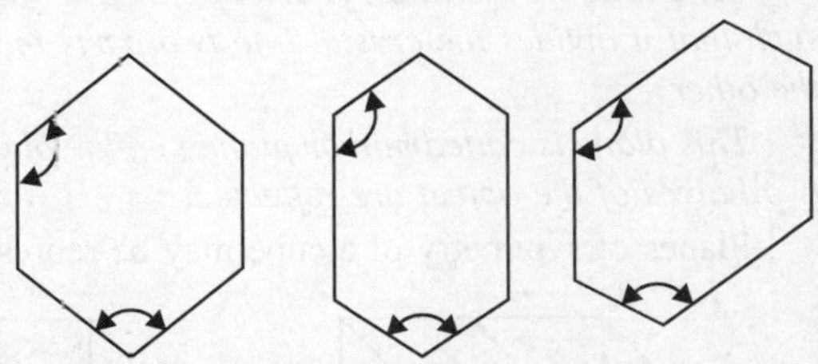

Fig. 4.23. Shapes of crystals (crystal habits) of a particular substance.

4.7.3 Symmetry elements of a crystal

The symmetry element is a geometrical entity such as a plane, a line or a point with respect to which one or more symmetry operation may be carried out.

Symmetry operation means movement of the body in such a way that after the movement has been carried out, every point of the body exactly coincides or is in coincidence with equivalent points of the body in the original orientation. This movement is called *symmetry operation.*

For example, when a molecule XYX is rotated in the plane of paper about an axis passing through the centre of Y through 180°, the new orientation has exactly the same apperance as the original one. In this case the symmetry element is axis of rotation and the symmetry operation is rotation (Fig. 4.24).

(*a*) Orginal orientation (*b*) Orientation after rotation through 180°

Fig. 4.24.

The symmetry properties can be expressed with the help of symmetry operation.

Different symmetry elements of a crystal are described as under :

(*i*) Centre of symmetry: *It is defined as an imaginary point within the crystal such that any line passing through this point intersects the opposite faces of the crystal at equal distances.* The centre of symmetry is also called the **centre of inversion** because if the crystal or the molecule is inverted through the centre of symmetry, it gives results equivalent to the original and hence indistinguishable from the original one. For example, the centre of inversion represented by solid dots of a few molecules are shown in Fig. 4.25.

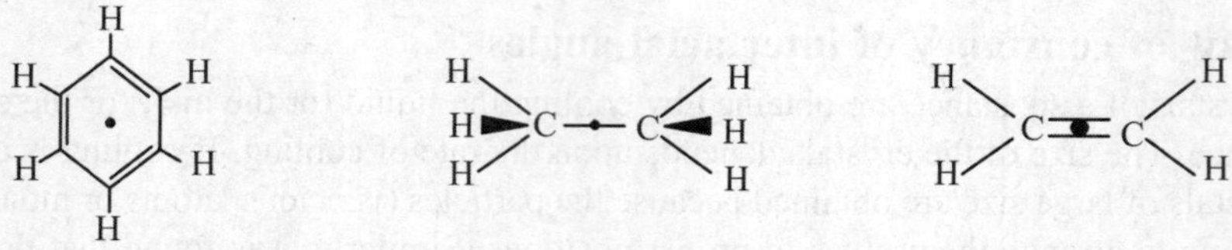

Fig. 4.25. Centres of inversion of a few molecules.

(*ii*) Plane of symmetry: *It is defined as an imaginary plane passing through, the crystal such that it divides the crystal into two parts in such a way that one part is the mirror image of the other.*

This plane is called mirror plane i.e., the plane may be regarded as mirror through which the two halves of the object are reflected.

Planes of symmetry of a cube may be represented as follows (Fig. 4.26).

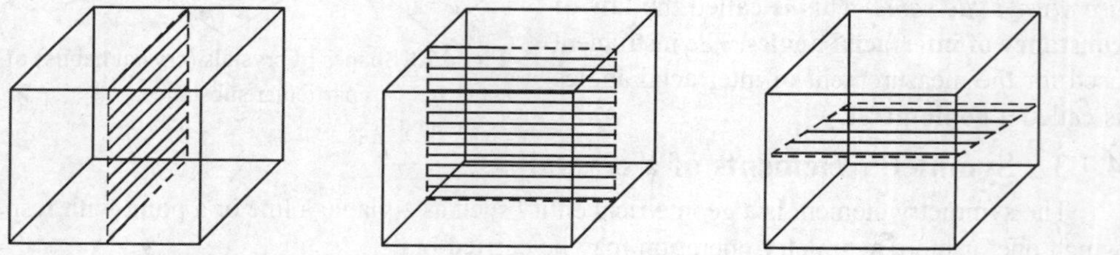

(a) Rectangular plane of Symmetry

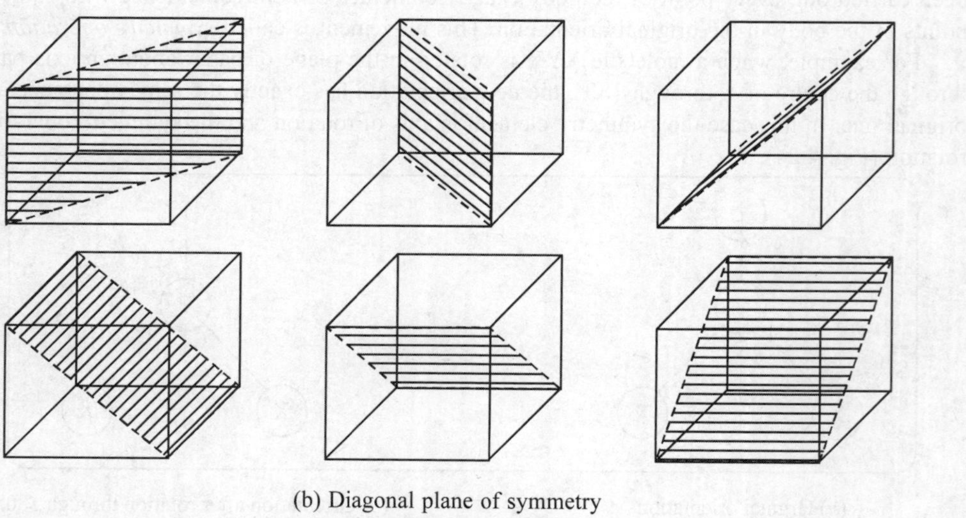

(b) Diagonal plane of symmetry

Fig. 4.26. Planes of symmetry of a cube.

There are nine planes of symmetry in a cube, six are diagonal planes and three rectangular planes.

(*iii*) Rotation axis of symmetry: *The rotation axis of symmetry or simply called the axis of symmetry is defined as the imaginary line passing through the crystal such that when the crystal is rotated about this line, exactly similar appearance occurs more than once in one complete revolution, i.e., in a rotation through 360°.* If similar appearance occurs twice in one complete revolution (*i.e.*, after every 180°), the axis is called *two-fold axis* of symmetry or a **diad axis.** Similarly the other possibilities found for similar appearance in one complete revolution are three, four times or six times (*i.e.*, after rotation through 120°, 90°, or 60°). The corresponding axes are called *three fold* (**triad**), *four-fold* (**tetrad**) and *six fold* (**hexad**) axes of symmetry.

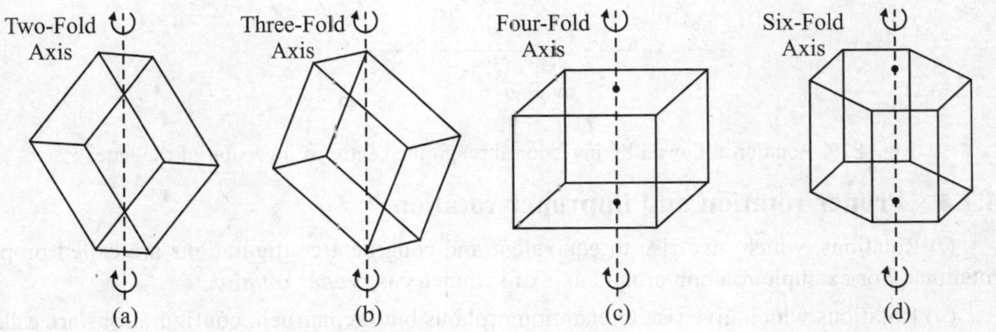

Fig. 4.27. Different axes of symmetry.

Hence in general, if the crystal presents the same appearance n times after rotation through $\frac{360°}{n}$, it is called *n*-fold axis of symmetry (Fig. 4.27).

It is important to mention that only 2-fold, 3-fold, 4-fold and 6-fold axes of symmetry are possible. 5-fold, 7-fold, 8-fold or higher-fold axes of symmetry are not possible because it is not possible to get compact structure with these shapes as some space is always left vacant. On the other hand it is possible to get compact structure of crystals with 2-fold, 3-fold, 4-fold and 6-fold axes of symmetry. The rotation about these axes gives identical appearance of various configurations.

(*iv*) Rotation-reflection axis of symmetry : When the rotation of a molecule or a crystal about an *n*-fold axis of proper rotation is followed by reflection through a plane perpendicular to the axis, the axis is called an *n*-fold rotation-reflection axis (Fig. 4.28). The new configuration obtained is not congruent but is enantiomorphous.

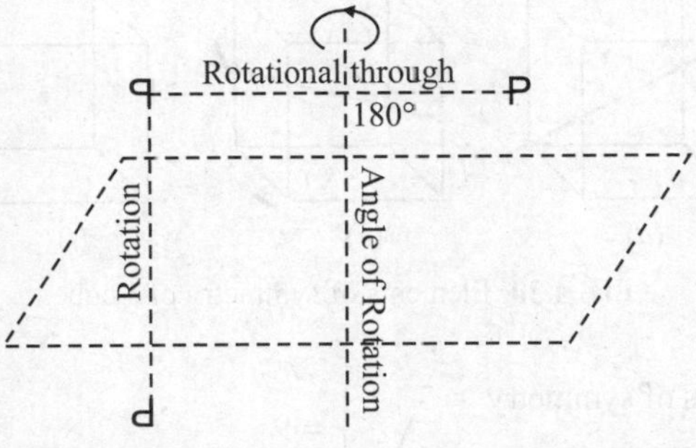

Fig. 4.28. Rotation followed by reflection of the letter P.

(*v*) **Rotation-inversion axis of symmetry:** When the rotation of a molecule or a crystal is followed by inversion through the centre of symmetry the axis is called an *n*-fold rotation-inversion axis (Fig. 4.29). The new configuration obtained is not congruent but is enantiomorphous, as in the above case.

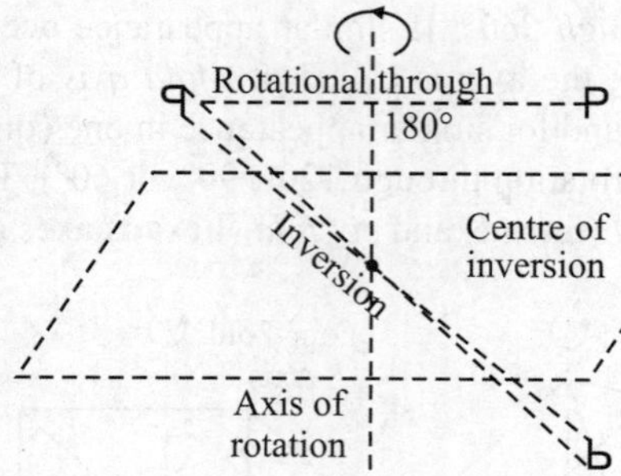

Fig. 4.29. Rotation followed by inversion through the centre of inversion of the letter P.

4.7.4 Proper rotation and improper rotation

(*i*) Rotations, which give rise to equivalent and congruent configurations are called proper rotations. For example rotation around axis of symmetry is *proper rotation.*

(*ii*) Rotations which give rise to enantiomorphous but incongruent configurations are called improper rotations. For example rotation around axis of symmetry followed by reflection or inversion leads to improper rotation.

4.7.5 Planes and axes of symmetry in a simple cube

The crystals of a particular substance always possess the same elements of symmetry. This is called **law of symmetry**. A cubic crystal always possesses the following 23 different elements of symmetry:

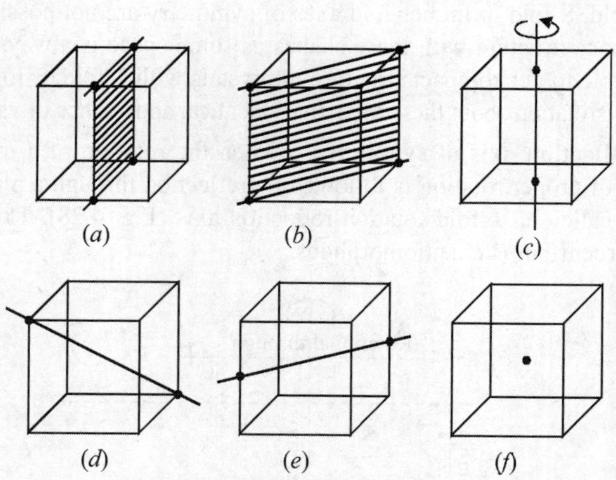

Fig. 4.30. Elements of symmetry of a cube.

$$\left.\begin{array}{ll}(a)\ \text{Rectangular planes of symmetry} & = 3 \\ (b)\ \text{Diagonal plane of symmetry} & = 6\end{array}\right\} = 9$$

(*c*) Four - fold axes of symmetry = 3
(*d*) Three-fold axes of symmetry = 4 } = 13
(*e*) Two-fold axes of symmetry = 6
(*f*) Centre of symmetry (located at the centre of gravity of the cube) = 1

An example of each type of symmetry possessed by a cube is shown in Fig. 4.30.

4.7.6 Crystallographic axes and axial ratio

Crystallographic Axes: In order to describe the relative directions and orientations of the faces and also the planes present within the crystal, three suitable axes are chosen which meet at a point, say *O*, (Fig. 4.31). These axes can be selected in a number of ways depending upon the symmetries of the crystal. The best choice is usually the lines, coinciding with or parallel to the edges between the principal faces. For example, in a cubic crystal, the best choice of the axes is the three lines mutually at right angles to each other. The three lines thus selected meeting at a point are called *crystallographic axes.*

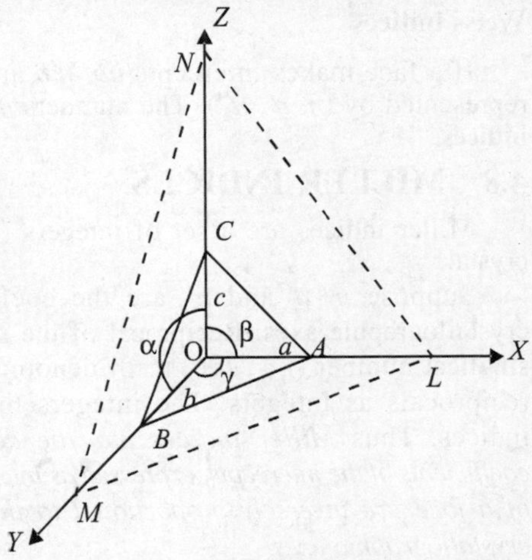

Fig. 4.31. Crystallographic axes and axial ratio.

The angles between the axes opposite to the intercepts *a, b* and *c* are represented by α, β and γ respectively. These angles are the same as the inter axial ratio. facial angles if the axes chosen are the lines coinciding with or parallel to the edges of the principal faces of the crystal.

Axial Ratio: To explain axial ratio, consider a plane called **unit plane** which meets the crystallo-graphic axes. For example, suppose the three axes are represented by *OX, OY* and *OZ* and the unit plane *ABC* cuts these axes at *A, B* and C respectively, making the intercepts *OA = a, OB = b, OC = c*. The ratio *a: b: c* is called the **axial ratio.** Thus *the axial ratio may by defined as the ratio of the intercepts made by the unit plane on the crystallographic axes.*

4.7.7 Law of rational indices

If the intercepts of any a face of the crystal on the three axes are measured, the following generalization is observed which is called law of rational indices:

The intercepts made by any face of the crystal on the crystallographic axes are either:

(*i*) *same as those of the unit plane, or*

(*ii*) *simple whole number multiples of those of the unit plane, or*

(*iii*) *one or two intercepts may be infinity, if the face is parallel to one or two axes, i.e., the face does not cut one or two axes.*

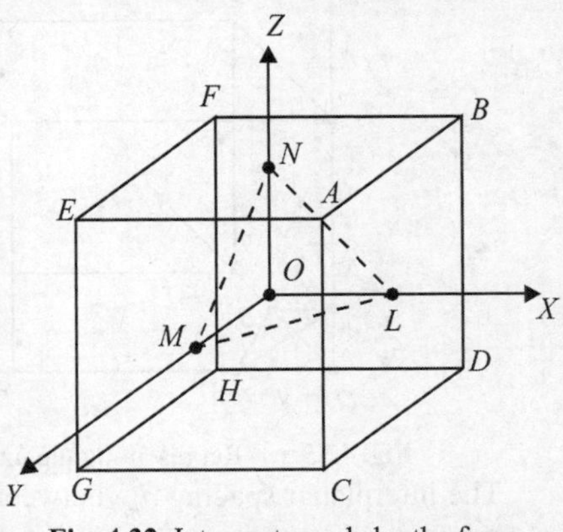

Fig. 4.32. Intercepts made by the faces of a cube on the axes.

For example, in Fig. 4.31, the intercepts made by the face *LMN* are 2*a*, 2*b* and 3*c* which are simple whole number multiples of those of the unit plane. Again consider the case of cube. Suppose the origin

O of the crystallographic axes lies at the centre of symmetry and the axes *OX, OY, OZ* are parallel to the edges as shown in Fig. 4.32. Suppose the axis *OX* cuts the face *ABCD* at the point *L*. Similarly, suppose the axes *OY* and *OZ* cut the faces at *M* and N. If *LMN* is taken as the unit plane, the intercepts made by the unit plane on the three axes are *OL, OM* and *ON*. The face *ABCD* cuts the *X*-axis at *L i.e.,* makes the same intercept as that made by the unit plane but it does not cut the Y-axis and Z-axis at all, because these two axes are parallel to this face. In other words, the intercepts made by the face *ABCD* on the *Y*-axis and *Z*-axis are *infinity.*

In general if a face *PQR* makes intercepts *p, q* and *r* on the three axes and the intercepts made by the unit plane are *a, b, c* then we have $p = na$, $q = n'b$ and $r = n''c$ where *n, n'* and *n''* are simple whole numbers or some out of these may be *infinity.*

Weiss indices

If a face makes intercepts *na, n′b* and *n″c* on the crystallographic axes, the face may be represented by (*n, n′, n″*). The numbers *n, n′, n″* thus used to represent a face are called **Weiss indices.**

4.8 MILLER INDICES

Miller indices are a set of integers (*h k l*) which are used to describe a given plane in a crystal.

Suppose *n, n′* and *n″* are the coefficients of the intercepts of a given plane on the crystallographic axes. Reciprocal of the numbers *n, n′* and *n″* are taken and multiplied by the smallest number (*i.e., L.C.M.* of denominators) so as to express all the reciprocals as integers. The integers thus obtained are called Miller indices. Thus *Miller indices may be defined as the reciprocals of the coefficients of the intercepts, expressed as integers.* Obviously, *the Miller indices of a face are inversely proportional to the intercepts of that face on the crystallographic axes.*

W.H. Miller

Alternatively, the intercepts made by any crystal plane such as DEF, in Fig. 4.33, are usually expressed in terms of the intercepts of the unit plane (*a, b, c*) as follows:

$$OD = \frac{a}{h},\ OE = \frac{b}{k} \text{ and } OF = \frac{c}{l}$$

Then the Miller indices of the plane are represented as (*hkl*)

The unit plane is represented as shown in Fig. 4.33 (*a*) and any other plane of the crystal may be represented as shown in Fig. 4.33 (*b*).

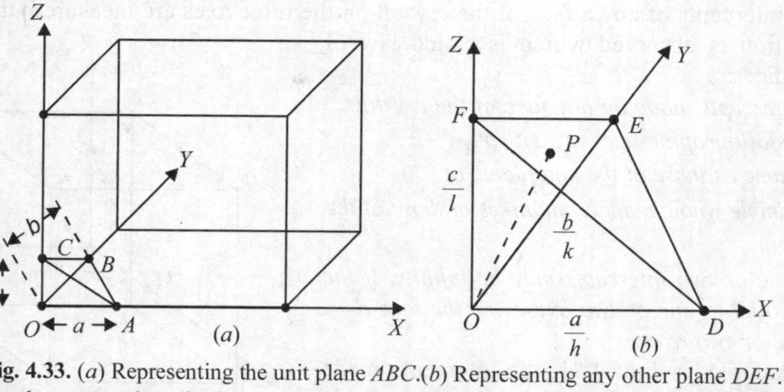

Fig. 4.33. (*a*) Representing the unit plane *ABC*.(*b*) Representing any other plane *DEF*.

The interplanar spacing d_{hkl} between the planes *hkl* in a cubic lattice ($a = b = c$) is given by

$$d_{hkl} = \frac{a}{\sqrt{h^2 + k^2 + l^2}}$$

where *a* is a side of the cube.

Examples of Miller Indices

(*i*) In Fig. 4.31, the intercepts made by the face *LMN* are $2a$, $2b$, $3c$. The reciprocals of the coefficients of *a, b, c* are $\frac{1}{2}, \frac{1}{2}, \frac{1}{3}$, *LCM* of 2, 2, 3 = 6. Multiplying the reciprocals by 6, we have 3, 3, 2. These are the Miller indices of the face *LMN* in Fig. 4.31. This face is represented as (332).

(*ii*) The intercepts made by the unit *ABC* [Fig. 4.31 or 4.33 (*a*)] plane are *a, b, c*. The coefficients of these intercepts are 1, 1, 1. Hence the unit plane is always represented as (111).

(*iii*) If some face does not cut some axis at all, the intercept on this axis is taken as infinity. For example, suppose a face makes intercepts $2a$ and $3b$ on the *X*-axis and *Y*-axis respectively and does not cut the *Z*-axis at all, the intercepts will be $2a$, $3b$ and ∞ (or $\infty\, c$). The reciprocals of the coefficients are 1/2, 1/3, 0. *LCM* of 2, 3 = 6. Multiplying by 6, the Miller indices will be 3, 2, 0, *i.e.*, the face will be represented as (320). Similarly, if a face makes intercepts *a* on the *X*-axis and does not cut the *Y*-axis and *Z*-axis at all (see Fig. 4.32, face *ABCD*), its Miller indices will be 1, 0, 0, *i.e.*, it will be represented as (100).

(*iv*) If some intercept made by a face is negative *i.e.*, the face is cut by the axis when extended in the opposite direction, it is represented by putting a bar over it. For example in Fig. 4.32, the intercept made by the face *EFGH* on the *X*-axis is $-a$ and this face does not cut the *Y*-axis and *Z*-axis at all. Hence this face is represented as $(\bar{1}00)$.

(*v*) **Miller Indices of the Faces of a Cube:** In Fig. 4.32, in which the origin of the crystallographic axes is the centre of symmetry of the cube, following the method as explained in (*iii*) and (*iv*) above, the various faces of the cube in terms of Miller indices may be represented as (100), (010), (001), $(\bar{1}00)$, $(0\bar{1}0)$ and $(00\bar{1})$.

***Example 7.* A crystal plane has intercepts on the three axes of crystal in the ratio of $\frac{3}{2}$: 2 : 1. What are the Miller indices of the plane?**

Solution. The reciprocal of the ratio of intercepts are

$$\frac{2}{3} : \frac{1}{2} : 1$$

Multiply by the L.C.M of 3 and 2 .*i.e.*, 6

$$4 : 3 : 6$$

The Miller indices of the plane are 436.

***Example 8.* Write Miller indices for any four faces of a Cubic unit Cell.**

Solution. Refer to Fig. 4.32.

If O is taken as the origin of crystallographic axes, then Miller indices of :

Plane *ABCD* are 100

Plane *ACGE* are 010

Plane *ABFE* are 001

Plane *EFGH* are $\bar{1}00$.

***Example 9.* The Weiss indices of a planes are $\frac{1}{2}\frac{1}{2}\frac{1}{2}$ What are the Miller indices?**

Solution. Miller indices are obtained by taking the reciprocal of Weiss indices and then by multiplying, if required, with a suitable number to obtain integral numbers.

Millar indices will be

$$\frac{1}{1/2}\;\frac{1}{1/2}\;\frac{1}{1/2} \text{ or } 222$$

Example 10. **Determine the interplanar spacing between the (221) planes of a cubic lattice of length 4·5 Å (450 pm).**

Solution. In a cubic lattice, the interplanar spacing d_{221} is given by

$$d_{221} = \frac{a}{\left(h^2 + k^2 + l^2\right)^{1/2}} \text{Å}$$

$$= \frac{4.5}{(4+4+1)^{1/2}} \text{Å} = 1.5 \text{ Å} = 150 \text{ pm}$$

Example 11. **Find the miller Indices of the plane that intersect the crystallographic axes at the distance a, 2b, ∞ c**

Solution. Miller indices of the plane are : $\frac{1}{1}\ \frac{1}{2}\ \frac{1}{\infty}$ or 210

4.8.1 Isomorphism and polymorphism

Certain crystalline substances which have similar chemical composition, (for example, sodium phosphate and sodium arsenate *i.e.*, Na_3PO_4 and Na_3AsO_4) are found to possess the same crystal shape. This phenomenon is known as **isomorphism** and the substances are said to be **isomorphous**.

There are certain substances which exist in more than one crystalline form which are formed under different conditions of temperature and pressure. This phenomenon is known as **polymorphism.** For example, carbon exists in two crystalline forms called graphite and diamond; calcium carbonate exists in two crystalline forms called calcite and aragonite. The polymorphism exhibited by the elements is commonly called **allotropy.**

4.8.2 Different crystallographic systems

Various crystallographic systems alongwith the axial characteristics and symmetry elements are given in the tabular form as shown below in Table 4.3

Table 4.3. Related Information Regarding Crystallographic Systems.

System	*Axial characteristics*	*Maximum Symmetry*	*Examples*
1. Cubic	$\alpha = \beta = \gamma = 90°$ $a = b = c$	Plane = 9 Axes = 13	NaC1, KC1, Alums, Diamond
2.Tetragonal	$\alpha = \beta = \gamma = 90°$ $a = b \neq c$	Planes = 5 Axes = 5	Sn, SnO_2, TiO_2
3.Orthorhombic or Rhombic	$\alpha = \beta = \gamma = 90°$ $a \neq b \neq c$	Planes = 3 Axes = 3	KNO_3, K_2SO_4, $BaSO_4$, Pb,CO_3 Rhombic sulphur
4.Monoclinic	$\alpha = \gamma = 90° \neq \beta$ $a \neq b \neq c$	Planes = 1 Axes = 1	$Na_2SO_4 \cdot 10H_2O$ $CaSO_4 \cdot 2H_2O$ Monoclinic sulphur
5. Triclinic	$\alpha \neq \beta \neq \gamma \neq 90°$ $a \neq b \neq c$	Planes = Nil Axes = Nil	$CuSO_4 \cdot 5H_2O$ $K_2Cr_2O_2$
6. Rhombohedral or Trigonal	$\alpha = \beta = \gamma \neq 90°$ $a = b = c$	Planes = 7 Axes = 7	$NaNO_3$,ICl,As,Sb,Bi
7. Hexagonal,	$\alpha = \beta = 90°$ $\gamma = 120°$ $a = b \neq c$	Planes = 7 Axes = 7	HgS, Ice Graphite, Mg, Zn, Cd

Example 12. **Identify the crystal system to which some solids having the following dimensions for their unit cell belong. Give examples of the solids.**

(*i*) $a \neq b \neq c$ $\alpha \neq \beta \neq \gamma \neq 90°$

(*ii*) $a \neq b \neq c$ $\alpha = \beta = \gamma = 90°$

Solution. (*i*) Solids having these characteristics are *triclinic.* Examples of this crystal system are $CuSO_4 \cdot 5H_2O$ and $K_2Cr_2O_7$.

(*ii*) Solids having these characteristics are *orthorhombic or rhombic.* Examples of this category of solids are KNO_3 and K_2SO_4.

Example 13. **Explain: We do not come across 5-fold, 7-fold, 8-fold or higher fold axis of rotation in crystals.**

Solution. We have only 2-fold, 3-fold, 4-fold and 6-fold axes of rotation and we do not have 5-fold, 7-fold, 8-fold or higher fold axes of rotation. This is due to the fact that the most essential requirement of a lattice is that it should be built up by the repetition of unit cells, the faces of each unit cell touching the faces of the surrounding unit cells without leaving any gap in-between. Such a close packing is possible if an object has 2-fold, 3-fold, 4-fold or 6-fold axis of symmetry but is not possible in case an object has 5-fold, 7-fold, 8-fold or higher-fold axis of symmetry. This is illustrated along side for objects having 6-fold axis of symmetry and 5-fold axis of symmetry. (Fig. 4.34). We find that a gap is created in a solid with 5-fold axis of symmetry.

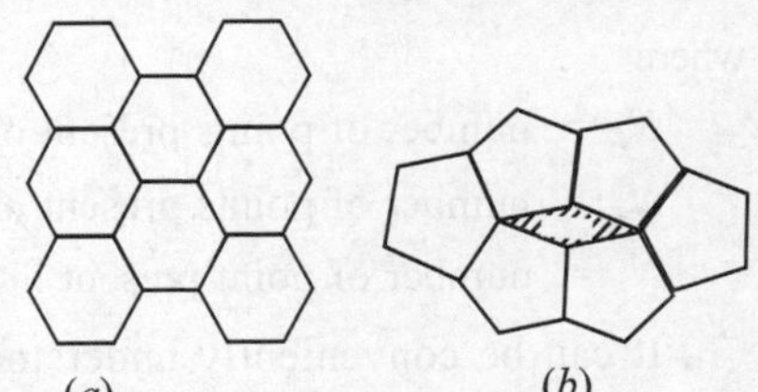

Fig. 4.34. (*a*) the close packing of hexagonal objects. (*b*) a gap created in the packing of pentagonal objects.

4.8.3 Bravais lattices

The detailed study of the internal structure by Bravais, has shown that the points (particles) may be present not only at the corners of unit cells but also at the centre of the faces or within the body of the unit cells. This gives rise to different types of lattices corresponding to the seven crystallographic systems. These 14 types of lattices are called **Bravais lattices.** The description of these lattices is given in Table 4.4 and the shapes are shown in the Fig. ahead The terms *simple, face-centresd* (*end-centred* and *body-centred* are explained as under:

(*i*) **Simple:** When the points are present only at the corners of the unit cell.

(*ii*) **Face-centred:** When in addition to the points at the corners, there are points at the centre of the each face of the unit cell.

(*iii*) **End-centred:** When there are points present at the centre of the end faces of the unit cell, in addition to points at the corners.

(*iv*) **Body-centred:** When in addition to the points at the corners, there is one point present within the body of each unit cell.

Table 4.4. Bravais lattices corresponding to different crystal systems.

System	*Number of Bravais lattices*	*Description of the lattices*
1. Cubic	3	Simple, Face-centred, Body-centred.
2. Tetragonal	2	Simple, Body-centred

System	*Number of Bravais lattices*	*Description of the lattices*
3. Orthorhombic or Rhombic	4	Simple, Face-centred End-centred, Body-centred.
4. Monoclinic	2	Simple, End-centred
5. Triclinic	1	Simple
6. Rhombohedral or trigonal	1	Simple
7. Hexagonal	1	Simple

Number of atoms per unit cell

The number of points per unit cell (N) is given by the expression

Auguste Bravais

$$N = \frac{N_c}{8} + \frac{N_f}{2} + N_i$$

where

N_c = number of points present on the corners of the unit cell

N_f = number of points present on the faces of the unit cell

N_i = number of point present in the interior of the unit cell.

It can be conveniently understood that each point present on the corner is shared by 8 unit cells, each point present on the face is shared by two unit cells and the point present in the interior is shared by no other unit cell.

For any simple lattice, $N = 8, N_f = 0, N_i = 0$, so that

$$N = \frac{8}{8} + 0 + 0 = 1$$

For a face-centred cubic lattice, $N_c = 8, N_f = 6, N_i = 0$ so that

$$N = \frac{8}{8} + \frac{6}{2} + 0 = 4$$

For body-centred orthorhombic lattice, $N_c = 8, N_f = 0, N_i = 1$ so that

$$N = \frac{8}{8} + 0 + 1 = 2.$$

For end-centred monoclinic lattice, $N_c = 8, N_f = 2, N_i = 0$ so that

$$N = \frac{8}{8} + \frac{2}{2} + 0 = 2.$$

4.8.4 Point groups and space groups

Theoretically, there are 32 different ways of arranging the elements of symmetry about a point. These different arrangements are called *points groups.* However, in a complete space lattice, because of the regular repetition of the pattern in the three dimensional space, two additional symmetry elements called **glide planes** and **screw axes** become possible. Because of these additional elements of symmetry, the number of arrangements possible for the elements of symmetry about a point of a space lattice becomes equal to 230. These 230 possible arrangements of the elements of symmetry about a lattice point are called **space groups.**

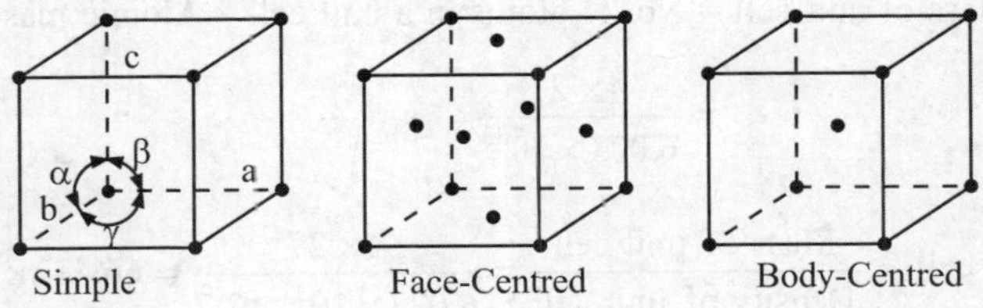

1. Cubic $\left\{\begin{array}{l}\alpha = \beta = \gamma = 90^\circ \\ a = b = c\end{array}\right\}$

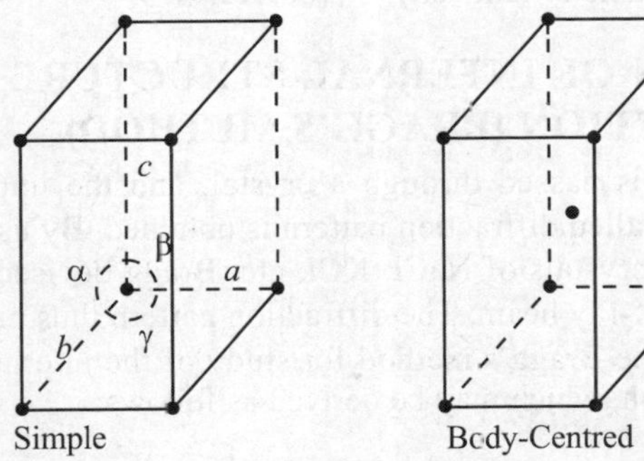

2. Tetragonal $\left\{\begin{array}{l}\alpha = \beta = \gamma = 90^\circ \\ a = b \neq c\end{array}\right\}$

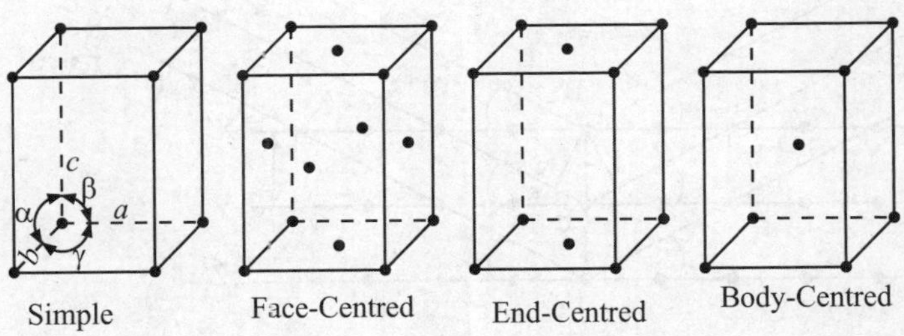

3. Orthorhombic $\left\{\begin{array}{l}\alpha = \beta = \gamma = 90^\circ \\ a \neq b \neq c\end{array}\right\}$

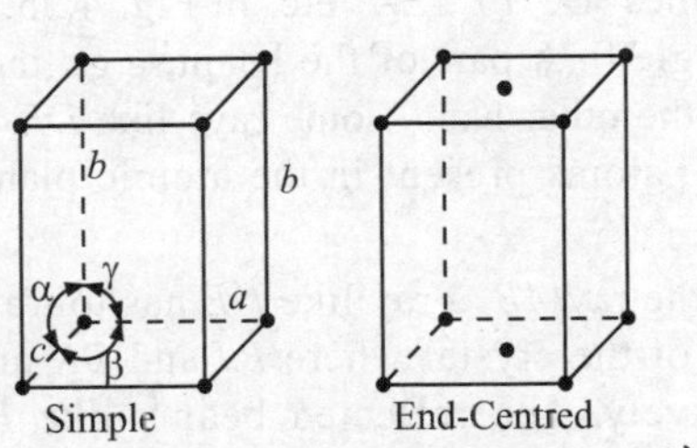

4. Monoclinic $\left\{\begin{array}{l}\alpha = \gamma = 90^\circ \neq \beta \\ a \neq b \neq c\end{array}\right\}$

Example 14. **A certain solid *X* (atomic mass 27) crystallises in a fcc structure. If the density of *X* is 2.7 *g* cm^{-3}, what is the length of the edge ?**

Solution. Mass of unit cell = No. of atoms in a unit cell × Atomic mass

$$= \frac{4 \times 27}{6.023 \times 10^{23}}$$

$$\text{Valume of unit cell} = \frac{\text{Mass of unit cell}}{\text{Density of unit cell}} = \frac{4 \times 27}{6.023 \times 10^{23} \times 2.7} = 66.412 \times 10^{-24}\ \text{cm}^3$$

$$\text{Length of edge} = (\text{volume of unit cell})^{\frac{1}{3}} = (66.412 \times 10^{-24})^{\frac{1}{3}} = 4.05 \times 10^{-8}\ \text{cm} = 4.05\text{Å}$$

4.9 INVESTIGATION OF INTERNAL STRUCTURE OF A SOLID BY X-RAY DIFFRACTION (BRAGG'S METHOD)

When a beam of X-ray is passed through a crystal, and the impression is taken on a photographic plate a pattern called diffraction pattern is obtained. By using this pattern, Bragg investigated the structure of crystals of NaCl, KCl, etc. Bragg devised a spectromter for the measurement of intensity of X-ray beam. The diffraction pattern thus produced can be used to study the crystal structure. The Bragg's method for study of the internal structure of crystals is based upon Bragg's equation, which may be derived as follows:

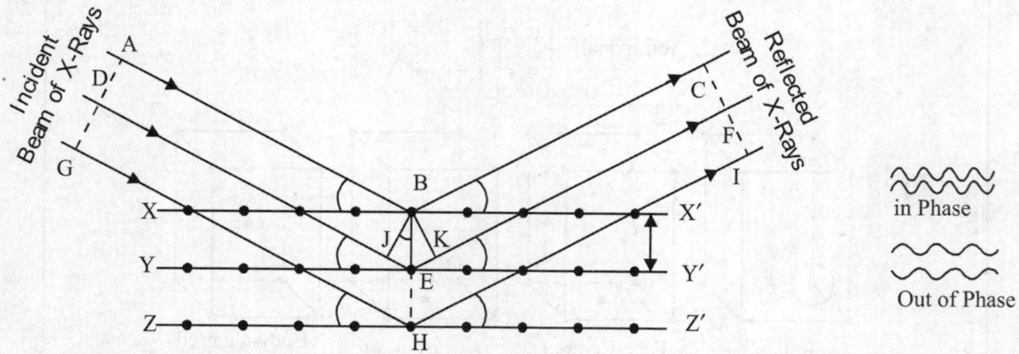

Fig. 4.35. Reflection of X-rays by equidistant parallel planes.

A crystal may be considered to be made up of a number of parallel equidistant atomic planes, as represented by the lines *XX′*, *YY′*, *ZZ″* etc. in Fig. 4.35. Suppose a beam of X-rays is incident on the crystal at angle θ. A part of the beam, *e.g.*, the ray *AB* is reflected at the point *B* along the path *BC*. On the other hand, some rays like *DE*, *GH*, etc., penetrate into the crystal and are reflected by the atoms present in the atomic planes *YY′*, *ZZ′*, etc., as shown below :

Evidently, as compared to the ray *AB*, a ray like *DE* has to travel a longer distance, equal to *JEK* in order to emerge out of the crystal where *BJ* and *BK* are perpendiculars drawn on the lines *DE* and *EF* respectively. The reflected beams like *BC*, *EF*, etc. then undergo interference with each other. If these reflected rays are in phase (*i.e.*, in the reflected rays, crests fall over the crests and troughs over the troughs), they reinforce each other and the intensity of the reflected rays is maximum. On the other hand, if the reflected rays are out of phase (*i.e.*, crests fall over the troughs), the intensity of the reflected beam is very low. If a photographic plate is placed to receive the reflected rays, *diffraction pattern* is obtained.

It is obvious that in order that the reflected rays BC and EF may be in phase, the extra distance JEK traversed by the ray DE should be an integral multiple of the wavelength λ of the X-rays, *i.e.*,

$$\text{Distance } JEK = n\lambda \qquad ...(i)$$

where n is an integer, *i.e.* 1, 2, 3, 4, etc. If d is the distance between the successive atomic planes, it is obvious from Fig. 4.35 that

$$JE = EK = d \sin \theta$$

so that

$$JEK = 2d \sin \theta \qquad ...(ii)$$

Putting this value in equation (i), we get

$$\boxed{2d \sin \theta = n\lambda} \qquad ...(iii)$$

W.L. Bragg (Father)

W.H. Bragg (Son)

This equation is called Bragg's equation. Using monochromatic X-rays, λ is constant. Also for a particular crystal when a particular face is facing the X-rays, d is constant. Thus by gradually increasing the value of θ, a number of positions are observed corresponding to n= 1, 2, 3, 4 etc., where the reflected beam will have maximum intensity. For other values of θ lying in between, the reflected beam will have intensity less than the maximum. Thus a diffraction pattern will be obtained in which the various maxima correspond to n= 1, 2, 3 etc. These are respectively called diffraction maxima of the first order, second order, third order etc.

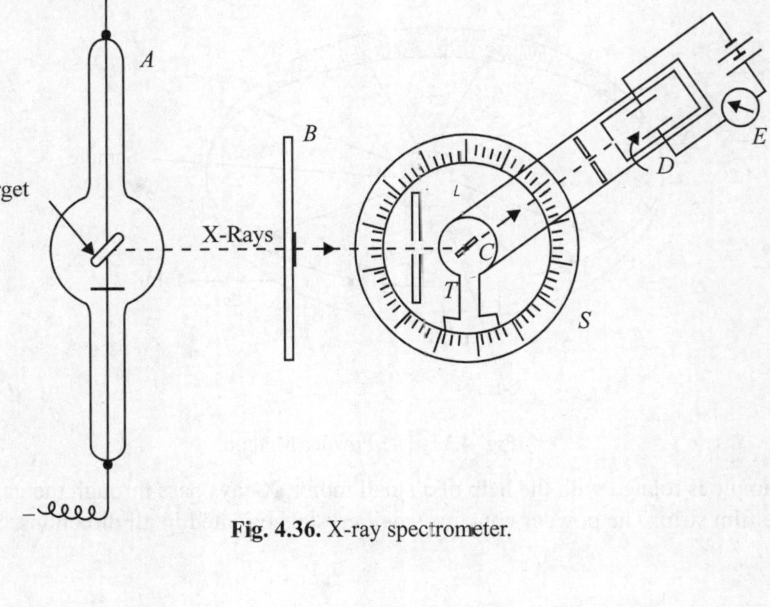

Fig. 4.36. X-ray spectrometer.

Measuring the angle θ at which the first maxima occurs (*i.e.*, first maximum intensity is observed) so that $n = 1$, and knowing the wavelength λ of X-rays used, the value of d can be calculated, using equation (*iii*).

Apparatus: The apparatus used by Bragg for studying the internal structure of crystals is called Bragg's X-ray spectometer and is shown in Fig. 4.36.

X-ray are generated in the *X*-ray tube *A*. These are made monochromatic *i.e.*, of single wavelength by passing through the absorbing screens *B*. Then they are passed through the slits *L*, to obtain the rays in a fine beam. This beam is then allowed to fall on some particular face of the crystal *C* which is mounted on a turn-table *T*. The turn-table can be moved over a circular scale *S*. The reflected beam is passed into the ionization chamber *D* which is connected to an electrometer *E*. The ionization chamber can also be moved on the circular scale *S* independent of the turn table *T*. The chamber contains some easily ionizable gas such as sulphur dioxide. Greater the intensity of the reflected beam, greater is the ionization produced in the gas and larger is the reading on the electrometer. When the crystal face is rotated by rotating the turn-table, the direction of the reflected beam changes. To receive the reflected beam into the ionization chamber every time, the ionization chamber is rotated through double the angle θ through which the crystal face is rotated. From these plots, the angles corresponding to the first, second, third maxima etc., are noted and the value of d calculated using Bragg's equation.

Calculation of *d*. For the face (100) of NaC1 crystal, when $n= 1$, $\theta = 5.9°$ so that $\sin\theta = 0.103$ Substituting these values in the Bragg equation, we get

$$2d \sin\theta = n\lambda$$

$$d = \frac{n\lambda}{2\sin\theta} = \frac{\lambda}{2\times 0.103} = 4.85\lambda$$

Thus knowing the wavelength of the X-ray used, distance '*d*' between the planes parallel to the face on which the X-rays are incident can be calculated.

4.9.1 Powder method for determination of crystal structure of a solid

When a single undistorted crystal of a solid is not available, we can use the powder method for crystal structural determination. In this method, the powdered crystalline substance is taken in a capillary tube. It is placed inside a camera containing a film strip (Fig. 4.37).

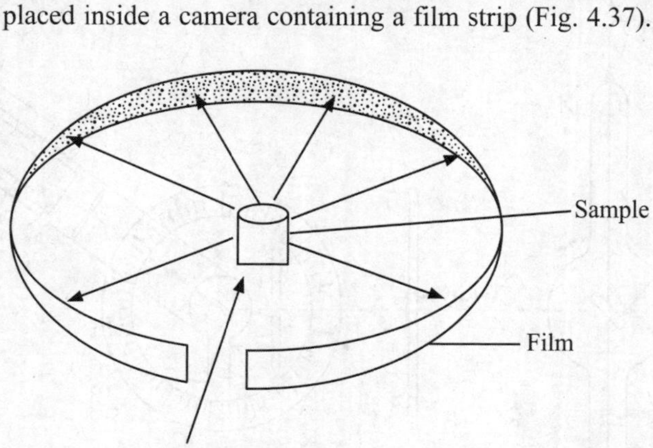

Fig. 4.37. The Powder Method.

The sample is rotated with the help of a small motor. X-rays pass through the gap between the ends of the film strip. The powder contains small crystals oriented in all directions. Some of these will reflect X-rays from each lattice plane simultaneously. The reflected X-rays will make an angle 2θ with the original direction. Lines of constant θ are obtained on the film. From the design of the camera, θ can be calculated for different crystal planes.

Fig. 4.38 illustrates the reflection of X-rays from the crystal planes

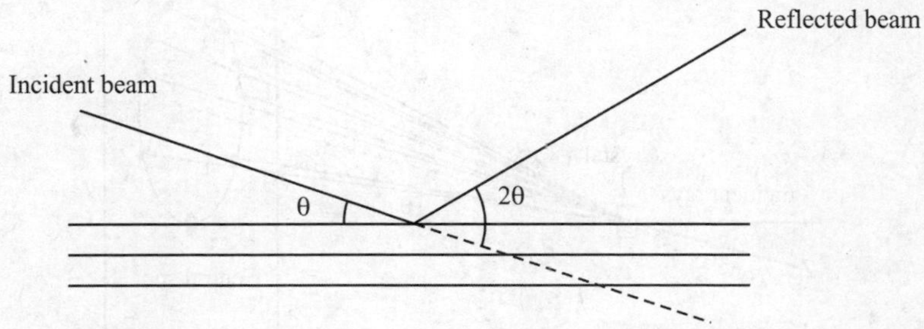

Fig. 4.38. Reflection of X-rays from crystal planes (powder method).

To calculate d, the interplanar distance, the Bragg's equation

$$n\lambda = 2d \sin \theta$$

as discussed in previous section is used.

4.9.2 Laue photographic method for the determination of structure of solids

The Laue method of examining the diffraction of X-rays by a crystal differs from other methods, in that a homogeneous beam of *X*-rays is used. The beam falls upon a small crystal and the diffracted beams are recorded on a photographic plate placed at a distance of a few centimeters from the crystal. The rays are limited to a narrow pencil by the slit system *A*, the crystal *C* is set on a holder which permits the adjustment of its orientation, and the photographic plate is placed at *S* (Fig. 4.39). A beam with a continuous range of wavelengths is conveniently obtained by using a bulb with a tungsten anticathode and a potential of about 65,000 volts.

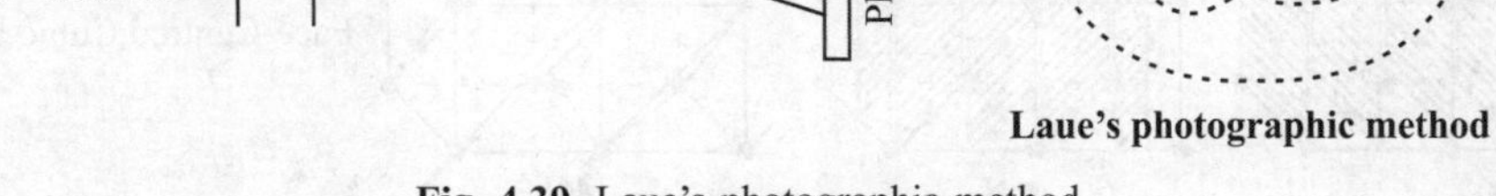

Fig. 4.39. Laue's photographic method.

A feature of the Laue photograph is the arrangement of the spots on ellipses which have one end of their major axis at the central spot. The same feature is shown more or less distinctly by all Laue photographs, the spots occurring where the ellipses intersect. All the spots which lie on any one ellipse are due to reflections by planes parallel to a common zone axis (Fig. 4.40). Incident rays fall upon a crystal at *C*, which has a zone axis parallel to *ZC*′, and we consider rays which are reflected by a plane of the corresponding zone. Suppose a reflecting plane is to be rotated about *ZC*′; the reflected beam will trace out a curve *EE*′ on the plate. If the incident rays make an angle θ with the zone axis, the reflected rays will always make the same angle with it, so that they lie on a cone of semi-vertical angle, θ around the zone axis. The ellipses in the Laue photograph are the intersections of such cones with the photographic plate. Each spot is at an intersection of ellipses, since a plane parallel to two zone axes is a crystal plane and can reflect the *X*-ray beam.

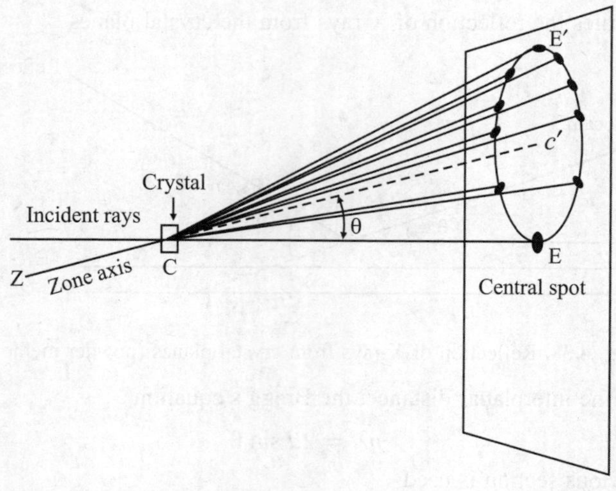

Fig. 4.40. Reflection of *X*-rays.

4.9.3 Structure of sodium chloride using Bragg's equation

There are three types of cubic lattices possible, namely, simple, face-centred and body-centred. Possible planes passing through the atoms are (100), (110) and (111), as shown in Fig. 4.41.

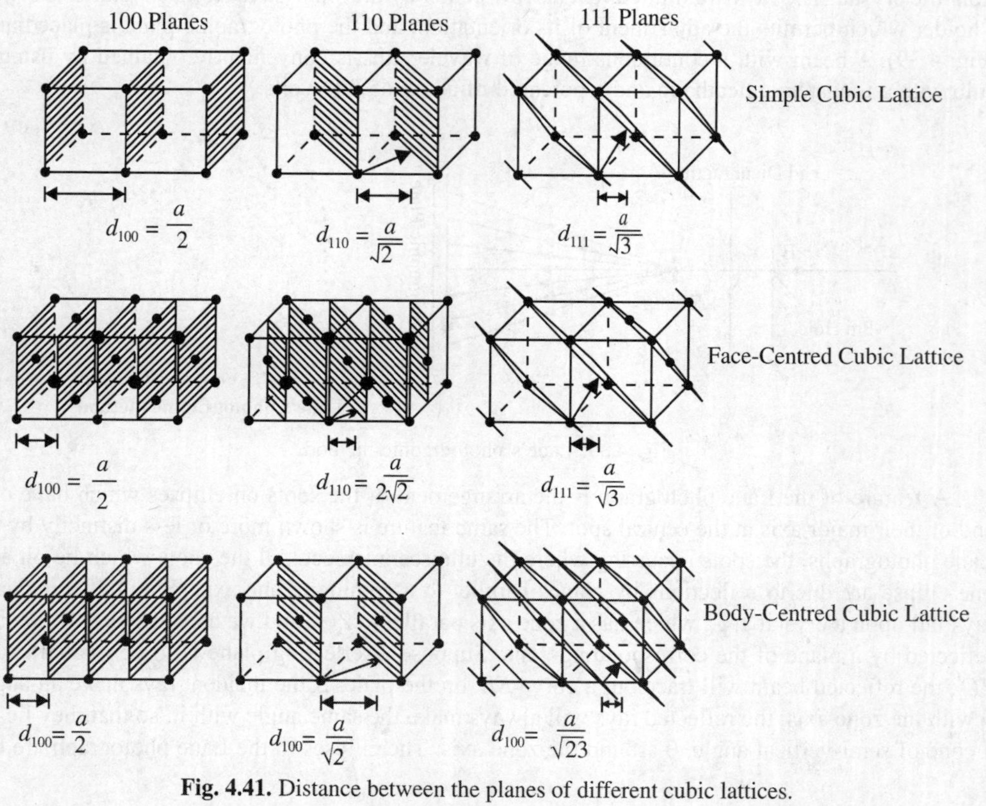

Fig. 4.41. Distance between the planes of different cubic lattices.

If a is the distance between the 100 planes of the simple cubic lattice, the distances between the planes of the planes of the three types of cubic lattices can be calculated theoretically and are given below each figure.

Thus the ratios $d_{100} : d_{110} : d_{111}$ for simple cubic, face-centred cubic and body-centred cubic lattices will be

Simple cube: $d_{100} : d_{110} : d_{111} = a : \frac{a}{\sqrt{2}} : \frac{a}{\sqrt{3}} = 1 : 0.707 : 0.577$

Face-centred cubic: $d_{100} : d_{110} : d_{111} = \frac{a}{2} : \frac{a}{2\sqrt{2}} : \frac{a}{\sqrt{3}} = 1 : 0.707 : 1.154$

Body-centred cubic : $d_{100} : d_{110} : d_{111} = \frac{a}{2} : \frac{a}{\sqrt{2}} : \frac{a}{2\sqrt{3}}$

$= 1 : 1.414 : 0.577$

In order to find out the type of the cubic lattice in the given crystal, the angles at which the first, second, or third order reflections, that take place from different planes are studied. The values of d_{100}, d_{110}, and d_{111} are calculated using Bragg's equation, *viz.*, $d = \frac{n\lambda}{2\sin\theta}$ where n is the order of reflection. The ratio $d_{100} : d_{110} : d_{111}$ is then calculated and compared with the ratios given above for different types of lattices.

Carrying out the study as above, it is found that NaC1 (as well as KC1) has face-centred cubic lattice as shown in Fig. 4.42.

The basis of this structure is Na^+ at (000) and Cl^- at $\left(\frac{1}{2}00\right)$

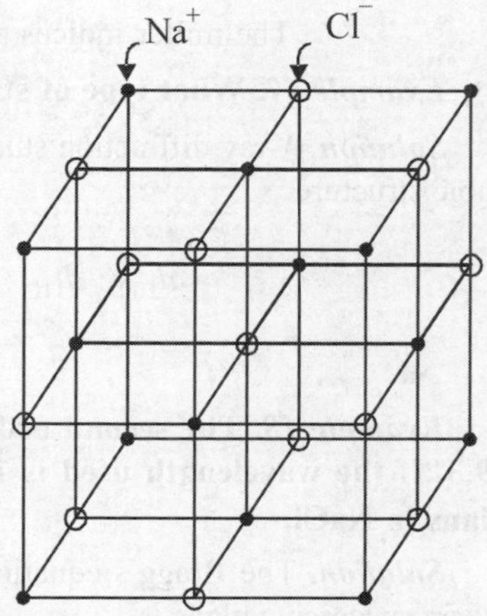

Fig. 4.42. Structure of NaCl (*fcc*).

Example 15. **The first order reflections from the 100, 110 and 111 planes of a given cubic crystal (NaCl crystal) were found to occur at angles 5.9°, 8.4° and 5.2° respectively. Determine the type of cubic lattice to which the crystal belongs.**

Solution. For first order reflections, $n = 1$ so that we have

$$d = \frac{n\lambda}{2\sin\theta} = \frac{\lambda}{2\sin\theta}$$

$$\therefore \quad d_{100} = \frac{\lambda}{2\,sin\,5.9^o} = \frac{\lambda}{2\times 0.103} = 4.845\lambda$$

$$d_{110} = \frac{\lambda}{2\sin 8.4^o} = \frac{\lambda}{2\times 0.146} = 3.425\lambda$$

$$d_{111} = \frac{\lambda}{2\sin 5.2^o} = \frac{\lambda}{2\times 0.094} = 5.319\lambda$$

$$\therefore \quad d_{100} : d_{110} : d_{111} = 4.845 : 3.425 : 5.319 = 1 : 0.706 : 1.096$$

This ratio is very close to that of the face-centred cubic. Hence the given crystal is face-centred cubic.

Example 16. **Answer the following:**

(*a*) $a \neq b \neq c, \alpha = \gamma = 90°, \beta \neq 90°$, identify the crystal system.

(*b*) The interplanar distance d_{hkl} ratio is 1 : 0.707: 0.577, identify the cubic lattice.

(*c*) The Weiss indices of a plane are 2, 2, ∞. What are its Miller indices ?

Solution. (*a*) The crystal system is monoclinic.

(*b*) It is a simple cubic system

(*c*) The Millar indices may be obtained as follows :

Take the reciprocal of Weiss indices

$$\frac{1}{2}\ \frac{1}{2}\ \frac{1}{\infty} \text{ or } \frac{1}{2}\ \frac{1}{2}\ 0$$

The miller indices are 110.

Example 17. **What type of structure is exhibited by KCl ?**

Solution. *X*-ray diffraction studies reveal that KCl possesses simple cubic structure. In simple cubic structure

$$d_{100} : d_{110} : d_{111} = a : \frac{a}{\sqrt{2}} : \frac{a}{\sqrt{3}}$$

$$= 1 : 0.707 : 0.577$$

Example 18. **The second order reflection of X-rays from (100) plans of NaCl occurs at 29.3°.If the wavelength used is 1.54Å, calculate the distance between two successive (100) plans in NaCl.**

Solution. The Bragg's equation relating the angle, wavelength of light and the distance between successive plans is

$$2d \sin\theta = n\lambda \text{ where } n \text{ is the order of reflection}$$

Substituting the values,we have

$$2d \sin 29.3° = 2 \times 1.54 \text{ or } d = \frac{1.54}{\sin 29.3} = 3.17Å$$

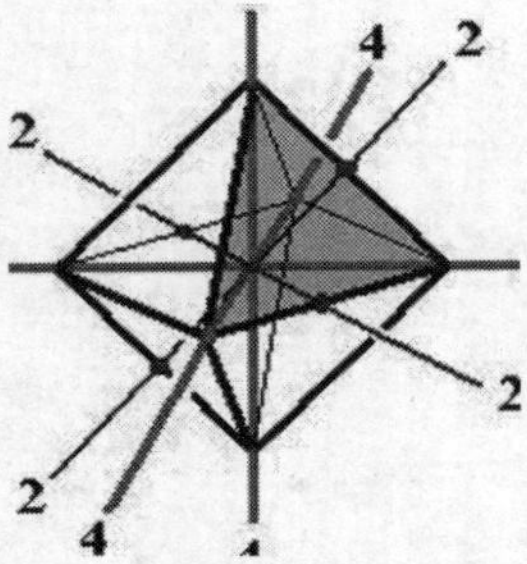

Symmetry elements

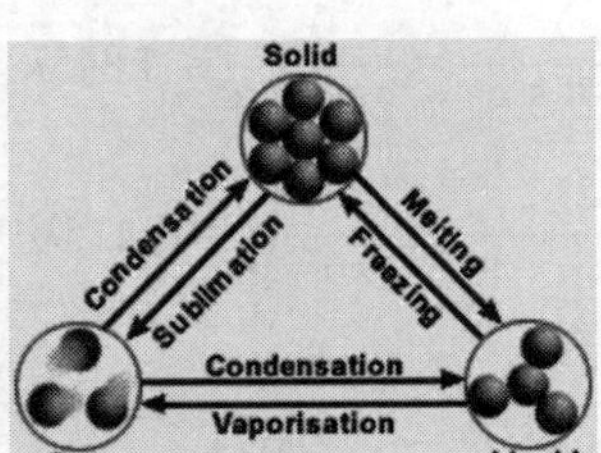

Interconversion of different states

Conversion of solid state into liquid

EXERCISES
(INCLUDING QUESTIONS FROM DIFFERENT UNIVERSITY PAPERS)

Objective Questions

Fill in the blanks

1. .... solids melt sharply at their melting point whereas .... solids melt gradually over a range of temperature.
2. Ionic solids are .... in polar solvents but .... in non-polar solvents
3. Coordination numbers of Na^+ and Cl^- in NaCl are ....
4. The ratio of the radius of the cation to the radius of the anion is called ....
5. The relation between atomic radius and the edge length in a body centred cubic soild is given by ....
6. Reciprocal of the coefficients of the intercepts on the axes, expreassed as integers are called ....
7. When there are atoms present at the centre of end faces of the unit cell, in addition to atom at the corners, the structure is called ....
8. The expression for Bragg's equation is ....
9. The methods for the determination of structure of a soild are : .... and ....
10. The equation for separation between the lattice planes is ....

Multiple choice QuestionS

Choose the correct option

1. Crystals can be classified into :
   (*a*) 7 systems (*b*) 9 systems (*c*) 10 systems
2. In Bragg's equation for diffraction of X-rays '*n*' represents :
   (*a*) no. of moles (*b*) quantum no (*c*) order of reflection (*d*) Avogadro no.
3. In a body centred cubic arrangement *A* ion occupies the centre while *B* ions occupy corners of cubic, the formula of solid is :
   (*a*) AB_2 (*b*) AB (*c*) A_2B (*d*) none of these
4. Closest packing is maximum in the crystal lattice of:
   (*a*) face centred cubic (*b*) bcc (*c*) simple centred cubic
5. Axis of symmetry are called "Hexad"
   (*a*) If original appearance is repeated twice
   (*b*) If original appearance is repeated 6 times
   (*c*) If original appearance is repeated after an angle of 90°
6. Radius ratio r_{Na+}/r_{Cl^-} in NaCl crystal is :
   (*a*) 0.524 (*b*) 0.414 (*c*) 0.732 (*d*) 0.80
7. Bragg's equation is :
   (*a*) $2n\lambda = d \sin\theta$ (*b*) $n\lambda = 2d \sin\theta$ (*c*) $\frac{n}{\lambda} = d \sin\theta$
8. The no. of atoms per unit cell in simple cubic, face centred cubic and body centred cubic are:
   (*a*) 1, 4, 2 (*b*) 4, 1, 2 (*c*) 2, 4,1 (*d*) 4, 8, 2

9. Which one of the crystal has the highest melting point ?
(*a*) Ionic crystal (*b*) Covalent crystal (*c*) Metallic crystal (*d*) Molecular crystal

10. Total no. of planes, axes and centre of symmetries in a crystal is termed as :
(*a*) elements of symmetry (*b*) symmetries
(*c*) symmetry operations

11. In a unit cell of NaCl lattice there are :
(*a*) 4 NaCl units (*b*) 3Na^+ ions (*c*) 6 Na^+ ions (*d*) 6Cl^- ions

12. Axis of symmetry in crystals is given by the general formula: '
(*a*) $\frac{360°}{n}$ (*b*) $n \times 360°$ (*c*) $n \times 180°$ (*d*) $n / 90°$

13. KC1 is a....... conductor of electricity :
(*a*) good (*b*) bad (*c*) semi (*d*) non

14. Name of equation related to X-ray diffraction by crystals is :
(*a*) Law equation (*b*) Bragg's equation (*c*) Lewis equation

15. Coordination no. for an atom in a primitive cubic unit cell is :
(*a*) 6 (*b*) 8 (*c*) 10 (*d*) 12

16. Interplanar distances in crystal can be determined by the equation :
(*a*) $n\lambda = 2d \sin\theta$ (*b*) $n = \lambda d \sin\theta$ (*c*) $\frac{n}{\lambda} = 2d \sin\theta$

17. The geometrical form consisting only of a regular array pf points in space is called:
(*a*) unit cell (*b*) lattice (*c*) crystal (*d*) amorphous solid

18. Miller indices in a crystal indicate :
(*a*) designation of planes incrystals (*b*) intercepts of plane on X axis
(*c*) reciprocals of fractional intercepts of that plane on various axes
(*d*) none

19. If a plane is parallel to an axis of crystal its Miller index will be :
(*a*) zero (*b*) infinity (*c*) one (*d*) none

20. How many different ways were suggested by Bravais in which similar points can be arranged in a three diamensional space :
(*a*) 14 (*b*) 10 (*c*) 7 (*d*) 9

21. If in a crystal intercepts are 1, ∞ and, ∞ the Miller indices are :
(*a*) (1 0 0) (*b*) (1 0 1) (*c*) (1 1 0) (*d*) (0 0 0)

ANSWERS

Fill in the blanks

1. Crystalline **2.** soluble **3.** 6 and 6 **4.** radius ratio

5. $r = \frac{\sqrt{3}}{4} a$ **6.** Miller indices **7.** end-centred **8.** $2d \sin\theta = n\lambda$

9. Powder method, Laue method **10.** $d_{hkl} = \frac{a}{\sqrt{h^2 + k^2 + l^2}}$

Multiple choice questions

1. (*a*)	**2.** (*c*)	**3.** (*b*)	**4.** (*a*)	**5.** (*b*)	**6.** (*a*)
7. (*b*)	**8.** (*a*)	**9.** (*c*)	**10.** (*a*)	**11.** (*a*)	**12.** (*a*)
13. (*b*)	**14.** (*b*)	**15.** (*a*)	**16.** (*a*)	**17.** (*b*)	**18.** (*c*)
19. (*a*)	**20.** (*a*)	**21.** (*a*)			

General Questions

1. What in meant by
(*i*) Rotation axis of symmetry :
(*ii*) Rotation–reflection axis of symmetry.
(*iii*) Rotation–inversion axis of symmetry.

2. (*a*) What is meant by space groups or point groups ?
(*b*) In how many space lattices, a total of 230 point groups are distributed ?

3. Define clearly the following terms :
(*i*) Axis of symmetry.
(*ii*) Mirror plane.
(*iii*) Inversion centre.
(*iv*) Improper rotation.

4. Describe the law of rational indices.

5. The crystallographic axes of a unit cell are *a*, *b* and *c*, draw a plane of which Miller indices are 221. Can there be more than one plane with the same Miller indices ?

6. What is meant by external feature of crystals ?
Explain clearly the various terms involved.

7. What is meant by total elements of symmetry ?
Explain by taking an example of a cubic crystal.

8. Explain the following terms with suitable examples :
(*i*) Plane of symmetry.
(*ii*) Axis of symmetry.
(*iii*) Centre of symmetry.

9. Draw the planes with Miller indices (112), (210).

10. Explain why, a five-fold axis of symmetry does not exist in any crystal.

11. What are Bravais lattice ?

12. What is meant by crystallographic elements and point goups ?

13. What is crystallography ?

14. In how many crystallographic groups the crystallographic elements are grouped ?

15. How will you calculate the number of particles per unit cell in the following :
(*i*) Simple cubic lattice.
(*ii*) Face centred cubic lattice.
(*iii*) Body centred cubic lattice.

16. Name different type of crystals. Name the units which occupy the lattice points with suitable Examples.

17. Show various axes of symmetry and planes of symmetry in a cubic crystal.

18. Show that for the given Miller indices in a crystal there is a whole set of equally spaced parrallel plane.

19. State and explain the laws of crystallography.

20. Represent diagramatically the Bravais lattices of

(*i*) Simple cubic lattice.

(*ii*) Face centered cubic lattice.

(*iii*) Body centered cubic lattice.

21. Differentiate between isomorphism and polymorphism.

22. What are the crystallographic dimensions of the unit cells of the following crystal systems :

(*i*) Triclinic.

(*ii*) Tetragonal.

23. What is plane of symmetry ? Show diagramatically various planes of symmetry in a cubic crystal.

24. Name various crystal systems with suitable examples.

25. What is meant by mirror plane and inversion centre ?

26. Calculate the number of the points per unit cell in

(*i*) a face-centred cubic lattice

(*ii*) a body-centred orthorhombic lattice

(*iii*) a simple cubic lattice.

27. Represent the following diagrammatically :

(*i*) Body - centred cubic lattice.

(*ii*) Face-centred orthorhombic lattice

(*iii*) Hexagonal lattice.

(*iv*) Simple tetragonal lattice.

28. What do you understand by space group ?

How many space groups are there in crystallography ?

29. Name the different crystal systems. Give their axial characteristic and one example of ech.

30. Briefly explain the terms :

(*i*) Unit cell.

(*ii*) Crystal lattice.

31. Differentiate between crystalline and amorphous solids. Justify that amorphous solid are liquids at all temperature.

32. (*a*) Show that for the given Miller indices in a crystal there is a whole set of equally spaced parallel planes.

(*b*) Draw the planes with Miller indices (112), (210).

33. (*a*) State and explain the law of constancy of interfacial angles.

(*b*) What is meant by mirror plane and inversion centre.

(*c*) Define improper rotation.

34. (*a*) Name the different crystal systems. Give their axial characteristic with one example of each.

(*b*) What are the crystallographic dimensions of the unit cells of the following systems :

(*i*) Triclinic.

(*ii*) Tetragonal.

35. (*a*) Explain the following terms with suitable examples :

(*i*) Plane of symmetry.

(*ii*) Axis of symmetry.

(*iii*) Centre of symmetry.

36. Identify for the crystal system to which each of the following belongs from the given data of their unit cells.

(*i*) $a = b = c = 9Å; \alpha = \beta = \gamma \neq 90°$

(*ii*) $a = b = c = 12Å; \alpha = \beta = \gamma = 90°$

(*iii*) $a \neq b \neq c; \alpha = \beta = \gamma = 90°$

(*iv*) $a = b \neq c; \alpha = \beta = \gamma = 90°$

(*v*) $a \neq b \neq c; \alpha = \beta = 90°; \gamma = 120°$

(*vi*) $a \neq b \neq c; \alpha = \beta \neq \gamma = 90°$

(*vii*) $a \neq b \neq c; \alpha = \gamma = 90° \neq \beta$

37. (*a*) Calculate the number of points per unit cell in

(*i*) a face centred cubic lattice

(*ii*) a body-centred orthorhombic lattice

(*iii*) a simple cubic lattice.

(*b*) What do you understand by space groups ?

How many space groups are there in crystallography ?

38. (*a*) Differentiate between isomorphism and polymorphism.

(*b*) What is meant by separation between lattice planes ? Derive the equation :

$$d_{hkl} = \frac{a}{\sqrt{h^2 + k^2 + l^2}}$$

in case of a cubic lattice.

39. (*a*) Name different types of crystal systems.

Draw different types of space lattices associated with cubic system.

(*b*) Represent the following diagramatically.

(*i*) Hexagonal lattice.

(*ii*) Simple tetragonal lattice.

40. Describe in detail the structure of sodium chloride. Show that the ratio $\frac{1}{d_{100}} : \frac{1}{d_{110}} : \frac{1}{d_{111}}$ $1 : \sqrt{2} : \frac{\sqrt{3}}{2}$ is obtained in sodium chloride crystal.

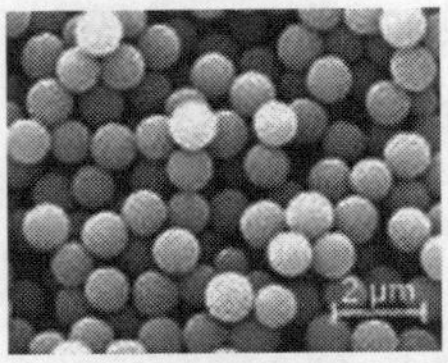

5

Colloidal State

5.1 INTRODUCTION

When water soluble substances like sodium chloride, copper sulphate, sugar etc., are put into water they dissolve and a homogeneous solution is obtained. The particles of the solute are not visible and their size is molecule size. Such mixtures are called *molecular solutions* or *true solutions.*

Now suppose we take muddy river water or an insoluble substance like lead sulphate or calcium sulphate is put into water. The particles of the solutes are visible even with the naked eye because their size is large. On keeping for some time, particles settle down. These mixtures are called *suspension.*

In between these two extremes, there are particles which are bigger than molecules but are too small to be seen even by a microscope. The colloidal state can thus be regarded as the intermediate stage between molecules and particles of a coarse suspension. A colloidal solution is also referred to as sol. These three states are illustrated in Fig. 5.1.

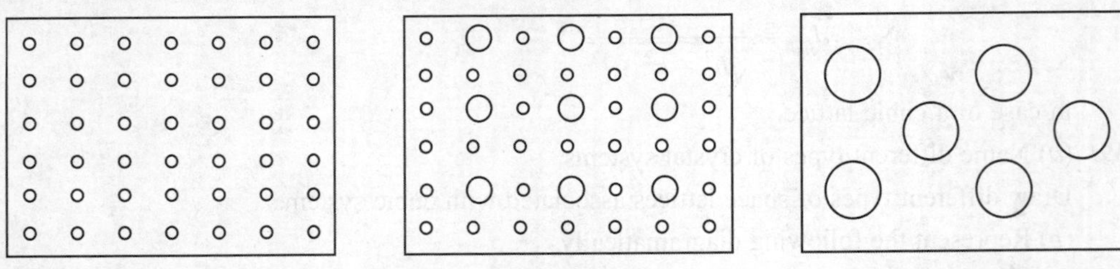

< 1 nm True solution — 1-100 nm Colloidal solution — > 100 nm Suspension

Fig. 5.1. Representation of true solution, colloidal solution and suspension.

The distinguishing characters of the three types of solutions are given in Table 5.1.

Table 5.1. Difference between a True Solution, Colloidal Solution and Suspension.

True Solution (1)	*Colloidal Solution* (2)	*Suspension* (3)
(*i*) The solute molecules in a true solution are not visible, even with a powerful microscope.	(*i*) The molecules of the dispersed phase cannot be seen with naked eye. However they can be made visible through a powerful microscope called Ultramicroscope.	(*i*) The particles are so big that they are visible even with naked eye. They sometimes settle at the bottom of the container, when the mixture is allowed to stand.

True Solution (1)	*Colloidal Solution* (2)	*Suspension* (3)
(*ii*) The particle size is below 1 nm in diameter.	(*ii*) The particle size is 1-100 nm in diameter.	(*ii*) The particle size is above 100 nm in diameter.
(*iii*) A true solution forms a homogeneous mixture.	(*iii*) A colloidal solution is heterogeneous.	(*iii*) Suspension is completely a heterogeneous mixture.
(*iv*) The particles of the solute in a true solution can pass through a parchment membrane.	(*iv*) The particles of a colloidal solution can not pass through a parchment membrane.	(*iv*) The suspension particle can not pass through the parchment membrane.
(*v*) The solute molecules in true solution can pass through even fine filter papers.	(*v*) The colloidal particles can also pass through ordinary filter papers, but can not pass through filter papers called Ultrafilters.	(*v*) The suspended particles are retained even by ordinary filter papers.
(*vi*) The osmotic pressure exerted by the solution is relatively more.	(*vi*) Colloidal solutions exert a very low osmotic pressure.	(*vi*) No osmotic pressure is observed.
(*vii*) When a strong beam of light is thrown on a true solution, the path of light is not illuminated.	(*vii*) The path of light is visible due to scattering of light, when a strong beam of light is passed through a colloidal solution (Tyndall effect).	(*vii*) No Tyndall effect is observed.
(*viii*) The solutions do not possess the property of adsorption (a surface phenomenon).	(*viii*) Colloidal solutions possess remarkable property of adsorption.	(*viii*) The phenomenon of adsorption is much less marked.
(*ix*) No electrical properties such as electroosmosis, cataphoresis and coagulation are observed.	(*ix*) Colloidal solutions exhibit these electrical properties to a great extent.	(*ix*) No such electrical properties are exhibited.

5.1.1 Various phases in a colloidal solution

A colloidal solution is a heterogeneous system consisting of two phases viz., the dispersed phase and the dispersion medium.

(*a*) Dispersed phase. It is just like the solute in a solution. It can be said that the component which is present in smaller proportion is the dispersed phase. For example in a colloidal solution of starch in water, starch is the dispersed phase.

(*b*) Dispersion medium. It is just like the solvent in a true solution. It is the component which is present in greater proportion. In the example of colloidal solution of starch in water, water is the dispersion medium.

It should be noted that particles of dispersed phase (colloidal particles) are bigger than those of dispersion medium (molecules or ions).

5.2. CLASSIFICATION OF COLLOIDS

Colloids can be classified on the basis of physical state of dispersed phase and dispersion medium, on the basis of appearance and on the basis of solvent affinity. Each of these is described below :

5.2.1 Classification of colloids on the basis of physical state of dispersed phase and dispersion medium

Each of the dispersed phase and the dispersion medium can be a solid or liquid or gas. There are eight different combinations possible. Table 5.2 describes the various colloids in which the dispersed phase and dispersion medium may have the same or different physical states. A colloidal solution with both the components in the gaseous state is not possible as according to Graham's Law of diffusion, they form a homogeneous mixture.

Table 5.2. Classification of colloids based on physical state of the phases.

S. No.	*Dispersed phase*	*Dispersion medium*	*Name*	*Examples*
1.	Solid	Solid	Solid sol	Coloured glass, alloys, gems.
2.	Solid	Liquid	Sol	Paints, inks, white of egg,colloidal Au, colloidal Pt.
3.	Solid	Gas	Aerosol	Smoke, dust.
4.	Liquid	Solid	Gel	Curds, cheese, jelly, pudding.
5.	Liquid	Liquid	Emulsion	Milk, butter, oil in water.
6.	Liquid	Gas	Aerosol	Mist, fog, clouds.
7.	Gas	Solid	Solid foam	Cake, bread, pumice stone.
8.	Gas	Liquid	Foam	Soap lather, aerated water.

5.2.2 Classification of colloids on the basis of appearance

On this basis, the colloids are divided into the following two main categories :

(*i*) **Sols.** When a colloidal solution appears a fluid, it is termed as sol. Sols are named after the dispersion medium. For example when dispersion medium is water, they are called **hydrosols**, if dispersion medium is alcohol they are called **alcosols** and if it is benzene, they are called **benzosols.**

(*ii*) **Gels.** When a colloid has a solid like rigid appearance it is called gel. The rigidity of a gel varies from substance to substance. Some substances may occur both as sols as well as **gels**. This depends upon the relative concentrations of the dispersed phase and dispersion medium.

The colloidal system in which the dispersed phase is a liquid and dispersion medium is solid is known as gel. A gel is formed by interlocking of the dispersed phase particles. Gelatin dissolves in warm water to form a colloidal solution which sets into a gel.

5.2.3 Classification of colloidal solutions on the basis of solvent affinity

When the dispersion medium exerts an attraction on the dispersed phase then we get **lyophilic Sol** (solvent loving). When the attraction between the dispersion medium and the dispersed phase is very little then the sol is called **lyophobic sol** (solvent hating).

Typical examples of lyophobic sols (non-reversible): Metals, sulphur, sulphides, silver halides, egg albumen, silicic acid and ferric hydroxide give non-reversible colloidal systems with water.

Examples of lyophilic (reversible) sols: Protein, starch, glue, gelatin and agar-agar give reversible colloidal system with water. Rubber in benzene is also a reversible colloidal system. Main points of difference between the two are given in Table 5.3.

Table 5.3. Difference between lyophilic and lyophobic colloids.

Lyophilic (Solvent loving)or Hydrophilic Colloids (water as solvent)	*Lyophobic (Solvent hating) or Hydrophobic Colloids (water as solvent)*
(*i*) Lyophilic colloids pass into the colloidal form readily when brought in contact with solvent.	(*i*) Lyophobic colloids do not form the colloidal solutions easily when treated with solvent.
(*ii*)They are called the reversible sols because they can be recovered from the colloidal solutions and can be converted into the colloidal form when desired.	(*ii*) They are called irreversible sols, because they can not be recovered from their colloidal form.
(*iii*) Small quantities of electrolytes do not effect precipitation, but only large quantities of electrolytes cause precipitation.	(*iii*) Even small quantities of the electrolytes can cause precipitation.
(*iv*) The particles are not easily detected in the ultra-microscope.	(*iv*) The particles are easily detected by an ultramicroscope.
(*v*) The particles may or may not migrate in an electric field. The migration may be in any direction.	(*v*) Particles migrate only in one direction in the presence of electric field.
(*vi*) Substances like starch, proteins, gums, soaps, and metasilicic acid are common examples.	(*vi*) Colloids of metals (Au, Pt), sulphur, assenius sulphide and silver iodide are common examples of this class.

5.2.4 Dispersion colloids, multimolecular or association colloids, macromolecular or molecular colloids

Dispersion colloids. In case of colloidal solutions in which the dispersed phase has little affinity for the dispersion medium (*i.e.* solid substance is insoluble in the dispersion medium), the size of the solid particles (dispersed phase) lies in the colloidal range. This system is called a dispersion colloid. For example, gold sol, hydrated ferric oxide sol, colloidal sulphur, a colloidal dispersion of As_2O_3 in water etc. Generally this type belongs to **lyophobic sols.**

Multimolecular or association colloids. Here an individual colloidal particle consists of an aggregate of a large number of small atoms or molecules. The size of such species lies in the colloidal range. The species thus formed are called associated colloids or multimolecular colloids, for example, soap solution, dyes, surface active agents etc. Generally in this type, there is occurrence of lyophobic and lyophilic portions in the same molecules. Thus the aggregates of ions or molecules with lyophilic as well as lyophobic parts are called **micelles.** Micelles are of colloidal dimension though the individual ions or molecules may not be of the colloidal dimensions.

Macromolecular or molecular colloids. In these systems each particle of the dispersed phase is of colloidal dimensions. Mostly this type consists of lyophilic sols. This resembles true solutions in some respects. Examples of this type include proteins, cellulose, the solutions of rubber and high polymers in organic solvents etc.

5.3 MICELLES

There are certain strong electrolytes which give a normal solution at low concentration but show colloidal nature at higher concentrations. These are called **micelles.** These substances may also be referred to as *Associated Colloids.* Soap is an example of this class. It forms a micelle. These substance are commonly named as *surface active substances.* They consist of lyophilic as well as lyophobic parts in the same molecule. A long hydrocarbon chain serves as lyophobic and an ionic group like carboxy or sulphonic serves as lyophilic. A micelle may contain as many as 100 molecules.

5.3.1 Cleansing action of soaps on the basis of micelle formation

Water is not a good medium for cleaning our body or clothes. This is because of the high surface tension of water. It does not spread evenly over the dirt particularly grease in the clothes or on the body. The exceptionally high value of surface tension of water could be explained in terms of hydrogen bonding. Without complete wetting of grease by water, we can not expect efficient cleaning. The function of soap or other synthetic detergents is to decrease the surface tension of water. This is made possible by micelle formation as shown in Fig. 5.2.

Ordinary soap is sodium stearate $C_{17}H_{35}COONa$. It gives $C_{17}H_{35}COO^-$ in solution. The hydrocarbon chain consisting of 17 carbon atoms is hydrophobic in nature. A number of such hydrocarbon chains are directed towards grease as shown in Fig. 5.2. the negative poles of the stearate ions then attract the positive ends of water (water is a polar molecule) forming a micelle.

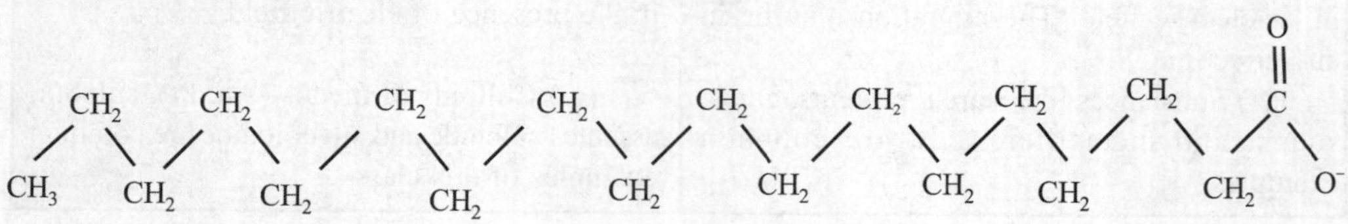

$(C_{17}H_{35}COO^-)$

The anion of a soap is represented as follows

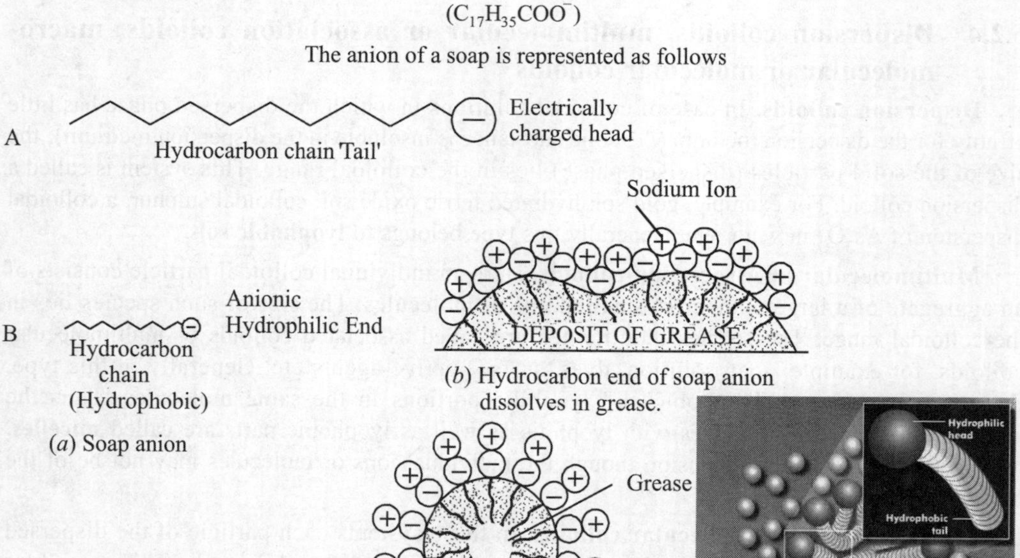

Fig. 5.2. A. Representation of a soap molecule, B. Formation of a micelle.

Dirt inside a micelle

The micelles are then washed away with more water. Thus soap helps in the cleaning of clothes or body. The function of soap is to reduce grease or oil to colloidal particle. therefore, soaps are emulsifying agents. Grease particles are prevented from coming closer because of the covering by soap molecules. Any synthetic detergent like sodium lauryl sulphate $C_{17}H_{25}SO^{-}_{4}$ Na^{+} would form a miscelle by a similar mechanism.

5.4 PREPARATION OF COLLOIDAL SOLUTIONS

A number of methods are available for the preparation of colloidal solutions. These are described as under.

5.4.1 Dispersion (size reduction) methods for the preparation of colloidal solution

Dispersion methods for the preparation of colloidal solution are described as under:

1. Mechanical dispersion or disintegration. It involves the following steps:

(*a*) First of all, the substance is powdered throughly.

(*b*) The finely ground substance is brought in the form of coarse suspension, by shaking with the dispersion medium.

(*c*) The suspension is then made to pass through a colloid mill (Disc Mill), which consists of two metallic discs almost touching each other and rotating with a terrific speed in opposite directions (Fig. 5.3). The suspension is thus subjected to a powerful shearing force, resulting in the breaking of suspended particles to form particles of colloidal dimensions.

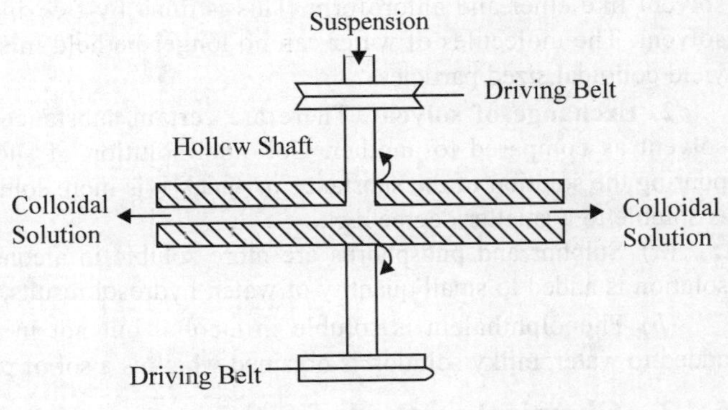

Fig. 5.3. Colloid Mill.

The colloid mills are now a days used in many industrial processes, such as the manufacture of dyestuffs, printing inks, varnishes, lacquers, paints, ointments, dental creams, oil and water colours, etc.

2. Electro-dispersion or Bredig's arc method. This method is employed for getting colloidal solutions of metals like gold, silver, copper, platinum etc. The metal rods (electrodes) are immersed in the dispersion medium say water. An arc is struck between the two electrodes (Fig. 5.4). As a result of powerful heat of arc, metal is converted into its vapours which get condensed at once (the vessel being placed in ice cold water bath), resulting in the formation of colloidal size particles. (To stabilize the solid, a small quantity of NaOH or K_2CO_3 is added).

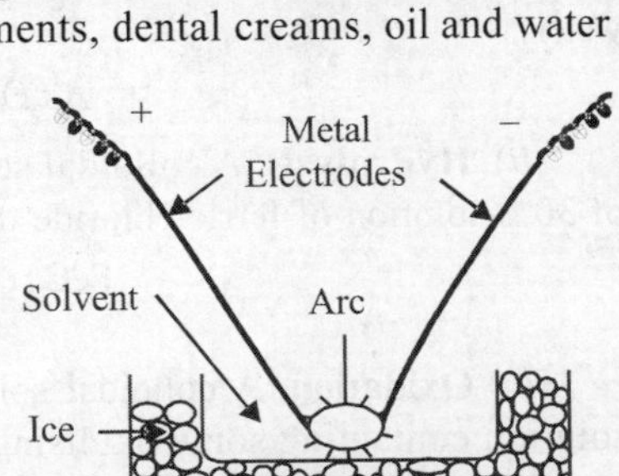

Fig. 5.4. Bredig's arc Method.

3. Peptization (Dispersion of a precipitate). It is the redispersion of the coagulated sol. Peptization *is defined as the process by which a colloidal solution can be obtained by dispersing the coagulated mass in which the colloidal particles pre-exist.* For example, ferric hydroxide is a coagulated sol and is a precipitate. On adding ferric chloride solution (Peptizing agent) to it, a reddish brown solution containing ferric hydroxide sol is obtained (Fig. 5.5)

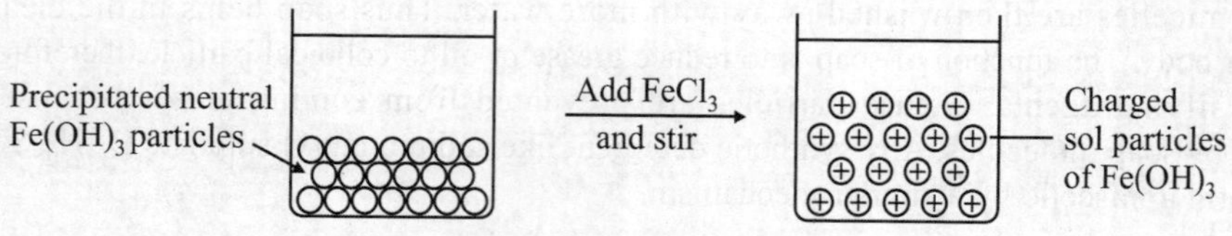

Fig. 5.5. Preparation of ferric hydroxide sol by peptization.

The excess of ferric chloride, if any, can be removed by dialysis. Similarly a precipitate of silver iodide can be peptised by shaking with a dilute solution of either KI or $AgNO_3$ when very dilute HCl acid is added to the freshly precipitated aluminium hydroxide, $Al(OH)_3$, a colloidal solution of aluminium hydroxide is obtained. The acid should be in small quantity.

5.4.2 Condensation or aggregation methods for the preparation of colloidal solutions

Condensation methods for the preparation of colloidal solutions are described as under:

1. Excessive cooling. This method is employed to get colloidal solution of ice in organic solvent like ether and chloroform. This is done by freezing a solution of water in the given solvent. The molecules of water can no longer be held in solution. They collapse together to yield colloidal sized particles.

2. Exchange of solvent. There are certain substances which are more soluble in one solvent as compared to another. Colloidal solution of such substances can be obtained by pouring the solution of the substance in which it is more soluble to the other solvent, in which it is soluble to a smaller degree *e.g.*,

(*a*) Sulphur and phosphorus are more soluble in alcohol than in water. If their alcoholic solution is added to small quantity of water, hydrosol results.

(*b*) Phenolphthalein is soluble in alcohol but not in water. If its alcoholic solution is added to water, milky solution is obtained which is a sol of phenolphthalein.

5.4.3 Chemical methods for the preparation of colloidal solutions

Chemical methods for preparing colloidal solutions are given below.

(*i*) **Double decomposition.** A colloidal solution of arsenious sulphide is obtained by adding water saturated with H_2S gas, to a solution of arsenius oxide (one gm. in 500 ml of water).

$$As_2O_3 + 3H_2S \longrightarrow As_2S_3 + 3H_2O$$

(*ii*) **Hydrolysis.** A colloidal solution of **ferric hydroxide** is obtained by adding 2 or 3 ml of 30% solution of ferric chloride to about 500 ml of hot boiling water.

$$FeCl_3 + 3H_2O \rightleftharpoons \underset{\text{(deep red)}}{Fe(OH)_3} + 3HCl$$

(*iii*) **Oxidation.** A colloidal solution of sulphur is obtained by passing H_2S gas through a solution containing some oxidising agent, say dilute HNO_3 or by passing H_2S into a solution of sulphur dioxide in water.

$$H_2S + [O] \longrightarrow H_2O + S$$

$$SO_3 + 2H_2S \longrightarrow 3S + 2H_2O$$

(*iv*) **Reduction.** A colloidal solution of gold (violet coloured) may be obtained by adding stannous chloride solution (reducing agent) to a solution of gold chloride.

$$2AuCl_3 + 2SnCl_2 \longrightarrow 3SnCl_4 + \underset{\text{(violet)}}{2Au}$$

5.4.4 Purification of colloidal solutions

A sol contains some electrolyte during its preparation. A small amount of electrolyte is essential for stability of the sol, but excess of it can cause coagulation. Therefore, it is essential to remove excess of electrolyte.

The purification of the sol is done by the following process.

Dialysis. The process of removing the dissolved electrolytes from the sol by means of a membrane is called dialysis.

A dialyser consists of a bag made of parchment or cellophane and suspended in running water in a trough. The mixture of the colloid and the crystalloid is taken in the bag and apparatus fitted as shown in Fig. 5.6 (*a*). After sometime, whole of the crystalloid in solution passes out while the colloid remains in the bag. It is further observed that the above process of separation can be accelerated if an electric field is applied around the parchment bag. The process is then called **electrodialysis.** This is shown in Fig. 5.6 (*b*).

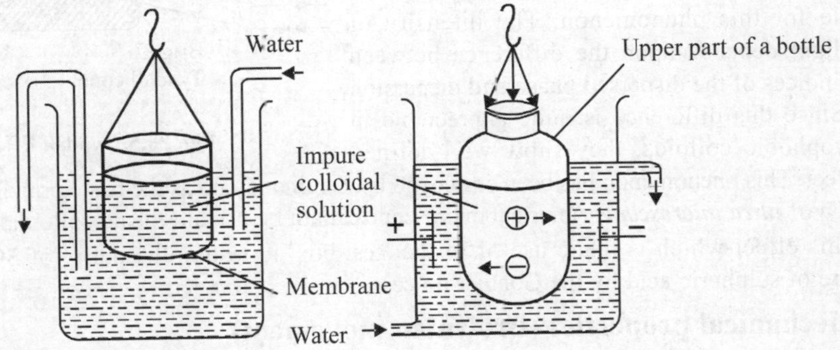

(*a*) Dialysis (*b*) Electrodialysis

Fig. 5.6. Dialysis and Electrodialysis

Explanation. Because of their bigger size, colloidal particles are unable to pass through the parchment bag while ions of the crystalloid are carried away by the running water in the dialyser. Further on applying the electric field, the movement of the ions is accelerated.

An important application of dialysis is in the purification of blood in the artificial kidney machine. The dialysis membrane permits small particles to pass through but colloidal haemoglobin does not pass through.

5.4.5 Ultrafiltration

Colloidal particles pass through an ordinary filter paper. This is because its pores are too big to retain colloidal particles. If the filter paper is impregnated with collodion, the pore size of the filter paper is much reduced. Such a modified filter paper is called **ultrafilter.**

Separation of colloidal particles from a solution by filtration through an ultrafilter is called ultrafiltration.

5.5 PROPERTIES OF SOLS (COLLOIDAL SOLUTIONS)

Different properties of colloidal solutions are discussed under the following headings :

5.5.1 Colour

Colour of a sol depends on the wavelength of the light scattered by the dispersed phase which, in turn, depends upon the size and nature of the particles. In the case of silver sol, the following data is available.

Colour of silver sol	Particle size (mm)
Orange-yellow	6×10^{-5}
Orange-red	9×10^{-5}
Purple	13×10^{-5}
Violet	15×10^{-5}

5.5.2 Optical properties (Tyndall effect)

A beam of light entering a dark room lights up the dust particles floating in air. Similarly *when a strong beam of light is concentrated on a colloidal solution the path of the beam is illuminated by a bluish light* (and becomes visible when observed from the side). This phenomenon is known as *Tyndall effect* (Fig. 5.7).

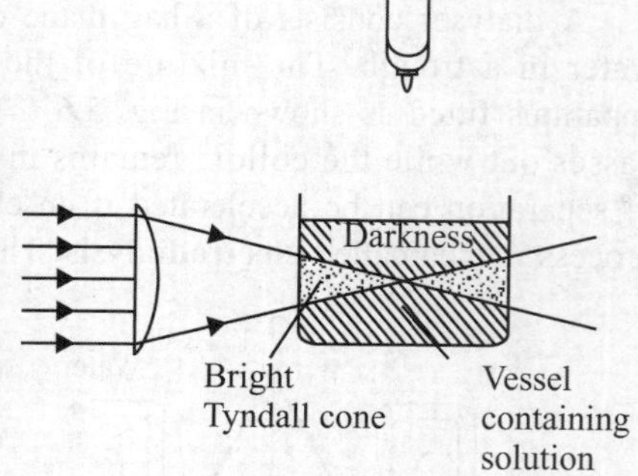

Fig. 20.5. Tyndall Effect.

The scattering of light by the colloidal particle is responsible for this phenomenon. The intensity of scattered light depends upon the difference between refractive indices of the dispersed phase and dispersion medium. Since this difference is quite appreciable in case of lyophobic colloids, they show well-defined Tyndall effect. This phenomenon has been employed by Zsigmondy on as the basic principle for the construction of ***ultra microscope***. A normal microscope cannot be used to locate colloidal particles.

It is this effect which is made use of in the 'test box' in testing the purity of SO_2 for the manufacture of sulphuric acid by the Contact process.

5.5.3 Mechanical properties (Brownian movement)

If we examine the colloidal solutions with an ultramicroscope, we find that suspended particles are in a constant rapid *zig-zag motion* (Fig. 5.8). This zig-zag motion is called *Brownian movement.* Robert Brown first noticed this movement while examining pollen grains suspended in water under a microscope.

Robert Brown

The kinetic theory gives a nice explanation of the phenomenon. The motion is due to unequal bombardment by the molecules of dispersion medium. With increase in size of particles, the probability of unequal bombardment diminishes and so Brownian movement decreases with increase of particle size.

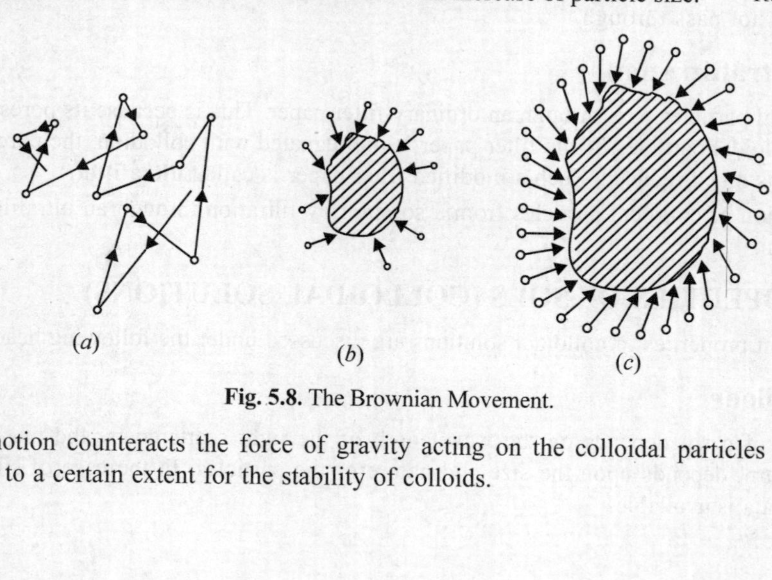

Fig. 5.8. The Brownian Movement.

This motion counteracts the force of gravity acting on the colloidal particles and is thus responsible to a certain extent for the stability of colloids.

Diffusion. Since the rates of diffusion decrease very rapidly with increase in the mass of the particles (Graham's law), colloidal solutions would diffuse much more slowly than ordinary molecular solutions. This property has been made use of in determining molecular weights of colloids. Since the equation to be used for the purpose also has a term for the *Avogadro's number* (the number of molecules in a gram molecule), the latter can also be determined from it, using solutes of known molecular masses.

Sedimentation of suspensoids. Although colloidal solutions are found to be fairly stable over considerable periods of time, they do very slowy settle down under the influence of gravity—even though some of them may take years to do so. The rates of settling [or sedimentation, as it is called] can be readily determined because the boundary line between the clear medium and the sol can be easily made out.

5.5.4 Electrical properties (Cataphoresis or Electrophoresis)

Colloidal particles carry an electric charge and therefore, move towards the oppositely charged electrode when the solution is placed under the influence of an electric field (Fig. 5.9). This *migration of colloidal particles under the influence of electric field is called cataphoresis.*

The type of charge on sol particle can be determined by noting the direction of their motion. The important applications of cataphoresis are:

(*a*) Removal of colloidal gases.

(*b*) Removal of dirt particles.

(*c*) Electro-deposition of rubber.

(*ii*) **Electro-osmosis.** A sol is electrically neutral. This reveals that dispersion medium carries an equal but opposite charge to the dispersed particles. If the dispersed phase is kept stationary by some suitable means, it is observed that the dispersion medium itself begins to move under the influence of an electric field. This motion of the dispersion medium under the influence of an electric field is **called electro-osmosis.**

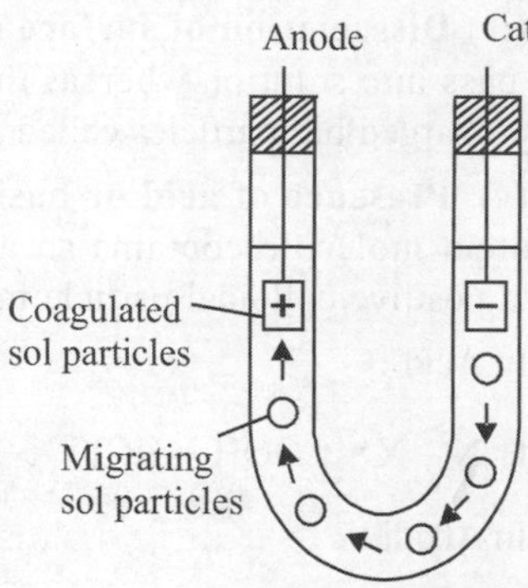

Fig. 5.9. Cataphoresis.

5.5.5. Stability of colloidal solutions

Colloidal solutions are stable, this is due to the fact that they are electrically charged. Colloidal particles carry either positive or negative charge. Since all the colloidal particles bear identical charge, they repel each other and do not precipitate easily under the effect of gravity.

The electrical charges carried by colloidal particles of various substances dispersed in water, as determined by electrophoresis experiments, are shown in table 5.4.

Tabld 5.4. Charge on the colloidal particles.

Positively Charged	*Negatively Charged*
1. Haemoglobin	1. Clay
2. Colloidal particles of Fe $(OH)_3$, Cr $(OH)_3$ and Al $(OH)_3$,	2. Metallic particles like those of gold, silver and platinum
3. Basic dye stuffs.	3. Sulphides of Cu, Pb, As, Sb, and Cd.
	4. Silicic acid, stannic acid and vanadium pentoxide.
	5. Gum-Arabic and soluble starch.
	6. Acid dye stuffs.

5.5.6 Origin of charge on colloidal particles

The electric charge is acquired by sol particles by one or more of the following means:

(*i*) **Friction.** The charge on the sol particles appears as a result of rubbing of the particles of dispersed phase against those of dispersion medium.

(*ii*) **Selective adsorption.** The sol particles have a tendency to adsorb anions or cations from the solution. For example, the colloidal particles of As_2S_3 can adsorb either As^{3+} or S^{2-} ions. Since during the precipitation of this sol, only sulphide ions from H_2S are in excess, therefore on a colloidal particle of As_2S_3, the S^{-2} ions get adsorbed and the particles become negatively charged.

Similarly, the sol of Fe $(OH)_3$ obtained by the hydrolysis of $FeC1_3$ is positively charged. In the medium, are present $Fe(OH)_3$ particles, Fe^{3+} and Cl^- ions. Fe^{3+} ions get adsorbed on the surface of Fe $(OH)_3$ and particles become positively charged.

The AgC1 particles can adsorb Cl^- ions from chloride solution and Ag^+ ions from solution containing Ag^+ ions. The colloidal particles becomes negatively charged in the first case and positively charged in the second case.

(*iii*) **Electron capture.** During electro-dispersion of the metals in Bredig's arc method.

(*iv*) **Dissociation of surface molecules.** Soap molecules dissociate to give ions. The cations pass into solution whereas the anions $(C_nH_{2n+1}COO^-)$ aggregate together to form a negatively charged big particles called micelle.

(*v*) **Presence of acid or basic group.** Protein molecules yield colloidal solutions. Since a protein molecule contains an acidic group—COOH and also a basic group—NH_2, it will give a positive colloidal particle in acidic medium and a negative particle in alkaline medium.

In Acid :

$$H_2N - X - COOH + HC1 \longrightarrow H^+{}_3N - X - COOH + Cl^-$$

In Alkali :

$$H_2N - X - COOH + NaOH \longrightarrow H_2N - XCOO^- + Na^+ + H_2O.$$

5.5.7 Theory of electrical double layer

The electrical properties of colloids can be explained satisfactorily by postulating the existence of an electrical double layer of opposite charges at the surface of separation between a solid and a liquid, *i.e.*, at a *solid-liquid interface.*

According to the modern views, when a solid is in contact with a liquid, a double layer of ions appears at the surface of separation, as shown in Fig. 5.10. One part of the double layer is fixed on the surface of the solid. This is known as the *fixed part* of the double layer. It consists of either positive ions or negative ions. The second part consists of a *diffuse* or a *mobile layer* of ions which extends into the liquid phase. This layer consists of ions of both the signs but its net charge is equal and opposite to that on the fixed part of the double layer. Such an arrangement of charges is shown in Fig. 5.10. In one case (Fig. 5.10 *a*), the fixed part of the double layer carries positive charge while in the second case (Fig. 5.10 *b*), the fixed part carries negative charge. It will be seen that in each case the net charge on the diffuse layer is equal and opposite to that on the fixed layer.

This theory is applicable to colloidal systems also. The ions which are preferentially adsorbed by the colloidal particles, are held in the fixed part of the double layer. These ions give the characteristic charge to the colloidal particles. The ions carrying the opposite charge, *i.e.*, the counter ions, are present more in the diffuse portion of the double layer, giving a net opposite charge to this layer.

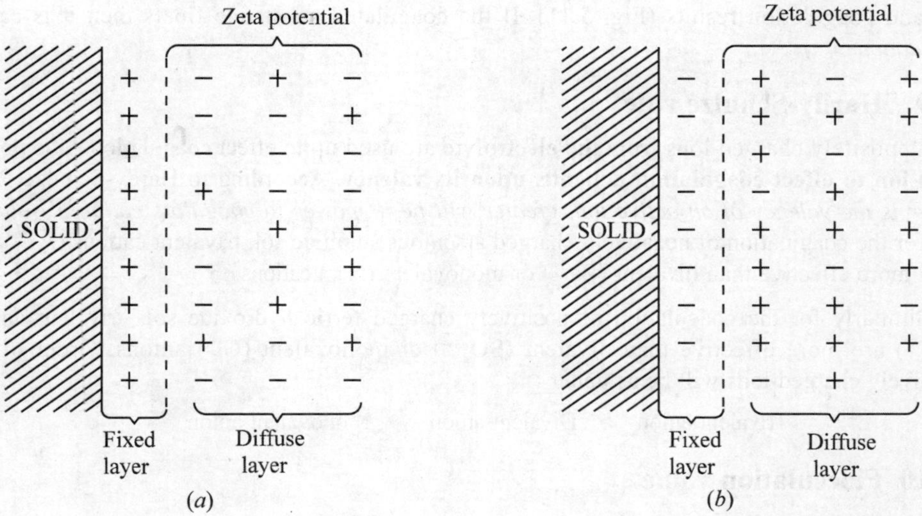

Fig. 5.10. The electrical double layer.

The existence of charges of opposite signs on the fixed and diffuse parts of the double layer leads to the appearance of a difference of potential between the two layers, as shown in Fig. This difference of potential between the two layers is known as **zeta potential** or **electrokinetic potential.** Thus this theory explains the charge and hence the stability of the colloidal particles.

5.5.8 Coagulation

It is a process in which the charge over colloidal particles is neutralised resulting in the precipitation of the colloidal particles. The process is brought about spontaneously by the addition of an electrolyte. For example, when barium chloride solution is added to arsenius sulphide sol it becomes turbid and after sometime a precipitate of arsenius sulphide separates out. The charge on the colloidal particles is neutralised. We know that arsenius sulphide is a negative

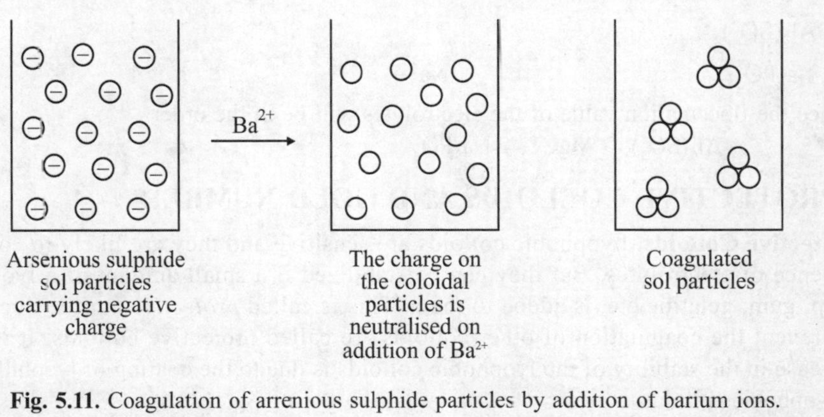

Fig. 5.11. Coagulation of arrenious sulphide particles by addition of barium ions.

sol. The colloidal particles carry negative charge. The sol is stable because of the mutual repulsion between the various colloidal particles having similar charge. But when barium chloride solution is added, the negative charge on the colloidal particles is neutralised by the +ve charge on barium ions. Thus the colloidal particles accumulate to form particles of larger size and coagulation results (Fig. 5.11). If the coagulated substance floats then it is called *flocculation.*

5.5.9 Hardy-Schulze rule

Oppositely charged ions from the electrolyte are used up to effect coagulation. **The power of an ion to effect coagulation depends upon its valency.** According to Hardy-Schulze Rule, *higher is the valency of an active ion, greater will be its power to coagulate colloidal solution e.g.,* for the coagulation of negatively charged arsenious sulphide sol, trivalent cations (Al^{+3}) are much more effective than bivalent (Ba^{2+}) or monovalent (Na^{+}) cations.

Similarly for the coagulation of positively charged ferric hydroxide sols trivalent anions (PO_4^{-3}) are more effective than divalent (SO_4^{-2}) or monovalent (Cl^{-}) anions. Order of the negatively charged ions will be as under :

Trivalent anion > Divalent anion > Monovalent anion

5.5.10 Flocculation value

The minimum concentration of an electrolyte in millimoles that must be added to one litre of a colloidal solution so as to bring about complete coagulation or flocculation is called the *flocculation value.*

Coagulation can also be brought about by mixing two oppositely charged colloidal solutions. For example, by mixing ferric hydroxide sol with arsenious sulphide sol, a mutual coagulation takes place.

***Example 1.* Arrange the following in increasing order of flocculation value for a negative sol.**

$$MgCl_2, Al_2(SO_4)_3, Na_3PO_4$$

Solution. For a negative sol, greater the charge on the cation, greater the flocculation value. The charges on the cation in the given electrolytes are as follows:

Electrolyte	*Cation*	*Charge*
$MgCl_2$	Mg^{2+}	+ 2
$Al_2(SO_4)_3$	Al^{3+}	+ 3
Na_3PO_4	Na^{+}	+ 1

Hence the flocculation value of the electrolytes will be in the order

$$Al_2(SO_4)_3 > MgCl_2 > Na_3PO_4$$

5.6 PROTECTIVE COLLOIDS AND GOLD NUMBER

Protective Colloids. Lyophobic colloids are sensitive and they are likely to coagulate in the presence of electrolytes. But they can be stabilized if a small amount of a lyophillic sol like soap, gum, gelatine etc. is added to them. This is called *protective action.* Such colloids which prevent the coagulation of other colloids are called protective colloids. It is believed that increase in the stability of the lyophobic colloids is due to the coating of lyophillic colloids on the lyophobic colloids.

Gold number. The protective action of the colloid is measured in terns of gold number. Gold number is defined *as the number of milligram of the protective colloid which just prevents the coagulation of 10 ml of gold solution when 1 ml of a 10 % solution of sodium chloride is added to it.* Gold number of a few protective colloids are given below :

Protective colloid	*Gold number*
Gelatin	0.005
Albumin	0.08
Starch	25

Example 2. **The coagulation of 100 ml of a colloidal solution of gold is completely prevented by addition of 0.25 g of starch to it after adding 1ml of 10% NaCl solution. Find out the gold number of starch.**

Solution. As per the definition of gold number, it is the number of milligrams of protective colloid (starch in this case), which prevents the coagulation of 10 ml of gold solution when 1ml of 10% NaCl solution is added to it.

In the present case

100 ml of colloidal solution requires to prevent coagulation = 0.25 g starch

= 250 mg starch

10 ml of colloided solution requires to prevent coagulation = 25 mg starch

Hence, gold number of starch = 25

Example 3. **Define gold number. The gold numbers of A, B, C and D are 0.005, 0.05, 0.5 and 5 respectively. Which of these has the greatest protective action? Explain your answer.**

Solution. For the definition of gold number, See Art. 5.6. *A* has the greatest protective action as we require the minimum quantity of *A* to prevent the coagulation of the gold solution when sodium chloride solution is added to it.

5.7 EMULSIONS

Colloidal solutions in which the dispersed phase as well as dispersion medium are liquids, are called emulsions. Emulsions are of two types.

1. Oil in water emulsions are those in which oil is the dispersed phase and water is the dispersion medium *e.g.*, milk is an emulsion of liquid fat in water.

2. Water in oil emulsions are those in which water is the dispersed phase and oil is the dispersion medium *e.g.*, butter, cod liver oil.

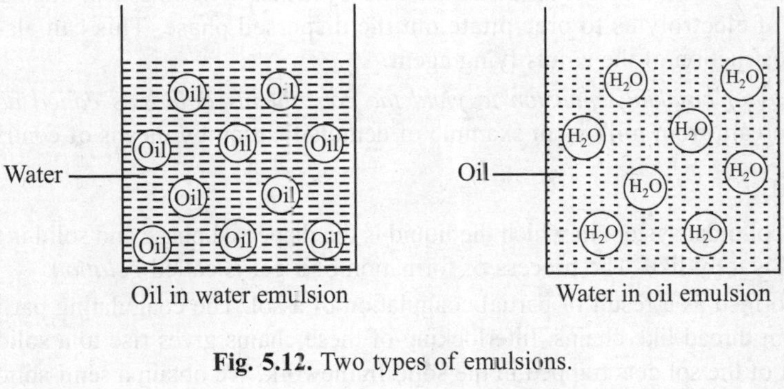

Fig. 5.12. Two types of emulsions.

Distinction between oil-in-water and water-in-oil emulsions

The two types of emulsions can be distinguished from each other by the following tests:

1. **Dye-test.** If on heating the emulsion with an oil soluble dye, emulsion gains the colour of the dye, it is a water-in-oil emulsion. If the emulsion does not catch the colour of the dye, it must be oil-in-water emulsion.

2. **Conductance method.** A small amount of an electrolyte is added to the emulsion. If the conductance increase, the emulsion is oil-in-water type. If there is no appreciable change, it is water-in-oil type.

5.7.1 Emulsifiers or Emulsifying agents

Normally emulsions are not stable colloidal systems. On standing, dispersed phase particles assemble together and form a separate layer. To stabilize emulsions, certain substances known as emulsifying agents or emulsifiers are added.

It is believed that the emulsifier surrounds the oil droplets, thus preventing them to come closer. It prevents them from coagulating. The function of the emulsifiers is to act as a protective layer around the oil droplets of the emulsion thus increasing its stability as shown in Figure 5.13.

Emulsions tend to coagulate because of the interfacial tension between the components of the emulsions. This effect is pronounced when there is a big difference between the surface tensions of the two components. In a bid to minimise the interfacial tension, they minimise the area of contact between the two components and thus the globules of the oil (*e.g.*, in oil-in-water emulsions) come together, thus destabilising the emulsion. An emulsifier prevents such a situation to occur.

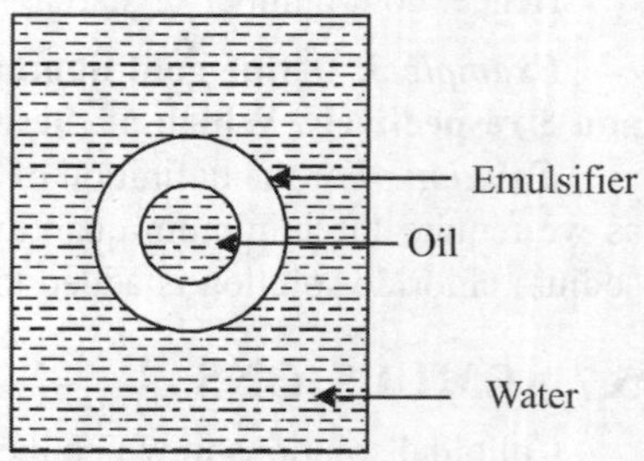

Fig. 5.13. Action of emulsifier.

Most commonly used emusifying agents are soaps and detergents. Other stabilizing agents are proteins and gums. Digestion of fats in the intestines is aided by emulsification. A little of the fat forms a sodium soap with the alkaline solution of the intestines and it emulsifies the rest of the fat thus making it easier for the digestive enzymes to do their work. Many of the lotions, creams and ointments are emulsions of oil-in-water or water-in-oil type. Emulsions find application in the concentration of ores also.

5.7.2 Demulsification

Emulsion can be broken to recover the constituents by means of heating, centrifuging or by adding large amount of electrolytes to precipitate out the dispersed phase. This can also be achieved by chemical destruction of the emulsifying agent

The process of breaking emulsion to yield the constituents liquids is called demulsification. Separation of cream from milk is an example of demulsification by means of centrifugation.

5.8 GELS

A gel is a colloidal system in which the liquid is the dispersed phase and solid is the dispersion medium. A gel is jelly-like. The process of formation of a gel is called *gelation.*

Gels are formed as a result of partial coagulation of a sol. The coagulating particles combine together to form thread like chains. Interlocking of these chains gives rise to a solid frame work. The liquid part of the sol gets trapped in the solid framework. We obtain a semi-solid porous mass called gel as shown in Fig. 5.14.

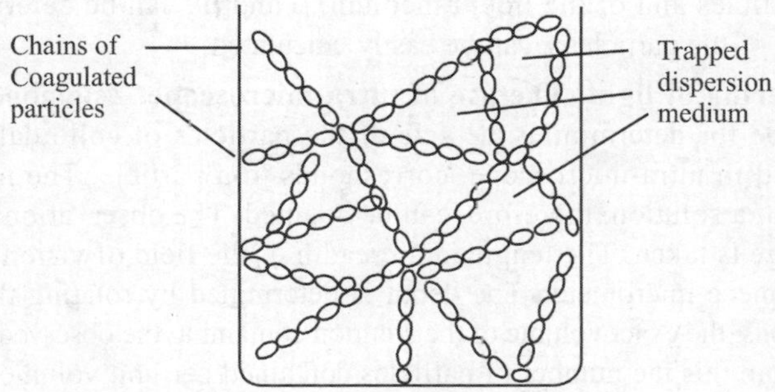

Fig. 5.14. Structure of a get.

5.8.1 Classification of gels

Gels may be classified into two types:

(*a*) Elastic gels: Gels which are elastic in nature are called elastic gels. Their shape changes when a force is applied on them. On releasing the force, the elastic gels regain their original shape. Gelatin, starch and soaps are examples of substances that can form elastic gels.

Such gels are formed by cooling concentrated solutions of lyophilic sols. There are no rigid forces of attraction between the particles of elastic gels.

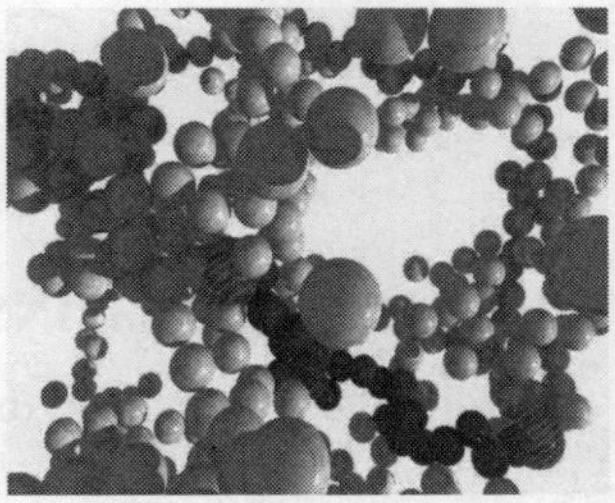

Drying and Cracking of gels

(*b*) Non-elastic gels: Gel which are rigid and do not show elasticity are called non-elastic gels. Silica gel is an example of non-elastic gel. Such gels are obtained by a specific chemical reaction. Silica gel, for example, is obtained by the addition of conc. hydrochloric acid to sodium silicate solution of a particular concentration, silicic acid is formed which polymerises to form silica gel. Particles of such a gel are linked by covalent bonds which give the gel a rigid structure.

5.9 VARIOUS METHODS TO DETERMINE THE SIZE OF COLLOIDAL PARTICLES

1. By using ultrafilters. An approximate idea about the size of particles in a colloidal system can be obtained by the use of ultrafilters. These are prepared by impregnating filter papers with collodion or gelatin which are subsequently hardened by immersing in formaldehyde. The pores can be made small enough to retain particles of colloidal dimensions. The size of the pores depends on the particular filter paper employed and the concentration of the collodion or gelatin solution used for impregnating it. It is thus possible to obtain a series of graded ultrafilters by means of which a colloidal solution may be separated into fractions containing particles of different sizes. An approximate estimate of the size of particles can be obtained from a knowledge of the dimensions of the pores of the ultrafilters. The latter parameter is determined from the pressure required to force air or water through the pores.

2. From Brownian movement. Colloidal particles suspended in a liquid medium are subjected to Brownian movement. They also tend to settle down due to gravitation. Under the influence of both these effects, the particles distribute themselves in a vertical column in accordance with the equation.

$$\frac{RT}{N_0}\ln\frac{n_1}{n_2} = \frac{4}{3}\pi r^3 (h_2 - h_1)(\rho-\rho')$$

Since N_0, the Avogadro's number, is known and since the number of particles n_1 and n_2 at two depths h_1 and h_2 of the vertical column can be counted with the help of an ultramicroscope and densities of the particles and of the liquid medium, ρ and ρ′, can be determined by the usual methods, the radius of the particles r can be easily calculated.

3. From scattering of light. The use of ultra-microscope. Zsigmondy and others used the ultra-microscope for determining the size of the particles of colloidal dimensions. Each spot of light viewed in ultra-microscope, corresponds to a particle. The number of particles in a given volume of a solution, therefore, can be counted. The observation is repeated several times and an average is taken. The length and breadth of the field of vision are measured with the help of an eye-piece micrometer. The depth is determined by rotating the slit through 90°. From these dimensions, the exact volume of the solution containing the observed number of particles can be obtained. From this the number of particles contained per unit volume of the solution can be determined. Let this number be *n*.

The next step is to evaporate a known volume of the colloidal solution to dryness. From the weight of the residue, the mass of colloidal particles per unit volume can easily be obtained. Let this be *m* gram. The particles are supposed to be spherical and the density ρ of the colloidal particles is supposed to be the same as that in the bulk state. Thus the volume of colloidal phase is m/ρ.

$$\therefore \quad m/\rho = \frac{4}{3}\pi r^3 \times n$$

or

$$r = \left(\frac{3m}{4\pi\rho n}\right)^{\frac{1}{3}}$$

5.10 APPLICATIONS OF COLLOIDS

Colloidal solution and the charge on colloids find extensive applications in our daily life and various other processes. A few important applications of colloids are summarised below.

1. Purification of drinking water. Ordinary water contains suspended impurities of colloidal size which do not settle down due to their charged nature. On adding a little of potash alum, the positive charge on aluminium ions neutralises the charge on colloidal particles, they settle down and water is cleaned for drinking.

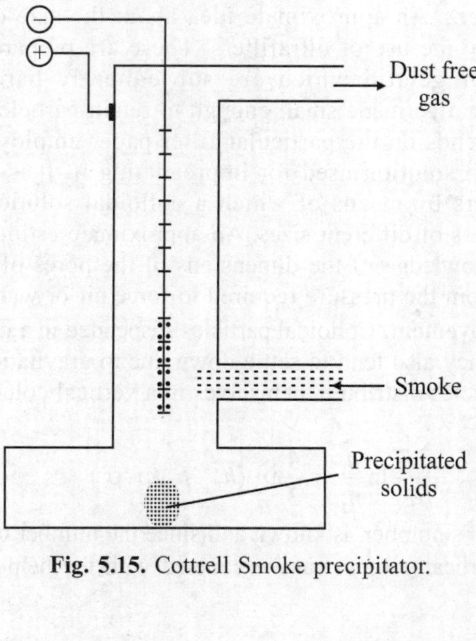

Fig. 5.15. Cottrell Smoke precipitator.

2. Pollution control. Chimneys of various industries bring out lot of smoke which acts as a health hazard. Smoke is a colloidal solution of carbon particles in air. This smoke is brought in contact with a conductor of opposite charge near the exit of the chimney. Smoke thus gets coagulated. Carbon particles settle down and carbonless fumes or air comes out of chimneys. The design is known as *Cottrell smoke precipitator* (Fig. 5.15).

3. Medicines. A large number of medicines are being produced in the form of sols these days. Such medicines are more effective as they have greater assimilation and adsorption qualities. It is found that ferric chloride solution checks excessive bleeding from a cut. It coagulates the blood due to it charged nature.

4. Sewage disposal. The sewage disposal carries dirt particles suspended in water. Separation of these particles from water is normally a difficult process. This difficulty is overcome by making the disposal water to come in contact with an electrode of opposite charge in a big tank. Dirt particles are precipitated and separated. They serve as a manure.

5. Photography. The photographic film contains a colloidal solution of silver bromide in gelatin. A photographic film can also be obtained on a glass plate by depositing a mixture of silver bromide and gelatin.

6. Formation of deltas. Delta is formed at a place where the river enters the sea. Sea water is rich in minerals whereas river water contains colloidal particles of clay. At the point of junction of the two waters, clay is deposited. This clay is fertile and is known as delta (Fig. 5.16).

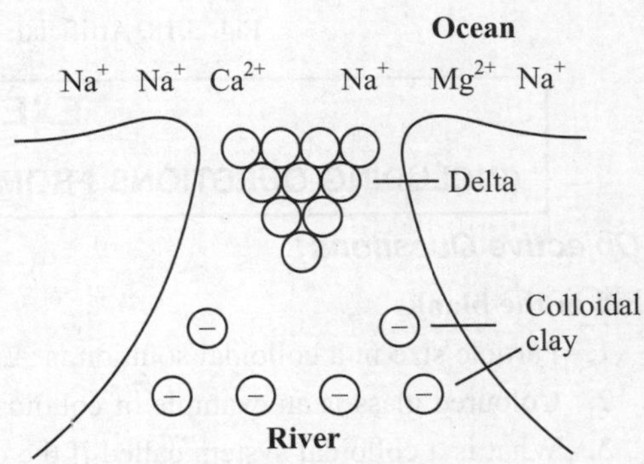

Fig. 5.16. Formation of Delta.

7. Blue colour of the Sky. Colour of a substance is the colour of the light transmitted by it. Colloidal particles of dust etc. scatter only the blue light and absorb the rest of the light. Hence blue light reaches us and the sky appears to be blue.(Fig. 5.17)

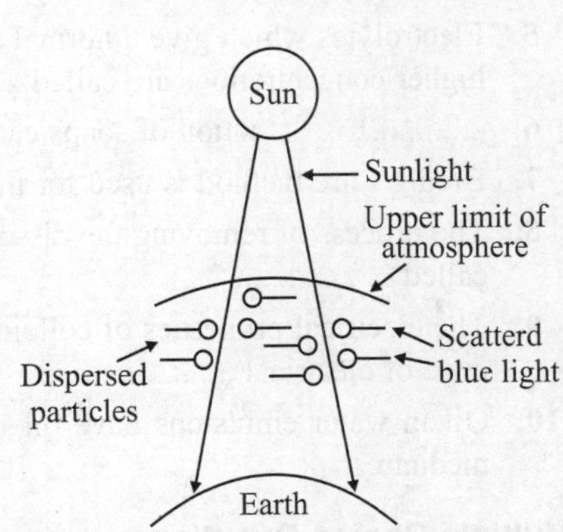

Fig. 5.17. Blue colour of sky (Illustration).

8. Chemical warfare. Smoke and mist screens are formed by the explosion of bombs. They are harmful to humans and gas masks which contain colloidal charcoal come to the rescue. They adsorb the toxic smoke.

9. Artificial kidney machines (dialysers). These machines which work on the principle of separation of colloidal particles from those of electrolytes have saved the lives of thousands of people who suffer from kidney failure. The function of the kidney is to separate blood (colloidal solution) from urea and uric acid (true solution). When the normal kidney stops working, it leads to accumulation of poisonous matter into the blood which causes death. In artificial kidney machine (dialyser) blood is made to pass through a series of cellophane tubes surrounded by a washing solution in water. The toxic chemicals diffuse into the washing solution. Purified blood is returned to the patient (Fig. 5.18).

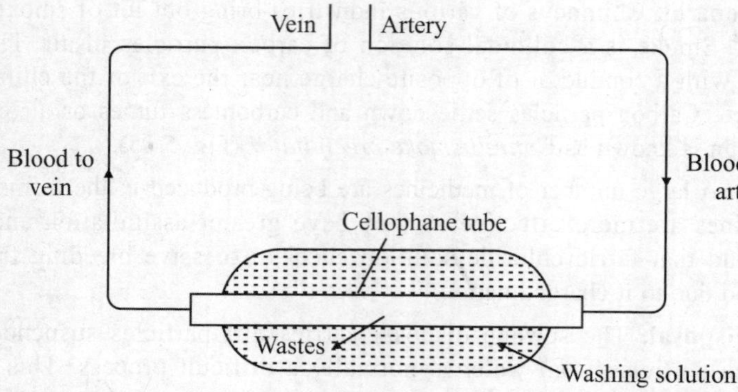

Fig. 5.18. Artificial kidney machine (dialyser).

EXERCISES

(INCLUDING QUESTIONS FROM DIFFERENT UNIVERSITY PAPERS)

Objective Questions

Fill in the blanks

1. Particle size in a colloidal solution is ....
2. Coloured glass is an example of colloidal solution of .... in ....
3. What is a colloidal system called if the dispersion medium is benzene
4. A small quantity of an electrolyte can cause precipitation of .... colloid
5. Electrolytes which give a normal solution at low concentration but show colloidal nature at higher concentrations are called ....
6. .... action of soaps can be explained on the basis of micelle formation
7. Bredig's arc method is used for the preparation of colloided solutions of ....
8. The process of removing the dissolved electrolytes from the sol by means of a membrane is called ....
9. The electrical properties of colloids can be explained satisfactorily by postulating the existence of electrical ....
10. Oil in water emulsions have oil in the .... phase and water in .... medium.

Multiple Choice Questions

Choose the correct option

1. Which substance gives positive sol ?
   (*a*) Gold (*b*) Starch (*c*) Ferric Hydroxide (*d*) As_2S_3
2. Sky looks blue due to :
   (*a*) dispersion effect (*b*) reflection
   (*c*) transmission (*d*) scattering
3. Which of the following is least effective in causing flocculation of ferric hydroxide sol ?
   (*a*) $K_3Fe(CN)_6$ (*b*) K_2CrO_4 (*c*) KBr (*d*) K_2SO_4
4. Which method is used to destroy a sol ?

(a) Condensation (b) Dialysis
(c) Diffusion (d) Addition of electrolyte

5. An emulsifer is a substance which :
(a) helps dispersion of liquid in liquid (b) stabilises the emulsion
(c) purifies the emulsion (d) coagulates the emulsion

6. Bleeding is stopped by applying $FeCl_3$ because :
(a) blood starts flowing in opposite direction
(b) $FeCl_3$ seals the blood vessels
(c) blood gets coagulated
(d) blood vessel is sealed

7. Kinetic activity of colloidal particles in dispersion medium is called :
(a) Electro-osmosis (b) Brownian movement (c) Cataphoresis

8. Who gave the rule ? "Greater the valency, higher in the coagulating power"
(a) Hardy-Schulze (b) Grahm (c) Faraday (d) Lewis

9. Gold number is a measure of :
(a) amount of gold present in colloidal solution
(b) amount of gold required to break the colloid
(c) amount of gold required to protect the colloidal
(d) none

10. Which one of following is + ve sol ?
(a) Blood (b) Clay soil
(c) Smoke (d) Gelatin in acidic solution

11. Tyndall phenomenon is shown by :
(a) dilute solutions (b) colloidal solutions
(c) suspension (d) true solution

12. Which of the following is emulsifier ?
(a) Soaps (b) Deteregents (c) Both (d) none

13. Migration of colloidal particles under the influence of electric field is called:
(a) Browian movement (b) Cataphoresis (c) Dialysis

14. Bredig's method is used for preparing :
(a) metal sol (b) organic compound
(c) Eutectic mixture (d) none

15. The potential difference between fixed charged layer and diffused layer with opposite charge is called :
(a) colloidal potential (b) electrode potential
(c) zeta potential (d) none

16. AgI is used for producing artificial rain because AgI
(a) has crystal structure similar to ice
(b) is easy to spray at high altitudes .
(c) is easy to synthesise
(d) is soluble in water

17. The Gold no.of sols A, B, C and D are 0.04, 0.002, 10 and 25 respectively. The protective powers are in the order:
(*a*) A > B > C > D (*b*) B > A > C > D (*c*) D > C > B > A (*d*) C > A > B > D

18. Stability of lyophilic sol is due to :
(*a*) Brownian motion only (*b*) electric charge only
(c) Both Brownian motion and charge (*d*) particle size

19. The emulsifying agent in milk is :
(*a*) Lactic acid (*b*) fat (*c*) Lactose (*d*) Casein

20. The minimum concentration of an electrolyte required to cause coagulation of a sol is called:
(*a*) flocculation value (*b*) gold no.
(*c*) coagulation no. (*d*) none

21. To coagulate gelatin sol, which is most effective ?
(*a*) NaCl (*b*) Na_3PO_4 (*c*) $AlCl_3$ (*d*) alcohol

22. A liquid is found to scatter a beam of light but passes through filter paper without leaving any residue, the liquid is a/an
(*a*) suspension (*b*) oil (*c*) sol (*d*) true solution

ANSWERS

Fill in the blanks

1. 1 – 100 nm	**2.** soild, solid	**3.** benzosol	**4.** lyophilic
5. gels	**6.** cleansing	**7.** metals	**8.** dialysis
9. double layer	**10.** dispersed, dispersion		

Multiple choice questions

1. (*c*)	**2.** (*d*)	**3.** (*c*)	**4.** (*d*)	**5.** (*b*)
6. (*c*)	**7.** (*b*)	**8.** (*a*)	**9.** (*d*)	**10.** (*a*)
11. (*b*)	**12.** (*c*)	**13.** (*b*)	**14.** (*a*)	**15.** (*c*)
16. (*a*)	**17.** (*b*)	**18.** (*c*)	**19.** (*d*)	**20.** (*a*)
21. (*c*)	**22.** (*c*)			

General Questions

1. What are colloids ? How are they classified ? How would you prepare the colloidal solution of
(*a*) gold; (*b*) arsenious sulphide; and (*c*) sulphur ?

2. (*a*) What is meant by peptization ? Give a suitable example.
(*b*) Describe a method for purifying colloidal solutions.
(*c*) Discuss the function of a protective colloid.

3. Discuss the origin of charge on colloidal particles. What is meant by electrical double layers ? What is meant by Zeta potential ?

4. What are emulsions and gels ? What distinguishes between elastic and non-elastic gels ?

5. What is electrophoresis ? How does this phenomenon provide information about the sign of charge on particles.

6. (*a*) What are lyophillic colloids ? Why are they called reversible colloids ?

(*b*) Why $MgCl_2$ is a better coagulant than KCl for As_2S_3 sol.

7. (*a*) Explain the stability of colloids.

(*b*) Write the difference between gels and emulsions.

(*c*) Explain the Schulze-Hardy rule for coagulation.

8. State different methods of preparation of colloidal dispersions. Describe Bredig's are method in detail.

9. What are emulsions ? How are they classified ? How are they prepared ?

10. (*a*) What are emulsions and what is emulsifying agent ?

(*b*) What are micelles ? Give examples.

11. (*a*) What is a colloid ? Discuss the essential difference between lyophillic and lyophobic colloids.

(*b*) Write short notes on :

(*i*) Tyndall effect and Brownian movement

(*ii*) Electrophoresis and its applications

12. (*a*) What is a protective colloid ? How does a hydrophillic colloid stablise a hydrophobic one ? Give an account of Gold number in this context.

(*b*) What do you understand by the coagulation value of an electrolyte ?

(*c*) What are gels ? Give examples.

13. Define gold number. The gold numbers of A, B, C and D are 0.005, 0.05, 0.5 and 5 respectively. Which of these has the greatest protective action. Explain your answer.

14. Give in details the methods of preparation of colloids.

15. (*a*) What is the size range of colloidal particles ? Distinguish between a molecular solution and a colloidal dispersion.

(*b*) Describe one method for the preparation of colloidal solutions.

16. Explain why :

(*a*) Alum is used in town water supply

(*b*) Alum is used in shaving

(*c*) Tyndall cone is formed when a beam of light is concentrated on a colloidal solution

17. Write a short note on 'origin of charge on colloidal particles'.

18. Give three points of difference between lyophobic and lyophillic colloids.

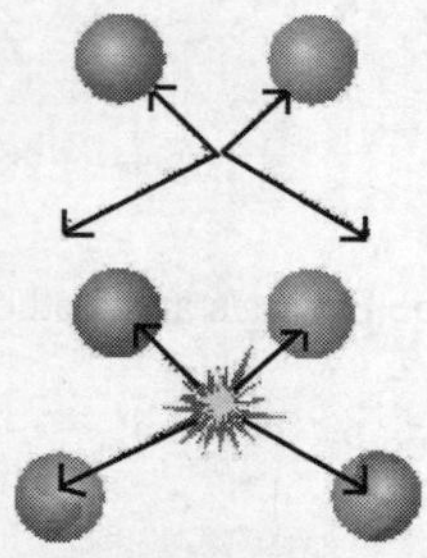

6

Chemical Kinetics

6.1 INTRODUCTION

The branch of chemistry that deals with the rates of reactions is called Chemical Kinetics. The following aspects are covered :

(*i*) Rate of a reaction and rate laws.

(*ii*) Effect of concentration, temperature, pressure and catalyst on the rate of reaction.

(*iii*) Mechanism of the reaction.

The knowledge of reaction kinetics is of great importance and utility in selecting optimum conditions for all industrial processes.

6.1.1 Rate of a reaction

Rate of a reaction may be defined as the change in the concentration of any one of the reactants or products per unit time. For example, consider the dissociation of PCl_5 to give PCl_3 and Cl_2.

$$PCl_5 \rightarrow PCl_3 + Cl_2$$

Rate of this reaction

$$= \frac{\text{Decrease in the conc. of } PCl_5}{\text{Time taken}}$$

or

$$= \frac{\text{Increase in the conc. } PCl_3 \text{ or } Cl_2}{\text{Time taken}}$$

Since the concentration is usually expressed in moles/litre and the time is taken in seconds or minutes, the units of the rate of reaction are therefore, moles litre^{-1} sec^{-1} or moles litre^{-1} min^{-1}.

Rate of a reaction does not remain constant throughout. Hence we can consider rate of a reaction at a particular instant which is defined as follows.

The rate of a reaction at any time is defined as the change in the concentration of any one of the reactants or products per unit time at that particular instant of time.

6.1.2 Expression of the rate of a reaction

Since the rate of a reaction is changing continuously throughout the course of the reaction, therefore to state the rate of reaction at any particular instant of time, a small interval of time dt may be taken during which the rate of reaction may be assumed to be almost constant. Suppose dx is the small change in concentration of any one of the reactants or products during the small interval of time dt (Fig. 6.1). Then the rate of reaction may be expressed as:

$$\text{Rate of reaction} = \frac{dx}{dt}$$

Further if the rate is expressed in terms of one of the reactants, a negative sign is attached to show that the concentration is decreasing.

For example, consider the reaction

$$A + B \longrightarrow P + Q$$

Then rate of this reaction may be expressed in any one of the following ways :

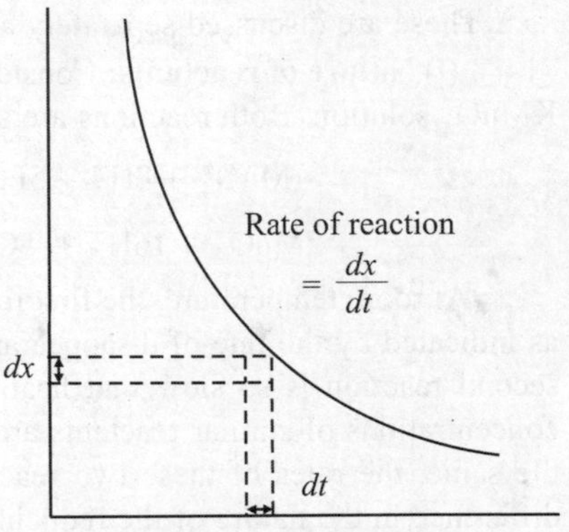

Fig. 6.1. Expressing rate of a reaction.

Rate of reaction

$$\left(\frac{dx}{dt}\right) = -\frac{dC_A}{dt} = -\frac{dC_B}{dt} = +\frac{dC_P}{dt} = +\frac{dC_Q}{dt}$$

where dC_A and dC_B represent small decrease in the concentrations of A and B respectively and dC_P and dC_Q represent small increase in the concentrations of P and Q respectively in the small interval of time dt.

For a more general reaction of the type

$$aA + bB \longrightarrow pP + qQ$$

a unique value of the rate of reaction can be obtained by expressing the rate of reaction as follows:

$$\text{Rate of reaction} = -\frac{1}{a}\frac{d[A]}{dt} = -\frac{1}{b}\frac{d[B]}{dt} = +\frac{1}{p}\frac{d[P]}{dt} = +\frac{1}{q}\frac{d[Q]}{dt}$$

6.1.3 Measurement of rate of a reaction

In order to measure the rate of a reaction, the concentration of one of the reactants or products is observed at different intervals of time. A common practice is to withdraw small amount of the reaction mixture (2 ml or 5 ml) at different intervals of time, cool it down immediately to nearly 0°C to arrest the reaction and then find out the concentration of the reactant or the product by suitable method, usually by titration against a suitable reagent. When some observable property like volume, pressure, optical rotation etc. change with time and can be observed directly at different intervals of time without stopping the reaction, it is preferably utilized. It is important to mention that factors like temperature etc. which affect the rate of the reaction are kept constant during the kinetic study of the reaction.

If concentration of one of the reactants is studied at different intervals of time and these values are plotted against the corresponding times, a graph of the type shown in Fig. 6.2 is obtained. The rate of reaction at any time may be obtained by the following procedure.

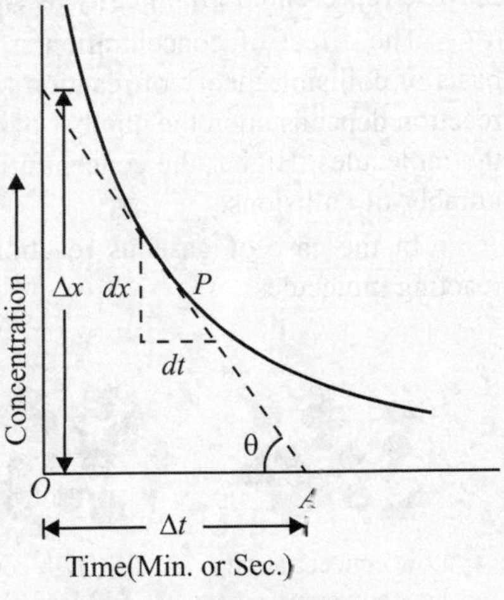

Fig. 6.2. Graphical method for measurement of the rate of reaction.

A tangent may be drawn on the curve at the point corresponding to the time at which the rate is required (Fig. 6.2). The slope of the tangent will then obviously give the rate of reaction at that time.

$$\text{rate} = \tan\theta = \frac{OB}{OA} = \frac{\Delta x}{\Delta t} = \frac{dx}{dt}$$

6.2 FACTORS AFFECTING RATE OF A REACTION

Following factors affect the rate of a reaction (*i*) Nature of the reactants (*ii*) Concentration and pressure of reactants (*iii*) Temperature (*iv*) Catalyst.

These are discussed separately as under :

(*i*) **Nature of reactants:** Consider the oxidation of ferrous ions and oxalate ions by acidified $KMnO_4$ solution. Both reactions are similar as represented below :

$$MnO_4^- + 8H^+ + 5Fe^{+2} \xrightarrow{\text{Fast}} Mn^{+2} + 5Fe^{+3} + 4H_2O$$

$$2MnO_4^- + 16H^+ + 5C_2O_4^{-2} \xrightarrow{\text{Slow}} 2Mn^{+2} + 10CO_2 + 8H_2O$$

At room temperature, the first reaction is very fast whereas the second reaction is quite slow as indicated by the rate of disappearance of pink colour of $KMnO_4$. In fact, in the titrations, the second reaction is so slow that heating is required. Thus in the above case, even if the same concentrations of similar reactants are taken in the two reactions and the temperature is also kept the same, the rates of these two reactions are still different. This is obviously only due to the difference in the nature of the reducing agents *i.e.*, Fe^{+2} ions and $C_2O_4^{-2}$ ions.

The difference in the rates may be ascribed to the fact that a reaction involves breaking of the bonds of the reacting molecules and the formation of new bonds in the product molecules. Energy is required for the breaking of bonds and energy is released when a bond is formed. The net energies involved are different for different reactions and hence the rates are different.

(*ii*) **Concentration of the reactants:** As a reaction proceeds, the reactants are consumed. Therefore, the concentration of the reactants decreases with the passage of time. It is observed experimentally that the rate of reaction also decreases as the time passes. This clearly indicates that the rate of reaction is directly proportional to the concentration of the reactants. The variation of the concentration of a reactant and hence that of the rate of reaction with time may be represented graphically as shown in Fig. 6.3.

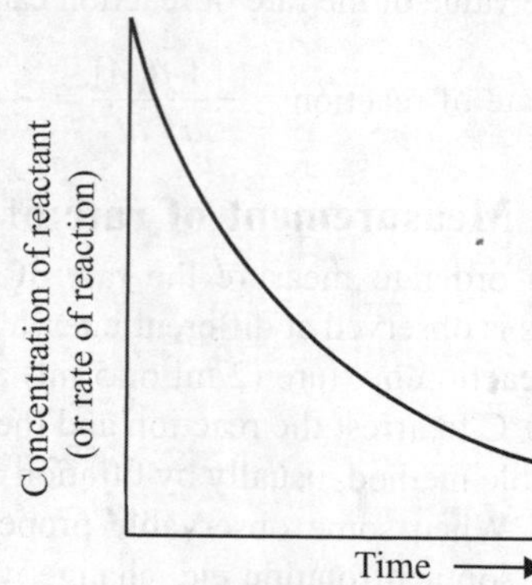

Fig. 6.3. Variation of concentration of a reactant (or the rate of reaction) with time.

The effect of concentration is explained on the basis of collision theory of reaction rates. The rate of a reaction depends upon the number of collisions between the molecules. Higher the concentrations, greater is the number of collisions.

In the case of gaseous reactions, the rate of reaction is proportional to the pressure of reacting molecules.

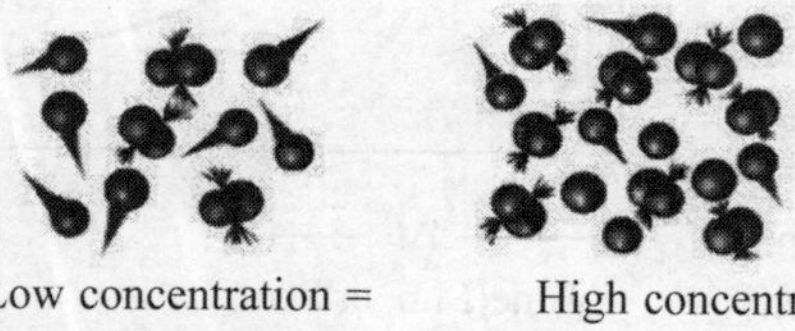

Concentration determines rate

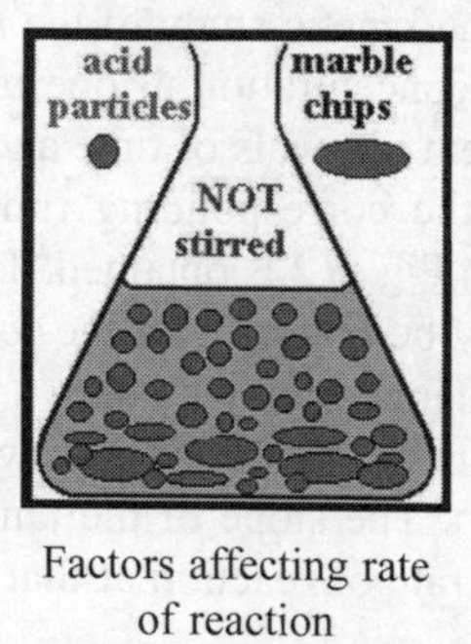

Factors affecting rate of reaction

(*iii*) **Effect of temperature on the rate of a reaction:** Temperature has a very significant effect on the rate of a reaction. A rise of temperature not only increases the speed of the reaction but in many cases it actually initiates the reaction. In the study of the effect of temperature on the reaction rates, two types of cases are observed:

(*a*) *For most of the reactions, the rate of reaction increases with rise of temperature.* As a rough rule, rate of a reaction becomes nearly double or sometimes even treble for every 10°C rise of temperature. This is generally expressed in the form of temperature coefficient as

$$\text{Temperature coefficient} = \frac{k_{(t+10)}}{k_t} \simeq 2 \text{ to } 3$$

where k_t and $k_{(t+10)}$ are the specific reaction rates at *t*°*C* and (*t*+ 10)°C respectively. It is really a wonderful observation because the number of collisions increase only by about 2% for 10° rise at ordinary temperatures whereas the reaction velocity may increase by 100 or 200 per cent or even more.

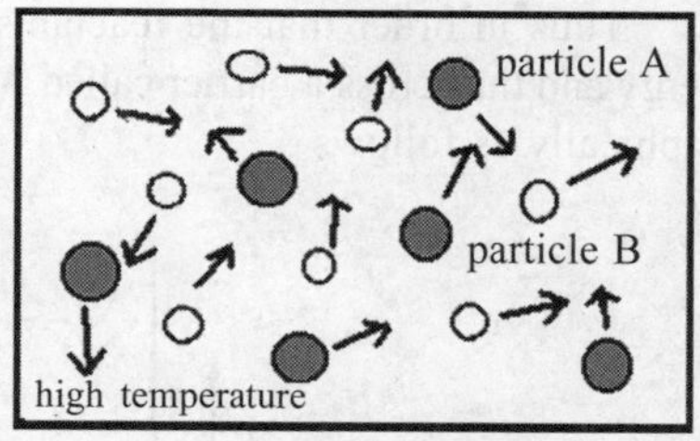

Rate of reaction increases with temperature

(*b*) *A large number of reactions are known which do not take place at room temperature but occur readily at higher temperature e.g.,* reaction between hydrogen and oxygen to form water takes place only when an electric spark is passed. Similarly, it is well known that coal can be kept in oxygen for years without burning but once the ignition is initiated, it goes on till the whole of coal burns away.

The above observations are explained on the basis of **"Collision Theory"**. According to this theory, substances react because their molecules collide with each other. All collisions are, however, not effective collisions. An **effective collision** is that collision which results into a chemical reaction. In order that a collision may be effective, the colliding molecules must possess a certain minimum amount of energy called threshold energy. Thus **threshold energy** *may be defined as the minimum amount of energy which the colliding molecules must possess in order that the collision between them may be an effective collision.*

The two types of cases may be explained as follows :

Explanation of the increase in reaction rate with rise of a temperature: The different molecules of the reactants do not possess the same energy. If the energies of the reacting molecules are plotted against the corresponding fractions, at a particular temperature *T,* a curve of the type shown in Fig. 6.4(*a*) is obtained (called *Maxwell's distribution of energies*). In the figure, E_t represents the threshold energy. The shaded area represents the fraction of molecules with energies equal to or greater than the threshold energy E_t. As the temperature is increased from *t* to (*t*+ 10)°C, the curve shifts as shown in Fig. 6.4(*b*). The shaded area beyond E_t now becomes almost double. This shows that the fraction of molecules possessing energy equal to or greater than the threshold energy becomes almost double. Hence the rate of the reaction also becomes almost double.

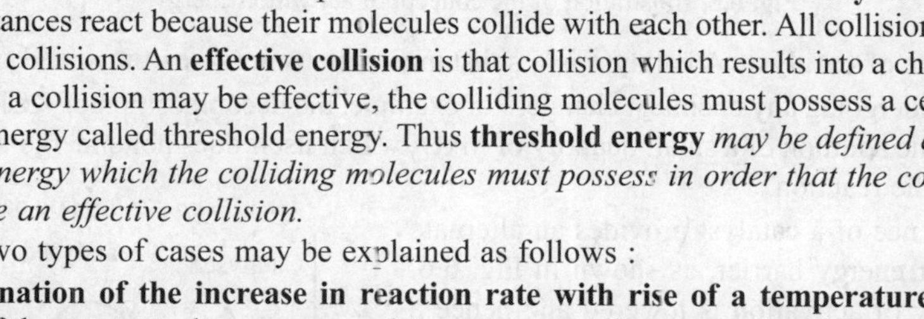

(*a*) (*b*)

Fig. 6.4. Distribution energies of reactants.

Explanation of the non-occurrence of reactions at room temperature and occurrence at higher temperature: At room temperature, there is no molecule possessing energy greater than the threshold energy and so there is no reaction. When temperature is increased, the molecules of the reactants absorb heat energy and thus a number of molecules start attaining energy equal to or greater than the threshold energy and hence the reaction starts.

The extra amount of energy which the molecules of the reactants have to absorb so that their energy becomes equal to the threshold energy is called the **Activation energy.** In other words,

Activation energy = Threshold energy – Energy actually possessed by the reacting molecules.

Thus in order that the reactants may change over to products, they have to absorb some energy and thus cross a barrier called **Activation Energy Barrier.** The situation may be represented graphically as follows :

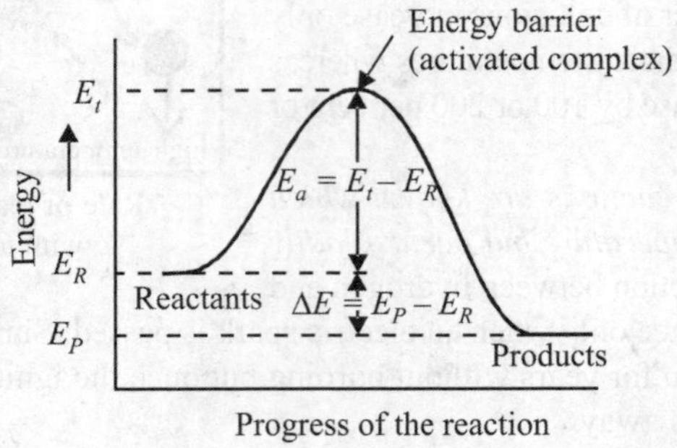

Fig. 6.5. Illustration of the concept of activation energy.

(*iv*) **Effect of catalyst:** A catalyst is a substance which increases the speed of a reaction without itself undergoing any chemical change. For example, the decomposition of $KClO_3$ is greatly accelerated by the addition of a small quantity of MnO_2 which itself does not undergo any chemical change during the reaction.

The presence of a catalyst provides an alternate path with lower energy barrier, as shown in Fig. 6.6. Thus the energy of activation is lowered and hence a greater number of molecules can cross the barrier and change over to products, thereby increasing rate of the reaction. In activation for the uncatalysed and catalysed reactions respectively E_t and $E_{t(ap)}$ represent total energy of the activated complex in the two paths.

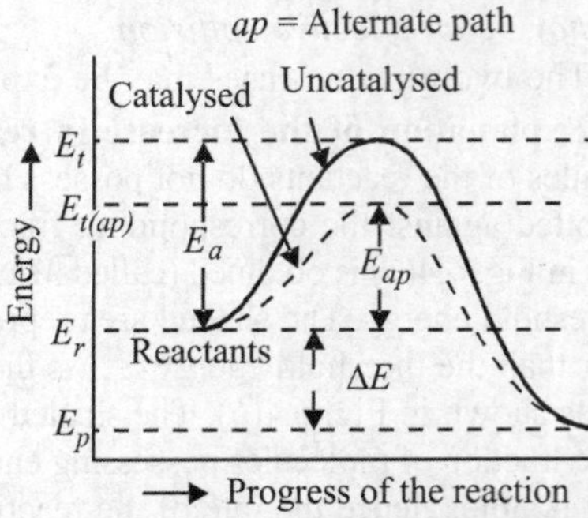

Fig. 6.6. Effect of addition of a catalyst.

Following observations are made from Fig. 6.6.

(*a*) ΔE $(= E_p - E_r)$ for the catalysed reaction is the same as ΔE for the uncatalysed reaction.

(*b*) For reversible reactions, the energy of activation for the reverse reaction is lowered to the same extent as for the direct reaction. Hence the presence of the catalyst in such reactions increases the speed of the forward reaction and that of the backward reaction to the same extent. Consequently, the equilibrium is not disturbed.

6.3 MOLECULARITY OF A REACTION

The number of reacting species (atoms, molecules or ions) which collide simultaneously in the elementary step to bring about a chemical reaction is defined as ***molecularity*** *of a reaction.* Molecularity of a reaction is always a whole number.

When the reaction involves decomposition of a single species, it will be ***unimolecular*** reaction. Decomposition of H_2O_2 is a unimolecular reaction

$$H_2O_2(l) \longrightarrow H_2O(l) + \frac{1}{2}O_2(g)$$

Similarly, when the reaction involves collision between ***two*** species, it will be a ***bimolecular*** reaction and if ***three*** species take part in the collision, it is called *trimolecular* or *termolecular reaction.*

Dissociation of HI into H_2 and I_2 is a ***bimolecular reaction***

$$2HI(g) \longrightarrow H_2(g) + I_2(g)$$

Reaction between NO and O_2 is an ***trimolecular reaction***

$$2NO(g) + O_2(g) \longrightarrow 2NO_2(g)$$

Reactions involving three or more molecules are uncommon. This is because occurrence of such reactions require simultaneous collisions of three or more molecules. The chances of occurrence of such collisions are very small. Reactions involving three or more molecules should be slow.

The reactions in many cases do not occur in single step but in a sequence of many steps. Each step involves the simultaneous collision of two or at the most three molecules. Such type of reactions take place through a sequence of two or more consecutive steps and are called complex reactions. Each elementary step has a molecularity of its own and none of the steps has a molecularity of more than three.

6.4 RATE LAW AND RATE CONSTANT

Consider the following general reaction

$$aA + bB \longrightarrow \text{Products}$$

Experimentally, it is observed that the rate of this reaction may not depend upon all the *a* concentration terms of *A* and all the *b* concentration terms of *B*. Suppose, experimentally the rate of the reaction is found to depend upon α concentration terms *A* and β concentration terms of *B*. Then we may write

$$\text{Rate} \propto [A]^{\alpha} \cdot [B]^{\beta}$$

$$\text{Rate} = k[A]^{\alpha}[B]^{\beta} \quad \text{... (i)}$$

Eq. (*i*) is called **rate law**.

Here $[A]$ and $[B]$ are the molar concentrations of *A* and *B* respectively and *k* is a constant called **velocity constant** or **rate constant**. If all concentrations are taken as unity.

i.e., $[A] = [B] = 1$ mol/litre, then

$$\text{Rate} = k.$$

Hence the *rate constant may be defined as the rate of the reaction when the concentration of each reactant is taken as unity.* That is why the rate constant is also called **specific reaction rate.**

6.4.1 Order of a reaction

Consider again the reaction :

$$\alpha A + \beta B \longrightarrow \text{Products}$$

According to rate law

$$\text{Rate} \propto [A]^{\alpha}[B]^{\beta}$$

The sum of the concentration terms on which the rate of a reaction actually depends is called the order of the reaction. Thus in the above case, order of reaction = α + β.

Hence the order of a reaction may be defined as the sum of the exponents to which the concentration terms in the rate law are raised to express the observed rate of reaction.

Depending upon whether $\alpha + \beta$ is equal to 0, 1, 2 or 3, the reactions are said to be of zero order, 1st order, 2nd order and 3rd order respectively. α and β are called the orders of the reaction with respect to *A* and *B* respectively.

Further it may be mentioned here that *the order of a reaction may not be always a whole number.* For example, at 450°C, acetaldehyde decomposes as

$$CH_3CHO \xrightarrow{450°C} CH_4 + CO$$

Experimentally, it is found that

$$\text{Rate of reaction} \propto [CH_3CHO]^{1.5}.$$

$\therefore$ Order of reaction = 1.5

The order of reaction is generally a small integer but under certain conditions it can be fraction or zero. Some reactions of different order are given below :

Reactions	Conc. Variable	Order
(*i*) Decomposition of NH_3 on W, MO	– –	Zero
(*ii*) Decomposition of HI on gold	– –	Zero
(*iii*) $N_2O_5\,(g) \longrightarrow N_2O_4\,(g) + \frac{1}{2}O_2\,(g)$	$C_{N_2O_5}$	One
(*iv*) $2HI\,(g) \longrightarrow H_2\,(g) + I_2\,(g)$	C^2_{HI}	Two
(*v*) $3KClO \longrightarrow KClO_3 + 2KCl$	C^2_{KClO}	Two
(*vi*) $2NO + Cl_2 \longrightarrow 2NOCl$	$C^2_{NO} \times C_{Cl_2}$	Three
(*v*) $CH_3CHO \xrightarrow{723\ k} CH_4 + CO$	$C^{1.5}_{CH_3CHO}$	1.5

6.4.2 Difference Between Molecularity and Order of Reaction

Order of reaction is sum of the powers to which the concentration terms are raised in the rate law. In other words, order of reaction is a number of reacting species whose concentration determines the rate of reaction. On the other hand, molecularity is the total number of reacting species which collide simultaneously to form the products.

The order of single step reaction or an elementary step of a multistep reaction is the same as its molecularity.

In some cases, one of the reactant is taken in such a large amount that its concentration does not change appreciably in the reaction and has no contribution in determining the order of reaction. For example

Hydrolysis of ethyl acetate in the presence of dilute mineral acid.

$$\underset{\text{Ethyl acetate}}{CH_3COOC_2H_5} + H_2O \xrightarrow{\text{Acid}} \underset{\text{Acetic acid}}{CH_3COOH} + \underset{\text{Ethyl alcohol}}{C_2H_5OH}$$

The molecularity of the above reaction appears to be **two** as it involves reaction between two reacting species. However, the concentration of only ethyl acetate changes during the reaction while that of water remains unchanged. Thus, rate of the reaction only depends upon the concentration of ethyl acetate.

The order of reaction, therefore, *is one*. The reaction appears to be of the second order but it follows first order kinetics. Such types of reaction are called *pseudo unimolecular reactions.*

Difference between molecularity and order of the reaction is given in the tabular form as under :

Molecularity	Order of Reaction
1. Molecularity is the number of reacting species which undergo simultaneous collisions in the elementary or simple reactions.	It is the sum of powers of the concentrations terms in the rate law equation.
2. It is a theoretical concept.	It can be determined experimentally.
3. It is always a whole number.	It can be whole number as well as fraction.
4. It can never be zero.	It can be zero.
5. It has no significance for a complex reaction and can be expressed for each elementary step in case of a complex reaction.	It is always the same for the overall reaction whether the reaction is complex or simple.

Example 1. **For the thermal decomposition of acetaldehyde at 800 K**

$$CH_3CHO \longrightarrow CH_4 + CO$$

following data is obtained :

Experiment	C_{CH_3CHO}	Rate of decomposition
(i)	0.1 mol L^{-1}	9.0×10^{-7} mol L^{-1} sec^{-1}
(ii)	0.2 mol L^{-1}	36×10^{-7} mol L^{-1} sec^{-1}
(iii)	0.4 mol L^{-1}	144×10^{-7} mol L^{-1} sec^{-1}

(*a*) What is the order of reaction ?

(*b*) Write rate equation for the reaction.

(*c*) Calculate the rate constant at 800 K.

(*d*) Determine the rate of decomposition of acetaldehyde at instant when C_{CH_3CHO} = 0.25 mol^{-1} L^{-1}

Solution. Let the rate law for the decomposition of acetaldehyde be

$$r = kC^a_{CH_3CHO}$$

(*a*) The following equation can be generated from experiments (*i*) and (*ii*).

$$9.0 \times 10^{-7} = k\,(0.1)^a \qquad \text{... (1)}$$

and $$36 \times 10^{-7} = k\,(0.2)^a \qquad \text{... (2)}$$

Dividing (2) by (1)

$$\frac{36 \times 10^{-7}}{9.0 \times 10^{-7}} = \left(\frac{0.2}{0.1}\right)^a$$

$$4 = 2^a$$

or $$2^a = 2^2 \quad \therefore \quad a = 2$$

∴ The order of reaction is 2 or the reaction is of the second order

(*b*) The rate equation for the reaction is

$$\text{Rate} = \frac{dx}{dt} = kC^2_{CH_3CHO}$$

(*c*) Rate constant *k* is given by

$$k = \frac{r}{C^2_{CH_3CHO}}$$

Substituting the data of experiment (*i*), we get

$$k = \frac{9.0\times10^{-7}}{(0.1)^2} = 9.0 \times 10^{-5}$$

$$\therefore \quad k = 9.0 \times 10^{-5} \text{ L mol}^{-1} s^{-1}$$

6.5 ZERO ORDER REACTION

A reaction is said to be of zero order if the rate of the reaction is independent of the concentration of all the reactants.

For such reactions the rate of the reaction remains constant throughout.

Following reactions are found to be of zero order :

(*i*) **Photochemical reactions:** Many such reactions have been found to be of zero order. For example, the combination between hydrogen and chlorine in presence of light may be studied by enclosing these gases in a tube and inverting the tube in a trough of water (Fig. 6.7). When the reaction takes place between H_2 and Cl_2, HCl is formed which dissolves into water and the level of water in the tube rises. Thus though the quantities of H_2 and Cl_2 decrease, their amounts per unit volume remain the same or the concentrations of H_2 and Cl_2 remain constant.

$$H_2\,(g) + Cl_2\,(g) \xrightarrow{h\nu} 2\,HCl\,(g)$$

(*ii*) **Heterogeneous reactions:** Two well known examples of this type are (*a*) the decomposition of hydrogen iodide on the surface of gold and (*b*) the decomposition of ammonia on the surface of tungsten, molybdenum or platinum at high temperatures. The gaseous molecules of the reacting substance get adsorbed on the surface of the catalyst. The surface is thus fully occupied with the molecules of the reactant. Some of these molecules react to form the products but immediately new molecules take up their position and thus the concentration of the reacting molecules remains constant on the surface and hence the rate of reaction remains constant irrespective of the total amount of the gaseous reactants taken initially.

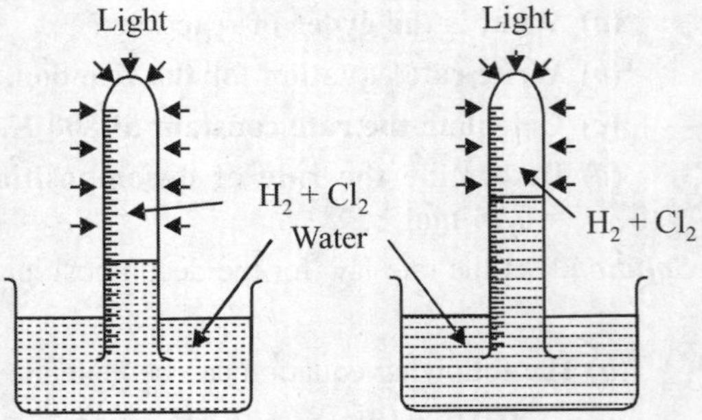

Fig. 6.7. Study of kinetics of a photochemical reaction.

6.5.1 Rate equation for zero order reaction

If *a* is the initial concentration, the expression for the rate of a reaction of zero order may be written as follows :

$$\frac{dx}{dt} \propto a^o$$

or $$\frac{dx}{dt} = k \qquad (a^o = 1)$$

or rate of reaction remains constant throughout.

where *k* is the rate constant for the zero order reaction.

The above expression may be written as

$$dx = kdt$$

Integrating this equation we get

$$x = kt + I$$

where I is constant of integration

When $t = 0$, $x = 0$; therefore $I = 0$

Hence $$\boldsymbol{x = kt} \quad \text{... } (i)$$

i.e., the amount of substance reacted ∝ time.

The expression for the half-life period (*i.e.*, the time taken for half the reaction to complete) may be obtained as follows :

When $x = \frac{a}{2}$, $t = t_{1/2}$

Substituting these values in eq. (*i*) above, we get

$$\frac{a}{2} = kt_{1/2}$$

or $$t_{1/2} = \frac{a}{2k}$$

i.e., the half-life period ∝ initial concentration.

***Example 2.* Determine the order of reaction and the rate constant of the decomposition of NH_3 on a tungsten wire at 857° C from the following data.**

Total Pressure p (torr)	**228**	**250**	**273**	**318**
Time t (sec)	**200**	**400**	**600**	**1000**

Solution. The rate of the reaction will be the rate of change of pressure

i.e. $$\text{Rate} = \frac{\Delta p}{\Delta t}$$

From the above data

Δp	22	45	90
Δt	200	400	800
$\frac{\Delta p}{\Delta t}$	0.1100	0.1125	0.1125

It is obvious that

$$\text{Rate} = \frac{\Delta p}{\Delta t} = \text{constant } (k) = \mathbf{0.1117}$$

It indicates that the rate of reaction does not depend on the partial pressure of the reactant.

∴ The reaction will be of zero order.

The specific reaction rate

$$k = 0.1117 \text{ torr } s^{-1}$$

6.6 FIRST-ORDER REACTION

A reaction is said to be of the first order if the rate of the reaction depends upon one concentration term only.

Let us consider a hypothetical case.

$$A \longrightarrow \text{Products}$$

Suppose we start with a moles per litre of the reactant A. After time t, suppose x moles per litre of it have decomposed. Therefore the concentration of A after time $t = (a - x)$ moles per litre. Then according to law of mass action,

Rate of reaction ∝ $(a - x)$

i.e., $$\frac{dx}{dt} \propto (a - x)$$

or $$\frac{dx}{dt} = k(a-x) \qquad \text{... (i)}$$

where k is called the *rate constant* or the *specific reaction rate* for the reaction of the first order.

The expression for the rate constant k may be derived as follows :

Equation (*i*) may be rewritten in the form

$$\frac{dx}{a-x} = kdt \qquad \text{...(ii)}$$

Integrating equation (*ii*), we get

$$\int\frac{dx}{a-x} = \int kdt$$

i.e. $$-\ln(a-x) = kt + I \qquad \text{...(iii)}$$

where I is a constant of integration.

In the beginning, when $t = 0$, $x = 0$ (as no substance has decomposed at the start of the reaction).

Putting these values in equation (*iii*), we get

$$-\ln(a-0) = k \times 0 + I$$

or $$-\ln a = I \qquad \text{...(iv)}$$

Substituting this value of I in equation (*iii*), we get

$$-\ln(a-x) = kt + (-\ln a)$$

or $$kt = \ln a - \ln(a-x)$$

$$= \ln\frac{a}{a-x}$$

or $$k = \frac{1}{t}\ln\frac{a}{a-x} \qquad \text{...(v)}$$

or $$\boldsymbol{k = \frac{2.303}{t}\log\frac{a}{a-x}} \qquad \text{... (vi)}$$

Equation (*vi*) is sometimes written in another form which is obtained as follows :

If the initial concentration is C_0 and the concentration after time t is C_t then putting $a = C_0$ and $(a-x) = C_t$, equation (*vi*) becomes

$$\boldsymbol{k = \frac{2.303}{t}\log\frac{C_0}{C_t}} \qquad \text{... (vii)}$$

When the initial concentration is not known, the value of k can be calculated as follows :

From equation (*vi*)

i.e. $$k = \frac{2.303}{t}\log\frac{a}{a-x}$$

or $$t = \frac{2.303}{k}\log\frac{a}{a-x}$$

or $$t = \frac{2.303}{k}\log a - \frac{2.303}{k}\log(a-x)$$

Let $(a - x_1)$ be the conc. of the reactant at time t_1 and $(a - x_2)$ be the conc. of the reactant time t_2.

$$t_2 = \frac{2.303}{k}\log a - \frac{2.303}{k}\log(a - x_2) \quad \text{... (viii)}$$

$$t_1 = \frac{2.303}{k}\log a - \frac{2.303}{k}\log(a - x_1) \quad \text{... (ix)}$$

Subtracting (*viii*) and (*ix*) we get,

$$(t_2 - t_1) = \frac{2.303}{k}\log(a - x_1) - \frac{2.303}{k}\log(a - x_2)$$

$$(t_2 - t_1) = \frac{2.303}{k}\log\frac{(a - x_1)}{(a - x_2)}$$

or $$k = \frac{2.303}{(t_2 - t_1)}\log\frac{(a - x_1)}{(a - x_2)}$$

6.6.1 Units of First-order Rate Constant (k)

As the value of k *i.e.*, the rate constant for the first order reaction depends upon the ratio $\frac{C_0}{C}$ or $\frac{a}{a-x}$ *i.e.*, the ratio of the two concentration terms, the value k will be independent of the units in which the concentration of the reactants is expressed. It has the dimension of reciprocal time *i.e.*, t^{-1}. If time is measured in seconds, k will be expressed in the unit sec^{-1}.

6.6.2 Some characteristics of first-order reactions

(*i*) *Any reaction of the first order must obey equation* (*vi*) or (*vii*). This may be tested by one of the following methods.

(*a*) *Substitution method.* Starting with a known concentration *a,* the concentration of the reactant $(a - x)$, at different intervals of time may be noted. For every value of t, the corresponding value of $(a - x)$ may be substituted in equation (*vi*). If the values of k thus obtained are nearly constant (within experimental error), the reaction is of the first order.

(*ii*) *The time taken for any fraction of the reaction to be completed is independent of the initial concentration.* For example, let us test this statement for *half* of the reaction to be completed. Writing the first-order equation.

$$t = \frac{2.303}{k}\log\frac{a}{a-x} \quad \text{... (i)}$$

When half of the reaction is completed, $x = a/2$. Representing the time taken for half of the reaction to be completed by $t_{1/2}$, above equation becomes

$$t_{1/2} = \frac{2.303}{k}\log\frac{a}{a - a/2}$$

$$= \frac{2.303}{k}\log 2 = \frac{0.693}{k} \quad \text{... (ii)}$$

Thus a does not appear in this equation so that $t_{1/2}$ is independent of a. Similarly, it can be seen that $t_{1/3}$, $t_{2/3}$ etc. will also be independent of the initial concentration.

The time taken for half of the reaction to be completed *i.e.*, the time in which the concentration of a reactant is reduced to half of its original value is called **half-life period** *of the reaction.*

(*iii*) *The units of k are independent of the units in which the concentrations are expressed.*

This is obvious from equation (*i*) because in this equation *k* depends upon the ratio of two concentrations *viz.* $\frac{a}{a-x}$ and the time *t*. Thus so long as both *a* and $(a-x)$ are expressed in the same concentration units, the value of *k* is not affected. The units of *k*, however, depend only upon the units of time *t*.

6.6.3 Examples of first order reactions

1. Decomposition of nitrogen peroxide
2. Decomposition of hydrogen peroxide
3. Inversion of cane sugar
4. Acid hydrolysis of ethyl acetate
5. Conversion of *N*-Chloroacetanilideinto *p*-chloroacetanide
6. Decomposition of benzene diazonium chloride
7. Thermal decomposition of Azo isopropane

These are discussed separately as under :

1. Decomposition of nitrogen pentoxide

The compound, nitrogen pentoxide, is a volatile solid which decomposes in the gaseous state as well as in the form of its solution in an inert solvent like carbon tetrachloride, chloroform etc. according to the equation

$$N_2O_5 \longrightarrow N_2O_4 + \frac{1}{2}O_2$$
$$N_2O_4 \rightleftharpoons 2NO_2$$

When the reaction is carried out in the solution, N_2O_4 and NO_2 remain in the solution and the volume of oxygen gas collected is noted at different intervals of time (Fig. 6.8). It can be observed that

$$\left.\begin{array}{l}\textit{Volume of oxygen gas collected}\\ \textit{at any time } (V_t)\end{array}\right\} \propto \textit{Amount of } N_2O_5 \textit{ decomposed } (x)$$

i.e., $x \propto V_t$

$$\left.\begin{array}{l}\textit{Volume of oxygen gas collected}\\ \textit{at infinite time } (V_\infty)\end{array}\right\} \propto \textit{Amount of } N_2O_5 \textit{ initially taken } (a)$$

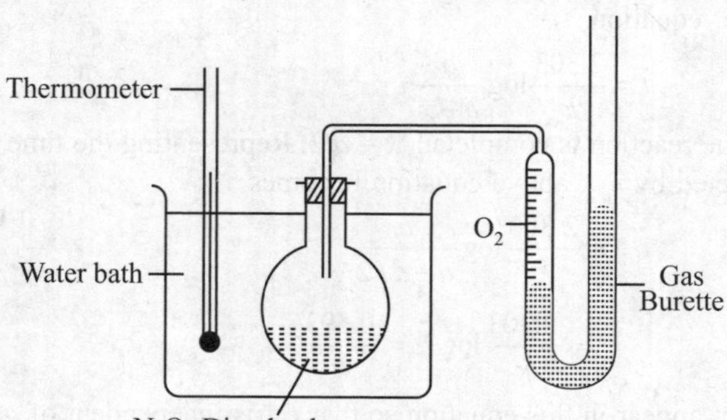

Fig. 6.8. An apparatus for monitoring the volume of O_2 evolved in the decomposition of N_2O_5 dissolved in carbnon tetrachloride.

i.e., $a \propto V_\infty$

Substituting these values in the first order equation *viz.*

$$k = \frac{2.303}{t} log \frac{a}{a-x}$$

we get

$$\mathbf{k = \frac{2.303}{t} \log \frac{V_\infty}{V_\infty - V_t}}$$

Example 3. **From the following data for the decomposition of N_2O_5 in carbon tetrachloride solution at 48°C, show that the reaction is of the first order and calculate the rate constant.**

Time (in minutes) :	**10**	**15**	**20**	**25**	**∞**
Vol. of O_2 evolved (in ml) :	**6.30**	**8.95**	**11.40**	**13.50**	**34.75**

Solution. If the reaction is of the first order, it must obey the equation

$$k = \frac{2.303}{t} \log \frac{a}{a-x}$$

$$= \frac{2.303}{t} \log \frac{V_\infty}{V_\infty - V_t}$$

In the present case, $V_\infty = 34.75$ ml

The value of k at each time can be calculated as follows :

Time (*min*)	V_t	$V_\infty - V_t$	$k = \frac{2.303}{t} \log \frac{V_\infty}{V_\infty - V_t}$ (min^{-1})
10	6.30	34.75 – 6.30 = 28.45	$k = \frac{2.303}{10} \log \frac{34.75}{28.45} = 0.01997$
15	8.95	34.75 – 8.05 = 25.80	$k = \frac{2.303}{15} \log \frac{34.75}{25.80} = 0.01985$
20	11.40	34.75 – 11.40 = 23.35	$k = \frac{2.303}{20} \log \frac{34.75}{23.35} = 0.01987$

Since the value of k comes out to be nearly constant, given reaction is of the first order. The *average* value of the rate constant = 0.01989 minute^{-1}.

Example 4. **The decomposition of an aqueous solution of ammonium nitrite was studied by placing the apparatus in a thermostat maintained at a particular temperature. The volume of nitrogen gas collected at different intervals of time was as follows :**

Time (minutes) :	**10**	**15**	**20**	**25**	**∞**
Vol. of N_2 (ml) :	**6.25**	**9.00**	**11.40**	**13.65**	**35.05**

From the above data prove that the reaction is of the first order.

Solution. Ammonium nitrite decomposes as follows : $NH_4\ NO_2 \longrightarrow 2H_2O + N_2$. Volume of nitrogen produced is observed after different intervals of time.

Vol of N_2 at any time $\propto x$.

Vol of N_2 after infinite time $\propto a$.

Substitute the values in the equation below :

$$k = \frac{2.303}{t} \log \frac{a}{a-x}$$

$$= \frac{2.303}{t} \log \frac{V_\infty}{V_\infty - V_t}$$

In the present case, $V_\infty = 35.05$ ml.

The value of k after each time can be calculated as follows:

Time (*min*)	V_t	$V_\infty - V_t$	$k = \frac{2.303}{t} \log \frac{V_\infty}{V_\infty - V_t}$ (min^{-1})
10	6.25	35.05 – 6.25 = 28.80	$k = \frac{2.303}{10} \log \frac{35.05}{28.80} = 0.0196$
15	9.00	35.05 – 9.00 = 26.05	$k = \frac{2.303}{15} \log \frac{35.05}{26.05} = 0.0198$
20	11.40	35.05 – 11.40 = 23.65	$k = \frac{2.303}{20} \log \frac{35.05}{23.65} = 0.0196$

Since the value of k comes out to be nearly constant, the given reaction is of the first order.

2. Decomposition of Benzene Diazonium chloride ($C_6H_5N = NCl$) in water :

Benzene diazonium chloride decomposes in water according to the following reaction :

$$\underset{\text{Benzene diazonium chloride}}{C_6H_5N = NCl} + H_2O \longrightarrow \underset{\text{Phenol}}{C_6H_5OH} + N_2 + HCl$$

The progress of the reaction can be studied by measuring volume of N_2 produced by the decomposition of benzene diazonium chloride at different intervals of time.

The volume of N_2 liberated is proportional to the decrease in concentration of benzene diazonium chloride.

Let V_∞ be the volume of N_2 measured at the end of the reaction and V_t be the volume at any time t, then

$$a \propto V_\infty$$

$$(a - x) \propto (V_\infty - V_t)$$

Substituting the values of a and $(a - x)$ in equation (*vi*), we get

$$k = \frac{2.303}{t} \log \frac{V_\infty}{V_\infty - V_t}$$

***Example 5.* Benzene diazonium chloride in aqueous solution decomposes according to the equation**

$$C_6H_5N_2Cl \longrightarrow C_6H_5Cl + N_2$$

Starting with an initial concentration of 10 g/litre, the volume of N_2 gas obtained at 50°C at different intervals of time was found to be as under :

t (min) :	**6**	**12**	**18**	**24**	**30**	**∞**
Vol. of N_2 (ml) :	**19.3**	**32.6**	**41.3**	**46.5**	**50.4**	**58.3**

Show that the above reaction follows the first order kinetics. What is the value of the rate constant?

Solution. If the given reaction follows first order kinetics, it must obey the equation

$$k = \frac{2.303}{t} \log \frac{a}{a - x}$$

In this case

Volume of N_2 obtained at infinite time $V_\infty \propto a$

Volume of N_2 obtained at any time (t) $V_t \propto x$

$$\therefore \qquad \frac{a}{a-x} \propto \frac{V_\infty}{V_\infty - V_t}$$

$$k = \frac{2.303}{t} \log \frac{V_\infty}{V_\infty - V_t}$$

In the present case, $V_\infty = 58.3$ ml.
The value of k after each time can be calculated as follows:

Time (*min*)	V_t	$V_\infty - V_t$	$k = \frac{2.303}{t} \log \frac{V_\infty}{V_\infty - V_t}$
6	19.3	58.3 – 19.3 = 39.0	$k = \frac{2.303}{6} \log \frac{58.3}{39.0} = 0.0670 \text{ min}^{-1}$
12	32.6	58.3 – 32.6 = 25.7	$k = \frac{2.303}{12} \log \frac{58.3}{25.7} = 0.0683 \text{ min}^{-1}$
18	41.3	58.3 – 41.3 = 17.0	$k = \frac{2.303}{18} \log \frac{58.3}{17.0} = 0.0685 \text{ min}^{-1}$
24	46.5	58.3 – 46.5 = 11.8	$k = \frac{2.303}{24} \log \frac{58.3}{11.8} = 0.0666 \text{ min}^{-1}$

Since the value of k comes out to be nearly constant, the reaction is of the first order. The mean value of $k = 0.0676 \text{ min}^{-1}$.

3. Decomposition of hydrogen peroxide

The decomposition of hydrogen peroxide in aqueous solution takes place according to the equation

$$H_2O_2 \xrightarrow{Pt} H_2O + \frac{1}{2}O_2$$

The kinetics of this reaction may be studied by making use of the fact that H_2O_2 solution can be titrated against $KMnO_4$ solution. Thus by withdrawing equal amounts of the solution (usually 5 ml) at regular intervals of time and titrating against the same $KMnO_4$ solution, the amount of H_2O_2 present can be found everytime. It is obvious that *for the same volume of the reaction solution withdrawn*

Volume of $KMnO_4$ *solution used before the commencement of the reaction i.e., at zero time* (V_0) $\propto$ *Initial concentration of* H_2O_2 (a)

i.e., $V_0 \propto a$

Volume of $KMnO_4$ *solution used at any instant of time t* (V_t) $\propto$ *Amount of* H_2O_2 *present at that instant i.e.,* $(a - x)$

i.e., $V_t \propto (a - x)$

Substituting these values in the first order equation, we get

$$k = \frac{2.303}{t} \log \frac{a}{a-x}$$

or $$k = \frac{2.303}{t} \log \frac{V_0}{V_t}$$

The decomposition of hydrogen peroxide, as tested by applying this equation, is found to be of the first order.

***Example 6.* From the following data, show that the decomposition of hydrogen peroxide is a reaction of the first order :**

***t* (min.)**	**0**	**10**	**20**
***V* (ml)**	**46.1**	**29.8**	**19.3**

where *t* is the time in minutes and *V* is the volume of standard $KMnO_4$ solution required for titrating the same volume of the reaction mixture.

Solution. Volume of $KMnO_4$ solution used ∝ amount of H_2O_2 present. Hence if the given reaction is of the first order, it must obey the equation

$$k = \frac{2.303}{t} \log \frac{a}{a-x}$$

$$k = \frac{2.303}{t} \log \frac{V_0}{V_t}$$

In the present case, $V_0 = 46.1$ ml

The value of k at each instant can be calculated as follows :

t (min)	V_t	$k = \frac{2.303}{t} \log \frac{V_0}{V_t}$
10	29.8	$k = \frac{2.303}{10} \log \frac{46.1}{29.8} = 0.0436 \text{ min}^{-1}$
20	19.3	$k = \frac{2.303}{20} \log \frac{46.1}{19.3} = 0.0435 \text{ min}^{-1}$

Thus, the value of k comes out to be nearly constant. Hence it is a reaction of the first order.

4. Conversion of *N*-chloroacetanilide into *p*-chloroacetanilide

The reaction can be represented as under.

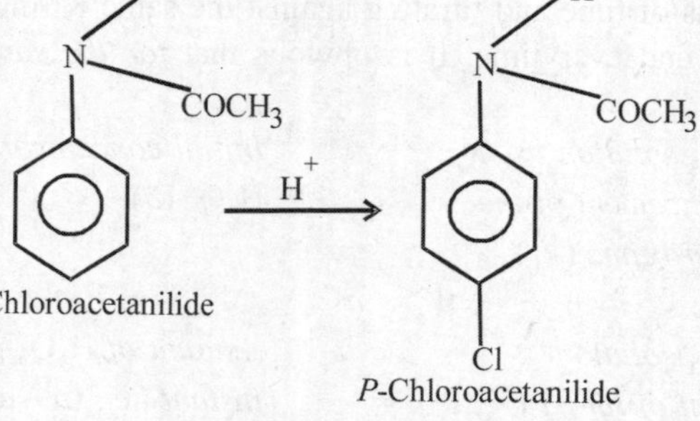

N-Chloroacetanilide *P*-Chloroacetanilide

The study of the kinetics of this reaction is based upon the fact that *N*-chloroacetanilide reacts with KI in presence of the acid to give iodine whereas *p*-chloroacetanilide does not. The reactions take place as follows:

$$2KI + H_2SO_4 \longrightarrow K_2SO_4 + 2HI$$

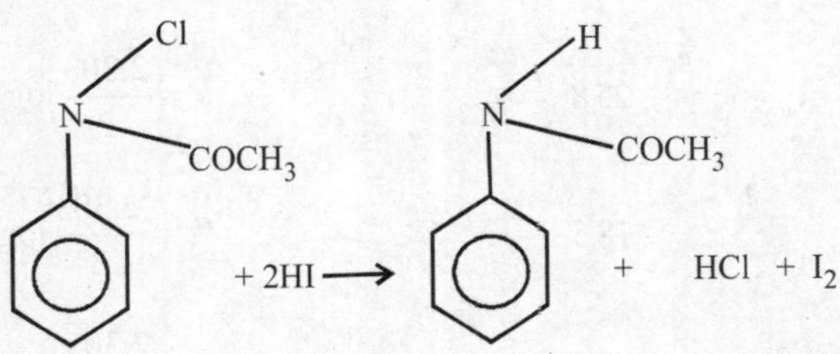

The iodine thus liberated can be found by titration against a standard hypo solution using starch as indicator. For the same volume of the reaction solution withdrawn at different times,

$$\left.\begin{array}{l}\textit{Volume of hypo solution}\\ \textit{used in the beginning}\\ \textit{i.e., at time } t = 0\ (V_0)\end{array}\right\} \propto \left\{\begin{array}{c}\text{Amount of } I_2\\ \textit{liberated at}\\ \textit{that stage}\end{array}\right\} \propto \left\{\begin{array}{c}\textit{Amount of}\\ \textit{N–chloroacetanilide}\\ \textit{present initially } (a)\end{array}\right.$$

or $\qquad V_0 \propto a$

$$\left.\begin{array}{l}\textit{Volume of hypo solution}\\ \textit{used at any instant of}\\ \textit{time } t\ (V_t)\end{array}\right\} \propto \left\{\begin{array}{l}\text{Amount of } I_2\\ \textit{liberated at}\\ \textit{that stage}\end{array}\right\} \propto \left\{\begin{array}{l}\text{Amount of } N\text{-chloroacetanilide}\\ \textit{present at that instant of time}\\ (a - x)\end{array}\right.$$

or $\qquad V_t \propto (a - x)$

Hence for the above reaction,

$$k = \frac{2.303}{t} \log \frac{a}{a - x}$$

$$k = \frac{2.303}{t} \log \frac{V_0}{V_t}$$

***Example 7.* In acidic solution *N*-chloroacetanilide changes into *p*-chloroacetanilide. The former can be estimated volumetrically by withdrawing a definite volume of the reaction mixture, adding KI and titrating the iodine liberated against hypo solution. The following results were obtained in a particular experiment :**

Time (hours) :	**0**	**1**	**2**	**3**	**4**	**∞**
Hypo sol. used (ml) :	**49.3**	**35.6**	**25.8**	**18.5**	**13.8**	**0.0**

Show that the above data conform to the first order reaction.

Solution. Volume of hypo solution used ∝ amount of *N*-chloroacetanilide present.

Hence if the reaction is of the first order, it must obey the equation

$$k = \frac{2.303}{t} \log \frac{a}{a - x}$$

As $V_0 \propto a$ and $V_t \propto a - x$, we have

$$k = \frac{2.303}{t} \log \frac{V_0}{V_t}$$

In the present case, $\qquad V_0 = 49.3$ ml

The value of k at different instants may be calculated as follows:

t (hrs)	V_t	$k = \frac{2.303}{t}\log\frac{V_0}{V_t}$
1	35.6	$k = \frac{2.303}{1}\log\frac{49.3}{35.6} = 0.3256\ \text{hr}^{-1}$
2	25.8	$k = \frac{2.303}{2}\log\frac{49.3}{25.8} = 0.3238\ \text{hr}^{-1}$
3	18.5	$k = \frac{2.303}{3}\log\frac{49.3}{18.5} = 0.3267\ \text{hr}^{-1}$
4	13.8	$k = \frac{2.303}{4}\log\frac{49.3}{13.8} = 0.3183\ \text{hr}^{-1}$

Since the value of k comes out to be nearly constant, the reaction is of the first order.

5. Hydrolysis of ethyl acetate in acidic medium

The reaction can be represented as under :

$$CH_3COOC_2H_5 + H_2O \xrightarrow{H^+} CH_3COOH + C_2H_5OH$$

In this reaction, acetic acid is one of the products the amount of which can be found by titration against standard NaOH solution. But being an acid-catalysed reaction, the acid present originally as a catalyst also reacts with NaOH solution. Hence *for the same volume of reaction mixture withdrawn* at different times.

$$\left.\begin{array}{l}\textit{Volume}\text{ of NaOH }\textit{solution}\\ \textit{used in the beginning i.e., at}\\ \textit{zero time }(V_0)\end{array}\right\} \propto \left\{\begin{array}{l}\textit{Amount of acid present only}\\ \textit{as catalyst (as no } CH_3COOH\\ \textit{is produced at } t = 0)\end{array}\right. \quad \ldots (i)$$

$$\left.\begin{array}{l}\textit{Volume}\text{ of NaOH}\\ \textit{solution used at any}\\ \textit{instant of time } t\ (V_t)\end{array}\right\} \propto \left\{\begin{array}{l}\textit{Amount of acid present}\\ \textit{as catalyst + Amount of}\\ CH_3COOH\ \textit{produced}\end{array}\right. \quad \ldots (ii)$$

Combining results (*i*) and (*ii*), we find that

$$\left.\begin{array}{l}\textit{Amount of } CH_3COOH\\ \textit{produced at any instant}\\ \textit{of time}\end{array}\right\} \propto \left\{(V_t - V_0)\right. \quad \ldots (iii)$$

$$\left.\begin{array}{l}\textit{But amount of } CH_3COOH\\ \textit{produced at any instant of time}\end{array}\right\} \propto \left\{\begin{array}{l}\textit{Amount of } CH_3COOC_2H_5\\ \textit{that has reacted } (x)\end{array}\right.$$

Hence $$(V_t - V_0) \propto x \quad \ldots (iv)$$

Further,

$$\left.\begin{array}{l}\textit{Volume of}\text{ NaOH }\textit{solution used after the}\\ \textit{reaction has taken place}\text{ for }\textit{a long time,}\\ \textit{called infinite time } (V_\infty)\end{array}\right\} \propto \left\{\begin{array}{l}\textit{Amount of acid present}\\ \textit{as catalyst + Max. amount}\\ \textit{of } CH_3COOH\ \textit{produced}\end{array}\right. \quad \ldots (v)$$

Combining results (*i*) and (*v*), we find that

$$\textit{Max. amount of } CH_3COOH\ \textit{produced} \propto (V_\infty - V_0)$$

$$\left.\begin{array}{l}\textit{But Max. amount of}\\ CH_3COOH\ \textit{produced}\end{array}\right\} \propto \left\{\begin{array}{l}\textit{Initial concentration of}\\ CH_3COOC_2H_5\ (a)\end{array}\right.$$

Hence $(V_\infty - V_0) \propto a$ . . . (*vi*)

From equations (*iv*) and (*vi*), we have

$$(a-x) \propto (V_\infty - V_0) - (V_t - V_0)$$

or $(a-x) \propto (V_\infty - V_t)$ . . . (*vii*)

Substituting the values of *a* and (*a* – *x*) from equations (*vi*) and (*vii*) in the first order equation,

$$k = \frac{2.303}{t} \log \frac{a}{a-x}$$

we have
$$k = \frac{2.303}{t} \log \frac{V_\infty - V_0}{V_\infty - V_t}$$

Example 8. **1.0 ml of ethyl acetate was added to 25 ml of *N*/2 HCl. 2 ml of the mixture were withdrawn from time to time during the progress of the hydrolysis of the ester and titrated against standard NaOH solution. The amount of NaOH required for titration at various intervals is given below :**

Time (min)	**0**	**20**	**75**	**119**	**183**	**∞**
NaOH used (ml) :	**20.24**	**21.73**	**25.20**	**27.60**	**30.22**	**43.95**

The value at infinite time was obtained by completing the hydrolysis on boiling. Show that it is a reaction of the first order and find the average value of the velocity constant.

Solution. Amount of NaOH used at $t = 0$ *i.e.*, $(V_0) \propto$ HCl present.

Amount of NaOH used at any time *t i.e.*, $(V_t) \propto$ HCl present + CH_3COOH formed.

CH_3COOH formed at any time ∝ ethyl acetate reacted

$$(V_t - V_0) \propto x$$

Amount of NaOH used at time $t = \infty$ *i.e.*, $(V_\infty) \propto$ HCl present + Max CH_3COOH formed.

Max. CH_3COOH formed ∝ Initial conc. of ethyl acetate

i.e., $(V_\infty - V_0) \propto a$

Hence if the given reaction is of the first order, it must obey the equation

$$k = \frac{2.303}{t} \log \frac{a}{a-x}$$

$$= \frac{2.303}{t} \log \frac{V_\infty - V_0}{V_\infty - V_t}$$

In the present case $V_0 = 20.24$ ml, $V_\infty = 43.95$ ml.

∴ $V_\infty - V_0 = 43.95 - 20.24 = 23.71$ ml.

The value of *k* at different instants can be calculated as follows :

t (min)	V_t	$V_\infty - V_t$	$k = \frac{2.303}{t} \log \frac{V_\infty - V_0}{V_\infty - V_t}$
20	21.73	43.95 – 21.73 = 22.22	$k = \frac{2.303}{20} \log \frac{23.71}{22.22} = 0.00324$
75	25.20	43.92 – 25.20 = 18.75	$k = \frac{2.303}{75} \log \frac{23.71}{18.75} = 0.00313$
119	27.60	43.95 – 27.60 = 16.35	$k = \frac{2.303}{119} \log \frac{23.71}{16.35} = 0.00312$
183	30.22	43.95 – 30.22 = 13.73	$k = \frac{2.303}{183} \log \frac{23.71}{13.73} = 0.00299$

Since the value of k comes out to be nearly constant, it is a reaction of the first order. The average value of $k = 0.00312\ \text{min}^{-1}$.

6. Inversion of cane sugar

The hydrolysis of sucrose in presence of a mineral acid takes place according to the equation

$$\underset{\substack{\text{Sucrose}\\ \text{(dextro-rotatory)}}}{C_{12}H_{22}O_{11}} + H_2O \xrightarrow{H^+} \underset{\substack{\text{Glucose}\\ \text{(dextro-rotatory)}}}{C_6H_{12}O_6} + \underset{\substack{\text{Fructose}\\ \text{(laevo-rotatory)}\\ \text{(laevo-rotatory)}}}{C_6H_{12}O_6}$$

An important characteristic of this reaction is that sucrose is dextro-rotatory whereas the products, glucose and fructose are dextro-rotatory and laevo-rotatory respectively. Further the laevo-rotation of fructose is more (– 92°) than the dextro rotation of glucose (+ 52.5°) so that the mixture as a whole is laevo-rotatory. Thus on hydrolysis, the dextro-rotatory sucrose gradually changes into the laevo-rotatory mixture. It is for this reason that the reaction is called **inversion of sucrose.**

The kinetics of the above reaction is studied by noting the angle of rotation at different intervals of time with the help of a polarimeter. Suppose

Reading of the polarimeter at zero time $= r_0$

Reading of the polarimeter at any time t $= r_t$

Reading of the polarimeter at infinite time $= r_\infty$

(*i.e.*, after 24 hours or more)

It is evident that the reading at zero time will be positive and it would decrease with the passage of time, pass through zero and ultimately become negative .

Angle of rotation at any instant of time, $(r_0 - r_t)$

$\propto$ *the amount of sucrose hydrolysed* (x)

i.e, $$x \propto (r_0 - r_t) \qquad ...(i)$$

Angle of rotation at infinite $(r_0 - r_\infty) \propto$ *the initial concentration of sucrose* (a)

i.e, $$a \propto (r_0 - r_\infty) \qquad ...(ii)$$

From results (*i*) and (*ii*), we have

$$(a - x) \propto (r_0 - r_\infty) - (r_0 - r_t)$$

i. e., $$(a - x) \propto (r_t - r_\infty) \qquad ...(iii)$$

Substituting the values of a and $(a - x)$ from equations (*ii*) and (*iii*) in the first order equation,

$$k = \frac{2.303}{t} \log \frac{a}{a-x}$$

We have $$k = \frac{2.303}{t} \log \frac{r_0 - r_\infty}{r_t - r_\infty}$$

***Example 9.* The inversion of cane sugar was studied in 0.9 *N* HCl at 25°C. The following polarimetric readings were obtained at different intervals of time :**

Time (minutes) :	**0.0**	**7.18**	**18.00**	**27.05**	**∞**
Reading (degrees) :	**+24.09**	**+21.41**	**+17.74**	**+15.00**	**– 10.74**

Show that the inversion of cane sugar is a unimolecular reaction.

Solution. If the reaction is of the first order, it must obey the equation

$$k = \frac{2.303}{t}\log\frac{a}{a-x}$$

$$= \frac{2.303}{t}\log\frac{r_0 - r_\infty}{r_t - r_\infty}$$

In the present case, $r_0 = 24.09$, $r_\infty = 10.74$

$\therefore$ $r_0 - r_\infty = + 24.09 - (- 10.74) = 34.83$.

The value of k at different instants may be calculated as under :

$t(min)$	r_t	$r_t - r_\infty$	$k = \frac{2.303}{t}\log\frac{r_0 - r_\infty}{r_t - r_\infty}$
7.18	+ 21.41	21.41– (– 10.74) = 32.15	$k = \frac{2.303}{7.18} \log \frac{34.83}{32.15} = 0.01116$
18.00	+ 17.74	17.74 – (– 10.74) = 28.48	$k = \frac{2.303}{18.00} \log \frac{34.83}{28.48} = 0.01118$
27.05	+ 15.00	15.00 – (– 10.74) = 25.74	$k = \frac{2.303}{27.05} \log \frac{34.83}{25.74} = 0.01119$

The constant values of k proves that the reaction is of the first order.

7. Thermal decomposition of Azo-isopropane

Azo-isopropane decomposes on heating to form hexane and N_2 according to the equation

$$\underset{\text{Azo-isopropane}}{(CH_3)_2CHN = NCH(CH_3)_2} \xrightarrow{(250 - 290^\circ C)} N_2 + \underset{\text{Hexane}}{C_6H_{14}}$$

The reaction is of the first order. It is a homogeneous reaction because all the reactants and the products are present in the gaseous state. The progress of the reaction can be followed by measuring the pressure of the reaction mixture at different intervals of time. From the values of the pressures at different intervals of time we can calculate the rate (k) constant for the reaction.

Let the initial pressure of Azo-isopropane be P_0 and P be the total pressure of the reaction system.

Let p be the decrease in pressure of azo-isopropane at time t and P_A be the pressure of azo-isopropane.

For every molecule of azo-isopropane decomposes, 1 molecule of N_2 and 1 molecule of hexane (C_6H_{14}) are produced. As a result the pressure increases with the passage of time.

At each stage of the reaction, pressure of azo-isopropane *i.e.* $P_A = (P_0 - p)$.

p is the decrease in pressure of azo-isopropane, the pressure of each of N_2 and C_6H_{14} increases and will be p. The total pressure of the system is given by the sum of partial pressures of azo-isopropane (P_A), nitrogen (P_{N_2}) and hexane ($P_{C_6H_{14}}$)

i.e. $$P = P_A + P_{N_2} + P_{C_6H_{14}}$$

$$= (P_0 - p) + p + p \quad \text{... (i)}$$

$$P = P_0 + p \quad \text{... (ii)}$$

$$[\because P_{N_2} = P_{C_6H_{14}} = p \text{ and } P_A = (P_0 - p)]$$

Now from (*ii*) $p = (P - P_0)$ ... (*iii*)

But $P_A = P_0 - p$

Substituting from (*iii*) the value of p we get

$$P_A = P_0 - (P - P_0) = 2P_0 - P$$

$$\therefore \quad P_A = 2P_0 - P \qquad \text{... } (iv)$$

The initial concentration of azo-isopropane
i.e. P_0 will be proportional to a.
and P_A *i.e.* conc. of azo-isopropane left after the time t will be proportional to $(a - x)$.
i.e. $a \propto P_0$

$$(a - x) \propto P_A \text{ or } (a - x) \propto (2P_0 - P)$$

Substituting the values of a and $(a - x)$ in the equation of first order

$$k = \frac{2.303}{t} \log \frac{P_0}{P_A}$$

or

$$\boxed{k = \frac{2.303}{t} \log \frac{P_0}{2(2P_0 - P)}}$$

***Example 10.* Decomposition of azo-isopropane is represented as**

$$\underset{\text{Azo-isopropane}}{(CH_3)_2CH\,N = NCH\,(CH_3)_2} \xrightarrow{270^\circ C} N_2 + \underset{\text{Hexane}}{C_6H_{14}}$$

was studied at 270°C. Following results were obtained :

Time (t) in seconds	0	200	400	1050
Pressure PA in mm of Hg	36.00	47.42	55.21	67.14

Show by using above data, that the thermal decomposition of azo-isopropane is a reaction of the first order.

Solution. For the thermal decomposition of azo-isopropane, the rate constant k is given by

$$k = \frac{2.303}{t} \log \frac{P_0}{2P_0 - P}$$

Here $P_0 = 36.00$, so that $2P_0 = 72.00$

The value of rate constant k at different intervals of time can be determined as follows :–

Time t (secs)	P (mm)	$2P_0 - P$	$k = \frac{2.303}{t} \log \frac{P_0}{2P_0 - P}$
200	47.42	24.50	$k = 1.908 \times 10^{-3}$
500	55.21	16.79	$k = 1.907 \times 10^{-3}$
1050	67.14	4.86	$k = 1.907 \times 10^{-3}$

The constancy in the value of k indicates that the thermal decomposition of azo-isopropane is a reaction of the first order.

MISCELLANEOUS PROBLEMS OF FIRST ORDER

***Example 11.* Show that in case of a first order reaction, the time required for 99.9% of the reaction to take place is about ten times than that required for half the reaction.**

Solution. Time for half-change is worked out as under

$$t_{1/2} = \frac{2.303}{k} \log \frac{a}{a - \frac{a}{2}}$$

$$= \frac{2.303}{k} \log 2 = \frac{2.303}{k}(0.3010)$$

Time for 99.9% change is worked out as under

$$t_{99.9\%} = \frac{2.303}{k} \log \frac{a}{a - 0.999a}$$

$$= \frac{2.303}{k} \log 10^3 = \frac{2.303}{k} \times 3$$

$$\frac{t_{99.9\%}}{t_{1/2}} = \frac{2.303}{k} \times 3 \times \frac{k}{2.303 \times 0.3010}$$

$$= \frac{3}{0.3010} \approx 10$$

Example 12. **A first order reaction is 40% complete in 50 minutes. Calculate the value of the rate constant. In what time will the reaction be 80% complete ?**

Solution. (*i*) For the first order reaction

$$k = \frac{2.303}{t} \log \frac{a}{a - x}$$

when $x = \frac{40}{100} a = 0.4\,a,$

$t = 50$ minutes (Given)

$$\therefore \quad k = \frac{2.303}{50} \log \frac{a}{a - 0.4a}$$

$$= \frac{2.303}{50} \log \frac{1}{0.6}$$

or $k = \mathbf{0.010216\ min^{-1}}$.

(*ii*) $t = ?,$

when $x = 0.8\,a$

$k = 0.010216\ \text{min}^{-1}$. (Calculated above)

$$\therefore \quad t = \frac{2.303}{k} \log \frac{a}{a - x}$$

$$= \frac{2.303}{0.010216} \log \frac{a}{a - 0.8a}$$

$$= \frac{2.303}{0.010216} \log \frac{1}{0.2}$$

$= \mathbf{157.58\ min.}$

Example 13. **The reaction**

$$SO_2Cl_2 \xrightarrow{k_1} SO_2 + Cl_2$$

is a first order reaction with $k_1 = 2.2 \times 10^{-5}$ sec^{-1} at 302°C. What percentage of SO_2Cl_2 will get decomposed in 90 minutes when the reaction is carried out at 302°C ?

Solution. Since the reaction is of the first order (given),

$$k = \frac{2.303}{t} \log \frac{a}{a - x}$$

Here we are given that

$k = 2.2 \times 10^{-5}\ \text{sec}^{-1}$

$t = 90$ minutes $= 90 \times 60 = 5400$ sec

$$\therefore \quad 2.2\times10^{-5} = \frac{2.303}{5400}\log\frac{a}{a-x}$$

or $$\log\frac{a}{a-x} = 0.0516$$

or $$\frac{a}{a-x} = \text{antilog } (0.0516) = 1.127$$

or $$a = 1.127\,a - 1.127x$$

or $$0.127\,a = 1.127x$$

$$\therefore \quad \frac{x}{a} = \frac{0.127}{1.127} = 0.113$$

$$= \mathbf{11.3\%}$$

Example 14. **At 100°C, the half-life period for the thermal decomposition of N_2O_5 is 4.6 sec and is independent of the initial pressure of N_2O_5. Calculate the specific rate constant at this temperature.**

Solution. Since the half-life period is independent of the initial pressure, this shows that the reaction is of the *first order.*

Hence it must obey the equation

$$k = \frac{2.303}{t}\log\frac{a}{a-x}$$

For half life period, $t = t_{1/2}$, $x = a/2$

$$\therefore \quad k = \frac{2.303}{t_{1/2}}\log\frac{a}{a-\frac{a}{2}}$$

$$= \frac{2.303}{t_{1/2}}\log 2$$

$$= \frac{2.303}{4.6}(0.3010) = \mathbf{0.1507\,sec^{-1}}$$

6.7 RADIOACTIVE DECAY AS A FIRST ORDER PHENOMENON

Radioactive substances emit alpha, beta and gamma particles. The mass of the substance decreases continuously as a result of decay. It has been found that radioactive decay follows first order kinetics. The rate constant in the case of nuclear decay is called **decay constant.**

Example 15. **The radioactive decay of atomic nucleus is a first order reaction. The half-life period of Radium i.e. $_{88}Ra^{226}$ is 1590 years. What is decay constant ?**

Solution. We know that for the first order reaction, half-life period is given by

$$t_{1/2} = \frac{0.693}{k}$$

$$k = \frac{0.693}{t_{1/2}} = \frac{0.693}{1590}$$

$$k = 0.000435849\ yr^{-1}$$

6.8 SECOND ORDER REACTION

A reaction is said to be of the second order if the rate of reaction depends upon two concentration terms.

Case I: When only one reactant is involved

The general equation for a reaction involving one reactant is

$$2A \longrightarrow \text{products}$$

Suppose the initial concentration of $A = a$ moles/litre.

Decrease in the concentration of A at any instant of time $= x$ moles/litre.

$\therefore$ Concentration of A at the instant of time $t = (a - x)$ mole/litre

Applying the law of mass action, the rate of reaction

$$\frac{dx}{dt} \propto [A]^2$$

$$\propto (a - x)^2$$

or
$$\frac{dx}{dt} = k\,(a - x)^2 \qquad ...(i)$$

where k is the rate constant for the second order reaction.

Equation (i) may be rewritten as

$$\frac{dx}{(a-x)^2} = kdt$$

Integrating this equation, we get

$$\int \frac{dx}{(a-x)^2} = \int kdt$$

i.e.,
$$\frac{(a-x)^{-1}}{-1} \times (-1) = kt + I$$

or
$$\frac{1}{(a-x)} = kt + I \qquad ...(ii)$$

where I is the constant of integration. In the beginning *i. e.,* at time $t = 0$, $x = 0$ (as no substance has reacted at zero time). Putting these values in equation (ii), we get

$$\frac{1}{a} = I$$

Substituting this value of I in equation (ii), we get

$$\frac{1}{a-x} = kt + \frac{1}{a}$$

$$kt = \frac{1}{a-x} - \frac{1}{a} \qquad ...(iii)$$

$$= \frac{a-(a-x)}{a(a-x)}$$

$$= \frac{x}{a(a-x)}$$

$$\boldsymbol{k = \frac{1}{t}\frac{x}{a(a-x)}} \qquad ...(iv)$$

If the initial concentration is written as C_0 and concentration at any instant of time t as C_t, then $a = C_0$, and $(a - x) = C_t$. Substituting these values in equation (iii), we get

$$k = \frac{\mathbf{1}}{\mathbf{t}}\left[\frac{\mathbf{1}}{\mathbf{C_t}} - \frac{\mathbf{1}}{\mathbf{C_0}}\right] \qquad ...(v)$$

This is another form of the expression for the rate constant of second order reaction.

Case II: When two different reactants with different initial concentration are involved

General equation for such reaction is

$$A + B \longrightarrow \text{Products}$$

Suppose the initial concentration of $A = a$ moles/litre, the initial concentration of $B = b$ moles/litre, amount of A that reacts in time $t = x$ moles/litre, then the amount of B that would react in the same time would also be x moles/litre

∴ At any instant of time t,

Concentration of $A = (a - x)$ moles/litre

and concentration of $B = (b - x)$ moles/litre

Applying the law of mass action, we get

$$\text{Rate of reaction } \frac{dx}{dt} \propto [A]\,[B]$$

$$\propto (a - x)\,(b - x)$$

or

$$\frac{dx}{dt} = k\,(a - x)\,(b - x) \qquad ...(i)$$

where k is the rate constant.

Equation (i) may be rewritten as

$$\frac{dx}{(a-x)(b-x)} = kdt \qquad ...(ii)$$

Resolving the left hand side into partial fractions, equation (ii) may be rewritten as

$$\frac{1}{(a-b)}\left[\frac{1}{(b-x)} - \frac{1}{(a-x)}\right]dx = kdt$$

Integrating this equation, we get

$$\frac{1}{(a-b)}\left(\int\frac{dx}{(b-x)} - \int\frac{dx}{(a-x)}\right) = \int kdt$$

or

$$\frac{1}{(a-b)}[-\ln\,(b - x) - \{-\ln(a - x)\}] = kt + I$$

or

$$\frac{1}{(a-b)}[\ln\,(a - x) - \ln\,(b - x)] = kt + I$$

or

$$\frac{1}{(a-b)}\left(\ln\frac{a-x}{b-x}\right) = kt + I \qquad ...(iii)$$

Where I is the constant of integration.

But at $t = 0$, $x = 0$, Putting these values in equation (iii), we get

$$\frac{1}{(a-b)}\left(\ln\frac{a}{b}\right) = I \qquad ...(iv)$$

Putting this value in equation (iii) , we get

$$\frac{1}{(a-b)}\ln\frac{a-x}{b-x} = kt + \frac{1}{(a-b)}\ln\frac{a}{b}$$

or

$$kt = \frac{1}{(a-b)}\ln\frac{a-x}{b-x} - \frac{1}{(a-b)}\ln\frac{a}{b}$$

$$= \frac{1}{(a-b)}\left[\ln\frac{a-x}{b-x} - \ln\frac{a}{b}\right]$$

$$= \frac{1}{(a-b)}\ln\left(\frac{a-x}{b-x}\times\frac{b}{a}\right)$$

or $$k = \frac{2.303}{t(a-b)} \log \frac{b(a-x)}{a(b-x)} \quad ...(v)$$

Example 16. **What would be the units of rate constant in the following rate law:**

rate = k [A] [B]

if the concentration is expressed in moles per litre and time in seconds.

Solution. For a second order reaction

$$k = \frac{1}{t}\frac{x}{a(a-x)}$$

$$= \frac{1}{\text{time}} \times \frac{\text{Concentration}}{\text{Concentration} \times \text{Concentration}}$$

or $$k = \frac{1}{\text{time} \times \text{concentration}}$$

Substituting for the units of time and concentration

$$k = \frac{1}{\text{s} \times \text{mol litre}^{-1}} = \text{litre mol}^{-1}\ \text{s}^{-1}$$

6.8.1 Some Characteristics of second order reaction

A second order reaction has the following characteristics:

(*i*) *Any reaction of the second order must obey the equation*

$$k = \frac{1}{t}\frac{x}{a(a-x)}$$

If on substituting the values of t, x, a and $(a - x)$, constant values of k are obtained, it means the reaction is of second order.

Graphical method can also be used.

We may use the second order equation in the form

$$kt = \frac{1}{C_t} - \frac{1}{C_0}$$

or $$\frac{1}{C_t} = kt + \frac{1}{C_0}$$

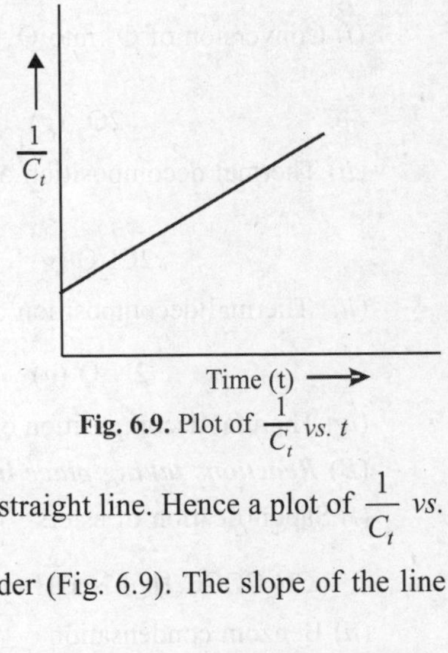

Fig. 6.9. Plot of $\frac{1}{C_t}$ *vs.* t

This equation is like $y = mx + c$ *i.e.*, the equation of a straight line. Hence a plot of $\frac{1}{C_t}$ *vs.* t should be a straight line if the reaction is of the second order (Fig. 6.9). The slope of the line will directly give the value of k.

(*ii*) *The time taken for any fraction of the reaction to be completed is inversely proportional to the initial concentration.* As an example, let us calculate the time taken for half of the reaction to be completed (*i.e.*, half life period). Time taken for half of the reaction to be completed is represented by $t_{1/2}$. Thus when $x = a/2$, $t = t_{1/2}$. Putting these values in equation for second-order reaction, we get

$$t_{1/2} = \frac{1}{k}\frac{a/2}{a(a-a/2)}$$

$$= \frac{1}{k}\cdot\frac{1}{a}$$

i.e, $$t_{1/2} \propto \frac{1}{a}$$

Similarly, it can be seen that for any fraction of the reaction to be completed, the time taken $\propto \frac{1}{a}$.

(*iii*) *The units of the rate constant k depend not only upon the units of time but also upon the units in which the concentrations are expressed.* This is obvious because there is one extra concentration term in the denominator in the equation for second order reaction.

$$k = \frac{1}{t}\frac{x}{a(a-x)}$$

$$= \frac{1}{\text{Time}} \cdot \frac{\text{Concentration}}{\text{Concentation} \times \text{Concentration}}$$

$$= \frac{1}{\text{Concentation} \times \text{Time}}$$

Thus the units of k for second order reaction are conc^{-1} time^{-1}.

6.8.2 Examples of second order reactions

(A) Gaseous Reactions.

There are many reactions which occur in the gaseous state and are kinetically of the second order *e.g.*

(*i*) Conversion of O_3 into O_2 at 100°C

i.e. $$2O_3\ (g) \xrightarrow{100^\circ\ C} 3O_2\ (g)$$

(*ii*) Thermal decomposition of Cl_2O to Cl_2 and O_2 at 200°C

$$2Cl_2O\ (g) \xrightarrow{200^\circ\ C} 2Cl_2\ (g) + O_2\ (g)$$

(*iii*) Thermal decomposition of N_2O

$$2N_2O\ (g) \longrightarrow 2N_2\ (g) + O_2\ (g)$$

(*iv*) Thermal decomposition of acetaldehyde at 518°C

(B) Reactions taking place in solutions

(*i*) Saponification of esters

$$CH_3COOC_2H_5 + NaOH \longrightarrow CH_3COONa + C_2H_5OH$$

(*ii*) Benzoin condensation

$$C_6H_5\underset{\substack{|\\H}}{C}=O + H-\overset{\substack{O\\||}}{C}-C_6H_5 \xrightarrow{KCN} C_6H_5-\underset{\substack{|\\H}}{\overset{\substack{OH\\|}}{C}}-\overset{\substack{O\\||}}{C}-C_6H_5$$

Benzoin

(*iii*) Conversion of ammonium cyanate to form urea

$$\underset{\text{Amm. cyanate}}{NH_4CNO} \longrightarrow \underset{\text{urea}}{H_2N-\overset{\substack{O\\||}}{C}-NH_2}$$

6.8.3 Kinetics of thermal decomposition of acetaldehyde

The thermal decomposition of acetaldehyde vapour takes place according to the equation

$$CH_3CHO \xrightarrow{791\,K} CH_4 + CO$$

From this equation, the reaction appears to be of the first order but experimentally it is found to be of second order and hence is preferably written as

$$2CH_3CHO \longrightarrow 2CH_4 + 2CO$$

As the reactants and the products are all gaseous, the progress of the reaction is followed by simply noting the pressure of the reaction mixture at different intervals of time. Further, as one molecule of acetaldehyde on decomposition produces two molecules (one of CH_4 and the other of CO), the pressure increases as the reaction progresses. It is obvious that

$$\left.\begin{array}{l}\textit{Pressure in the beginning} \\ \textit{i.e., at zero time}, (P_0)\end{array}\right\} \propto \left\{\begin{array}{l}\textit{Intital concentration} \\ \textit{of acetaldehyde}\,(a)\end{array}\right.$$

i.e. $\quad a \propto P_0 \quad$ ...(*i*)

At any other instant of time, *t*, the total pressure of the reaction mixture (P_t) will be given by

$$P_t = P_{CH_3CHO} + P_{CH_4} + P_{CO} \qquad ...(ii)$$

If at this instant of time, *p* is the decrease in pressure of CH_3CHO then *p* will be increase in the pressure of each of CH_4 and CO so that we have

$$P_{CH_3CHO} = P_0 - p \qquad ...(iii)$$

$$P_{CH_4} = p$$

$$P_{CO} = p$$

Substituting these values in equation (*ii*), we get

$$P_t = (P_0 - p) + p + p$$

$$= P_0 + p$$

or $\quad p = P_t - P_0 \quad$ ...(*iv*)

$$\left.\begin{array}{l}\textit{But decrease in the} \\ \textit{Pressure of} \\ \textit{acetaldelyde}\ (P)\end{array}\right\} \propto \left\{\begin{array}{l}\textit{Acetaldehyde} \\ \textit{decomposed}(x)\end{array}\right.$$

i.e. $\quad x \propto p \quad$ ...(*v*)

Combining this result with equation (*iv*), we get

$$x \propto P_t - P_0 \qquad ...(vi)$$

From equation (*i*),

$$a \propto P_0$$

$\therefore \quad a - x \propto P_0 - (P_t - P_0)$

or $\quad (a - x) \propto 2P_0 - P_t$

Substituting these values in the second order equation,

$$k = \frac{1}{t}\,\frac{x}{a(a-x)}$$

we have $$k = \frac{1}{t}\frac{P_t - P_0}{P_0(2P_0 - P_t)}$$

Example 17. **The thermal decomposition of acetaldehyde was studied by Hinshelwood and Hutchison at 518°C. Starting with an initial pressure of 363 mm of Hg, the following results were obtained at different intervals of time.**

***t* (sec)**	**: 42**	**73**	**105**	**190**
Increase in pressure, Δ*p* (mm)	**: 34**	**54**	**74**	**114**

Show that the reaction is of the second order.

Solution. If the reaction is of second order, it must obey the equation.

$$k = \frac{1}{t}\frac{P_t - P_0}{P_0(2P_0 - P_t)}$$

Here we are given that

$$P_0 = 363 \text{ mm of Hg}$$

Δp means $P_t - P_0$

The value of k at different intervals may be calculated as follows :

t (sec)	$P_t - P_0$	$2P_0 - P_t = P_0 - (P_t - P_0)$	$k = \frac{1}{t}\frac{P_t - P_0}{P_0(2P_0 - P_t)}$
42	34	363 – 34 = 329	$k = \frac{1}{42}\times\frac{34}{363\times329} = 6.79\times10^{-6}$
73	54	363 – 54 = 309	$k = \frac{1}{73}\times\frac{54}{363\times309} = 6.59\times10^{-6}$
105	74	363 – 74 = 289	$k = \frac{1}{105}\times\frac{74}{363\times289} = 6.71\times10^{-6}$
190	114	363 – 114= 249	$k = \frac{1}{190}\times\frac{114}{363\times249} = 6.64\times10^{-6}$

The constancy in the value of k shows the reaction to be of second order.

6.8.4 Saponification (alkaline hydrolysis) of esters

The reaction of esters with sodium hydroxide solution is called saponification of esters. For example, the reaction of ethyl acetate with sodium hydroxide solution takes place as follows:

$$\underset{\text{Ethyl acetate}}{CH_3COOC_2H_5} + NaOH \longrightarrow \underset{\text{Sodium acetate}}{CH_3COONa} + \underset{\text{Ethyl alcohol}}{C_2H_5OH}$$

The reaction involves NaOH as one of the reactants, the concentration of which decreases as the reaction progresses. Starting with known concentration of ethyl acetate and sodium hydroxide (preferably equal concentrations), the concentration of NaOH at any instant of time can be found by withdrawing a definite amount of the reaction mixture and titrating against a standard acid. It is obvious that for the same volume of the reaction mixture withdrawn at different intervals of time,

$$\left.\begin{array}{l}\textit{Volume of the standard}\\ \textit{acid used at the beginning}\\ \textit{of the experiment i.e. at}\\ \textit{zero time,}(V_0)\end{array}\right\} \propto \left\{\begin{array}{l}\textit{Initial concentration}\\ \textit{of } \mathrm{NaOH}\,(a)\end{array}\right.$$

i.e. $$a \propto V_0 \qquad \text{....}(i)$$

$$\left(\begin{array}{l}\textit{Volume of standard acid}\\ \textit{solution used at any}\\ \textit{instant of time } t,\ (V_t)\end{array}\right) \propto \left\{\begin{array}{l}\textit{Concentration of}\\ \mathrm{NaOH}\ \textit{at that}\\ \textit{instant i.e.}\,(a-x)\end{array}\right.$$

i.e., $$(a - x) \propto V_t \qquad \text{...}(ii)$$

From equations (*i*) and (*iii*), we get

$$x \propto V_0 - V_t \qquad \text{...}(iii)$$

Substituting the values of *a*, *x* and (*a* – *x*) from equations (*i*), (*ii*) and (*iii*) in the second order equation,

$$k = \frac{1}{t}\frac{x}{a(a-x)}$$

we have $$k = \frac{1}{t}\frac{(V_0 - V_t)}{V_0 V_t}$$

Example 18. **Starting with equal concentration of ethyl acetate and sodium hydroxide, the saponification of ethyl acetate was carried out. Same volume of the reaction mixture was withdrawn at different intervals of time and titrated with the same acid. The following data were obtained :**

Time (min.) :	**0**	**5**	**15**	**25**
Acid used (ml) :	**16**	**10.24**	**6.13**	**4.32**

Show that the reaction is of the second order.

Solution. If the reaction is of the second order, it must obey the equation

$$k = \frac{1}{t}\frac{x}{a(a-x)} = \frac{1}{t}\frac{(V_0 - V_t)}{V_0 \cdot V_t}$$

In the present case, $V_0 = 16$ ml

The value of *k* at different instants of time can be calculated as follows :

t (min.)	V_t	$(V_0 - V_t)$	$k = \frac{1}{t}\frac{(V_0 - V_t)}{V_0 V_t}$
5	10.24	5.76	$k = \frac{1}{5} \times \frac{5.76}{16 \times 10.24} = 0.00703$
15	6.13	9.87	$k = \frac{1}{15} \times \frac{9.87}{16 \times 6.13} = 0.00671$
25	4.32	11.68	$k = \frac{1}{25} \times \frac{11.68}{16 \times 4.32} = 0.00676$

Thus the value of *k* is fairly constant. The reaction is of the second order.

MISCELLANEOUS EXAMPLES OF SECOND ORDER

Example 19. **A second order reaction in which the initial concentrations of both the reactions are same is 25% complete in 600 sec. How long will it take for the reaction to go to 75% completion?**

Solution. For a second order reaction

$$k = \frac{1}{t}\frac{x}{a(a-x)}$$

For the first case, $x = \frac{25}{100}a = 0.25\ a$

$$t = 600 \text{ sec.}$$

$$\therefore \quad k = \frac{1}{200} \times \frac{0.25a}{a(a-0.25a)}$$

$$= \frac{1}{600} \times \frac{0.25}{0.75a} = \frac{1}{1800a}$$

For the second case, $x = \frac{75}{100}a = 0.75a$

$$t = \text{To be calculated}$$

$$\therefore \quad k = \frac{1}{t} \times \frac{0.75a}{a(a-0.75a)}$$

$$= \frac{1}{t} \times \frac{0.75a}{0.25a} = \frac{3}{ta}$$

But $k = \frac{1}{1800a}$ (from the first case)

$$\therefore \quad \frac{1}{1800a} = \frac{3}{ta}$$

or $t = 3 \times 1800 =$ **5400 sec.**

Example 20. **Decomposition of a gas is of second order. It takes 40 minutes for 40% of a gas to be decomposed when its initial concentration is 4×10^{-2} mole/litre. Calculate the specific reaction rate.**

Solution. For a second order reaction

$$k = \frac{1}{t}\frac{x}{a(a-x)}$$

Here, we are given that

$$a = 4 \times 10^{-2} \text{ mole/litre}$$

$$t = 40 \text{ minutes}$$

$$x = \frac{40}{100} \times a = 0.4a$$

Substituting the values in the above formula, we get

$$k = \frac{1}{40} \times \frac{0.4a}{a(a-0.4a)}$$

$$= \frac{1}{40} \times \frac{0.4}{0.6a} = \frac{1}{40} \times \frac{0.4}{0.6 \times 4 \times 10^{-2}}$$

$$= \mathbf{0.4167 \text{ litre mol}^{-1} \text{ min}^{-1}}$$

6.9 THIRD ORDER REACTION

A reaction is said to be of the third order if the rate of reaction depends upon three concentration terms.

The simplest case of this type can be represented as

$$3A \longrightarrow \text{Products}$$

The expression for the rate constant of such a reaction may be derived as follows :

Suppose a is the initial concentration of A and x is the decrease in the concentration of A in time t. Then the concentration of A after time $t = (a - x)$. Now, according to the law of mass action, the rate of the above reaction at the instant of time t will be given by

$$\frac{dx}{dt} \propto [A]^3$$

i.e.

$$\frac{dx}{dt} \propto (a-x)^3$$

or

$$\frac{dx}{dt} = k\,(a-x)^3 \qquad ...(i)$$

where k is the rate constant for the third order reaction.

Equation (i) may be rewritten as

$$\frac{dx}{(a-x)^3} = kdt$$

Integrating this equation, we get

$$\int \frac{dx}{(a-x)^3} = \int kdt$$

i.e.

$$\frac{(a-x)^{-2}}{-2} \times (-1) = kt + I$$

or

$$\frac{1}{2(a-x)^2} = kt + I \qquad ...(ii)$$

where I is the constant of integration.

Now at $t = 0$, $x = 0$. Putting these values in equation (ii), we get

$$\frac{1}{2a^2} = I \qquad ...(iii)$$

Substituting this value in equation (ii), we get

$$\frac{1}{2(a-x)^2} = kt + \frac{1}{2a^2}$$

or

$$kt = \frac{1}{2(a-x)^2} - \frac{1}{2a^2}$$

$$= \frac{1}{2}\left[\frac{1}{(a-x)^2} - \frac{1}{a^2}\right]$$

$$= \frac{1}{2}\left[\frac{a^2-(a-x)^2}{a^2(a-x)^2}\right]$$

$$= \frac{1}{2}\left[\frac{a^2-(a^2+x^2-2ax)}{a^2(a-x)^2}\right]$$

$$= \frac{1}{2}\left[\frac{2ax - x^2}{a^2(a-x)^2}\right]$$

$$= \frac{1}{2}\left[\frac{x(2a - x)}{a^2(a-x)^2}\right]$$

or $$\boldsymbol{k = \frac{1}{2t}\left(\frac{x(2a - x)}{a^2(a-x)^2}\right)} \qquad \text{... } (iv)$$

Replacing a by C_0 (*i.e.* initial concentration of A) and x by $C_0 - C_t$ (*i.e.* decrease in the concentration of A), equation (*iv*) may be rewritten as

$$k = \frac{1}{2t}\left[\frac{(C_0 - C_t)(2C_0 - C_0 + C_t)}{C_0^2(C_0 - C_0 + C_t)^2}\right]$$

$$= \frac{1}{2t}\frac{(C_0 - C_t)(C_0 + C_t)}{C_0^2 C_t^2}$$

$$= \frac{1}{2t}\frac{C_0^2 - C_t^2}{C_0^2 C_t^2}$$

or $$\boldsymbol{k = \frac{1}{2t}\left[\frac{1}{C_t^2} - \frac{1}{C_0^2}\right]} \qquad \text{...} (v)$$

6.9.1 Some characteristics of third order reactions

Third order reactions have the following characteristics.

(*i*) A reaction is of the third order, if it obeys the equation

$$k = \frac{1}{2t}\left[\frac{x(2a - x)}{a^2(a-x)^2}\right] \qquad \text{...} (i)$$

Thus on substituting the values of t, x and a, constant values of k must be obtained.

(*ii*) *The time taken for any fraction of the reaction to be completed is inversely proportional to the square of the initial concentration.* For example, the time taken for half of the reaction to *be* completed (*i.e. half period*) may be calculated as follows :

When $x = \frac{a}{2}$, $t = t_{1/2}$

Putting these values in equation (*i*), we get

$$t_{1/2} = \frac{1}{2k}\left\{\frac{\frac{a}{2}\left(2a - \frac{a}{2}\right)}{a^2\left(a - \frac{a}{2}\right)^2}\right\}$$

or $$t_{1/2} = \frac{1}{2k}\left\{\frac{\frac{a}{2}\cdot\frac{3a}{2}}{a^2\,\frac{a^2}{4}}\right\}$$

or $$t_{1/2} = \frac{1}{2k}\frac{3}{a^2}$$

i.e. $$t_{1/2} \propto \frac{1}{a^2}$$

(*iii*) *The units of the rate constant k depend upon the units of the concentration.* This may be seen by calculating the dimensions of k with the help of equation for third order.

$$k = \frac{1}{2t}\left[\frac{x(2a-x)}{a^2(a-x)^2}\right]$$

$$= \frac{1}{\text{Time}}\left[\frac{\text{Concentration} \times \text{Concentration}}{(\text{Concentration})^2 \times (\text{Concentration})^2}\right]$$

$$= \frac{1}{\text{Time}}\frac{1}{(\text{Concentration})^2}$$

Thus any change in the unit of concentration will change the value of k.

6.9.2 Some examples of third order reaction

Five reactions in gaseous phase, each involving nitric oxide as one of the reactions are known to be third order reactions.

(*i*) $2NO + Cl_2 \longrightarrow 2NOCl$
(*ii*) $2NO + Br_2 \longrightarrow 2NOBr$
(*iii*) $2NO + O_2 \longrightarrow 2NO_2$
(*iv*) $2NO + H_2 \longrightarrow N_2O + H_2O$
(*v*) $2NO + D_2 \longrightarrow N_2O + D_2O$

Besides, reduction of ferric chloride by stannous chloride is an example of third order reaction in solution.

$$2FeCl_3 + SnCl_2 \longrightarrow 2FeCl_2 + SnCl_4$$

Why are the reactions of higher order rare ?

The reason for this may be explained on the basis of **collision theory.** According to this theory, in order that a reaction may take place between certain molecules, these molecules must come together simultaneously to collide with each other. As the probability for three molecules to come together simultaneously and collide are less than those for two molecules, the reactions of third order are fewer than those of second order. Similarly, the chances for four or more molecules to come together simultaneously and collide are still less (Fig. 6.10), hence the reactions of fourth or higher order are very rare.

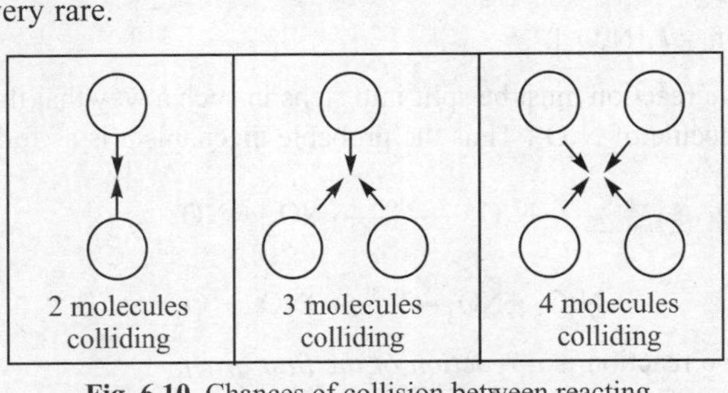

Fig. 6.10. Chances of collision between reacting molecules decrease as the molecularity increases.

6.9.3 Relations between half-life period and order of a reaction

The relations are as under :

(*i*) *For reactions of zero order*

$$t_{1/2} \propto a$$

i.e. half-life period is directly proportional to the initial concentration.

(*ii*) *For reactions of the first order*

$$t_{1/2} = \frac{2.303}{k}\log 2$$

$$= \frac{2.303}{k}(0.3010) = \frac{0.693}{k}$$

i.e. half-life period is independent of the initial concentration.

(*iii*) *For reactions of the second order*

$$t_{1/2} \propto \frac{1}{a}$$

i.e. half-life period is inversely proportional to the initial concentration.

(*iv*) *For reactions of third order*

$$t_{1/2} \propto \frac{1}{a^2}$$

i.e. half-life period is inversely proportional to the square of the initial concentration.

In general, for a reaction of the *n*th order

$$t_{1/2} \propto \frac{1}{a^{n-1}}$$

6.10 FORMULATION OF MECHANISM OF A REACTION

A series of step reactions, proposed to account for the overall reaction is called the mechanism of the reaction.

The writing of step reactions is usually based upon experimental evidence. *Slowest step (called the rate determining step) must involve the molecules on which the rate of reaction actually depends as observed experimentally.* The following examples will illustrate the point more clearly:

(*i*) *Thermal decomposition of dinitrogen pentoxide. The balanced equation may be written as*

$$2N_2O_5 \longrightarrow 4NO_2 + O_2$$

However, according to the rate law, it is found that

Rate of reaction = $k\,[N_2O_5]$.

Hence the above reaction must be split into steps in such a way that the slowest step should involve only one molecule of N_2O_5. Thus the probable mechanism is as follows:

$$N_2O_5 \xrightarrow{\text{Slow}} NO_2 + NO_3$$

$$N_2O_5 + NO_3 \xrightarrow{\text{Fast}} 3NO_2 + O_2$$

Hence the above reaction is *a reaction of the first order.*

(*ii*) *Bromination of propanone* (*acetone*). The bromination of propanone in dilute aqueous solution is represented by the equation

$$CH_3COCH_3 + Br_2 \longrightarrow CH_3COCH_2Br + HBr$$

Experimentally, it is found that

Rate of reaction $\propto$ $[CH_3COCH_3]$ only (and is independent of the concentration of bromine).

Hence the rate determining step (slow step) would involve a molecule of acetone only. The suggested steps are as follows :

(*a*) CH_3COCH_3 (Ketonic from) $\xrightarrow{\text{Slow}}$ $CH_3-\underset{\underset{\text{OH}}{|}}{C}=CH_2$ (enolic from)

(*b*) $CH_2-\underset{\underset{\text{OH}}{|}}{C}=CH_2 + Br_2 \xrightarrow{\text{Fast}} CH_3COCH_2Br + HBr$

(*iii*) *Combination of* NO_2 *and* F_2 *to form* NO_2F. The reaction may be represented as

$$2NO_2 + F_2 \longrightarrow 2NO_2F$$

Experimentally, it is found that

Rate of reaction = $k\,[NO_2]\,[F_2]$

i.e. the rate law equation involves only one molecule of NO_2 and one molecule of F_2. Hence the probable mechanism is as follows :

$$NO_2 + F_2 \xrightarrow{\text{Slow}} NO_2F + F$$

$$NO_2 + F \xrightarrow{\text{Fast}} NO_2F$$

Hence the above reaction must be a *reaction of the second order.*

(*iv*) *Oxidation of hydrogen bromide with oxygen.* The oxidation of hydrogen bromide with oxygen in the temperature range 400 – 600°C takes place rapidly as follows :

$$4HBr(g) + O_2(g) \longrightarrow 2H_2O(g) + 2Br_2\,(g)$$

Experimentally, it is found that

Rate of reaction $\propto$ [HBr] $[O_2]$

This implies that the slowest step of the mechanism of the above reaction should involve only one molecule of HBr and one molecule of O_2. Hence the probable mechanism is as follows :

$$HBr + O_2 \xrightarrow{\text{Slow}} HOOBr$$

$$HOOBr + HBr \xrightarrow{\text{Fast}} 2HOBr$$

$$HOBr + HBr \xrightarrow{\text{Fast}} H_2O + Br_2] \times 2$$

Overall reaction : $$4HBr + O_2 \longrightarrow 2H_2O + 2Br_2$$

(*v*) *Reduction of nitric oxide with hydrogen.* The reaction

$$2NO + 2H_2 \longrightarrow N_2 + 2H_2O$$

The rate law equation is found to be

$$\text{Rate}\left(\frac{dx}{dt}\right) = k[NO]^2[H_2]$$

This suggests that the reaction is of the third order and the rate determining step (slow and step) should involve two molecules of NO and one molecule of H_2. Thus the following mechanism has been proposed:

$$2NO + H_2 \xrightarrow{\text{Slow}} N_2 + H_2O_2$$

$$H_2O_2 + H_2 \xrightarrow{\text{Fast}} 2H_2O$$

Example 21. **Order and molecularity of a reaction may not be the same. Explain**

Solution. *Order* of a reaction may be defined as the sum of exponents to which the concentration terms in the rate law are raised to express the observed rate of reaction.

Molecularity, in the case of simple reactions, is the sum of molecules of the reactants as represented in the balanced equation. In complex reactions, it is the number of molecules involved in the slowest step.

Order and molecularity may not be the same.

Consider the hydrolysis of an ester with a mineral acid

$$CH_3COOC_2H_5 + H_2O \xrightarrow{H^+} CH_3COOH + C_2H_5OH$$

This reaction has the molecularity equal to 2, while its order has been found to 1.

Consider the decomposition of acetaldehyde at 450°C

$$CH_3CHO \xrightarrow{450°C} CH_4 + CO$$

It is a unimolecular reaction (molecularity = 1) but experimentally its order has been found to be 1.5.

6.11 METHODS FOR THE DETERMINATION OF ORDER OF A REACTION

A number of methods are available for the determination of order of a reaction. Some of them are described as under:

1. Integral Rate Equation Method

In this method, the initial concentrations of the reactants and the concentration of any one of the reactants or the products at different instants of time are determined experimentally. Then either of the following two methods may be employed :

(*i*) *Substitution method* (*Hit and Trial Method*). The appropriate values of a and x (or C_0 and C_t) are substituted into the integrated forms of the rate equations for the first order, second order and third order one by one. The rate equation which gives the constant value of k gives the order of the reaction.

(*ii*) *Graphical method.* In this method, a suitable function is plotted against time.

(*a*) If the plot of log $(a - x)$ against time is a straight line, the reaction is of the first order.

(*b*) If the plot of $\frac{1}{a-x}$ against time is a straight line, the reaction is of the second order.

Example 22. **Following data is obtained for the conversion of ammonium cyanate into urea**

Time in minutes	**0**	**45**	**72**	**157**	**312**	**600**
Conc. of Amm. cyanate unchanged in mol l^{-1}	**0.0916**	**0.0740**	**0.0656**	**0.0512**	**0.0348**	**0.0228**

What is the order of reaction ?

Solution. Substituting the data in the first order rate equation, if the value of k obtained is not constant, the data is then substituted in the second order rate equation and so on. The rate equation which gives constant value of k gives the correct order of reaction.

In the present examples

$$a = 0.0916 \; mol \; l^{-1}$$

Time (t) in minutes	$(a-x)$	$k = \frac{2.303}{t} \log \frac{a}{a-x}$	$k = \frac{1}{2t}\left[\frac{1}{(a-x)} - \frac{1}{a}\right]$
45	0.0740	5.74×10^{-3}	5.77×10^{-2}
72	0.0656	4.63×10^{-3}	5.99×10^{-2}
157	0.0512	3.70×10^{-3}	5.48×10^{-2}
312	0.0348	3.08×10^{-3}	5.51×10^{-2}
600	0.0228	2.31×10^{-3}	5.49×10^{-2}

Since the value of k comes out to be almost constant by using rate equation for the second order reaction. Therefore, it is evident that the reaction is of ***second order.***

2. Differential rate equation method

This method is based upon the fact that the rate of a reaction of the nth order is proportional to the nth power of its concentration, *i.e.*

$$\text{Rate of reaction} \left(\frac{dx}{dt}\right) \propto C^n$$

or
$$r = kC^n$$

where k is the rate constant for the reaction and C is the concentration at the instant of time t when the rate is r Hence for two different concentrations,

$$r_1 = kC_1^n$$

$$r_2 = kC_2^n$$

$$\therefore \quad \frac{r_1}{r_2} = \left(\frac{C_1}{C_2}\right)^n$$

or
$$\log \frac{r_1}{r_2} = n \log \frac{C_1}{C_2}$$

or
$$\log r_1 - \log r_2 = n (\log C_1 - \log C_2)$$

or
$$n = \frac{\log r_1 - \log r_2}{\log C_1 - \log C_2}$$

Thus to calculate the value of the order n, we require to know the values of the rates r_1 and r_2 at two different concentrations C_1 and C_2 respectively.

Concentration of the reactant may be plotted against time and the rate at any concentration may be found from the slope of the tangent at that point.

Example 23. **For the reaction.**

$$\mathbf{A + 2\,B \longrightarrow Product}$$

It is found that if the concentration of A is doubled the rate increases four times but when the concentration of B is doubled, the rate becomes doubled. What is the order of reaction with respect to A and B and what is the over all order of reaction ?

Solution. Let the rate law expression for the reaction

$$A + 2B \longrightarrow \text{Product be}$$

$$r = k\,(C_A)^a \cdot (C_B)^b \qquad \dots (i)$$

Case (*i*) When the conc. of A alone is doubled, the rate will become 4 times. The rate is given by

$$4r = k\,(2C_A)^a \cdot (C_B)^b \qquad \text{... } (ii)$$

Case (*ii*) When the conc. of B alone is doubled, the rate becomes ***double.*** The rate is given by

$$2r = k\,(C_A)^a \cdot (2\,C_B)^b \qquad \text{... } (iii)$$

Dividing (*i*) by (*iii*)

$$\frac{r}{2r} = \frac{k(C_A)^a \cdot (C_B)^b}{k(C_A)^a \cdot (2C_B)^b}$$

or $$\left(\frac{1}{2}\right)^1 = \left(\frac{1}{2}\right)^b$$

∴ $$b = 1$$

Order w.r.t B = 1

Now dividing (*i*) by (*ii*), we get

$$\frac{r}{4r} = \frac{k(C_A)^a \cdot (C_B)^b}{k(2C_A)^a \cdot (C_B)^b}$$

$$\frac{1}{4} = \left(\frac{1}{2}\right)^a$$

or $$\left(\frac{1}{2}\right)^2 = \left(\frac{1}{2}\right)^a$$

∴ $$a = 2$$

Hence order w.r.t. $A = 2$

∴ Total order of reaction $= a + b = 2 + 1 = 3$

Example 24. **For the reaction between gaseous chlorine and nitric oxide**

$$2NO + Cl_2 \longrightarrow 2NOCl$$

it is found that doubling the concentration of both reactants increases the rate eight times but doubling the chlorine concentration alone doubles the rate. What is the order of reaction with respect to nitric oxide and chlorine ?

Solution. Suppose the initial concentration of NO is a moles/litre and that of Cl_2 is b moles/litre. Suppose the order of reaction w.r.t. NO is n_1 and w.r.t. Cl_2 is n_2. Then initially the rate of reaction will be given by

$$r = k\,(a)^{n_1}\,(b)^{n_2} \qquad \text{...}(i)$$

In the first case when the concentration of both the reactants is doubled, we will have

$$8r = k\,(2a)^{n_1}\,(2b)^{n_2} \qquad \text{...}(ii)$$

In the second case when the concentration of only Cl_2 is doubled, we have

$$2r = k\,(a)^{n_1}\,(2b)^{n_2} \qquad \text{...}(iii)$$

Now to get the values of n_1 and n_2, we have to solve equations (*i*), (*ii*) and (*iii*)

From equations (*i*) and (*iii*), we get

$$2k\,[a]^{n_1}\,[b]^{n_2} = k\,[a]^{n_1}\,[2b]^{n_2}$$

or $$2[b]^{n_2} = [2b]^{n_2}$$

or $$2 = [2]^{n_2}$$

which gives $$n_2 = 1$$

Again, from equations (*ii*) and (*iii*), we get

$$2k\,[2a]^{n_1}\,[2b]^{n_2} = 4k\,[a]^{n_1}\,[2b]^{n_2}$$

or
$$[2a]^{n_1} = 4[a]^{n_1}$$

or
$$2^{n_1} = 4 \text{ or } n_1 = 2$$

Thus $n_1 = 2$ and $n_2 = 1$. This means that the order of reaction *w.r.t* NO is 2 and *w.r.t.* Cl_2 is 1. The total order of reaction is, therefore, 2 + 1 = 3.

3. Fractional change (half-life period) method

In general, we can say that $t_{1/2} \propto \dfrac{1}{a^{n-1}}$

where *n* is the order of the reaction.

The time taken for half of the reaction to be completed is called half-life period *and is represented by* $t_{1/2}$.

For two different initial concentrations, we may write

$$(t_{1/2})_1 \propto \frac{1}{a_1^{\,n-1}}$$

and
$$(t_{1/2})_2 \propto \frac{1}{a_2^{\,n-1}}$$

Dividing one equation by other, we get

$$\frac{(t_{1/2})_1}{(t_{1/2})_2} = \left(\frac{a_2}{a_1}\right)^{n-1}$$

Taking logarithm of both the sides, we get

$$\log\,(t_{1/2})_1 - \log(t_{1/2})_2 = (n-1)\,[\log a_2 - \log a_1].$$

This on simplification gives

$$n = 1 + \frac{\log(t_{1/2})_1 - \log(t_{1/2})_2}{\log a_2 - \log a_1}$$

Thus starting with two different initial concentrations and finding the half life period in each case, the value of *n* can be calculated.

***Example 25.* The half-life period of a substance is 50 minutes at a certain concentration. When the concentration is reduced to one half of the initial concentration, the half life period is 25 minutes. Calculate order of the reaction.**

Solution. Suppose the initial concentration is *a* moles/litre.

Then $(t_{1/2})_1 = 50$ minutes, $a_1 = a$

$$(t_{1/2})_2 = 25 \text{ minutes}, \; a_2 = \frac{a}{2}$$

Substituting the values in the expression

$$\frac{(t_{1/2})_1}{(t_{1/2})_2} = \left(\frac{a_2}{a_1}\right)^{n-1}$$

we get
$$\frac{50}{25} = \left(\frac{a/2}{a}\right)^{n-1}$$

or
$$2 = \left(\frac{1}{2}\right)^{n-1}$$

or $$\log 2 = (n-1) \log \frac{1}{2}$$

$$= -(n-1) \log 2$$

or $$1 = -(n-1)$$

or $$n = 0$$

Hence the reaction is of the *zero order.*

4. Ostwald's isolation method

This method can be used for finding the order of those reactions which involve two or more reactants. The method is based upon the fact that the order of a reaction does not depend upon the concentration of those reactants which are taken in excess. Turn by turn, all reactants except one are taken in excess. The order thus observed is with respect to the reactant that has been *isolated* from the others. The total or the overall order of the reaction is the sum of the orders found with respect to the different reactants one by one by isolation.

For example, suppose a reaction involves three reactants *A, B* and *C*. In the first case *B* and *C* are taken in excess (so that *A* is isolated). The order thus observed is the order with respect to *A*. Similarly by taking *A* and *C* in excess, the order with respect to *B* can be found and by taking *A* and *B* in excess, the order with respect to *C* can be found. Suppose the orders with respect to *A, B* and *C* are found to be *a, b* and *c* respectively. Then the overall order of the reaction is $a + b + c$.

6.12 COMPLICATIONS IN THE DETERMINATION OF ORDER OF REACTION

There are many reactions which do not occur strictly in accordance with the rate equation. These reactions generally involve two or more side reactions in addition to the main reaction. The kinetics of such reactions becomes more complicated. The overall rate of such types of reactions is the resultant of several different rate expression. The various complications are due to the following disturbing factors :

(*i*) Reversible or Opposing Reactions

(*ii*) Consecutive Reactions

(*iii*) Side Reactions or Parallel Reactions

(*i*) **Opposing or Reversible reactions:** *If in a reaction, the reactants react to form the products and the products also react to give back the original substances under the same conditions, it is called a reversible reaction.* In other words, the reaction takes place not only in the forward direction but also in the backward direction (*i. e.* the forward reaction is opposed by the reverse reaction). Let us consider a general reversible reaction in which both the forward and the backward reactions are of the first order

$$A \underset{k_b}{\overset{k_f}{\rightleftharpoons}} B$$

If the velocity constant for the backward reaction (k_b) is very small as compared to that of the forward reaction (k_f), the former can be neglected in comparison to the latter. However, if the forward and the backward reactions take place at comparable rates, the rate equation has to be modified.

Some examples of the reversible reactions whose kinetics have been studied are as follows:

(*a*) $2HI \rightleftharpoons H_2 + I_2$ (*Both forward and backward reactions are of second order*)

(*b*) $2NO + O_2 \rightleftharpoons 2NO_2$ (*Forward reaction is of third order, backward reaction is of second order*)

(*c*) $CH_3COOH + C_2H_5OH \rightleftharpoons CH_3COOC_2H_5 + H_2O$

(*ii*) **Consecutive reactions:** In case of complex reactions, the order of the reaction is usually not equal to the molecularity of the reaction as observed from the stoichiometry of the equation. Such reactions are, therefore, supposed to take place in a number of steps. The slowest step is the rate determining step. *The various step reactions thus written for the overall reaction are called consecutive reactions.* For example, let us consider the general complex reaction,

$$2A + B \longrightarrow C$$

Suppose by experiment, the rate of reaction $\propto$ $[A]$ $[B]$ *i.e.* the rate of reaction depends upon one molecule of A and one molecule of B. The slow step should, therefore, involve only one molecule of A and one molecule of B. The step reactions (or the consecutive reactions) may, therefore, written as

(*i*) $$A + B \xrightarrow{\text{Slow } (k_1)} D$$

(*ii*) $$A + D \xrightarrow{\text{Fast } (k_2)} C$$

(*iii*) **Parallel or Side reactions:** *If in a reaction, the reactants react to form two different products by two different paths such that one product is in larger amount than the other, then the reaction giving the larger amount is called main reaction while the other giving the smaller amount of the product is called* **side reaction or parallel reaction** *i.e.*

$$A \xrightarrow{k_1} B\ (90\%)\ \ldots\ \textit{Main reaction}$$

$$A \xrightarrow{k_2} C\ (10\%)\ \ldots\ \textit{Side reaction}$$

If the side reaction takes place only to a very small extent, it can be neglected in comparison to the main reaction. However, if the two reactions take place in comparable amounts, the rate equation has to be modified.

A few examples of parallel or side reactions are given below :

(*a*) *Nitration of phenol*

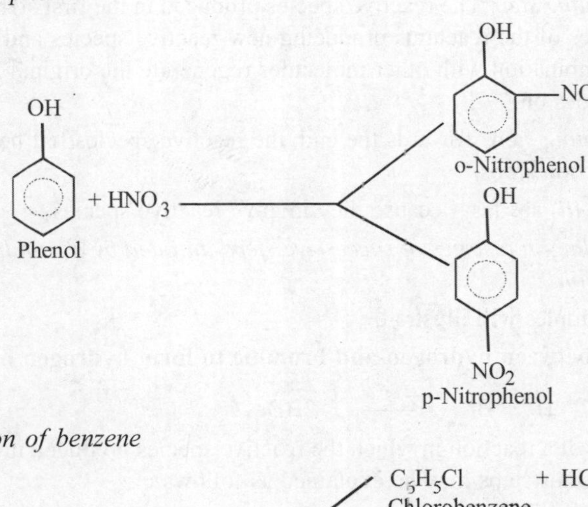

(*b*) *Chlorination of benzene*

$$C_6H_6 + Cl_2 \longrightarrow C_6H_5Cl + HCl \quad \text{(Chlorobenzene)}$$

$$C_6H_6 + Cl_2 \longrightarrow C_6H_6Cl_6 \quad \text{(Benzene hexachloride)}$$

6.13 SURFACE REACTIONS

Gaseous reactions which take place exclusively on the walls of the container are called surface reactions. The rate of these reactions is affected in a number of ways as follows:

(*i*) *If the surface is sparsely covered* (as in the case at high temperatures) the rate of reaction depends only upon the number of molecules colliding per unit time on the surface *i.e.* on the pressure (p) of the gaseous reactants. We may write

$$\frac{dx}{dt} = kp$$

Hence all such reactions are kinetically the reactions of the first order. Examples of such reactions are the decomposition of phosphine on the surface of glass, silica or tungsten, and decomposition of hydrogen iodide on platinum.

(*ii*) *If the surface is almost completely covered, it* is found that the rate of reaction is inde pendent of pressure *i.e.*

$$\frac{dx}{dt} = \text{constant}$$

Such reactions are then kinetically the reactions of the zero order. Examples of such reactions are decomposition of ammonia on the surface of tungsten and decomposition of hydrogen iodide on the surface of gold.

(*iii*) *If some of the molecules of the reactants are firmly held on the surface of the solid,* the rate of the reaction is obviously retarded.

6.14 CHAIN REACTIONS

A chain reaction takes place through the following steps:

(*i*) *Chain initiating step.* This is the slowest step in which the reactive species are produced.

(*ii*) *Chain propagating step.* The reactive species produced in the first step react immediately with the other molecules of the reactants producing new reactive species and products. The new reactive species on combination with other molecules regenerate the original reactive species. In this way the reaction goes on.

(*iii*) *Chain terminating step.* Towards the end, the reactive species left behind combine with each other and the reaction stops.

The steps (*ii*) and (*iii*) are fast because they involve reactive species.

A reaction proceeding in a series of successive steps initiated by a suitable primary process is called a chain reaction.

Following two examples will illustrate :

(1) Combination between hydrogen and bromine to form hydrogen bromide :

$$H_2 + Br_2 \longrightarrow 2HBr$$

This is an example of a reaction in which the reactive species produced in the chain initiating step are atoms. The various steps may be explained as follows :

(*i*) *Chain initiating step.* In the presence of light, bromine molecules dissociate to form the highly reactive bromine atoms.

$$Br_2 \xrightarrow{h\nu} 2\dot{B}r \text{ (Photochemical reaction)}$$

(*ii*) *Chain propagating step.* The above reaction is immediately followed by the following reactions :

$$\dot{Br} + H_2 \longrightarrow HBr + \dot{H}$$

$$\dot{H} + Br_2 \longrightarrow HBr + \dot{Br}$$

The bromine atom thus produced reacts with more of H_2 and the reaction goes on till whole of hydrogen or bromine present is used up.

(*iii*) *Chain terminating step.* The left off atoms then combine with each other or with the molecule of the product and the reaction stops.

$$\dot{H} + HBr \longrightarrow H_2 + \dot{Br}$$

$$\dot{Br} + \dot{Br} \longrightarrow Br_2$$

(2) Thermal decomposition of acetaldehyde at high temperature :

$$CH_3CHO \longrightarrow CH_4 + CO$$

This is an example of a reaction which proceeds through the free radical mechanism. The measured order of the reaction is 1.5 and the possible mechanism is as follows :

(*i*) *Chain initiating step*

$$CH_3CHO \longrightarrow \underbrace{\dot{C}H_3 + \dot{C}HO}_{\text{Free radicals}}$$

(*ii*) *Chain propagating step*

$$\dot{C}H_3 + CH_3CHO \longrightarrow CH_4 + CH_3\dot{C}O$$

$$\dot{C}HO \longrightarrow CO + \dot{H}$$

$$\dot{H} + CH_3CHO \longrightarrow H_2 + CH_3\dot{C}O$$

(*iii*) *Chain terminating Step*

$$\dot{C}H_3 + \dot{C}H_3 \longrightarrow C_2H_6$$

$$\dot{C}H_3 + CH_3\dot{C}O \longrightarrow CH_3COCH_3$$

$$2CH_3\dot{C}O \longrightarrow CH_3CO.COCH_3$$

6.15 TEMPERATURE DEPENDENCE AND ARRHENIUS EQUATION

The effect of temperature on the rate of a reaction was studied by Arrhenius. The equation, called **Arrhenius equation,** is usually written in the form

$$k = Ae^{-E_a/RT} \quad ...(i)$$

where A is a constant called **frequency factor** (because it gives the frequency of collisions of the reacting molecules), E_a is called **energy of activation,** R is gas constant and T is the absolute temperature.

The energy of activation (E_a) is an important quantity as it is characteristic of the reaction. Using the above equation, its value can be calculated as follows :

Taking logarithm of both side of equation (*i*), we get

$$\ln k = \ln A - \frac{E_a}{RT} \qquad ...(ii)$$

If the values of the rate constant at temperatures T_1 and T_2 are k_1 and k_2 respectively, then we have .

$$\ln k_1 = \ln A - \frac{E_a}{RT_1} \qquad ...(iii)$$

and

$$\ln k_2 = \ln A - \frac{E_a}{RT_2} \qquad ...(iv)$$

Subtracting equation (*iii*) from equation (*iv*), we get

$$\ln k_2 - \ln k_1 = -\frac{E_a}{RT_2} - \left(-\frac{E_a}{RT_1}\right)$$

$$= \frac{E_a}{RT_1} - \frac{E_a}{RT_2}$$

Svante Arrhenius

or

$$\ln \frac{k_2}{k_1} = \frac{E_a}{R}\left(\frac{1}{T_1} - \frac{1}{T_2}\right)$$

$$= \frac{E_a}{R}\left(\frac{T_2 - T_1}{T_1\, T_2}\right)$$

or

$$\log \frac{k_2}{k_1} = \frac{E_a}{2.303R}\left(\frac{T_2 - T_1}{T_1\, T_2}\right) \qquad ...(v)$$

Thus knowing the values of the rate constants k_i and k_2 at two different temperatures T_1 and T_2, the value of E_a can be calculated. Alternatively, knowing the rate constant at any one temperature, its value at another temperature can be calculated provided the value of E_a is known for that reaction.

Equation (*ii*), may be written as

$$\ln k = -\frac{E_a}{RT} + \ln A$$

$$\log k = -\frac{E_a}{2.303\,RT} + \log A \qquad ...(vi)$$

This equation is of the form $y = mx + c$ *i.e.* the equation of a straight line. Thus a plot of log k *vs* $\frac{1}{T}$ is a straight line as shown in Fig. 6.11.

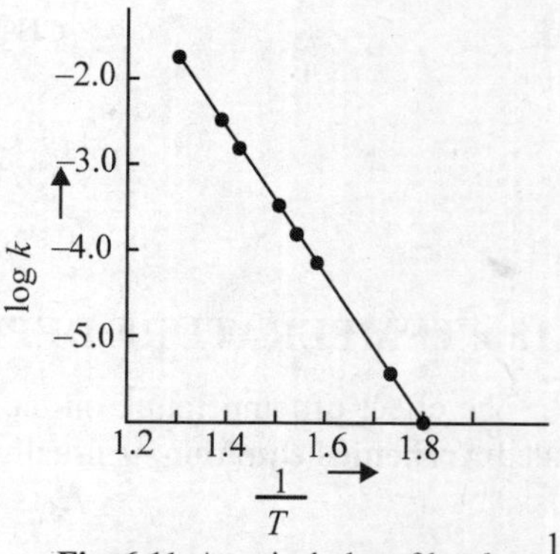

Fig. 6.11. A typical plot of log k *vs* $\frac{1}{T}$.

Further from equation (*vi*), we have

Slope of the line

$$= -\frac{E_a}{2.303\,R}$$

Thus by measuring the slope of the line, the value of E_a can be calculated.

Example 26. **ln k versus 1/T graph was plotted to calculate the activation energy of a reaction using Arrhening equation for the effect of temperature on reaction rate. The slope of the straight line was found to be – 2.55 × 10⁴. Calculate the activation energy of the reaction.**

Solution. Slope of the graph straight line is given by :

$$\text{Slope} = -\frac{E_a}{2.303 \times R}$$

or
$$-\frac{E_a}{2.303 \times R} = -2.55 \times 10^4$$

Remove negative sign from both sides and put the value of R in joule

$$\frac{E_0}{2.303 \times 8.314} = 2.55 \times 10^4$$

or
$$E_a = 2.55 \times 10^4 \times 2.303 \times 8.314$$

or
$$E_a = 4.88 \times 10^5 \text{ J.}$$

Example 27. **The activation energy of a reaction is 225 kcal mol^{-1} and the value of rate constant at 40°C is 1.8 × 10^{-5} sec^{-1}. Calculate the frequency factor, A.**

Solution. Here; we are given that

$$E_a = 22.5 \text{ kcal mol}^{-1} = 22500 \text{ cal mol}^{-1}$$
$$T = 40°\text{ C} = 40 + 273 = 313 \text{ K}$$
$$k = 1.8 \times 10^{-5} \text{ sec}^{-1}$$

Substituting the values in the equation

$$\log k = -\frac{E_a}{2.303\, RT} + \log A$$

or
$$\log A = (1.8 \times 10^{-5}) + \frac{22500}{2.303 \times 1.987 \times 313}$$
$$= (\log 1.8) - 5 + 15.7089$$
$$= 0.2553 - 5 + 15.7089 = 10.9642$$
$$A = \text{antilog}\,(10.9642)$$
$$= \mathbf{9.208 \times 10^{10}} \textbf{ collisions/sec.}$$

Example 28. **Calculate the activation energy of a reaction whose reaction rate at 27°C gets doubled for 10°C rise in temperature.**

Solution. Here we are given that

when $T_1 = 27 + 273 \text{ K} = 300 \text{ K},$

$k = k$ (say)

When $T_2 = 27 + 10°\text{C} = 310 \text{ K}$

$k_2 = 2k$

Substituting these values in the equation

$$\log\frac{k_2}{k_1} = \frac{E_a}{2.303R}\left(\frac{T_2 - T_1}{T_1 T_2}\right)$$

we get $$\log\frac{2k}{k} = \frac{E_a}{2.303\times8.314}\times\frac{310-300}{300\times310}$$

or $$\log 2 = \frac{E_a}{2.303\times8.314}\times\frac{10}{300\times310}$$

or **E_a= 53598.6 J mol^{-1} or 53.599 kJ mol^{-1}.**

THEORIES OF REACTION RATES

Two theories which are used to predict the rates of reactions are :

1. Collision Theory

2. Transition state theory or theory of absolute reaction rates or theory of activated complex formation

6.16 COLLISION THEORY OF REACTION RATES

This theory is based upon the kinetic theory of gases according to which the molecules of a gas are continuously moving and hence colliding with each other. Since for a collision to take place, at least two molecules must be involved, therefore the simple case to consider is that of *bimolecular* reactions. Hence in the discussion that follows, we shall first discuss the collision theory for bimolecular reactions and then for unimolecular reactions.

(*a*) **Collision theory for bimolecular reactions:** In a bimolecular reaction, in order that two molecules may react, they should not only collide but the collision must be associated with sufficient energy so that the molecules are activated. This means that all collisions are not effective. The number of effective collisions is only a fraction of the total number of collisions. Taking the *specific reaction rate* (k) *as equal to the number of effective collisions per cc per second,* if Z is the number of molecules per cc taking part in the collisions and f is the fraction of those molecules that are activated, we can write

$$k = Z \times f \qquad ...(i)$$

Now, according to kinetic theory of gases, the fraction of the molecules in a gas possessing energy greater than a particular value E, is given by the equation

$$f = \frac{\Delta N}{N} = e^{-E/RT}$$

where ΔN = number of molecules possessing energy greater than E

and N = total number of molecules.

If E corresponds to E_a, the energy of activation, then the fraction of the activated molecules will be given by

$$f = \frac{\Delta N}{N} = e^{-E_a/RT} \qquad ...(ii)$$

Putting the value of f from equation (*ii*) in equation (*i*), we get

$$\boldsymbol{k = Ze^{-Ea/RT}} \qquad ...(iii)$$

Example 29. **The rate constant k_r for the second order decomposition of N_2O follows the equation**

$$\mathbf{k_r = 5 \times 10^{11} \times e^{-29,000/T}}$$

calculate the energy of activation and pre-exponential factor with reaction.

Solution. Arrhenius equation for the reaction is

$$k_r = A\, e^{-Ea/RT}$$

Comparing this equation with the given equation

$$k_r = 5 \times 10^{11}\, e^{-29,000/T}$$

$$\frac{E_a}{RT} = \frac{29,000}{T}$$

$$E_a = 29,000 \times 8.314 = 241106 \text{ J mol}^{-1} = 241.1 \text{ kJ mol}^{-1}$$

The frequency factor $\mathbf{A = 5 \times 10^{11}\ L\ mol^{-1}\ s^{-1}}$

6.16.1 Failure of simple collision theory

The collision theory of reaction rate predicts the value of rate constant very accurately for atomic species or simple and small molecules. There are many reactions for which the predicted and the experimental value of rate constant differ widely. Major points of failure of the theory are:

(*i*) Polymerisation of 1, 3-butadiene or combination of two molecules of ethene, the theoretical value of k_r is 10^4 times greater than the experimentally calculated value.

(*ii*) For the reaction between ethyl alcohol and acetic anhydride at 75°C, the theoretically calculated rate constant k_r is 10^5 times greater than the experimental value.

(*iii*) The theory is only applicable to homogeneous gas phase bimolecular reactions which are generally of the second order.

(*iv*) The plot of $ln\ k_r$ against 1/T is not strictly a straight line as predicted by Arrhenius equation. This is because frequency factor is not completely independent of temperature rather it has square root dependence of temperature *i.e.*

$$k_r = \text{Constant} \times \sqrt{T} \times e^{-E_a/RT}$$

(*v*) The theory totally ignores the structure and properties of the reacting molecule. The bigger or complex molecules do not behave as rigid spheres.

(*vi*) The theory does not account for the higher rates of chain reactions.

(*vii*) The theory fails to account for the first order kinetics of bimolecular gaseous reaction at high pressure.

(*viii*) There is no way to determine E_a from the theory itself.

6.16.2 Causes of failure of collision theory

(1) The main cause of the failure of collision theory for reactions involving complex or large molecule is due to over simplification in the assumption that the entire energy of the molecules is the translation kinetic energy which is true only in case of simple molecules which behave as rigid spheres. The complex molecules do not behave as such and these molecules possess vibrational and rotational kinetic energies in addition to translational kinetic energy.

(2) Another cause of the failure of collision theory is that the proper activation energy by the colliding molecules is not the only pre-requisite for the reaction to occur. For effective collision, suitable orientation of the colliding molecules is also essential for the reaction to occur.

The effect of orientation of molecules in space for a simple reaction

$$2HI \longrightarrow H_2 + I_2$$

can be represented as follows (Fig. 6.12)

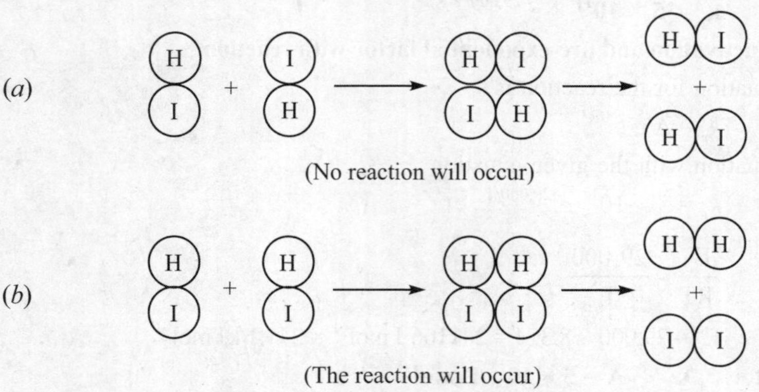

Fig. 6.12. Proper orientation of molecules during collision is necessary for the reaction to take place.

It can be concluded that collisions of only properly oriently molecules possessing energy greater than E_a will result into the reaction.

6.16.3 Modification of the equation of collision theory

A good agreement between the calculated value and the observed value is found for a number of bimolecular reactions in the gaseous phase as well as in the solution. Still there are many reactions for which the calculated value is found to be much higher than the observed value. Hence equation (*iii*) is modified as

$$k = pZe^{-E_a/RT} \qquad ...(iv)$$

where the factor *p,* introduced in the above equation, is called **probability factor** or **steric factor**. It accounts for the fact that all collisions between the activated molecules may not result into a reaction; only those collisions between the activated molecules result into reaction which collide with proper orientations. The value of *p* varies from 1 to 10^{-9}. When $p = 1$, the reaction obeys collision theory.

6.16.4 Thermodynamic interpretation of steric factor

Consider a reversible reaction

$$A_2 + B_2 = 2AB$$

Rate constant k_f (for forward reaction) $= p_1 z_1\, e^{-E_1/RT}$

Rate constant k_b (for backward reaction) $= p_2 z_2\, e^{-E_2/RT}$

The value of equilibrium constant K is given by

$$K = \frac{k_f}{k_b} = \frac{p_1 z_1\, e^{-E_1/RT}}{p_2 z_2\, e^{-E_2/RT}} \qquad ...(i)$$

From the knowledge of thermodynamics

$$\Delta G^* = -RT\, ln\, K \qquad ...(ii)$$

or $$K = e^{-\Delta G^*/RT} \qquad ...(iii)$$

But $$\Delta G^* = \Delta H^* - T\,\Delta s^*$$

∴ $$K = e^{-\frac{1}{RT}(\Delta H^* - T\Delta S^*)}$$

or $$K = e^{-\Delta H^*/RT} \cdot e^{\Delta S/R} \qquad ...(iv)$$

where ΔG^* = Free energy change

$$\Delta H^* = \text{Enthalpy change}$$
$$\Delta S^* = \text{Entropy change}$$

for the activation process *i.e.*, in going from reactants to the activated complex state.

From (*i*) $$K = \frac{p_1 z_1}{p_2 z_2} \cdot e^{-(E_1 - E_2)/RT}$$

Since, $e^{-(E_1 - E_2)/RT} = e^{-\Delta H/RT}$

$$\therefore \quad K = \frac{p_1 z_1}{p_2 z_2} \cdot e^{-\Delta H/RT} \quad \text{... (v)}$$

Comparing (*iv*) and (*v*)

$$\frac{p_1 z_1}{p_2 z_2} \cong e^{\Delta S^*/R} \quad \text{... (vi)}$$

If the molecules A_2, B_2 and AB are of comparable dimensions, then $z_1 \cong z_2$ so that

$$e^{\Delta S^*/R} \cong \frac{p_1}{p_2} \quad \text{... (vii)}$$

The steric factor '*p*' should be interpreted in terms of entropy of activation Δ S* rather than the probability of proper orientation of molecules.

(*i*) When **Δ S* is + ve, *p*** will large and reaction will be fast.

(*ii*) When **Δ S* is – ve, *p*** will be small and reactions will be slower than that predicted from the collision theory.

6.16.5 Collision theory for unimolecular reactions (Lindemann hypothesis)

A unimolecular reaction, in general, means a reaction of the type

$$A \longrightarrow \text{Products}$$

In such reactions, only one molecule of the reactant takes part in the reaction. Consequently, the question arises : How do molecules in a unimolecular reaction attain their energy of activation ? The answer to this question was first suggested by Lindemann in 1922. He pointed out that the behaviour of unimolecular reactions can be explained on the basis of **reaction.** During this time, the activated molecules may either react or be deactivated to ordinary molecules. Consequently, the rate of reaction will not be proportional to all the molecules activated but only to those which remain active. This concept may be illustrated as follows :

$$\underset{\text{Ordinary Molecules}}{A + A} \underset{k_2}{\overset{k_1}{\rightleftharpoons}} \underset{\text{Activated Molecules}}{A^*} + A$$
$$A^* \xrightarrow{k_3} \text{Products}$$

where k_1 k_2, k_3 are the rate constants of the different processes involved as shown above.

6.17 TRANSITION STATE THEORY OR THEORY OF ABSOLUTE REACTION RATES OR THEORY OF ACTIVATED COMPLEX FORMATION

According to this theory, it is believed that before the reacting molecules change into products, they form an intermediate, called activated complex which has higher energy than both the reactants and the products. It is for this reason that the theory is called Transition state theory. The activated complex is in equilibrium with the reacting molecules and the rate of reaction is given by the rate of decomposition of the complex to form the products. Thus we may represent the situation as follows :

$$\underset{\text{Reactants}}{A + B} \rightleftharpoons \underset{\substack{\text{Activated}\\\text{complex}}}{X^*} \xrightarrow{k} \text{Products}$$

By making use of some fundamental properties of the reacting molecules, it was shown by Eyring that the rate constant (k) for any reaction (irrespective of its molecularity or order) is given by

$$k = \frac{RT}{Nh} K \qquad ...(i)$$

where R = Gas constant
h = Planck's constant
N = Avogadro's number
T = Absolute temperature
and K = Equilibrium constant for the formation of the activated complex from the reacting molecules, that is

$$K = \frac{[X^*]}{[A]\,[B]}$$

Now, from thermodynamics, we know that

$$\Delta G^* = -RT \ln K$$

or $$\ln K = -\Delta G^* / RT$$

or $$K = e^{-\Delta G^* / RT} \qquad ...(ii)$$

where ΔG^* is the free energy of activation for the reaction $A + B \rightarrow X^*$

Substituting this value of K in equation (i), we get

$$k = \frac{RT}{Nh} e^{-\Delta G^* / RT} \qquad ...(iii)$$

Further substituting for $\Delta G^* = \Delta H^* - T\Delta S^*$, where ΔH^* and ΔS^* are the changes in enthalpy and entropy of activation respectively, we get

$$k = \frac{RT}{Nh} e^{-(\Delta H - T\Delta S^*)/RT}$$

$$= \frac{RT}{Nh} e^{\Delta S^*/R} . e^{-\Delta H^*/RT} \qquad ...(iv)$$

Significance of the equation of the transition state theory: The equation (iii) involves two factors *i.e.* $= \frac{RT}{Nh}$ and $e^{-\Delta G^*/RT}$. The first factor is constant at constant temperature. Hence the rate of reaction depends only upon the second factor. This factor involves ΔG^* with a negative sign. This means that at particular temperature, *greater the value of the free energy of activation for a reaction, the slower will be the reaction.*

Comparison with Arrhenius equation: Comparing equation (*iv*) with the Arrhenius equation *viz.* $k = Ae^{-E_a/RT}$ we observe that we should have

$$A = \frac{RT}{Nh} e^{-\Delta S^*/R}$$

and $$E_a = \Delta H^*$$

(*i.e.* enthalpy of activation is equal to the energy of activation).

Comparison with the collision theory equation: Comparing equation (*iv*) *with* the collision theory equation *viz.*

$$k = PZe^{-E_a/RT}$$

we observe that, $$PZ = \frac{RT}{Nh} e^{\Delta S^*/R}$$

Thus the probability factor P is related to the entropy of activation ΔS^*.

6.17.1 Advantages of transition state theory over collision theory

The transition state theory is better than the collision theory in a number of ways.

For example,

(*i*) In collision theory, the correction factor P was introduced arbitrarily whereas in the transition state theory, its inclusion has been justified in terms of entropy of activation ΔS^*.

(*ii*) The concept of the formation of an activated complex seems to be more appropriate than assuming that the molecules just collide together and change into products.

6.18 EXPERIMENTAL METHODS OF CHEMICAL KINETICS

The underlying principle behind kinetic studies of reactions is to follow the progress of reaction with passage of time. In other words, we need to know how much of the reactant changes into the product at different intervals of time. The technique to be employed to make kinetic studies will depend upon the nature, medium and the physical state of the reactants and products. If the reactant or product is gaseous, barometric method may be employed. If the reactant or product contains some acid or base, titration methods are handy to obtain the concentration of the reactants at different intervals of time. Other methods which involve the use of instruments are described below.

1. Conductometric methods. A conductometer is employed to study the progress of a reaction involving either ionic reactants or product. With the passage of time, conductance is measured at different interval of time. This gives us the concentration of reactants at different times, after which it is possible to find rate of the reaction.

2. Potentiometric methods. A potentiometer is employed to study the progress of the reaction involving a redox couple in which one component gets oxidised and the other gets reduced. The e.m.f. of the redox couple is observed after different intervals of time. From these values, we can calculate the concentration of reactants after different intervals of time. This enables us to arrive at the rates of reaction.

3. Polarimetric methods. This method is employed to study reactions in which either the reactant or product or both are optically active *i.e.* rotate the plane of polarised light. For example, the inversion of cane sugar (sucrose) can be studied by this method. Sucrose is dextrorotatory. On hydrolysis, it is broken down to glucose which is dextrorotatory and fructose which is laevorotatory. But the laevorotation of fructose is greater than the dextrorotation of glucose. Thus the solution obtained after hydrolysis is found to be laevorotatory. The instrument used in this method is called polarimeter.

4. Spectrophotometric method. This method is employed to study reaction which involve either coloured reactants or products. Let us say the reactant is coloured. As the reaction proceeds, the solution becomes less and less coloured because the product is colourless. The intensity of transmitted light is noted after different intervals of time. Thereafter, we can obtain the concentration of reactant after different intervals of time and calculate the rate constant.

EXERCISES
(INCLUDING QUESTIONS FROM DIFFERENT UNIVERSITY PAPERS)

Objective Questions

Fill in the blanks

1. If the concentration is expressed in mol dm^{-3} and time in seconds, the unit of rate constant for a second order reaction will be ....
2. Hydrolysis of ester in the presence of a base is a .... order reaction.
3. To start a reaction, a minimum amount of energy known as .... energy should be possessed by the reacting molecules
4. Temperature coefficent of most of the reactions lies between ....
5. Reactions of higher order are ....
6. A series of step reactions, proposed to account for overall reaction is called the .... of the reaction
7. The middle step in a chain reaction is known as chain .... step
8. Arrhenius equation is given by ....
9. Two theories to explain reaction rates are (*i*) .... (*ii*) Transition state theory
10. In a third order reaction, the half– life period is inversely proportional to .... of the unitial concentration.

Multiple Choice Questions

Choose the correct option

1. The rate constant of a reaction is equal to its rate at constant temperature when the concentration of reactants is :
   (*a*) unity (*b*) zero (*c*) two (*d*) three
2. The rate of a reaction, on increasing concentration of the reactants :
   (*a*) increases (*b*) decreases (*c*) remains unaltered (*d*) none of these
3. The half life period of a reaction is independent of the initial concentration of the reactants. The reaction is of
   (*a*) first order (*b*) zero order (*c*) second order (*d*) third order
4. According to activated complex theory, the rate of a reaction depends on :
   (*a*) the intensity of collision
   (*b*) initial velocity of molecules
   (*c*) energy of activation of the molecules
   (*d*) the velocity to molecules with which they cross the energy barrier.
5. According to the collision theory the rate of a reaction depends on :
   (*a*) the average collision of molecules
   (*b*) the total number of molecules

(*c*) the number of colloiding molecules per ml per unit time
(*d*) None of these

6. The essential condition for a reaction to occur, according to collision theory is that the
(*a*) molecules should dissociate after collision
(*b*) molecules should acquire activation energy
(*c*) molecules should become deactivated
(*d*) volume of molecules should decrease

7. The rate equation of a reaction is $dx/dt = k[A]^a [B]^b$ for which the order of the reaction is :
(*a*) $a + b$ (*b*) $a - b$ (*c*) a (*d*) b

8. The one which is a unimolecular reaction is :
(*a*) $2HI \rightarrow H_2 + I_2$ (*b*) $N_2O_5 \rightarrow N_2O_5 + (1/2)\,O_2$
(*c*) $H_2 + Cl_2 \rightarrow 2HCl$ (*d*) $PCl_3 + Cl_2 \rightarrow PCl_5$

9. The specific rate constant of a first order reaction depend on
(*a*) concentration of the reactants (*b*) time
(*c*) concentration of the products (*d*) temperature

10. On increasing temperature rate of a reaction increases, because:
(*a*) the number of activated molecules increases.
(*b*) the number of activated molecules decreases.
(*c*) the number of activated molecules becomes zero.
(*d*) the activated molecules move with a low speed.

11. The statement which is true in respect of order of a reaction is :
(*a*) it is always a whole number
(*b*) it is always a fraction.
(*c*) it may be zero, a whole number or a fraction
(*d*) it can never be a fraction.

12. The half life period $t_{1/2}$ of a reaction varies with the initial concentration of a reaction as $t_{1/2} \propto \frac{1}{a^{1/2}}$, the order of reaction is
(*a*) 0 (*b*) 1 (*c*) 2 (*d*) none of these.

13. The equation which expresses the effect of temperature on the velocity constant of reaction is :
(*a*) Faraday equation (*b*) Arrhenius equation
(*c*) Avogadro's equation (*d*) none of these

14. The ratio of the velocity constant for a reaction at two temperature differing by 10°C is called its :
(*a*) rate constant (*b*) temperature coefficient
(*c*) activation energy(*d*) specific reaction rate

15. For the chemical change $A \rightarrow B$
it is found that the rate reaction doubles when the concentration is increased four times. The order in A for this reaction is
(*a*) two (*b*) one (*c*) zero (*d*) half

16. The half life period of a first order of reaction is 10 minutes. The time required for the concentration the reactant to change from 0.08 M to 0.02 M is

(*a*) 10 minutes (*b*) 20 minutes (*c*) 30 minutes (*d*) 40 minutes

17. The rate of which a substance reacts is proportional to its :

(*a*) active mass (*b*) concentration in g/litre

(*c*) mole fraction (*d*) none of these

18. The rate of a reaction usually

(*a*) increases with increase in temperature.

(*b*) decreases with increase in temperature.

(*c*) remains unaffected with increase in temperature.

(*d*) none of these.

19. It the concentration is expressed moles/litre, the unit of rate constant for a first order reaction is

(*a*) mole litre^{-1} sec.$^{-1}$ (*b*) mole litre^{-1}

(*c*) sec.$^{-1}$ (*d*) mole^{-1}

20. In first order reaction, the conc. of reactant decreases from 1. 0M to 0.25 M in. 20 minutes. The value of *k* is :

(*a*) 0.6932 (*b*) 6.932 (*c*) 0.06932 (*d*) None of these

21. The log k for a reaction is plotted against 1/ T. The slope of the straight line will give.

(*a*) frequency factor (*b*) number of collisions

(*c*) energy of activation (*d*) none of these

22. The active mass of 2g hydrogen kept in a 2 litre flask is :

(*a*) mole / litre (*b*) 0.5 mole/litre (*c*) 2 mole / litre (*d*) none of these

ANSWERS

Fill in the blanks

1. dm^3 mol^{-1} sec^{-1} **2.** second **3.** threshold **4.** 2 and 3

5. rare **6.** mechanism **7.** propagating **8.** $k = Ae^{-Ea/RT}$

9. collision theory **10.** square

Multiple choice questions

1. (*a*) **2.** (*a*) **3.** (*a*) **4.** (*a*) **5.** (*c*)

6. (*a*) **7.** (*a*) **8.** (*b*) **9.** (*d*) **10.** (*b*)

11. (*c*) **12.** (*d*) **13.** (*b*) **14.** (*b*) **15.** (*d*)

16. (*c*) **17.** (*a*) **18.** (*a*) **19.** (*c*) **20.** (*c*)

21. (*c*) **22.** (*b*)

General Questions

1. Explain the term rate of reaction. How it can be determined ? Why the rate of a reaction cannot be measured by giving the total amount of the reactant coasumed by the total time taken ? How we can express the rate of a reaction ?

2. Define chemical kinetics. What is meant by rate of a reaction ? Why do the rates of chemical reaction differ from one another ?

3. Discuss the importance of chemical kinetics in the study of mechanism of reactions. Illustrate your answer by giving suitable examples.
4. (*a*) What is meant by pseudo-unimolecular reactions? Explain with examples.
   (*b*) Derive an expression for the time required for comletion of one third of a first order reaction.
5. (*a*) How does the value of the rate constant of a second order reaction depend upon the units of concentrations of the reaction ?
   (*b*) What do you understand by the terms :
   (*i*) Order of reaction
   (*ii*) Molecularity of a reaction ?
   Explain with examples.
6. (*a*) Deduce the expression for the specific rate constant for a second order reaction of the type: 2A $\longrightarrow$ product.
   (*b*) In a reaction 2R $\longrightarrow$ product
   the concentration of the reactant R after time t second was A moles/liter. Derive an expression for the rate constant if the initial concentration was
   A_0 moles/litre. What will be the units of the rate constant ?
7. (*a*) Show that for a first order reaction the half-life period is independent of the initial concentration.
   (*b*) Give two examples of zero order reactions.
8. (*a*) On what factors does the rate of a chemical reaction depend ? Explain with examples.
   (*b*) What is meant by the term molecularity of a reaction ? Why the molecularity of a reaction seldom exceeds three ?
9. (*a*) What is meant by the terms rate law and rate constant ?
   (*b*) Distinguish between order of a reaction and molecularity of a reaction.
10. (*a*) Write two examples in which the order of reaction is fractional.
    (*b*) Give two examples of first order reactions.
    (*c*) Write down the units of specific rate constant for first order reaction.
11. (*a*) Why the reactions of higher order are rare ?
    (*b*) A first order reaction is 40% complete at the end of 50 minutes. What is the value of rate constant in sec^{-1} ? In how many minutes will the reaction be 80% complete ?
    [**Ans.** 1.7×10^{-4} sec^{-1}; 157.8 min]
12. (*a*) Define and explain the term order of reaction and rate constant of a reaction.
    (*b*) Differentiate between rate constant and rate of a reaction.
13. (*a*) Derive mathematical expression for the rate constant of a second order reaction.
    A + B $\longrightarrow$ Product
    (*b*) Derive mathematical expression for half-life period of a second order reaction.
14. (*a*) Give a brief account of the methods for determining the order of reaction.
    (*b*) The optical rotation of sucrose in 1N-HCl at various time was given below :

Time in minutes	0	7.8	18.00	27.05	∞
Rotation in degree	+ 24.0	+ 21.7	+ 17.7	+ 15.0	– 10.74

(*i*) Show that inversion of sucrose is a first order reaction.

(*ii*) When is the solution optically inactive ? [**Ans.** 109 minutes]

15. (*a*) Name the methods used for the determination of order of reaction.

(*b*) Discuss one method in detail.

(*c*) How will you show experimentally that hydrolysis of ethyl acetate by alkali when alkali is not used in excess is a bimolecular reaction ?

16. (*a*) Discuss the kinetics of the first order parallel reaction.

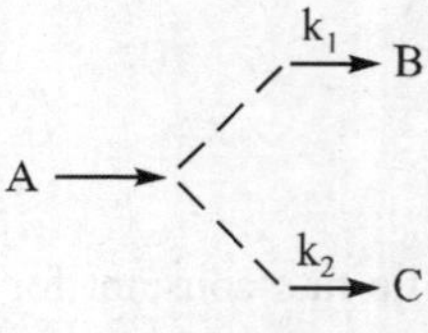

Explain how the values of k_1 and k_2 can be determined separately. Show that the reaction as a whole is of first order.

17. Give a brief account of the temperature dependence of reaction rates.

18. (*a*) What is a second order reaction ? Derive an equation given the velocity constant of a second-order reaction ?

(*b*) Show that :

(*i*) for a first order reaction

$$A \longrightarrow \text{product}$$

$$t_{1/2} = \frac{0.693}{K}$$

(*ii*) for a second order reaction

$$2A \longrightarrow \text{Product}$$

$$t_{1/2} = \frac{1}{k_a}$$

(*iii*) for a third order reaction

$$3A \longrightarrow \text{Product}$$

$$t_{1/2} = \frac{3}{2} \times \frac{1}{k_a^2}$$

19. (*a*) What are consecutive reactions ? Discuss the kinetics of the first order consecutive reaction :

$$A \xrightarrow{k_1} B \xrightarrow{k_2} C$$

Show graphically the variation of concentration of A, B and C with time.

20. (*a*) The rate constant of a certain reaction increases two fold with increase in temperature by 10 degree. Explain clearly.

(*b*) What is meant by activation energy ? How it can be determined ?

21. Discuss clearly the kinetics of first order reaction.

$$A \underset{k_2}{\overset{k_1}{\rightleftharpoons}} B$$

Show how the values of k_1 and k_2 are determined in reversible reactions.

22. (*a*) Comment on the statement :

The molecularity of the slowest step in a chemical reaction gives the order of the overall reactions.

(*b*) Discuss various methods for the determination of order of reaction.

23. (*a*) What are the chain reactions ? Differentiate between stationary chain reactions and branching chain reactions.

(*b*) Discuss the kinetics of Hydrogen-Bromine reaction.

24. (*a*) 'The magnitude of the activation energy depends upon the nature of the reacting species.' Support this statement with suitable examples.

(*b*) Explain activation energy, potential energy barrier and Arrhenius law.

25. Give an account of Arrhenius equation for temperature dependence of the reaction rates. Show how it is possible to determine the value of activation energy of the reaction.

26. (*a*) State and explain rate law. How does it differ from law of mass action ?

(*b*) What are psuedo-unimolecular reactions ? Give two examples in support of your answer ?

27. Write short notes on the following :

(*a*) Order of a reaction

(*b*) Molecules of a reaction

(*c*) Collision theory of reaction rate.

28. (*a*) What is activation energy ? On the basis of the concept of activation energy explain why the reaction having the same concentration of the reactions differ so much in their rates.

(*b*) Derive mathematical expression for the rate constant of a reaction

$$A + B \longrightarrow \text{Products}$$

(*c*) Show that in case of first order reaction the time required for 99.9% of the reaction to take place is about ten times than that required for half the reaction.

29. (*a*) Explain the term energy of activation. How is the activation energy of a reaction determined ?

(*b*) The values of the rate constant observed for the reaction.

$$2HI \longrightarrow H_2 + I_2$$

at 356°C and 443°C are 3.02×10^{-5} mol^{-1} dm^2 mol^{-1} and 2.53×10^{-3} mol^{-1} dm^3 s^{-1} respectively. Calculate the energy of activation for the forward and backward reaction. $\Delta H = 16.32$ k J mol^{-1} [**Ans.** $E_{(Forward)}$ = 190.613 kJ : E_a(Backward) = 174.293 kJ]

30. Explain :

(*a*) Why :

(1) Rate constant of a reaction is generally coubled for 10°C in temperature.

(2) Reactions of higher molecularity are rare.

(*b*) With a suitable example show that how chemical kinetics helps in determining the mechanism of reactions.

(*c*) A second order reaction in which initial concentration of both the reactants is same complete in 20 minutes. How long will it take for the reaction to complete to 75% ?

[**Ans.** 180 minutes]

31. (*a*) Give the units of specific rate constant for second order reaction.

(*b*) What is energy of activation ? Justify the statement that for a fast reaction energy has a low value whereas for slow reaction it has a high value.

(*c*) In a second order reaction, half-life period is 60 minutes when the initial concentration is 0.02 mole per litre. Calculate the value of specific reaction rate.

(*d*) What is the order of reaction

$$CH_3COOC_2H_5 + H_2O \xrightarrow{H^+} CH_3COOH + C_2H_5OH$$

[**Ans.** 0.8333 mol^{-1} lt min^{-1}]

32. Derive the expression for the time required for completion of 1/3 of a first order reaction and write down the units of specific rate constant and also show that in first order reaction.

(*a*) Half-life period is independent of the initial concentration of the reactants.

(*b*) The rate constant is independent of the change of concentration units.

33. (*a*) What are initial and instantaneous rates ? How do they differ from average rate ?

(*b*) The optical rotation of sucrose in 0.5 M – HCl at various intervals are given below :

Time (minutes)	0	10	20	40	∞
Rotation (degree)	+ 32.4	+ 28.4	+ 25.5	+ 19.6	– 1.1

Show that the reaction is of first order.

34. (*a*) Derive an expression for the velocity constant of reaction.

$$CH_3COOC_2H_5 + NaOH \longrightarrow CH_3COONa + C_2H_5OH$$

when alkali is used in large excess.

(*b*) The half-life period of a substance is an hour at a certain concentration. When concentration is reduced to one half of the initial concentration, the half-life period is half an hour. Calculate the order of reaction.

35. (*a*) What is the order of a reaction ? Comment on the statement that order of a reaction is an experimental quantity. Describe the half-life period method for determining the order of a reaction ?

(*b*) What are chain reactions ? Study the rate law for the chain reaction between hydrogen and bromine ?

(*c*) The half-life period of a first order reaction A $\longrightarrow$ B is 10 minutes. What percentage of A remains after 1 hour ? [**Ans.** 1.5651%]

36. (*a*) Deduce the expression for the rate constant of a second order reaction of the type

$$2A \longrightarrow \text{Product}$$

(*b*) A second order reaction in which the initial concentration of both the reactants are same is 25% complete in 600 seconds. How long will it take for the reaction to go to 70% completion ? [**Ans.** 5400 seconds]

37. (*a*) Explain any one of the methods used for the determination of the order of reaction.

(*b*) Give an example of the chemical reaction where order and molecularities are different.

(*c*) Why do reaction rates vary from reaction to reaction even if the concentrations of reactants and the temperatures are kept constant ?

(*d*) Explain how will you determine the activation energy of a reaction by using Arrhenius equation.

38. (*a*) At 298.4°C azomethane decomposes mainly according to the reaction

$$CH_3 - N = N - CH_3\ (g) \longrightarrow C_2H_6\ (g) + N_2\ (g)$$

If this be the only reaction obeying first order kinetics with a rate constant, *k*, as 2.50×10^{-4} sec^{-1}, what will be the partial pressure of reactants and products when azomethane initially at 200 mm Hg decomposes for 30.0 min ?

(b) Deduce a general rate equation for a second order bimolecular reaction when the initial concentrations of the reactants are different.

39. (a) What is the difference between Energy of activation and Threshold energy ?

(b) Explain Arrhenius equation.

(c) An artificially produced radio isotope decomposes according to the first order law with a half life period of 15 minutes. In what time will 80% of the sample be decomposed ?

40. (a) Explain the concept of energy of activation and potential energy barrier. Discuss the effect of temperature on the rate of a reaction.

(b) What are the advantages of transition state theory over the collision theory ?

41. (a) The activation energy of a non-catalysed reaction at 37°C is 200 k cal mol^{-1} and the activation energy of the same reaction when catalysed by an enzyme is 6.0 k cal mol^{-1}. Calculate the ratio of the rate constants of the catalysed and non-catalysed reactions. Assume frequency factor to be the same in both the cases.

(b) Calculate the activation energy of a reaction whose rate constant doubles when the temperature is raised from 22°C to 32°.

42. Write briefly on the following :

(1) Explosive reactions

(2) Influence of temperature on rates of reactions.

43. (a) What is energy of activation ? Justify the statement :

'For a fast reaction energy of activation for the reaction has a low value whereas for a slow reaction it has a high value'.

(b) In a second order reaction, half-life period is 60 minutes when the initial concentration is 0.02 mole per litre. Calculate the value of specific reaction rate.

44. (a) What are consecutive and parallel reactions ? Discuss the kinetics of the first order consecutive reaction :

$$A \xrightarrow{k_1} B \xrightarrow{k_2} C$$

Represent graphically the concentration changes in case of A and B with time.

45. (a) N_2 (g) + O_2 (g) $\longrightarrow$ 2NO (g)

Express the rate of this reaction in terms of reactants and products. What are the units of rate ?

(b) The rate of a reaction depends upon concentration of reactants. Explain.

(c) Show that for a first order reaction

$$k = \frac{2.303}{t} \log \frac{C_0}{C_t}$$

where the symbols have their usual meanings.

(d) A second order reaction in which the initial concentration of both the reactants is 0.01 mol/dm^3 is 25 % complete in 300 seconds. How long it takes for the reaction to go to 75% completion?

46. (i) What is the order of reaction ? Describe the half-life period method for determining the order of reaction.

(ii) What are chain reactions ? Study the rate law for chain reaction between hydrogen and bromine.

47. (*a*) Write a note on stationary state hypothesis.

(*b*) Calculate the activation energy of a reaction whose reaction rate at 27°C gets doubled for 10°C rise in temperature.

(*c*) Discuss the kinetic of the first order consecutive reaction :

$$A \xrightarrow{k_1} B \xrightarrow{k_2} C$$

Deduce expression for the concentration of A, B and C at any time. Show graphically the variation of concentration of A, B and C with time.

48. (*a*) Derive the equation :

$$k = \frac{2.303}{t} \log \frac{a}{a - x}$$

for a first order reaction what are the units of *k*.

(*b*) State and explain the following :

(*i*) Consecutive reactions.

(*ii*) Parallel reactions.

49. (*a*) Derive a second order rate equation for the reaction.

$$2A \longrightarrow \text{Products}$$

(*b*) Explain that half-life period in the second order reaction depends upon the initial concentration of the reactions.

(*c*) Explain the following :

(*i*) Some reactions take place at higher temperature but not at room temperature.

(*ii*) Rate of forward reaction is greater than the rate of reverse reaction in case of an endothermic reaction.

(*d*) Why reactions of higher order rare ?

50. (*a*) Differentiate order and molecularity of a reaction.

(*b*) Define a second order reaction and deduce the expression for the rate constant of second order reaction of the type.

$$2A \longrightarrow \text{Product}$$

(*c*) Hydrolysis of ethyl acetate in the presence of H^+ ions is of.......order (*Fill in the blank*)

(*d*) Optical rotation of sucrose in 1 N – HCl at various times was given as shown below :

Time (minutes)	0	7.18	18.00	∞
Rotation (degrees)	+24.9	+21.7	+17.7	–10.74

Show that inversion of sucrose is a first order reaction.

51. (*a*) Enumerate the methods employed for determining the order of reaction. Discuss one method in detail.

(*b*) Outline the basic principles involved in the theories of reaction rates.

(*c*) Derive third order rate equation for the reaction

$$3A \longrightarrow \text{Product}$$

52. (*a*) Give one example of a third order reaction.

(*b*) Define : (*i*) Parallel reaction (*ii*) Consecutive reactions.

(*c*) Define half-life time. The half-life time for a chemical reaction at a particular concentration is 50 minutes. When the concentration is doubled the half-life time becomes 100 minutes. Calculate the order of the reaction.

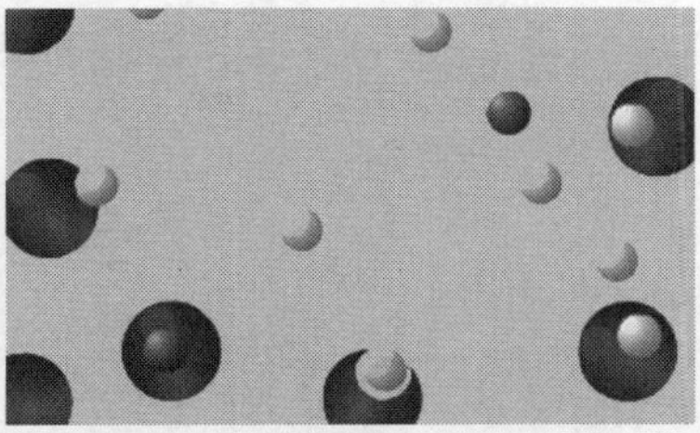

7

Catalysis

7.1 INTRODUCTION

It was Berzelius who realised in 1836 that there are some substances which increase the rate of a reaction without themselves getting consumed. He named such substances as **catalysts** (Greek *kata* = wholly, *lein* = to loosen). He believed that catalysts functioned by loosening the bonds in reacting molecules.

A catalyst is a substance that changes (usually increases) the rate of a reaction but is recovered chemically unchanged at the end of the reaction and the phenomenon is called catalysis.

Positive and negative catalysis. *If a catalyst increases the rate of a reaction, it is called a* positive catalyst *and the phenomenon is called positive catalysis.* A few examples of positive catalysis are given below :

(*i*) *Decomposition of potassium chlorate in presence of* MnO_2 *as catalyst*

Jacob Berzelius

$$2KClO_3(s) \xrightarrow{MnO_2(s)} 2KCl(s) + 3O_2(g)$$

(*ii*) *Oxidation of* SO_2 *to* SO_3 *using* NO *as catalyst*

$$SO_2(g) + \frac{1}{2}O_2(g) \xrightarrow{NO(g)} SO_3(g)$$

(*iii*) *Manufacture of ammonia by Haber's process using finely divided iron as catalyst*

$$N_2(g) + 3H_2(g) \xrightarrow{Fe} 2NH_3(g)$$

If a catalyst decreases the rate of a reaction, it is called a **negative catalyst** *and the phenomenon is called negative catalysis.* A few examples of negative catalysis are given below:

(*i*) *Oxidation of chloroform is retarded in the presence of a small quantity of ethyl alcohol*

$$\underset{\text{Chloroform}}{2CHCl_3} + O_2 \xrightarrow{C_2H_5OH} \underset{\text{Phosgene}}{2COCl_2} + 2HCl$$

(*ii*)*Decomposition of* H_2O_2 *is retarded in the presence of traces of acetanilide*

$$2H_2O_2 \xrightarrow{\text{Acetanilide}} 2H_2O + O_2$$

7.1.1 Activation energy diagram for catalysed and uncatalysed reactions

A catalyst decreases the activation energy of the reactants. It lowers the energy barrier and thus makes the reaction faster, as shown in Fig. 7.1. However in a reversible reaction, the catalyst decreases the activation energy of the forward reaction as well as that of the backward reaction by the same amount. Thus both the forward and the backward reactions are accelerated to the same extent. As a result, the equilibrium is attained quickly but does not change position.

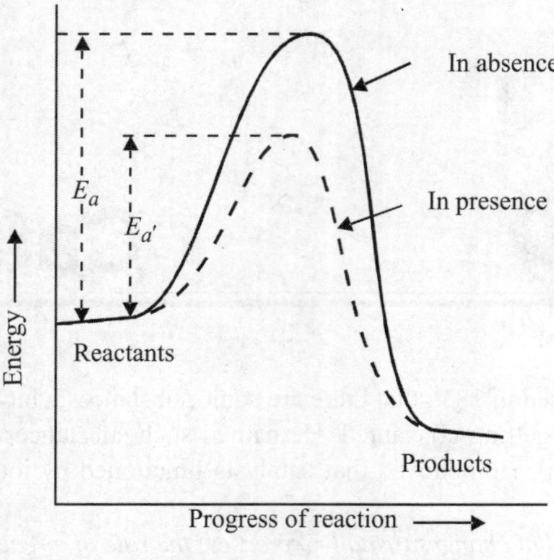

Fig. 7.1. Activation energy diagram for uncatalysed and catalysed reactions.

7.1.2 Promoters (activators)

Sometimes the presence of some other substances which are not necessarily catalysts increases the efficiency of a catalyst. Such substances thus act as catalyst for the catalyst and are called promoters or activators. Molybdenum acts as a promoter for the iron catalyst in the manufacture of ammonia by Haber's process.

Explanation of promoter action

There is no clear explanation of promoter action. It is assumed that the following changes take place on adding the promoter to the catalyst.

(*i*) *Change in lattice spacing*. The lattice spacing is increased thus enhancing the space between the catalyst particles. The bonds between the molecule of the reactant (say H_2) are weakened and cleaved. This makes the reaction go faster (Fig. 7.2).

This explains how Mo acts as a promoter in the manufacture of NH_3 with Fe as catalyst in the Haber's process.

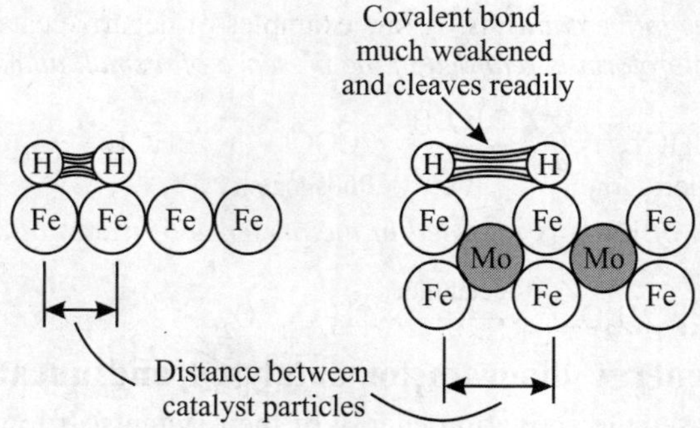

Fig. 7.2. Increase of crystal lattice spacing of catalyst makes the reaction go faster.

(*ii*) *Increase of active sites*. The number of peaks and cracks on the catalyst surface is increased on the addition of promoter to the catalyst. This increases the concentration of reactants (based on adsorption theory) and rate of the reaction.

7.1.3 Inhibitors (catalytic poisons)

In some cases, the presence of small quantities of impurities in the reacting substances makes the catalyst inactive. Such substances are called *catalytic poisons* or *inhibitors*. Catalytic poisons are usually also poisonous to organisms. Arsenious oxide and hydrogen cyanide are two of the most powerful catalytic poisons. Arsenious oxide poisons platinized asbestos used as catalyst in the manufacture of H_2SO_4 by contact process.

Similarly platinum catalyst used in the oxidation of H_2 is poisoned by CO.

$$2H_2 + O_2 \xrightarrow{Pt} 2H_2O \qquad \text{(Poisoned by the presence of CO)}$$

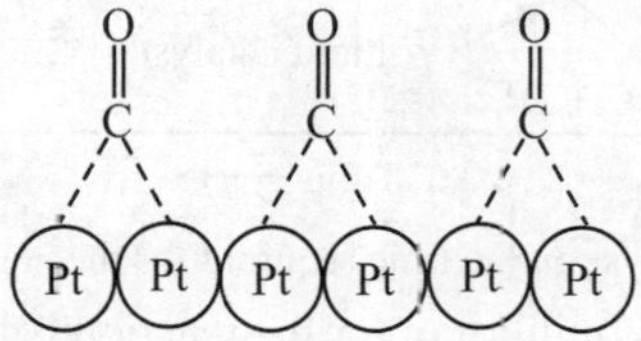

Fig. 7.3. Poisoning of platinum catalyst by carbon monoxide.

Explanation of the action of inhibitors (poisoning)

1. The poison is adsorbed on the catalyst surface in preference to the reactants. The surface of catalyst becomes unavailable to the reactant and the rate of reaction falls. Poisoning by As_2O_3 or CO appears to be of this kind (Fig. 7.3)
2. The catalyst may combine chemically with the poison. Poisoning of iron by H_2S falls in this category (Fe is a catalyst in the manufacture of NH_3 by Haber's process.

$$Fe + H_2S \longrightarrow FeS + H_2$$

7.2 GENERAL CHARACTERISTICS OF CATALYSED REACTION

Some characteristic properties of catalysed reactions are given below :

(i) The catalyst remains unchanged chemically at the end of the reaction. The amount of the catalyst recovered at the end of the reaction is found to be the same as the amount taken at the beginning of the reaction. Thus a catalyst does not undergo any chemical change although its physical state such as size, colour etc. may change. For example, MnO_2 which is used as a catalyst in the decomposition $KClO_3$ changes its physical state from coarse powder to fine powder at the end of the reaction.

Smooth platinum gauze which is used for the oxidation of NH_3 may turm rough at the end of the reaction.

(ii) Only a small quantity of the catalyst is needed for carrying out the reaction. There are reactions where very small amount of catalyst is required to convert very large amounts of reactants into the products. For example,

One gram of platinum is quite sufficient for the decomposition of 10^8 litres of hydrogen peroxide. However, there are some reactions especially homogeneous catalysed where the rate increases with the increase in concentration of the catalyst. In the inversion of cane-sugar in the presence of dil. HCl as a catalyst, the rate of the reaction increases with the increase in concentration of HCl. On the other hand there are reactions in which the rate of the reaction increases proportionally with the increase in the surface area of the catalyst.

(iii) A catalyst does not affect the state of equilibrium in a reversible reaction. It does not alter the value of equilibrium constant. From thermodynamic point of view, for a reversible reaction whether it occurs in the presence or absence of a catalyst, the free energy of the process is the same. A catalyst accelerates both the forward and backward reaction and helps to attain the equilibrium quickly (Fig. 7.4).

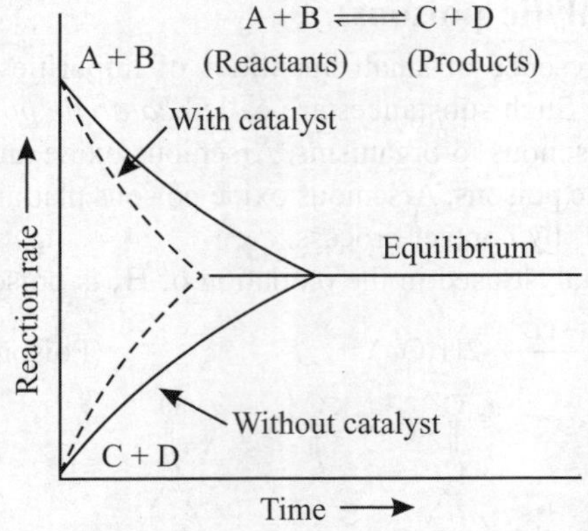

Fig. 7.4. The effect of a catalyst on the time required for the equilibrium to be established.

The figure indicates that the equilibrium for the reaction between A + B → C + D is attained in a shorter time in the presence of a catalyst.

(*iv*) *A catalyst does not initiate a reaction, it simply increases or decreases the speed of the reaction.* However, this is not always the case because there are large number of reactions which are initiated in the presence of a catalyst. The function of the catalyst is to follow an alternative path for the reaction which requires lower activation energy.

(*v*) *A catalyst is specific in its action.* A given catalyst can catalyst only a specific reaction. For example, MnO_2 can catalyst the decomposition of $KClO_3$ but fails for other reactions. Similarly enzymes are also specific in their action.

(*vi*) *A catalyst has an optimum temperature.* A catalyst can act effectively at a particular temperature for a specified reaction. This temperature is known as optimum temperature. For example,

The optimum temperature for the oxidation of SO_2 to SO_3 in the presence of Pt. catalyst is 450° to 500°C.

(*vii*) *A catalyst is poisoned by the presence of certain substances even in small amounts. These are called catalystic poisons.* Some powerful catalytic poisons are As_2O_3.HCN.CO etc. Presence of these substances makes the catalyst inactive.

(*viii*) *Activity of a catalyst is increased by the presence of substances known as promoter.* For example, in the manufacture of ammonia by Haber's process, finally divided iron is supported over molybdenum. Here molybdenum (Mo) acts as a promoter for the catalyst iron.

7.3 THEORIES OF CATALYSIS

Different types of mechanism have been proposed by which the catalyst can act. The major requirement for such type of reactions to occur is that the reactants must acquire some minimum amount of energy called threshold energy. *The energy required to activate the reactant species to the level of threshold energy is known as **activation energy.***

In the presence of a catalyst, the magnitude of activation energy required is less than in the uncatalysed reaction because the reaction follows a different path which has low activation energy. Due to lower value of activation energy of the catalysed reaction greater number of reactant species participate in the reaction and thereby the rate of reaction is increased.

There are two main theories which explain the role of catalyst in the catalytic reactions. These are :

7.3.1 Intermediate compound formation theory

According to this theory, a catalyst must chemically combine with one or more of the reactants to form an ***unstable intermediate product.*** It decomposes or combines later with other reactant to form the product and the catalyst is regenerated.

Consider a reaction

$$A + B \xrightarrow{X} AB \text{ (X is catalyst)}$$

The reaction is supposed to occur by the following steps :

(*i*) $A + X \longrightarrow AX$ (Unstable intermediate)

(*ii*) $AX + B \longrightarrow AB + X$

OR

(*i*) $A + B + X \longrightarrow ABX$ (Unstable intermediate)

(*ii*) $ABX \longrightarrow AB + X$

In many cases the intermediates have been isolated. For example,

(1) During the manufacture of sulphuric acid by lead chamber process, an intermediate compound nitrososulphuric acid ($HOSO_2NO_2$) is actually formed if the amount of water vapours is small.

$$\underbrace{2SO_2 + H_2O + NO + NO_2}_{\text{catalyst}} + O_2 \longrightarrow \underset{\text{Nitro-sulphuric acid}}{2HO - SO_2 - NO_2}$$

$$\underset{\substack{\text{Nitrososulphuric acid}\\\text{(Intermediate)}}}{2HO - SO_2 - NO_2} + H_2O \longrightarrow 2\,H_2SO_4 + \underbrace{NO + NO_2}_{\text{Catalyst}}$$

(2) In the formation of diethyl ether by the action of conc. H_2SO_4 on ethyl alcohol, an intermediate ethyl hydrogen sulphate is formed which reacts with another molecules of ethyl alcohol to form diethyl ether and H_2SO_4 is regenerated.

$$\underset{\text{(Reactant)}}{C_2H_5OH} + \underset{\text{(Catalyst)}}{H_2SO_4} \longrightarrow \underset{\text{(Ethyl hydrogen sulphate)}}{C_2H_5HSO_4} + H_2O$$

$$\underset{\text{(Intermediate)}}{C_2H_5HSO_4 + C_2H_5OH} \xrightarrow{140^\circ} \underset{\text{(Product)}}{C_2H_5 - O - C_2H_5} + \underset{\text{(Catalyst)}}{H_2SO_4}$$

The intermediate compound formation theory satisfactorily accounts for homogeneous catalysis but it fails to explain the gaseous reaction taking place on the surface of solids, *i.e.* heterogeneous catalysis. Also this theory does not explain deactivation by a catalyst poison and activation by a catalyst promoter.

7.3.2 Adsorption Theory

Heterogeneous reactions generally occur through adsorption of the reactants on the surface of the catalyst. The catalytic action of a surface depends on the extent of absorption of the reactants. The molecules of the reacting gases are adsorbed at different sites on the surface of a catalyst. These adsorbed molecules may then combine to form ***adsorbed activated complex*** which then decomposes to form the products which are desorbed from the surface. Following steps are involved in the whole process.

(*i*) *Diffusion of reactants to the surface.*

(*ii*) *Adsorption of reactant molecules on to the surface.*

(*iii*) *Chemical reaction on the surface.*

(*iv*) *Desorption of products from the surface.*

(*v*) *Diffusion of products away from the surface.*

In general, step (*iii*) determines the rate of reaction as it is the slowest step.

The stronger forces are operative in chemisorption which tend to bring about a partial loosening of the bonds in the adsorbed reacting molecules and a very small amount of energy is required to form the activated complex for the reaction.

Adsorption takes place on *active centres* on the surface of a catalyst. X-rays studies of the catalyst show that there are free valencies present on the surface of a catalyst because of a number of edges, corners, cracks and peaks on the rough surface of the catalyst. *The surface atoms at these sites have free valencies which can be used to attach molecules. The reacting gas molecules are adsorbed at these sites to form surface activated complex. These sites are called active centres.*

This explains how the catalytic activity of the finely divided catalyst is increased.

The surface of nickel crystal may be represented in Fig. 7.5. The dotted lines represent free valencies of nickel atoms. At the peak there will be greater number unsatisfied valencies. Isolated nickel atom would be the centres of great activity.

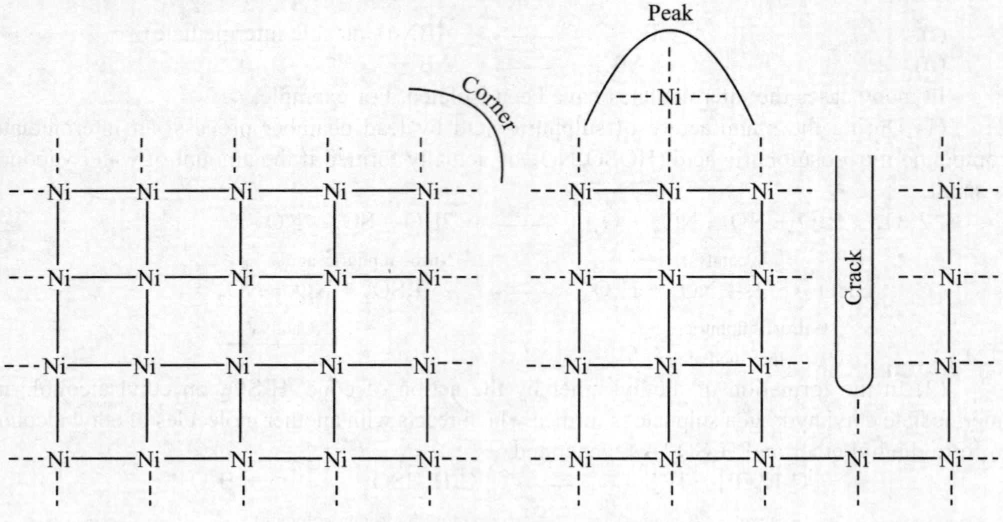

Fig. 7.5. Active centres on catalyst surface.

Function of Surface in Heterogeneous Catalysis.

A molecule is activated when it gets adsorbed on two or more active sites. Molecules can react more probably when they are held side by side on the surface than during the collisions in the bulk of the gas. Hence spacing of the active sites and the mode of attachment of the gas molecules are most important.

Examples.

When alcohol vapours are passed over reduced copper at 623K, the alcohol undergoes dehydrogenation while it gets dehydrated on alumina.

The possible orientation of the molecules are shown below in Fig. 7.6

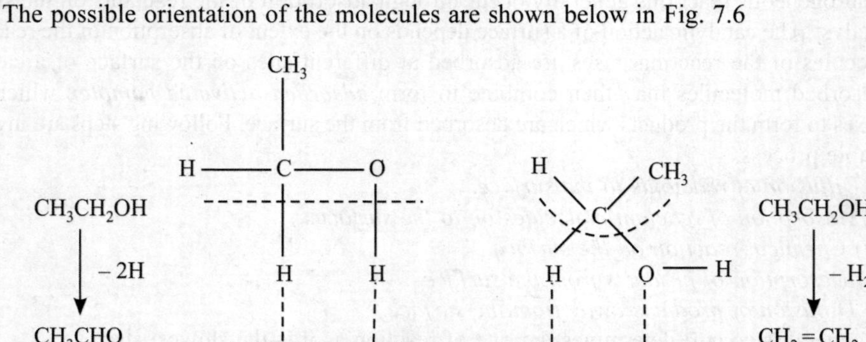

Fig. 7.6. Adsorption of C_2H_5OH molecule on different catalytic surface.

In Fig. 7.6 (*a*), the C–H and O–H bonds get weakened and dehydrogenation is favoured. In Fig. 7.6 (*b*), C–O and C–H bonds get weakened or stretched and dehydration is favoured.

7.4 HOMOGENOUS CATALYSIS

When in a reaction the catalyst is present in the same phase as the reactants, it is called homogeneous catalysis.

Examples of homogeneous catalysis are given below :

A. Gaseous phase

(*a*) *Reaction between* SO_2 *and* O_2 *to form* SO_3 *in presence of nitric oxide* (NO)

$$2SO_2(g) \xrightarrow{NO(g)} 2SO_3(g)$$

(*b*) *Decomposition of acetaldehyde in the presence of iodine vapours*

$$CH_3CHO(g) \xrightarrow{I_2(g)} CH_4(g) + CO(g)$$

(*c*) *Decomposition of vapours of* N_2O *at 1000 K in presence of traces of chlorine gas*

$$2N_2O(g) \xrightarrow[1000\ K]{Cl_2(g)} 2N_2(g) + O_2(g)$$

(*d*) *Decomposition of ozone in presence of* NO or N_2O_5

$$2O_3(g) \xrightarrow{NO\ (g)} 3O_2\ (g)$$

B. Liquid phase

(*a*) **Acid-Base catalysis**

(*i*) *Hydrolysis of cane-sugar (sucrose) in the presence of a mineral acid.* (*I*nversion of cane-sugar).

$$\underset{}{C_{12}H_{22}O_{11}(aq)} + H_2O(l) \xrightarrow{H^+(aq)} \underset{\text{Glucose}}{C_6H_{12}O_6(aq)} + \underset{\text{Fructose}}{C_6H_{12}O_6(aq)}$$

(*ii*) *Hydrolysis of methyl acetate by an acid or a base*

$$CH_3COOCH_3(aq) + H_2O(l) \xrightarrow[\text{or } OH^-(aq)]{H^+(aq)} CH_3COOH(aq) + CH_3OH(aq)$$

(*iii*) *Hydrolysis of starch to glucose in presence of an acid*

$$\underset{\text{Starch}}{(C_6H_{10}O_5)_n} + nH_2O \xrightarrow{H^+} \underset{\text{Glucose}}{nC_6H_{12}O_6}$$

(*b*) *Decomposition of* H_2O_2 *in the presence of* I^- *ion or* Br_2

$$2H_2O_2(aq) \xrightarrow[\text{or } Br_2]{I^-(aq)} 2H_2O\ (l) + O_2(g)$$

7.4.1 Mechanism of homogeneous catalytic reactions

A homogeneous catalytic usually takes place in two steps as follows :

1st step: The catalyst reacts with the reactant to form an intermediate complex

$$A + C \longrightarrow AC$$

2nd step: The complex then decomposes to form the product regenerating the catalyst or reacts with a second reactant to give the product and regenerating the catalyst *i.e.*

$$AC \longrightarrow \text{Product} + C$$

or

$$AC + B \longrightarrow \text{Product} + C$$

This is illustrated with examples given below :

(1) *Mechanism of the reaction* $2SO_2 + O_2 \xrightarrow{NO} 2SO_3$

1st step: $2NO + O_2 \longrightarrow 2NO_2$

2nd step: $2NO_2 + 2SO_2 \longrightarrow 2NO + 2SO_3$

Both the above steps are fast. Hence the overall reaction is also fast whereas in the absence of catalyst the reaction is slow.

(2) *Mechanism of the reaction* $CH_3CHO \xrightarrow{I_2} CH_4 + CO$

1st step: $CH_3CHO + I_2 \longrightarrow CH_3I + HI + CO$

2nd step: $CH_3I + HI \longrightarrow CH_4 + I_2$

(3) *Mechanism of the reaction* $2N_2O \xrightarrow{Cl_2} 2N_2 + O_2$

1st step: $Cl_2 \longrightarrow 2Cl$

2nd step: $N_2O + Cl \longrightarrow N_2 + ClO$

3rd step: $2ClO \longrightarrow Cl_2 + O_2$

(4) *Mechanism of the reaction* $2O_3 \xrightarrow{N_2O_5} 3O_2$

1st step: $2N_2O_5 \longrightarrow 2N_2O_4 + O_2$

2nd step: $O_3 + N_2O_4 \longrightarrow O_2 + N_2O_5$

(5) *Mechanism of the reaction* $2H_2O_2 \longrightarrow 2H_2O + O_2$

1st step: $H_2O_2 + I^- \longrightarrow H_2O + IO^-$

2nd step: $H_2O_2 + IO^- \longrightarrow H_2O + O_2 + I^-$

7.4.2 Kinetics of homogeneous catalytic reactions

The rate equation in the presence of catalyst is found out experimentally. This tells the type and number of the different molecules taking part in the rate determining step (slowest step), which then helps to write the probable mechanism of the reaction. This is illustrated with the help of following examples.

(1) *Kinetics of the decomposition of acetaldehyde in the presence of iodine vapours as catalyst.* The decomposition of a acetaldehyde takes place as

$$CH_3CHO(g) \longrightarrow CH_4(g) + CO(g)$$

The observed rate equation for the this reaction is found to be

$$\text{Rate} = \frac{d[CH_3CHO]}{dt} = k_1 [CH_3CHO]^2$$

which suggests the mechanism as follows :

$$CH_3CHO + CH_3CHO \longrightarrow 2CH_4 + 2CO$$

In the presence of iodine vapours (as catalysts), the rate equation is observed to be

$$\text{Rate} = \frac{d[CH_3CHO]}{dt} = k_1 [CH_3CHO][I_2]$$

This suggests the probable mechanism of the catalysed reaction as follows :

$$CH_3CHO + I_2 \xrightarrow{\text{slow}} CH_3I + HI + CO \quad \text{...Rate determining step}$$

$$CH_3I + HI \xrightarrow{\text{fast}} CH_4 + I_2$$

(2) *Kinetics of the dehydration of tert-butyl alcohol in the presence of HBr.* The dehydration of tert-butyl alcohol takes place as

$$\underset{\text{tert-butyl alcohol}}{(CH_3)_3\,C-OH} \longrightarrow \underset{\text{2-Methylpropene}}{(CH_3)_2C=CH_2} + H_2O$$

In the absence of catalyst, the alcohol does not decompose at any appreciable rate upto about 700 K. On adding a small amount of HBr, rapid decomposition occurs at 500 K. The rate equation for the uncatalysed reaction is

$$\text{Rate} = -\frac{d[t\text{–BuOH}]}{dt} = k_1[t-\text{BuOH}]$$

i.e. it is a first order reaction with very low value of k_1.

In the presence of HBr, the rate equation is observed to be

$$\text{Rate} = -\frac{d[t\text{–BuOH}]}{dt} = k_2[t-\text{BuOH}][\text{HBr}]$$

i.e. it is a second order reaction with high value of k_2.

This suggests that a possible mechanism for the catalysed reaction is

$$(CH_3)_3\,C-OH + HBr \xrightarrow{\text{slow}} (CH_3)_3\,C-Br + H_2O \quad \text{...Rate determining step}$$

$$(CH_3)_3\,C-Br \xrightarrow{\text{fast}} (CH_3)_2\,C=CH_2 + HBr$$

7.4.3 Rate of an acid catalysed reaction

In an acid-catalysed reaction; the first step is the transfer of a proton (H^+ ion) from the acid HA to the substrate S (one of the reactants)

$$S + HA \underset{k_1'}{\overset{k_1}{\rightleftharpoons}} SH^+ + A^- \qquad \text{... }(i)$$

This is then followed by one of the two possibilities *i.e.* either proton transfer to the solvent *i.e.*,

$$SH^+ + H_2O \xrightarrow{k_2} P + H_3O^+ \qquad (P = \text{Product}) \qquad \text{...}(ii)$$

or proton transfer to the base A^- *i.e.*,

$$SH^+ + A^- \longrightarrow P + HA \qquad \text{...}(iii)$$

Taking reaction (*i*) as the first step and reaction (*ii*) as the second step, then taking the latter as the rate determining step, the reaction rate is given by

$$-\frac{d[S]}{dt} = k_2[SH^+] \qquad \text{...}(iv)$$

But the reaction of $[SH^+]$ with time is given by

$$\frac{d[SH^+]}{dt} = \text{Rate of formation of } SH^+ \text{ from S and HA}$$

$$- \text{Rate of reaction of } SH^+ \text{ with } H_2O$$

$$= \{k_1[S][HA] - k_1'[SH^+][A^-]\} - \{k_2\,[SH^+]\}$$

But steady state condition implies that the rate of change of concentration of $[SH^+]$ with time should be zero.

i.e.
$$\frac{d[SH^+]}{dt} = 0$$

Thus k_1 [S] [HA] – k_1' [SH^+] [A^-] – k_2 [SH^+] = 0

or
$$[SH^+]\ \{k_1'\ [A^-] + k_2\} = k_1[S]\ [HA]$$

or
$$[SH^+] = \frac{k_1[S][HA]}{k_1'[A^-] + k_2} \qquad ...(v)$$

Putting this value in equation (*iv*), we get

$$-\frac{d[S]}{dt} = k_1k_2 \times \frac{[S][HA]}{k_1'[A^-] + k_2} \qquad ...(vi)$$

If $k_2 << k_1'$ [A^-], equation (*vi*) becomes

$$-\frac{d[S]}{dt} = \frac{k_1k_2[S][HA]}{k_1'[A^-]} \qquad ...(vii)$$

7.5 HETEROGENEOUS CATALYSIS

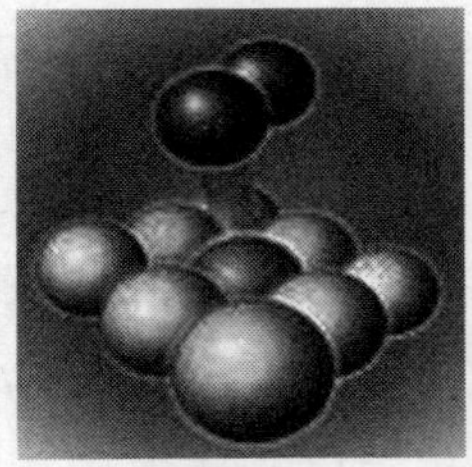

Heterogeneous catalysis

If the catalyst is present in a different phase from that of the reactants, in a reaction, it is called heterogeneous catalysis. The catalysts are usually solids whereas the reactants are gaseous or liquid. The reactions occur at the surface of the catalyst. That is why this type of catalysis is also called *surface catalysis* or *contact catalysis.* Evidently, greater the surface area of the catalyst, greater is the efficiency of the catalyst. For this reason, finely divided solids are usually used as catalysts.

Examples of heterogeneous catalytic reactions

(*i*) *Manufacture of ammonia from nitrogen and hydrogen using finely divided iron as catalyst*

$$N_2(g) + 3H_2(g) \xrightarrow[\text{divided Fe}]{\text{Finely}} 2NH_3(g)$$

(*ii*) *Formation of* SO_3 *from* SO_2 *and* O_2 *in contact process, using finely divided platinum or vanadium pentoxide as catalyst for manufacture of sulphuric acid*

$$2SO_2\ (g) + O_2\ (g) \xrightarrow[\text{or } V_2O_5]{\text{Pt}} 2SO_3\ (g)$$

(*iii*) *Manufacture of methyl alcohol from* CO *and* H_2 *using* ZnO *and* Cr_2O_3 *as catalysts*

$$CO(g) + 2H_2(g) \xrightarrow{ZnO/Cr_2O_3} CH_3OH(g)$$

7.5.1 Heterogeneous catalysis is surface catalysis

Consider the bromination of ethylene

$$CH_2 = CH_2(g) + Br_2(g) \longrightarrow BrCH_2 - CH_2Br(g)$$

This reaction takes place readily in a glass vessel as 470 K. It was thought to be an ordinary homogeneous catalytic reaction. However it was found that the rate is higher in a smaller reaction vessel or in a vessel packed with glass beads but the rate is reduced when the vessel is coated on the inside with paraffin wax. This shows that the reaction is not occurring in the gas phase but is occurring on the glass surface of the vessel. The glass acts as a catalyst.

7.5.2 Heterogeneous catalysts are specific

Consider the decomposition of formic acid. If the acid vapour is passed through a heated glass tube, the reaction is about 50% dehydration and 50% dehydrogenation *i. e.*

$$(a)\quad HCOOH \longrightarrow H_2O + CO$$

$$(b)\quad HCOOH \longrightarrow H_2 + CO_2$$

If the tube is packed with Al_2O_3, only reaction (*i*) occurs but if it is packed with ZnO, only reaction (*ii*) occurs. This illustrates specificity of a heterogeneous catalyst.

7.5.3 Mechanism of heterogeneous catalytic reactions

A heterogeneous catalytic reaction takes place through the following sequence of steps. (See Fig. 7.7)

(*i*) Diffusion of the reactant molecules towards the surface of the catalyst.

(*ii*) Adsorption of the reactant molecules on the surface of the solid.

(*iii*) Chemical reaction of the adsorbed molecules with the surface of the catalyst resulting into stretching and weakening of bonds forming an activated complex with the catalyst which then decomposes or even breaking of the bonds of reactant molecules.

Steps (*ii*) and (*iii*) are jointly called *chemisorption.*

(*iv*) Desorption of the products from the surface.

(*v*) Diffusion of the products away from the surface.

The various steps are illustrated with the following examples .

(*a*) *Hydrogenation of ethylene in presence of finely divided Ni, Pt or Pd*

It is illustrated as under

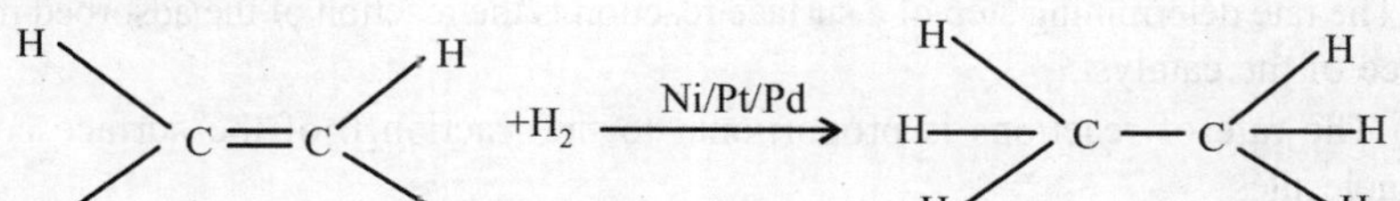

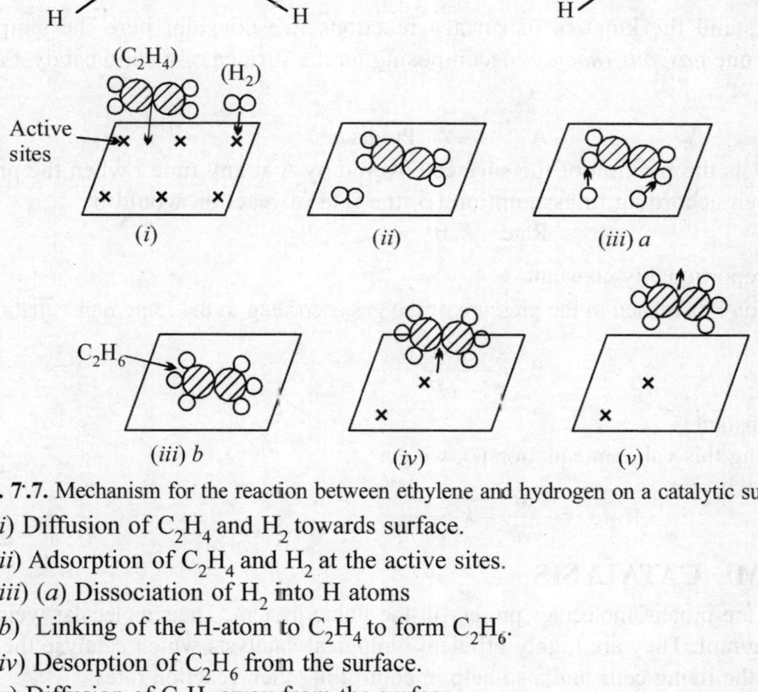

Fig. 7.7. Mechanism for the reaction between ethylene and hydrogen on a catalytic surface.

(*i*) Diffusion of C_2H_4 and H_2 towards surface.

(*ii*) Adsorption of C_2H_4 and H_2 at the active sites.

(*iii*) (*a*) Dissociation of H_2 into H atoms

(*b*) Linking of the H-atom to C_2H_4 to form C_2H_6.

(*iv*) Desorption of C_2H_6 from the surface.

(*v*) Diffusion of C_2H_6 away from the surface.

(*b*) *Decomposition of* HI *in the presence of platinum.*

$$2HI\,(g) \longrightarrow H_2(g) + I_2(g)$$

The mechanism of this reaction may be represented as follows :

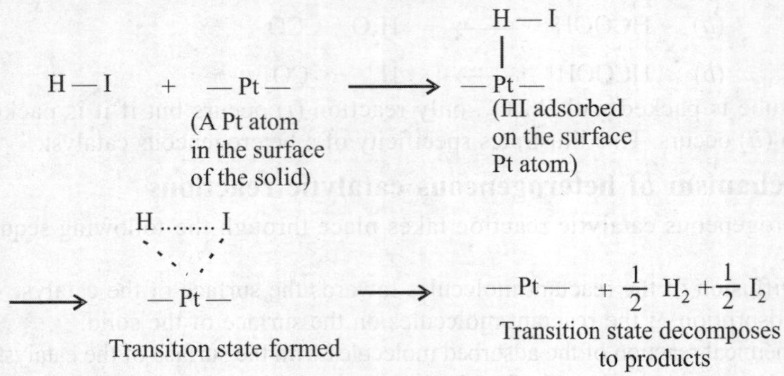

7.5.4 Kinetics of heterogeneous (surface) catalytic reactions

A heterogeneous catalytic reaction takes place in a number of consecutive steps. Although any of these steps may be slowest and hence rate determining, yet from the point of view of the study of the kinetics of these reactions, the reaction of the adsorbed molecules is considered to be the rate of determining. Further the rate of reaction is supposed to be proportional to the fraction θ of the surface covered by the reactant molecules.

Thus the two *assumptions* in the study of kinetics of the surface reactions are :

(*i*) The rate determining step of a surface reaction is the reaction of the adsorbed molecules with the surface of the catalyst.

(*ii*) The rate of reactions is proportional to the fraction θ of the surface covered by the reactant molecules.

To understand the kinetics of surface reactions, we consider here the simplest case involving only one *gaseous reactant* decomposing on the surface of a solid catalyst according to the reaction

$$A \longrightarrow \text{Products}$$

Then if θ_A is the fraction of the surface covered by A at any time *t* when the pressure of the gas is *P,* then according to assumption (*i*), the rate of reaction would be

$$\text{Rate} = k'\,\theta_A \qquad ...(i)$$

where *k'* is a proportionality constant.

The fraction θ is related to the pressure of the gas according to the Langmuir equation, *viz.*

$$\theta = \frac{bP}{1+bP} \qquad ...(ii)$$

where *b* is a constant.

Substituting this value in equation (*i*), we get

$$\text{Rate} = -\frac{dP}{dt} = k'\frac{bP}{1+bP} \qquad ...(iii)$$

7.6 ENZYME CATALYSIS

Enzymes are protein molecules present in the living system. Their molecular weights range from 10^4– 10^6 g/mol. They are highly efficient biological catalysts which catalyse the reactions taking place in the living cells and also help in controlling their reaction rates.

The reaction in enzyme catalysis occurs at a specific site on the protein molecule. This is called the **active site.** The reactants in an enzyme reaction are referred to **as substrates**.

A few examples of enzyme-catalysed laboratory reactions are given below :

$$(i)\ \text{Starch} \xrightarrow{\text{Diastase}} \text{Maltose} \xrightarrow{\text{Maltase}} \text{Glucose} \xrightarrow{\text{Zymase}} \text{Alcohol} + CO_2$$

$$(ii)\ \text{Cane sugar} \xrightarrow[+H_2O]{\text{Invertase}} \underset{(\text{Glucose} + \text{Fructose})}{\text{Invert sugar}} \xrightarrow{\text{Zymase}} \text{Alcohol} + CO_2$$

The names of the enzymes are written on the arrows.

Enzymes possess certain characteristics which are different from other types of catalysts. A few of these are briefly described below :

(*i*) *Their efficiency is very high i.e.* many of them are so powerful catalysts that they increase the rate of reaction by many orders of magnitude even at quite low concentrations.

(*ii*) *They are highly specific in their catalytic action.* For example, urease even in very low concentrations catalyses the hydrolysis of urea but it has no detectable effect on the hydrolysis rate of substituted ureas such as methyl urea. There are many such examples to quote. However there are some exceptions to this generalization.

(*iii*) *They are highly sensitive to temperature.* The activity of the enzyme is usually found to be maximum at the normal temperature of the organism in which the enzyme is found. Above this temperature, usually the activity is first found to increase temporarily and then drops to zero when the protein is denatured.

(*iv*) *They are highly sensitive to the acidity or basicity of the solution.* It is found that the enzymes show a sharp change in activity if the *p*H of the solution is changed.

7.6.1 Mechanism of enzyme reactions (Michaelis Menten equation)

The mechanism of an enzymes reaction as proposed by Michaelis and Menten is as follows:

The long chains of the enzyme (protein) molecules are coiled to make a rigid colloidal particle with cavities on its surface. These cavities which are of characteristic shape and rich in active groups ($-NH_2$, $-COOH$, $-SH$, $-OH$) are termed **active centres.** The molecules of substrate which have complementary shape fit into these cavities just as a key fits into a lock (**Lock-and-Key theory**). With the help of active groups, the enzyme forms an activated complex with the substrate which decomposes to yield the products. Thus the substrate molecule enters the cavities, forms complex and reacts, and the products move out of the cavities (Fig. 7.8)

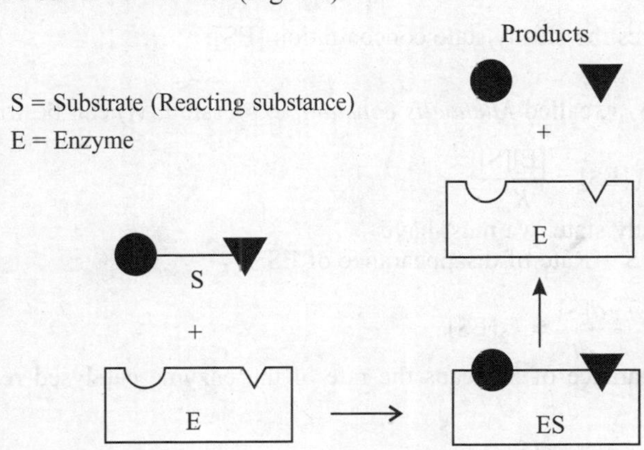

Fig. 7.8. Illustration of the lock-and-key model of enzyme catalysis.

The first step in the enzyme reaction is the reaction between the enzyme E (at the active site) and the substrate S (*i.e.* reactants of the enzyme reaction) to form the complex ES.

$$E + S \underset{k_1'}{\overset{k_1}{\rightleftharpoons}} ES \quad ...(i)$$

This reaction is very fast and the equilibrium is rapidly established.

The complex then dissociates forming one or more products P and the enzyme is regenerated

$$ES \underset{k_2'}{\overset{k_2}{\rightleftharpoons}} E + P \quad ...(ii)$$

This reaction usually goes to completion, so the rate coefficient k_2' is effectively zero. The steps (*i*) and (*ii*) can be written together as

$$E + S \underset{k_1'}{\overset{k_1}{\rightleftharpoons}} ES \underset{k_2'}{\overset{k_2}{\rightleftharpoons}} E + P \quad ...(iii)$$

Thus the enzyme-substrate complex ES is an intermediate and the net reaction is S → P. As a result, the rate of appearance of the product must be equal to the rate of disappearance of the substrate S, *i.e.*

$$\frac{d[P]}{dt} = -\frac{d[S]}{dt} \quad ...(iv)$$

This implies that the concentration of the complex ES, represented by [ES] must be in a *steady state.* In other words, the rate of change of concentration of ES with time must be equal to zero. But rate of change of concentration of ES with time will be given by

$$\frac{d[ES]}{dt} = \text{Rate of formation of ES from E and S} - \text{Rate of decomposition of ES into E and P}$$

$$= \{k_1[E][S] - k_1'[ES]\} - \{k_2[ES]\} \quad ...(v)$$

For steady state condition, $d[ES]/dt = 0$. Hence

$$k_1[E][S] - k_1'[ES] - k_2[ES] = 0$$

or $$[ES]\{k_1' + k_2\} = k_1[E][S]$$

or $$[ES] = \frac{k_1[E][S]}{k_1' + k_2} \quad ...(vi)$$

This expression gives the steady state concentration [ES].

Putting $\frac{k_1' + k_2}{k_1} = K_{m'}$, called *Michaelis constant,* expression (*vi*) can be written as

$$[ES] = \frac{[E][S]}{K_m} \quad ...(vii)$$

Since [ES] is a steady state, we must have

Rate of disappearance of S = Rate of disappearance of ES.

i.e $$-\frac{d[S]}{dt} = k_2[ES] \quad ...(viii)$$

But rate of disappearance of S means the rate of the enzyme catalysed reaction, usually represented by v. Hence

$$v = -\frac{d[S]}{dt} = k_2[ES] \quad ...(ix)$$

Substituting the value of [ES] from equation (*vii*) in equation (*ix*), we get

$$v = -\frac{d[S]}{dt} = k_2[\mathrm{ES}] = k_2\,\frac{[\mathrm{E}][\mathrm{S}]}{K_m} \quad ...(x)$$

The equation is, however, not useful in this form since it includes the concentration of free enzyme [E] whereas the experimentally measurable quantity is $[E_0]$, the total concentration of the enzyme *i.e.* free as well as combined in the complex ES. But we have

$$[\mathrm{E_0}] = [\mathrm{E}] + [\mathrm{ES}]$$

so that

$$[\mathrm{E}] = [\mathrm{E_0}] - [\mathrm{ES}]$$

Substituting this value in equation (*vii*), we get

$$[\mathrm{ES}] = \frac{\{[\mathrm{E_0}] - [\mathrm{ES}]\}[\mathrm{S}]}{K_m}$$

or $$K_m\,[\mathrm{ES}] = [\mathrm{E_0}][\mathrm{S}] - [\mathrm{ES}][\mathrm{S}]$$

or $$K_m\,[\mathrm{ES}] + [\mathrm{ES}][\mathrm{S}] = [\mathrm{E_0}][\mathrm{S}]$$

or $$[\mathrm{ES}]\,\{K_m + [\mathrm{S}]\} = [\mathrm{E_0}][\mathrm{S}]$$

or $$[\mathrm{ES}] = \frac{\{[\mathrm{E_0}][\mathrm{S}]}{K_m + [\mathrm{S}]} \quad ...(xi)$$

Substituting this value in equation (*ix*), we get

$$v = -\frac{d[S]}{dt} = k_2[\mathrm{ES}] = \frac{k_2[\mathrm{E_0}][\mathrm{S}]}{K_m + [\mathrm{S}]} \quad ...(xii)$$

This equation is known as *Michaelis-Menten equation* which contains the measurable quantities, $[E^0]$ and [S].

Equation (*xii*) is usually written in terms of maximum velocity v_m of the enzyme-catalysed reaction which is reached when [S] becomes so large that $[S] >> K_m$ so that K_m can be neglected in comparison to [S]. Then equation (*xii*) takes the form

$$v_m = k_2[E_0] \quad ...(xiii)$$

Hence $$\frac{v}{v_m} = \frac{[\mathrm{S}]}{K_m + [\mathrm{S}]} \quad ...(xiv)$$

From this equation, we find that

$$\text{when } [\mathrm{S}] = K_m,\ \frac{v}{v_m} = \frac{1}{2}$$

i.e. $$v = \frac{1}{2}v_m.$$

Thus *Michaelis constant is equal to that concentration of the substrate at which the rate of the enzyme-catalysed reaction falls to one-half of its maximum value.* Thus the values of v_m and K_m can be obtained by plotting rate of reaction, *v* versus the concentration of the substrate S.

7.7 AUTOCATALYSIS

When one of the products of a reaction acts as a catalyst for that reaction, the phenomenon is known as autocatalysis.

Examples of Autocatalysis

1. Oxidation of oxalic acid. When oxalic acid is oxidised by acidified potassium permanganate, mangamous sulphate or Mn^{2+} ions) produced during the reaction acts as a catalyst for the reaction.

$$2\ KMnO_4 + 5\ H_2C_2O_4 + 3H_2SO_4 \longrightarrow 2MnSO_4 + K_2SO_4 + 8\ H_2O + 10CO_2$$

2. Hydrolysis of an ester. Hydrolysis of ethyl acetate forms acetic acid and ethanol. Acetic acid acts as a catalyst for the reaction

$$CH_3COOC_2 + H_2O \xrightarrow[\text{Catalyst}]{H+} CH_3COOH + C_2H_5OH$$

Whenever autocatalysis occurs the initial rate of reaction rises as the catalytic product is formed unlike in a normal reaction where the rate of the reaction decreases with the passage of time. The curve shown in Fig. 7.9 shows a maximum when the reaction is complete.

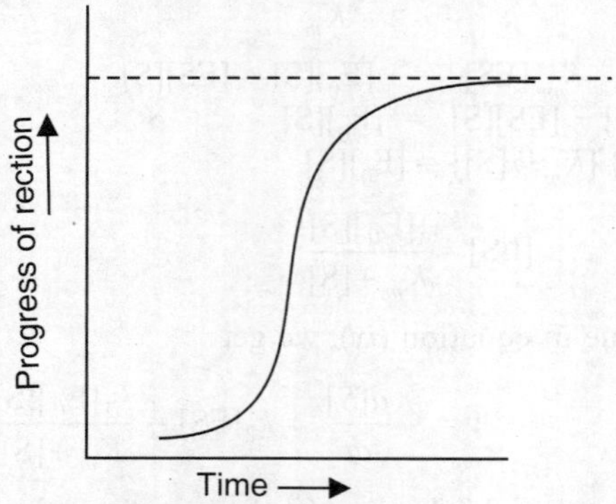

Fig. 7.9. Curve showing the rise of rate of reaction with time.

Example 1. **Write down important criteria of a catalyst.**

Solution. Important features (also known as criteria) of a catalyst are given below.

1. *A catalyst remains unchanged in mass and chemical composition at the end of the reaction.*
   Qualitative and quantitative analysis of the catalyst shows that it undergoes no change in mass or chemical nature. It might undergo a physical change.
2. *A small quantity of catalyst is generally sufficient to complete a reaction.*
   It has been observed that in most of the reactions, only a very small amount of the catalyst can produce huge quantities of the products. Only in exceptional cases, like in Friedel Crafts reaction, we require large quantities of the catalyst to obtain good yield.
3. A catalyst is more effective when finely divided powdered than in the coarse state. Finely divided nickel is a better catalyst than lumps of solid nickel.
4. *A catalyst is specific in action*

   A catalyst which works for a reaction need not work for another reaction. The same reactant may yield different products with different catalysts. For example,

$$C_2H_5OH \xrightarrow{Al_2O_3} \underset{\text{Ethene}}{C_2H_4} + H_2O \qquad C_2H_5OH \xrightarrow{Cu} \underset{\text{Ethanal}}{CH_3\ CHO} + H_2$$

Example 2. **The activity of a catalyst increases on decrease in particle size. Explain.**

Solution. It has been observed that catalysts in fine powder are more effective than in lump form. This is because the free valencies, which are responsible for catalytic activity increase as the particle size decreases. This is illustrated in Fig. 7.10.

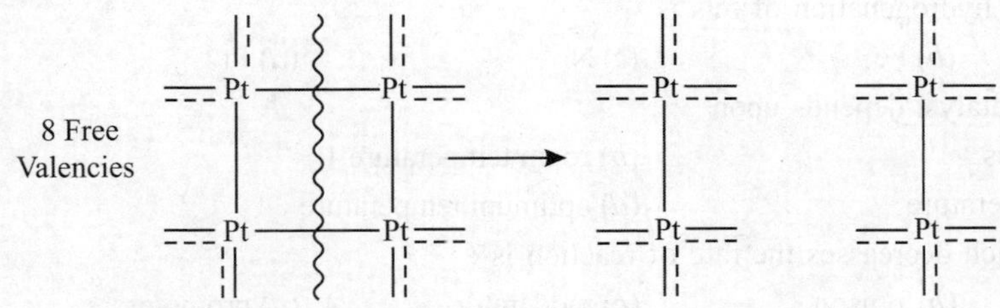

Fig. 7.10. Subdivision of a catalyst makes it more efficient due to increase of free valence bonds. Dotted lines indicate valencies

One big particle of Pt catalyst has 8 free valencies. On division into two smaller particles. The number of free valencies increases to 12. The number of free valencies will go increasing on further subdivision of the catalyst particles.

EXERCISES

(INCLUDING QUESTIONS FROM DIFFERENT UNIVERSITY PAPERS)

Objective Questions

Fill in the blanks

1. Subdivision of a catalyst makes it more efficient due to increase of ....
2. A .... remains unchanged in mass and chemical composition at the end of the reaction.
3. When one of the products of a reaction acts as a catalyst for that reaction, the phenomenon is know as ....
4. .... theory explains the mechanism of enzyme reactions
5. The enzyme .... converts glucose into alcohol.
6. If the catalyst is present in a different phase from that of the reactants, it is called .... catalyst
7. Hydrolysis of cane sugar in the presence of a mineral acid is an example of .... catalysis
8. A catalyst does not .... a reaction, it simply increases or decreases the rate of the reaction.
9. Platinum catalyst used in the oxidation of H_2 is poisoned by ....
10. The promoter change the .... in the catalyst which help in weakening the bond between reactant molecules.

Multiple Choice Questions

Choose the correct option

1. The name catalyst was proposed by

   (*a*) Rutherford (*b*) Langmuir (*c*) Graham (*d*) Berzelius

2. Half life of a radioactive element depends upon

   (*a*) amount of the element (*b*) the temperature
   (*c*) the pressure (*d*) none of the above

3. Which of the following are most efficient catalysts ?

   (*a*) Transition metals (*b*) Alkali metals
   (*c*) Radioactive metals (*d*) Alkaline earth metals

4. Catalyst used for hydrogenation of oils is
(*a*) V_2O_5 (*b*) Fe (*c*) Ni (*d*) Pt
5. Efficiency of a catalyst depends upon
(*a*) molecular mass (*b*) room temperature
(*c*) absolute temperature (*d*) optimum temperature
6. A substances which decreasses the rate of reaction is :
(*a*) inhibitor (*b*) poison (*c*) moderator (*d*) promoter
7. A catalyst is a substance which
(*a*) increases the equilibrium concentration of the products
(*b*) changes the equilibrium constant of the reaction
(*c*) shortens time to reach equilibrium
(*d*) supplies energy to the reaction
8. In the reaction $KMnO_4 + H_2SO_4 + H_2C_2O_4 \longrightarrow$ products, Mn^{2+} ions act as
(*a*) positive catalyst (*b*) negative catalyst
(*c*) autocatalyst (*d*) enzyme catalyst
9. Which is not an example of homogeneous catalysis ?
(*a*) Formation of SO_3 in chamber process
(*b*) Formation of SO_3 in contact process
(*c*) Hydrolysis of methyl acetate in the presence of an acid
(*d*) Decomposition of $KClO_3$ in the presence of MnO_2
10. Which of the following statements is not correct ?
(*a*) Catalyst action is specific
(*b*) Catalyst does not alter the equilibrium
(*c*) A small amount of catalyst is sufficient to catalyse the reaction
(*d*) The catalyst initiates the reaction

ANSWERS

Fill in the blanks

1. free valence bonds **2.** catalyst **3.** autocatalysis
4. lock and key **5.** zymase **6.** heterogeneous
7. homogeneous **8.** initiate **9.** CO
10. lattice spacing

Multiple choice questions

1. (*d*) **2.** (*d*) **3.** (*a*) **4.** (*c*) **5.** (*d*)
6. (*a*) **7.** (*c*) **8.** (*c*) **9.** (*b*) **10.** (*d*)

General Questions

1. What do you mean by the term catalysis ? What are the general characteristics of catalysed reaction ?
2. Explain giving examples :
(*i*) Catalytic poisons
(*ii*) Action of promoters

(*iii*) Active centres

(*iv*) The specific nature of catalyst.

3. Discuss the characteristics of a catalysed reaction. What are catalytic inhibitors ?

4. What are catalytic inhibitors ?

5. What is enzyme catalysis ? How will you account for the fact that enzyme catalysed reactions have maximum activity at a definite *p*H. Give the mechanism of enzyme catalysed reaction.

6. Differentiate clearly between homogeneous and heterogeneous catalysis with suitable examples.

7. For the reaction :

$$E + S \underset{k_2}{\overset{k_1}{\rightleftharpoons}} ES$$

$$ES \xrightarrow{k_3} E + P$$

where E is enzyme

S is substrate

ES is enzyme substrate complex

And P is product

Following the Michaelis-Menten mechanism, deduce the equation for the rate of reaction.

8. Write notes on the following :

(*a*) Intermediate compound formation theory.

(*b*) Adsorption theory for heterogeneous catalysis.

9. Define acid base catalysis. Explain the acid-base catalysis by giving suitable examples.

10. (*a*) Derive Michaelis–Menten equation.

(*b*) What is Michaelis constant ? How can it be determined ?

11. Show that for a unimolecular surface reaction

$$r = \frac{k_1 k_2 p_A}{k_1 p_A + k_{-1} + k_2}$$

where p_A is the partial pressure of reactant A,

k_1 = Rate constant for adsorption process

k_{-1} = Rate constant for desorption

k_2 = Rate constant for decomposition.

12. What are enzymes ? How do they act as catalyst ? Explain in terms of Michaelis–Mention mechanism.

13. Give the kinetics of unimolecular surface reaction.

14. (*a*) What are enzymes ? How do they act as catalyst ?

(*b*) What is acid-base catalyst ? Give examples.

15. (*a*) What is catalyst ? How does it work ?

(*b*) Write short notes on the following :

(*i*) Acid-base catalysis.

(*ii*) Enzyme catalysis.

(*c*) Explain the application of adsorption in catalysis

16. Write short notes on

(*i*) Catalytic poison of

(*ii*) Enzyme catalysis.

17. Discuss the mechanism of enzyme catalysed reactions. Deduce Michaelis-Menten equation and give conditions under which an enzyme catalysed changes its order.

18. Derive Michaelis-Menten equation for enzyme catalysis.

19. (*a*) Discuss the mechanism of an enzyme reaction as proposed by Michaelis-Menten equation.

(*b*) Derive a general expression for the rate of reaction of an acid catalysed reaction.

8

Thermodynamics–I*

(First Law of Thermodynamics and Thermochemistry)

8.1. THERMODYNAMICS

Thermo means heat and *dynamics* means motion or mechanical work. Hence thermodynamics means that branch of science which deals with the conversion of heat into mechanical work and vice versa. In a wider sense *thermodynamics is a branch of science that deals with quantitative relationships between heat and other forms of energy*. Study of thermodynamics is based on three laws *viz.* first, second and third laws of thermodynamics. There is no theoretical proof of the laws, but nothing contrary to the laws has been reported so far.

Objectives of Thermodynamics

(*i*) **To predict the feasibility of a process:** Whether or not, a process will occur under the given set of conditions, can be predicted by applying the principles of thermodynamics.

(*ii*) **To estimate the yield of the products:** Thermodynamic relations help to predict the yield of the products obtainable in a process or a reaction. It is possible to know the extent upto which the reaction takes place before attaining equilibrium.

(*iii*) **To deduce some important relationships in physical chemistry:** It has been possible to deduce some important and useful results such as Raoult's law of lowering of vapour pressure and expressions for depression in freezing point, elevation in boiling point, distribution law and phase rule from the study of thermodynamics.

8.8.1. Limitations of Thermodynamics

Thermodynamics suffers from certain limitations. These are :

(*i*) It deals with the properties like temperature, pressure *etc.* of the matter in bulk *i.e.* with macroscopic quantities and not with the microscopic quantities. Thus, it deals with large groups of atoms, molecules and ions rather than individual atoms, molecules or ions.

(*ii*) It helps to predict the feasibility of a process but does not tell anything about the time taken for the process to complete or the rate at which the process would proceed.

(*iii*) It concerns itself only with the initial and final states of a system and is not concerned with the path by which the change is brought about *i.e.* it does not reveal the mechanism of a process.

8.2. DEFINITION OF CERTAIN THERMODYNAMICS TERMS

1. System and surroundings: *The part of the universe chosen for thermodynamic consideration (i.e. to study the effect of temperature, pressure etc.) is called a system.*

The remaining portion of the universe, excluding the system, is called *surroundings*.

2. Open, closed and isolated systems

(*a*) **Open system:** *A system is said to be an open system if it can exchange both matter and energy with the surroundings.* For example, if some water is kept in an open vessel (Fig. 8.1), exchange of both matter and energy takes place between the system and the surroundings.

*Not in UGC syllabus for B.Sc. first year.

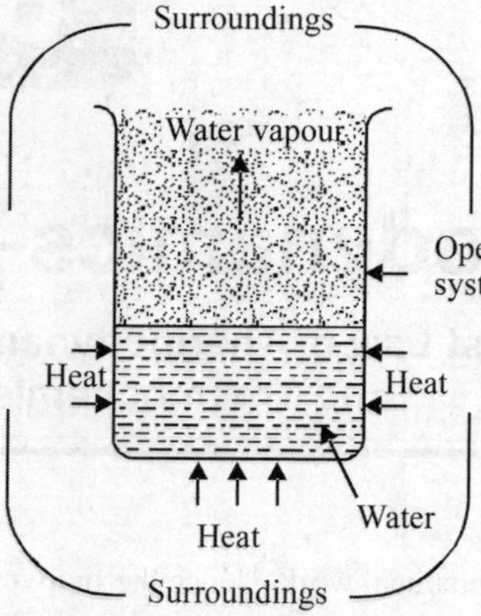

Fig. 8.1. Example of open system: Water kept in an open vessel.

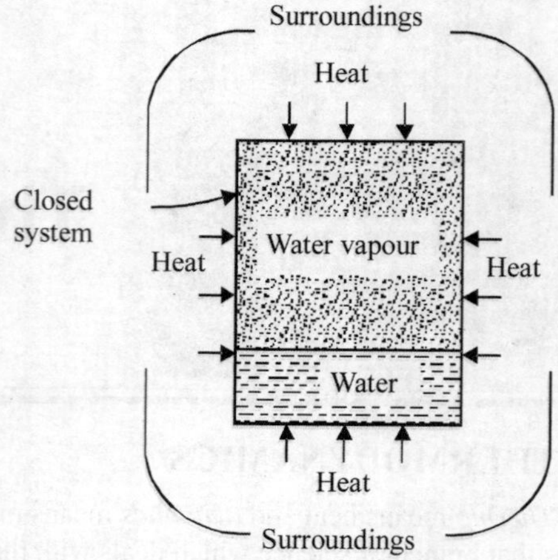

Fig. 8.2. Water placed in an closed vessel.

(*b*) **Closed system:** *If a system can exchange only energy with the surroundings but not matter, it is called a closed system.* For example, if some water is placed in a closed metallic vessel (Fig. 8.2), then as the vessel is closed, no exchange of matter between the system and the surroundings can take place. But, as the vessel has conducting walls, exchange of energy can take place between the system and the surroundings.

(*c*) **Isolated system:** *If a system can exchange neither matter not energy with the surroundings, it is called an isolated system.* For example, if water is placed in a vessel which is closed as well as insulated, no exchange of matter or energy is possible between the system and the surroundings. This constitutes an isolated Fig. 8.3.

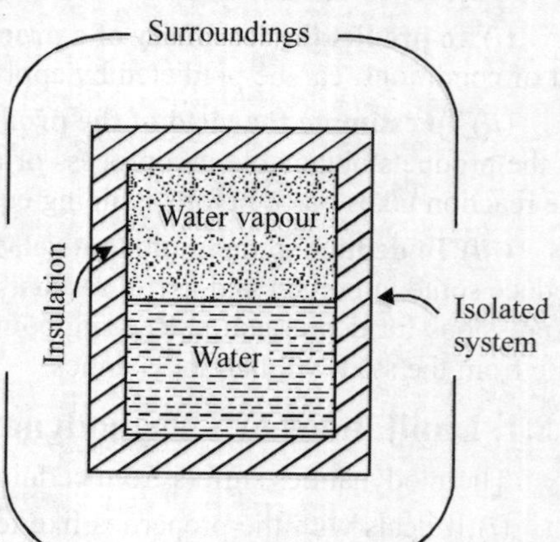

Fig. 8.3. An example of an isolated system.

3. Homogeneous system: *A system is said to be homogeneous if it is uniform throughout.* For example, a gas or a mixture of gases or a pure liquid or a pure solid or a solution of a solid in a liquid are all homogeneous systems. Such a system has one phase only.

4. Heterogeneous system: *If a system is not uniform throughout, it is said to be heterogeneous.* It then consists of two or more phases which are separated from each other by definite bounding surfaces. A few common examples of the heterogeneous systems are (*a*) a system consisting of a liquid and its vapour, (*b*) two or more immiscible liquids, (*c*) a mixture of two or more solids.

5. Macroscopic system: *Macroscopic system means a system containing a large amount of the substance.* In other words, a macroscopic system is one in which there are large number of particles *i.e.* atoms, ions or molecules.

6. Macroscopic property: *A property associated with the collective behaviour of particles in a macroscopic system is called a macroscopic property.* A few examples of these properties are pressure, volume, temperature, surface tension, viscosity, density, refractive index etc.

7. State of the system and state variables

State of system: The state of a system means a condition of the system when the various macroscopic properties like pressure, volume, temperature etc. of the system have definite values. The first and the last state of the system are called *initial state* and *final state* respectively.

State variables: Macroscopic properties are known as state variables. The four most common macroscopic properties which are sufficient to define the state of a system are *composition, pressure, volume* and *temperature.* If these properties are fixed, all other physical properties of the system are automatically fixed.

8. Thermodynamic equilibrium: A system is said to be in thermodynamic equilibrium if the macroscopic properties of the system in various phases do not undergo any change with time.

In fact, the thermodynamic equilibrium denotes that three different types of equilibria which must exist simultaneously. These are :

(*i*) **Thermal equilibrium:** This equilibrium demands that there should be no flow of *heat* from one portion of the system to another. This can happen if the temperature of the system remains constant.

(*ii*) **Mechanical equilibrium:** This equilibrium implies that no work should be done by one part of the system over another *i.e.* no macroscopic movement of matter should occur within the system or of the system with respect to its surroundings. This is possible if the pressure of the system remains constant.

(*iii*) **Chemical equilibrium:** This equilibrium demands that no change in composition should take place in any part of the system with passage of time.

9. Extensive and intensive properties

(*i*) **Extensive properties:** These are the properties which depend upon the quantity of the matter contained in the system. Examples of extensive properties are *mass, volume, energy, heat capacity* etc. The total value of an extensive property is equal to the sum of the values for the separate parts into which the system may be divided for the sake of convenience.

(*ii*) **Intensive properties:** These are those properties which depend only upon the nature of the substance and are independent of the amount of the substance present in the system. The common examples of these properties are *temperature, pressure, refractive index, viscosity, density, surface tension, specific heat* etc. It is because pressure and temperature are intensive properties independent of the quantity of the matter present that they are frequently used as variables to describe the state of a system.

Interestingly an extensive property may become intensive property by specifying unit amount of the substance concerned. Thus mass and volume are extensive properties but density is intensive property, heat capacity is an extensive property but specific heat is intensive.

10. Thermodynamic processes

(*i*) **Isothermal process:** *When a process is carried out in such a manner that the temperature remains constant throughout the process, it is called an isothermal process.* Obviously, when such a process occurs, heat can flow from the system to the surroundings and vice versa in order to keep the temperature of the system constant.

(*ii*) **Adiabatic process:** *When a process is carried out in such a manner that no heat can flow from the system to the surroundings* or *vice versa, i.e., the system is completely insulated from the surroundings, it is called an adiabatic process.*

(*iii*) **Isochoric process:** *It is a process during which the volume of the system is kept constant.*

(*iv*) **Isobaric process:** *It is a process during which the pressure of the system is kept constant.*

11. Reversible and irreversible processes

A reversible process is a process which is carried out infinitesimally slowly so that all changes occurring in the direct process can be exactly reversed and the system remains almost in a state of equilibrium at all times. In other words, a reversible process may be defined as a process which is conducted in such a manner that at every stage, driving force is only infinitesimally greater than the opposing force and which can be reversed by increasing the opposing force by an infinitesimal amount.

Strictly speaking, a reversible process is almost impossible as it would require an infinite time for completion. A process can be made very nearly reversible in some cases. For example, if an opposing E.M.F. is applied to a cell, then no current flows if the opposing E.M.F is exactly equal to that of the cell. If the opposing E.M.F. is very slightly (infinitesimally) less than that of the cell, a very small current is given out by the cell. If the opposing E.M.F. is slightly greater than that of the cell, a very small current flows in the opposite direction.

A reversible process can also be demonstrated with the help of the following experiment as shown in Fig. 8.4.

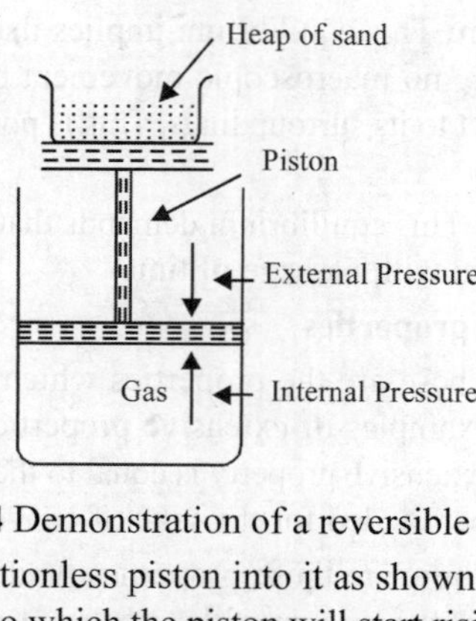

Fig. 8.4 Demonstration of a reversible process

Take a cylinder and fit a frictionless piston into it as shown in Fig. 8.4. The cylinder contains a gas at a moderate pressure, due to which the piston will start rising upwards. Balance it by placing a petri dish containing fine sand over the piston. In the balanced position, the piston will be static.

Now if we add a very small sand grain to the petri dish, the piston will start moving ownwards very slowly. On the other hand, if we take out a very small sand grain from the petri dish, the piston will start moving upwards very slowly. In both cases, it demonstrates reversible process.

Irreversible process. A process which does not meet the above requirements is called an **irreversible process.** In other words, an irreversible process is defined as that process which is not carried out infinitesimally slowly (instead, it is carried out rapidly) so that the successive steps of the direct process cannot be retraced and any change in the external conditions disturbs the equilibrium.

It is interesting to note that all naturally occurring processes like the flow of water down a hill or the flow of heat from a hot end to a cold end of an iron bar etc. are thermodynamically irreversible.

8.3. INTERNAL ENERGY AND CHANGE IN INTERNAL ENERGY

Internal energy: The evolution or absorption of energy in different processes clearly shows that every substance must be associated with some definite amount of energy, the actual value of which depends upon the nature of the substances (*i.e.* arrangement of atoms and electrons within the molecules) and the conditions of temperature, pressure, volume and composition. The energy associated with a substance is called its *internal energy* and is usually denoted by the symbol E or U.

Change of internal energy: It is not possible to find the absolute value of internal energy of a substance because it involves certain quantities (like translational, vibrational and rotational kinetic energies) which cannot be measured.

However, it is not required to know the absolute value of internal energy possessed by any substance (or a system). What is required in different processes is simply the *change of internal energy* when the reactants change into products or when system changes from initial state to the final state. This is easily measurable and is represented by ΔE (or ΔU).

Thus if the internal energy of a system in the initial state is E_1 (or U_1) and in the final state, it is E_2 (or U_2) then the change of internal energy (ΔE or ΔU) may be given by

$$\Delta E \text{ (or } \Delta U) = E_2 - E_1$$

Similarly, in a chemical reaction if E_R is the internal energy of the reactants and E_p is the internal energy of the products, then energy change accompanying the process would be

$$\Delta E = E_P - E_R \text{ or } \Delta U = U_P - U_R$$

Sign conventions in energy: Obviously, if $E_1 > E_2$ (or $E_R > E_P$), the extra energy possessed by the system in the initial state (or the reactants) would be given out, and ΔE will be negative according to the above equations. Similarly if $E_1 < E_2$ (or $E_R < E_P$), energy will be absorbed in the process and ΔE will be positive.

Hence ΔE (or ΔU) *is negative* if energy *is evolved*

and ΔE (or ΔU) *is positive* if energy *is absorbed*

Units of U. The unit of energy is erg or joule, 1 Joule = 10^7 ergs

8.4. EXPRESSION FOR WORK OF EXPANSION AGAINST CONSTANT PRESSURE

The expression for such a work may be derived as follows :

Consider a gas enclosed in a cylinder fitted with a frictionless piston. (Fig.8.5).

Area of cross-section of the cylinder = a sq cm

Pressure on the piston (which is slightly less than internal pressure of the gas so that the gas can expand) = P.

Distance through which gas expands = dl cm

As pressure is force per unit area, force (f) acting on the piston will be

$$f = P \times a$$

$\therefore$ Work done by the gas (*i.e.* the system)

$$= \text{Force} \times \text{Distance} = f \times dl$$

$$= P \times a \times dl$$

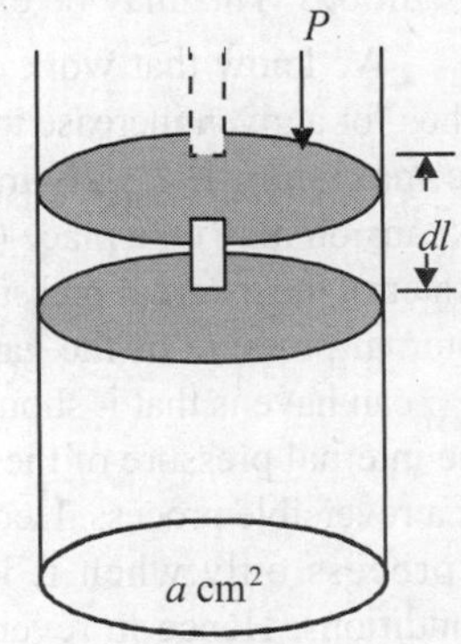

Fig. 8.5. Expansion of a gas.

But $a \times dl = dV$ is a small increase in the volume of the gas. Hence the small amount of work (δW) done by the gas can be written as

$$\delta W = P\, dV$$

If the gas expands from initial volume V_1 to the final volume V_2, then the total work done (w) will be given by

$$w = \int_{V_2}^{V_1} P\, dV$$

If the external pressure P against which the gas expands remains almost constant throughout the process, the above result may be written as

$$w = P\int_{V_2}^{V_1} dV = P(V_2 - V_1)$$

$$= P.\Delta V$$

where $\Delta V = (V_2 - V_1)$ is the total change in the volume of the gas.

If the external pressure (P) is slightly greater more than the pressure of the gas, the gas will contract *i.e.* the work will be done by the surroundings on the system. However, the same formulae will apply for the work done.

It may be mentioned here that P is the *external* pressure and hence is sometimes written as P_{ext} so that

$$w = P_{ext} \times \Delta V.$$

Note: According to the latest SI conventions of sign, w is taken as positive if work is done on the system (*i.e.* $w_{contraction}$ is positive) and it is taken as negative if work is done by the system (*i.e.* $w_{expansion}$ is negative). Thus $w_{exp} = -P\Delta V$ and $w_{contraction} = + P\Delta V$.

Electrical work. The electrical work done is expressed as under:

Electrical work done = Electromotive force (emf) × Quantity of electricity

Gravitational work. Gravitational work done when an object of mass (m) moves against the force of gravitation (g) and the displacement of the object is h, is given by:

Work done = $m \times g \times h$

8.4.1. Work done in a reversible process is the maximum work obtainable

The work done by a system is the maximum work if the process is carried out under reversible conditions. This may be explained as follows :

We know that work done by the system = $P_{ext} \times \Delta V$. Thus for a given increase in volume (ΔV), the work done can be maximum if P_{ext} is maximum. But in order that the expansion may take place (*i.e.* the work may be done by the system), the external pressure should always be less than the internal pressure of the gas. Thus the maximum value that P_{ext} can have is that it should be *infinitesimally smaller* than the internal pressure of the gas. But these are the conditions of a reversible process. Hence the maximum work is done in a process only when it is carried out under reversible conditions. Hence in reversible processes, w is sometimes replaced by w_{max}.

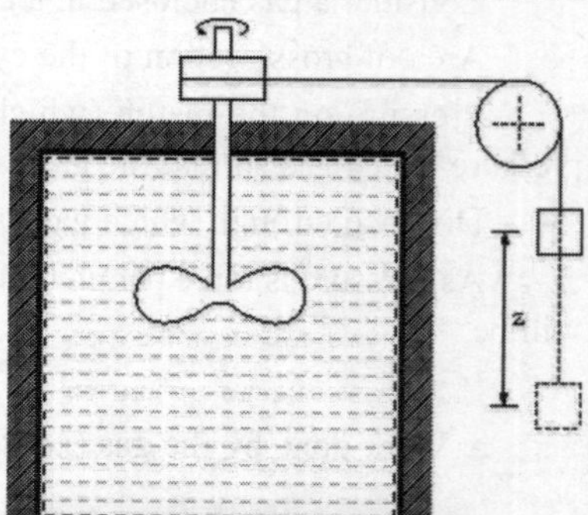

Joule's apparatus

8.5. HEAT

Heat is a mode of energy exchanged between the system and the surroundings as a result of the difference of temperature between them. It is usually represented by the letter q.

Sign of 'q'. When heat is given by the system to the surroundings, it is **given a negative** sign. When heat is **absorbed by the system** from the surroundings, it is given a **positive** sign.

Units of 'q'. Heat is usually measured in terms of calories. *A calorie is defined as the quantity of heat required to raise the temperature of one gram of water through 1°C.*

In the S.I. units, heat is expressed in terms of **joules**. The two units are related to each other as under:

1 calorie = 4.184 joules

8.5.1. Zeroth law of thermodynamics

The zeroth law of thermodynamics states: *When two bodies have equality of temperature with a third body, they in turn have equality of temperature with each other.*

We are so familiar with this fact that it does not need any further experiments for its support. However, since this fact is not derivable from other laws and since it logically prececes the first and the second law of thermodynamics, it has been called as the zeroth law of thermodynamics.

Heat flows in the direction of decreasing temperature....

8.6. FIRST LAW OF THERMODYNAMICS

First law of thermodynamics can be stated in a number of ways. Some statements of the law are:

(*i*) Energy can neither be created nor destroyed, although it may be converted from one form to another.

(*ii*) The total energy of an isolated system remains constant, although it may undergo transformation from one form to the other.

(*iii*) Whenever certain quantity of some form of energy disappears, an exactly equivalent amount of some other form of energy must be produced.

(*iv*) It is impossible to construct a perpetual motion machine *i.e.* a machine which would produce work continuously without consuming energy.

Einstein in 1905 showed that the energy could be created by the destruction of mass. The energy produced (E) by the destruction of mass m is given by the equation

$$E = mc^2 \text{ (called Einstein equation)}$$

where c is the velocity of light. In the light of this idea, the law of conservation of energy has been modified and now it is called *law of conservation of mass and energy.* It may be stated as follows:

The total mass the energy of an isolated system remains constant in chemical thermodynamics.

8.6.1. Mathematical formulation of the first law of thermodynamics

If heat q is supplied to a system, it may be used up partly to increase the internal energy of the system and partly to do some mechanical work. If ΔU is the increase in internal energy of the system and w is the work done by the system, then we have

$$\Delta U = q + w \quad \text{... } (i)$$

If the work done is work of expansion,

$$w = -P\Delta V$$

or

$$q = \Delta U + P\Delta V$$

If the changes are very small, it may be written as

$$\delta q = dU + PdV \quad \text{... } (ii)$$

8.6.2. Internal energy is a state function but work and heat are not

Internal energy: The internal energy is a state function, that is, when a process occurs, the change in its value depends only upon its value in the initial state and that in the final state and is independent of the path by which the change is brought about. For example, consider a system having a particular pressure, volume and temperature represented by the point A in Fig. 8.6. Suppose the pressure, volume and temperature are changed in such a way that the system is brought to the point B by path I. If the system is returned to the point A by path II, then the energy change involved in path I must be equal in magnitude to that involved in path II. If the energy changes involved in paths I and II are not the same, then suppose that the increase in internal energy by path I is greater than the decrease in internal energy along path II. Thus by carrying out the process $A \rightarrow B$ by path I and the reverse process $B \rightarrow A$ by path II, though the original state has been restored, some energy would have been created. This is contrary to the first law of thermodynamics. Thus it may be inferred that the energy change involved in the process $A \rightarrow B$ must be equal in magnitude to that involved in the reverse process $B \rightarrow A$ by path II. In other words, *the energy change accompanying a process is independent of the path followed and depends only upon the initial and the final states.* Thus if U_A is the internal energy of the system in the initial state and U_B that in the final state, then the change in internal energy accompanying the process would be given by

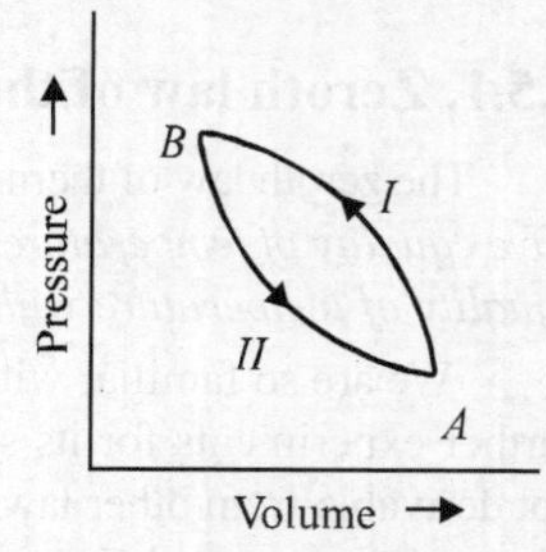

Fig. 8.6 Energy changes in direct and reverse paths.

$$\Delta U = U_B - U_A$$

Work and heat: Unlike internal energy, the work done during a process does not depend merely upon the initial and the final states of the system but it depends upon the path followed. Hence work is not a state function but is a path function. This is explained as under :

Suppose we want to change the system from state A to state B. This may be done by following different paths as shown in the different figures below (Fig. 8.7). Confining ourselves to the work of expansion only, we know that the work done for a small change in volume dV is PdV where P is the external pressure. Thus the total work done is the sum of PdV terms and this is equivalent to the area under the curve in the P–V diagrams. From the figures given below it is obvious that the areas under the curves are different and hence the work done is different when different paths are followed.

That heat is also not a state function follows directly from the mathematical formulation of the first law of thermodynamics *viz.*,

$$q = \Delta U - w$$

Thus the heat q is the sum of two quantities *i.e.* the internal energy change, ΔU and the work done, w. Now ΔU is a state function whereas w is not. Now since w depends upon the path followed, therefore, q will also depend upon the path followed. In other words, q is also not a state function but is a path function.

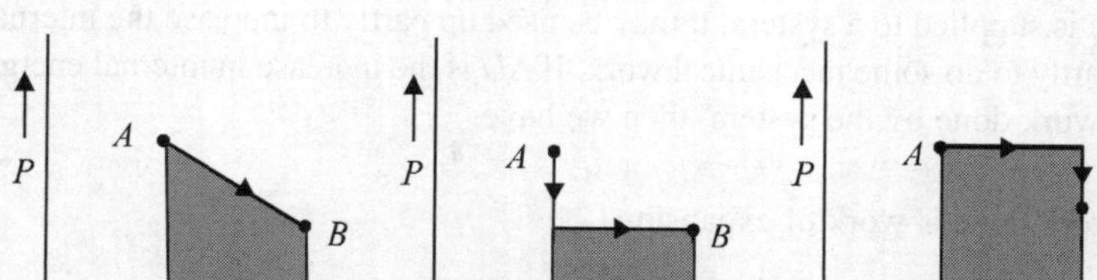

Fig 8.7 Different paths followed in going from state A to state B.

8.7. ENTHALPY OR HEAT CONTENT

If a process is carried out at constant pressure, the work of expansion is given by

$$w = P\Delta V \qquad \text{... }(i)$$

where ΔV is the increase in volume and P is the constant pressure.

According to first law of thermodynamics, we know that

$$q = \Delta U - w \quad \text{... (ii)}$$

where q is the heat absorbed by the system, ΔU is the increase in internal energy of the system and w is the work done by the system.

Under condition of constant pressure, putting $w = -P\Delta V$ and representing the heat absorbed by q_p, we get

$$q_p = \Delta U + P\Delta V \quad \text{... (iii)}$$

Let us suppose when the system absorbs q_p calories of heat, its internal energy increases from E_1 to E_2 and the volume increases from V_1 to V_2. Then we have

$$\Delta U = U_2 - U_1 \quad \text{... (iv)}$$

and $$\Delta V = V_2 - V_1 \quad \text{... (v)}$$

Putting these values in equation (*iii*) above, we get

$$q_p = (U_2 - U_1) + P(V_2 - V_1)$$

$$q_p = (U_2 + PV_2) - (U_1 + PV_1) \quad \text{... (vi)}$$

Now as U, P and V are the functions of state, therefore, the quantity $U + PV$ must also be a state function. The thermodynamic quantity $U + PV$ is called the **heat content or enthalpy** of the system and is represented by the symbol H *i.e.* the enthalpy may be defined mathematically by the equation

$$H = U + PV$$

Thus if H_2 is the enthalpy of the system in the final state and H_1 is the value in the initial state, then

$$H_2 = U_2 + P\,V_2$$

and $$H_1 = U_1 + PV_1$$

Putting these values in equation (*vi*), we get

$$q_p = H_2 - H_1$$

or $$q_p = \Delta H \quad \text{... (vii)}$$

where $\Delta H = H_2 - H_1$ is the enthalpy change of the system. Hence **enthalpy change** *of a system is equal to the heat absorbed by the system at constant pressure.*

Further, putting the value of q_p, from equation (*vii*) in equation (*iii*) we get

$$\Delta H = \Delta U + P\Delta V \quad \text{... (viii)}$$

Hence *the enthalpy change accompanying a process may also be defined as the sum of the increase in internal energy of the system and the pressure-volume work done i.e., the work of expansion.*

$P\Delta V$ can be replaced by ΔnRT, so that eq. (*viii*) can be written as

$$\Delta H = \Delta U + \Delta nRT$$

SOLVED EXAMPLE ON HEAT OF FORMATION

Example. The heat of formation of methane at 27°C is –19.3 kcal when the measurements are made at constant pressure. What will be the heat of formation at constant volume?

Solution. Apply the relationship

$$\Delta H = \Delta U + \Delta n \,.\, RT \quad \text{... (i)}$$

ΔH is the heat of reaction at constant pressure whereas ΔU is the heat of reaction at constant volume.

For the reaction

$$\Delta H = -19.3 \text{ kcal}$$

$$C(s) + 2H_2(g) \longrightarrow CH_4(g)$$

No. of moles of the gaseous reactants = 2

No. of moles of the gaseous products = 1

Thus $\Delta n = 1-2 = -1$

Substituting the values in eq. (*i*) above

$$-19.3 \times 1000 \text{ cal} = \Delta U + (-1) \times (1.987 \text{ cal}) \times 300$$

or $\Delta U = -19300 - 5961$ cal

or $\Delta U = -13339$ cal or -13.339 kcal.

8.8. HEAT CAPACITY

Heat capacity of a system is defined as the amount of heat required to raise the temperature of the system through 1° C.

If q is the amount of heat supplied to a system to raise the temperature from T_1 to T_2, then the heat capacity of the system is given by

$$C = \frac{q}{T_2 - T_1} = \frac{q}{\Delta T} \quad \text{... } (i)$$

The value of C has to be considered in a narrow range because the heat capacity varies with the temperature.

If δq is the small amount of heat absorbed by a system which raises the temperature of the system by a small amount dT, then the heat capacity of the system is given by

$$C = \frac{\delta q}{\delta T} \quad \text{... } (ii)$$

There are two types of heat capacities, *viz.*,

(*i*) Heat capacity at constant volume (C_v)

(*ii*) Heat capcity at constant pressure (C_p)

(*i*) Heat capacity at constant volume (C_v)

According to first law of Thermodynamics

$$\delta q = dU + PdV \quad \text{... } (iii)$$

Substituting for δq from eq. (*iii*) into eq. (*ii*)

$$C = \frac{dU + PdV}{dT} \quad \text{... } (iv)$$

When the volume is kept $dV = 0$ and the eq. (*iv*) becomes

$$C_v = \left(\frac{\partial U}{\partial T}\right)_v \quad \text{...} (v)$$

or *for an ideal gas,* this equation may simply be written as

$$C_v = \frac{dU}{dT} \quad \text{... } (vi)$$

C_v is the heat capacity at constant volume.

Thus *the heat capacity at constant volume may be defined as the rate of change of internal energy with temperature at constant volume.*

(*iii*) Heat capacity at constant pressure, C_p

When the pressure is kept constant during the absorption of heat, equation (*iv*) becomes

$$C_p = \left(\frac{\partial U}{\delta T}\right)_p + P\left(\frac{\partial V}{\partial T}\right)_p \quad \text{... } (vii)$$

We know that the heat content or enthalpy of a system is given by

$$H = E + PV$$

Differentiating w.r.t. T at constant P, we get

$$\left(\frac{\partial H}{\partial T}\right)_p = \left(\frac{\partial U}{\partial T}\right)_p + P\left(\frac{\partial V}{\partial T}\right)_p \quad \text{...} (viii)$$

Combining equations (*vii*) and (*viii*), we get

$$C_p = \left(\frac{\partial H}{\partial T}\right)_p \quad \text{... } (ix)$$

or for any ideal gas, this equation may simply be put in the form

$$C_p = \frac{dH}{dT} \quad \text{.. } (x)$$

C_p is the heat capacity at constant pressure.

Thus *the heat capacity at constant pressure may be defined as the rate of change of enthalpy with temperature at constant pressure.*

8.8.1. Relation between C_p and C_v

If the volume of the system is kept constant when the heat is supplied to a system, then no work is done by the system. Thus the heat absorbed by the system is used up completely to increase the internal energy of the system. And if the pressure of the system is kept constant when the heat is supplied to the system, then some work of expansion is also done by the system in addition to the increase in internal energy. Thus if at constant pressure, the temperature of the system is to be raised through the same value as at constant volume, then some extra heat is required for doing the work of expansion. Thus $C_p > C_v$.

We have the following equations for C_v and C_p.

$$C_v = \frac{dU}{dT} \quad \text{... } (i)$$

$$C_p = \frac{dH}{dT} \quad \text{... } (ii)$$

The difference between the heat capacities of an ideal gas can be obtained by subtracting equation (*i*) from equation (*ii*). So we have

$$C_p - C_v = \frac{dH}{dT} - \frac{dU}{dT} \quad \text{... } (iii)$$

But $H = U + PV$ (by definition)

and $PV = RT$ (for 1 mole of an ideal gas)

$\therefore$ $H = U + RT.$

Differentiating this equation w.r.t T, we get

$$\frac{dH}{dT} = \frac{dU}{dT} + R \quad \text{... } (iv)$$

or

$$\frac{dH}{dT} - \frac{dU}{dT} = R \quad \text{... } (v)$$

Combining equation (*iii*) and (*v*),

$$\mathbf{C_p - C_v = R} \quad \text{(for 1 mole of an ideal gas)}$$

Thus C_p is greater then C_v by the gas constant R, *i.e.*, approximately 2 calories or 8.314 joules.

The ratio of specific heats (C_p/C_v) is represented by γ

Thus $$\gamma = \frac{C_p}{C_v}$$

It is found that the value of γ comes out to be about 1.67 for monoatomic gases, 1.40 for diatomic gases and 1.30 for triatomic gases.

8.8.2. Joule's Law

The law can be stated as: *The change of internal energy of an ideal gas with volume at constant temperature is zero.*

Mathematically, $$\left(\frac{\partial U}{\partial V}\right)_T = 0$$

This result has been obtained from the Joule Thomson experiment as explained hereunder.

A combination of bulbs A and B connected through a stop cock S is immersed in water taken in a trough (Fig. 8.8 *a*). Bulb A contains as ideal gas at certain pressure with the stop cock closed. There is complete vaccum in the flask B. The bulbs reach a thermal equilibrium after some time. The temperature of water is noted.

The stop cock S is now opened and the gas is allowed to move to bulb B. The bulbs reach a thermal equilbrium after some time. The temperture of water is again noted.

We find that there is no change in temperature. On opening the stop cock, the gas from the bulb A expands into the evacuated bulb B *i.e* the opposing force $P_{\text{opposing}} = 0$. Thus work of expansion

$$dW = -P_{\text{opposing}}\, dV = 0$$

From the first law of thermodynamics, $dv = dq + dw$

As $dW = 0$, we have $dV = dq$

As the temperature of water has not changed, $dq = 0$. Hence $dU = 0$. The mathematical form of Joule's law can be arrived at as follows:

Taking U as a function of volume and temperature

$$U = f(V, T)$$

Differentiating

$$dU = \left(\frac{\partial U}{\partial T}\right)_V dT + \left(\frac{\partial U}{\partial V}\right)_T dV \quad \text{... } (i)$$

We have found earlier that in such a situation $dU = 0$ and $dT = 0$.

Substituting for dU and dT in eq. (i) above, we have

$$\left(\frac{\partial U}{\partial V}\right)_T dV = 0 \qquad \text{but } dV \neq 0$$

$$\therefore \quad \left(\frac{\partial U}{\partial V}\right)_T = 0$$

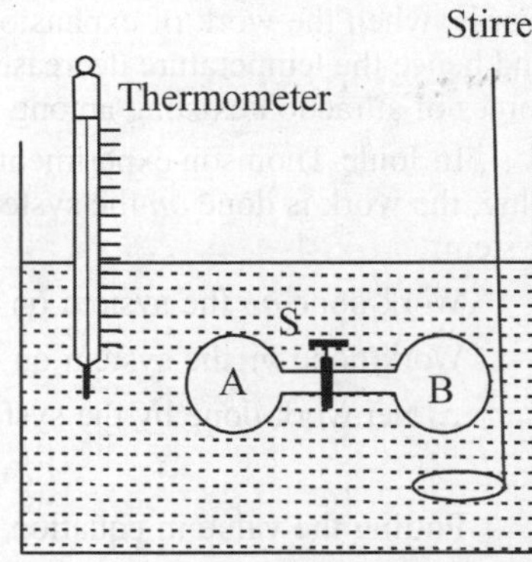

Fig. 8.8a. Joule Thomson effect

James P. Joule

8.8.3. Joule-Thomson effect

It was observed by Joule and Thomson that *when a real gas at a certain pressure expands adiabatically through a porous plug or a fine hole into a region of low pressure, it is accompanied by cooling (except for hydrogen and helium which get warmed up)*. This phenomenon is known as **Joule-Thomson effect.**

Fig. 8.8*b*. shows the experimental set-up of Joule and Thomson experiment.

A tube thoroughly insulated is fitted with a porous plug in the middle and pistons on either side as shown in the figure. The pressures of the gas are kept different on the two sides of the plug, being P_1 on the left side and P_2 on the right side where $P_1 > P_2$. By applying pressure on the piston on the left slowly enough (so as not to change the pressure P_1), a volume of the gas equal to V_1 is forced slowly through the porous plug and then allowed to expand to the pressure P_2 and volume V_2 by moving the piston on the right outward. When the process is being carried out as above, the temperature on the two sides of the porous plug is recorded accurately. The experiment is repeated with a different gases. It is observed that a fall in temperature of the gas takes place in case of all gases except hydrogen and helium when the experiments are carried out at ordinary temperature *i.e.* at room temperature.

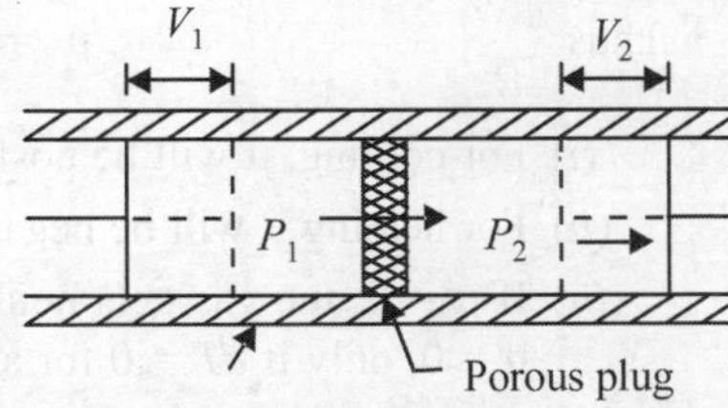

Fig. 8.8b Joule-Thomson experiment (Expansion of real gas through a porous plug).

As the process is carried out adiabatically,

$$q = 0$$

$\therefore$ By first law of thermodynamics *viz.* $\Delta U = q + w$

$$\Delta U = w$$

or $\quad w = \Delta U \quad$ ... (i)

It can be said that the work done, during the expansion of the gas under adiabatic conditions, is at the cost of the internal energy. In other

Sir William Thomson

words, when the work of expansion is done adiabatically, the internal energy of the gas decreases and hence the temperature decreases. In fact, a part of the internal energy is used up to overcome the forces of attraction existing among the molecules (van der Waals' forces).

In Joule-Thomson experiment, which is an example of adiabatic process, on the left side of the plug, the work is done *on* the system whereas on the right side of the plug, the work is done by the system.

Work done *on* the system on the left side = $P_1 V_1$

Work done *by* the system on the right side = $-P_2 V_2$

$\therefore$ Net work done by the system is

$$w = P_1 V_1 - P_2 V_2 \qquad \text{... } (ii)$$

Putting the value in equation (*i*), we get

$$\Delta U = P_1 V_1 - P_2 V_2$$

Further putting $\quad \Delta U = U_2 - U_1$

we get $\quad U_2 - U_1 = P_1 V_1 - P_2 V_2$

or $\quad U_2 + P_2 V_2 = U_1 + P_1 V_1$

or $\quad H_2 = H_1$

i.e. $\quad \Delta H = H_2 - H_1 = 0$

Thus it leads to the conclusion that *when the expansion of a gas takes place adiabatically through a porous plug of a fine hole, the enthalpy of the system remains constant.*

8.8.4. Joule-Thomson coefficient

Joule-Thomson coefficient μ may be defined as the temperature change in degrees produced by a drop of one atmospheric pressure when the gas expands under conditions of constant enthalpy. It is expressed as :

Thus
$$\mu = \left(\frac{\partial T}{\partial P}\right)_H$$

(*i*) For cooling, μ will be positive (because dT and dP both will be negative).

(*ii*) For heating μ will be negative (because dT is positive while dP is negative).

(*iii*) If $\mu = 0$, the gas gets neither heated up not cooled on adiabatic expansion (because $\mu = 0$, only if $dT = 0$ for any value *of* dP).

8.8.5. Inversion temperature

Every gas has a definite temperature (at a particular pressure) at which $\mu = 0$. Below this temperature μ is positive and above this temperature μ is negative. *This temperature (at a particular pressure) at which μ = 0 i.e. the gas neither cooled down nor heated up on adiabatic expansion is called the* **inversion temperature.**

8.8.6. Helium and hydrogen show heating on adiabatic expansion in Joule-Thomson experiment

The inversion temperatures of most of the gases are very high (much higher than the room temperature). That is why these gases undergo cooling on adiabatic expansion under ordinary conditions of temperature. On the other hand, the inversion temperatures of hydrogen and helium are very low (being much below 0°C). Hence these gases undergo heating on adiabatic expansion under ordinary conditions of temperature. But, if these gases are first cooled to temperatures below their inversion temperatures and then allowed to expand adiabatically, these gases also show cooling effect.

8.9. RELATIONSHIP OF μ WITH OTHER THERMODYNAMIC QUANTITIES

Enthalpy H is a state function. Taking temperature T and pressure P as the independent variables.

$$H = f(T, P) \qquad \text{... } (i)$$

$$\therefore \quad dH = \left(\frac{\partial H}{\partial P}\right)_T dP + \left(\frac{\partial H}{\partial T}\right) dT \qquad \text{... } (ii)$$

In Joule-Thomson effect, the enthalpy of the system remains constant *i.e.*,

$$dH = 0$$

Putting this value in the above equation, we get

$$\left(\frac{\partial H}{\partial P}\right)_T dP + \left(\frac{\partial H}{\partial T}\right)_p dT = 0 \qquad \text{... } (iii)$$

Dividing this equation throughout by dP and imposing the condition of constant enthalpy, we get

$$\left(\frac{\partial H}{\partial P}\right)_T + \left(\frac{\partial H}{\partial T}\right)_P \left(\frac{\partial T}{\partial P}\right)_H = 0 \qquad \text{... } (iv)$$

But $\left(\frac{\partial H}{\partial T}\right)_P = C_p$, heat capacity at constant pressure,

and $\left(\frac{\partial T}{\partial P}\right)_H = \mu$, Joule-Thomson coefficient ... (v)

Putting these values in equation (iv), we get

$$\left(\frac{\partial H}{\partial P}\right)_T - \mu C_p = 0$$

or

$$\mu = -\frac{1}{C_P}\left(\frac{\partial H}{\partial P}\right)_T \qquad \text{...}(vi)$$

Thus μ can be calculated from $\left(\frac{\partial H}{\partial P}\right)_T$ and *vice versa.*

SOLVED EXAMPLE ON JOULE-THOMSON COEFFICIENT

Example . Explain why an ideal gas shows neither heating nor cooling in Joule-Thomson experiment ?

OR

Why is Joule-Thomson coefficient of an ideal gas zero ?

Solution. An ideal gas, undergoes neither cooling nor heating on adiabatic expansion in Joule-Thomson experiment. This is expected as explained below:

The intermolecular forces of attraction (*i.e.* van der Waals forces) in an ideal gas are negligible. Hence no energy is used up in overcoming these forces of attraction when the gas expands adiabatically. Thus the internal energy of the gas does not fall and therefore the temperature also does not fall.

8.10. EXPRESSION FOR WORK DONE IN ISOTHERMAL REVERSIBLE EXPANSION OF A GAS

A general expression for the work of expression in an isothermal reversible process may be derived as follows :

Let P be the pressure of the gas within a system undergoing an isothermal expansion by a reversible process. Then in order that the expansion may take place very slowly (*i.e.* reversibly) the external pressure should be $P - dP$, where dP is very small quantity. The small amount of work (δw) done by the system when its volume increases by an infinitesimal amount, dV, is equal to the product of the external pressure and the volume change *i.e.*,

$$\delta w = -(P - dP)\, dV \quad \text{... (i)}$$

Neglecting the term $dP \times dV$ which is very small (being the product of two very small quantities), equation (*i*) becomes

$$\delta w = -P.\, dV \quad \text{... (ii)}$$

Hence the total work *(w)* done by the system when its volume increases from V_1 to V_2 will be given by

$$w = -\int_{V_1}^{V_2} PdV \quad \text{... (iii)}$$

Now, for n moles of an ideal gas, the equation of state is

$$PV = nRT$$

so that

$$P = \frac{nRT}{V} \quad \text{... (iv)}$$

Substituting this value is equation (*iii*), we get

$$w = -\int_{V_1}^{V_2} \frac{nRT}{V} dV \quad \text{... (v)}$$

For an isothermal process, T is constant. Also n and R are constant quantities. Hence equation (*v*) may be written as

$$w = -nRT \int_{V_1}^{V_2} \frac{dV}{V}$$

$$= -nRT\, ln \frac{V_2}{V_1} \quad \text{...(vi)}$$

or

$$w = -2{\cdot}303\, nRT \log \frac{V_2}{V_1} \quad \text{... (vii)}$$

A relation can also be obtained in terms of pressure.

As the process is isothermal *i.e.* the temperature remains constant, therefore by Boyle's law

$$P_1 V_1 = P_2 V_2 \quad \text{... (viii)}$$

where P_1, V_1 are the initial pressure and volume and P_2, V_2 are the final pressure and volume of the system.

Equation (*viii*) can be written as

$$\frac{V_2}{V_1} = \frac{P_1}{P_2}$$

Substituting this value in equation (*vii*), we get

$$w = -2{\cdot}303\ nRT \log \frac{P_1}{P_2} \qquad ...(ix)$$

Work done in a *reversible process* is the *maximum* work obtainable from a system. Hence w in equations (*vii*) and (*ix*) may be replaced by w_{max}. Expressions (*vii*) and (*ix*) may thus be written as

$$w_{max} = -2{\cdot}303\ nRT \log \frac{V_2}{V_1}$$

$$= -2{\cdot}303\ nRT \log \frac{P_1}{P_2}$$

Internal energy of a system is a function of temperature. As the temperature remains constant in an isothermal process, internal energy does not change or $\Delta U = 0$.

Expression for q (heat absorbed): According to first law of thermodynamics

$$q = \Delta U - w$$

But $\Delta U = 0$ (*as explained above*)

$\therefore$ $q = -w$

Thus all the heat absorbed is used up completely in doing the work of expression only. Hence the expressions for the heat absorbed in an isothermal reversible expansion of an *ideal gas* are same as for the work done *i.e.*

$$q = -w = 2{\cdot}303\ nRT \log \frac{V_2}{V_1} = 2{\cdot}303\ nRT \log \frac{P_1}{P_2}$$

8.10.1. Values for ΔU: and ΔH in an isothermal expansion process

Expression for ΔU : Since the internal energy of an *ideal gas* is function of temperature only, therefore, in an isothermal process as the temperature remains constant, the internal energy of the system (containing an ideal gas) remains constant

i.e $\Delta U = 0$

Expression for ΔH

$$\Delta H = \Delta (U + PV)$$

$$= \Delta U + \Delta(PV)$$

$$= \Delta U + \Delta(nRT)$$

$$= \Delta U + \Delta nR\ \Delta T$$

$$= 0 + 0 \quad \begin{cases} \Delta U = 0 \text{ as explained above and} \\ \Delta T = 0 \text{ as the process is isothermal} \end{cases}$$

i.e. $\Delta H = 0$

8.10.2. Values of q, w, ΔU and ΔH in case of isothermal irreversible expansion when

(*i*) opposing pressure is zero (*ii*) opposing pressure is constant and less than P_2 but greater than zero.

The expressions for w, q, ΔU and ΔH for the above expansion may be obtained as follows :

(*i*) **For free expansion:** Here $P_{opp} = 0$

$$\therefore \quad w = -\int P_{opp}\, dV = 0$$

Hence

$$q = w = 0$$

$$\Delta U = \Delta H = 0$$

(as the gas is *ideal* and expansion is *isothermal*)

(*ii*) **For intermediate expansion:** Here P_{opp} = constant and lies in the range $0 < P_{opp} < P_2$. Hence

$$w = -\int_{V_1}^{V_2} P_{opp}\, dV = -P_{opp}(V_2 - V_1)$$

$$\therefore \quad q = -w = P_{opp}(V_2 - V_1)$$

$$\Delta U = \Delta H = 0$$ (as the gas is *ideal* and expansion is *isothermal*)

It may be noted that the work done in this process will be less than the work done in the reversible process.

8.11. SOLVED EXAMPLES ON ISOTHERMAL REVERSIBLE EXPANSION OF GAS

Example 1. Ten moles of an ideal gas at the initial pressure of one atmosphere at 0°C were expanded reversibly under isothermal conditions to a final pressure of 0·1 atmosphere. Calculate the work done by the gas, the change in internal energy and the heat absorbed by the system (R = 8·314 J K^{-1} deg^{-1} mol^{-1}).

Solution. Here we are given that

$$n = 10 \text{ moles}$$

$$P_1 = 1 \text{ atm}$$

$$P_2 = 0{\cdot}1 \text{ atm}$$

$$T = 0°\text{C} = 273 \text{ K}$$

$$R = 8.314 \text{ JK}^{-1}\text{ mol}^{-1}$$

$$\therefore \quad w_{max} = -2.303\, nRT \log \frac{P_1}{P_2}$$

$$= -2.303 \times 10 \times 8{\cdot}314 \times 273 \times \log \frac{1}{0.1}$$

$$= \mathbf{-52271.7\ J} \qquad (\because \log 10 = 1.0)$$

For an ideal gas during isothermal process

$$\Delta U = 0$$

$\therefore$ By first law of thermodynamics ($q = \Delta U - w$)

$$q = -w_{max} = 52271.7 \text{ J}$$

Example 2. One mole of an ideal gas is heated at constant pressure from 0°C to 100°C (*i*) Calculate the work involved (*ii*) If the gas were expanded isothermally and reversibly at 0°C from 1 atmosphere to some other pressure *P*, what must be the final pressure if the isothermal work is equal to the work involved in (*i*)?

Solution. Work involved when the gas is heated at constant pressure from 0°C to 100°C is obtained as follows :

(*i*)
$$
\begin{aligned}
w &= -P\Delta V = -P(V_2 - V_1)\\
&= -(PV_2 - PV_1)\\
&= -(nRT_2 - nRT_1)\\
&= -nR(T_2 - T_1)\\
&= 1\times 8.314\,(373 - 273)\\
&= \mathbf{831{\cdot}4\ J}
\end{aligned}
$$

(*ii*) Final pressure if the process is carried out isothermally involving the work as in step (*i*) is obtained as under

$$w = -2.303\, nRT \log \frac{P_1}{P_2}$$

Put $P_1 = 1$ atm, $T = 273$ *K*, $R = 8.314$ JK^{-1} and $n = 1$

This on solving gives

$$\mathbf{P = 0.6933\ atm}$$

Example 3. Calculate the pressure-volume work performed by the system during reversible isothermal expansion of two moles of an ideal gas from 2 litres to 10 litres at 20°C.

Solution. Here we are given that

$$
\begin{aligned}
n &= 2 \text{ moles}\\
V_1 &= 2 \text{ litres}\\
V_2 &= 10 \text{ litres}\\
T &= (20 + 273)\text{ K} = 293\text{ K}
\end{aligned}
$$

Taking $R = 8.314$ Joules degree^{-1} mol^{-1}

$$
\begin{aligned}
w_{max} &= -2{\cdot}303\, nRT \log \frac{V_2}{V_1}\\
&= -2.303 \times 2 \times 8.314 \times 293 \times \log \frac{10}{2}\\
&= -11220.2 \times \log 5\\
&= -112202 \times 0.6990\\
&= \mathbf{7842.9\ joules}
\end{aligned}
$$

PROBLEMS FOR PRACTICE

1. Find the work required to compress reversibly 1 mole of a perfect gas from 1 to 100 atmospheres at 27°C. **[Ans.** 11488 J]

2. Calculate the work done during isothermal expansion of one mole of an ideal gas from 10 atmoshpheres to 1 atmosphere at 27°C. **[Ans.** – 5.74 kJ]

3. Ten moles of an ideal gas are compressed isothermally and reversibly at 100°C from pressure of 1 atmosphere to a pressure of 10 atmosphere. Calculate the following quantities:

(*i*) *W* and (*ii*) ΔH **[Ans.** $W = 71419$ J, $\Delta H = 0$]

4. Calculate the work done by 10 litres of an ideal gas at 0°C and 10 atm in an isothermal reversible expansion to a final pressure of 1 atm (R = 8.314 joules).

[**Hint.** (*i*) Reduce the given volume to volume at S.T.P.]

$$\frac{P_1 V_1}{T_1} = \frac{P_2 V_2}{T_2}$$

$$\frac{10\times10}{273} = \frac{1\times V_2}{273}$$

or $$V_2 = 100 \text{ litres at S.T.P.}$$

(*ii*) Calculate the number of moles as follows:

22.4 litres at S.T.P. = 1 mole

100 litres at S.T.P = 4.46 moles

$$w = -2.303\, nRT \log \frac{P_1}{P_2}$$

or $$w = -23.3 \text{ kJ}]$$

5. Calculate the maximum work per mole when 5 moles of an ideal gas expand isothermally and reversibly from 20 litres to 100 litres at 27°C. [R = 8.314 × 10^7 ergs mol^{-1} deg^{-1}]

[**Ans.** – 200.76 × 10^5 ergs]

6. Estimate the work done in joules when the volume of 16.0 g of oxygen at 300 K changes isothermally from 5.0 to 25.0 litres [R = 8.3143 Jk^{-1} mol^{-1}] [**Ans.** – 2007.6 J]

7. One mole of an ideal gas is heated at constant pressure from 0°C to 100°C.

(*i*) Calculate the work involved (*ii*) If the gas were expanded isothermally and reversibly at 0°C from 1 atmosphere to some other pressure P, what must be the final pressure if the isothermal work is equal to the work involved in (*i*)

[**Hint.** (*i*) $w = -P\Delta V = -P(V_2 - V_1) = (PV_2 - PV_1) = -(nRT_2 - nRT_1)$
$= -nR\,(T_2 - T_1) = -1 \times 8.314\ (373 - 273) = -831.4$ J]

(*ii*) $w = 2.303 - nRT \log \frac{P_1}{P_2}$. Put P_1= 1 atm, T = 273 K, R = 8.314 JK^{-1} mol^{-1}, n = 1

On solving, we get P_2 = 0.6933 atm]

8. Calculate the value of w, ΔU and ΔH in the isothermal and reversible expansion of 64 grams of oxygen at 410 K from an initial volume of 20 litres to a final volume of 60 litres. Assume oxygen behaves ideally. [**Ans.** w = –7490.8J, $\Delta H = \Delta U$ = zero]

9. Estimate the maximum work which can be obtained by the isothermal expansion of 1 mole of an ideal gas from 2.24 litres to 22.4 litres at 273 K (R = 8.3143 JK^{-1} mol^{-1}).

[**Ans.** – 5227.4 J]

10. Calculate the work done when 5 moles of an ideal gas at 27°C expand from an initial pressure of 5 atm to 1 atm reversibly.

8.12. CALCULATION OF VALUES OF *w*, *q*, ΔU AND ΔH IN ADIABATIC EXPANSION OF AN IDEAL GAS

An adiabatic process is the one in which no heat enters or leaves the system of any stage. For every infinitesimal change of the process,

$$\delta q = 0$$

Putting this value in the equation of the first law *viz.*

$$dU = \delta q + \delta w, \text{ we get}$$

or $$dU = \delta w \quad \text{... } (i)$$

If the work involved is the work of expansion only, then if dV is the small increase in volume and P is the pressure of the gas, then

$$\delta w = -PdV \quad \text{... } (ii)$$

Putting this value in equation (*i*), we get

$$dU = -PdV \quad \text{... } (iii)$$

For an ideal gas, $$C_v = \frac{dE}{dT}$$

or $$dE = C_v dT$$

Putting this value in equation (*iii*) we get

$$C_v dT = -PdV \quad \text{... } (v)$$

Also
$$\begin{aligned} \Delta H &= \Delta (U + PV) \\ &= \Delta U + \Delta(PV) \\ &= \Delta U + \Delta(RT) \\ &= \Delta U + R\,\Delta T \\ &= C_v \Delta T + R\,\Delta T \\ &= (C_v + R)\,\Delta T \\ &= (C_v + R)\,(T_2 - T_1) \\ &= C_p\,(T_2 - T_1) \quad \text{... } (vi) \end{aligned}$$

In a nutshall,

For any type of adiabatic process (reversible or irreversible)

$$\begin{aligned} q &= 0 \\ w &= \Delta E \\ \Delta U &= C_v\,\Delta T = C_v\,(T_2 - T_1) \\ \Delta H &= C_p\,\Delta T = C_p\,(T_2 - T_1) \end{aligned}$$

8.12.1. Reversible adiabatic expansion

I. Relationship between temperature and volume in reversible adiabatic expansion

Consider 1 mol of the *ideal gas.* Then according to the gas equation

$$PV = RT$$

or $$P = \frac{RT}{V} \quad \text{... } (i)$$

Also $$C_v\,dT = -PdV \quad \text{... } (ii)$$

Substituting for P from eq. (*i*) in (*ii*)

$$C_v dT = -\frac{RT}{V}\, dV$$

or

$$C_v \frac{dT}{T} = -R\frac{dV}{V} \qquad \text{... }(iii)$$

Now suppose the volume of the gas changes from V_1 to V_2 when the temperature changes from T_1 to T_2. Then assuming C_v to be independent of temperature, and integrating equation (*iii*) between the limits T_1, T_2 and V_1, V_2, we get

$$C_v \int_{T_1}^{T_2} \frac{dT}{T} = -R\int_{V_1}^{V_2} \frac{dV}{V}$$

$$C_v [\ln T]_{T_1}^{T_2} = -R[\ln V]_{V_1}^{V_2}$$

$$C_v \ln\frac{T_2}{T_1} = -R\ln\frac{V_2}{V_1}$$

$$C_v \ln\frac{T_2}{T_1} = R\ln\frac{V_1}{V_2}$$

Now, for 1 mol of an ideal gas we know that

$$C_p - C_v = R$$

Putting this value of R in equation (*iv*), we get

$$C_v \ln\frac{T_2}{T_1} = (C_p - C_v)\ln\frac{V_1}{V_2}$$

Dividing throughout by C_v we get

$$\ln\frac{T_2}{T_1} = \left(\frac{C_p}{C_v} - 1\right)\ln\frac{V_1}{V_2}$$

Putting $\dfrac{C_p}{C_v} = \gamma$, the ratio of the two heat capacities, we get

$$\ln\frac{T_2}{T_1} = (\gamma - 1)\ln\frac{V_1}{V_2}$$

or

$$\ln\frac{T_2}{T_1} = \ln\left(\frac{V_1}{V_2}\right)^{\gamma-1}$$

Taking antilogarithm of both sides, we get

$$\frac{T_2}{T_1} = \left(\frac{V_1}{V_2}\right)^{\gamma-1} \quad \frac{T_2}{T_1} = \left(\frac{V_1}{V_2}\right)^{\gamma-1}$$

or

$$T_1 V_1^{\gamma-1} = T_2 V_2^{\gamma-1}$$

or

$$\boxed{\mathbf{TV^{\gamma-1} = constant}}$$

II. Relationship between temperature and pressure in adiabatic expansion of an ideal gas

For an ideal gas, $P_1V_1 = RT_1$

and $P_2V_2 = RT_2$

i.e. $$V_1 = \frac{RT_1}{P_1}$$

and $$V_2 = \frac{RT_2}{P_2}$$

$$\therefore \quad \frac{V_1}{V_2} = \frac{\dfrac{RT_2}{P_1}}{\dfrac{RT_2}{P_2}} = \frac{T_1P_2}{T_2P_1} \quad \text{... }(i)$$

Also $$\frac{T_2}{T_1} = \left(\frac{V_1}{V_2}\right)^{\gamma-1} \quad \text{... }(ii)$$

(The relation between T and V in case of adiabatic expansion)

From (*i*) and (*ii*)

$$\frac{T_2}{T_1} = \left(\frac{T_1P_2}{T_2P_1}\right)^{\gamma-1}$$

or $$\frac{T_2}{T_1}\times\left(\frac{T_2}{T_1}\right)^{\gamma-1} = \left(\frac{P_2}{P_1}\right)^{\gamma-1}$$

or $$\left(\frac{T_2}{T_1}\right)^{\gamma} = \left(\frac{P_2}{P_1}\right)^{\gamma-1}$$

or $$\left(\frac{T_2}{T_1}\right)^{\gamma} = \left(\frac{P_1}{P_2}\right)^{1-\gamma}$$

Taking the γth root of both sides, we get ... (*iii*)

$$\frac{T_2}{T_1} = \left(\frac{P_1}{P_2}\right)^{(1-\gamma)/\gamma}$$

or $$T_1P_1^{(1-\gamma)/\gamma} = T_2P_2^{(1-\gamma)/\gamma}$$

or $$\boxed{TP^{(1-\gamma)/\gamma} = \text{constant}}$$

III. Relationship between pressure and volume in adiabatic expansion of an ideal gas

Consider the relations between temperature and volume and temperature and pressure

$$\frac{T_2}{T_1} = \left[\frac{V_1}{V_2}\right]^{\gamma-1} \qquad \text{... (i)}$$

$$\frac{T_2}{T_1} = \left[\frac{P_1}{P_2}\right]^{(1-\gamma)/\gamma} \qquad \text{... (ii)}$$

From (*i*) and (*ii*)

$$\left(\frac{P_1}{P_2}\right)^{(1-\gamma)/\gamma} = \left(\frac{V_1}{V_2}\right)^{\gamma-1}$$

or
$$\left(\frac{P_1}{P_2}\right)^{(1-\gamma)/\gamma} = \left(\frac{V_2}{V_1}\right)^{1-\gamma}$$

Taking (1 – γ)th root of both sides, we have

$$\left(\frac{P_1}{P_2}\right)^{1/\gamma} = \frac{V_2}{V_1}$$

Raising both sides to the power γ, we have

$$\frac{P_1}{P_2} = \left(\frac{V_2}{V_1}\right)^{\gamma}$$

or
$$P_1V_1^{\gamma} = P_2V_2^{\gamma}$$

$$PV^{\gamma} = \text{constant}$$

8.12.2. Work done in an adiabatic reversible expansion of an ideal gas

The expression for the work done in an adiabatic reversible expansion of an ideal gas may be obtained as under:

For adiabatic process, we have

$$PV^{\gamma} = \text{constant}$$

Differentiating this equation, we get

$$P(\gamma V^{\gamma-1})\, dV + V^{\gamma}\, dP = 0$$

or
$$\gamma P V^{\gamma-1}\, dV + V^{\gamma}\, dP = 0$$

Dividing throughout by $V^{\gamma-1}$, we get

$$\gamma PdV + VdP = 0$$

or
$$VdP = -\gamma PdV \qquad \text{... (i)}$$

For n moles of an ideal gas, $PV = nRT$

Differentiating this equation completely, we get

$$PdV + VdP = nRdT \qquad \text{... (ii)}$$

Substituting for VdP from equation (*i*) in this equation we get

$$PdV - \gamma PdV = nRdT$$

or $(1-\gamma)\,PdV = nRdT$

or $$PdV = \frac{nRdT}{1-\gamma} \quad \text{... (iii)}$$

Work done in the adiabatic process is given by general equation

$$w = -\int_{V_1}^{V_2} PdV \quad \text{... (iv)}$$

Substituting for PdV from equation (*iii*) in this equation, and changing the limits from volumes to temperatures, we get

$$w = -\int_{T_1}^{T_2} \frac{nRdT}{1-\gamma}$$

$$= -\frac{nR}{1-\gamma}\int_{T_1}^{T_2} dT$$

$$= -\frac{nR}{1-\gamma}(T_2 - T_1) \quad \text{... (v)}$$

Hence $$\boldsymbol{w = -\frac{nR(T_2 - T_1)}{1-\gamma}}$$

Expression for work done may be obtained in a different form.

Putting $R = C_p - C_v$ and $\gamma = C_p/C_v$ equation (*v*) becomes

$$w = -\frac{n(C_p - C_v)(T_2 - T_1)}{1 - \dfrac{C_p}{C_v}}$$

$$= -\frac{n(C_p - C_v)(T_2 - T_1)}{\dfrac{(C_v - C_p)}{C_p}}$$

$$= -\frac{n(C_p - C_v)(T_2 - T_1)}{\dfrac{-(C_p - C_v)}{C_v}}$$

$= nC_v\,(T_2 - T_1)$, (for n moles of the ideal gas)

$= C_v\,(T_2 - T_1)$, (for 1 mol of the ideal gas) ... (*vi*)

For n moles of an ideal gas,

$P_1 V_1 = nRT_1$ (for the initial state)

and $P_2 V_2 = nRT_2$ (for the final state)

Substituting these values in equation (*v*), we get

$$\boldsymbol{w = -\frac{P_2 V_2 - P_1 V_1}{1-\gamma}} \quad \text{... (vii)}$$

SOLVED EXAMPLES ON ADIABATIC REVERSIBLE EEXPANSION OF AN IDEAL GAS

Example 1. For a certain gas C_p = 8.58J. Two moles of the gas are expanded adiabatically from an initial temperature of 20°C to a final temperature of – 45.4°C. Calculate the work done in the above case.

Solution. $C_p = 8.58\text{J}, C_v = 8.58 - 2 = 6.58\text{J}$

Therefore $r = \dfrac{C_p}{C_v} = \dfrac{8{\cdot}58}{6{\cdot}58} = 1.3$

$$n = 2, R = 8.314\text{J K}^{-1}\text{ mol}^{-1}$$

Work done
$$w = -\frac{nR(T_2 - T_1)}{1-\gamma}$$
$$= -\frac{2\times 8.314\times(-65.4)}{(1-1.3)}$$
$$= -\frac{2\times 8.314\times 65.4}{0.3}\text{J}$$
$$= \mathbf{-\ 3624.9\ J}$$

Example 2. Three moles of a perfect gas with C_v = 5.0 cal/mol/degree at 1.0 atmosphere are to be compressed adiabatically and reversibly from a volume of 75 litre at 1·0 atmosphere to a pressure of 100 atmospheres. Predict (*a*) the final volume of the gas (*b*) the final temperature of the gas.

Solution. Here we are given that

$$n = 3 \text{ moles}$$
$$C_v = 5.0 \text{ cal/mol/degree}$$
$$P_1 = 1.0 \text{ atm}$$
$$P_2 = 100 \text{ atm}$$
$$V_1 = 75 \text{ litres}$$

For a perfect gas
$$C_p = C_v + R$$
$$= 5{\cdot}0 + 2.0 = 7.0 \text{ cal } (R \cong 2 \text{ cal})$$

$\therefore$
$$\gamma = \frac{C_p}{C_v} = \frac{7.0}{5.0} = 1.4$$

(*a*) *To calculate the final volume*

$$P_1V_1^\gamma = P_2V_2^\gamma$$

or
$$\left(\frac{V_1}{V_2}\right)^\gamma = \frac{P_2}{P_1}$$

or
$$\left(\frac{75}{V_2}\right)^{1{\cdot}4} = \frac{100}{1.0} = 100$$

$$\therefore \quad 1.4\,[\log 75 - \log V_2] = \log 100$$

$$\text{or} \quad 1.4\,(1.8751 - \log V_2) = 2$$

$$\text{or} \quad 1.4 \log V_2 = 0.62514$$

$$\text{or} \quad \log V_2 = 0.4465$$

$$V_2 = \text{antilog}\,(0.4465)$$

$$= \mathbf{2{\cdot}796\ litres}$$

(*b*) *To calculate the final temperature*

For a perfect gas,

$$PV = nRT$$

When

$$P = 100 \text{ atm}$$

$$V = 2.796 \text{ litres (calculated above)}$$

$$n = 3 \text{ moles (given)}$$

$$R = 0{\cdot}0821 \text{ litre atm deg}^{-1} \text{ mol}^{-1}$$

$$\therefore \quad T = T = \frac{P \times V}{n \times R} = \frac{100 \times 2.796}{3 \times 0.0821} = \mathbf{1135.2\ K}$$

Example 3. Two moles of H_2 are compressed adiabatically from S.T.P. conditions to occupy a volume of 4.48 litres. Calculate the final temperature (γ for H_2= 1.41).

Solution. We are given that $n = 2$

so that

$$\left.\begin{aligned} V_1 &= 2 \times 22.4 = 44.8 \text{ litres} \\ T_1 &= 273 \text{ K} \end{aligned}\right\} \text{at S.T.P.}$$

$$V_2 = 4.48 \text{ litres}$$

$$T_2 = ?$$

$$\gamma = 1.41$$

Substituting the values in the formula

$$T_1 V_1^{\gamma-1} = T_2 V_2^{\gamma-1}$$

$$\text{or} \quad \frac{T_2}{T_1} = \left(\frac{V_1}{V_2}\right)^{\gamma-1}$$

We get

$$\frac{T_2}{273} = \left(\frac{44.8}{4.48}\right)^{1{\cdot}41-1} = (10)^{0{\cdot}41}$$

$$\text{or} \quad T_2 = 273 \times (10)^{0{\cdot}41}$$

$$\therefore \quad \log T_2 = \log 273 + 0.41 \log 10$$

$$= 2{\cdot}4362 + 0.41$$

$$= 2{\cdot}8462$$

$$\text{or} \quad T_2 = \text{antilog } 2{\cdot}8462 = \mathbf{701.8\ K}$$

Example 4. A dry gas at NTP is expanded adiabatically from 1 litre to 5 litre. Calculate the final temperature assuming ideal behaviour ($C_p / C_v = 1.4$)

Solution. Apply the relation $TV^{\gamma-1}$ = constant

$$T_1 V_1^{\gamma-1} = T_2 V_2^{\gamma-1} \quad \text{or} \quad \frac{T_1}{T_2} = \left(\frac{V_2}{V_1}\right)^{\gamma-1}$$

$$T_1 = 273 \text{ K}, \quad T_2 = ? \quad V_1 = 1 \text{ litre}, \quad V_2 = 5 \text{ litre}$$

Substituting the values in the equation above,

$$\frac{273}{T_2} = 5^{1.4-1} \quad \text{or} \quad \frac{273}{T_2} = 5^{0.4}$$

Taking logarithm of both sides

$$\log 273 - \log T_2 = 0.4 \log 5$$

or $$\log T_2 = \log 273 - 0.4 \log 5$$

or $$\log T_2 = 2.4362 - (0.4 \times 00.6990)$$

$$= 2.1566$$

or $$T_2 = 143.4 \text{ K}$$

PROBLEMS FOR PRACTICE

1. Two moles of hydrogen at S.T.P are compressed adiabatically to a volume of 10 litres. Calculate the final pressure and temperature of the gas. Given that γ for hydrogen is 1.41.

   [**Ans.** Final Pressure = 8.30 atm, Final temperature = 505 K]

   [**Hint.** $V_1 = 2 \times 22.4 = 44.8$ litres, $V_2 = 10$ litres]

2. 14 grams of N_2 at 290 K are compressed adiabatically from 8 to 5 litres. Calculate the final temperature and the work done on the gas. Assume $C_p = \frac{7}{2}R$.

   [**Ans.** $T = 349.9$ K, $w = 622.5$ J]

3. A quantity of air ($\gamma = 1.4$) at 300 K is compressed adiabatically to one third of its volume. Find the change in its temperature. [**Ans.** 465.5 K]

4. Calculate the change in energy for 2 moles of H_2 warmed at constant volume from 25° to 50°C. Given that for the gas near room temperature, C_v is constant and is about 5 cal/deg

   [**Ans.** 250 ergs.]

5. For a certain ideal gas $C_p = 8.58$ cal mol^{-1}. What will be the final volume and temperature when 2 moles of the gas at 20°C and 15 atm are allowed to expand adiabatically and reversibly to 5 atm pressure? Also calculate the work done in the above process.

   [**Ans.** $V = 7.44$ litres $t = -45.4$°C, $w = -3625$ J]

   [**Hint.** First calculate T_2, then find out w using the formular $w = -\dfrac{nR(T_2 - T_1)}{1-\gamma}$

   $T_1\gamma = 293$ K, $\qquad T_2 = 227.6$ K

   $\gamma = \dfrac{8.58}{6.58} = 1.30$, $n = 2$, $\qquad R = 8.314$ JK^{-1} mol^{-1}]

8.12.3. Comparison of isothermal and adiabatic expansion of a real gas

For a reversible adiabatic process, pressure and volume are related as

$$PV^{\gamma} = \text{constant}$$

While *for a reversible isothermal process,* the relationship is simply given by Boyle's law, *viz.*

$$PV = \text{constant}$$

Since C_p is always greater than C_v,

therefore the ratio $\frac{C_p}{C_v} = \gamma$ is greater than unity.

Hence the increase of volume for a given decrease of pressure will be less in an adiabatic than in an isothermal expansion. Therefore the plot of pressure against volume *i.e.,* pressure volume curve will be steeper for an adiabatic than for an isothermal process, starting at the same point, as shown is Fig. 8.9.

Further, since the area under the pressure-volume curve gives the work of expansion, from Fig. 8.9, it may be observed that for the same increase of volume, the work done in an adiabatic process is less than in an isothermal process.

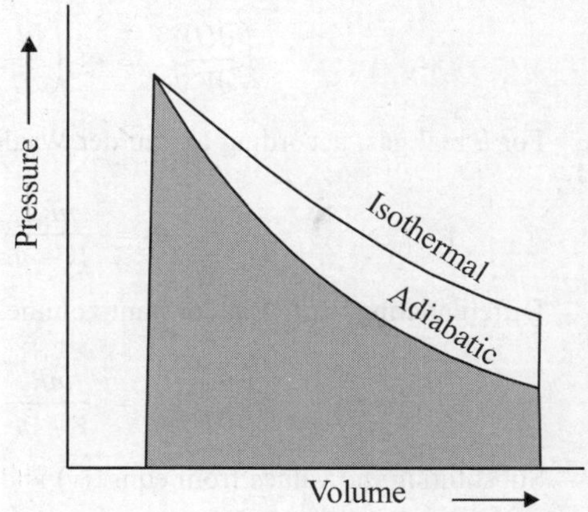

Fig .8.9 Isothermal and adiabatic curves

8.13. CALCULATION OF *w*, *q*, Δ*U* AND Δ*H* IN THE EXPANSION OF A REAL GAS IN ISOTHERMAL REVERSIBLE EXPANSION

(*i*) **Expression for *w*,**

$$dw = -P_{ext}\, dV$$

For the reversible expansion

$$P_{ext} = P_{int} - dP \simeq P_{int}$$

where P_{int} is the pressure (internal pressure) of the gas

Hence $$dw = -P_{int}\, dV \quad \text{... (i)}$$

As the gas is real, P_{int} can be obtained from van der Waal's equation *i.e.*

$$\left(P + \frac{an^2}{V^2}\right)(V - nb) = nRT$$

We get

$$P_{int} = \frac{nRT}{V - nb} - \frac{an^2}{V^2} \quad \text{... (ii)}$$

Putting this value in eqn (*i*), we get

$$-dw = \left(\frac{nRT}{V-nb} - \frac{an^2}{V^2}\right) dV$$

$$\therefore \quad -w = \int_{V_1}^{V_2} \left(\frac{nRT}{V-nb} - \frac{an^2}{V^2}\right) dV$$

$$= nRT\left[\ln(V - nb)\right]_{V_1}^{V_2} - an^2\left[-\frac{1}{V}\right]_{V_1}^{V_2}$$

or $$-w = nRT \ln\frac{V_2 - nb}{V_1 - nb} + an^2\left(\frac{1}{V_2} - \frac{1}{V_1}\right) \quad \text{... (iii)}$$

(*ii*) **Expression for Δ*U* :** According to thermodynamic equation of state

$$\left(\frac{\partial U}{\partial V}\right)_T = T\left(\frac{\partial U}{\partial T}\right)_V - P \quad \text{... (iv)}$$

For a real gas, according to van der Waal's equation

$$P = \frac{nRT}{V-nb} - \frac{an^2}{V^2} \quad \text{... (v)}$$

Differentiating w.r.t. T at constant volume, we get

$$\therefore \quad \left(\frac{\partial P}{\partial T}\right)_V = \frac{nR}{V-nb} \quad \text{... (vi)}$$

Substituting the values from eqns. (*v*) and (*vi*) in eqn. (*iv*) we get

$$\left(\frac{\partial U}{\partial V}\right)_T = \frac{nRT}{V-nb} - \left(\frac{nRT}{V-nb} - \frac{an^2}{V^2}\right) = \frac{an^2}{V^2} \quad \text{... (vii)}$$

For dU, we can write

$$dU = \left(\frac{\partial U}{\partial T}\right)_V dT + \left(\frac{\partial U}{\partial V}\right)_T dV \quad \text{... (viii)}$$

For the isothermal process, $dT = 0$

$$\therefore \quad dU = \left(\frac{\partial U}{\partial V}\right)_T dV = \frac{an^2}{V^2} dV \quad \text{[Substituting value from eqn (vii)}$$

Hence for a finite process,

$$\Delta U = \int_{V_1}^{V_2} \frac{an^2}{V^2} dV$$

or

$$\boldsymbol{\Delta U = -an^2\left(\frac{1}{V_2} - \frac{1}{V_1}\right)} \quad \text{... (ix)}$$

(*iii*) **Expression for *q* :** According to first law of thermodynamics

$$q = \Delta U - w$$

Substituting the values of w and ΔE from eqns. (*iii*) and (*ix*), we get

$$q = -an^2\left(\frac{1}{V_2} - \frac{1}{V_1}\right) + nRT \ln\frac{V_2 - nb}{V_1 - nb} + an^2\left(\frac{1}{V_2} - \frac{1}{V_1}\right)$$

or

$$\boldsymbol{q = nRT \ln\frac{V_2 - nb}{V_1 - nb}} \quad \text{... (x)}$$

(*iv*) **Expression for ΔH :** We have

$$\Delta H = \Delta(U + PV)$$
$$= \Delta U + \Delta(PV)$$

Substituting the value of ΔU and P from eqns (*ix*) and (*ii*), we get

$$\Delta H = -an^2\left(\frac{1}{V_2}-\frac{1}{V_1}\right)+\Delta\left[\left(\frac{nRT}{V-nb}-\frac{an^2}{V^2}\right)V\right]$$

$$= -an^2\left(\frac{1}{V_2}-\frac{1}{V_1}\right)+nRT\left(\frac{V_2}{V_2-nb}-\frac{V_1}{V_1-nb}\right)-an^2\left(\frac{1}{V_2}-\frac{1}{V_1}\right)$$

$$= -2an^2\left(\frac{1}{V_2}-\frac{1}{V_1}\right)+nRT\left[\frac{V_2-nb+nb}{V_2-nb}-\frac{V_1-nb+nb}{V_1-nb}\right]$$

$$= -2an^2\left(\frac{1}{V_2}-\frac{1}{V_1}\right)+nRT\left[\left(1+\frac{nb}{V_2-nb}\right)-\left(1+\frac{nb}{V_1-nb}\right)\right]$$

$$\boldsymbol{\Delta H = -2an^2\left(\frac{1}{V_2}-\frac{1}{V_1}\right)+n^2bRT\left(\frac{1}{V_2-nb}-\frac{1}{V_1-nb}\right)} \quad \text{... (xi)}$$

8.13.1. *q, w,* ΔE and ΔH in adiabatic reversible expansion of a real gas

(*i*) Value for *q*: For the adiabatic process, as already explained,

$$q = 0$$

(*ii*) Expressions for *w* and ΔU: By first law of thermodynamics,

$$\Delta U = q + w$$

As $q = 0$, therefore $\Delta U = w$

The expression for ΔU and hence for w can be obtained as follows :

$$dU = \left(\frac{\partial U}{\partial T}\right)_v dT + \left(\frac{\partial U}{\partial V}\right)_T dV$$

Putting $(\partial U/\partial T)_v = nC_v$ (for n moles) and $\left(\frac{\partial U}{\partial V}\right)_T = \frac{an^2}{V^2}$, we get

$$dU = nC_v dT + \frac{an^2}{V^2}dV \quad \text{... (i)}$$

Hence for a finite process,

$$\Delta U = \int_{T_1}^{T_2} nC_v dT + \int_{V_1}^{V_2}\frac{an^2}{V^2}dV$$

Taking C_v to be independent of temperature, we have

$$\boldsymbol{\Delta U = nC_v(T_2-T_1) - an^2\left(\frac{1}{V_2}-\frac{1}{V_1}\right)} \quad \text{... (ii)}$$

This expression can be used for calculation of ΔU if T_1, T_2, V_1 and V_2 are known. However, if T_2 is not given, it is first calculated as follows :

$$dU = dw = -PdV$$

or

$$dU = -PdV \quad \text{... (iii)}$$

For a real gas $$P = \frac{nRT}{V-nb} - \frac{an^2}{V^2}$$

and $$dU = nC_v dT + \frac{an^2}{V^2} dV$$ Refer Eqn. (*i*) above

Substituting these values in eqn. (*iii*), we get

$$nC_v dT + \frac{an^2}{V^2} dV = -\frac{nRT}{V-nb} dV + \frac{an^2}{V^2} dV$$

$$nC_v dT = -\frac{nRT}{V-nb} dV$$

$$C_v dT = -\frac{nRT}{V-nb} dV$$

Separating the variables, it can be rewritten as

$$C_v \frac{dT}{T} = -R \frac{dV}{V-nb}$$

Integrating between suitable limits, we get

$$\int_{T_1}^{T_2} C_v \frac{dT}{T} = -R \int_{V_1}^{V_2} \frac{dV}{V-nb}$$

or $$C_v [\ln T]_{T_1}^{T_2} = -R \left[\ln(V-nb)\right]_{V_1}^{V_2}$$

or $$C_v \ln \frac{T_2}{T_1} = -R \ln \frac{V_2-nb}{V_1-nb}$$

or $$\ln\left(\frac{T_2}{T_1}\right)^{C_v} = \ln\left(\frac{V_1-nb}{V_2-nb}\right)^R$$

or $$\left(\frac{T_2}{T_1}\right)^{C_v} = \left(\frac{V_1-nb}{V_2-nb}\right)^R$$

or $$\frac{T_2}{T_1} = \left(\frac{V_1-nb}{V_2-nb}\right)^{R/C_v}$$

or $$\boldsymbol{T_2 = T_1 \left(\frac{V_1-nb}{V_2-nb}\right)^{R/C_v}} \quad \text{... (iv)}$$

Knowing V_1, V_2 and T_1, T_2 (temperature after expansion) can be calculated. This may be substituted in eqn. (*ii*) to get ΔU and hence *w*.

(*iii*) **Expression for ΔH**

$$\Delta H = \Delta(U + PV) = \Delta U + \Delta(PV)$$

$$= \Delta U + \Delta\left[\left(\frac{nRT}{V-nb} - \frac{an^2}{V^2}\right)V\right]$$

$$= \Delta U + nR\left(\frac{V_2 T_2}{V_2 - nb} - \frac{V_1 T_1}{V_1 - nb}\right) - an^2\left(\frac{1}{V_2} - \frac{1}{V_1}\right)$$

Substituting the value of ΔU from eqn. (*ii*), we get

$$\Delta H = nC_v(T_2 - T_1) - 2an^2\left(\frac{1}{V_2} - \frac{1}{V_1}\right) + nR\left(\frac{V_2 T_2}{V_2 - nb} - \frac{V_1 T_1}{V_1 - nb}\right) \quad ...(v)$$

8.14. THERMOCHEMISTRY

The branch of chemistry which deals with heat changes accompanying chemical reactions is called **thermochemistry.**

A chemical reaction involves rearrangement of atoms, *i.e.*, some bonds within the reactant molecules are broken while some new bonds in molecules of the products are formed. Energy is consumed in the dissociation of bonds while energy is evolved when the bonds are formed. If the total energy evolved is greater than the total energy consumed, the net result is the *release of energy.* On the other hand, if the energy consumed is greater than the energy evolved, the net result is the *absorption of energy.*

All chemical reactions are usually accompanied by energy changes. These energy changes appear in the form of heat, light, work, electricity, etc. These energy changes are of great practical utility both for domestic and industrial purposes. A few reactions alongwith the form in which the accompanying energy changes appear and the use to which they are put are listed below :

(*i*) Burning of coal in air produces *heat.* That is why it is used as a fuel.

$$C\,(s) + O_2\,(g) \longrightarrow CO_2\,(g) + \text{heat}$$

(*ii*) *Digestion and metabolism of carbohydrates* in biological systems produce *heat* which maintains the body temperature.

(*iii*) *Burning of candle* (*wax*) *in air* produces heat and light. It is, therefore, used as a source of light. Wax is a hydrocarbon.

(*iv*) *Reaction that takes place in a galvanic cell* produces electrical energy which may be used to run an electrical motor or to ring a bell, etc.

$$Zn\,(s) + Cu^{2+}\,(aq) \longrightarrow Zn^{2+}\,(aq) + Cu\,(s)$$

Units of Heat and Work. The heat changes are measured in calories, kilocalories, joules or kilojoules.

1 calorie = 4.184 J   1 kcal = 4.184 kJ

The S.I. unit of heat is joule or kilojoule.

Work is measured in ergs or joules. The S.I. unit of work is joule.   1 Joule = 10^7 ergs

8.14.1. Enthalpy and enthalpy change

When a process is carried out at constant pressure and temperature, there is a change in volume.

If the volume increases, some mechanical work is done by the system so that the internal energy decreases. Similarly, if the volume decreases, some mechanical work is done on the system so that the internal energy increases. In all such processes occurring at constant temperature when pressure remains constant, the total energy of a system is given by a thermodynamic quantity called **heat**

content or **enthalpy** of the system and not by internal energy. It is denoted by H and is defined mathematically by the equation

$$H = E + PV$$

where E is the internal energy, P and V are the pressure and volume of the system respectively.

The **enthalpy** *of a system may be defined as the sum of the internal energy and the product of pressure and volume of the system.*

The *physical significance of enthalpy* may be put as under :

The **enthalpy** *of a substance is the amount of energy stored within the substance, i.e., available for conversion into heat.*

Like internal energy, *the heat content or enthalpy of a substance is also a state function, i.e.,* depends only on the state of the system and is independent of the manner by which the state has been attained.

Enthalpy change. Like internal energy, the absolute value of enthalpy of a substance cannot be determined. However, the change in the enthalpy when a system undergoes a change can be determined experimentally accurately and that serves our purpose.

A relation between enthalpy change and internal energy change exists as follows :

$$\Delta H = \Delta E + P\,\Delta V \qquad \dots (ii)$$

where $\Delta E = E_2 - E_1$ represents the change in internal energy

and $\Delta V = V_2 - V_1$ represents the change in volume of the system.

Thus, **enthalpy change** *of a system may be defined as the sum of internal energy change and the pressure-volume work done.*

If q_p is the heat absorbed at constant pressure, then $q_p = \Delta H$ ... (*iii*)

Thus, enthalpy change *is the heat absorbed or evolved at constant pressure.*

Standard enthalpy change. The enthalpy change of a reaction when all the reactants and the products are in their standard states, *i.e.*, at 298 K and under a pressure of one atmosphere is known as the standard enthalpy change. It is usually represented by ΔH°.

As a chemical reaction involves breaking of bonds of the reactants and the formation of new bonds in the products, some net energy change (called the enthalpy of reaction) takes place.

$$\text{Enthalpy change of a reaction} = \begin{bmatrix}\text{Energy consumed to}\\ \text{break the bonds of}\\ \text{the reactants}\end{bmatrix} - \begin{bmatrix}\text{Energy released in the}\\ \text{formation of the bonds in}\\ \text{the products}\end{bmatrix}$$

If energy consumed > energy released, the net result is the absorption of energy and the reaction is *endothermic.*

If energy released > energy consumed, the net result is the evolution of energy and the reaction is *exothermic.*

LATEST SIGN CONVENTIONS FOR *q* OR ΔH, ΔU AND *w*.

(*i*) When heat is absorbed by the system, *q* is positive.
(*ii*) When heat is given out by the system, *q* is negative.
(*iii*) When energy is absorbed by the system, ΔU is positive.
(*iv*) When energy is given out by the system, ΔU is negative.
(*v*) When work is done on the system, *w* is positive.
(*vi*) Work done by the system, *w* is negative.

8.14.2. Exothermic and endothermic reactions

Exothermic Reactions *are those reactions which are accompanied by the evolution of heat.*

The quantity of heat produced is shown alongwith the products with a 'plus' sign. A few examples of exothermic reactions are given below :

$$C(s) + O_2(g) \longrightarrow CO_2(g) + 393.5 \text{ kJ}$$

$$H_2(g) + \frac{1}{2} O_2(g) \longrightarrow H_2O(g) + 248.8 \text{ kJ}$$

$$CH_4(g) + 2O_2(g) \longrightarrow CO_2(g) + 2H_2O(g) + 890.4 \text{ kJ}$$

Endothermic Reactions *are those reactions which are accompanied by absorption of heat.*

Heat absorbed can be written with the products with a 'minus' sign. A few examples of endothermic reactions are given below :

$$N_2(g) + O_2(g) \longrightarrow 2NO(g) - 180.7 \text{ kJ}$$

$$C(s) + H_2O(g) \longrightarrow CO(g) + H_2(g) - 131.4 \text{ kJ}$$

$$C(s) + 2S(s) \longrightarrow CS_2(l) - 92.0 \text{ kJ}$$

The enthalpy change (ΔH) accompanying a reaction is given by

$$\Delta H = \text{Heat content of products} - \text{Heat content of reactants} = H_P - H_R$$

A reaction is exothermic, if the total heat content of the reactants is more than that of the products, i.e., $H_R > H_P$. Thus, ΔH will be negative for an exothermic process.

Thus, the exothermic reactions given above may be written in terms of ΔH as :

$$C(s) + O_2(g) \longrightarrow CO_2(g), \quad \Delta H = -393.5 \text{ kJ}$$

$$H_2(g) + \frac{1}{2} O_2(g) O(l) \longrightarrow H_2O(l), \quad \Delta H = -285.8 \text{ kJ}$$

$$CH_4(g) + 2O_2(g) \longrightarrow CO_2(g) + 2H_2O(l), \quad \Delta H = -890.4 \text{ kJ}$$

A reaction is endothermic if the total heat content of reactants is less than that of the products, i.e.,

$$H_R < H_P$$

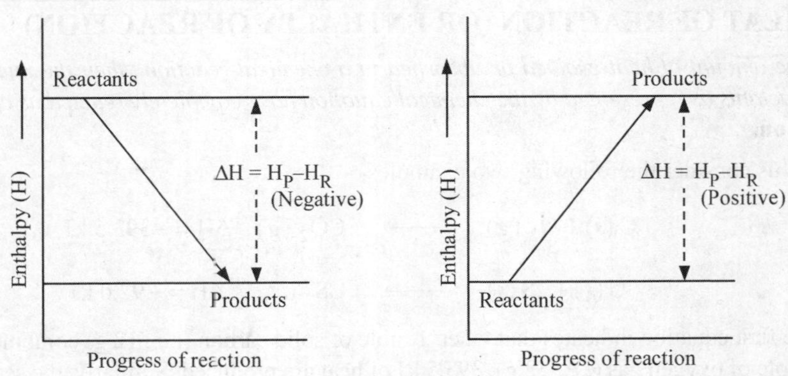

(*a*) (*b*)

Fig. 8.10 (*a*) Exothermic reaction (*b*) Endothermic reaction.

Thus, ΔH will be positive for an endothermic reaction.

Endothermic reactions given above may be written in terms of ΔH as :

$$N_2(g) + O_2(g) \longrightarrow 2NO(g), \quad \Delta H = +180.7 \text{ kJ}$$

$$C(s) + H_2O(g) \longrightarrow CO(g) + H_2(g), \quad \Delta H = +131.4 \text{ kJ}$$

$$C(s) + 2S(s) \longrightarrow CS_2(l), \quad \Delta H = +92.0 \text{ kJ}$$

Exothermic and endothermic reactions may be represented graphically as shown in Fig. 8.10(*a*) and 8.10(*b*).

8.14.3. Thermochemical equation

A balanced chemical equation not only indicating the quantities of the different reactants and products but also indicating the amount of heat evolved or absorbed is called a **thermochemical equation.**

Fractional coefficients are allowed in writing a thermochemical equation. For example, the formation of water is written as

$$H_2(g) + \frac{1}{2}O_2(g) \longrightarrow H_2O(l), \quad \Delta H = -285.8 \text{ kJ}$$

Thus, 285.8 kJ of heat is produced when 1 mole of hydrogen reacts with 0.5 mole of oxygen. If the quantities of reactants are doubled, the heat produced will also be doubled. For example, in the above case, we may write

$$2H_2(g) + O_2(g) \longrightarrow 2\ H_2O(l), \quad \Delta H = -571.6 \text{ kJ}$$

SOME POINTS TO REMEMBER ABOUT THERMOCHEMICAL EQUATIONS

(*i*) Unless otherwise mentioned, ΔH values are for the standard state of the substances (*i.e.*, 298 K and 1 atm pressure).

(*ii*) The physical state (*s, l, g*, etc.) of the different substances must be mentioned as the heat evolved or absorbed depends upon the physical state of the substances.

(*iii*) If the coefficients of the substances are multiplied or divided by some number, the value of ΔH is also multiplied or divided by the same number.

(*iv*) If the reaction is reversed, the sign of ΔH changes but the magnitude remains the same.

8.15. HEAT OF REACTION (OR ENTHALPY OF REACTION)

The amount of heat evolved or absorbed in a chemical reaction when the number of moles of the reactants as represented by the chemical equation have completely reacted, is called the **heat of reaction.**

Let us consider the following two examples :

$$C(s) + O_2(g) \longrightarrow CO_2(g) \quad \Delta H = -393.5 \text{ kJ}$$

$$C(s) + 2S(s) \longrightarrow CS_2(l), \quad \Delta H = +92.0 \text{ kJ}$$

The first equation indicates that when 1 mole of solid carbon (*i.e.*, 12 g) combines completely with 1 mole of oxygen gas (*i.e.*, 32 g), 393.5 kJ of heat are produced. Similarly, the second equation tells that when 1 mole of solid carbon (*i.e.*, 12 g) reacts completely with 2 moles of sulphur (64 g), 92 kJ of heat are absorbed.

Heat of reaction, ΔH = Heat content of products – Heat content of reactants

If the reaction is reversed, the sign of ΔH is also reversed.

8.15.1. Factors Influencing Heat of Reaction

The heat of a reaction depends upon a number of factors as explained below :

(i) Quantities of the reactants involved. If the quantities of the reactants are changed, the heat of reaction changes proportionally.

(ii) Physical state of the reactants and products. Since latent heat is involved in the change of state, therefore, the physical state of the reactants and products affects the heat of reaction. For example, when hydrogen and oxygen gases combine to form liquid water, the heat of reaction is different from the value when they combine to form water vapours.

$$H_2\,(g) + \frac{1}{2}O_2\,(g) \longrightarrow H_2O\,(l), \quad \Delta H = -\,285.8 \text{ kJ}$$

$$H_2\,(g) + \frac{1}{2}O_2\,(g) \longrightarrow H_2O\,(g), \quad \Delta H = -\,248.8 \text{ kJ}$$

It is, therefore, essential to write the symbols *s, l, g* or *aq* to indicate whether a particular substance is solid, liquid, gaseous or an aqueous solution.

(iii) Allotropic modifications. Heat of reaction changes if a different allotropic form is involved in the reaction. For example,

$$C\,(\text{diamond}) + O_2\,(g) \longrightarrow CO_2\,(g), \quad \Delta H = -\,395.4 \text{ kJ}$$

$$C\,(\text{graphite}) + O_2\,(g) \longrightarrow CO_2\,(g), \quad \Delta H = -\,393.5 \text{ kJ}$$

(iv) Concentration of solutions. If the solutions are involved in a reaction, their concentrations affect the heat of reaction.

(v) Temperature. The heat of reaction depends upon the temperature at which the reactants and products are taken. Generally, the heat of reaction increases with temperature.

(vi) Conditions of constant pressure or constant volume, i.e., whether the reaction takes place at constant pressure or at constant volume.

If q_p and q_v represent the heats of reaction at constant pressure and volume respectively, then the following equation is applicable.

$$q_p = q_v + \Delta n_g\, RT$$

where $$\Delta n_g = n_p - n_r$$

n_p and n_r represent the number of moles of gaseous reactants and products respectively. R and T have the usual meanings.

8.15.2. Heat (or enthalpy) of combustion

The heat of combustion of a substance is defined as the heat change (usually the heat evolved) when 1 mole of substance is completely burnt or oxidised in oxygen. For example,

$$C\,(s) + O_2\,(g) \longrightarrow CO_2\,(g), \Delta H = -\,393.5 \text{ kJ}$$

When one mole of carbon (12 g) burns completely in oxygen, 393.5 kJ of heat are evolved.

Thus, heat of combustion of carbon is 393.5 kJ.

The significance of the words *burns completely* must be clearly understood. For example, carbon may be oxidised to carbon monoxide (CO) or carbon dioxide (CO_2). The heats evolved in the two cases are different, *viz.*,

$$C(s) + \frac{1}{2}O_2(g) \longrightarrow CO(g), \quad \Delta H = -110.5 \text{ kJ}$$

$$C(s) + O_2(g) \longrightarrow CO_2(g), \quad \Delta H = -393.5 \text{ kJ}$$

Complete oxidation means oxidation to CO_2 and not to CO. Hence, the heat of combustion of carbon is 393.5 kJ mol^{-1}.

8.15.3. Calorific value of foods and fuels

For the working of the human machine, we eat carbohydrates, fats, etc., in the form of food. The carbohydrates are first decomposed in our body by the enzymes to form glucose which is then oxidised by the oxygen that we inhale to produce energy as given by the following equation.

$$C_6H_{12}O_6(s) + 6O_2(g) \longrightarrow 6CO_2(g) + 6H_2O(g) + 2900 \text{ kJ}$$

This oxidation reaction is usually called '*combustion of food*'.

Different foods have different heats of combustion. These are usually expressed in terms of their calorific values.

The calorific value of a fuel or food is the amount of heat in calories or joules produced from the complete combustion of one gram of the fuel or the food.

Thus, according to the above reaction, 1 mole of glucose, *i.e.*, 180 g of glucose produce 2900 kJ of heat.

$$\text{Hence, calorific value of glucose} = \frac{2900}{180} = 16.1 \text{ kJ g}^{-1}$$

Calorific values of some foods and fuels are given in Tables 8.1 and 8.2.

Table 8.1: Calorific Values of Foods per Gram

Food	Calorific value kJ / gram	Food	Calorific value kJ / gram
Ghee	37.65	Fish	5.2
Butter	30.50	Meat	12.0
Cheese	14.56	Egg (yolk)	5.4
Milk (buffalo's)	4.90	Egg (white)	14.0
Milk (cow's)	2.20	Sugar	16.0
Curd	2.51	Rice	16.7
Soyabean	18.07	Potatoes	4.0

Table 8.2 : Calorific Values of some Common Fuels in kJ per kg

Fuel	kJ per kg
Wood	16,000
Charcoal	29,000
Natural gas	46,000
Kerosene	38,000

A normal youngman needs 10,000 kJ of energy per day for normal functioning.

SOLVED EXAMPLE ON CALORIFIC VALUE

Example . A cylinder of indane gas contains 11.2 kg of butane. A normal family needs 25000 kJ energy per day for cooking. How long will the cylinder of gas last assumping 100% combustion?

If the efficiency of combustion is 85%, how long will the cylinder last? Heat of combustion of butane is 2658 kJ mol^{-1}.

Solution. Mol. mass of butane is 58 g mol^{-1}.

Thus, 58 g butane on complete combustion produces = 2658 kJ

$$\therefore \text{ 11200 g butane on complete combustion produces } = \frac{2658 \times 11200}{58} \text{ kJ} = 513268.96 \text{ kJ.}$$

$$\text{The cylinder will last } = \frac{513268.96}{25000} = 20.5 \text{ days}$$

If the efficiency of combustion is only 85%, the cylinder will last = 0.85 × 20.5 ≈ 17.5 days.

8.16. STANDARD STATE

The standard molar enthalpy of every element at 1 atm pressure and 298.15 K in the most stable state is taken as zero. These conditions of 1 atm pressure and 298.15 K are called standard state conditions. It is not possible to determine the absolute value of molar enthalpy, but we can use the above convention to find out the standard molar enthalpy of a substance. The most stable form of hydrogen is gaseous H_2, that of bromine is liquid Br_2 while that of iodine is solid I_2. Their molar standard enthalpies are taken as zero at 1 atmospheric pressure and 298.15 K temperature. In case of elements exhibiting allotrophy, we consider the most stable form under the conditions of standard state.

8.16.1. Enthalpy of formation and standard enthalpy of formation

Heat of Formation (or Enthalpy of Formation) : *The enthalpy of formation of a substance is defined as the heat change that takes place when 1 mole of the substance is formed from its elements under given conditions of temperature and pressure. It is usually represented by ΔH_f.*

Standard enthalpy of formation *of a substance is defined as the heat change accompanying the formation of 1 mole of the substance in the standard state from its elements, also taken in the standard state (i.e., 298 K, and 1 atmospheric pressure). It is usually represented by $\Delta H^o{}_f$.*

For example, in the reaction

$$C\,(s) + O_2\,(g) \longrightarrow CO_2\,(g), \quad \Delta H^o{}_f = -\,393.5 \text{ kJ}$$

When 1 mole of CO_2 (*g*) is formed from its elements, viz., C (*s*) and O_2 (*g*) (all substances being taken in the standard state), 393.5 kJ of heat is produced. Hence, the standard heat of formation of gaseous CO_2 is 393.5 kJ mol^{-1}.

Importance of standard enthalpies of formation. Knowing the standard enthalpies of formation of the different compounds involved in a chemical reaction, the standard enthalpy change of the given reaction can be obtained using the formula.

$$\Delta H^o{}_{\text{reaction}} = \begin{bmatrix}\text{Sum of the standard heats}\\ \text{of formation of products}\end{bmatrix} - \begin{bmatrix}\text{Sum of the standard heats}\\ \text{of formation of reactants}\end{bmatrix}$$

i.e.,

$$\Delta H^o{}_{\text{reaction}} = \Sigma\,\Delta H^o{}_f \text{ (Products)} - \Sigma\,\Delta H^o{}_f \text{ (Reactants)}$$

Thus, for a general reaction

$$aA + bB \longrightarrow cC + dD$$

$$\Delta H^o{}_{\text{reaction}} = [c\,\Delta H^o{}_f(\text{C}) + d\,\Delta H^o f\,(\text{D})] - [a\,\Delta H^o{}_f(\text{A}) + b\,\Delta H^o{}_f(\text{B})]$$

Remember ΔH^o_f is taken as zero for the elements in a reaction. Suppose we are interested in finding the heat of combustion of methane. The reaction is represented as :

$$CH_4\,(g) + 2O_2(g) \longrightarrow CO_2\,(g) + 2H_2O\,(l)$$

$$\Delta H^o_{reaction} = [\Delta H^o_f \text{ for } CO_2 + (2 \times \Delta H^o_f \text{ for } H_2O)] - [\Delta H^o_f \text{ for } CH_4 + (2 \times \Delta H^o_f \text{ for } O_2)]$$

Putting the values of heats of formation in the standard state

$$\Delta H^o_{reaction} = [-393.5 + 2 \times (-285.8)] - [-74.9 + 0] = -965.1 + 74.9 = -890.2 \text{ kJ}$$

SOLVED EXAMPLES ON ENTHALPIES OF FORMATION

Example 1. Enthalpies of formation of CO_2 (g) and H_2O (l) under standard conditions are 394.65 kJ and 285.84 kJ per mole. If the standard enthalpy of combustion of acetaldehyde (CH_3CHO) be 1167.62 kJ per mole, find its enthalpy of formation.

Solution. From the given data

(a) $$C + O_2 \longrightarrow CO_2\,(g); \qquad \Delta H = -394.65 \text{ kJ}$$

(b) $$H_2 + \frac{1}{2}O_2 \longrightarrow H_2O\,(l); \qquad \Delta H = -285.84 \text{ kJ}$$

(c) $$CH_3CHO + \frac{5}{2}O_2 \longrightarrow 2CO_2 + 2H_2O\,(g); \qquad \Delta H = -1167.62 \text{ kJ}$$

Our aim is to find out the ΔH value for the reaction

$$2C + 2H_2 + \frac{1}{2}O_2 \longrightarrow CH_3CHO$$

Using standard enthalpy values

ΔH^o_f for $CO_2 = -394.65$ kJ, ΔH^o_f for $H_2O = -285.84$ kJ and ΔH^o_f for $O_2 = 0$

($\because$ Enthalpy of formation of elements is taken as zero).

Let ΔH^o_f for CH_3CHO be x kJ.

ΔH value for any reaction = Enthalpies of the products – Enthalpies of the reactants.

From the equation

$$CH_3CHO + \frac{5}{2}O_2 \longrightarrow 2CO_2\,(g) + 2H_2O \quad ; \quad \Delta H = -1167.62 \text{ kJ}$$

$$\therefore -1167.62 = [2 \times \Delta H^o_f\,(CO_2) + 2 \times \Delta H^o_f(H_2O)] - [\Delta H^o_f(CH_3CHO) + \frac{5}{2} \times \Delta H^o_f(O_2)]$$

$$= 2 \times (-394.65) + 2 \times (-285.84) - x + 0 = -789.30 - 571.68 - x$$

$$-1167.62 = -1360.98 - x \quad \text{or} \quad x = -193.36 \text{ kJ}$$

Hence, enthalpy of formation of CH_3CHO is **–193.36 kJ.**

Example 2. Calculate ΔH° for the reaction : $CO_2(g) + H_2(g) \longrightarrow CO(g) + H_2O(g)$ given that ΔH^o_f for $CO_2(g)$, $CO(g)$ and $H_2O(g)$ are – 393.5, – 111.3 and – 241.8 kJ mol^{-1} respectively.

Solution. ΔH^o for the reaction $CO_2(g) + H_2(g) \longrightarrow CO(g) + H_2O(g)$ is

$$\Delta H^o = [H^o{}_{f\,(products)} - H^o{}_{f\,(reactants)}]$$

$$= [(-111.3) + (-241.8)] - [(-393.5) + 0] = (-111.3 - 241.8) - (-393.5)$$

$$= -353.1 + 393.5$$

ΔH^o = **+ 40.4 kJ.** [Enthalpy of every element in standard state is assumed as zero].

Example 3. Enthalpies of formation of C_2H_5OH (*l*), CO_2 (*g*) and H_2O (*l*) are –277.0, – 393.5 and – 285.8 kJ mole^{-1} respectively. Calculate enthalpy change for reaction.

$$C_2H_5OH\,(l) + 3O_2\,(g) \longrightarrow 2CO_2(g) + 3H_2O\,(l)$$

Solution. ΔH^o for the reaction $C_2H_5OH\,(l) + 3O_2\,(g) \longrightarrow 2CO_2\,(g) + 3H_2O\,(l)$ is

$$\Delta H^o = \Delta H^o_{f(products)} - \Delta H^o_{f(reactants)}$$

$$= [2 \times \Delta H^o{}_f(CO_2) + 3 \times \Delta H^o{}_f(H_2O)] - [\Delta H^o{}_f(C_2H_5OH) + 3 \times \Delta H^o{}_f(O_2)]$$

$$= [2 \times (-393.5) + 3 \times (-285.8)] - [-277 + 0] = -787.0 - 857.4 + 277$$

$$= -1644.4 + 277 = \mathbf{-1367.4\ kJ.}$$

8.16.2. Heat (or enthalpy) of neutralization

The heat of neutralization of an acid by a base is defined as the heat change when one gram equivalent of the acid is neutralized by a base, the reaction being carried out in dilute aqueous solution.

The heat of neutralization of a base by an acid is defined in a similar manner.

For example, when one gram equivalent of HCl is neutralized by NaOH or one gram equivalent of NaOH is neutralized by HCl, in dilute and aqueous solution, 57.1 kJ of heat is produced.

$$NaOH + HCl \longrightarrow NaCl + H_2O, \quad \Delta H = -57.1\ kJ$$

Hence, the heat of neutralization of HCl with NaOH or NaOH with HCl is 57.1 kJ.

The heat of neutralization of any strong acid (HCl, HNO_3, H_2SO_4) with a strong base (NaOH, KOH) or vice versa, is always the same, *i.e.*, 57.1 kJ. This is because the strong acids, strong bases and the salts that they form are all completely ionized in dilute aqueous solution. Thus, the reaction between a strong acid and a strong base may be written as

$$Na^+ + OH^- + H^+ + Cl^- \longrightarrow Na^+ + Cl^- + H_2O, \quad \Delta H = -57.1\ kJ$$

Cancelling out common ions on the two sides, we have

$$H^+\,(aq) + OH^-\,(aq) \longrightarrow H_2O\,(l), \quad \Delta H = -57.1\ kJ$$

Thus, neutralization may be regarded as a reaction between the H^+ ions given by the acid with the OH^- ions given by the base. Since strong acids and strong bases ionize completely in dilute aqueous solution, the number of H^+ ions and OH^- ions produced by one gram equivalent of the strong acid and the strong base is always the same. Hence, the heat of neutralization between a strong acid and a strong base is always the same. *i.e.*, –57.1 kJ.

If the acid or the base is weak, the heat of neutralization is usually less than 57.1 kJ. Consider the neutralization of a weak acid like acetic acid with a strong base like NaOH. Acetic acid ionizes to a small extent whereas NaOH ionizes completely as

$$(i)\ CH_3COOH \rightleftharpoons CH_3COO^- + H^+ \qquad (ii)\ NaOH \longrightarrow Na^+ + OH^-$$

When H^+ ions given by the acid combine with the OH^- ions given by the base, the equilibrium (*i*) shifts to the right, *i.e.*, more of acetic acid dissociates. A part of the heat produced during the combination of H^+ ions and OH^- ions is used up for the complete dissociation of acetic acid. Hence, the net heat evolved in the above reaction is less than 57.1 kJ.

Compared to a strong acid, a weak acid produces a smaller number of H^+, hence, heat of neutralization is smaller.

Example 1. Determine the amount of heat released when :

(*a*) 0.75 mole of nitric acid is neutralised by 0.75 mole of sodium hydroxide.

(*b*) 0.75 mole of hydrochloric acid is neutralised by 0.5 mole of potassium hydroxide.

(*c*) 900 ml of 0.5 M hydrochloric acid solution is mixed with 500 ml of 0.4 M sodium hydroxide solution. Assuming the specific heat of water as 4.18 J K^{-1} g^{-1}, also calculate the rise in temperature.

(*d*) 200 ml of 0.2 M sulphuric acid is mixed with 300 ml of 0.1 M sodium hydroxide solution. Calculate the rise in temperature.

Solution. (*a*) The reaction is :

$$(0.75 \text{ moles})\ HNO_3 + (0.75 \text{ moles})\ NaOH \longrightarrow NaNO_3 + H_2O.$$

When one mole of HNO_3 is neutralised by 1 mole of NaOH, heat released is 57.1 kJ. Therefore, when 0.75 mole of HNO_3 is neutralised by 0.75 mole of NaOH, heat released will be

$$57.1 \times 0.75 = \mathbf{42.8\ kJ.}$$

(*b*) Out of 0.75 mole of HCl, 0.5 mole will react with 0.5 mole of KOH and 0.25 mole of HCl will be left unreacted.

$$\text{Heat released} = 57.1 \times 0.5 = \mathbf{28.5\ kJ.}$$

(*c*) $\quad$ 1000 ml of 1 M HCl = 1 mole

$$\therefore \quad 900 \text{ ml of } 0.5 \text{ M HCl} = \frac{900 \times 0.5}{1000} = 0.45 \text{ mole}$$

Similarly,

$$500 \text{ ml of } 0.4 \text{ M NaOH} = \frac{500 \times 0.4}{1000} = 0.2 \text{ mole}$$

Out of 0.45 mole of HCl, 0.2 mole of HCl reacts with 0.2 mole of NaOH

$$\therefore \quad \text{Heat evolved} = 57.1 \times 0.2 = 11.4 \text{ kJ.}$$

To calculate the rise in temperature, use the relation

Mass × Sp. heat × Rise in temp.= Heat evolved

Heat evolved is 11.4 kJ or 11400 J, mass of the mixture is 900 g + 500 g (assuming the densities of the solution as unity) and sp. heat is 4.18 J.

$$\therefore \quad \text{Rise in temp.} = \frac{\text{Heat evolved}}{\text{Mass} \times \text{Sp. heat}} = \frac{11.4 \times 1000}{1400 \times 4.18} = 1.95^\circ$$

(*d*) 1 molecule of H_2SO_4 releases two H^+ ions

$$200 \text{ ml. of } 0.2M\ H_2SO_4 = \frac{2\times200\times0.2}{1000} \text{ moles of } H^+ = 0.08 \text{ moles of } H^+$$

$$300 \text{ ml of } 0.1 \text{ M NaOH } = 0.03 \text{ mole } OH^-$$

Out of 0.08 mole of H^+, 0.03 mole H^+ will react with 0.03 mole OH^-.

$\therefore$ Heat evolved $= 57.1 \times 0.03 = 1.7$ kJ.

$\therefore$ Rise in temperature $= \dfrac{1.7\times1000}{500\times4.18} = 0.8^o$.

Example 2. Calculate the amount of heat evolved when :

(*i*) 600 cm^3 of 0.1 M hydrochloric acid is mixed with 200 cm^3 of 0.2 M sodium hydroxide solution.

(*ii*) 250 cm^3 of 0.2 M sulphuric acid is mixed with 400 cm^3 of 0.5 M potassium hydroxide solution.

Assuming that the specific heat of water is 4.18 J K^{-1} g^{-1} and ignoring the heat absorbed by the container, thermometer, stirrer, etc., what would be the rise in temperature in each of the above cases?

Solution. (*i*)

$$600 \text{ cm}^3 \text{ of } 0.1M \text{ HCl} = \frac{0.1}{1000}\times600 \text{ mole of HCl}$$

$$= 0.06 \text{ mole of HCl} = 0.06 \text{ mole of } H^+ \text{ ions}$$

$$200 \text{ cm}^3 \text{ of } 0.2M \text{ NaOH} = \frac{0.2}{1000}\times200 \text{ mole of NaOH}$$

$$= 0.04 \text{ mole of NaOH} = 0.04 \text{ mole of } OH^- \text{ ions}$$

Thus, 0.04 mole of H^+ ions will combine with 0.04 mole of OH^- ions to form 0.04 mole of H_2O and 0.02 mole of H^+ ions will remain unreacted.

Heat evolved when 1 mole of H^+ ions combines with 1 mole of OH^- ions = 57.1 kJ.

$\therefore$ Heat evolved when 0.04 mole of H^+ ions combines with 0.04 mole of OH^- ions.

$= 57.1 \times 0.04 =$ **2.284 kJ.**

(*ii*)
$$250 \text{ cm}^3 \text{ of } 0.2M\ H_2SO_4 = \frac{0.2}{1000}\times250 \text{ mole of } H_2SO_4$$

$$= 0.05 \text{ mole of } H_2SO_4 = 0.10 \text{ mole of } H^+ \text{ ions}$$

$$400 \text{ cm}^3 \text{ of } 0.5M \text{ KOH} = \frac{0.5}{1000}\times400 \text{ mole of KOH}$$

$$= 0.2 \text{ mole of KOH} = 0.2 \text{ mole of } OH^- \text{ ions}$$

Thus, 0.10 mole of H^+ ions will neutralize 0.10 mole of OH^- ions, (out of 0.2 mole of OH^- ions) to form 0.10 mole of H_2O.

Hence, heat evolved $= 57.1 \times 0.1 =$ **5.71 kJ.**

In case (*i*), heat produced $= 2.284$ kJ $= 2284$ J

Total mass of the solution $= 600 + 200 = 800$ g

Specific heat $= 4.18$ JK^{-1}g^{-1}

$$Q = m \times s \times \Delta t$$

$$\therefore \quad \Delta t = \frac{Q}{m \times s} = \frac{2284}{800 \times 4.18} = \mathbf{0.68^o.}$$

In case (*ii*), heat produced = 5710 J

Total mass of the solution = 250 + 400 = 650 g

$$\therefore \quad \Delta t = \frac{Q}{m \times s} = \frac{4568}{650 \times 4.18} = \mathbf{1.68^o.}$$

8.17. DIFFERENT KINDS OF HEAT (OR ENTHALPY) OF REACTION

1. Heat of Solution (or Enthalpy of solution) : *The heat of solution of a substance in a particular solvent is defined as the heat change (i.e., the amount of heat evolved or absorbed) when 1 mole of the substance is dissolved in such a large volume of the solvent that further addition of the solvent does not produce any more heat change.*

Thus, the thermochemical equations for the dissolution of KCl and $CuSO_4$ in water may represented as

$$KCl\,(s) + aq \longrightarrow KCl\,(aq), \quad \Delta H = +18.6 \text{ kJ}$$

$$CuSO_4\,(s) + aq \longrightarrow CuSO_4\,(aq), \quad \Delta H = -66.5 \text{ kJ}$$

Thus,

the heat of solution of KCl = + 18.6 kJ mol^{-1}

and heat of solution of $CuSO_4$ = – 66.5 kJ mol^{-1}

It is interesting to note that the salts like copper sulphate, calcium chloride, etc., when present in the hydrated state (*i.e.,* $CuSO_4.5H_2O$, $CaCl_2.6H_2O$, etc.) dissolve with the absorption of heat. For example,

$$CuSO_4.5H_2O + aq \longrightarrow CuSO_4\,(aq), \Delta H = +11.7 \text{ kJ}$$

Thus, it can be generalized that the process of dissolution is usually endothermic for :

(*i*) salts which do not form hydrates like NaCl, KCl, KNO_3, etc.

(*ii*) hydrated salts like $CuSO_4.5H_2O$, $CaCl_2.6H_2O$, etc.

2. Heat of Hydration. *The amount of heat change (i.e., the heat evolved or absorbed) when one mole of the anhydrous salt combines with the required number of moles of water so as to change into the hydrated salt, is called the heat of hydration.*

For example, the heat of hydration of copper sulphate is –78.2 kJ mol^{-1}. This is represented as follows:

$$CuSO_4\,(s) + 5H_2O \longrightarrow CuSO_4.5H_2O\,(s), \quad \Delta H = -78.2 \text{ kJ}$$

3. Heat of Fusion. *Heat of fusion is the heat change that takes place when one mole of a solid substance changes into its liquid state at its melting point.*

Heat of fusion (ΔH_{fus}) of ice (m.p. = 273 K) is 6.0 kJ mol^{-1}. It may be represented as follows :

$$\underset{\text{Ice}}{H_2O\,(s)} \longrightarrow \underset{\text{Water}}{H_2O\,(l)}, \quad \Delta H = +6.0 \text{ kJ}$$

4. Heat of Vaporisation. *It is the heat change that takes place when one mole of a liquid changes into its gaseous state at its boiling point.*

For example, the heat of vaporisation ($\Delta H_{vap.}$) of water into its gaseous state at the boiling point (373 K) is 40.7 kJ. It may be represented as follows :

$$\underset{\text{Water}}{H_2O\,(l)} \longrightarrow \underset{\text{Steam}}{H_2O\,(g)}, \quad \Delta H = +\,40.7 \text{ kJ}$$

5. Heat of Sublimation. *Heat of sublimation of a substance is the heat change that takes place when 1 mole of a solid changes directly into vapour phase at a given temperature below its melting point.*

Sublimation is a process in which a solid on heating changes directly into gaseous state below its melting point. For example, the heat of sublimation of iodine is 62.39 kJ mol^{-1}. It is represented as follows :

$$I_2\,(s) \longrightarrow I_2\,(g), \quad \Delta H = +\,62.39 \text{ kJ}$$

Most solids, that sublime readily are molecular solids, e.g., iodine and naphthalene, etc.

8.18. HESS'S LAW OF CONSTANT HEAT SUMMATION

G. H. Hess

G.H. Hess, a Russian chemist, in 1840, gave a law about the heats of reactions which states as follows :

The total amount of heat evolved or absorbed in a reaction is the same whether the reaction takes place in one step or in a number of steps. In other words, the total amount of heat change in a reaction depends only upon the nature of the initial reactants and the nature of the final products and not upon the path or the manner by which this change is brought about.

Hess's law may be illustrated with the help of following examples :

(1) Sulphur (Rhombic) burns to form SO_3 to evolve 395.4 kJ of heat.

$$S_R + \frac{3}{2} O_2\,(g) \longrightarrow SO_3\,(g), \quad \Delta H = -395.4 \text{ kJ}$$

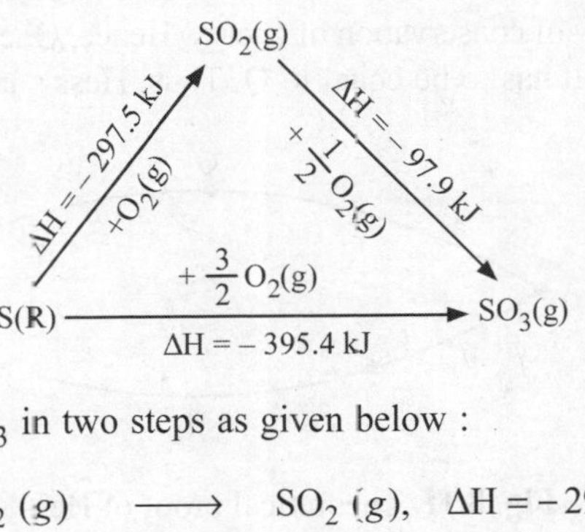

Sulphur may change to SO_3 in two steps as given below :

$$S_R + O_2\,(g) \longrightarrow SO_2\,(g), \quad \Delta H = -297.5 \text{ kJ}$$

$$SO_2(g) + \frac{1}{2} O_2\,(g) \longrightarrow SO_3\,(g), \quad \Delta H = -97.9 \text{ kJ}$$

(See accompanying figure)

Total heat evolved in the two steps is $-\,297.5 + (-\,97.9) = -\,395.4$ kJ.

This is the same as for the direct reaction in one step.

(2) When carbon (graphite) burns to form carbon dioxide directly in one step, 393.5 kJ mol^{-1} of heat is evolved.

$$C\,(s) + O_2\,(g) \longrightarrow CO_2\,(g), \quad \Delta H = -393.5 \text{ kJ}$$

If carbon burns to form carbon monoxide first which then burns to form carbon dioxide, the heats evolved in the two steps are as follows (see accompanying figure).

$$C\,(s) + \frac{1}{2}O_2\,(g) \longrightarrow CO\,(g), \quad \Delta H = -111.5 \text{ kJ}$$

$$CO\,(g) + \frac{1}{2}O_2\,(g) \longrightarrow CO_2\,(g), \quad \Delta H = -282.0 \text{ kJ}$$

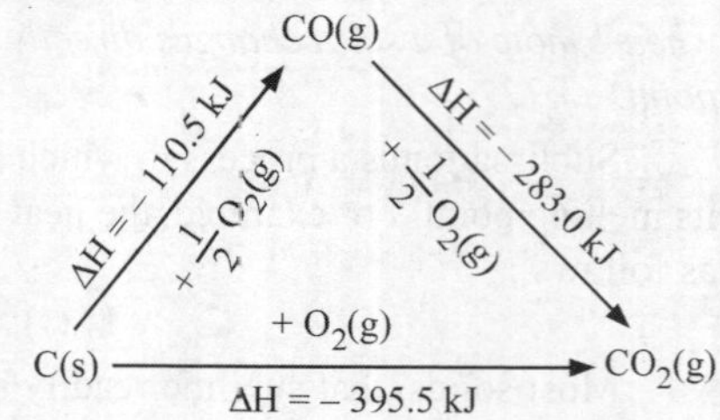

Thus, the total heat evolved in the two steps comes out to be (–111.5) + (–282.0) = –393.5 kJ, which is the same when the reaction takes place directly in one step.

Proof of Hess's Law. Consider the general reaction : A $\longrightarrow$ D

Suppose the heat evolved in this reaction directly is Q joules.

Now suppose the same reaction takes place in three steps as follows :

$$A \xrightarrow{\text{Step 1}} B \xrightarrow{\text{Step 2}} C \xrightarrow{\text{Step 3}} D$$

Suppose the heats evolved in these three steps are q_1, q_2 and q_3 joules respectively (Fig. 8.11).

Thus, the total heat evolved from A to D = $q_1 + q_2 + q_3 = Q'$ joules (say)

According to Hess's law, we must have Q = Q′

Suppose Hess's law is not correct. In that case either Q′ < Q or Q′ > Q. Let us suppose Q′ > Q. We shall now complete the cycle, *i.e.,* we shall go from A to D through three steps and return to A in one step and we shall see what heat changes take place.

Heat change from D to A in one step is reverse of heat change from A to D in one step.

In the direct process from A to D, Q joules of heat are evolved. Therefore, in direct process from D to A, Q joules of heat are absorbed.

Now if we move from A to D in steps and return to A directly. Q′ – Q joules of heat is produced. Thus, by repeating the cyclic process a number of times, a large amount of heat can be created. This is, however, against the law of conservation of energy. Hence, Q′ cannot be greater than Q. Similarly. Q′ cannot be less than Q. It has to be equal to Q. Thus, Hess's law is proved.

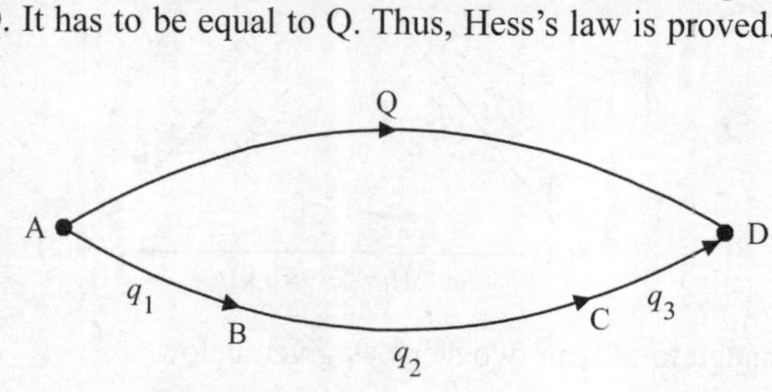

Fig. 8.11. Theoretical proof of Hess's law.

8.18.1. Applications of Hess's Law

Hess's law is very useful in determining the heats (enthalpies) of many reactions, which cannot otherwise be measured by direct methods. The thermochemical equations are added, subtracted, multiplied or divided just like other algebraic equations. Some important applications of Hess's law are :

***(i)* Determination of enthalpy of formation.** The heat of formation of some compounds can not be determined directly by experiment. This can be calculated by the application of Hess's law.

Suppose we want to determine the heat of formation of CO

$$C + \frac{1}{2}O_2 \longrightarrow CO; \quad \Delta H = ?$$

By direct experiment, it is not possible because in the reaction between carbon and oxygen, we may get a mixture of CO and CO_2. But we require only CO. Therefore, from the experiment it is not possible to determine the heat of formation.

We know experimentally, the heat of combustion of pure carbon and carbon monoxide. It is represented by the following equations :

(*a*) $$C + O_2 \longrightarrow CO_2 : \quad \Delta H = -393.5 \text{ kJ}$$

(*b*) $$CO + \frac{1}{2}O_2 \longrightarrow CO_2 : \quad \Delta H = -283.0 \text{ kJ}$$

Now, applying Hess's law on (*a*) and (*b*), we get the required equation for the heat of formation of CO. This can be achieved by subtracting (*b*) from (*a*), *i.e.*, by carrying out (*a* – *b*).

$$C + \frac{1}{2}O_2 \longrightarrow CO; \quad \Delta H = -110.5 \text{ kJ}$$

Therefore, enthalpy of formation of CO is – 110.5 kJ.

***(ii)* Determination of heat (enthalpy) of transition.** Transition of an element from one allotropic form to another is a slow process and requires heating of one allotropic form of the element for a long time at constant temperature. Hence, enthalpies of transition of elements cannot be determined experimentally. They are, however, determined indirectly by means of Hess's law. Thus, Hess's law is useful in determining the enthalpies of transition of the allotropic forms such as rhombic sulphur to monoclinic sulphur, yellow phosphorus to red phosphorus, graphite to diamond, etc.

Suppose we want to determine the enthalpy of transition of sulphur (rhombic) into sulphur (monoclinic). Each variety is combustioned and heats of combustion are found out. This is represented by the following equations :

(*a*) $$S(R) + O_2 \longrightarrow SO_2; \quad \Delta H = -295.1 \text{ kJ}$$

(*b*) $$S(M) + O_2 \longrightarrow SO_2 : \quad \Delta H = -296.4 \text{ kJ}$$

Subtracting (*b*) from (*a*), we get

$$S(R) \longrightarrow S(M); \quad \Delta H = -295.1 - (-296.4) = +1.3 \text{ kJ}$$

Thus, transition of one gram atom of rhombic sulphur to monoclinic sulphur absorbs 1.3 kJ of heat.

***(iii)* Calculation of enthalpies from given data.** Thermochemical equations can be added, subtracted, multiplied or divided by any numerical factor like algebraic equations. Hess's law is, thus, used for the calculation of enthalpy changes for the reactions for which direct experimental measurements are not available.

Example 1. Calculate the heat of formation of acetic acid if its heat of combustion is 867 kJ mol^{-1}. The heats of formation of CO_2 (g) and H_2O (*l*) are –393.5 kJ mol^{-1} and –285.9 kJ mol^{-1} respectively.

Solution. The equations given are :

(*a*) $CH_3COOH + 2O_2 \longrightarrow 2CO_2 + 2H_2O;$ $\Delta H = -867$ kJ

(*b*) $C + O_2 \longrightarrow CO_2;$ $\Delta H = -393.5$ kJ

(*c*) $H_2 + \frac{1}{2}O_2 \longrightarrow H_2O;$ $\Delta H = -205.9$ kJ

Our aim is to find out ΔH value for the reaction

$$2C + 2H_2 + O_2 \longrightarrow CH_3COOH; \quad \Delta H = ?$$

To obtain this equation, multiply equations (*b*) and (*c*) by 2 and add. We get equation (*d*)

$$2C + 2O_2 + 2H_2 + O_2 \longrightarrow 2CO_2 + 2H_2O$$

or

(*d*) $2C + 2H_2 + 3O_2 \longrightarrow 2CO_2 + 2H_2O;\ \Delta H = -1358.8$ kJ

Subtracting (*a*) from (*d*), we get

$$2C + 2H_2 + 3O_2 \longrightarrow 2CO_2 + 2H_2O;\ \Delta H = -1358.8 \text{ kJ}$$

$$CH_3COOH + 2O_2 \longrightarrow 2CO_2 + 2H_2O;\ \Delta H = -867 \text{ kJ}$$

$$- \quad - \qquad - \quad -$$

$$2C + 2H_2 + O_2 \longrightarrow CH_3COOH \quad \Delta H = -491.8 \text{ kJ}$$

∴ Heat of formation of acetic acid = **– 491.8 kJ.**

Example 2. Find the heat of formation of SO_2 from the following thermochemical equations :

(*a*) $S(s) + \frac{3}{2}O_2(g) \longrightarrow SO_3(g) + X$ **kJ**

(*b*) $SO_2(s) + O_2(g) \longrightarrow SO_3(g) + Y$ **kJ**

Solution. Our aim is to find out ΔH value for the reaction

$$S + O_2 \longrightarrow SO_2 = \Delta H = ?$$

This can be obtained by subtracting (*b*) from (*a*), *i.e.,* by carrying out (*a* – *b*)

(*a*) $S(s) + \frac{3}{2}O_2(g) \longrightarrow SO_3(g);$ $\Delta H = -X$ kJ

(*b*) $SO_2(g) + \frac{1}{2}O_2(g) \longrightarrow SO_3(g);$ $\Delta H = -Y$ kJ

$$- \qquad - \qquad -$$

$$S + O_2 \longrightarrow SO_2(g) \qquad \Delta H = -X - (-Y) = -X + Y$$

$$= -(X - Y) \text{ kJ}$$

Example 3. Calculate enthalpy of formation of anhydrous Al_2Cl_6 from the following data :

(*a*) $2Al\ (s) + 6\ HCl\ (aq) \longrightarrow Al_2Cl_6\ (aq) + 3H_2\ ;\ \Delta H = 1004.4$ kJ

(*b*) $H_2\ (g) + Cl_2\ (g) \longrightarrow 2HCl\ (g)\ ;\ \Delta H = 754.74$ kJ

(*c*) $HCl\ (g) + aq \longrightarrow HCl\ (aq)\ ;\ \Delta H = -73.24$ kJ

(*d*) $Al_2Cl_6\ (s) + aq \longrightarrow Al_2Cl_6\ (aq)\ ;\ \Delta H = -\ 642.69$ kJ

Solution. Thermochemical equation for the formation of anhydrous Al_2Cl_6 is

$$2Al + 3Cl_2 \longrightarrow Al_2Cl_6\ (s);\quad \Delta H = ?$$

To obtain ΔH value for this equation from the given data we perform as

$$[a + (3 \times b) + (6 \times c)] - d$$

Rearranging the given equations, we get

$$2Al\ (s) + 6HCl\ (aq) + 3H_2 + 3Cl_2 + 6\ HCl\ (g) + aq - Al_2Cl_6\ (s) - aq \longrightarrow Al_2Cl_6\ (aq) + 3H_2\ (g) + 6HCl\ (g) + 6HCl\ (aq) - Al_2Cl_6\ (aq)$$

$$2Al\ (s) + 3Cl_2\ (g) \longrightarrow Al_2Cl_6\ (s),\ \Delta H = 1004.4 + 3 \times 754.74 + 6 \times (73.24) - (-\ 642.69) = 3471.87 \text{ kJ}$$

Example 4. Calculate enthalpy of formation of methane. Given :

(*a*) $C\ (graphite) + O_2\ (g) \longrightarrow CO_2(g)\ ;\ \Delta H = -\ 393.5$ kJ

(*b*) $2H_2\ (g) + O_2\ (g) \longrightarrow 2H_2O\ (l);\ \Delta H = -\ 571.8$ kJ

(*c*) $CH_4\ (g) + 2O_2\ (g) \longrightarrow CO_2\ (g) + 2H_2O\ (l);\ \Delta H = -\ 890.3$ kJ

Solution. We want to find out the value of ΔH for the reaction

$$C\ (graphite) + 2H_2\ (g) \longrightarrow CH_4;\quad \Delta H = ?$$

This can be obtained by the operation $(a + b) - c$, we get

$$C\ (graphite) + O_2 + 2H_2 + O_2 \longrightarrow CO_2 + 2H_2O \quad .... (a + b)$$

or $C\ (graphite) + 2O_2 + 2H_2 \longrightarrow CO_2 + 2H_2O\ ;\ \Delta H = -\ 965.3$ kJ

$$CH_4 + 2O_2 \longrightarrow CO_2 + 2H_2O\ ;\ \Delta H = -890.3 \text{ kJ}$$

$$-\quad -\quad\quad -\quad -$$

$$C\ (graphite) + 2H_2 \longrightarrow CH_4 \quad \Delta H = -\ 75.0 \text{ kJ} \quad ... (a + b) - c$$

Example 5. From the thermochemical equations :

(*a*) $N_2\ (g) + 3H_2\ (g) \longrightarrow 2NH_3\ (g);\ \Delta H = -\ 92.40$ kJ

(*b*) $2H_2\ (g) + O_2\ (g) \longrightarrow 2H_2O;\ \Delta H = -\ 483.79$ kJ

Find ΔH_f for NH_3 (*g*) and ΔH_f for H_2O (*g*).

Solution. Heat of formation (ΔH_f) is for one mole

$$\therefore \quad \Delta H_f \text{ for } NH_3 = \frac{-92.4}{2} = -46.2 \text{ kJ}$$

$$\therefore \quad \Delta H_f \text{ for } H_2O = \frac{-483.79}{2} = \mathbf{-241.895\,kJ.}$$

Example 6. Calculate the heat of formation of carbon disulphide given that the heats of combustion of carbon disulphide, carbon and sulphur are –1109 kJ, –394.6 kJ and –298.7 kJ respectively.

Solution. The thermochemical equations for the combustion of CS_2, C and S are

(*a*) $CS_2 + 3O_2 \longrightarrow CO_2 + 2SO_2$; $\Delta H = -1109$ kJ

(*b*) $C + O_2 \longrightarrow CO_2$; $\Delta H = -394.6$ kJ

(*c*) $S + O_2 \longrightarrow SO_2$; $\Delta H = -298.7$ kJ

Our aim is to find out the value of ΔH for the following equation,

$C + 2S \longrightarrow CS_2$; $\Delta H = ?$

Multiplying (*c*) by 2 and adding to (*b*), we get

$$2S + 2O_2 + C + O_2 \longrightarrow 2SO_2 + CO_2$$

or

(*d*) $2S + C + 3O_2 \longrightarrow 2SO_2 + CO_2$; $\Delta H = (-298.7) \times 2 + (-394.6) = -992.0$ kJ

Subtracting (*a*) from (*d*), we get

$2S + C + 3O_2 \longrightarrow 2SO_2 + CO_2$; $\Delta H = -992.0$ kJ

$CS_2 + 3O_2 \longrightarrow CO_2 + 2SO_2$; $\Delta H = -1109$ kJ

$\quad - \quad - \qquad\quad - \quad -$

$2S + C \longrightarrow CS_2$; $\Delta H = -992 - (-1109) = +117$ kJ

$$\therefore \quad \text{Heat of formation of } CS_2 = +117 \text{ } \mathbf{kJ\ mol^{-1}}$$

Example. 7. The enthalpy of combustion of ethylene (C_2H_4) is – 1410 kJ. What will be calorific value of ethylene per gram?

Solution. The thermochemical equation for the combustion of C_2H_4 is

$$\underset{28 \text{ grams}}{C_2H_4} + O_2 \longrightarrow 2CO_2 + 2H_2O\ ;\ \Delta H = -1410 \text{ kJ}$$

It is evident from the above equation that

28 grams of C_2H_4 give heat = –1410 kJ

$\therefore$ 1 gram of C_2H_4 will give heat $= \dfrac{-1410}{28} = -50.36$ kJ

$\therefore$ Calorific value of ethylene $= -$ 50.36 kJ.

Example 8. One gram of an organic liquid A (mol. wt. = 128) liberates 150 joules of heat on solidification. Calculate the enthalpy of fusion of A.

Solution. Molecular weight of liquid A = 128 g

Heat evolved when 1 gram of liquid A solidifies = 150 J

$\therefore$ Heat evolved when 128 grams (1 mole) of liquid A solidify = 150 × 128 = 19200 J

$$\text{Liquid (A)} \longrightarrow \text{Solid (A)} ; \Delta H = -19200 \text{ J}$$

Rearranging this equation, we get

$$\text{Solid (A)} \longrightarrow \text{Liquid (A)} ; \Delta H = +\ 19200$$

$\therefore$ Enthalpy of fusion of **A = + 19200 joules.**

Example 9. Calculate the heat of formation of KCl from the following data :

$$(i) \quad \text{KOH}\ (aq) + \text{HCl}\ (aq) \longrightarrow \text{KCl}\ (aq) + \text{H}_2\text{O}\ (l); \quad \Delta H = -57.3 \text{ kJ}$$

$$(ii) \quad \text{H}_2\ (g) + \frac{1}{2}\text{O}_2\ (g) \longrightarrow \text{H}_2\text{O}\ (l); \quad \Delta H = -286.2 \text{ kJ}$$

$$(iii) \quad \frac{1}{2}\text{H}_2\ (g) + \frac{1}{2}\text{Cl}_2\ (g) + aq \longrightarrow \text{HCl}\ (aq); \quad \Delta H = -164.4 \text{ kJ}$$

$$(iv) \quad \text{K}\ (s) + \frac{1}{2}\text{O}_2\ (g) + \frac{1}{2}\text{H}_2\ (g) + aq \longrightarrow \text{KOH}\ (aq); \quad \Delta H = -487.4 \text{ kJ}$$

$$(v) \quad \text{KCl}\ (s) + aq \longrightarrow \text{KCl}\ (aq); \quad \Delta H = +\ 18.4 \text{ kJ}$$

Solution. Our aim is to find out ΔH for the following reaction.

$$(vi) \quad \text{K}(s) + \frac{1}{2}\text{Cl}_2\ (g) \longrightarrow \text{KCl}\ (s), \Delta H = ?$$

In order to get this thermochemical equation, we follow the following two steps :

Step 1. Adding Eqns. (*iii*) and (*iv*) and subtracting Eqn. (*v*), we have

$$(vii) \quad \text{K}(s) + \frac{1}{2}\text{Cl}_2\ (g) + \text{H}_2(g) + \frac{1}{2}\text{O}_2\ (g) \longrightarrow \text{KCl}\ (s) + \text{HCl}\ (aq) + \text{KOH}\ (aq) - \text{KCl}\ (aq)$$

$$\Delta H = -487.4 + (-164.4) - (18.4) = \mathbf{-670.2\ kJ.}$$

Step 2. To cancel out the terms of this equation which do not appear in the required equation (*vi*), add Eqn. (*i*) to Eqn. (*vii*) and subtract Eqn. (*ii*) from their sum. This gives

$$\text{K}\ (s) + \frac{1}{2}\text{Cl}_2\ (g) \longrightarrow \text{KCl}\ (s) ; \Delta H = -670.2 + 57.3 - (-286.2) = \mathbf{-441.3\ kJ.}$$

Example 10. Calculate the heat of hydration of anhydrous copper sulphate ($CuSO_4$) into hydrated copper sulphate ($CuSO_4.5H_2O$). Given that the heats of solutions of anhydrous copper sulphate and hydrated copper sulphate are – 66.5 and +11.7 kJ mol^{-1} respectively.

Solution. Data provided is :

$$(i) \quad CuSO_4\,(s) + aq \longrightarrow CuSO_4\,(aq); \quad \Delta H = -66.5 \text{ kJ}$$

$$(ii) \quad CuSO_4.5H_2O\,(s) + aq \longrightarrow CuSO_4\,(aq); \quad \Delta H = +\,11.7 \text{ kJ}$$

Our aim is to find out heat of the following reaction.

$$CuSO_4\,(s) + 5H_2O\,(l) \longrightarrow CuSO_4.5H_2O\,(s); \quad \Delta H = ?$$

Equation (*i*) can be written in two steps as :

$$(iii) \quad CuSO_4\,(s) + 5H_2O\,(l) \longrightarrow CuSO_4.5H_2O\,(s); \quad \Delta H = q_1 \text{ kJ}$$

$$(iv) \quad CuSO_4.5H_2O\,(s) + aq \longrightarrow CuSO_4\,(aq); \quad \Delta H = q_2 \text{ kJ}$$

According to Hess's law $q_1 + q_2 = -\,66.5$ kJ ....(*v*)

Further, equations (*ii*) and (*iv*) are same $\therefore\ q_2 = +\,11.7$ kJ

Putting this value in eq. (*v*) above, we get

$$q_1 + 11.7 = -\,66.5 \text{ or } q_1 = -\,66.5 - 11.7 \text{ kJ} = -78.2 \text{ kJ}$$

Thus, equation (*iii*) may be written as

$$CuSO_4\,(s) + 5H_2O\,(l) \longrightarrow CuSO_4.5H_2O\,(s)\,; \Delta H = -78.2 \text{ kJ}$$

This is what we aimed at. Hence, the required value of the heat of hydration is $\Delta H = -\,78.2$ kJ.

Example 11. Calculate the heat of formation of carbon monoxide (CO) from the following data :

$$(i) \quad C\,(s) + O_2\,(g) \longrightarrow CO_2\,(g)\,; \quad \Delta H = -\,393 \text{ kJ}$$

$$(ii) \quad CO\,(g) + \frac{1}{2}\,O_2\,(g) \longrightarrow CO_2\,(g)\,; \quad \Delta H = -\,282 \text{ kJ}$$

Solution. Our aim is to find out ΔH for the following reaction :

$$(i) \quad C\,(s) + \frac{1}{2}O_2\,(g) \longrightarrow CO\,(g)\,;\ \Delta H = ?$$

Subtracting equation (*ii*) from (*i*), we get

$$C\,(s) + \frac{1}{2}O_2\,(g) - CO\,(g) \longrightarrow 0; \quad \Delta H = -393.3 - (-282.8) = -110.5 \text{ kJ}$$

$$\text{or} \quad C\,(s) + \frac{1}{2}O_2\,(g) \longrightarrow CO\,(g); \quad \Delta H = -110.5 \text{ kJ}$$

$\therefore$ Heat of formation of CO, ΔH_f = **110.5 kJ.**

Example 12. Calculate the heat change accompanying the transformation of C (graphite) to C (diamond). Given that the heats of combustion of graphite and diamond are 393.5 and 395.4 kJ mol^{-1} respectively.

Solution. Data provided is :

$$(i) \quad C\,(\text{graphite}) + O_2\,(g) \longrightarrow CO_2\,(g); \quad \Delta H = -393.5 \text{ kJ}$$

$$(ii) \quad C\,(\text{diamond}) + O_2\,(g) \longrightarrow CO_2\,(g); \quad \Delta H = -395.4 \text{ kJ}$$

Our aim is to find out ΔH for the following reaction

$$C\text{ (graphite)} \longrightarrow C\text{ (diamond)}; \qquad \Delta H = ?$$

Subtracting equation (*ii*) from (*i*), we get

$$C\text{ (graphite)} - C\text{ (diamond)} \longrightarrow 0; \qquad \Delta H = -393.5 - (-395.4) = +1.9 \text{ kJ}$$

or
$$C\text{ (graphite)} \longrightarrow C\text{ (diamond)}; \qquad \Delta H = +1.9 \text{ kJ.}$$

Q. 13. The heat of combustion of ethyl alcohol (C_2H_5OH) is 1380.7 kJ mol^{-1}. If the heats of formation of CO_2 and H_2O are 394.5 and 286.6 kJ mol^{-1} respectively, calculate the heat of formation of ethyl alcohol.

Solution. Data provided is :

(*i*) $$C_2H_5OH + 3O_2 \longrightarrow 2CO_2 + 3H_2O; \qquad \Delta H = -1380.7 \text{ kJ}$$

(*ii*) $$C + O_2 \longrightarrow CO_2; \qquad \Delta H = -394.5 \text{ kJ}$$

(*iii*) $$H_2 + \frac{1}{2}O_2 \longrightarrow H_2O; \qquad \Delta H = -286.6 \text{ kJ}$$

Our aim is to determine ΔH for the following reaction

$$2C + 3H_2 + \frac{1}{2}O_2 \longrightarrow C_2H_5OH$$

In order to get this thermochemical equation multiply Eqn. (*ii*) by 2 and Eqn. (*iii*) by 3 and substract Eqn. (*i*) from their sum, *i.e.,* operating : 2 × Eqn. (*ii*) + 3 × Eqn. (*iii*) – Eqn. (*i*), we get

$$2C + 3H_2 + \frac{1}{2}O_2 \longrightarrow C_2H_5OH; \qquad \Delta H = 2(-394.5) + 3(-286.6) - (-1380.7) = -268.1 \text{ kJ}$$

Thus, the heat of formation of ethyl alcohol = ΔH_f = **–268.1 kJ mol^{-1}.**

Example 14. Calculate the standard enthalpy of formation of n-butane given that the standard enthalpies of combustion of n-butane (g), C (graphite) and H_2(g) are – 2878.5 kJ mol^{-1}, 393.5 kJ mol^{-1} and – 285.3 kJ mol^{-1} respectively.

Solution. We aim at the equation

$$4C + 5H_2 \longrightarrow C_4H_{10}; \qquad \Delta H = ?$$

The given data can be expressed in the form of the following equations.

(*i*) $$C_4H_{10} + 6\frac{1}{2}O_2 \longrightarrow 4CO_2 + 5H_2O; \qquad \Delta H = -2878.5 \text{ kJ}$$

(*ii*) $$4C + 4O_2 \longrightarrow 4CO_2; \qquad \Delta H = 4 \times (-393.5) \text{ kJ}$$

(*iii*) $$5H_2 + \frac{5}{2}O_2 \longrightarrow 5H_2O; \qquad \Delta H = 5 \times (-285.3) \text{ kJ}$$

Add (*ii*) and (*iii*) and subtract (*i*)

$$(-1574.0) + (-1426.5) - (-2878.5)$$
$$= -122.0 \text{ kJ}$$

Example 15. Heats of solution (ΔH) for $BaCl_2.2H_2O$ and $BaCl_2$ are 8.8 and –20.6 kJ respectively. Calculate the heat of hydration of $BaCl_2$ to $BaCl_2.2H_2O$.

Solution. We are given

$$BaCl_2.2H_2O\ (s) + aq \longrightarrow BaCl_2\ (aq); \quad \Delta H = 8.8\ kJ \quad ...\ (i)$$

$$BaCl_2\ (s) + aq \longrightarrow BaCl_2\ (aq); \quad \Delta H = -20.6\ kJ \quad ...\ (ii)$$

We aim at :

$$BaCl_2\ (s) + 2H_2O \longrightarrow BaCl_2.2H_2O\ (s); \quad \Delta H = ? \quad ...\ (iii)$$

Equation (*ii*) may be written in two steps as

$$BaCl_2\ (s) + 2H_2O \longrightarrow BaCl_2.2H_2O\ (s) \quad ; \Delta H = \Delta H_1\ (say) \quad ...\ (iv)$$

$$BaCl_2.2H_2O\ (s) + aq \longrightarrow BaCl_2\ (aq) \quad ; \Delta H = \Delta H_2\ (say) \quad ...\ (v)$$

Then according to Hess's law

$$\Delta H_1 + \Delta H_2 = -20.6\ kJ$$

But $\Delta H_2 = 8.8\ kJ$ [∵ Equation (*i*) = Equation (*v*)]

∴ $\Delta H_1 = -20.6 - 8.8 = -29.4\ kJ$

But Equation (*iii*) = Equation (*iv*)

Hence, the heat of hydration of $BaCl_2$ = **–29.4 kJ.**

Example 16. Given the following thermochemical equations :

$$(i) \quad \textbf{S (rhombic)} + \mathbf{O_2}\ (g) \longrightarrow \mathbf{SO_2}\ (g)\ ; \quad \mathbf{\Delta H = -\ 297.5\ kJ}$$

$$(ii) \quad \textbf{S (monoclinic)} + \mathbf{O_2}\ (g) \longrightarrow \mathbf{SO_2}\ (g)\ ; \quad \mathbf{\Delta H = -300.0\ kJ}$$

Calculate ΔH for the transformation of one gram atom of rhombic sulphur into monoclinic sulphur.

Solution. We aim at :

$$S\ (rhombic) \longrightarrow S\ (monoclinic); \quad \Delta H = ?$$

Equation (*i*) – Equation (*ii*) gives

$$S\ (rhombic) - S\ (monoclinic) \longrightarrow 0; \quad \Delta H = 297.5 - (-300.0) = 2.5\ kJ$$

or

$$S\ (rhombic) \longrightarrow S\ (monoclinic); \quad \Delta H = \mathbf{-2.5\ kJ.}$$

Thus, for the transformation of one gram atom of rhombic sulphur into monoclinic sulphur, 2.5 kJ of heat is absorbed.

Example 17. Calculate the enthalpy of formation of $N_2O_5(g)$:

$$\mathbf{N_2(g) + \frac{5}{2}\,O_2(g) \longrightarrow N_2\,O_5(g)}$$

The following information is given :

$$\textbf{(i)} \quad \mathbf{N_2\ (g) + O_2\ (g) \longrightarrow 2NO\ (g);} \quad \mathbf{\Delta H^\circ = +\ 180\ kJ}$$

$$\textbf{(ii)} \quad \mathbf{4\ NO_2\ (g) + O_2\ (g) \longrightarrow 2N_2\,O_5\ (g);} \quad \mathbf{\Delta H^\circ = -\ 120.6\ kJ}$$

$$\textbf{(iii)} \quad \mathbf{2\ NO\ (g) + O_2\ (g) \longrightarrow 2NO_2\ (g);} \quad \mathbf{\Delta H^\circ = -\ 114.0\ kJ}$$

Solution. Perform the operation (*i*) $+ \frac{1}{2}$ (*iii*) to get the desired equation as

$$N_2(g) + \frac{5}{2}O_2(g) \longrightarrow N_2O_5\ (g)$$

Applying Hess's law, perform the same operation on enthalpies to obtain the enthalpy of formation of N_2O_5 (g)

$$\Delta H_f = [180 - \frac{1}{2}(120.6) - 114.0] \text{ kJ}$$
$$= 5.7 \text{ kJ}$$

PROBLEMS FOR PRACTICE

1. Calculate the enthalpy of formation of acetic acid from the following data:

(*i*) $C(s) + O_2(g) \longrightarrow CO_2(g), \Delta H = -393.7 \text{kJ}$

(*ii*) $H_2 + (g) + \frac{1}{2}O_2(g) \longrightarrow H_2O(l), \Delta H = -285.8 \text{ kJ}$

(*iii*) $CH_3COOH(l) + 2O_2(g) \longrightarrow 2CO_2(g) + 2H_2O(l), \Delta H = -873.2 \text{ kJ}$

[**Ans.** – 485.8 kJ mol^{-1}]

2. Calculate the standard enthalpy of formation of SO_3 at 298 K using the following reactions and enthalpies.

$S_8(s) + 8O_2(g) \longrightarrow 8SO_2(g), \Delta H° = -2775 \text{ kJ mol}^{-1}$

$2SO_2(g) + O_2(g) \longrightarrow 2SO_3(g), \Delta H° = -198 \text{ kJ mol}^{-1}$

[**Ans.** – 445.9 kJ mol^{-1}]

3. From the following thermochemical equations, calculate the standard enthalpy of formation of HCl (*g*).

(A) $H_2(g) \longrightarrow 2H(g), \quad \Delta H = +436.0 \text{ kJ mol}^{-1}$

(B) $Cl_2(g) \longrightarrow 2Cl(g), \quad \Delta H = +242.7 \text{ kJ mol}^{-1}$

(C) $HCl(g) \longrightarrow H(g) + Cl(g), \quad \Delta H = +431.8 \text{ kJ mol}^{-1}$

[**Ans.** – 92.45 kJ mol^{-1}]

4. Calculate the enthalpy of formation of acetic acid if the enthalpy of combustion to CO_2 (g) and H_2O (*l*) is –867.0 kJ mol^{-1} and enthalpies of formation of CO_2 (*g*) and H_2O (*l*) are respectively –393.5 and –285.9 kJ mol^{-1} [**Ans.** –491.8 kJ mol^{-1}]

5. Calculate the enthalpy of formation of sucrose ($C_{12}H_{22}O_{11}$) from the following data:

(*i*) $C_{12}H_{22}O_{11} + 12O_2 \longrightarrow 12CO_2 + 11H_2O; \quad \Delta H = -5200.7 \text{ kJ mol}^{-1}$

(*ii*) $C + O_2 \longrightarrow CO_2, \quad \Delta H = -394.5 \text{ kJ mole}^{-1}$

(*iii*) $H_2 + \frac{1}{2}O_2 \longrightarrow H_2O, \quad \Delta H = -285.8 \text{ kJ mol}^{-1}$

[**Ans.** – 2677.1 KJ mol^{-1}]

6. Ethylene on combustion gives carbon dioxide and water. Its enthalpy of combustion is 1410.0 kJ mol^{-1}. If the enthalpy of formation of CO_2 and H_2O are 393.3 kJ and 286.2 kJ respectively. Calculate the enthalpy of formation of ethylene. [**Ans.** + 51.0 kJ mol^{-1}]

7. Calculate the enthalpy of formation of benzene, given that enthalpies of combustion of benzene, carbon and hydrogen are –3281.5 kJ, –394.9 kJ and –286.1 kJ/mol, respectively. [**Ans.** +53.8 kJ mol^{-1}]

8. Calculate the enthalpy of formation of *n*-butane from the following data:

(*i*) $2C_4H_{10}(g) + 13O_2(g) \longrightarrow 8CO_2(g) + 10H_2O(l), \Delta H = -5757.2$ kJ mol^{-1}

(*ii*) $C(s) + O_2(g), \longrightarrow CO_2(g), \quad \Delta H = -405.4$ kJ mol^{-1}

(*iii*) $2H_2(g) + O_2(g) \longrightarrow 2H_2O(l), \quad \Delta H = -572.4$ kJ mol^{-1}

On what law are your calculations based? [**Ans.** –174 kJ mol^{-1}]

9. Calculate the enthalpy of formation of anhydrous Al_2Cl_6 from the following data:

(*i*) $2Al(s) + 6HCl(aq) \longrightarrow Al_2Cl_6(aq) + 3H_2(g) + 1004.2$ kJ mol^{-1}

(*ii*) $H_2(g) + Cl_2(g) \longrightarrow 2HCl(g) + 184.1$ kJ mol^{-1}

(*iii*) $HCl(g) + aq \longrightarrow HCl(aq) + 73.2$ kJ mol^{-1}

(*iv*) $2Al_2Cl_6(s) + aq \longrightarrow Al_2Cl_6(aq) + 643.1$ kJ mol^{-1}

[**Ans.** –1352.6 kJ mol^{-1}]

10. Calculate the enthalpy of formation of carbon disulphide given that the enthalpy of combustion of carbon disulphide is 110.2 kJ mol^{-1} and those of sulphur and carbon are 297.4 kJ and 394.5 kJ/g atom respectively. [**Ans.** –879.1 kJ mol^{-1}]

8.19. BOND ENTHALPY OR BOND ENERGY

The bond energy of a particular bond is defined as the average amount of energy released when one mole of bonds are formed from isolated gaseous atoms or the amount of energy required when 1 mole of bonds are broken so as to get the separated gaseous atoms.

For diatomic molecules (like H_2, HCl, etc.), the bond energy is equal to the dissociation energy of the molecules but for a *polyatomic molecule* like CH_4, the bond dissociation energies of the four C — H bonds are different. Hence, an average value is taken. Bond energies of some common bonds are given in Table 8.3.

Table 8.3 Bond energies of some common bonds

Bond	Bond energy (kJ mol^{-1})	Bond	Bond energy (kJmol^{-1})
H — H	436	C — H	414
H — F	565	O — H	463
H — Cl	431	N — H	389
H — Br	364	C — C	347
H — I	297	C = C	619
F — F	155	C ≡ C	812
Cl — Cl	242	C — Cl	326
Br — Br	190	C — O	335
I — I	149	C = O	707
O = O	494	C — N	293
N ≡ N	941	C = N	616
		C ≡ N	879

Heat of reaction can be calculated from bond energy data. The following formula can be used directly in these calculations.

$\Delta H_{reaction} = \Sigma$ Bond energies of reactants $- \Sigma$ Bond energies of products

PROBLEMS ON CALCULATION OF BOND ENERGY

Example 1. Calculate the enthalpy change for the reaction

$$H_2\,(g) + Br_2\,(g) \longrightarrow 2HBr\,(g)$$

Given that the bond energies of H — H, Br — Br and H — Br are 435, 192 and 364 kJ mol^{-1} respectively.

Solution. First method. *By calculating the total energy absorbed and released.*

Energy absorbed in the dissociation of 1 mole of H – H bonds = 435 kJ

Energy absorbed in the dissociation of 1 mole of Br – Br bonds = 192 kJ

Total energy absorbed = 435 + 192 = 627 kJ

Energy released in the formation of 1 mole of H – Br bonds = 364 kJ

∴ Energy released in the formation of 2 moles of H – Br bonds = 2 × 364 = 728 kJ

Energy released is greater than energy absorbed.

Hence, net result is the release of energy.

Energy released = 728 kJ – 627 kJ = 101 kJ

Thus, for the given reaction,

ΔH = **–101 kJ.**

Second method. *Using the formula directly.*

$\Delta H_{reaction}$ = Σ Bond energies of reactants – Σ Bond energies of products

= [Bond energy (H — H) + Bond energy (Br — Br)] –[2 × Bond energy (H— Br)]

= 435 + 192 – (2 × 364) = 627 –128 = **–101 kJ.**

Hess's law can also be applied to solve the problem.

Example 2. Calculate the bond energy of C — H bond, given the heat of combustion of methane (CH_4), graphite and hydrogen are 891 kJ, 394 kJ and 286 kJ respectively while the heat of sublimation of graphite is 717 kJ and the heat of dissociation of hydrogen molecule is 436 kJ.

Solution. The data given is :

$$CH_4\,(g) + 2O_2\,(g) \longrightarrow CO_2\,(g) + 2H_2O\,(g);\ \Delta H = -891 \text{ kJ} \quad ...(i)$$

$$C\,(s) + O_2\,(g) \longrightarrow CO_2\,(g);\ \Delta H = -394 \text{ kJ} \quad ...(ii)$$

$$H_2\,(g) + \frac{1}{2}O_2\,(g) \longrightarrow H_2O\,(g);\ \Delta H = -286 \text{ kJ} \quad ...(iii)$$

$$C\,(s) \longrightarrow C\,(g);\ \Delta H = +717 \text{ kJ} \quad ...(iv)$$

$$H_2\,(g) \longrightarrow 2H\,(g);\ \Delta H = +436 \text{ kJ} \quad ...(v)$$

We want ΔH for the following reaction

$$\frac{1}{4}\,[CH_4\,(g)] \longrightarrow C\,(g) + 4H\,(g)]; \qquad \Delta H = ? \qquad \ldots (vi)$$

The problem is solved in two steps.

First we calculate ΔH for $CH_4\,(g) \longrightarrow C\,(g) + 4H\,(g)$

and then divide the result by 4.

Operating Eqn. (*i*) + Eqn. (*iv*) + 2 × Eqn. (*v*), we get

$$CH_4\,(g) + 2O_2\,(g) + C\,(s) + 2H_2\,(g) \longrightarrow CO_2\,(g) + 2H_2O\,(g) + C\,(g) + 2H\,(g) \quad \ldots (vii)$$

To cancel out the terms not needed now operate Eqn. (*vii*) – 2 × Eqn. (*iii*) – Eqn. (*ii*).

$$CH_4\,(g) \longrightarrow C\,(g) + 4H\,(g);$$

$$\Delta H = -891 + 717 + 2 \times 436 - 2 \times (-286) - (-394) = +1664 \text{ kJ.}$$

This is the energy required for the dissociation of 4 moles of C—H bonds.

Bond dissociation energy for C—H bond $= \frac{1664}{4}$ kJ = **416 kJ mol⁻¹.**

PROBLEMS FOR PRACTICE

1. ΔH for the reaction

H — C ≡ N (g) + 2 H_2 (g) ⟶ 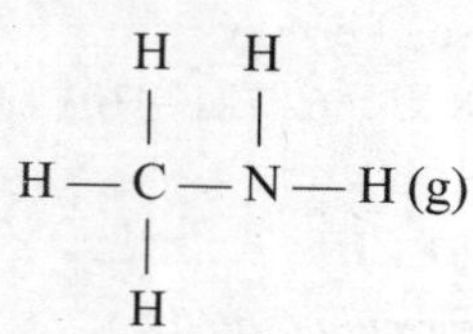 (g)

is – 150 kJ. Calculate the bond energy of C ≡ N bond.

[Given bond energies of C — H = 414 kJ mol^{-1}; H — H = 435 kJ mol^{-1}; C — N = 293 kJ mol^{-1}, N — H = 396 kJ mol^{-1}] **[Ans.** 893 kJ mol^{-1}]

2. Calculate the enthalpy of hydrogenation of C_2H_2 (*g*) to C_2H_4 (*g*).

Given bond energies. C — H = 414.0 kJ mol^{-1}, C ≡ C = 827.6 kJ mol^{-1}, C = C = 606.0 kJ mol^{-1}, H — H = 430.5kJ mol^{-1}) **[Ans.** – 175.9 kJ mol^{-1}]

3. Calculate the C — C bond energy from the following data:

(*i*) $2C\text{ (graphite)} + 3H_2\,(g) \longrightarrow C_2H_6\,(g),\ \Delta H = -84.67$ kJ

(*ii*) $C\text{ (graphite)} \longrightarrow C\,(g),\ \Delta H = 716.7$ kJ

(*iii*) $H_2\,(g) \longrightarrow 2\,H\,(g),\ \Delta H = 435.9$ kJ

Assume the C — H bond energy as 416 kJ. **[Ans.** – 964 kJ mol^{-1}]

4. Propane has the structure $H_3C - CH_2 - CH_3$. Calculate the change in enthalpy for the following reaction:

$$C_3H_8\,(g) + 5O_2\,(g) \longrightarrow 3\,CO_2\,(g) + 4\,H_2O\,(g)$$

Given that average bond enthalpies are :

C — C	C — H	C = O	O = O	O — H
347	414	741	498	464 kJ mol^{-1}

[Ans. – 1662 kJ mol^{-1}]

5. Calculate ΔH° for the reaction

$$CH_2 = CH_2 + 3\,O_2 \longrightarrow 2\,CO_2 + H_2O$$

Given that the average bond energies of the different bonds are

Bond	C — H	O = O	C = O	O — H	C = C
Bond energy (kJ mol^{-1})	414	499	724	460	619

[**Ans.** – 964 kJ mol^{-1}]

6. Calculate the enthalpy change for the following reaction

$$H_2\,(g) + Cl_2\,(g) \longrightarrow 2HCl\,(g)$$

Given that the bond dissociation energies of H — H, Cl — Cl and H — Cl are 437 kJ, 244 kJ and 433 kJ mol^{-1} respectively. [**Ans.** – 185 kJ]

MISCELLANEOUS EXAMPLES

Example 1. If a man takes a diet which gives him energy equal to 9500 kJ per day and he expends energy in all forms to a total of 12000 kJ per day, what is the change in internal energy per day? If the energy lost was stored as sucrose (1632 kJ per 100 g), how many days should it take to lose 1 kg? Ignore water loss.

Solution. Loss of energy per day = 12000 – 9500 = 2500 kJ

For a loss of 1632 kJ of energy, sucrose ($C_{12}H_{22}O_{11}$) lost = 100 g (given)

$\therefore$ For a loss of 2500 kJ of energy, sucrose lost $= \dfrac{100}{1632} \times 2500\text{ g} = 153.2\text{ g}$

Thus, loss of 153.2 g in weight takes place in 1 day.

$\therefore$ Time taken to lose 1 kg or 1000 g glucose $= \dfrac{1}{153.2} \times 1000 =$ **6.5 days.**

Example 2. From the thermochemical equation

$$C_6H_6\,(l) + 7\frac{1}{2}\,O_2\,(g) \longrightarrow 3H_2O\,(l) + 6CO_2\,(g)\,; \qquad \Delta H = -3264.64\text{ kJ mol}^{-1}$$

Calculate the energy evolved when 39 g of C_6H_6 are burnt in an open container.

Solution. 1 mole of benzene = 78 g

When 78 g of C_6H_6 burn, heat evolved = 3264.64 kJ

$\therefore$ When 39 g of C_6H_6 burn, heat evolved $= \dfrac{3264.64}{78} \times 39 =$ **1632.32 kJ.**

Example 3. The heat evolved in the combustion of methane is given by the following equation :

$$CH_4\,(g) + 2O_2\,(g) \longrightarrow CO_2\,(g) + 2H_2O\,(l); \qquad \Delta H = -890.3\text{ kJ}$$

(*a*) How many grams of methane would be required to produce 445.15 kJ of heat on combustion?

(*b*) How many grams of carbon dioxide would be formed when 445.15 kJ of heat are evolved?

(c) What volume of oxygen at STP would be used in the combustion process (a) or (b)?

Solution. (a) From the given equation,

Heat produced from 1 mole of CH_4, *i.e.*, 16 g of CH_4 = 890.3 kJ

Heat produced from 8 g of CH_4 = **445.15 kJ.**

(b) From the given equation,

When 890.3 kJ of heat are evolved, CO_2 formed = 44 g

∴ When 445.15 kJ of heat are evolved, CO_2, formed = **22 g.**

(c) O_2 used in the production of 890.3 kJ of heat = 44.8 litres at STP.

Hence, O_2 used in the production of 445.15 kJ of heat = **22.4 litres at STP.**

Example 4. The heat evolved in the combustion of glucose is shown in the following equation :

$$\mathbf{C_6H_{12}O_6 + 6O_2\,(g) \longrightarrow 6CO_2\,(g) + 6H_2O\,(g);\ \Delta H = -\,2840\ kJ}$$

What is the energy requirement for production of 0.36 g of glucose by the reverse reaction?

Solution. Data provided is

$$C_6H_{12}O_6 + 6O_2\,(g) \longrightarrow 6CO_2\,(g) + 6H_2O\,(g); \qquad \Delta H = -2840\ kJ$$

Writing the reaction in reverse direction

$$6CO_2\,(g) + 6H_2O\,(g) \longrightarrow C_6H_{12}O_6\,(s) + 6O_2\,(g); \qquad \Delta H = +\,2840\ kJ$$

For production of 1 mole of $C_6H_{12}O_6$ (180 g) heat required (absorbed) = 2840 kJ

∴ For production of 0.36 g of glucose, heat absorbed = $\frac{2840}{180} \times 0.36 =$ **5.68 kJ.**

Example 5. The thermochemical equation for solid and liquid rocket fuel are given below :

$$\mathbf{2Al\,(s) + 1\tfrac{1}{2}O_2\,(g) \longrightarrow Al_2O_3\,(s); \qquad \Delta H = -1667.8\ kJ}$$

$$\mathbf{H_2\,(g) + \tfrac{1}{2}O_2\,(g) \longrightarrow H_2O\,(l); \qquad \Delta H = -285.9\ kJ}$$

(a) If equal masses of aluminium and hydrogen are used, which is better rocket fuel?

(b) Determine ΔH for the reaction $Al_2O_3\,(s) \longrightarrow 2Al\,(s) + 1\frac{1}{2}O_2\,(g)$

Solution.

2 moles of Al = 54 g

54 g of Al on combustion give heat = 1667.8 kJ

∴ 1 g of Al on combustion gives heat = $\frac{1667.8}{54} = 30.9$ kJ

1 mole of H_2 = 2 g

2 g of H_2 on combustion give = 285.9 kJ of heat

∴ 1 g of H_2 on combustion will give = $\frac{285.9}{2} = 142.95$ kJ of heat

Thus, H_2 is a better rocket fuel.

(b) Writing the reverse of the first reaction, we have

$$Al_2O_3\,(s) \longrightarrow 2Al\,(s) + 1\frac{1}{2}O_2\,(g);\ H = +\,1667.8\text{ kJ}$$

Thus, for the reaction given in part (*b*) of the problem,

$$\Delta H = +\,\mathbf{1667.8\ kJ.}$$

8.20. VARIATION OF HEAT OF REACTION WITH TEMPERATURE KIRCHOFF'S EQUATION

Kirchoffs

Heat evolved or absorbed depends upon the temperature at which the reaction is carried out. The relationship showing the variation of the heat of reaction with temperature was given by Kirchoff in 1856.

Kirchoff's equation gives the relationship between variation in heat of reaction with temperature. It can be stated as :

The change in the heat of reaction at constant pressure for every degree change of temperature is equal to the change in the heat capacity at constant pressure, accompanying the reaction.

It can be derived as under.

Consider the general reaction

$$A \longrightarrow B$$

as shown in Fig.8.11

Suppose the heat of reaction (heat evolved) at temperature $T_1 = \Delta H_1$ and the heat of reaction (heat evolved) at temperature $T_2 = \Delta H_2$.

Starting from A at temperature T_1, we can reach B at temperature T_2 by two different paths :

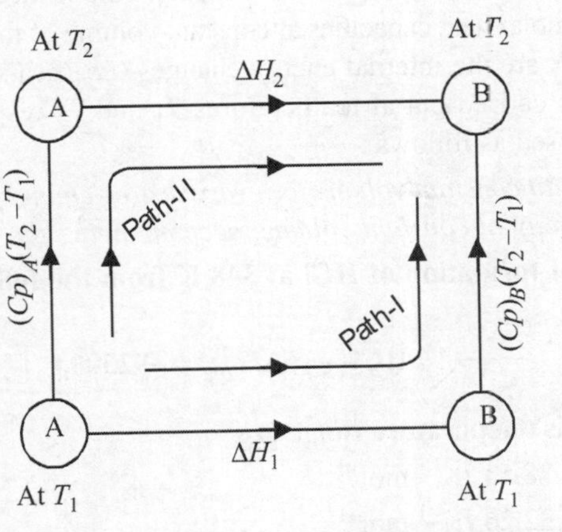

Fig 8.11 Changing A to T_1 to B at T_2 by two different paths.

Path I : We perform the reaction at temperature T_1 to obtain B at the same temperature and then heat this up to the temperature T_2. Then

Heat **evolved** in the first stage = ΔH_1

Heat **absorbed** in the second stage = $(C_P)_B\,(T_2 - T_1)$

where $(C_P)_B$ is the average molar heat capacity of the products.

$\therefore$ Net heat evolved $= \Delta H_1 - (C_p)_n(T_2 - T_1)$

Path II: We first heat up the reactants to the temperature T_2 and then perform the reaction to give the products at the same temperature. Then if $(C_P)_A$ is the average molar heat capacity of the reactant *A*,

Heat **absorbed** in the first stage $= (C_p)_A\,(T_2 - T_1)$

Heat **evolved** in the second stage $= \Delta H_2$

$\therefore$ Net heat evolved $= \Delta H_2 - (C_p)_A\,(T_2 - T_1)$.

Since the initial and the final states are the same in both the above cases, therefore, by law of conservation of energy, we must have

Net heat evolved by path I = Net heat evolved by path II

i.e. $\Delta H_1 - (C_p)_B\ (T_2 - T_1) = \Delta H_2 - (C_p)_A\,(T_2 - T_1)$

or $$\frac{\Delta H_2 - \Delta H_1}{T_2 - T_1} = \Delta C_p$$

When the reaction is carried out at constant volume: We have the thermodynamic relation

$$\Delta H = \Delta U + P\Delta V$$

Under conditions of constant volume, $\Delta V = 0$

$$\therefore \quad \Delta H = \Delta U$$

Hence equation derived above takes the form

$$\frac{\Delta U_2 - \Delta U_1}{T_2 - T_1} = \Delta C_v$$

where $\Delta C_v = (C_v)_B - (C_v)_A$ is the change in the heat capacity at constant volume accompanying the reaction, $(C_v)_B$ and $(C_v)_B$ are the mean molar heat capacities at constant volume of the reactants and the products respectively, ΔU_1 and ΔU_2 are the internal energy changes (*i.e.*, heats of reaction at constant volume) when the reaction is carried out at temperatures T_1 and T_2 respectively. Thus Kirchoff's equation may also be expressed as follows :

The change in the heat of reaction at constant volume for every degree change of temperature is equal to the change in the heat capacity at constant volume, accompanying the reaction.

Example 1. Calculate the heat of formation of HCl at 348 K from the following data :

$$\frac{1}{2}\mathbf{H_2}(g) + \frac{1}{2}\mathbf{Cl_2}(g) \longrightarrow \mathbf{HCl}(g);\ \Delta H^0_{298} = -92300\,\mathbf{J}$$

The mean heat capacities over this temperature range are

H_2 (*g*); C_p = 28.53 $JK^{-1}\ mol^{-1}$

Cl_2 (*g*); C_p = 32.26 $JK^{-1}\ mol^{-1}$

HCl (*g*); C_p = 28.49 $JK^{-1}\ mol^{-1}$

Solution. The reaction under consideration is

$$\frac{1}{2}H_2(g) + \frac{1}{2}Cl_2(g) \longrightarrow HCl\ (g)$$

For this reaction,

$$\Delta C_p = (C_p)_{HCl} - \frac{1}{2}(C_p)_{H_2} - \frac{1}{2}(C_p)_{Cl_2}$$

$$= 28.49 - \frac{1}{2}(28.53) - \frac{1}{2}(32.26)$$

$$= 1.91\ JK^{-1}\ mol^{-1}$$

We are given that

$$\Delta H_1 = -92300\ J$$
$$T_1 = 298\ K$$
$$\Delta H_2 = ?$$

According to Kirchoff's equation

$$\frac{\Delta H_2 - \Delta H_1}{T_2 - T_1} = \Delta C_p$$

or

$$\Delta H_2 = \Delta H_1 + \Delta C_p\ (T_2 - T_1)$$
$$= -92300 + (-1{\cdot}91)\ (348 - 298)$$
$$= -92300 - 95{\cdot}5$$
$$= \mathbf{-92395.5\ J}$$

PROBLEMS FOR PRACTICE

1. Calculate the heat of formation of H_2O (liquid) at 383 K, if ΔH for the reaction

$$H_2\ (g) + \frac{1}{2}\ O_2\ (g) \longrightarrow H_2O\ (l)$$

at 298 K is –298.06 kJ. The average value of heat capacities between the two temperature for H_2 (*g*), O_2 (*g*) and H_2O (*l*) are 27.61, 29.50 and 75.31 $JK^{-1}\ mol^{-1}$ respectively.

[**Ans.** 283.92 kJ]

2. If the standard heat of formation of HCl gas at 25° C is –22.66 kcal mol^{-1}, calculate the value at 60° C. C_p for H_2 (*g*) = 6.82 cal mol^{-1} deg^{-1}

3. The heat of solution of ammonia at 25° C was found to be –11.04 k cal. Calculate the heat of solution at 348 K. Given that the mean heat capacities of N_2, H_2 and NH_3 are 6.80, 6.77 and 8.86 cal/degree/mol respectively.

[**Hint.** $\frac{1}{2}\ N_2 + \frac{3}{2}\ H_2 \longrightarrow NH_3$

$$\Delta C_p = (C_p)\ NH_3 - \frac{1}{2}\ (C_P)\ N_2 - \frac{3}{2}(C_p)H_2$$

$$T_1 = 25 + 273 = 298\ K,\ T_2 = 348\ K]$$

4. The molar heat capacities at constant pressure of H_2 (*g*), Cl_2 (*g*) and HCl (*g*) are respectively 29.3, 34.7 and 28.9 JK^{-1}. If the heat of formation of HCl (*g*) at constant pressure at 293 K is –91.2 kJ, what will be its heat of formation at 313 K?

[**Ans.** –91.27 kJ]

5. Consider the reaction

$$H_2O\ (g) \longrightarrow H_2 + \frac{1}{2}\ O_2\ ;\ \Delta H° = +57780\ cal\ at\ 18°\ C.$$

What would be ΔH° value at 25° C? The C_p values are

$C_p\ (H_2O) = 8.02$, $C_p\ (H_2) = 6.89$ and $C_p\ (O_2) = 7.96$ cal deg^{-1} mol^{-1}

[**Ans.** 57796.45 kcal]

SOME SOLVED CONCEPTUAL PROBLEMS

Problem 1. Under what conditions, an extensive property may become intensive property? Give an example.

Solution. An extensive property might become intensive when unit amount of the substance is specified. For example, mass is extensive, but mass per unit volume (*i.e.* density) is intensive. Similarly, heat capacity is extensive but specific heat (which is heat capacity per unit mass) is intensive.

Problem 2. State the variables which are kept constant in the following processes

(*a*) isothermal (*b*) Isobaric (*c*) Isocaloric

Solution. (*i*) Temperature (*ii*) Pressure (*iii*) Volume

Problem 3. Which of the following expressions represent work done and under what conditions?

(*a*) $P\Delta V$ (*b*) $\Delta (PV)$ (*c*) $P_{ext} \int_{V_1}^{V_2} dV$

Solution

(*i*) $P\Delta V$ denotes work done against constant external pressure, P when the volume changes from V_1 to V_2 such that $\Delta V = V_2 - V_1$

(*ii*) It does not stand for work done

(*iii*) It represents the work done when there is reversible expansion of the gas from the volume V_1 to V_2 against extend pressure P_{ext}

Problem 4. What are the conditions under which *q* and *w* become state functions. Discuss briefly.

Solution.

(*i*) At constant volume, there is no work of expansion, hence $w = 0$. By first law of thermodynamics $\Delta U = q_v$. As ΔU is a state function, q_v is also a state function.

(*ii*) At constant external pressure $w = - P\Delta V$. Hence w becomes a state function.

(*iii*) $\Delta U = q_p + w$. As ΔU is a state function, w under constant pressure is also a state function [see (*ii*) above]. Therefore q_p is also a state function.

Problem 5. What are standard state conditions? Which form of carbon is assigned a value of zero for molar enthalpy in the standard state and why?

Solution. A substance is considered to be in standard state if it is at a temperature of 298.15 K and a pressure of one atmosphere. Graphite is assigned a value of zero for enthalpy in the standard state because it is the stable form.

Problem 6. In the case of polyatomic molecules, why is the bond energy of a particular type of bond, the average value?

Solution. Bond energy for the dissociation of a particular bond is not the same in different molecules. It is not the same even in the same molecule like CH_4. Bond energies of four C — H are not equal because after the dissociation of every C — H bond, the electronic environments change. Hence the average value is taken.

Problem 7. What is the Thermodynamic basis of Hess's law?

Solution. Enthalpy is a state function. That means enthalpy change during a reaction does not depend upon the reaction path in which the change is brought about. It depends only upon the nature of the reactants and the final products.

EXERCISES
(INCLUDING QUESTIONS FROM DIFFERENT UNIVERSITY PAPERS)

Multiple Choice Questions

Choose the correct option

1. First law of thermodynamics is the law of

(*a*) entropy (*b*) free energy

(*c*) conservation of energy (*d*) none of these

2. The mathematical form of first law of thermodynamics is

(*a*) $dq = dU + dw$ (*b*) $dq = dU - dw$

(*c*) $dU = dq + dw$ (*d*) none of these

3. Standard state refers to

(*a*) One atmosphere pressure and 25° C temperature

(*b*) One atmosphere pressure and 100° C temperature

(*c*) One atmosphere pressure and 0° C temperature

(*d*) none of these

4. Endothermic reaction is one in which

(*a*) heat is converted into electricity (*b*) heat is absorbed

(*c*) heat is given out (*d*) heat is converted into mechanical work

5. Thermochemistry is the study of

(*a*) heat changes accompanying chemical reaction

(*b*) net entropy change in a reaction

(*c*) net free energy change in a chemical reaction

(*d*) none of these

6. A process during which there is no heat change is called

(*a*) an adiabatic process (*b*) reversible process

(*c*) irreversible process (*d*) none of these

7. A process in which a system comes to its initial position after a series of operations is called

(*a*) reversible (*b*) cyclic

(*c*) adiabatic (*d*) none of these

8. A reaction whose heat of reaction shows the bond energy of HCl is

(*a*) $HCl\,(g) \longrightarrow H\,(g) + Cl\,(g)$ (*b*) $2\,HCl\,(g) \longrightarrow H_2\,(g) + Cl_2\,(g)$

(*c*) $HCl\,(g) \longrightarrow \frac{1}{2}H_2\,(g) + \frac{1}{2}Cl_2\,(g)$ (*d*) $HCl\,(g) \longrightarrow H^+\,(g) + Cl^-(g)$

9. A process which proceeds infinitesimally slowly is called

(*a*) irreversible (*b*) reversible

(*c*) isothermal (*d*) adiabatic

10. The equations showing the effect of temperature change on the heat of reaction is known as

(*a*) Arrhenius equation (*b*) Kirchoff's equation

(*c*) Ostwald's equation (*d*) none of these

11. The heat of neutralisation is maximum for the reaction

(*a*) NaOH and HCl (*b*) NaOH and CH_3COOH

(*c*) HCl and NH_4OH (*d*) NH_4OH and CH_3COOH

12. "The resultant heat change in a chemical reaction is the same whether it takes place in one or several stages". This statement is called

(*a*) Le-Chatelier's principle (*b*) Hess's law

(*c*) Lavoisier-Laplace law (*d*) Joule Thomson effect

13. Standard heat of formation of an element is

(*a*) infinite (*b*) zero

(*c*) 100 kJ (*d*) 200 kJ

14. The apparatus used for measuring heat of reaction at constant volume is called

(*a*) calorimeter (*b*) colourimeter

(*c*) Pyrometer (*d*) pyknometer

15. The enthalpies of elements in their standard state are taken as zero. Thus, the enthalpy of formation of a compound

(*a*) will always be positive (*b*) will always be negative

(*c*) may be positive or negative (*d*) will always be zero

16. The heat content of a system is called

(*a*) internal energy (*b*) enthalpy

(*c*) free energy (*d*) entropy

17. For the reaction $N_2 + 3H_2 \rightleftharpoons 2NH_3$, q_p is equal to

(*a*) $q_v - 2RT$ (*b*) $q_v + 2RT$

(*c*) $q_v + RT$ (*d*) $q_v - RT$

18. The least random state of water system is

(*a*) ice (*b*) liquid water

(*c*) steam (*d*) equal randomness in all

19. Identify the intensive quantities from the following

(*a*) Enthalpy (*b*) Temperature

(*c*) Volume (*d*) Refractive Index

20. The heat of neutralisation of HCl and NaOH is 13.7 kcal and that of NaOH and CH_3COOH is 13.2 kcal. Heat of ionisation of CH_3COOH is

(*a*) 13.3 kcal (*b*) 13.7 kcal

(*c*) 0.5 kcal (*d*) none of these

ANSWERS

Multiple choice questions

1. (*c*)	**2.** (*c*)	**3.** (*a*)	**4.** (*b*)	**5.** (*a*)	**6.** (*a*)
7. (*b*)	**8.** (*a*)	**9.** (*b*)	**10.** (*b*)	**11.** (*a*)	**12.** (*b*)
13. (*b*)	**14.** (*a*)	**15.** (*d*)	**16.** (*b*)	**17.** (*a*)	**18.** (*a*)
19. (*d*), (*a*)	**20.** (*c*)				

Short Answer Questions

1. Point out the main difference between
   (*a*) Open, closed and isolated system
   (*b*) Reversible and irreversible processes
2. What do you understand by the state of a system? What are state variables? Why should it be sufficient to specify the values of a few variables to define the state of a system instead of specifying the values of all the variables?
3. Derive expression for the work done when a gas expands from volume V_1 to V_2 against external pressure, P.
4. Define Thermodynamics. What are its main objectives?
5. Distinguish between
   (*i*) State function and path function.
   (*ii*) Isothermal and adiabatic process.
   (*iii*) Dependent and independent variables.
6. Explain that the work done during reversible expansion is the maximum work for a given increase of volume and minimum to compress the gas for a given decreases of volume.
7. What are extensive and intensive properties? Give three examples of each of them.
8. Give in brief the objectives and limitations of thermodynamics.
9. What is the relationship between Heat and Mechanical work? Define Joule's mechanical equivalent of heat. What is its value?
10. State First law of thermodynamics in two different ways. Derive its mathematical formulation.
11. Briefly explain the terms Enthalpy and Enthalpy change. Is it intensive or extensive property?
12. Prove that the heat capacity at constant volume (C_v) for an ideal gas having translational kinetic energy only is constant and equal to 3 cal mol^{-1} and is independent of temperature.
13. State and explain Joule's law about the variation of internal energy of an ideal gas with volume.
14. What do you understand by Inversion temperature ?
15. Define Zeroth law of thermodynamics. What is its most important application?
16. Explain the fact that internal energy is a state function but work and heat are path functions.
17. Derive expression for molar heat capacities C_v and C_P in terms of internal energy change and enthalpy change and hence show that $C_p - C_v = R$ for one mole of an ideal gas.
18. Prove that C_p for an ideal gas is constant and nearly equal to 5 cal degree^{-1} mol^{-1}.
19. Write a short note on Joule-Thomson Effect.
20. Derive expression for Joule Thomson coefficient and inversion temperature in terms of van der Waals' constants.
21. What generalization do you draw about the relationship between the atomicity of the gases and the ratio (γ) of their specific heats?

General Questions

1. State and explain Joule's law.
2. State First Law of Thermodynamics in at least three different ways. Derive the mathematical expression for it. Explain the fact that internal energy is a state function but heat and work are not.

3. State and explain the following terms with suitable examples:
   (*i*) Open, closed and isolated systems
   (*ii*) Thermodynamic equilibrium
   (*iii*) Thermodynamics Reversible and Irreversible processes
   (*iv*) Extensive and Intensive properties
   (*v*) State function and path function
4. What is Joule-Thomson effect? Justify that during this process, enthalpy of the system remains constant. Also explain what you understand by Joule-Thomson coefficient and Inversion temperature. Derive that Joule-Thomson coefficient is given by

$$\mu = -\frac{1}{C_p}\left(\frac{\partial H}{\partial P}\right)_T$$

5. Derive expressions for w, q, ΔU and ΔH when an ideal gas undergoes (*i*) isothermal reversible expansion (*ii*) isothermal irreversible expansion.
6. Derive expressions for w, ΔU, q and ΔH for isothermal reversible expansion of a real gas.
7. Define 'Bond energy'. How is bond energy of a bond calculated for a polyatomic molecule? How is it used to calculate the enthalpy change of a reaction?
8. Derive that for reversible adiabatic expansion of an ideal gas:
   (*i*) PV^γ = constant (*ii*) work done, $w = C_v\,(T_2 - T_1)$.
9. Define standard enthalpy of reaction, standard enthalpy of formation and standard enthalpy of combustion. What is Hess's law? What is the thermodynamic basis of Hess's law?
10. Briefly explain the terms internal energy, work and heat. Derive the expression for the work of expansion. Explain that work done by a system in a reversible process is the maximum work for a given increases of volume and for a given decreases of volume, it is the minimum work to reverse the process.
11. Explain the terms (*i*) 'Enthalpy' and 'Enthalpy change' (*ii*) Heat capacity. Derive expression for heat capacity at constant volume and that at constant pressure. Derive the relationship between them.
12. Prove thermodynamically that Joule-Thomson coefficient for an ideal gas is zero. Also deduce expression for Joule-Thomson coefficient for real gases in terms of van der Waals constants '*a*' and '*b*'
13. What do you understand by Heat of reaction at constant volume and that at constant pressure. Derive the relationship between them. Under what conditions, the two are equal?
14. Prove that for a monatomic gas (having transtational kinetic energy only), $C_v = \frac{3}{2}R$.
15. Derive expressions for ΔU and ΔH for adiabatic reversible expansion of real gas.
16. Briefly explain the terms enthalpy of neutralisation. Comment on their values if both are strong or one of them is weak or both are weak?
17. Derive thermodynamically Kirchoff's equation giving the variation of heat of reaction with temperature.
18. Derive thermodynamically Kirchoff's equation.

Or

Show that the temperature dependence of heat of reaction is given by the relation

$$\left(\frac{\partial \Delta H}{\partial T}\right)_P = \Delta C_p.$$

Name the equation.

19. (*a*) Distinguish between reversible isothermal expansion and reversible adiabatic expansion.

(*b*) Show that for adiabatic expansion of an ideal gas,

$$PV^{\gamma} = \text{constant}$$

20. Derive the following expression for the reversible adiabatic expansion of a real gas.

$$T_2 = T_1\left(\frac{V_1 - nb}{V_2 - nb}\right)^{R/C_p}$$

21. Define 'Bond energy' for a diatomic molecule and for a polyatomic molecule. How bond energy data helps to calculate the enthalpy change of a reaction?

22. How can the reversible iscthermal expansion of an ideal gas be brought about? Derive an expression for the work of expansion of such a process.

23. What are the values/expressions for q, w, ΔU and ΔH for

(*i*) irreversible isothermal free expansion of an ideal gas?

(*ii*) irreversible isothermal intermediate expansion of an ideal gas?

Out of reversible and irreversible expansion which one will absorb less heat and why?

24. What are the values/ expressions, for q, w, ΔU and ΔH for

(*i*) irreversible adiabatic free expansion of an ideal gas?

(*ii*) irreversible adiabatic intermediate expansion of an ideal gas?

25. How is the final temperature of irreversible adiabatic expansion of an idea gas calculated?

26. Derive the expression for the work done in an adiabatic expansion. Will the expression be different for adiabatic expansion along a reversible and an irreversible path ?

27. Derive relationship between heat of reaction at constant pressure and that at constant volume.

28. How maximum amount of the work during isothermal expansion of a gas can be obtained? Derive an expression for the maximum work obtainable from the isothermal expansion of n moles of a perfect gas.

29. The work done in a reversible isothermal expansion of an ideal gas is greater than the work done in reversible adiabatic expansion. Explain.

30. Show the maximum work is produced in a reversible isothermal expansion of a gas.

31. Define 'Standard enthalpy of formation'. Taking a suitable example, prove that the standard enthalpy of a compound is equal to its standard enthalpy of formation.

32. Compare reversible isothermal expansion with reversible adiabatic expansion of an ideal gas.

33. Define standard enthalpy of combustion. Explain with a suitable example.

34. State Hess's law of constant heat summation. Describe any two of its applications.

35. Write notes on the following

(*a*) Heat of combustion (*b*) Heat of formation

(*c*) Heat of neutralisation

APPENDIX 1

GREEK APLPHABET TABLE

Capital	*Small*	*Pronunication*	*English equivalent*
Α	α	alpha	a
Β	β	beta	b
G	γ	gamma	g, n
Δ	δ	delta	d
Ε	ε	eosilon	e
Ζ	ζ	zeta	z
Η	η	eta	$\bar{e}$
Θ	θ	theta	th
Ι	ι	iota	i
Κ	κ	kappa	k
Λ	λ	lamba	l
Μ	μ	mu	m
Ν	ν	nu	n
Ξ	ξ	xi	x
Ο	ο	omricon	o
Π	π	pi	p
Ρ	ρ	rho	r, rh
Σ	σ	sigma	s
Τ	τ	tau	t
Υ	υ	upsilon	y, u
Φ	φ	phi	ph
Χ	χ	chi	ch
Ψ	ψ	psi	ps
Ω	ω	omega	$\bar{o}$

APPENDIX 2

SOME USEFUL TABLES

Table 1. Seven base units

Physical Quantity	*Name of Unit*	*Official S.I. symbol*
Length	metre	m
Mass	kilogram	kg
Time	second	s
Electric current	ampere	A
Temperature	kelvin	K
Luminous intensity	candela	cd
Amount of substance	mole	mol

Table 2. Some commonly used derived units

Physical Quantity	*Name of Unit*	*Symbol*
Area	Square metre	m^2
Volume	Cubic metre	m^3
Density	Kg per cubic metre	$kg\ m^{-3}$
Velocity	Metre per second	ms^{-1}
Acceleration	Metre per sec per sec	ms^{-2}
Force	Newton	$N = kg\ ms^{-2}$
Power	Watt	$W = Js^{-1}$
Pressure	Pascals (Newtons per sq. meter)	$Pa = Nm^{-2} = kg\ m^{-1}s^{-2}$
Work or Energy (Heat etc)	Joule	$J = Nm = kg\ m^2\ s^{-2}$
Magnetic flux density	Tesla	$T = kg\ s^{-2}\ A^{-1}$
Frequency	Hertz	$Hz = s^{-1}$
Electric charge	Coulomb	$C = As$
Potential difference	Volt	V
Electric resistance	Ohms	$\Omega = VA^{-1}$
Electric conductance	$Ohms^{-1}$	$\Omega^{-1} = AV^{-1}$